HANDBOOK OF IMAGING IN BIOLOGICAL MECHANICS

HANDBOOK OF IMAGING IN BIOLOGICAL MECHANICS

Edited by
Corey P. Neu
Guy M. Genin

CRC Press
Taylor & Francis Group
Boca Raton London New York

CRC Press is an imprint of the
Taylor & Francis Group, an **informa** business

CRC Press
Taylor & Francis Group
6000 Broken Sound Parkway NW, Suite 300
Boca Raton, FL 33487-2742

Printed on acid-free paper
Version Date: 20140915

International Standard Book Number-13: 978-1-4665-8813-4 (Pack - Book and Online)

Library of Congress Cataloging-in-Publication Data

Handbook of imaging in biological mechanics / editors, Corey P. Neu and Guy M. Genin.
p. ; cm.
Includes bibliographical references and index.
ISBN 978-1-4665-8813-4 (alk. paper)
I. Neu, Corey P., editor. II. Genin, Guy M., editor.
[DNLM: 1. Biomechanical Phenomena--physiology. 2. Diagnostic Imaging--methods. WN 180]

RC78.7.D53
616.07'54--dc23 2014015960

Visit the Taylor & Francis Web site at
http://www.taylorandfrancis.com

and the CRC Press Web site at
http://www.crcpress.com

Contents

Section V Tissue and Organ Imaging Methods: High Throughput and Image Quantification

Section VI Tissue and Organ Imaging Methods: Multiscale Imaging and Modeling

Section VII Cellular and Molecular Imaging Methods: Microenvironments—Native ECM

Section VIII Cellular and Molecular Imaging Methods: Microenvironments—In Vitro Culture Systems

Section XII Cellular and Molecular Imaging Methods: Surfaces and Single Molecule Measurements

Preface

Predictive models require data for validation. Acquiring these data has been a central challenge in the field of biological mechanics since its earliest days. Planning for this CRC handbook stretches back not quite that long but nevertheless spans a number of years. This handbook arose from a need for a concise presentation of the range of imaging protocols for biomechanics across hierarchies, as identified in a series of workshops and conference sessions that we and others have organized on the subject. The handbook presents a range of primers on the most successful techniques used at each of a number of length scales. Our goals were to provide a comprehensive resource and guide that is accessible to students and researchers and therefore facilitate access to modern imaging methods used to nondestructively or noninvasively measure mechanics and biological responses of tissues, cells, and molecules. An emphasis has been placed on providing an overview of practical method implementation, in addition to directions and challenges facing further development and expanded utility of each method.

This handbook was a major undertaking that could not have succeeded without hard work from a great many people. We thank our many colleagues who encouraged us to assemble this handbook and who provided their criticism and feedback on the project as it developed. We are indebted to Leong LiMing and David Fausel from CRC Press and Taylor & Francis, who saw from the very start the importance of this handbook for those navigating a rapidly emerging field. We are grateful to all of the authors of the chapters, all leaders in their fields, for assembling this fantastic set of primers. Each chapter went through multiple rounds of anonymous peer review by a great number of leading scholars, and we are indebted to all of those who assisted in that exhaustive and exhausting enterprise. Most of all, we are grateful to our wives and children for allowing us time away to make all of this come together—thank you Nancy, Anika, Signe, Li, Jacob, and Samuel!

Editors

Corey P. Neu directs the Soft Tissue Bioengineering Laboratory at Purdue University and holds an appointment on the faculty in Purdue's Weldon School of Biomedical Engineering. He studies force transfer in cells and tissues in the context of disease and regeneration. His current research focuses on structural and mechanical changes to articular cartilage in the pathogenesis of osteoarthritis, and multiscale biomechanics and mechanobiology of collagen-rich tissues. A primary theme of this research is the development of methods to noninvasively quantify the mechanics of cells and tissues *in vivo*, requiring advancement of a broad range of imaging modalities. Professor Neu earned his ScB and ScM in mechanical and biomedical engineering from Brown University, his PhD in biomedical engineering from the University of California (UC) at Davis, and postdoctoral training in cell biology and tribology from UC Davis and Berkeley. He is the recipient of a Faculty Early Career Development (CAREER) Program award from the National Science Foundation.

Guy M. Genin holds appointments on the faculty of mechanical engineering and materials science at Washington University in St. Louis, and of neurological surgery at the Washington University School of Medicine. He studies interfaces and adhesion in nature, physiology, and engineering. His current research focuses on interfaces between tissues at the attachment of tendon to bone, between cells in cardiac fibrosis, and between cell membranes and the cell wall in plant defenses. A central challenge in all of this work is identifying noninvasively how mechanical fields are perturbed by interfaces, and a primary focus of his lab is developing noninvasive imaging tools to quantify this. Professor Genin earned his BSCE in civil engineering and MS in engineering mechanics from Case Western Reserve University, and his SM in applied mechanics and PhD in solid mechanics from Harvard. He received post-doctoral training at Cambridge and Brown. He is the recipient of numerous awards for engineering design, teaching, and research, including a Research Career Award from the National Institutes of Health and the Skalak Award from the American Society of Mechanical Engineers.

Contributors

Khaled Z. Abd-Elmoniem
Biomedical and Metabolic Imaging Branch
National Institute of Diabetes and Digestive and Kidney Diseases
National Institutes of Health
Bethesda, Maryland

Steven G. Adie
Department of Electrical and Computer Engineering
and
Beckman Institute for Advanced Science and Technology
University of Illinois at Urbana-Champaign
Champaign, Illinois

Antonia Antoniou
Department of Mechanical Engineering
Georgia Institute of Technology
Atlanta, Georgia

Leon Axel
Department of Radiology
New York University School of Medicine
New York, New York

Russell J. Bailey
Department of Materials
School of Engineering and Materials Science
Queen Mary University of London
London, United Kingdom

Jeffrey C. Bamber
Institute of Cancer Research
Surrey, United Kingdom

Asa H. Barber
Department of Materials
School of Engineering and Materials Science
Queen Mary University of London
London, United Kingdom

Paul E. Barbone
Department of Mechanical Engineering
Boston University
Boston, Massachusetts

Ashraf Bastawros
Department of Aerospace Engineering
Iowa State University
Ames, Iowa

Philip V. Bayly
Department of Mechanical Engineering and Materials Science
and
Department of Biomedical Engineering
Washington University in St. Louis
St. Louis, Missouri

James S. Bell
Department of Physics
College of Engineering Mathematics and Physical Science
University of Exeter
Exeter, United Kingdom

Gearóid P. Berry
Institute of Cancer Research
Surrey, United Kingdom
and
UBS AG
Singapore

Agata Blasiak
Bionanoscience Group
Conway Institute for Biomolecular and Biomedical Sciences
and
School of Chemistry and Chemical Biology
University College Dublin
Dublin, Ireland

Stephen A. Boppart
Department of Bioengineering
and
Department of Electrical and Computer Engineering
and
Beckman Institute for Advanced Science and Technology
University of Illinois at Urbana-Champaign
Champaign, Illinois

Kent Butz
School of Mechanical Engineering
Purdue University
West Lafayette, Indiana

Amnon Buxboim
Molecular and Cell Biophysics Lab
University of Pennsylvania
Philadelphia, Pennsylvania

Deva Chan
Weldon School of Biomedical Engineering
Purdue University
West Lafayette, Indiana

Sohae Chung
Department of Radiology
New York University School of Medicine
New York, New York

Erik H. Clayton
Department of Mechanical Engineering and Materials Science
Washington University in St. Louis
St. Louis, Missouri

Benjamin R. Coleman
Department of Mechanical Engineering
University of Washington
Seattle, Washington

Charles G. Cranfield
Victor Chang Cardiac Research Institute
and
School of Medical and Molecular Biosciences
University of Technology Sydney
Sydney, New South Wales, Australia

Joseph J. Crisco III
Bioengineering Laboratory
Department of Orthopedics
Alpert Medical School of Brown University
and
Rhode Island Hospital
and
Center for Biomedical Engineering and School of Engineering
Brown University
Providence, Rhode Island

Michelle R. Dawson
School of Chemical and Biomolecular Engineering
and
Parker H. Petit Institute for Bioengineering and Biosciences
Georgia Institute of Technology
Atlanta, Georgia

Dennis E. Discher
Molecular and Cell Biophysics Lab
University of Pennsylvania
Philadelphia, Pennsylvania

Henry J. Donahue
Division of Musculoskeletal Sciences
Department of Orthopaedics and Rehabilitation
Pennsylvania State University College of Medicine
Hershey, Pennsylvania

Jean-François Dord
Computational Combustion Laboratory
General Electric Global Research Center
Niskayuna, New York

Carlos Drummond
Centre de Recherche Paul Pascal (Centre National de la Recherche Scientifique [CNRS])–Université Bordeaux 1
Pessac, France

Randall L. Duncan
Departments of Biological Sciences and Biomedical Engineering
University of Delaware
Newark, Delaware

Elliot L. Elson
Department of Mechanical Engineering and Materials Science
Washington University in St. Louis
and
Department of Biochemistry and Molecular Biophysics
Washington University School of Medicine
St. Louis, Missouri

Brian C. Fabien
Department of Mechanical Engineering
University of Washington
Seattle, Washington

Xiaoxi Fan
Key Laboratory of Biomedical Information Engineering of the Ministry of Education
School of Life Science and Technology
and
Bioinspired Engineering and Biomechanics Center
Xi'an Jiaotong University
Xi'an, China

Xiaoyao Fan
Thayer School of Engineering
Dartmouth College
Hanover, New Hampshire

Genevieve Farrar
Department of Mechanical Engineering
University of Washington
Seattle, Washington

Elizabete Rodrigues Ferreira
Department of Civil Engineering
University of Minnesota
Minneapolis, Minnesota

Eric A. Galburt
Department of Biochemistry and Molecular Biophysics
Washington University in St. Louis
St. Louis, Missouri

Brian S. Garra
Center for Devices and Radiological Health
U.S. Food and Drug Administration
Silver Spring, Maryland

and

Washington DC Veterans Affairs Medical Center
Washington, District of Columbia

Damian C. Genetos
Department of Anatomy, Physiology, and Cell Biology
School of Veterinary Medicine
University of California
Davis, California

Guy M. Genin
Department of Mechanical Engineering and Materials Science
Washington University in St. Louis
and
Department of Neurological Surgery
Washington University School of Medicine
St. Louis, Missouri

Niloufar Ghoreishi
Department of Biomedical Engineering
Washington University in St. Louis
St. Louis, Missouri

Sevan Goenezen
Department of Mechanical Engineering
Texas A&M University
College Station, Texas

Robert E. Guldberg
Parker H. Petit Institute of Bioengineering and Biosciences
and
George W. Woodruff School of Mechanical Engineering
Georgia Institute of Technology
Atlanta, Georgia

Roi Gurka
School of Coastal and Marine Systems Science
Coastal Carolina University
Conway, South Carolina

Candice Haase
Department of Biomedical Engineering
Texas A&M University
College Station, Texas

Eni Halilaj
Bioengineering Laboratory
Department of Orthopedics
Alpert Medical School of Brown University
and
Rhode Island Hospital
and
Center for Biomedical Engineering and School of Engineering
Brown University
Providence, Rhode Island

Timothy J. Hall
Medical Physics Department
University of Wisconsin
Madison, Wisconsin

Bumsoo Han
Weldon School of Biomedical Engineering
and
School of Mechanical Engineering
College of Engineering
Purdue University
West Lafayette, Indiana

Yu Long Han
Key Laboratory of Biomedical Information Engineering of the Ministry of Education
School of Life Science and Technology
and
Bioinspired Engineering and Biomechanics Center
Xi'an Jiaotong University
Xi'an, China

Alex Hartov
Thayer School of Engineering
Dartmouth College
Hanover, New Hampshire

and

Norris Cotton Cancer Center
Dartmouth College
Lebanon, New Hampshire

Jonathan T. Henderson
Weldon School of Biomedical Engineering
Purdue University
West Lafayette, Indiana

Michael J. Higgins
ARC Centre of Excellence for Electromaterials Science
Intelligent Polymer Research Institute
and
Australian Institute for Innovative Materials Faculty
Innovation Campus
University of Wollongong
Sydney, New South Wales, Australia

Guoyou Huang
Key Laboratory of Biomedical Information Engineering of the Ministry of Education
School of Life Science and Technology
and
Bioinspired Engineering and Biomechanics Center
Xi'an Jiaotong University
Xi'an, China

Michael F. Insana
Department of Bioengineering
and
Department of Electrical and Computer Engineering
and
Beckman Institute for Advanced Science and Technology
University of Illinois at Urbana-Champaign
Champaign, Illinois

Jerome Irianto
Molecular and Cell Biophysics Lab
University of Pennsylvania
Philadelphia, Pennsylvania

Jacob N. Israelachvili
Department of Chemical Engineering
and
Materials Department
University of California Santa Barbara
Santa Barbara, California

Krasimir Ivanov
Bionanoscience Group
Conway Institute for Biomolecular and Biomedical Sciences
and
School of Chemistry and Chemical Biology
University College Dublin
Dublin, Ireland

Irena L. Ivanovska
Molecular and Cell Biophysics Lab
University of Pennsylvania
Philadelphia, Pennsylvania

K. Venkatesan Iyer
Mechanobiology Institute
National University of Singapore
Singapore

Songbai Ji
Thayer School of Engineering
Dartmouth College
Hanover, New Hampshire

and

Geisel School of Medicine
Dartmouth College
Lebanon, New Hampshire

Ines Jimenez-Palomar
Department of Materials
School of Engineering and Materials Science
Queen Mary University of London
London, United Kingdom

Roland Kaunas
Department of Biomedical Engineering
Texas A&M University
College Station, Texas

Devrim Kilinc
Bionanoscience Group
Conway Institute for Biomolecular and Biomedical Sciences
and
School of Chemistry and Chemical Biology
University College Dublin
Dublin, Ireland

Michal Klosowski
Department of Materials
Imperial College London
London, United Kingdom

Brian Koons
Department of Mechanical Engineering
Virginia Tech
Blacksburg, Virginia

William Lai
Department of Aerospace Engineering
Iowa State University
Ames, Iowa

Yu-Hui Lai
Department of Biomedical Engineering
Indiana University–Purdue University Indianapolis
Indianapolis, Indiana

Gil U. Lee
Bionanoscience Group
Conway Institute for Biomolecular and Biomedical Sciences
and
School of Chemistry and Chemical Biology
University College Dublin
Dublin, Ireland

Jerry S.H. Lee
Center for Strategic Scientific Initiatives
National Cancer Institute
National Institutes of Health
Bethesda, Maryland

and

Department of Chemical and Biomolecular Engineering
Johns Hopkins University
Baltimore, Maryland

Wesley R. Legant
Janelia Farm Research Campus
Howard Hughes Medical Institute
Ashburn, Virginia

Fei Li
Bioinspired Engineering and Biomechanics Center
and
Department of Chemistry
School of Sciences
Xi'an Jiaotong University
Xi'an, China

Peng Li
Bionanoscience Group
Conway Institute for Biomolecular and Biomedical Sciences
and
School of Chemistry and Chemical Biology
University College Dublin
Dublin, Ireland

Alex Liberzon
School of Mechanical Engineering
Tel Aviv University
Tel Aviv, Israel

Angela Lin
Parker H. Petit Institute of Bioengineering and Biosciences
and
George W. Woodruff School of Mechanical Engineering
Georgia Institute of Technology
Atlanta, Georgia

Baoyu Liu
Coulter Department of Biomedical Engineering
Georgia Institute of Technology
Atlanta, Georgia

Keng-Ku Liu
Department of Mechanical Engineering and Materials Science
Washington University in St. Louis
St. Louis, Missouri

Tian Jian Lu
Bioinspired Engineering and Biomechanics Center
Xi'an Jiaotong University
Xi'an, China

Alexander Macmillan
Biomedical Imaging Facility
Lowy Cancer Research Centre
The University of New South Wales
Sydney, New South Wales, Australia

Ekta Makhija
Mechanobiology Institute
National University of Singapore
Singapore

Jessica C. Mansfield
Department of Physics
College of Engineering Mathematics and Physical Science
University of Exeter
Exeter, United Kingdom

Boris Martinac
Victor Chang Cardiac Research Institute
and
St Vincent's Clinical School
The University of New South Wales
Sydney, New South Wales, Australia

Elena Martines
Bionanoscience Group
Conway Institute for Biomolecular and Biomedical Sciences
and
School of Chemistry and Chemical Biology
University College Dublin
Dublin, Ireland

Kathleen M. McAndrews
School of Chemical and Biomolecular Engineering
Georgia Institute of Technology
Atlanta, Georgia

David W. McComb
Department of Materials Science and Engineering
Center for Electron Microscopy and Analysis
The Ohio State University
Columbus, Ohio

Julian Moger
Department of Physics
College of Engineering, Mathematics and Physical Science
University of Exeter
Exeter, United Kingdom

Douglas C. Moore
Bioengineering Laboratory
Department of Orthopedics
Alpert Medical School of Brown University
and
Rhode Island Hospital
Providence, Rhode Island

Peter R. Mouton
Department of Pathology and Cell Biology
University of South Florida College of Medicine
Tampa, Florida

Raja Muthupillai
Department of Diagnostic and Interventional Radiology
St Luke's Episcopal Hospital
Houston, Texas

Sungsoo Na
Department of Biomedical Engineering
Indiana University–Purdue University Indianapolis
Indianapolis, Indiana

Amrinder Nain
School of Biomedical Engineering and Sciences
and
Department of Mechanical Engineering
Virginia Tech
Blacksburg, Virginia

Eric A. Nauman
School of Mechanical Engineering
and
Weldon School of Biomedical Engineering
and
Department of Basic Medical Sciences
Purdue University
West Lafayette, Indiana

Ali Nekouzadeh
Department of Biomedical Engineering
Washington University in St. Louis
St. Louis, Missouri

Corey P. Neu
Weldon School of Biomedical Engineering
Purdue University
West Lafayette, Indiana

Assad A. Oberai
Department of Mechanical, Aerospace and Nuclear Engineering
Rensselaer Polytechnic Institute
Troy, New York

Ruth J. Okamoto
Department of Mechanical Engineering and Materials Science
Washington University in St. Louis
St. Louis, Missouri

Vijay Parthasarathy
Philips Research North America
Briarcliff Manor, New York

Tarpit K. Patel
Bioengineering Laboratory
Department of Orthopedics
Alpert Medical School of Brown University
and
Rhode Island Hospital
Providence, Rhode Island

Keith D. Paulsen
Thayer School of Engineering
Dartmouth College
Hanover, New Hampshire

and

Norris Cotton Cancer Center
Dartmouth College
Lebanon, New Hampshire

Alexandra E. Porter
Department of Materials
Imperial College London
London, United Kingdom

Jerry L. Prince
Department of Electrical and Computer Engineering
Johns Hopkins University
and
Russell H. Morgan Department of Radiology and Radiological Science
Johns Hopkins University School of Medicine
Baltimore, Maryland

David Reece
Parker H. Petit Institute of Bioengineering and Biosciences
and
Wallace H. Coulter Department of Biomedical Engineering
Georgia Institute of Technology
Atlanta, Georgia

David W. Roberts
Geisel School of Medicine
Hanover, New Hampshire

and

Norris Cotton Cancer Center
Dartmouth College
Lebanon, New Hampshire

Marina Ruths
Department of Chemistry
University of Massachusetts Lowell
Lowell, Massachusetts

Timothy J. Schaewe
Medtronic, Inc.
Surgical Technologies
Louisville, Colorado

W. Paul Segars
Department of Radiology
Duke University
Durham, North Carolina

Jin-Yu Shao
Department of Biomedical Engineering
Washington University in St. Louis
St. Louis, Missouri

Puja Sharma
School of Biomedical Engineering and Sciences
Virginia Tech
Blacksburg, Virginia

Kevin Sheets
School of Biomedical Engineering and Sciences
Virginia Tech
Blacksburg, Virginia

Sandra J. Shefelbine
Department of Mechanical and Industrial Engineering
Northeastern University
Boston, Massachusetts

Jae-Won Shin
Molecular and Cell Biophysics Lab
University of Pennsylvania
Philadelphia, Pennsylvania

G.V. Shivashankar
Mechanobiology Institute
and
Department of Biological Sciences
National University of Singapore
Singapore

David A. Simon
Medtronic, Inc.
Surgical Technologies
Louisville, Colorado

Srikanth Singamaneni
Department of Mechanical Engineering and Materials Science
Washington University in St. Louis
St. Louis, Missouri

Kyle Spinler
Molecular and Cell Biophysics Lab
University of Pennsylvania
Philadelphia, Pennsylvania

Joe Swift
Molecular and Cell Biophysics Lab
University of Pennsylvania
Philadelphia, Pennsylvania

Sirimuvva Tadepalli
Department of Mechanical Engineering and Materials Science
Washington University in St. Louis
St. Louis, Missouri

Shefali Talwar
Mechanobiology Institute
National University of Singapore
Singapore

Zachary J. Taylor
School of Mechanical Engineering
Tel Aviv University
Tel Aviv, Israel

Abhishek Tondon
Department of Biomedical Engineering
Texas A&M University
College Station, Texas

Yiider Tseng
Department of Chemical Engineering
and
Institute for Cell Engineering and Regenerative Medicine
University of Florida
Gainesville, Florida

Alexander I. Veress
Department of Mechanical Engineering
University of Washington
Seattle, Washington

Sherry L. Voytik-Harbin
Weldon School of Biomedical Engineering
College of Engineering
and
Department of Basic Medical Sciences
College of Veterinary Medicine
Purdue University
West Lafayette, Indiana

Qiaoqiao Wan
Department of Biomedical Engineering
Purdue University
West Lafayette, Indiana

and

Department of Biomedical Engineering
Indiana University–Purdue University Indianapolis
Indianapolis, Indiana

Hui Wang
MEMC Electronic Materials, Inc.
St. Peters, Missouri

Lin Wang
Bioinspired Engineering and Biomechanics Center
Xi'an Jiaotong University
Xi'an, China

Pengfei Wang
Bioinspired Engineering and Biomechanics Center
Xi'an Jiaotong University
Xi'an, China

Yue Wang
Department of Bioengineering
and
Beckman Institute for Advanced Science and Technology
University of Illinois at Urbana-Champaign
Champaign, Illinois

C. Peter Winlove
Department of Physics
College of Engineering, Mathematics and Physical Science
University of Exeter
Exeter, United Kingdom

Haguy Wolfenson
Department of Biological Sciences
Columbia University
New York, New York

Feng Xu
Key Laboratory of Biomedical Information Engineering of the Ministry of Education
School of Life Science and Technology
and
Bioinspired Engineering and Biomechanics Center
Xi'an Jiaotong University
Xi'an, China

Jiming Zhang
Physics Department and Texas Center for Superconductivity
University of Houston
and
Department of Diagnostic and Interventional Radiology
St Luke's Episcopal Hospital
Houston, Texas

Section I

Introduction

1

Imaging in Biological Mechanics

Corey P. Neu and Guy M. Genin

Advances in biological mechanics have progressed hand in hand with advances in imaging since the field's earliest days. Even the early biological mechanics work of da Vinci could be called a product of advances in the production of architectural images. The architectural image production techniques of the Renaissance were a natural fit for da Vinci's favored biomechanical models of pulleys and levers.

However, more recent advances in imaging have required much more adaptation to make an impact on biological mechanics. The oldest known photograph, *View from the Window at Le Gras,* dates to circa 1826 and was created by inventor Nicéphore Niépce at Saint-Loup-de-Varennes, a commune in central France. The contrast of a commonly known version (Figure 1.1) was manually enhanced from the original plate, to highlight main features in the field of view, including portions of buildings and the surrounding countryside. A very long exposure time, over hours to days, was needed to form the image, focused by a camera obscura, on a small pewter plate coated with a naturally occurring light-sensitive material (bitumen of Judea). Long exposure times were typical of photographs in following years, including static buildings and still human portraits, where a distinct *lack* of motion was needed to prevent blurring artifacts and render a clean image depicting sharp features and morphologies. Early photography was a very poor fit for the study of biological mechanics in particular and of motion in general.

The breakthrough was forthcoming, with imaging of biological mechanics as its focus. Early photographs of motion, and precursors to the modern film industry, were captured in *The Horse in Motion*, 1878, by Eadweard Muybridge, in Palo Alto, California. Muybridge took advantage of advances in faster shutter speeds and film emulsions to capture the images (Figure 1.2) with multiple cameras, initially set up around a race track and triggered by a thread as the horse passed. The image series resolved a debate popular in the era, that is, that during trotting, all four feet of the horse were off the ground at the same time.

In a broader sense, the images of *The Horse in Motion* represent a significant step forward to address a fundamental biological question—how can we maximally observe, record, and analyze our dynamic world? Over logarithmic length scales, from individual molecules to whole organs and animals, movement and the influence of mechanical forces are ubiquitous. Especially, at small length scales, approaching and exceeding the cellular scale, spatiotemporal resolution becomes a premium, with one (e.g., temporal resolution) often maximized at the expense of the other (e.g., spatial resolution). A diverse array of new methods have been developed over decades of research that allow us to now peer inside of animals, organs, tissues, and cells to reveal new information that was not previously measured. Modern methods apply fundamentals from physics and biology, alone and in combination, based on x-rays (Figure 1.3), magnetic resonance (MR), genetic engineering, and microelectromechanical systems, to name only a few. The refinement of these methods has increasingly allowed the field to move beyond the qualitative interpretation of biological mechanics that was the focus of Muybridge and toward the fitting of quantitative models of biological mechanics.

With this book, we hope to provide a comprehensive resource and guide for students and researchers to access modern imaging methods used to nondestructively or noninvasively measure mechanics and biological responses of tissues, cells, and molecules. We specifically targeted key methods at multiple (tissue, cell, molecular) length scales, presented by leading authorities. An emphasis has been placed on providing an overview of practical method implementation, in addition to directions and challenges facing further development and expanded utility of each method.

The handbook is divided into two parts, with technologies sorted by the hierarchical scales at which they have enjoyed the most widespread usage. The first part is devoted to methods that have been most important to our current understanding of tissue- and organ-level biomechanics. These methods encompass what might be viewed as the 21st-century sequel to Muybridge's pioneering work: the range of techniques available to quantify noninvasively the motion and deformation of entire organisms and organs and the tissues that comprise

FIGURE 1.1
Earliest known photograph, *View from the Window at Le Gras*, 1826. Early photographs showed a distinct *lack* of motion, owning to the long exposure times typically required for image formation.

FIGURE 1.2
Early example of motion pictures, *The Horse in Motion*, 1878. Motion capture was possible using advances in shutter speeds and film emulsions, allowing observers to see a midstride instant when all four hooves were off the ground.

them. A great number of the primary tools we currently have available were designed initially for imaging only static organisms. The pioneers who have made some of the most significant advances in the imaging of biological mechanics have done so by, like Muybridge, finding ways to strobe these static imaging modalities to make them suitable for applications completely different from any that their original inventors seem to have imagined. We are particularly pleased that a number of these pioneers are in fact the authors of chapters in this handbook.

The first part begins with a selection of chapters describing noninvasive techniques that are based upon

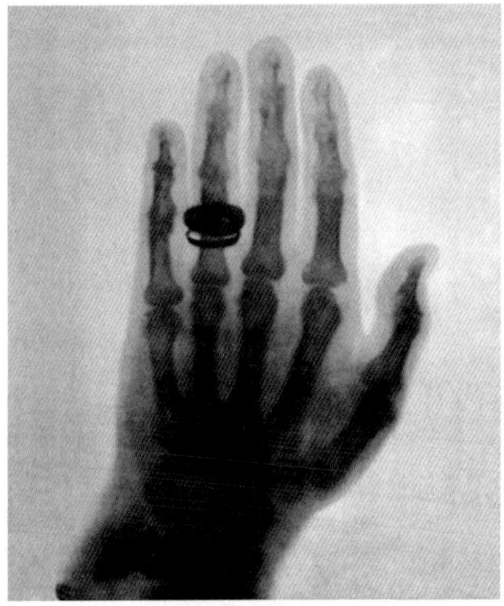

FIGURE 1.3
Early image revealing the bony structures inside the hand. Modern methods allow us to peer inside the body, revealing structure and function of organs, tissues, cells, and molecules.

MR imaging, a modality that was at first as ill suited to the measurement of motion as was early photography. The first major breakthrough in our opinion was the MR tagging technology described by Axel and Chung in Chapter 2, a technology that enables noninvasive estimation of the deformation of the human heart. This technology has been made more accurate and accessible through the harmonic phase (HARP) and zHARP algorithms of Abd-Elmoniem and coworkers (Chapter 3). Additional advances for measuring waves resulting from very subtle vibrations using MR elastography techniques have provided a tool for noninvasive estimation of viscoelastic material properties (see Chapter 4, by Zhang and Muthupillai, and Chapter 5, by Bayly et al.). An important extension of this is recent methods to use the displacement fields associated with these waves for the study of material properties (dualMRI, as described in Chapter 6).

The adaptation of technologies for static imaging to the study of biomechanics is a theme that continues through the next two sections of the handbook. The section on ultrasound techniques in biomechanics highlights how wave motion associated with what was originally a static imaging modality has been harnessed to estimate mechanical properties of tissues noninvasively (Chapter 7, by Wang and coworkers), and how this mechanical information has subsequently developed back into another mechanism for medical imaging to detect stiffness changes associated with tumors (Chapter 8, by Garra). The section on computed tomography describes similar successes in converting the highly static modality of x-ray imaging into a tool capable of quantifying joint kinematics in vivo (Chapter 9, by Moore and coworkers), and also describes progress in challenges associated with x-ray dosages, spatial resolution, and structure–function relationships (Chapter 10, by Reece and coworkers).

The handbook then shifts to optical and high-throughput techniques for estimating strain and stress fields from images of a deforming biological specimen. The first algorithms to automate this process and to extend beyond tracking fiducial markers by hand were a range of texture correlation algorithms. As described by Wang et al. in Chapter 11, even these well-established methods pose challenges when applied to biological tissues. The large strains that can be associated with the testing of biological tissues have motivated the development of hyperelastic warping methods (see Chapter 12, by Veress and coworkers), and needs specific to the imaging of fluid mechanics have motivated the development of broadly applicable particle image velocimetry tools (see Chapter 13, by Taylor et al.). The application of this and other modalities to high-throughput screening is often a requirement for biological specimens due to sample-to-sample variability and warrants special consideration. These considerations are described for a number of imaging modalities in the chapter by Han and coworkers (Chapter 14).

The end goal of applying estimated deformation fields to the prediction of mechanical properties has spawned a field of its own, and the final section of the first part of the handbook is devoted to challenges and successes in this area. The intertwining of imaging techniques into constitutive models is central to the success of these efforts, as can be seen in Chapter 15, on poroelastic and hyperelastic imaging, by Barbone and coworkers. The handbook then continues with a chapter detailing the state-of-the-art stereological methods that are so central to these efforts (Chapter 16, by Mouton). The section is then rounded out with a chapter on the challenges of recursively bridging length scales to estimate deformations in real time in critical applications (see Chapter 17, by Ji et al., on real-time model updating during neurosurgical procedures) and a chapter on techniques for the broader challenge of fitting multiscale mechanical models to MRI data (Chapter 18, by Butz and coworkers).

The second part of the book brings together methods that have been most effective for imaging at the level of cells and individual molecules. The challenges associated with these imaging techniques are intertwined with challenges in preparing a natural or at least realistic environment for the specimen that is being imaged. The first section presents chapters describing methods to translate and extend into three dimensions the vast body of imaging-derived biomechanics data acquired for cells cultured upon two-dimensional substrata. At the heart of this is the need to develop models that are based upon quantitative stereological data for tissues and engineered culture environments. The development of nonlinear optical methods such as coherent Raman scattering, two-photon fluorescence imaging, and second-harmonic generation have been central to these efforts and are described in the chapter by Mansfield and coworkers (Chapter 19). The section then continues with two chapters on ways that cell—extracellular matrix (ECM) interactions (Chapter 20, by Harbin and Han) and cellular mechanics (Chapter 21, by Genin and Elson) can be estimated through integrated modeling, imaging, and mechanical testing.

A challenge with all of these three-dimensional culture systems based upon native ECM proteins is that the models required to estimate cellular mechanics from the results of tests on a tissue construct involve much uncertainty in the modeling of the native ECM proteins. Therefore, as described in the next section, in vitro culture systems that enable cell mechanics to be estimated from the deformation of materials whose mechanics are much better defined have been central to our understanding of cell biomechanics. The three-dimensional cell traction assay of Legant (Chapter 22) is a prime example of this, and much of our basic understanding of how cells respond to steric cues (Chapter 23, by Sheets and coworkers) and cyclic stretching (Chapter 24, by Tondon and coworkers) has been derived from these integrated mechanical testing and imaging systems.

Extending down to the level of individual cells and proteins has required still greater integration of imaging and mechanical testing, and these are described in a section presenting chapters on imaging techniques for cellular and subcellular biophysics. One of the earliest systems for this, dating back to the 1950s, is the aspiration of cells by micropipettes, involving cells drawn into a glass tube at a prescribed level of pressure. The chapter by Liu and Shao (Chapter 25) describes the integration of imaging and mechanics in these systems, and modern extensions that enable the measurement of subcellular biophysics through enhanced imaging. Another early system, whose application to cells dates to the 1970s, is atomic force microscopy (AFM, Chapter 26, by Tadepalli and coworkers). Although AFM is used heavily as an imaging tool, this chapter describes the ways in which AFM's roots in the mechanical testing of cells and proteins are still producing exciting breakthroughs. Mechanical tests at the mesoscale that lies between the length scale of AFM and that of the macroscale techniques of the first part of the handbook, however, have remained elusive until recently. Chapter 28 by Jimenez-Palomar and coworkers describes a recent breakthrough in extending these AFM-like techniques to the testing of bundles of proteins through the combination of AFM-like mechanical testing and other imaging modalities.

The remaining imaging modalities at this length scale involve tracking of fluorescence and analysis of fluctuations or variations in fluorescence that can occur randomly or in response to perturbations. These methods all have their roots in the early work on fluorescence correlation spectroscopy (FCS) and fluorescence photobleaching recovery (FPR, now commonly called *FRAP*, with the *A* standing for *after*) in the 1970s; a historical perspective and an overview of many of these techniques is found in Chapter 40, by Elson, in the last section of the handbook. This section, however, focuses on extensions of such technologies specifically for the understanding of cellular and subcellular biophysics. Chapter 27, by Wolfenson, highlights innovative tools that have been developed for combining FPR with biochemical assays to quantify the mechanics of cellular adhesion. The development of special probes that use Förster resonance energy transfer (FRET) has been central in development of the ability to quantify the mechanisms and rates of mechanotransduction of living cells. This is described in the chapter by Wan and coworkers (Chapter 29). Finally, tracking of the motion of fluorescent particles within cells has led to the first measurements of intracellular viscosity, as described by Dawson and coworkers (Chapter 30; see also Chapter 40, by Elson, for an additional perspective).

Extending another length scale down, the next section in the handbook focuses on the biomechanics and biophysics of the cell nucleus, an area that has blossomed as evidence has emerged that mechanical distortion of living cells may alter gene expression and regulation. A tremendous challenge at this level is imaging without obstructing the function of a cell. The first chapter in this section describes techniques for overcoming challenges in imaging chromatin structure and dynamics. The name "chromatin" was in fact assigned by Flemming in 1881 owing to its special "affinity to dyes" and highlights the challenge in visualizing chromatin in vivo. The chapter by Makhija and coworkers (Chapter 31) describes another descendent of FCS, fluorescence anisotropy imaging, which has recently shed important light on chromatin dynamics. A host of other flurorescence microscopy techniques have recently been developed and brought to bear on the problem of imaging the relationship between nuclear mechanics and DNA transcription, and these are summarized in the chapter by Ivanovska and coworkers (Chapter 32). Recent advances in imaging and strain mapping have enabled such work to extend down into the nuclues itself, as described in the chapter by Henderson and Neu (Chapter 33).

Thereafter, the next section of the handbook progresses still further down the length hierarchy to individual ion channels. Chapter 34, by Macmillan and coworkers, describes the range of techniques available for combined mechanics and imaging study of mechanosenstitive channels, from electrophysiological voltage- and patch-clamp techniques to fluorescence methods such as FRET. The work is challenging, and many unanswered questions persist at this hierarchical level about even the most elementary questions of mechanobiology. Chapter 35, by Donahue and coworkers, summarizes one very important set of mechanobiological questions about the nature of mechanotransduction in bone and the imaging challenges and success in ongoing efforts to implicate specific ion channel in skeletal mechanotransduction.

The final section of the handbook describes the suite of technologies that exist for probing the mechanics of surfaces, membranes, and single molecules. At this level, only a handful of imaging modalities exist, with choices limited largely to scanning probe microscopy, electron microscopy, and fluorescence-based methods. The section starts with a chapter from Lee and coworkers (Chapter 36) on challenges associated with locating, grasping, and stretching of single molecules using AFM. At this scale, magnetic probing and actuation of proteins and surfaces has also been highly effective, and key techniques, successes, and challenges are summarized in the chapter by Galburt (Chapter 38). However, structure and interactions are also important, as many central problems in biological mechanics of molecules involve large assemblages of molecules interacting with each other and with surfaces and interfaces. A suite of techniques exists for quantifying these interactions and is summarized in the chapter on surface force microscopy (Chapter 37) by Ruths and coworkers. The application of high-energy imaging techniques to biological mechanics at the length scale of molecules has been quite limited, largely due to challenges in specimen preparation. However, a series of recent breakthroughs has enabled the application of transmission electron microscopy to the study of bone and partially mineralized tissues. Some of this work is described in the chapter by Klosowski and coworkers (Chapter 39).

The section concludes with two chapters related to fluorescence methods for quantifying the dynamics of proteins. The first is a chapter by Elson (Chapter 40) on FCS and related methods such as FPR and single-particle tracking that take data on random fluctuations or directed changes in fluorescence to derive mechanical parameters that describe convection, diffusion, and viscosity in living systems and on membranes. The second is a chapter by Nekouzadeh and Ghoreishi (Chapter 41) on stereological techniques to quantify fluorescence microscopy images of evolving distributions of protein structures and relate these to mechanical interactions.

This handbook presents a broad-ranging set of topics in a rapidly evolving area. Our goal in bringing these topics together in a single handbook is to provide readers with an appreciation for the usefulness of methods for fundamental studies in biomechanics and biology in the context of normal physiology, growth, disease, damage, regeneration, and repair. We hope that in doing so, the handbook will facilitate the continued extension of these methods to applications across hierarchies.

We thank the numerous individuals who contributed to the completion of this handbook. We are grateful to editors and staff at CRC Press, including Li-Ming Leong. There are countless reviewers who helped in the peer review of the included chapters. Finally, we thank the contributing authors for excellent contributions in their areas of specialization.

Section II

Tissue and Organ Imaging Methods

Magnetic Resonance Imaging Methods

2

MRI Tagging of the Heart

Leon Axel and Sohae Chung

CONTENTS

Magnetic resonance imaging (MRI) with magnetization tagging permits direct imaging of motion patterns within the heart wall, through the motion of the MRI-visible tags. The tag motion can also be revealed through corresponding changes in the phase of the images. Analysis of the tagged images can provide information on the motion and deformation of the heart wall over time, in two or three dimensions.

2.1 Introduction

Magnetic resonance imaging (MRI) is a noninvasive imaging technique, which does not require the use of potentially harmful ionizing radiation; it uses the physical phenomenon of nuclear magnetic resonance (NMR). In NMR, the magnetic moments of certain nuclei, such as hydrogen, which is ubiquitous in the human body, cause them to tend to align with an external magnetic field, leading to a net macroscopic magnetization. This net magnetization is very weak, even in the strong magnetic fields used in conventional MRI systems (typically 1.5 T, on the order of 30,000 times the earth's magnetic field), and would not be readily detectable by conventional methods. However, these same nuclei also have a property, called spin, which causes them to exhibit a resonance phenomenon at a frequency proportional to the external magnetic field. While the equilibrium condition of the nuclear magnetization is aligned with the external magnetic field, when a suitable oscillating magnetic field is applied at the resonance frequency, we can cause the direction of the nuclear magnetization to rotate away from the equilibrium orientation; the amount of this rotation depends on the strength and duration of the pulse of the oscillating radio-frequency (RF) excitation field. If the final orientation of the nuclear magnetization is aligned at an angle to the external magnetic field, the collective nuclear magnetization will precess about the direction of the external field, producing a weak but detectable signal (proportional to the component of the nuclear magnetization perpendicular to the external field) in a suitable receiver; this is the NMR signal. This signal will not persist indefinitely but will decay away exponentially at a rate characterized by a time constant, the "T_2" relaxation time. This time depends on the state of the tissue being imaged, providing a mechanism that can be used for generating image contrast by acquiring the signal at times when there is partial signal decay, with different resulting relative signal intensities. While magnetic field inhomogeneity may cause the signal to apparently decay away more rapidly than the underlying T_2 time, manipulation of the field homogeneity or application of additional RF excitation pulses may cause the signal to transiently reappear, as an "echo." Meanwhile, the component of the nuclear magnetization along the external field exponentially approaches its equilibrium value at a rate characterized by another time constant, the "T_1" relaxation time; the state of incomplete recovery of the magnetization along the external field at times on the order of the T_1 relaxation time is called "saturation." This time also depends on the state of the tissue being imaged, thus providing another mechanism that can be used for image contrast by acquiring the signal at times when there is partial saturation.

To create images from the detected signal, one can encode information on the position of the signal sources by applying pulses of a superimposed magnetic field varying linearly with position (gradients) while the signal is being detected; the resulting changes in the local magnetic field lead to corresponding changes in the resonant frequency of the local signal sources, which can be decoded through the Fourier transform of the net signal for image reconstruction. Multiple such signal detections, each with different frequency encoding, are typically needed for imaging. Motion during the imaging process, such as that related to cardiac motion or respiration, can result in blurring of the images. However, synchronization of the image data acquisition with the cardiac cycle ("gating") can be used to create sharp images of different phases of the cardiac cycle from data acquired over multiple cardiac cycles (assuming that they are all similar). Keeping the imaging time short enough for the subject to suspend respiration allows for elimination of the potentially blurring effects of respiratory motion; otherwise, a similar respiratory gating process can be used to handle it.

Cardiac MRI can provide high-quality images of the moving heart, and it is becoming a valuable tool for clinical and research evaluation of the heart. However, there is a paucity of natural landmarks seen within the heart wall in conventional MRI of the heart, leading to difficulty in directly assessing the motion patterns within the myocardium. A unique aspect of MRI is the ability to noninvasively create MRI-visible synthetic marks or "tags" within tissue, by producing a local perturbation of the tissue magnetization. These tags will persist for times on the order of the relaxation time of the tissue, and they will move with the underlying tissue when it moves. Such regional magnetization tagging was even used prior to the development of MRI, to demonstrate the effects of flow in blood vessels.[1] The motion of the tags, seen in serial MRI images after creation of the tags, will thus directly reveal the corresponding motion of the surrounding tissue. In the case of the heart, the relaxation time of the myocardium is on the order of the cardiac cycle length. This permits study of the whole cardiac cycle by tagging at a reference phase of the cycle (e.g., end diastole); this can be done either with single-shot or real-time imaging, on a beat-to-beat basis, or with segmented cine imaging, combining imaging data acquired from multiple heart cycles to reconstruct an "average" cardiac cycle. Such tags can be created either individually, with selective excitation, or simultaneously, with nonselective excitation (e.g., with spatial modulation of magnetization [SPAMM[2]]). We will here provide an overview of MRI tagging of the heart, citing some representative basic references; we will not attempt to provide an exhaustive review of the (now extensive) related literature (recent reviews include Refs. 3–5; see also Chapters 3–6).

While much can be learned qualitatively about cardiac motion from a simple visual inspection of the motion of the tags in the tagged images, quantitative analysis can provide additional potentially useful information. While manual tracking of tag positions in conventional serial magnitude images is generally too time intensive for most purposes, the tracking process can be speeded up in a variety of ways, working in the image or the Fourier domains. As an alternative approach to tag tracking, one can use modified imaging methods to convert the motion of the tags into corresponding signal phase changes; these phase-extended tagging methods can be used either for a direct phase-based readout of tag motion (e.g., displacement encoding with stimulated echoes [DENSE][6]) or for phase-related magnitude changes due to motion (e.g., strain-encoded MRI [SENC][7]).

The motion data recovered from tagged images can be used to reconstruct 2-D or even 3-D motion patterns within the heart wall. Furthermore, one can separate the reconstructed motion field into both rigid body and deformation motion components. One can then use these data to study the effects of different cardiac states on the heart motion.

2.2 Magnetization Tagging in MRI

2.2.1 Magnetization Tagging

The simplest approach to producing magnetization tags in MRI is to use selective excitation of a single slice perpendicular to the imaged slice, prior to the actual imaging.

In the case of a direct imaging of the excited spins, when using the excited magnetization as the tissue tag (with appropriate refocusing and imaging magnetic field gradients), motion of the tagged spins out of the originally excited plane between the time of excitation and imaging position encoding may occur, for example, due to flow. This will be seen as a corresponding imaged displacement of the bright moving spins away from the bright band of signal representing the stationary spins that remain within the originally excited region.[8] Such excitation tags will show the flow profiles of blood moving through the excited plane within vessels; the amount of time available for the displacement will be limited by the T_2 relaxation time of the moving spins. One can similarly use the selective excitation to produce saturation of the magnetization within the tagging plane.[9] In this case, subsequent images will show the stationary tissue in the tagged plane as a darker band within the corresponding location in the image, due to incomplete recovery of the local magnetization. Any motion of the tagged tissue between the time of tag creation and image acquisition will be seen as an associated motion of the darker tissue band or "stripe" within the image; the amount of time available for the displacement will be limited by the T_1 relaxation time of the moving spins, as well as the specific imaging parameters used.

While the use of selective excitation for producing magnetization tagging is conceptually straightforward, the need for explicit prescription of the location to be tagged is an additional time-consuming step in the image acquisition process. In addition, the relatively small number of such tags that can be readily produced, even with multiplexing of the excitation process, is a limiting factor for the ability to use selective tagging for some potential applications. As an alternative, one can use nonselective excitation methods, such as SPAMM,[2] for producing magnetization tagging. This permits one to efficiently create a whole family of tags at once, without the need to specify individual tagging locations. The excitation process can be adjusted to control the width and shape of the tagged regions. In the basic "1-1" SPAMM tag creation method (Figure 2.1), one can use a minimum set of two RF pulses and a gradient pulse for creating the tags. First, in the preparation phase, an initial excitation RF pulse is used to nonselectively create a uniform excitation within the tissue (Figure 2.1, time B). Next, a gradient pulse is applied along the direction perpendicular to the desired tag plane locations to produce an evolving phase variation of the excited magnetization along its direction (Figure 2.1, time C). The strength and duration of the gradient pulse will determine the spacing (wavelength) of the resulting periodic phase variations (the magnitude of the magnetization will still be uniform). Finally, a second excitation RF pulse is used to detect the new periodic phase variation of the excited spins, by converting it into a corresponding periodic variation in their magnetization (Figure 2.1, time D). One can now detect any interval motion of the tissue by imaging it after a delay and looking

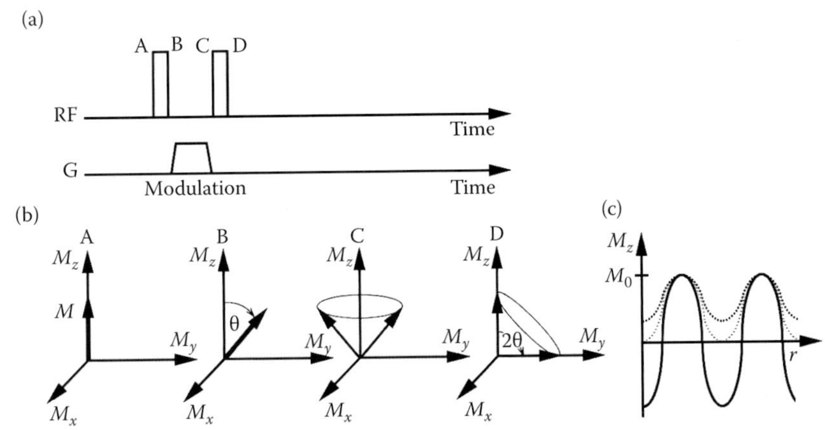

FIGURE 2.1
SPAMM tagging. (a) The timing diagram of a pulse sequence for 1-1 SPAMM shows a set of two RF pulses and a gradient pulse (G) for creating the tags. Letters A–D indicate the corresponding times in (b). (Redrawn from Axel, L. and Dougherty, L., *Radiology*, 171, 1989.) (b) The state of magnetization (M) at different times in the pulse sequence in (a) is diagrammed, as follows: Time A shows the magnetization prior to initiation of the modulation sequence (in [a]). The magnetization (M) is at its equilibrium state along the main magnetic field (in the z direction). Time B shows the magnetization after the first RF pulse. The RF pulse has flip angle θ, with phase along the x-axis in the rotating frame. Time C shows the magnetization after the modulating gradient, with a uniform magnitude but a spatially modulated phase. Time D shows the magnetization after a second RF pulse with the same values as the first, now with a spatially modulated magnitude. (Redrawn from Axel, L. and Dougherty, L., *Radiology*, 171, 1989.) (c) The resulting sinusoidal magnetization pattern along the direction of the gradient, r, is shown, with different possible degrees of modulation of M_z, ranging from full inversion (solid line) to partial saturation (dotted lines). The different initial states could reflect either different flip angles or different degrees of relaxation after an initial inverting modulation. (Redrawn from Axel, L. and Dougherty, L., *Radiology*, 171, 1989.)

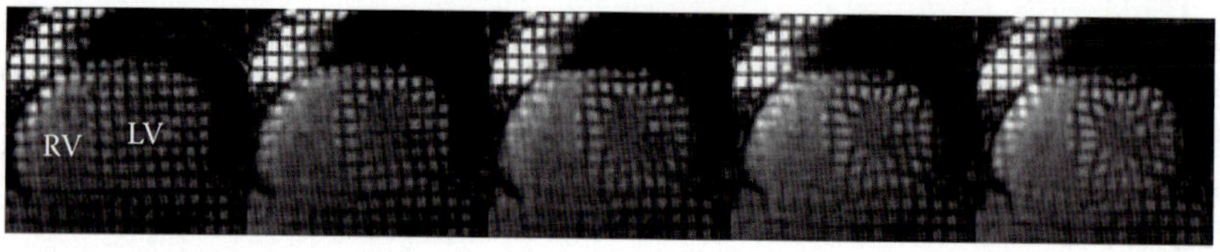

FIGURE 2.2
Grid tagging. This representative example of serial short-axis SPAMM grid-tagged images of the right ventricle (RV) and left ventricle (LV) shows interval motion of the tissue by imaging it with incremental 49 ms increases in delay between SPAMM sequence and imaging.

for corresponding displacement in the images of the tagged (darker) regions (Figure 2.2). The tags will fade according to the T_1 time with the delay after tagging. In addition to the intrinsic fading of the tags due to the T_1 relaxation time, if there is continuous imaging of the tag motion (i.e., cardiac "cine" imaging), the imaging process itself can also contribute to the apparent fading. In such a case, keeping the imaging RF flip angles relatively small can help to reduce this fading,[10] although at a potential cost in signal-to-noise ratio (SNR). Imaging at higher magnetic fields leads to longer myocardial T_1 times, with associated greater tag persistence. The blood within the cavities of the heart and great vessels will also show tags on the initial images, but the relatively large and complex motion of the blood will generally result in rapid loss of the tag images within the blood.

The simple nonselective SPAMM tagging approach described in the previous paragraph leads to sinusoidal profile linear tags, which are somewhat "fuzzy" for visual tracking (although not necessarily undesirable for more automated approaches to tracking); one can use various modifications of the tagging process to alter the tags, if desired. To make sharper tag profiles, one can use a longer train of alternating RF and gradient pulses. In particular, using pulses distributed according to a binomial series can lead to sharper tags (like a power of a sinusoid).[11] Approaches based on digital filter design (e.g., the Shinnar–LeRoux algorithm[12]) can be used for further optimization of the tag shapes. Acquiring two sets of simple SPAMM-tagged images with alternating polarity, and then subtracting the images (complementary spatial modulation of magnetization [CSPAMM][13]), can produce difference images that have a net greater amount of intensity modulation with an associated greater resistance to tag fading but at the cost of longer data acquisition. Producing images with other amounts of relative offset of the tags can produce a greater effective density of spatial sampling with the tags; however, any inconsistency of the positions of the underlying tissue between acquisitions will reduce the utility of such an approach. Other sorts of tagging patterns (e.g., radial tagging,[14–15] ring tagging[16]), which may adapt them more closely to the geometry of the heart, can be produced with more complex combinations of RF and gradient pulses. However, such more complex excitation sequences require more imaging time to play out, and they also require more operator time to set up. In addition, as the heart is only approximately symmetrical about its axis, these tags may still not be locally aligned with the local natural axes of the heart wall.

2.2.2 Magnetization Tag Imaging

With conventional magnetization tagging, one relies on tracking the tags from frame to frame to study the underlying motion. The tagging density achievable in tagged MRI is only limited by the ability of the imaging step to resolve the tags; one can easily create tags that are more finely spaced than the pixel resolution of conventional MRI. Increasing the resolution of the imaging may come at the cost of increased imaging time and decreased SNR. One way to get around this limitation, for effectively denser sampling of the motion, is to use a phase-sensitive approach to detecting the tags. Specifically, in the DENSE[6] imaging approach, if one applies an additional gradient pulse during the imaging step, like the one used for creating the original phase variation in the tagging process, one will be effectively producing a stimulated echo whose local net signal phase will reflect any interval motion between the original tagging gradient and the subsequent imaging "unwrapping" gradient (Figure 2.3). This is like the Stejskal–Tanner pulse sequence,[17] which is used for diffusion MRI, except that in this case, one is interested in the coherent motion component, rather than the incoherent motion of diffusion. Thus, we can see that there is

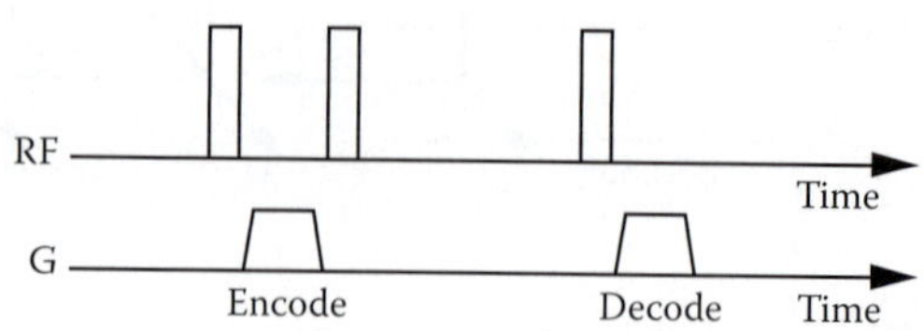

FIGURE 2.3
DENSE displacement encoding. Similarly to the SPAMM sequence in Figure 2.1, in DENSE, the gradient pulse (G) between the first two RF pulses is used to encode position through a modulation of magnetization. In this case, the third RF pulse, together with the decoding gradient pulse, is used to create a stimulated echo. Displacement of the tissue along the direction of the gradient between the second and third RF pulses will produce a corresponding shift of the phase of the stimulated echo.

a strong connection between these apparently different ways to study motion with MRI. Since the motion is being encoded into a single variable, the signal phase, one can only recover one motion component at a time; one also needs to acquire two images in order to find the net phase change due to motion. Thus, DENSE imaging can be relatively time consuming, compared to simple magnitude-tagged MRI. Note that the associated effective phase dispersion within each pixel produced with this approach will result in a net loss of the signal strength of a square root of 2. Similarly, any significant dispersion of the motion within the imaging voxel, for example, due to shearing or rotation, will also result in a further corresponding phase dispersion and loss of signal. In another approach to the use of phase-sensitive imaging to detect tag motion, one can use a controlled refocusing of the phase dispersion within a voxel to detect changes in the spacing of the tags through the associated changes in the net signal magnitude. In particular, in the SENC approach to detecting through-plane components of strain, one can use tagged image acquisitions with the magnetization tags created within the slice, parallel to the plane of the image (Figure 2.4). Imaging the slice with different strengths of the unwrapping gradient will result in different net signal strengths, depending on the amount of residual phase dispersion within the voxel: a decreased tag spacing due to a motion that compresses the slice perpendicular to its face will result in a signal that is higher with a stronger unwrapping pulse, while an increased tag spacing due to a stretching of the slice thickness will result in a signal that is higher with a weaker unwrapping pulse. Note that a rigid body rotation of the slice relative to its initial orientation will also result in an effective increase in the tag spacing in the direction perpendicular to its face, which could falsely look like a deformation.

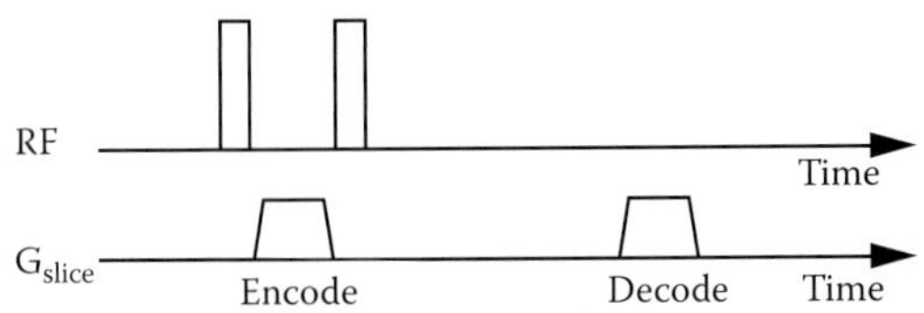

FIGURE 2.4
SENC through-plane deformation encoding. Similarly to the SPAMM sequence in Figure 2.1, in SENC, the gradient pulse between the first two RF pulses is used to encode position through a modulation of magnetization, in this case along the direction of the slice thickness. The "tuning" decoding gradient pulse then produces a periodic phase modulation along this direction. The magnitude of the resulting signal will be strongest when the periodicities match. The difference in the signals found when using a weaker "low tune" or a stronger "high tune" decoding gradient pulse can be used to calculate the amount of compression or stretching of the tissue in the slice between the times of the encoding and decoding pulses.

2.3 Analysis of Tagged MRI

2.3.1 Image-Domain Tag Tracking

With conventional magnitude image tagged MRI, one wants to extract the implicit point correspondence information from the tagged images, by tracking the positions of the tags over time. Note that a linear tag only provides position information along the direction component perpendicular to itself. However, if one has images of two orthogonal families of tags (either acquired in separate images or combined in one image), one can combine the two corresponding sets of serial position information to recover information on the within-plane motion field (fully at the points of intersection of the tags, and by interpolation between them). One approach to automatically finding the positions of the tags within the image is to use a suitable set of matched filters, for example, templates that look like the tag intersections, which can be correlated with the image to look for local maxima that should (ideally) correspond to the locations of such intersections.[18] Another approach to tag tracking is to use "active contours,"[19–20] for example, tag-like smooth lines that can be attracted to local areas of lower intensity in the image.[21] Optical flow-type motion tracking techniques can also be used to look for corresponding areas in serial images[22–23]; note that the tag fading may make this approach more difficult to use here than in other applications. A different sort of automated tag image-tracking approach is to use Gabor filters to search for the tags and their positions within the images.[24–26] A Gabor filter is a sinusoidally modulated Gaussian function (or a Gaussian windowed sinusoid).[27] A pair of such orthogonal filters (e.g., incorporating a sinusoid and a cosinusoid) with the same wavelength as the initial underlying tagged image can be convolved with the image; the local response to the filters can be used to estimate the local phase of the underlying sinusoid, that is, the local position within the quasi-periodic tagging pattern. This local phase will change with the motion of the tags; the change in phase will provide information on the corresponding motion of the tags over time. It may be necessary to do some phase unwrapping of the phase difference images to recover unambiguous positions for the tags. For images with orthogonal tag families, one can use corresponding orthogonally oriented sets of Gabor filters. In general, the tags will change their orientation and spacing over time, as well as their local position, due to the complex motion of the underlying heart wall. One can use a family of Gabor filters, with a suitable range of sinusoidal wavelengths and orientations, to better capture the corresponding tag positions. One can also use the magnitude of the response to the Gabor filters to produce an image with the image of the tags suppressed, along with suppression of the untagged regions; this can be useful for segmentation of the tagged cardiac regions from the image.[28] In fact, the segmentation of the heart wall is often the most time-consuming part of the analysis of tagged MRI data.

2.3.2 Fourier Domain Tag Tracking

As an alternative approach to tracking the tags with the image, one can use a filtering operation carried out in the Fourier domain, for example, harmonic phase MRI (HARP).[29–31] Here, we apply a mask to the Fourier transform of the tagged image, which is centered on the expected location of the strong signal associated with the quasi-periodic tags. An inverse Fourier transform of this masked region will yield a complex image whose local phase depends on the corresponding local relative position of the tag pattern; changes in the position of the underlying tag pattern will be reflected in corresponding changes of the phase of the filtered image. The HARP method will be discussed at greater length in another chapter of this book (Chapter 3).

The use of Gabor filters, described in Section 2.3.1, for a direct image-domain implementation, can also be carried out in the Fourier domain (Figure 2.5); this is generally a more computationally efficient implementation. In the Fourier domain, the use of Gabor filters resembles the HARP approach. However, the use of a bank of Gabor filters, with suitable corresponding sizes and locations in the Fourier domain,[25] can provide more robustness to possible changes in tag spacing and orientation due to motion than the usual use of a fixed filter mask in HARP analysis.

2.3.3 Phase-Modulated Tag Tracking

In the analysis of DENSE image data, one will generally have to provide some unwrapping of the phase information, to deal with the possibility of aliasing in the images. One will also need to deal with relatively low SNR in the images; additional phase dispersion within the voxels will result in further loss of SNR, as mentioned in Section 2.2.2. One can acquire two sets of DENSE images with different relative phases of the refocusing pulses, to produce different relative polarity of the desired and undesired echoes, and then subtract the images to eliminate the undesired echoes, similar to CSPAMM with the use of cosine and sine modulation to eliminate (CANSEL[32]). While this approach can produce difference images that have a net reduction in artifacts, this comes at the cost of longer data acquisition.

In the analysis of SENC data, one will generally need to compare the signal in two images acquired with different values of the unwrapping gradient ("high tune" and "low tune") in order to calculate the corresponding difference in the within-slice tag spacing between the time of tag creation and imaging. Thus, there will be a greater loss of signal in the more poorly unwrapped image, with an associated decrease in the SNR. While one can, in principle, combine the SENC within-slice tagging for assessment of through-plane changes in length with conventional tagging for in-plane motion assessment, the associated loss of signal in the tagged

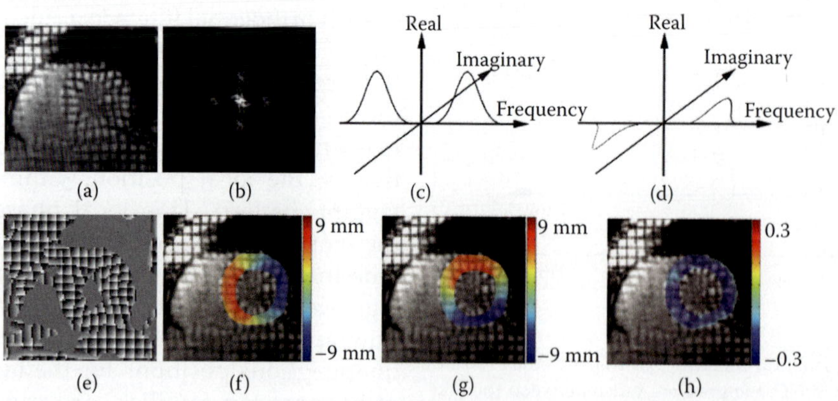

FIGURE 2.5
Fourier domain tag tracking using Gabor filters. (a) A representative cardiac grid-tagged magnetic resonance (MR) image is shown along with (b) the corresponding Fourier transform domain equivalent of the tagged image. A Gabor filter consists of (c) even- and (d) odd-symmetric functions. It can be represented as a shifted Gaussian function in the Fourier domain. A family of Gabor filters can be applied to cover the first harmonic peak efficiently during the motions of the cardiac cycle, by choosing suitable values of relevant parameters, such as magnitude of center frequency, orientation, and the extent of corresponding Gaussian functions in the Fourier domain. (e) The phase image from applying Gabor filters is shown, together with (f) the corresponding displacement maps of the vertical tags and (g) the horizontal tags. (h) The corresponding circumferential strain map is also shown.

regions will impair the ability to recover the through-plane motion information there.

2.4 Tissue Motion Analysis from Magnetization Tagging

2.4.1 Motion Analysis

There are two general frames of reference used in the description of motion, the "Eulerian," referring to a frame of reference fixed in the space through which an object is moving, and the "Lagrangian," referring to a frame of reference fixed to the material of the moving object. One can mathematically relate these frames to each other, transforming back and forth between them, if necessary. When considering the motion of an object like the heart, one is generally more interested in the Lagrangian approach, for example, tracking material points within the heart wall. The use of magnetization tagging lends itself naturally to such an approach, as the tag images move with the underlying tagged material points. However, when carrying out an analysis of the motion from tagged images, one must account for the fact that the images are acquired at fixed locations in space, which is more like the Eulerian picture. As mentioned in Section 2.1, a general motion field can be decomposed into an equivalent combination of rigid body motion and deformation. The rigid body motion can be further decomposed into translation and rotation components. Deformation is commonly described quantitatively through the associated "strain," or local fractional change in length; one can also calculate the "stretch," or ratio of initial and final lengths. In addition to looking at the evolution of these quantities over time, one can look at their corresponding rates of change, that is, velocity and strain rate. Note that while the position change and the associated velocity are vector quantities, the deformation strain and strain rate are tensor quantities, requiring more components to fully describe them. That is, at a given location at a given time, the velocity will have a well-defined magnitude and direction, whereas the tissue at that location may be increasing in length along one direction while it is simultaneously decreasing in length along another direction. The strain is commonly described in terms of the corresponding fractional changes in length along the components of a given local reference frame, along with the associated shears between those directions. Alternatively, it can be described as the "principal strains," that is, the magnitudes and (orthogonal) directions of the greatest and least values of the strain at that location, as well as of the strain along the direction perpendicular to those directions.

2.4.2 Motion Reconstruction

As mentioned in Section 2.3.1, one can use the information from serial images of two orthogonal families of tags to reconstruct the 2-D motion field in the heart wall within the image plane. If one has acquired suitable sets of tagged image data in different locations and orientations, one can similarly reconstruct the 3-D motion field within the heart wall in the imaged volume. One way to approach this is with the finite element method (FEM), wherein one represents the heart as composed of a suitable set of discrete polyhedral elements. If the motion can be considered to vary smoothly within those elements, one need only specify the values of the motion at the vertices of the elements to describe the motion field. If one then adjusts these values so as to minimize the discrepancies between the predicted and observed motion at the imaged locations of the tags, one can recover an estimate of the full motion within the heart wall.[33–35] Note that one may also need to adjust the data to account for possible cardiac position inconsistencies between different image acquisitions. Meshless methods are similar to the FEM approach but do not require an explicit division of the heart into finite elements. Here, the motion of each point is considered to depend on the motion of its closest neighbors. Given a suitable set of specified points within the heart wall, one can adjust their motion to best match the observed and predicted tag motion; the motion of the rest of the heart wall can then be similarly recovered from their motion.[36–37]

2.4.3 Motion Characterization/Classification

In evaluating the motion of the heart that is recovered from tagged MRI, one has the problem of how to analyze a large amount of complex motion data, which are distributed over both space and time. In conventional approaches to evaluating cardiac motion, one typically seeks to reduce cardiac function to a set of simple characteristic numbers, such as the ejection fraction, or assign a simple qualitative motion description, such as hypokinesis, to a set of regions of the heart wall. In contrast, here one is faced with an effective "embarrassment of riches," with a dense set of position (and velocity) vectors (with magnitude and direction values) and shear/strain (and strain rate) tensors within the wall, which may evolve over time in complex and spatially varying ways. Thus, to make this mass of potentially useful motion data more clinically meaningful and accessible, one will likely need to develop new computer-based tools for the automatic analysis and classification of these motion patterns,[38] along with new means to visualize both the "raw" motion data and the results of this analysis.

2.5 Applications of Magnetization Tagging in MRI

2.5.1 Cardiac Motion

The applications of tagged MRI have primarily focused on the left ventricle, reflecting its normally thicker wall and simpler structure (making for an easier tag tracking task), as well as the generally greater level of clinical interest in the left ventricle (Figure 2.6). However, one can also apply tagged MRI for the study of the right ventricle, especially when it is hypertrophied. In carrying out tagged studies of the heart motion, one is somewhat hampered by the relatively limited amount of information available on the expected normal values for the detailed motion of the heart, as well as the relatively time-consuming and cumbersome nature of the existing analysis tools. However, there has already been limited use of tagged MRI for the study of some representative clinical conditions, such as hypertrophic cardiomyopathy,[39–47] myocardial infarction,[48–63] right ventricle hypertrophy,[35,64–66] and dyssynchrony.[67–69] It is to be expected that the development of improved analysis tools and the accumulation of more experience with studying the detailed motion of normal and abnormal hearts will lead to a greater use of tagged MRI for the evaluation of heart disease.

2.5.2 Cardiac-Induced Liver Motion

In performing tagged MRI studies of the heart, it can be seen that the motion of the heart induces related motion

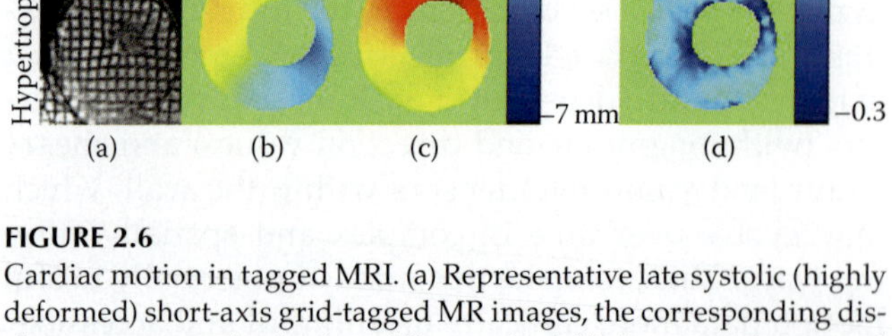

FIGURE 2.6
Cardiac motion in tagged MRI. (a) Representative late systolic (highly deformed) short-axis grid-tagged MR images, the corresponding displacement maps (scales in millimeters) of (b) the vertical tags and (c) the horizontal tags, and (d) the circumferential strain map are shown for, from top to bottom, a representative normal heart (26 years old, female); the heart of a dilated cardiomyopathy patient (51 years old, male); and the heart of a hypertrophied patient (63 years old, female). Significant decreases in motion were found in both patient hearts.

of the left lobe of the liver, on the other side of the diaphragm. The presence of stiffening of the liver due to fibrosis associated with cirrhosis produces qualitative and quantitative changes in the motion observed with tagged MRI[70] (Figure 2.7). One can use the same sorts of tag motion analysis methods for the study of the liver motion as have been applied to the study of the heart, such as Gabor filters. This approach provides a promising new means to noninvasively assess the stiffening effects of liver disease, but there is still very limited experience with its clinical application.

2.5.3 Noncardiac Applications of Tagged MRI

While the principal area of application of tagged MRI has been the cardiovascular system, the use of MRI with magnetization tagging can be applied to many other areas where it is desired to use magnetization tagging for tracking the motion of deforming or flowing structures that can be imaged with MRI, whether biological, industrial, or otherwise. In the case of tagged MRI of cardiovascular structures, one has the natural periodicity of the cardiac cycle driving the motion, allowing one to use data acquired from multiple cardiac cycles, with the tags applied at a particular consistent phase of the cycle (e.g., end diastole), to construct high-resolution "cine" images of the tag motion in a composite "average" cycle. In extending the use of tagged MRI beyond the cardiovascular system, if one has a periodically driven motion, one can use a similar approach for making tagged MRI movies of the motion. Note that the tags only persist for times on the order of the T_1 relaxation time of the images' structure, which limits the reasonably achievable duration of such tagged movies. Note also that it is difficult to precisely repeat voluntary motions many times and that fatigue may rapidly set in, further limiting reproducibility. When studying nonperiodic kinds of motions, one is restricted to "single-shot" types of image acquisitions, with a resulting limitation in the achievable spatial and temporal resolution. In this case, one needs to synchronize the tagging with a particular time in the motion being studied, for example, initiation of an active or passive motion of a musculoskeletal structure. The sound of the gradient pulses associated with the tagging process can provide a convenient auditory cue to initiate a motion. Alternatively, a suitable sensor can detect the initiation of a motion and trigger the tagging and imaging processes.

2.5.4 Potential Future Directions and Technical Challenges to Be Overcome

Potential future directions for tagged MRI in the heart include technical improvements in the imaging and analysis methods, as discussed further at the end of

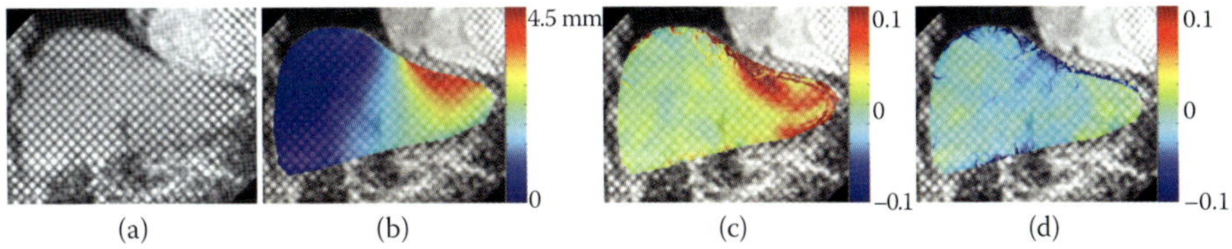

FIGURE 2.7
Cardiac-induced liver motion in tagged MRI. (a) Representative healthy liver grid-tagged MR image with the following superimposed: (b) displacement map (scales in millimeters), (c) first principal strain map, and (d) second principal strain map. (Redrawn from Chung, S. et al., *Magn. Reson. Med.*, 65, 2011.)

this section, and establishing better criteria for both the normal range to be expected for variables describing cardiac motions and the kinds of alterations of these variables found in different clinical conditions. There has been much recent interest in the recently developed "compressed sensing" or sparsity-based approaches to MRI image data acquisition and reconstruction. These have still been relatively little explored in the context of tagged MRI, but they have the potential for both improving the effective spatial and temporal resolution of tagged MRI and providing improved means for extracting the underlying motion field from the tagged images. Although tagged MRI provides the capability of extracting high-dimensional motion variables from the images, such as the tensor variables related to strain and strain rate, these are usually only dealt with in a more reduced form, such as through vector or scalar quantities computed from them. While less intuitive to handle, working more directly with the tensor variables may provide additional useful information on the motion being studied. Similarly, while most tagged MRI data are considered primarily on a frame-by-frame basis, a more fully integrated spatiotemporal approach to the analysis of the tagged images may provide a better understanding of the evolution of the motion and its alteration in different conditions. Machine learning methods may provide useful tools for such analysis and characterization of the very large and complex data sets that tagged MRI can produce; this area is still underexplored. As one develops such improved methods for data analysis, one will also need to acquire and analyze well-controlled image data sets of normal subjects, in order to establish the expected normal ranges of patterns of motion, as well as of well-characterized patients with different clinical conditions, to establish the correlations between the images and the underlying pathophysiology.

Some of the principal remaining technical challenges to be overcome for improving and extending the use of tagged MRI, in the heart and elsewhere, are related to the development of improved methods for both imaging and image analysis. As in other kinds of imaging, there is always a need to speed up the imaging process and increase the temporal and spatial resolution; sparsity-based methods may provide a means to achieve further gains in imaging speed and resolution. There is also always a need to speed up the image analysis process and make it more robust; sparsity-based methods may also be of help here. In particular, a sparsity-based approach to a joint process of image reconstruction and motion characterization may provide a powerful means of improving tagged MRI studies of motion. Providing effective means for visualization of both the complex motions revealed by tagged MRI and the high-dimensional results of its analysis is an inherently difficult challenge, but it will be important to provide such tools, in order to be able to get the most useful information from these kinds of studies.

2.6 Conclusion

Tagged MRI provides a potentially very powerful means of investigating the motion of the heart in health and disease. However, its clinical impact has still been relatively limited. This largely reflects the limitations of the currently available analysis tools. With improvements in these tools and the accumulation of more clinical experience with their use, it is likely that tagged MRI will play an increasing role in the evaluation of the heart.

References

1. Singer, J. R. Blood flow rates by nuclear magnetic resonance measurements. *Science* **130**, 1652–1653 (1959). doi: 10.1126/science.130.3389.1652.
2. Axel, L. & Dougherty, L. MR Imaging of motion with spatial modulation of magnetization. *Radiology* **171**, 841–845 (1989).
3. Axel, L., Montillo, A. & Kim, D. Tagged magnetic resonance imaging of the heart: A survey. *Med Image Anal* **9**, 376–393 (2005). doi:10.1016/j.media.2005.01.003. S1361-8415(05)00016-2 [pii].

4. Simpson, R. M., Keegan, J. & Firmin, D. N. MR assessment of regional myocardial mechanics. *J Magn Reson Imaging* **37**, 576–599 (2013). doi:10.1002/Jmri.23756.
5. Ibrahim, E.-S. H. Myocardial tagging by cardiovascular magnetic resonance: Evolution of techniques—pulse sequences, analysis algorithms, and applications. *J Cardiovasc Magn Reson* **13**(36), 1–40 (2011).
6. Aletras, A. H., Ding, S. J., Balaban, R. S. & Wen, H. DENSE: Displacement encoding with stimulated echoes in cardiac functional MRI. *J Magn Reson* **137**, 247–252 (1999). doi:10.1006/jmre.1998.1676.
7. Osman, N. F., Sampath, S., Atalar, E. & Prince, J. L. Imaging longitudinal cardiac strain on short-axis images using strain-encoded MRI. *Magnet Reson Med* **46**, 324–334 (2001). doi:10.1002/Mrm.1195.
8. Axel, L., Shimakawa, A. & MacFall, J. A time-of-flight method of measuring flow velocity by magnetic resonance imaging. *Magn Reson Imaging* **4**, 199–205 (1986).
9. Zerhouni, E. A., Parish, D. M., Rogers, W. J., Yang, A. & Shapiro, E. P. Human heart: Tagging with MR imaging-a method for noninvasive assessment of myocardial motion. *Radiology* **169**, 59–63 (1988).
10. Reeder, S. B. & McVeigh, E. R. Tag contrast in breath-hold CINE cardiac MRI. *Magn Reson Med* **31**, 521–525 (1994).
11. Axel, L. & Dougherty, L. Heart wall motion: Improved method of spatial modulation of magnetization for MR imaging. *Radiology* **172**, 349–350 (1989).
12. Axel, L. & Shinnar, M. Method for generating optimal pulses for magnetic resonance imaging using spatial modulation of magnetization. *US Patent 5,217,016* (1993).
13. Fischer, S. E., McKinnon, G. C., Maier, S. E. & Boesiger, P. Improved myocardial tagging contrast. *Magn Reson Med* **30**, 191–200 (1993).
14. Bolster, B. D., Mcveigh, E. R. & Zerhouni, E. A. Myocardial tagging in polar coordinates with use of striped tags. *Radiology* **177**, 769–772 (1990).
15. Bosmans, H. et al. Left ventricular radial tagging acquisition using gradient-recalled-echo techniques: Sequence optimization. *Magn Reson Mater Phy* **4**, 123–133 (1996). doi:10.1007/Bf01772519.
16. Spiegel, M. A., Luechinger, R., Schwitter, J. & Boesiger, P. RingTag: Ring-shaped tagging for myocardial centerline assessment. *Invest Radiol* **38**, 669–678 (2003). doi:10.1097/01.rli.0000084888.79706.3a.
17. Stejskal, E. O. & Tanner, J. E. Spin diffusion measurements: Spin echoes in the presence of a time-dependent field gradient. *J Chem Phys* **42**, 288–292 (1965). doi:10.1063/1.1695690.
18. Kraitchman, D. L., Young, A. A., Chang, C. N. & Axel, L. Semi-automatic tracking of myocardial motion in MR tagged images. *IEEE Trans Med Imaging* **14**, 422–433 (1995). doi:10.1109/42.414606.
19. Guttman, M. A., Prince, J. L. & McVeigh, E. R. Tag and contour detection in tagged MR images of the left ventricle. *IEEE Trans Med Imaging* **13**, 74–88 (1994). doi:10.1109/42.276146.
20. Kumar, S. & Goldgof, D. Automatic tracking of SPAMM grid and the estimation of deformation parameters from cardiac MR images. *IEEE Trans Med Imaging* **13**, 122–132 (1994). doi:10.1109/42.276150.
21. Amini, A. A., Chen, Y., Curwen, R. W., Mani, V. & Sun, J. Coupled B-snake grids and constrained thin-plate splines for analysis of 2-D tissue deformations from tagged MRI. *IEEE Trans Med Imaging* **17**, 344–356 (1998). doi:10.1109/42.712124.
22. Dougherty, L., Asmuth, J. C., Blom, A. S., Axel, L. & Kumar, R. Validation of an optical flow method for tag displacement estimation. *IEEE T Med Imaging* **18**, 359–363 (1999). doi:10.1109/42.768845.
23. Prince, J. L. & Mcveigh, E. R. Motion estimation from tagged MR image sequences. *IEEE T Med Imaging* **11**, 238–249 (1992). doi:10.1109/42.141648.
24. Qian, Z., Metaxas, D. N. & Axel, L. Extraction and tracking of MRI tagging sheets using a 3D Gabor filter bank. *Conf Proc IEEE Eng Med Biol Soc* **1**, 711–714 (2006). doi:10.1109/IEMBS.2006.259542.
25. Chung, S., Chen, T. & Axel, L. Evaluation of heart wall motion from tagged MRI using Gabor filter bank. *Proceedings of the International Society for Magnetic Resonance in Medicine* 3598 (2007).
26. Chen, T., Wang, X. X., Chung, S., Metaxas, D. & Axel, L. Automated 3D motion tracking using Gabor filter bank, robust point matching, and deformable models. *IEEE T Med Imaging* **29**, 1–11 (2010). doi:10.1109/Tmi.2009.2021041.
27. Gabor, D. Theory of communications. *IEEE Proc* **93**, 429–457 (1946).
28. Huang, J., Qian, Z., Huang, X., Metaxas, D. & Axel, L. Tag separation in cardiac tagged MRI. *Med Image Comput Comput Assist Interv* **11**, 289–297 (2008).
29. Osman, N. F. & Prince, J. L. Visualizing myocardial function using HARP MRI. *Phys Med Biol* **45**, 1665–1682 (2000). doi:10.1088/0031-9155/45/6/318.
30. Osman, N. F., Kerwin, W. S., McVeigh, E. R. & Prince, J. L. Cardiac motion tracking using CINE harmonic phase (HARP) magnetic resonance imaging. *Magnet Reson Med* **42**, 1048–1060 (1999). doi:10.1002/(Sici)1522-2594(199912)42:6<1048::Aid-Mrm9>3.0.Co;2-M.
31. Osman, N. F., McVeigh, E. R. & Prince, J. L. Imaging heart motion using harmonic phase MRI. *IEEE T Med Imaging* **19**, 186–202 (2000). doi:10.1109/42.845177.
32. Epstein, F. H. & Gilson, W. D. Displacement-encoded cardiac MRI using cosine and sine modulation to eliminate (CANSEL) artifact-generating echoes. *Magn Reson Med* **52**, 774–781 (2004). doi:10.1002/mrm.20232.
33. Young, A. A., Kraitchman, D. L., Dougherty, L. & Axel, L. Tracking and finite element analysis of stripe deformation in magnetic resonance tagging. *IEEE Trans Med Imaging* **14**, 413–421 (1995). doi:10.1109/42.414605.
34. Young, A. A. Model tags: Direct three-dimensional tracking of heart wall motion from tagged magnetic resonance images. *Med Image Anal* **3**, 361–372 (1999). doi:10.1016/S1361-8415(99)80029-2.
35. Hu, Z., Metaxas, D. & Axel, L. In vivo strain and stress estimation of the heart left and right ventricles from MRI images. *Med Image Anal* **7**, 435–444 (2003). doi:10.1016/S1361-8415(03)00032-X.
36. Chen, T., Wang, X., Metaxas, D. & Axel, L. Fast motion tracking of tagged MRI using angle-preserving meshless registration. *Med Image Comput Comput Assist Interv* **11**, 313–320 (2008).

37. Wang, X., Chen, T., Zhang, S., Metaxas, D. & Axel, L. LV motion and strain computation from tMRI based on meshless deformable models. *Med Image Comput Comput Assist Interv* **11**, 636–644 (2008).
38. Qian, Z., Liu, Q., Metaxas, D. N. & Axel, L. Identifying regional cardiac abnormalities from myocardial strains using nontracking-based strain estimation and spatio-temporal tensor analysis. *IEEE Trans Med Imaging* **30**, 2017–2029 (2011). doi:10.1109/TMI.2011.2156805.
39. Maier, S. E. et al. Evaluation of left ventricular segmental wall motion in hypertrophic cardiomyopathy with myocardial tagging. *Circulation* **86**, 1919–1928 (1992).
40. Young, A. A., Kramer, C. M., Ferrari, V. A., Axel, L. & Reichek, N. Three-dimensional left ventricular deformation in hypertrophic cardiomyopathy. *Circulation* **90**, 854–867 (1994).
41. Dong, S. J. et al. Left ventricular wall thickness and regional systolic function in patients with hypertrophic cardiomyopathy. A three-dimensional tagged magnetic resonance imaging study. *Circulation* **90**, 1200–1209 (1994).
42. Kramer, C. M. et al. Regional heterogeneity of function in hypertrophic cardiomyopathy. *Circulation* **90**, 186–194 (1994).
43. Beache, G. M. et al. Intramural mechanics in hypertrophic cardiomyopathy. Functional mapping with strain-rate MR imaging. *Radiology* **197**, 117–124 (1995).
44. Bergey, P. D. & Axel, L. Focal hypertrophic cardiomyopathy simulating a mass: MR tagging for correct diagnosis. *AJR Am J Roentgenol* **174**, 242–244 (2000). doi:10.2214/ajr.174.1.1740242.
45. Ennis, D. B. et al. Assessment of regional systolic and diastolic dysfunction in familial hypertrophic cardiomyopathy using MR tagging. *Magn Reson Med* **50**, 638–642 (2003). doi:10.1002/mrm.10543.
46. Soler, R., Rodriguez, E., Monserrat, L., Mendez, C. & Martinez, C. Magnetic resonance imaging of delayed enhancement in hypertrophic cardiomyopathy: Relationship with left ventricular perfusion and contractile function. *J Comput Assist Tomogr* **30**, 412–420 (2006). doi:10.1097/00004728-200605000-00011.
47. Kim, Y. J. et al. Delayed enhancement in hypertrophic cardiomyopathy: Comparison with myocardial tagging MRI. *J Magn Reson Imaging* **27**, 1054–1060 (2008). doi:10.1002/jmri.21366.
48. Kramer, C. M. et al. Regional differences in function within noninfarcted myocardium during left-ventricular remodeling. *Circulation* **88**, 1279–1288 (1993).
49. Lima, J. A. C. et al. Segmental motion and deformation of transmurally infarcted myocardium in acute postinfarct period. *Am J Physiol-Heart C* **268**, H1304–H1312 (1995).
50. Kramer, C. M. et al. Remote noninfarcted region dysfunction soon after first anterior myocardial infarction—A magnetic resonance tagging study. *Circulation* **94**, 660–666 (1996).
51. Marcus, J. T. et al. Myocardial function in infarcted and remote regions early after infarction in man: Assessment by magnetic resonance tagging and strain analysis. *Magnet Reson Med* **38**, 803–810 (1997). doi:10.1002/mrm.1910380517.
52. Geskin, G. et al. Quantitative assessment of myocardial viability after infarction by dobutamine magnetic resonance tagging. *Circulation* **98**, 217–223 (1998).
53. Croisille, P. et al. Differentiation of viable and nonviable myocardium by the use of three-dimensional tagged MRI in 2-day-old reperfused canine infarcts. *Circulation* **99**, 284–291 (1999).
54. Gotte, M. J. W. et al. Recognition of infarct localization by specific changes in intramural myocardial mechanics. *Am Heart J* **138**, 1038–1045 (1999). doi:10.1016/S0002-8703(99)70068-2.
55. Nagel, E. et al. Cardiac rotation and relaxation after anterolateral myocardial infarction. *Coron Artery Dis* **11**, 261–267 (2000). doi:10.1097/00019501-200005000-00009.
56. Gerber, B. L. et al. Microvascular obstruction and left ventricular remodeling early after acute myocardial infarction. *Circulation* **101**, 2734–2741 (2000).
57. Kramer, C. M. et al. Global alterations in mechanical function in healed reperfused first anterior myocardial infarction. *J Cardiov Magn Reson* **2**, 33–41 (2000). doi:10.3109/10976640009148671.
58. Gotte, M. J. W. et al. Quantification of regional contractile function after infarction: Strain analysis superior to wall thickening analysis in discriminating infarct from remote myocardium. *J Am Coll Cardiol* **37**, 808–817 (2001). doi:10.1016/S0735-1097(00)01186-4.
59. Garot, J. et al. Alterations of systolic left ventricular twist after acute myocardial infarction. *Am J Physiol-Heart C* **282**, H357–H362 (2002).
60. Heijman, E., Strijkers, G. J., Habets, J., Janssen, B. & Nicolay, K. Magnetic resonance imaging of regional cardiac function in the mouse. *MAGMA* **17**, 170–178 (2004). doi:10.1007/s10334-004-0082-4.
61. Gilson, W. D., Yang, Z., French, B. A. & Epstein, F. H. Measurement of myocardial mechanics in mice before and after infarction using multislice displacement-encoded MRI with 3D motion encoding. *Am J Physiol Heart Circ Physiol* **288**, H1491–H1497 (2005). doi:10.1152/ajpheart.00632.2004. 00632.2004 [pii].
62. Ashikaga, H. et al. Electromechanical analysis of infarct border zone in chronic myocardial infarction. *Am J Physiol-Heart C* **289**, H1099–H1105 (2005). doi:10.1152/ajp.00423.2005.
63. Ivancevic, M. K. et al. High-resolution complementary spatial modulation of magnetization (CSPAMM) rat heart tagging on a 1.5 Tesla Clinical Magnetic Resonance System: A preliminary feasibility study. *Invest Radiol* **42**, 204–210 (2007). doi:10.1097/01.rli.0000255646.58831.4b.
64. Young, A. A., Fayad, Z. A. & Axel, L. Right ventricular midwall surface motion and deformation using magnetic resonance tagging. *Am J Physiol* **271**, H2677–H2688 (1996).
65. Fayad, Z. A. et al. Right ventricular regional function using MR tagging: Normals versus chronic pulmonary hypertension. *Magn Reson Med* **39**, 116–123 (1998).
66. Haber, I., Metaxas, D. N. & Axel, L. Three-dimensional motion reconstruction and analysis of the right ventricle using tagged MRI. *Med Image Anal* **4**, 335–355 (2000). doi:10.1016/S1361-8415(00)00028-1.

67. Curry, C. W. et al. Mechanical dyssynchrony in dilated cardiomyopathy with intraventricular conduction delay as depicted by 3D tagged magnetic resonance imaging. *Circulation* **101**, E2 (2000).
68. Helm, R. H. et al. Cardiac dyssynchrony analysis using circumferential versus longitudinal strain: Implications for assessing cardiac resynchronization. *Circulation* **111**, 2760–2767 (2005). doi:10.1161/Circulationaha.104.508457.
69. Lardo, A. C., Abraham, T. P. & Kass, D. A. Magnetic resonance imaging assessment of ventricular dyssynchrony: Current and emerging concepts. *J Am Coll Cardiol* **46**, 2223–2228 (2005). doi:10.1016/j.jacc.2005.09.015.
70. Chung, S., Breton, E., Mannelli, L. & Axel, L. Liver stiffness assessment by tagged MRI of cardiac-induced liver motion. *Magn Reson Med* **65**, 949–955 (2011). doi:10.1002/mrm.22785.

3

Frequency-Domain Analysis of Tagged MRI and Imaging Strategies

Khaled Z. Abd-Elmoniem, Vijay Parthasarathy, and Jerry L. Prince

CONTENTS

Measuring myocardial motion and strain using tagged magnetic resonance imaging provides valuable diagnostic indicators of heart disease. For over a decade, the clinical adoption of magnetic resonance (MR) tagging was limited due to the lengthy image acquisitions involved, multiple long patient breath-holds, and manually intensive postprocessing. The implementations of the harmonic phase (HARP) magnetic resonance imaging (MRI) methodology provided fast and efficient postprocessing tools for tagged images. This chapter presents a review of HARP-tagged MRI analysis and a brief survey of related MR pulse sequences that exploit HARP concepts to provide accelerated imaging, volumetric imaging, and improved quantification of myocardial motion and strain from tagged MR images.

3.1 Introduction

Magnetic resonance (MR) tagging was invented in 1988 as a method to encode displacement in myocardial tissue. For over two decades since that invention, various methods have been proposed to extract displacement and strain information from these images (see also Chapters 2 and 4–6). Most of these methods are based on spatial-domain pixel-based image analysis techniques such as optical flow motion tracking and tag line tracking. In 1999, Osman et al.[1,2] developed the harmonic phase (HARP) method, the first frequency-domain technique for tagged MR image analysis. HARP exploits the fact that displacement information in tagged MRI is encoded in the Fourier domain of the images similar to the way that information is encoded in frequency-modulated (FM) signals in communication theory. This communication theory framework of HARP has changed the way researchers acquire and process tagged MR data. In particular, the HARP framework has been exploited in new MRI techniques for accelerated, three-dimensional acquisition and computation of displacement and strain in moving tissues. In this chapter, we discuss HARP and other imaging and image processing techniques that emerged based on HARP and the FM concept of tagged MRI.

3.2 Spatial Modulation of Magnetization

3.2.1 Spatial Modulation of Magnetization Contrast Generation

The intensities that appear in an MR image depend on both the type and timing of the variable magnetic fields used. Both the type and timing of the variable magnetic fields can be controlled using a set of instructions called a pulse sequence. A pulse sequence is a timed set of instructions that determine the spatially variable magnetic field at any given time, and is usually drawn in the form of a timing diagram as shown in Figure 3.1. This figure shows the pulse sequence to produce a tagging pattern called 1-1 spatial modulation of magnetization (SPAMM).[3,4] The topics discussed in this chapter are best described using the concept of SPAMM tagging, which is one of several ways to produce tag lines in an MR image. Details of other methods of tag production can be found in the literature.[5–7]

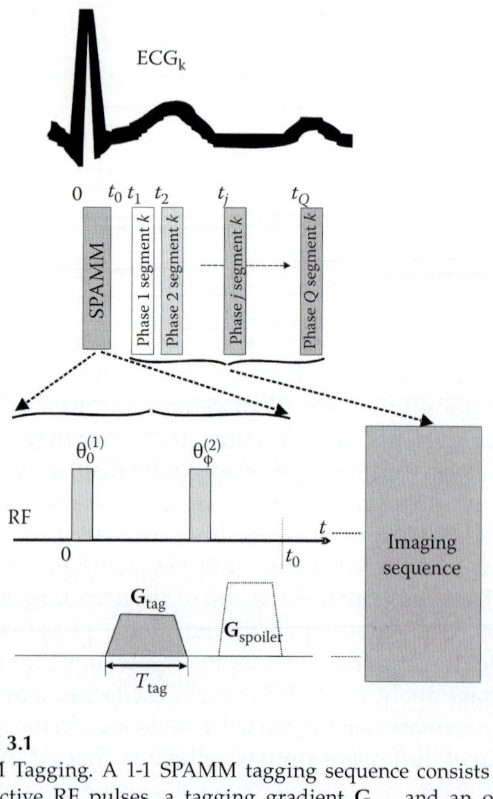

FIGURE 3.1
SPAMM Tagging. A 1-1 SPAMM tagging sequence consists of two nonselective RF pulses, a tagging gradient $\mathbf{G}_{tag}$, and an optional spoiler gradient $\mathbf{G}_{spoiler}$. The RF flip angles $\theta_0^{(1)}$ and $\theta_\varphi^{(2)}$ are independent from each other.

The unique aspect of 1-1 SPAMM is that it produces a sinusoidal pattern of tags, which we will later see is ideally suited for HARP-MRI. As shown in Figure 3.1, two nonselective radio-frequency (RF) pulses are triggered by the *R*-wave of the echocardiogram (ECG) and are played out with flip angles $\theta_0^{(1)}$ and $\theta_\phi^{(2)}$, respectively. The RF pulses are separated in time by a temporally and spatially varying magnetization gradient vector $\mathbf{G}_{tag}$. The first RF pulse flips the tissue magnetization vector $\boldsymbol{M}$ by $\theta_0^{(1)}$ degrees. The tagging gradient $\mathbf{G}_{tag}$ creates phase dispersion in the transverse magnetization component M_{xy}; that is, the magnetization vector fans out to a degree that depends on its spatial location along the direction of $\mathbf{G}_{tag}$. The second RF pulse rotates M_{xy} by $\theta_\phi^{(2)}$ degrees about the *x*-axis with an initial phase of ϕ. Consequently, the component of M_{xy} along the $\pm y$-axes are rotated to the $\pm z$-axes. The longitudinal magnetization M_z at a spatial point $\mathbf{p}$ is now modulated by a sinusoidal pattern. In the most common implementation, the RF pulses have flip angles of 90°, and M_z can be represented mathematically by

$$M_z(\mathbf{p},t_0) = M_z^0 \cos\left(\gamma \int_0^{T_{tag}} \mathbf{G}_{tag}^T \mathbf{p} dt + \phi\right), \quad (3.1)$$

where t_0 is the time immediately after the application of the tagging sequence.

Tag contrast fades with time due to magnetization relaxation. Consequently, tagged magnetization decays, and untagged magnetization recovers. In general, if $\mathbf{G}_{tag}$ is not function of spatial position, the measured SPAMM magnetization at $t = t_k$ is given by

$$M_z(\mathbf{p},t_k) = M_z^0\left[A(t_k)\cos\left(\omega_0^T\mathbf{p} + \phi\right) + B(t_k)\right], \quad (3.2)$$

where $\omega_0 = \left(\gamma \int_0^{T_{tag}} \mathbf{G}_{tag}^T\, dt\right)$ is the frequency of the tag sinusoid, and $A(t_k)$ and $B(t_k)$ are the time-varying coefficients representing the exponential decay of the tagged magnetization and the recovery of the untagged magnetization, respectively. Complementary SPAMM (CSPAMM) is a technique that can reduce the effect of tag fading by preventing the buildup of untagged magnetization $B(t_k)$. In CSPAMM, two SPAMM images are acquired according to Equation 3.2 with the special condition that the first image has $\phi = 0$ and the second image has $\phi = \pi$. Subtraction of these two images removes the signal component caused by untagged magnetization.

Whether SPAMM or CSPAMM is used, image analysis methods for tagged MRI are generally concerned only with the tagged magnetization, so the following assumption can be used:

$$M_z(\mathbf{p},t_k) \approx \rho(\mathbf{p},t_k)\cos\left(\omega_0^T\mathbf{p}\right), \tag{3.3}$$

where $\rho(\mathbf{p},t_k) = M_z^0 A(t_k)$.

3.2.2 Encoding Tissue Motion Using Tagging

A tagged image can be written as $I(x, y, t; z_0)$ where z_0 is the z coordinate specified by slice selection and spatial encoding. Since heart tissue is moving, the magnetization residing at $\mathbf{x}_{z_0} = [x, y, z_0]^T$ at current time t must have moved from its original location $\mathbf{p}(\mathbf{x}_{z_0})$ at time t_0; $\mathbf{p}(\mathbf{x}_{z_0})$ is known as the reference map. The tagged image $I(x, y, t_k; z_0)$ is a map of the longitudinal magnetization of $M_z(\mathbf{x}_{z_0}, t_k)$. Longitudinal magnetization is a material property of the tissue so that when the tissue moves, the longitudinal magnetization moves along with it. Therefore, the longitudinal magnetization satisfies

$$M_z(\mathbf{x}_{z_0},t_k) = M_z(\mathbf{p}(\mathbf{x}_{z_0},t_k),t_k), \tag{3.4}$$

$$M_z(\mathbf{x}_{z_0},t_k) = \rho\left(\mathbf{p}(\mathbf{x}_{z_0},t_k)\right)\cos\left(\omega_0^T\mathbf{p}(\mathbf{x}_{z_0},t_k)\right). \tag{3.5}$$

This magnetization can be expressed in terms of the displacement of the tissue $\mathbf{u}(\mathbf{x}_{z_0}, t_k)$, which is defined as the vector difference between the current position and its reference position:

$$\mathbf{u}(\mathbf{x}_{z_0},t_k) = \mathbf{x}_{z_0} - \mathbf{p}(\mathbf{x}_{z_0},t_k). \tag{3.6}$$

Therefore, using Equations 3.5 and 3.6, the map of longitudinal magnetization on the imaging slice can be rewritten in terms of its displacement:

$$M_z(\mathbf{x}_{z_0},t_k) = \rho(\mathbf{x}_{z_0},t_k)\cos\left(\omega_0^T(\mathbf{x}_{z_0} - \mathbf{u}(\mathbf{x}_{z_0},t_k))\right), \tag{3.7}$$

or, simply,

$$M_z(\mathbf{x},t_k) = \rho(\mathbf{x},t_k)\cos\left(\omega_0^T(\mathbf{x} - \mathbf{u}(\mathbf{x},t_k))\right). \tag{3.8}$$

3.3 HARP Model of MR Tagging

3.3.1 Frequency Modulation of Tissue Motion

A vertically tagged image as shown in Figure 3.2a can be considered a product of two spatial signals:

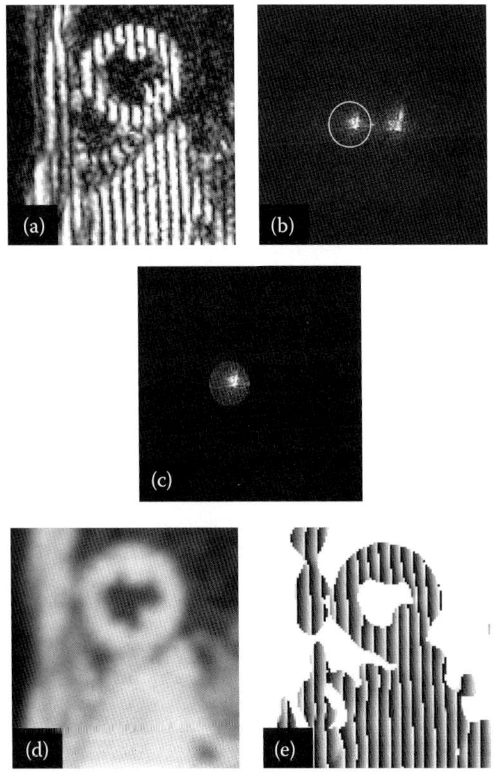

FIGURE 3.2
Harmonic phase filtering. (a) A vertically tagged CSPAMM image. (b) Image Fourier spectrum with the HARP band-pass filter as white circle. (c) The Fourier spectrum of one harmonic peak. (d) HARP magnitude image, $\rho(\mathbf{x}, t_k)$. (e) HARP phase $\phi^v(\mathbf{x}, t_k)$ masked using the magnitude image.

the anatomy signal $\rho(\mathbf{x}, t_k)$, Figure 3.2d, and the tagging signal $\cos(\omega_x x - \omega_x u_x(\mathbf{x}, t_k))$. The multiplication of these two signals in the spatial domain is analogous to the convolution of their Fourier transforms. Figure 3.2b shows the spectrum of a vertically tagged image. It consists of two harmonic spectral peaks, whose centers are located at the first harmonic frequencies of the tagging sinusoid. The premise of HARP-MRI is that most of the spectral energy corresponding to the motion of the tissue is localized around either one of the harmonic spectral peaks. A band-pass filter (see circle in Figure 3.2b) is used to filter one of the harmonic peaks, and the rest of the spectrum is zero padded. The zero-padded spectrum, Figure 3.2c, is an asymmetric spectrum. The inverse Fourier transform of this one-sided spectrum results in a complex image called the vertical harmonic image I_{HARP}^v, which can be expressed as

$$I_{\text{HARP}}^v(\mathbf{x},t_k) = \rho(\mathbf{x},t_k)e^{j\phi^v(\mathbf{x},t_k)}, \tag{3.9}$$

where the vertical HARP map is given by

$$\phi^v(\mathbf{x}, t_k) = \omega_x x - \omega_x u_x(\mathbf{x}, t_k). \tag{3.10}$$

A similar analysis can be carried out for the horizontally tagged images, yielding the horizontal HARP map

$$\phi^h(\mathbf{x}, t_k) = \omega_y y - \omega_y u_y(\mathbf{x}, t_k). \quad (3.11)$$

These two equations represent the basic concept of HARP—the HARP is linearly related to 2-D tissue displacement. This concept can be visually corroborated by comparing the tagged image in Figure 3.2a and the HARP image in Figure 3.2e. Note how the phase wraps in the HARP image exactly coincide with the troughs of the tagging pattern. The phase wrap occurs because the phase of a complex image is computed between $-\pi$ and π. As a result, the actual extracted phase images are wrapped versions of ϕ^v and ϕ^h:

$$\varphi^h = \mathcal{W}(\phi^h), \quad \varphi^v = \mathcal{W}(\phi^v), \quad (3.12)$$

where the nonlinear function $\mathcal{W}(.)$ is defined by

$$\mathcal{W}(\phi) = \mathrm{mod}(\phi + \pi, 2\pi) - \pi. \quad (3.13)$$

The displacement $\mathbf{u}(\mathbf{x}, t_k)$ can be used for various indices of motion like trajectories of tissue points, Eulerian strain, velocities, and Lagrangian strains.[1]

It is important to understand that while it is apparent that the motion information is encoded in a small portion of *k*-space, higher-spatial-frequency information encoded in *k*-space represents high-resolution details of the anatomical boundaries in the image. However, choosing a larger filter size to include this high-resolution detail will also include interfering information from the other harmonic peaks. Therefore, typically, a compromise must be made between higher-resolution anatomy and strain mapping and between an acceptable amount of artifacts in the resulting strain maps.

3.3.2 HARP In-Plane Motion Tracking

HARP motion tracking[2] is based on the fact that each material point $\mathbf{p}$ has a locally unique HARP vector $\boldsymbol{\varphi} \stackrel{\text{def}}{=} [\varphi^h \; \varphi^v]^T$, that is,

$$\boldsymbol{\varphi}(\mathbf{X}(\mathbf{p}, t_k), t_k) = \boldsymbol{\varphi}(\mathbf{X}(\mathbf{p}, t_{k+1}), t_{k+1}) \quad (3.14)$$

where $\mathbf{X}(\mathbf{p}, t_k)$ is the position of the point $\mathbf{p}$ at time t_k. Therefore, as a tissue point moves, its HARP vector value moves with it. An in-plane path line or trajectory can be generated by tracking that phase value throughout the cine images (see Figure 3.3). Starting from an arbitrarily selected point $\mathbf{p}$ at time t_k, $\boldsymbol{\varphi}(\mathbf{X}(\mathbf{p}, t_k), t_k)$ is evaluated. In the next time frame at t_{k+1}, a local iterative search is carried out to find $\mathbf{x}$ that satisfies

$$\boldsymbol{\varphi}(\mathbf{x}, t_{k+1}) - \boldsymbol{\varphi}(\mathbf{X}(\mathbf{p}, t_k), t_k) = 0. \quad (3.15)$$

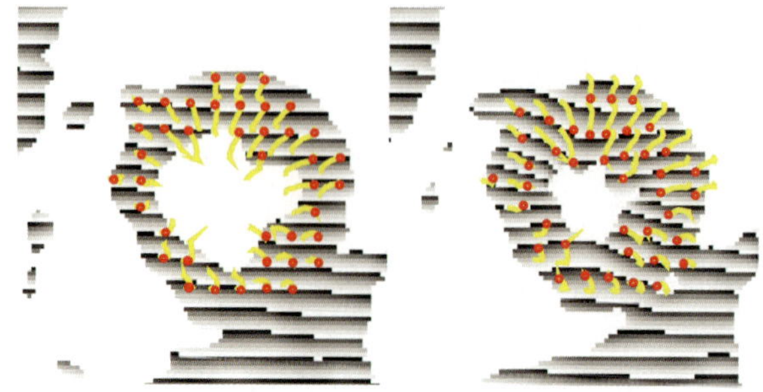

FIGURE 3.3
HARP tracking. Trajectories of arbitrary pooled points superimposed on φ^h at two different cardiac phases. Yellow curves represent the path lines throughout the systole period of the cardiac cycle. Red points represent $\mathbf{X}(\mathbf{p}, t)$, the location of the material point, at the specific cardiac phase in view.

3.3.3 Eulerian Strain Measurement

The deformation gradient tensor $\boldsymbol{F}$ completely characterizes the local deformation of the tissue from one configuration to another.[8] If $\mathbf{dX}$ is the size of an infinitesimally small piece of tissue that deforms to $\mathbf{dx}$ at a later time, then

$$\mathrm{d}\mathbf{X} = \boldsymbol{F}^{-1}\,\mathrm{d}\mathbf{x}. \quad (3.16)$$

It can be shown that

$$\boldsymbol{F}^{-1} = \boldsymbol{\Omega}^{-T}\nabla^*\boldsymbol{\varphi}, \quad (3.17)$$

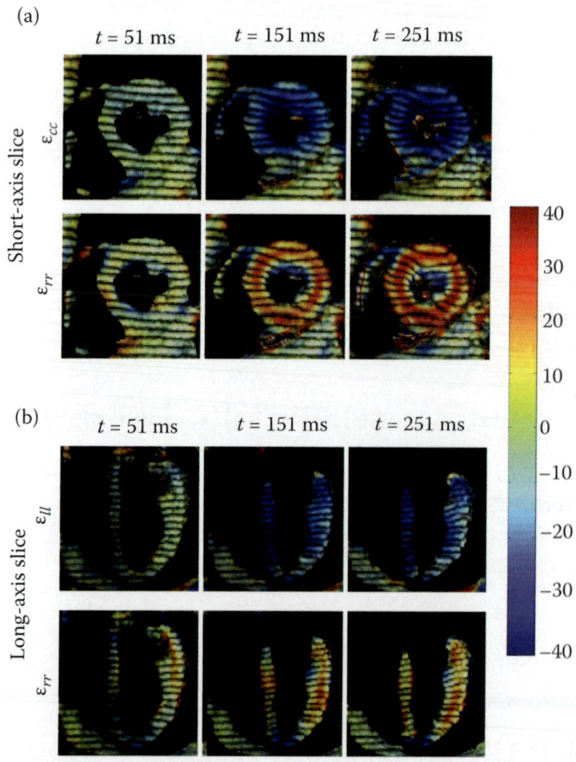

FIGURE 3.4
Eulerian strain maps at three different cardiac phases. (a) Circumferential strain ε_{cc} and radial strain ε_{rr} of a short-axis slice. (b) Longitudinal strain ε_{ll} and radial strain ε_{rr} calculated on a long-axis slice.

where $\Omega = \begin{pmatrix} \omega_x & 0 \\ 0 & \omega_y \end{pmatrix}$, and $\nabla^*\varphi = \begin{pmatrix} \nabla^*\varphi^h \\ \nabla^*\varphi^v \end{pmatrix}$ and the modified gradient operator is defined by

$$\nabla^*\varphi = \begin{cases} \nabla\varphi & |\nabla\varphi| \le |\nabla\mathcal{W}(\varphi+\pi)|, \\ \nabla\mathcal{W}(\varphi+\pi) & \text{otherwise.} \end{cases} \quad (3.18)$$

A measure of apparent strain in the unit direction **dr** is given by

$$\varepsilon(\mathbf{x}, t_k; \mathbf{dr}) = 1 - F^{-1}\,\mathbf{dr}. \quad (3.19)$$

Due to the prolate spheroid shape of the left ventricle, the most relevant strains in the short-axis slices are the circumferential strain ε_{cc} and the radial strain ε_{rr}, which are calculated in the circumferential and radial directions, respectively. In long-axis slices, radial strain ε_{rr} and longitudinal strain ε_{ll} (also called shortening) are commonly calculated in the radial and longitudinal directions, respectively (see Figure 3.4).

3.4 Real-Time Monitoring of Cardiac Regional Function

3.4.1 Motivation

HARP analysis of tagged cardiac MRI requires information from a single *k*-space harmonic peak, while conventional imaging unnecessarily acquires large regions in *k*-space. This suggests that a substantial reduction in the imaging time can be achieved by restricting the acquisition in *k*-space to a small region centered on the desired harmonic peak. That amount of data is sufficient for quantification of myocardial motion and strain.

3.4.2 Two Heartbeat HARP-MRI

Sampath et al.[9] and Abd-Elmoniem et al.[10] developed the FastHARP imaging sequence and myocardial regional function monitoring system to exploit the small amount of Fourier information that is required by HARP postprocessing. As depicted in Figure 3.5, the image acquisition windows during a typical FastHARP scan is centered on the rightmost and topmost harmonic peaks when the imaged slice is tagged in the vertical and horizontal directions, respectively. Using this scheme, full visualization of the 2-D in-plane myocardial mechanical function is now possible from just two cardiac cycles. An accelerated algorithm for image reconstruction and strain computation and visualization in less than 10 ms was also developed in order to enable real-time strain monitoring.[10]

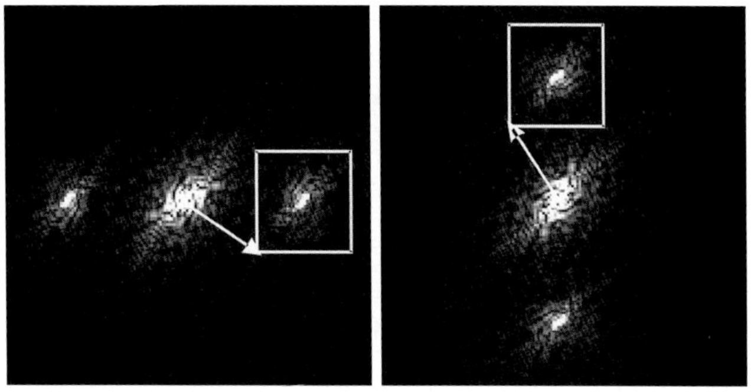

FIGURE 3.5
SPAMM image in the Fourier domain. Typical FastHARP Fourier-space harmonic peak acquisition windows with (left) vertical tags and (right) horizontal tags.

The FastHARP sequence as shown in Figure 3.6 is optimized for the real-time acquisition of the *k*-space harmonic peaks. A train of 15 tagged MR images is acquired every heartbeat at a frame rate of 25 frames per second with orthogonally oriented tagging directions every successive heartbeat. The data from the two most recent heartbeats are combined using HARP analysis to yield images showing the evolution in the in-plane regional motion and strain in the heart. The real-time recording of regional myocardial function with only one heartbeat delay was demonstrated to be significantly valuable in a canine stress testing experiment involving balloon

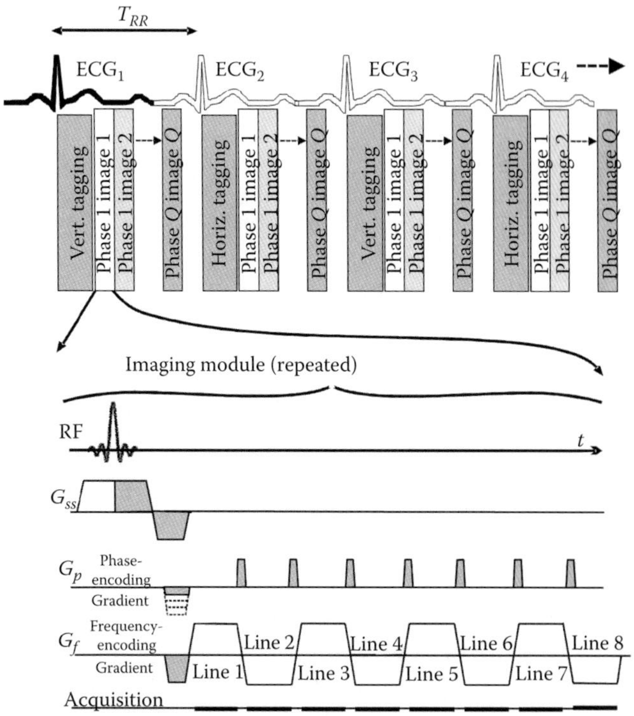

FIGURE 3.6
The FastHARP pulse sequence. Vertical and horizontal tags are applied in successive heartbeats, and the corresponding spectral peaks in Fourier space are acquired in a corresponding alternating fashion.

inflation in the left anterior descending coronary artery.[11] The study demonstrated that the earliest possible detection of ischemia was by HARP-MRI (9.5 s) as compared to cine MRI (33 s) and ECG alterations (63 s).

3.5 Analysis of Multiple Spectral Peaks

3.5.1 Peak-Combination HARP

Due to magnetic field inhomogeneity, there is always a residual phase φ^e superimposed on the image. Therefore, in the presence of field inhomogeneity artifacts, the phase images in Equations 3.10 and 3.11 (corresponding to the positive harmonic peaks) should be replaced by Equations 3.20 and 3.21.

As well, the phase images from the negative harmonic peaks are obtained from Equations 3.22 and 3.23

$$\phi^{v+}(\mathbf{x}, t_k) = \varphi^e(\mathbf{x}, t_k) + \omega_x x - \omega_x u_x(\mathbf{x}, t_k), \quad (3.20)$$

$$\phi^{h+}(\mathbf{x}, t_k) = \varphi^e(\mathbf{x}, t_k) + \omega_y y - \omega_y u_y(\mathbf{x}, t_k). \quad (3.21)$$

$$\phi^{v-}(\mathbf{x}, t_k) = \varphi^e(\mathbf{x}, t_k) + \omega_x x - \omega_x u_x(\mathbf{x}, t_k), \quad (3.22)$$

$$\phi^{h-}(\mathbf{x}, t_k) = \varphi^e(\mathbf{x}, t_k) + \omega_y y - \omega_y u_y(\mathbf{x}, t_k). \quad (3.23)$$

One way to suppress these artifacts and improve the quality of strain maps is to utilize both positive and negative harmonic peaks in strain computations.[12] The phase-combination images are

$$\bar{\phi}^v(\mathbf{x}, t_k) = (\phi^{v+} - \phi^{v-})/2, \quad (3.24)$$

$$\bar{\phi}^h(\mathbf{x}, t_k) = (\phi^{h+} - \phi^{h-})/2. \quad (3.25)$$

Studies have demonstrated that with peak combination, the artifactual phase is suppressed, thus improving the accuracy and reproducibility in HARP evaluation.

3.6 Total Removal of Unwanted Harmonic Peaks

3.6.1 Motivation

Consider a single SPAMM data set with line tags. The Fourier transform of that data set will show three peaks, as shown in Figure 3.5: one peak in the center of the spectrum and two conjugate peaks due to the sinusoidal tagging. The central peak does not carry any motion information and simply interferes with the other peaks. A typical solution for this problem is to use a small HARP filter size, which reduces interference but also reduces the dynamic range and worsens the resolution. In another solution, the phase images corresponding to the negative and positive harmonic peaks in *k*-space are combined before HARP evaluation to eliminate any spurious phase and to reduce ringing artifacts. On the other hand, CSPAMM addresses these limitations by removing the center peak, as previously shown in Equation 3.3 and Figure 3.2b.

A more general approach is to isolate all the acquired harmonic peaks individually, suppress the interference from all the harmonic peaks on each other, and utilize the motion information in each separated peak in calculating a more robust estimate of displacement and strain. The realization of these ideas has been proposed via the total removal of unwanted harmonic peaks (TruHARP).[13,14]

3.6.2 Theory

Vertical and horizontal cosine-modulated SPAMM images are given by

$$I_{A_v}(\mathbf{x}, t) = \rho e^{j\varphi^e}[1 + \cos(\omega x - \omega u_x)] \quad (3.26)$$

and

$$I_{A_h}(\mathbf{x}, t) = \rho e^{j\varphi^e}[1 + \cos(\omega y - \omega u_y)]. \quad (3.27)$$

In TruHARP, in addition to I_{A_h} and I_{A_v}, the following three specially tagged SPAMM images are also acquired:

$$I_{B_v}(\mathbf{x}, t) = \rho e^{j\varphi^e}[1 - \cos(\omega x - \omega u_x)], \quad (3.28)$$

$$I_{C_v}(\mathbf{x}, t) = \rho e^{j\varphi^e}[1 + \sin(\omega x - \omega u_x)], \quad (3.29)$$

$$I_{C_h}(\mathbf{x}, t) = \rho e^{j\varphi^e}[1 + \sin(\omega y - \omega u_y)]. \quad (3.30)$$

From these five image sequences, the central peak image I_{dc}, the positive and negative vertical harmonic images $I^{v\pm}_{\text{HARP}}$, and the horizontal harmonic image harmonic $I^{h\pm}_{\text{HARP}}$ are extracted as follows (without the use of band-pass filtering):

$$I_{dc} = (I_{A_v} + I_{B_v})/2, \quad (3.31)$$

$$I^{v\pm}_{\text{HARP}} = (I_{A_v} - I_{B_v})/4 \mp j(I_{C_v} - I_{dc})/2, \quad (3.32)$$

$$I^{h\pm}_{\text{HARP}} = (I_{A_h} - I_{dc})/2 \mp j(I_{C_h} - I_{dc})/2. \quad (3.33)$$

The TruHARP framework has been implemented in a single breath-hold MR tagging protocol permitting

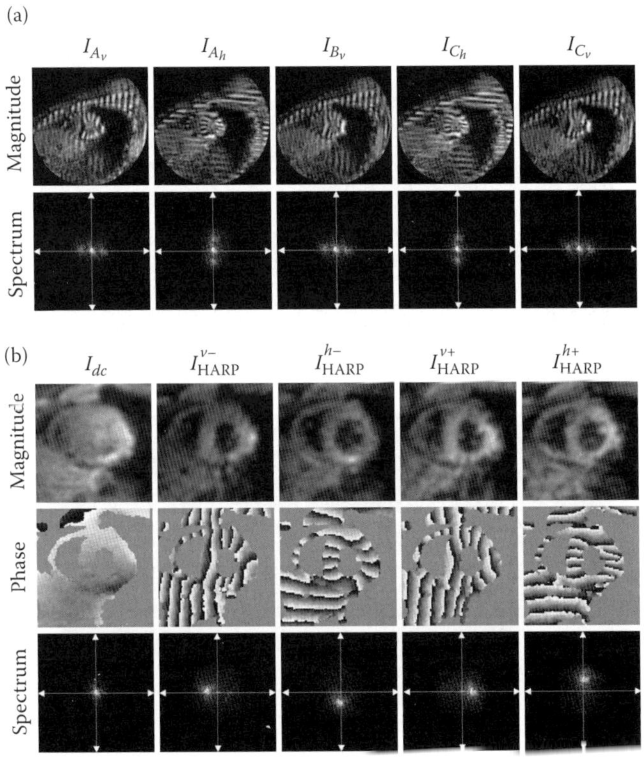

FIGURE 3.7
An example of (a) a TruHARP data set and (b) the calculated harmonic images over the part of the image containing the heart as marked by the white box in (a).

two-dimensional strain resolution of 4.37 mm that is only limited by the resolution and quality of the acquired images instead of being limited by the tag spacing or the proximity of harmonic peaks. An example data set as shown in Figure 3.7 demonstrates the concept and the ability to resolve the individual harmonic peak images. Images can then be used for a high-resolution displacement and strain quantification.[13,14]

3.7 Three-Dimensional Motion Encoding

3.7.1 Motivation

So far in this chapter, the discussion has been focusing on the measurement of apparent 2-D motion and strain in an imaged plane that is fixed during the acquisition. That setup ignores the fact that cardiac motion and strain are three-dimensional properties. This fixed-plane imaging approach has two limitations. First, due to the through-plane motion, the slice imaged at that plane location differs from one time point to another. Second, the motion and strain measured only demonstrate a partial incomplete picture of the actual myocardial mechanical function. Extending 2-D tagging to 3-D has been problematic for a long time due to the lengthy acquisition time. To quantify 3-D strain, methods were proposed to acquire multiple orthogonal image slices covering the volume of interest.[15–19] Aside from patient discomfort due to the long multiple breath-holds required for data acquisition, long imaging times often cause image artifacts and inaccuracies due to image registration errors.

zHARP is a fast tagged MRI pulse sequence and post-processing method that can image and track the 3-D displacement of all points in a single slice (requiring only two breath-holds).[20,21] zHARP also computes and tracks the 3-D strain tensor at any material point in a stack of parallel zHARP images.

3.7.2 Theory

zHARP is an MRI tagging pulse sequence that tracks the imaged slice during the cardiac cycle similarly to slice-following CSPAMM[22] (SF-CSPAMM). It extends SF-CSPAMM to encode the through-plane displacement, as shown in Figure 3.8, by adding a through-plane phase-encoding (z-encoding) gradient G_{enc} to the refocusing of the slice excitation directly before data acquisition. As a result, the acquired vertically and horizontally tagged data are now formulated as follows:

$$I_v(\mathbf{x},t) = \rho e^{j\varphi^e} e^{+j\kappa u_z} \cos(\omega x - \omega u_x), \qquad (3.34)$$

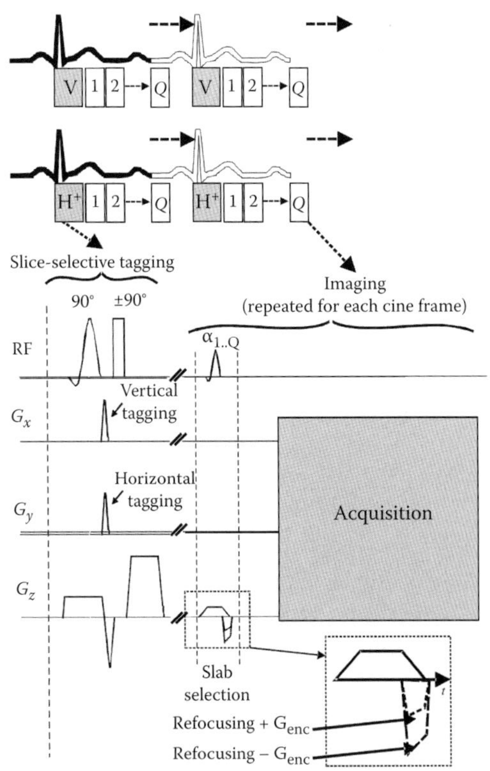

FIGURE 3.8
Single-slice zHARP pulse sequence.

and

$$I_{\mathrm{h}}(\mathbf{x},t) = \rho e^{j\varphi^e} e^{-j\kappa u_z} \cos(\omega y - \omega u_y), \quad (3.35)$$

where $\kappa = \gamma G_{enc}$ is the through-plane or z-encode frequency and γ is the gyromagnetic ratio. After harmonic peak filtering, four harmonic images are obtained:

$$I_{zHARP}^{v\pm} = \frac{\rho}{2} e^{j(\varphi^e + \kappa u_z \pm \omega u_x)}, \quad (3.36)$$

$$I_{zHARP}^{h\pm} = \frac{\rho}{2} e^{j(\varphi^e - \kappa u_z \pm \omega u_y)}, \quad (3.37)$$

From the phases in Equations 3.36 and 3.37, the three phases related to motion in 3-D can be extracted and arranged as follows:

$$\mathbf{\Phi}(x,y,t) = \begin{pmatrix} \phi_x \\ \phi_y \\ \phi_z \end{pmatrix} = \begin{pmatrix} \mathcal{W}(2\omega u_x) \\ \mathcal{W}(2\omega u_y) \\ \mathcal{W}(4\kappa u_z) \end{pmatrix}$$

$$= \begin{pmatrix} \angle\left(I_{zHARP}^{v+} \times \overline{I_{zHARP}^{v-}}\right) \\ \angle\left(I_{zHARP}^{h+} \times \overline{I_{zHARP}^{h-}}\right) \\ \angle\left(I_{zHARP}^{v+} \times I_{zHARP}^{v-} \times \overline{I_{zHARP}^{h+}} \times \overline{I_{zHARP}^{h-}}\right) \end{pmatrix}. \quad (3.38)$$

The 3-D displacement vectors (u_x, u_y, u_z) can be extracted from these phases, and then the 3-D path line of each material point in an imaged slice can be extracted.[20] In this way, 3-D motion is computed, although only 2-D slice data are acquired.

If a stack of two or more slices is available, the 3-D displacement gradient $\nabla\mathbf{u}(\mathbf{x}, t)$ can be calculated as follows:

$$\nabla\mathbf{u}(\mathbf{x},t) = \begin{pmatrix} \nabla u_x \\ \nabla u_y \\ \nabla u_z \end{pmatrix}$$

$$= \begin{pmatrix} \frac{\omega^{-1}}{2\Delta x} & 0 & 0 \\ 0 & \frac{\omega^{-1}}{2\Delta y} & 0 \\ 0 & 0 & \frac{\kappa^{-1}}{2\Delta z} \end{pmatrix} \nabla^{*}\mathbf{\Phi}(\mathbf{x},t). \quad (3.39)$$

where Δx, Δy, and Δz are the physical spacings between the adjacent points in the x, y, and z directions, respectively. The 3-D displacement gradient $\nabla\mathbf{u}(\mathbf{x}, t)$ can then be used to calculate the 3-D strain tensor.[21] An example of 3-D displacement path lines and 3-D strain tensor components are illustrated in Figure 3.9.

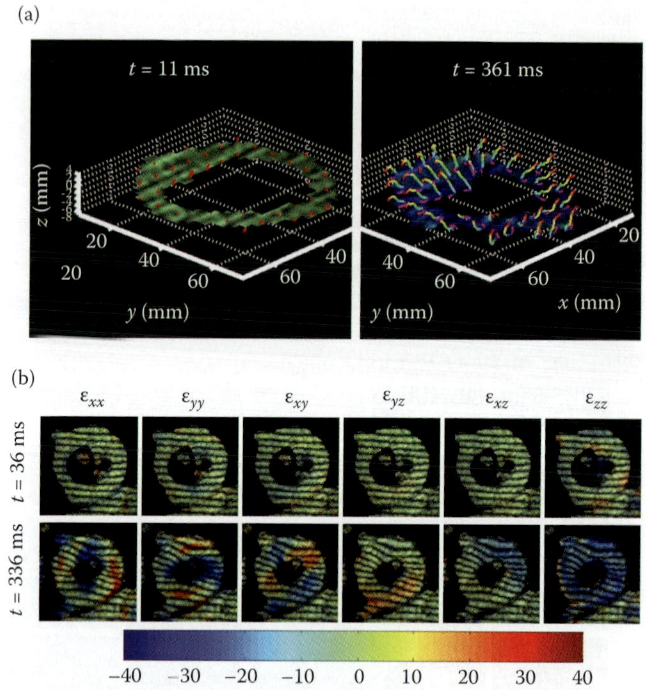

FIGURE 3.9
(a) 3-D representation of ε_{zz} in a short-axis slice and the motion path lines of selected points. (b) 3-D strain tensor components superimposed on their corresponding tagged short-axis slice at two time points during the cardiac cycle.

3.8 Related Development

Several techniques have been developed to improve various aspects of HARP-MRI, including accelerated imaging, 3-D coverage, model-based reconstruction, and so forth. For example, to capture the complex 3-D cardiac motion of the whole heart throughout the cardiac cycle, multiple slices are typically acquired at multiple orientations. This approach is prone to misregistration and requires long acquisition times. Multiple approaches have been proposed to encode the 3-D cardiac displacement of the whole heart and acquire the 3-D data or the 3-D harmonic peaks.[23,24] Meanwhile, strain-encoding (SENC) imaging encodes through-plane strain within the imaged slice and directly provides quantitative strain measurements with minimum data processing.[25,26] In other work, Sampath et al.[27] combined HARP and SENC in one acquisition technique that acquires in only six heartbeats both HARP and SENC images and provides dense measurements of radial, circumferential, and longitudinal strains within a single short-axis slice.

In addition to tagging-framework techniques, cardiac strain can be quantified using displacement encoding with stimulated echoes (DENSE) imaging.[28] HARP and DENSE share many similarities.[29] Instead of postprocessing filtering of the harmonic peaks, DENSE suppresses unwanted peaks during acquisition. Therefore, tissue displacement is readily encoded in the phase of the DENSE MR signal. HARP was originally developed to accelerate fast low to moderate resolution analysis of MR tagging data. DENSE, on the other hand, was intended for high-resolution imaging of a single temporal frame in the cardiac cycle. However, later developments in both techniques led to comparable implementations and strain quantification results from both HARP and DENSE.

3.9 Closing Remarks

In this chapter, we presented methods for measuring the regional cardiac function from tagged MR images based on the principle of HARP imaging and analysis. HARP mathematically characterizes the appearance of the spectra of tagged MR images and utilizes harmonic images for computing the motion of the heart. This allows fast imaging of cardiac regional function, which is feasible in real time with no breath-holding, and accelerated measuring of 3-D motion and 3-D strain tensors. Strain quantification of the left ventricle has been successfully demonstrated in the systolic phase of the cardiac cycle. Challenging hurdles and opportunities exist in optimization of imaging to cover the diastolic phase that associated with diastolic heart failure. In addition, understanding subtle details of the strain profile requires further improvement in temporal resolution during acquisition. Accelerated whole-heart coverage and strain quantification in the right ventricle are necessary for analysis of diseases such as pulmonary hypertension.

References

1. Osman, N. F. & Prince, J. L. Visualizing myocardial function using HARP MRI. *Phys Med Biol* **45**, 1665–1682 (2000).
2. Osman, N. F., Kerwin, W. S., McVeigh, E. R. & Prince, J. L. Cardiac motion tracking using CINE harmonic phase (HARP) magnetic resonance imaging. *Magn Reson Med* **42**, 1048–1060 (1999).
3. Axel, L. & Dougherty, L. Heart wall motion: Improved method of spatial modulation of magnetization for MR imaging. *Radiology* **172**, 349–350 (1989).
4. Axel, L. & Dougherty, L. MR imaging of motion with spatial modulation of magnetization. *Radiology* **171**, 841–845 (1989).
5. Axel, L., Montillo, A. & Kim, D. Tagged magnetic resonance imaging of the heart: A survey. *Med Image Anal* **9**, 376–393 (2005).
6. Zerhouni, E. A., Parish, D. M., Rogers, W. J., Yang, A. & Shapiro, E. P. Human heart: Tagging with MR imaging—a method for noninvasive assessment of myocardial motion. *Radiology* **169**, 59–63 (1988).
7. Mosher, T. J. & Smith, M. B. A DANTE tagging sequence for the evaluation of translational sample motion. *Magn Reson Med* **15**, 334–339 (1990).
8. Lai, W. M., Rubin, D. & Krempl, E. *Introduction to Continuum Mechanics*, 3rd edn (Pergamon Press, 1993).
9. Sampath, S., Derbyshire, J. A., Atalar, E., Osman, N. F. & Prince, J. L. Real-time imaging of two-dimensional cardiac strain using a harmonic phase magnetic resonance imaging (HARP-MRI) pulse sequence. *Magn Reson Med* **50**, 154–163 (2003).
10. Abd-Elmoniem, K. Z., Sampath, S., Osman, N. F. & Prince, J. L. Real-time monitoring of cardiac regional function using FastHARP MRI and region-of-interest reconstruction. *IEEE Trans Biomed Eng* **54**, 1650–1658 (2007).
11. Kraitchman, D. L. et al. Quantitative ischemia detection during cardiac magnetic resonance stress testing by use of FastHARP. *Circulation* **107**, 2025–2030 (2003).
12. Ryf, S., Tsao, J., Schwitter, J., Stuessi, A. & Boesiger, P. Peak-combination HARP: A method to correct for phase errors in HARP. *J Magn Reson Imaging* **20**, 874–880 (2004).
13. Agarwal, H. K., Prince, J. L. & Abd-Elmoniem, K. Z. Total removal of unwanted harmonic peaks (TruHARP) MRI for single breath-hold high-resolution myocardial motion and strain quantification. *Magn Reson Med* **64**, 574–585 (2010).
14. Agarwal, H. K., Abd-Elmoniem, K. Z. & Prince, J. L. Truharp: Single breath-hold Mri for high resolution cardiac motion and strain quantification. *Proc IEEE Int Symp Biomed Imaging* **2009**, 458–461 (2009).
15. O'Dell, W. G., Moore, C. C., Hunter, W. C., Zerhouni, E. A. & McVeigh, E. R. Three-dimensional myocardial deformations: Calculation with displacement field fitting to tagged MR images. *Radiology* **195**, 829–835 (1995).
16. Denney, T. S. Jr. & McVeigh, E. R. Model-free reconstruction of three-dimensional myocardial strain from planar tagged MR images. *J Magn Reson Imaging* **7**, 799–810 (1997).
17. Denney, T. S. Jr., Gerber, B. L. & Yan, L. Unsupervised reconstruction of a three-dimensional left ventricular strain from parallel tagged cardiac images. *Magn Reson Med* **49**, 743–754 (2003).
18. Pan, L., Prince, J. L., Lima, J. A. & Osman, N. F. Fast tracking of cardiac motion using 3D-HARP. *IEEE Trans Biomed Eng* **52**, 1425–1435 (2005).

19. Sampath, S. & Prince, J. L. Automatic 3D tracking of cardiac material markers using slice-following and harmonic-phase MRI. *Magn Reson Imaging* **25**, 197–208 (2007).
20. Abd-Elmoniem, K. Z., Osman, N. F., Prince, J. L. & Stuber, M. Three-dimensional magnetic resonance myocardial motion tracking from a single image plane. *Magn Reson Med* **58**, 92–102 (2007).
21. Abd-Elmoniem, K. Z., Stuber, M. & Prince, J. L. Direct three-dimensional myocardial strain tensor quantification and tracking using zHARP. *Med Image Anal* **12**, 778–786 (2008).
22. Fischer, S. E. et al. True myocardial motion tracking. *Magn Reson Med* **31**, 401–413 (1994).
23. Stoeck, C. T., Manka, R., Boesiger, P. & Kozerke, S. Undersampled cine 3D tagging for rapid assessment of cardiac motion. *J Cardiovasc Magn Reson* **14**, 60 (2012).
24. Rutz, A. K., Ryf, S., Plein, S., Boesiger, P. & Kozerke, S. Accelerated whole-heart 3D CSPAMM for myocardial motion quantification. *Magn Reson Med* **59**, 755–763 (2008).
25. Pan, L. et al. Real-time imaging of regional myocardial function using fast-SENC. *Magn Reson Med* **55**, 386–395 (2006).
26. Osman, N. F. Detecting stiff masses using strain-encoded (SENC) imaging. *Magn Reson Med* **49**, 605–608 (2003).
27. Sampath, S., Osman, N. F. & Prince, J. L. A combined harmonic phase and strain-encoded pulse sequence for measuring three-dimensional strain. *Magn Reson Imaging* **27**, 55–61 (2009).
28. Aletras, A. H., Ding, S., Balaban, R. S. & Wen, H. DENSE: Displacement encoding with stimulated echoes in cardiac functional MRI. *J Magn Reson* **137**, 247–252 (1999).
29. Kuijer, J. P. et al. DENSE and HARP: Two views on the same technique of phase-based strain imaging. *J Magn Reson Imaging* **24**, 1432–1438 (2006).

4

Magnetic Resonance Elastography

Jiming Zhang and Raja Muthupillai

CONTENTS

There has been a longstanding scientific interest in the noninvasive evaluation of the biomechanical properties of soft tissues. Magnetic resonance imaging (MRI) has emerged as a useful tool for assessing tissue response to mechanical loading. In this chapter, we briefly summarize the theoretical basis of elasticity imaging. We describe various types of stress employed to cause tissue deformation and MRI methods that are used to measure the resulting strain, and we outline some of the common algorithms used to extract metrics reflecting tissue elasticity.

4.1 Introduction

For centuries, physicians have palpated tissues to use tissue stiffness as a marker of health [1,2]. Palpation is a qualitative method of characterizing tissue stiffness. In a nonsurgical setting, palpation is restricted to superficial regions of the body, and its reach is determined by the patient's body habitus. Over the years, there have been several attempts to extend the reach of the palpating hand of the physician deep within the body. Nearly three decades of active development by a number of researchers has resulted in the evolution of noninvasive imaging methods that may be broadly referred to as elasticity imaging techniques [3–77].

Notwithstanding the intrinsic scientific attraction in developing an imaging modality based on elasticity—a fundamentally new contrast parameter for imaging—the primary reason for the sustained interest in developing elasticity imaging methods is the recognition that the underlying contrast parameter, such as tissue shear elastic modulus, can span six orders of magnitude, from a few kilopascal for glandular breast tissue to a few hundred megapascal for cartilage and bone [78]. This is in sharp contrast to conventional imaging modalities such as ultrasound, x-ray computed tomography (CT), or magnetic resonance imaging (MRI), where the underlying contrast parameters span one or two orders of magnitude (Figure 4.1) [79].

From a technical development point of view, it is desirable to have an elasticity imaging method that is noninvasive, quantitative, and accessible to remote areas of the body. Elasticity imaging involves the following three basic steps: (1) Impose an internally generated or externally applied source of mechanical stress on the tissue of interest. (2) Measure tissue response to mechanical stress using a suitable technique such as ultrasound, MRI, or optical imaging methods. (3) Reconstruct quantitative (or semiquantitative or qualitative) maps of tissue stiffness with the knowledge (full or limited) of source stress distribution and tissue deformation. Elasticity imaging methods are broadly classified using the type of stress applied, the imaging method used for measuring tissue response, and the reconstruction approach used to estimate indices of tissue elasticity.

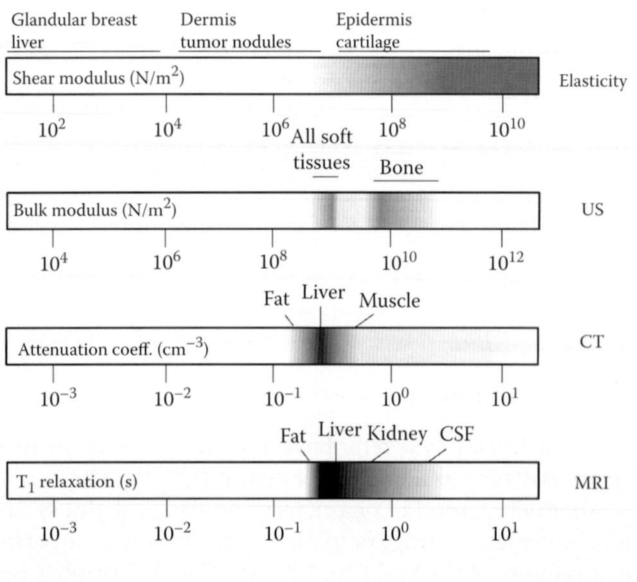

FIGURE 4.1
The bulk elastic modulus, x-ray attenuation coefficient, and MR relaxation parameters that form the basis of tissue contrast in ultrasound, CT, and MR span about an order of magnitude in soft tissues. In contrast, the shear elastic modulus spans nearly six orders of magnitude between different tissue types.

There has been a long-standing scientific interest in evaluating the biomechanical properties of soft tissues, and initial work was mostly done on ex vivo tissue specimens [3,4,7,13,57,77,80–83]. Many ultrasound-based imaging techniques have been used to measure tissue response to static or quasi-static stress [5,6,8,9,12,14,16,18,78,84,85], and these methods paved the way for newer imaging modalities such as MRI for assessment of the tissue response. *The scope of this review will be restricted to a survey of elasticity imaging methods that use MRI as the tool for the measurement of tissue response to a mechanical stimulus.*

4.2 Theory of Elasticity Imaging

The mechanical properties of a medium can be characterized from its response to an applied force. The response of the medium to the applied force can be either rigid-body motion (i.e., translation and rotation) or deformation. The deformation can be reversible (elastic) or irreversible (viscous, plastic, or flow) or a combination of both. Soft tissues are generally considered to be viscoelastic. The linear relationship between applied force (stress tensor) and the resulting deformation (strain tensor) is described by a simple constitutive equation (under the assumption of infinitesimal small deformation) [86,87]:

$$\tau_{ij} = C_{ijkl}\,\varepsilon_{kl}\; i, j, k, l = 1,2,3 \tag{4.1}$$

where τ_{ij} and ε_{kl} are the stress and strain tensors, respectively, and C_{ijkl} is the elastic modulus tensor (rank 4) of 81 components. However, not all of these 81 components are independent. Since both stress and strain tensors are symmetric tensors ($C_{ijkl} = C_{jikl}$ and $C_{ijkl} = C_{ijlk}$), the elastic modulus tensor is reduced to only 36 independent constants. This can be further reduced to two well-known independent Lame′ constant parameters (λ and μ, which are related to longitudinal and shear deformation, respectively) under the assumption of an isotropic medium, and Equation 4.1, can be simplified as

$$\tau_{ij} = 2\mu\varepsilon_{ij} + \lambda\varepsilon_{nn}\delta_{ij} \tag{4.2}$$

where δ_{ij} is the Kronecker delta (δ_{ij} is 1 if $i = j$, otherwise 0), where the strain tensor (ε_{ij}) is defined in terms of the displacement u_{ij} as

$$\varepsilon_{ij} = \frac{1}{2}\left(\frac{\partial u_i}{\partial x_j} + \frac{\partial u_j}{\partial x_i}\right) = \frac{1}{2}(u_{i,j} + u_{j,i}). \tag{4.3}$$

The Young's modulus (E) and Poisson ratio (σ) are often used in ultrasound elastography and are related to the two Lame′ constants as

$$E = \frac{\mu(3\lambda + 2\mu)}{\lambda + \mu} \text{ and } \sigma = \frac{\lambda}{2(\lambda + \mu)}. \tag{4.4}$$

Equation 4.2 can be inserted into the equation of motion to calculate the mechanical properties [87–89]:

$$\sum_{j=1}^{3} \frac{\partial \tau_{ij}}{\partial x_j} + f_i = \rho \frac{\partial^2 u_i}{\partial t^2},\ i = 1,2,3 \tag{4.5}$$

where f_i is force per unit volume applied on the body in the x_i direction and ρ is the density of medium.

4.2.1 Static or Quasi-Static Elasticity Imaging

In the case of static or quasi-static stress, the right term of Equation 4.5 is zero, and plugging Equations 4.2 and 4.3 into Equation 4.5 gives

$$\sum_{j=1}^{3} \frac{\partial(\mu(u_{i,j} + u_{j,i}) + \lambda u_{n,n}\delta_{ij})}{\partial x_j} + f_i = 0,\ i = 1,2,3 \tag{4.6}$$

which can be rewritten in a more compact vector form as

$$\mu\nabla^2 \boldsymbol{u} + (\lambda + \mu)\nabla(\nabla \cdot \mathbf{u}) + \boldsymbol{f} = 0 \tag{4.7}$$

where $\boldsymbol{u}$ and $\boldsymbol{f}$ are the displacement and force vectors and ∇^2 and $\nabla\cdot$ represent the Laplacian and divergence operators, respectively. It can be seen that the estimation of the mechanical properties relies on the accurate measurements of the displacement vector and also on the accurate estimation of its higher-order derivatives. The unique solution of this system is also determined by the boundary conditions and the applied force.

4.2.2 Dynamic Elasticity Imaging

The theoretical basis of elasticity imaging in the case of dynamic stress or time-varying stress has been derived in detail elsewhere [45]. In brief, in an isotropic, homogeneous, linearly elastic medium, dynamic stress initiates a mechanical wave within the medium. By plugging Equation 4.2 into Equation 4.5 and setting the applied force (f) to zero during the wave propagation, one can obtain

$$\sum_{j=1}^{3} \frac{\partial(2\mu\varepsilon_{ij} + \lambda\varepsilon_{nn}\delta_{ij})}{\partial x_j} = \rho \frac{\partial^2 u_i}{\partial t^2},\ i = 1,2,3. \tag{4.8}$$

Substituting the strain expression (Equation 4.3) into Equation 4.8 gives

$$\sum_{j=1}^{3} \frac{\partial(\mu(u_{i,j} + u_{j,i}) + \lambda u_{n,n}\delta_{ij})}{\partial x_j} = \rho \frac{\partial^2 u_i}{\partial t^2},\ i = 1,2,3. \tag{4.9}$$

Rewriting this equation in the vector form under the assumption of local homogeneity of medium results in

$$\rho \frac{\partial^2 \boldsymbol{u}}{\partial t^2} = \mu\ \nabla^2 \boldsymbol{u} + (\lambda + \mu)\nabla(\nabla \cdot \mathbf{u}) \tag{4.10}$$

where ρ is the medium density, μ is the shear modulus (the second Lame′ parameter), and λ is the first Lame′ parameter.

For biological elasticity imaging, a number of simplifying assumptions can be made. The contributions from both longitudinal and shear waves are taken into account in Equation 4.10. As discussed previously, in elasticity imaging, the relevant parameter of interest is the shear-wave contribution, given its large dynamic range. In soft tissues, bulk elastic modulus is much greater than the shear elastic modulus (by as much as four orders of magnitude), and many elasticity reconstruction methods exploit this large difference to separate the contributions of these two components. Alternatively, longitudinal wave contribution can be removed from consideration by taking the curl of the displacement vector $\boldsymbol{u}$ ($\nabla \times \boldsymbol{u}$), and the wave Equation 4.10 may be further simplified as

$$\rho \frac{\partial^2 \boldsymbol{u}}{\partial t^2} = \mu\nabla^2 \boldsymbol{u} \tag{4.11}$$

$$\frac{\partial^2 \boldsymbol{u}}{\partial t^2} - v_s^2 \nabla^2 \boldsymbol{u} = 0 \tag{4.12}$$

$$v_s = \sqrt{\frac{\mu}{\rho}} \tag{4.13}$$

where v_s is the shear-wave speed within the medium. It should be noted that the shear elastic modulus thus computed is a real quantity only for purely elastic materials. For a more detailed discussion about the theoretical basis of elasticity imaging, we direct the reader elsewhere [31,47,90–92]. For a viscoelastic soft tissue, the shear-wave speed and attenuation are frequency dependent, and this dispersion could potentially serve as an additional contrast parameter. In many experimental settings, often, an *effective* shear modulus is calculated by only measuring the local shear-wave speed at a particular frequency.

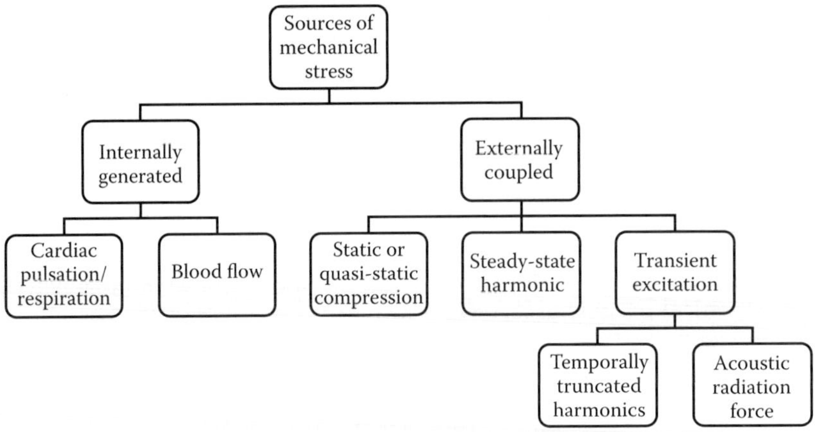

FIGURE 4.2
Broad classification of sources of stress used in various MRE imaging approaches.

4.3 Sources of Mechanical Stimulation

In broad terms, tissue deformation can be initialized by internally generated stress within the body by physiologic activity such as blood flow [93], cardiac pulsation [61], and respiration [94], or by externally applied stress that is transmitted to the organ of interest (Figure 4.2).

4.3.1 Internally Generated Stress

Physiologic phenomena such as cardiac pulsation, blood flow, and respiration can serve as sources of stress. The measurement of tissue deformation caused by such internally generated sources of stress can yield information about local tissue stiffness. MRI-based cardiac tagging techniques are often used to measure intrinsic tissue deformation associated with cardiac pulsation and respiration. For a detailed review of tissue tagging methods, the reader is referred elsewhere [95] (see also Chapters 1 and 2). Similarly, the expansion and contraction of large blood vessels such as the aorta during the cardiac cycle could be used to estimate vascular compliance (Figure 4.3), when coupled with external cuff-based pressure measurements as a surrogate for local blood pressure [93].

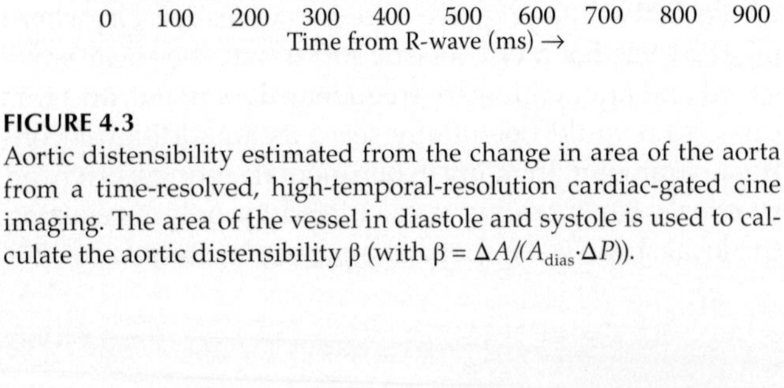

FIGURE 4.3
Aortic distensibility estimated from the change in area of the aorta from a time-resolved, high-temporal-resolution cardiac-gated cine imaging. The area of the vessel in diastole and systole is used to calculate the aortic distensibility β (with $\beta = \Delta A/(A_{dias} \cdot \Delta P)$).

A surrogate for the functional state of the aorta is the measurement of pulse-wave velocity (PWV)—a metric that can be measured using MRI. The principle behind the estimation of PWV using MRI is given in Figure 4.4.

Limitations of internally generated, physiologic sources of stress while the lack of a need for any external devices to impart stress within the body is the primary advantage of such internally generated sources of stress, the primary limitation of using internally generated physiologic sources of stress comes from the lack of a quantifiable estimate of local stress distribution to permit absolute estimates of tissue elasticity. Another key limitation is that elasticity imaging using internal physiologic sources is typically limited to organs within the body that are close to the heart, lungs, and major blood vessels. Last, when using physiological sources of stress, the amplitude and periodicity of applied stress are not controllable and are typically low-frequency motion (a few hertz). This lack of control over the motion frequency and amplitude of stress hampers the estimation of viscous properties of tissue.

Unlike internally generated sources of stress, the control over amplitude and frequency of externally applied stress can provide greater flexibility for elasticity imaging. In fact, the bulk of magnetic resonance (MR) elasticity imaging methods use an externally applied stress.

4.3.2 Externally Induced Stress

4.3.2.1 Static or Quasi-Static Stress

In static stress mode, a constant force is applied and sustained at the surface of the tissue, causing tissue deformation. The extent and type of deformation of the

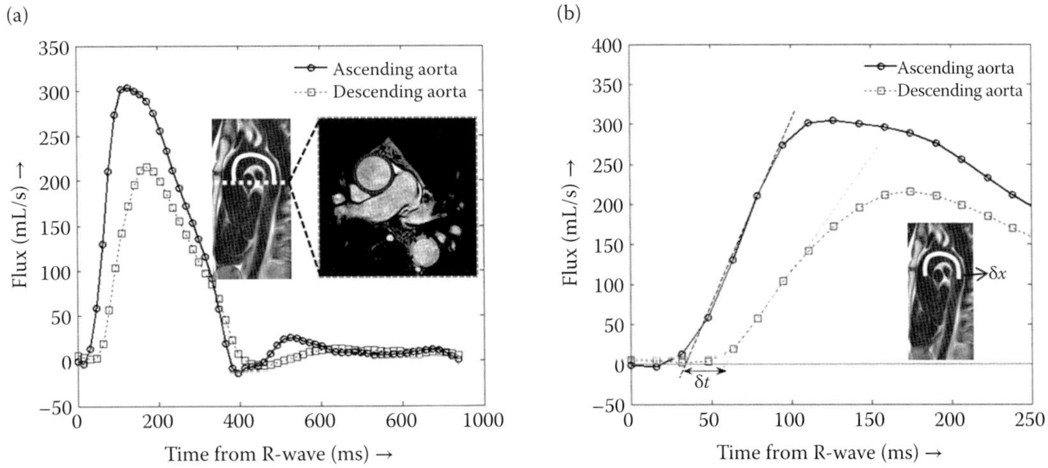

FIGURE 4.4
(a) Estimation of pulse-wave velocity (PWV) using MR phase contrast method. Blood flow across the ascending and descending aorta is measured by positioning a transverse slice (dotted white line) dissecting both sections. The flow across the ascending (thick line, circles) and descending aorta (dotted lines, squares) is shown. (b) The three-dimensional distance (δx) between the location of the intersection of the transverse slice on the ascending and descending aorta is measured, and the time of arrival of the foot of the pulse wave at each location is measured from the flow curves (δt). PWV is calculated as the ratio of $\delta x/\delta t$, and it is related to distensibility, $\beta = 1/(\rho \cdot \text{PWV}^2)$, by the Bramwell–Hill model [93], where ρ is tissue density.

tissue is reflective of the tissue's mechanical property [33,67,96–99]. If an accurate knowledge of the applied stress distribution as well as the resulting strain is available, then it is possible to estimate tissue mechanical properties such as the Young's modulus [100]. The basic principle behind static elastography is described in Figure 4.5.

The use of static or quasi-static stress as a means of tissue deformation has some limitations. Precise quantitative knowledge about the static stress distribution within the body is difficult to estimate in vivo, and therefore,

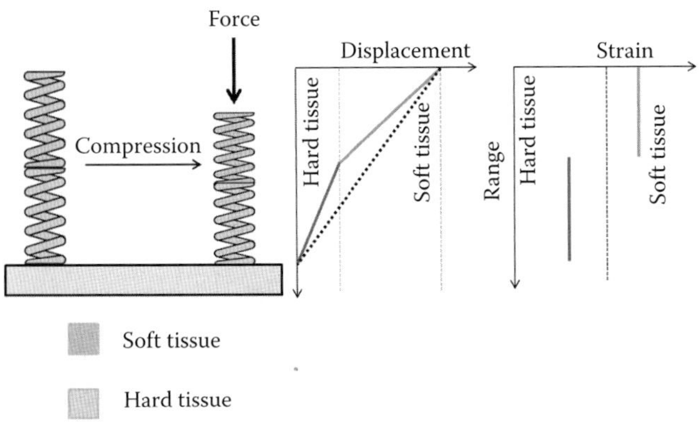

FIGURE 4.5
Principle of estimating tissue elasticity using static compression methods. As shown, the stiffer tissue undergoes less displacement than the softer tissue for a given force, and this relative difference in tissue response is displayed as a strain map. (Adapted from the presentation by Emelianov, S. in 50th AAPM Meeting: 35-9888-35854-789, 2008.)

quantitative estimation of the elastic modulus is challenging. Second, even if the source distribution can be modeled accurately, the measured elastic modulus may be different for different amounts of strain (strain hardening), potentially complicating interpretation. It has been suggested that such strain hardening could potentially serve as a characteristic tissue mechanical property [33,67,96–99]. Last, static stress elastography cannot be used to estimate time-dependent tissue properties like viscosity.

4.3.2.2 Steady-State Harmonic Stress

An alternative to static or quasi-static stress is the application of a periodic steady-state harmonic stress (see also Chapters 5 and 6). It injects mechanical waves within the body, and the estimation of the propagation speed of these waves can provide information regarding tissue mechanical properties. Mechanical waves can be either longitudinal (compression) or transverse (shear). Soft tissues are nearly incompressible, and in most soft tissues within the body, the longitudinal mechanical wave speed is close to 1540 m/s. In contrast, the low-frequency (<300 Hz) shear-wave speed of propagation within tissues is on the order of a few meters per second and has a large variation across different tissue types. Therefore, shear waves are more prominently used for the purposes of elasticity imaging for their potential for generating contrast between tissue types. In this approach, a mechanical driver (such as a vibrating drum) imparts low-frequency (<300 Hz) steady-state harmonic mechanical vibrations on the skin surface [21,31–33,39,40,45,47–49,51,52,72–74,92,101–107]. The spectral content of these vibrations can be pure monotones or a combination of frequencies. These mechanical vibrations propagate within the body as both longitudinal and transverse waves. For reasons discussed in Section 4.2.2, we focus on shear-wave propagation within the tissue. Minute tissue displacements (on the order of a few hundred nanometers) caused by shear-wave propagation within the body are measured using an imaging device. These displacement measurements capture snapshots of propagating mechanical waves. The spatiotemporal behavior of the shear waves can then be used to estimate tissue mechanical properties. For example, in an isotropic, homogenous, elastic medium, the shear-wave speed and the elastic modulus are related by a simple expression (Equation 4.13):

$$\mu = \rho v_s^2. \tag{4.14}$$

The advantages of using steady-state harmonic mechanical stress to generate shear waves include the following: first, such harmonic stress can be readily applied to most organs within the human body with a proper combination of mechanical transducers and coupling mechanisms. Published results show such steady-state shear waves have been injected into the liver [47,108,109], skeletal muscle [35,41,110,111], heart [58,112,113], brain [44,51,114,115], breast [40,45,116], lungs [117,118], uterine cavity [119,120], and so forth. Second, unlike static or quasi-static compression methods, steady-state harmonic stress-based methods do not require knowledge of the precise stress distribution within the body and can then be used to evaluate the time-dependent mechanical properties of soft tissue, that is, viscosity.

The main disadvantages of using such steady-state harmonic shear waves are as follows: first, shear-wave attenuation is high, and therefore, this approach may have a limited depth of penetration within the body and may require sophisticated transducers to permit uniform acoustic illumination across the entire organ of interest. The displacement amplitude signal-to-noise ratio (SNR_d) diminishes as the wave propagates within the body, and low SNR_d could lead to spurious results. Second, the shear-wave field can be complex, due to diffraction phenomena such as reflection, refraction, and mixture of longitudinal and shear-wave components. This requires sophisticated postprocessing methods to estimate tissue elasticity from displacement measures of this complex wave behavior within tissue. Last, it should be noted that accurate estimation of shear elasticity from these wave fields necessitates both volumetric acquisition (to accurately estimate the wave vector), and three-dimensional motion encoding (to fully measure the displacement vector). The combination of these requirements prolongs the acquisition time substantially. These issues will be discussed in more detail in subsequent sections.

4.3.2.3 Transient Induced Stress

Temporally truncated harmonic stress: Instead of steady-state harmonic mechanical excitation that results in complex wave propagation within tissue, it is possible to truncate the harmonic mechanical excitation to a few cycles. Such truncation minimizes the possibility of complex wave behavior within tissue. For example, in a homogeneous medium, local wave speed may be calculated by directly measuring the distance traveled by the wave front at two time points. McCracken et al. [44] found that stiffness values estimated from such temporally truncated harmonic excitation were comparable to those obtained using steady-state harmonic stress.

Acoustic radiation force (ARF)-induced stress: It is possible to generate shear waves within the body by focusing ultrasound waves within the body and by exploiting the mechanical force imparted at the focal point by acoustic waves. The so-called ARF can cause localized mechanical stress at the focal point and set off a propagating mechanical wave from this point. An imaging device can measure the local tissue displacement in the path of the propagating mechanical wave (Figure 4.6).

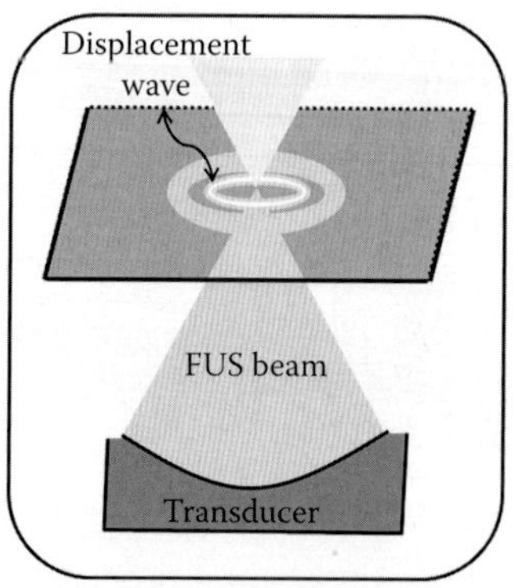

FIGURE 4.6
Mechanism of using acoustic radiation force (ARF) to initiate local shear wave propagation deep within the tissue. A high-intensity focused ultrasound transducer concentrates ultrasound energy at a focal point several centimeters deep into the tissue, resulting in highly localized stress. Measurement of tissue displacement at the plane orthogonal to the beam direction captures the propagation of shear waves within the medium.

The ARF from focused ultrasound has been proposed to remotely generate localized stress in tissue, and the resultant displacement was measured by ultrasound correlation method [121]. However, clinical ultrasound measures only the component of motion that occurs along the axis of the ultrasound beam, making it difficult to estimate true tissue deformation. In this regard, MRI is an attractive imaging modality to measure tissue displacement caused by ARF due to its inherent ability to measure displacement in any direction (so-called Magnetic Resonance-Acoustic Radiation Force Imaging [MR-ARFI] methods).

However, MR-ARFI-based methods need a complicated setup that includes MRI-compatible high-intensity focused ultrasound transducers incorporated within the MRI scanner. In addition, the focused ultrasound beam path is limited by the ultrasound acoustic windows within the body and the ability to probe elasticity is restricted to the reach of the focused ultrasound transducer. Stated differently, the combined use of ultrasound and MRI methods to measure tissue elasticity imposes dual constraints intrinsic to both MRI and focused ultrasound. Last, the shear waves attenuate rapidly within tissues, and as a result, only localized information about tissue elasticity in the region surrounding the ultrasound focus can be obtained. While the ability to obtain local estimates of shear elasticity is an advantage, it entails long scan times, when elasticity information over a larger region is desired.

4.4 MR Imaging of Tissue Deformation

Tissue deformation caused by the applied stress can be directly measured using imaging methods such as MRI and ultrasound. In this section, we will review the MRI methods used for measuring tissue deformation caused by static, quasi-static, and dynamic stress. Here, we use the term "spin" to refer to the intrinsic property of hydrogen nuclei.

Relationship between MR signal and spin motion: It is well known that the precession frequency of hydrogen nuclei within the body is proportional to the applied magnetic field ($\boldsymbol{B}_0$) and is given the following expression: $\omega_0 = \gamma B_0$, where ω_0 is the angular frequency of precession and γ is the gyromagnetic ratio characteristic of the nuclei. For protons, $\gamma/2\pi = 42.57$ MHz/T. The spatial position of spins ($\boldsymbol{r}$) is accomplished by the introduction of additional magnetic field gradients ($\boldsymbol{G}$) ($\omega_0 = \gamma(B_0 + \boldsymbol{G}.\boldsymbol{r})$). Furthermore, specific motion-encoding gradients (MEGs) can be designed to impart sensitivity to spin motion. A commonly used approach is to apply a bipolar MEG in the direction of spin motion to impart sensitivity to the velocity of spins in the direction of MEG. Because the phase information is affected by a number of non–motion-related factors, such as background gradients, those induced by air–tissue interfaces, eddy currents, gradient imperfections, and so forth, typical phase contrast sequence acquisitions involve two acquisitions. For example, spin velocity can be encoded in the phase difference image obtained with two otherwise identical acquisitions with different sensitivities to spin velocity values, to minimize the contribution from such nonmotion-related effects. Such acquisitions are commonly referred to as phase contrast methods and have found widespread clinical use for evaluating blood flow [122] and for reducing motion-induced artifacts in clinical imaging [123,124].

4.4.1 MRI of Quasi-Static Motion

A modified MRI phase contrast sequence can be used to evaluate tissue deformation under quasi-static or static compression. Quasi-static MR elastography (MRE) was first utilized to visualize mechanical properties of tissues [125]. This method compresses an object at low frequencies between 0 and 5 Hz [99]. At these low deformation frequencies, wave propagation can be neglected, and the object can be approximated to be in a state of static stress [33].

In this approach, MEG is applied when the tissue is at a state of compression and when the tissue is at an uncompressed state. The presence of motion in the direction of the applied MEG induces a phase shift in the measured MR signal. Typically, in the second acquisition, all pulse sequence parameters are kept identical, but tissue is left uncompressed. Subtraction of the phase of the two acquisitions yields information about the relative deformation of the tissue under applied compression.

Plewes et al. [125] first demonstrated this approach using a spin-echo sequence by interlacing an MEG between the intervals of 90° and 180° radio-frequency (RF) pulses, and in the gap between 180° and the readout (Figure 4.7) [125]. A tissue-mimicking phantom consisting of materials with different stiffness was compressed during the first half of the MEG (between 90° and 180° RF pulses) and was allowed to get back to its uncompressed state during the second half of the MEG (between the 180° and the readout). A reference acquisition without compression was also acquired. The subtraction of phase information from the two acquisitions yielded information about local deformation. A limitation of using a simple spin-echo method was that the tissue might not have been entirely in the uncompressed state by the time the second half of the MEG was applied after the refocusing pulse. This problem was alleviated by the use of stimulated echoes—a technique that uses three 90° RF pulses, as proposed by Chenevert et al. [97].

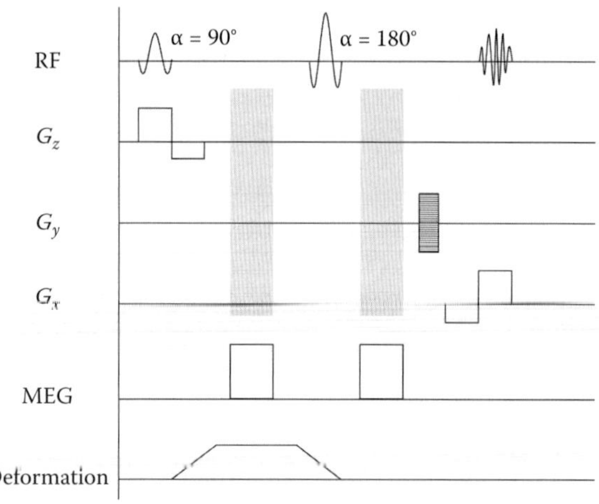

FIGURE 4.7
Spin-echo-based MRE pulse sequence used in static MRE. Tissue displacement caused by static compression is encoded in the phase of the received MR signal with MEG. The MEG can be superimposed along any desired direction (indicated by gray bars) to encode tissue motion.

Local tissue displacement (r) in the presence of MEG of strength G and duration τ due to applied low-frequency stress results in an additional phase shift φ in the received MR signal:

$$\varphi = \gamma G\tau \cdot r. \quad (4.15)$$

The displacement sensitivity (β: rad · m^{-1}) resulting from the spin-echo or stimulated spin-echo sequence is defined as follows:

$$\beta = \frac{\phi}{r} = \gamma G\tau. \quad (4.16)$$

The spatial derivative of the measured displacement yields local strain and generates a strain-based contrast. Except in the case of most simple geometries, in quasi-static MRE, it is difficult to estimate the local stress distribution. As a result, most quasi-static MRE methods provide qualitative strain images. It should be noted that the level of strain imparted could result in strain hardening of tissues, and some groups have suggested this as another potential mechanism to generate tissue contrast based on mechanical properties of tissues. A number of studies have demonstrated the potential for using quasi-static MRE for clinical diagnosis [33,98,99]. Recently, McGrath et al. [67] have applied the quasi-static MRE method to study tissue biomechanical properties involved in pathology fixation.

4.4.2 MRI of Steady-State Harmonic Motion

The propagation of harmonic shear waves within tissue induces small cyclic displacement of spins, and encoding cyclic motion information in the phase of the MR signal requires the development of necessary mathematical formulation and experimental methodologies [21]. In this section, we will consider the case of encoding cyclic motion occurring at a single frequency. However, the theoretical principles can be directly expanded to include multifrequency or multiband mechanical excitation [44,52].

Cyclic displacement of spins in conjunction with shear-wave propagation is of the form

$$\boldsymbol{r}(t) = \boldsymbol{r}_0 + \xi_0 \exp[j(\boldsymbol{k} \cdot \boldsymbol{r} - \omega t + \alpha)] \quad (4.17)$$

where r_0 is the mean position of the spin and ξ_0 is the peak displacement of the spin from its mean position. ω is the angular frequency of mechanical wave excitation, $\boldsymbol{k}$ is the wave number, and α is the initial phase offset of the wave.

As in the case of static/quasi-static MRE methods, it is possible to encode this cyclic motion with an appropriate MEG waveform that is interlaced between the excitation and the readout as phase in the received MR signal.

Then the dynamic phase shift of the moving magnetization can be rewritten as follows:

$$\phi(r,\alpha) = \gamma \int G(r,t') \cdot (r_0 + \xi_0 e^{j(k \cdot r - \omega t' + \alpha)}) dt'. \quad (4.18)$$

One potential MEG to encode cyclic motion is a gradient waveform at the same frequency of mechanical excitation (Figure 4.8) [21].

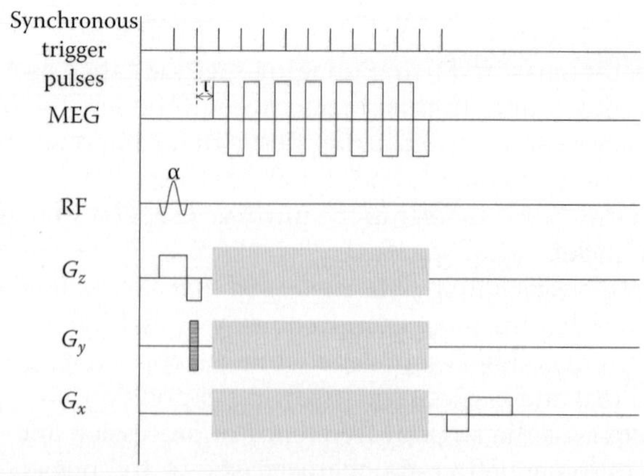

FIGURE 4.8
Dynamic MRE method for encoding cyclic motion. The frequency spectrum of the MEG was chosen to match the frequency spectrum of the mechanical excitation applied. The application of the MEG is synchronized with mechanical excitation using triggers generated from the MRI scanner. The gray bars indicate that the MEG can be applied along any direction to encode three-dimensional tissue displacements.

For example, consider the simplest case of an MEG with N bipolar pairs, switched at the same frequency ($f = \omega/2\pi$) as the spin motion:

$$G(r,t) = \begin{cases} +G_0;\ t \in [nT, (2n+1)T/2] \\ -G_0;\ t \in [(2n+1)T/2, (n+1)T] \end{cases} \tag{4.19}$$

where $n = 0, 1, 2, 3,..., N - 1$. G_0 is the gradient strength amplitude in Tesla per meter. To simplify the analysis, the ramp times of MEG are considered negligible. Substituting this MEG into Equation 14.18, the phase shift induced by cyclic motion in the received MR signal is given by [21]

$$\phi(r,\alpha) = \frac{2\gamma NT(G_0 \cdot \xi_0)}{\pi} \sin(k \cdot r + \alpha). \tag{4.20}$$

The following observations can be made from Equation 4.20. For a mechanical wave at steady state,

1. The measured phase shift is proportional to the component of the displacement amplitude vector in the direction of the MEG, MEG amplitude, and number of cycles of MEG. This makes measuring the micron and sub-micron level motion feasible by increasing either the gradient strength of MEG or number of cycles of MEG or both.
2. The phase difference image depicts a single snapshot of the mechanical wave (wave vector k, displacement amplitude vector ξ_0, and initial wave phase α) propagating within the tissue.
3. The three-dimensional displacement vector (r) can be fully characterized by repeating measurements by superimposing MEG on orthogonal directions.
4. The wave vector (k) can be fully characterized by acquiring a three-dimensional volumetric acquisition.
5. By repeating the measurements at different temporal offsets α by adjusting synchronizing trigger pulses between the MEG and cyclic tissue motion, one can capture a series of snapshots of the mechanical wave propagating within the tissue. This information regarding the temporal behavior of wave propagation within tissue can be used to resolve complex wave propagation caused by reflection, refraction, and so forth within tissue.

Thus, dynamic shear-wave MRE provides a comprehensive means to evaluate three-dimensional, time-resolved, shear-wave propagation within tissue, at the expense of prolonged acquisition times.

While Equation 4.20 provides the specific case of encoding cyclic motion occurring at single frequency, the theoretical principles can be extended to include a more general case by considering the sensitivity of the gradient waveform to harmonic motion [21].

The MEG waveform, $G_r(t)$, may be considered to be a continuous waveform that is temporally truncated by the windowing function $w(t)$, and then the accumulated phase of received MR signal by motion can be written as

$$\phi(\tau) = \int_0^{\tau} \gamma[G_r(t) \cdot w(t)] \cdot r(t)\mathrm{d}t \tag{4.21}$$

where

$$w(t) = \begin{cases} 1,\ 0 < t < \tau \\ 0,\ \text{otherwise.} \end{cases} \tag{4.22}$$

If $G_r(f)$ is the Fourier transformation (FT) of $G_r(t)$, then Equation 4.21 can be rewritten as

$$\begin{aligned} \phi(\tau) &= \gamma \int_0^{\tau} \left[\int_{-\infty}^{\infty} G_r(f)\exp(-j2\pi ft)\mathrm{d}f \cdot w(t) \right] \cdot r(t)\mathrm{d}t \\ &= \gamma \int_{-\infty}^{\infty} G_r(f) \cdot \left[\int_0^{\tau} w(t) \cdot r(t)\exp(-j2\pi ft)\mathrm{d}t \right] \mathrm{d}f \end{aligned} \tag{4.23}$$

and further simplified as

$$\phi(\tau) = \gamma \int_{-\infty}^{\infty} G_r(f) \cdot [w(-f) \otimes r(-f)]\mathrm{d}f \tag{4.24}$$

where $w(f)$ and $r(f)$ are the FT of $w(t)$ and $r(t)$, respectively, and $\otimes$ is the convolution operator. Equation 4.24 represents a Fourier-based generalization of the phase shift.

In other words, the MEG function could be considered as a "filter" that selectively measures components of introduced mechanical motion. In its simplest implementation, to measure a single harmonic frequency, as discussed in the previous section, the frequency of MEG is chosen to match the frequency of mechanical excitation. More generally, the spectral response of the MEG waveform could be tailored to selectively extract single or multiple harmonic components of imparted mechanical motion [126].

4.4.3 MRI of Transient Shear Waves by ARF

The ARF can be used to generate a localized transient stress deep within the tissue [127–129]. This stress initiates a transient mechanical wave emanating from the ultrasound focus. The mechanical transient wave propagation is governed by the tissue mechanical properties (shear modulus and viscosity). The MRE imaging approach for ARF-induced stress employs a combination of strategies used in the case of quasi-static MRE and harmonic MRE [130,131].

In this MR-ARFI approach, the displacement of tissue caused by ARF is encoded into the phase of the received MR signal in a manner analogous to the approach used in quasi-static MRE by using bipolar MEGs (Figure 4.9). Furthermore, in a manner analogous to steady-state harmonic MRE, transient wave propagation within the tissue is captured by acquiring a series of images at varying temporal delays between the ARF-induced displacement and the MEG.

The ARF excitation is synchronized with the predetermined time delay (τ) with respect to the onset of the MEG. Two interleaved phase images were acquired at same time delay (τ) with opposite polarity of MEG. The phase shift introduced by transient motion $r(t)$ can be generally written as follows:

$$\phi = \int_0^{\delta} \gamma G(r,\tau) \cdot r(\tau) \mathrm{d}\tau. \tag{4.25}$$

For the sake of simplicity, if the MEG is assumed to be a rectangular gradient waveform (with no ramps) of gradient strength G_0 and duration δ (period: 2δ), this MEG encodes a phase shift that is proportional to the

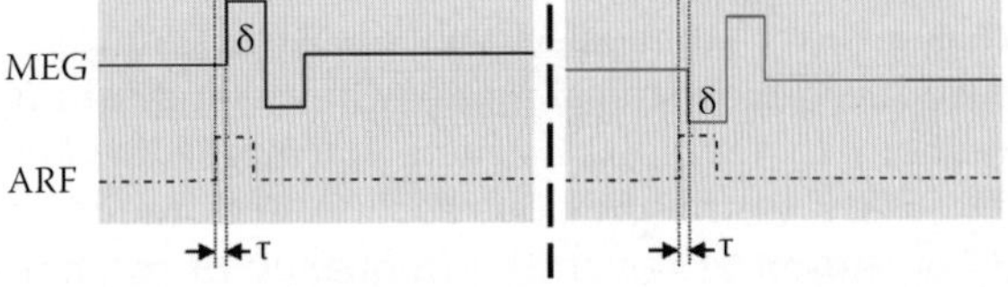

FIGURE 4.9
A 2-D gradient-echo–based technique to capture transient tissue motion caused by acoustic radiation force (ARF) is shown. As in the case of static MRE, a bipolar MEG gradient encodes tissue displacement caused by ARF into the phase of the received MR signal and can be applied along any preferred direction. As in the case of dynamic MRE, phase offset (τ) between the MEG and the applied ARF can be adjusted to capture temporal propagation of transient shear wave within the medium. The gradient applied at both z and x directions after readout was used to destroy the magnetization history.

average displacement $\bar{r}$ that occurs during the motion-encoding time as [121]

$$\phi = \gamma \delta \boldsymbol{G}_0 \cdot \bar{\boldsymbol{r}}. \tag{4.26}$$

By acquiring another reference phase acquisition with identical acquisition parameters with the mechanical excitation turned off, one can eliminate spurious, nonmotion-related phase contributions. The resulting phase difference image captures a snapshot of the propagating transient mechanical wave emanating from the ultrasound focus. By varying the time delay between ARF and MEG, multiple snapshots of the propagating mechanical wave can be obtained, and in a homogeneous medium, the distance traveled by the wave as a function of time can reveal the mechanical wave speed and, therefore, elasticity.

While the bulk of MR-based ARF imaging methods rely on monitoring the propagation of a transient impulse within the medium, by applying repeated ARF-induced mechanical stimulations, and encoding the resulting steady-state motion using cyclic MEG, one can obtain a pseudo–steady-state condition mimicking harmonic MRE [34]. However, repeated application of an ARF-induced stimulus using a focused ultrasound transducer can result in local heating.

4.5 Tissue Mechanical Parameter Reconstruction

The raw output from all MRE imaging methods is a measurement of local tissue displacement. Local mechanical properties of tissue need to be estimated from these displacement measurements by making a number of simplifying assumptions, with respect to tissue isotropy, homogeneity, and compressibility.

4.5.1 Compression-Based Methods

In compression-based MRE methods, the imaging of tissue under the states of compression and noncompression results in a phase difference image that reflects local tissue displacement. The magnitude of the spatial gradient of the displacement map provides a map of local strain [33,97]. The gradient operations tend to amplify the noise, and therefore, these techniques demand high $\mathrm{SNR_d}$ acquisitions [96]. While strain maps provide qualitative information regarding the local tissue stiffness, there are a number of challenges associated with generating quantitative information on tissue elasticity using static MRE methods.

The calculation of quantitative tissue elasticity information from the strain maps without explicit knowledge of the applied stress distribution is difficult. Some groups have attempted to solve this inverse problem by using iterative reconstruction methods [67]. One approach is to combine the experimentally measured strain information with the stress information provided by a finite-element model of the experimental setup, to calculate regional tissue mechanical properties [67]. McGrath et al. [67] have demonstrated the utility of this approach in estimating tissue mechanical properties of ex vivo tissue specimens, before and after fixation. This is an area of active research (see also Chapter 18).

4.5.2 Reconstruction of Steady-State Harmonic Motion Image Processing

A broad range of postprocessing methods have been described in the literature to extract tissue mechanical properties from the so-called wave images, which capture a snapshot of the propagating mechanical wave within the tissue. In dynamic MRE images, it is possible to capture a series of snapshots that depict temporal behavior of wave propagation within tissue (Figure 4.10) [33,50,92,101,131].

Unlike the static/quasi-static MRE, the ability to acquire temporal information of the data provides additional flexibility in postprocessing. We will briefly summarize some of the common reconstruction algorithms used to extract tissue mechanical properties from dynamic steady-state MRE images.

4.5.2.1 Phase Gradient Method

The simplest and the most direct reconstruction method for postprocessing steady-state dynamic MRE images is

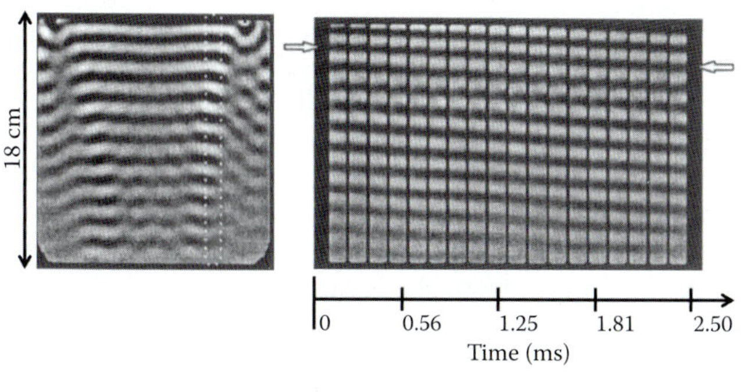

FIGURE 4.10
Single frame of a steady-state MRE image depicting wave propagation within a homogenous tissue-mimicking phantom is shown on the left. A thin strip (dotted white lines) extracted from a series of images acquired at progressively increasing temporal delays between the MEG and the mechanical excitation captures wave propagation. From the spatial progress of the wave front (indicated by the white arrows) over the 2.5 ms duration, one can estimate the shear wave speed. The ability to capture the three-dimensional spatiotemporal behavior of wave propagation within tissue provides additional flexibility in extracting mechanical properties during reconstruction.

the phase gradient method [92]. In this approach, a simple harmonic wave propagating through the material is captured at multiple time points. For each pixel, the temporally resolved displacement data are filtered along the time domain to extract the harmonic frequency component of interest. The resulting amplitude and phase of the displacement vector can provide information regarding the mechanical properties of the tissue.

For example, the spatial gradient of phase estimated along the direction of a single-frequency (f) planar wave propagating within a homogenous medium gives a direct measure of local frequency (k, radians per unit distance) of wave propagation, which can be related to the shear elastic modulus as follows:

$$\mu = \rho v_s^2 = \rho(\lambda f)^2 = \rho\left(\frac{2\pi}{k} f\right)^2. \tag{4.27}$$

Furthermore, the peak amplitude of displacement at each spatial location can be used to estimate local wave attenuation.

The advantages of the phase gradient method are its simplicity and the ability to estimate the local frequency with high spatial resolution. However, the derivative operation used in the analysis is intrinsically sensitive to noise. Second, this approach fails when the tissue displacement is not a pure harmonic and consists of superposition of harmonics, arising from reflections and diffractions—a common scenario in steady-state acquisitions. Third, this technique also requires information about the direction of wave propagation, which may be hard to estimate in vivo. Last, the shear modulus estimated from the wave speed (ρv_s^2) does not explicitly describe the frequency-dependent viscous behavior of the tissue. So, such shear modulus estimates should be considered as an "equivalent" or "effective" shear modulus or shear stiffness of a purely elastic material that supports shear-wave propagation at the measured velocity (v_s) at a given mechanical frequency (f).

4.5.2.2 Multiscale Local Frequency Estimation Algorithm

Some of the limitations of the phase gradient method are eliminated by an image processing approach, commonly referred to as the local frequency estimation (LFE) algorithm. Knutsson et al. [132] described a multiscale algorithm for estimating the local frequency of a signal, over a large number of scales. Manduca et al. [92] further developed LFE algorithms suitable for processing steady-state MRE wave images.

In brief, the LFE algorithm estimates the instantaneous or local frequency from quadrature filters, which are a product of a lognormal radial frequency function and a directional function. The net response of the lognormal filter is the product of the radial response functions and the directional response function at multiple scales. By taking the ratios of such filter responses from two filters with different center frequencies, one can arrive at an estimate for the local frequency. Several such estimates can be obtained from filters with a range of center frequencies. These estimates can be weighted according to the energy in the filtered output to obtain a robust estimate of local frequency. The bandwidth of the estimate provides a measure of the variance of the estimate and is often used as a quality measure or a certainty measure. Manduca et al. [92] have shown that the results from the LFE approach are equivalent to solving the Helmholtz equation under the assumptions of local homogeneity, incompressibility, and no attenuation.

Unlike the phase gradient method, the LFE algorithm does not make assumptions about the direction of wave propagation and is relatively robust against complications arising from wave diffraction. Second, this algorithm has proven to be robust for its sophisticated multiscale data averaging in the estimation and simultaneously provides a confidence map of the LFE estimation. Third, LFE is essentially an image processing approach and therefore can estimate shear elastic modulus from a single wave image.

The main drawback of the LFE approach is that it does not directly provide information about shear viscosity or other time-dependent mechanical properties. Similar to the phase gradient method, the LFE approach also estimates an *effective* shear modulus of a material (equivalent to that of a purely elastic material that supports shear-wave propagation at the measured velocity) as it solely relies on the estimation of local spatial frequency of wave propagation within tissue. Furthermore, the LFE method does not provide as high spatial resolution as the spatial gradient method, and the LFE images are blurred at object boundaries. This imposes an intrinsic limit to the resolvable size of the object for the LFE algorithm (Figure 4.11). Manduca et al. have shown that while the LFE algorithm can *detect* the presence of a "hard" inclusion in the background of homogeneous soft medium that is as small as one-eighth of the wavelength of the propagating shear wave, the accuracy of the shear elasticity of the embedded hard object improves only when the object size is about half the size of the wavelength. For a more extensive description of the LFE method, the reader is referred to the Manduca et al. [92].

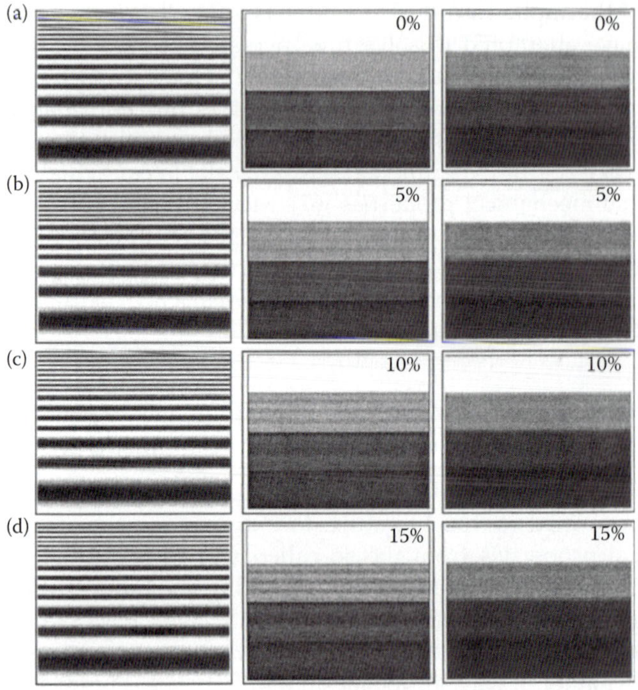

FIGURE 4.11
The digital phantoms and the spatial frequency estimates using the phase gradient method (middle) and the LFE method (right) are shown. While the phase gradient method preserves sharp edges compared to the LFE approach, it degrades more rapidly with increasing levels of noise (rows a through d), compared to the LFE method.

4.5.2.3 Inversion Algorithm

An attractive feature of steady-state, dynamic MRE methods is the ability to acquire extensive information regarding the spatiotemporal behavior of shear-wave propagation within tissue, and this has spawned a broad spectrum of approaches to extract tissue mechanical properties by inverting the measured displacement data. Direct inversion of the measured acoustic wave field within tissue is complex and computationally intensive, and inversion often involves a number of simplifying assumptions. Most inversion algorithms assume that tissue density is uniform for all tissues (ρ = 1.0 kg/m^3) and that the material is homogenous (at least locally), isotropic, and linear [92].

Under the assumption of local homogeneity, Oliphant et al. [36] rewrote the wave Equation 4.10 as an algebraic matrix expression (as opposed to a differential equation) and directly calculated elastic modulus by estimating local derivatives from the local neighborhood. The disadvantage of such algebraic inversion methods is the amplification of noise due to the derivative operations. Oliphant et al. proposed smoothing of the displacement data to minimize such noise amplification. Romano et al. [133] proposed taking analytical derivatives of appropriately chosen test functions instead of numerical derivatives to estimate the Lamé constants [133].

The analysis can be further simplified by eliminating information from longitudinal wave propagation. Fortuitously, in soft tissues, the speed of longitudinal wave propagation is more than two orders greater than shear-wave speed (1540 m/s as opposed to 1–10 m/s). As a result, in soft tissues, $\lambda \gg \mu$, and these two parameters are effectively decoupled. At typical frequencies used in dynamic MRE (<300 Hz), the wavelength of longitudinal waves in soft tissues is on the order of several meters, and therefore, the displacements caused by longitudinal waves change gradually over space, and the longitudinal wave contribution is neglected in some reconstruction approaches. Some groups have proposed taking the curl of the displacement wave field to entirely remove displacement contribution from longitudinal waves [45,92] to simplify the Equation 4.10 further to a simple Helmholtz equation

$$\rho\omega^2 \boldsymbol{u} + \mu\nabla^2 \boldsymbol{u} = 0 \tag{4.28}$$

$$\mu = -\frac{\rho\omega^2 \boldsymbol{u}}{\nabla^2 \boldsymbol{u}} \tag{4.29}$$

However, taking the curl of the displacement vector requires the measurement of all three orthogonal components of displacement, and the curl operation itself can amplify noise. Given the difference of two orders of magnitude in the relative wavelengths of longitudinal and shear waves, some groups have suggested using a high-pass filter to suppress the low spatial frequency contribution from longitudinal waves as an alternative to taking the curl of the displacement wave field [92]. The quality of the separation of the longitudinal wave field from the displacement data will depend on the spectral frequency response of the chosen filter. In any event, a careful accounting of the longitudinal wave contribution is necessary for accurate extraction of tissue shear elastic modulus [45].

It should be noted that the shear elastic modulus thus computed is a complex quantity (as noted in Section 4.2), and in a typical viscoelastic material, shear elastic modulus is frequency dependent. The frequency-dependent behavior of tissue can be evaluated by repeating dynamic MRE measurements at multiple frequencies. The complex shear elastic modulus estimated at these frequencies can be fitted to a suitable rheological model (e.g., Voigt, Maxwell, Zener, etc.) to more comprehensively characterize the tissue mechanical property [49]. Sinkus et al. [134] suggest that, within the clinically used range of mechanical frequencies, complex shear elastic modulus of breast tissue follows a power-law behavior and that it is possible to extract the exponent of the frequency power law from the complex shear elastic modulus. It remains to be seen if such analysis could shed further insights into the physical state of tissue.

A number of other approaches, including direct inversion of the differential equations without assumptions regarding homogeneity, have also been proposed [37], at the cost of increased computational complexity. Some finite-element-based approaches attempt to minimize the measured displacements against numerically computed tissue displacements in an iterative fashion [36,37,90,133]. These models require additional geometric information and are computationally intensive. The degree of sophistication of the reconstruction algorithm has to be balanced against the practical constraints in data acquisition in a clinical setting. Comprehensive data acquisition for MRE is very time consuming. The complete resolution of the displacement vector requires motion encoding along three orthogonal directions. The unambiguous resolution of the wave vector requires a volumetric acquisition with isotropic voxels. The complete resolution of the temporal behavior of the displacement field requires sampling the propagating wave field, at an adequate sampling rate. The characterization of the dispersive behavior of the tissue requires reacquiring all of the data at multiple frequencies. Therefore, while dynamic MRE, in principle, can provide all these data, the time involved to get this comprehensive characterization of the wave field is often clinically unrealistic. Developing methods to reduce the total acquisition time of dynamic MRE is an area of active research.

From a strictly reconstruction point of view, it is important to recognize that the clinical choice of reconstruction algorithm may ultimately depend more on the ability of the algorithm to provide clinically relevant distinctions between pathological states of tissue in a robust manner, with least discomfort to the patient at the time of data acquisition, than on the ability to provide an accurate estimate of elasticity based on a comprehensive set of high-SNR, dynamic MRE data. This recognition spurs the continued development of acquisition methods and tailoring reconstruction algorithms with judicious assumptions to extract clinically relevant mechanical properties of tissues.

4.5.3 Reconstruction of Transient Wave Motion Image Processing

A transient mechanical excitation may be initiated either by an actuator imparting a single cycle of mechanical excitation at a given frequency [44] or by an impulse pulse delivered by methods such as ARF using a focused ultrasound transducer [127,130,131]. Unlike single-frequency or multi-frequency harmonic excitation associated with steady-state harmonic excitation, it should be noted that the transient wave excitation is broadband in nature by virtue of the apodization of the mechanical excitation. If the spectral sensitivity of the MEG is also broad, then tissue motion at all these frequencies will be encoded in the received MR signal. Therefore, for a viscoelastic material, any shear elastic modulus calculation based on the wave speed should be considered as a weighted average of the stiffness values at different frequencies that are encoded by the MEG.

Reconstruction algorithms for transient mechanical wave propagation rely primarily on two forms of analysis, that is, the time-of-arrival method and local inversion of the propagating transient pulse.

4.5.3.1 Time of Arrival

Estimation of tissue mechanical properties using the time-of-arrival method relies on the ability of the acquisition technique to capture a propagating transient shear wave at multiple time points (Figure 4.12). For example, in a homogeneous medium, the distance traveled by the wave front from the source of mechanical disturbance at different time points can be used to calculate the wave speed (v_s). In lieu of such wave front tracking, it is possible to calculate the time of arrival for each pixel, by fitting the temporal displacement to a higher-order polynomial [44]. The wave speed on each pixel can then be accurately calculated by performing a one-dimensional gradient along the direction of wave traveling on these time-of-arrival images. For a shear wave, then, the shear modulus μ at each pixel of the material is calculated from the wave speed image:

$$\mu = \rho v_s^2. \tag{4.30}$$

As discussed in the previous section, this shear elastic modulus estimated from the wave speed should be considered as a *weighted average* of all spectral components of the transient wave encoded by the MEG. There are some key differences between transient wave images acquired from externally applied transient excitation (one cycle of harmonic motion) and from internally initiated transient excitation by ARF.

In the case of transient wave images generated using *externally applied transient excitation*, the mechanical waves have to propagate from the skin surface to the target anatomy of interest. As a result, by the time the waves reach the target anatomy, there could be substantial attenuation, and the SNR_d may be too small to permit accurate evaluation of time of arrival. Second, this technique is sensitive to wave diffraction effects such as reflection, refraction, and so forth, which could confound a clear estimation of the direction of wave propagation (Figure 4.13).

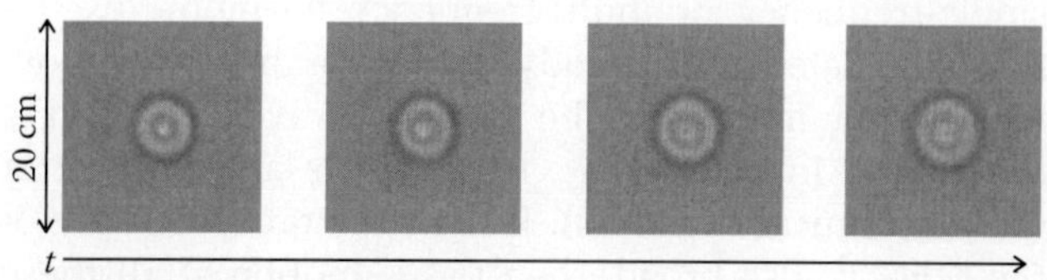

FIGURE 4.12
Shear wave initiated by ARF is captured at 0.4 ms intervals in a homogenous tissue-mimicking phantom. Note the spatial progress of the shear wave front at increasing temporal delays. The effective shear wave speed is calculated as 4.01 ± 0.06 m/s by tracking the wave front.

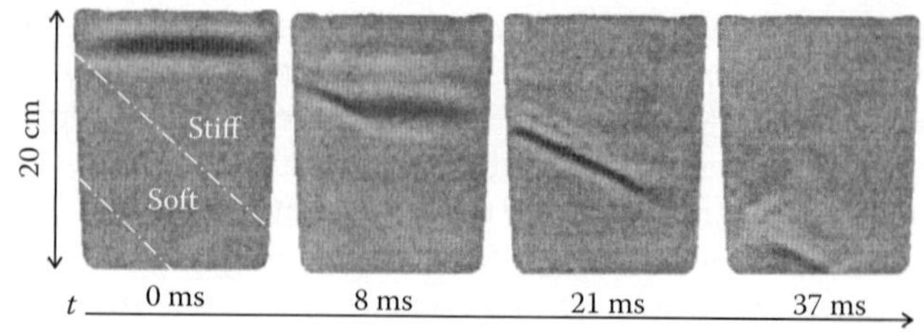

FIGURE 4.13
A single period of a harmonic mechanical excitation applied at the surface of a tissue-mimicking phantom is captured at progressively increasing temporal offsets between the application of the mechanical excitation and the MEG. The phantom consists of materials at two different stiffness values; notice the effect of reflections at the interfaces, as well as the attenuation of the propagating wave deep within the tissue.

In the case of transient impulses generated deep within the body via ARF, the reduction of displacement SNR_d due to wave propagation from the surface to the region of interest is not as significant an issue. ARF-based methods also provide the flexibility to selectively inject mechanical waves at specific regions of interest with phased-array focused ultrasound transducers. However, methods relying on the ARF for transient wave generation are constrained by the availability of suitable acoustic windows.

4.5.3.2 Transient Direct Inversion

This method is fundamentally analogous to the inversion techniques described in the case of steady-state mechanical excitation. Unlike the steady-state case, the shear modulus is calculated by local inversion of the wave equation as [44]

$$\mu = \rho \frac{\partial^2 u / \partial t^2}{\nabla^2 u}. \tag{4.31}$$

The elasticity reconstruction of this transient direct inversion algorithm relies on the proper calculation of local spatial and second temporal derivatives of displacement at each location, and thus, the elasticity can be estimated through time. A suitable spatiotemporal window must be chosen to reliably estimate these spatiotemporal derivatives.

Similar to the steady-state inversion-based reconstruction methods, inversion of transient waves is also noise sensitive due to the need for spatiotemporal derivative operations to estimate shear-wave velocity. Reliable calculation of local derivatives requires sufficient displacement amplitude in the spatiotemporal domain; for example, if the spatiotemporal window encompasses regions where the shear waves are yet to arrive, then the local estimates of shear-wave velocity will be inaccurate. To mitigate this problem, McCracken et al. [44]

proposed a final stiffness estimate that is weighted by the displacement amplitude of the underlying wave images.

4.6 Current Clinical Applications and MRE Challenges

Among the MRE methods, the dynamic steady-state MRE method has been investigated widely for potential clinical applications. Steady-state MRE is attractive as a clinical elasticity imaging tool because of its noninvasiveness, ability to evaluate elasticity over a large area encompassing entire organs, and operator independence. More importantly, dynamic steady-state MRE provides a unique opportunity to fully characterize the displacement vector (3-D motion encoding) and wave vector (volumetric data acquisition), as well as the ability to repeat measurements at multiple frequencies (to evaluate dispersive properties of tissue). This tremendous potential to provide a wealth of information about the mechanical properties of tissue has prompted researchers to investigate its clinical value in the evaluation of a number of organs, including the breast [40,45,116], liver [47,108], brain [44,51,114,115], heart [58,112,113,135], lung [117,118], cartilage [53], skeletal muscle [35,41,111], prostate [55], spleen [136], thyroid [137], and vasculature [138] (see also Chapters 1–2 and 4–5).

Of these investigations, arguably the most mature clinical applications are in the evaluation of hepatic fibrosis (diffuse disease) and characterization of focal breast lesions [40,116,134]. We will provide a brief overview of these two clinical applications; for other emerging clinical applications, the reader is referred elsewhere [79,91,139].

4.6.1 Hepatic Fibrosis Assessment with MRE

An array of pathologic processes such as autoimmune diseases, hepatitis C and other infections, fatty liver disease, alcohol use, and metabolic syndrome can cause hepatic fibrosis. An accurate estimate of the degree of fibrosis is a useful tool for risk stratification and appropriate patient management. The current gold standard for estimating liver fibrosis is invasive needle biopsies with connective tissue staining. Needle biopsies involve patient morbidity, a small but nonnegligible risk of patient complications (0.01%–0.1%), and inherent sampling errors associated with biopsy [140,141]. MRE has been widely investigated for the diagnosis of hepatic diseases [48]. Liver stiffness measured using dynamic steady-state hepatic MRE (at 60 Hz) progressively increased with the stage of liver fibrosis [54,106,108] (Figure 4.14). Based on receiver operating characteristic (ROC) analysis, a cutoff of 2.93 kPa

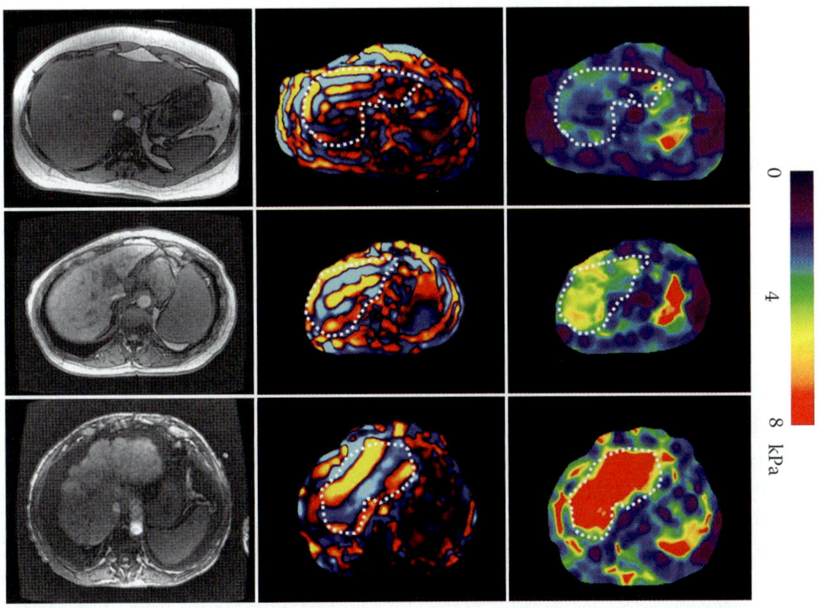

FIGURE 4.14
Conventional MRI magnitude (left), shear-wave (middle), and shear stiffness (right) images obtained in three clinical subjects. Images from a 29-year-old female with a fatty liver and history of elevated liver tests are shown in top row. The liver stiffness was estimated to be 2.7 kPa (range: 2.6–2.7 kPa). The images in the middle row were obtained in a 64-year-old female with primary biliary cirrhosis, and stiffness of the liver in this subject was estimated to be 5.1 kPa (range: 4.7–5.4 kPa), reflected by the longer wavelength. Images in the bottom row were obtained in a 65-year-old male with portal hypertension and advanced-stage hepatocellular carcinoma (alcohol related) with portal invasion, and the liver stiffness was substantially higher and estimated to be 10.2 kPa (range: 6.9–12.0 kPa). (Courtesy of Dr. Richard L. Ehman, Mayo Clinic, Rochester, Minnesota.)

was found to be an optimal threshold for distinguishing healthy livers from fibrotic ones, with sensitivity and specificity values of 98% and 99%, respectively.

In addition to the estimation of diffuse liver diseases such as fibrosis, MRE is also being investigated as a means to characterize hepatic tumors. Venkatesh et al. [142] have shown that malignant liver tumors have significantly greater mean shear stiffness than benign tumors and normal liver tissue, and a cutoff value of 5 kPa can differentiate malignant tumors from benign tumors and normal liver parenchyma.

4.6.2 Breast Cancer Assessment with MRE

One in eight women are at risk for developing breast cancer in their lifetime, and in fact, breast cancer is the most commonly diagnosed cancer in women [143,144]. If breast cancer is detected at a localized state, the 5-year survival rate in women is over 90%, in contrast to the poor prognosis when the cancer is detected at a later stage [145]. Dynamic contrast-enhanced MR imaging (DCE-MRI) has near-perfect sensitivity to detect breast cancer but has poor specificity, resulting in a number of unnecessary biopsies. Over the years, a number of ultrasound-based methods have been used to exploit the inherent difference in stiffness between normal and pathologic states of breast tissue [146,147] with varying degrees of diagnostic performance. Early dynamic MRE studies, in a small number of subjects, also found that the breast lesions were stiffer than normal breast tissue [40,116]. More recently, in a larger cohort of patients, Sinkus et al. [134] found that the specificity of MR-based detection of breast lesions can be improved from 40% to 60%, at 100% sensitivity, by the addition of MRE information to conventional DCE-MRI-based clinical decision making. These promising results might pave way for MRI-based screening in selective patient populations at higher risk for breast cancer.

4.6.3 MRE Challenges

Despite its tremendous promise and demonstrated clinical utility in some areas, MRE is still in its infancy with vibrant research in all aspects, such as inducing mechanical excitation within the body, developing appropriate MR pulse sequences for data acquisition, and image reconstruction algorithms. As an imaging modality, MRE has all the restrictions associated with MR as an imaging modality, for example, patient claustrophobia, inability to image patients with non–MR-compatible implants, and so forth. MRE is difficult to apply in tissues with very short T_2 or T_2^*, for example, in liver tissue with significant iron deposition. Further refinement of existing pulse sequences such as ultrashort TE sequences may mitigate this current limitation of MRE methods in its application to the evaluation of tissues with short T_2 or T_2^*. As is typical of most emerging imaging modalities, the lack of standardization on all aspects of MRE is a significant barrier to its routine clinical adoption. MRE techniques range from estimating elasticity from 2-D slices, with motion sensitization along a single direction, to more comprehensive data acquisition methods that acquire data volumetrically with multiaxis motion encoding [38,114,148]. Reconstruction algorithms range from the basic phase gradient methods to more complicated variants of inversion approaches. Although MRE is beginning to be used by several groups, there have been few published multicenter studies to demonstrate clinical validity of MRE in the hands of nonexpert users. Such multicenter studies are essential for widespread acceptance of MRE in clinical practice and help address reimbursement issues in this cost-conscious health care environment.

References

1. Bynum, W.F., and Porter, R. 2005. *Medicine and the Five Senses*. Cambridge University Press, Cambridge, UK.
2. Elias, D., Sideris, L., Pocard, M., de Baere, T., Dromain, C., Lassau, N., and Lasser, P. 2005. Incidence of unsuspected and treatable metastatic disease associated with operable colorectal liver metastases discovered only at laparotomy (and not treated when performing percutaneous radiofrequency ablation). *Ann Surg Oncol* 12: 298–302.
3. Lolsky, H. 1963. *Stress Waves in Solids*. New York: Dover.
4. Truong, X.T., Jarrett, S.R., and Nguyen, M.C. 1978. Method for deriving viscoelastic modulus for skeletal-muscle from transient pulse-propagation. *IEEE Trans Biomed Eng* 25:382–384.
5. Dickinson, R.J., and Hill, C.R. 1980. An ultrasonic method of analyzing tissue motion. *Br J Radiol* 53:626–627.
6. Wilson, L.S., and Robinson, D.E. 1982. Ultrasonic measurement of small displacements and deformations of tissue. *Ultrason Imaging* 4:71–82.
7. Eisenscher, A., Schweg-Toffler, E., Pelletier, G., and Jacquemard, P. 1983. Rhythmic echographic palpation. Echosismography. A new technic of differentiating benign and malignant tumors by ultrasonic study of tissue elasticity. *J Radiol* 64:255–261.
8. Tristam, M., Barbosa, D.C., Bamber, J.C., Cosgrove, D.O., and Hill, C.R. 1986. Ultrasonic analysis of soft-tissue movement in clinical-assessment of malignant-tumors. *Ultrasonic Imaging* 8:45–46.
9. Krouskop, T.A., Dougherty, D.R., and Vinson, F.S. 1987. A pulsed Doppler ultrasonic system for making noninvasive measurements of the mechanical properties of soft tissue. *J Rehabil Res Dev* 24:1–8.

10. Zerhouni, E.A., Parish, D.M., Rogers, W.J., Yang, A., and Shapiro, E.P. 1988. Human heart: Tagging with MR imaging—A method for noninvasive assessment of myocardial motion. *Radiology* 169:59–63.
11. Mosher, T.J., and Smith, M.B. 1990. A DANTE tagging sequence for the evaluation of translational sample motion. *Magn Reson Med* 15:334–339.
12. Parker, K.J., Huang, S.R., Musulin, R.A., and Lerner, R.M. 1990. Tissue response to mechanical vibrations for "sonoelasticity imaging." *Ultrasound Med Biol* 16:241–246.
13. Lewa, C.J. 1991. Magnetic-resonance-imaging in the presence of mechanical waves. *Spectrosc Lett* 24:55–67.
14. Ophir, J., Cespedes, I., Ponnekanti, H., Yazdi, Y., and Li, X. 1991. Elastography: A quantitative method for imaging the elasticity of biological tissues. *Ultrason Imaging* 13:111–134.
15. Skovoroda, A.R., Lubinski, M.A., Shapo, B.M., and O'Donnell, M. 1992. Ultrasound elasticity imaging using Fourier based speckle tracking algorithm. *Ultrason Symp, 1992. Proc, IEEE* 2:1065–1068.
16. Cespedes, I., Ophir, J., Ponnekanti, H., and Maklad, N. 1993. Elastography: Elasticity imaging using ultrasound with application to muscle and breast in vivo. *Ultrason Imaging* 15:73–88.
17. O'Donnell, M., Skovoroda, A.R., Shapo, B.M., and Emelianov, S.Y. 1994. Internal displacement and strain imaging using ultrasonic speckle tracking. *IEEE Trans Ultrason Ferroelectr Freq Control* 41:314–325.
18. Emelianov, S.Y., Lubinski, M.A., Weitzel, W.F., [illegible], and O'Donnell, M. 1995. Elasticity imaging for early detection of renal pathology. *Ultrasound Med Biol* 21:871–883.
19. Fowlkes, J.B., Emelianov, S.Y., Pipe, J.G., Skovoroda, A.R., Carson, P.L., Adler, R.S., and Sarvazyan, A.P. 1995. Magnetic-resonance-imaging techniques for detection of elasticity variation. *Med Phys* 22:1771–1778.
20. Ponnekanti, H., Ophir, J., Huang, Y., and Cespedes, I. 1995. Fundamental mechanical limitations on the visualization of elasticity contrast in elastography. *Ultrasound Med Biol* 21:533–543.
21. Muthupillai, R., Lomas, D.J., Rossman, P.J., Greenleaf, J.F., Manduca, A., and Ehman, R.L. 1995. Magnetic resonance elastography by direct visualization of propagating acoustic strain waves. *Science* 269:1854–1857.
22. Levinson, S.F., Shinagawa, M., and Sato, T. 1995. Sonoelastic determination of human skeletal-muscle elasticity. *J Biomech* 28:1145–1154.
23. Rubens, D.J., Hadley, M.A., Alam, S.K., Gao, L., Mayer, R.D., and Parker, K.J. 1995. Sonoelasticity imaging of prostate cancer: In vitro results. *Radiology* 195:379–383.
24. Gao, L., Parker, K.J., Alam, S.K., and Lernel, R.M. 1995. Sonoelasticity imaging: Theory and experimental verification. *J Acoust Soc Am* 97:3875–3886.
25. Smith, J.A., Muthupillai, R., Greenleaf, J.F., Rossman, P.J., Hulshizer, T.C., and Ehman, R.L. 1996. Tissue characterization with MR elastography. *Radiology* 201:362–362.
26. Lubinski, M.A., Emelianov, S.Y., Raghavan, K.R., Yagle, A.E., Skovoroda, A.R., and O'Donnell, M. 1996. Lateral displacement estimation using tissue incompressibility. *IEEE Trans Ultrason Ferroelectr Freq Control* 43:247–256.
27. Garra, B.S., Cespedes, E.I., Ophir, J., Spratt, S.R., Zuurbier, R.A., Magnant, C.M., and Pennanen, M.F. 1997. Elastography of breast lesions: Initial clinical results. *Radiology* 202:79–86.
28. Greenleaf, J.F., Dutt, V., Muthupillai, R., Manduca, A., and Ehman, R.L. 1996. Measurement of shear waves in tissue. *Engineering in Medicine and Biology Society, 1996. Bridging Disciplines for Biomedicine. Proceedings of the 18th Annual International Conference of the IEEE* 5:2172–2173.
29. Andreev, V.G., Dmitriev, V.N., Pishchalnikov, Y.A., Rudenko, O.V., Sapozhnikov, O.A., and Sarvazyan, A.P. 1997. Observation of shear waves excited by focused ultrasound in a rubber-like medium. *Acoust Phys* 43:123–128.
30. Walker, C.L., Foster, F.S., and Plewes, D.B. 1998. Magnetic resonance imaging of ultrasonic fields. *Ultrasound Med Biol* 24:137–142.
31. Bishop, J., Samani, A., Sciarretta, J., and Plewes, D.B. 2000. Two-dimensional MR elastography with linear inversion reconstruction: Methodology and noise analysis. *Phys Med Biol* 45:2081–2091.
32. Kruse, S.A., Smith, J.A., Lawrence, A.J., Dresner, M.A., Manduca, A., Greenleaf, J.F., and Ehman, R.L. 2000. Tissue characterization using magnetic resonance elastography: Preliminary results. *Phys Med Biol* 45:1579–1590.
33. Plewes, D.B., Bishop, J., Samani, A., and Sciarretta, J. 2000. Visualization and quantification of breast cancer biomechanical properties with magnetic [illegible]. *Phys Med Biol* 45:1591–1610.
34. Wu, T., Felmlee, J.P., Greenleaf, J.F., Riederer, S.J., and Ehman, R.L. 2000. MR imaging of shear waves generated by focused ultrasound. *Magn Reson Med* 43:111–115.
35. Dresner, M.A., Rose, G.H., Rossman, P.J., Muthupillai, R., Manduca, A., and Ehman, R.L. 2001. Magnetic resonance elastography of skeletal muscle. *J Magn Reson Imaging* 13:269–276.
36. Oliphant, T.E., Manduca, A., Ehman, R.L., and Greenleaf, J.F. 2001. Complex-valued stiffness reconstruction for magnetic resonance elastography by algebraic inversion of the differential equation. *Magn Reson Med* 45:299–310.
37. Van Houten, E.E., Miga, M.I., Weaver, J.B., Kennedy, F.E., and Paulsen, K.D. 2001. Three-dimensional subzone-based reconstruction algorithm for MR elastography. *Magn Reson Med* 45:827–837.
38. Weaver, J.B., Van Houten, E.E.W., Miga, M.I., Kennedy, F.E., and Paulsen, K.D. 2001. Magnetic resonance elastography using 3D gradient echo measurements of steady-state motion. *Med Phys* 28:1620–1628.
39. Wu, T., Felmlee, J.P., Greenleaf, J.F., Riederer, S.J., and Ehman, R.L. 2001. Assessment of thermal tissue ablation with MR elastography. *Magn Reson Med* 45:80–87.
40. McKnight, A.L., Kugel, J.L., Rossman, P.J., Manduca, A., Hartmann, L.C., and Ehman, R.L. 2002. MR elastography of breast cancer: Preliminary results. *AJR Am J Roentgenol* 178:1411–1417.
41. Sack, I., Bernarding, J., and Braun, J. 2002. Analysis of wave patterns in MR elastography of skeletal muscle using coupled harmonic oscillator simulations. *Magn Reson Imaging* 20:95–104.

42. Calle, S., Remenieras, J.P., Matar, O.B., Hachemi, M.E., and Patat, F. 2005. Temporal analysis of tissue displacement induced by a transient ultrasound radiation force. *J Acoust Soc Am* 118:2829–2840.
43. Chen, Q., Ringleb, S.I., Manduca, A., Ehman, R.L., and An, K.N. 2005. A finite element model for analyzing shear wave propagation observed in magnetic resonance elastography. *J Biomech* 38:2198–2203.
44. McCracken, P.J., Manduca, A., Felmlee, J., and Ehman, R.L. 2005. Mechanical transient-based magnetic resonance elastography. *Magn Reson Med* 53:628–639.
45. Sinkus, R., Tanter, M., Catheline, S., Lorenzen, J., Kuhl, C., Sondermann, E., and Fink, M. 2005. Imaging anisotropic and viscous properties of breast tissue by magnetic resonance-elastography. *Magn Reson Med* 53:372–387.
46. Glaser, K.J., Felmlee, J.P., and Ehman, R.L. 2006. Rapid MR elastography using selective excitations. *Magn Reson Med* 55:1381–1389.
47. Huwart, L., Peeters, F., Sinkus, R., Annet, L., Salameh, N., ter Beek, L.C., Horsmans, Y., and Van Beers, B.E. 2006. Liver fibrosis: Non-invasive assessment with MR elastography. *NMR Biomed* 19:173–179.
48. Rouviere, O., Yin, M., Dresner, M.A., Rossman, P.J., Burgart, L.J., Fidler, J.L., and Ehman, R.L. 2006. MR elastography of the liver: Preliminary results. *Radiology* 240:440–448.
49. Klatt, D., Hamhaber, U., Asbach, P., Braun, J., and Sack, I. 2007. Noninvasive assessment of the rheological behavior of human organs using multifrequency MR elastography: A study of brain and liver viscoelasticity. *Phys Med Biol* 52:7281–7294.
50. Rump, J., Klatt, D., Braun, J., Warmuth, C., and Sack, I. 2007. Fractional encoding of harmonic motions in MR elastography. *Magn Reson Med* 57:388–395.
51. Xu, L., Lin, Y., Xi, Z.N., Shen, H., and Gao, P.Y. 2007. Magnetic resonance elastography of the human brain: A preliminary study. *Acta Radiol* 48:112–115.
52. Asbach, P., Klatt, D., Hamhaber, U., Braun, J., Somasundaram, R., Hamm, B., and Sack, I. 2008. Assessment of liver viscoelasticity using multifrequency MR elastography. *Magn Reson Med* 60:373–379.
53. Lopez, O., Amrami, K.K., Manduca, A., and Ehman, R.L. 2008. Characterization of the dynamic shear properties of hyaline cartilage using high-frequency dynamic MR elastography. *Magn Reson Med* 59:356–364.
54. Talwalkar, J.A., Gross, J.B., Venkatesh, S.K., Yin, M., Glockner, J., Takahashi, N., Watt, K.D., Charlton, M.R., Kamath, P.S., and Ehman, R. 2008. Magnetic resonance elastography accurately detects moderate to severe hepatic fibrosis in obese individuals with Nafld. *Hepatology* 48: 810a.
55. Chopra, R., Arani, A., Huang, Y., Musquera, M., Wachsmuth, J., Bronskill, M., and Plewes, D. 2009. In vivo MR elastography of the prostate gland using a transurethral actuator. *Magn Reson Med* 62:665–671.
56. Mariappan, Y.K., Rossman, P.J., Glaser, K.J., Manduca, A., and Ehman, R.L. 2009. Magnetic resonance elastography with a phased-array acoustic driver system. *Magn Reson Med* 61:678–685.
57. Oestreicher, H.L. 1951. Field and impedance of an oscillating sphere in a viscoelastic medium with an application to biophysics. *J Acoust Soc Am* 23:701–714.
58. Sack, I., Rump, J., Elgeti, T., Samani, A., and Braun, J. 2009. MR elastography of the human heart: Noninvasive assessment of myocardial elasticity changes by shear wave amplitude variations. *Magn Reson Med* 61:668–677.
59. Asbach, P., Klatt, D., Schlosser, B., Biermer, M., Muche, M., Rieger, A., Loddenkemper et al. 2010. Viscoelasticity-based staging of hepatic fibrosis with multifrequency MR elastography. *Radiology* 257:80–86.
60. Cheng, S., Gandevia, S.C., Green, M., Sinkus, R., and Bilston, L.E. 2011. Viscoelastic properties of the tongue and soft palate using MR elastography. *J Biomech* 44:450–454.
61. Chung, S.H., Breton, E., Mannelli, L., and Axel, L. 2011. Liver stiffness assessment by tagged MRI of cardiac-induced liver motion. *Magn Reson Med* 65:949–955.
62. McGarry, M.D., Van Houten, E.E., Perrinez, P.R., Pattison, A.J., Weaver, J.B., and Paulsen, K.D. 2011. An octahedral shear strain-based measure of SNR for 3D MR elastography. *Phys Med Biol* 56:N153–N164.
63. Okamoto, R.J., Clayton, E.H., and Bayly, P.V. 2011. Viscoelastic properties of soft gels: Comparison of magnetic resonance elastography and dynamic shear testing in the shear wave regime. *Phys Med Biol* 56:6379–6400.
64. Olsen, D.A., Song, P., Glaser, K.J., and Ehman, R.L. 2011. Cardiac-gated hepatic MR elastography with intrinsic transient waveforms. *Proc Int Soc Magn Reson Med* 19.
65. Warner, L., Yin, M., Glaser, K.J., Woollard, J.A., Carrascal, C.A., Korsmo, M.J., Crane, J.A., Ehman, R.L., and Lerman, L.O. 2011. Noninvasive in vivo assessment of renal tissue elasticity during graded renal ischemia using MR elastography. *Invest Radiol* 46:509–514.
66. Garteiser, P., Doblas, S., Daire, J.L., Wagner, M., Leitao, H., Vilgrain, V., Sinkus, R., and Van Beers, B.E. 2012. MR elastography of liver tumours: Value of viscoelastic properties for tumour characterisation. *Eur Radiol* 22(1): 2169–2177.
67. McGrath, D.M., Foltz, W.D., Al-Mayah, A., Niu, C.J., and Brock, K.K. 2012. Quasi-static magnetic resonance elastography at 7 T to measure the effect of pathology before and after fixation on tissue biomechanical properties. *Magn Reson Med* 68:152–165.
68. Othman, S.F., Curtis, E.T., Plautz, S.A., Pannier, A.K., Butler, S.D., and Xu, H. 2012. MR elastography monitoring of tissue-engineered constructs. *NMR Biomed* 25: 452–463.
69. Rustogi, R., Horowitz, J., Harmath, C., Wang, Y., Chalian, H., Ganger, D.R., Chen, Z.E., Bolster, B.D. Jr., Shah, S., and Miller, F.H. 2012. Accuracy of MR elastography and anatomic MR imaging features in the diagnosis of severe hepatic fibrosis and cirrhosis. *J Magn Reson Imaging* 35:1356–1364.
70. Hirsch, S., Guo, J., Reiter, R., Papazoglou, S., Kroencke, T., Braun, J., and Sack, I. 2014. MR elastography of the liver and the spleen using a piezoelectric driver, single-shot wave-field acquisition, and multifrequency dual parameter reconstruction. *Magn Reson Med* 71(1):267–277.

71. Feng, Y., Clayton, E.H., Chang, Y., Okamoto, R.J., and Bayly, P.V. 2013. Viscoelastic properties of the ferret brain measured in vivo at multiple frequencies by magnetic resonance elastography. *J Biomech* 46:863–870.
72. Green, M.A., Geng, G., Qin, E., Sinkus, R., Gandevia, S.C., and Bilston, L.E. 2013. Measuring anisotropic muscle stiffness properties using elastography. *NMR Biomed* 26(11):1387–1394.
73. Clayton, E.H., Okamoto, R.J., and Bayly, P.V. 2013. Mechanical properties of viscoelastic media by local frequency estimation of divergence-free wave fields. *J Biomech Eng* 135:021025.
74. Weaver, J.B., Pattison, A.J., McGarry, M.D., Perreard, I.M., Swienckowski, J.G., Eskey, C.J., Lollis, S.S., and Paulsen, K.D. 2012. Brain mechanical property measurement using MRE with intrinsic activation. *Phys Med Biol* 57:7275–7287.
75. Serai, S.D., Towbin, A.J., and Podberesky, D.J. 2012. Pediatric liver MR elastography. *Dig Dis Sci* 57(10): 2713–2719.
76. Ichikawa, S., Motosugi, U., Ichikawa, T., Sano, K., Morisaka, H., Enomoto, N., Matsuda, M., Fujii, H., and Araki, T. 2012. Magnetic resonance elastography for staging liver fibrosis in chronic hepatitis C. *Magn Reson Med Sci* 11:291–297.
77. Sarvazyan, A.P., Pasechni, V.I., and Shnol, S.E. 1968. Low velocity of sound in gels and protoplasmatic structures. Possible biological significance of this phenomenon. *Biofizika* 13:587–594.
78. Sarvazyan, A.P., Skovoroda, A.R., Emelianov, S.Y., Fowlkes, J.B., Pipe, J.G., Adler, R.S., Buxton, R.B., and Carson, P.L. 1995. Biophysical bases of elasticity imaging. *Acoust Imaging* 21:223–240.
79. Mariappan, Y.K., Glaser, K.J., and Ehman, R.L. 2010. Magnetic resonance elastography: A review. *Clin Anat* 23:497–511.
80. Fung, Y.C. 1993. *Biomechanics: Mechanical Properties of Living Tissues*. 2. New York: Springer-Verlag.
81. Birnholz, J.C., and Farrell, E.E. 1985. Fetal lung development: Compressibility as a measure of maturity. *Radiology* 157:495–498.
82. Dejong, P.G.M., Arts, T., Hoeks, A.P.G., and Reneman, R.S. 1990. Determination of tissue motion velocity by correlation interpolation of pulsed ultrasonic echo signals. *Ultrasonic Imaging* 12:84–98.
83. Parker, K.J., and Lerner, R.M. 1992. Sonoelasticity of organs: Shear waves ring a bell. *J Ultrasound Med* 11:387–392.
84. Lerner, R.M., Huang, S.R., and Parker, K.J. 1990. "Sonoelasticity" images derived from ultrasound signals in mechanically vibrated tissues. *Ultrasound Med Biol* 16:231–239.
85. Yamakoshi, Y., Sato, J., and Sato, T. 1990. Ultrasonic imaging of internal vibration of soft tissue under forced vibration. *IEEE Trans Ultrason Ferroelectr Freq Control* 37:45–53.
86. Auld, B.A. 1990. *Acoustic Fields and Waves in Solids*. Malabar, FL: Krieger.
87. Tschoegl, N.W. 1989. *The Phenomenological Theory of Linear Viscoelastic Behavior: An Introduction*. Berlin Heidelberg, German: Springer-Verlag.
88. Timoshenko, S., and Goodier, J. 1951. *Theory of Elasticity*. New York: McGraw-Hill.
89. Landau, L., and Lifshitz, E. 1987. *Fluid Mechanics*. New York: Pergamon.
90. Sinkus, R., Tanter, M., Xydeas, T., Catheline, S., Bercoff, J., and Fink, M. 2005. Viscoelastic shear properties of in vivo breast lesions measured by MR elastography. *Magn Reson Imaging* 23:159–165.
91. Sarvazyan, A., Hall, T.J., Urban, M.W., Fatemi, M., Aglyamov, S.R., and Garra, B.S. 2011. An overview of elastography—An emerging branch of medical imaging. *Curr Med Imaging Rev* 7:255–282.
92. Manduca, A., Oliphant, T.E., Dresner, M.A., Mahowald, J.L., Kruse, S.A., Amromin, E., Felmlee, J.P., Greenleaf, J.F., and Ehman, R.L. 2001. Magnetic resonance elastography: Non-invasive mapping of tissue elasticity. *Med Image Anal* 5:237–254.
93. Westenberg, J.J., van Poelgeest, E.P., Steendijk, P., Grotenhuis, H.B., Jukema, J.W., and de Roos, A. 2012. Bramwell–Hill modeling for local aortic pulse wave velocity estimation: A validation study with velocity-encoded cardiovascular magnetic resonance and invasive pressure assessment. *J Cardiovasc Magn Reson* 14:2.
94. Watanabe, H., Kanematsu, M., Kitagawa, T., Suzuki, Y., Kondo, H., Goshima, S., Kajita, K. et al. 2010. MR elastography of the liver at 3 T with cine-tagging and bending energy analysis: Preliminary results. *Eur Radiol* 20:2381–2389.
95. Ibrahim el, S.H. 2011. Myocardial tagging by cardiovascular magnetic resonance: Evolution of techniques—Pulse sequences, analysis algorithms, and applications. *J Cardiovasc Magn Reson* 13:36.
96. Bishop, J., Samani, A., Sciarretta, J., Luginbuhl, C., and Plewes, D.B. 2001. A signal/noise analysis of quasi-static MR elastography. *IEEE Trans Med Imaging* 20:1183–1187.
97. Chenevert, T.L., Skovoroda, A.R., O'Donnell, M., and Emelianov, S.Y. 1998. Elasticity reconstructive imaging by means of stimulated echo MRI. *Magn Reson Med* 39:482–490.
98. Fu, Y.B., Chui, C.K., Teo, C.L., and Kobayashi, E. 2011. Motion tracking and strain map computation for quasi-static magnetic resonance elastography. *Med Image Comput Comput Assist Interv* 14:428–435.
99. Samani, A., Bishop, J., and Plewes, D.B. 2001. A constrained modulus reconstruction technique for breast cancer assessment. *IEEE Trans Med Imaging* 20:877–885.
100. Skovoroda, A.R., Lubinski, L.A., Emelianov, S.Y., and O'Donnell, M. 1999. Reconstructive elasticity imaging for large deformations. *IEEE Trans Ultrason Ferroelectr Freq Control* 46:523–535.
101. Muthupillai, R., and Ehman, R.L. 1996. Magnetic resonance elastography. *Nat Med* 2:601–603.
102. Hamhaber, U., Grieshaber, F.A., Nagel, J.H., and Klose, U. 2003. Comparison of quantitative shear wave MR-elastography with mechanical compression tests. *Magn Reson Med* 49:71–77.
103. Van Houten, E.E., Doyley, M.M., Kennedy, F.E., Weaver, J.B., and Paulsen, K.D. 2003. Initial in vivo experience with steady-state subzone-based MR elastography of the human breast. *J Magn Reson Imaging* 17:72–85.

104. Sack, I., McGowan, C.K., Samani, A., Luginbuhl, C., Oakden, W., and Plewes, D.B. 2004. Observation of nonlinear shear wave propagation using magnetic resonance elastography. *Magn Reson Med* 52:842–850.
105. Othman, S.F., Xu, H., Royston, T.J., and Magin, R.L. 2005. Microscopic magnetic resonance elastography (microMRE). *Magn Reson Med* 54:605–615.
106. Venkatesh, S.K., Yin, M., and Ehman, R.L. 2013. Magnetic resonance elastography of liver: Technique, analysis, and clinical applications. *J Magn Reson Imaging* 37:544–555.
107. Kim, D., Kim, W.R., Talwalkar, J.A., Kim, H.J., and Ehman, R.L. 2013. Advanced fibrosis in nonalcoholic fatty liver disease: Noninvasive assessment with MR elastography. *Radiology* 268(2):411–419.
108. Yin, M., Woollard, J., Wang, X., Torres, V.E., Harris, P.C., Ward, C.J., Glaser, K.J., Manduca, A., and Ehman, R.L. 2007. Quantitative assessment of hepatic fibrosis in an animal model with magnetic resonance elastography. *Magn Reson Med* 58:346–353.
109. Salameh, N., Peeters, F., Sinkus, R., Abarca-Quinones, J., Annet, L., Ter Beek, L.C., Leclercq, I., and Van Beers, B.E. 2007. Hepatic viscoelastic parameters measured with MR elastography: Correlations with quantitative analysis of liver fibrosis in the rat. *J Magn Reson Imaging* 26:956–962.
110. Bensamoun, S.F., Ringleb, S.I., Littrell, L., Chen, Q., Brennan, M., Ehman, R.L., and An, K.N. 2006. Determination of thigh muscle stiffness using magnetic resonance elastography. *J Magn Reson Imaging* 23:242–247.
111. Jenkyn, T.R., Ehman, R.L., and An, K.N. 2003. Noninvasive muscle tension measurement using the novel technique of magnetic resonance elastography (MRE). *J Biomech* 36:1917–1921.
112. Elgeti, T., Rump, J., Hamhaber, U., Papazoglou, S., Hamm, B., Braun, J., and Sack, I. 2008. Cardiac magnetic resonance elastography. Initial results. *Invest Radiol* 43:762–772.
113. Robert, B., Sinkus, R., Gennisson, J.L., and Fink, M. 2009. Application of DENSE-MR-elastography to the human heart. *Magn Reson Med* 62:1155–1163.
114. Hamhaber, U., Sack, I., Papazoglou, S., Rump, J., Klatt, D., and Braun, J. 2007. Three-dimensional analysis of shear wave propagation observed by in vivo magnetic resonance elastography of the brain. *Acta Biomater* 3:127–137.
115. Sack, I., Beierbach, B., Hamhaber, U., Klatt, D., and Braun, J. 2008. Non-invasive measurement of brain viscoelasticity using magnetic resonance elastography. *NMR Biomed* 21:265–271.
116. Lorenzen, J., Sinkus, R., Lorenzen, M., Dargatz, M., Leussler, C., Roschmann, P., and Adam, G. 2002. MR elastography of the breast: Preliminary clinical results. *Rofo* 174:830–834.
117. Goss, B.C., McGee, K.P., Ehman, E.C., Manduca, A., and Ehman, R.L. 2006. Magnetic resonance elastography of the lung: Technical feasibility. *Magn Reson Med* 56:1060–1066.
118. McGee, K.P., Hubmayr, R.D., Levin, D., and Ehman, R.L. 2009. Feasibility of quantifying the mechanical properties of lung parenchyma in a small-animal model using (1)H magnetic resonance elastography (MRE). *J Magn Reson Imaging* 29:838–845.
119. Gennisson, J.L., Muller, M., Ami, O., Kohl, V., Gabor, P., Musset, D., and Tanter, M. 2012. Shear wave elastography in obstetrics: Quantification of cervix elasticity and uterine contraction. *2011 IEEE International Ultrasonics Symposium (IUS)* 2094–2097.
120. Preis, K., Swiatkowska-Freund, M., and Pankrac, Z. 2010. Elastography in the examination of the uterine cervix before labor induction. *Ginekologia Polska* 81:757–761.
121. Sarvazyan, A.P., Rudenko, O.V., Swanson, S.D., Fowlkes, J.B., and Emelianov, S.Y. 1998. Shear wave elasticity imaging: A new ultrasonic technology of medical diagnostics. *Ultrasound Med Biol* 24:1419–1435.
122. Moran, P.R. 1982. A flow velocity zeugmatographic interlace for NMR imaging in humans. *Magn Reson Imaging* 1:197–203.
123. Haacke, E.M., and Lenz, G.W. 1987. Improving MR image quality in the presence of motion by using rephasing gradients. *AJR Am J Roentgenol* 148:1251–1258.
124. Gullberg, G.T., Wehrli, F.W., Shimakawa, A., and Simons, M.A. 1987. MR vascular imaging with a fast gradient refocusing pulse sequence and reformatted images from transaxial sections. *Radiology* 165:241–246.
125. Plewes, D.B., Betty, I., Urchuk, S.N., and Soutar, I. 1995. Visualizing tissue compliance with MR imaging. *J Magn Reson Imaging* 5:733–738.
126. Muthupillai, R., Rossman, P.J., Lomas, D.J., Greenleaf, J.F., Riederer, S.J., and Ehman, R.L. 1996. Magnetic resonance imaging of transverse acoustic strain waves. *Magn Reson Med* 36:266–274.
127. Nightingale, K., McAleavey, S., and Trahey, G. 2003. Shear-wave generation using acoustic radiation force: In vivo and ex vivo results. *Ultrasound Med Biol* 29:1715–1723.
128. Cho, S.H., Lee, J.Y., Han, J.K., and Choi, B.I. 2010. Acoustic radiation force impulse elastography for the evaluation of focal solid hepatic lesions: Preliminary findings. *Ultrasound Med Biol* 36:202–208.
129. Cheng, Y., Li, R., Li, S.N., Dunsby, C., Eckersley, R.J., Elson, D.S., and Tang, M.X. 2012. Shear wave elasticity imaging based on acoustic radiation force and optical detection. *Ultrasound Med Biol* 38:1637–1645.
130. Zhang, J., Hor, P.H., and Muthupillai, R. 2012. Rapid magnetic resonance imaging of displacement wave propagation introduced by a transient high intensity focused ultrasound using a gradient echo sequence. *Proceedings, 2012 Conference, International Society for Magnetic Resonance in Medicine* 20:2922.
131. Souchon, R., Salomir, R., Beuf, O., Milot, L., Grenier, D., Lyonnet, D., Chapelon, J.Y., and Rouviere, O. 2008. Transient MR elastography (t-MRE) using ultrasound radiation force: Theory, safety, and initial experiments in vitro. *Magn Reson Med* 60:871–881.
132. Knutsson, H., Westin, C.J., and Granlund, G. 1994. Local multiscale frequency and bandwidth estimation. *Proceedings of the IEEE International Conference on Image Processing* 1:36–40.
133. Romano, A.J., Bucaro, J.A., Ehman, R.L., and Shirron, J.J. 2000. Evaluation of a material parameter extraction algorithm using MRI-based displacement measurement. *IEEE Trans Ultrason Ferroelectr Freq Control* 47:1575–1581.

134. Sinkus, R., Siegmann, K., Xydeas, T., Tanter, M., Claussen, C., and Fink, M. 2007. MR elastography of breast lesions: Understanding the solid/liquid duality can improve the specificity of contrast-enhanced MR mammography. *Magn Reson Med* 58:1135–1144.
135. Kolipaka, A., McGee, K.P., Araoz, P.A., Glaser, K.J., Manduca, A., Romano, A.J., and Ehman, R.L. 2009. MR elastography as a method for the assessment of myocardial stiffness: Comparison with an established pressure-volume model in a left ventricular model of the heart. *Magn Reson Med* 62:135–140.
136. Talwalkar, J.A., Yin, M., Venkatesh, S., Rossman, P.J., Grimm, R.C., Manduca, A., Romano, A., Kamath, P.S., and Ehman, R.L. 2009. Feasibility of in vivo MR elastographic splenic stiffness measurements in the assessment of portal hypertension. *AJR Am J Roentgenol* 193:122–127.
137. Bahn, M.M., Brennan, M.D., Bahn, R.S., Dean, D.S., Kugel, J.L., and Ehman, R.L. 2009. Development and application of magnetic resonance elastography of the normal and pathological thyroid gland in vivo. *J Magn Reson Imaging* 30:1151–1154.
138. Woodrum, D.A., Herrmann, J., Lerman, A., Romano, A.J., Lerman, L.O., and Ehman, R.L. 2009. Phase-contrast MRI-based elastography technique detects early hypertensive changes in ex vivo porcine aortic wall. *J Magn Reson Imaging* 29:583–587.
139. Glaser, K.J., Manduca, A., and Ehman, R.L. 2012. Review of MR elastography applications and recent developments. *J Magn Reson Imaging* 36:757–774.
140. Thampanitchawong, P., and Piratvisuth, T. 1999. Liver biopsy: Complications and risk factors. *World J Gastroenterol* 5:301–304.
141. Mendez, P., Regev, A., Molina, E., Berho, M., Bejarano, P., Ladino, M., and Schiff, E. 2003. Sampling error and reliability of liver biopsy in patients with non-alcoholic fatty liver disease (NAFLD). *Hepatology* 38:673a.
142. Venkatesh, S.K., Yin, M., Glockner, J.F., Takahashi, N., Araoz, P.A., Talwalkar, J.A., and Ehman, R.L. 2008. MR elastography of liver tumors: Preliminary results. *AJR Am J Roentgenol* 190:1534–1540.
143. Feuer, E.J., Wun, L.M., Boring, C.C., Flanders, W.D., Timmel, M.J., and Tong, T. 1993. The lifetime risk of developing breast-cancer. *J Natl Cancer Inst* 85:892–897.
144. Howlader, N., Noone, A.M., and Krapcho, M. 2012. *SEER Cancer Statistics Review. 1975–2009 (Vintage 2009 Populations).* Bethesda, MD: National Cancer Institue. Retrieved September 7, 2012.
145. Jemal, A., Siegel, R., Ward, E., Hao, Y.P., Xu, J.Q., and Thun, M.J. 2009. Cancer Statistics. *CA Cancer J Clin* 59: 225–249.
146. Itoh, A., Ueno, E., Tohno, E., Kamma, H., Takahashi, H., Shiina, T., Yamakawa, M., and Matsumura, T. 2006. Breast disease: Clinical application of US elastography for diagnosis. *Radiology* 239:341–350.
147. Schaefer, F.K.W., Heer, I., Schaefer, P.J., Mundhenke, C., Osterholz, S., Order, B.M., Hofheinz, N., Hedderich, J., Heller, M., Jonat, W. et al. 2011. Breast ultrasound elastography-results of 193 breast lesions in a prospective study with histopathologic correlation. *Eur J Radiol* 77:450–456.
148. Muthupillai, R., Dutt, V., Rossman, P.J., Hulshizer, T.C., Manduca, A., and Ehman, R.L. 1997. Three dimensional magnetic resonance elastography with multi-axis motion encoding. *Proc Int Soc Magn Reson Med* 5:460.

5

Magnetic Resonance Elastography of the Brain

Philip V. Bayly, Erik H. Clayton, Guy M. Genin, and Ruth J. Okamoto

CONTENTS

Magnetic resonance elastography (MRE) allows noninvasive measurement of the mechanical properties of the living, intact brain. Quantitative measurements of the brain's material properties and structure are critical to understanding traumatic brain injury. In addition, mechanical properties of brain tissue change due to development, disease, and degeneration; such changes may lead to new methods for diagnosis or therapy. MRE studies have been performed recently in the human brain and in the brains of mice and other animals. In this chapter, we review the experimental methods used in these studies and the techniques to estimate mechanical properties from MRE data. We discuss the assumptions, technical choices, and limitations specific to MRE of the mammalian brain. Particular attention is given to safe and effective actuation; the relationships among frequency, sampling, and length scale; and the choice of an appropriate material model of brain tissue, which in general is nonlinear, dissipative, anisotropic, and heterogeneous.

5.1 Introduction

Magnetic resonance elastography (MRE) has emerged as an important technique to extract information about the mechanics of the brain (see also Chapters 2–4 and 6). In MRE, magnetic resonance images (MRIs) of waves propagating in soft tissue are acquired noninvasively and analyzed to estimate the tissue's mechanical properties (Muthupillai et al. 1995). Because the brain is well hidden and delicate, its mechanical behavior in vivo remains incompletely characterized.

MRE measurement of mechanical properties and their variations can illuminate the processes of injury, development, and disease in the human brain. Traumatic brain injury (TBI) is a leading cause of death and disability in children and young adults (Corrigan et al. 2010; Langlois et al. 2006), and TBI due to explosive blast is a common and debilitating injury among military personnel in combat (Hoge et al. 2008). In most

TBIs, the direct cause of injury is the large, fast deformation of brain tissue caused by acceleration of the skull. Brain deformation depends on tissue stiffness (shear modulus) as well as on the brain's anatomical structure and connections to the skull. The magnitudes, locations, and directions of deformation remain largely unknown, however, due to the difficulty of measuring or modeling the brain's mechanical response. Reliable computer simulations of TBI would be valuable, but such models require (1) accurate estimates of material parameters, (2) knowledge of mechanical interactions between the brain and skull, and perhaps most importantly, (3) data with which to compare model predictions. MRE can address all three requirements in vivo.

The process of brain development is also a topic of great interest and sustained research effort. Cortical folding occurs during the third trimester in humans and is critical to normal cognitive and emotional function. Van Essen (Van Essen 1997) has hypothesized that tensile stresses in axons cause folding. Others (Bayly et al. 2013; Richman et al. 1975; Xu et al. 2010) suggest that compressive stresses due to growth drive the process. Here, also, knowledge of the brain's mechanical properties in the early stages of development is essential to model and understand cortical folding.

Changes in the brain's material properties are thought to accompany normal aging (Sack et al. 2009, 2011). These changes may be amplified in degenerative disorders like Alzheimer's disease (Murphy et al. 2011, 2012). Brain tumors, like tumors in superficial tissue, also exhibit differences in stiffness relative to the healthy brain (Murphy et al. 2013; Xu et al. 2007). These changes in mechanical properties might be used for clinical purposes, for example, as techniques for diagnosis or evaluation of therapy. Macroscopic mechanical changes, because they are related to cellular and microstructural properties, might also elucidate the underlying biology of disease and degeneration.

The previous chapter outlines the basic principles of MRE. MRE involves three basic steps: (1) excitation of waves in tissue, (2) acquisition of images of these waves, and (3) analysis of the measured wave images to estimate material properties. Each of these three steps, *excitation*, *acquisition*, and *inversion*, must be performed carefully and evaluated critically to obtain useful and reliable information about brain mechanics.

5.2 Excitation

MRE requires the excitation of propagating waves in tissue. In the brain, this is accomplished by mechanical vibration applied to the skull; vibrations are transmitted to the brain through anatomical pathways. The location, frequency, amplitude, and direction of vibration will all affect the resulting wave field. In this section, we discuss the basis for the choice of excitation parameters and review examples from recent studies.

5.2.1 Actuators and Transmission Mechanisms

An electromechanical actuator is typically the core of any excitation system for brain MRE. Recent MRE studies have used the following actuators:

1. A self-contained electromagnetic (voice coil) actuator (Clayton et al. 2012; Johnson et al. 2013; Sack et al. 2008). These actuators are not magnetic resonance (MR) compatible; they contain magnetic components. They must therefore be placed outside the scanner's magnetic field. If used in the scanner room, extreme care is required to make sure that the actuator is not catastrophically drawn into the magnet.
2. A simple current-carrying coil inside the magnetic field of the scanner (Atay et al. 2008; Green et al. 2008; Hamhaber et al. 2007, 2010). The coil experiences (oscillatory) torque $\mathbf{M} = NiA\mathbf{n} \times \mathbf{B}_0$, where N is the number of turns, i is the (oscillatory) current in the coil, A is the area of the coil cross section, $\mathbf{n}$ is the axis of the coil, and $\mathbf{B}_0$ is the scanner's magnetic field. The coil itself is usually copper and thus safe to use in the scanner. The coil and any other metal components must be far enough from the imaged volume to avoid susceptibility artifacts.
3. A piezoelectric actuator (Clayton et al. 2011; Uffmann and Ladd 2008).

These three general actuation approaches have been used in both human and animal studies.

Human Studies

1. External electromagnetic actuation: Sack et al. (2008) and Johnson et al. (2013) coupled the actuator to the skull by a long rod connected to a cradle supporting the skull (Figure 5.1a). Clayton et al. (2012) used a commercially available system (Resoundant, Rochester, Minnesota) that transmits pressure waves through long tubes from an actuator in an adjacent room. Oscillatory pressure loading is delivered to the skull by passive "paddles," which are preloaded against the skull (Figure 5.1b). Murphy et al. (2011) used a similar device in a study of the effect of Alzheimer's disease on brain mechanical properties.

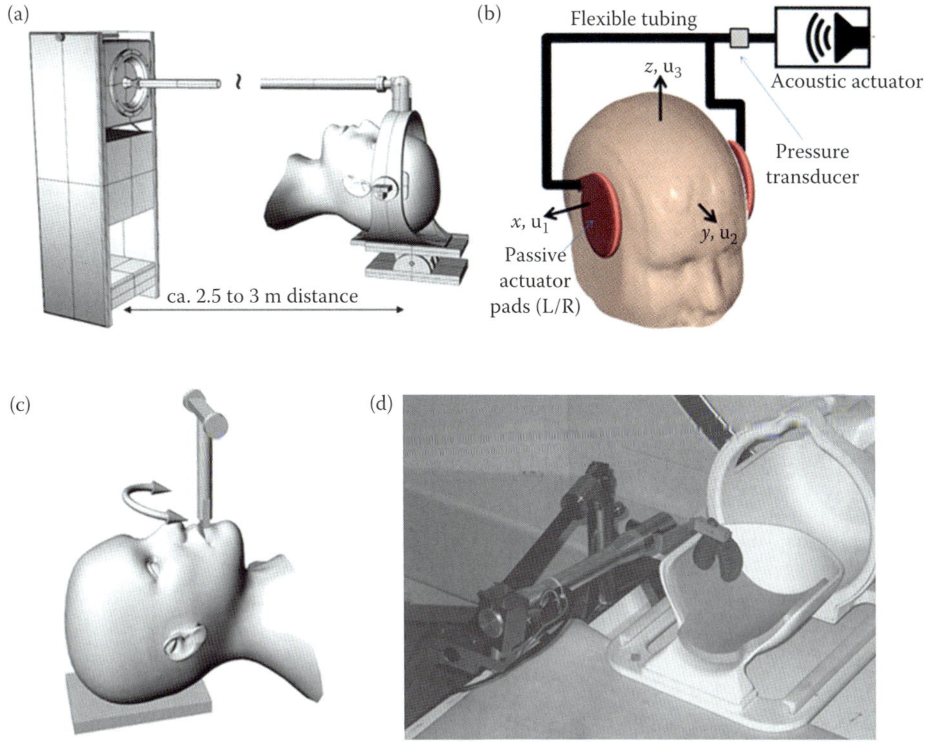

FIGURE 5.1
Actuation schemes for human brain MRE. (a) Head rocking device connected by a long rod to an electromagnetic actuator (speaker, or voice coil) outside the strong magnetic field of the scanner. (Reproduced from Sack, I. et al., *Neuroimage*, 46, 2009.) (b) Passive drumlike pads excited by acoustic pressure transmitted via flexible tubing from an electromagnetic voice coil actuator (Resoundant). (Reproduced from Clayton, E.H. et al., *J. R. Soc. Interface.*, 9, 2012.) (c) "Bite bar" actuated by alternating current passing through a coil in the magnetic field of the scanner. (Reproduced from Hamhaber, U. et al., *J. Magn. Reson. Imaging*, 32, 2010.) (d) Piezoelectrically actuated bite bar. (Reproduced from Uffmann, K. and Ladd, M.E., *IEEE Eng. Med. Biol. Mag.*, 27, 2008.)

2. Current-carrying coils inside the scanner magnetic field have been used to drive either a bite bar (Green et al. 2008; Hamhaber et al. 2010) (Figure 5.1c) or a cradle supporting the head (Kruse et al. 2008).
3. A piezoelectric actuator coupled to a bite bar was designed by Uffmann and Ladd (2008) (Figure 5.1d) for human brain MRE.

Animal and ex vivo Studies

1. An external electromagnetic actuator was used by Vappou et al. (2007) to drive a long rod and excite waves in the porcine brain ex vivo (Figure 5.2a) and rat brain in vivo (Vappou et al. 2008). Schregel et al. (2012) used an external actuator to vibrate a long rod inside the bore of a 7 T scanner, in order to study the effects of demyelination in the mouse brain by MRE. Pattison et al. (2010a) used a subwoofer (electromagnetic speaker) to provide acoustic pressure loading at 85 Hz in an MRE study of the feline brain.
2. A current-carrying coil inside an 11.7 T scanner was used by Atay et al. (2008) to induce waves in the mouse brain for MRE. To maximize transmission, the tip of a vibrating plastic arm was attached to the exposed skull.
3. Piezoelectric actuators are well suited to small-animal MRE studies, since they are compact, MR compatible, and capable of high frequencies. Clayton et al. (2011) and Feng et al. (2013a) used piezoelectric excitation in studies of the mouse and ferret brain, respectively (Figure 5.2b).

5.2.2 Amplitude and Frequency of Excitation

In studies involving human subjects, the most important concerns are the safety and comfort of the subject. Humans routinely experience transient skull accelerations due to normal activity (e.g., running or jumping) as well as vibrations due to driving or operating machinery. Dynamic mechanical strain amplitudes of 0.02–0.05

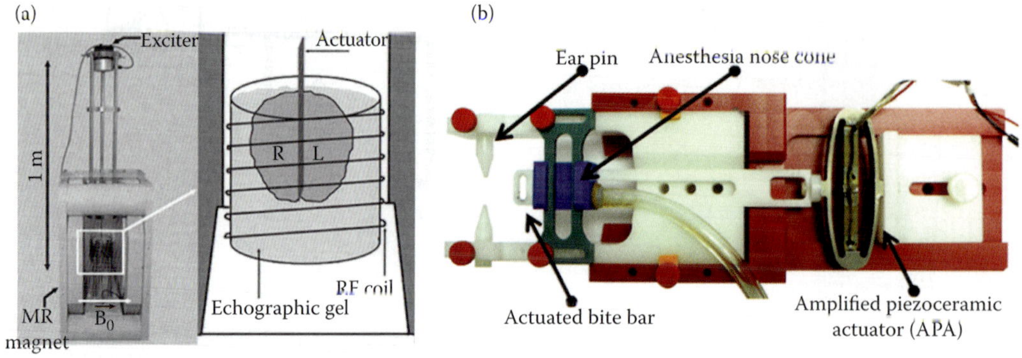

FIGURE 5.2
Example excitation systems for brain MRE studies in animal models. (a) Electromagnetic excitation system for MRE of the porcine brain in vitro. RF, radio frequency. (Reproduced from Vappou, J. et al., *MAGMA*, 20, 2007. With permission.) (b) MR-compatible piezoelectric actuator (APA-100, Cedrat Technologies) coupled to the incisor bar of a head-holding apparatus for MRE studies in the mouse, in vivo. (Reproduced from Clayton, E.H. et al., *Phys. Med. Biol.*, 56, 2011. With permission.) Finally, the intrinsic motion of the brain due to pulsatile blood flow has been used to estimate mechanical properties. (From Pattison, A.J. et al., Poroelastic MRE reconstructions of brain acquired with intrinsic activation, *Proceedings of the ISMRM*, 2010b, and Weaver, J.B. et al., *Phys. Med. Biol.*, 57, 2012.) The obvious benefit is elimination of the necessity for any external excitation; the cost is the lack of control of excitation parameters (amplitude, frequency, location, or direction).

(dimensionless) at strain rates on the order of 1/s have been observed at skull accelerations typical of normal activity (1G–3G or 10–30 m/s^2). Brain MRE studies typically involve harmonic excitation at frequencies between 25 and 100 Hz with tissue displacements of 5–50 μm (Ehman et al. 2008). Such displacement fields have wavelengths of several centimeters. In the worst-case scenario, if the highest displacements (50 μm) occurred at the highest frequencies (100 Hz) in shear waves with the shortest (2 cm) wavelength, maximal strain ≈ 0.0015 and strain rate ≈ 0.9/s would be observed. In short, the strains experienced in MRE are several orders of magnitude less than those associated with injury and significantly lower than those experienced in normal activity. Ehman et al. (2008) have also found the accelerations typical of MRE to be well within the European Union's recommended guidelines for vibration tolerance.

Strain amplitude also determines how well the small-strain approximation of linear elasticity describes the response of brain tissue during MRE. Most soft tissue exhibits nonlinear behavior at strain magnitudes from physiological levels (0.02–0.05) to injury levels (>0.20). However, due to the inherently small strain amplitudes noted in the previous paragraph, MRE only probes the linear behavior of brain tissue near its reference state. It is possible to apply MRE to tissue undergoing large quasi-static deformations, in order to investigate nonlinear mechanical properties.

Excitation frequency, *f*, is a critical parameter as it affects the wavelength and attenuation (and thus penetration depth) of waves. The material parameters of brain tissue, like other viscoelastic materials, depend on frequency. All other factors being equal, shorter wavelengths allow more waves to exist in a given domain, thus enabling higher accuracy and resolution in material parameter estimates. For a given set of material properties, the wavelength, λ, is inversely proportional to frequency: $\lambda = c/f$, where *c* is propagation speed (meters per second) and *f* is frequency in hertz. The propagation speed of shear waves in an isotropic linear elastic material depends on shear modulus, μ, and density, ρ: $c = \sqrt{\mu/\rho}$.

The frequency–wavelength relationship motivates researchers to increase excitation frequency, but excitation frequency is limited by several factors. First, all MRI data are spatially discrete. To satisfy basic Nyquist–Shannon sampling requirements (i.e., to avoid aliasing), the wavelength must be at least twice the voxel size (Therrien 2004). Experience suggests, however, that accurate material parameter estimates require closer to 10 voxels per wavelength (Papazoglou et al. 2008). Maximum frequency is thus limited by voxel size, which in turn is constrained by requirements for signal-to-noise ratio and acquisition time. It is also physically more difficult to produce measurable displacements at higher frequencies in deep tissue. High-frequency waves dissipate more rapidly in soft tissue (Clayton et al. 2012), and less displacement is provided by mechanical actuators at high frequencies (inertial forces increase as the square of excitation frequency). Finally, the inverse relationship between frequency and wavelength is confounded in brain tissue by the frequency dependence of material parameters. In a linear viscoelastic material, the parameter μ is replaced by a complex shear modulus,

*G**, the amplitude of which typically increases with frequency (Clayton et al. 2011).

The frequencies used in human brain MRE have ranged between 25 and 100 Hz (Clayton et al. 2012; Green et al. 2008; Hamhaber et al. 2010; Kruse et al. 2008; Sack et al. 2008). These frequencies provide wave penetration into deep brain tissue and wavelengths on the order of 2–4 cm. Klatt et al. (2007), Sack et al. (2009), and Hamhaber et al. (2010) pioneered the use of multifrequency MRE to acquire data at four frequencies (25, 37.5, 50, and 62.5 Hz) simultaneously. In brain MRE studies of small animals (Atay et al. 2008; Clayton et al. 2011; Feng et al. 2013a; Schregel et al. 2012), excitation frequencies from 400 to 1800 Hz have been used. These frequencies provide wavelengths in the range of [illegible] mm, comparable to the dimensions of typical anatomical structures. Attenuation is not a critical limiting factor in small-animal MRE, but frequency-dependent stiffening of brain tissue is clear (Clayton et al. 2011).

5.3 Image Acquisition

In brain MRE, like other MR imaging techniques, sequence design is critical to obtaining data with sufficient resolution, accuracy, field of view, contrast, and signal-to-noise ratio, to obtain accurate estimates of material parameters. All these objectives must be balanced with the requirement for reasonably short imaging times.

As described in the previous chapter, the basic MRE sequence consists of a gradient-echo, spin-echo, or echo-planar imaging (EPI) sequence, augmented with oscillatory motion-sensitizing gradients that are synchronized with the electromechanical actuator. The resulting phase images are proportional to the harmonic displacement in the direction of the motion-sensitizing gradient. In general, brain motion is fully three-dimensional (3-D). To fully characterize the displacement field, MRE thus requires motion-sensitizing gradients applied in three different directions. Typically, gradients are applied in both positive and negative directions, and the resulting phase images are subtracted to eliminate the effects of background field inhomogeneity. To obtain data at N different time points, the gradient oscillations are delayed by small time increments, $\Delta t_n = T/N$, relative to the mechanical vibrations, where T is the vibration period. Total scan time is lengthened both by the repetitions required to obtain multiple displacement components and the repetitions required to acquire data at multiple time points. To accommodate the trade-offs in signal-to-noise ratio, resolution, coverage, and scan time, brain MRE is often done in a limited number of slices with voxels that are larger than those typical of anatomical scans.

5.3.1 Human Brain MRE Sequences and Parameters

Several recently used MRE sequences are reviewed briefly as follows. The field strength and scanner make are noted for each study. McCracken et al. (2005) and Kruse et al. (2008) used modified gradient-echo sequences for human brain MRE (1.5 T GE), an approach that was later followed by Clayton et al. (2012) with 3 × 3 × 3 mm^3 voxels (1.5 T Siemens). Green et al. (2008) and Zhang et al. (2011) used modified spin-echo [illegible] with [illegible] voxels 3 × 3 × 3 mm^3 (3 T Phillips). Klatt et al. (2007), Hamhaber et al. (2007, 2010), and Sack et al. (2009) have published a number of studies based on modified single-shot EPI; typical resolution is 1.5 × 1.5 × 6 mm^3 (1.5 T Siemens). Recently, Romano and coauthors (2012) published human brain MRE data obtained with this sequence in 30 slices with 2 × 2 × 2 mm^3 voxels (1.5 T Siemens). Johnson et al. (2013) have developed a multishot spiral MRE sequence that provides high resolution (2 × 2 × 2 mm^3 voxels) and short imaging times (Figure 5.3). They acquired 20 slices with all three directions of motion encoding in under 10 min (3 T Siemens).

5.3.2 Animal Brain MRE Sequences and Parameters

Atay et al. (2008) and Clayton et al. (2011) have used a modified spin-echo sequence to perform MRE studies in the mouse brain with a resolution of 0.25 × 0.25 ×

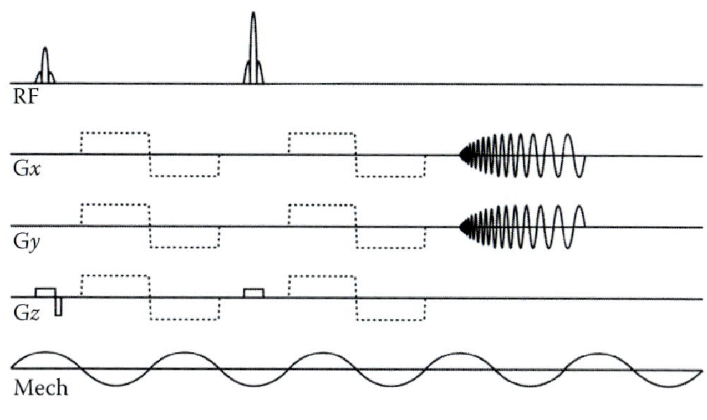

FIGURE 5.3
Diagram for an MRE pulse sequence with multishot, variable-density, spiral readout gradients, designed to achieve high spatial resolution (2 × 2 × 2 mm^3) and short imaging times (<10 min). The solid trace at the bottom depicts the mechanical excitation; dashed traces superimposed on the imaging gradient channels show the motion-sensitizing gradients. Images of the displacement field at different time points are obtained by shifting the gradients by a fraction of a cycle, relative to the motion. (Reproduced from Johnson, C.L. et al., *Magn. Reson. Med.*, 70, 404–12, 2013. With permission.)

0.4 mm^3 (15 slices, 11.7 T Varian scanner) and 0.25 × 0.25 × 0.25 mm^3 (29 slices, 4.7 T Varian), respectively. Schregel et al. (2012) also used a modified spin-echo sequence to image the mouse brain (0.3 × 0.3 × 0.3 mm^3, 10 slices, 7 T Bruker). Pattison and coauthors (2010a) report results in the feline brain using a modified spin-echo sequence with 1.2 × 1.2 × 1.8 mm^3 voxels (12 slices, 3 T Phillips). Vappou et al. (2008) used a modified spin-echo sequence at low magnetic field (0.1 T Bouhnik) to perform MRE in one slice with large voxel size (0.55 × 0.55 × 7 mm^3). Feng et al. (2013a) used a modified gradient-echo sequence with 0.5 × 0.5 × 0.5 mm^3 voxels for MRE of the ferret brain (11 slices, 4.7 T Varian).

5.4 Material Models and Inversion Algorithms

5.4.1 Material Models of Brain Tissue

MRE provides estimates of parameters for models of brain tissue. Brain tissue is a complex material, which can be modeled with varying degrees of simplicity and accuracy. The simplest model is a linear elastic, isotropic, incompressible, locally homogeneous material. While brain tissue is not completely incompressible, the effects of compressibility may be removed by analyzing the curl of the displacement field, which is not affected by the contributions of longitudinal waves (Sinkus et al. 2005a). The assumption of local homogeneity allows each subdomain to be analyzed as if the material parameters were constant within that subdomain, which simplifies considerably the governing equations (Manduca et al. 2001). Isotropic, linear elastic material behavior was assumed for the human brain by Kruse et al. (2008), for the feline brain by Pattison et al. (2010a), and for the mouse brain by Atay et al. (2008).

The most obvious deficiency of the linear elastic material model is that it does not capture dissipative effects, driven by the dependence of stress on strain rate. Viscoelastic material models of brain tissue represent a significant improvement to linear elastic models, at relatively little cost. To describe the response of a linear, isotropic, viscoelastic material to harmonic loading, the elastic shear modulus μ is replaced by a complex shear modulus $G^* = G' + iG''$. The real part of the complex modulus, the storage modulus G', represents the elastic component, and the imaginary part, the loss modulus G'', represents viscous effects. Linear, isotropic, viscoelastic material models of brain tissue have been used in most recent MRE studies of the human (Green et al. 2008; Hamhaber et al. 2007; Papazoglou et al. 2008; Sack et al. 2009; Sinkus et al. 2005a) and animal (Clayton et al. 2011; Feng et al. 2013a; Schregel et al. 2012) brain. Rayleigh, or proportional, damping models, in which both shear modulus and density are complex valued, have been proposed (Van Houten et al. 2011) and applied to brain MRE (Johnson et al. 2013).

Viscoelastic and Rayleigh models are based on observations of the dissipative behavior of tissue—thus, they are phenomenological rather than mechanistic. In contrast, in poroelastic models, brain tissue is explicitly modeled as a porous elastic solid containing fluid, which more accurately describes the underlying physics. Poroelastic models have also been used to interpret brain MRE data in several recent studies (Pattison et al. 2010b; Perrinez et al. 2009; Weaver et al. 2012).

Mechanical anisotropy in brain tissue is gaining attention among MRE researchers. White matter in the brain is largely composed of aligned axonal fibers, and the resulting microstructural anisotropy has been convincingly demonstrated by diffusion tensor imaging (DTI) (Figures 5.4 and 5.5). Anisotropy may be important to the mechanism of TBI; axonal injury in white matter and at the interface of white and gray matter is thought to underlie many of the cognitive effects. Several recent papers have begun to focus specifically on anisotropy using MRE. In an MRE study of breast tissue, Sinkus et al. (2005a) introduced a set of equations that included two different shear moduli. Namani et al. (2009) demonstrated that MRE could identify two different shear moduli in magnetically aligned fibrin gels. Romano colleagues (2012) used an orthotropic, viscoelastic model, guided by DTI estimates of axon fiber direction, to estimate anisotropic parameters of white matter in the human brain. Qin et al. (2013) combined DTI and MRE to estimate anisotropic shear moduli in an anisotropic

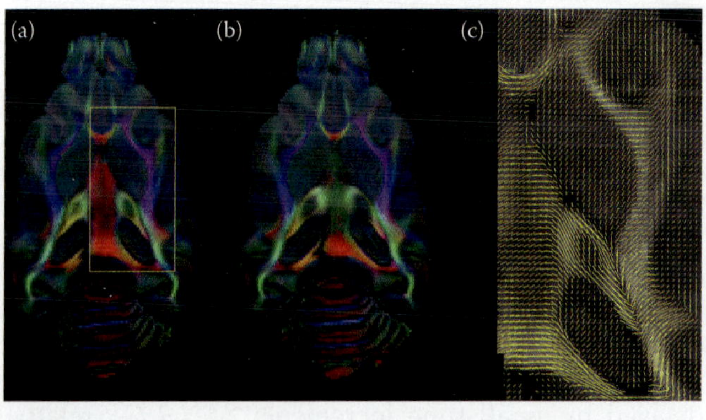

FIGURE 5.4
Diffusion tensor images (DTIs) of structural anisotropy in the ferret brain. (a–b) Diffusion is faster along the dominant fiber direction indicated by color (red = left–right; green = anterior–posterior; blue = inferior–superior). Brightness indicates the magnitude of anisotropy. (c) Vector field of dominant fiber direction inferred from the direction of maximum diffusivity. (Courtesy of C.D. Kroenke [unpublished].)

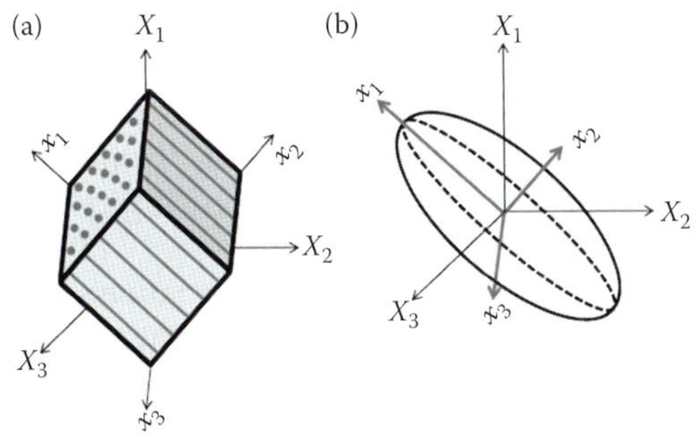

FIGURE 5.5
Connection of structural anisotropy to diffusion anisotropy in a fibrous material. (a) The local coordinate system (x_1, x_2, x_3) is aligned with the fiber axis. Displacements are measured in the laboratory frame (X_1, X_2, X_3). (b) The fiber axis coincides with the direction of maximal diffusivity—the major axis of the diffusion ellipsoid. (Reproduced from Clayton, E. H., Magnetic resonance elastography of the brain: From phantom to mouse to man. PhD Thesis, Saint Louis, MO, Washington University, 2012. With permission.)

gel phantom and compared their estimates to values from direct mechanical tests.

Mechanical anisotropy (Figure 5.6) introduces significant complexity into modeling and validation of brain tissue. It is not simple to completely characterize even the simplest anisotropic case, a transversely isotropic material. A recent study by Feng and coworkers (2013b) showed that white matter exhibits anisotropy in both

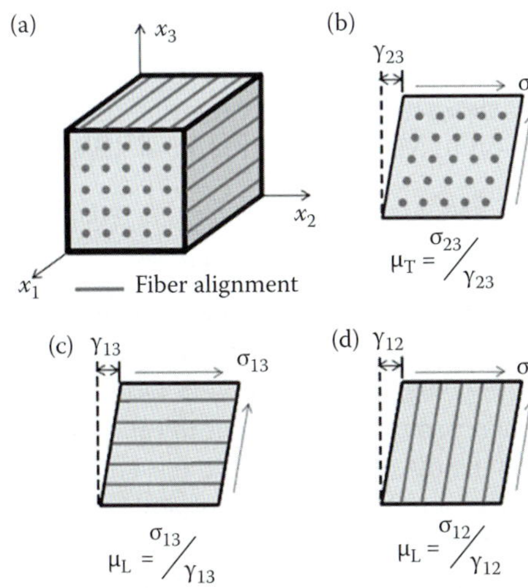

FIGURE 5.6
Mechanical anisotropy in shear. The response to shear loading in a transversely isotropic material depends on the relationship between the direction of loading and the symmetry axis of the material. (a) A schematic diagram of an element of fibrous soft material with symmetry axis determined by fiber direction. (b–d) Depiction of shear in the plane of isotropy and in a perpendicular plane. Shear moduli will differ in these cases due to fiber–matrix interactions. Shear deformations also involve fiber stretch if displacements are not perfectly parallel or perpendicular to the fiber axis. (Reproduced from Clayton, E.H. et al., *J. Biomech. Eng.*, 135, 021025. 2013. With permission.)

tension and shear. It is important to note that anisotropy in tensile properties may affect shear wave propagation, because shear deformations in general involve stretch along the fiber axis (Parker 1984). In transversely isotropic materials, the effect of tensile anisotropy is seen in the dependence of the "fast" shear wave on direction, while the behavior of the "slow" shear wave only reflects anisotropy in shear moduli (Parker 1984).

It should be clear that brain tissue is difficult to model and that increasing the complexity of the material model increases the difficulty of estimating and interpreting material parameters. In fact, each model is likely to be inaccurate under some conditions. We advocate incorporating into every inversion algorithm an assessment of how well (or poorly) the material model fits the data. In recent work by Clayton et al. (2011), Okamoto et al. (2011), and Feng et al. (2013a), a normalized residual error (NRE) estimate was used to assess whether a linear, viscoelastic, isotropic, locally homogeneous model provided a reasonable fit to MRE data in the neighborhood of each voxel.

5.4.2 Inversion Algorithms

A variety of different algorithms for estimating material parameters have been used in brain MRE studies, even for identical material models. The problem may be posed as follows: given (a) measurements of dynamic, 3-D displacement in a 3-D volume and (b) a material model with spatially varying parameters, what distribution of these material parameters is most consistent with the data? Four general approaches are in common use:

1. Local (spatial) frequency estimation (LFE)
2. Fitting of measured data to the partial differential equation (PDE) of shear wave motion
3. Substituting data into the weak form of the equation of motion and solving for the parameters (the "virtual fields method" [VFM])
4. Performing inverse finite element analysis (iteratively modifying the parameters of a finite element model until simulated displacement fields match measured data)

Each of these approaches, and their application to brain MRE, are described in more detail in the following.

5.4.2.1 *Local Frequency Estimation*

LFE provides a quick estimate of parameters for a linear, isotropic, elastic or viscoelastic material model (Manduca et al. 2001). At a given frequency, the local wavelength is proportional to propagation speed ($\lambda = c/f$), which, in elastic materials, depends on shear modulus and density: $c = \sqrt{\mu/\rho}$. In viscoelastic materials, the wavelength depends on the storage modulus G' rather than μ. A fast and robust method for estimating 2-D local spatial frequency was published by Knutsson et al. (1994) and has been used recently in human brain MRE (Clayton et al. 2012, 2013).

5.4.2.2 Fitting Data to the PDE

The PDE governing displacement in an isotropic, homogeneous, linear elastic continuum is

$$\rho\frac{\partial^2\mathbf{u}(\mathbf{X},t)}{\partial t^2} = \mu\nabla^2\mathbf{u}(\mathbf{X},t) + (\lambda+\mu)\nabla(\nabla\cdot\mathbf{u}(\mathbf{X},t)). \quad (5.1)$$

The displacement, $\mathbf{u}$, of a material element depends on its reference position, $\mathbf{X}$, and time, t. It can be separated into dilatational ($\mathbf{u}_d$) and isochoric or shear ($\mathbf{u}_s$) components: $\mathbf{u}(\mathbf{X}, t) = \mathbf{u}_d(\mathbf{X}, t) + \mathbf{u}_s(\mathbf{X}, t)$. The equations for the dilatational and isochoric components of the displacement field can be decoupled using the identities $\nabla\cdot\mathbf{u}_s(\mathbf{X}, t) = 0$ and $\nabla\times\mathbf{u}_d(\mathbf{X}, t) = 0$. If the shear wave displacement field arises from harmonic wave motion $\mathbf{u}_s(\mathbf{X}, t) = \mathbf{U}(\mathbf{X})\exp(i\omega t)$, the equation for $\mathbf{U}(\mathbf{X}) = \mathbf{U}'(\mathbf{X}) + i\mathbf{U}''(\mathbf{X})$, the complex, spatially varying coefficient of $\exp(i\omega t)$, is

$$-\rho\omega^2\mathbf{U}(\mathbf{X}) = \mu\nabla^2\mathbf{U}(\mathbf{X}) \quad (5.2a)$$

$$-\rho\omega^2\mathbf{\Gamma}(\mathbf{X}) = \mu\nabla^2\mathbf{\Gamma}(\mathbf{X}) \quad (5.2b)$$

where $\mathbf{\Gamma} = \nabla\times\mathbf{U}$. The viscoelastic analog of this equation is simply obtained, according to the correspondence principle (Flügge 1975), by replacing the (real) shear modulus with the complex modulus G^*.

The first task in finding the shear modulus by fitting data to the shear wave equation is to remove the effects of dilatational (longitudinal) waves. Since brain tissue is nearly incompressible, the bulk modulus is several orders of magnitude larger than the shear modulus. The shear modulus is the mechanical parameter of interest; shear modulus governs deformation in TBI and changes with age and disease. The most direct approach is to take the curl of the displacement field as in the work of Sinkus et al. (2005b) and Clayton et al. (2013). The curl of displacement satisfies the same wave equation as the transverse displacement field and contains no contributions from the longitudinal component. Another approach is to spatially filter the displacement field with a high-pass filter (Glaser et al. 2012), since longitudinal waves will be much longer than shear waves. However, the choice of a cutoff wavelength for such a filter is seldom obvious and can affect parameter estimates. Finally, a frequency-domain Helmholtz filter (Romano et al. 2005) can be used to eliminate longitudinal components.

Estimates of the curl and Laplacian are obtained numerically and substituted into the shear wave equation of motion (Equation 5.2a or 5.2b). Almost always, some low-pass filtering, spatial averaging, or smoothing is performed to reduce the effects of noise, which are amplified by numerical differentiation. At each voxel, or in a small neighborhood surrounding the voxel, the PDE (Equation 5.2a or 5.2b) is then solved for the unknown material parameters. Variations of this approach have been used to estimate viscoelastic properties of the human and animal brain (Clayton et al. 2011; Feng et al. 2013a; Green et al. 2008; Hamhaber et al. 2007, 2010; Klatt et al. 2007; Sack et al. 2008; Zhang et al. 2011).

5.4.2.3 VFM

In the VFM, the measured data are used in the weak form of the governing equations, rather than the PDE itself. Briefly, the strain and displacement fields are multiplied by corresponding "virtual fields" (discretized test functions) and integrated over a subdomain, and the resulting virtual work equations are solved for the unknown material parameters. This technique avoids one numerical differentiation step. The general VFM approach is described by Pierron and Grediac (2012) and its application to MRE by Pierron et al. (2010).

5.4.2.4 Inverse Finite Element Analysis

The use of finite element models to estimate mechanical parameters from MRE data was pioneered by Van Houten et al. (1999, 2001, 2003) and Doyley et al. (2003). This approach provides the distribution of material parameters that produces the closest match between measured displacement fields and the results of finite element simulations. Van Houten et al. (1999) developed an "overlapping subzone" approach, in which the optimization is performed in subsets of the domain. This approach has recently been used with 3-D, high resolution, human brain MRE data to estimate parameters of an isotropic, Rayleigh-damped material model (Johnson et al. 2013).

In each of the approaches described in this section, it is possible to provide some estimate of the degree of confidence that should be placed in local parameter estimates. The LFE algorithm of Knutsson et al. (1994) provides a "certainty" parameter as well as the local frequency estimate at each pixel. McGarry et al. (2011) proposed an octahedral shear strain (OSS) signal-to-noise ratio as a criterion for confidence in estimated material parameters. Local fits to the PDE of wave propagation can be evaluated based on a residual error criterion (Atay et al. 2008; Clayton et al. 2011; Feng et al. 2013a). Residual error is also appropriate for inverse finite element analysis, as it captures the difference between experimental measurements and model predictions. If the residual error is small relative to the variance in the data in the vicinity of a given voxel, then the material model is likely to be appropriate in that region. Residual error comparisons could be used to compare multiple material models on a given data set.

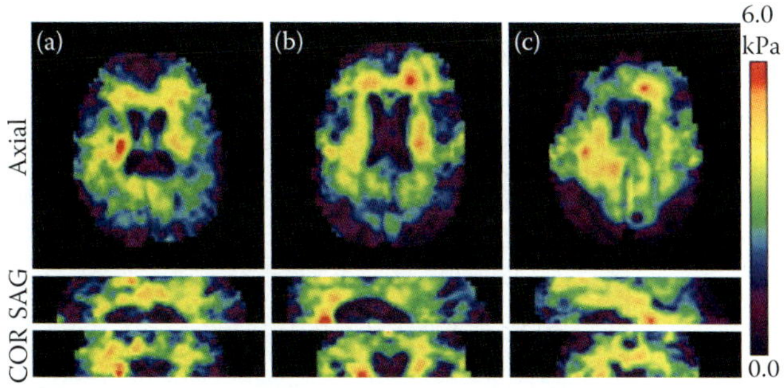

FIGURE 5.8
Example results from an MRE study in the human brain. Distribution of the storage modulus, G′, in axial, sagittal (SAG), and coronal (COR) image planes in each of three subjects (a, b, c). (Reproduced from Johnson, C.L. et al., *Magn. Reson. Med.*, 70, 404–12, 2013.)

5.5 MRE Estimates of Material Parameters in the Animal and Human Brain

Figures 5.7 through 5.10 show example displacement fields and estimated fields of material parameters from human and animal brain MRE studies, respectively. Tables 5.1 and 5.2 summarize recent estimates of viscoelastic moduli in the human and animal brains. Direct comparison between studies is complicated by the differences in material models (e.g., elastic, viscoelastic,

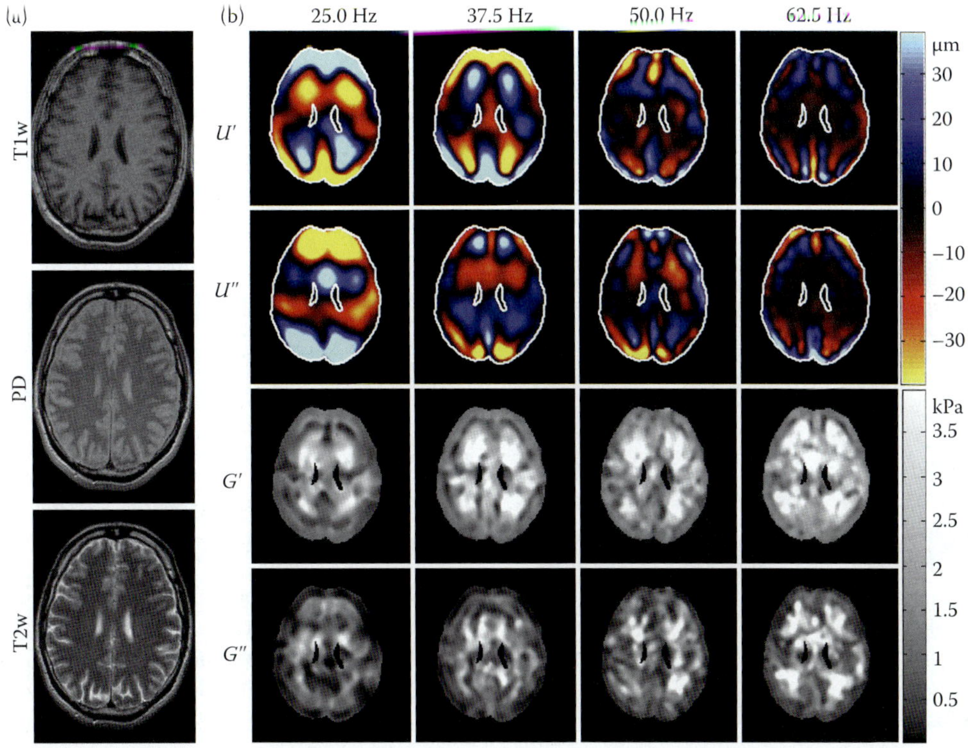

FIGURE 5.7
Example MRE study in the human brain. (a) Standard anatomical MR images: T1-weighted (T1w), proton density (PD), and T2-weighted (T2w). (b) MRE data. The variables U' and U'' represent the real and imaginary components of the complex harmonic displacement field (Equation 5.2). The parameters G' and G'' are the storage and loss moduli of an isotropic viscoelastic material model. (Reproduced from Sack, I. et al., *Neuroimage*, 46, 2009.)

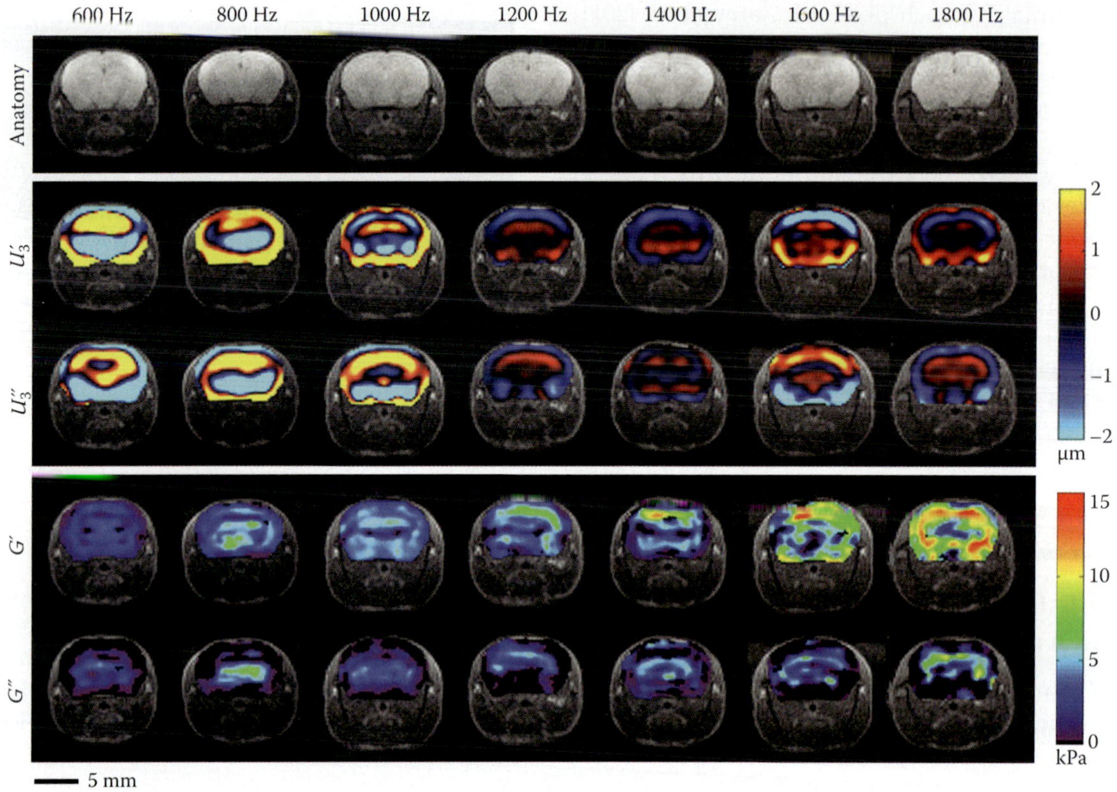

FIGURE 5.9
MRE in the mouse brain at multiple frequencies. (Top row) Anatomical images. (Middle rows) Images of the through-plane displacement field. The variables U' and U'' represent the real and imaginary components of the complex harmonic displacement field (Equation 5.2a). (Bottom rows) The storage (G') and loss (G'') moduli estimated by MRE in these image planes. (Reproduced from Clayton, E.H. et al., *Phys. Med. Biol.*, 56, 2011.)

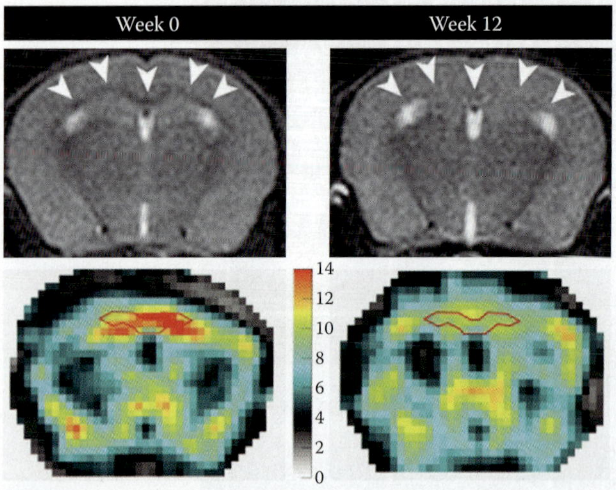

FIGURE 5.10
MRE of the normal and myelin-deficient mouse brain. (Top row) T2-weighted MRI. (Bottom row) Amplitude of the complex shear modulus $|G^*|$ in (left) a healthy control mouse and (right) a mouse deficient in myelin after 12 weeks of continuous cuprizone diet. (Reproduced from Schregel, K. et al., *Proc. Natl. Acad. Sci. USA*, 109, 2012.)

Rayleigh damped). Results are shown in terms of the parameters of the isotropic, viscoelastic model: G' and G''.

5.6 Summary, Discussion, and Future Challenges

The variation in modulus estimates in Tables 5.1 and 5.2 is notable. Some of the variation in parameter estimated may be explained by the frequency range, which spans 25–1800 Hz. It is quite plausible that brain tissue exhibits a much higher effective stiffness at high frequencies. In general, as techniques mature, the variation in estimates shrinks.

The fundamental challenges in MRE are to increase spatial resolution, to improve material models, and to increase the accuracy and robustness of parameter estimates. It is also important to assess the degree of

TABLE 5.1

Estimates of Viscoelastic Parameters from MRE Studies of the Human Brain

		G′ (kPa)			G″ (kPa)				
Study	***f*** **(Hz)**	**Gray**	**White**	**Global Avg.**	**Gray**	**White**	**Global Avg.**	**Field (T)**	**Voxel (mm³)**
(1)	25	—	—	1.2 ± 0.02	—	—	0.5 ± 0.04	1.5	3 × 3 × 6
(1)	25	—	—	1.3 ± 0.02	—	—	0.5 ± 0.01	7.0	3 × 3 × 6
(2)	25	—	—	1.1 ± 0.2	—	—	0.5 ± 0.1[a]	1.5	1.5 × 1.5 × 6
(1)	37.5	—	—	1.6 ± 0.03	—	—	0.5 ± 0.01	1.5	3 × 3 × 6
(1)	37.5	—	—	1.5 ± 0.03	—	—	0.6 ± 0.01	7.0	3 × 3 × 6
(2)	37.5	—	—	1.3 ± 0.3	—	—	0.6 ± 0.1[a]	1.5	1.5 × 1.5 × 6
(3)	45	2.8 ± 0.5	3.7 ± 0.8	—	0.8 ± 0.2	1.3 ± 0.4	—	1.5	3 × 3 × 3
(1)	50	—	—	1.9 ± 0.03	—	—	0.6 ± 0.02	1.5	3 × 3 × 6
(1)	50	—		1.8 ± 0.03		—	0.5 ± 0.01	7.0	3 × 3 × 6
(4)	50	1.7 ± 1.0	2.2 ± 1.0	—	0.7 ± 0.5	1.1 ± 0.6	—	3.0	2 × 2 × 2
(4)	50	1.9 ± 1.0	2.5 ± 1.0	—	0.8 ± 0.6	1.2 ± 0.8	—	3.0	2 × 2 × 2
(5)	50	—	4.0 ± 0.5[b]	—	—	—	0.9 ± 1.0[b]	3.0	2 × 2 × 2
(2)	50	—	—	1.5 ± 0.2	—	—	0.6 ± 0.1[a]	1.5	1.5 × 1.5 × 6
(3)	60	3.1 ± 0.3	3.3 ± 0.1	—	1.7 ± 0.3	2.0 ± 0.1	—	1.5	3 × 3 × 3
(1)	62.5	—	—	2.5 ± 0.08	—	—	0.7 ± 0.02	1.5	3 × 3 × 6
(1)	62.5	—	—	2.2 ± 0.02	—	—	0.6 ± 0.01	7.0	3 × 3 × 6
(2)	62.5	—	—	2.0 ± 0.2	—	—	0.8 ± 0.1[a]	1.5	1.5 × 1.5 × 6
(3)	80	4.4 ± 0.3	4.7 ± 0.6	—	2.3 ± 0.2	2.4 ± 0.5	—	1.5	3 × 3 × 3
(6)	80	5.3 ± 1.3	10.7 ± 1.4	—	—	—	—	1.5	4 × 4 × 5
(7)	80	2.3 ± 0.2	2.4 ± 0.2		1.1 ± 0.0	1.2 ± 0.2	—	3.0	3 × 3 × 3
(8)	83.3	—	—	3.5 ± 1.3	—	—	—	1.5	1.5 × 1.5 × 6
(9)	90	3.0 ± 0.1	2.7 ± 0.1	—	2.5 ± 0.2	2.5 ± 0.2	—	3.0	3 × 3 × 3
(10)	100	5.2 ± 0.4	13.6 ± 1.3	—	—	—	—	1.5	3.1 × 3.1 × 5
(6)	Transient	7.5 ± 1.6	11.6 ± 2.4	—	—	—	—	1.5	4 × 4 × 5

Note: Studies: 1, Hamhaber et al. (2010); 2, Sack et al. (2009); 3, Clayton et al. (2012); 4, Johnson et al. (2013); 5, Romano et al. (2012); 6, McCracken et al. (2005); 7, Zhang et al. (2011); 8, Hamhaber et al. (2007); 9, Green et al. (2008); 10, Kruse et al. (2008).

[a] Loss modulus G'' calculated from shear viscosity η (i.e., $G'' = \omega\eta$).

[b] Smallest of three unequal shear moduli in orthotropic model.

confidence and precision in these estimates. For both research and clinical purposes, it would be ideal to detect and characterize fine anatomical structures and small lesions in the brain. Detection and characterization of small features are easier with high imaging resolution (small voxel size) and short MRE wavelength. It is essential to understand and account for the factors that lead to differences in estimated parameters, and to continue to improve inversion algorithms. Anisotropy and nonlinearity are key properties of brain tissue that have not yet been significantly illuminated by MRE.

These objectives are related by a several constraints and trade-offs. Resolution and signal-to-noise ratio could be improved by increasing imaging time, but scan durations longer than 30 min could become uncomfortable and costly. To increase resolution and accuracy, shorter MRE wavelengths are preferred, but these require higher excitation frequencies, which are more strongly attenuated. The most highly resolved and accurate inversion techniques, which are those based on inverse finite element methods, are also the most time consuming and computationally demanding. Robust measurement of anisotropy may require excitation in multiple locations to produce waves with different polarization and propagation directions. The combination of MRE with DTI appears promising in the study of anisotropy. The ability to use MRE to probe nonlinear mechanical behavior of brain tissue remains an open question.

Ongoing and future research should confront these problems. Improved sequence design (Johnson et al. 2013) can help to overcome some of these trade-offs in image resolution and imaging time. In terms of inversion, a hybrid approach may be warranted, in which a relatively fast, approximate analysis is performed immediately, followed by a high-resolution, more precise inversion if needed. A simple and intuitive approach is needed to handle anisotropy, but such an

TABLE 5.2

Estimates of Viscoelastic Parameters from MRE Studies of Animal Brain

Animal (study)	f (Hz)	G' (kPa)			G'' (kPa)			Field (T)	Voxel (mm³)
		Gray	White	All	Gray	White	All		
Mouse (1)	600	—	—	1.6 ± 1.0	—	—	1.0 ± 0.8	4.7	0.25 × 0.25 × 0.25
Mouse (1)	800	—	—	2.5 ± 1.5	—	—	1.7 ± 1.5	4.7	0.25 × 0.25 × 0.25
Mouse (2)	1000	—	—	9.3 ± 0.6[a]	—	—	4.2 ± 0.2[a]	7.0	0.30 × 0.30 × 0.30
Mouse (3)	1200	—	—	13.8 ± 1.5	—	—		11.7	0.25 × 0.25 × 0.40
Mouse (1)	1200	—	—	4.1 ± 1.6	—	—	1.7 ± 1.3	4.7	0.25 × 0.25 × 0.25
Mouse (4)	1500	—	—	25.0 ± 6.4	—	—	—	3.0	0.47 × 0.47 × 3
Mouse (1)	1800	—	—	7.8 ± 3.3	—	—	3.0 ± 2.4	4.7	0.25 × 0.25 × 0.25
Rat (5)	180	—	—	8.5 ± 0.4	—	—	7.1 ± 0.6	0.1	0.73 × 0.55 × 7
Ferret (6)	400	3.1 ± 0.3	1.1 ± 0.2	—	3.1 ± 0.3	1.0 ± 0.3	—	4.7	0.50 × 0.50 × 0.50
Ferret (6)	600	5.3 ± 0.4	1.9 ± 0.2	—	5.2 ± 0.3	1.9 ± 0.2	—	4.7	0.50 × 0.50 × 0.50
Ferret (6)	800	6.9 ± 0.5	7.3 ± 0.2	—	2.3 ± 0.6	2.2 ± 0.5	—	4.7	0.50 × 0.50 × 0.50
Cat (7)	85	8.3 ± 3.7	7.1 ± 2.8	—	—	—	—	3.0	1.2 × 1.2 × 1.8
Pig (8)	80	—	—	1.2 ± 0.2	—	—	0.9 ± 0.2	0.1	0.70 × 0.94 × 10
Pig (8)	100	—	—	1.1 ± 0.1	—	—	0.9 ± 0.2	0.1	0.70 × 0.94 × 10
Pig (8)	120	—	—	1.2 ± 0.1	—	—	1.1 ± 0.1	0.1	0.70 × 0.94 × 10
Pig (8)	140	—	—	1.2 ± 0.1	—	—	1.1 ± 0.2	0.1	0.70 × 0.94 × 10

Note: Studies: 1, Clayton et al. (2011); 2, Schregel et al. (2012); 3, Atay et al. (2008); 4, Murphy et al. (2012); 5, Vappou et al. (2008); 6, Feng et al. (2013a); 7, Pattison et al. (2010a); 8, Vappou et al. (2007).

[a] G' and G'' computed from reported parameters magnitude $|G^*|$ and phase $\angle G^*$.

approach must also be accurate, robust, and useful. Important steps in this direction have been taken (Feng et al. 2013b; Qin et al. 2013; Romano et al. 2012) (Figure 5.11), but more work remains before anisotropic properties can be extracted with confidence. Finally, imaging of shear waves can provide information other than estimates of tissue mechanical parameters. The directions of wave propagations and energy transfer illuminate the effects of anatomical structure, including stiff membranes, vessels, and other attachments, on brain behavior (Figure 5.12).

In conclusion, MRE has proven to be a powerful tool with which to characterize the material and structural properties of the brain. Its continued development will lead to improved understanding of TBI, as well as the development of new scientific and clinical techniques.

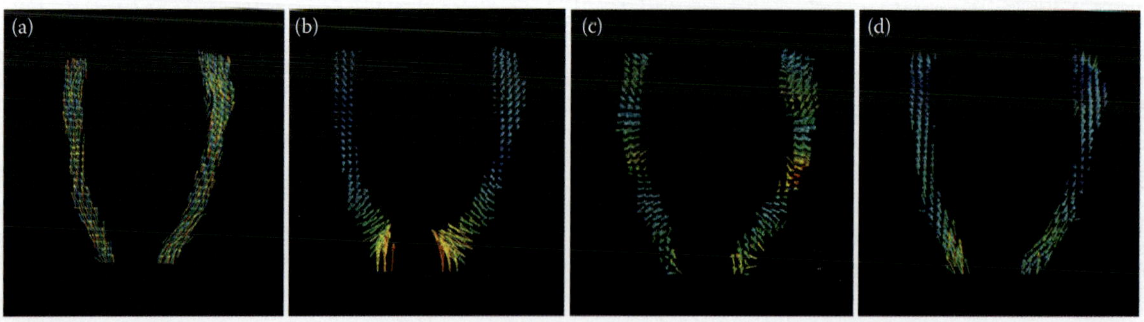

FIGURE 5.11
Estimation of anisotropic material parameters in human brain MRE begins with identification of displacement components with respect to fiber direction. (a) Image of the right and left corticospinal tracts (CSTs) of one volunteer obtained using DTI. Arrows show the direction of maximal diffusivity, which is taken to be the fiber direction. (b) Displacement field in the CSTs at one time point. (c) Displacement components associated with transverse waves propagating up the CSTs. (d) Displacement components corresponding to longitudinal waves propagating up the CSTs. Arrows in (b), (c), and (d) represent scaled displacement vectors. (Reproduced from Romano, A. et al., *Magn. Reson. Med.*, 68, 2012. With permission.)

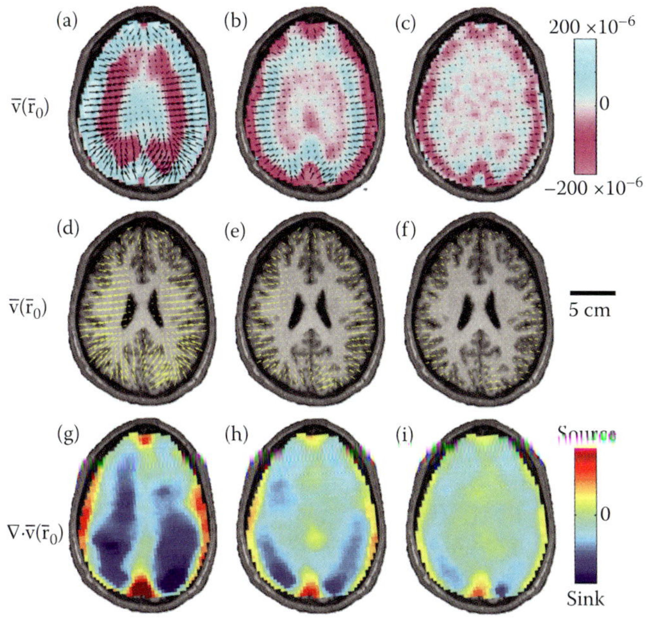

FIGURE 5.12
Imaging of shear waves can illuminate structural behavior as well as material properties. Vector fields of average propagation direction superimposed on (a–c) estimated curl fields and (d–f) high-resolution anatomical images in one human subject at three frequencies. Shear waves in the brain appear to propagate inward from the brain–skull boundary toward the ventricles. (g–i) Divergence of the propagation direction field in the anterior/posterior falx and above the tentorium indicates that these membranous structures may act like sources of shear wave excitation. (a, d, g) 45 Hz; (b, e, h) 60 Hz; and (c, f, i) 80 Hz. (Reproduced from Clayton, E.H. et al., *J. R. Soc. Interface.*, 9, 2012. With permission.)

References

Atay, S. M., Kroenke, C. D., Sabet, A. & Bayly, P. V. 2008. Measurement of the dynamic shear modulus of mouse brain tissue in vivo by magnetic resonance elastography. *J Biomech Eng*, 130, 021013.

Bayly, P. V., Okamoto, R. J., Xu, G., Shi, Y. & Taber, L. A. 2013. A cortical folding model incorporating stress-dependent growth explains gyral wavelengths and stress patterns in the developing brain. *Phys Biol*, 10, 016005.

Clayton, E. H. 2012. Magnetic resonance elastography of the brain: From phantom to mouse to man. PhD Thesis, Saint Louis, MO, Washington University.

Clayton, E. H., Garbow, J. R. & Bayly, P. V. 2011. Frequency-dependent viscoelastic parameters of mouse brain tissue estimated by MR elastography. *Phys Med Biol*, 56, 2391–406.

Clayton, E. H., Genin, G. M. & Bayly, P. V. 2012. Transmission, attenuation and reflection of shear waves in the human brain. *J R Soc Interface*, 9, 2899–910.

Clayton, E. H., Okamoto, R. J. & Bayly, P. V. 2013. Mechanical properties of viscoelastic media by local frequency estimation of divergence-free wave fields. *J Biomech Eng*, 135, 021025.

Corrigan, J. D., Selassie, A. W. & Orman, J. A. 2010. The epidemiology of traumatic brain injury. *J Head Trauma Rehabil*, 25, 72–80.

Doyley, M. M., Weaver, J. B., Van Houten, E. E., Kennedy, F. E. & Paulsen, K. D. 2003. Thresholds for detecting and characterizing focal lesions using steady-state MR elastography. *Med Phys*, 30, 495–504.

Ehman, E. C., Rossman, P. J., Kruse, S. A., Sahakian, A. V. & Glaser, K. J. 2008. Vibration safety limits for magnetic resonance elastography. *Phys Med Biol*, 53, 925–35.

Feng, Y., Clayton, E. H., Chang, Y., Okamoto, R. J. & Bayly, P. V. 2013a. Viscoelastic properties of the ferret brain measured in vivo at multiple frequencies by magnetic resonance elastography. *J Biomech*, 46, 863–70.

Feng, Y., Okamoto, R. J., Namani, R., Genin, G. M. & Bayly, P. V. 2013b. Measurements of mechanical anisotropy in brain tissue and implications for transversely isotropic material models of white matter. *J Mech Behav Biomed Mater*, 23, 117–32.

Flügge, W. 1975. *Viscoelasticity*. Berlin; New York, Springer-Verlag.

Glaser, K. J., Manduca, A. & Ehman, R. L. 2012. Review of MR elastography applications and recent developments. *J Magn Reson Imaging*, 36, 757–74.

Green, M. A., Bilston, L. E. & Sinkus, R. 2008. In vivo brain viscoelastic properties measured by magnetic resonance elastography. *NMR Biomed*, 21, 755–64.

Hamhaber, U., Klatt, D., Papazoglou, S., Hollmann, M., Stadler, J., Sack, I., Bernarding, J. & Braun, J. 2010. In vivo magnetic resonance elastography of human brain at 7 T and 1.5 T. *J Magn Reson Imaging*, 32, 577–83.

Hamhaber, U., Sack, I., Papazoglou, S., Rump, J., Klatt, D. & Braun, J. 2007. Three-dimensional analysis of shear wave propagation observed by in vivo magnetic resonance elastography of the brain. *Acta Biomater*, 3, 127–37.

Hoge, C. W., McGurk, D., Thomas, J. L., Cox, A. L., Engel, C. C. & Castro, C. A. 2008. Mild traumatic brain injury in U.S. Soldiers returning from Iraq. *N Engl J Med*, 358, 453–63.

Johnson, C. L., McGarry, M. D., Van Houten, E. E., Weaver, J. B., Paulsen, K. D., Sutton, B. P. & Georgiadis, J. G. 2013. Magnetic resonance elastography of the brain using multishot spiral readouts with self-navigated motion correction. *Magn Reson Med*, 70, 404–12.

Klatt, D., Hamhaber, U., Asbach, P., Braun, J. & Sack, I. 2007. Noninvasive assessment of the rheological behavior of human organs using multifrequency MR elastography: A study of brain and liver viscoelasticity. *Phys Med Biol*, 52, 7281–94.

Knutsson, H., Westin, C. F. & Granlund, G. 1994. Local multiscale frequency and bandwidth estimation. *Proceedings of the IEEE International Conference on Image Processing*.

Kruse, S. A., Rose, G. H., Glaser, K. J., Manduca, A., Felmlee, J. P., Jack, C. R. Jr. & Ehman, R. L. 2008. Magnetic resonance elastography of the brain. *Neuroimage*, 39, 231–7.

Langlois, J. A., Rutland-Brown, W. & Wald, M. M. 2006. The epidemiology and impact of traumatic brain injury: A brief overview. *J Head Trauma Rehabil*, 21, 375–8.

Manduca, A., Oliphant, T. E., Dresner, M. A., Mahowald, J. L., Kruse, S. A., Amromin, E., Felmlee, J. P., Greenleaf, J. F. & Ehman, R. L. 2001. Magnetic resonance elastography: Non-invasive mapping of tissue elasticity. *Med Image Anal*, 5, 237–54.

McCracken, P. J., Manduca, A., Felmlee, J. & Ehman, R. L. 2005. Mechanical transient-based magnetic resonance elastography. *Magn Reson Med*, 53, 628–39.

McGarry, M. D., Van Houten, E. E., Perrinez, P. R., Pattison, A. J., Weaver, J. B. & Paulsen, K. D. 2011. An octahedral shear strain-based measure of SNR for 3D MR elastography. *Phys Med Biol*, 56, N153–64.

Murphy, M. C., Curran, G. L., Glaser, K. J., Rossman, P. J., Huston, J. III, Poduslo, J. F., Jack, C. R. Jr., Felmlee, J. P. & Ehman, R. L. 2012. Magnetic resonance elastography of the brain in a mouse model of Alzheimer's disease: Initial results. *Magn Reson Imaging*, 30, 535–9.

Murphy, M. C., Huston, J. III, Glaser, K. J., Manduca, A., Meyer, F. B., Lanzino, G., Morris, J. M., Felmlee, J. P. & Ehman, R. L. 2013. Preoperative assessment of meningioma stiffness using magnetic resonance elastography. *J Neurosurg*, 118, 643–8.

Murphy, M. C., Huston, J. III, Jack, C. R. Jr., Glaser, K. J., Manduca, A., Felmlee, J. P. & Ehman, R. L. 2011. Decreased brain stiffness in Alzheimer's disease determined by magnetic resonance elastography. *J Magn Reson Imaging*, 34, 494–8.

Muthupillai, R., Lomas, D. J., Rossman, P. J., Greenleaf, J. F., Manduca, A. & Ehman, R. L. 1995. Magnetic resonance elastography by direct visualization of propagating acoustic strain waves. *Science (New York, N Y)*, 269, 1854–7.

Namani, R., Wood, M. D., Sakiyama-Elbert, S. E. & Bayly, P. V. 2009. Anisotropic mechanical properties of magnetically aligned fibrin gels measured by magnetic resonance elastography. *J Biomech*, 42, 2047–53.

Okamoto, R. J., Clayton, E. H. & Bayly, P. V. 2011. Viscoelastic properties of soft gels: Comparison of magnetic resonance elastography and dynamic shear testing in the shear wave regime. *Phys Med Biol*, 56, 6379.

Papazoglou, S., Hamhaber, U., Braun, J. & Sack, I. 2008. Algebraic Helmholtz inversion in planar magnetic resonance elastography. *Phys Med Biol*, 53, 3147–58.

Parker, D. F. 1984. Elastic wave propagation in strongly anisotropic solids. In: Spencer, A. J. M. (ed.) *Continuum Theory of the Mechanics of Fibre-Reinforced Composites*. New York, Springer-Verlag.

Pattison, A. J., Lollis, S. S., Perrinez, P. R., Perreard, I. M., McGarry, M. D., Weaver, J. B. & Paulsen, K. D. 2010a. Time-harmonic magnetic resonance elastography of the normal feline brain. *J Biomech*, 43, 2747–52.

Pattison, A. J., Perreard, I. M., Weaver, J. B. & Paulsen, K. D. 2010b. Poroelastic MRE reconstructions of brain acquired with intrinsic activation. *Proceedings of the ISMRM*. Stockholm, Sweden, 3404.

Perrinez, P. R., Kennedy, F. E., Van Houten, E. E., Weaver, J. B. & Paulsen, K. D. 2009. Modeling of soft poroelastic tissue in time-harmonic MR elastography. *IEEE Trans Biomed Eng*, 56, 598–608.

Pierron, F., Bayly, P. V. & Namani, R. 2010. Application of the virtual fields method to magnetic resonance elastography data. *Proceedings of the Society for Experimental Mechanics Annual Conference*. Indianapolis, IN, Springer.

Pierron, F. & Grediac, M. 2012. *The Virtual Fields Method: Extracting Constitutive Mechanical Parameters from Full-Field Deformation Measurements*. New York, Springer.

Qin, E. C., Sinkus, R., Geng, G., Cheng, S., Green, M., Rae, C. D. & Bilston, L. E. 2013. Combining MR elastography and diffusion tensor imaging for the assessment of anisotropic mechanical properties: A phantom study. *J Magn Reson Imaging*, 37, 217–26.

Richman, D. P., Stewart, R. M., Hutchinson, J. W. & Caviness, V. S. 1975. Mechanical model of brain convolutional development. *Science*, 189, 18–21.

Romano, A., Scheel, M., Hirsch, S., Braun, J. & Sack, I. 2012. In vivo waveguide elastography of white matter tracts in the human brain. *Magn Reson Med*, 68, 1410–22.

Romano, A. J., Abraham, P. B., Rossman, P. J., Bucaro, J. A. & Ehman, R. L. 2005. Determination and analysis of guided wave propagation using magnetic resonance elastography. *Magn Reson Med*, 54, 893–900.

Sack, I., Beierbach, B., Hamhaber, U., Klatt, D. & Braun, J. 2008. Non-invasive measurement of brain viscoelasticity using magnetic resonance elastography. *NMR Biomed*, 21, 265–71.

Sack, I., Beierbach, B., Wuerfel, J., Klatt, D., Hamhaber, U., Papazoglou, S., Martus, P. & Braun, J. 2009. The impact of aging and gender on brain viscoelasticity. *Neuroimage*, 46, 652–7.

Sack, I., Streitberger, K. J., Krefting, D., Paul, F. & Braun, J. 2011. The influence of physiological aging and atrophy on brain viscoelastic properties in humans. *PLoS One*, 6, e23451.

Schregel, K., Wuerfel, E., Garteiser, P., Gemeinhardt, I., Prozorovski, T., Aktas, O., Merz, H., Petersen, D., Wuerfel, J. & Sinkus, R. 2012. Demyelination reduces brain parenchymal stiffness quantified in vivo by magnetic resonance elastography. *Proc Natl Acad Sci U S A*, 109, 6650–5.

Sinkus, R., Tanter, M., Catheline, S., Lorenzen, J., Kuhl, C., Sondermann, E. & Fink, M. 2005a. Imaging anisotropic and viscous properties of breast tissue by magnetic resonance-elastography. *Magn Reson Med*, 53, 372–87.

Sinkus, R., Tanter, M., Xydeas, T., Catheline, S., Bercoff, J. & Fink, M. 2005b. Viscoelastic shear properties of in vivo breast lesions measured by MR elastography. *Magn Reson Imaging*, 23, 159–65.

Therrien, C. W. 2004. *Discrete Random Signals and Statistical Signal Processing*. Englewood Cliffs, NJ, Prentice Hall.

Uffmann, K. & Ladd, M. E. 2008. Actuation systems for MR elastography: Design and applications. *IEEE Eng Med Biol Mag*, 27, 28–34.

Van Essen, D. C. 1997. A tension-based theory of morphogenesis and compact wiring in the central nervous system. *Nature*, 385, 313–8.

Van Houten, E. E., Doyley, M. M., Kennedy, F. E., Weaver, J. B. & Paulsen, K. D. 2003. Initial in vivo experience with steady-state subzone-based MR elastography of the human breast. *J Magn Reson Imaging*, 17, 72–85.

Van Houten, E. E., Miga, M. I., Weaver, J. B., Kennedy, F. E. & Paulsen, K. D. 2001. Three-dimensional subzone-based reconstruction algorithm for MR elastography. *Magn Reson Med*, 45, 827–37.

Van Houten, E. E., Paulsen, K. D., Miga, M. I., Kennedy, F. E. & Weaver, J. B. 1999. An overlapping subzone technique for MR-based elastic property reconstruction. *Magn Reson Med*, 42, 779–86.

Van Houten, E. E., Viviers, D., McGarry, M. D., Perrinez, P. R., Perreard, I. I., Weaver, J. B. & Paulsen, K. D. 2011. Subzone based magnetic resonance elastography using a Rayleigh damped material model. *Med Phys*, 38, 1993–2001.

Vappou, J., Breton, E., Choquet, P., Goetz, C., Willinger, R. & Constantinesco, A. 2007. Magnetic resonance elastography compared with rotational rheometry for in vitro brain tissue viscoelasticity measurement. *MAGMA*, 20, 273–8.

Vappou, J., Breton, E., Choquet, P., Willinger, R. & Constantinesco, A. 2008. Assessment of in vivo and post-mortem mechanical behavior of brain tissue using magnetic resonance elastography. *J Biomech*, 41, 2954–9.

Weaver, J. B., Pattison, A. J., McGarry, M. D., Perreard, I. M., Swienckowski, J. G., Eskey, C. J., Lollis, S. S. & Paulsen, K. D. 2012. Brain mechanical property measurement using MRE with intrinsic activation. *Phys Med Biol*, 57, 7275–87.

Xu, G., Knutsen, A. K., Dikranian, K., Kroenke, C. D., Bayly, P. V. & Taber, L. A. 2010. Axons pull on the brain, but tension does not drive cortical folding. *J Biomech Eng*, 132, 071013.

Xu, L., Lin, Y., Han, J. C., Xi, Z. N., Shen, H. & Gao, P. Y. 2007. Magnetic resonance elastography of brain tumors: Preliminary results. *Acta Radiol*, 48, 327–30.

Zhang, J., Green, M. A., Sinkus, R. & Bilston, L. E. 2011. Viscoelastic properties of human cerebellum using magnetic resonance elastography. *J Biomech*, 44, 1909–13.

6

Displacements under Applied Loading by MRI in Soft Biomaterials and Tissues

Corey P. Neu

CONTENTS

Large deformations can be noninvasively quantified and visualized at high spatiotemporal resolution within a broad range of biomaterials and tissues using displacements under applied loading by magnetic resonance imaging (dualMRI). The synchronous timing of MRI acquisitions with external mechanical loading has enabled the detailed description of internal three-dimensional (3D) displacements and strains in hydrogels, explanted cartilage and intervertebral disc, and complex load-bearing joints of the body, providing a noninvasive means to characterize both tissue regeneration and disease in longitudinal studies. Here, we introduce the basic implementation of dualMRI and discuss current applications that visualize the mechanics of biomaterials and tissues, highlighting recent developments for in vivo imaging and the use of complimentary MRI methods that improve the study of relationships among tissue structure, disease, and biomechanical function.

6.1 Introduction

Imaging modalities offer an ideal approach to noninvasively study the mechanics of biomaterials and tissues. Imaging allows for the nondestructive measurement and three-dimensional (3-D) visualization of material deformation and changes in bulk morphology, which are not typically available by standard mechanical testing methods. Invasive techniques are numerous in the literature and encompass the use of external or internal gages [1] or other transducers [2] that may alter the mechanical behavior of the material. In contrast, imaging-based methods can reveal internal mechanics using speckle and texture tracking in blood vessels [3] and cells [4] or x-ray diffraction in bone [5].

Magnetic resonance imaging (MRI) is a noninvasive modality that is commonly used to visualize the morphology of biomaterials and tissues and can be further configured to measure multiple phenomena including flow and diffusion [6] (see also Chapters 2–5). The sensitivity of magnetic resonance (MR) data to motion has been known for over a half century [7], and most modern techniques exploit the motion sensitivity of the phase (and not magnitude) data from the complex MR signal. Under many implementations, the "phase" information is directly proportional to displacements [8], while the "magnitude" image is familiar to clinicians and scientists interested to observe tissue morphology, typically with excellent soft tissue contrast compared to other techniques like optical imaging or ultrasound [9] (Figure 6.1). The internal mechanics of biomaterials and tissues remain a challenge to measure and requires postprocessing of image data using texture correlation [10], tracking of imposed fiducial markers like tag lines [11,12], or the implementation of techniques that exploit the spatiotemporally rich phase data of the MR signal [13,14].

For load-bearing tissues like articular cartilage, displacements under applied loading by MRI (dualMRI) is a suitable method for the noninvasive assessment of the internal tissue biomechanics. Common methods to study in vivo cartilage mechanics, often by MRI or radiography [15–18], rely on measurements of nominal thickness (Figure 6.1b) or volume changes that, alone, are too insensitive to describe the internal mechanical behavior of the heterogeneous response [13,14,19]. MR elastography is a powerful technique that has been used to compute material properties for cartilage ex vivo [20–23], but the (small) load magnitudes and (high) frequencies required for this technique are not easily adaptable in a broad range of applications, including imaging in vivo, where high magnitude loading (i.e., body weight) at low frequency (i.e., walking cadence) is desired to mimic normal daily activities [24]. Elastography is also challenging for the measurement of the thin (2–3 mm thick) cartilage embedded deep within the joint space. In contrast, dualMRI was developed to measure internal strains under the aforementioned physiologically relevant conditions and at high precision [13,14] to capture intrinsic heterogeneity of the tissue response.

In this chapter, we provide an overview of dualMRI for the measurement of displacements and strain in soft biological tissues and biomaterials. The essential steps for the method are described, inclusive of MRI pulse sequence design, synchronization of MRI actions with exogenous mechanical loading, and the use of postprocessing methods to obtain spatial maps of deformation. We further describe initial implementations of dualMRI for the measurement of tissue degeneration and repair in model systems.

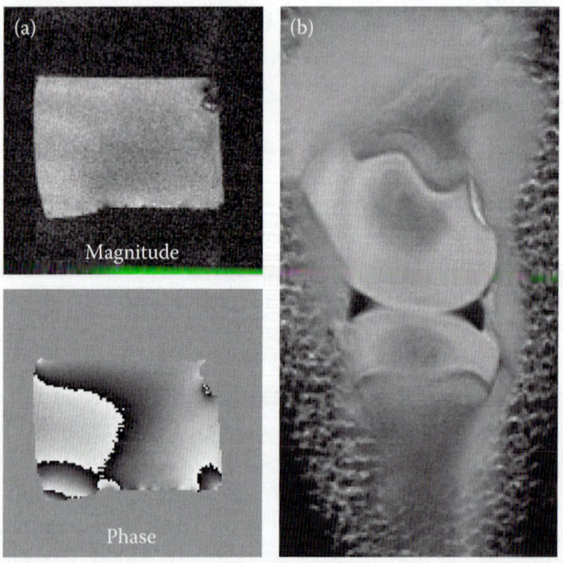

FIGURE 6.1
Magnetic resonance magnitude and phase images. Magnitude images render (a) morphology of silicone phantom materials and (b) tissue contrast in tibiofemoral joints. Phase data are sensitive to displacements and can be used to calculate three-dimensional strain in tissues and biomaterials. (The tibiofemoral joint image in (b) was adapted from Chan, D.D. et al., *Magn. Reson. Med.*, 61, 2009.)

6.2 dualMRI: Theory and Method

6.2.1 Overview

dualMRI is a highly versatile method (Figures 6.1 through 6.3) that can measure biomaterial and tissue deformation during the time course of loading. The principle underlying dualMRI is the sensitivity of MRI phase to motion. While dualMRI broadly encompasses more general motion-sensitive methods based on tag registration [12,19,25], most modern implementations incorporate phase-contrast methods [13,14], which rely on sensitizing the change in phase data, $\Delta\phi$, of the spin system to changes in displacement, Δx, according to $\Delta\phi = \gamma_H t_{enc}(G_{de} - G'_{de})\Delta x$, with the gyromagnetic ratio, γ_H, duration of the encoding gradient, t_{enc}, encoding gradient in any arbitrary xyz direction, G_{de}, and a common gradient magnitude used to eliminate other phase contributions, G'_{de}. Phase-based dualMRI methods are very precise, with displacement and strain precisions on the order of 11 μm and 0.1%, respectively [13].

Measurement of deformation in biomaterials and tissues by dualMRI is accomplished in three primary steps: (1) design of an appropriate MRI pulse sequence, (2) application of synchronous MR imaging and mechanical loading actions, and (3) postprocessing of image data to extract patterns of displacements and strain. While

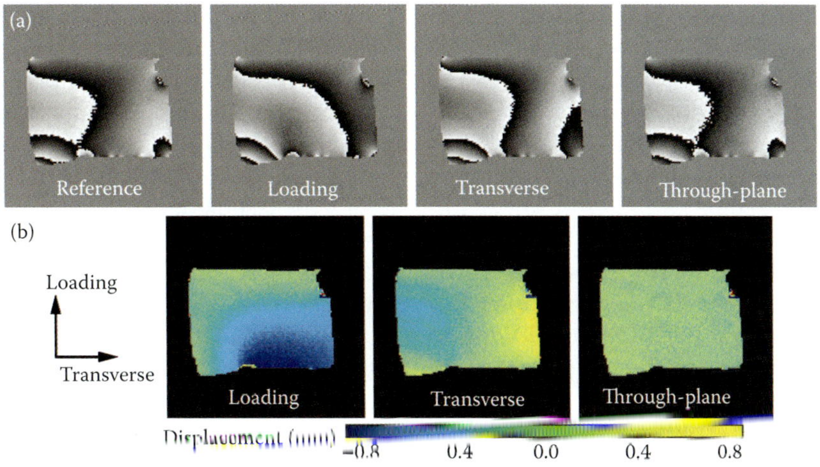

FIGURE 6.2
MRI phase data are used to compute displacements. (a) Reference phase data acquired without the use of a displacement-encoding gradient, G'_{de}, is subtracted from data acquired with G_{de} applied in loading, transverse, through-plane, or arbitrary (not shown) directions. (b) Phase data are converted to displacement fields in two or three dimensions and further used for strain calculations.

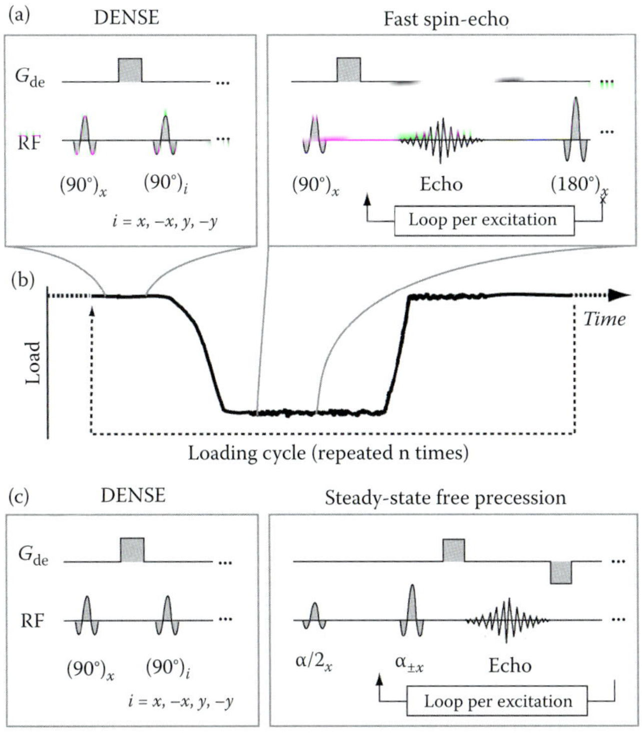

FIGURE 6.3
dualMRI synchronizes a displacement encoding with stimulated echoes (DENSE) preparation, image acquisition sequence, and exogenous mechanical loading. (a–b) DENSE is typically applied prior to mechanical loading and involves a displacement-encoded gradient, G_{de}, and two radiofrequency pulses. During ramp or arbitrary loading [13], data are acquired using a second MRI pulse sequence, for example, (a) fast spin-echo (FSE) or (c) steady-state free precession (SSFP) readout. (Deva Chan, PhD, is gratefully acknowledged for her assistance in the creation of this figure.)

the method is readily adaptable to tissues with intrinsic motion, for example, cardiac tissue [8], here we focus on systems that require exogenous loading, including load-bearing musculoskeletal tissues of the body. dualMRI can be configured to measure motion at low frequency (e.g., that mimic the walking cycle) and over a range in load magnitudes, allowing for the measurement of large (finite) strains.

6.2.2 MRI Pulse Sequences

Strain measures through the 3-D volume of articular cartilage were based on MRI tagging [12] and phase-contrast displacement encoding [14]. Typical implementation of phase-contrast displacement-encoding sequences involves a displacement encoding with stimulated echoes (DENSE) preparation of the tissue in a reference or initial configuration, followed by image acquisition at a later time in a current or deformed configuration (Figure 6.3). For the DENSE preparation, a displacement-encoded gradient (e.g., G_{de}) is applied between two 90° radio-frequency (RF) pulses, allowing for tissue positions to be encoded and then stored prior to movement during a variable mixing time (TM). Phase cycling of the second pulse is used to eliminate image artifacts [26]. Following the TM, the stored echo is returned to the imaging plane, and G_{de} is used to decode the current configuration of the magnetization and phase.

The choice of the image acquisition sequence provides a flexible framework for deformation measurements in a variety of applications. In particular, DENSE with a fast spin-echo readout (DENSE-FSE) was used to measure strain in cartilage explants [14], whole joints [27–29], and cartilage defect models [30]. The strain data were further used to estimate stresses and material properties using mathematical modeling [31]. Although FSE is appropriate for tissues and biomaterials with midrange to long T_2 values, DENSE can be paired with MR acquisition sequences better suited for short and ultrashort T_2 values. More recently, DENSE with a steady-state free precession readout (DENSE-SSFP) was developed [13] (based on ref. 32) for intact joints [28,29]. In addition to enabling easy visualization of cartilage, ligaments, and menisci in joints, the total acquisition time for dualMRI in intact knees using DENSE-SSFP was approximately 16 min, inclusive of cyclic loading with a 2.5/5.0 s loading/total period, whereas a comparable DENSE-FSE acquisition required >120 min.

6.2.3 Synchrony of Imaging and Mechanical Loading

Precise timing of MRI and mechanical loading actions is required for dualMRI. The biomaterial or tissue is placed inside an MRI-compatible loading apparatus, located within an RF coil, magnetic field gradient system, and permanent magnet of the MRI system. Imaging and loading are typically synchronized using a digital trigger from software (e.g., C++ or LabView), enabling millisecond timing control of MRI and load apparatus actions. MRI data are normally acquired over multiple loading cycles because of three competing factors: (1) time required for image acquisition, which typically spans minutes for 2-D and 3-D data; (2) viscoelasticity, for example, tissue creep, which can lead to motion artifacts and biased displacement and strain results; and (3) decay of the DENSE preparation signal, which is diminished according to the T1 of the material and on the order of seconds for cartilage in most field strengths. Consequently, a portion of the image data is obtained during each loading cycle, and only after the material has been preconditioned to a steady-state load-displacement response [33]. Pneumatic actuation is used for rapid material loading (e.g., compression), with targeted, full-scale loading (e.g., 0.5 times body weight) occurring in a few hundred milliseconds [34]. The MRI data can be acquired during ramp-and-hold loading or over arbitrary loading waveforms, with spatial resolutions at or exceeding 100 μm in-plane, and a high temporal resolution of ~2–3 ms (limited by the repetition time [TR] of the DENSE-SSFP sequence), at the cost of increased total imaging time.

6.2.4 Image Postprocessing

The pipeline for image processing follows several defined steps. First, recombination of phase-cycled data is accomplished prior to application of a 2-D inverse Fourier transform. Phase corrections (of zeroth and first order) and unwrapping are applied to transformed data, and displacements (e.g., Δx) are computed using a difference of phase ($\Delta\phi$) computed in the reference phase data from the displacement-encoded phase data. Displacements are smoothed using a 5 × 5 pixel² Gaussian kernel [35] and used to determine the deformation gradient tensor, F, using a maximum likelihood estimation [14,36]. Strain is then typically expressed at each pixel of interest as components of the Green–Lagrange strain tensor, $E(= 1/2(F^TF - I))$, or as principal strains.

6.3 Current Applications of dualMRI

6.3.1 Articular Cartilage

Finite strains have been measured in articular cartilage explants and intact joints. In the early development of dualMRI, a cartilage deformation by tag registration (*CDTR*) technique was used to characterize displacements and strain using a specialized tagging-based MRI pulse sequence [12,19,25]. The advantage of the CDTR technique was the relatively rapid MRI acquisition time, but the technique suffered severely due to the availability of relatively few tag lines to track motion in the thin tissue coupled with the need to extrapolate strains near the tissue boundary. Because of these limitations, dualMRI was inspired by the stimulated echo acquisition mode (STEAM) sequence [20] for the phase-based measurement of strain (Figure 6.4). In recent studies of intact joints, strains were consistently found to be heterogeneous through the depth of tibiofemoral cartilage during compressive loading, with the largest strains in the loading direction located in the middle zone of the tissue and tensile strains documented in the direction transverse to loading [9,13,27–29].

In an effort to translate dualMRI for use in humans in vivo, two new sequences were implemented that reduced the overall acquisition time of images. DENSE was paired with an SSFP readout [13] and then with a single-shot fast spin-echo (SSFSE) readout [34], allowing a complete acquisition of displacement-encoded data in minutes. In the case of DENSE-SSFP, strains have been documented not only in the articular cartilage of intact human joints but, additionally, in the ligament and meniscus, other joint tissues that are typically difficult to visualize using sequences like FSE alone [14]. We have also demonstrated that strains are a sensitive measure of joint degeneration in human tissues [37,38]. For DENSE-SSFSE, we recently documented direct

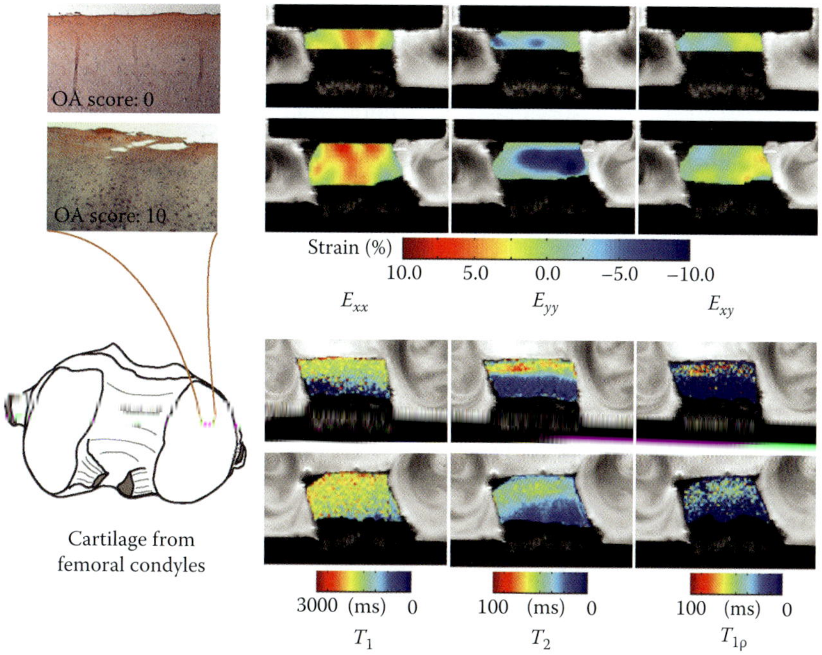

FIGURE 6.4
Spatially dependent multicontrast MRI measures were acquired in explanted human cartilage and compared to histological assessment of OA severity. Using explanted cartilage from the condyles of tibiofemoral joints of osteoarthritic patients, finite strains (E_{xx}, E_{yy}, E_{xy}) were determined by dualMRI. Relaxivity measures were also acquired, allowing for pixel-by-pixel registration of MRI data. (Adapted from Griebel, A.J. et al., *Magn. Reson. Med.*, 2013.)

measurement of in vivo cartilage deformation in human volunteers throughout the tissue interior for the first time [39], opening the possibility of longitudinal studies in multiple applications, including preclinical and clinical trials of treatment for joint disease.

6.3.2 Intervertebral Disc

Measurement of mechanical strains in the intervertebral disc (IVD) allows for characterization of normal function [34,40] and the progression of degeneration [27,41]. In previous studies, dualMRI quantified displacements and strain in a puncture model of IVD degeneration [27,42,43]. IVDs between the fourth and fifth lumbar vertebrae (L4/L5) of rabbits were punctured with a 16-gauge needle, sacrificed after 28 days, and isolated with adjacent L3/L4 IVDs to include the full disc and the attached superior and posterior vertebral bodies. IVDs were placed in a loading device and 9.4 T vertical bore MRI system. dualMRI was acquired during compressive loading at 0.33 Hz, followed by in-plane deformation calculations. Displacements of the IVDs, in particular, each nucleus pulposus, were altered in most untreated compared to punctured discs. dualMRI was also registered to another independent MRI measure, dGEMRIC-derived GAG concentrations [27], further extending the use of multiple MRI measures (i.e., "multicontrast MRI" [37]) for the best characterization of tissue damage. IVD deformation was most recently implemented in human tissues [34], demonstrating the potential for in vivo imaging at clinical (e.g., 3 T) field strengths.

6.3.3 Tissues Engineering, Phantom Materials, and Hydrogels

The measurement of strain in a broad range of biomaterials is important as a functional measure and primary end point for the successful regeneration of many biological tissues such as articular cartilage, spine, and heart. We have routinely imaged materials including silicone gels, with excellent elastomeric properties [25], and hydrogels like agarose [30,44] used as engineered constructs. In one study, dualMRI computed detailed deformation fields throughout the interior of native and engineered cartilage using an articular defect model as a study system [30]. Strains were highest in the tissue constructs compared to surrounding cartilage, revealing discontinuities in the tissue surface that altered the distribution of strain near the indentor–construct interface, in addition to locations in tissue–construct boundaries that lacked integration. The sensitivity of dualMRI for engineered tissues was not evaluated over time, but we expect that changes (e.g., lower strains) would be observed in the construct as matrix is deposited by embedded cells.

6.3.4 Estimation of Internal Stresses and Material Properties

With the appropriate choice of a constitutive relationship, it is possible to compute stress distributions and material properties in biomaterials and tissues based on dualMRI finite strains. In one study, finite strains were determined in the medial tibiofemoral cartilage of cyclically loaded joints [29,31] and coupled to joint geometry and (plane stress) mathematical modeling, defined by linear elastic and nonlinear constitutive relations, to compute stresses and elastic moduli. In a plane stress model, we observed nonuniform stress distributions throughout the cartilage [31]. Large differences in stresses and material properties were found between the tibial and femoral cartilage. Extension of the approach to include other forms of MRI contrast (e.g., relaxivity) allows for the inverse calculation of constitutive relation parameters with increasing complexity and degrees of freedom.

6.4 Future Applications of dualMRI

6.4.1 dualMRI in Models of Disease

Osteoarthritis (OA) is a degenerative disease of articular cartilage in synovial joints, often leading to pain, disability, and total joint arthroplasty. OA reflects a complex interplay of biomechanical, biochemical, metabolic, and genetic factors, which are often triggered by injury, and mediated by inflammation, catabolic cytokines (e.g., IL-1β), and enzymes (e.g., aggrecanase) [45–47]. A critical barrier in cartilage imaging is the use of methods that are unfortunately best suited to visualize cartilage damage in advanced OA [48,49]. dualMRI is one technology that may allow for the noninvasive detection of early OA using biomechanics as an imaging biomarker, especially when combined with complementary imaging biomarkers [37].

Cartilage strain may be used as a functional indicator of joint damage and repair. dualMRI computes cartilage displacements and strain, quantities that are indirectly related to tissue stiffness and that change with OA, since the stiffness of cartilage is known to decrease in OA [50–52]. Changes in cartilage strain patterns from baseline values, or changes over time, would indicate tissue regions with altered stiffness and degenerative changes. Importantly, cartilage defects, ACL tears and reconstruction, and meniscus injuries all lead to cartilage changes and OA in the long term [53–55], suggesting that cartilage strain and dualMRI may be useful to assess a broad range of musculoskeletal joint problems and treatments.

One possible approach to visualize cartilage damage involves the combined mapping of spatial patterns of mechanical strain by dualMRI with measures of magnetic relaxation rates and cartilage morphology by quantitative MRI (qMRI). We demonstrated the use of multiple MRI acquisitions, termed "multicontrast MRI" [37], inclusive of dualMRI and qMRI, using explanted human osteochondral tissues exhibiting a wide range of structural integrity from patients with diagnosed OA undergoing elective total knee replacement surgery (Figure 6.4). Osteochondral specimens were obtained from standardized regions within the joint, representing a distribution of wear and OA severity. Explants bathed in PBS were scanned using a 14.1 T MRI system (Bruker Medical GMBH). For dualMRI, the explants were cyclically and intermittently loaded in unconfined compression with a flat platen at 235 kPa and 0.33 Hz until they reached a quasi–steady-state load-deformation behavior. A DENSE-SSFP imaging sequence was used with a displacement-encoding gradient area of 2.13 π/mm. SSFP parameters were identical to those described previously. MRI-based displacement fields were used to compute finite strains (transverse [E_{xx}], axial [E_{yy}], and shear [E_{xy}]). Spatial (pixel-by-pixel) qMRI maps of T_1, T_2, and $T_{1\rho}$ relaxation time measurements were acquired.

Multicontrast MRI better predicted histologically assessed OA compared to any individual parameter alone. We regressed statistical model-predicted OA scores with measured OA scores, attaining an R^2 of 0.482 with high significance ($P < 0.001$) [37]. We demonstrated that including the relaxivity parameters in a predictive model along with dualMRI allows for a greater explanation of the variability seen in the data. Interestingly, our predictive model treated a patient as a random effect, so in future studies, the development and application of patient-specific information (e.g., age, gender, body mass index [BMI], activity level, limb alignment, family history) may improve the multicontrast MRI predictive ability for early OA.

6.4.2 dualMRI in Tissue Regeneration and Repair

In vivo imaging biomarkers are needed to noninvasively characterize the biomechanical function of regeneration and repair in musculoskeletal tissues. Emerging biological therapies and management strategies, including gene therapy and cell implantation [56–61], may ultimately be effective therapies for musculoskeletal tissues following injury and disease. Critical to evaluate the success of emerging therapies is the need for tools that can assess health, disease, and damage in musculoskeletal tissues, ideally using noninvasive, in vivo methods that are suited for longitudinal assessment in animals and humans. Imaging biomarkers that determine the biomechanical function of cartilage are especially important given the normal load-bearing action of the tissue [62] and the sensitivity of cartilage biomechanics to joint injury and disease [51,52]. For example, standard-of-care treatment for chondral defects inevitably leads to posttraumatic OA. Cartilage defects are a common result of joint trauma and sports injuries, approaching 36% of athletic populations [63]. Microfracture is a gold-standard surgical procedure used to treat focal cartilage defects and involves creating small fractures in the underlying subchondral bone to access progenitor cells in the marrow, fill the defect with a blood clot, and accelerate repair [64–66]. Unfortunately, fibrocartilage forms in the defect space, which is mechanically inferior to normal hyaline cartilage [67–69]. dualMRI may help us distinguish differences in the repair quality of hyaline versus fibrocartilage and evaluate new strategies for cartilage repair in the long term.

6.5 Conclusion

The sensitivity of MR to motion, described decades ago [7], has led to a broad range of specialized techniques that provide detailed information about the mechanics of biomaterials and tissues. MRI traditionally provides excellent contrast and morphology of soft tissues compared to other medical imaging modalities like ultrasound or radiography. The ability of MRI to provide information on other phenomena, like motion, expands capability and enables the possibility for characterization of disease and regeneration. dualMRI, in particular, provides a flexible framework for the characterization of large deformations in biomaterials and tissues that are typically subjected to exogenous mechanical loading, like musculoskeletal tissues.

Numerous challenges remain on the horizon to best adapt dualMRI for improved clinical utility. Spatiotemporal resolution remains a major limitation of dualMRI, especially in vivo, where it is desirable to obtain detailed time course information of small tissue structures like cartilage, which may exhibit only microscale cracking or damage in the early stages of disease or trauma. Future implementations that address the need to provide phase-contrast data with high signal-to-noise ratio will likely best be suited for disease and regeneration models. A second challenge is the design of exogenous mechanical loading modes (e.g., compression, shear) that best reveal changes in tissues over time. There is evidence that bending of the IVD may be a better measure than compression to reliably describe strains with minimal variability over patient populations [70]. Finally, controlled bench-side experiments, with defined application of disease (e.g., cytokine-induced [71]) or damage (e.g., trauma [72,73]) models, are required to help define the sensitivity of dualMRI to longitudinal changes in biomaterials in tissues in relation to concomitant biochemical and biosynthesis changes at the tissue and cellular levels.

Acknowledgments

This work was supported in part by grants from the National Science Foundation (NSF) (CMMI 1100554) and the National Institutes of Health (NIH) (R01 AR063712 and R21 AR064178). Deva Chan, PhD, is gratefully acknowledged for her assistance in the creation of Figure 6.3.

References

1. Monahan, J.J. et al., In vivo strain patterns in the four major canine knee ligaments. *J Orthop Res*, 1984. **2**(4): pp. 408–18.
2. McNally, D.S., and M.A. Adams, Internal intervertebral disc mechanics as revealed by stress profilometry. *Spine*, 1992. **17**(1): pp. 66–73.
3. Zhang, J. et al., Strain and mechanical behavior measurements of soft tissues with digital speckle method. *J Biomed Opt*, 2005. **10**(3): p. 034021.
4. Lim, K.Y., J.T. Henderson, and C.P. Neu, Cell and tissue deformation measurements: Texture correlation with third-order approximation of displacement gradients. *J Biomech*, 2013. **46**(14): pp. 2490–6.
5. Almer, J.D., and S.R. Stock, Internal strains and stresses measured in cortical bone via high-energy X-ray diffraction. *J Struct Biol*, 2005. **152**(1): pp. 14–27.
6. Haacke, E.M. et al., *Magnetic Resonance Imaging: Physical Principles and Sequence Design*. New York: John Wiley & Sons, Inc., 1999.
7. Hahn, E.L., Detection of sea-water motion by nuclear precession. *Geophys Res*, 1960. **65**: p. 776.
8. Aletras, A.H. et al., DENSE: Displacement encoding with stimulated echoes in cardiac functional MRI. *J Magn Reson*, 1999. **137**(1): pp. 247–52.
9. Chan, D.D., and C.P. Neu, Probing articular cartilage damage and disease by quantitative magnetic resonance imaging. *J R Soc Interface*, 2013. **10**(78): p. 20120608.
10. O'Connell, G.D. et al., Human internal disc strains in axial compression measured noninvasively using magnetic resonance imaging. *Spine*, 2007. **32**(25): pp. 2860–8.
11. Bayly, P.V. et al., Deformation of the human brain induced by mild acceleration. *J Neurotrauma*, 2005. **22**(8): pp. 845–56.
12. Neu, C.P. et al., MRI-based technique for determining nonuniform deformations throughout the volume of articular cartilage explants. *Magn Reson Med*, 2005. **53**(2): pp. 321–8.

13. Chan, D.D., and C.P. Neu, Transient and microscale deformations and strains measured under exogenous loading by noninvasive magnetic resonance. *PLos One*, 2012. **7**(3): p. e33463.
14. Neu, C.P., and J.H. Walton, Displacement encoding for the measurement of cartilage deformation. *Magn Reson Med*, 2008. **59**: pp. 149–55.
15. Coleman, J.L. et al., Diurnal variations in articular cartilage thickness and strain in the human knee. *J Biomech*, 2013. **46**(3): pp. 541–7.
16. Eckstein, F. et al., In vivo cartilage deformation after different types of activity and its dependence on physical training status. *Ann Rheum Dis*, 2005. **64**(2): pp. 291–5.
17. Stammberger, T. et al., Determination of 3D cartilage thickness data from MR imaging: Computational method and reproducibility in the living. *Magn Reson Med*, 1999. **41**(3): pp. 529–36.
18. Yao, J. et al., Magnetic resonance image analysis of meniscal translation and tibio-menisco-femoral contact in deep knee flexion. *J Orthop Res*, 2008. **26**(5): pp. 673–84.
19. Neu, C.P., M.L. Hull, and J.H. Walton, Heterogeneous three-dimensional strain fields during unconfined cyclic compression in bovine articular cartilage explants. *J Orthop Res*, 2005. **23**(6): pp. 1390–8.
20. Hardy, P.A. et al., Imaging articular cartilage under compression—Cartilage elastography. *Magn Reson Med*, 2005. **53**(5): pp. 1065–73.
21. Lopez, O. et al., Characterization of the dynamic shear properties of hyaline cartilage using high-frequency dynamic MR elastography. *Magn Reson Med*, 2008. **59**(2): pp. 356–64.
22. Lopez, O. et al., Developments in dynamic MR elastography for in vitro biomechanical assessment of hyaline cartilage under high-frequency cyclical shear. *J Magn Reson Imaging*, 2007. **25**(2): pp. 310–20.
23. Tervonen, O. et al., MR imaging of knee hyaline cartilage: Evaluation of two- and three-dimensional sequences. *J Magn Reson Imaging*, 1993. **3**(4): pp. 663–8.
24. Winter, D.A., *Biomechanics and Motor Control of Human Movement*. John Wiley & Sons, Inc., 1990.
25. Neu, C.P., M.L. Hull, and J.H. Walton, Error optimization of a three-dimensional magnetic resonance imaging tagging-based cartilage deformation technique. *Magn Reson Med*, 2005. **54**(5): pp. 1290–4.
26. Epstein, F.H., and W.D. Gilson, Displacement-encoded cardiac MRI using cosine and sine modulation to eliminate (CANSEL) artifact-generating echoes. *Magn Reson Med*, 2004. **52**(4): pp. 774–81.
27. Chan, D.D. et al., Mechanical deformation and glycosaminoglycan content changes in a rabbit annular puncture disc degeneration model. *Spine*, 2011. **36**(18): pp. 1438–45.
28. Chan, D.D., C.P. Neu, and M.L. Hull, Articular cartilage deformation determined in an intact tibiofemoral joint by displacement-encoded imaging. *Magn Reson Med*, 2009. **61**(4): pp. 989–93.
29. Chan, D.D., C.P. Neu, and M.L. Hull, In situ deformation of cartilage in cyclically loaded tibiofemoral joints by displacement-encoded MRI. *Osteoarthritis Cartilage*, 2009. **17**(11): pp. 1461–8.
30. Neu, C.P. et al., Characterization of engineered tissue construct mechanical function by magnetic resonance imaging. *J Tissue Eng Regen Med*, 2009. **3**(6): pp. 477–85.
31. Butz, K.D., D.D. Chad, E.A. Nauman, and C.P. Neu, Stress distributions and material properties determined in articular cartilage from MRI-based finite strains. *J Biomech*, 2011. **13;44**(15):2667–72.
32. Lin, A.P. et al., Circumferential strain int the wall of the common carotid artery: Comparing displacement-encoded and cine MRI in volunteers. *Magn Reson Med*, 2008. **60**(1): pp. 8–13.
33. Neu, C.P., and M.L. Hull, Toward an MRI-based method to measure non-uniform cartilage deformation: An MRI-cyclic loading apparatus system and steady-state cyclic displacement of articular cartilage under compressive loading. *J Biomech Eng*, 2003. **125**(2): pp. 180–8.
34. Chan, D.D., and C.P. Neu, Intervertebral disc internal deformation measured by displacements under applied loading with MRI at 3T. *Magn Reson Med*, 2013. **71**(3):1231–7.
35. Chan, D.D., D. Toribio, and C.P. Neu, Displacement smoothing for the precise MRI-based measurement of strain in soft biological tissues. *Comput Methods Biomech Biomed Eng*, 2013. **16**(8): pp. 852–60.
36. Geers, M.G.D., R. de Borst, and W.A.M. Brekelmans, Computing strain fields from discrete displacement fields in 2D-solids. *Int J Solids Struct*, 1996. **33**(29): pp. 4293–307.
37. Griebel, A.J., S.B. Trippel, N.C. Emery, and C.P. Neu, Noninvasive assessment of osteoarthritis severity in human explants by multicontrast MRI. *Magn Reson Med*, 2013. **71**(2):807–14.
38. Griebel, A.J., S.B. Trippel, and C.P. Neu, Noninvasive dualMRI-based strains vary by depth and region in human osteoarthritic articular cartilage. *Osteoarthritis Cartilage*, 2013. **21**(2): pp. 394–400.
39. Chan, D.D. et al., In vivo displacements and strains measured in human knee articular cartilage using dualMRI. In *Transactions of the Annual Meeting of the Orthopaedic Research Society*, New Orleans, LA, 2014.
40. An, H.S., K. Masuda, and N. Inoue, Intervertebral disc degeneration: Biological and biomechanical factors. *J Orthop Sci*, 2006. **11**(5): pp. 541–52.
41. Wang, C. et al., Advances in magnetic resonance imaging for the assessment of degenerative disc disease of the lumbar spine. *Semin Spine Surg*, 2007. **19**(2): pp. 65–71.
42. Masuda, K., T.R. Oegema Jr., and H.S. An, Growth factors and treatment of intervertebral disc degeneration. *Spine*, 2004. **29**(23): pp. 2757–69.
43. Sobajima, S. et al., A slowly progressive and reproducible animal model of intervertebral disc degeneration characterized by MRI, X-ray, and histology. *Spine*, 2005. **30**(1): pp. 15–24.
44. Griebel, A.J., M. Khoshgoftar, T. Novak, C.C. van Donkelaar, and C.P. Neu, Direct noninvasive measurement and numerical modeling of depth-dependent strains in layered agarose constructs. *J Biomech*, 2014. **27;47**(9):2149–56.
45. Moos, V. et al., Immunohistological analysis of cytokine expression in human osteoarthritic and healthy cartilage. *J Rheumatol*, 1999. **26**(4): pp. 870–9.

46. Tetlow, L.C., D.J. Adlam, and D.E. Woolley, Matrix metalloproteinase and proinflammatory cytokine production by chondrocytes of human osteoarthritic cartilage: Associations with degenerative changes. *Arthritis Rheum*, 2001. **44**(3): pp. 585–94.
47. van den Berg, W.B. et al., Role of tumour necrosis factor alpha in experimental arthritis: Separate activity of interleukin 1beta in chronicity and cartilage destruction. *Ann Rheum Dis*, 1999. **58 Suppl 1**: pp. 140–8.
48. Duvvuri, U. et al., T1rho-relaxation in articular cartilage: Effects of enzymatic degradation. *Magn Reson Med*, 1997. **38**(6): pp. 863–7.
49. Regatte, R.R. et al., T1rho relaxation mapping in human osteoarthritis (OA) cartilage: Comparison of T1rho with T2. *J Magn Reson Imaging*, 2006. **23**(4): pp. 547–53.
50. Kleemann, R.U. et al., Altered cartilage mechanics and histology in knee osteoarthritis: Relation to clinical assessment (ICRS Grade). *Osteoarthritis Cartilage*, 2005. **13**(11): pp. 958–63.
51. Setton, L.A., D.M. Elliott, and V.C. Mow, Altered mechanics of cartilage with osteoarthritis: Human osteoarthritis and an experimental model of joint degeneration. *Osteoarthritis Cartilage*, 1999. **7**(1): pp. 2–14.
52. Setton, L.A. et al., Mechanical properties of canine articular cartilage are significantly altered following transection of the anterior cruciate ligament. *J Orthop Res*, 1994. **12**(4): pp. 451–63.
53. Friel, N.A., and C.R. Chu, The role of ACL injury in the development of posttraumatic knee osteoarthritis. *Clin Sports Med*, 2013. **32**(1): pp. 1–12.
54. Jiang, Y.Z. et al., Cell transplantation for articular cartilage defects: Principles of past, present, and future practice. *Cell Transplant*, 2011. **20**(5): pp. 593–607.
55. Petty, C.A., and J.H. Lubowitz, Does arthroscopic partial meniscectomy result in knee osteoarthritis? A systematic review with a minimum of 8 years' follow-up. *Arthroscopy*, 2011. **27**(3): pp. 419–24.
56. Diekman, B.O. et al., Cartilage tissue engineering using differentiated and purified induced pluripotent stem cells. *Proc Natl Acad Sci U S A*, 2012. **109**(47): pp. 19172–7.
57. Shi, S. et al., Growth factor regulation of growth factor production by multiple gene transfer to chondrocytes. *Growth Factors*, 2013. **31**(1): pp. 32–8.
58. Trippel, S. et al., Gene therapy for articular cartilage repair. *Proc Inst Mech Eng H*, 2007. **221**(5): pp. 451–9.
59. Trippel, S.B., S.C. Ghivizzani, and A.J. Nixon, Gene-based approaches for the repair of articular cartilage. *Gene Ther*, 2004. **11**(4): pp. 351–9.
60. Valonen, P.K. et al., In vitro generation of mechanically functional cartilage grafts based on adult human stem cells and 3D-woven poly(epsilon-caprolactone) scaffolds. *Biomaterials*, 2010. **31**(8): pp. 2193–200.
61. Wu, L. et al., Regeneration of articular cartilage by adipose tissue derived mesenchymal stem cells: Perspectives from stem cell biology and molecular medicine. *J Cell Physiol*, 2013. **228**(5): pp. 938–44.
62. Mow, V.C., A. Ratcliffe, and A.R. Poole, Cartilage and diarthrodial joints as paradigms for hierarchical materials and structures. *Biomaterials*, 1992. **13**(2): pp. 67–97.
63. Flanigan, D.C. et al., Prevalence of chondral defects in athletes' knees: A systematic review. *Med Sci Sports Exerc*, 2010. **42**(10): pp. 1795–801.
64. Dorotka, R. et al., Repair of articular cartilage defects treated by microfracture and a three-dimensional collagen matrix. *Biomaterials*, 2005. **26**(17): pp. 3617–29.
65. Erggelet, C. et al., Formation of cartilage repair tissue in articular cartilage defects pretreated with microfracture and covered with cell-free polymer-based implants. *J Orthop Res*, 2009. **27**(10): pp. 1353–60.
66. Hoemann, C.D. et al., Chitosan-glycerol phosphate/blood implants improve hyaline cartilage repair in ovine microfracture defects. *J Bone Joint Surg Am*, 2005. **87**(12): pp. 2671–86.
67. Furukawa, T. et al., Biochemical studies on repair cartilage resurfacing experimental defects in the rabbit knee. *J Bone Joint Surg Am*, 1980. **62**(1): pp. 79–89.
68. Peterson, L. et al., Two- to 9-year outcome after autologous chondrocyte transplantation of the knee. *Clin Orthop Relat Res*, 2000. **374**: pp. 212–34.
69. Wegener, B. et al., Matrix-guided cartilage regeneration in chondral defects. *Biotechnol Appl Biochem*, 2009. **53**(Pt 1): pp. 63–70.
70. Gossett, P.C. et al., Noninvasive mapping of strain fields in a human L4-L5 intervertebral disc under physiologically-relevant axial and bending loads. In *ASME Summer Bioengineering Conference*, Sunriver, OR, 2013.
71. DuRaine, G. et al., Regulation of the friction coefficient of articular cartilage by TGF-beta1 and IL-1beta. *J Orthop Res*, 2009. **27**(2): pp. 249–56.
72. Lee, J.H. et al., Mechanical injury of cartilage explants causes specific time-dependent changes in chondrocyte gene expression. *Arthritis Rheum*, 2005. **52**(8): pp. 2386–95.
73. Sui, Y. et al., Mechanical injury potentiates proteoglycan catabolism induced by interleukin-6 with soluble interleukin-6 receptor and tumor necrosis factor alpha in immature bovine and adult human articular cartilage. *Arthritis Rheum*, 2009. **60**(10): pp. 2985–96.

Section III

Tissue and Organ Imaging Methods

Ultrasound

7

Ultrasound and Optical Methods for Dynamic Viscoelastic Imaging

Yue Wang, Steven G. Adie, Stephen A. Boppart, and Michael F. Insana

CONTENTS

This chapter summarizes the approaches to tissue elasticity imaging based on ultrasonic and/or optical modalities that have been developed during the last three decades. We describe a wide range of techniques, from simply mapping soft-tissue stiffness to accurately estimating a complex modulus of viscoelastic tissues, in vivo. We describe how experimental parameters determine the spatiotemporal properties of the force that then couple to the properties of the imaging system used to track deformations and to specific tissue geometries. These experimental-design parameters determine the resolution and noise properties of the image, and they can also be set to select which elements of tissue structure contribute to the object contrast. Parameter selection depends on measurement goals, which might include anything from a basic-science investigation into mechanobiology of disease processes to diagnosis of focal diseases in a patient. Although there are many coupled parameters to consider, they provide a broad pallet of opportunities for discovery and diagnosis related to changes in the mechanical properties of tissues.

7.1 Introduction

It is well known that pathological changes in tissues are often accompanied by changes in the mechanical properties of tissue. For example, liver stiffness was found to predict the 5-year survival of patients chronically infected with hepatitis B virus (de Ledinghen et al. 2013). Arterial stiffness was found to positively correlate with several inflammatory markers in essential hypertension (Mahmud and Feely 2005). Also, the tissues surrounding malignant lesions often stiffen as cancer cells conscript wound healing mechanisms to remodel the extracellular matrix (ECM) in ways favorable to malignant tumor growth (Lu et al. 2011). Cancer is associated with a range of tissue changes affecting the cellular mechanoenvironment that result in a disease-promoting reactive stroma. The density and cross-linking of ECM collagen increase as fluid pressure builds from lymphatic collapse, in part from epithelial hyperplasia (Butcher et al. 2009; Lu et al. 2012; Lunt et al. 2008). Compared to baseline elastic modulus values for normal fibroglandular breast tissues, fibrocystic disease and low-grade malignant mammary tumors exhibit threefold to sixfold stiffness increases. High-grade invasive ductal carcinomas can be an average of 13 times stiffer (Samani et al. 2007). By imaging tissue mechanical properties, one can provide a sensitive indication of early inflammatory processes that can improve patient diagnosis and prognosis, as well as assist with treatment planning and the evaluation of therapeutic responses (Hayashi et al. 2012).

The terms *elastography* and *elasticity imaging* are used interchangeably to describe methods for imaging mechanical properties of tissue. An imaging modality is used to track the deformation response of a tissue to an applied mechanical force. Phase-sensitive imaging modalities (ultrasound, MRI [see Chapters 2–6], optical) are most often used because of their high sensitivity to sub-millimeter-scale movements. The large object contrast in mechanical properties associated with several diseases provides specific advantages over alternative imaging techniques, provided that the applied forces and the induced deformations can be quickly and accurately tracked over time. Ultrasonic and optical coherence tomography (OCT)-based elastography methods offer safe, low-cost, and high-speed methods for tracking tissue movement in two or more dimensions. All major manufacturers of sonographic systems now offer some form of elasticity imaging because of the growing number of clinical applications. For example, clinical trials involving breast cancer patients show that the number of unnecessary biopsies can be significantly reduced in low-risk populations by applying ultrasonic elastography to patients with a positive sonographic exam (Sadigh et al. 2012). Ultrasonic elasticity imaging has also been used intraoperatively to monitor thermal ablation of hepatic lesions (van Vledder et al. 2011). Optical tracking of tissue deformation is matched to the spatial scales required of several clinical applications involving skin (Kennedy et al. 2012a) and eye (Ford et al. 2011) diseases, and for monitoring engineered tissue growth (Liang et al. 2009).

We now summarize several common elastographic methods involving ultrasonic and optical modalities. Each method has distinguishing features that influence the diagnostic information provided: the source, spatial extent, and frequency content of tissue force excitation; the dimensionality of the motion tracking; and the type of data reconstruction that yields strain or modulus images. Despite differences, the goal of all elasticity imaging techniques is to describe some vital aspect of the cellular mechanoenvironment that enhances clinical diagnosis or improves understanding of the role of mechanobiology in disease progression and treatment. We classify methods based on whether they measure the instantaneous deformation or time-dependent deformation.

- **Stiffness imaging methods**: When the position of tissue anatomy is recorded immediately before and after force application, we can map the instantaneous stiffness using strain images. Strain images are computed from local gradients of tissue displacements (Chaturvedi et al. 1998). If the force is broadly applied to the tissue, as in compression elastography, where a roughly uniform stress is assumed, then strain approximates stiffness (Sridhar et al. 2007; Kennedy et al. 2012b; McGrath et al. 2012). Because small forces (<10 N) are applied slowly (>1 s), there is time for the deformations to propagate throughout the tissue, and therefore, strains at each point in the image are influenced to some degree by properties mechanically coupled throughout the tissue, which influences spatial resolution in unpredictable ways. Alternatively, an acoustic radiation force can be applied locally and quickly (Palmeri et al. 2009) so that the localized deformation describes local tissue stiffness. In either case, stiffness contrast described by a strain image is capable of differentiating benign from malignant tumors in the breast (Itoh et al. 2006; Cho et al. 2010), prostate (Pallwein et al. 2007), and thyroid (Lyshchik et al. 2007; Wang et al. 2012a), and of guiding interventional procedures near nerve bundles (Palmeri et al. 2009), among many other applications.
- **Dynamic imaging methods**: When force is applied to tissues and held constant, pulsed, or varied harmonically as deformations are tracked in a time series of images, we can form images of the viscoelastic (time-dependent) tissue properties. A complex modulus is often estimated (Greenleaf et al. 2003), where the real part is the elastic modulus that can be closely related to tissue stiffness, as described in the previous paragraph. The complex modulus also describes tissue viscosity, which is proportional to the rate of change of deformation. Viscoelastic properties measured over a broad range of applied forces reveal many details about the role of mechanobiology in disease progression (Paszek et al. 2005; Butcher et al. 2009). They also convey important diagnostic information about breast (Fatemi et al. 2002; Alizad et al. 2004; Qiu et al. 2008; Tanter et al. 2008; Cho et al. 2010), liver (Sandrin et al. 2003; Huwart et al. 2006; Bavu et al. 2011), thyroid (Bhatia et al. 2012), and arterial (Schmitt et al. 2011) disease conditions.

Dynamic methods apply force frequencies as low as 0.1 Hz and as high as 10 kHz. By selecting the excitation frequency and spatial distributions of the applied force, we can examine different components of tissue structure. In stiffness imaging methods, image contrast arises primarily from the elastic compliance of the ECM. In dynamic imaging methods at low force frequencies, we probe the viscoelastic properties of the solid ECM (Sridhar et al. 2007), but, at high force frequencies, viscous movement of interstitial fluids moving through solid components of tissue dominates mechanical responses (Greenleaf et al. 2003; Sridhar et al. 2007). Once we understand the mechanobiology of specific diseases and disease states, selection of the excitation frequency can be used to "tune in" the most diagnostic tissue components. Research into how diagnostic information can be derived from the force-frequency landscape is ongoing.

All elasticity imaging research has two coupled goals: improved differential diagnosis and increased understanding of the role of mechanobiology in disease onset, progression, and response to treatment. This chapter provides a review of the field of ultrasound and optical elastography and then focuses on two promising methods of dynamic elasticity imaging studied by the authors.

7.2 Development of Ultrasound and Optical Elastography

7.2.1 Ultrasonic Methods

The long and diverse history of ultrasonic elasticity imaging development can be found in review papers (Gao et al. 1995; Nightingale 2011; Parker et al. 2012; Sarvazyan et al. 2013) and special issues (e.g., Insana and Bamber 2000). Among the first dynamic ultrasonic elasticity imaging methods was the method of Sato and colleagues (1984). They measured the nonlinear propagation of a high-intensity, low-frequency, transient sound pulse (pumping wave) using a low-intensity, high-frequency, harmonic sound wave (probing wave) to quantify tissue properties that can be related to mechanical moduli. They later simplified the measurement by imaging induced vibration waves in the frequency range of 50–250 Hz. The complex modulus was found from the wave amplitude and phase (Yamakoshi et al. 1990), which is an approach commonly used today. Realizing the diagnostic significance of imaging the complex modulus, many joined the search for simpler and more accurate measurements based on Doppler imaging of shear waves induced at the skin surface (e.g., see the review by Parker et al. 2012). Above 20 kHz, tissues behave like fluids and do not support significant shear-wave propagation. At lower frequencies, however, shear waves can propagate many millimeters. There are simple equations relating the speed and absorption of shear waves to the corresponding complex shear modulus (Greenleaf et al. 2003).

One family of dynamic shear-wave imaging methods (e.g., Deffieux et al. 2009; Orescanin 2010a; Sarvazyan et al. 2013) applies focused, harmonic force patterns (50–1000 Hz) to tissues to measure wave speed and absorption from the amplitude and phase of the shear wave generated (Figure 7.1). These properties can be used to assess the viscoelastic, anisotropic, and nonlinear properties of tissues. However, the phase of shear waves measured during harmonic excitation and low-attenuation conditions can be strongly affected by reflections at interfaces of the heterogeneities we intend to image (Wang and Insana 2013).

Replacing harmonic forces with transient forces minimizes reflections at the cost of lowered signal-to-noise ratio (SNR) for shear-wave imaging. One type of transient elastography (commercial available as Fibroscan) (Sandrin et al. 2003) involves a series of broadband force excitations (20–200 Hz) generated by high-intensity acoustic pulses tracked by recording echoes from subsequent imaging pulses. Moduli are estimated from the time of arrival of the shear wave. Acoustic radiation force impulse (ARFI) imaging transmits a single short-duration, high-intensity acoustic pulse, and the displacement at the "pushing spot" is tracked using subsequent short-duration, low-intensity imaging pulses. This approach provides a spatial resolution comparable to B-mode imaging (Nightingale et al. 2001, 2003; Nightingale 2011). Other dynamic methods include vibro-acoustography, where confocal, harmonic acoustic beams of slightly different frequencies induce audible vibrations where the beams intersect. The intersection beam volume and its steered location define the spatial resolution and position of the signal within the body; the audible vibration signal is detected using a hydrophone (Fatemi and Greenleaf 1998; Fatemi et al. 2002).

7.2.2 Optical Methods

Optical imaging can, in principle, provide higher resolution and improved sensitivity to tissue displacements compared to ultrasound, owing to the shorter wavelength of light. Laser speckle imaging has been utilized to evaluate the mechanical properties of tissue phantoms and various biological samples, including cortical bone and arteries, as well as natural and engineered normal and cancerous skin tissues (Duncan et al. 1994; Kirkpatrick and Brooks 1998; Chiang 2008). Recently, digital holographic techniques have been developed to provide wide-field imaging (over tens of millimeters) of surface acoustic waves, with the ability to reconstruct 3-D depth-resolved mechanical properties from the frequency-dependent velocities (dispersion curve) of these waves (Li et al. 2011; Mohan and Oldenburg 2012).

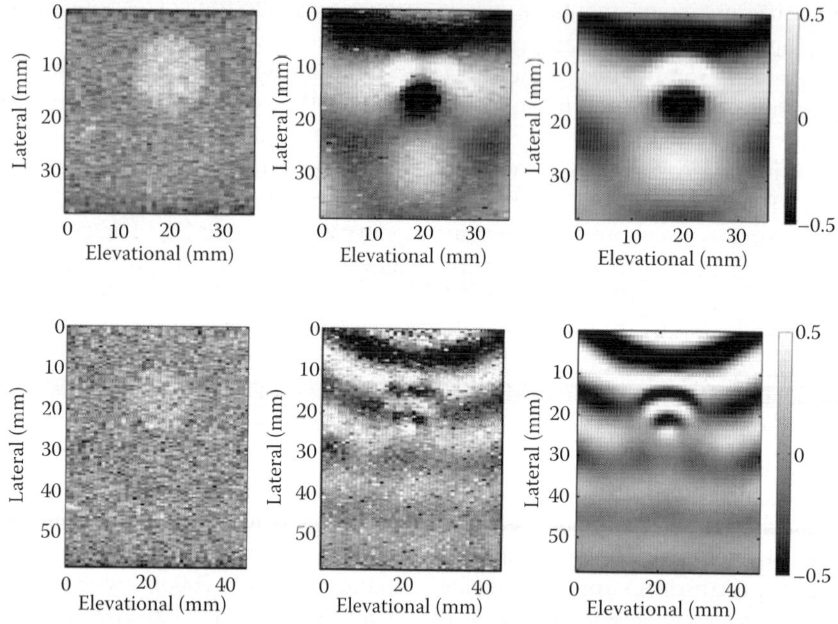

FIGURE 7.1
C-scan sonograms (top view of a 3-D scan) of a tissue-like gelatin phantom containing a soft 15 mm hyperechoic inclusion (left column). Also shown are measured (middle column) and simulated (right column) ultrasonic shear-wave images of the phantom. The shear-wave frequency is 100 Hz (top row) and 150 Hz (bottom row). Images show waves entering from the top and propagating downward. Shortening of the shear wavelength within the soft inclusion and the uneven SNR from wave absorption and bending at interfaces are both apparent from the shear-wave images. (From Orescanin, M. et al. *IEEE Trans. Ultrason. Ferroelectr. Freq. Control*, 58, 2011.)

Optical coherence elastography (OCE), a method developed in the late 1990s (Schmitt 1998), has received much attention due to its ability to provide high-resolution tomographic images of strain in soft highly scattering tissues. OCE measures tissue displacements using OCT (Huang et al. 1991; Fercher et al. 2003), an optical analogue of ultrasonic imaging, which provides an improvement of one to two orders of magnitude in spatial resolution over ultrasound. However, the penetration depth of OCT is only 2–3 mm in highly scattering tissues, which could limit some of the noninvasive applications. As in ultrasound elastography, OCE has transitioned from utilizing static or quasi-static (<20 Hz) excitation, with the associated microscopic displacement fields detected by speckle tracking methods (Kirkpatrick et al. 2006), to a method capable of imaging dynamic audio frequency (20 Hz to 20 kHz) mechanical excitations with nanometer-scale displacement sensitivity using phase-sensitive OCT (Liang et al. 2008; Kennedy et al. 2011). The ability to detect oscillations over a wide range of mechanical excitation frequencies has facilitated OCT-based complex mechanical spectroscopy (imaging of both the amplitude and mechanical phase response), thus providing an extra dimension over which to characterize the mechanical properties of tissue (Adie et al. 2010). Recent advancements in ultra-compact needle-based OCT has also enabled deep tissue OCE (Kennedy et al. 2012c), enabling the use of OCE for image-guided biopsy.

Various methods have been explored in OCE. The studies mentioned in the preceding paragraphs have predominantly utilized external actuators in contact with the sample surface to apply bulk quasi-static compression or induce dynamic oscillations. Magnetomotive OCE (MM-OCE) has utilized magnetic actuation (magnetomotion) of iron oxide particles infused or distributed within the sample to internally induce dynamic oscillations. MM-OCE has been used to perform optical rheology of viscoelastic materials via step excitation (Crecea et al. 2009) or a frequency sweep (Oldenburg and Boppart 2010). Acoustic radiation force has recently been explored in OCE, either utilizing single-point excitation with focused ultrasound to generate shear waves (Liang et al. 2009; Razani et al. 2012; Li et al. 2012), or using wide-area excitation with an unfocused ultrasound transducer to perform 3-D OCE of arterial plaques ex vivo (Qi et al. 2012). OCE based on noncontact excitation of shear waves has been demonstrated using an air-puff method (Alonso-Caneiro et al. 2011) or the absorption of pulsed laser light.

Applications of OCE have sought to leverage the high-resolution and/or nanometer-scale displacement sensitivity afforded by OCT. It has the potential to examine very heterogeneous tissues such as cancerous lesions. Several ex vivo studies have shown early promise for the detection of arterial plaques (Chau et al. 2004). OCE has also been investigated for imaging the biomechanical properties of breast and other cancers ex vivo (Wang et al. 2012b), with potential applications in tumor-margin detection. One ex vivo study of human breast tissues utilized static compression to measure the stress–strain curves of resected human breast tissues, finding that Young's moduli of benign fibroadenoma and invasive ductal carcinoma were, respectively, a factor of 2 and a factor of 4 higher than normal breast tissue (Srivastava et al. 2011). Due to the direct access for optical imaging, the skin is a prime target for in vivo OCE applications, motivating the development of several methods for imaging the elastic response of skin (Liang and Boppart 2010). OCE has also shown promise for studying the role of mechanics in developmental biology and in engineered tissue constructs in vitro (Ko et al. 2006). In a recent in vitro study, spectroscopic MM-OCE was utilized to investigate the elastic properties of blood clots, using blood infused with superparamagnetic iron oxide nanoparticles (Oldenburg et al. 2012). The noncontact (air-puff and laser pulse absorption) methods mentioned in the previous paragraph have been developed for the measurement of corneal biomechanics.

7.3 Dynamic Approaches to Quantitative Viscoelastic Properties

One advantage of dynamic elasticity imaging over other approaches is the ability to provide information necessary to quantitatively estimate a complex modulus. Necessary components of a dynamic elastography system are tissue excitation, motion tracking, and property field reconstruction.

7.3.1 Tissue Excitation Mechanisms

Tissue excitation is arguably the most important aspect of quantitative elasticity from an image science perspective. The bandwidth and spatiotemporal pattern of the applied force determines object and image contrast, overall spatial resolution, SNR, and the artifacts that must be managed during reconstruction. Except for passive elastography methods that use physiological motion as the motion source (Mai and Insana 2002; Gallot et al. 2011), most dynamic methods apply a force stimulus. There are several excitation mechanisms to generate forced motion of tissue. Some make direct contact with the tissue, while others use waves to apply forces remotely. Noninvasive contact methods include surface vibration with a plate or a rod (Li et al. 2012) driven by an acoustic or piezoelectric actuator (Tse et al. 2011). Invasive contact methods based on a needle inserted into tissue have also been used (Orescanin et al. 2010b; DeWall et al. 2011). Remote application methods include forces generated by the application of acoustic (Nightingale et al. 2001) or magnetic (Crecea et al. 2009) waves. Selection of the force applicator depends on the application.

Three types of forces are commonly applied, with the goal of generating deformations identified as bulk, local, and surface wave. Bulk motion is generated when the entire sample volume is excited simultaneously. For example, in one study, breast tissue samples were placed on a plate that compressed or sheared the sample volume uniformly (Liang et al. 2008). Another study used a magnetic field to harmonically excite nanoparticles embedded in samples at frequencies up to 400 Hz to create standing waves in cylindrical tissue samples (Oldenburg and Boppart 2010). From the resonance modes generated by the sample geometry, Young's modulus of the sample was estimated. Finally, MRI was used to track the deformation-wave modes in cylindrical samples (Yasar et al. 2013). Bulk methods can generate accurate modulus estimates for homogeneous samples with well-defined geometries.

In heterogeneous tissues measured in vivo, there are major resolution advantages to exciting and tracking deformations locally. We discussed ARFI (Nightingale et al. 2001) and vibro-acoustography methods (Fatemi and Greenleaf 1998) that use localized acoustic radiation force, and the cylindrical waves generated around needles or wires inserted into tissue are vibrated along their axis (Orescanin and Insana 2010a; DeWall et al. 2011). Techniques like thermoacoustic and photoacoustic imaging (Ke et al. 2012) apply microwave and near-infrared optical pulses, respectively, to induce local acoustic waves internally from the thermal stress that occurs wherever the electromagnetic (EM) energy is absorbed. The emerging sound waves are focused and mapped into images by ultrasound arrays located at the skin surface or by an intravascular ultrasound probe. Object contrast is generated by differential absorption of the EM waves, while spatial resolution is controlled ultrasonically (Kruger et al. 1995). Viscoelastic properties are estimated from the phase delay of the photoacoustic shock wave as it travels through tissues to the transducer (Gao et al. 2011). Elasticity images from impulse forces are more difficult to reconstruct quantitatively than those from harmonic forces, so spatial resolution may be traded off for improved modulus accuracy when quantitation is important for the clinical task.

Rayleigh waves generated at the surface of a volume and Lamb waves generated within platelike media have both compressional- and shear-wave components that behave differently from bulk shear waves. The distinction is important using our acoustic radiation force–OCE (ARF-OCE) method, where forces are generated at surfaces of abrupt changes in acoustic impedance. Other investigators have described the elastic properties generated from these surface waves to characterize cardiovascular tissues (Nenadic et al. 2011).

7.3.2 Imaging Systems for Dynamic Motion Tracking

The major techniques for imaging tissue deformations in response to mechanical forces can be grouped by the spatial coverage afforded by each modality (see Figure 7.2). On the macroscopic scale, magnetic resonance elastography (see Chapters 2–6) and ultrasound methods have fields of view capable of imaging whole organs. At the microscopic scale, atomic force microscopy and optical microscopy provide nanoscale to micron-scale imaging of cells and subcellular structures (see Chapters 22, 25–26, 36–37, and 40). OCT and ultrasound describe material properties at spatial scales most relevant to assessing the state of soft-tissue lesions (Paszek et al. 2005; Wirtz et al. 2011). Together, these modalities provide a multiscale description of soft-tissue mechanics.

Ultrasonic elastography, which employs radiofrequency (RF) echo signals, offers echo phase to track tissue displacements on the order of micrometers (1/100 of the wavelength) even though the spatial resolution for displacement is hundreds of micrometers (spatial extent of the tracking pulse). Conversely, the use of B-mode image data (demodulated RF echoes) is most sensitive to movements on the order of the spatial resolution for displacement estimation (pulse dimensions). Thus, we usually find ultrasonic RF signals used in oncological elastography and B-mode signals used in cardiac elastography.

In OCE, low-coherence interferometry is used to measure the time of flight of backscattered light, which is analogous to ultrasonic pulse-echo imaging methods. The shorter optical wavelength means that OCE is most sensitive to nanometer-scaled displacements, with a spatial resolution of 1 μm being feasible (Povazay et al. 2002). Relative to ultrasonic methods, OCE can sense significantly smaller movements (nanometers) at much higher spatial resolution (micrometers) but at the cost of significantly less depth of penetration (millimeters). Smaller displacements improve the validity of the linear tissue deformation assumptions often made in modulus reconstructions. Light from a broadband optical source is split into sample and reference arms, and the interference of the back-reflected light from each arm is detected by a spectrometer in a configuration known as spectral-domain OCT (SD-OCT) (Linder et al. 2002). Each spectrum measured by a line-scan camera in the spectrometer is used to compute a single depth trace, or A-scan. Galvanometer mirrors in the sample arm raster scan the beam along one or two transverse dimensions to perform two- or three-dimensional tomographic imaging, respectively, over millimeter-scale fields of view.

Phase-sensitive SD-OCT has become the method of choice to quantify dynamic motion because no mechanical scanning is needed in the optical source or the detection arm of the OCT interferometer (Wang et al. 2006; Liang et al. 2010). This method was used to demonstrate 3-D OCE of human skin in vivo (Kennedy et al. 2011). The ability to tune mechanical excitation frequencies appropriate for the task has enabled mechanical spectroscopic tomography (Adie et al. 2010) with shear-wave frequencies at 1 kHz to be imaged at a 5 kHz A-scan rate. In the last 5 years, high-speed SD-OCT (with A-scan rates around 100 kHz or higher) has become widespread. High-speed acquisition has yet to be exploited for phase-sensitive OCE above 1 kHz.

7.3.3 Estimation of Viscoelastic Parameters

Strain images indicate tissue stiffness but cannot reveal the intrinsic constitutive properties of tissues. For this, one must measure stress and strain and interpret them within the context of the measurement geometry. Dynamic methods can provide sufficient local information to estimate intrinsic viscoelastic properties, such as the complex modulus, although these too can be affected by measurement artifacts and assumptions. Estimation of viscoelastic parameters generally requires that we adopt a physical model for mechanical wave motion and a rheological model that relates the stress and strain fields.

Typical physical models for continua include a second-order partial differential equation, where lumped parameters are found from mass-spring model systems (Equation 7.1) (Liang et al. 2008), a simple stress–strain relationship (Equation 7.2) (Ponnekanti et al. 1994), and the more general Navier–Stokes equation (Equation 7.3) (Fox and McDonald 1992). These correspond to bulk, local, and wave motions, respectively.

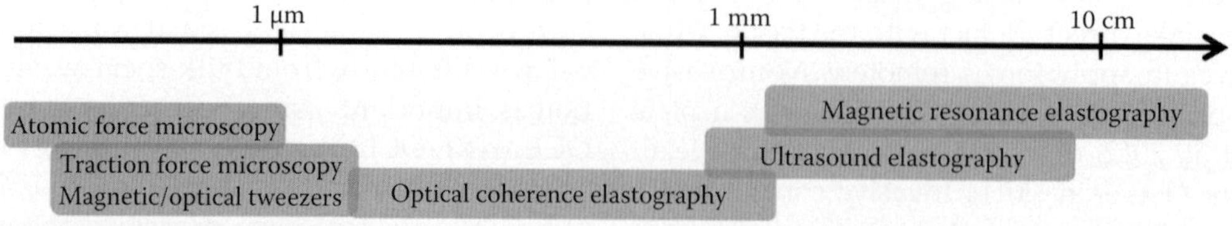

FIGURE 7.2
Elastographic modalities characterized by their range of spatial coverage.

$$m\partial_{tt}^{2}u_{z} = Gu_{z} + F \tag{7.1}$$

$$\vec{\sigma} = G\vec{u} \tag{7.2}$$

$$\nabla[G(\nabla\vec{u})] + \vec{F} = \rho\partial_{tt}^{2}\vec{u} \tag{7.3}$$

In Equations 7.1–7.3, F is the excitation force, m is the total mass on which the force acts, ρ is the local tissue density, $\vec{u}$ is the strain vector, $\vec{\sigma}$ is the stress vector, and G is the mechanical modulus that we wish to estimate. Forward and inverse techniques were developed to estimate G based on physical models. Forward techniques include an iterative finite-element method (Miga 2003) and finite-difference methods (Sinkus et al. 2000) that provide the highest spatial resolution for the estimation of displacement. They also account for known boundary conditions, but the computational loads are often too high to permit real-time imaging. Inverse techniques are simpler and widely applied. They include direct inversion (Papazoglou et al. 2008) and iterative parameter-optimizing inversion (Doyley et al. 2007). These estimates can suffer from high levels of estimation noise.

The rheological model relates stress to strain and thus can be used to estimate medium properties. In the physical model, G appears as an operator representing the rheological model. Tissues can be anisotropic, inhomogeneous, nonlinear, and poroviscolastic, so G can be complex, with many parameters that are all candidates for imaging. The parameter space is reduced by assuming locally isotropic, homogeneous, and linear tissue responses in the applied force range. Thus, G can be simplified to a few modulus parameters as candidates for imaging. Examples of viscoelastic rheological models are Kelvin–Voigt, Maxwell, and Zener (Orescanin 2010a; Schmitt et al. 2011). One could also report G as a complex matrix to avoid invoking a rheological model. This approach is referred to as model-independent modulus estimation (Schmitt et al. 2011).

The choice of viscoelastic parameter and its estimation technique depends on factors such as the excitation method, the diagnostic feature of the tissue, the dimension of strain data, and the minimum spatial resolution required.

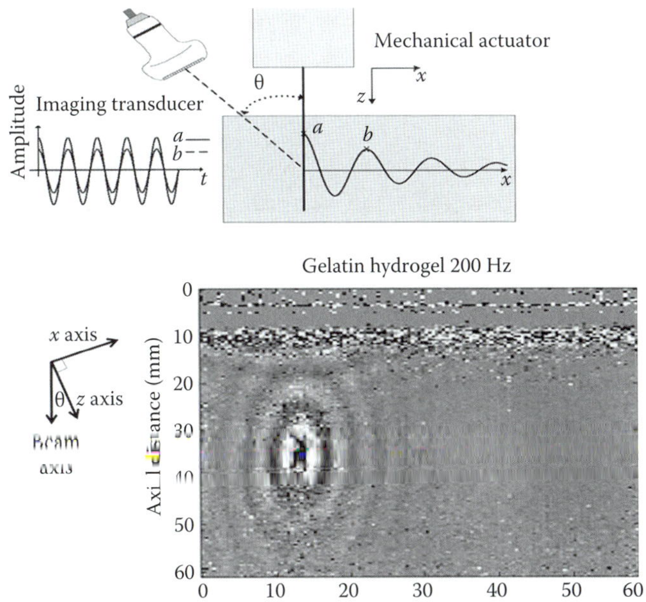

FIGURE 7.3
A vibrating needle generates patterns of 200 Hz cylindrical shear waves inside a gelatin block that are imaged using standard ultrasonic Doppler techniques. (From Orescanin 2010a.)

7.4 Ultrasonic Shear-Wave Imaging Using Vibrating-Needle Stimulation

Ultrasonically guided breast needle biopsy is commonly used. Needle insertion provides an opportunity to vibrate the needle and excite harmonic cylindrical waves for shear-wave imaging with a well known shear-wavefield geometry (Figure 7.3).

Another advantage of harmonic needle vibration is the large shear-wave amplitudes generated, as compared with other force-excitation techniques. The needle body surface should be roughened to increase the friction between the tissue and the needle, which avoids slippage. The rough surfaces also enhance ultrasonic scattering of the needle at arbitrary incidences to aid in visualizing needle position. Needles are driven sinusoidally along the long axis by an actuator with an upper frequency response limited at about 500 Hz. The frequency limit is determined by the mechanical response of the actuator, slippage, and the minimum motion that an ultrasonic Doppler system can detect (Orescanin 2010a). Tissue motion can be tracked by commercial scanners between 5 and 13 MHz operating in spectral Doppler mode. Pulse repetition frequency (PRF) of the Doppler pulses should be large enough to avoid aliasing, usually 5–10 kHz. The ultrasound transducer is placed next to the needle on the skin surface and tipped at a 15°–30° angle to locate the needle in the tissue (Figure 7.4).

Displacements are estimated from the temporal phase shift between pulses at a fixed location using a standard lag-one autocorrelation estimator (Kasai 1986) or a modified lag-k estimator (Orescanin et al. 2010). Lag-k estimators significantly reduce velocity estimation variance for narrow-band pulses when the echo SNR <30 dB, but at the cost of a lower imaging frame rate. Phase-wrapping errors occur when displacements

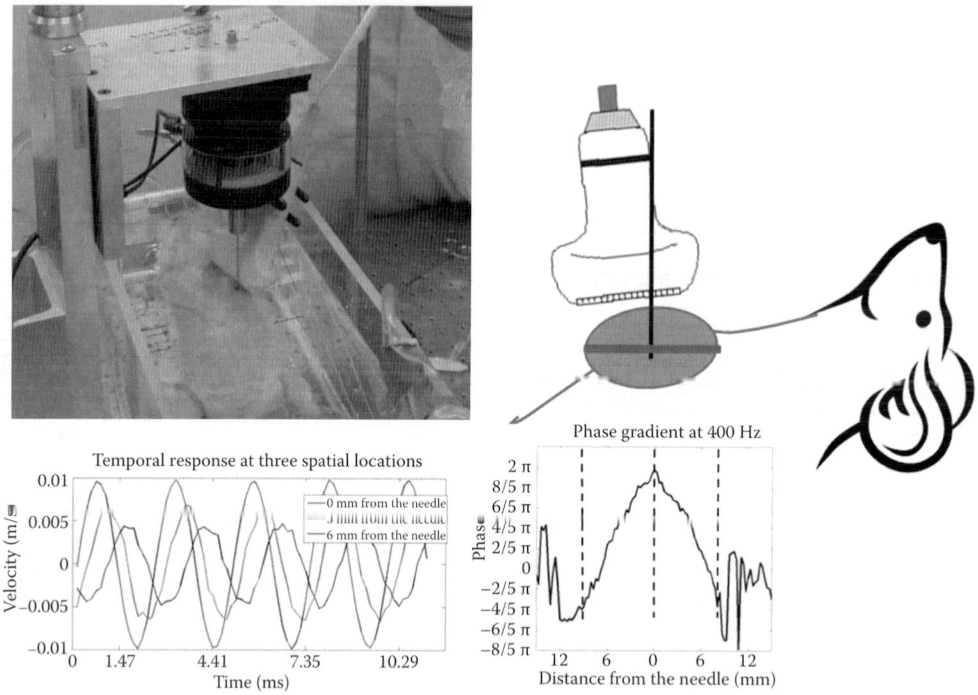

FIGURE 7.4
Illustration of in vivo ultrasonic shear-wave imaging methods using vibrating-needle stimulation. (From Wang, Y. and Insana, M.F., *Ultrason. Imaging*, 35, 2013.)

are larger than one-half of the ultrasound wavelength. This requires operator control of actuator voltage to avoid phase ambiguity. Usually, only microscale displacement amplitudes are generated in tissue. During in vivo imaging, cardiopulmonary motions are much larger and can complicate displacement estimation for harmonic stimuli. As in blood flow imaging, it is helpful to apply a high-pass filter to the displacement data to remove unwanted tissue and transducer motions.

Once the wave field is imaged, quantitative mechanical information can be estimated. The Helmholtz inversion and phase-gradient techniques are two common inversion techniques frequently applied to estimate the complex modulus:

$$G = \frac{-\rho\omega^2 U_z}{\nabla^2 U_z}. \tag{7.4}$$

The Helmholtz inversion (Equation 7.4) can be derived from Equation 7.3 once it is expressed in the frequency domain. Equation 7.4 is used for modulus reconstruction from a 3-D displacements field. Helmholtz inversion requires heavy filtering before and after inversion since the Laplacian operator amplifies any high-frequency shear-wave noise. Figure 7.5 shows the real part of G from 3-D Helmholtz inversion on the same inclusion phantom shown in Figure 7.1. The soft 1 kPa cylindrical inclusion embedded in the 2.5 kPa background can be accurately imaged in 3-D using this technique. There is some spatial distortion of the cylinder near the ends caused by the upper boundary. The phase-gradient method is typically employed since it is more tolerant to noise. It calculates the spatial phase shift of the wave along the wave-propagation direction and relates the wave number k to G using

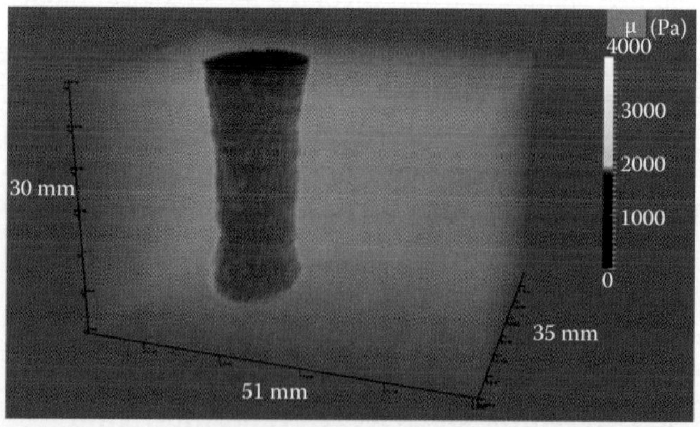

FIGURE 7.5
3-D elastic modulus image of a soft cylindrical inclusion in a stiff gelatin gel block imaged using ultrasonic shear-wave methods with vibrating-needle stimulation. (From Orescanin, M. et al. *IEEE Trans. Ultrason. Ferroelectr. Freq. Control*, 58, 2011.)

$$k = \sqrt{\frac{\rho\omega^2}{G}}. \tag{7.5}$$

However, a decision needs to be made on how many spatial locations will be used to estimate the local shear-wave speed assuming local homogeneity. A trade-off exists between spatial resolution and accuracy of shear-wave phase estimation depending on the degree of tissue heterogeneity. Under the assumption of global homogeneity, a single shear-wave speed at each frequency is derived for the sample using linear regression on the phase-gradient curve. Measurable shear-wave dispersion is found in most tissues (Figure 7.6).

Rheological models: Viscoelastic models describe the dispersive behavior of shear-wave speeds over the measurement frequencies. The Kelvin–Voigt model is the most widely used rheological model (Figure 7.6). Table 7.1 shows the estimated elastic and viscous moduli of gelatin samples using the Kelvin–Voigt model. The elastic modulus is verified through independent rheometry measurements. However, some researchers have shown that it often does not describe the experimental data as well as alternative models, such as the Zener model, in viscous tissues (Schmitt et al. 2011; Wang and Insana 2013). Model-independent estimation of G requires estimation of the real and imaginary parts of the complex wave number k. That is, speed and attenuation of the shear wave must be estimated, and the latter is highly susceptible to errors when computed over small distances (Schmitt et al. 2011).

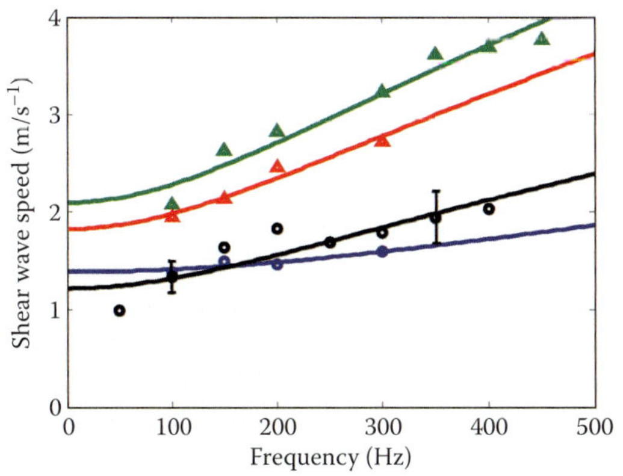

FIGURE 7.6
Shear-wave dispersion curves of four rat fibroadenoma tumors (shear-wave speeds measured from 50 to 450 Hz). Triangles denote measurements from two stiff tumors; circles denote data from two soft tumors. Each dispersion curve is fit to Kelvin–Voigt models to derive elastic and viscous moduli. (From Wang, Y. and Insana, M.F., *Ultrason. Imaging*, 35, 2013.)

TABLE 7.1
Estimated Viscoelastic Parameters from Kelvin–Voigt Model

Gelatin	Elastic Modulus (Pa)	Viscous Modulus (Pa s)	Rheometer Modulus (Pa)
4%	570	0.21	571
8%	2919	1.0	2286

Shear-wave reflections: Reflections from the boundaries of a small lesion will bias toward lower values the speed estimated from phase-gradient methods (Wang and Insana 2013). We showed using simulations and experiments that the diameter of an inclusion must be larger than two wavelengths when the medium is weakly attenuating (less for tissue-like attenuation) to minimize the influence of reflected waves. Directional filters can help reduce the influence of reflected shear waves of dynamic elasticity images (Manduca et al. 2003).

Overall, needle-based tissue excitation has many imaging advantages over other methods when tissue invasiveness is not a significant concern. It is a useful tool when carrying out basic science research and when incorporated into breast needle biopsy procedures.

7.5 ARF-OCE

Needle-excitation provides cylindrical shear waves from which accurate modulus estimates can be obtained in homogeneous tissues several centimeters in extent and at frequencies less than 800 Hz. Acoustic radiation force provides excitation frequencies as high as 10 kHz for high-spatial-resolution imaging within heterogeneous tissues, but often with an unknown spatial force pattern, making accurate modulus estimates difficult to achieve. The larger force bandwidth results in much smaller shear-wave amplitudes that can be too small for ultrasound detection but are perfectly matched to the sensing capabilities of OCT.

The radiation force from a focused acoustic beam applied to the propagation medium is proportional to the absorption coefficient of the tissue and the time-average beam intensity (Nightingale 2011). Thus, the applied force is largest at the beam focus within an attenuating medium. If, however, the sound beam crosses a strongly reflecting boundary, surface acoustic waves are formed at the boundary, in addition to bulk acoustic waves within the medium. Since the penetration of OCT in scattering samples is limited to about 2 mm in highly scattering tissues, the acoustic beam was focused at the top sample–air interface to generate shear waves that propagate near the surface of the sample (Figure 7.7). We selected a circular, single-element, 8 MHz, *f*/1 transducer to deliver the acoustic radiation force. The small *f*-number keeps the force excitation focused to an area about the size of the sound wavelength, in this case, 0.2 mm. The magnitude of the acoustic force was easily controlled by the transducer driving voltage. Harmonic-wave excitation was achieved by modulating the continuous-wave 8 MHz amplitude anywhere in the range of 100 to 10,000 Hz. Because OCT offers high phase stability and SNR, tissue movements as small as 10 nm can be readily measured. The low driving force of the transducer allowed forces to be applied continuously without heating. Acoustic radiation exerts forces in only one direction; thus, harmonic oscillations generate a net displacement in tissue that depends on frequency and tissue properties.

OCT is ideal for detecting low-amplitude, high-frequency oscillations at high spatial resolution, and therefore, it is well suited to tomographic imaging near a sample surface. Although the limited depth of penetration limits clinical applications, OCE is ideal for basic science investigations into the viscoelastic properties of tissues over a broad force-frequency bandwidth.

Shear-wave modulation frequencies from 200 Hz to 5 kHz are easily achievable at an A-scan rate of 65,000 lines per second. We have imaged shear waves with frequencies as high as 10 kHz using this method. The analytic signal was computed from the raw optical data acquired in M-mode and was used to calculate the phase difference between adjacent A-lines at each axial position as a function of time. As in ultrasound displacement detection, phase wrapping was avoided. Figure 7.8 illustrates

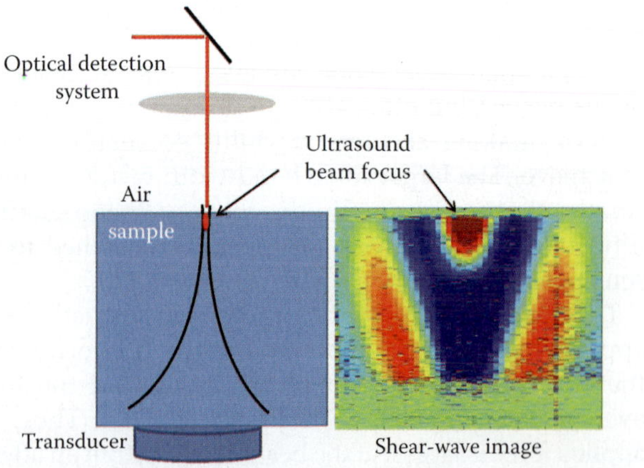

FIGURE 7.7
The acoustic radiation force generated at the top sample–air surface (left) generates a combination of surface and bulk shear waves that form a cone-shaped geometry (right). These waves are imaged using OCT with a beam that is incident on the top surface.

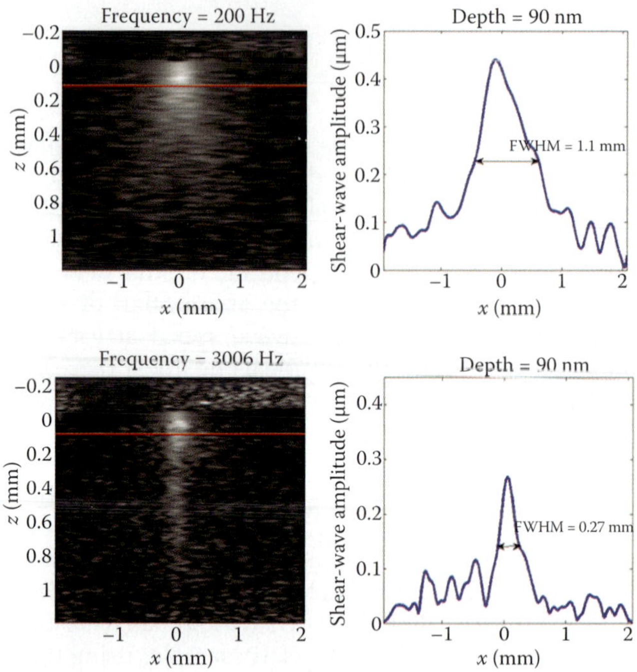

FIGURE 7.8
(Left) OCE displacement maps are shown at acoustically generated shear-wave frequencies of 200 Hz (top row) and 3000 Hz (bottom row) for an 8% (stiff, elastic) gelatin gel phantom. The top of the sample is located at depth $z = 0$. The brightest area (left) corresponds to the largest displacements (right) at the ultrasound focus, indicated by the red horizontal lines in the OCE displacement maps. The mechanical interaction length, indicated by the full-width-half-maximum (FWHM) values, affects the mechanical spatial resolution.

spatial patterns of displacement when a harmonic acoustic force is applied at the top surface of the sample. The largest displacements appear at the ultrasound-beam focus. Both the magnitude and the width of the displacement pattern depend on the carrier frequency of the sound waves. This demonstrates the frequency-dependent mechanical interaction length that can be measured with ARF-OCE.

Parameter estimation: Methods for computing the shear modulus in ARF-OCE images are the same as in ultrasound. The difference is the high sensitivity and spatial resolution of OCT detection. As shown in Figure 7.9(a), the acoustic radiation force at the surface generates shear waves with much wider bandwidth than needle excitation. Figure 7.9(b) shows the dispersion curves of a homogeneous gelatin sample and an ex vivo porcine liver sample. Tissues are known to be more dispersive than hydrogels because of higher shear-wave attenuation. Shear-speed dispersion over such wide bandwidth may reflect viscosity, nonlinearity, and geometry information of the

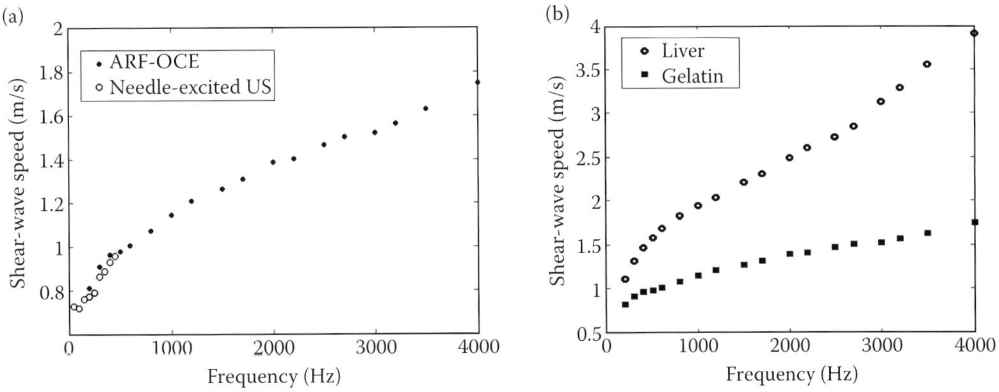

FIGURE 7.9
(a) Comparison of dispersion curve bandwidths obtained from ultrasonic (US) shear-wave imaging and OCE methods. (b) Dispersion curves for a homogeneous 4% gelatin sample and fresh ex vivo porcine liver.

tissue and thus should be interpreted carefully to improve our understanding about the rheological behavior of tissues and for choosing the most appropriate rheological models.

7.6 Conclusion and Future Outlook

Dynamic elasticity imaging methods have been developed and translated into a clinically viable imaging modality over the last two decades. Used to complement the structural information provided by ultrasound and optical imaging of tissues, the mechanical properties derived from these methods have been clearly shown to be sensitive to pathology-related tissue structure and have significantly added to the diagnostic utility of these techniques. These biomechanical parameters may also serve as novel biomarkers that have the potential to provide important diagnostic information. It has long been appreciated that palpation of tissues by physicians during physical exams, and the detection of altered mechanical properties such as stiffening during inflammatory processes or during tumor growth, is diagnostically useful. With advances in ultrasound and optical techniques for dynamic viscoelastic imaging, it is possible to examine or "palpate" tissues and cells at increasingly higher resolution, with greater sensitivity, and with increased diagnostic value. To promote the reliability and value of these methods in both clinical and scientific applications, researchers in this field will need to further investigate the mechanical resolution of each elastography imaging system and to find the suitable driving schemes and rheological models that could magnify the mechanical contrast of each type of biological tissue. For shear-wave frequencies below 1000 Hz, shear-wave dispersion behaviors in normal and diseased tissue states have been widely studied by currently available methods. Shear-wave frequencies higher than 1000 Hz, however, have yet to be exploited. The primary hurdle before applying these methods in vivo is to mount the tissue precisely on the delicately assembled excitation and tracking devices so that the desired motion pattern could be generated and collected without distortion. As advances in ultrasound elastography and new technologies such as ARF-OCE emerge, our understanding of cell and tissue mechanics will undoubtedly improve, as will our ability to utilize this new information for potentially earlier or more accurate diagnoses.

References

Adie SG, X Liang, BF Kennedy, R John, DD Sampson, and SA Boppart. "Spectroscopic Optical Coherence Elastography." *Optics Express* 18, no. 25 (2010): 25519–34.

Alizad A, M Fatemi, DH Whaley, and JF Greenleaf. "Application of Vibro-Acoustography for Detection of Calcified Arteries in Breast Tissue." *Journal of Ultrasound in Medicine* 23, no. 2 (2004): 267–73.

Alonso-Caneiro D, K Karnowski, BJ Kaluzny, A Kowalczyk, and M Wojtkowski. "Assessment of Corneal Dynamics with High-Speed Swept Source Optical Coherence Tomography Combined with an Air Puff System." *Optics Express* 19, no. 15 (2011): 14188–99.

Bavu E, JL Gennisson, M Couade, J Bercoff, V Mallet, M Fink, A Badel, A Vallet-Pichard, B Nalpas, M Tanter, and S Pol. "Noninvasive in vivo Liver Fibrosis Evaluation Using Supersonic Shear Imaging: A Clinical Study on 113 Hepatitis C Virus Patients." *Ultrasound in Medicine and Biology* 37, no. 9 (2011): 1361–73.

Bhatia KSS, CSL Tong, CCM Cho, EHY Yuen, YYP Lee, and AT Ahuja. "Shear Wave Elastography of Thyroid Nodules in Routine Clinical Practice: Preliminary Observations and Utility for Detecting Malignancy." *European Radiology* 22, no. 11 (2012): 2397–406.

Butcher DT, T Alliston, and VM Weaver. "A Tense Situation: Forcing Tumour Progression." *Nature Reviews Cancer* 9, no. 2 (2009): 108–22.

Chaturvedi P, MF Insana, and TJ Hall. "2-D Companding for Noise Reduction in Strain Imaging." *IEEE Transactions on Ultrasonics Ferroelectrics and Frequency Control* 45, no. 1 (1998): 179–91.

Chau AH, RC Chan, M Shishkov, B MacNeill, N Iftimiia, GJ Tearney, RD Kamm, BE Bouma, and MR Kaazempur-Mofrad. "Mechanical Analysis of Atherosclerotic Plaques Based on Optical Coherence Tomography." *Annals of Biomedical Engineering* 32, no. 11 (2004): 1494–503.

Chiang F-P. "Micro-/Nano-Speckle Method with Applications to Materials, Tissue Engineering and Heart Mechanics." *Strain* 44, no. 1 (2008): 27–39.

Cho N, WK Moon, HY Kim, JM Chang, SH Park, and CY Lyou. "Sonoelastographic Strain Index for Differentiation of Benign and Malignant Nonpalpable Breast Masses." *Journal of Ultrasound in Medicine* 29, no. 1 (2010): 1–7.

Crecea, V, AL Oldenburg, X Liang, TS Ralston, and SA Boppart. "Magnetomotive Nanoparticle Transducers for Optical Rheology of Viscoelastic Materials." *Optics Express* 17, no. 25 (2009): 23114–22.

de Ledinghen V, J Vergniol, C Barthe, J Foucher, F Chermak, B Le Bail, W Merrouche, and PH Bernard. "Non-Invasive Tests for Fibrosis and Liver Stiffness Predict 5-Year Survival of Patients Chronically Infected with Hepatitis B Virus." *Alimentary Pharmacology and Therapeutics* 37, no. 10 (2013): 979–88.

Deffieux T, G Montaldo, M Tanter, and M Fink. "Shear Wave Spectroscopy for in vivo Quantification of Human Soft Tissues Visco-Elasticity." *IEEE Transactions on Medical Imaging* 28, no. 3 (2009): 313–22.

DeWall RJ, T Varghese, and EL Madsen. "Shear Wave Velocity Imaging Using Transient Electrode Perturbation: Phantom and ex vivo Validation." *IEEE Transactions on Medical Imaging* 30, no. 3 (2011): 666–78.

Doyley MM, Q Feng, JB Weaver, and KD Paulsen. "Performance Analysis of Steady-State Harmonic Elastography." *Physics in Medicine and Biology* 52, no. 10 (2007): 2657–74.

Duncan DD, SJ Kirkpatrick, FF Mark, and LW Hunter. "Transform Method of Processing for Speckle Strain-Rate Measurements." *Applied Optics* 33, no. 22 (1994): 5177–86.

Fatemi M, and JF Greenleaf. "Ultrasound-Stimulated Vibro-Acoustic Spectrography." *Science* 280, no. 5360 (1998): 82–5.

Fatemi M, LE Wold, A Alizad, and JF Greenleaf. "Vibro-Acoustic Tissue Mammography." *IEEE Transactions on Medical Imaging* 21, no. 1 (2002): 1–8.

Fercher AF, W Drexler, CK Hitzenberger, and T Lasser. "Optical Coherence Tomography—Principles and Applications." *Reports on Progress in Physics* 66, no. 2 (2003): 239–303.

Ford MR, WJ Dupps Jr, AM Rollins, AS Roy, and Z Hu. "Method for Optical Coherence Elastography of the Cornea." *Journal of Biomedical Optics* 16, no. 1 (2011): 016005.

Fox RW and AT McDonald. *Introduction to Fluid Mechanics*. 4th ed. New York: J. Wiley, 1992.

Gallot T, S Catheline, P Roux, J Brum, N Benech, and C Negreira. "Passive Elastography: Shear-Wave Tomography from Physiological-Noise Correlation in Soft Tissues." *IEEE Transactions on Ultrasonics Ferroelectrics and Frequency Control* 58, no. 6 (2011): 1122–6.

Gao G, S Yang, and D Xing. "Viscoelasticity Imaging of Biological Tissues with Phase-Resolved Photoacoustic Measurement." *Optics Letters* 36, no. 17 (2011): 3341–3.

Gao L, KJ Parker, SK Alam, and RM Lerner. "Sonoelasticity Imaging—Theory and Experimental-Verification." *Journal of the Acoustical Society of America* 97, no. 6 (1995): 3875–86.

Greenleaf JF, M Fatemi, and MF Insana. "Selected Methods for Imaging Elastic Properties of Biological Tissues." *Annual Review of Biomedical Engineering* 5 (2003): 57–78.

Hayashi M, Y Yamamoto, M Ibusuki, S Fujiwara, S Yamamoto, S Tomita, M Nakano, K Murakami, K Iyama, and H Iwase. "Evaluation of Tumor Stiffness by Elastography Is Predictive for Pathologic Complete Response to Neoadjuvant Chemotherapy in Patients with Breast Cancer." *Annals of Surgical Oncology* 19, no. 9 (2012): 3042–9.

Huang D, EA Swanson, CP Lin, JS Schuman, WG Stinson, W Chang, MR Hee, T Flotte, K Gregory, CA Puliafito, and JG Fujimoto. "Optical Coherence Tomography." *Science* 254, no. 5035 (1991): 1178–81.

Huwart L, F Peeters, R Sinkus, L Annet, N Salameh, LC ter Beek, Y Horsmans, and BE Van Beers. "Liver Fibrosis: Non-Invasive Assessment with MR Elastography." *NMR in Biomedicine* 19, no. 2 (2006): 173–9.

Insana MF, and JC Bamber. "Editorial: Tissue Motion and Elasticity Imaging." *Physics in Medicine and Biology* 45, no. 6 (2000).

Itoh A, E Ueno, E Tohno, H Kamma, H Takahashi, T Shiina, M Yamakawa, and T Matsumura. "Breast Disease: Clinical Application of US Elastography for Diagnosis." *Radiology* 239, no. 2 (2006): 341–50.

Kasai C. "Real-Time Two-Dimensional Blood-Flow Imaging Using an Autocorrelation Technique." *IEEE Transactions on Ultrasonics Ferroelectrics and Frequency Control* 33, no. 1 (1986): 94.

Ke H, TN Erpelding, L Jankovic, C Liu, and LV Wang. "Performance Characterization of an Integrated Ultrasound, Photoacoustic, and Thermoacoustic Imaging System." *Journal of Biomedical Optics* 17, no. 5 (2012): 056010-1–056010-6.

Kennedy BF, SH Koh, RA McLaughlin, KM Kennedy, PRT Munro, and DD Sampson. "Strain Estimation in Phase-Sensitive Optical Coherence Elastography." *Biomedical Optics Express* 3, no. 8 (2012a): 1865–79.

Kennedy BF, X Liang, SG Adie, DK Gerstmann, BC Quirk, SA Boppart, and DD Sampson. "*In Vivo* Three-Dimensional Optical Coherence Elastography." *Optics Express* 19, no. 7 (2011): 6623–34.

Kennedy BF, M Wojtkowski, M Szkulmowski, KM Kennedy, K Karnowski, and DD Sampson. "Improved Measurement of Vibration Amplitude in Dynamic Optical Coherence Elastography." *Biomedical Optics Express* 3, no. 12 (2012b): 3138–3152.

Kennedy KM, BF Kennedy, RA McLaughlin, and DD Sampson. "Needle Optical Coherence Elastography for Tissue Boundary Detection." *Optics Letters* 37, no. 12 (2012c): 2310–2.

Kirkpatrick SJ and BW Brooks. "Micromechanical Behavior of Cortical Bone as Inferred from Laser Speckle Data." *Journal of Biomedical Materials Research* 39, no. 3 (1998): 373–9.

Kirkpatrick, SJ, RK Wang, and DD Duncan. "OCT-Based Elastography for Large and Small Deformations." *Optics Express* 14, no. 24 (2006): 11585–97.

Ko HJ, W Tan, R Stack, and SA Boppart. "Optical Coherence Elastography of Engineered and Developing Tissue." *Tissue Engineering* 12, no. 1 (2006): 63–73.

Kruger RA, PY Liu, YR Fang, and CR Appledorn. "Photo-acoustic Ultrasound (PAUS)—Reconstruction Tomography." *Medical Physics* 22, no. 10 (1995): 1605–9.

Li C, G Guan, X Cheng, Z Huang, and RK Wang. "Quantitative Elastography Provided by Surface Acoustic Waves Measured by Phase-Sensitive Optical Coherence Tomography." *Optics Letters* 37, no. 4 (2012): 722–4.

Li S, KD Mohan, WW Sanders, and AL Oldenburg. "Toward Soft-Tissue Elastography Using Digital Holography to Monitor Surface Acoustic Waves." *Journal of Biomedical Optics* 16, no. 11 (2011): 116005-1–116005-7.

Liang X, M Orescanin, KS Toohey, MF Insana, and SA Boppart. "Acoustomotive Optical Coherence Elastography for Measuring Material Mechanical Properties." *Optics Letters* 34, no. 19 (2009): 2894–6.

Liang X, and SA Boppart. "Biomechanical Properties of in vivo Human Skin from Dynamic Optical Coherence Elastography." *IEEE Transactions on Biomedical Engineering* 57, no. 4 (2010): 953–9.

Liang X, SG Adie, R John, and SA Boppart. "Dynamic Spectral-Domain Optical Coherence Elastography for Tissue Characterization." *Optics Express* 18, no. 13 (2010): 14,183–14,190.

Liang X, AL Oldenburg, V Crecea, EJ Chaney, and SA Boppart. "Optical Micro-Scale Mapping of Dynamic Biomechanical Tissue Properties." *Optics Express* 16, no. 15 (2008): 11052–65.

Linder MW, P Andretzky, F Kiesewetter, and G Hausler. "Spectral Radar: Optical Coherence Tomography in the Fourier Domain." In: Bouma B, Tearney GJ, editors. *Handbook of Optical Coherence Tomography*. New York: Marcel Dekker Inc., (2002): 335–358.

Lu P, VM Weaver, and Z Werb. "The Extracellular Matrix: A Dynamic Niche in Cancer Progression." *Journal of Cell Biology* 196, no. 4 (2012): 395–406.

Lu PF, K Takai, VM Weaver, and Z Werb. "Extracellular Matrix Degradation and Remodeling in Development and Disease." *Cold Spring Harbor Perspectives in Biology* 3, no. 12 (2011): 24.

Lunt SJ, A Fyles, RP Hill, and M Milosevic. "Interstitial Fluid Pressure in Tumors: Therapeutic Barrier and Biomarker of Angiogenesis." *Future Oncology* 4, no. 6 (2008): 793–802.

Lyshchik A, T Higashi, R Asato, S Tanaka, J Ito, M Hiraoka, MF Insana, AB Brill, T Saga, and K Togashi. "Cervical Lymph Node Metastases: Diagnosis at Sonoelastography—Initial Experience." *Radiology* 243, no. 1 (2007): 258–67.

Mahmud A, and J Feely. "Arterial Stiffness Is Related to Systemic Inflammation in Essential Hypertension." *Hypertension* 46, no. 5 (2005): 1118–22.

Mai JJ, and MF Insana. "Strain Imaging of Internal Deformation." *Ultrasound in Medicine and Biology* 28, no. 11–12 (2002): 1475–84.

Manduca A, DS Lake, SA Kruse, and RL Ehman. "Spatio-Temporal Directional Filtering for Improved Inversion of MR Elastography Images." *Medical Image Analysis* 7, no. 4 (2003): 465–73.

McGrath DM, WD Foltz, A Al-Mayah, CJ Niu, and KK Brock. "Quasi-Static Magnetic Resonance Elastography at 7 T to Measure the Effect of Pathology before and After Fixation on Tissue Biomechanical Properties." *Magnetic Resonance in Medicine* 68, no. 1 (2012): 152–65.

Miga MI. "A New Approach to Elastography Using Mutual Information and Finite Elements." *Physics in Medicine and Biology* 48, no. 4 (2003): 467–80.

Mohan KD, and AL Oldenburg. "Elastography of Soft Materials and Tissues by Holographic Imaging of Surface Acoustic Waves." *Optics Express* 20, no. 17 (2012): 18887–97.

Nenadic, IZ, MW Urban, S Aristizabal, SA Mitchell, TC Humphrey, and JF Greenleaf. "On Lamb and Rayleigh Wave Convergence in Viscoelastic Tissues." *Physics in Medicine and Biology* 56, no. 20 (2011): 6723–38.

Nightingale K. "Acoustic Radiation Force Impulse (ARFI) Imaging: A Review." *Current Medical Imaging Reviews* 7, no. 4 (2011): 328–39.

Nightingale K, S McAleavey, and G Trahey. "Shear-Wave Generation Using Acoustic Radiation Force: In vivo and ex vivo Results." *Ultrasound in Medicine and Biology* 29, no. 12 (2003): 1715–23.

Nightingale KR, ML Palmeri, RW Nightingale, and GE Trahey. "On the Feasibility of Remote Palpation Using Acoustic Radiation Force." *Journal of the Acoustical Society of America* 110, no. 1 (2001): 625–34.

Oldenburg AL, and SA Boppart. "Resonant Acoustic Spectroscopy of Soft Tissues Using Embedded Magnetomotive Nanotransducers and Optical Coherence Tomography." *Physics in Medicine and Biology* 55, no. 4 (2010): 1189–201.

Oldenburg AL, G Wu, D Spivak, F Tsui, AS Wolberg, and TH Fischer. "Magnetomotive Optical Coherence Tomography and Labeled Platelets." *IEEE Journal of Selected Topics in Quantum Electronics* 18, no. 3 (2012): 1100–9.

Orescanin M, and MF Insana. "Shear Modulus Estimation with Vibrating Needle Stimulation." *IEEE Transactions on Ultrasonics Ferroelectrics and Frequency Control* 57, no. 6 (2010a): 1358–67.

Orescanin, M, MA Qayyum, KS Toohey, and MF Insana. "Dispersion and Shear Modulus Measurements of Porcine Liver." *Ultrasonic Imaging* 32, no. 4 (2010b): 255–66.

Orescanin, M, Y Wang, and MF Insana. "3-D FDTD Simulation of Shear Waves for Evaluation of Complex Modulus Imaging." *IEEE Transactions on Ultrasonics Ferroelectrics and Frequency Control* 58, no. 2 (2011): 389–98.

Pallwein L, M Mitterberger, P Struve, G Pinggera, W Horninger, G Bartsch, FR Aigner, A Lorenz, F Pedross, and F Frauscher. "Real-Time Elastography for Detecting Prostate Cancer: Preliminary Experience." *BJU International* 100, no. 1 (2007): 42–6.

Palmeri ML, JJ Dahl, DB Macleod, SA Grant, and KR Nightingale. "On the Feasibility of Imaging Peripheral Nerves Using Acoustic Radiation Force Impulse Imaging." *Ultrasonic Imaging* 31, no. 3 (2009): 172–82.

Papazoglou S, U Hamhaber, J Braun, and I Sack. "Algebraic Helmholtz Inversion in Planar Magnetic Resonance Elastography." *Physics in Medicine and Biology* 53, no. 12 (2008): 3147–58.

Parker KJ, MM Doyley, and DJ Rubens. "Imaging the Elastic Properties of Tissue: The 20 Year Perspective." *Physics in Medicine and Biology* 57, no. 16 (2012): R1–R29.

Paszek MJ, N Zahir, KR Johnson, JN Lakins, GI Rozenberg, A Gefen, CA Reinhart-King et al. "Tensional Homeostasis and the Malignant Phenotype." *Cancer Cell* 8, no. 3 (2005): 241–54.

Ponnekanti H, J Ophir, and I Cespedes. "Ultrasonic-Imaging of the Stress-Distribution in Elastic Media Due to an External Compressor." *Ultrasound in Medicine and Biology* 20, no. 1 (1994): 27–33.

Povazay B, K Bizheva, A Unterhuber, B Hermann, H Sattmann, AF Fercher, W Drexler et al. "Submicrometer Axial Resolution Optical Coherence Tomography." *Optics Letters* 27, no. 20 (2002): 1800–2.

Qi W, R Chen, L Chou, G Liu, J Zhang, Q Zhou, and Z Chen. "Phase-Resolved Acoustic Radiation Force Optical Coherence Elastography." *Journal of Biomedical Optics* 17, no. 11 (2012): 110505-1–110505-3.

Qiu Y, M Sridhar, JK Tsou, KK Lindfors, and MF Insana. "Ultrasonic Viscoelasticity Imaging of Nonpalpable Breast Tumors: Preliminary Results." *Academic Radiology* 15, no. 12 (2008): 1526–33.

Razani, M, A Mariampillai, C Sun, TWH Luk, VXD Yang, and MC Kolios. "Feasibility of Optical Coherence Elastography Measurements of Shear Wave Propagation in Homogeneous Tissue Equivalent Phantoms." *Biomedical Optics Express* 3, no. 5 (2012): 972–80.

Sadigh, G, RC Carlos, CH Neal, and BA Dwamena. "Ultrasonographic Differentiation of Malignant from Benign Breast Lesions: A Meta-Analytic Comparison of Elasticity and BIRADS Scoring." *Breast Cancer Research and Treatment* 133, no. 1 (2012): 23–35.

Samani A, J Zubovits, and D Plewes. "Elastic Moduli of Normal and Pathological Human Breast Tissues: An Inversion-Technique-Based Investigation of 169 Samples." *Physics in Medicine and Biology* 52, no. 6 (2007): 1565–76.

Sandrin L, B Fourquet, JM Hasquenoph, S Yon, C Fournier, F Mal, C Christidis et al. "Transient Elastography: A New Noninvasive Method for Assessment of Hepatic Fibrosis." *Ultrasound in Medicine and Biology* 29, no. 12 (2003): 1705–13.

Sarvazyan AP, MW Urban, and JF Greenleaf. "Acoustic Waves in Medical Imaging and Diagnostics." *Ultrasound in Medicine and Biology* 39, no. 7 (2013): 1133–46.

Sato T, A Fukushima, N Ichida, H Ishikawa, H Miwa, Y Igarashi, T Shimura, and K Murakami. "Nonlinear Parameter Tomography System Using Counter-Propagating Probe and Pump Waves." *Ultrasonic Imaging* 6, no. 2 (1984): 237–37.

Schmitt C, AH Henni, and G Cloutier. "Characterization of Blood Clot Viscoelasticity by Dynamic Ultrasound Elastography and Modeling of the Rheological Behavior." *Journal of Biomechanics* 44, no. 4 (2011): 622–9.

Schmitt JM. "OCT Elastography: Imaging Microscopic Deformation and Strain of Tissue." *Optics Express* 3, no. 6 (1998): 199–211.

Sinkus R, J Lorenzen, D Schrader, M Lorenzen, M Dargatz, and D Holz. "High-Resolution Tensor MR Elastography for Breast Tumour Detection." *Physics in Medicine and Biology* 45, no. 6 (2000): 1649–64.

Sridhar M, J Liu, and MF Insana. "Elasticity Imaging of Polymeric Media." *Journal of Biomechanical Engineering-Transactions of the ASME* 129, no. 2 (2007): 259–72.

Srivastava A, Y Verma, KD Rao, and PK Gupta. "Determination of Elastic Properties of Resected Human Breast Tissue Samples Using Optical Coherence Tomographic Elastography." *Strain* 47, no. 1 (2011): 75–87.

Tanter M, J Bercoff, A Athanasiou, T Deffieux, JL Gennisson, G Montaldo, M Muller, A Tardivon, and M Fink. "Quantitative Assessment of Breast Lesion Viscoelasticity: Initial Clinical Results Using Supersonic Shear Imaging." *Ultrasound in Medicine and Biology* 34, no. 9 (2008): 1373–86.

Tse ZT, YJ Chan, H Janssen, A Hamed, I Young, and M Lamperth. "Piezoelectric Actuator Design for MR Elastography: Implementation and Vibration Issues." *International Journal of Medical Robotics and Computer Assisted Surgery* 7, no. 3 (2011): 353–60.

van Vledder MG, EM Boctor, LR Assumpcao, H Rivaz, P Foroughi, GD Hager, UM Hamper, TM Pawlik, and MA Choti. "Intra-Operative Ultrasound Elasticity Imaging for Monitoring of Hepatic Tumour Thermal Ablation." *HPB* 12, no. 10 (2010): 717–23.

Wang H-L, S Zhang, X-J Xin, L-H Zhao, C-X Li, J-L Mu, and X-Q Wei. "Application of Real-Time Ultrasound Elastography in Diagnosing Benign and Malignant Thyroid Solid Nodules." *Cancer Biology Medicine* 9, no. 2 (2012a): 124–7.

Wang RK, Z Ma, and SJ Kirkpatrick. "Tissue Doppler Optical Coherence Elastography for Real Time Strain Rate and Strain Mapping of Soft Tissue." *Applied Physics Letters* 89, no. 14 (2006): 144103-1–144103-3.

Wang S, J Li, RK Manapuram, FM Menodiado, DR Ingram, MD Twa, AJ Lazar, DC Lev, RE Pollock, and KV Larin. "Noncontact Measurement of Elasticity for the Detection of Soft-Tissue Tumors Using Phase-Sensitive Optical Coherence Tomography Combined with a Focused Air-Puff System." *Optics Letters* 37, no. 24 (2012b): 5184–6.

Wang Y and MF Insana. "Viscoelastic Properties of Rodent Mammary Tumors Using Ultrasonic Shear-Wave Imaging." *Ultrasonic Imaging* 35, no. 2 (2013): 126–45.

Wirtz D, K Konstantopoulos, and PC Searson. "The Physics of Cancer: The Role of Physical Interactions and Mechanical Forces in Metastasis." *Nature Reviews Cancer* 11, no. 7 (2011): 512–22.

Yamakoshi Y, J Sato, and T Sato. "Ultrasonic-Imaging of Internal Vibration of Soft-Tissue under Forced Vibration." *IEEE Transactions on Ultrasonics Ferroelectrics and Frequency Control* 37, no. 2 (1990): 45–53.

Yasar TK, D Klatt, RL Magin, and TJ Royston. "Selective Spectral Displacement Projection for Multifrequency MRE." *Physical in Medicine and Biology* 58, no. 16 (2003): 5771–81.

8

Ultrasound Imaging of Mechanical Properties of Cancers

Brian S. Garra

CONTENTS

The increased stiffness of cancers relative to normal tissues and benign tumors has long been important in physical diagnosis. Recently, it has become possible to image this characteristic using strain imaging in ultrasound (US) or shear wave imaging in both US and magnetic resonance scanning. US stiffness imaging has been applied to a number of tumors including breast cancer, prostate cancer, thyroid cancer, and lymph node metastasis. By far, most research has been done on breast cancer diagnosis with reported ROC curve areas (AUC) of 0.89 to 0.95 for distinguishing benign from cancerous lesions for strain elastography when used with standard B-mode US. Widespread clinical adoption has been slow, however, due to difficulties in training both operators and readers to properly acquire and interpret the elastograms. Shear wave elastography has shown promise, but the potential for falsely low shear wave speed estimates in cancers has recently been demonstrated. Lymph node strain elastography has given good results with AUC values of 0.87 to 0.91. Shear wave evaluation of lymph nodes, thyroid and prostate is still at an early stage with prostate results showing the most promise. Thyroid nodule evaluation using strain elastography has been challenging with lower sensitivity and specificity values than in breast or lymph nodes. Better results have been reported when strain ratios are calculated instead of using a visual scoring system. Similar results in the prostate have been shown where strain ratio calculation may yield better sensitivity and specificity than simple subjective grading.

8.1 Introduction

The use of tissue hardness or stiffness in medical diagnosis dates from ancient times. The Edwin Smith Papyrus dating from about 1500 BCE[1] and attributed by some[2] to physician, architect, and priest Imhotep, who lived between 3000 and 2500 BCE, mentions palpation of organs and tumors as part of a standard patient examination. Chinese medical texts from the third century BCE mention palpation, and the techniques were probably known much earlier. Tumors, both benign and malignant, have long been known to be firmer than the surrounding tissues in which they arise, and palpable hardness and lack of mobility are classic signs of a malignancy versus a benign lesion.[3]

Soft tissue imaging has augmented and, in many situations, replaced palpation for the diagnosis of tumors and cancer, beginning with ultrasound (US) in the 1950s and 1960s, computed tomography (CT) in the 1970s, and magnetic resonance imaging (MRI) in the 1980s. Even with these advanced imaging methods, there is still considerable overlap in the appearance of cancers versus benign tumors, so the imaging of tumor stiffness/hardness has been welcomed by clinicians as an additional way to diagnose both cancers and various benign conditions such as liver cirrhosis.[4,5]

8.2 Mechanisms of Increased Stiffness in Cancers

The question of why cancers are stiffer than benign lesions is an area of active investigation. Possible mechanisms of increased stiffness include packing the extracellular matrix (ECM) with additional cells,[6] tension placed on the ECM by attached tumor cells,[7] increased interstitial pressure caused by leaky capillaries, the presence of microcalcifications, and increased abnormal collagen in the ECM induced by dysregulation of collagen deposition in cancers.[8] Recent work has also associated the abnormal stiffened ECM with increased incidence of metastasis[9] and malignant behavior of breast cancer cells.[10,11] On the other hand, there is evidence that the tumor cells themselves are less stiff then their normal counterparts and that lower stiffness of the cells themselves correlates with malignant potential.[12] Also, tumor necrosis centrally may result in decreased tumor stiffness and even liquefaction of the central portion of a tumor. So a hallmark of many cancers is heterogeneity of internal mechanical properties.

A cancer typically induces changes in surrounding tissues, including increased blood flow, increased tissue edema, infiltration by inflammatory cells, and actual formation of fibrosis (desmoplasia). These effects may lead to stiffening of surrounding tissues, even before the cancer invades those tissues, and are likely responsible for cancers feeling larger on palpation than a benign lesion of similar size. As noted in the next section, the stiffening of adjacent tissues can be observed on tissue elasticity imaging and may be used to help identify cancers.

8.3 US Imaging of Tumor Stiffness

Imaging of the mechanical properties of tissues using US can be performed using standard grayscale imaging. The simplest method that can be used for superficial tissues is to compress a lesion using the US transducer, noting the amount of flattening or compression that the lesion undergoes and comparing it with the measured flattening of the surrounding tissue (Figure 8.1). This method can be used to help distinguish benign from malignant breast masses[13] but never came into widespread use, possibly because the breast imaging community was focused at the time on integrating grayscale US into the workup of solid breast tumors, an area that had been previously dominated by mammography.

A more subjective method known as echopalpation can be used to evaluate lesion stiffness when a palpable lesion is felt. In this method, a finger or paperclip is placed between the transducer and the skin and is slid along the length of the transducer. The position of the finger or paperclip is noted when the edge of the lesion is felt. This helps to identify the margins of the palpable lesion and can help to alleviate patient anxiety when a normal structure or benign-appearing lesion is identified as the cause of the palpable lump. The compressibility of the lesion can also be easily evaluated with soft lesions deforming like a sponge and hard lesions showing little if any focal deformation under the finger or paperclip (Figure 8.2).

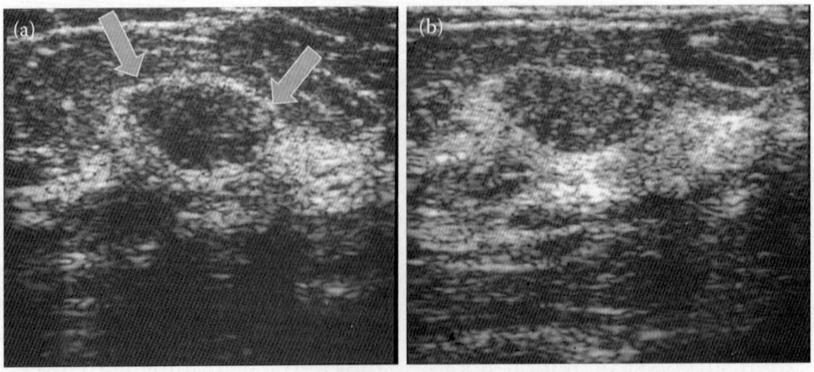

FIGURE 8.1
Ultrasound of a benign fibroadema (a) before compression and (b) during compression. The arrows point to the lesion. Note the flattening of the lesion during compression.

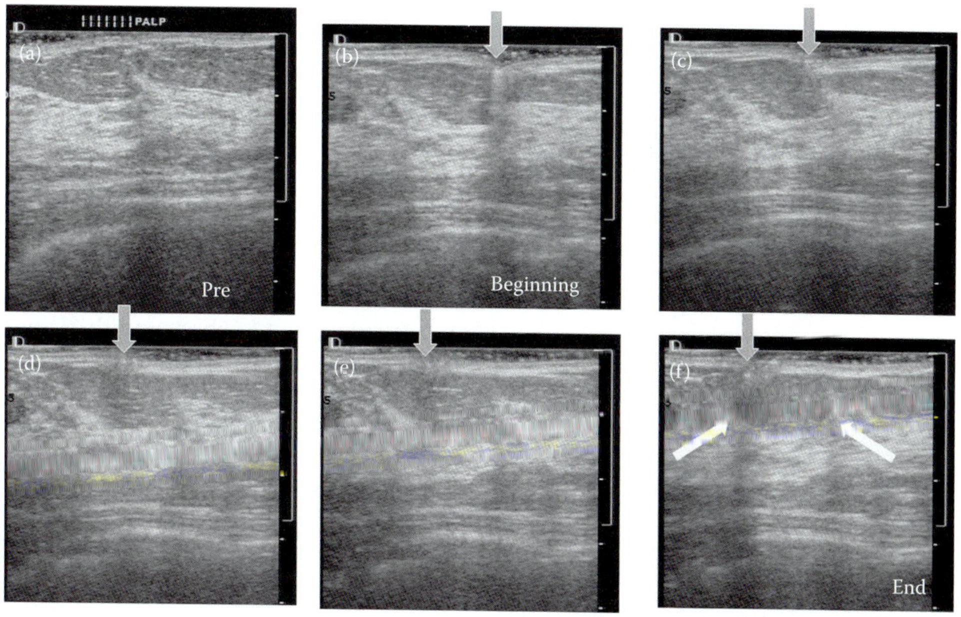

FIGURE 8.2
Scans of the breast showing echopalpation using a paper clip. (a) Breast prior to compressing locally with the paperclip. (b–e) Paper clip being slid across the breast from right to left. The paper clip is the small bright focus with a weak shadow behind it. Dark gray arrows mark the spot of the clip. The mass can be felt by the operator holding the paper clip. Figure 8.2b shows the clip at the right edge of the palpable lump, and Figure 8.2f shows it at the left edge. The mass corresponding to the palpable lump is marked by the white arrows in Figure 8.2f. Doppler color signal is marked by white arrows

Another simple method for evaluating tissue stiffness uses vibrations transmitted into the body from the patient's vocal chords. In this method (called "fremitus" after the physical examination test, having a patient hum), color or power Doppler imaging is used to detect the vibrations in tissues such as the breast caused by having the patient hum (Figure 8.3). This method, although useful, never became widely popular because variability in detectability of lesions caused by both variations in Doppler sensitivity and variations in the pitch at which different patients hum.

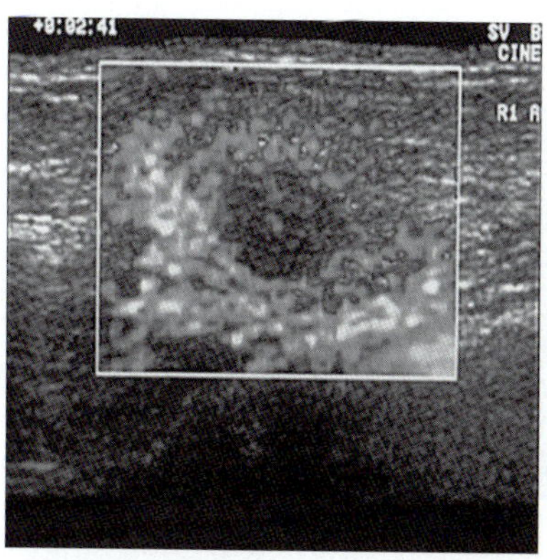

FIGURE 8.3
Fremitus. Power Doppler image of a breast lesion, which shows no color when the patient hums because of the different stiffness of the lesion compared to the surrounding tissue.

True tissue elasticity imaging developed out of tissue motion research conducted in England in the mid-1980s.[14,15] The early techniques used M-mode US* to monitor tissue position changes over time with subsequent Fourier analysis of the waveforms. M-mode was eventually supplemented with Doppler-based tissue motion detection in 1987, and this subsequently evolved at the University of Rochester into a color Doppler-based tissue stiffness imaging technique called sonoelasticity imaging[16,17] (Figure 8.4).

The imaging of tissue stiffness became more practical with the development of strain imaging, now known as strain elastography, first reported by Cespedes and Ophir in 1991.[18] In this method, radio-frequency backscatter data are acquired from tissue prior to compression and then again after compression by between 0.2% and 1.5% usually by pressing with the US transducer itself or by compression from internal body movements such as breathing or vascular pulsation. The displacement of the tissue at each point in the image caused by the compression is estimated using cross-correlation

* M-mode refers to an ultrasound method of tracking change in axial position of a structure over time. Ultrasound pulses are sent repeatedly in one direction and echoes from each structure along that direction (A-line) are plotted on the vertical axis vs. time on the horizontal axis. Straight horizontal lines correspond to structures that are not moving towards or away from the transducer and wavy lines indicate movement towards and away from the transducer over time. B-mode ultrasound refers to normal grayscale imaging ultrasound.

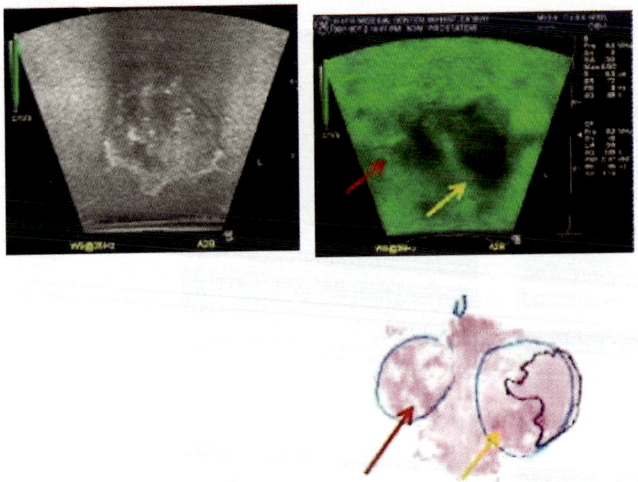

FIGURE 8.4
Sonoelasticity imaging. The top left image is a prostate gland embedded in gelatin. Top right is the sonoelasticity image showing no color in the areas of increased stiffness, which correspond to cancers in the bottom right image (arrows). (Courtesy of B. Casteneda.)

methods, and the displacement *change* at each depth (the strain) is displayed as the strain elastogram (Figure 8.5). Strain elastography is becoming a mature technology, with most US manufacturers offering this capability on their systems. Since strain is dependent on the amount of compression applied, which varies, the ratio of strain in a lesion is often compared to that in adjacent "normal" tissue. Such strain ratios may be less operator dependent that pure strain images or strain values.

A newer method of imaging tissue stiffness involves estimation of shear wave speed, which increases with increasing tissue hardness. Shear waves are transverse waves (not compressional waves like acoustic waves), and since shear waves travel much more slowly than acoustic waves, an US scanner can track the tissue displacements generated by shear waves and thus estimate shear wave speed. Shear waves may be induced in tissue by applying an external mechanical vibrator or by a beam of acoustic waves. When an acoustic beam (compression waves) is transmitted into tissue, shear waves are generated at the margins of the beam and propagate at right angles to the acoustic wave. A higher-intensity or specially formed acoustic beam must be used to generate shear waves of sufficient amplitude to be detected.[19,20]

Shear wave speed is dependent on both the elastic properties of tissue and tissue viscosity, which cause the shear wave speed to change with frequency. Since shear waves are typically generated at frequencies of 50–500 Hz by clinical systems, different systems may give different shear wave speeds for the same tissue. This and other possible system dependencies are being addressed by the radiological society of north america (RSNA) Quantitative Imaging Biomarker Alliance Ultrasound Technical Committee so that results from different systems may be compared.[21]

The key features of cancers on strain elastograms are increased stiffness relative to benign tumors and surrounding tissues, increased size of the tumor on the elastogram versus the sonogram, and tumor attachment to the surrounding tissues. The increased stiffness and relative size of tumors were reported in the initial clinical paper on breast elastography[22] and were confirmed in later publications.[23,24] The attachment of cancers to surrounding tissues is manifest on palpation by lack of mobility of the mass and can be imaged using a newer technique known as axial shear strain elastography (ASSE)[25] (Figure 8.6). Almost all cancers exhibit increased stiffness, with the other features being more variable. Cancers also exhibit greater nonlinearity in stiffness with increased precompression,[26,27] a factor that some are attempting to exploit as an additional feature to help distinguish cancers.[28,29]

Less extensive results are available for shear wave speed estimates in cancers, but initial results have been

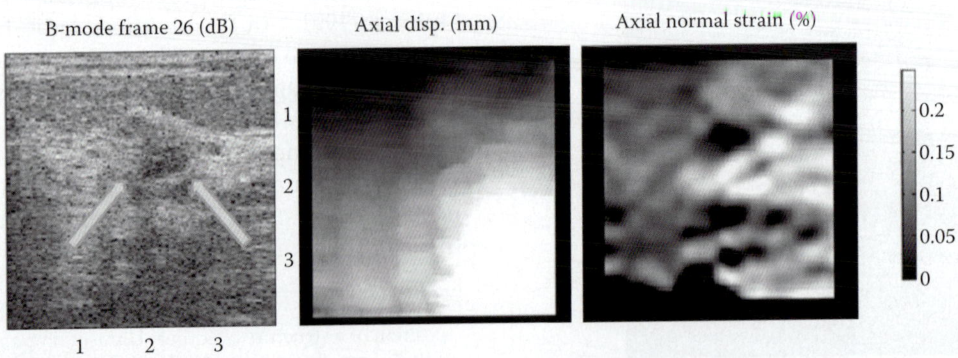

FIGURE 8.5
Strain elastogram. The arrows point to a hypoechoic breast lesion on the B-mode image. The middle image is the displacement map, and the left image is the resulting strain elastogram showing the lesion as dark (stiff) and slightly larger than on the sonogram. This is a characteristic of breast cancers.

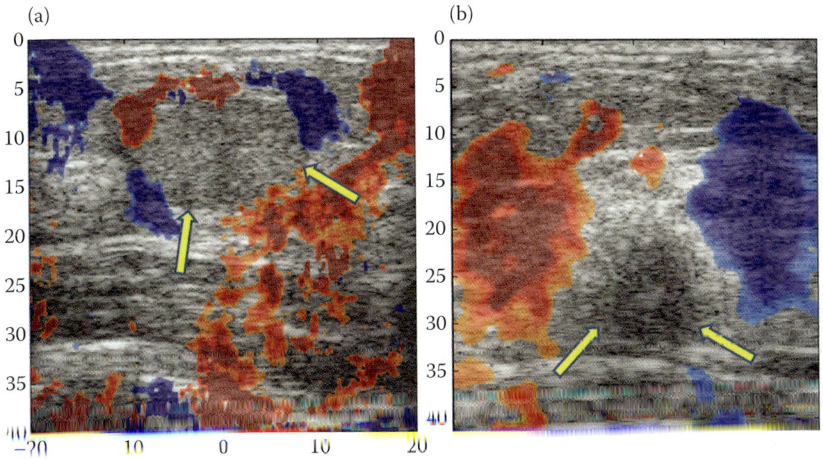

FIGURE 8.6
Axial shear strain elastogram (ASSE). (a) Axial shear strain elastogram for a fibroadenoma (arrows). The encircling color bands are narrow because a slip boundary is present between the lesion and surrounding tissue, causing all the strain to be concentrated into a narrow zone (the slip boundary) at the lesion edge. (b) Invasive ductal cancer (arrows). Note the large thick color bands at the lesion margin showing that the shear strain is spread over a large area because the lesion is anchored to the surrounding tissue. (Courtesy of Dr. Arun Thitai-Kumar, University of Texas Medical School, Houston.)

promising. Both direct estimation of elastic modulus using shear wave speed and the ratio of shear wave speed in a lesion to that in surrounding tissue have been used. These results are discussed in Section 8.4.1.

8.4 Elastography Applications in Oncology

Strain elastography has been tried in most malignancies with variable success, and now that shear wave elastographic devices are becoming more available, that method is being attempted in many other tumors as well. The most commonly reported uses are for evaluation of breast cancer, prostate cancer, thyroid cancer, and lymph node metastasis.

8.4.1 Breast Cancer

Evaluation of breast masses for possible cancer was the first systematically studied application of elastography. The original work[22] was performed by compressing the breast parallel to the chest wall against a mammography cassette, but this quickly evolved into compression against the chest wall without any degradation in image quality. A combination of lesion stiffness and size (relative to the sonogram) was used to demonstrate good separability of benign from malignant lesions. Later work confirmed these findings in larger numbers of patients with reported receiver operating characteristic (ROC) curve areas (AUC) of 0.89–0.95 for distinction of benign from cancerous lesions.[30–34] A standard method of grading breast lesions qualitatively using the color display of strain elastograms on Hitachi systems (Figure 8.7) was developed,[35] and multiple studies have shown this grading to be useful for diagnosis of cancers.[36] More quantitative methods using the ratio of strain in the lesion to that of surrounding tissue (strain ratios) have been also found to be useful and possibly more accurate than the five-point grading scale method.[37–39]

Lesion size is a factor in the accuracy of breast elastography. While lesions as small as 3 mm have been studied using elastography,[40] problems in the assessment of small lesions exist that are related to the data window sizes used for gathering and analyzing strain data and shear wave speed data. The minimum size of radiofrequency data windows for displacement tracking and regions of interest for shear wave speed is typically 3–10 mm, resulting in averaging of the lesion with surrounding tissue and increased variability in estimates for small lesions. A recent meta-analysis of the effects of breast mass size on elastography results[41] showed similar sensitivity and specificity for strain elastography across three size ranges (5–10, 10–19, >19 mm), with sensitivity ranging from 74% to 82% and specificity ranging from 90% to 94%. However, the number of very small lesions (5–6 mm) in the meta-analysis may have been quite small.

For strain elastography, a method of distinguishing high-quality from low-quality elastograms is important since it is often difficult for observers to judge which of the many elastograms obtained accurately portrays tissue strain values. For example, an elastogram containing mostly decorrelation noise may still show a lesion clearly

Score	Classification standard	Typical image
1	Strain is seen in the entire hypoechoic area (the entire lesion is shown in green similar to the surrounding tissue).	
1*	BGR (blue–green–red) three-layer pattern—typical artefact seen in a cystic lesion.	
2	Strain is seen within most of the hypoechoic area, but some areas show no strain (the lesion is a mixture of green and blue).	
3	Strain appears only in the periphery with no strain in the center of the lesion (the center of the lesion is shown as blue with the periphery in green).	
4	No strain is measured within the lesion (the entire lesion is shown in blue).	
5	No strain is measured within the lesion nor in the surrounding tissues (the lesion and the surrounding tissues are blue).	

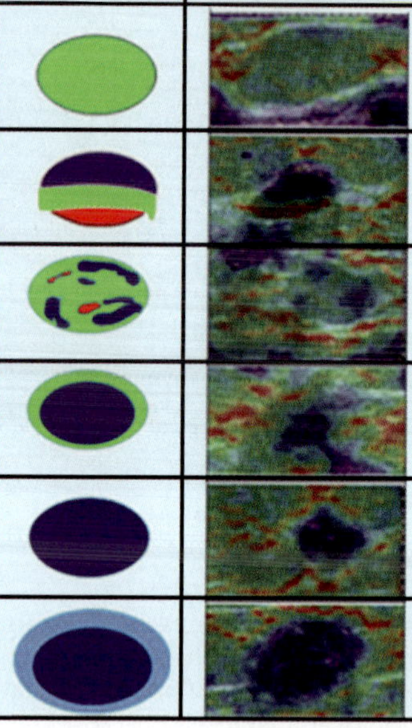

FIGURE 8.7
Tsukuba elasticity score. The table shows the 1–5 scoring system popularized by Itoh et al. and used on the Hitachi strain elastography systems. Note that when the stiff lesion is larger than the corresponding ultrasound lesion, the diagnosis of cancer can be made confidently. (From http://www.hitachi-medical-systems.eu/products-and-services/ultrasound/hitachi-real-time-tissue-elastography-hi-rte/clinical-applications.html, August 2013.)

because the lesion compresses less than the surrounding tissue, resulting in lower strain and less decorrelation within the lesion (Figure 8.8). Alternatively, an elastogram created with very low compression may not show a lesion if the displacements and strain values are very low throughout the image. Most manufacturers provide an indicator of elastogram quality to help the observer select the "best" elastograms for visual evaluation. Such indicators may flag low-quality elastograms but usually do not indicate for the user what problem caused the low-quality elastogram or how to fix the problem. Problems with strain elastograms fall into four categories:

1. Too much precompression was applied, making lateral and out-of-plane movement more pronounced and stiffening all tissues.
2. The rate of compression is too low or too high, making the total displacement between images either too small or too large for tissue displacements to be estimated accurately by cross-correlation.
3. Lateral motion occurs during compression caused by either uneven pressure across the transducer face or curvature of the surface against which the tissue is being compressed (usually the chest wall for breast elastography).
4. Out-of-plane motion is occurring during compression, caused by either too much precompression or curvature/tilt of the surface against which the tissue is being compressed.

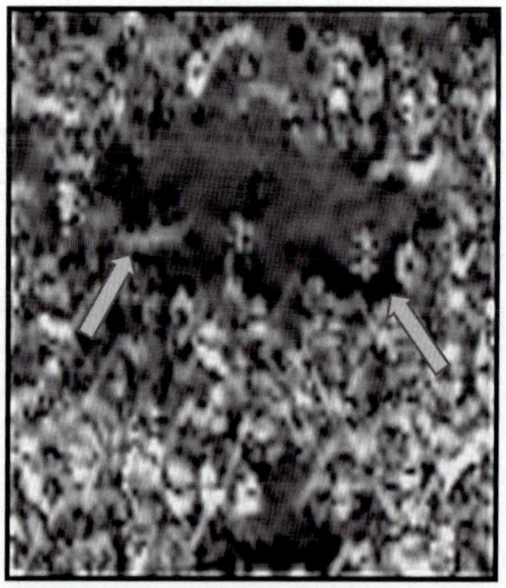

FIGURE 8.8
Lesion visualized in noisy elastogram. The image has the speckled pattern of noise due to large displacements and a failure of the cross-correlation algorithm to accurately estimate those displacements. The breast lesion, however, appears as a noise free-area (arrows) probably, due to increased stiffness and resulting decreased tissue movement, resulting in more accurate displacement estimates.

In my experience, careful evaluation of the B-mode image during compression and relaxation can provide valuable information about which of these problems is occurring. There is a significant learning curve for operators wanting to perform high-quality strain elastography, and there is a need for standardized approaches to training operators to perform and interpret strain elastograms.

A newer method of quantification of breast lesion stiffness involves the estimation of shear wave speed within the breast lesion. Initial clinical results using this method have been promising,[42,43] and a more extensive multicenter trial has shown a significant increase in AUC from an already high value of 0.950 to an even higher value of 0.962 at $P = .005$ when using B-mode US features alone versus B-mode US features plus shear wave elastographic features.[44] Table 8.1 lists the elastic moduli computed from shear wave speed measurements for a number of studies.[44,45–49]

Shear wave elastography is not perfect, however, and the images may be misleading unless the interpreter is properly trained on interpretation. Initial published studies tended to show the "classic" appearance of a cancer—that of a very stiff tumor (usually displayed in red) surrounded by softer surrounding tissue (Figure 8.9)—but later studies demonstrated lesions with an apparently low shear wave speed within the lesion but foci of increased speed in the peritumoral tissues (Figure 8.10). Initially, the explanation for cancers with low internal shear wave speeds was that central necrosis was present but failure of pathological correlation and no hint of softness on corresponding strain images suggest a problem with shear wave imaging itself. Possible explanations for erroneous shear wave speeds in the center of lesions include presence of a weak noisy shear wave due to tissue heterogeneity in cancers and reflected shear waves at the boundary of tumors causing errors in shear wave speed estimators.[50–52] Despite these problems, accurate identification of cancers seems achievable if a lesion is graded as suspicious for cancer if any focus of increased shear wave speed is present either in the lesion or in the adjacent surrounding tissues.

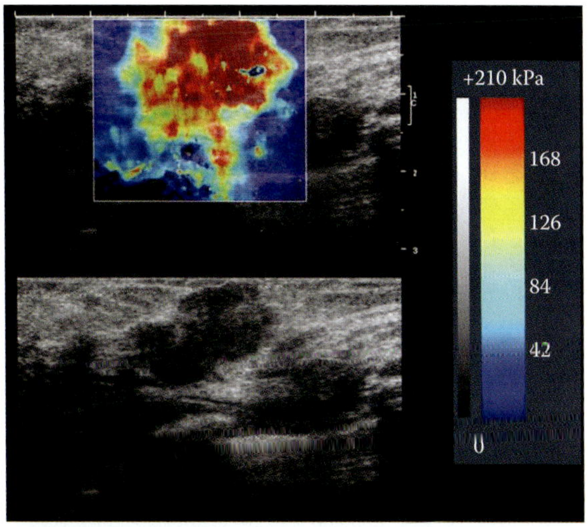

FIGURE 8.9
Invasive ductal cancer on shear wave speed imaging. Lobulated invasive ductal carcinoma grade III showing high stiffness (red in top image) throughout the tumor mass. The grayscale ultrasound is shown in the bottom image. (From http://www.supersonicimagine.com/Community/Scientific-papers/Breast2/Ductal-carcinoma, accessed May 28, 2013.)

TABLE 8.1
Elastic Moduli for Breast Lesions

Reference	Lesion Type	Number	Equipment	Mean Young's modulus (Mean SWS)	Range/IQR	Std. Dev.
Berg et al. 2012[44]	Benign	527	SSI	43	25–90 IQR	
Awad 2013[45]	Benign	55	SSI	50	10–99	
Tozaki et al. 2011[46]	Benign	13	S2000	22.4 (2.73)	9.4–42.2	0.91
Bai et al. 2012[47]	Benign	102	S2000	15.2 (2.25)	4.8–71	1
Evans et al. 2010[48]	Benign	23	SSI	28	18–44	
Berg et al. 2012[44]	Cancer	277	SSI	177	137–180 IQR	
Awad 2013[45]	Cancer	61	SSI	134	110–297	
Tozaki et al. 2011[46]	Cancer	17	S2000	96.8 (5.68)	17.9–166	23.7
Bai et al. 2012[47]	Cancer	15	S2000	107 (5.96)	4.1–248	26
Evans et al. 2010[48]	Cancer	28	SSI	140	29–293	
Chang et al. 2013[49]	Cancer	337	SSI	146	12–300	57
Evans et al. 2010[48]	DCIS	1	SSI	76	—	
Evans et al. 2010[48]	LCIS	1	SSI	82	—	
Bai et al. 2012[47]	Normal Breast	143	S2000	8.5 (1.68)	2–26	8.5

Note: DCIS, ductal carcinoma in situ; IQR, 25th percentile to 75th percentile range; LCIS, lobular carcinoma in situ; S2000, Siemens S2000; SSI, Aixplorer from Supersonic Imagine; SWS, shear wave speed. Young's modulus is listed in kPa. SWS is listed in meters per second. Note the lower Modulus values for S2000 versus SSI, which may be due in part to differences in shear wave frequency between the two systems and also in part to some normal breast tissue being included in the region of interest on the Siemens system.[53]

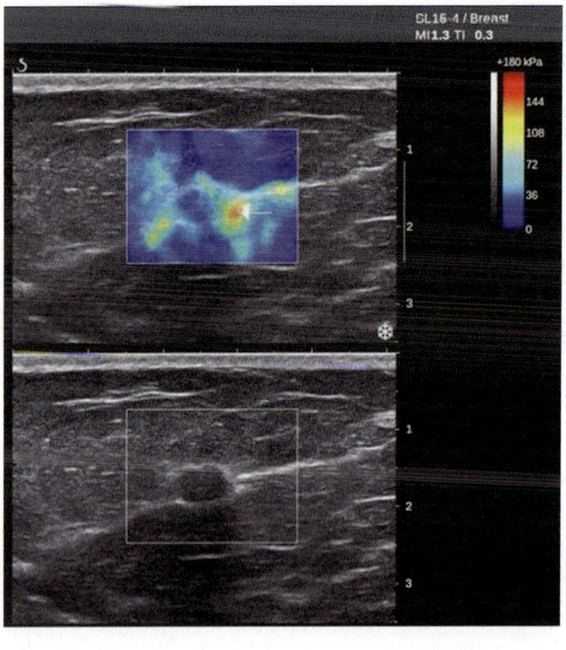

FIGURE 8.10
Invasive ductal cancer with benign features on grayscale image and no increase in stiffness within the tumor—only two foci of increased stiffness adjacent to the tumor mass (upper image, arrow). (From Evans A, Whelehan P, Thomson K et al. *Breast Cancer Res.*, 2010;12(6): R104 and http://www.medscape.com/viewarticle/735153_3, accessed May 28, 2013.)

8.4.2 Lymph Node Evaluation

Enlarged or palpable firm lymph nodes are a common diagnostic problem. Nodes most often are abnormal due to the presence of inflammation or in response to nearby infection or inflammation, but the possibility of metastatic cancer must always be considered. A reliable method for determining whether an enlarged lymph node is involved in cancer or not would be extremely useful for selecting nodes for biopsy in patients with suspected or confirmed malignancy.

A number of studies using strain elastography for lymph node evaluation have appeared[54–58] using both strain ratios and the color display-based grading system originally developed for breast masses. This work has shown that similar to primary cancers, metastatic disease causes lymph nodes to become stiffer than normal or benign-reactive lymph nodes.

The first report to appear[54] found a specificity of 98% using strain ratios in cervical nodes, and Alam et al.[57] achieved a specificity of 100% using strain elastography pattern analysis. ROC analysis in this study yielded AUC values of 0.87 for elastography alone and 0.90 for B-mode US. Choi et al., on the other hand, showed little improvement by using elastographic grading (modified Itoh–Tsukuba scale) over grayscale US alone, with B-mode AUC = 0.81 and elastography AUC = 0.78.[59] Wojcinski reported on 180 axillary lymph nodes evaluated using color strain elastography using a scoring system that separated the cortex of nodes from the medullary regions and found that elastographic score plus a cortical thickness >3 mm gave high specificity (99.3%) but low sensitivity for metastatic disease.[60] More recently, a meta-analysis of nine studies and 835 nodes was reported showing that strain ratio estimation performed better than elasticity scoring methods, with hierarchical summary ROC (HSROC) of 0.88 for elasticity scoring and 0.91 for strain ratio methods.[61]

Shear wave elastography results for lymph nodes are just beginning to appear. A pilot study in cervical lymph nodes[62] demonstrated a relatively low AUC of 0.77 for classification of benign versus metastatic nodes, with malignant nodes showing uniformly low shear wave speeds or a heterogeneous appearance. The discrepancy between sheer wave speed imaging and strain imaging may be the result of reflected shear waves or tissue heterogeneity, as was discussed for breast lesion imaging in Section 8.4.1. Choi et al. also demonstrated heterogeneity of shear wave speed in some metatstatic nodes but showed increased sensitivity (91% vs. 88%) and specificity (97% vs. 94%) for shear wave speed imaging as compared to grayscale evaluation.[63] Further clinical results may help to clarify the situation and better define the criteria for shear wave speed classification of metastatic disease in lymph nodes.

8.4.3 Thyroid Nodule Evaluation

Increasing numbers of thyroid nodules are being detected using US and during CT examinations of the thorax and neck. Guidelines for biopsy of nodules based on size are trending downward, with recent guidelines by the American Association of Clinical Endocrinologists suggesting that nodules larger than 5 mm should be biopsied in certain subsets of patients.[64] Since most thyroid glands containing nodules have many more small nodules than large, a reliable method of selecting smaller nodules for biopsy is needed. US examination has been the key examination for detection and characterization of thyroid nodules for a number of years, and criteria for distinguishing benign and malignant nodules have been published.[65,66] Hypoechogenicity, irregular margins, lack of a hypoechoic halo, and noncystic nature have all been reported as features, but the most useful single feature is the presence of microcalcifications or macrocalcifications, which, in a solid nodule, increase the likelihood of cancer to as high as 59%.[67] On the other hand, a large number of benign nodules (24%) will also contain calcifications.[68] Clearly, better methods for distinguishing benign from cancerous nodules are needed.

Strain elastography was initially shown to be potentially useful for classification of nodules as benign versus malignant[69–73] by both external compression and use of carotid artery pulsations. Sensitivities of 88%–94% and specificities of 81%–90% were achieved. As with other malignancies, thyroid cancers were found to be stiffer than benign nodules. More recently, however, results have been more mixed, with Kagoya et al. reporting only 73% sensitivity and 64% specificity[74] and Trimboli et al. reporting 81% sensitivity and 62% specificity for elastography alone[75] in a large multicenter trial. However, in the latter study, when elastography was combined with grayscale features, sensitivity jumped to 97%. Moon,[73] on the other hand, concluded that elastography was inferior to grayscale US alone, with a sensitivity of 91.7% and a specificity of 88% for US compared to 65.4% and 58.2% for strain elastography alone. In this study, elastographic performance declined in smaller lesions, with a sensitivity and specificity of 46% and 48.7% for lesions ≤5 mm. They also concluded that elastography plus US was inferior to US alone, but their sensitivity and specificity values for combined US and elastography (94.5% and 47.5%) are not convincing. The variability seen in elastography results is not surprising given the very subjective nature of the four- or five-point grading systems used. Also, thyroid strain elastography can be very challenging technically due to pulsations from the adjacent common carotid artery and the steeply sloping neck contour, which promotes lateral movement of the thyroid, which degrades the strain image. The initial work by Lyschik[69] and later work by Bae[76] made use of strain ratios, which might at least remove some of the subjectivity in interpretations. At least one study has shown the superiority of strain ratios to a visual scoring system for characterization of thyroid nodules, with an AUC of 0.907 for strain ratio versus 0.829 for a visual scoring system.[77]

Because shear wave speed imaging does not require external compression, the problems created by applying compression in the complex and heterogeneous neck can be avoided. Thyroid evaluation results using shear wave speed are now just appearing in the literature. Although the studies clearly show that malignant thyroid nodules are stiffer than benign ones, discriminability of benign versus malignant nodules has been disappointing, with reported AUC values of 0.58–0.74.[78,79] For comparison, conventional US gave an AUC of 0.792 in one study.[79] Several studies have reported an improvement in performance when conventional US and shear wave speed are used together, but the magnitude of the improvement is variable.[79,80] One publication has suggested a threshold of 66 kPa for distinguishing benign from malignant nodules, which gave a sensitivity of 80% and a specificity of 90.5%,[81] but given that the goal is to not miss any cancers, a higher sensitivity at the cost of lower specificity may be preferable. Hopefully, as the technique of shear wave speed estimation is refined, these clinical results will improve.

8.4.4 Prostate Evaluation

Prostate cancer was responsible for 14% of new cancers in men in 2008 worldwide[82] and is the second most common cancer in men after lung cancer. Current treatment protocols suggest active surveillance for clinically "insignificant" disease and more aggressive treatment only for "significant" disease. Given that the current criteria for distinguishing significant from nonsignificant disease include a tumor volume of 0.5–1.3 mL, detection of early prostate cancer is clearly of great interest. The detection and evaluation of prostate cancer was the very first application proposed for elastography when the technique was developed in the early 1990s and was also the application suggested for sonoelasticity imaging when it appeared about the same time.[83] Unlike breast cancer, where many lesions are visible and the main task is to correctly classify them, many prostate cancers are not visible sonographically, making detection the most important task. Since then, more than 30 articles reporting performance of elasticity imaging in vivo have appeared, and a thorough systematic review was recently published.[84] Both magnetic resonance (MR) and US elasticity imaging have been tried but the in vivo MR papers have been either case reports or small series of 6–19 patients.[85,86] In those studies, prostate cancer was stiffer than benign tissue, although the results from the different studies were not comparable. For US elastography, by far the most common technique used has been strain elastography with free hand compression with color display of stiffness, accounting for 13 of 18 papers published with pathology correlation between 2002 and 2012.[87–99] The vast majority of the work has been performed using one US system (Hitachi EUB 8500) with a color map overlay on the grayscale image where red = soft and blue = stiff. Exact criteria for classification of a lesion as malignant vary, as do the results achieved, with sensitivities varying from 51% to 86% and specificities from 62% to 86%. Some improvement over simple classification schemes is suggested in one paper when strain ratios are used (AUC = .90 for cancer vs. benign, sensitivity = 74.5%, specificity 83.3%).[100] Additional work is needed to confirm this possibility.

Work on the use of shear wave elastography in the prostate has just begun. The initial results are very encouraging. An exhibit by one group demonstrated a sensitivity of 90% and a specificity of 100% in 18 patients with an elevated prostate specific antigen (PSA).[101] The authors used the stiffness ratio between the nodule and adjacent parenchyma, and a cutoff stiffness ratio of 1.5 was found to give the best discrimination. Another group demonstrated even better sensitivity and specificities of 96.2% when a cutoff of 37 kPa was used for distinguishing benign from malignant.[102] Using a cutoff of 40 kPa, sensitivity was 100%, and a reduction in false positives from 79% (for US without shear wave elastography) to 50% would have been achieved. More recently, another study using shear wave imaging achieved good results as well in 50 patients with sensitivity and specificity of 0.90 and 0.88, respectively.[102] For patients with an elevated PSA, sensitivity and specificity improved somewhat to 0.93 for each. If these results are confirmed in additional studies with careful correlation between shear wave imaging and pathology, a new and useful tool for prostate cancer detection and biopsy guidance will have been found. There is even a suggestion in a more recent paper[103] as well as some other publications that a correlation exists between Gleason score and stiffness, which would be helpful in determining the significance of a cancer, if found. Newer methods employing the nonlinear behavior of stiffness in cancers and the viscosity of the tumor (using a viscoelastic model) are still in the early research stages.

Tumors of several other organs have been examined using elasticity imaging and quantification. Efforts to diagnose pancreatic cancer using elastography have focused on endoscopic US,[104,105] though there is at least one publication describing transcutaneous evaluation of pancreatic carcinoma using elastography.[106] Early results were promising, but a more recent study[107] was less optimistic, reporting an AUC of 0.72, with the problem primarily being one of low specificity (sensitivity of 100% and specificity of only 16.7%). Evaluation of focal liver lesions to distinguish hepatocellular carcinoma (HCC) from other tumors has also been studied, though most of the focus in the liver has been the detection of liver fibrosis. Preliminary work[108,109] has been promising, with AUC values as high as 0.94. More recently, a study has shown similar performance (sensitivity = 92.5%, specificity = 88.8%) using both endoscopic US and transabdominal US.[110] Another study[111] suggests that elastography may be used to distinguish HCC from other malignancies such as cholangiocarcinoma and metastatic disease.

Testicular lesions have been evaluated using real-time elastography with promising initial results.[112] In recent study using the scoring system of Itoh, the sensitivity and specificity for benign versus malignant lesions were again promising at 87.5% and 98.2%, respectively.[113] Single reports on the evaluation of skin cancers[114] and salivary gland tumors[115] have appeared, but results are too preliminary to draw any conclusions.

8.4.5 Treatment Monitoring

A potentially important future application of elasticity imaging is for the monitoring of tumor ablations and other therapies. Tumor ablations typically result in increased tissue stiffness.[116,117] The increased stiffness is easily imaged using strain elastography, making it possible to readily identify the margins of the ablation, which are not visible on sonography. Quantification of the stiffness change may be useful for determining the adequacy of the ablative treatment. Several studies have successfully demonstrated treatment monitoring in the liver and prostate gland using elastography.[118–120] Recently, a study of rectal carcinomas using shear wave elastography was conducted in 31 patients to determine if the method would be effective for treatment monitoring.[121] The study demonstrated a softening of the tumor manifested by a significant decline in mean shear wave speed from 3.13 to 2.11 m/s. Tumors confined to the rectal wall also had lower stiffness than tumors that had invaded through the wall. A correlation was seen between shear wave speed decline and tumor regression grading by other methods.

Elasticity imaging has also been used to monitor neoadjuvant chemotherapy response in breast cancer patients.[122] The authors found a significant decrease in strain ratios in treatment responders versus those who did not respond.

8.5 Magnetic Resonance Elastography and US Elastography

MRI (see Chapters 2–6) may also be used to estimate tissue stiffness by measuring shear wave speed resulting from vibrations induced in tissue by an external mechanical vibrator. This method (magnetic resonance elastography [MRE]) can be used for diffuse disease and for focal lesions, both benign and malignant. One significant advantage of MRE is that it can be applied to organs not generally accessible to US, such as the brain and lungs.[123] The results of MRE cannot be directly compared with US shear wave estimates because of the difference in shear wave frequencies used by the two methods. The use of MRE is likely to be more limited than US shear wave elastography because of the higher cost of MRE. The cost of MRE equipment and scanner time is also a likely reason why studies of clinical efficacy for MRE are proceeding at a much slower pace than for US.

8.6 Future Trends

In the future, further work on grading of cancers using elasticity imaging alone and in combination with other US analysis methods such as radio-frequency backscatter analysis will help to increase the clinical value of elasticity imaging. Also, better training methods to make image acquisition and interpretation easier for physicians and technologists will help to move elasticity estimation further into the mainstream of US diagnosis. Further work to confirm the value of elastographic imaging as a means to monitor treatment is also likely. In addition, it is important to note that many key applications for elasticity imaging are related to detection of fibrosis and inflammation, which helps to identify patients at risk for cancer development. Finally, continuing development of strain and shear wave detection and analysis algorithms should improve the accuracy of the quantitative stiffness estimates obtainable from elasticity imaging systems. This includes improved quality monitoring tools during acquisition and shear wave speed estimation algorithms.

8.7 Conclusion

The imaging of elasticity has evolved considerably since it was first developed and now includes several methods for quantification that greatly increase its usefulness. Increased size and stiffness are seen in breast cancers, whereas most other cancers appear to exhibit only increased stiffness. Elasticity imaging for breast cancer is here, with thyroid, lymph node, and prostate cancer application following close behind. Further refinements in technique and technology will make elasticity imaging a key tool for the detection and treatment monitoring of cancers.

References

1. Allen JP. *The Art of Medicine in Ancient Egypt*. (New York: The Metropolitan Museum of Art, 2005), 70.
2. Breasted JH. *The Edwin Smith Surgical Papyrus*: Published in facsimile and hieroglyphic transliteration with translation and commentary in two volumes. (Chicago: University of Chicago Press, 1991), 9.
3. Fu KL, Fu YS, Bassett LW, Cardall LW, Lopen JK. Invasive malignancies. In: *Diagnosis of Diseases of the Breast*, 2nd Ed. Edited by LW Bassett (Philadelphia, PA: Saunders, 2005), 483.
4. Friedrich-Rust M, Ong M-F, Martens S, Sarrazin C, Bojunga J, Zeuzem S, Herrmann E. Performance of transient elastography for the staging of liver fibrosis: A meta-analysis. *Gastroenterology* 2008;134:960–974.e968.
5. Fierbinteanu-Braticevici CAD, Usvat R, Cretoiu D, Baicus C, Marinoschi G. Acoustic radiation force imaging sonoelastography for noninvasive staging of liver fibrosis. *World J Gastroenterol* 2009;15:5525–5532.
6. Katira P, Bonnecase RT, Zaman MH. Modeling the mechanics of cancer; effect of changes in cellular and extracellular mechanical properties. *Front Oncol*. 2013;3:145. doi: 10.3389/fonc.2013.00145. Published online June 11, 2013. Prepublished online. March 18, 2013.
7. Pasek MJ, Zahir N, Johnson KR, Lakins JN, Rozenberg GI, Gefen A, Reinhard-King CA, Margulles SS, Dembo M, Boettiger D, Hammer DA, Weaver VM. Tensional homeostasis and the malignant phenotype. *Cancer Cell* 2005;8:241–254.
8. Butcher DT, Alliston T, Weaver VM. A tense situation: Forcing tumour progression. *Nat Rev Cancer* 2009;9: 108–122.
9. Ramaswamy S, Ross KN, Lander ES, Golub TR. A molecular signature of metastasis in primary solid tumors. *Nat Genet* 2003;33:49–54.
10. Levental KR, Yu H, Kass L, Lakins JN, Egeblad M, Erler M, Erle JT, Fong SF, Csiszar K, Giacca A, Weninger W, Yamauchi M, Gasser DL, Weaver VM. Matrix crosslinking forces tumor progression by enhancing integrin signaling. *Cell* 2009;139(5):891–906.
11. Lu P, Weaver VM, Werb Z. The extracellular matrix: A dynamic niche in cancer progression. *J Cell Biol* 2012;196(4):395–406.
12. Swaminathan V, Mythreye K, O'Brien ET, Berchuck A, Blobe GC, Superfine R. Mechanical stiffness grades metastatic potential in patient tumor cells and in cancer cell lines. *Cancer Res* 2011;71(15):5075–5080.
13. Kelly KM. Breast ultrasound. *Crit Rev Diagn Imaging* 1996;37:79–161.
14. Tristam M, Barbosa DC, Cosgrove DO, Nassiri DK, Bamber JC. Ultrasound study of in vitro kenetic characteristics of human tissues. *Ultrasound Med Biol* 1986;12:927–937.
15. Tristam M, Barbosa DC, Cosgrove DO, Bamber JC, Hill CR. Application of Fourier analysis to clinical study of patterns of tissue movement. *Ultrasound Med Biol* 1988;14:695–707.
16. Lerner RM, Parker KJ, Holen J, Gramiak R, Waag RC. Sono-elasticity: Medical elasticity images derived from ultrasound signals in mechanically vibrated targets. *Acoust Imaging* 1988;16:317–327.
17. Parker KJ, The evolution of vibration sono-elastography. *Current Medical Imaging Reviews* 2011;7:283–291.
18. Ophir J, Céspedes I, Ponnekanti H, Yazdi Y, Li X. Elastography: A quantitative method for imaging the elasticity of biological tissues. *Ultrason Imaging* 1991;13(2):111–134.
19. Bercoff J, Tanter M, Fink M. Supersonic shear imaging: A new technique for soft tissue elasticity mapping. *IEEE Trans Ultrason Ferroelectr Freq Control* 2004;51(4): 396–409.
20. Nightingale K, McAleavey S, Trahey G. Shear-wave generation using acoustic radiation force: *In vivo* and *ex vivo* results. *Ultrasound Med Biol* 2003;29(12):1715–1723.
21. Available from: http://qibawiki.rsna.org/index.php?title=Ultrasound_SWS_tech_ctte.
22. Garra BS, Ophir J, Spratt SR, Zuurbier RA, Magnant CM, Pennanen M. Elastography of breast lesions: Initial clinical results. *Radiology* 1997;202:79.

23. Hall TJ, Zhu Y, Spalding CS. In vivo real-time freehand palpation imaging. *Ultrasound Med Biol* 2003;29(3): 427–435.
24. Barr RG. Real–Time ultrasound elasticity of the breast: Initial clinical results. *Ultrasound Q* 2010;26:61–66.
25. Thitaikumar A, Mobbs LM, Kraemer-Chant CM, Garra BS, Ophir J. Breast tumor classification using axial shear strain elastography: A feasibility study. *Phys Med Biol* 2008;53:4809–4823.
26. Krouskop TA, Wheeler TM, Kallel F, Garra BS, Hall TJ. Elastic moduli of breast and prostate tissues under compression. *Ultrason Imaging* 1998;20(4):260–274.
27. Barr RG, Zhang Z. Effects of precompression on elasticity imaging of the breast. *JUM* 2012;31(6):895–902.
28. Oberai AA, Gokhale NH, Goenezen S, Barbone PE, Hall TJ, Sommer AM, Jiang J. Linear and nonlinear elasticity imaging of soft tissue *in vivo*: Demonstration of feasibility. *Phys Med Biol* 2009;54:1191–1207.
29. Goenezen S, Dord J-F, Sink Z, Barbone PE, Jiang J, Hall TJ, Oberai AA. Linear and nonlinear elastic modulus imaging: An application to breast cancer diagnosis. *IEEE Trans Med Imaging* 2012;31(8):1628–1637.
30. Scaperrotta G, Ferranti C, Costa C, Mariani L, Marchesini M, Suman L, Folini C, Bergonzi S. Role of sonoelastography in non-palpable breast lesions. *Eur Radiol* 2008;18(11):2381–2389. Available from: http://dx.doi.org/10.1007/s00330-008-1032-8.
31. Schaefer FKW, Heer I, Schaefer PJ, Mundhenke C, Osterholz S, Order BM, Hofheinz N, Hedderich J, Heller M, Jonat W, Schreer I. Breast ultrasound elastography—Results of 193 breast lesions in a prospective study with histopathologic correlation. *Eur J Radiol* 2011;77(3): 450–456. Available from: http://www.ejradiology.com/article/S0720-048X(09)00510-5/abstract (cited May 24, 2013).
32. Sohn Y-M, Kim MJ, Kim E-K, Kwak JY, Moon HJ, Kim SJ. Sonographic elastography combined with conventional sonography: How much is it helpful for diagnostic performance? *J Ultrasound Med* 2009;28:413–420.
33. Burnside ES, Hall TJ, Sommer AM, Hesley GK, Sisney GA, Svensson WE, Fine JP, Jiang J, Hangiangreou NJ. Differentiating benign from malignant solid breast masses with US strain imaging. *Radiology* 2007;245: 401–410.
34. Chung SY, Moon WK, Choi JW, Cho N, Jang M, Kim KG. Differentiation of benign from malignant nonpalpable breast masses: A comparison of computer-assisted quantification and visual assessment of lesion stiffness with the use of sonographic elastography. *Acta Radiol (Stockholm, Sweden: 1987)* 2010;51(1):9–14.
35. Itoh A, Ueno E, Tohno E, Kamma H, Takahashi H, Shiina T, Yamakawa M, Matsumura T. Breast disease: Clinical application of US elastography for diagnosis. *Radiology* 2006;239:341–350.
36. Carlsen JF, Ewertsen C, Lönn L, Nielsen MB. Strain elastography ultrasound: An overview with emphasis on breast cancer diagnosis. *Diagnostics* 2013;3:117–125.
37. Zhi H, Xiao X-Y, Yang H-Y, Ou B, Wen Y-L, Luo B-M. Ultrasonic elastography in breast cancer diagnosis: Strain ratio vs. 5-point scale. *Acad Radiol* 2010;17(10):1227–1233.
38. Cho N, Moon WK, Kim HY, Chang JM, Park SH, Lyou CY. Sonoelastographic strain index for differentiation of benign and malignant nonpalpable breast masses. *J Ultrasound Med* 2010;29(1):1–7.
39. Gheonea IA, Stoica Z, Bondari S. Differential diagnosis of breast lesions using ultrasound elastography. *Indian J Radiol Imaging* 2011;21(4):301–305.
40. Kumm TR, Szabunio MM. Elastography for characterization of breast lesions: Initial clinical experience. *Cancer Control* 2010;17:156–161.
41. Sadigh G, Carlos RC, Neal CH, Wojcinski S, Dwamena BA. Impact of breast mass size on accuracy of ultrasound elastography vs. conventional B-mode ultrasound: A meta-analysis of individual participants. *Eur Radiol* 2013;23:1006–1014.
42. Tanter M, Bercoff J, Athanasiou A, Deffieux T, Gennisson JL, Montaldo G, Muller M, Tardivon A, Fink M. Quantitative assessment of breast lesion viscoelasticity: Initial clinical results using supersonic shear imaging. *Ultrasound Med Biol* 2008;34:1373–1386.
43. Athanasiou A, Tardivon A, Tanter M, Sigal-Zafrani B, Bercoff J, Deffieux T, Gennisson J-L, Fink M, Neuenschwander S. Breast lesions: Quantitative elastography with supersonic shear imaging—preliminary results. *Radiology* 2010;256:297–303.
44. Berg WA, Cosgrove DO, Doré CJ, Schäfer FK, Svensson WE, Hooley RJ, Ohlinger R, Mendelson EB, Balu-Maestro C, Locatelli M, Tourasse C, Cavanaugh BC, Juhan V, Stavros AT, Tardivon A, Gay J, Henry JP, Cohen-Bacrie C; BE1 Investigators. Shear-wave elastography improves the specificity of breast US: The BE1 multinational study of 939 masses. *Radiology* 2012;262(2):435–449.
45. Awad FM. Role of supersonic shear wave imaging quantitative elastography (SSI) in differentiating benign and malignant solid breast masses. *Egyptian J Rad Nuc Med* 2013;44(3):681–685.
46. Tozaki M, Isobe S, Fukuma E. Preliminary study of ultrasonographic tissue quantification of the breast using the acoustic radiation force impulse (ARFI) technology. *Eur J Rad* 2011;80:e182–e187.
47. Bai M, Du L, Gu J, Li F, Jia X. Virtual touch tissue quantification using acoustic radiation force impulse technology initial clinical experience with solid breast masses. *J Ultrasound Med* 2012;31:289–294.
48. Evans A, Whelehan P, Thomson K, McLean D, Brauer K, Purdie C, Jordan L, Baker L, Thompson A. Quantitative shear wave ultrasound elastography: Initial experience in solid breast masses. *Breast Cancer Res* 2010;12(6): R104. Available from: http://breast-cancer-research.com/content/12/6/R104.
49. Chang JM, Park IA, Lee SH, Kim WH, Bae MS, Koo HR, Yi A, Kim SJ, Cho N, Moon WK. Stiffness of tumours measured by shear-wave elastography correlated with subtypes of breast cancer. *Eur Radiol* 2013;23:2450–2458.
50. Wang Y. Quantitative estimate of mechanical properties from in vivo tumors of rodent animals. Ph.D. Thesis, Univ of Illinois, Urbana-Champaign, IL, 2012.
51. Nightingale K. Personal communication. Duke University.
52. Barr RG. Shear wave imaging of the breast: Still on the learning curve. *JUM* 2012;31(3):347–350.

53. Milkowski A. Personal communication. Siemens Medical Systems, November 2013.
54. Lyshchik A, Higashi T, Asato R, Tanaka S, Ito J, Hiraoka M, Insana MF, Brill AB, Saga T, Togashi K. Cervical lymph node metastases: Diagnosis at sonoelastography—initial experience. *Radiology* 2007;243:258–267.
55. Furukawa MK, Kubota A, Hanamura H, Furukawa M. Clinical application of real-time tissue elastography to head and neck cancer: Evaluation of cervical lymph node metastasis with real-time tissue elastography. *Nippon Jibiinkoka Gakkai Kaiho* 2007;110:503–505.
56. Zhang Y, Lv Q, Yin Y, Xie M, Xiang F, Lu C, Yan T, Li W, Xu H, Huang Y. The value of ultrasound elastography in differential diagnosis of superficial lymph nodes. *Front Med China* 2009;3(3):260–271.
57. Alam F, Naito K, Horiguchi J, Fukuda H, Tachikake T, Ito K. Accuracy of sonographic elastography in the differential diagnosis of enlarged cervical lymph nodes: Comparison with conventional B-mode sonography. *Am J Roentgenol* 2008;191:604–610.
58. Teng DK, Wang H, Lin YQ, Sui GQ, Guo F, Sun LN. Value of ultrasound elastography in assessment of enlarged cervical lymph nodes. *Asian Pac J Cancer Prev* 2012;13(5):2081–2085.
59. Choi JJ, Kang BJ, Kim SH, Lee JH, Jeong SH, Yim HW, Song BJ, Jung SS. Role of sonographic elastography in the differential diagnosis of axillary lymph nodes in breast cancer. *J Ultrasound Med* 2011;30(4):429–436.
60. Wojcinski S, Dupont J, Schmidt W, Cassel M, Hillemanns P. Real-time ultrasound elastography in 180 axillary lymph nodes: Elasticity distribution in healthy lymph nodes and prediction of breast cancer metastases. *BMC Med Imaging* 2012;12:35–45.
61. Ying L, Hou Y, Zheng H-M, Lin X, Xie Z-L, Hu Y-P. Real-time elastography for the differentiation of benign and malignant superficial lymph nodes: A meta-analysis. *Eur J Radiol* 2012;81(10):2576–2584.
62. Bhatia KS, Cho CC, Tong CS, Yuen EH, Ahuja AT. Shear wave elasticity imaging of cervical lymph nodes. *Ultrasound Med Biol* 2012;38(2):195–201.
63. Choi YJ, Lee JH, Lim HK, Kim SY, Han MW, Cho K-J et al. Quantitative shear wave elastography in the evaluation of metastatic cervical lymph nodes. *Ultrasound Med Biol* 2013;39(6):935–940.
64. Gharib H, Papini E, Paschke R, Valcavi R, Hegedüs L, Vitti P. AACE/AME/ETA thyroid nodule guidelines. *Endocr Pract* 2010;16(suppl 1):1–43.
65. Frates MC, Benson CB, Charboneau JW, Cibas ES, Clark OH, Coleman BG, Cronan JJ, Doubilet PM, Evans DB, Goellner JR, Hay ID, Hertzberg BS, Intenzo CM, Jeffrey RB, Langer JE, Larsen PR, Mandel SJ, Middleton WD, Reading CC, Sherman SI, Tessler FN. Society of Radiologists in Ultrasound. Management of thyroid nodules detected at US: Society of Radiologists in Ultrasound consensus conference statement. *Radiology* 2005;237:794–800.
66. Cooper DS, Doherty GM, Haugen BR, Kloos RT, Lee SL, Mandel SJ, Mazzaferri EL, McIver B, Pacini F, Schlumberger M, Sherman SI, Steward DL, Tuttle RM. Revised American Thyroid Association management guidelines for patients with thyroid nodules and differentiated thyroid cancer. *Thyroid* 2009;19(11):1–48.
67. Sieberling KA, Dutra JC, Grant T, Barjamovic S. Role of N-trathyroidal cacifications detected on ultrasound as a maker of malignancy. *The Laryngoscope* 2004; 114(10):1753–1757.
68. Wienke JR, Chong WK, Fielding JR, Zou KH, Mittelstaedt CA. Sonographic features of benign thyroid nodules: Interobserver reliability and overlap with malignancy. *J Ultrasound Med* 2003;22:1027–1031.
69. Lyshchik A, Higashi T, Asato R, Tanaka S, Ito J, Mai J, Pellot-Barakat C, Insana MF, Brill AB, Saga T, Hiraoka M, Togashi K. Thyroid gland tumor diagnosis at US elastography. *Radiology* 2005;237(1):202–211.
70. Dighe M, Bae U, Richardson ML, Dubinsky TJ, Minoshima S, Kim Y. Differential diagnosis of thyroid nodules with US elastography using carotid artery pulsation. *Radiology* 2008;248(2):662–669.
71. Asteria C, Giovanardi A, Pizzocaro A, Cozzaglio L, Morabito A, Somalvico F, Zoppo A. US-elastography in the differential diagnosis of benign and malignant thyroid nodules. *Thyroid* 2008;18(5):523–531.
72. Hong Y, Liu X, Li Z, Zhang X, Chen M, Luo Z. Real-time ultrasound elastography in the differential diagnosis of benign and malignant thyroid nodules. *J Ultrasound Med* 2009;28(7):861–867.
73. Moon HJ, Sung JM, Kim EK, Yoon JH, Youk JH, Kwak JY. Diagnostic performance of gray-scale US and elastography in solid thyroid nodules. *Radiology* 2012;262(3):1002–1013.
74. Kagoya R, Monobe H, Tojima H. Utility of elastography for differential diagnosis of benign and malignant thyroid nodules. *Otolaryngol Head Neck Surg* 2010;143:230–234.
75. Trimboli P, Guglielmi R, Monti S, Misischi I, Graziano F, Nasrollah N, Amendola S, Morgante SN, Deiana MG, Valabrega S, Toscano V, Papini E. Ultrasound sensitivity for thyroid malignancy is increased by real-time elastography: A prospective multicenter study. *J Clin Endocrinol Metab* 2012;97(12):4524–4530.
76. Bae U, Dighe M, Dubinsky T, Minoshima S, Shamdasani V, Kim Y. Ultrasound thyroid elastography using carotid artery pulsation: Preliminary study. *J Ultrasound Med* 2007;26(6):797–805.
77. Wang H, Brylka D, Sun LN, Lin YQ, Sui GQ, Gao J. Comparison of strain ratio with elastography score system in differentiating malignant from benign thyroid nodules. *Clin Imaging* 2013;37(1):50–55.
78. Bhatia KS, Tong CL, Cho CC, Yuen EH, Lee YY, Ahuja AT. Shear wave elastography of thyroid nodules in routine clinical practice: Preliminary observations and utility for detecting malignancy. *Eur Radiol* 2012;22(11):2397–2406.
79. Kwak JY, Kim EK. Diagnostic performance of quantitative shear wave ultrasound elastography for thyroid cancer. *J Korean Thyroid Assoc* 2011;4(2):109–113.
80. Kim H, Kim JA, Son EJ, Youk JH. Quantitative assessment of shear-wave ultrasound elastography in thyroid nodules: Diagnostic performance for predicting malignancy. *Eur Radiol* 2013 (Epub ahead of print).
81. Veyrieres JB, Albarel F, Lombard JV, Berbis J, Sebag F, Oliver C, Petit P. A threshold value in Shear Wave

elastography to rule out malignant thyroid nodules: A reality? *Eur J Radiol* 2012;81(12):3965–3972.

82. Ferlay JSH, Bray F, Forman D, Mathers C, Parking DM. GLOBOCAN 2008 v1.2, Cancer Incidence and Mortality Worldwide: IARC CancerBase No. 102010. Available from: http://globocan.iarc.fr.
83. Parker KJ. The evolution of vibration sonoelastography. *Current Medical Imaging Reviews* 2011;7:283–291.
84. Good DW, Stewart GD, Hammer S, Scanlan P, Shu W, Phipps S, Reuben R, McNeill AS. Elasticity as a biomarker for prostate cancer: A systematic review. *BJU Int* 2014;113(4):523–534.
85. Sahebjavaher RS, Baghani A, Honarvar M, Sinkus R, Salcudean SE. Transperineal prostate MR elastography: Initial in vivo results. *Magn Reson Med* 2013;69:411–420.
86. Li S, Chen M, Wang W, Zhao W, Wang J, Zhao X, Zhou C. A feasibility study of MR elastography in the diagnosis of prostate cancer at 3.0T *Acta Radiol* 2011;52:354–358.
87. Cochlin DL, Ganatra RH, Griffisths DF. Elastography in the detection of prostatic cancer. *Clin Radiol* 2002;57:1014–1020.
88. Pallwein L, Mitterberger M, Pinggera G, Aigner F, Pedross F, Gradl J, Pelzer A, Bartsch G, Frauscher F. Sonoelastography of the prostate: Comparison with systematic biopsy findings in 492 patients. *Eur J Radiol* 2008;65:304–310.
89. Nelson ED, Slotoroff CB, Gomella LG, Halpern EJ. Targeted biopsy of the prostate: The impact of color Doppler imaging and elastography on prostate cancer detection and Gleason score. *Urology* 2007;70:1136–1140.
90. Kamoi K, Okihara K, Ochiai A, Ukimura O, Mizutani Y, Kawauchi A, Miki T. The utility of transrectal real-time elastography in the diagnosis of prostate cancer. *Ultrasound Med Biol* 2008;34:1025–1032.
91. Ferrari FS, Megliola A, Drudi FM, Trovarelli S, Ponchietti R. Real-time elastography in the diagnosis of prostate cancer. *JUM* 2009;12:22–31.
92. Romagnoli A, Autieri G, Centrella D, Gastaldi C, Pedaci G, Rivolta L, Pozzi E, Anghileri A, Cerabino M, Bianchi CM, Roggia A. Real-time elastography in the diagnosis of prostate cancer: Personal experience. *Urologia* 2010;77:248–253.
93. Giurgiu CR, Manea C, Crisan N, Bungardean C, Coman I, Dudea SM. Real-time sonoelastography in the diagnosis of prostate cancer. *Med Ultrason* 2011;13:5–9.
94. Miyagawa T, Tsutsumi M, Matsumura T, Kawazoe N, Ishikawa S, Shimokama T, Miyanaga N, Akaza H. Real-time elastography of rhthe diagnosis of prostate cancer: Evaluation of elastographic moving images. *Jpn J Clin Oncol* 2009;39:394–398.
95. Pallwein L, Mitterberger M, Struve P, Pinggera G, Horninger W, Bartsch G, Aigner F, Lorenz A, Pedross F, Frauscher F. Real-time elastography for detecting prostate cancer: Preliminary experience. *BJU Int* 2007;100:42–46.
96. Sumura M, Shigeno K, Hyuga T, Yoneda T, Shiina H, Igawa M. Initial evaluation of prostate cancer with real-time elastography based on step-section pathologic analysis after radical prostatectomy: A preliminary study. *Int J Urol* 2007;14:811–816.
97. Salomon G, Köllerman J, Thederan I, Chun FK, Budäus L, Schlomm T, Isbarn H, Heinzer H, Huland H, Graefen M. Evaluation of prostate cancer detection with ultrasound real-time elastography: A comparison with step section pathological analysis after radical prostatectomy. *Eur Urol* 2008;54:1354–1362.
98. Walz J, Marcy M, Maubon T, Brunelle S, Laroche J, Gravis G, Salem N, Bladou F. Real time elastography in the diagnosis of prostate cancer: Comparison of preoperative imaging and histology after radical prostatectomy. *Prog Urol* 2011;21:925–931.
99. Brock M, von Bodman C, Palisaar RJ, Löppenberg B, Sommerer F, Deix T, Noldus J, Eggert T. The impact of real-time elastography guiding a systematic prostate biopsy to improve cancer detection rate: Prospective study of 353 patients. *J Urol* 2012;187:2039–2043.
100. Zhang Y, Tang J, Li YM, Fei X, Lv FQ, He EH, Li QY, Shi HY. Differentiation of prostate cancer from benign lesions using strain index of transrectal real-time tissue elastography. *Eur J Radiol* 2012;81:857–862.
101. Correas JMKA, Tissier AM, Vassiliu V, Eiss D, Helenon O. Trans-rectal quantitative shear wave elastography: Application to prostate cancer-A feasibility study. Poster No C-1748. Congress: ECR, 2011.
102. Barr RG, Memo R, Schaub CR. Shear wave ultrasound elastography of the prostate: Initial results. *Ultrasound Q* 2012;28(1):13–20.
103. Ahmad S, Cao R, Varghese T, Bidaut L, Nabi G. Transrectal quantitative shear wave elastography in the detection and characterisation of prostate cancer. *Surg Endosc* 2013; 27(9):3280–3287.
104. Adrian S, Peter V, Florin G, Gheonea DI, Gorunescu M, Ciurea T, Popescu GL, Iordache A, Hassan H, Iordache S. Neural network analysis of dynamic sequences of EUS elastography used for the differential diagnosis of chronic pancreatitis and pancreatic cancer. *Gastrointest Endosc* 2008;68:1086–1094.
105. Giovannini MB, Botelberge T, Bories E, Pesenti C, Caillol F, Esterni B, Monge G, Arcidiacono P, Deprez P, Yeung R, Schimdt W, Schrader H, Szymanski C, Dietrich C, Eisendrath P, Van Laethem J-L, Deviere J, Vilmann P, Saftoiu A. Endoscopic ultrasound elastography for evaluation of lymph node and pancreatic masses: A multicenter study. *World J Gastroenterol* 2009;15:1587–1593.
106. Uchida H, Hirooka Y, Itoh A, Kawashima H, Hara K, Nonogaki K, Kasugai T, Ohno E, Ohmiya N, Niwa Y, Katano Y, Ishigami M, Goto H. Feasibility of tissue elastography using transcutaneous ultrasonography for the diagnosis of pancreatic diseases. *Pancreas* 2009;38:17–22.
107. Dawwas MF, Taha H, Leeds JS, Nayar MK, Oppong KW. Diagnostic accuracy of quantitative EUS elastography for discriminating malignant from benign solid pancreatic masses: A prospective, single-center study. *Gastrointest Endosc* 2012;76(5):953–961.
108. Gheorghe L, Iacob S, Iacob R, Dumbrava M, Becheanu G, Herlea V, Gheorghe C, Lupescu I, Popescu I. Real time elastography—A non-invasive diagnostic method of small hepatocellular carcinoma in cirrhosis. *J Gastrointestin Liver Dis* 2009;18:439–446.

109. Fahey BJ, Nelson RC, Bradway DP, Hsu SJ, Dumont DM, Trahey GE. In vivo visualization of abdominal malignancies with acoustic radiation force elastography. *Phys Med Biol* 2008;53:279.
110. Sandulescu L, Padureanu V, Iulia C, Dumitrescu NB, Streba CT, Gheonea DI, Cazacu S, Ciurea T, Rogoveanu I, Saftoiu A. A pilot study of real time elastography in the differentiation of focal liver lesions. *Curr Health Sci J* 2012;38(1):1–4 (on line).
111. Masuzaki R, Tateishi R, Yoshida H, Sato T, Ohki T, Goto T, Yoshida H, Sato S, Sugioka Y, Ideda H, Shiina S, Kawabe T, Omata M. Assessing liver tumor stiffness by transient elastography. *Hepatol Int* 2007;1:394–397.
112. Pallwein L, Pallwein F, Schurich M, Fischbach V, Steiner H, Frauscher F. Sonoelastography of the testicles: Preliminary results in the diagnosis of different pathological processes. Poster at the 5th International Conference on the Ultrasonic Measurement and Imaging of Tissue Elasticity. Snowbird UT, Oct 8–11, 2006. Available from: http://www.elasticityconference.org/prior_conf/2005/2005Proceedings.pdf#page=28.
113. Goddi F, Sacchi A, Magistretti G, Almolla J, Salvadore M. Real-time tissue elastography for testicular lesion assessment. *Eur Radiol*. 2012; 22(4):721–730. doi: 10.1007/s00330-011-2312-2. Online first. October 16, 2011.
114. Dasgeb B, Siegel E. Elastographic ultrasound quantitative analysis combined with high frequency imaging for characterization of benign and malignant skin lesions. Presented at the 95th Scientific Assembly of the Radiological Society of North America, Chicago, IL, November 29–December 4, 2009.
115. Bhatia KS, Rasalkar DD, Lee YP, Wong KT, King AD, Yuen HY, Ahuja AT. Evaluation of real-time qualitative sonoelastography of focal lesions in the parotid and submandibular glands: Applications and limitations. *Eur Radiol* 2010;20:1958–1964.
116. Bharat S, Techavipoo U, Kiss MZ, Liu W, Varghese T. Monitoring stiffness changes in lesions after radiofrequency ablation at different temperatures and durations of ablation. *Ultrasound Med Biol* 2005;31:415–422.
117. Curiel L, Souchon R, Rouvière O, Gelet A, Chapelon JY. Elastography for the follow-up of high-intensity focused ultrasound prostate cancer treatment: Initial comparison with MRI. *Ultrasound Med Biol* 2005;31(11):1461–1468.
118. Rivaz H, Fleming I, Assumpcao L, Fichtinger G, Hamper U, Choti M, Hager G, Boctor E. Ablation monitoring with elastography: 2D in-vivo and 3D ex-vivo studies. *MICCAI*, Lecture Notes in Computer Science 2008;5242: 458–466.
119. Varghese T, Zagzebski JA, Lee FT. Elastographic imaging of thermal lesions in the liver in vivo following radiofrequency ablation: Preliminary results. *Ultrasound Med Biol* 2002;28:1467–1473.
120. Souchon R, Rouviere O, Gelet A, Detti V, Srinivasan S, Ophir J, Chepelon J. Visualization of HIFU lesions using elastography of the human prostate in vivo: Preliminary results. *Ultrasound Med Biol* 2003;29(7):1007–1015.
121. Rafaelsen SR, Vagn-Hansen C, Sørensen T, Lindebjerg J, Pløen J, Jakobsen A. Ultrasound elastography in patients with rectal cancer treated with chemoradiation. *Eur J Radiol* 2013;82(6):913–917. doi: 10.1016/j.ejrad.2012.12.030. Epub. February 11, 2013.
122. Falou O, Sadeghi-Naini A, Prematilake S, Sofroni E, Papanicolau N, Iradji S et al. Evaluation of neoadjuvant chemotherapy response in women with locally advanced breast cancer using ultrasound elastography. *Transl Onc* 2013;6:17–24.
123. Di Ieva A, Grizzi F, Rognone E, Tse ZTH, Parittotokkaporn T, Rodriguez F, Matula C, Trattnig S, Rodriguez Y, Baena R. Magnetic resonance elastography: A general overview of its current and future applications in brain imaging. *Neurosurg Rev* 2010;33:137–145.

Section IV

Tissue and Organ Imaging Methods

Radiography and Computed Tomography

9

Computed Tomography Image-Based Kinematic Analysis: An Overview

Douglas C. Moore, Eni Halilaj, Tarpit K. Patel, and Joseph J. Crisco III

CONTENTS

Accurate measurement of skeletal kinematics in vivo is essential for understanding normal and pathological joint function, disease progression, and treatment. However, making measurements noninvasively in living subjects, and reporting what can be very complex joint motion, is challenging and attainable with only a few methodologies. This chapter provides an overview of computed tomography–based kinematic analysis using markerless bone registration. First described in the late 1980s, computed tomography (CT)–based kinematic analysis has been used to evaluate the function of numerous joints, including the wrist and hand, forearm, elbow, knee, foot and ankle, and spine. It is ideally suited to the investigation of complex, multibone articulations, where out-of-plane motion is the norm. The conceptual basis for the technique is described, as are details of the principal steps in the analytical pipeline, including image acquisition, segmentation and bone model generation, markerless registration, coordinate system selection, and data presentation. Emphasis is placed on the more well-established static imaging methodology, though dynamic CT-based kinematic analysis is touched on briefly. The goal is to provide a resource for groups interested in implementing the technique, as well as those with modest experience.

9.1 Introduction

The in-depth understanding of normal joint function—and the changes associated with aging, injury, and disease—requires the ability to quantitatively evaluate skeletal kinematics. Historically, a broad range of techniques have been employed to perform kinematic analysis, from the early, elegant work of descriptive anatomists to the more recent, computationally sophisticated in vivo 3-D dynamic analyses by investigators with access to biplane videoradiography systems. Important advances have been made using cadaver testing, surface marker-based motion tracking, single-plane and stereoradiogrammetry, fluoroscopy, and volume imaging systems (i.e., computed tomography [CT] and magnetic resonance imaging [MRI]). The choice of the best technique for a given study depends on the nature of the specific research question, acceptable level of invasiveness, requisite accuracy, and available resources.

The purpose of this chapter is to provide an overview of current CT-based kinematic analysis. One of few techniques available for the noninvasive analysis of skeletal motion in vivo, CT-based kinematic analysis- has been used to evaluate joint motion in the hand and wrist,[1–16] forearm,[17–19] elbow,[20–22] spine,[23,24] foot/ankle,[25] and knee.[26] The popularity of the technique emanates from its conceptual simplicity and ease of implementation: serial 3-D image volumes are acquired using standard CT imaging systems, and rigid body transforms are generated directly via registration of bone models segmented from the acquired images. Much of the software required for analysis is commercially available or easily programmed.

CT-based kinematic analysis is most applicable in situations where bone overlap precludes the use of planar imaging techniques, such as in the wrist and hand, and where the motion of interest can be accommodated within the bore of the CT scanner. The chief limitations of CT-based kinematic analysis are that it requires exposure to ionizing radiation and it is not suitable for evaluating high-speed dynamic or impact-related events. To date, most CT-based kinematic analyses have been performed using static imaging protocols. Dynamic (4-D) CT is emerging as an intriguing potential alternative to serial static CT for kinematic analysis,[13–15,27–30] though with the scanners used to date, image artifacts have been an issue due to limited detector width and/or temporal resolution.[13–15,30] It is anticipated that advances in CT technology will improve image quality and broaden application of the technology.

Regardless of the imaging methodology—static or dynamic CT (or MRI)—the analysis pipeline is similar: image segmentation and bone model generation, model registration, coordinate system selection, and data presentation. The primary outcome variables include bone posture at each scanned position, and transformation matrices reflecting the relative motion between positions. Importantly, the image data sets allow the exploration of secondary outcome measures, such as bone morphology, joint spacing, and ligament lengths.

9.2 CT Imaging

CT volume images form the basis for CT-based skeletal kinematic analysis. Accordingly, care must be taken that the acquired images reflect the important features of the anatomy and task of interest and that sufficient positions are imaged to facilitate reasonable estimates of joint motion. Imaging of the upper extremity and axial skeleton can typically be performed using standard clinical scanners, where the table and gantry aperture are oriented horizontally. Analysis of joints in the lower extremity is more challenging, however, due to the potential need to accommodate weight bearing. Regardless of the anatomy, custom fixtures are helpful to standardize positioning and limit motion during imaging (Figure 9.1). Static CT-based kinematic analysis was

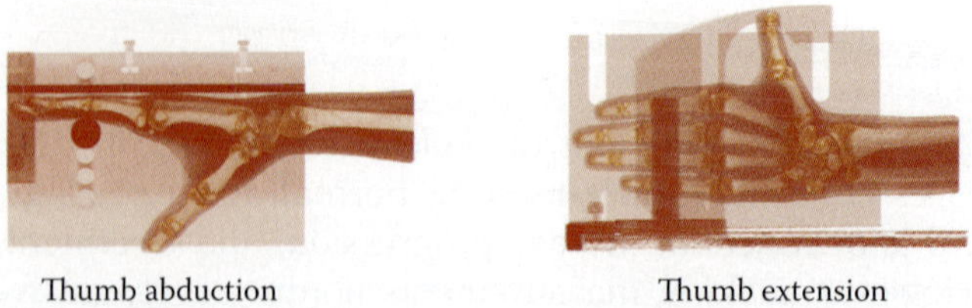

FIGURE 9.1
Sequential image acquisition. Sequential CT volume images form the basis for CT-based kinematic analysis, in this case, two poses for an analysis of thumb carpometacarpal joint kinematics. Custom-designed radiolucent fixtures are used to standardize positioning and limit motion during imaging.

developed using singleslice scanners, but it is much more efficient with multislice scanners.[31] Multislice scanners are capable of rapid image acquisition over long coverage intervals and the generation of thin (submillimeter) slice images. Quantitative kinematic analysis using dynamic CT requires high-speed, high-resolution, multidetector (≥64-slice) scanners and the use of gated CT imaging techniques developed for cardiac applications.[13,15,30]

9.2.1 Imaging Parameters

9.2.1.1 General Comments

Skeletal kinematics (and other derived outcome measures) are obtained directly via registration of bone models from the 3-D image data sets (or grayscale image data in the case of voxel-based registration). Accordingly, each bone of interest must be located within the scanner's acquisition field of view (AFOV) (or scan field of view [SFOV]) at each scan position, and the images must be of sufficient resolution and quality to capture any bony features of interest. In general, segmentation of bone from surrounding cartilage, ligament, and muscle is straightforward, given the contrast between the relatively radiopaque bone and radiolucent soft tissues. However, imaging artifacts can obscure the bone edges and reduce the accuracy of the segmented bone models and calculated outcome variables.

9.2.1.2 Image Resolution

Acquired image data must have spatial resolution sufficient to adequately capture features of the bones required for segmentation and registration. CT resolution is a function of the x-ray focal spot and detector sizes, and the reconstruction algorithm.[32] During reconstruction, the raw image data are discretized to yield 3-D digital image volumes composed of smaller rectangular subunits (voxels). The in-plane (x–y) dimension of the image voxels are defined by the reconstruction field of view (RFOV, or display field of view [DFOV]), and the longitudinal dimension is determined by the slice thickness. The RFOV is the width (and height) of the AFOV mapped to the square tomographic image matrix. The standard slice image matrix for clinical CT scanners is 512 × 512, so the nominal in-plane pixel size is RFOV/512; slice width is limited by the scanner's collimated detector width and interpolation algorithms.[33]

Absent published studies evaluating the influence of CT image resolution on calculated kinematic outcome measures, the tendency is to generate images with the highest possible spatial resolution by scanning at machine-limited z-axis intervals (typically 0.5 to 0.62 mm) and reconstructing with the smallest RFOV that completely captures the anatomy of interest. For the wrist and hand, RFOVs of 15 to 25 cm are common, yielding in-plane resolutions of 0.29 to 0.49 mm.[5,13,15,34] An advantage of capturing high-resolution image data is that it can be used for the analysis of bone morphology, for example, bone size or articular facet shape, which may define subpopulations of individuals or correlate with the development of pathology.[35] The presumption is that higher-resolution image data better reproduce subtle bone surface features than lower-resolution image data.

9.2.1.3 CT Artifacts

Artifacts are commonly encountered in clinical CT images, from a number of intrinsic and extrinsic factors.[36] The most common and/or problematic for CT-based kinematic analysis are motion artifacts and metal artifacts. Motion artifacts are caused by movement of the imaged anatomy during scanning, result in blurring and double images, and lead to step-off artifacts in the segmented bone models. The risk of motion artifacts is reduced with high-speed scanners and can be largely eliminated with the use of positioning fixtures that stabilize the region of interest with respect to the patient table. Motion artifacts are inherent in dynamic (4-D) CT, but decrease as motion velocity decreases and scanner speed (temporal velocity) increases.[13] Metal leads to starburst-shaped bright and dark streaking in the CT images that obscures image content, rendering accurate bone segmentation challenging. While metal artifact reduction is available on some scanners, its effectiveness depends on the algorithm used and the size and density of the metal object. The best solution is to avoid including metal in the first place, either by removing it where possible (i.e., jewelry) or via positioning that keeps it outside the axial slices of interest.

Image noise, which can also adversely affect CT-based kinematic analysis, is the result of statistical error due to low photon counts at the CT detector. In general, noise increases as slice thickness and/or tube current × time (milliampere seconds) is decreased. There is a balance between resolution, radiation dose, and image noise. Slice thickness must be decreased to increase resolution, but to attain the highest z-axis resolutions (0.5 to 0.625 mm) postpatient collimators are employed, which partially occlude the detectors, mechanically limiting photon density.[31] Radiation dose is directly proportional to the tube current × time product (milliampere seconds), so there is incentive to minimize this value to limit dose. In general, high-contrast objects such as bone are less affected by image noise,[37] so analysis protocols that employ serial segmentation are more noise tolerant; noise may be an issue with analysis protocols that rely on automated methods to perform segmentation and registration simultaneously. Regardless of the method used, it is useful to consult with a radiologist to optimize image quality and dose.

9.2.2 Radiation Exposure

An important consideration when contemplating CT-based skeletal kinematics is exposure to ionizing radiation, which is commonly accepted to increase the risk of certain types of cancer, even at low doses.[38–40] In general, CT imaging involves more radiation exposure than other x-ray imaging techniques (such as plain films or fluoroscopy), and the repeated scanning required for CT-based skeletal kinematic analysis compounds the problem. However, because musculoskeletal tissues are relatively insensitive to radiation exposure,[41] investigations of skeletal kinematics in the extremities are generally considered to pose manageable risk, as evidenced by the numerous extremity-related publications.[1–26] The analysis of more centrally located skeletal elements, such as the spine,[23,24] poses more risk, as more radiosensitive tissue is irradiated. Regardless, it is important to include input from radiologists and/or radiation physicists during imaging protocol development to minimize exposure. It goes without saying that any research on human subjects must be reviewed and approved by an institutional review board (IRB).

The most common method for defining radiation exposure is "effective dose," which is an estimate of whole-body cancer risk resulting from partial-body radiation exposure. Effective dose is calculated as the sum of the radiation absorbed by the tissues being imaged (i.e., skin, muscle, bone, bone marrow for extremity imaging) weighted by the sensitivities of the tissues to radiation (0.01, 0.05, 0.01, and 0.12, respectively).[41] The effective dose for clinical CT scans of the wrist and hand is ~0.03 mSv, in contrast to chest and lumbar spine CT scans, which are ~5.3 mSv and 19.2 mSv, respectively.[41] For kinematic studies of the wrist and hand, we can perform thirteen ~170 mm CT scans with a total effective dose of less than 4 mSv (at 80 kV, 40 mAs, and 0.625 mm pitch). This is significantly less than the exposures for many common clinical exams and only slightly more than a year's worth of background radiation, which is on the order of 3 mSV.[41] Studies of more proximal articulations, such as the shoulder and hip, would involve higher effective doses.

Easily modifiable determinants of radiation exposure (i.e., effective dose) are the amount of energy used for imaging (tube voltage and current × time) and the radiosensitivity of the tissue being imaged. Accordingly, tube voltage should be reduced to the point where image contrast is optimized, and the current × time product should be titrated against image noise. Patient positioning and shielding should be optimized as well, to limit exposure to radiosensitive tissues. Our approach to experimental design in this context tends to be iterative. We work closely with our radiation physicist and radiology colleagues to determine exposure risk and balance that with our experimental protocol. We find the reference by Biswas et al.[41] relevant and extremely helpful.[3]

9.3 Image Segmentation and Model Generation

9.3.1 General Comments

Accurate bone models are required for CT-based kinematic analysis, regardless of the method ultimately used for registration, which aligns the models and yields the transformation matrix reflecting motion between sequential scan positions (Figure 9.2). The bone models are generated via standard image segmentation techniques, which are used to define the bone-soft tissue interfaces in the individual CT slice images and group them appropriately. There are an increasing number of image processing programs with sophisticated, built-in tools that facilitate the process of image segmentation and surface rendering, such as Image J (National Institutes of Health, Bethesda, Maryland, USA) and Osirix (Pixmeo, Bernex, Switzerland), both of which are open source, and commercial products such as Mimics® (Materialize, Leuven, Belgium), Amira (FEI Visualization Sciences Group, Burlington, Massachusetts, USA), and Simpleware (Simpleware Ltd., Exeter, United Kingdom).

Given the broad applicability and use of segmentation in medical image processing, numerous methods have been developed that range in complexity and

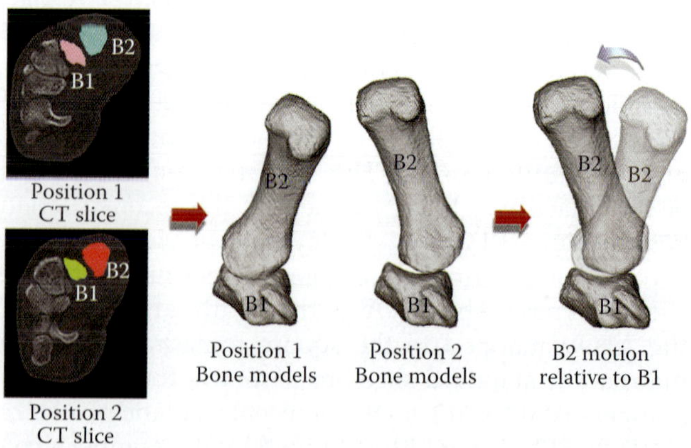

FIGURE 9.2
Overview of markerless registration. In CT image-based kinematic analysis, motion is determined via registration of bone surfaces (B1 and B2) from sequential images, based on the features and shape of the bones of interest. The analytical pipeline includes image segmentation, bone model generation, and registration. Relative motion is calculated from the transforms that reflect the motion of individual bones in the global coordinate system.

labor intensiveness.[42–44] Fully automated methods for segmenting large bones can be effective, but they have remained elusive for small bones due to the limits of clinical CT image quality and normal variations in bone density. While the surface of dense cortical bone is relatively easy to identify computationally, the problem is more challenging around joints (i.e., in the metaphysis and epiphysis) and the small bones in the wrist, hands, and feet. Here, the bones are in close proximity, and the thinness of the subchondral bone and the vascular penetrations reduce contrast, resulting in leakage through weak edges and the merging of neighboring bones.[45] Accordingly, most bone segmentation for CT-based kinematic analysis of small bones is done using manual, or manually assisted, methods.

9.3.2 Manual Segmentation

Manual segmentation is conceptually the simplest method for the generation of bone models from CT image data and historically has been considered the gold standard against which other methodologies are compared. Essentially, it involves tracing of the individual bone contours on sequential slice images and then grouping the related contours into their respective bones. Disadvantages of manual segmentation are that it is labor intensive and subjective, which can lead to intraobserver and interobserver variability. However, when performed carefully, manual segmentation can be accurate and effective. A recent comparison of phalangeal bone models generated via manual segmentation of CT scans and surface laser scanning indicates that manual segmentation can be performed with subvoxel accuracy. Specifically, the average distance between bone models generated from CT scans acquired at 0.34 × 0.34 × 0.40 mm and models generated via 0.2 mm resolution surface laser scanning was 0.20 mm.[46]

9.3.3 Semiautomated Segmentation

A step up from manual segmentation, semiautomated segmentation involves limited manual editing of images segmented computationally. Semiautomated segmentation allows for the incorporation of high-level expert knowledge, especially when local image contrast is low or where poor image resolution or pathology causes adjacent bones to merge, such as in osteoarthritis.

Any of the methods described for the automated segmentation of bone from CT scans can be implemented for semiautomated segmentation, either alone or in combination with one another. Several of the most common include global and local thresholding,[47,48] edge detection,[48] region growing,[15,49,50] and deformable models.[51] The choice of method, or sequence of methods, will depend on image quality and the anatomy of interest. Each has strengths and weaknesses. For example, global thresholding is simple and computationally efficient, but error increases as density varies within the structure of interest. Region growing has similar limitations. Edge detection works well where there are large intensity gradients at the bone margins but can lead to merging of bones where contrast is low. Regardless of the method or methods used, the results obtained via automatic segmentation must be reviewed slice by slice and manually edited by a trained operator. In the end, we recommend optimizing the segmentation workflow using representative image data sets and then implementing it consistently in the study scans.

9.3.4 Bone Surface Generation and Export

Once the margins of the bones of interest are defined via segmentation, 3-D bone surface models are generated for export and registration. Early methods used binary edge detection to generate bone contours, which were then exported before being grouped and analyzed using custom-written code.[11,52] Surface rendering is largely automated in the more sophisticated commercial and open-source software packages. For example, in Mimics, 3-D bone volumes are created from fully segmented image stacks with a single function ("Calculate 3D"). The results are tessellated surface renderings consisting of triangular mesh elements connecting an underlying point cloud. Similarly, export is typically a single-click process, to any of several usable file formats (i.e., STL, VRML, etc.).

9.3.5 Example Segmentation: Hand and Wrist

In the wrist and hand, where the anatomy is complex and the images are often low contrast, we use Mimics software and a segmentation sequence that involves global thresholding, region growing, and both manual and automated editing.

The first step is global thresholding, which produces a single 3-D mask that includes all of the bones in the image. The global threshold value is selected based on histograms of the image tomograms. The wrist contains few tissues with distinct densities, so the image histogram is essentially bimodal, with overlapping peaks for soft tissue and bone.[11] Numerous criteria have been proposed for selecting optimal thresholds.[53] We have successfully thresholded at a value equal to the mean of the peak bone intensity and the intersection of the soft tissue and bone intensities,[11] as well as Mimics' predefined bone threshold, with manual adjustment depending on imaging parameters.[6,54,55] Individual carpal bones are then separated via manual erasure of areas where neighboring bones have merged, followed by region growing to identify individual bone masks, and Boolean subtraction. Dilation and erosion are used to fill small (one- to two-pixel) discontinuities in the perimeter of the bone mask, and larger bounded holes are filled automatically to yield homogeneous masks that match the original bone boundaries.

9.3.6 Smoothing

The raw segmented bone volumes typically appear pixelated because the bone contours are generated via binary assignment of the image voxels during the segmentation process. For visualization purposes, some measure of smoothing is often performed to remove noise.[46] Smoothing is not necessary for kinematic calculations, however, as the common registration algorithms accommodate unsmoothed images. Although there are no published studies evaluating the effect of smoothing on kinematic calculations, in our experience, the effect of modest smoothing is minimal. Modest smoothing yields only small changes in bone centroid location and inertia axis orientation (unpublished data). Aggressive smoothing can, however, affect bone volume and shape calculations and can obscure anatomic features that may be of interest, such as the shape and margins of the articular surfaces, and potential ligament insertion sites. These features are important for modeling and estimation of joint contact and ligament strains. If smoothing is used in the analytic pipeline, then its effects on the outcome variables of interest should be thoroughly evaluated.

Smoothing may be introduced at any step in the segmentation process, deliberately or inadvertently. Deliberate smoothing of the binary image can be performed before the surface models are generated,[46] as the models are created (Mimics incorporates smoothing into its "Calculate 3D" tool), or on the exported models using mesh editing tools such as Geomagic Studio® (Geomagic USA, Morrisville, North Carolina). Smoothing can also occur as an inadvertent by-product of the automated or manual segmentation process. Certain operations, such as the creation of high-resolution contours from thresholded image data or contour filling operations, may involve interpolation, which by definition has a smoothing effect. It is important to understand how smoothing affects important bone features and how these changes might influence any outcome measures of interest, for example, with a parametric analysis of the unsmoothed and the range of smoothing values under consideration.

9.4 Registration—Calculating Kinematics

9.4.1 Markerless Registration for Sequential Analysis

Image-based markerless bone registration (MBR) was first described in 1988 as a noninvasive method for rigorously evaluating motion of the carpal bones in the wrist[56] and then was refined in the late 1990s for practical implementation.[3,11] In MBR, the unique shape and features of the bones are the registered elements, as opposed to fiducial points employed for marker-based and radiostereometric analysis, and motion is defined as the relative difference in pose positions (Figure 9.2). Surface fitting is a commonly used method for bone registration.[57,58] Alternative methods include inertia registration,[11,56] which is more computationally efficient, and voxel-based methods, which have been shown to provide better accuracy with less segmentation effort.[59]

Regardless of the method employed for registration, the output is a series of transforms (T) that map the bone surface models between positions:

$$x_n = T(x_A) \tag{9.1}$$

where $x_n = (x_n, y_n, z_n)$ and $x_A = (x_A, y_A, z_A)$ correspond to the same points on the surface but at different positions, n and A. For skeletal kinematics, bone deformations are assumed to be negligible, and the transformations can be expressed in terms of rigid body rotations and translations:

$$x_n = \boldsymbol{R}_{An} x_A + \boldsymbol{t}_{An} \tag{9.2}$$

where $\boldsymbol{R}_{An}$ is the 3 × 3 rotation matrix and $\boldsymbol{t}_{An}$ is 1 × 3 translation vector describing motion from position A to position n. It is important to note that these transforms are expressed in terms of the global coordinate system defined by the imaging system.

9.4.2 Surface Registration

The implementation of surface registration for skeletal kinematics has its roots in the more general problem of medical image registration, where the spatial integration of complementary information from different modalities (i.e., CT or MRI and PET) can be useful for diagnosis and treatment.[57,60,61] For skeletal kinematics, point or model-based methods are used for registration, as all (or most) of the surface points are available to drive the alignment. Essentially, surface-based registration seeks the values of $\boldsymbol{R}$ and $\boldsymbol{t}$ that minimize:

$$\min_{R,t} \sum_{i=1}^{N} \left\| x_{n_i} - (\boldsymbol{R}x_{A_i} + \boldsymbol{t})^2 \right\| \tag{9.3}$$

where there is one-to-one correspondence between x_{A_i} and x_{n_i}, and N is the number of points on the surface.

In practice, there is not one-to-one correspondence between points on the bone surface models. Accordingly, registration is performed by minimizing a global metric of the distance between the surfaces. There are numerous methods for rigid model-based registration.[60,61] Two common and widely implemented methods are those based on the Iterative Closest Point (ICP) algorithm described by Besl and McKay[58] and the "hat-and-head" method described by Pelizzari et al.[57] With the ICP methods, the mean-square distance between the closest point pairs is minimized,[58,62,63] while with the hat-and-head method, the sum of the distances between the surfaces of a "hat" object and "head" object along a radial from the head centroid is minimized.[57,64]

Surface registration is robust and widely used for skeletal kinematic analysis.[4,11,15,65] It is a common function available in commercial data visualization and modeling software (e.g., Amira, Geomagic Studio [Geomagic], Analyze [Biomedical Imaging Resource, Rochester, Minnesota]), -although the specific algorithm used is rarely described. Surface registration is also available as a feature in several open-source medical image analysis products, such as 3D Slicer, Osirix, and Amide (see also http://www.uiowa.edu/mihpclab/mias.html), and it can also be implemented fairly easily in custom-written code. A significant disadvantage of surface registration is that it requires segmentation of the bones of interest at each scan position.

9.4.3 Inertia Registration

An alternative to surface registration is inertia registration.[11,56,60] Inertia registration is less computationally intensive than surface fitting, as it involves registration of the centroids and principal axes of inertia to yield the rotation and translation matrices, as opposed to the full point cloud used to represent the bone surfaces. Bone volumes, centroids, and mass moments of inertia can be efficiently calculated using discretized versions of Green's theorem, which yields the area and centroid of plane figures via integration of the perimeter.[52] Disadvantages of inertia registration are that it requires segmentation of all of the bones of interest at each position, and its use is limited to bones that are completely imaged in all positions.

9.4.4 Voxel-Based Registration

The most advanced registration methods currently used for skeletal kinematics are voxel based, in that they use the full image data set for registration, rather than derived objects generated via segmentation.[15,59,66] As with surface registration, there are numerous methods for full-image-content-based voxel registration.[60] However, few have been implemented for skeletal kinematic analysis. Conceptually, the simplest among these is gray value matching, which minimizes intensity differences voxel by voxel across the volume of interest.[15,66] Depending on the implementation, the bones may be segmented in each image[65] or, for more automated methods, only on a single index scan.[66] An additional full-image voxel-based method that has been used for CT-based skeletal kinematics is chamfer matching.[66]

We currently use tissue-classified distance fields for carpal bone registration, which is a hybrid approach that combines the strengths of segmentation-based and voxel property methods.[59] In our method, the bones are extracted from a high-quality reference scan (typically anatomic neutral), and superresolution distance fields are generated from each image volume using unsupervised tissue classification. The distance fields and bone models are then used to automatically register the bones in each of the CT volumes for the motion sequence. The advantages of our method are that it requires limited segmentation (only a single position), it is largely automated, and it yields results with subvoxel accuracy. A significant limitation of the method is that it is computationally expensive, so to this point it has only been implemented to run on multimode computer clusters.

9.4.5 Calculating Kinematics

The outcome of the registration process is a series of transformation matrices that describe the motion of each bone from its reference position to each of its subsequent positions, in terms of the CT scanner's global coordinate system. Joint kinematics reflects the *relative* change in location and posture of the bones at each scanned position. Accordingly, once the global motions for the bones and positions of interest have been determined, transforms for relative motion ($\boldsymbol{M}$) are computed. This is done by multiplying the motion transform for one bone with the inverse of the transform of the other:

$$\boldsymbol{M}_{An} = \boldsymbol{T}_{b1An}^{-1}\boldsymbol{T}_{b2An} \tag{9.4}$$

where $\boldsymbol{M}_{An}$ is a 4 × 4 matrix representing the motion of bone 2 with respect to bone 1 between positions A and n in the global reference frame, and $\boldsymbol{T}_{b1An}$ and $\boldsymbol{T}_{b2An}$ are the transforms for bone 1 and bone 2 in the global (CT) reference frame. Motion ($\boldsymbol{M}$) is calculated in terms of the global reference frame and must be transformed to a local coordinate system for presentation and interpretation. Typically, motion of a distal segment is calculated with respect to a proximal segment and reported in terms of an anatomically based coordinate system.

9.5 Data Presentation

9.5.1 General Comments

To be of practical use in describing kinematics at a specific joint, the data must be presented in terms of an anatomically meaningful local reference frame and parameterized in a clinically interpretable way. There are several common systems for reporting kinematic data, which differ from one another in complexity and interpretability (most commonly Cardan/Euler angles, helical axes, and joint coordinate system [JCS]). The choice of reporting methods is influenced by the joint and motion being studied, the intended audience (clinical vs. basic science), and the rigor demanded by the research question being asked. Six-degree-of-freedom motion occurs in most, if not all, synovial joints, though not all questions require rigorous 3-D analysis.

9.5.2 Coordinate Systems

Clinically, joint motion is described in terms of rotations (and translations) in the sagittal (flexion/extension), frontal (abduction/adduction), and transverse planes of the body, and with few exceptions—notably the thumb carpometacarpal (CMC) joint, which is oblique to the palm—kinematic data are reported in the same general reference frame. This requires transformation of the kinematic data from the global CT reference frame to a joint-specific reference frame (Figure 9.3). This is done by embedding coordinate systems in the bone (or bones) of interest and then transforming the global motion data into the coordinate system for that bone (or those bones). Typically, motion of a distal segment is reported with respect to a mathematically fixed proximal segment (i.e., tibia with respect to femur), so it is most convenient to embed the coordinate system in the proximal segment. Cartesian coordinate systems (x, y, z) are used almost exclusively, consistent with the standard anatomic coordinate system defined by the intersections of the frontal, midsagittal, and transverse planes.[67–69]

Ideally, the embedded coordinate systems are aligned with the principal rotation axes of the joint, to simplify interpretation of the output rotations and translations. Typically, anatomic features on the bone, and/or the anatomic planes, are used to locate the first two orthogonal axes, and the third is generated via completion of the right-handed coordinate system.[4,55,70–72] The anatomic features used to define the coordinate system can be selected manually,[73,74] or the coordinate systems can be generated using semiautomated feature-fitting algorithms.[4,55,72,75] An alternate strategy uses joint motion to define the coordinate system.[76] Regardless of the method selected, accurate coordinate system definition is important. Failure to align the segment coordinate systems with anatomical axes of a joint can result in "kinematic cross talk," or the incorrect reporting of components of rotation about one axis as rotation about another (e.g., components of knee extension reported as external rotation).[77,78]

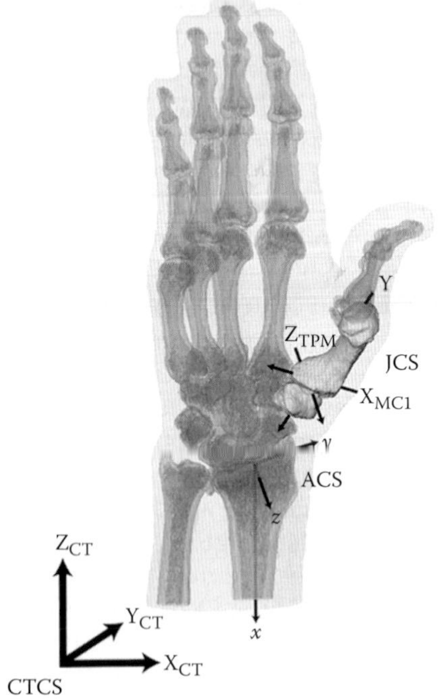

FIGURE 9.3
Coordinate system selection. The registration of the bones in sequential CT images yields motion transforms in the CT coordinate system (CTCS). To be of use in describing kinematics at a specific joint, the data must transformed before they can be presented in terms of an anatomically meaningful local reference frame, such as the anatomic coordinate system (ACS) embedded here in the distal radius. The ISB-recommended joint coordinate system (JCS) utilizes two segment-fixed axes (here in the trapezium and base of the first metacarpal) and a third floating axis that is perpendicular to both segment-fixed axes.

9.5.3 Kinematic Paramaterization: Euler Angles, JCS, and Helical Axis of Motion

The most commonly used method for describing three-dimensional joint attitude (rotation) is with the use of Euler angles. Essentially, Euler angles are an ordered sequence of rotations (typically α, β, γ) about the three axes of the selected anatomic coordinate system that reflect the change in position of the distal segment. The advantages of Euler angles are that they are conceptually simple and easy to understand. However, they have several limitations that make them best suited for describing relatively simple motions largely aligned with the coordinate system. Chief among them is the fact that Euler angles are sequence dependent; that is, the magnitudes of the angular rotations can differ depending on the order in which they are decomposed.[79–81] Practically, this means that for any given joint rotation, there are six potential correct solutions for the sequential rotations about the x, y, and z coordinate axes (x-y-z, x-z-y, y-x-z, y-z-x, z-x-y, and z-y-x). If the rotation is primarily about one coordinate axis, then the six solutions are essentially identical. However, if the rotation is about an oblique axis, the calculated rotations can vary significantly. For example, for a total rotation of 30° about an axis that lies halfway between the y and z coordinate axes, R_x ranges from −4.1° to 5.12°, R_y ranges from −19.4° to 22.2°, and R_z ranges from 22.2° to 20.7°, depending on the sequence selected. If Euler angles are used, it is crucial that the sequence used to decompose the transformation matrix also be reported. The additional limitations with Euler angles are that translations must be described separately, they contain no information on the center or axis of rotation, and they are susceptible to gimbal lock.[81,82] Because of these limitations, in many cases, it is preferable to use the JCS or helical axis parameters when reporting 3-D kinematics.

The Standardization and Terminology Committee (STC) of the International Society of Biomechanics (ISB) has recommended use of the JCS for the upper extremity, lower extremity, and spine,[68,69] based on the approach first described by Grood and Suntay in 1983.[70] With the Grood and Suntay method, the two principal rotations of the joint are reported as rotations about two segment-fixed axes, while the third, more minor rotation is reported in terms of a floating axis that is perpendicular to both segment-fixed axes[70] (Figure 9.3). Advantages of the method are that it yields results that can be readily communicated in terms of commonly used clinical reference frames and it incorporates both rotations *and* translations. The disadvantages of the Grood and Suntay method are that the JCS is not mutually orthogonal and that it is susceptible to gimbal lock and kinematic cross talk.[78] Nevertheless, it has been widely adopted, especially where complex kinematic data need to be communicated to a clinical audience.

In our experience, the reporting of complete 3-D kinematics is most efficiently accomplished using helical-axis-of-motion (HAM) parameters.[79,83] HAM parameters completely describe 3-D motion in terms of a single vector that is both the axis of rotation and direction of translation for an object as it moves from one position to the next. HAM axes are vector representations of motion, providing a general solution usable for all joints, and it is not susceptible to gimbal lock. HAM parameters include the position and direction of the axis in 3-D space, the angle of rotation about the axis (ϕ), and the amount of translation along the axis (d). Not only do the HAM parameters completely describe the rigid body transform, but they also can be readily reduced to describe simple motions. For example, when motion is planar, the intersection of the rotation axis with the plane of the motion is the instantaneous center of rotation. For complex motions, this axis will move in space, and the set of axes computed is often referred to as the envelope of helical axes.[84] A benefit to using the HAM rotation axis is that it describes the physical location of the rotation axes and is readily visualized. The disadvantages of HAM axes are that they can be difficult to interpret anatomically or reduce to the simple clinical descriptions of motion, and they are prone to error for very small rotations (<3°).

9.5.4 Joint Space and Ligament Elongations

As noted in the introduction, the image data sets obtained during CT-based kinematic analysis allow for the exploration of secondary outcome measures, such as bone morphology, joint spacing, and ligament lengths. The measurement of joint space from weight-bearing x-rays of the knee remains the only validated assessments of osteoarthritis progression.[85] Joint space can be visualized and quantified accurately using images obtained from CT-based kinematic analysis, even in complex joints,[86] and if enough positions are imaged, reasonable models of the cartilage surfaces can be computed.[87] Similarly, once ligament insertions are identified on the bone surface models, ligaments can be modeled, and ligament elongations can be computed from the relative bone kinematics.[88] Ligament elongations can yield important insight into ligament and joint function, although at this point, information on ligament strains cannot be determined. Without a priori knowledge of ligament strain in at least one position, these calculated elongations cannot directly yield corresponding strain values.

9.6 Accuracy

In general, the accuracy of CT-based kinematic analysis is comparable to—or slightly better than—the resolution of the CT images on which the analysis is based. Accuracy has been assessed via the direct comparison of kinematic results generated via markerless registration of cadaver wrist bones and clusters of radiopaque spherical markers affixed to those bones.[59,89] In the first, which evaluated two surface registration methods, as well as inertia registration, translation errors were generally less than 1 mm, and rotation errors were less than 2° for 1-mm-thick images acquired at 12 cm RFOV (0.234 × 0.234 × 1.0 mm^3 image voxel dimension). In the second, which used the same experimental construct to evaluate the accuracy of tissue-classified distance fields, translation errors were less than 0.3 mm, and rotation errors were less than 0.4° for 0.31 × 0.31 × 1 mm^3 images (and less than 0.4 mm and 0.7° for 0.94 × 0.94 × 1 mm^3 images).[59] The average rotational and translational errors for 4-D analysis of thumb CMC motion have been reported to be less than 0.6° and 0.7 mm (though they vary slightly by bone).[15] Despite these figures, accuracy should be evaluated by each research group, for each of their different potential applications (i.e., wrist vs. knee vs. foot/ankle). Accuracy predictably varies as a function of bone and specific technique (i.e., segmentation technique and registration method), and what is acceptable for one application (kinematics) may be unsuitable for another (i.e., joint space analysis).

9.7 Conclusion

The ability to measure complex skeletal motion noninvasively using CT image-based markerless registration techniques has led to important advances in the understanding of the anatomy and function of joints in the upper extremity, lower extremity, and spine. Although biplane videoradiography has emerged as a powerful tool for the dynamic assessment of in vivo skeletal kinematics,[90–92] the infrastructure for CT image-based kinematic analysis is more broadly available, and it is still superior for the investigation of complex, overlapping, multibone articulations. Going forward, advances in CT and image processing technology should make it increasingly easy to implement.

Acknowledgments

Over the years, funding for the refinement of the MBR techniques implemented in our lab has been via grants from the National Institutes of Health (National Institute of Child Health and Human Development [NICHD] HD052127, National Institute of Arthritis and Musculoskeletal and Skin Diseases [NIAMS] AR053648 and AR059185). Partial salary support for the writing of this chapter was provided by NIH AR059185.

References

1. Belsole, R. J. et al. Scaphoid orientation and location from computed, three-dimensional carpal models. *Orthop Clin North Am* **17,** 505–510 (1986).
2. Blankenhorn, B. D. et al. Carpal kinematics after proximal row carpectomy. *J Hand Surg* **32,** 37–46 (2007).
3. Feipel, V. & Rooze, M. Three-dimensional motion patterns of the carpal bones: An in vivo study using three-dimensional computed tomography and clinical applications. *Surg Radiol Anat SRA* **21,** 125–131 (1999).
4. Crisco, J. J. et al. Carpal bone postures and motions are abnormal in both wrists of patients with unilateral scapholunate interosseous ligament tears. *J Hand Surg Am* **28,** 926–937 (2003).
5. Crisco, J. J. et al. In vivo radiocarpal kinematics and the dart thrower's motion. *J Bone Joint Surg Am* **87,** 2729–2740 (2005).
6. Leventhal, E. L., Moore, D. C., Akelman, E., Wolfe, S. W. & Crisco, J. J. Carpal and forearm kinematics during a simulated hammering task. *J Hand Surg Am* **35,** 1097–1104 (2010).
7. Leventhal, E. L., Moore, D. C., Akelman, E., Wolfe, S. W. & Crisco, J. J. Conformational changes in the carpus during finger trap distraction. *J Hand Surg Am* **35,** 237–244 (2010).
8. Leventhal, E. L. et al. Interfragmentary motion in patients with scaphoid nonunion. *J Hand Surg Am* **33,** 1108–1115 (2008).
9. Rainbow, M. J., Crisco, J. J., Moore, D. C. & Wolfe, S. W. Gender differences in capitate kinematics are eliminated after accounting for variation in carpal size. *J Biomech Eng* **130,** 041003 (2008).
10. Rainbow, M. J. et al. In vivo kinematics of the scaphoid, lunate, capitate, and third metacarpal in extreme wrist flexion and extension. *J Hand Surg* **38,** 278–288 (2013).
11. Crisco, J. J., McGovern, R. D. & Wolfe, S. W. Noninvasive technique for measuring in vivo three-dimensional carpal bone kinematics. *J Orthop Res Off Publ Orthop Res Soc* **17,** 96–100 (1999).
12. Neu, C. P., Crisco, J. J. & Wolfe, S. W. In vivo kinematic behavior of the radio-capitate joint during wrist flexion-extension and radio-ulnar deviation. *J Biomech* **34,** 1429–1438 (2001).
13. Tay, S.-C. et al. Four-dimensional computed tomographic imaging in the wrist: Proof of feasibility in a cadaveric model. *Skeletal Radiol* **36,** 1163–1169 (2007).
14. Leng, S. et al. Dynamic CT technique for assessment of wrist joint instabilities. *Med Phys* **38 Suppl 1,** S50 (2011).
15. Goto, A., Leng, S., Sugamoto, K., Cooney, W. P. 3rd, Kakar, S. & Zhao, K. In vivo pilot study evaluating the thumb carpometacarpal joint during circumduction. *Clin Orthop Relat Res* **472,** 1106–1113 (2014).
16. Shores, J. T., Demehri, S. & Chhabra, A. Kinematic "4 dimensional" CT imaging in the assessment of wrist biomechanics before and after surgical repair. *Eplasty* **13,** e9 (2013).
17. Moore, D. C. et al. Three-dimensional in vivo kinematics of the distal radioulnar joint in malunited distal radius fractures. *J Hand Surg Am* **27,** 233–242 (2002).
18. Crisco, J. J. et al. Effects of distal radius malunion on distal radioulnar joint mechanics—An in vivo study. *J Orthop Res Off Publ Orthop Res Soc* **25,** 547–555 (2007).
19. Tay, S. C. et al. A method for in-vivo kinematic analysis of the forearm. *J Biomech* **41,** 56–62 (2008).
20. Miyake, J., Moritomo, H., Kataoka, T., Murase, T. & Sugamoto, K. In vivo three-dimensional motion analysis of chronic radial head dislocations. *Clin Orthop* **470,** 2746–2755 (2012).
21. Baeyens, J. P. et al. In vivo 3D arthrokinematics of the proximal and distal radioulnar joints during active pronation and supination. *Clin Biomech* **21 Suppl 1,** S9–S12 (2006).
22. Kim, H.-J. et al. Influence of forearm rotation on proximal radioulnar joint congruency and translational motion using computed tomography and computer-aided design technologies. *J Hand Surg* **36,** 811–815 (2011).
23. Salem, W., Lenders, C., Mathieu, J., Hermanus, N. & Klein, P. In vivo three-dimensional kinematics of the cervical spine during maximal axial rotation. *Man Ther* **18,** 339–344 (2013).

24. Salem, W. & Klein, P. In vivo 3D kinematics of the cervical spine segments during pre-manipulative positioning at the C4/C5 level. *Man Ther* **18,** 321–326 (2013).
25. Ledoux, W. R., Rohr, E. S., Ching, R. P. & Sangeorzan, B. J. Effect of foot shape on the three-dimensional position of foot bones. *J Orthop Res Off Publ Orthop Res Soc* **24,** 2176–2186 (2006).
26. Kawashima, K., Tomita, T., Tamaki, M., Murase, T., Yoshikawa, H. & Sugamoto, K. In vivo three-dimensional motion analysis of osteoarthritic knees. *Mod Rheumatol* **23,** 646–652 (2013).
27. Carelsen, B. et al. 4D rotational X-ray imaging of wrist joint dynamic motion. *Med Phys* [illegible], [illegible]–2776 (2005).
28. Carelsen, B. et al. Detection of in vivo dynamic 3-D motion patterns in the wrist joint. *IEEE Trans Biomed Eng* **56,** 1236–1244 (2009).
29. Foumani, M. et al. In-vivo three-dimensional carpal bone kinematics during flexion-extension and radio-ulnar deviation of the wrist: Dynamic motion versus step-wise static wrist positions. *J Biomech* **42,** 2664–2671 (2009).
30. Tay, S. C. et al. Understanding the relationship between image quality and motion velocity in gated computed tomography: Preliminary work for 4-dimensional musculoskeletal imaging. *J Comput Assist Tomogr* **32,** 634–639 (2008).
31. Rydberg, J., Liang, Y. & Teague, S. D. Fundamentals of multichannel CT. *Radiol Clin North Am* **41,** 465–474 (2003).
32. Sprawls, P. AAPM tutorial. CT image detail and noise. *Radiogr Rev Publ Radiol Soc North Am Inc* **12,** 1041–1046 (1992).
33. Flohr, T. G. et al. Multi-detector row CT systems and image-reconstruction techniques. *Radiology* **235,** 756–773 (2005).
34. Halilaj, E., Rainbow, M. J., Got, C., Schwartz, J. B., Moore, D. C., Weiss, A.-P. C., Ladd, A. L. & Crisco, J. J. In vivo kinematics of the thumb carpometacarpal joint during three isometric functional tasks. *Clin Orthop Relat Res* **472,** 1114–1122 (2014).
35. Moore, D. C., Crisco, J. J., Trafton, T. G. & Leventhal, E. L. A digital database of wrist bone anatomy and carpal kinematics. *J Biomech* **40,** 2537–2542 (2007).
36. Boas, E. & Fleischmann, D. CT artifacts: Cases and reduction techniques. *Imaging Med* **4,** 229–240 (2012).
37. Oka, K. et al. Accuracy analysis of three-dimensional bone surface models of the forearm constructed from multidetector computed tomography data. *Int J Med Robot Comput Assist Surg MRCAS* **5,** 452–457 (2009).
38. Cardis, E. et al. The 15-country collaborative study of cancer risk among radiation workers in the nuclear industry: Estimates of radiation-related cancer risks. *Radiat Res* **167,** 396–416 (2007).
39. Cardis, E. et al. Risk of cancer after low doses of ionising radiation: Retrospective cohort study in 15 countries. *BMJ* **331,** 77 (2005).
40. Brenner, D. J. & Hall, E. J. Computed tomography—An increasing source of radiation exposure. *N Engl J Med* **357,** 2277–2284 (2007).
42. Biswas, D. et al. Radiation exposure from musculoskeletal computerized tomographic scans. *J Bone Joint Surg Am* **91,** 1882–1889 (2009).
42. Pal, N. R. & Pal, S. K. A review on image segmentation techniques. *Pattern Recognit* **26,** 1277–1294 (1993).
43. Pham, D. L., Xu, C. & Prince, J. L. Current methods in medical image segmentation. *Annu Rev Biomed Eng* **2,** 315–337 (2000).
44. Heimann, T. & Meinzer, H.-P. Statistical shape models for 3D medical image segmentation: A review. *Med Image Anal* **13,** 543–563 (2009).
45. Calder, J., Tahmasebi, A. M. & Mansouri, A.-R. A variational approach to bone segmentation in CT images. In *SPIE Med. Imaging 2011 Image Process* March 9, 2011, 79620B, 15 pp. (2011). doi:10.1117/12.877355.
46. DeVries, N. A. et al. Validation of phalanx bone three-dimensional surface segmentation from computed tomography images using laser scanning. *Skeletal Radiol* **37,** 35–42 (2008).
47. Zhang, J., Yan, C.-H., Chui, C.-K. & Ong, S.-H. Fast segmentation of bone in CT images using 3D adaptive thresholding. *Comput Biol Med* **40,** 231–236 (2010).
48. Rathnayaka, K., Sahama, T., Schuetz, M. A. & Schmutz, B. Effects of CT image segmentation methods on the accuracy of long bone 3D reconstructions. *Med Eng Phys* **33,** 226–233 (2011).
49. Kang, Y., Engelke, K. & Kalender, W. A. A new accurate and precise 3-D segmentation method for skeletal structures in volumetric CT data. *IEEE Trans Med Imaging* **22,** 586–598 (2003).
50. Bryce, C. D. et al. Validation of three-dimensional models of in situ scapulae. *J Shoulder Elbow Surg* **17,** 825–832 (2008).
51. Sebastian, T. B., Tek, H., Crisco, J. J. & Kimia, B. B. Segmentation of carpal bones from CT images using skeletally coupled deformable models. *Med Image Anal* **7,** 21–45 (2003).
52. Crisco, J. J. & McGovern, R. D. Efficient calculation of mass moments of inertia for segmented homogeneous three-dimensional objects. *J Biomech* **31,** 97–101 (1998).
53. Batenburg, K. J. & Sijbers, J. Optimal threshold selection for tomogram segmentation by projection distance minimization. *IEEE Trans Med Imaging* **28,** 676–686 (2009).
54. Kamal, R. N., Rainbow, M. J., Akelman, E. & Crisco, J. J. In vivo triquetrum-hamate kinematics through a simulated hammering task wrist motion. *J Bone Joint Surg Am* **94,** e85 (2012).
55. Halilaj, E., Rainbow, M. J., Got, C. J., Moore, D. C. & Crisco, J. J. A thumb carpometacarpal joint coordinate system based on articular surface geometry. *J Biomech* **46,** 1031–1034 (2013).
56. Belsole, R. J. et al. Mathematical analysis of computed carpal models. *J Orthop Res* **6,** 116–122 (1988).
57. Pelizzari, C. A., Chen, G. T., Spelbring, D. R., Weichselbaum, R. R. & Chen, C. T. Accurate three-dimensional registration of CT, PET, and/or MR images of the brain. *J Comput Assist Tomogr* **13,** 20–26 (1989).
58. Besl, P. J. & McKay, N. D. A method for registration of 3-D shapes. *IEEE Trans Pattern Anal Mach Intell* **14,** 239–256 (1992).
59. Marai, G. E., Laidlaw, D. H. & Crisco, J. J. Super-resolution registration using tissue-classified distance fields. *IEEE Trans Med Imaging* **25,** 177–187 (2006).

60. Maintz, J. B. & Viergever, M. A. A survey of medical image registration. *Med Image Anal* **2,** 1–36 (1998).
61. Audette, M. A., Ferrie, F. P. & Peters, T. M. An algorithmic overview of surface registration techniques for medical imaging. *Med Image Anal* **4,** 201–217 (2000).
62. Zhang, Z. Iterative point matching for registration of free-form curves and surfaces. *Int J Comput Vis* **13,** 119–152 (1994).
63. Lavallee, S. & Szeliski, R. Recovering the position and orientation of free-form objects from image contours using 3D distance maps. *IEEE Trans Pattern Anal Mach Intell* **17,** 378–390 (1995).
64. Chen, G. T. & Pelizzari, C. A. Image correlation techniques in radiation therapy treatment planning. *Comput Med Imaging Graph Off J Comput Med Imaging Soc* **13,** 235–240 (1989).
65. Goto, A. et al. In vivo three-dimensional wrist motion analysis using magnetic resonance imaging and volume-based registration. *J Orthop Res* **23,** 750–756 (2005).
66. Snel, J. G. et al. Quantitative in vivo analysis of the kinematics of carpal bones from three-dimensional CT images using a deformable surface model and a three-dimensional matching technique. *Med Phys* **27,** 2037–2047 (2000).
67. Wu, G. & Cavanagh, P. R. ISB recommendations for standardization in the reporting of kinematic data. *J Biomech* **28,** 1257–1261 (1995).
68. Wu, G. et al. ISB recommendation on definitions of joint coordinate system of various joints for the reporting of human joint motion—part I: Ankle, hip, and spine. International Society of Biomechanics. *J Biomech* **35,** 543–548 (2002).
69. Wu, G. et al. ISB recommendation on definitions of joint coordinate systems of various joints for the reporting of human joint motion—part II: Shoulder, elbow, wrist and hand. *J Biomech* **38,** 981–992 (2005).
70. Grood, E. S. & Suntay, W. J. A joint coordinate system for the clinical description of three-dimensional motions: Application to the knee. *J Biomech Eng* **105,** 136–144 (1983).
71. Kobayashi, M. et al. Normal kinematics of carpal bones: A three-dimensional analysis of carpal bone motion relative to the radius. *J Biomech* **30,** 787–793 (1997).
72. Miranda, D. L., Rainbow, M. J., Leventhal, E. L., Crisco, J. J. & Fleming, B. C. Automatic determination of anatomical coordinate systems for three-dimensional bone models of the isolated human knee. *J Biomech* **43,** 1623–1626 (2010).
73. Conti, G., Cristofolini, L., Juszczyk, M., Leardini, A. & Viceconti, M. Comparison of three standard anatomical reference frames for the tibia–fibula complex. *J Biomech* **41,** 3384–3389 (2008).
74. Cooney, W. P. III, Lucca, M. J., Chao, E. Y. & Linscheid, R. L. The kinesiology of the thumb trapeziometacarpal joint. *J Bone Joint Surg Am* **63,** 1371–1381 (1981).
75. Amadi, H. O., Hansen, U. N. & Bull, A. M. J. A numerical tool for the reconstruction of the physiological kinematics of the glenohumeral joint. *Proc Inst Mech Eng [H]* **223,** 833–837 (2009).
76. Ferreira, L. M., King, G. J. W. & Johnson, J. A. Motion-derived coordinate systems reduce inter-subject variability of elbow flexion kinematics. *J Orthop Res Off Publ Orthop Res Soc* **29,** 596–601 (2011).
77. Piazza, S. J. & Cavanagh, P. R. Measurement of the screw-home motion of the knee is sensitive to errors in axis alignment. *J Biomech* **33,** 1029–1034 (2000).
78. Cheze, L., Dumas, R., Comtet, J. J., Rumelhart, C. & Fayet, M. A joint coordinate system proposal for the study of the trapeziometacarpal joint kinematics. *Comput Methods Biomech Biomed Eng* **12,** 277–282 (2009).
79. Woltring, H. J. Representation and calculation of 3D joint movement. *Hum Mov Sci* **10,** 603–616 (1991).
80. Cole, G. K., Nigg, B. M., Ronsky, J. L. & Yeadon, M. R. Application of the joint coordinate system to three-dimensional joint attitude and movement representation: A standardization proposal. *J Biomech Eng* **115,** 344–349 (1993).
81. Karduna, A. R., McClure, P. W. & Michener, L. A. Scapular kinematics: Effects of altering the Euler angle sequence of rotations. *J Biomech* **33,** 1063–1068 (2000).
82. Woltring, H. J. 3-D attitude representation of human joints: A standardization proposal. *J Biomech* **27,** 1399–1414 (1994).
83. Panjabi, M. M., Krag, M. H. & Goel, V. K. A technique for measurement and description of three-dimensional six degree-of-freedom motion of a body joint with an application to the human spine. *J Biomech* **14,** 447–460 (1981).
84. Blankevoort, L., Huiskes, R. & de Lange, A. Helical axes of passive knee joint motions. *J Biomech* **23,** 1219–1229 (1990).
85. Strand, V. & Hochberg, M. C. *Osteoarthr Diagn Manag,* 639–649 (WB Saunders Co, Philadelphia, PA, 2001).
86. Marai, G. E. et al. Estimating joint contact areas and ligament lengths from bone kinematics and surfaces. *IEEE Trans Biomed Eng* **51,** 790–799 (2004).
87. Marai, G. E., Crisco, J. J. & Laidlaw, D. H. A kinematics-based method for generating cartilage maps and deformations in the multi-articulating wrist joint from CT images. *Conf Proc Annu Int Conf IEEE Eng Med Biol Soc IEEE Eng Med Biol Soc Conf* **1,** 2079–2082 (2006).
88. Rainbow, M. J. et al. Elongation of the dorsal carpal ligaments: A computational study of in vivo carpal kinematics. *J Hand Surg Am* **37,** 1393–1399 (2012).
89. Neu, C. P., McGovern, R. D. & Crisco, J. J. Kinematic accuracy of three surface registration methods in a three-dimensional wrist bone study. *J Biomech Eng* **122,** 528–533 (2000).
90. Tashman, S. & Anderst, W. In-vivo measurement of dynamic joint motion using high speed biplane radiography and CT: Application to canine ACL deficiency. *J Biomech Eng* **125,** 238–245 (2003).
91. Bey, M. J., Zauel, R., Brock, S. K. & Tashman, S. Validation of a new model-based tracking technique for measuring three-dimensional, in vivo glenohumeral joint kinematics. *J Biomech Eng* **128,** 604–609 (2006).
92. Brainerd, E. L. et al. X-ray reconstruction of moving morphology (XROMM): Precision, accuracy and applications in comparative biomechanics research. *J Exp Zool Part Ecol Genet Physiol* **313,** 262–279 (2010).

10

Contrast-Enhanced MicroCT Imaging

David Reece, Angela Lin, and Robert E. Guldberg

CONTENTS

The use of contrast agents with microcomputed tomography (microCT) allows for the quick, nondestructive, and quantifiable 3D visualization of a wide variety of organs and tissues. Rigorous morphometric and attenuation density analyses of microCT images have facilitated improved understanding of structure–function relationships in organs such as the heart, bone, and kidneys while also providing quantitative metrics for evaluating therapeutics for diseases such as cancer, osteoarthritis, and pulmonary disorders. This chapter introduces a range of contrast enhancement techniques and their application in combination with microCT imaging to visualizing various physiological systems.

10.1 Introduction

Understanding the normal or pathological function of biological tissues requires a detailed knowledge of the underlying tissue structure. Computed tomography (CT) provides 3-D, high-resolution imaging and is a useful tool for allowing researchers to gain more complete structural information about anatomical systems. Additionally, by using this powerful imaging modality to analyze physiological systems during developmental or pathological states, structural changes due to these processes can be correlated to relevant functional changes. These imaging methods also provide quantitative metrics that can be utilized to assess the effectiveness of potential therapeutics to retain or restore native tissue and organ structure and function.

CT was first introduced in the early 1970s[1] as a more complete way of using x-rays for medical diagnostics. CT allows radiologists to build 3-D images of the scanned anatomical region from sequential cross sections along one of the region's axes. Each cross section is reconstructed from multiple cone- or fan-beam x-ray scans taken from several incremental angles around the center axis. Because x-ray attenuation is proportional to density, reconstructed CT images provide both morphology and, with calibration, local density information for the scanned anatomical region.[2] These images can be used to three-dimensionally quantify morphological and compositional parameters such as volume and average density, respectively.

CT is heavily used in clinical settings[3]; however, as clinical CT scanners are generally restricted to voxel sizes on the order of several hundreds of microns,[4,5] their use in the laboratory has been limited. MicroCT, referring to CT imaging at resolutions on the scale of 1 to 100 microns, has thus been developed to use enhanced hardware (microfocus x-ray sources, higher-resolution detectors) and software (improved reconstruction and morphometric analysis algorithms) to visualize and analyze smaller anatomical features.[6–9] These systems and techniques provide more applicability in the preclinical setting, including small-animal imaging.

As most soft tissues attenuate x-rays at a similar level to water, contrast agents are utilized to increase the radiodensity and, therefore, visibility of the tissue.[10–12] Certain contrast agents, such as ionic contrast agents, can also be used at equilibrium as indicators of the composition of the tissue.[13,14] Contrast-enhanced microCT imaging allows for relatively quick, high-resolution imaging with the additional benefit of generating quantitative 3-D measures of tissue structures. These quantitative capabilities are not easily achieved with other methodologies such as serial histology,[15] which is the gold standard and validation method of choice for many of these applications. Magnetic resonance imaging (MRI), relying on the interactions of atomic nuclei with magnetic fields and radio waves,[16] has also been adapted for high-resolution imaging (microMRI) of various organs and tissues.[17–22]

This chapter presents various applications of contrast-enhanced microCT and describes contributions to fields that benefit from the added quantitative capabilities the technique can provide. Common contrast agent categories are first described, followed by examples of contrast-enhanced microCT applications in different physiological systems.

10.2 Contrast Agents

10.2.1 Immersion Contrast Agents

The most straightforward method of utilizing contrast agents is to immerse the sample in the agent, allowing molecules containing contrast elements to diffuse into the tissue. Contrast agents such as osmium tetroxide, ionic and nonionic iodinated solutions, lead-uranyl acetate, and phosphotungstic acid (PTA) have been used in this manner for imaging various tissues.[12,19,23–36] The majority of these studies have been limited to ex vivo imaging of tissue structures, though contrast agents can also be injected directly into the region or organ of interest in vivo.[37–42]

For ionic contrast agents, charged molecules containing contrast-enhancing elements diffuse into the tissue and partition themselves relative to the native ionic components depending on the charge of the contrast agent and the tissue. The distribution of contrast agent enhances visibility and dictates x-ray attenuation variations within

TABLE 10.1

Referenced Immersion Contrast Agents

Contrast Agent	Trade Name	Referenced Uses	References	Infiltration Time	Charge	Doses/Concentration
Gallocyanin-chromalum		General tissue stain	11	16 h	N/A	5% (w/v) in water
Iodine		General tissue stain	11,12,43	30 min–25 h	–	3–8 mg/mL
Iodine potassium iodide		General tissue stain	11,12,23	30 min–2 days	N/A	1%–7.5% (w/v) in water
Phosphotungstic acid		General tissue stain	11,12	2–16 h	N/A	0.3% (w/v) in 70% ethanol
Osmium tetroxide		General tissue stain	11,12,24,31,36	1–16 h	N/A	1%–2% (w/v) in buffer
Ioxaglate	Hexabrix	Cartilage	13,14,44–46,47–50, 51–54	20 min–36 h	–	12–320 mg I/mL
Iothalamate	Cysto Contray II	Cartilage	14,51,55,56	1–24 h	–	27–80 mg I/mL
Cationic contrast agents		Cartilage	14,51,52,56,58	1–24 h	+	8–27 mg I/mL
Iopamidol	Iopamiron	Spinal cord	19,30	14 days	N/A	100–150 mg I/mL
Lead-uranyl acetate		Bone microdamage	27	21 days	N/A	20% lead acetate and 7% uranyl acetate in 70% acetone
Diatrizoate	Hypaque	Brain	34	5 days	N/A	17–62 mg I/mL
Iohexol	Omnipaque	Brain	35	5 days	N/A	9–300 mg I/mL

Note: Table of immersion contrast agents referenced in chapter with trade name, referenced uses, referenced papers, infiltration time, charge, and dose or concentration.

the tissue. This technique has been used in cartilage to quantify tissue morphology,[13,44–49,57,58] assess the distribution of ionic molecules within the cartilage when the contrast is at equilibrium,[13,14,47–53,55,56] and detect diffusivity differences due to degeneration[43,54] (Table 10.1).

10.2.2 Perfusion Contrast Agents

Agents such as barium sulfate and lead-chromate-based silicon rubber (Microfil, Flow Tech Inc.) are used to investigate the vasculature of entire organisms or organs ex vivo.[15,59–81] Immediately postmortem, the vascular anatomy of the entire animal or an organ of interest is flushed with saline (sometimes with the addition of a vasodilator), perfused with a fixative, and, finally, perfused with the injectable contrast agent. The agent is then allowed to cure before scanning. Care needs to be taken during the injection process to avoid bursting small vessels with pressure generated in perfusing the contrast agent. Viscosity and particle size of the contrast agent can also affect the extent and uniformity of perfusion into the microvasculature.[82,83]

10.2.3 In Vivo Intravascular Contrast Agents

Intravascular delivery of iodinated contrast agents has been heavily used to image various organs in vivo.[17,18,21,22,39,42,84–108] While the animal is anesthetized, the contrast agent is introduced directly into the bloodstream, usually through the tail or jugular vein. Depending on the specific application, the agent is delivered either continuously throughout the scanning time or as a bolus dose at a set time point before the scan. After the scan, the contrast agent is naturally cleared from the blood. For many applications, the vascular residence time of iodinated contrast agents such as iopamidol (Isovue 370, Bracco Diagnostics) or iohexol (Omnipaque, GE Healthcare or Daiichi-Sankyou) is very short. Blood-pool contrast agents such as iodinated triglyceride (glyceryl-2-oleoyl-1,3-di-7-(3-amino-2,3,6-tri-iodophenyl)-hepanoate) (ITG-DHOG) and polyethylene glycol (PEG)-modified ITG-DHOG lipid emulsions[84] (Fenestra VC and LC, Advanced Research Technologies Inc.) and nanoscale liposomes[85,92] have been developed to improve bloostream residence time. These contrast agents are typically injected in the same manner as other intravascular agents, though some researchers introduce the contrast agent intraperitoneally for abdominal organ imaging.[39] All of these contrast agents have been shown to be well tolerated in model animals, and many are cleared for human use (Table 10.2).

10.3 Circulatory and Respiratory System Applications

10.3.1 Reduction of Motion Artifacts

Because the heart and lungs continue to function when animals are anesthetized, the movement of these organs can cause distortions within the microCT images. One method of managing motion during imaging acquisition is prospective gating, where the animal is intubated and respiration is controlled by a ventilator.[109] The scanner is programmed to collect images at the same point in each breathing cycle. Retrospective gating is another method of minimizing effects of lung movement and involves free-breathing animals whose respiratory rates are tracked while the scans are performed.[87] After scanning, images taken at a specific point in the breathing cycle are isolated and used for analyses.

TABLE 10.2

Referenced In Vivo Contrast Agents

Contrast Agent	Trade Name	Referenced Uses	References	Doses
ITG-DHOG lipid emulsions	Fenestra LC	Liver, spleen	17,39,84,95,98,100,105–108	1–20 mL/kg
Liposomes containing iopamidol		Liver	18,88,91–94	10–16 mL/kg
PEG-modified ITG-DHOG lipid emulsions	Fenestra VC	Vasculature, liver, spleen, pancreas	21,39,86,87,89,98,99	2.5–20 mL/kg
Iomeprol	Iomeron	Vasculature	22,103,104	2–14 mL/kg
Ioxaglate	Hexabrix	Cartilage	57	70 μL/knee
Liposomes containing iohexol		Vasculature	85	13 mL/kg
Iopamidol	Iopamiron or Isovue	Vasculature	86,90,99	5–25 mL/kg
Liposomes containing iomeprol		Liver	96,97	1–30 mL/kg
Iodine-based contrast agent	Exia	Liver, spleen	98	2.5–5 mL/kg
Iohexol	Omnipaque	Vasculature	101	1–6 mL/hr
Alkaline earth-based nanoparticles	ExiTron nano	Vasculature, liver	102	4 mL/kg

Note: Table of in vivo contrast agents referenced in chapter with trade name, referenced uses, referenced papers, and doses.

10.3.2 Heart

Contrast-enhanced microCT has been shown to be useful in imaging the coronary vasculature in animal models and was first used by Beighley et al.[59] to investigate the branching patterns in the arterial tree of rat hearts ex vivo. Prior to heart excision and microCT scanning, the vasculature of the rats was perfused with Microfil to provide sufficient contrast such that the vasculature could be differentiated from the surrounding soft tissue during image analysis. This technique was later utilized for the first visualization and quantification of the 3-D structure of the vasa vasorum, a network of smaller blood vessels that surround larger vessels such as the coronary artery.[62,63] Differences in vascular density and vessel diameter in the vasa vasorum in hypercholesterolemic[62] and balloon-injured arteries were also quantified.[63] Rodriguez-Porcel et al.[64] then applied this quantitative technique to detect significant changes in the entire vasculature of hypercholesterolemic versus normal pig hearts. This was the first study to establish that contrast-enhanced microCT was sensitive enough to detect and quantify 3-D pathophysiological changes in the entire heart. These papers, along with others detailing improved automatic segmentation and mapping algorithms,[65] demonstrate the utility of contrast-enhanced microCT for investigating the structure and pathophysiology of coronary vasculature.

Contrast-enhanced microCT can also be used to investigate cardiac function in vivo, as first shown by Badea et al.[86] In this study, Isovue 370 or Fenestra VC was injected into mice to provide the blood with sufficient contrast such that blood could be distinguished from surrounding soft tissue in image processing. The mice were scanned using prospective gating techniques to reduce artifacts caused by breathing and cardiac motion. Functional parameters (left ventricle diastolic and systolic volumes, ejection fraction, stroke volume, and cardiac output) were calculated from the 3-D microCT images. In further work that used a novel liposomal contrast agent[92] to increase both the intensity and the residence time of blood-pool contrast, Badea et al.[88] showed that this technique was sensitive enough to detect changes in the same functional parameters caused by dobutamine-induced cardiac stress. These functional differences agreed with previous results measured with MRI,[20,110] validating the approach of using contrast-enhanced microCT to provide quantitative assessments of cardiac function.

Contrast-enhanced imaging techniques combined with retrospective gating have also been used to quantify similar cardiac functional changes resulting from myocardial infarction (MI) in mice, as shown by Detombe et al.[89] Figure 10.1 shows long-axis (a, b) and short-axis (c, d) microCT images of mouse hearts with sham operation (a, c) or induced MI (b, d) during the diastole phase. The functional parameters calculated from the reconstructed 3-D images agreed well with microMRI assessments of MI in small animals.[20] Taken together, these studies by Badea et al. and Detombe et al. demonstrate the ability of contrast-enhanced microCT to quantify changes in cardiac functionality in small-animal models in vivo. These techniques could potentially be utilized in studies of cardiac genetic modifications and cardiac toxicity of new drugs.

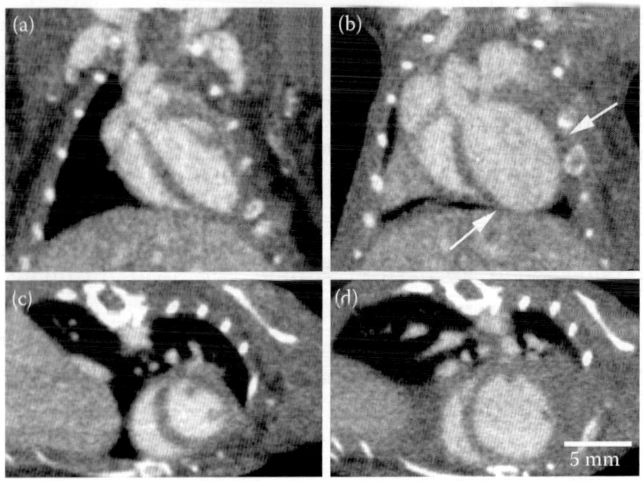

FIGURE 10.1
(a, b) Long-axis and (c, d) short-axis images of mouse hearts with sham operation (a, c) or induced MI (b, d) during the diastole phase. Quantitative analysis confirmed the qualitative observations that the entire heart of the MI mouse was enlarged, indicating the global effects of the infarct. The infarct is evident in the thinned wall in the apex of the left ventricle, below the arrows. (From Detombe, S.A. et al., *Invest. Radiol.*, 43, 2008.)

10.3.3 Vasculature

Neovascularization is associated with a number of physiological and pathological phenomena, including bone healing, atherosclerosis, and cancer. Contrast-enhanced microCT provides a rapid method of imaging, quantifying, and investigating the role of neovascularization in these various processes.

Duvall et al.[15] used contrast-enhanced microCT in the mouse hind-limb ischemia model to develop a quantitative methodology for evaluating compensatory collateral vessel formation. The vasculature of the mice was perfused with Microfil to provide sufficient contrast such that the vasculature could be differentiated from the surrounding soft tissue during image analysis. The authors applied established microCT trabecular bone morphometry analysis methods to the perfused vascular network to obtain quantitative measures of vasculature volume, connectivity, number, thickness, separation, and anisotropy. Figure 10.2 shows a 3-D microCT image at 14 days postsurgery with pseudo-color mapping of vessel thicknesses (red = larger, blue/green = smaller). The authors also showed that changes in voxel size and segmentation values used in the analyses could significantly affect the resulting quantitative measures, thus emphasizing the importance of selecting appropriate values for each. This was the first time that quantitative bone morphometry analysis methods were applied to contrast-enhanced microCT scans of vasculature and showed that this method could be utilized to investigate neovascular processes in pathological situations.

As tumors grow, new vasculature is formed, but these vessels possess lower structural integrity and greater permeability compared to native vasculature.[111,112] Using a liposomal-iodine blood-pool contrast agent[92] to increase vascular contrast, Ghaghada et al.[93] visualized the neovascularization of tumors in mice and quantified the increase in x-ray attenuation within the tumors in vivo. Figure 10.3 demonstrates the use of these liposomal-iodine nanoparticles in this application. Vessel attenuation increased immediately after nanoparticle injection, while tumor tissue attenuation increased at later time points. This suggested that nanoparticles leaked into the extravascular space, and the authors speculated that this phenomenon could be quantified as a parameter of vascular permeability. Using the same nanoparticle contrast agent and a dual-energy microCT scanner, which scans the specimen at two different energy levels simultaneously, Moding et al.[94] demonstrated that contrast-enhanced microCT could quantify vascular permeability in mouse sarcomas. The authors reported increased vessel volume and vascular permeability in radiation treated tumors 1 day after treatment, compared to untreated controls, and speculated that this increase in vessel leakiness could be utilized in delivering cancer therapies. These studies demonstrate that in vivo contrast-enhanced microCT can be used to quantify vasculature functional parameters such as vasculature permeability in addition to morphometric parameters.

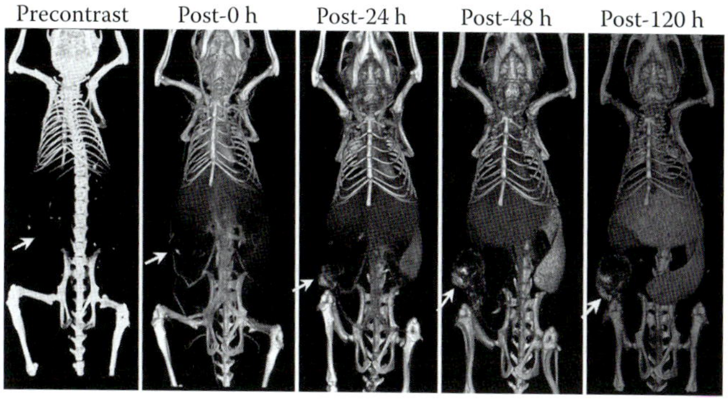

FIGURE 10.3
Dynamics of tumor signal [illegible] three-dimensional volume-rendered images demonstrating the [illegible] and accumulation of nanoparticle contrast agent within the tumor (arrow). Immediately after administering the nanoparticle contrast agent, overall body and tumor vasculatures are easily visualized. Tumor accumulation of nanoparticle contrast agent was observed at as early as 24 h. No image-detectable nanoparticle contrast agent signal was observed from the blood-pool at post-120 h (day 5) as confirmed by the absence of any vessels or heart signal on post-120 h image. However, the tumor was clearly enhanced. Liver (directly below ribs) and spleen (right) were also enhanced, likely because of contrast agent clearance. (From Ghaghada, K.B. et al., *Acad. Radiol.*, 18, 2011.)

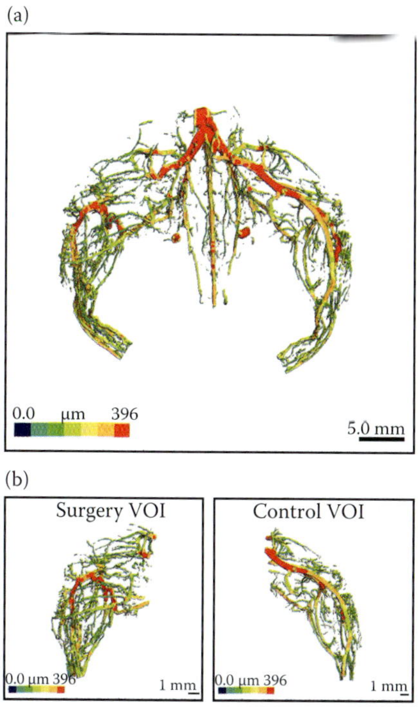

FIGURE 10.2
Representative 14-day postsurgery contrast-enhanced microCT images of rat hind-limb ischemia with a color-coded vessel diameter scale mapped to the 3-D image surface. (a) 3-D rendering of entire hind-limb vasculature. (b) Volume of interest (VOI) analyzed in the surgical ligation and control limbs. Color scale represents vessel size. Red is mapped to the surface of vessels with diameter ≥396 μm, and the color scale is linear down to zero. (From Duvall, C.L. et al., *Am. J. Physiol. Heart Circ. Physiol.*, 287, 2004.)

10.3.4 Lungs

Detailed knowledge of lung structure–function relationships is of considerable interest for studying pulmonary disease, including pulmonary hypertension, pulmonary cancers, and chronic obstructive pulmonary disease, and in testing the cardiopulmonary safety of new drugs or therapeutics. Badea et al.[90] developed a contrast-enhanced microCT technique for imaging lung perfusion at the temporal resolution of one heartbeat in small-animal models. In this context, lung perfusion is defined as the process by which blood perfuses through the lung vasculature after leaving the heart. Multiple small-dose injections of Isovue 370 separated by at least 2 min were delivered into the jugular vein to increase the contrast of the blood such that the blood could be differentiated from the surrounding soft tissue. Multiple injections were needed to achieve 3-D images at the required temporal resolution. Prospective gating techniques were used to reduce motion artifacts during the scans. Figure 10.4 shows sequential microCT perfusion images taken at 140 ms temporal resolution (one rat heartbeat) at one location in the lungs. A second paper by the same group demonstrated a similar technique in the mouse.[91] 4-D images (three spatial dimensions and one temporal dimension) of lung perfusion had not been achieved at such high resolutions prior to these studies. These techniques could be used to provide valuable blood flow information in small-animal models of pulmonary diseases or in testing cardiopulmonary drug safety.

Multiple studies have also investigated the use of microCT to image and quantify lung functional parameters. Lam et al.[113] investigated the use of contrast-enhanced microCT for imaging and measuring in vivo ventilation, or the rate of air intake into the lungs, in rats. Xenon was utilized to provide contrast to the air breathed by the rats during imaging, and the increase in contrast over time within the lungs was used in a theoretical model to calculate ventilation. Further validation is needed, however, to show if this method is sensitive enough to detect ventilation differences in various pulmonary diseases.

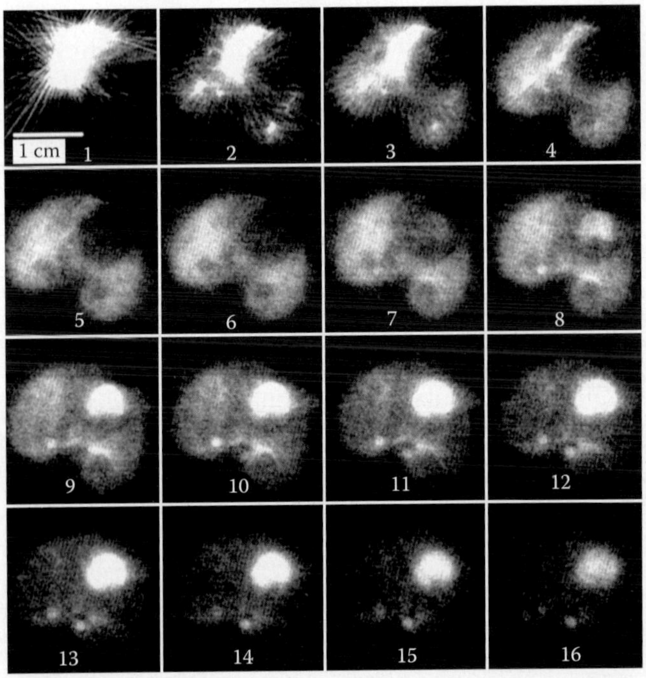

FIGURE 10.4
Sixteen successive microCT perfusion images at an axial slice cutting through the lung at heartbeat temporal resolution (140 ms). (From Badea, C.T. et al., *Med. Phys.*, 37, 2010.)

10.4 Musculoskeletal System Applications

10.4.1 Cartilage

Cartilage defects and diseases, such as osteoarthritis, are the most common cause of disability in the United States, and presently, there are few options to help prevent cartilage degeneration or to replace cartilage after complete loss. One major limitation in the development of effective cartilage treatments has been the inability to quantitatively measure 3-D changes in the articular cartilage. MicroCT imaging techniques using ionic immersion contrast agents, as highlighted in the following studies, demonstrate that contrast-based methodologies can provide high-throughput 3-D analyses for cartilage degeneration, regeneration, and potentially, evaluation of therapeutics.

Palmer et al.[13] introduced equilibrium partitioning of an anionic contrast agent for microCT imaging (EPIC-microCT) to evaluate the morphology and proteoglycan (PG) content of cartilaginous tissues. PGs are a key component of articular cartilage, playing an important functional role in cartilage matrix by regulating the osmotic pressure within the tissue, and PG loss is an early indicator of cartilage damage.[114] As PGs have a negative charge, areas of high PG content have a high overall negative charge density, and areas of low PG content have low negative charge density. An anionic contrast agent diffused into cartilage until equilibrium would therefore be expected to distribute inversely proportional to PG content. Indeed, when cartilage is equilibrated with the contrast agent Hexabrix (ioxaglate meglumine and ioxaglate sodium, Guerbet) and is then scanned with microCT, areas of lower x-ray attenuation indicate lower contrast agent concentration, which correlate to regions of greater PG content and healthy cartilage. Conversely, areas of higher x-ray attenuation indicate higher contrast agent concentration and correspond to regions of lower PG content, indicative of cartilage degradation.[13]

Xie et al.[44] utilized EPIC-microCT to extract 3-D cartilage morphological data for distal femoral cartilage in the rat. Cartilage thickness measurements were repeatable and agreed well with experimental needle probe thickness measurements and 2-D histomorphometry estimates. Significant differences were shown in cartilage volume, thickness, and surface area for rats of different ages. A later study demonstrated that EPIC-microCT was also capable of quantifying significant changes in PG content (as measured by average volumetric EPIC-microCT attenuation and correlated to histological safranin-O optical density of a central 8 μm section) resulting from chondroitinase ABC digestion.[50] These studies demonstrate that EPIC-microCT can be used to quantify morphological and compositional properties in healthy and degraded articular cartilage in intact joints, suggesting that the technique may be useful for evaluating therapeutics in small-animal models of cartilage degeneration. EPIC-microCT has since been used to assess cartilage damage[illegible] and regeneration.[45,46]

Cationic contrast agents have also been utilized for cartilage imaging.[14,51,52,55,56] These solutions contain positively charged molecules; therefore, contrast agent molecules can permeate readily into tissues of high negative charge or PG content, such as healthy cartilage. Use of a cationic contrast agent has been shown to provide higher contrast enhancement and increased correlation with PG content within nondegraded cartilage when compared to the same concentration of anionic contrast agents.[51,56] Hayward et al.[58] used this technique to calculate cartilage volume in calluses of a mouse femoral fracture model. They showed that cartilage volumes matched in contrast-enhanced microCT and histomorphometric measurements. This suggests that cationic contrast-enhanced microCT can measure cartilage volume in callus formation and can provide better understanding of the fracture healing process.

The major assumption with EPIC-microCT is that the contrast agent distribution is most significantly dependent on the fixed charge density of the cartilage. However, a recent paper by Kokkonen et al.[53] showed that increased collagen cross-linking significantly decreased contrast agent diffusion into the cartilage while leaving the PG content unaltered. These results may influence the interpretation of studies utilizing equilibrium-based ionic contrast agent techniques. Other studies have also shown that cartilage degeneration can affect contrast agent diffusion, and this change in diffusion coefficient can therefore be used to identify cartilage defects.[43,54]

To measure cartilage changes in joint degeneration models in vivo, Siebelt et al.[57] employed microCT arthrography (microCTa) whereby Hexabrix was injected into the knee joints of rats immediately prior to imaging.[57] In three different joint degeneration models, cartilage volume decreased and attenuation increased compared to baseline, signifying an increase in cartilage damage, and this correlated well with histological scoring. Figure 10.5 shows histological and microCT images of the degenerative models and controls. The authors acknowledged that joints were scanned before contrast equilibrium was

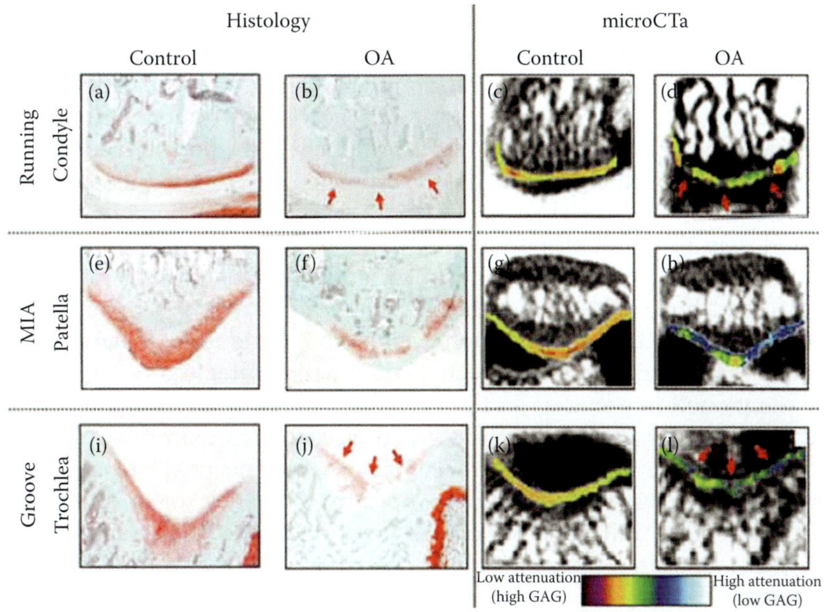

FIGURE 10.5
(a, b, e, f, i, j) Histology sections stained with safranin-O showing glycosaminoglycan (GAG) distribution (red indicates high GAG concentration) (c, d, g, h, k, l) and microCTa images for (a–d) condylar cartilage in running animals, (e–h) patellar cartilage in monoiodoacetate injected animals, and (i–l) animals with grooved trochlear cartilage. The red arrows indicate small focal lesions in the running model seen on histology and detected with microCTa. For the groove model, focal defects were created by the tipped point of a surgical tool, and comparable changes were seen on microCTa scans (red arrows). (From Siebelt, M. et al., *J. Orthop. Res.*, 29, 2011.)

reached. Thus, attenuation changes could not be directly correlated with PG content and instead were likely due to combined changes in PG content and matrix properties. These factors, along with differences in diffusion properties, may have limited the sensitivity in detecting degradative changes. However, this study importantly showed that cartilage degradation changes were detectable in vivo, thus enabling longitudinal analyses.

10.4.2 Bone Healing

Due to its mineralized nature, bone does not typically require contrast agents for microCT analysis. However, to elucidate mechanisms involved in bone formation or remodeling, contrast agents have been used to enhance visibility of surrounding relevant tissues, such as vasculature, as well as bone itself in analyses of microdamage.[25–29]

Bone microdamage consists of microscopic fractures in bone and accumulates with normal aging and repetitive mechanical loading. This microdamage results in reduced bone elastic modulus and fracture toughness, thereby increasing the likelihood of fractures. In the work of Landrigan et al.,[29] bone samples were loaded in tension and then soaked in barium chloride and sodium sulfate to allow barium sulfate to precipitate into the microdamage on the bone surface, increasing the contrast of the microdamage for microCT imaging. Microdamage accumulation was quantified and seen to increase with increased mechanical loading. Similar results were observed with basic fuchsin histology, and there was good correlation between microcrack density as measured by histology and microdamage density measured with microCT. However, the microCT resolution and limited contrast absorption prevented the quantification of microdamage morphologies, such as crack length, with microCT.

Vasculature has been closely linked with new bone growth, and improved understanding of the relationship between vasculature and bone growth is critical to the development of regenerative therapies for bone defects or malformations. Vasculature perfusion techniques similar to those mentioned in previous sections were first used to study bone and vasculature interactions by Moore et al.[66] in investigating neovasculature associated with distraction osteogenesis. Contrast-enhanced microCT has since been utilized to investigate vascular and bone interactions in osteonecrosis[67–69] and bone tissue engineering.[70–72]

MicroCT provides new quantitative measures for evaluating tissue engineering strategies that affect bone vascularization. Zhang et al.[70] utilized this technique to investigate the role of periosteal progenitor cells in bone grafts. Two weeks after the bone graft procedures in rat femora, Microfil was injected into the vasculature of the rats to increase contrast such that the vasculature could be differentiated from the surrounding tissue. Previously described morphometric analyses[15] showed that devitalizing the bone grafts prior to transplantation significantly reduced vascular volume around the graft and decreased bone formation. Adding bone morphogenetic protein-producing bone marrow stromal cells to the devitalized grafts significantly attenuated this reduction in vascular volume and bone formation. Raines et al.[73] later used contrast-enhanced microCT to assess the effects of hyaluronic acid and demineralized bone matrix on bone marrow regeneration in a rat bone marrow ablation model, and the results indicated that demineralized bone matrix and hyaluronic acid altered the vasculature of the regenerated marrow.

Contrast-enhanced microCT can also provide insights into contributions of the mechanical environment on vasculature and bone regeneration. Boerckel et al.[74] used microCT to examine the effects of functional loading on the vascular and bone healing response in a rat bone segmental defect model. Rats underwent surgery to create a large segmental defect in the femur, and rigid fixation plates were used to limit mechanical loading of the defects. The fixation plates could be unlocked to allow physiologic loads to be transferred to the defect site at early (at time of surgery) or late (4 weeks after surgery) time points. Three weeks after the plates were unlocked, the rat vasculature was perfused to enhance vessel contrast, and the defect site was harvested and scanned. Vascular morphometric analyses showed that early loading of the defect site hindered vascular ingrowth and resulted in decreased bone regeneration when compared to unloaded defect sites. Delayed loading, however, promoted an increase in vessel thickness and increased bone regeneration when compared to unloaded controls, demonstrating that mechanical loading affected both bone neovascularization and regeneration. These studies indicate that contrast-enhanced microCT can be used to image and quantify bone vascularization changes in multiple animal models, and this can provide new insights into injury mechanics and treatment strategies.

In studies where contrast-enhanced vascular structures are in proximity to bone, segmentation of the tissues can be challenging. When imaging Microfil-perfused tissues, most applications require demineralization of the bone to analyze the vasculature. New formulations of barium sulfate compounds have been developed to allow the simultaneous analysis of bone and vasculature without decalcification.[61] Regardless, perfusion techniques to enhance contrast in microCT have been utilized for the detailed imaging and quantification of neovascularization and its relationship to the formation of new bone and marrow. Furthermore, these techniques can be used to facilitate evaluation of novel therapeutics to alter the neovascularization in bone regeneration applications.

10.5 Gastrointestinal and Urinary System Applications

The majority of contrast-enhanced microCT studies investigating abdominal organs involve the kidney and liver. These organs will be the focus of this section. Other notable studies have used contrast-enhanced microCT to image the spleen,[95,97,98] colon,[40,115] and pancreas[99] and to measure the peritoneal fluid surface area.[57,41]

10.5.1 Liver

Contrast-enhanced microCT provides imaging ability and quantifiable parameters for tracking multiple liver pathologies, including cysts,[100] ischemic lesions,[101] and tumors.[39,42,84,102,105–108] Bakan et al.[84] developed a novel liver-specific contrast agent (Fenestra LC) to enhance the contrast of healthy liver and was the first to utilize contrast-enhanced microCT to detect liver tumors. Fenestra LC is taken up primarily by healthy hepatocytes, increasing the x-ray attenuation of healthy tissue, while cancerous or diseased hepatocytes do not absorb the contrast agent. As such, healthy liver has high contrast in microCT images, and diseased tissue appears as areas of low contrast. This technique was later used by Aprahamian et al.[39] to test the anticancer potential of *myo*-inositol trispyrophosphate (ITPP) by tracking tumor growth in a rat orthotopic hepatocellular carcinoma (HCC) model. Rats received saline (negative control), doxorubicin (conventional therapeutic), or ITPP over a period of 11 weeks. Fenestra VC was injected intravenously 6 h before each imaging session to enhance vascular contrast, and Fenestra LC was injected intraperitoneally immediately prior to imaging for increased liver contrast. Figure 10.6 illustrates the development or prevention of tumors within the liver in response to treatment. Quantification of tumor volumes showed that rats treated with ITPP had significantly smaller tumor volumes than doxorubicin-treated rats. These results correlated well with animal survival index and histology, demonstrating that microCT can provide accurate tracking of liver tumor growth in therapeutic testing.

Masyuk et al.[75] utilized contrast-enhanced microCT to image and quantify parameters describing the biliary tree in livers. This study investigated differences

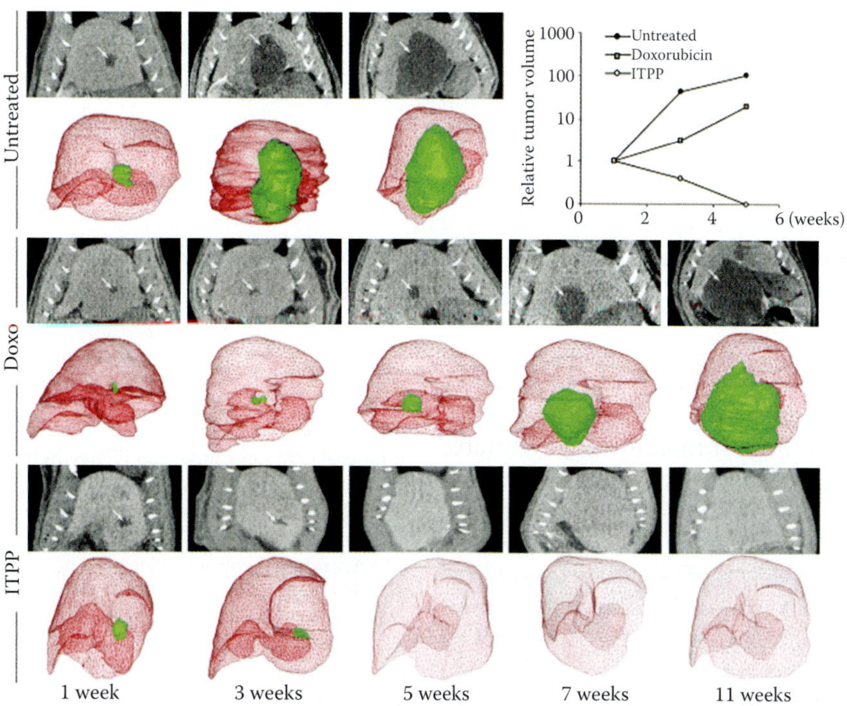

FIGURE 10.6
Liver tumor progression as detected by microCT imaging of untreated rats (top), rats treated with doxorubicin (Doxo, center), and rats treated with *myo*-inositol trispyrophosphate (ITPP, bottom). In the 2-D slices, tumor is indicated by an arrow (monochrome images). Tumor volumes were determined via 3-D transparent simulations created by 3-D virtual-reality and rendering software (color images) and plotted as a ratio to the initial volume (graph, upper right). (From Aprahamian, M. et al., *ChemBioChem.*, 12, 2011.)

between the biliary trees of rats fed a diet containing α-naphthylisothiocyanate (ANIT) and those of rats fed a normal diet. ANIT intake results in increased proliferation of cholangiocytes (the epithelial cells that line the biliary tree) and increased remodeling of the biliary tree, which is a common pathological change in liver disease.[116,117] Microfil was injected into the common bile duct to increase the contrast such that the biliary tree could be distinguished from the surrounding soft tissue in imaging. Quantitative analysis showed that ANIT-fed rats exhibited more biliary tree branching. This resulted in large increases in both total biliary tree volume and surface area in ANIT-fed rats and showed that perfusion contrast-enhanced microCT could effectively be used to image and quantify remodeling of the biliary tree.

10.5.2 Kidney

Visualizing renal vascular 3-D structure is critical to understanding both normal and pathological relationships between vascular structure and tubular function. Researchers have utilized perfusion contrast agents combined with microCT imaging techniques to study small-animal renal vasculature ex vivo in normal[76] and pathological[77–79] conditions. Fortepiani et al.[79] demonstrates the benefit of using contrast-enhanced microCT to investigate renal vasculature in a study on the effects of nitrous oxide (NO) synthesis inhibition on renal microvasculature structure. Rats were provided with untreated water, water with an NO synthesis inhibitor, or water with the NO synthesis inhibitor and an angiotensin II type-1 (Ang II AT_1) inhibitor. Plasma and urinary protein assays have shown that chronic inhibition of NO synthesis decreases renal function while Ang II AT_1 inhibition prevents those changes.[118] The rats were perfused with Microfil to provide sufficient contrast such that the kidney vasculature could be differentiated from the surrounding soft tissue during image analysis. Afterwards, the kidneys were excised and scanned. Significant reductions in vascular volume fraction in response to the NO synthesis inhibitor were detectable from the contrast-enhanced microCT images, while the Ang II AT_1 inhibitor attenuated those reductions. It was also noted that the different sections of the kidneys underwent different levels of vascular remodeling. These results, in conjunction with companion physiological and histological results, provided improved understanding of the effects of NO synthesis inhibition on kidney structure/function. This study establishes the utility of contrast-enhanced microCT for quantifying differences in renal vasculature parameters and demonstrates the benefit of whole-organ 3-D quantification that microCT can provide.

10.6 Nervous System Applications

In comparison to other organs and tissues, there have been fewer studies using contrast-enhanced microCT to image neurological and cranial features. Brain imaging will be the focus of this section. Researchers have also imaged the spinal cord,[19,30] inner ear nerves,[31] and ocular drug delivery,[32,33] but few quantitative analyses have been performed.

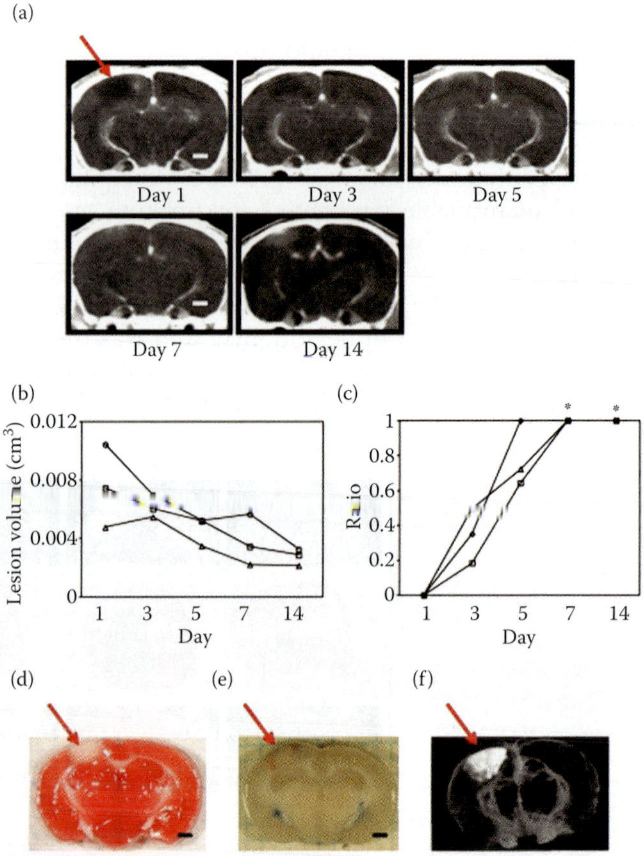

FIGURE 10.7
Temporal changes in the infarct area in the photochemically induced thrombotic brain damage (PIT-BD) mouse model. (a) Serial contrast-enhanced microCT images taken from a representative PIT-BD mouse brain. (b) Time-dependent changes in % high attenuation area (HAA) and total volume of lesion. Changes in lesion volumes of three middle cerebral artery occlusion (MCAO) mice through days 1 to 14 are shown. (c) Temporal changes in %HAA in the same PIT-BD mice shown in (b). Significant differences were seen in %HAA between day 1 and days 7 and 14 (Kruskal–Wallis with Dunn post hoc tests, $^*P < .05$). (d) TTC-stained coronal section of a PIT-BD mouse brain, day 7. The arrow indicates the ischemic lesion. (e) Leakage of Evans blue indicating elevated vascular permeability in the infarct area. (f) 3-D image of the PIT-BD lesion (white area) reconstructed from sequential microCT images. Scale bars: 1 mm (a, d, e). (From Hayasaka, N. et al., *PloS One*, 7, 2012.)

10.6.1 Brain

Most of the neurological contrast-enhanced microCT work has focused on the brain, especially tumor detection[34,103] and vasculature.[80,81,104] One unique application of contrast-enhanced microCT has been in furthering the understanding of brain ischemia and stroke, as implemented by Hayasaka et al.[101] In this study, contrast-enhanced microCT was used to detect tissue damage in vivo in two mouse models of brain ischemia and a rat model of spontaneously occurring stroke. Omnipaque was intravenously injected into the vasculature of anesthetized animals prior to and during the brain microCT scans. Areas of reduced contrast within the brain indicated lower contrast perfusion from the vasculature, most likely due to low blood flow. Areas of high contrast indicated increased contrast perfusion from the vasculature due to a lesion caused by stroke or ischemia. Using these contrast differences, lesion growth was followed longitudinally. MicroCT measurements of lesion growth correlated well with whole mount staining (see Figure 10.7). In the rat studies, both microCT and microMRI were used to calculate lesion volume, and there were no significant differences between the volumes obtained from the two methodologies. In another study, Dobrivojevic et al.[35] utilized contrast-enhanced microCT to measure ischemic lesion volume in mouse brains ex vivo. After excision, the brains were immersed in Omnipaque to increase the contrast of the lesions. Lesion volumes measured by microCT correlated well with volumes calculated from Nissl staining, a histological stain traditionally used to detect ischemic nervous tissue lesions. These results demonstrate that contrast-enhanced microCT provides sufficiently detailed quantitative information to detect and track ischemic brain lesions both in vivo and ex vivo and has potential for use in treatment and pathophysiologic studies.

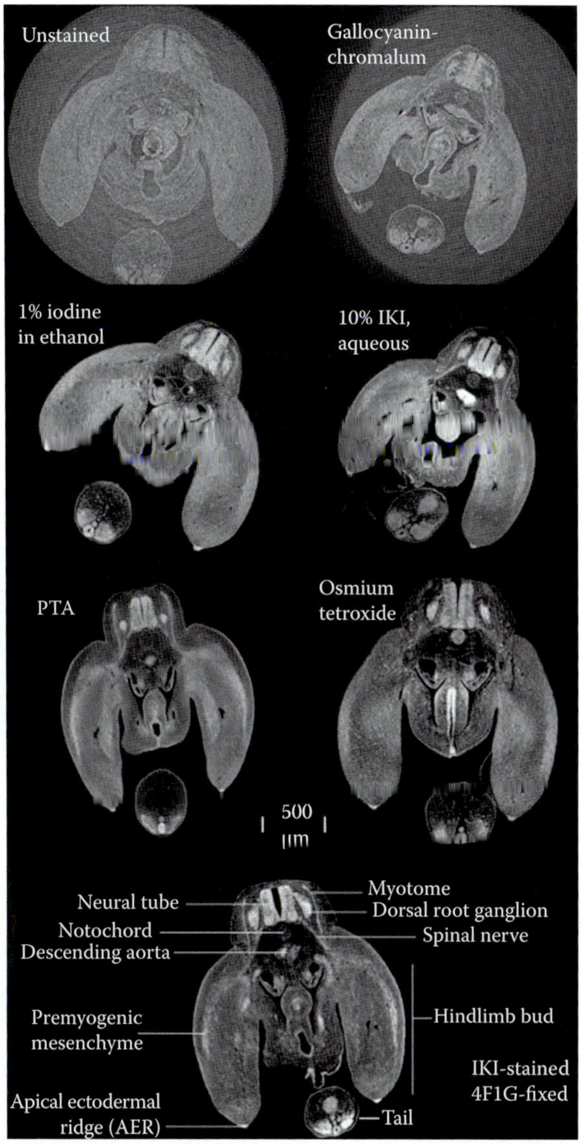

FIGURE 10.8
MicroCT cross sections of chick embryos unstained or stained with gallocyanin-chromalum, 1% iodine, 10% iodine potassium iodide, phosphotungstic acid, or osmium tetroxide. The bottom-most figure contains labels of the different tissue features seen in the scans. (From Metscher, B.D., *Dev. Dyn.*, 238, 2009.)

10.7 Embryonic Development Applications

Accurate, detailed images providing distinction between different tissues and organs in embryos are necessary to understand developmental processes and potentially as a tool for assessing drug toxicity. Metscher[12] applied contrast-enhanced microCT to obtain such images of chick embryos. The embryos were stained with osmium tetroxide, gallocyanin-chromalum, 1% iodine in alcohol, iodine potassium iodide (IKI), or PTA. Figure 10.8 contains microCT cross sections of the embryos demonstrating the different staining intensities achieved with the various stains. For all the stains, neural tissue and denser tissue such as the apical ectodermal ridge and premyogenic mesenchyme exhibited the greatest contrast. Although osmium tetroxide had been more commonly used for x-ray attenuation enhancement, PTA and IKI provided comparable or better contrast,[12] were simpler, and were less toxic. IKI also had much faster tissue penetration compared to osmium tetroxide and PTA. These results establish the use of contrast-enhanced microCT to visualize embryonic morphology, and in a later paper, Metscher[11] demonstrated that these same techniques and stains could be used to visualize multiple other model species.[11]

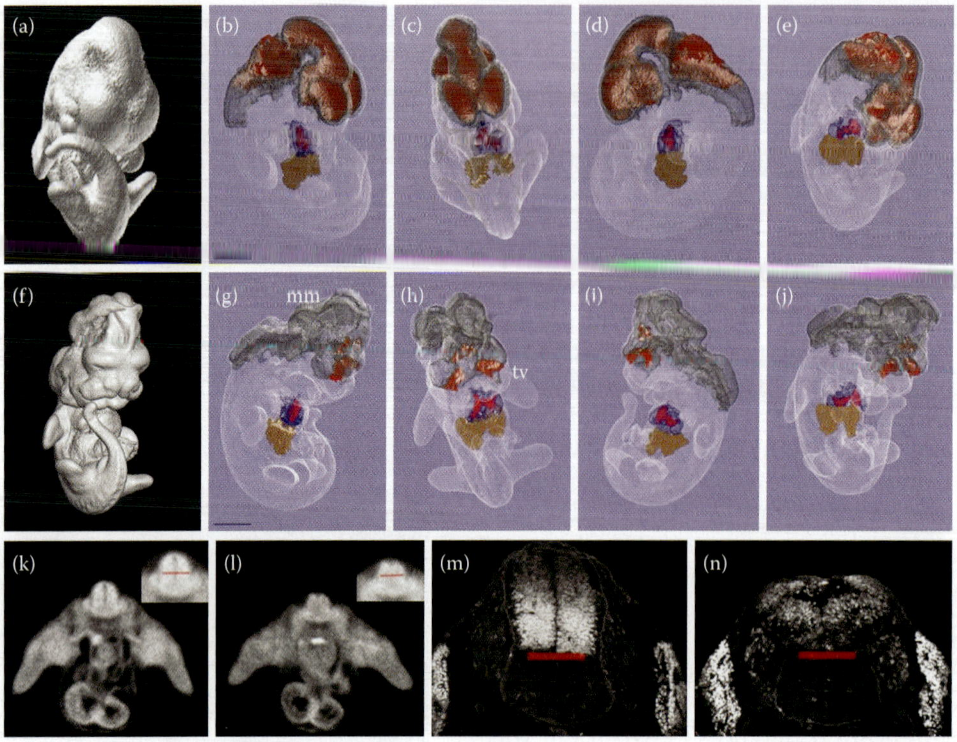

FIGURE 10.9
(a–e) Wild-type E11.5 embryo compared to (f–j) a mutant littermate that expresses the *Pax3:Fkhr* oncogene in neural crest and myogenic tissue. (a, f) Isosurface renderings. (b–e, g–j) Segmentation renderings of cranial neural crest (grey), brain vesicles (red), liver (brown), cardiac muscle (blue), and the conotruncal cavity (red within the blue cardiac muscle). From the segmented renderings, the mutant embryo is noted to have partial neural tube closure failure, resulting in absence of the hindbrain vesicle, overgrown midbrain mesenchyme (mm), as well as severely hypomorphic telencephalic vesicles (tv, shown in red). (k, l) MicroCT axial sections of the neural tube at the level of the forelimbs (dorsal surface is up) from wild-type and control embryos, respectively. Magnification of the neural tube is shown in the upper right insets of each panel. The sulcus limitans, which separates the dorsal and ventral neural tube, is shown as a red line. Severe blunting of the dorsal neural tube is evident in the mutant embryo. (m, n) Axial confocal microscopy sections from embryos of the same genotype and developmental stage as the corresponding axial cross section as in panels (k) and (l), respectively. Sections were stained with the Pax7 immunohistochemical marker of the dorsal neural tube. Pax7-stained cells of the dorsal neural tube and flanking dermomyotome are white. The sulcus limitans is shown as a red line, confirming the dorsal neural tube blunting seen in the microCT sections. Scale bars in panels (b) and (f) are each 1.2 mm. (From Johnson, J.T. et al., *PLoS Genet.*, 2, 2006.)

To show that contrast-enhanced microCT can detect phenotypic changes in mutated embryos, Johnson et al.[36] used contrast-enhanced microCT to image neural development in the embryos of Pax3:Fkhr and wild-type mice. If the Pax3:Fkhr gene is expressed in place of the Pax3 gene, incomplete neural tube closure is expected. Mouse embryos were stained with osmium tetroxide prior to microCT scanning to provide contrast between various tissues in the embryos. Figure 10.9 clearly shows partial neural tube closure in the mutant mouse embryos. These results were confirmed with confocal microscopy, suggesting that contrast-enhanced microCT was able to detect phenotype alterations caused by genetic mutations. In addition, contrast-enhanced microCT could potentially be utilized to screen for embryonic or fetal abnormalities in developmental toxicity studies, similar to the work of Winkelmann and Wise[38] with detecting fetal skeletal abnormalities with standard microCT.

10.8 Conclusions

Since microCT was first introduced, it has become increasingly useful as an imaging tool for small-animal research; however, one major limitation of contrast-enhanced microCT is the use of ionizing radiation. In large doses, this can negatively affect animal health during in vivo scans.[119,120] This makes contrast-enhanced microCT less attractive for longitudinal studies where temporal proximity is required for scans. However, improvements in scanner technology and reconstruction algorithms have allowed for the reduction of scan time and radiation dose.[91,121] New studies using phase contrast microCT also show promise for reducing radiation dose and increasing resolution, both with[122] and without[123] the use of contrast agents.

MicroMRI is often used in similar applications as microCT. Advantages of using microMRI are that contrast agents may not be required and no ionizing radiation is used. Advantages of using contrast-enhanced microCT are greater system availability, faster scan times, and higher resolutions at lower costs. A comparative study of the two imaging modalities concluded that while MRI may be preferable for longitudinal studies to eliminate repeated radiation exposure, contrast-enhanced microCT may be more suitable for studies requiring higher resolution and throughput.[18]

While initially designed to quantify bone morphology and density, the utility of microCT has been expanded to further application areas through the use of contrast agents. Examples described in this chapter demonstrate the use of immersion, perfusion, and in vivo intravascular contrast agents to enhance visibility and characterization of soft tissues. Contrast-enhanced microCT can be used to detect both structural and functional changes within biological materials in a wide variety of organs and tissues. These techniques provide powerful additions to histological analyses as they are faster, require less processing, provide 3-D images and quantitative data, and are often nondestructive. Further augmentation of contrast-enhanced microCT capabilities, including innovations in phase contrast microCT imaging, will continue to increase the utility of contrast-enhanced microCT for both ex vivo and longitudinal in vivo studies, making these techniques more available and advantageous for researchers studying biological tissue structure and function.

References

1. Hounsfield, G. N. Computerized transverse axial scanning (tomography): Part 1. Description of system. *Br J Radiol* **46,** 1016–22 (1973).
2. ASTM Standard E1441-11. *Standard Guide for Computed Tomography (CT) Imaging: E1441-11*, ASTM International, West Conshohocken, PA (2011). doi:10.1520/E1441-11. www.astm.org.
3. Mettler, F. A. et al. Radiologic and nuclear medicine studies in the united radiation and worldwide: Frequency, radiation dose, and comparison with other radiation sources—1950–2007. *Radiology* **253,** 520–31 (2009).
4. Yang, W. J. et al. High-definition computed tomography for coronary artery stents imaging compared with standard-definition 64-row multidectector computed tomography. *J Comput Assist Tomogr* **36,** 295–300 (2012).
5. Kazakauskaite, E. et al. Image quality in low-dose coronary computed tomography angiography with a new high-definition CT scanner. *Int J Cardiovasc Imaging* **29,** 471–7 (2013).
6. Feldkamp, L. A., Davis, L. C. & Kress, J. W. Practical cone-beam algorithm. *J Opt Soc Am A* **1,** 612–9 (1984).
7. Flannery, B. P., Deckman, H. W., Roberge, W. G. & Amico, K. L. D. Three-dimensional X-ray microtomography. *Science* **237,** 1439–44 (1987).
8. Rüegsegger, P., Koller, B. & Müller, R. A microtomographic system for the nondestructive evaluation of bone architecture. *Calcif Tissue Int* **58,** 24–9 (1996).
9. Hildebrand, T. & Rüegsegger, P. A new method for the model-independent assessment of thickness in three-dimensional images. *J Microsc* **185,** 67–75 (1997).
10. Paulus, M. J., Gleason, S. S., Kennel, S. J., Hunsicker, P. R. & Johnson, D. K. High resolution X-ray computed tomography: An emerging tool for small animal cancer research. *Neoplasia* **2,** 62–70 (2000).
11. Metscher, B. D. MicroCT for comparative morphology: Simple staining methods allow high-contrast 3D imaging of diverse non-mineralized animal tissues. *BMC Physiol* **9,** (2009).
12. Metscher, B. D. MicroCT for developmental biology: A versatile tool for high-contrast 3D imaging at histological resolutions. *Dev Dyn* **238,** 632–40 (2009).
13. Palmer, A. W., Guldberg, R. E. & Levenston, M. E. Analysis of cartilage matrix fixed charge density and three-dimensional morphology via contrast-enhanced microcomputed tomography. *Proc Natl Acad Sci USA* **103,** 19255–60 (2006).
14. Joshi, N. S., Bansal, P. N., Stewart, R. C., Snyder, B. D. & Grinstaff, M. W. Effect of contrast agent charge on visualization of articular cartilage using computed tomography: Exploiting electrostatic interactions for improved sensitivity. *J Am Chem Soc* **131,** 13234–5 (2009).
15. Duvall, C. L., Taylor, W. R., Weiss, D. & Guldberg, R. E. Quantitative microcomputed tomography analysis of collateral vessel development after ischemic injury. *Am J Physiol Hear Circ Physiol* **287,** H302–10 (2004).
16. Pykett, I. L. et al. Principles of nuclear magnetic resonance imaging. *Radiology* **143,** 157–68 (1982).
17. Martiniova, L. et al. Noninvasive monitoring of a murine model of metastatic pheochromocytoma: A comparison of contrast-enhanced microCT and nonenhanced MRI. *J Magn Reson Imaging* **29,** 685–91 (2009).
18. Pandit, P. et al. The utility of micro-CT and MRI in the assessment of longitudinal growth of liver metastases in a preclinical model of colon carcinoma. *Acad Radiol* **20,** 430–9 (2013).
19. Saito, S., Mori, Y., Yoshioka, Y. & Murase, K. High-resolution ex vivo imaging in mouse spinal cord using micro-CT with 11.7T-MRI and myelin staining validation. *Neurosci Res* **73,** 337–40 (2012).
20. Wiesmann, F. et al. Dobutamine-stress magnetic resonance microimaging in mice: Acute changes of cardiac geometry and function in normal and failing murine hearts. *Circ Res* **88,** 563–9 (2001).
21. Choukèr, A. et al. Comparison of Fenestra VC contrast-enhanced computed tomography imaging with gadopentetate dimeglumine and ferucarbotran magnetic resonance imaging for the in vivo evaluation of murine liver damage after ischemia and reperfusion. *Invest Radiol* **43,** 77–91 (2008).
22. Schambach, S. J., Bag, S., Groden, C., Schilling, L. & Brockmann, M. A. Vascular imaging in small rodents using micro-CT. *Methods* **50,** 26–35 (2010).

23. Stephenson, R. S. et al. Contrast enhanced micro computed tomography resolves the 3-dimensional morphology of the cardiac conduction system in mammalian hearts. *PLoS One* **7,** e35299 (2012).
24. Pai, V. M. et al. Coronary artery wall imaging in mice using osmium tetroxide and micro-computed tomography (micro-CT). *J Anat* **220,** 514–24 (2012).
25. Parkesh, R., Lee, T. C., Gunnlaugsson, T. & Gowin, W. Microdamage in bone: Surface analysis and radiological detection. *J Biomech* **39,** 1552–6 (2006).
26. Parkesh, R., Gowin, W., Lee, T. C. & Gunnlaugsson, T. Synthesis and evaluation of potential CT (computer tomography) contrast agents for bone structure and microdamage analysis. *Org Biomol Chem* **4,** 3611–7 (2006).
27. Tang, S. Y. & Vashishth, D. A non-invasive in vitro technique for the three-dimensional quantification of microdamage in trabecular bone. *Bone* **40,** 1259–64 (2007).
28. Turnbull, T. L., Gargac, J. A., Niebur, G. L. & Roeder, R. K. Detection of fatigue microdamage in whole rat femora using contrast-enhanced micro-computed tomography. *J Biomech* **44,** 2395–400 (2011).
29. Landrigan, M. D. et al. Contrast-enhanced micro-computed tomography of fatigue microdamage accumulation in human cortical bone. *Bone* **48,** 443–50 (2011).
30. Saito, S. & Murase, K. Visualization of mouse spinal cord microscopic structures by use of ex vivo quantitative micro-CT images. *Radiol Phys Technol* **6,** 7–13 (2013).
31. Lareida, A. et al. High-resolution X-ray tomography of the human inner ear: Synchrotron radiation-based study of nerve fiber bundles, membranes and ganglion cells. *J Microsc* **234,** 95–102 (2009).
32. Patel, S. R., Lin, A. S. P., Edelhauser, H. F. & Prausnitz, M. R. Suprachoroidal drug delivery to the back of the eye using hollow microneedles. *Pharm Res* **28,** 166–76 (2011).
33. Smith, C. A. et al. A framework for modeling ocular drug transport and flow through the eye using micro-CT. *Phys Med Biol* **57,** 6295–307 (2012).
34. De Crespigny, A. et al. 3D micro-CT imaging of the postmortem brain. *J Neurosci Methods* **171,** 207–13 (2008).
35. Dobrivojević, M., Bohaček, I., Erjavec, I., Gorup, D. & Gajović, S. Computed microtomography visualization and quantification of mouse ischemic brain lesion by nonionic radio contrast agents. *Croat Med J* **54,** 3–11 (2013).
36. Johnson, J. T. et al. Virtual histology of transgenic mouse embryos for high-throughput phenotyping. *PLoS Genet* **2,** e61 (2006).
37. Bergua, L. et al. Peritoneal membrane recruitment in rats: A micro-computerized tomography (muCT) study. *Pediatr Nephrol* **23,** 2179–84 (2008).
38. Winkelmann, C. T. & Wise, L. D. High-throughput micro-computed tomography imaging as a method to evaluate rat and rabbit fetal skeletal abnormalities for developmental toxicity studies. *J Pharmacol Toxicol Methods* **59,** 156–65 (2009).
39. Aprahamian, M. et al. Myo-InositolTrisPyroPhosphate treatment leads to HIF-1α suppression and eradication of early hepatoma tumors in rats. *Chembiochem* **12,** 777–83 (2011).
40. Boll, H. et al. Double-contrast micro-CT colonoscopy in live mice. *Int J Color Dis* **26,** 721–7 (2011).
41. Breton, E. et al. In vivo peritoneal surface area measurement in rats by micro-computed tomography (microCT). *Perit Dial Int* **28,** 188–94 (2008).
42. Graham, K. C. et al. Contrast-enhanced microcomputed tomography using intraperitoneal contrast injection for the assessment of tumor-burden in liver metastasis models. *Invest Radiol* **43,** 488–95 (2008).
43. Kokkonen, H. T., Jurvelin, J. S., Tiitu, V. & Töyräs, J. Detection of mechanical injury of articular cartilage using contrast enhanced computed tomography. *Osteoarthr Cartil* **19,** 295–301 (2011).
44. Xie, L., Lin, A. S. P., Levenston, M. E. & Guldberg, R. E. Quantitative assessment of articular cartilage morphology via EPIC-microCT. *Osteoarthr Cartil* **17,** 313–20 (2009).
45. Moyer, H. R. et al. A new animal model for assessing cartilage repair and regeneration at a nonarticular site. *Tissue Eng Part A* **16,** 2321–30 (2010).
46. Lee, C. S. et al. Adipose stem cells can secrete angiogenic factors that inhibit hyaline cartilage regeneration. *Stem Cell Res Ther* **3,** 35 (2012).
47. Xie, L. et al. Quantitative imaging of cartilage and bone morphology, reactive oxygen species, and vascularization in a rodent model of osteoarthritis. *Arthritis Rheum* **64,** 1899–908 (2012).
48. Kotwal, N., Li, J., Sandy, J., Plaas, A. & Sumner, D. R. Initial application of EPIC-μCT to assess mouse articular cartilage morphology and composition: Effects of aging and treadmill running. *Osteoarthr Cartil* **20,** 887–95 (2012).
49. Thote, T. et al. Localized 3D analysis of cartilage composition and morphology in small animal models of joint degeneration. *Osteoarthr Cartil* **21,** 1132–41 (2013).
50. Xie, L., Lin, A. S. P., Guldberg, R. E. & Levenston, M. E. Nondestructive assessment of sGAG content and distribution in normal and degraded rat articular cartilage via EPIC-microCT. *Osteoarthr Cartil* **18,** 65–72 (2010).
51. Bansal, P. N., Stewart, R. C., Entezari, V., Snyder, B. D. & Grinstaff, M. W. Contrast agent electrostatic attraction rather than repulsion to glycosaminoglycans affords a greater contrast uptake ratio and improved quantitative CT imaging in cartilage. *Osteoarthr Cartil* **19,** 970–6 (2011).
52. Stewart, R. C. et al. Contrast-enhanced CT with a high-affinity cationic contrast agent for imaging ex vivo bovine, intact ex vivo rabbit, and in vivo rabbit cartilage. *Radiology* **266,** 141–50 (2013).
53. Kokkonen, H. T. et al. Computed tomography detects changes in contrast agent diffusion after collagen cross-linking typical to natural aging of articular cartilage. *Osteoarthr Cartil* **19,** 1190–8 (2011).
54. Kulmala, K. A. M. et al. Contrast-enhanced micro-computed tomography in evaluation of spontaneous repair of equine cartilage. *Cartilage* **3,** 235–44 (2012).
55. Bansal, P. N., Joshi, N. S., Entezari, V., Grinstaff, M. W. & Snyder, B. D. Contrast enhanced computed tomography can predict the glycosaminoglycan content and biomechanical properties of articular cartilage. *Osteoarthr Cartil* **18,** 184–91 (2010).
56. Bansal, P. N. et al. Cationic contrast agents improve quantification of glycosaminoglycan (GAG) content by

contrast enhanced CT imaging of cartilage. *J Orthop Res* **29,** 704–9 (2011).
57. Siebelt, M. et al. Quantifying osteoarthritic cartilage changes accurately using in vivo microCT arthrography in three etiologically distinct rat models. *J Orthop Res* **29,** 1788–94 (2011).
58. Hayward, L. N. M., de Bakker, C. M. J., Gerstenfeld, L. C., Grinstaff, M. W. & Morgan, E. F. Assessment of contrast-enhanced computed tomography for imaging of cartilage during fracture healing. *J Orthop Res* **31,** 567–73 (2013).
59. Beighley, P. E., Thomas, P. J., Jorgensen, S. M. & Ritman, E. L. 3D Architecture of myocardial microcirculation in intact rat heart: A study with micro-CT. *Adv Exp Med Biol* **430,** 165–75 (1997).
60. Jackowski, M., Papademetris, X., Dobrucki, L. W., Sinusas, A. J. & Staib, L. H. Characterizing vascular connectivity from microCT images. *Med Image Comput Comput Assist Interv* **8,** 701–8 (2005).
61. Sider, K. L., Song, J. & Davies, J. E. A new bone vascular perfusion compound for the simultaneous analysis of bone and vasculature. *Microsc Res Tech* **73,** 665–72 (2010).
62. Kwon, H. M. et al. Enhanced coronary vasa vasorum neovascularization in experimental hypercholesterolemia. *J Clin Invest* **101,** 1551–6 (1998).
63. Kwon, H. M. et al. Adventitial vasa vasorum in balloon-injured coronary arteries: Visualization and quantitation by a microscopic three-dimensional computed tomography technique. *J Am Coll Cardiol* **32,** 2072–9 (1998).
64. Rodriguez-Porcel, M. et al. Altered myocardial microvascular 3D architecture in experimental hypercholesterolemia. *Circulation* **102,** 2028–30 (2000).
65. Lee, J., Beighley, P., Ritman, E. & Smith, N. Automatic segmentation of 3D micro-CT coronary vascular images. *Med Image Anal* **11,** 630–47 (2007).
66. Moore, D. C., Leblanc, C. W., Muller, R., Crisco, J. J. & Ehrlich, M. G. Physiologic weight-bearing increases new vessel formation during distraction osteogenesis: A micro-tomographic imaging study. *J Orthop Res* **21,** 489–96 (2003).
67. Qin, L. et al. Multiple bioimaging modalities in evaluation of an experimental osteonecrosis induced by a combination of lipopolysaccharide and methylprednisolone. *Bone* **39,** 863–71 (2006).
68. Zhang, G. et al. Epimedium-derived phytoestrogen exert beneficial effect on preventing steroid-associated osteonecrosis in rabbits with inhibition of both thrombosis and lipid-deposition. *Bone* **40,** 685–92 (2007).
69. Zhang, G. et al. A novel semisynthesized small molecule icaritin reduces incidence of steroid-associated osteonecrosis with inhibition of both thrombosis and lipid-deposition in a dose-dependent manner. *Bone* **44,** 345–56 (2009).
70. Zhang, X. et al. Periosteal progenitor cell fate in segmental cortical bone graft transplantations: Implications for functional tissue engineering. *J Bone Min Res* **20,** 2124–37 (2005).
71. Bolland, B. J. R. F., Kanczler, J. M., Dunlop, D. G. & Oreffo, R. O. C. Development of in vivo muCT evaluation of neovascularisation in tissue engineered bone constructs. *Bone* **43,** 195–202 (2008).
72. Shen, X. et al. Prolyl hydroxylase inhibitors increase neoangiogenesis and callus formation following femur fracture in mice. *J Orthop Res* **27,** 1298–305 (2009).
73. Raines, A. L. et al. Hyaluronic acid stimulates neovascularization during the regeneration of bone marrow after ablation. *J Biomed Mater Res A* **96,** 575–83 (2011).
74. Boerckel, J. D., Uhrig, B. A., Willett, N. J., Huebsch, N. & Guldberg, R. E. Mechanical regulation of vascular growth and tissue regeneration in vivo. *Proc Natl Acad Sci USA* **108,** E674–80 (2011).
75. Masyuk, T. V., Ritman, E. L. & LaRusso, N. F. Quantitative assessment of the rat intrahepatic biliary system by three-dimensional reconstruction. *Am J Pathol* **158,** 2079–88 (2001).
76. Garcia-Sanz, A., Rodriguez-Barbero, A., Bentley, M. D., Ritman, E. L. & Romero, J. C. Three-dimensional microcomputed tomography of renal vasculature in rats. *Hypertension* **31,** 440–4 (1998).
77. Ortiz, M. C. et al. Microcomputed tomography of kidneys following chronic bile duct ligation. *Kidney Int* **58,** 1632–40 (2000).
78. Bentley, M. D. et al. Enhanced renal cortical vascularization in experimental hypercholesterolemia. *Kidney Int* **61,** 1056–63 (2002).
79. Fortepiani, L. A. et al. Effect of losartan on renal microvasculature during chronic inhibition of nitric oxide visualized by micro-CT. *Am J Physiol Ren Physiol* **285,** F852–60 (2003).
80. Plouraboue, F. et al. X-ray high-resolution vascular network imaging. *J Microsc* **215,** 139–48 (2004).
81. Abruzzo, T. et al. Microscopic computed tomography imaging of the cerebral circulation in mice: Feasibility and pitfalls. *Synapse* **62,** 557–65 (2008).
82. Müller, B., Fischer, J., Dietz, U., Thurner, P. J. & Beckmann, F. Blood vessel staining in the myocardium for 3D visualization down to the smallest capillaries. *Nucl Instrum Methods Phys Res B* **246,** 254–61 (2006).
83. Grabherr, S. et al. Angiofil-mediated visualization of the vascular system by microcomputed tomography: A feasibility study. *Microsc Res Tech* **71,** 551–6 (2008).
84. Bakan, D. A. et al. Imaging efficacy of a hepatocyte-selective polyiodinated triglyceride for contrast-enhanced computed tomography. *Am J Ther* **8,** 359–65 (2001).
85. Kao, C., Hoffman, E. A., Beck, K. C., Bellamkonda, R. V. & Annapragada, A. V. Long-residence-time nano-scale liposomal iohexol for X-ray-based blood pool imaging. *Acad Radiol* **10,** 475–83 (2003).
86. Badea, C. T., Fubara, B., Hedlund, L. W. & Johnson, G. A. 4-D micro-CT of the mouse heart. *Mol Imaging* **4,** 110–6 (2005).
87. Drangova, M., Ford, N. L., Detombe, S. A., Wheatley, A. R. & Holdsworth, D. W. Fast retrospectively gated quantitative four-dimensional (4D) cardiac micro computed tomography imaging of free-breathing mice. *Invest Radiol* **42,** 85–94 (2007).
88. Badea, C. T., Hedlund, L. W., Cook, J., Berridge, B. R. & Johnson, G. A. Micro-CT imaging assessment of dobutamine-induced cardiac stress in rats. *J Pharmacol Toxicol Methods* **63,** 24–9 (2011).
89. Detombe, S. A. et al. Longitudinal follow-up of cardiac structure and functional changes in an infarct mouse model using retrospectively gated micro-computed tomography. *Invest Radiol* **43,** 520–9 (2008).

90. Badea, C. T. et al. Lung perfusion imaging in small animals using 4D micro CT at heartbeat temporal resolution. *Med Phys* **37,** 54–62 (2010).
91. Badea, C. T., Johnston, S. M., Qi, Y. & Johnson, G. A. 4D micro-CT for cardiac and perfusion applications with view under sampling. *Phys Med Biol* **56,** 3351–69 (2011).
92. Mukundan, S. et al. A liposomal nanoscale contrast agent for preclinical CT in mice. *AJR Am J Roentgenol* **186,** 300–7 (2006).
93. Ghaghada, K. B. et al. Evaluation of tumor microenvironment in an animal model using a nanoparticle contrast agent in computed tomography imaging. *Acad Radiol* **18,** 20–30 (2011).
94. Moding, E. J. et al. Dual-energy micro-computed tomography imaging of radiation-induced vascular changes in primary mouse sarcomas. *Int J Radiat Oncol Biol Phys* **85,** 1353–9 (2013).
95. Almajdub, M. et al. In-vivo high-resolution X-ray microtomography for liver and spleen tumor assessment in mice. *Contrast Media Mol Imaging* **2,** 88–93 (2007).
96. Fouillet, X. et al. Enhancement of computed tomography liver contrast using and detection of small liver tumors in rats. *Acad Radiol* **2,** 576–83 (1995).
97. Montet, X. et al. Improved visualization of vessels and hepatic tumors by micro-computed tomography (CT) using iodinated liposomes. *Invest Radiol* **42,** 652–8 (2007).
98. Willekens, I. et al. Time-course of contrast enhancement in spleen and liver with Exia 160, Fenestra LC, and VC. *Mol Imaging Biol* **11,** 128–35 (2009).
99. Kitahashi, T. et al. Imaging study of pancreatic ductal adenocarcinomas in Syrian hamsters using X-ray micro-computed tomography (CT). *Cancer Sci* **101,** 1761–6 (2010).
100. Stroope, A. et al. Hepato-renal pathology in pkd2ws25/- mice, an animal model of autosomal dominant polycystic kidney disease. *Am J Pathol* **176,** 1282–91 (2010).
101. Hayasaka, N. et al. In vivo diagnostic imaging using micro-CT: Sequential and comparative evaluation of rodent models for hepatic/brain ischemia and stroke. *PLoS One* **7,** e32342 (2012).
102. Boll, H. et al. Micro-CT based experimental liver imaging using a nanoparticulate contrast agent. A longitudinal study in mice. *PLoS One* **6,** e25692 (2011).
103. Schwarz, M. et al. In vivo imaging of MSCT and micro-CT: A comparison. *Rofo* **182,** 322–6 (2010).
104. Schambach, S. J. et al. Ultrafast high-resolution in vivo volume-CTA of mice cerebral vessels. *Stroke* **40,** 1444–50 (2009).
105. Weber, S. M. et al. Imaging of murine liver tumor using microCT with a hepatocyte-selective contrast agent: Accuracy is dependent on adequate contrast enhancement. *J Surg Res* **119,** 41–5 (2004).
106. Ohta, S. et al. MicroCT for high-resolution imaging of ectopic pheochromocytoma tumors in the liver of nude mice. *Int J Cancer* **119,** 2236–41 (2006).
107. Henning, T. et al. Imaging characteristics of DHOG, a hepatobiliary contrast agent for preclinical microCT in mice. *Acad Radiol* **15,** 342–9 (2008).
108. Suckow, C. E. & Stout, D. B. MicroCT liver contrast agent enhancement over time, dose, and mouse strain. *Mol Imaging Biol* **10,** 114–20 (2008).
109. Badea, C., Hedlund, L. W. & Johnson, G. A. Micro-CT with respiratory and cardiac gating. *Med Phys* **31,** 3324 (2004).
110. Daire, J.-L. et al. Cine and tagged cardiovascular magnetic resonance imaging in normal rat at 1.5 T: A rest and stress study. *J Cardiovasc Magn Reson* **10,** (2008).
111. Pasqualini, R., Arap, W. & McDonald, D. M. Probing the structural and molecular diversity of tumor vasculature. *Trends Mol Med* **8,** 563–71 (2002).
112. Bergers, G. & Benjamin, L. E. Tumorigenesis and the angiogenic switch. *Nat Rev Cancer* **3,** 401–10 (2003).
113. Lam, W. W. et al. Micro-CT imaging of rat lung ventilation using continuous image acquisition during xenon gas contrast enhancement. *J Appl Physiol* **103,** 1848–56 (2007).
114. Sandy, J. D. & Verscharen, C. Analysis of aggrecan in human knee cartilage and synovial fluid indicates that aggrecanase (ADAMTS) activity is responsible for the catabolic turnover and loss of whole aggrecan whereas other protease activity is required for C-terminal processing in vivo. *Biochem J* **358,** 615–26 (2001).
115. Durkee, B. Y., Weichert, J. P. & Halberg, R. B. Small animal micro-CT colonography. *Methods* **50,** 36–41 (2010).
116. Larusso, N. F. Morphology, physiology, and biochemistry of biliary epithelia. *Toxicol Pathol* **24,** 84–9 (1996).
117. Forbes, S. J. & Parola, M. Liver fibrogenic cells. *Best Pr Res Clin Gastroenterol* **25,** 207–17 (2011).
118. Verhagen, A. M. et al. Losartan-sensitive renal damage caused by chronic NOS inhibition does not involve increased renal angiotensin II concentrations. *Kidney Int* **56,** 222–31 (1999).
119. Kouzmanova, M., Ivanov, S., Nankova, V. & Traykov, L. Effects of ionizing and nonionizing radiation on rat hemopoiesis. *Electromagn Biol Med* **16,** 161–8 (1997).
120. Boone, J. M., Velazquez, O. & Cherry, S. R. Small-animal X-ray dose from micro-CT. *Mol Imaging* **3,** 149–58 (2004).
121. Nett, B. E., Brauweiler, R., Kalender, W., Rowley, H. & Chen, G.-H. Perfusion measurements by micro-CT using prior image constrained compressed sensing (PICCS): Initial phantom results. *Phys Med Biol* **55,** 2333–50 (2010).
122. Lundström, U. et al. X-ray phase contrast for CO2 micro-angiography. *Phys Med Biol* **57,** 2603–17 (2012).
123. Moosmann, J. et al. X-ray phase-contrast in vivo microtomography probes new aspects of Xenopus gastrulation. *Nature* **497,** 374–7 (2013).

Section V

Tissue and Organ Imaging Methods

High Throughput and Image Quantification

11

Application of Digital Image Correlation for Multiscale Biomechanics

Hui Wang, William Lai, Antonia Antoniou, and Ashraf Bastawros

CONTENTS

Applications of digital image correlation (DIC) to explore the multiscale mechanical properties of biological and biological-like tissues are presented. A highlight of DIC principles and DIC's mathematical framework is reviewed. Different methods of sample decoration to generate high-contrast speckle patterns for both soft and hard tissue are discussed. The relation between different length scales of DIC techniques is presented for highly localized deformation field, especially near bounding surfaces and interfaces. Careful consideration is required for the imaging magnification and the scale of the measured phenomena, the scale of the decorating speckles, and the correlation window size to yield high-fidelity deformation field. Two examples of DIC applications to hard and soft biological materials are presented to elucidate the DIC technique potential. The evolution of localized inhomogeneous deformation in hard tissues of porous trabecular bone-like materials is summarized. The extension of the technique to measure the three-dimensional interaction forces and deformation in soft tissue representing cell interaction with their extracellular environment are also summarized. The continued development of DIC and its extension to 3D volume measurements will provide viable tools to probe the interactions between biochemical and mechanical signaling in various cellular processes.

11.1 Introduction

Digital image correlation (DIC) has been developed for many years as a noncontact method to evaluate the full-field deformation measurements by analyzing the acquired images of an object subjected to an external stimulus (e.g., load, temperature, or electric and magnetic field). It was first proposed in the early 1980s to compute the two-dimensional in-plane surface strains and displacements [1–3]. It has been used either as a quantitative tool for average and homogeneous strains of macroscopically uniform deformation [2,4] or as a whole-field deformation mapping tool to assess plastic deformation patterns [5], measure highly localized deformation fields such as localization during tensile loading [6], monitor the collapse of aluminum alloy foams under compressive loads [7], characterize strain bifurcation in foams [8], measure the deformation field surrounding a crack tip in cellular media [9], as well as measure microscopic large strain from scanning electron microscopy (SEM) topography images [10]. The DIC technique is further developed to measure the full three-dimensional surface deformation employing the principles of stereo vision using a two-camera system [11,12] to measure the full three-dimensional shapes and deformation of planar or curved surfaces.

As a noncontact method, DIC provides a noninvasive tool to study the biomechanics of both hard and soft tissues. A wide range of hard tissues were explored, covering different classes of bone tissue [13–17], as well as hard biological tissues such as hoof horn [18], lobster cuticle [19], and beetle wing membrane [20]. Soft tissues ranging from arterial tissue [21], skin tissue under dynamic loading [22], liver tissue [23], to the cell-level deformation and interaction with the environment [24–26] were also explored. DIC has been employed as an average strain measuring technique to derive the tissue average mechanical properties and the appropriate material constants, under the assumption of some constitutive framework. Mechanical properties of a diverse group of tissues were investigated, such as anisotropic elasticity of blood vessel [18,21,27,28], incompressibility of brain tissue [29], human skin mechanical properties [30], mechanical properties of the femoral head [31], role of moisture on the elastic properties of hoof horn in cows [18], role of moisture on the elastoplastic properties of lobster shells [19], as well as the general nonlinear mechanical response of soft tissue [32]. The DIC technique has been also employed to study the local deformation mechanisms and interaction with the environment, such as the evolution of surface strain distribution on a mouse tibia under loading [14], fracture of proximal femurs [15], local strain distribution within early bone callus [16] and the associated evolution of damage and fracture [33], loosening of cemented total hip replacements [18], as well as cement–bone interface and local relative displacement [34]. Details of inelastic and localized deformation of bovine fibrolamellar bone were also explored [17]. DIC strain measurements have been also extended to a three-dimensional tissue volume, elaborating the details of the deformation field of the inner points and connecting ligaments within such volume. The technique relies on a volumetric imaging technique such as high-resolution x-ray computed tomography imaging, with the ability to measure the strains within a trabecular bone strut [13]. Three-dimensional imaging by laser scanning confocal microscopy with fluorescent particles provides in situ monitoring for cell-level interaction with the environment. When combined with digital volume correlations (DVCs), cell-mediated deformation of the extracellular matrix can be quantified [24–26].

This chapter provides relevant details of the DIC technique for relevant applications in biomechanics investigations. The article highlights the principles of the DIC technique, proper sample preparation, and proper gage length. Detailed mathematical formulation can be found elsewhere [3,8,28]. Two examples of DIC applications are presented here: an example of hard tissue for porous trabecular bone-like materials, and the other example for soft tissue representing cell interaction with their extra cellular environment. Fluorescent nanospheres are viable options, especially for 3D volume imaging of cell–cell interactions.

11.2 DIC Principles

11.2.1 Technique Basics

As a deformation measurement technique, DIC provides an estimate of the displacement field on the surface of the specimen or within its bulk. The DIC technique relies on correlating an array of features in a pair of digital images taken before and after deformation, via a mathematically well-defined function, as depicted in Figure 11.1. The DIC technique has different correlation algorithms, such as estimating discrete displacement data by local optimal match of intensity pattern in subsets (subset method) [35], or obtaining continuous displacement field function by full-field optimal match of the intensity pattern (continuous method) [36]. Different intensity value interpolation schemes are employed, including

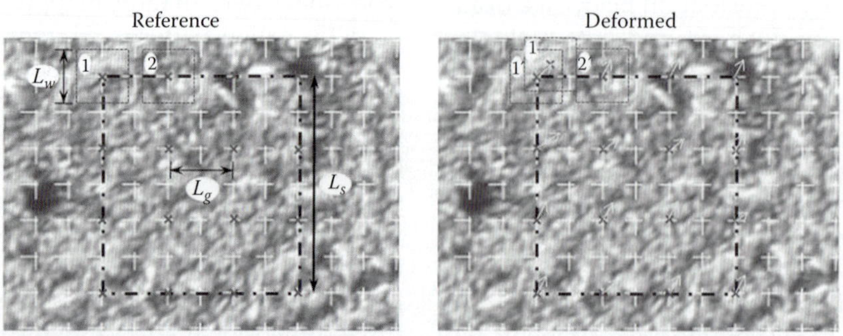

FIGURE 11.1
Principles of full-field digital image correlations; (1) coarse correlation search to find initial guess in unit pixel, and (2) fine search with subpixel resolution.

the most common polynomial or B-spline for 2D surface [37] and tricubic interpolator based on 1D Hermitian cardinal basis functions for 3D volume [38]. The continuous method normally predicts accurate displacements and their gradients, when applied to a deformation field without localization [36]. However, it would smear out localizations, for highly inhomogeneous deformation fields, especially with discontinuities.

A typical DIC analysis involves a two-step process of coarse and fine search as shown in Figure 11.1. The coarse search step is based on maximizing the intensity-based correlation coefficient between pairs of subwindows to an integer number of pixels. The fine search utilizes the coarse search results as an initial guess for a series of 2D Newton–Raphson iterations that resolve displacements and their gradients within corresponding subwindows with subpixel resolution [3]. Since the image intensity is spatially sampled with pixel periodicity, an interpolation scheme is utilized to provide the subpixel accuracy. Different interpolation schemes have been explored such as bilinear, bicubic, high-order polynomials and the B-spline method [37]. The acquired nodal displacements from both coarse and fine search represent the average displacement of each subwindow. Despite the accuracy of the numerical algorithm, the perceived resolution is about 0.01 pixel [39]. It should be noted that the acquired nodal displacement fields, while in themselves have high accuracy, possess strong fluctuation to render a compatible displacement field. A low-pass filter might be utilized on the acquired displacement field with a minimum wavelength greater than twice the grid spacing L_g to eliminate the high-frequency digital noise, without obscuring any displacement localization. Utilizing the filtered nodal displacements, the Lagrangian strain components on original configuration, or the Almansi strain component on the deformed configuration can be evaluated.

11.2.2 Imaging Scale and Magnification

The core assumption of the DIC technique is finding a representative region with different variation in the intensity level so that accurate matching can be performed to define the relative local subwindow motion. This region should have almost uniform deformation. While a feature on the image requires a minimum of 2 pixels to be represented (the Nyquist frequency limit), a practical limit of the feature size within the subwindow is about 5–10 pixels to ensure its unique characteristics (see also Chapter 33). Accordingly, a correlation window of about $L_w \sim 15$ pixels or greater is required. The L_w scale represents the gage length for strain calculation. In addition, the L_w scale controls the magnification factor between the object and the image plane, wherein local deformation uniformity assumption prevails. Therefore, the highest magnification that might be used is controlled by the uniformity of the deformation that could be mapped to an area of about 15–20 pixels. Beyond such a limit, a redundancy of the nodal displacement field might occur.

A full representation of the displacement field could be acquired by performing the correlation analysis on a structured or unstructured grid. Structured grids are typically used when a continuous, space-filling dense material is tested. Unstructured analysis grids are typically implemented for cellular and porous material, following local topography, like the cancellous bone structure. Figure 11.2 shows the unstructured grid analysis of 2D bone-like material, a honeycomb. One layer of honeycomb cells is assembled into a double shear-lap to understand the local cell interaction with the attached skin, which is a scenario similar to the cancellous bone tissue interaction with implants. A skeletal tracking algorithm is utilized to locate the analysis points within the cell ligaments. Figure 11.2b shows the details of the measured displacement components parallel to the bounding skin. Figure 11.2c shows a full field distribution of the measured engineering shear strain within the cellular structure. The role of the bounding skin can be seen in locking the deformation of the attached cell ligament to the skin, leading to a stiffening effect within the first core cell near the face sheet. The deformation is focused and localized as a band within the core of the cellular layer.

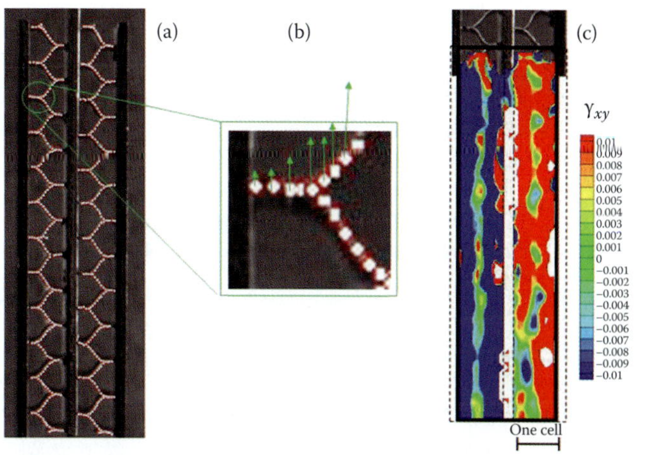

FIGURE 11.2
DIC deformation analysis of bone-like structure at the cell level (a honeycomb) using unstructured mesh to follow local topological features. (a) Skeletal tracking to identify the analysis points. (b) Cell level displacement vector, parallel to the attached skin. (c) Calculated engineering shear strain within the cellular structure, showing a boundary layer cell-locking effect with the attaching skin and localized deformation within the core.

11.2.3 Mapping Localized Deformation

Large deformation of soft tissues or highly porous hard tissues is highly nonuniform, especially near interfaces with different stiffness and elastic moduli. Resolving these types of deformation heterogeneity requires careful selection of the image magnification factor and the analysis grid periodicity. To illustrate the interaction of the different length scale, a parametric study is presented in Figure 11.3. A mathematical one-dimensional localization is imbedded into the image of Figure 11.1, with a displacement jump of 1 pixel, over a different localization length scale L_o. A range of L_w and L_g was employed to resolve this localization. The correlation window size is typically around L_w ~ 15 pixels, independent of the image size or magnification. In practice, L_o is dictated by the image size and the magnification. Figure 11.3a shows a range of localization: $0.2L_w > L_o > 2L_w$. It is clearly shown that the apparently measured localization scale, L_{app}, approaches the true values of L_o, when L_o approaches twice the correlation window size L_w. Additionally, when the scale of localization is much smaller than L_w, errors of many orders of magnitude in estimating L_o will occur. For example, for $L_o/L_w = 0.18$, the error in estimating L_o is about 670%, a three orders of magnitude error. There is a small range for changing the correlation window as shown in Figure 11.3b. L_w can be as small as 10 pixels for images with high dynamic range. It remains true that when $L_o < 2L_w$, true size of the localization band cannot be estimated. Finally, the role of the grid spacing L_g is shown in Figure 11.3c. The extremely dense grid does not enhance the accuracy of a uniform deformation field, as can be seen away from the localization domain for all the grid spacing range. However, for an inhomogeneous deformation field, the grid spacing has to be around L_g ~ 10%–20%L_o. Accordingly, L_w is typically set around 15 pixels or higher based on the uniformity and contrast of the image. Lower contrast images require a larger correlation window. Then, the imaging magnification has to be changed so that any localization within the field of view should map to $L_o > 2L_w$. The grid spacing is adjusted in such a case to be around 0.1–0.2L_o. As a final word of caution, resolving localized deformation especially near boundaries and interfaces requires careful cross-examination with higher topographic images revealing the true localization length scale. Otherwise, both the spatial scale of the DIC measured localization and its amplitude will be biased with significant errors.

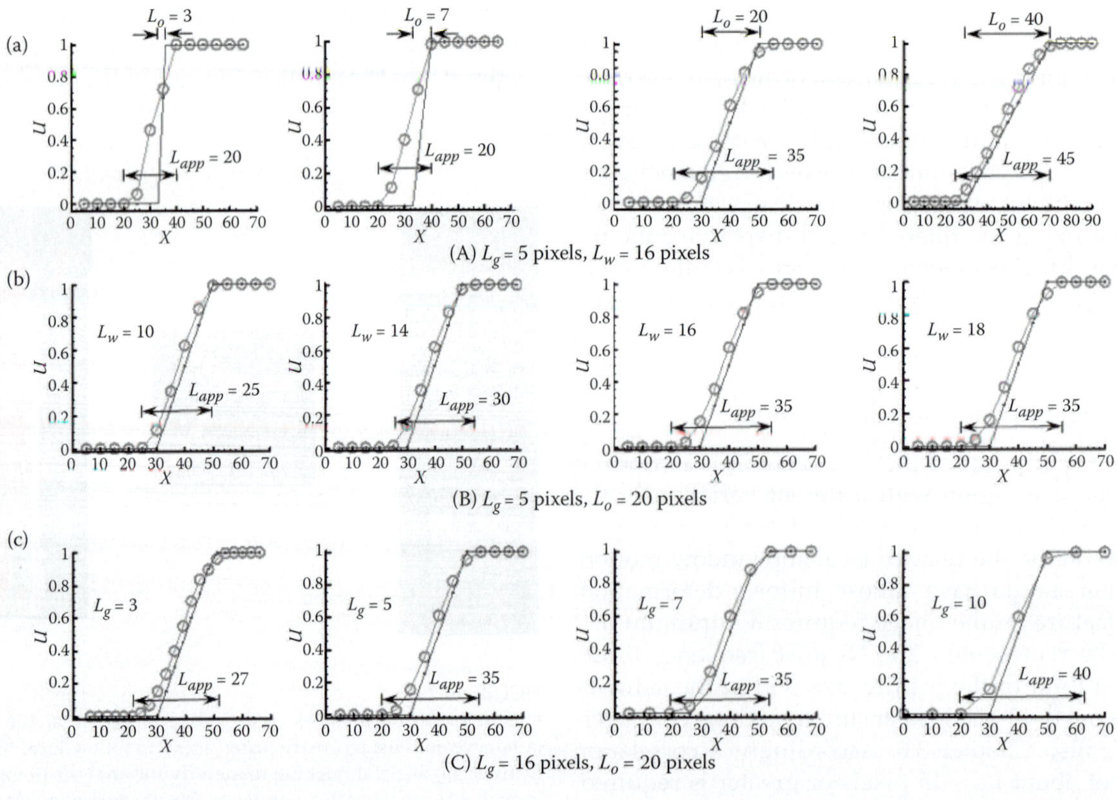

FIGURE 11.3
Role of different length scales affecting DIC measurements of localized deformation. (a) Role of localization length scale, L_o, compared to the correlation window L_w. (b) Role of the correlation window, L_w. (c) Role of grid spacing, L_g. DIC measurements marked with hollow circles and the band mathematical representation marked by a solid line. Coordinate x and localization amplitude U are in pixels.

11.2.4 Sample Preparation

Sample preparation for the DIC technique entails generating local variation of the intensity level within the correlation window. The natural surface texture of the sample, such as the case of cancellous bone, may be sufficient. For hard specimens, surface decoration may be utilized by depositing random features on the surface utilizing acrylic black and white spray paint, and physical vapor deposition, or e-beam sputtering of many conductive elements such as graphite, gold, silver, or chromium. Electrophoretic deposition for a wide range of microcolloidal and nanocolloidal suspensions can be used to generate a wide range of feature sizes over several orders of length scales, spanning from nanoscale to microscale. The key issue in selecting the proper particle size is to map the resulting cluster features to at least 5–10 pixels of the imaging system in order to provide unique and distinctive features. Thus, the proper size of the decorating feature is tied to the imaging scale and magnification and the uniformity of the interrogated deformation field.

The surface or bulk decoration process is quite difficult for soft tissues, since the stiffness of the painted or deposited layer may interfere with the total stiffness of the tissue, or greatly affect its local behavior, especially in a bioactive environment. Fluorescent microspheres are typically used for laser scanning confocal microscopy as a staining agent. They represent also a great contrast enhancer for DIC, either for surface or volumetric measurements [24–26].

11.3 Application of DIC Technique to Biological and Biological-Like Materials

11.3.1 Compression of a Closed Cell Foam, A Biological-Like Material

The deformation of a biological-like material, a closed cell aluminum foam (ALPORAS), with a 7.5% relative density is illustrated here under compression. The cells are relative equiaxed polyhedra (Figure 11.4). The mechanical isotropic response is approximately isotropic [7,40]. The compressive Young's modulus upon unloading is E = 0.7–1.0 GPa, and the nominal plateau stress, designated as the first peak on the compression stress–strain curve of Figure 11.4c, is around 1.2 MPa. The deformation is highly nonuniform with a compaction band spanning a single cell as shown in Figure 11.4b, especially following the peak stress. Details of the deformation mechanism can be found elsewhere [7]. Here, the details of the strain localization between stages A and B, following the nominal plateau stress, will be discussed.

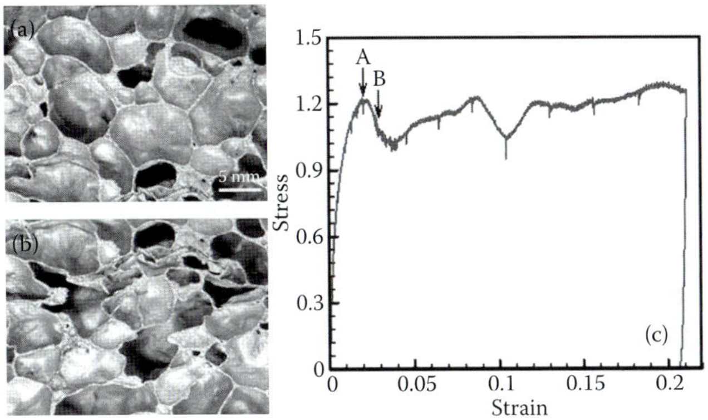

FIGURE 11.4
Example of biological like material; compression of closed cell aluminum foam. (a) Optical image of the microstructure before deformation. (b) Localized compression band after deformation, spanning a single row of cells. (c) Uniaxial compression stress–strain curve, showing the analysis points. (Images a and b are from Bastawros, A. F., and Evans, A. G., *Advanced Engineering Materials* 4:210–214, 2000.)

Surface images were recorded by a commercial video camera with a charge-coupled device array of 1024 × 1528 pixels. A wide aperture lens (F1.4) with an extended depth of field was used with the camera. Two fiber optic light sources were used to provide oblique white light illumination and shadow reduction. The deformation process of the foam is recorded by a sequence of images with a resolution of 60 μm/pixel. The captured images are analyzed by the DIC subset method, with L_g = 10 pixels and L_w = 16 pixels. The compaction band is mapped to about 40–50 pixels. The resulting overall maximum in-plane shear strain increment $\left(\varepsilon_{eff} = \left|\varepsilon_{22} - \varepsilon_{11}\right| = \sqrt{(\varepsilon_{xx} - \varepsilon_{yy})^2 + 4\varepsilon_{xy}^2}\right)$ between stages A and B is shown in Figure 11.5. An excessive localized deformation can be seen. To elaborate such localization, Figure 11.6 provides details of the deformation for the marked rectangular zone within the localized band in Figure 11.5 in the form of the total maximum in-plane

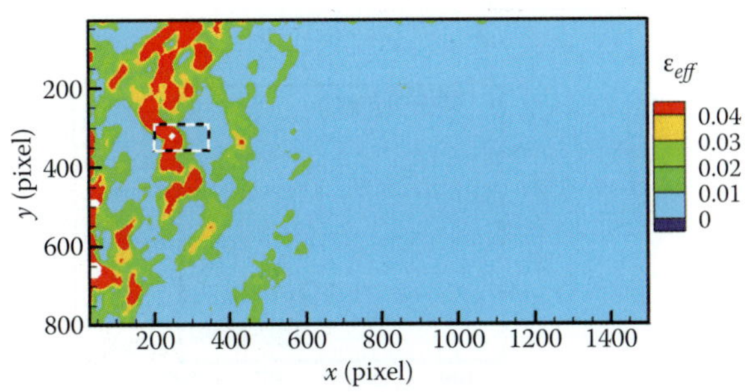

FIGURE 11.5
DIC-derived strain map for the in-plane incremental maximum shear strain of loading stage B relative to A. Exclusive localization can be seen, having a round one cell in thickness. Coordinates are in pixels.

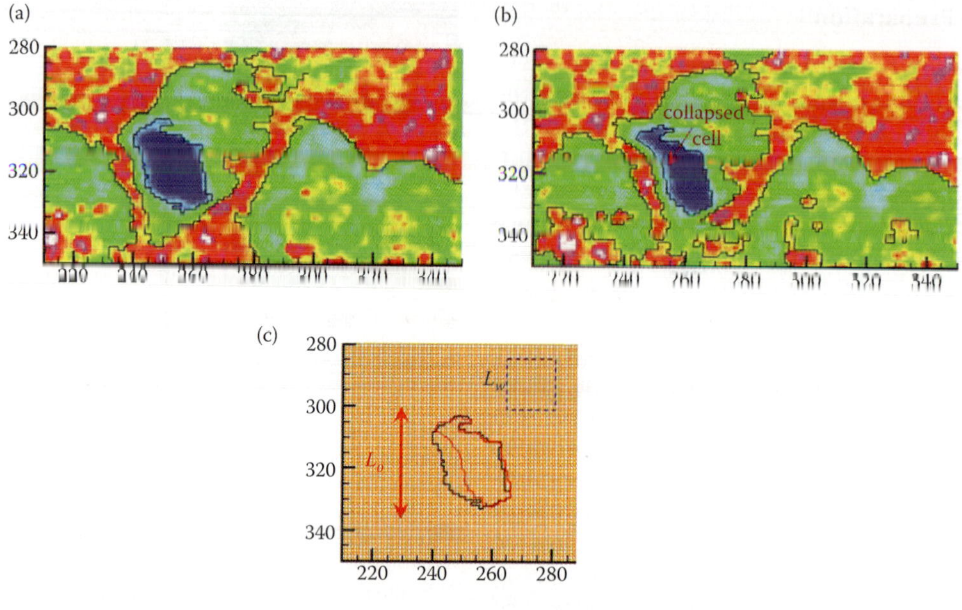

FIGURE 11.6
Details of cell crushing and localization as derived from the DIC analysis, with cell-level resolution, showing total strains for (a) stage "A" and (b) stage "B." (c) Superposition of the marked region as a collapsed cell, raced from the DIC strain contours. Also marked is the relative size of L_o and L_w in pixels. The color represents the magnitude of ε_{eff}.

shear strain, relative to the original configuration. The edges of the current traces of the cell walls are marked on both Figure 11.6a and b. Figure 11.6c shows an overlay of the cell shape at stage A, with its deformed shape at stage B, highlighting the process of cell collapse. The zone of strain localization (highlighted in blue) coincides with the collapsed cell. The grid scale of Figure 11.6c is in pixel number. The relative scale of the correlation window L_w and the scale of the localization band L_o are also highlighted to ensure the condition of localized deformation imaging of $L_o > 2L_w$. Figure 11.7 shows the distribution of the measured displacement components across the localized band with the corresponding derived ε_{eff} across the band. Figure 11.7c shows both the effective compression bandwidth (~40 pixels) and the amplitude of the localized strain within the band (~14%).

The presented case highlights the DIC potential to measure deformation localization over a diverse length scale for biological materials. However, it is critical to map the localization length scale to a proper correlation window. The DIC technique would provide both the spatial dimensions of these inhomogeneous deformation bands as well as quantifiable magnitudes of these deformations.

11.3.2 Cell Interaction with Their Extracellular Environment

Cells interact with their extracellular environment with mechanical forces exerted by their cytoskeleton. These forces influence cell–cell interaction, adhesion, and migration. Such cell interactions typically show different characteristics in a 2D and 3D environment. The general

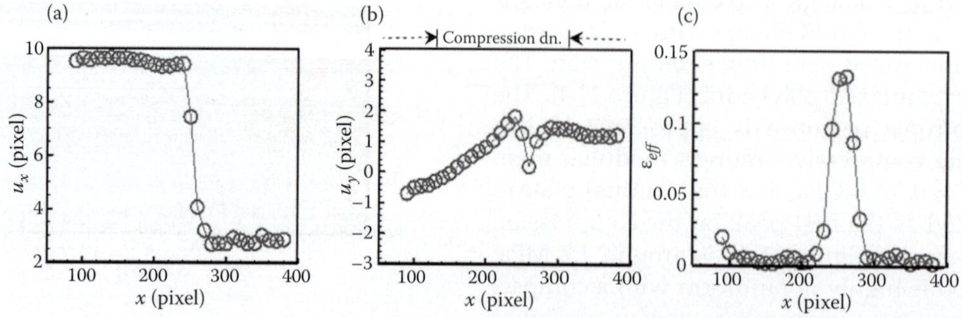

FIGURE 11.7
Distribution of the DIC-derived displacement and effective strain distribution across the localization band. The strain distribution approximately shows the bandwidth.

framework of DIC can be applied to measure these interactions and the associated cell forces and deformation. A 3D traction force microscopy technique [24–26] utilizing DIC principles was developed to track and quantify the cell-mediated deformation of their extracellular matrix. It was also used to investigate cell–substrate interaction in order to map the traction field during different cellular processes. The technique utilizes the 3D imaging capability of laser scanning confocal microscopy and the extension of DIC to the DVC method [41] to measure the 3D displacement and traction field from a volumetric confocal image stack. Substrates, being fibronectin-modified poly acrylamide gels, are typically decorated with fluorescent microspheres during gel formation to provide the proper spatial contrast variation for the correlation algorithm. Cells are decorated with the proper fluorescent proteins. The use of a different laser wavelength and imaging filter would enable the tracking of both the substrate and cell deformation. While the spatial imaging resolution of the system is controlled by the confocal microscope to be around 1 μm, the measured displacement change (the subpixel resolution limit) utilizing DVC is about 0.12 μm.

The pull–push interaction of a 3T3 fibroblast cell on a fibronectin substrate, resolved by the DVC [25], is shown in Figure 11.8. The figure highlights the three-dimensional interaction of the tracked cell with the gel substrate. The cell-mediated deformation in the gel shows a simultaneous pushing and pulling on the substrate during migration. During locomotion, the cell exerts extensive polarization and unbalanced adhesion forces. Figure 11.8b shows a 3D rendering of the pattern and extent of counteracting forces in the substrate. A simultaneous strong upward pull near the end of the cell is accompanied with forward motion of the cell and pushing on the material near the front of the cell.

The ability of the DVC to provide a full three-dimensional displacement field would enable the quantification of the traction forces, developed by migrating cells [24], as shown in Figure 11.9. The full strain tensor $\varepsilon(\mathbf{x}, t)$ can be calculated from the measured displacement field at every material point as a function of the spatial material coordinate $\mathbf{x}$ and time t. For a simple linear elastic incompressible material, the full tensorial stress state at each material point $\sigma(\mathbf{x}, t)$ can be calculated through the shear modulus μ, such that $\boldsymbol{\sigma}(\mathbf{x}, t) = \mu \cdot \varepsilon(\mathbf{x}, t)$. Knowing the full stress state in the substrate, the calculation of the cell traction force $\mathbf{T}(\mathbf{x}, t)$ is straightforward, $\mathbf{T}(t) = \boldsymbol{\sigma}(\mathbf{x}, t) \cdot \mathbf{n}$, where $\mathbf{n}$ is the normal to the substrate surface. Figure 11.9 shows the time evolution of cell-induced traction forces for migration of a 3T3 fibroblast cell on a fibronectin substrate as a function of depth (x_3) over 70 min along an arbitrary slice of the substrate below the cell's long axis. The cell traction force distributions show that the cell's leading edge (left) applies a downward force on the substrate. Concurrently, the trailing edge (right) applies an upward force. This detailed traction measurement is an evidence of a cellular detachment peeling mechanism during locomotion [24].

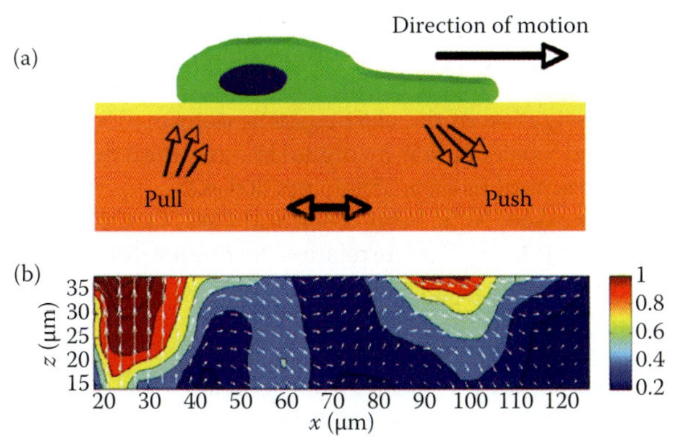

FIGURE 11.8
Results of DVC for 3T3 fibroblast cell adhesion on fibronectin-modified polyacrylamide gels, showing the 3D push–pull phenomenon of the cell-mediated deformation in the gel. (a) Schematic of a cell simultaneously pushing and pulling the underlying material during migration. (b) Displacement profile in the z-plane under the cell illustrates the push–pull hypothesis at t = 35 min. A strong pull upward near the rear of the cell coordinates with the forward motion of the cell and the pushing of the material near the front of the cell. The color bar represents the magnitude of the total 3D displacement vectors, and the arrows represent the displacement vectors in the x–z-plane. (After Maskarinec, S. A. et al., *Proceedings of the National Academy of Sciences of the United States of America*, 106(52), pp. 22108–22113, 2009.)

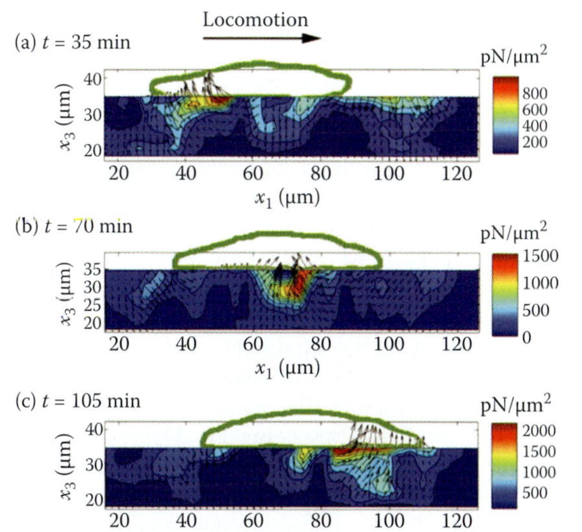

FIGURE 11.9
Results of DVC measurement for the evolution of traction forces due to 3T3 fibroblast cell migration on fibronectin-modified polyacrylamide gels. The contour plots show the magnitude of the three-dimensional traction force vector in pN = mm². The black arrows represent the in-plane traction forces as magnitude and direction. The cell outline, shown in green, is recorded by confocal imaging. (After Franck, C. et al., *PLoS ONE* 6(3): e17833. doi:10.1371/journal.pone.0017833, 2011.)

11.4 Conclusion

The several case studies presented here are indicative of the capabilities of the DIC technique to enable multiscale studies and biological materials. However, a reliable full 3D deformation field requires careful consideration of the scales of the measured phenomena, the scale of the decorating speckles, and the correlation window size. The extension of the technique to interfacial traction measurements of the cell and its environment will enable the understanding of the interactions between biochemical processes and the mechanical signaling in various cellular processes.

References

1. Peters, W. H., and Ranson, W. F., 1982, "Digital imaging techniques in experimental stress-analysis," *Optical Engineering*, 21(3), pp. 427–431.
2. Chu, T. C., Ranson, W. F., Sutton, M. A., and Peters, W. H., 1985, "Applications of digital-image-correlation techniques to experimental mechanics," *Experimental Mechanics*, 25(3), pp. 232–244.
3. Bruck, H. A., McNeill, S. R., Sutton, M. A., and Peters, W. H., 1989, "Digital image correlation using Newton-Raphson method of partial-differential correlation," *Experimental Mechanics*, 29(3), pp. 261–267.
4. Zhao, W. Z., and Jin, G. C., 1996, "An experimental study on measurement of Poisson's ratio with digital correlation method," *Journal of Applied Polymer Science*, 60(8), pp. 1083–1088.
5. Tong, W., 1997, "Detection of plastic deformation patterns in a binary aluminum alloy," *Experimental Mechanics*, 37(4), pp. 452–459.
6. Wattrisse, B., Chrysochoos, A., Muracciole, J. M., and Nemoz-Gaillard, M., 2001, "Analysis of strain localization during tensile tests by digital image correlation," *Experimental Mechanics*, 41(1), pp. 29–39.
7. Bastawros, A. F., Bart-Smith, H., and Evans, A. G., 2000, "Experimental analysis of deformation mechanisms in a closed-cell aluminum alloy foam," *Journal of the Mechanics and Physics of Solids*, 48(2), pp. 301–322.
8. Wang, Y., and Cuitino, A. M., 2002, "Full-field measurements of heterogeneous deformation patterns on polymeric foams using digital image correlation," *International Journal of Solids and Structures*, 39(13–14), pp. 3777–3796.
9. Antoniou, A., Onck, P. R., and Bastawros, A. F., 2004, "Experimental analysis of compressive notch strengthening in closed-cell aluminum alloy foam," *Acta Materialia*, 52(8), pp. 2377–2386.
10. Kang, J., Schmidt, T., Jain, M., and Wilkinson, D., June 4–7, 2006, "Microscopic material characterization using SEM topography images correlation," *SEM Annual Conference and Exposition on Experimental and Applied Mechanics 2006*, Society for Experimental Mechanics Inc., Bethel, CT, pp. 1389–1395.
11. Luo, P. F., Chao, Y. J., Sutton, M. A., and Peters, W. H., 1993, "Accurate measurement of 3-dimensional deformations in deformable and rigid bodies using computer vision," *Experimental Mechanics*, 33(2), pp. 123–132.
12. Luo, P. F., Chao, Y. J., and Sutton, M. A., 1994, "Application of stereo vision to 3-dimensional deformation analysis in fracture experiments," *Optical Engineering*, 33(3), pp. 981–990.
13. Verhulp, E., van Rietbergen, B., and Huiskes, R., 2004, "A three-dimensional digital image correlation technique for strain measurements in microstructures," *Journal of Biomechanics*, 37(9), pp. 1313–1320.
14. Sztefek, P., Vanleene, M., Olsson, R., Collinson, R., Pitsillides, A. A., and Shefelbine, S., 2010, "Using digital image correlation to determine bone surface strains during loading and after adaptation of the mouse tibia," *Journal of Biomechanics*, 43(4), pp. 599–605.
15. Vaananen, S. P., Yavari, S. A., Weinans, H., Zadpoor, A. A., Jurvelin, J. S., and Isaksson, H., 2013, "Repeatability of digital image correlation for measurement of surface strains in composite long bones," *Journal of Biomechanics*, 46(11), pp. 1928–1932.
16. Thompson, M. S., Schell, H., Lienau, J., and Duda, G. N., 2007, "Digital image correlation: A technique for determining local mechanical conditions within early bone callus," *Medical Engineering and Physics*, 29(7), pp. 820–823.
17. Benecke, G., Kerschnitzki, M., Fratzl, P., and Gupta, H. S., 2009, "Digital image correlation shows localized deformation bands in inelastic loading of fibrolamellar bone," *Journal of Materials Research*, 24(2), pp. 421–429.
18. Zhang, D. S., and Arola, D. D., 2004, "Applications tissues of digital image correlation to biological," *Journal of Biomedical Optics*, 9(4), pp. 691–699.
19. Sachs, C., Fabritius, H., and Raabe, D., 2006, "Experimental investigation of the elastic-plastic deformation of mineralized lobster cuticle by digital image correlation," *Journal of Structural Biology*, 155(3), pp. 409–425.
20. Jin, T., Goo, N. S., Woo, S.-C., and Park, H. C., 2009, "Use of a digital image correlation technique for measuring the material properties of beetle wing," *Journal of Bionic Engineering*, 6(3), pp. 224–231.
21. Zhang, D. S., Eggleton, C. D., and Arola, D. D., 2002, "Evaluating the mechanical behavior of arterial tissue using digital image correlation," *Experimental Mechanics*, 42(4), pp. 409–416.
22. Jacquemoud, C., Bruyere-Garnier, K., and Coret, M., 2007, "Methodology to determine failure characteristics of planar soft tissues using a dynamic tensile test," *Journal of Biomechanics*, 40(2), pp. 468–475.
23. Gao, Z., Lister, K., and Desai, J. P., 2010, "Constitutive modeling of liver tissue: Experiment and theory," *Annals of Biomedical Engineering*, 38(2), pp. 505–516.
24. Franck, C., Maskarinec, S. A., Tirrell, D. A., and Ravichandran, G., 2011, "Three-dimensional traction force microscopy: A new tool for quantifying cell-matrix interactions," *PLoS ONE* 6(3): e17833. doi:10.1371/journal.pone.0017833.

25. Maskarinec, S. A., Franck, C., Tirrell, D. A., and Ravichandran, G., 2009, "Quantifying cellular traction forces in three dimensions," *Proceedings of the National Academy of Sciences of the United States of America*, 106(52), pp. 22108–22113.
26. Notbohm, J., Kirn, J. H., Asthagiri, A. R., and Ravichandran, G., 2012, "Three-dimensional analysis of the effect of epidermal growth factor on cell-cell adhesion in epithelial cell clusters," *Biophysical Journal*, 102(6), pp. 1323–1330.
27. Zhou, J., and Fung, Y. C., 1997, "The degree of nonlinearity and anisotropy of blood vessel elasticity," *Proceedings of the National Academy of Sciences of the United States of America*, 94(26), pp. 14255–14260.
28. Sutton, M. A., Ke, X., Lessner, S. M., Goldbach, M., Yost, M., Zhao, F., and Schreier, H. W., 2008, "Strain field measurements on mouse carotid arteries using microscopic three-dimensional digital image correlation," *Journal of Biomedical Materials Research Part A*, 84A(1), pp. 178–190.
29. Libertiaux, V., Pascon, F., and Cescotto, S., 2011, "Experimental verification of brain tissue incompressibility using digital image correlation," *Journal of the Mechanical Behavior of Biomedical Materials*, 4(7), pp. 1177–1185.
30. Evans, S. L., and Holt, C. A., 2009, "Measuring the mechanical properties of human skin in vivo using digital image correlation and finite element modelling," *Journal of Strain Analysis for Engineering Design*, 44(5), pp. 337–345.
31. Wang, Q., Xie, H., Tang, P., Yao, Q., Huang, P., Chen, P., and Huang, F., 2009, "A study on the mechanical properties of beagle femoral head using the digital speckle correlation method," *Medical Engineering and Physics*, 31(10), pp. 1228–1234.
32. Moerman, K. M., Holt, C. A., Evans, S. L., and Simms, C. K., 2009, "Digital image correlation and finite element modelling as a method to determine mechanical properties of human soft tissue in vivo," *Journal of Biomechanics*, 42(8), pp. 1150–1153.
33. Yao, X., Wang, P., and Dai, R., 2008, "Experimental study of damage and fracture of cancellous bone using a digital speckle correlation method," *Journal of Biomedical Optics*, 13(3), pp. 034026-1–034026-7.
34. Mann, K. A., Miller, M. A., Cleary, R. J., Janssen, D., and Verdonschot, N., 2008, "Experimental micromechanics of the cement–bone interface," *Journal of Orthopaedic Research*, 26(6), pp. 872–879.
35. Sutton, M. A., Wolters, W. J., Peters, W. H., Ranson, W. F., and McNeill, S. R., 1983, "Determination of displacements using an improved digital correlation method," *Image and Vision Computing*, 1(3), pp. 133–139.
36. Cheng, P., Sutton, M. A., Schreier, H. W., and McNeill, S. R., 2002, "Full-field speckle pattern image correlation with B-spline deformation function," *Experimental Mechanics*, 42(3), pp. 344–352.
37. Schreier, H. W., Braasch, J. R., and Sutton, M. A., 2000, "Systematic errors in digital image correlation caused by intensity interpolation," *Optical Engineering*, 39(11), pp. 2915–2921.
38. Bay, B. K., Smith, T. S., Fyhrie, D. P., and Saad, M., 1999, "Digital volume correlation: Three-dimensional strain mapping using X-ray tomography," *Experimental Mechanics*, 39(3), pp. 217–226.
39. Wang, H., 2010, *Experimental Study of Inhomogeneous Deformation in Bulk Metallic Glass and Composite During Wedge-Like Cylindrical Indentation*. Iowa State University, Ames, IA.
40. Bastawros, A. F., and Evans, A. G., 2000, "Deformation heterogeneity in cellular Al alloys," *Advanced Engineering Materials*, 4, pp. 210–214.
41. Franck, C., Hong, S., Maskarinec, S. A., Tirrell, D. A., and Ravichandran, G., 2007, "Three-dimensional full-field measurements of large deformations in soft materials using confocal microscopy and digital volume correlation," *Experimental Mechanics*, 47, pp. 427–438.

12

Image-Based Estimation of Passive Myocardial Properties Using Finite Element Modeling

Alexander I. Veress, Benjamin R. Coleman, Genevieve Farrar, W. Paul Segars, and Brian C. Fabien

CONTENTS

Patient-specific cardiac finite element (FE) models, which are used to estimate the stress state of the heart, are highly dependent upon the assumed material properties defined in the models. A method is developed for determining cardiac material properties from computed tomography imaging data. Custom optimization software is developed to interact with a nonlinear large-deformation FE solver in order to perform nonlinear optimization on the material parameters of a patient-specific FE left ventricular (LV) model. The optimization method is tested using synthetic diastolic images based on an FE model with known material parameters as well as in clinical images for a normal male patient. The optimization technique produced material parameter estimates that are in good agreement with the parameters of the FE/LV model that is used to produce the synthetic images. Sensitivity studies show that the optimization process is relatively insensitive to the parameter starting values but highly sensitive to the diastolic loading pressures.

12.1 Introduction

Cardiac mechanical performance depends on myocardial tissue elongation and contraction during the filling and ejection phases of the cardiac cycle. Deformation, stress, and strain in the myofibers supply valuable information about the many factors affecting cardiac physiology and pathophysiology.[1] Estimates of the tissue stresses provide information about factors driving potential tissue adaptation.[7] This is evident in pathologies that are marked by tissue remodeling such hypertension-induced hypertrophy, hypertrophic cardiomyopathy, dilated cardiomyopathy, fibrosis, and infarction-based scar formation.

Finite element (FE) computational models of the heart have been developed to gain a greater understanding of the regional stresses and strains present in both the normal and pathological heart.[3–8] Wall stress, in particular, is believed to be one of the driving forces behind remodeling of the myocardium, which can been seen in hypertension-induced hypertrophy[9,10] and in the remodeling of the remote myocardium associated with myocardial infarction.[11,12] Characterizing the stress state of the heart has proven to be difficult in practice, as there are no direct methods for measuring the in vivo stress state within the human heart.[13] FE models of the heart provide the best estimate of wall stress in the normal heart as well as hearts subjected to disease processes.[14–17] Accurate modeling of the normal and diseased cases require patient-specific models of the heart, which are commonly defined from clinical cardiac imaging.[18,19] FE model analyses have studied pathologies including ischemia,[19,20] left ventricular aneurysm,[21] and infarction.[20,22] FE models have also been used to study cardiac pacing and rhythm disturbances.[23,24] The accuracy of cardiac FE analysis depicting deformations over the cardiac cycle is dependent upon the accuracy of the model geometry, the applied boundary conditions, and the assigned material properties. Highly accurate subject-specific model geometries can be generated from high-resolution imaging such as computed tomography (CT) and magnetic resonance imaging (MRI). Another factor that influences the accuracy of the output of the FE model is the definition of subject-specific boundary conditions such as external loads and displacement conditions. Because it is extremely rare for subject-specific LV pressure data to be available due to the inherent complication risks associated with in vivo pressure measurements, the boundary conditions assigned to FE models are usually based on literature values or through the coupling of a numerical circulatory system to the FE models.[23,25,26]

The final component of the model that affects the accuracy of the stress predictions is the assigned material properties representing the myocardium. In practice, these are difficult to estimate due to the complex, three-dimensional nature of the myocardial tissue structure (Figure 12.1). In order to characterize the material properties experimentally, an ex vivo triaxial test would be

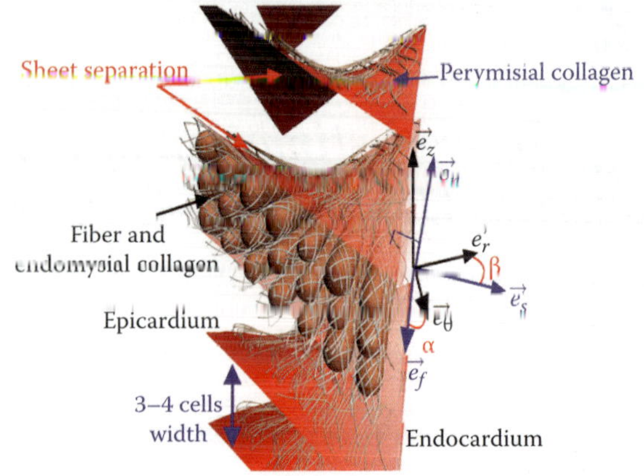

FIGURE 12.1
Cardiac tissue structure. The myocardium is composed of fibers arranged into sheets that are held together by collagen. (Courtesy of Grant T. Gullberg. With permission.)

required. However, triaxial tests are currently difficult to perform in practice.[13] Instead of triaxial testing, sequential ex vivo biaxial tests have been used as the standard for determining the material properties. This type of ex vivo experimental testing cannot be applied to living patients, and ex vivo conditions may result in different material parameter values than in vivo conditions. Furthermore, an accurate model must capture the significant variation in material properties between individuals, especially in disease cases.

For these reasons, noninvasive image-based techniques for patient-specific analysis are an important area of research. Computational models have been developed to study the in vivo properties of the human heart. Several material optimization studies have been done comparing FE model strain output to measurements of the strain field from tagged MRI imaging.[27] These studies are limited to image modalities that can provide deformation and strain information, such as tagged MRI. A general optimization method would be able to estimate material properties using a variety of clinical imaging modalities.

The motivation for the following work was to develop an optimization technology that could be used with a variety of imaging modalities. Optimization technology was developed to determine passive material properties. The optimization technique uses a patient-specific FE model of the left ventricle (LV) and gated CT image data sets depicting diastolic filling to estimate the material properties. This technique has an advantage over the studies based on tagged MRI in that it does not rely on measurement of the strain field and can be applied to any number of imaging modalities in order to obtain in vivo material parameter estimates. In order to assess this method's applicability to clinical imaging data, the optimization software was applied to a synthetic CT image data set. The deformations depicted in these images were based on an FE model in which the material properties were exactly defined, thus providing a true "gold standard."

12.2 FE Model

12.2.1 Model Geometry

An FE model of the LV was developed based on geometry from a high-resolution CT image data set of a normal male patient who was imaged using a Siemens dual-source multislice CT (MS CT) scanner (Siemens AG, Erlangen, Germany). One hundred gated image volumes were obtained over two cardiac cycles. Each image volume was 512 × 512 × 371 pixels with an in-plane resolution of 0.322 mm/pixel and a slice thickness of 0.4 mm/pixel. An image volume approximately one-third of the way into diastole[20] was selected to define the reference model geometry, as the wall stress at this point in the cycle is believed to be at a minimum. Complete details of the image acquisition may be found in the work of Veress et al.[19,20]

The image was scaled to be isotropic, with a voxel dimension of 0.4 mm/pixel, and was rotated into short and long-axis orientations, using cubic B-spline interpolation.[28] A sequence of slight Gaussian blurring (radius, 0.5 voxel) was then applied to reduce the intensity variation within the myocardium and the blood pool. Next, a power law transformation, with a value of 1.3,[29] was used to increase the differentiation between the myocardium and the blood pool. Finally, the brightness and contrast were adjusted. The semiautomatic active contour segmentation algorithm in ITK-SNAP[30] was used to define the myocardium. Surfaces (Figure 12.2a) were obtained from the segmentation data through the application of a MATLAB® implementation of the marching cubes algorithm. A linear hexahedral element mesh was then projected between the smoothed epicardial and endocardial surfaces using the preprocessor TrueGrid.[31] The final mesh (Figure 12.2b) contained 4720 nodes and 3771 elements. The FE model used in this study was previously tested and verified to produce realistic LV deformations over the entire cardiac cycle.[19,20]

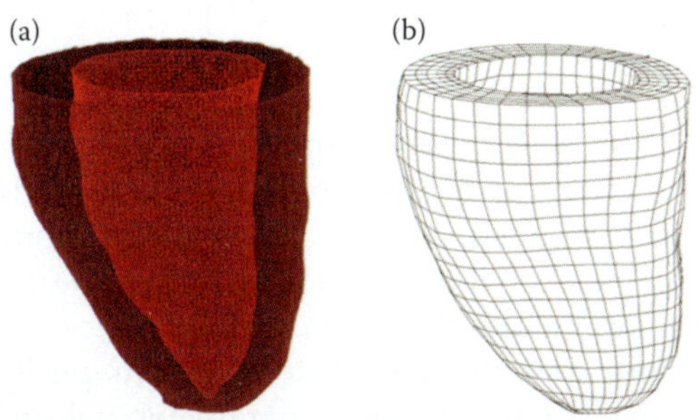

FIGURE 12.2
FE model geometry. (a) Smoothed endocardial and epicardial surfaces. (b) Finite element mesh.

12.2.2 Boundary Conditions

The boundary conditions that may be considered in a forward model of the LV include intraventricular (lumen) pressure, thoracic pressure, pericardial pressure, and base displacement (Figure 12.3). Because the external pressure boundary conditions, thoracic and pericardial pressures, are not yet well quantified in the literature and are likely to have a minimal contribution, the primary loading that is commonly included in FE models is the intraventricular diastolic pressure. A lumen pressure of 10 mm Hg (1.333 kPa) was used as the end-diastolic pressure value. This pressure boundary condition was applied to the element faces representing the endocardial surface. The base surface nodes were allowed to move freely in the in-plane directions, while the long-axis deformation was prescribed by measurements made for the aortic valve translation over this part of the cardiac cycle (approximately 1 cm).

12.2.3 Constitutive Model

The LV myocardium was represented as a slightly compressible transversely isotropic hyperelastic material.[19,20,32] The fiber distribution was varied linearly from

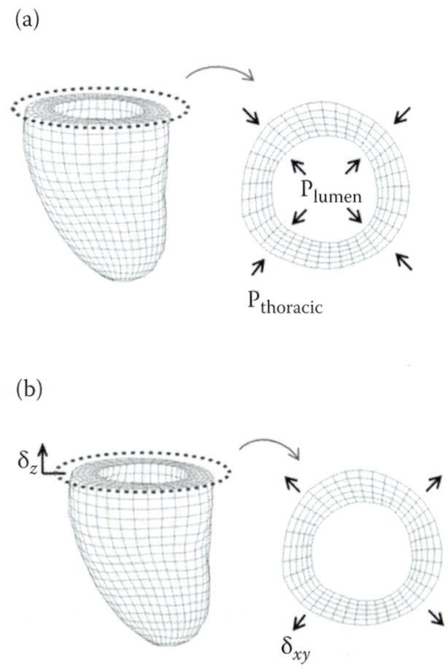

FIGURE 12.3
Forward FE model boundary conditions. (a) Pressure boundary conditions. (b) Displacement boundary conditions.

−82° at the epicardium to 75° at the endocardium. The strain energy function of the constitutive model represents the passive myocardium as an isotropic material with a nonlinear fiber. Complete details of this material model as well as the FE implementation can be found in the work of Weiss et al.[32]

$$W = C_1(\tilde{I}_1 - 3) + C_2(\tilde{I}_2 - 3) + \frac{1}{2}K[ln(J)]^2 + F(\tilde{\lambda}) \quad (12.1)$$

$$\frac{\partial F}{\partial \tilde{\lambda}} = \begin{cases} 0, & \tilde{\lambda} < 1 \\ \frac{C_3}{\tilde{\lambda}}\left(\exp(C_4(\tilde{\lambda} - 1)) - 1\right), & \tilde{\lambda} \geq 1 \end{cases} \quad (12.2)$$

C_1 and C_2 scale the behavior of the isotropic matrix, modeled as a Mooney–Rivlin material. K represents the bulk modulus. J is the Jacobian, and $F(\tilde{\lambda})$ determines the shape of the exponential rise, a characteristic of soft tissue. For the fiber component, $\tilde{\lambda}$ is the deviatoric fiber stretch ratio, C_3 is a scaling coefficient for the exponential, and C_4 controls the shape of the exponential. The Cauchy stress tensor is given by

$$\boldsymbol{T}^{(p)} = p\mathbf{1} + \frac{2}{J}\left[(\tilde{W}_1\tilde{\boldsymbol{B}} + \tilde{\lambda}W_{\tilde{\lambda}}\boldsymbol{a}\otimes\boldsymbol{a}) - \frac{1}{3}(\tilde{W}_1\tilde{I}_1 + \tilde{\lambda}W_{\tilde{\lambda}})\mathbf{1}\right] \quad (12.3)$$

where $\tilde{W}_1$ and $W_{\tilde{\lambda}}$ are strain energy derivatives with respect to $\tilde{I}_1$ and $\tilde{\lambda}$, $\tilde{\boldsymbol{B}}$ is the deviatoric left deformation tensor, p is a pressure term that enforces incompressibility, **1** is the identity tensor, and "⊗" represents the outer product operation. NIKE3D,[33] a nonlinear, large-deformation FE analysis package, contains an implementation of this constitutive model. The finite elasticity problem was solved using this material model to govern the response of the ventricular myocardium to the applied loads and boundary conditions. The passive parameters (Table 12.1) for this material model were obtained through curve fits to biaxial testing data.[34]

TABLE 12.1

Passive Myocardial Material Parameters Determined from Curve Fits to Biaxial Test Data

C_1 (dynes/cm²)	C_2 (dynes/cm²)	C_3 (dynes/cm²)	C_4	K (dynes/cm²)
2.5×10^4	2700	300	16	1.95×10^6

12.3 Optimization

Nonlinear optimization was used to estimate the parameters governing the passive myocardium by matching the deformation of the left ventricular FE model to the deformation documented in clinical cardiac images.

12.3.1 Nonlinear Optimization Using Gradient Descent

The first step in any optimization problem is to define a function that relates the variables to the desired outcome. This function is referred to as a cost function (also known as an objective or loss function). Unconstrained optimization problems, involving a scalar cost function, can be expressed as

$$\text{find } \boldsymbol{x}^* \text{ such that } f(\boldsymbol{x}^*) \ \forall \boldsymbol{x} \in \mathcal{R}^n \quad (12.4)$$

where $\boldsymbol{x}^*$ is a n-dimensional vector of parameters that minimizes the scalar function $f(\boldsymbol{x})$.

An overview of the theory of nonlinear optimization using gradient descent methods is presented in the following work. The methodology involves two steps: first finding the direction in the parameter space that leads toward the minimum and then finding the distance in that direction that reduces the value of the cost function. These two steps are repeated in an iterative process until a convergence criterion is satisfied.

Gradient descent methods will find a set of parameters that yield a local minimum. Constraints can be placed on the parameter values to limit the solutions to the optimization problem to values that are appropriate to the application. The general form of these methods includes equality and inequality constraints:[35]

$$\text{find } \boldsymbol{x}^* \text{such that } f(\boldsymbol{x}^*) \leq f(\boldsymbol{x})$$
$$\forall \boldsymbol{x} \in \mathcal{R}^n \text{ subject to } \begin{cases} c_i(x) = 0, \ i \in \mathcal{B}_1 \\ c_i(x) \geq 0, \ i \in \mathcal{B}_2 \end{cases} \quad (12.5)$$

where $\mathcal{B}_1$ and $\mathcal{B}_2$ are two distinct sets of indices and the c_i's are functions on a subset of $\mathcal{R}^n$. For this study, simple inequality constraints are necessary to ensure that the parameters of the material model of the myocardium remain greater than zero.

To ensure that the parameters remain positive, as is necessary for the material parameter values being sought in the optimization, the process can be performed with the squares of the parameters so the cost function $f(\boldsymbol{x}^2)$ is minimized instead of $f(\boldsymbol{x})$.

12.3.2 Steepest Descent

The normalized gradient was used to determine the steepest descent direction,

$$\boldsymbol{d} = -\nabla f(x)/\|\nabla f(x)\|. \tag{12.6}$$

Some advantages of this method are simplicity and low storage requirements. During the search, each direction will be orthogonal to the previous search direction. For cost functions that have nearly circular level sets around the minimum, this method works well and will converge very quickly.[36] However, if the level sets of the cost function are elliptical, resulting in a long narrow valley surrounding the minimum, there is a potential for the optimization requiring a high number of iterations to reach convergence (Figure 12.4). More sophisticated methods of determining the search direction have been developed to overcome this limitation, such as the conjugate gradient, simultaneous perturbation algorithm for stochastic optimization, as well as various Newton methods. For the class of problems in which the material properties are being estimated the gradient descent method was shown to be very robust.[37]

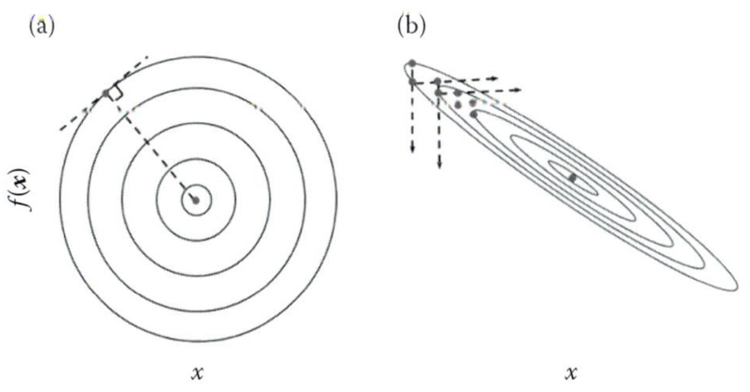

FIGURE 12.4
Steepest descent optimization method. The gradient descent direction is perpendicular to the level sets of $f(x)$. (a) The gradient direction of a circular level set will indicate the direction of the minimum. (b) Elliptical topography requires many iterations because the steepest direction will not point to the local minimum.

The gradients used in steepest descent can be determined through successive perturbation in which finite difference approximations are used to determine each partial derivative. A forward finite difference approximation is given as follows:

$$\frac{\partial f_+(x)}{x_i} = \frac{f(x+\Delta x_i{}^* e_i) - f(x)}{\Delta x_i} \tag{12.7}$$

where Δx_i is the step length for the ith parameter and $\boldsymbol{e}_i$ is the ith column of the identity matrix.

12.4 Software

Custom software was developed to perform material parameter estimation for the passive myocardium. The software is capable of executing sensitivity studies of the cost function parameter space as well as performing the nonlinear optimization. The optimization software interfaces with the FE solver, the template and target images, a target segmentation image, and a control file (Figure 12.5). This software is composed of three main components: interaction with the FE solver, image processing, and nonlinear optimization (Figure 12.6). Interaction with the FE solver involves reading and writing formatted text files specific to NIKE3D, initiating the FE analyses, and reading the FE analysis results. The image processing components were developed to work with clinical cardiac images and labeled segmentation images. The image processing used for the segmentation has been implemented through interaction with the Insight Toolkit (ITK) libraries[38–40] and uses the concept of generic programming based on C++ templates.

12.4.1 Cost Functions

Scalar-valued cost functions were developed to compare the model deformation to the image deformation. The method for measuring the image deformation varies by

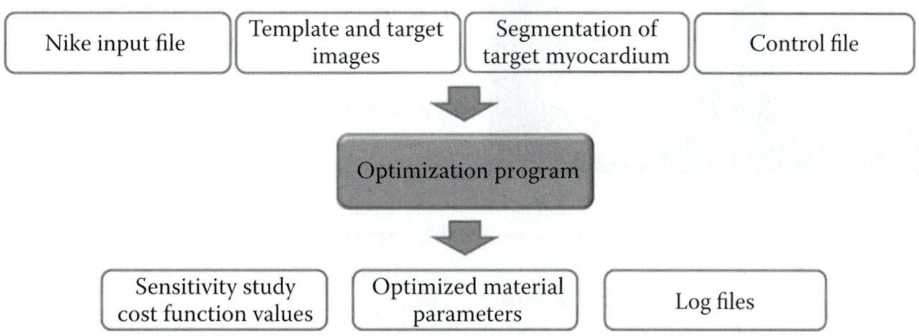

FIGURE 12.5
FE model optimization software overview.

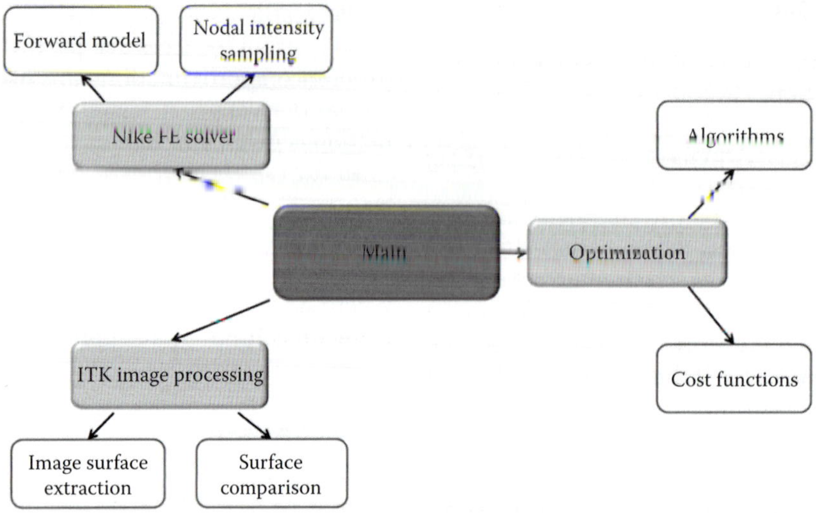

FIGURE 12.6
Custom software organization diagram showing the three main components.

cost function and depends on a template (reference) and a target (deformed) image.

12.4.2 Image Intensity Cost Function

An image intensity cost function was developed to compare the model deformation to the image deformation. This function measures the change in image intensity values at each element of the FE mesh (Figure 12.7). This cost function required a deforming template image and the target image so that the image intensities could be sampled at each integration point within each element in the mesh. To produce the deforming template, the template image is sampled at each integration point of the FE mesh of the LV, thus associating these template image intensities with the integration points. In this manner, the FE mesh then becomes a discretized representation of the template image capable of undergoing deformation.[41] The intensity values for the discretized template and sampled target images were averaged on

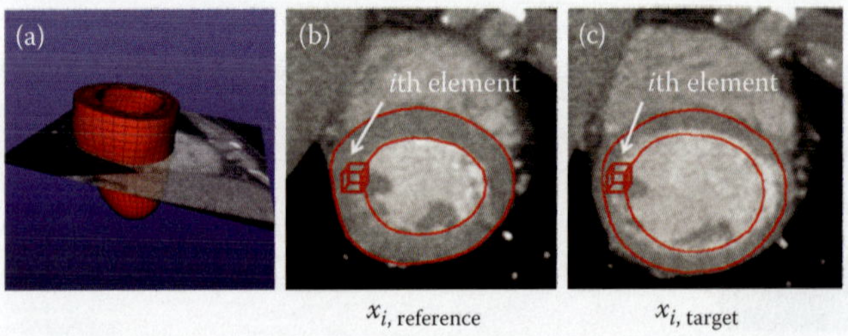

FIGURE 12.7
Image intensity sampling diagram. The intensity difference was calculated as the difference in element intensity between the template and target image. (a) Finite element mesh showing image slice location. (b) Template image short-axis slice showing a sample element. (c) Target image short-axis slice showing sample element.

an element basis. Measuring the change in intensities at each element in the FE mesh yields a vector of comparison values:

$$\Delta x_i = x_{i,\text{target}} - x_{i,\text{reference}} \tag{12.8}$$

where x_i represents the image intensity value for the ith element. A scalar-valued cost function is obtained through a measure of this vector. The root mean square (RMS) error (intensity) is given by

$$\text{RMS}_I(x) = \sqrt{\frac{1}{n}\sum_{1}^{n}(x_{i,\text{target}} - x_{i,\text{reference}})^2}. \tag{12.9}$$

The image intensity sampling at each element and measurement of the RMS error were implemented within NIKE3D. The software determines the element image intensities at the beginning and at the end of the simulation and exports the RMS error and the absolute average of the differences to a text file. The primary optimization software then reads this text file to obtain the cost function value.

12.4.3 Surface Comparison Cost Function

A surface comparison cost function was developed to compare the endocardial and epicardial surface of the FE simulation model results to the endocardial and epicardial surfaces from a segmentation of the target image (Figure 12.8). This cost function measures the smallest Euclidean distance between points on a discretization of each surface. A scalar value for the cost function is obtained as a measure of the minimum Euclidean

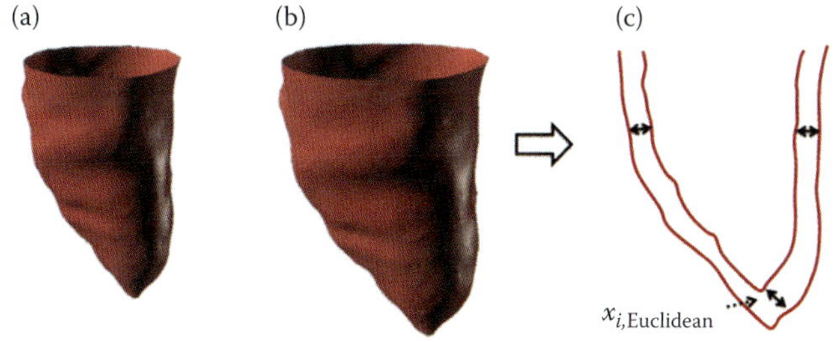

FIGURE 12.8
Minimum distance surface comparison cost function diagram. (a) Model endocardial surface compared with the (b) corresponding segmentation surface. (c) Slice through surfaces showing distance measurements.

distance array. The RMS measure of this vector is given by

$$\text{RMS}_S(x) = \sqrt{\frac{1}{n}\sum_{1}^{n}(x_{i,\text{Euclidean}})^2} \tag{12.10}$$

where n is the number of points in the discretization of the boundary surface and x is the vector containing the Euclidean distances at each point on the surface. Other measures that were investigated were the L2 and the infinity vector norms. The vector of minimum Euclidean distances between surfaces is determined by calling the ITK image processing libraries. The image processing to determine these distances occurs in two stages.

The first step in the evaluation of the surface-based cost function is performed at the start of the program, prior to running a sensitivity study or optimization. A Euclidean distance map image for both the epicardial and endocardial boundaries is calculated. These images are used as distance lookup tables when comparing the surfaces from the FE model to the surfaces from the target image segmentation. Since the distance map calculation is computationally intensive, it is performed for the boundaries of the target myocardial segmentation, which remain constant over the duration of the program. Calculating the distance maps for these boundaries rather than for the boundaries of the surfaces from each FE simulation greatly reduced the overall computational load. The distance map images are retained in memory for computational speed as well as written to disk.

The image processing pipeline for the computation of the distance maps involves several steps (Figure 12.9). The program begins by reading the target segmentation image. The intensity label for the myocardium is then determined from the program input file and used to extract the myocardium through binary thresholding. The binary image of the myocardium is then separated into two images, one representing the endocardial region and the other representing the epicardial region. To accomplish this task, the blood pool is filled in by applying a morphological hole-filling algorithm on each short-axis plane in the image. This results in a volume bounded by the epicardial surface. A volume representing the endocardial surface is then obtained by applying an exclusive disjunction (XOR) filter to the myocardial image and the filled epicardial image. Once the epicardial and endocardial volume images have been obtained, the next step is to generate boundary images that can be used in generating the distance map images. This stage is accomplished through 3-D morphological boundary extraction. The resulting images contain a one-pixel-thick border representing the endocardial and epicardial surfaces. The boundary pixel surface images are then used to

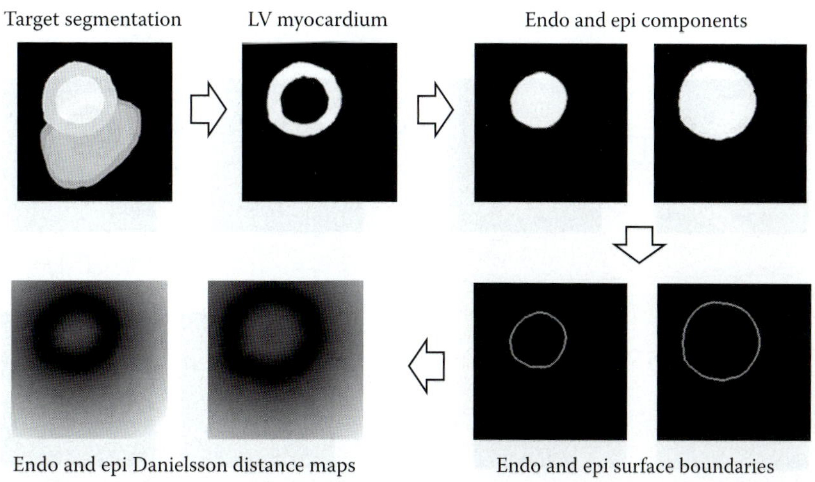

FIGURE 12.9
Distance map image processing flow diagram. All of the image processing steps are 3-dimensional except for the surface boundary extraction step. 2-D slices from each image volume are shown to demonstrate the result of each step.

generate a distance map of the minimum Euclidean distances from each point in the image to a boundary pixel. This is accomplished through the computation of a 3-D Danielsson distance map.[42]

The second step of the surface comparison occurs later during the sensitivity study analysis or optimization run. The software computes a boundary pixel image for the endocardial and epicardial surfaces of the left ventricular geometry defined by the FE solution for comparison to the distance maps calculated at the beginning of the run. The processing pipeline (Figure 12.10) for these boundary images is very similar to the distance map pipeline, differing in the initial and final step.

A voxel mapping program is used to obtain a binary image of the myocardium from the FE model solution. This software takes an image of the FE model reference state and applies the deformation map obtained at a specified time point during the FE simulation to generate a deformed image representing the intensity values located within the deformed FE model geometry. Once the deformed image has been generated, it is converted to a binary image of the myocardium through binary thresholding. This image is then processed to generate one-pixel-thick boundary images for the endocardial and epicardial surfaces.

Once the boundary images for the endocardial and epicardial FE solution surfaces have been obtained, they are compared to the image segmentation surfaces by multiplying the boundary image with the corresponding distance map (Figure 12.11). This is accomplished by counting the nonzero pixels in the boundary image to determine the size of the distance array. An ITK iterator was then used to visit each nonzero pixel in the boundary image and place the value at the corresponding location in the distance image into the distance array. Once the distance arrays for the endocardial and epicardial surface have been obtained, they are used to generate a scalar value, using Equation 12.10. This cost function returns a scalar measure of the surface boundary distances. Since there are two surfaces to compare, the measures of each distance vector are summed together to obtain the cost function value.

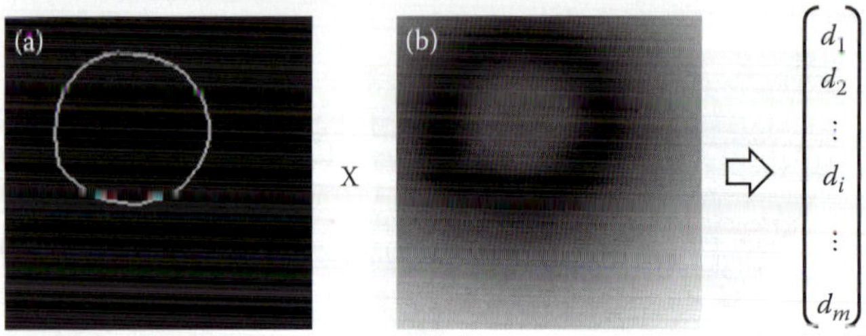

FIGURE 12.11
Depiction of the epicardial surface distance calculation. (a) A representative slice from epicardial boundary pixel image and (b) the corresponding slice from the epicardial distance map.

12.4.4 Gradient Descent Implementation

A custom gradient descent implementation was developed for this study[43] (Algorithm 12.1) where the descent direction was determined based on the gradient of the cost functions. The normalized gradient vector (Equation 12.11) was used to determine the descent direction.

$$d_2 = -\nabla J(x)/\|\nabla J(x)\| \tag{12.11}$$

A simple backtracking line search method (Figure 12.12 and Algorithm 12.2) was implemented rather than a more robust approach, such as the Armijo rule line search,[35] due to limited precision of the material parameters. The default values for the inputs to Algorithm 12.2 are $\gamma = 1$, $\alpha = 0.5$. Additionally, Algorithm 12.2 defaults to a line

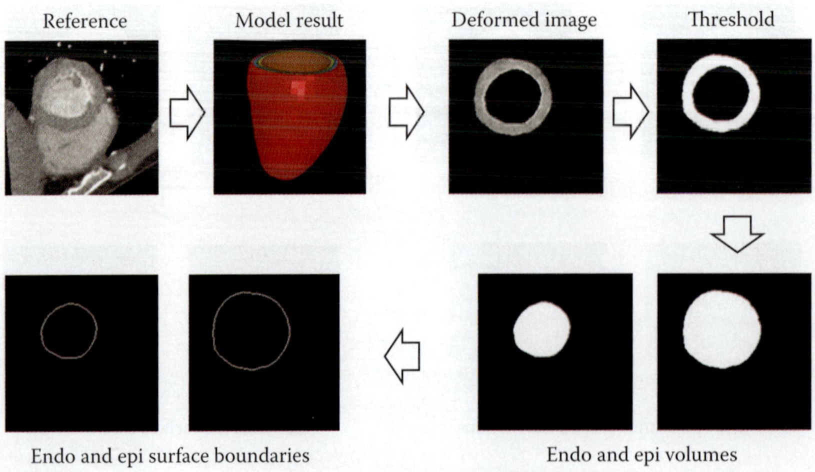

FIGURE 12.10
Model image surface extraction flow diagram.

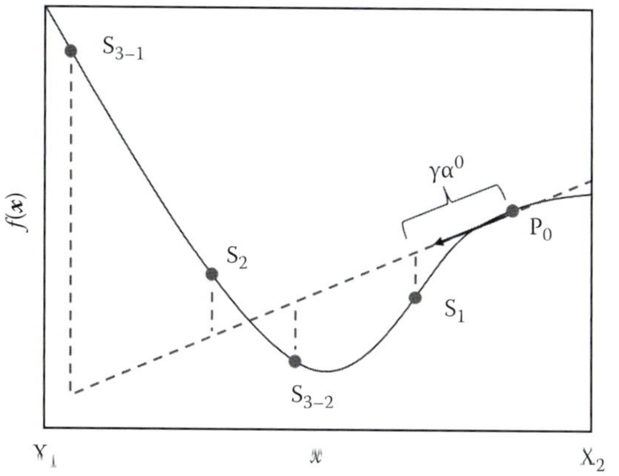

FIGURE 12.12
Backtracking line search. Starting at point P_0 and moving in the direction d, we wish to find a set of parameters that minimize $f(x)$. S_1, S_2, and S_3 represent three different values of γ. S_1 and S_2 result in an immediate reduction in the function value. The initial step for S_{3-1} results in an increase of the function value. Backtracking using α = 1/2 results in S_{3-2}, which reduces the function value by the second iteration.

search that does not remember the previous line search step size $\alpha_{previous}$. Enabling this feature can greatly reduce the number of line search iterations and, consequently, the number of FE simulations that must be run during this stage of the optimization.

Algorithm 12.1: Steepest Descent Algorithm

Input: x_0, initial parameters, ε_t, convergence tolerance, k_{max}, maximum number of iterations
Output: x such that $\|\nabla f(x^k)\| \leq \varepsilon_t$, StopCode, reason for termination
for $k = 0,1,\ldots, k_{max}$ do
(1a) If $\|\nabla f(x^k)\| \leq \varepsilon_t$ break
(1b) Compute $d^{(k)}$ such that $\nabla f(x^{(k)})^T d^{(k)} < 0$
(2) Perform a backtracking line search to find $\alpha = \alpha^* > 0$ that minimizes $f(x^{(k)} + \alpha^* d^{(k)})$
(3) $x^{(k+1)} = x^{(k)} + \alpha^* d^{(k)}$
end for
$x = x^{(k)}$

Algorithm 12.2: Backtracking Line Search

Input: x, current optimization parameters, $d^{(k)}$, descent direction,
$\gamma > 0$, step scale factor, $0 < \alpha < 1$, backtracking factor,
j_{max}, maximum number of line iterations,
$\alpha_{previous}$, the step length from the previous line search (*optional*)
Output: x such that $f(x^{(k)} + \alpha d^{(k)}) < f(x^{(k)})$, StopCode, reason for termination

for $j = 0,1,\ldots, j_{max}$ **do**
if $\alpha_{previous}$ set $\alpha_0 = \alpha_{previous}$
else $\alpha_0 = 1$
$\alpha^{(j+1)} = \gamma\alpha^j\alpha_0$
$x^{(j)} = x^{(k)} + \alpha^{(j+1)} d^{(k)}$
if $x^{(j)} = x^{(k)}$ break due to insufficient parameter precision to register change
if $f(x^j) < f(x^{(k)})$ break
end for
$x^{(k)} = x^{(j4)}$

12.5 Synthetic Cardiac Study to Test Parameter Estimation

A synthetic image data set was generated to validate the performance of the optimization technique for a case where all of the FE model parameters were known. This synthetic image data set provided a true gold standard and was used to determine the applicability of the technology to cardiac imaging. A sensitivity study was conducted to determine how the results were affected by the initial values of the material parameters as well as the loading pressure.

12.5.1 Synthetic Image Data Study

A synthetic diastolic image was generated based on a forward FE model of end-diastole. The deformation of the FE model was applied to the CT image of the reference configuration (one-third of the way to diastole) to generate the synthetic diastolic image (Figure 12.13)

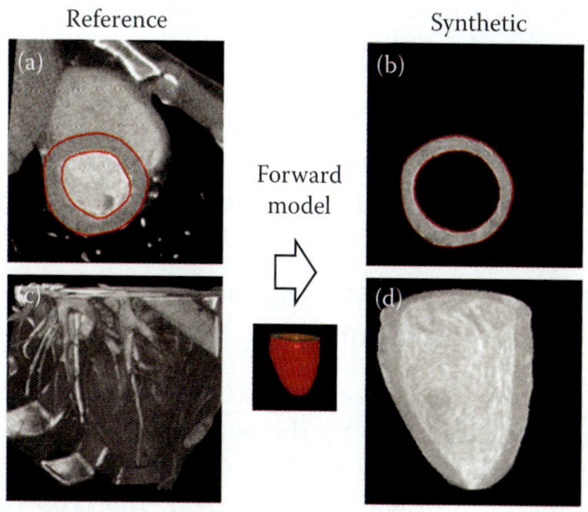

FIGURE 12.13
Synthetic diastolic image based on FE forward model. Voxel mapping program applies the deformation map to (c) the reference image, generating (d) a 3-D image of the left ventricle. (a, b) A 2-D slice near the base and (c, d) a 3-D rendering are shown for both images. The 3-D rendering of the reference image shows surrounding tissue such as the vasculature and ribcage, whereas the synthetic image contains only the left ventricular myocardium.

TABLE 12.2

C_1–C_4 Gradient Descent Synthetic Optimization Results

		Image	Surface
Forward parameters	C_1	2.500E + 04	2.500E + 04
	C_2	2.700E + 03	2.700E + 03
	C_3	3.000E + 02	3.000E + 02
	C_4	1.600E + 01	1.600E + 01
Optimized parameters	C_1	2.461E + 04	2.100E + 04
	C_2	2.957E + 03	2.946E + 03
	C_3	3.137E + 02	3.747E + 02
	C_4	1.659E + 01	1.719E + 01

TABLE 12.3

C_1–C_4 Gradient Descent Synthetic Optimization Error Results

	Run	Image	Surface
Parameters	C_1	−1.56%	−4.00%
	C_2	9.52%	9.11%
	C_3	4.57%	24.90%
	C_4	3.69%	7.44%

using the mapping program that is utilized in the surface cost function. The synthetic image is composed of the voxels of the reference image that are deformed to align with the FE forward model solution. Regions outside of the FE forward model mesh solution are set to zero as there is no mesh to sample these locations.

The optimization was carried out using both the surface and intensity cost functions. The final material parameter estimations can be found in Table 12.2 for the C_1–C_4 parameters using the gradient descent optimization. The optimization results show relatively good correspondence with the material parameters used to create the forward end-diastolic image. The error values ranged from 2% to 10% for the image-based cost function optimization and 4% to 25% for the surface-based cost function optimization (Table 12.3).

12.5.2 Sensitivity to Material Property Initial Conditions

To investigate the range of perturbations that could be successfully estimated by the software, C_1 and C_4 were optimized for models with a wide variety of initial parameters, ranging up to 50% from the true parameters (Table 12.4). A gradient descent optimization using the intensity cost function was performed for each set of starting parameters. The C_1 estimates produced by the optimization were all within 3% of the true solution. The estimates for the C_4 were less accurate. Optimization estimates when the initial parameters were within 20% of the true solution were all within 10% of the true solution

TABLE 12.4

Synthetic C_1 and C_4 Initial Conditions

C_1	C_4
12,500	8
37,500	24
37,500	8
12,500	24
20,000	12.8
30,000	19.2
22,500	14.4

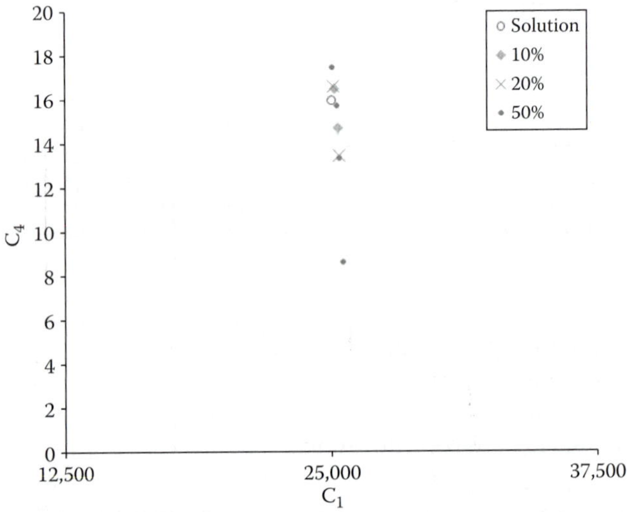

FIGURE 12.14
Synthetic C_1 and C_4 parameter estimation results for several initial parameter sets. The initial conditions were varied by percentage of the solution parameters. The parameters are correctly estimated within a 20% variation. For a 50% variation, one of the test runs significantly underestimated C_4.

with the exception of a single C_4 estimate (Figure 12.14). Initial parameters that were overestimated seemed to yield better optimization results.

12.5.3 Sensitivity to End-Diastolic Pressure Loading Initial Conditions

The effect of variation of the end-diastolic pressure was carried out in order to determine the relative sensitivity of the optimization results to the loading. The end-diastolic pressure loading was varied over the range considered physiologically normal (Table 12.5). No other optimization parameters were altered. The expectation for performing this test was that the material properties would directly depend on the applied load.

The results of the parameter optimizations using the synthetic images showed that the C_1–C_4 parameter estimations were directly affected by the end-diastolic loading (2% to 64% of the true parameters). As was found in the

TABLE 12.5

Sensitivity to Changes in End-Diastolic Pressure

Pressure (mm Hg)	C_1	C_2	C_3	C_4
10.00[a]	25,000.0	2700.0	300.0	16.0
10.00	24,610.4	2957.0	313.7	17.2
12.50	32,393.5	3145.0	292.9	14.0
8.50	20,177.6	2825.1	328.6	17.1
4.75	9113.7	2761.7	304.9	16.2
10.00	24,945.4	Fixed	Fixed	15.4
12.50	32,785.8	Fixed	Fixed	13.6
8.50	20,375.8	Fixed	Fixed	15.7
4.75	9153.3	Fixed	Fixed	16.2
	% ΔC_1	% ΔC_2	% ΔC_3	% ΔC_4
10.00	−1.56%	9.52%	4.57%	7.50%
12.50	29.57%	16.48%	−2.37%	−12.50%
8.50	−19.29%	4.63%	9.53%	6.88%
4.75	−63.55%	2.29%	1.63%	1.25%
10.00	−0.22%	–	–	−3.75%
12.50	31.14%	–	–	−15.00%
8.50	−18.50%	–	–	−1.88%
4.75	−63.39%	–	–	1.25%

[a] Forward model values used to generate synthetic image data set. The variation of the LV pressure.

previous results, all of the optimized models had excellent qualitative alignment with the deformation observed in the synthetic image. These results highlight the fact that accurate determination of the initial pressure loading is a necessary condition for successful parameter estimation.

12.5.4 Applications to Clinical Human Imaging Data

Parameter estimation was performed for the passive material properties of a human LV depicted in the reference CT data image set that was used to generate the FE model and the associated end-diastolic image data set. A sensitivity study was performed to study the parameter space as well as parameter optimizations on all four parameters, as well as on C_1 and C_4 only.

12.5.4.1 Sensitivity Studies

A sensitivity analysis was performed to gain an understanding of the behavior of the intensity cost function with respect to the most sensitive parameters C_1 and C_4. The parameters C_1 and C_4 were varied over a wide range of values (Table 12.6) surrounding the literature values,

TABLE 12.6

C_1, C_4 Sensitivity Study Parameter Range

C_1 (dynes/cm^2)	C_4
1000–10,000	2–30

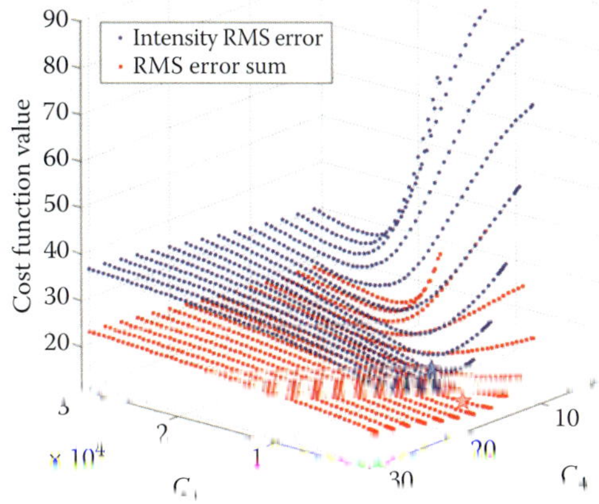

FIGURE 12.15
Optimization surface for the most sensitive parameters C_1 and C_4. Stars represent the location of the minimum values.

in which C_2 and C_3 were held constant at their literature values. The cost function responses were graphed to create a surface representation of the sensitivity study. Previous work showed that C_2 and C_3 were directly affected by the selection of the parameters' initial values rather than by the optimization process.[37]

The cost function response was reasonably smooth and revealed a single minimum within the range of parameters selected (Figure 12.15). Some of the simulations for low values of the parameters resulted in divergent solutions that were not included in the analysis. The large cost function values at the lower end of the parameter range indicate that the mesh is expanding well beyond the deformation observed in the image. Under these extreme deformations, the mesh is stretching beyond its limit, resulting in inverted elements and cessation of the FE analysis.

12.5.4.2 Optimization of C_1–C_4 Using Image Intensity Cost Function

The gradient descent optimization software was initially run for the C_1–C_4 parameter space optimizing all four parameters, as well as for C_1 and C_4 only using the image intensity cost function. The optimization runs were performed starting from initial parameters that were used in the synthetic optimization studies.

The optimized solutions showed excellent qualitative alignment. A typical short-axis slice for the optimization results can be found in Figure 12.16, shown in comparison to the LV model analysis that used literature values. The parameter estimates varied from the literature values (Table 12.7). One of the likely causes of the deviation is that the end-diastolic pressure loading was not measured during the image acquisition. The results

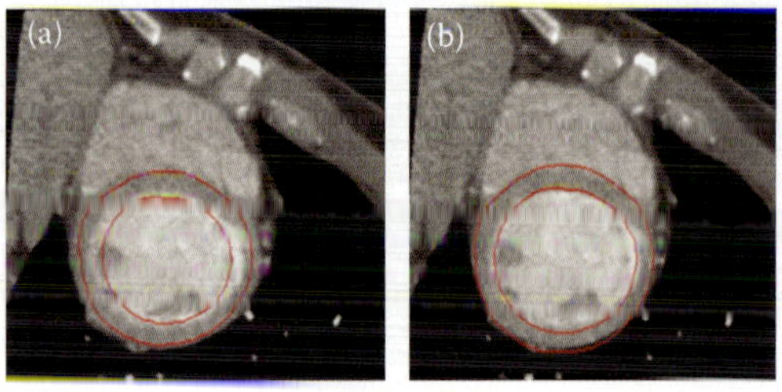

FIGURE 12.16
Gradient descent optimization short-axis alignment using human data. (a) FE Model epicardial and endocardial surface alignment using literature values for the material properties. (b) FE model results using C_1 and C_4 values determined from optimization using intensity cost function.

TABLE 12.7
Optimization Results with 10 mm Hg End-Diastolic Pressure Loading

	C_1	C_2	C_3	C_4
Initial	25,000.0	2700.0	300.0	16.0
4 parameter	5502.4	1708.0	205.7	28.0
C_1 and C_4 only	5194.2	Fixed	Fixed	27.8

from Section 12.5.3 indicate that the values used for the end-diastolic loading pressure used in the FE analyses have a profound effect upon the estimates of material parameters. It is unlikely that the loading pressure was the assumed 10 mm Hg for the subject depicted in the CT image used in the analysis.

12.6 Discussion

The application of the steepest descent optimization program on synthetic images with known displacements was found to provide relatively accurate estimations of the material parameters. The image intensity-based cost functions consistently provided more accurate estimates of the parameters than the surface-based ones. Sensitivity studies conducted on these image data sets suggested that variation of the starting values for the parameters had little effect on the final optimized parameter values. However, as expected, the variation of the loading pressure used in the FE model was found to have a profound effect upon the estimated parameters and so is necessary for accurate parameter estimation.

The determination of in vivo estimates of the parameters governing the passive behavior of the myocardium has been the subject of several studies, all of which used strain-based objective functions. Most of these studies utilized gradient-based optimization methodologies[27,44–47] versus other optimization methodologies.[47,48] It has been suggested that nongradient methods were more likely to find the global minimum values.[49,50] However, the cost function surface analysis (described in Section 12.5.4.1 and in previous work[37]) suggests that the parameter spaces associated with the transversely isotropic material model associated with the present study have a single global minimum without any apparent local minima for the C_1 and C_4 optimizations, thus ensuring convergence. However, the final values for C_2 and C_3 appear to be more influenced by the initial value assignment than the optimization process.[37] Therefore, the values of C_2 and C_3 should be specified through other means, such as by least-squares fitting of biaxial test data.[19,20]

This image-based optimization could be applied to the problem of determining active contractile properties of the LV. The active contractile properties could be obtained by using the end-systolic image instead of the end-diastolic image. This type of optimization was carried out by Sun et al.[21] to study of the contractile properties of the myocardium by determining estimates for two contractile parameters. Measurements of strain and ventricular volumes were obtained from an ovine heart with a left ventricular aneurysm. The predicted strains from the FE models were compared to the strain fields measured by tagged MRI. The passive properties of the FE models and some of the active contraction parameters were defined using literature values or values from previous studies. The active material property optimization was based on the successive response surface method. The results of this study found a significant decrease in the active contractile properties of the tissue in the infarcted regions. The material property estimates in the infarct region were found to be close in value to those of the remote regions. They conducted a series of sensitivity studies to determine how various assumptions for the passive material parameters affected the active contraction parameter estimates. In the worst case, a 10% increase in one of the passive material properties resulted in a 5.6% change in one of the active contractile properties. Deviations in the active contraction parameters such as the stress-free sarcomere length had a profound effect on the estimated optimization parameters. For example, a 5% decrease in stress-free sarcomere length resulted in a 195% increase in one of the active contraction parameters. Many of these parameters, such as stress-free sarcomere length or intracellular calcium concentrations, cannot be determined using noninvasive means, so one must rely on literature values. This type of analysis could be applied to both the passive and active contraction parameters provided that both systolic and diastolic tagged MRI acquisitions were obtained.

The image-based optimization described in this chapter represents the only optimization methodology that does not require strain measurements to obtain parameter estimates. All of the previous studies used cost functions based on the differences in model-predicted strains and strains obtained from tagged MRI. Therefore, this methodology can be applied to many different imaging modalities rather than being restricted to tagged MRI as long as the modality is of sufficient resolution. Image-based optimization could be used to derive parameter estimates for the entire myocardium as was done in this study or applied to specific regions (e.g., ischemia or infarction). For example, gated, delayed enhanced MRI images could be used to identify infarcted regions and be used to determine the passive material properties within these regions. One need only specify a set of elements, those within the infarcted region, to be subject to the optimization process.

All of the optimization methods described in this chapter require reasonable estimates for the loading pressures in order to obtain meaningful parameter estimates. The gold standard for determining end-diastolic pressure is through the utilization of micromanometer conductance catheters.[51] However, the risk of complications during this procedure is significant, precluding common use of this type of procedure. Another invasive method for estimation of left ventricular filling pressures is the measurement of the pulmonary wedge pressure. While still invasive, this procedure is commonly performed in the clinical setting and provides an accurate estimate of left ventricular end-diastolic pressure for normal function.[52,53] In cases of diastolic dysfunction, the wedge pressure wave provides a close estimate of left ventricular end-diastolic pressure.[53–55] In cases of pulmonic and aortic regurgitation, pulmonary wedge pressure provides very poor estimates for LV filling pressures.[56]

Several studies[57–60] have found that estimations of end-diastolic pressure can be made using transmitral diastolic filling and atrial contraction velocity measurements. While these studies used Doppler echocardiography, 2-D and 3-D vec MRI could also be used to determine these parameters.[61] Another predominantly echocardiographic-based methodology uses pulmonary artery deceleration time to predict end-diastolic pressure in cases where there is diastolic and systolic dysfunction.[62] In cases where systolic function is preserved, the estimations produced by pulmonary artery deceleration time do not provide accurate end-diastolic pressure estimates.[63,64] Yamamoto et al.[64] found that the pulmonary vein deceleration time corrected by the pulmonary artery deceleration time can give more accurate predictions for cases without systolic dysfunction. These studies suggest that noninvasive imaging techniques have the potential to be used to obtain loading pressure estimates, which could then be combined with optimization techniques to obtain patient-specific material parameter estimates. However, it is unknown whether these Doppler-based pressure estimates are accurate enough for parameter estimation. The efficacy of using of noninvasive pressure measurements remains to be determined.

12.7 Limitations

Several assumptions were made in constructing the model for this project. First, the scope of the model was restricted to structural mechanics. Other levels of multiscale cardiac models, such as coupling with circulatory or fluid models, were beyond the scope of this work. Second, the assumption of a stress-free state at one-third the way into diastole[65] neglects the residual stresses that are present in the myocardial wall. While the use of the reference configuration one-third into diastole is common in cardiac FE modeling, some studies have been done to try to quantify the residual stresses using an inverse approach.[66] This type of analysis was also beyond the scope of the current work. Third, the myocardium was assumed to be homogenous for this study both in material definitions as well as for the fiber definitions. These are common simplifications in cardiac FE modeling. The extent to which material parameters may vary spatially has not been well documented due to the difficulties in determining the quantifying heterogeneity, and the material properties of normal myocardium are generally assumed to be homogenous for FE modeling of the heart.[24,67–74] It remains to be seen if an image-based material property optimization such as described in this chapter could be applied to the determination of regional material properties (septal, posterior, lateral, and anterior). However, the determination of element-level material properties would not be possible, because the optimization would likely not converge and the definition of approximately 3700 separate material property definitions in the case of our models would simply not be practical. Additionally, it has been shown that there is significant fiber angle variation circumferentially around the ventricle.[75] Diffusion tensor MRI has been used to determine myofiber distributions in formalin-fixed hearts.[76–79] A recent study by Nielles-Vallespin et al.[80] has indicated that the imaging sequence they developed could provide a reliable method for determining fiber distributions in vivo. However, to date, it has only been applied to a limited number of patients.

Another limitation of this study is the use of only the diastolic phase of the cardiac cycle to determine the parameters. Passive inflation may not be appropriate loading for the estimation of all of the parameters. In particular, the estimate of the shear modulus C_1 may not be correct, because during passive inflation, the ventricle experiences only limited indirect shear forces, so the diastolic loading may not be fully testing this parameter. Active contraction results in significant shear forces that are generated due to contraction of the helical fiber structure. The systolic contraction phase of the cardiac cycle could be included in the optimization to provide an estimate of C_1.

12.8 Future Work

This technology could easily be applied to the study of a population of normal hearts to determine the variation in properties. This would provide an effective range of material parameters for the normal population. This type of population-based work could then be expanded to study the effects of pathologies (e.g., ischemia, infarction, and heart failure) on the passive and active contraction material parameters. However, these studies would be contingent upon the availability of loading pressure estimates.

Patient-specific FE models have been applied to a variety of pathologies, as discussed in this chapter, and are based on clinical medical imaging. Beyond the specifying the heart geometry, clinical imaging can provide valuable information on the perfusion state (CT,[81–84] MRI,[85–88] positron emission tomography [PET],[82,89–91] and single photon emission tomography [SPECT][92–96]) and the underlying metabolism[92,97,98] of the myocardial tissue (SPECT). The development of minimally invasive and noninvasive technologies to supplement the information available in clinical imaging is necessary to further the utility of patient-specific models. Toward this end, nonlinear optimization techniques have been applied to determine the material parameters of an FE model as described in this book chapter. This allows the model to reproduce the gross deformations documented in the images.

This methodology has the potential to be applied to the determination regional material properties in the LV. A similar synthetic model process could be used with assumed material property variations. The accuracy of the optimization could be tested using a range of material property variations. The uniqueness of the estimated material property values as well as the convergence characteristics would need to be determined. The technology could also be applied to infarcted hearts to determine the properties of the compromised tissue. The optimization would be applied to the affected tissue (infarcted and border zone) and the remote tissue. Similarly, the methodology could be used to estimate the fiber distribution of the LV. This would take the form of an additional step following the determination of the passive material properties of the LV and would involve optimization of the model to the end-systolic image data sets.

12.9 Conclusions

The use of an image-based optimization method can give reasonable estimates of passive material parameters using standard CT imaging. Two cost functions were evaluated for estimating material parameters of the LV based on an FE model of cardiac diastole. One of the cost functions was based on the change of image intensities at points within the FE mesh, while the other was based on a comparison between the surfaces generated from CT image segmentation. This study found that the intensity-based cost function provided better parameter estimates than the surface-based cost function. This type of image-based optimization technique could potentially be applied to other clinical imaging modalities such as MRI. The technique can be applied retrospectively to archived data sets to investigate the material parameters of normal and pathological hearts provided a reasonable estimation of the diastolic filling pressure is available. Without a pressure estimate, the optimization could be carried out using the limits of normal/pathological filling pressures. This would provide a range of material parameters within which the true values lie.

Acknowledgments

We thank Roy Kerckhoffs at the University of California, San Diego, for his timely responses to our questions concerning cardiac mechanics. We are indebted to his thoughtful communication. This work was supported in part by the National Institutes of Health under 1R01HL091036. This work benefited from the use of the Insight Segmentation and Registration Toolkit (ITK), an open-source software developed as an initiative of the US National Library of Medicine and available at http://www.itk.org. In addition, the multiplatform configuration tool CMake was used for configuring ITK and facilitating its use. CMake was partially funded by the US National Library of Medicine as part of the Insight Toolkit project. CMake is an open-source system, and it is freely available at http://www.cmake.org.

References

1. McCulloch, A. D. In *The Biomedical Engineering Handbook Electrical Engineering Handbook Series* (ed. J. D. Bronzino) (CRC Press/Taylor & Francis, Boca Raton, FL, 2006).
2. Geerts, L., Kerckhoffs, R., Bovendeerd, P. & Arts, T. *Lecture Notes in Computer Science* 81–90 (Springer-Verlag, Berlin, 2003).

3. McCulloch, A. D. *Biomedical Engineering Handbook The Electrical Engineering Handbook Series* 418–439 (IEEE Press, Boca Raton, FL, 1995).
4. Hunter, P. J., Nash, M. & Sands, G. *Computational Biology of the Heart John Wiley Series on Nonlinear Science* 173–217 (Wiley, New York, 1996).
5. Hunter, P. J. & Smaill, B. H. The analysis of cardiac function: A continuum approach. *Prog. Biophys. Mol. Biol.* **52**, 101–164 (1989).
6. Le Grice, I. *A Finite Element Model of the Myocardial Structure: Implications for Electrical Activiation in the Heart* (New Zealand: University of Auckland, 1992).
7. Wong, A. & Rautaharju, P. Stress distribution within the left ventricular wall approximated as a thick ellipsoidal shell. *Am. Heart J.* **75**, 649–662 (1968).
8. Hunter, P., Smaill, B., Nielsen, P. & Le Grice, I. *Computational Biology of the Heart* 171–215 (Wiley, 1997).
9. Ganau, A. et al. Patterns of left ventricular hypertrophy and geometric remodeling in essential hypertension. *J. Am. Coll. Cardiol.* **19**, 1550–1558 (1992).
10. Owens, G. K. Influence of blood pressure on development of aortic medial smooth muscle hypertrophy in spontaneously hypertensive rats. *Hypertension* **9**, 178–187 (1987).
11. Sutton, M. G. S. J. & Sharpe, N. N. Left ventricular remodeling after myocardial infarction pathophysiology and therapy. *Circulation* **101**, 2981–2988 (2000).
12. Parachuri, V. R. & Adhyapak, S. M. Ventricular Geometry in Post-Myocardial Infarction Aneurysms: Implications for Surgical Ventricular Restoration (Springer Publishing, London, 2012).
13. Costa, K. D., Holmes, J. W. & McCulloch, A. D. Modelling cardiac mechanical properties in three dimensions. *Philos. Trans. Math. Phys. Eng. Sci.* **359**, 1233 (2001).
14. McCulloch, A. D. & Mazhari, R. Regional myocardial mechanics: Integrative computational models of flow–function relations. *J. Nucl. Cardiol.* **8**, 506–519 (2001).
15. Mazhari, R., Omens, J. H., Pavelec, R. S., Covell, J. W. & McCulloch, A. D. Transmural distribution of three-dimensional systolic strains in stunned myocardium. *Circulation* **104**, 336–341 (2001).
16. Mazhari, R. & McCulloch, A. D. Integrative models for understanding the structural basis of regional mechanical dysfunction in ischemic myocardium. *Ann. Biomed. Eng.* **28**, 979–990 (2000).
17. Mazhari, R., Omens, J. H., Covell, J. W. & McCulloch, A. D. Structural basis of regional dysfunction in acutely ischemic myocardium. *Cardiovasc.* Res. **47**, 284–293 (2000).
18. Neal, M. L. & Kerckhoffs, R. Current progress in patient-specific modeling. *Brief Bioinform.* **11**, 111–126 (2010).
19. Veress, A. I., Segars, W. P., Weiss, J. A., Tsui, B. M. & Gullberg, G. T. Normal and pathological NCAT image and phantom data based on physiologically realistic left ventricle finite-element models. *IEEE Trans. Med. Imaging* **25**, 1604–1616 (2006).
20. Veress, A. I., Segars, W. P., Tsui, B. M. & Gullberg, G. T. Incorporation of a left ventricle finite element model defining infarction into the XCAT imaging phantom. *IEEE Trans. Med. Imaging* **30**, 915–927 (2011).
21. Sun, K. et al. A computationally efficient formal optimization of regional myocardial contractility in a sheep with left ventricular aneurysm. *J. Biomech. Eng.* **131**, 111001 (2009).
22. Wenk, J. F. et al. Regional left ventricular myocardial contractility and stress in a finite element model of posterobasal myocardial infarction. *J. Biomech. Eng.* **133**, 044501 (2011).
23. Kerckhoffs, R. C., McCulloch, A. D., Omens, J. H. & Mulligan, L. J. Effects of biventricular pacing and scar size in a computational model of the failing heart with left bundle branch block. *Med. Image. Anal.* **13**, 362–369 (2009).
24. Kerckhoffs, R. C., Narayan, S. M., Omens, J. H., Mulligan, L. J. & McCulloch, A. D. Computational modeling for bedside application. *Heart Fail. Clin.* **4**, 371–378 (2008).
25. Wenk, J. F., Ge, L., Zhang, Z., Soleimani, M., Potter, D. D., Wallace, A. W., Tseng, E., Ratcliffe, M. B., & Guccione, J. M. A coupled biventricular finite element and lumped-parameter circulatory system model of heart failure. *Comput. Methods Biomech. Biomed. Eng.* **16**, 807–818 (2012). doi: 10.1080/10255842.2011.641121.
26. Veress, A. I., Raymond, G. M., Gullberg, G. T. & Bassingthwaighte, J. B. B. Left ventricular finite element model bounded by a systemic circulation model. *J. Biomech. Eng.* **135**, 54502 (2013). doi: 10.1115/1.4023697.
27. Augenstein, K. F., Cowan, B. R., LeGrice, I. J. & Young, A. A. Estimation of cardiac hyperelastic material properties from MRI tissue tagging and diffusion tensor imaging. *Med. Image Comput. Comput. Assist. Interv.* **9**, 628–635 (2006).
28. Meijering, E. H., Niessen, W. J. & Viergever, M. A. Quantitative evaluation of convolution-based methods for medical image interpolation. *Med. Image. Anal.* **5**, 111–126 (2001).
29. Katouzian, A., Prakash, A. & Konofagou, E. A new automated technique for left-and right-ventricular segmentation in magnetic resonance imaging. *Conf. Proc. IEEE Eng. Med. Biol. Soc.* **1**, 3074–3077 (2006).
30. Yushkevich, P. A. et al. User-guided 3D active contour segmentation of anatomical structures: Significantly improved efficiency and reliability. *Neuroimage* **31**, 1116–1128 (2006).
31. Brower, R. W., ten Katen, H. J. & Meester, G. T. Direct method for determining regional myocardial shortening after bypass surgery from radiopaque markers in man. *Am. J. Cardiol.* **41**, 1222–1229 (1978).
32. Weiss, J. A. Ph.D. Dissertation 00131 (The University of Utah, Department of Bioengineering, 1994).
33. Maker, B. N., Ferencz, R. M. & Hallquist, J. O. *NIKE3D: A Nonlinear, Implicit, Three-Dimensional Finite Element Code for Solid and Structural Mechanics* (Lawrence Livermore Nat. Lab, 1990).
34. Novak, V. P., Yin, F. C. & Humphrey, J. D. Regional mechanical properties of passive myocardium. *J. Biomech.* **27**, 403–412 (1994).
35. Nocedal, J. & Wright, S. J. *Numerical Optimization*, 2nd edn. (Springer: New York, 2006).
36. LeVeque, R. J. *Finite Difference Methods for Ordinary and Partial Differential Equations: Steady-State and Time-Dependent Problems* (Society for Industrial and Applied Mathematics: Philadelphia, PA, 2007).

37. Coleman, B. R. Three dimensional estimation of passive myocardial parameters through finite element model optimization. Masters of Science Thesis (University of Washington, 2011).
38. Ibanez, L., Schroeder, W., Ng, L. & Cates, J. *The ITK Software Guide*, 2nd edn. (Kitware, Inc.: New York, 2005).
39. Yoo, T. S. et al. Engineering and algorithm design for an image processing Api: A technical report on ITK—The Insight Toolkit. *Stud. Health Technol. Inform.* **85**, 586–592 (2002).
40. The insight segmentation and registration toolkit. Available at http://www.itk.org.
41. Veress, A., Phatak, N. & Weiss, J. Deformable image registration with hyperelastic warping. In *Handbook of Biomed. Image Anal.*, Vol. 3 (eds. Suri, J. S., Wilson, D. L. & Laxminarayan, S.) 487–534 (Springer: New York, 2005).
42. Danielsson, P. E. Euclidean distance mapping. *Comp. Graph. Image Proc.* **14**, 227–248 (1980).
43. Agrawal, S. K. & Fabien, B. C. *Optimization of Dynamic Systems*. (Kluwer Academic Publishers, Dordrecht, Germany, 1999).
44. Moulton, M. J. et al. An inverse approach to determining myocardial material properties. *J. Biomech.* **28**, 935–948 (1995).
45. Moulton, M. J. et al. Myocardial material property determination in the in vivo heart using magnetic resonance imaging. *Int. J. Card. Imaging* **12**, 153–167 (1996).
46. Okamoto, R. J. et al. Epicardial suction: A new approach to mechanical testing of the passive ventricular wall. *J. Biomech. Eng.* **122**, 479–487 (2000).
47. Augenstein, K. F., Cowan, B. R., LeGrice, I. J., Nielsen, P. M. & Young, A. A. Method and apparatus for soft tissue material parameter estimation using tissue tagged Magnetic Resonance Imaging. *J. Biomech. Eng.* **127**, 148–157 (2005).
48. Wang, V. Y. et al. Modelling passive diastolic mechanics with quantitative MRI of cardiac structure and function. *Med. Image Anal.* **13**, 773–784 (2009).
49. Einstein, D. R., Freed, A. D., Stander, N., Fata, B. & Vesely, I. Inverse parameter fitting of biological tissues: A response surface approach. *Ann. Biomed. Eng.* **33**, 1819–1830 (2005).
50. Chen, K., Fata, B. & Einstein, D. R. Characterization of the highly nonlinear and anisotropic vascular tissues from experimental inflation data: A validation study toward the use of clinical data for in-vivo modeling and analysis. *Ann. Biomed. Eng.* **36**, 1668–1680 (2008).
51. Burkhoff, D., Mirsky, I. & Suga, H. Assessment of systolic and diastolic ventricular properties via pressure-volume analysis: A guide for clinical, translational, and basic researchers. *Am. J. Physiol. Heart Circ. Physiol.* **289**, H501–H512 (2005).
52. Teplick, R. S. Measuring central vascular pressures: A surprisingly complex problem. *Anesthesiology.* **67**, 289–291 (1987).
53. Falicov, R. E. & Resnekov, L. Relationship of the pulmonary artery end-diastolic pressure to the left ventricular end-diastolic and mean filling pressures in patients with and without left ventricular dysfunction. *Circulation* **42**, 65–73 (1970).
54. Rahimtoola, S. H. et al. Relationship of pulmonary artery to left ventricular diastolic pressures in acute myocardial infarction. *Circulation* **46**, 283–290 (1972).
55. Stott, D. K., Marpole, D. G., Bristow, J. D., Kloster, F. E. & Griswold, H. E. The role of left atrial transport in aortic and mitral stenosis. *Circulation* **41**, 1031–1041 (1970).
56. Mark, J. B. *Predicting Left Ventricular End-Diastolic Pressure in Atlas of Cardiovascular Monitoring* (Churchill Livingstone, 1998).
57. Mulvagh, S., Quinones, M. A., Kleiman, N. S., Cheirif, J. & Zoghbi, W. A. Estimation of left ventricular end-diastolic pressure from Doppler transmitral flow velocity in cardiac patients independent of systolic performance. *J. Am. Coll. Cardiol.* **20**, 112–119 (1992).
58. Appleton, C. P., Galloway, J. M., Gonzalez, M. S., Gaballa, M. & Basnight, M. A. Estimation of left ventricular filling pressures using two-dimensional and Doppler echocardiography in adult patients with cardiac disease. Additional value of analyzing left atrial size, left atrial ejection fraction and the difference in duration of pulmonary venous and mitral flow velocity at atrial contraction. *J. Am. Coll. Cardiol.* **22**, 1972–1982 (1993).
59. Giannuzzi, P. et al. Doppler-derived mitral deceleration time of early filling as a strong predictor of pulmonary capillary wedge pressure in postinfarction patients with left ventricular systolic dysfunction. *J. Am. Coll. Cardiol.* **23**, 1630–1637 (1994).
60. Vanoverschelde, J. L. et al. Noninvasive estimation of pulmonary arterial wedge pressure with Doppler transmitral flow velocity pattern in patients with known heart disease. *Am. J. Cardiol.* **75**, 383–389 (1995).
61. Nordmeyer, S. et al. Four-dimensional velocity-encoded magnetic resonance imaging improves blood flow quantification in patients with complex accelerated flow. *J. Magn. Reson. Imaging* **37**, 208–216 (2013).
62. Olariu, A., Wellnhofer, E., Grafe, M. & Fleck, E. Noninvasive estimation of left ventricular end-diastolic pressure by pulmonary venous flow deceleration time. *Eur. J. Echocard.* **4**, 162–168 (2003).
63. Pozzoli, M., Traversi, E. & Roelandt, J. R. Non-invasive estimation of left ventricular filling pressures by Doppler echocardiography. *Eur. J. Echocardiogr.* **3**, 75–79 (2002).
64. Yamamoto, K. et al. Determination of left ventricular filling pressure by Doppler echocardiography in patients with coronary artery disease: Critical role of left ventricular systolic function. *J. Am. Coll. Cardiol.* **30**, 1819–1826 (1997).
65. Kerckhoffs, R. C. et al. Coupling of a 3D finite element model of cardiac ventricular mechanics to lumped systems models of the systemic and pulmonic circulation. *Ann. Biomed. Eng.* **35**, 1–18 (2007).
66. de Putter, S. et al. Patient-specific initial wall stress in abdominal aortic aneurysms with a backward incremental method. *J. Biomech.* **40**, 1081–1090 (2007).
67. Dang, A. B. et al. Effect of ventricular size and patch stiffness in surgical anterior ventricular restoration: A finite element model study. *Ann. Thorac. Surg.* **79**, 185–193 (2005).
68. Walker, J. C. et al. Helical myofiber orientation after myocardial infarction and left ventricular surgical

restoration in sheep. *J. Thorac. Cardiovasc. Surg.* **129**, 382–390 (2005).
69. Walker, J. C. et al. MRI-based finite-element analysis of left ventricular aneurysm. *Am. J. Physiol. Heart Circ. Physiol.* **289**, H692–H700 (2005).
70. Walker, J. C. et al. Magnetic resonance imaging-based finite element stress analysis after linear repair of left ventricular aneurysm. *J. Thorac. Cardiovasc. Surg.* **135**, 1094–1102, 1102.e1091–e1092 (2008).
71. Wenk, J. F. et al. A coupled biventricular finite element and lumped-parameter circulatory system model of heart failure. *Comput. Methods Biomech. Biomed. Eng* **16**, 807–818 (2013).
72. Kerckhoffs, R. C. et al. Multi-scale modeling of excitation-contraction coupling in the normal and failing heart. *Conf Proc IEEE Eng Med. Biol. Soc.* 2009, 4281–4282 (2009).
73. Kerckhoffs, R. C. et al. Cardiac resynchronization: Insight from experimental and computational models. *Prog. Biophys. Mol. Biol.* **97**, 543–561 (2008).
74. Kerckhoffs, R. C., Omens, J. H. & McCulloch, A. D. Mechanical discoordination increases continuously after the onset of left bundle branch block despite constant electrical dyssynchrony in a computational model of cardiac electromechanics and growth. *Europace* **14 Suppl 5**, v65–v72 (2012).
75. Rohmer, D., Sitek, A. & Gullberg, G. T. Reconstruction and visualization of fiber and laminar structure in the normal human heart from ex vivo diffusion tensor magnetic resonance imaging (DTMRI) data. *Invest. Radiol.* **42**, 777–789 (2007).
76. Chen, J. et al. Regional ventricular wall thickening reflects changes in cardiac fiber and sheet structure during contraction: Quantification with diffusion tensor MRI. *Am. J. Physiol. Heart Circ. Physiol.* **289**, H1898–H1907 (2005).
77. Chen, J. et al. Remodeling of cardiac fiber structure after infarction in rats quantified with diffusion tensor MRI. *Am. J. Physiol. Heart Circ. Physiol.* **285**, H946–H954 (2003).
78. Frindel, C., Robini, M., Croisille, P. & Zhu, Y. M. Comparison of regularization methods for human cardiac diffusion tensor MRI. *Med. Image Anal.* **13**, 405–418 (2009).
79. Gullberg, G. T., Defrise, M., Panin, V. Y. & Zeng, G. L. Efficient cardiac diffusion tensor MRI by three-dimensional reconstruction of solenoidal tensor fields. *Magn. Reson. Med.* **19**, 233–256 (2001).
80. Nielles-Vallespin, S., Mekkaoui, C., Gatehouse, P., Reese, T. G., Keegan, J., Ferreira, P. F., Collins, S., Speier, P., Feiweier, T., de Silva, R., Jackowski, M. P., Pennell, D. J., Sosnovik, D. E. & Firmin, D. In vivo diffusion tensor MRI of the human heart: Reproducibility of breath-hold and navigator-based approaches. *Magn. Reson. Med.* **70**, 454–465 (2013).
81. de Roos, A. Myocardial perfusion imaging with multidetector CT: Beyond lumenography. *Radiology* **254**, 321–323 (2010).
82. Heller, G. V., Calnon, D. & Dorbala, S. Recent advances in cardiac PET and PET/CT myocardial perfusion imaging. *J. Nucl. Cardiol.* **16**, 962–969 (2009).
83. Perrone-Filardi, P. et al. Cardiac computed tomography and myocardial perfusion scintigraphy for risk stratification in asymptomatic individuals without known cardiovascular disease: A position statement of the Working Group on Nuclear Cardiology and Cardiac CT of the European Society of Cardiology. *Eur. Heart J.* **32**, 1986–1993, 1993a, 1993b (2011).
84. Thomas, G. S., Martinez, C. L. & Rice, D. L. Sequential myocardial perfusion imaging and cardiac CT: What to do with incidental CT findings? *Am. Heart Hosp. J.* **4**, 71–73 (2006).
85. Epstein, F. H. et al. Multislice first-pass cardiac perfusion MRI: Validation in a model of myocardial infarction. *Magn. Reson. Med.* **47**, 482–491 (2002).
86. Health Quality Ontario. Functional cardiac magnetic resonance imaging (MRI) in the assessment of myocardial viability and perfusion: An evidence-based analysis. *Ont. Health Technol. Assess Ser.* **3**, 1–82 (2003).
87. Lubbers, D. D. et al. The additional value of first pass myocardial perfusion imaging during peak dose of dobutamine stress cardiac MRI for the detection of myocardial ischemia. *Int. J. Card. Imag.* **24**, 69–76 (2008).
88. Misko, J. et al. Co-registration of cardiac MRI and rest gated SPECT in the assessment of myocardial perfusion, function and viability. *J. Nucl. Cardiol.* **8**, 389–397 (2006).
89. Blanksma, P. K. et al. Quantitative myocardial mapping of perfusion and metabolism using parametric polar map displays in cardiac PET. *J. Nucl. Med.* **36**, 153–158 (1995).
90. Schelbert, H. R. PET contributions to understanding normal and abnormal cardiac perfusion and metabolism. *Ann. Biomed. Eng.* **28**, 922–929 (2000).
91. Wu, Y. W. et al. PET assessment of myocardial perfusion reserve inversely correlates with intravascular ultrasound findings in angiographically normal cardiac transplant recipients. *J. Nucl. Med.* **51**, 906–912 (2010).
92. Alexanderson, E., Ricalde, A., Siu, S. & Meave, A. SPECT perfusion abnormalities in patients with noncompacted myocardium: A comparative image with cardiac magnetic resonance. *J. Nucl. Cardiol.* **11**, 634–636 (2004).
93. Cacciabaudo, J. M. & Szulc, M. Gated cardiac SPECT: Has the addition of function to perfusion strengthened the value of myocardial perfusion imaging? *J. Nucl. Med.* **42**, 1050–1052 (2001).
94. Cavusoglu, Y. et al. Regional distribution and extent of perfusion abnormalities, and the lung to heart uptake ratios during exercise thallium-201 SPECT imaging in patients with cardiac syndrome X. *Can. J. Cardiol.* **21**, 57–62 (2005).
95. Dorbala, S. et al. Prognostic value of SPECT myocardial perfusion imaging in patients with elevated cardiac troponin I levels and atypical clinical presentation. *J. Nucl. Cardiol.* 14, 53–58 (2007).
96. Sciagra, R. et al. Myocardial perfusion imaging using gated SPECT in heart failure patients undergoing cardiac resynchronization therapy. *J. Nucl. Med.* **45**, 164–168 (2004).
97. Bateman, T. M., Kolobrodov, V. V., Vasin, A. P. & O'Keefe, J. H. Jr. Extended acquisition for minimizing attenuation artifact in SPECT cardiac perfusion imaging. *J. Nucl. Med.* **35**, 625–627 (1994).
98. Chen, J. et al. Onset of left ventricular mechanical contraction as determined by phase analysis of ECG-gated myocardial perfusion SPECT imaging: Development of a diagnostic tool for assessment of cardiac mechanical dyssynchrony. *J. Nucl. Cardiol.* **12**, 687–695 (2005).

13

Particle Image Velocimetry for Biological Mechanics

Zachary J. Taylor, Roi Gurka, and Alex Liberzon

CONTENTS

Particle image velocimetry (PIV) is an optical-based technique for flow measurement. The technique enables the measurement of the velocity distribution within a given flow field with a high degree of accuracy over a large spatial extent. This chapter overviews the use of PIV in biological applications with particular focus on extensions to the technique that are often required in biological flows including measuring the time dependence of various phenomena and extracting accurate flow information near immersed boundaries. In addition, a case study is provided demonstrating the use of PIV in the measurement of forces in animal propulsion and locomotion.

13.1 Introduction

Particle image velocimetry (PIV) is a nonintrusive, state-of-the-art technique for flow measurements.[1,2] The principles of the technique are based on acquiring images of illuminated seed particles and correlating the displacement of a localized collection of particles over time, which yields a velocity vector. The experimental setup of a PIV system typically consists of several subsystems. In most applications, tracers (particles, droplets, etc.) are added to the investigated flow, and a light sheet is formed from a coherent light source (commonly a laser), which illuminates the particles. The light scattered by the tracers is then recorded on a sequence of frames, and the mean displacement of the particle images is determined through evaluation of pairs of PIV images by applying a spatial cross-correlation function. Assuming a suitable choice of tracers,[3] the local two-component flow velocity between two consecutive illuminations can be determined with a high degree of accuracy.[4]

The accuracy of the measurement results strongly depends on a large number of parameters such as particle concentration, size distribution, and shape[5]; illumination source; recording device[6,7]; and synchronization between the illumination, acquisition, and recording systems.[4] An appropriate choice of the different parameters of the cross-correlation analysis (e.g., interrogation area, time between pulses, scaling) will also influence the accuracy of the results. These themes arise across a number of techniques related to tracking of particles and textures to map deformation (see, for example, Chapters 11, 12, 15, 22, 30, and 40). Detailed descriptions of various implementations of the PIV method are given in the review of Adrian.[2] Although several components are required in a typical PIV setup, it is likely that less expensive systems will support many applications.[8] Furthermore, open-source software packages such as OpenPIV (www.openpiv.net) are available in numerous interfaces and programming languages like MATLAB, Python, or C++.[9]

The use of PIV in biological mechanics is considerably widespread at present and has led to many important discoveries. Aquatic animals, similar to avian ones, exhibit unique fluid dynamics, which are of interest to biologists and engineers. PIV has been used extensively in the study of aquatic propulsion of fish[10–13] and other swimmers.[14,15] In addition to propulsion, the study of fish feeding and the hydrodynamics of the predator–prey relationship have been investigated using PIV.[16,17] Additional topics using PIV but not covered here include hemodynamics,[18,19] respiratory systems,[20] microorganisms,[21] and marine invertebrates.[22] Although outside the scope of the current review, PIV techniques are also currently being applied in microscopy,[23] with different approaches being evaluated, including light-emitting diode (LED) illumination[24] as well as confocal techniques.[25] At much larger scales, the mysteries of bird flight have been quantitatively explored using PIV. One of the first studies in this regard was done by Spedding et al.,[26] where PIV was used to measure the wake of a thrush nightingale (*Luscinia luscinia*). Through this study, and subsequent studies,[27,28] it has been suggested that birds create a smoothly varying vortex structure over a range of flight speeds and that this structure may be a general feature of bird flight. Furthermore, bat wakes were shown to be similar to bird wakes using PIV even though they contain several unique features.[29,30] Recently, Kirchhefer et al.[31] demonstrated similarity features between the wakes of bats and starlings (*Sturnus vulgaris*) through a detailed study using time-resolved PIV. Other flow features resolved by PIV in different species of birds, such as the hummingbird, were addressed by Warrick et al.,[32] who suggest that these birds may operate at Reynolds numbers sufficiently low to exploit a key mechanism typical of insect hovering.

One of the inherent challenges when studying the interaction of fluid dynamics in many biological applications is the variability of experimental conditions when exploring living subjects. For example, fish can change swimming mode depending on flow conditions and their context (swimming, feeding, escaping). Thus, in addition to the stochastic nature of turbulent flows, the variability in the behavior of the test subject may require certain extensions to the classical PIV technique in order to provide useful insights on the biological mechanics. In this chapter, the state-of-the-art add-ons to the PIV technique are discussed, including how to account for high- and low-frequency temporal variations (Section 13.2.1) and methods to improve the accuracy of flow near immersed boundaries (Section 13.2.2). Techniques to account for variations in the frame of reference and extensions to three dimensions are also discussed (Section 13.2.3) along with how to couple mass transport measurements with velocity data (Section 13.2.4). Furthermore, a case study on the measurement of forces using PIV data is presented in the context of animal locomotion and propulsion in Section 13.3.

13.2 Extensions to the PIV Technique

13.2.1 Time Dependence

Many physical phenomena vary with time. Historically, it has been necessary to make assumptions about the nature of the time dependence of the particular phenomenon. For example, in the study of flapping flight, it is often assumed that quasi–steady-state conditions exist, implying that the process behaves in an equivalent series of time-invariant (steady) flows. However, unsteady aerodynamics are known to be important to both insect and bird flight,[33,34] thus suggesting the need for increased time resolution. Likewise, there are many scenarios in biological flows where the time variation is one of the dominant features that cannot be ignored or assumed to be negligible. In addition to unsteady fluid dynamics, time resolution is also of practical interest for experimentalists as a way to cope with the inherent variability in the actions of living test subjects.[35]

Until recently, the measurement of time dependence using optical techniques, such as PIV, has been limited. On one hand, the spatial resolution was set by the camera sensor size to provide high spatial resolution, yet on the other hand, the fastest CCD cameras with high spatial resolution were limited to sampling frequencies on the order of 10 Hz. Once complementary metal-oxide semiconductor (CMOS) devices were introduced,[36] the potential of resolving the time dependence of the velocity field became feasible. With the advent of high-speed imaging, with sufficient quality and spatial resolution, sampling frequencies on the order of 1000 Hz are now common. The increase in the image sampling frequency enforced the use of a high-energy source for illumination that can deliver a sufficient amount of energy, over a short period of time, at a high frequency. The current choices that are available are continuous lasers, high-repetition-rate lasers, and LEDs.[37] The choice of which light source to use depends mainly on the application and available resources. For air flow applications, a high-repetition-rate laser would be required, while a continuous laser or LED is most often sufficient in water. However, it is important to note that, unlike lasers, there are currently no reliable methods to focus LED light into a thin light sheet for large measurement volumes. As previously discussed, research is underway to develop methods for the use of LED light in smaller measurement volumes—particularly in the field of microscopy.[24]

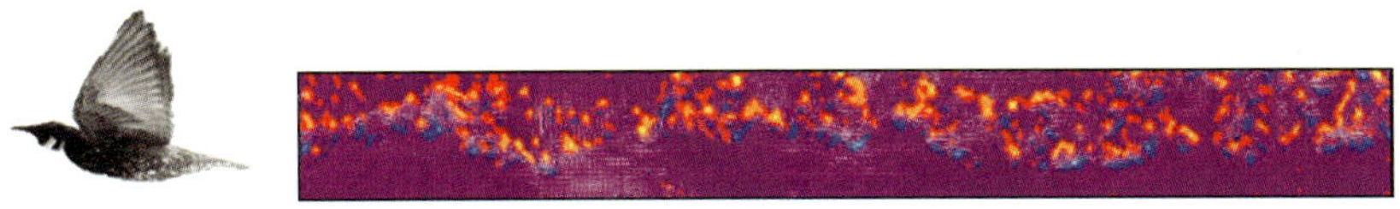

FIGURE 13.1
Reconstructed wake of a European starling measured using long-duration time-resolved PIV. The arrows are fluctuating velocities, and the contours represent the vorticity. (From Ben-Gida, H. et al., *PLoS ONE*, doi:0.1371/journal.pone.0080086.)

In addition to the laser requirements, data transfer from the imaging device to a computer or hard disk becomes increasingly complex. High-speed cameras can save images to an on-board memory card. However, their size is limited, and the amount of storage usually corresponds to only a few seconds of data acquisition. Due to the variability of biological systems and the existence of turbulent fluid dynamics, it is often desirable to treat the data statistically. The high-frequency turbulent events can be resolved if the sampling frequency is sufficiently high; however, low-frequency events (such as flapping wings) require a longer sampling length for statistical analysis. The length of the data record can be increased by using frame grabbers that stream data directly to an array of solid-state hard disks at rates in excess of 1 GB/s. Such a solution was proposed by Taylor et al.,[9] where long-duration capabilities were utilized at the Advanced Facility for Avian Research at the University of Western Ontario. The system is capable of sampling 1000 images per second, per camera, for 25 min continuously. This system allowed for the characterization of the unsteady wake behind a freely flying European starling[31] and the assessment of the contribution of the unsteady drag during flapping flight.[38] The wake data were reconstructed to visualize complete wing-beat cycles,[38] and one of these complete wing-beat cycles is shown in Figure 13.1.

13.2.2 PIV near Immersed Boundaries

Much of the important fluid dynamics in biological systems occurs near bodies immersed in fluid, such as the boundary layer on swimming fish.[39] In the study of flow near immersed boundaries, one of the most significant challenges is the layer of light—or laser glaze—created at the interface between the body and the fluid due to reflections. In conventional applications, an antireflective coating can be applied; however, these coatings are often toxic, rendering them inapplicable to living systems. Moreover, in many studies, the location of the surface may be moving (e.g., a moving animal or a compliant wall), and such cases often require high accuracy of the flow data in the proximity of the surface. In this section, techniques to enhance the accuracy of the PIV data near the surface are discussed.

In the study of the boundary layer on a freely swimming fish, the relatively sparse seeding density in the boundary layer limited the accuracy of the data in this region.[39] This limitation was resolved by manually tracking particles in two dimensions close to the surface (the reader is referred to Section 13.2.3 for a review of an automatic tracking algorithm). Another category of PIV applications in which precise definition of immersed object boundaries is crucial is in the studies of two-body or multibody interactions, such as contact, settlement, deposition, and prey–predator interactions. For example, Zilman et al.[40] studied the probability of contact between marine larvae and a cylindrical collector. In this case, it was shown that surface glaze on the cylindrical collector can drastically change the quality and the statistics of the measurements. A typical solution of wavelength filtering (e.g., 532 nm Nd:YAG laser sheet, fluorescent particles, and a band-pass light filter on a camera lens) cannot be applied because of the toxicity of the fluorescent particles and the presence of the larvae. The solution in this case was to measure both the flow around the cylinder and the motion of larvae using Lagrangian tracking of flow tracers and of the moving objects (as discussed in more detail in Section 13.2.3). Furthermore, the acceleration of the swimming organism and/or sharp changes of the trajectory curvature, when measured accurately, can assist in identifying the critical time of contact or settlement.[40]

Another example when the distance between two objects is crucial is in the fluid dynamics of feeding fish. PIV has been used extensively in this field of study,[23,24,32,41] and a typical pair of PIV images is shown in Figure 13.2. The accuracy of the PIV measurements

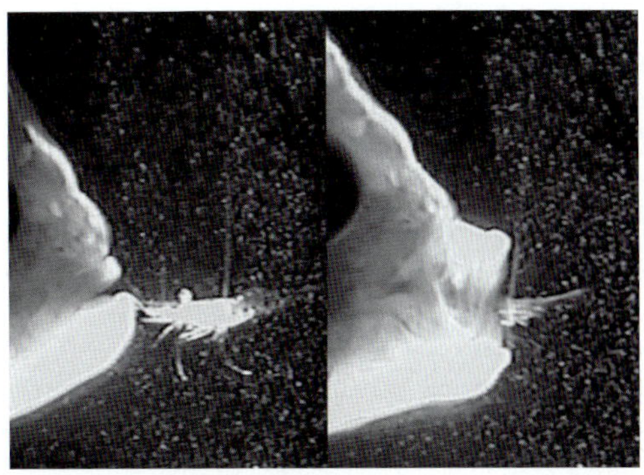

FIGURE 13.2
Bluegill sunfish feeding on a ghost shrimp event as recorded by PIV cameras at two time instances 12 ms apart. The mouth size is approximately 10 mm in diameter. (Courtesy of Holzman, R. et al. *J. Exp. Biol.*, 211, 2008.)

and of the distance between the objects is strongly influenced by the laser glaze on both the predator and the prey. Rather than a conventional surface treatment, however, the problem can be addressed by modification to the existing algorithms.

Theunissen et al.[42] highlighted the following problems that arise in the classical PIV cross-correlation algorithm applied to regions close to solid boundaries: (1) truncation of the particle displacement signal at the wall, (2) spurious light speckles due to reflections, and (3) insufficient resolution in the direction normal to the boundary. One of the major sources for these problems in conventional cross-correlation analysis is the use of the fast Fourier transform (FFT), which assumes periodicity in all directions. This assumption is violated in the interrogation windows near surfaces. The proposed solutions[42] include correlation using windows of an arbitrary shape that also take into account a mask over the reflecting surfaces. Furthermore, the algorithm is combined with an automatic image preprocessing routine that removes bright spots above a certain threshold. Another challenge that arises when estimating the velocity close to an immersed surface is that the vector location is inherently shifted toward the solid boundary (or its mask), as demonstrated in Figure 13.3. In this case, an adaptive method for boundary compensation should be used.[42,43]

Further algorithmic solutions have been suggested[44,45] for cases in which the wall normal particle image displacements near a body surface are smaller than the tracer diameters themselves. The main concept of these algorithms is to perform a series of preprocessing steps starting from boundary identification, followed by a conformal transformation of the image to a regular PIV grid. It is then suggested to split the cross-correlation steps into the tangential and normal directions of motion, performing two 1-D correlations rather than the standard 2-D correlation.[42]

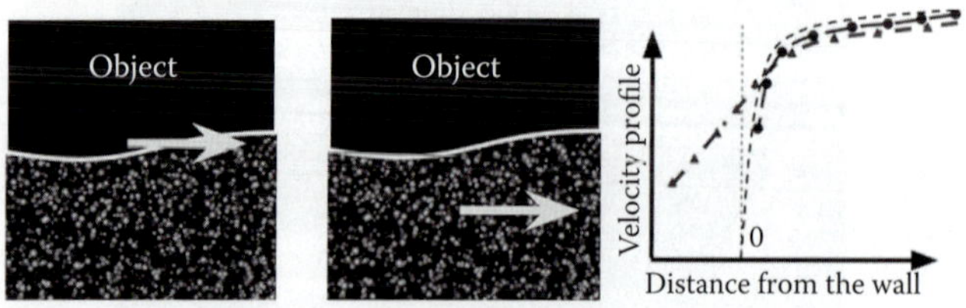

FIGURE 13.3
Schematic demonstrating the differences between the classic PIV algorithm and the boundary-compensated algorithms proposed in ref. 42. In the left panel, the standard PIV algorithm determines the incorrect location of the vector, and in the center panel, improved estimation using the boundary-compensated method is shown. The effect of the boundary-compensated method on the velocity profile is shown in the right panel by comparing the standard method (▲) and the boundary-compensated method (●) with the theoretical result (dashed line). (From Theunissen, R. et al., *Exp. Fluids*, 45, 2008.)

The work of Adhikari and Longmire[46] offers a solution based on experimental equipment rather than modifications to the algorithm. In this case, an infrared tomographic three-dimensional measurement system was employed to measure flow fields and the motion of the objects simultaneously in a predator–prey scenario. Through the use of a near-infrared laser (wavelength of 808 nm), concerns over the organism being affected by the sight of the laser pulses was removed. However, invisible high-power laser light could lead to severe safety concerns for the researchers. There is also increased absorption of the illumination light in water at these long wavelengths as opposed to visible wavelength illumination. Nevertheless, the ability to track two interacting animals and the continuously changing flow field in between the two surfaces was achieved.[46] The importance of three-dimensional measurements was also shown since the full three-dimensional vortical flow structures in the wake differed noticeably from reconstructions based on multiple two-dimensional planar PIV recordings.[46]

13.2.3 Lagrangian Frame of Reference: Particle-Tracking Methods

In some instances, it is desirable to measure the flow features as they move in the flow field in space and time (the Lagrangian frame of reference) rather than measuring changes in the flow at a specific location with time (the Eulerian frame of reference). PIV strictly operates in a Eulerian framework. To partially adapt PIV data into a Lagrangian framework, one typically has to track an object (e.g., a fish or a vortex) within the PIV images. For example, Anderson et al.[39] used manual two-dimensional particle tracking in order to address the flow in the near-surface layer of a fish. However, using similar components from a classical PIV system, it is possible to perform three-dimensional particle-tracking experiments.

Three-dimensional particle-tracking velocimetry (3D-PTV[47]) utilizes the concept that every fluid point, if it is tagged at an instant in time, will move along its own unique trajectory. Although it is a nontrivial task, tracking this object does not involve further theoretical or analytical approximations. The result, if continued for a long time, or repeated for a large number of fluid tracers (small seeding tracers similar to those used in PIV), provides an exact flow field similar to that measured using PIV at all possible cross-sections. Furthermore, there is an advantage in terms of the obtained data since it is not fixed in space and allows one to analyze the motion of the fluid and of the organisms (or their surfaces) in time and space.

A typical 3D-PTV setup is shown in Figure 13.4. The required components are volumetric illumination and at least three cameras. The light source need not

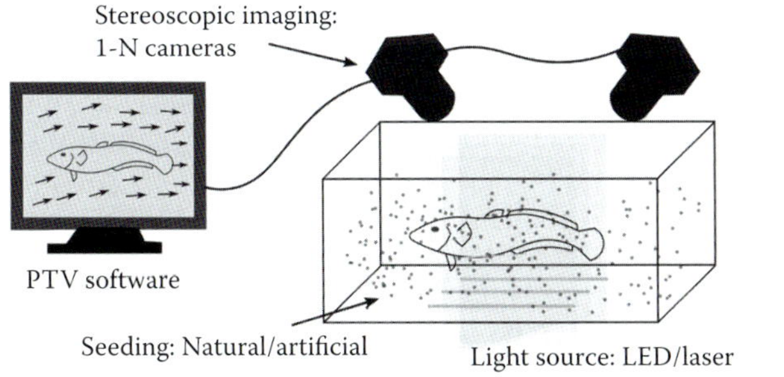

FIGURE 13.4
Typical 3D-PTV setup: volumetric illumination of the volume of interest and multiple digital video cameras.

be monochromatic, coherent, or collimated; therefore, while lasers can be used, less expensive options such as white light, LEDs, or stroboscopes are sufficient as long as the flow tracers and/or the moving objects reflect and diffract the light. The cameras capture the 3-D position of each flow tracer in the volume of interest, and it is recommended to use three or more cameras in order to resolve ambiguities. Possible ambiguities can arise due to the high density of flow tracers, optical occlusions, and the variable index of refraction at the inter-media interfaces (e.g., air–glass or air–water). For more details the reader is referred to Dracos[47] in addition to the open-source software solution OpenPTV (http://www.openptv.net), which allows easy adoption of the 3D-PTV method.

3D-PTV has been applied to a vast variety of flows from cardiovascular hemodynamics to the motion of bioorganisms. For example, Boutsianis et al.[48] used the Lagrangian description of laminar-turbulent transitions in an anatomically accurate abdominal aortic aneurysm model while the pulsating thoracic aorta flow was also measured using this technique.[49] By tracking passive flow particles and marine larvae (*Bugula neritina*), the flow and contact past a cylindrical collector were measured by Zilman et al.,[40] who demonstrate the advantages of this technique in multibody experiments. In a study of animal locomotion, Gharib et al.[50] have used similar techniques to measure the flow around a swimming boxfish. These studies demonstrate one of the additional advantages of Lagrangian methods: the possibility to obtain the shape and the motion of the compliant surface (e.g., the aorta walls or the fish body). Typical results are shown in Figure 13.5, where Lagrangian trajectories of the flow tracers are plotted in the pulsating aortic flow at different time instances. Also shown in Figure 13.5 is the trajectory of the actively swimming marine larva in the flow passing a cylindrical collector.

13.2.4 Mass and Momentum Transport

Transport phenomena in biological systems play a major role in various physical, chemical, and biological processes. For example, PIV has been used to study the hydrodynamics of krill (*Euphasia pacifica*).[51] Through these measurements, it was shown that a volume 18 times larger than the krill's own volume is affected by the jet from the swimming krill. It was suggested that the other krill may use the momentum and mass transport of this jet to track each other.[51] Denissenko et al.[52] examined how the red swamp crayfish (*Procambarus clarkii*) uses outward directed currents to generate an inflow that could serve for odor acquisition. These crayfish inhabit rice fields and other

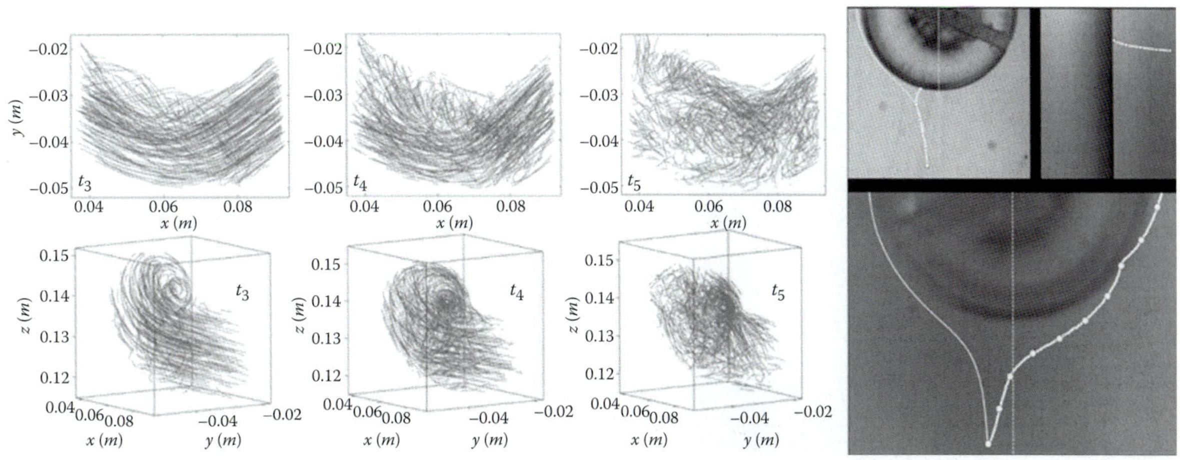

FIGURE 13.5
Typical results of 3D-PTV. (Left panel) Multiple trajectories representing the developing vortical structure in an aortic flow field at different phase instances of a cardiac pulse. (From Gülan, U. et al., *Exp. Fluids*, 53, 2012.) (Right panel) Motion of the flow tracers and of a marine larvae (*Bugula neritina*) in the proximity of a cylinder collector (two of the views show the top, while the upper right view shows the side of the cylinder). (From Zilman, G. et al., *J. Exp. Biol.*, 216, 2013.)

stagnant water bodies; therefore, they utilize fanning as an active olfactory mechanism. These studies demonstrate the use of PIV to develop hypotheses about how the flow carries odor to chemosensory receptors. In order to confirm these hypotheses, it is advantageous to simultaneously measure the mass transport of the odor field with the momentum transport measured by PIV. In this section, an extension to PIV known as planar laser-induced fluorescence (PLIF) is presented for the purpose of measuring momentum and mass transport. The utilization of PLIF combined with PIV in a biological environment is important when scalar transport plays a dominant role in the investigated phenomena.

PLIF is a nonintrusive technique for measuring the spatial scalar concentration field. Conceptually, PLIF is similar to flow visualization, yet it provides a quantitative description of the transported scalar. Essentially, PLIF instrumentation is similar to PIV, and the user does not have to significantly change the PIV setup. However, for simultaneous capture of PIV and PLIF, it is typical to use two cameras: one for PIV and the other for PLIF. The PLIF technique relies on the relationship between the fluorescent light emission, the concentration of tracer dye, and the intensity of the laser. At low concentration levels, there is a linear relationship between the dye concentration and the emitted light intensity. Thus, calibration for many experiments is relatively straightforward, as detailed by Crimaldi.[53] Since the activation light source (the laser) has a different wavelength than the fluoresced light, filters can be fitted to the cameras so that the PIV camera measures only the light reflected from the particles, while the PLIF camera measures only the fluoresced light. Figure 13.6 demonstrates the utilization of PLIF in environmental research where a contaminant plume is captured in a water tunnel. PIV yields simultaneous velocities (arrows), revealing the interaction between the velocity field and the concentration of contaminants.[54,55]

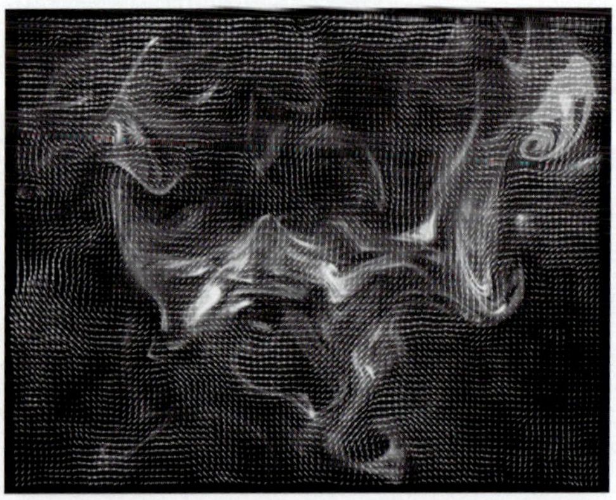

FIGURE 13.6
Measurements of pollutant dispersion from a point source in turbulence using simultaneous PIV/PLIF. The image of the fluoresced light from the scalar (Rhodamine 6G) is overlaid with the fluctuating velocity vectors. (From Sarathi, P. et al., *Exp. Fluids*, 52, 2012.)

The PIV/PLIF technique was employed by Weissburg et al.[56] In their study on the odor navigation of blue crabs (*Callinectes sapidus*) in turbulent flows. They examined the potentially competing influences of hydrodynamic forces and odor transport to explain blue crab behavioral responses to the simultaneous demands of efficient locomotion and acquisition of sensory cues. They found that, rather than adopting a fixed orientation that minimizes the drag force, blue crabs decrease their relative angle (increase drag) when odor is present in low-speed flow, while assuming a drag minimizing posture under other conditions. Another example of the use of PIV/PLIF in a biological system is in the release of a fluorescent flow tracer from coral skeletons in the laboratory experiment of Koehl and Reidenbach.[57] They showed how the chemical cues released from corals can cause larvae to settle at particular locations. Thus, by using PIV coupled with PLIF, insights can be learned about how animals interact with the normally invisible chemical tracers in the flow. These findings can lead to explanations in behavioral responses ranging from locomotion to mating and can aid in the understanding of how animals navigate in nature based on cues other than vision.

13.3 Case Study: Measurement of Forces in Animal Propulsion

In many biological applications, it is desirable to know the forces acting on a body immersed in a fluid. While relatively small species can struggle against viscosity,[58] birds and fish have evolved to move through the air and water with great speed and efficiency. Realistic force measurements are a significant challenge in biological propulsion experiments. For example, it is unlikely that realistic loads are obtained by tethering or gluing a specimen to a force balance. Thus, nonintrusive flow measurements such as PIV have been exploited to measure forces. There are two main approaches to measuring forces using PIV: (1) through approximation of the terms in the momentum integral equation and (2) through measuring vortex circulation in the wake and relating it to the force.[59] In this section, the use of the momentum integral equation with PIV data is demonstrated, followed by a discussion on using vortex identification methods to calculate circulation.

13.3.1 Force Estimation by the Momentum Integral Equation

In the majority of cases, the animal to be studied is fully immersed in fluid (typically either air or water). Thus, a control volume can be established around the animal assuming constant density, ρ, as shown in Figure 13.7. In this section, common assumptions regarding animal locomotion and propulsion are discussed with an emphasis on using PIV to estimate terms in the momentum integral equation. For a more general discussion on force measurement using PIV, the reader is referred to Noca,[60] and for simplicity the notation here is kept consistent with this reference.

The force, F, on a body immersed in an incompressible fluid is expressed through the momentum integral equation:

$$F = -\frac{d}{dt}\int_{V(t)} \rho u \, dV + \oint_{S(t)} n \cdot [-pI - (u - u_s)\rho u + \mu \nabla u] dS - \oint_{S_b(t)} n \cdot (u - u_s)\rho u \, dS \quad (13.1)$$

where u is a flow velocity vector; u_s is the surface velocity; pI is the pressure multiplied by the identity matrix; $\mu\nabla u$ are the viscous stresses; $V(t)$, $S(t)$, and $S_b(t)$ are the overall control volume, control surface, and immersed body surface, respectively; and n being the unit vector to the surface.

For most purposes, it is plausible to assume the following boundary conditions on the control surface enveloping the body, $S_b(t)$: (1) no through flow, $(u - u_s) \cdot n|_b = 0$, and (2) no slip, $(u - u_s)|_b = 0$. While these assumptions are generally accepted in bird aerodynamics and fish hydrodynamics, it should be noted that there are still many unsolved questions about the efficiency of bird flight, including possible flow through the wing.[61] For rigid body motion, $u_s|_b = f(t) + x_b \times g(t)$, where $f(t)$ is a translational velocity vector and $g(t)$ is a rigid body angular velocity vector. By applying these assumptions, expanding the term containing the time derivative, and neglecting viscous stresses, the force on the immersed body is described by

$$F = -\rho \int_{V(t)} \frac{\partial u}{\partial t} dV - \oint_{S(t)} n \cdot pI \, dS - \rho \oint_{S(t)} (n \cdot u) u \, dS - \rho \oint_{S_b(t)} (n \cdot u_s) u_s \, dS. \quad (13.2)$$

As an example of Equation 13.2, consider the control surface shown in Figure 13.7. Since PIV is a planar measurement tool, the typical control surface is of unit depth providing an estimate of the sectional force. The control surface is chosen such that the inflow conditions are uniform ($u = U_\infty i$, $p = p_\infty$) and the lines connecting $(x_1, y_1) \rightarrow (x_4, y_4)$ and $(x_2, y_2) \rightarrow (x_3, y_3)$ are streamlines with pressure $p = p_\infty$. An example force calculation is now carried out for the force in the x direction, F_x, with the velocity vector $u = ui + vj$ obtained from the PIV measurement plane shown in Figure 13.7.

Since biological flows are rarely steady in either an instantaneous or an average sense, one of the challenges in force estimation is computing the volume integral of the time derivative. As a result, this term is often neglected; however, an approximation of this term[38] has shown its potential importance in bird propulsion. The pressure term is also often neglected due to experimental limitations. However, the importance of this term in unsteady aerodynamics remains an open question, and although outside the scope of the current discussion, there have been notable developments in techniques aimed at obtaining an estimate of the pressure field directly from PIV measurements.[60,62,63]

Because there is no flow across a streamline ($n \cdot u = 0$), and because the inlet is assumed to be uniform (i.e., $u = U_\infty i$), the following expression can be found from the surface integral of the velocity:

$$\rho \int_{y_3}^{y_4} u(U_\infty - u) dy, \quad (13.3)$$

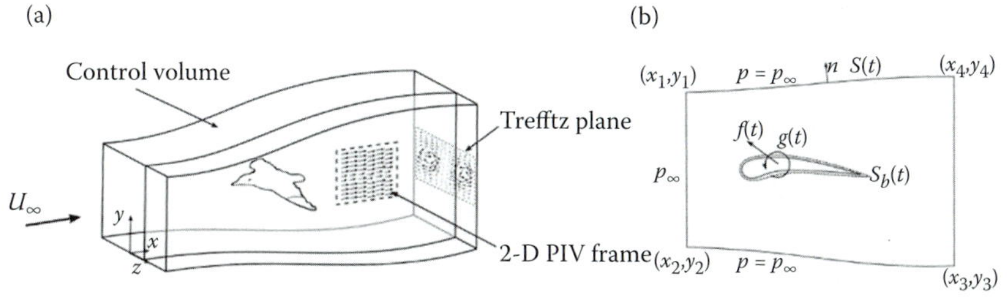

FIGURE 13.7
(a) Example of a control volume around an immersed, freely flying bird. (b) A control surface resulting from the cross-sectional cut plane.

incorporating the integral form of the conservation of mass. In classical aerodynamics, Equation 13.3 is referred to as the profile drag. The integral of the body surface is zero for the steady case; however, in cases such as flapping flight, the velocity of the bird's wing can be nonnegligible. The importance of this term is especially evident when attempting to compute lift in a similar fashion. Thus, while drag/thrust forces are often evaluated in this way,[38,59] different methods are used to estimate the total lift force using PIV, which are discussed in the following section.

13.3.2 Measurement of Circulation by Vortex Identification

There is a direct connection between lift and vortex circulation in classical aerodynamics, and many studies have shown the link between vortex structure and the propulsion of fliers and swimmers. For example, Dabiri[64] uses vortex ring structures to estimate the thrust for jellyfish propulsion,[65] while elliptical vortex loops have been suggested as a model for the lift generated by bird wakes.[66–68] By using PIV measurements, the measured circulation in these loops has been used to approximate the total lift and drag forces.[26] Likewise, a reverse Kármán vortex street is one of the characteristics in the thrust-generating wake of swimming fish.[13,69] Thus, the identification of vortices from PIV data is particularly important for estimates of circulation.

To illustrate some of the difficulties in rigorously identifying a vortex, a potential flow scenario is constructed by superposition of an idealized Rankine vortex, a shear flow, and uniform flow in Figure 13.8a. From an initial observation of Figure 13.8a, it appears that the center of the vortex is located at $(x, y) = (0, 5)$, yet, it is actually located at $(x, y) = (0, 0)$. In order to identify a vortex "by eye," the speed of the vortex should be known a priori.[70] Since this limitation is often impractical, another method used to identify vortices is to compute the vorticity, $\omega = \nabla \times \boldsymbol{u}$, which is independent of the reference frame and shown in Figure 13.8b. Figure 13.8b demonstrates that vorticity reveals not only regions of strong rotation but also regions where there is high shear. Thus, a different set of criteria is necessary to identify vortices.

In a three-dimensional flow field, several different vortex identification methods have been proposed.[71–73] However, in a two-dimensional flow, these three methods are all equivalent.[73] Even though most natural flows are three-dimensional, many implementations of PIV obtain two-dimensional velocity fields (see Section 13.2.3 for extensions to three dimensions). As a result, it is only possible to identify vortices aligned approximately normal to the imaging plane. The criterion for a vortex in two dimensions is the same regardless of the aforementioned methods: a complex eigenvalue pair for the two-dimensional velocity gradient tensor $\nabla \boldsymbol{u}$. The complex portion of the complex pair is typically referred to as the swirling strength,[74,75] λ_{ci}, and is defined as

$$\lambda_{ci} = \sqrt{\left[\frac{1}{2}\left(\frac{\partial u}{\partial x} + \frac{\partial v}{\partial y}\right)\right]^2 - 2\left(\frac{\partial u}{\partial x}\frac{\partial v}{\partial y} - \frac{\partial u}{\partial y}\frac{\partial v}{\partial x}\right)}. \quad (13.4)$$

The swirling strength has been computed for the potential flow case and is shown in Figure 13.8c. In Figure 13.8d, the uniform flow has been subtracted so that the observer is traveling at the reference frame of the vortex. Comparing the swirling strength in Figure 13.8c with the vortex in Figure 13.8d reveals the capabilities of the swirling strength in identifying the vortex. Furthermore, the swirling strength contours allow for unambiguous definition of vortex boundaries for further analyses.

The circulation of a vortex is

$$\Gamma = \oint_C \boldsymbol{u} \cdot d\boldsymbol{l} = \iint_A \omega \cdot \boldsymbol{n}\, dA. \quad (13.5)$$

Thus, the boundary of the vortex must be identified to calculate the circulation associated with the individual

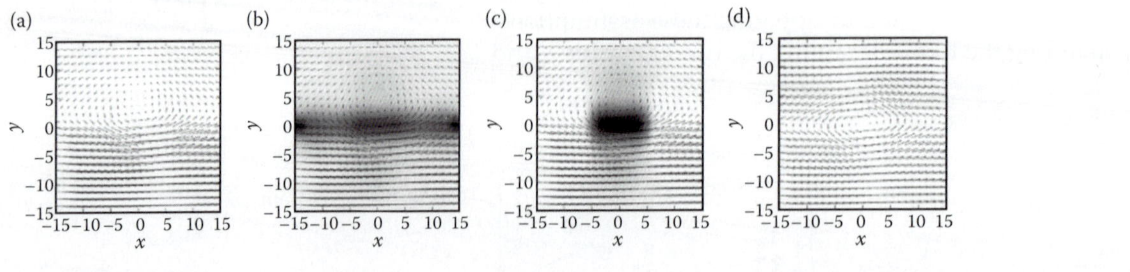

FIGURE 13.8
A potential flow of the superposition of a Rankine vortex (radius = 6), uniform flow, and shear. (a) Velocity vectors of the potential flow field. (b) Velocity vectors in addition to a grayscale map of the vorticity. (c) Vectors in addition to a grayscale map of the swirling strength. (d) Velocity vectors with the uniform flow subtracted.

vortex. This task is often performed by (1) identifying the center of the vortex and (2) defining the radius of the vortex by a threshold related to the maximum vorticity in the center of the vortex (e.g., <5%).[76,77] Once the boundaries of a vortex are identified, the circulation of the vortex can be computed, and by using an appropriate model, the forces can be approximated.[26,68] Increasing the accuracy in estimates of vortex circulation can be used to refine assumptions about the vortex wake topology in addition to the relationship between a given wake structure and the propulsive forces.

13.4 Summary

PIV is an established tool in fluid measurements, and it has remained state-of-the-art due to the continual refinements and extensions to the technique. In the study of biological mechanics, it has been used in a wide range of applications ranging from internal flows to predator–prey relationships to avian/aquatic propulsion, a sampling of which has been reviewed here.

While there are many extensions to the classical PIV technique, four of the main extensions have been addressed in this chapter: (1) long-duration time-resolved PIV to resolve both short and long time scales; (2) methods to resolve the velocity close to immersed surfaces; (3) 3D-PTV for a Lagrangian description in moving frames of reference; and (4) the ability to couple PIV with PLIF for momentum and mass transport. These extensions allow for a greater depth of insight to the flow in many biological systems.

As an example of the possible insights obtained from PIV, a case study on force measurements was presented. The spatial extent and the high spatial resolution from PIV have been shown to offer significant insight into the fluid forces governing animal propulsion and locomotion. Though significant progress has already been made in the understanding of fluid dynamics in biological systems because of PIV to date, continued work to overcome some of the remaining technological hurdles will lead to even greater understanding in the future. The scales over which PIV is applicable are continually being expanded as accuracy increases at the micro scale and resolution is increased at the larger scales. Likewise, the available time resolution is only expected to increase. As already discussed, one of the most exciting current and future developments is the extension of the technique to include the full three-dimensional flow. Insight into this additional dimension is expected to bring many new discoveries in the interaction of fluids and biological systems.

Acknowledgments

Z.J. Taylor is grateful to the Azrieli Foundation for the award of an Azrieli Fellowship. The authors acknowledge the support of the Bi-National U.S.–Israel Science Foundation under grant no. 2008051. A. Liberzon's work was partially supported by the Israel Science Foundation under grant no. 782/08. R. Gurka was partially supported by the Israel Science Foundation under grant no. 426/12.

References

1. Raffel, M., Willert, C.E., Wereley, S.T. & Kompenhans, J. *Particle Image Velocimetry—A Practical Guide*, 2nd Edition. Springer-Verlag, Heidelberg (2007).
2. Adrian, R.J. Particle-imaging techniques for experimental fluid mechanics. *Annu. Rev. Fluid Mech.* **23**, 261–304 (1991).
3. Melling, A. Tracer particles and seeding for Particle Image Velocimetry. *Meas. Sci. Technol.* **8**, 1406–1416 (1997).
4. Huang, H., Dabiri, D. & Gharib, M. On errors in digital Particle Image Velocimetry. *Meas. Sci. Technol.* **8**, 1427–1440 (1997).
5. Hadad, T. & Gurka, R. Effects of particle size, concentration and surface coating on turbulent flow properties obtained using PIV/PTV. *Exp. Therm. Fluid Sci.* **45**, 203–212 (2013).
6. Liberzon, A., Gurka, R. & Hetsroni, G. XPIV—Multi-plane stereoscopic Particle Image Velocimetry. *Exp. Fluids* **36**, 355–362 (2004).
7. Hain, R. & Kahler, C.J. Fundamentals of multiframe Particle Image Velocimetry (PIV). *Exp. Fluids* **42**, 575–587 (2007).
8. Ryerson, W.G. & Schwenk, K. A simple, inexpensive system for digital Particle Image Velocimetry (DPIV) in biomechanics. *J. Exp. Zool.* **317**, 127–140 (2011).
9. Taylor, Z.J., Gurka, R., Kopp, G.A. & Liberzon, A. Long-duration time-resolved PIV to study unsteady aerodynamics. *IEEE Trans. Instrum. Meas.* **59**, 3262–3269 (2010).
10. Nauen, J.C. & Lauder, G.V. Quantification of the wake of rainbow trout (*Oncorhynchus mykiss*) using three-dimensional stereoscopic digital Particle Image Velocimetry. *J. Exp. Biol.* **205**, 3271–3279 (2002).
11. Lauder, G.V. & Madden, P.G.A. Fish locomotion: Kinematics and hydrodynamics of flexible foil-like fins. *Exp. Fluids* **43**, 641–653 (2007).
12. Lauder, G.V. & Madden, P.G.A. Advances in comparative physiology from high-speed imaging of animal and fluid motion. *Annu. Rev. Physiol.* **70**, 143–163 (2008).
13. Taylor, Z.J., Liberzon, A., Gurka, R., Holzman, R., Reesbeck, T. & Diez, F.J. Experiments on the vortex wake of a swimming knifefish. *Exp. Fluids* **54**, 1588 (2013).

14. Catton, K.B., Webster, D.R., Brown, J. & Yen, J. Quantitative analysis of tethered and free-swimming copepodid flow fields. *J. Exp. Biol.* **210**, 299–310 (2007).
15. Dabiri, J.O., Colin, S.P., Katija, K. & Costello, J.H. A wake-based correlate of swimming performance and foraging behavior in seven co-occurring jellyfish species. *J. Exp. Biol.* **213**, 1217–1227 (2010).
16. Holzman, R., Collar, D.C., Day, S.W., Bishop, K.L. & Wainwright, P.C. Scaling of suction-induced flows in bluegill: Morphological and kinematic predictors for the ontogeny of feeding performance. *J. Exp. Biol.* **211**, 2658–2668 (2008).
17. Holzman, R. & Wainwright, P.C. How to surprise a copepod: Strike kinematics reduce hydrodynamic disturbance and increase stealth of suction feeding fish. *Limnol. Oceanogr.* **54**, 2201–2212 (2009).
18. Steinman, D.A. & Taylor, C.A. Flow imaging and computing: Large artery hemodynamics. *Annu. Biomed. Eng.* **33**, 1704–1709 (2005).
19. Poelma, C., Vennemann, P., Lindken, R. & Westerweel, J. In vivo blood flow and wall shear stress measurements in the vitelline network. *Exp. Fluids* **45**, 703–713 (2008).
20. Kleinstreuer, C. & Zhang, Z. Airflow and particle transport in the human respiratory system. *Annu. Rev. Fluid Mech.* **42**, 301–334 (2010).
21. Malkiel, E., Sheng, J., Katz, J. & Strickler, J.R. The three-dimensional flow field generated by a feeding calanoid copepod measured using digital holography. *J. Exp. Biol.* **206**, 3657–3666 (2003).
22. Mass, T., Genin, A., Shavit, U., Grinstein, M. & Tchernov, D. Flow enhances photosynthesis in marine benthic autotrophs by increasing the efflux of oxygen from the organism to the water. *Proc. Natl. Acad. Sci. U.S.A.* **107**, 2527–2531 (2010).
23. Nesbitt, W.S., Westein, E., Tovar-Lopez, F.J., Tolouei, E., Michell, A., Fu, J., Carberry, J., Fouras, A. & Jackson, S.P. A shear gradient-dependent platelet aggregation mechanism drives thrombus formation. *Nat. Med.* **15**, 665–675 (2009).
24. Hagsäter, S.M., Westergaard, C.H., Bruus, H. & Kutter, J.P. Investigations on LED illumination for micro-PIV including a novel front-lit configuration. *Exp. Fluids* **44**, 211–219 (2008).
25. Malone, M.H., Sciaky, N., Stalheim, L., Hahn, K.M., Linney, E. & Johnson, G.L. Laser-scanning velocimetry: A confocal microscopy method for quantitative measurement of cardiovascular performance in zebrafish embryos and larvae. *BMC Biotechnol.* **7**, 40 (2007).
26. Spedding, G.R., Rosen, M. & Hedenström, A. A family of vortex wakes generated by a thrush nightingale in free flight in a wind tunnel over its entire natural range of flight speeds. *J. Exp. Biol.* **206**, 2313–2344 (2003).
27. Hedenström, A., Rosén, M. & Spedding, G.R. Vortex wakes generated by robins *Erithacus rubecula* during free flight in a wind tunnel. *J. R. Soc. Interface* **3**, 263 (2006).
28. Rosén, M., Spedding, G.R. & Hedenström, A. Wake structures and wingbeat kinematics of a house-martin *Delichon urbica*. *J. R. Soc. Interface* **4**, 659 (2007).
29. Hedenström, A., Johansson, L.C., Wolf, M., von Busse, R., Winter, Y. & Spedding, G.R. Bat flight generates complex aerodynamic tracks. *Science* **316**, 894–897 (2007).
30. Hubel, T.Y., Hristov, N.I., Swartz, S.M. & Breuer, K.S. Time-resolved wake structure and kinematics of bat flight. *Exp. Fluids* **46**, 933–943 (2009).
31. Kirchhefer, A.J., Kopp, G.A. & Gurka, R. The near wake of a freely flying European starling. *Phys. Fluids* **25**, 051902 (2013).
32. Warrick, D.R., Tobalske, B.W. & Powers, D.R. Aerodynamics of the hovering hummingbird. *Nature* **435**, 1094–1097 (2005).
33. Muijres, F.T., Johansson, L.C., Barfield, R., Wolf, M., Spedding, G.R. & Hedenström, A. Leading-edge vortex improves lift in slow-flying bats. *Science* **319**, 1250–1253 (2008).
34. Hubel, T.Y. & Tropea, C. Experimental investigation of a flapping wing model. *Exp. Fluids* **46**, 945–961 (2009).
35. Day, S.W., Higham, T.E., Cheer, A.Y. & Wainwright, P.C. Spatial and temporal patterns of water flow generated by suction-feeding bluegill sunfish *Lepomis macrochirus* resolved by Particle Image Velocimetry. *J. Exp. Biol.* **208**, 2661–2671 (2005).
36. Okamoto, K. *Particle Image Velocimetry: Recent Improvements*, chapter Dynamic PIV: A strong tool to resolve the unsteady phenomena, pp. 17–27. Springer, Berlin, Heidelberg (2004).
37. Geoghegan, P.H., Buchmann, N.A., Soria, J. & Jermy, M.C. Time-resolved PIV measurements of the flow field in a stenosed, compliant arterial model. *Exp. Fluids* **54**, 1528 (2013).
38. Ben-Gida, H., Kirchhefer, A., Taylor, Z.J., Bezner-Kerr, W., Guglielmo, C.G. et al. Estimation of unsteady aerodynamics in the wake of a freely flying European starling (*Sturnus vulgaris*). *PLoS One* **8 Suppl 11,** e80086 (2013). doi:0.1371/journal.pone.0080086.
39. Anderson, E.J., McGillis, W.R. & Grosenbaugh, M.A. The boundary layer of swimming fish. *J. Exp. Biol.* **204**, 81–102 (2001).
40. Zilman, G., Novak, J., Liberzon, A., Perkol-Finkel, S. & Benayahu, Y. The hydrodynamics of contact of a marine larvae, *bugula neritina*, with a cylinder. *J. Exp. Biol.* **216**, 2789–2797 (2013).
41. Staab, K.L., Holzman, R., Hernande, P. & Wainwright, P.C. Independently evolved upper jaw protrusion mechanisms show convergent hydrodynamic function in teleost fishes. *J. Exp. Biol.* **215**, 1456–1463 (2012).
42. Theunissen, R., Scarano, F. & Riethmuller, M.L. On improvement of PIV image interrogation near stationary interfaces. *Exp. Fluids* **45**, 557–572 (2008).
43. Nogueira, J., Lecuona, A. & Rodriguez, P.A. Local field correction PIV, implemented by means of simple algorithms and multigrid versions. *Meas. Sci. Technol.* **12**, 1911–1921 (2001).
44. Nguyen, C.V., Nguyen, T.D., Wells, J.C. & Nakayama, A. Interfacial PIV to resolve flows in the vicinity of curved surfaces. *Exp. Fluids* **48**, 577–587 (2010).
45. Jeon, Y.J. & Sung, H.J. PIV measurement of flow around an arbitrarily moving body. *Exp. Fluids* **50**, 787–798 (2011).
46. Adhikari, D. & Longmire, E.K. Infrared tomographic PIV and 3D motion tracking system applied to aquatic predator-prey interaction. *Meas. Sci. Technol.* **24**, 024011 (2013).

47. Dracos, T., editor. *Three-Dimensional Velocity and Vorticity Measuring and Imaging Analysis Techniques*. Kluwer, Dordrecht, Netherlands (1996).
48. Boutsianis, E., Guala, M., Olgac, U., Wildermuth, S., Hoyer, K., Ventikos, Y. & Poulikakos, D. CFD and PTV steady flow investigation into anatomically accurate abdominal aortic aneurysm. *J. Biomech. Eng.* **131**, 011008 (2009).
49. Gülan, U., Lüthi, B., Holzner, M., Liberzon, A., Tsinober, A. & Kinzelbach, W. Experimental study of aortic flow in the ascending aorta via particle tracking velocimetry. *Exp. Fluids* **53**, 1469–1485 (2012).
50. Gharib, M., Pereira, F., Dabiri, D., Hove, J.R. & Modarress, D. Quantitative flow visualization: Toward a comprehensive flow diagnostic tool. *Integr. Comp. Biol.* **42**, 964–970 (2002).
51. Yen, J., Brown, J. & Webster, D.R. Analysis of the flow field of the krill. *Euphausia Pacifica*. *Mar. Freshw. Behav. Physiol.* **36**, 307–319 (2003).
52. Denissenko, P., Lukaschuk, S. & Breithaupt, T. The flow generated by an active olfactory system of the red swamp crayfish. *J. Exp. Biol.* **210**, 4083–4091 (2007).
53. Crimaldi, J.P. Planar laser induced fluorescence in aqueous flows. *Exp. Fluids* **44**, 851–863 (2008).
54. Gurka, R., Liberzon, A., Sarathi, P. & Sullivan, P.J. Diffusion of scalar concentration from localized sources in turbulent flows. *Rev. Chem. Eng.* **26**, 13–27 (2010).
55. Sarathi, P., Gurka, R., Kopp, G.A. & Sullivan, P.J. A calibration scheme for quantitative concentration measurements using simultaneous PIV and PLIF. *Exp. Fluids* **52**, 247–259 (2012).
56. Weissburg, M.J., James, C.P., Smee, D.L. & Webster, D.R. Fluid mechanics produces conflicting constraints during olfactory navigation of blue crabs. *Callinectus Sapidus*. *J. Exp. Biol.* **206**, 171–180 (2003).
57. Koehl, M.A.R. & Reidenbach, M.A. Swimming by microscopic organisms in ambient water flow. *Exp. Fluids* **43**, 755–768 (2007).
58. van Duren, L.A. & Videler, J.J. Escape from viscosity: The kinematics and hydrodynamics of copepod foraging and escape swimming. *J. Exp. Biol.* **206**, 269–279 (2003).
59. Spedding, G.R. & Hedenström, A. PIV-based investigations of animal flight. *Exp. Fluids* **46**, 749–763 (2009).
60. Noca, F. On the evaluation of time-dependent fluid-dynamic forces on bluff bodies. PhD thesis, California Institute of Technology, Pasadena, CA (1997).
61. Tucker, V.A. Gliding birds: Reduction of induced drag by wing tip slots between the primary feathers. *J. Exp. Biol.* **180**, 285–310 (1993).
62. Gurka, R., Liberzon, A., Hefetz, D., Rubinstein, D. & Shavit, U. Computation of pressure distribution using PIV velocity data. In *PIV03 Symp. Particle Image Velocimetry*, pp. 671–676, Santa Barbara, CA (1999).
63. Khodarahmi, I., Shakeri, M., Sharp, M.K. & Amini, A.A. Using PIV to determine relative pressures in a stenotic phantom under steady flow based on the pressure-Poisson equation. In *32nd Annu. Int. Conf. IEEE EMBS*, Buenos Aires, Argentina (2010).
64. Dabiri, J.O. On the estimation of swimming and flying forces from wake measurements. *J. Exp. Biol.* **208**, 3519–3532 (2005).
65. Dabiri, J.O., Colin, S.P. & Costello, J.H. Fast swimming hydromedusae exploit velar kinematics to form an optimal vortex wake. *J. Exp. Biol.* **209**, 2025–2033 (2006).
66. Rayner, J.M.V. A new approach to animal flight mechanics. *J. Exp. Biol.* **80**, 17–54 (1979).
67. Rayner, J.M.V. A vortex theory of animal flight. Part 1. The vortex wake of a hovering animal. *J. Fluid Mech.* **91**, 697–730 (1979).
68. Rayner, J.M.V. A vortex theory of animal flight. Part 2. The forward flight of birds. *J. Fluid Mech.* **91**, 731–763 (1979).
69. Triantafyllou, M.S., Triantafyllou, G.S. & Yue, D.K.P. Hydrodynamics of fishlike swimming. *Annu. Rev. Fluid Mech.* **32**, 33–53 (2000).
70. Perry, A.E. & Steiner, T.R. Large-scale vortex structures in turbulent wakes behind bluff bodies. Part 1. Vortex formation process. *J. Fluid Mech.* **174**, 233–270 (1987).
71. Hunt, J.C.R., Wray, A.A. & Moin, P. Eddies, stream, and convergence zones in turbulent flows. Technical report, Center for Turbulence Research CTR-S88, Stanford University, Stanford, CA (1988).
72. Chong, M.S., Perry, A.E. & Cantwell, B.J. A general classification of three-dimensional flow fields. *Phys. Fluids* **2**, 765–777 (1990).
73. Jeong, J. & Hussain, F. On the identification of a vortex. *J. Fluid Mech.* **285**, 69–94 (1995).
74. Zhou, J., Adrian, R.J., Balachandar, S. & Kendall, T.M. Mechanisms for generating coherent packets of hairpin vortices in channel flow. *J. Fluid Mech.* **387**, 353–396 (1999).
75. Adrian, R.J., Christensen, K.T. & Liu, Z.-C. Analysis and interpretation of instantaneous turbulent velocity fields. *Exp. Fluids* **29**, 275–290 (2000).
76. Cantwell, B. & Coles, D. An experimental study of entrainment and transport in the turbulent near wake of a circular cylinder. *J. Fluid Mech.* **136**, 321–374 (1983).
77. Taylor, Z.J., Palombi, E., Gurka, R. & Kopp, G.A. Features of the turbulent flow around symmetric elongated bluff bodies. *J. Fluids Struct.* **27**, 250–265 (2011).

14

High-Throughput Imaging Methodologies for Biomechanical Testing

Yu Long Han, Guoyou Huang, Lin Wang, Xiaoxi Fan, Fei Li, Pengfei Wang, Tian Jian Lu, and Feng Xu

CONTENTS

High-throughput strategies involving arrays of artificial tissue constructs offer tremendous promise for clinical diagnostics and fundamental cell biomechanics research. These tissue constructs consist of cells cultured in a well-defined three-dimensional environment, and are becoming increasingly useful tools as researchers begin to overcome the challenges of statistical sample-to-sample variability through these high-throughput techniques. While the cell culture techniques for high-throughput screening of these tissue constructs are well established, many challenges persist in high-throughput strategies for probing changes to biomechanics induced, for example, in response to a specific drug. However, advances are being made in applying many of the wide range of imaging techniques described in this handbook to high-throughput testing, and these hold great potential. In this chapter, we review three major imaging approaches that have begun to show potential for high-throughput biomechanical testing, including optical imaging, mechanical imaging, and ion-conductance imaging.

14.1 Introduction

Cells in vivo respond to and modify a complex microenvironment that presents multiple stimuli including chemical, electrical, and mechanical factors.[1,2] Accumulating evidence has demonstrated that mechanical cues such as stiffness, stress, and strain play an important role in determining cell behavior[3,4] and that living cells alter extracellular matrix (ECM) actively in response to these mechanical cues.[5,6] Such cooperative evolution of cell function and ECM properties are hallmarks of tissue growth and disease progression.[7] For example, some cancers present distinct changes to tissue and ECM mechanical properties, enabling initial clinical diagnosis through palpation and also affecting cell growth and further tumor development.[8] Bone remodeling,[9,10] lung development,[11] and fibrous skin callus formation[12] are also regulated by the dynamic mechanical microenvironment. Studies of the mechanics of cells, ECM and tissues are therefore important for deciphering the fundamental mechanisms underlying these significant physiological and pathophysiological processes.

A range of well-developed approaches exists, including tensile and compressive testing, for characterizing the mechanical properties of centimeter-scale biological tissues.[13–16] These have significantly advanced the understanding of the effect of mechanical cues on tissue development by establishing connections between mechanical properties and biological functions of a range of tissues or organs.[17–19] However, a difficult balance must be struck when extending these approaches to high-throughput testing of tissue constructs. On the one hand, extending the spatial resolution of many of these approaches below the tissue (centimeter) level is often difficult, favoring large specimens. On the other-hand, diffusion barriers for soluble factors exist in large specimens, and large perturbations are possible due to sample handling during mechanical testing,[20] favoring small specimens. These problems make high-throughput testing of larger specimens challenging.

Smaller-scale (e.g., micrometer) tests are promising to address the aforementioned challenges across disciplines including cell biology, biomechanics, and clinical diagnostics.[21] Although tissue-level biomechanical imaging can image centimeter-level changes to biomechanics,[22–24] critical techniques such as magnetic resonance (MR) elastography are not practical for high-throughout screening of smaller specimens (see Chapters 2–6). Atomic force microscopy (AFM) can reach the required length scales to distinguish, for example, cancer cells from normal cells by mechanical testing.[25,26] More recently, cancers at different stages have been identified from stiffness mapping of tumor biopsies.[27] However, challenges exist with probing cell and tissue mechanics using AFM (cf. Chapters 21, 26, and 36), and widespread application of these techniques will require advances in experimental methodologies for microscale biomechanical testing.

Numerous approaches have been developed for mechanical testing of biological materials at levels of single cells, cell aggregates, and microscale tissues ("microtissues").[28] However, most of these methods are currently low throughput as well. In this chapter, we summarize emerging imaging technologies that we believe to show the most potential for high-throughput biomechanical testing, including optical imaging, mechanical imaging and ion-conductance imaging (Table 14.1). For each, we describe some basic principles and potential applications and then discuss their current advantages and disadvantages in terms of spatial resolution, temporal resolution, and high-throughput capability. We also highlight directions of future research that we believe to be essential to achieving high-throughput biomechanical testing.

TABLE 14.1

Comparison of Optical, Mechanical, and Ion-Conductance Imaging Approaches for Biomechanical Testing

Approach	Optical	Mechanical	Ion-Conductance
Spatial resolution	Low 0.1 μm	Middle ~50–100 nm	High 10–20 nm
Temporal resolution	Real time	Real time	Real time
Throughput	High ~2000 cells/s	Low ~1 cell/60 s	Low ~1 cell/56 s

14.2 Optical Imaging

14.2.1 Principles

Optical imaging in biomechanical testing often requires optical quantification of deformation fields during mechanical loading, often with the goal of eliminating the need for a mechanical force sensor. Mechanical properties of interest are then estimated from these deformation fields using a simulation specific to the experimental setup. The procedure thus includes mechanical loading, visualization, and simulation. In this section, we introduce principles of optical estimation of deformation fields. We then describe applications involving specific mechanical loading systems and associated computational simulations.

Optical imaging is well developed across many areas of engineering and medicine.[29] The basic principle is to take advantage of interactions of light, from infrared to visible to ultraviolet, with biological materials[29] (Figure 14.1a). Interactions with tissues and biological materials, such as absorption, scattering, and fluorescent emission, can vary tremendously over this range (Figure 14.1b). For instance, photons with wavelengths of 300–600 nm are strongly absorbed by living tissues, while those with wavelengths in the near-infrared (700–1000 nm) are minimally attenuated by living tissues[30] (Figure 14.1c). In this section, we review some optical imaging approaches that are promising for high-throughput biomechanical testing of cells, cell aggregates, and microtissues. We refer the reader to recent reviews of other promising modalities such as near-infrared optical imaging.[31–33]

Optical imaging with visible light is especially common for in vitro biomechanical testing, taking advantage of the absorption and reflection of light when interacting with biological materials. Mechanical properties such as shear modulus, Poisson ratio, and viscoelastic responses of biological materials can be derived from optical images of time-dependent deformation of cells/tissues using the classical theories of mechanics[34–37] (see also Chapters 20–24). Taking the micropipette method as an example[38] (Figure 14.2c, see also Chapter 25), the elastic properties of tested samples can be specified using the following theory:

$$L = \frac{3\phi}{2\pi E}\Delta P \tag{14.1}$$

where L is the length of material aspirated into a micropipette under a pressure drop ΔP, ϕ is a parameter depending on the geometry of the pipette (i.e., the inner and outer diameters), and E is the local Young's modulus of the materials to be specified on the basis of the experimental data.

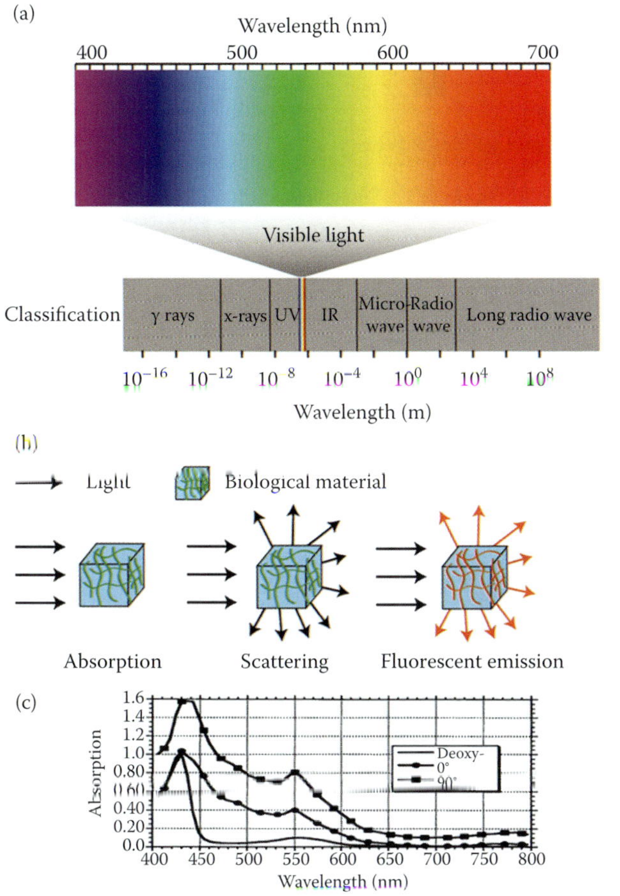

FIGURE 14.1
Principles of optical imaging for biological materials. (a) Classification of light by wavelength. (b) Interaction of light with biological materials including absorption, scattering, and fluorescent emission. (c) Absorption of light with different wavelengths by living tissues. UV, ultraviolet light; IR, infrared light. (From Marquez, G. et al., *Appl. Opt.*, 37, 1998.)

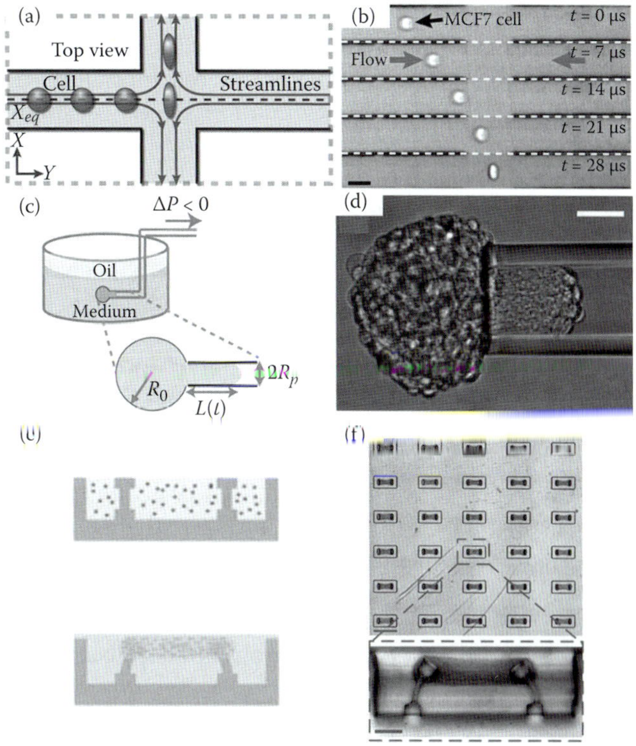

FIGURE 14.2
Optical imaging for biomechanical testing. (a, b) High-throughput measurement of single-cell mechanical properties using a microfluidics platform. (From Gossett, D.R. et al. Proceedings of the National Academy of Sciences, 2012.) (c, d) Measurement of cell aggregate mechanical properties using aspiration. (From Guevorkian, K. et al., *Phys. Rev. Lett.*, 104, 2010.) (e, f) High-throughput measurement of microtissue mechanical properties using microelectromechanical systems. (From Gossett, D.R. et al. Proceedings of the National Academy of Sciences, 2012, and Legant, W.R. et al., *Proc. Natl. Acad. Sci.*, 106, 2009.)

Optical imaging systems are usually amenable to high-throughput testing because light can be processed and recorded in a high-throughput manner by optical components (e.g., reflector and camera). Thus, optical imaging can be integrated with other platforms such as microfluidic or microelectromechanical systems (MEMS) and finds widespread application in biomechanical testing owing to its convenience, relatively low cost, and high throughput capacity.

14.2.2 Applications

At the cell level, a successful example is testing the mechanical properties of large populations of living cells by combining a microfluidic chip and a high-speed camera.[39] In the microfluidic chip, cells were stretched hydrodynamically by inertial focusing, automatically imaged, and analyzed (Figure 14.2a). The system can achieve single-cell mechanical measurements at a throughput of approximately 2000 cells per second, which provides a high-throughput platform and lays the foundation of performing clinical assays using mechanical biomarkers.

At the cell aggregate level, the micropipette aspiration technique has been used for studying the mechanical properties of cell aggregates (e.g., cancer aggregates).[40] In this approach, the cell aggregate is partially aspirated into a pipette (diameter of 40–70 µm) with controlled pressure. A CCD camera accompanied by an inverted microscope records movies of the deformation, and mechanical properties of cell aggregates can be calculated by Equation 14.1 or models in Ref. 40. The whole testing process may take several hours, depending upon the movement of cell aggregates inside the micropipette in response to the aspiration. Although this approach has been applied only to cell aggregates in series, it has the potential to be applied in a high-throughput fashion through parallel aspiration of cell aggregates.

At the microtissue level, advances in MEMS technologies enable microscale tissue engineering such as beating heart microtissues. MEMS devices incorporating micropillar arrays enable estimation of the mechanical properties of cell populations and microtissues which combined with real-time optical monitoring of the deformations of the micropillars induced by active contractile forces.[39,41] More recently, the incorporation of magnetic beads into MEMS devices has enabled application of external mechanical loading to such samples, thereby enabling quantification of tissue stiffness as well.[20] In this setup, a magnetic bead is affixed to a micropillar, forming a "magnetic pillar" that can be actuated by an external magnetic field or magnetic force microscope (see Chapter 38), thereby applying an external mechanical load. The mechanics of the arrays of micropillars are tunable by adjusting their material properties and dimensions during fabrication. The ability to create chips with large arrays of these micropillars and defined wells for the molding of tissue constructs lends much promise for high-throughput application of these technologies.

In conclusion, optical imaging is the most direct and developed approach for high-throughput imaging of biomechanics. Well-developed optical image sensors such as CCD cameras allow real-time high-throughput imaging from the macroscale to the submicroscale. Open challenges include deriving information about three-dimensional (3-D) phenomena from planar images, often stacks of confocal fluorescence microscopy images. An additional challenge is that direct optical imaging of cells within tissues relies largely on immunostaining of specific molecules within cells or upon insertion of fluorescent markers into living cells (cf. Chapters 21, 34, and 40). The former poses a challenge because it requires fixation of tissues, and the latter because such markers can affect cell function in ways that can be variable and that are difficult to control.

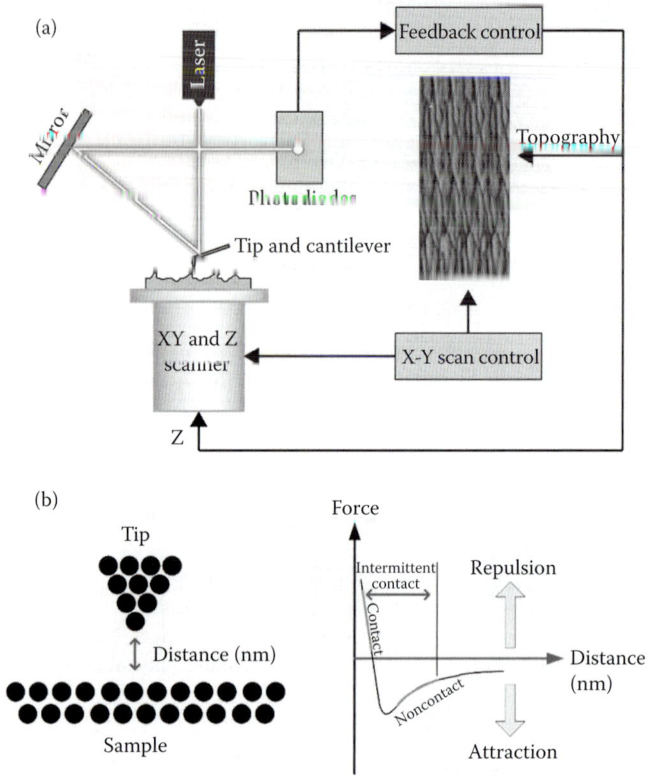

FIGURE 14.3
Principles of AFM. (a) The AFM probe consists of a sharp tip attached to a flexible cantilever. When the tip approaches the sample surface closely, the cantilever deflects due to interaction forces. The resulting flexure of the cantilever is measured from the deflection of a laser reflected onto an array of photodiodes. (b) The force between the tip and the surface of the sample can shift from repulsion to attraction.

14.3 AFM Imaging

14.3.1 Principles

The earliest AFM-like instruments were developed in the 1970s specifically for the purpose of probing biological mechanics (see Chapters 21 and 26 for a historical perspective). AFM continues to enjoy widespread application for scanning of surface morphologies and for testing of biomechanics. The AFM probe measures the interaction force between a substrate and a sharp tip that is attached to a flexible cantilever, and the sharp tip can be scanned over a sample surface (Figure 14.3). Force-displacement data may consist of a contact region where the probe contacts the sample surface and a noncontact region where the probe does not (Figure 14.3b). The probe can be pulled up and down by a piezoelectric controller to approach sample surface. When the tip approaches the sample surface, the cantilever deflects upward or downward due to attractive or repulsive interaction forces and does so with a magnitude that is proportional to the interaction force and can therefore be estimated with Euler–Bernoulli beam theory.[42] These interaction forces arise from a combination of a number of factors, including contact, capillary, magnetic, van der Waals, and electrostatic interactions. As in the earliest AFM-like machines from the 1970s, the resulting flexure of the cantilever is measured using a reflected laser beam striking a quadrant detector (Figure 14.3). In a modern device, AFM enables force resolution down to 10 piconewtons.

These force measurements enable a broad range of surface assays. The stiffness of a cell or tissue can be estimated through analysis of the dynamic force spectrum obtained in the force spectroscopy experiments. Height contours of a surface can be revealed by the deflection of the cantilever as it scans over a sample. AFM can thus provide both morphological and mechanical information on the surface of a wide range of materials, including biological specimens, for instance, single cells and microtissues. With modern piezo stages, nanometer spatial resolution and piconewton force resolution are possible.

14.3.2 Applications

AFM has been used to measure biomechanical properties, including the mechanics of a wide range of living cells, such as microbial cells and mammalian cells, at nanoscale.[43] Testing is possible in an aqueous environment, permitting in situ imaging of cells.

AFM has provided unique insight into the interaction of microbes with their environment. However, AFM is slow compared to timescales associated with the inherent mobility of microbes, and chemical fixation and mechanical entrapment are usually required.[44,45] After immobilization, microbial cells can be imaged at a molecular level or whole-cell level. For molecular-level imaging, an AFM tip can be coated with molecular ligands that interact with complementary molecules on microbial membranes. AFM can be used for localized and conformational analysis of such membrane molecules. For instance, Francius et al.[46] investigated the type and disposition of polysaccharides on bacteria membrane using lectin-modified AFM tip. They found that these polysaccharides differ with the bacterium's genotype in terms their polymeric, surface elasticity, and morphological properties (Figure 14.4a). At the level of entire cells, surface compressibility, homogeneity, and structure have been studied using AFM (Figure 14.4b),[47] and 3-D morphological changes have been probed over the entire life cycle to provide insight into structural mechanisms underlying cytokinesis.[48]

AFM analysis of mammalian cell and molecular biomechanics has promise for the detection of diseases, including sickle cell anemia, malaria, and cancer.[49,50] AFM assays of the mechanical mechanisms of the development of certain cancers provide a possible means for detection and characterization of these cancers. For instance, cells of certain cancerous cell lines are more complaint than their healthy counterparts (Figure 14.4c through d), including bladder cancer,[51] cervical cancer,[26] and breast cancer,[52] indicating promise for mechanical properties as effective diagnostic markers. Mechanical efforts to detect cancerous cells from clinical samples yield results that are consistent with those of established methods, such as immunohistochemical staining.[25] However, these AFM mechanical measurements must be performed one cell at a time and are thus poor candidates for high-throughput screening of cancerous

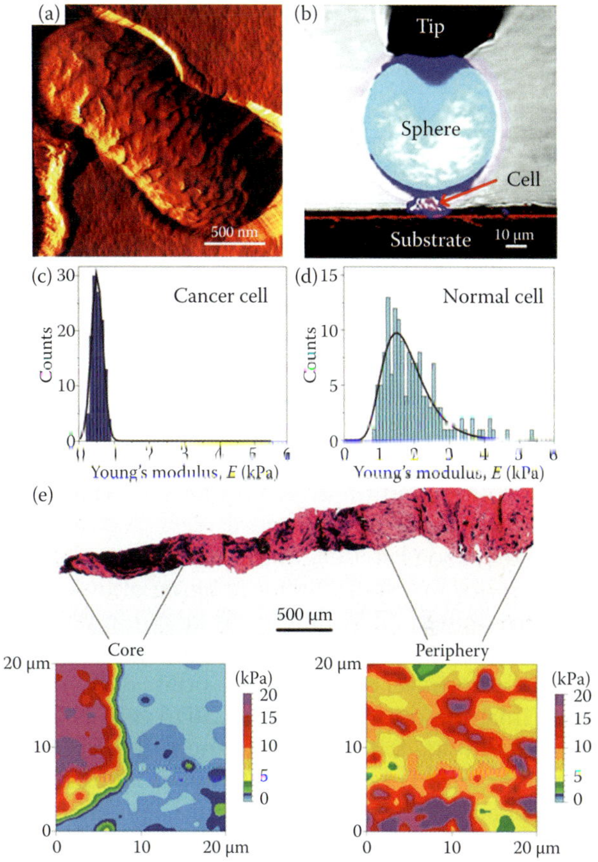

FIGURE 14.4
AFM imaging for biomechanical testing. (a) Molecular imaging of polysaccharides on the surface of a bacterium. (From Francius, G. et al., *ACS Nano*, 2, 2008.) (b) Compresion of a single cell using AFM. (From Lulevich, V. et al., *Langmuir*, 22, 2006.) Comparison of the mechanical properties of (c) normal and (d) cancerous cells. (From Iyer, S. et al., *Nat. Nano*, 4, 2009.) (e) AFM stiffness mapping of cancerous specimens to identify the different stages of cancer. (From Plodinec, M. et al., *Nat. Nano*, 7, 2012.)

cells when compared to immunohistochemical testing, which probes entire slides of cells at a time.

An additional challenge with translating AFM of cells to the clinic for cancer diagnostics is the fact that cells in vivo are surrounded by ECM. Discriminating the mechanical responses of cells from that of the surrounding ECM within a native tumor tissue requires intricate mechanical modeling and can rarely be achieved with adequate spatial/structural resolution and force sensitivity (cf. Chapter 21). A potential pathway to overcoming this challenge, and to simultaneously achieving high-throughput AFM measurement, is the mechanical mapping of microtissues (Figure 14.4e).[27] AFM tests on samples of malignant breast tissue show significant mechanical heterogeneity and a distribution of peaks, while tests performed on normal tissue show a single peak. Furthermore, with the development of a tumor and progression through the various stages of cancer, the mechanical profiles of malignant tissue evolve, potentially enabling assessment of the stage of a particular cancer.

In summary, AFM provides a platform that is capable of imaging both morphological and mechanical properties of biological materials and of providing biomechanical data useful to cell biology and medical diagnostics. Despite being versatile and powerful, challenges persist for high-throughput biomechanics testing using AFM, especially given that AFM can test only one sample at a time.[53] This low throughput has highly limited practical use in biomedical applications where a large number of cells need to be tested or screened. However, the development of AFM techniques for microtissues offers promise as a way of testing large populations of cells simultaneously, and preliminary data for screening of breast cancer show much promise.

For morphological imaging, temporal resolution is another problem, especially for imaging of living cells. A high-resolution AFM image can take minutes, especially for a highly corrugated cell surface, while many biological processes of interest can occur over seconds. Spatial imaging of unfixed mammalian cells is limited to 50–100 nm due to the dynamic and soft properties of cell membranes. This is further complicated by the sensitivity of an AFM tip to contamination by macromolecules on a cell membrane. However, this problem can be addressed by a noncontact ion-conductance microscopy technique, as discussed in the next section.

14.4 Ion-Conductance Imaging

14.4.1 Principles

Scanning ion-conductance microscopy (SICM) is an emerging tool for imaging the morphology of biological specimens and for dynamically measuring their mechanical properties.[54–57] SICM resembles AFM in that it uses a nanometer-size tip to scan over a sample surface to obtain a topographical map of and functional information about a sample. In contrast to AFM, SICM employs a cone-shaped glass nanopipette, with a typical tip opening diameter of 50–100 nm after pulling from a glass capillary, that is filled with electrolytic solution as the probe. SICM operates exclusively in a noncontact mode.[58]

The measurement principle of SICM is as follows (Figure 14.5). As the nanopipette tip scans over a sample surface, bathed in an electrolyte solution, change of an ionic current (or ionic conductance) between the tip opening and the sample surface is monitored and recorded. This conductance can reflect the sample morphology and function, and SICM is therefore

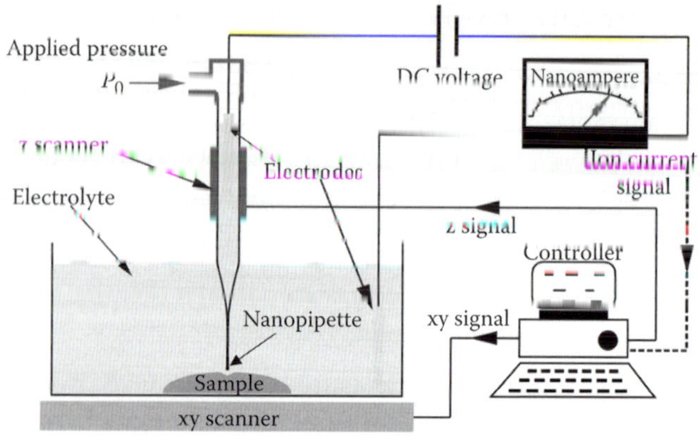

FIGURE 14.5
Schematic of a typical SICM setup for measuring sample morphology and mechanical properties. For mechanical measurements, an external pressure P_0 is applied to the upper nanopipette opening. DC, direct current. (From Schäffer, T.E., *Anal. Chem.*, 85, 2013.)

an "ion-conductance imaging" technique for sample analysis.

A typical SICM setup is shown in Figure 14.5,[59] and it involves one Ag/AgCl electrode placed inside the nanopipette filled with an electrolyte solution, and another placed in an external electrolyte solution containing the sample. Voltage applied across the electrolyte via the two Ag/AgCl electrodes generates an ion current that is highly dependent upon the tip–sample distance. The magnitude of generated ion current, *I*, is determined by the applied potential, *U*, and the overall resistance of the system, R_T[56]:

$$I = \frac{U}{R_T} \tag{14.2}$$

The total resistance R_T of the system arises from the sum of the resistance R_p of the nanopipette and the access resistance R_{ac} between the nanopipette and the sample surface, where

$$R_p = \frac{l}{\kappa \pi r_p r_i} \tag{14.3}$$

$$R_{ac} \approx \frac{3}{2}\frac{1}{\kappa \pi d}\ln(r_o/r_i). \tag{14.4}$$

Here, r_i and r_o are the inner and outer radii of the nanopipette opening, respectively; r_p is the inner radius of the nanopipette base; l is the nanopipette length; κ is the conductivity of the electrolyte in the nanopipette; and d is the distance between the opening of the nanopipette and the surface of the sample. As is evident from Equations 14.3 and 14.4, R_p is constant for nanopipettes with the same geometry, whereas R_{ac} is strongly influenced by the tip–sample distance d. Combining Equations 14.2 through 14.4 yields a relation showing a strong dependence of the measured ionic current I on the tip–sample distance d:

$$I(d) = \frac{U}{R_p + R_{ac}} \approx \frac{U}{R_p}\left(1 + \frac{3}{2}\ln\left(\frac{r_o}{r_i}\right)\frac{r_p r_i}{ld}\right)^{-1}. \quad (14.5)$$

As the nanopipette tip approaches the sample surface and the tip–sample surface gap that ions can flow through becomes smaller, the electrical resistance of the nanopipette–sample system increases, and the ionic current decreases. For probing of surface morphology, changes in ionic current measured by an amplifier are used as a feedback signal: by adjusting the height of the nanopipette during scanning to keep this current constant and recording this height at the position of each measurement, a 3-D map of surface features can be estimated.

One method to further measure the mechanical properties of a sample by SICM is to measure the sample response to a perturbations of local pressure applied through the SICM tip. A study by Sánchez et al.[60] involved exerting an external pressure of P_0 (typically between P_{ATM} − 100 kPa and P_{ATM} + 100 kPa, where P_{ATM} is atmospheric pressure) from the upper opening to generate a local pressure on the sample surface. The nanopipette is held close to the sample surface in this experiment and is maintained at a constant tip–sample distance using SICM feedback control. As the pressure applied to the upper nanopipette opening is slowly increased, the local pressure on the sample surface below the nanopipette tip increases, increasingly indenting the sample and pushing the sample surface further away from the tip. The nanopipette is then moved downward to maintain a constant ionic current, thereby enabling an estimate of the sample deformation: the more compliant the sample and the larger the applied pressure, the larger the displacement. The relationship between the vertical nanopipette displacement Z_{avg} and the applied pressure P_0 can be used to estimate the mechanical properties of the surface[60]:

$$Z_{avg} = \frac{16(1-v^2)r_i}{3\pi E}P_0 \quad (14.6)$$

where, for an isotropic specimen, E is the local Young's modulus and v is the local Poisson ratio. By recording the displacement Z_{avg} as a function of pressure P_0, estimating Poisson ratio, and fitting to Equation 14.6, an estimate of the local Young's modulus of sample is obtained.

14.4.2 Applications

Since SICM can simultaneously apply local pressures to and measure local mechanical properties of samples, an important application of SICM is biomechanical testing of cell mechanics. Compared to optical microscopy, SICM and AFM both have the advantage of providing local information with subcellular, even nanometer-scale, resolution, depending upon the probe used. A specific advantage of SICM is the ability to apply well-defined noncontact mechanical forces to soft or fragile biological samples under physiological conditions.

Since the first study of cell mechanics using SICM was reported by Sánchez et al.,[60] SICM has been applied to measure mechanical properties of many living cells, including cardiomyocytes, epithelial cells, and neurons. For example, the distribution of the mechanical properties of live dorsal root ganglion neurons was studied using SICM (Figure 14.6). Just below the membrane was an initial compliant region with $E \approx 0.5$ kPa (Figure 14.6a,

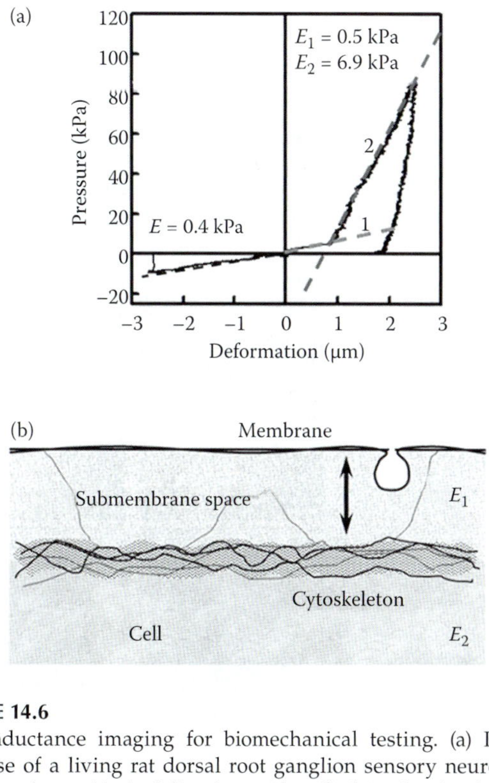

FIGURE 14.6
Ion-conductance imaging for biomechanical testing. (a) Pressure response of a living rat dorsal root ganglion sensory neuron. Two different slopes (dashed lines) are identified in the displacement-versus-pressure curve, giving two different values of sample Young's moduli, one below and one above a displacement of ≈ 1 μm. (b) Model of the region below the plasma membrane. The presence of a 1-μm-thick submembrane space between the membrane and the cytoskeleton explains the occurrence of the two different slopes in the pressure response. (From Sánchez, D. et al., *Biophys. J.*, 95, 2008.)

dashed line 1), and further below was a stiffer region with $E \approx 6.9$ kPa (Figure 14.6a, dashed line 2).[59] A negative pressure applied to the nanopipette pushed the membrane away from the cell by several micrometers, exhibiting a Young's modulus of $E \approx 0.4$ kPa (Figure 14.6a, dashed black line), similar to the region just below the membrane. This result suggests the presence of an ~1-μm-thick submembrane region above the cytoskeleton (Figure 14.6b), which might provide a general mechanism for osmoregulation via an interaction of the cytoskeleton with sodium channels. In another study, Pellegrino et al.[61] indented a living NIH3T3 fibroblast cell with a range of applied pressures using SICM and obtained a map of local Young's moduli.

These SICM methods are slow and unsuitable for the rapid measurements needed for high-throughput screening. However, Rheinlaender and Schaffer[62] developed an SICM method for fast quantitative mapping of the stiffness of a living fibroblast cell. In this method, a nanopipette was repetitively advanced and retracted over several positions on a fibroblast while constant applied pressure was applied. Rheinlaender and Schaffer found that as the compliance of the area below the nanopipette increases, deformed in response to an applied pressure, and hence, the effective tip–sample separation increases. The current change in response to a pressure perturbation is thus sensitive to the compliance of the surface. The map of recorded SICM current–distance curves could therefore be analyzed using

$$E = \frac{AP_0}{S_\infty / S - 1} \tag{14.7}$$

where A is a constant depending on the pipette geometry (relative wall thickness and inner half-cone angle α), S is the slope of an SICM current–distance curve, and S_∞ is the slope that would be obtained for an infinitely stiff sample region. From Equation 14.7, the sample topography (height) can be simultaneously obtained by recording the nanopipette position at 99% ionic current.

A key study supporting the feasibly of the Rheinlaender–Schaffer SICM method involved cells containing a green fluorescence protein to label the actin cytoskeleton (Figure 14.7).[63] The actin fibers could be located from the images of stiffness. Effects of cell migration due to osmotic swelling could also be observed. This highlights the potential of SICM for characterization of cytoskeleton structure and dynamics, intracellular processes, and interactions between the cell and its environment.[63]

In summary, compared to other technologies for biomechanical imaging, SICM offers several advantages, such as noncontact imaging; high spatial and fast temporal resolution; simple sample preparation; compatibility with cell

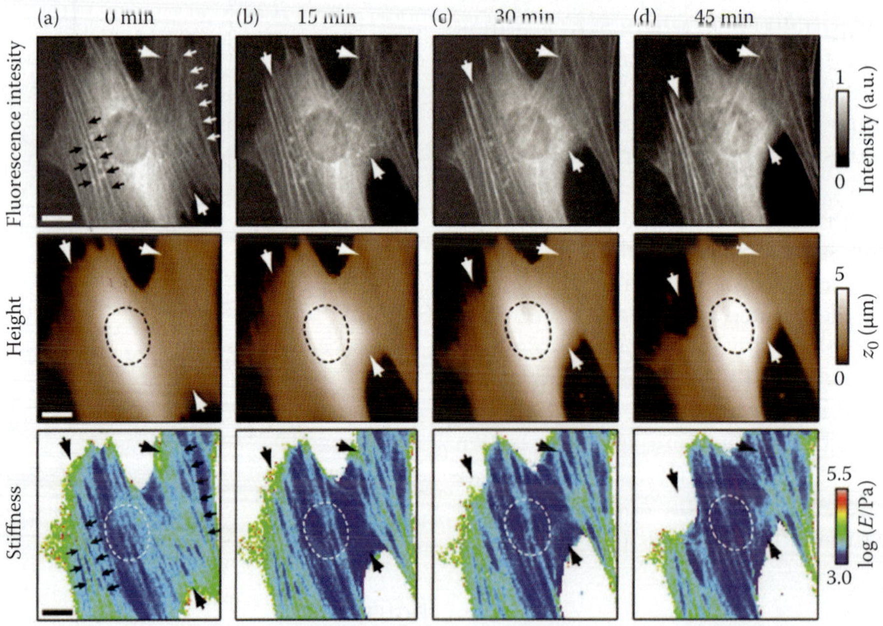

FIGURE 14.7
Visualization of the dynamics of individual cytoskeleton fibers in a living NIH3T3 fibroblast cell using SICM (a–d). The image sequence shows actin fluorescence (top row), SICM height (middle row), and SICM stiffness in terms of Young's modulus (bottom row). The fluorescence images show a dense network of actin fibers. In the surface morphology images, the cell appears smooth, with no substructure. The highest region of the cell is where the nucleus is located (dashed curve). The stiffness images show individual actin fibers whose dynamics are clearly resolved. Images were taken at four time points: (a) 0 min, (b) 15 min, (c) 30 min, and (d) 45 min. Some fibers correspond to fibers in the fluorescence images (marked with small arrows). Slow (upper large arrows) and fast (lower large arrow) modes of cytoskeletal reorganization are visible. Scale bars: 15 μm. (From Rheinlaender, J. and Schäffer, T.E., *Soft Matter*, 9, 2013.)

culture conditions; and real-time, in vivo mapping of cell surface microstructure.[56,57,64,65] Although the first SICM techniques were too slow for this, recent advances indicate promise for relatively high-throughput screening of cells and microtissues. SICM shows potential for in vivo and in vitro investigation of biological samples at the micrometer and nanometer scales, and has been demonstrated for cell imaging,[66] dynamic monitoring of living cells,[67] and mechanical stimulation of living cells.[68] SICM analysis of cell mechanics might be applied to investigate rheological properties of cells and probe important aspects of cell development, growth, differentiation, motility, and malignancy in a high throughout fashion. SICM-based study of cellular responses to locally applied pressures may help elucidate force-transduction pathways.

To truly harness the high throughput of SICM, imaging speed must be further enhanced. Although it is much faster than AFM, SICM is not yet sufficiently fast for several important phenomena, even at the level of single cells. For example, it is not sufficiently fast to capture reentrant surfaces and membrane folds. Developments involving higher-speed scanning, application of multi-barreled nanopipettes such as SICM tips, and combination with other techniques (such as AFM, scanning electrochemical microscopy, and patch-clamp methods) may open further avenues of inquiry.

14.5 Conclusion

High-throughput screening is needed to overcome challenges associated with the intrinsic mechanical heterogeneity of biological samples. There are several methods for high-throughput screening work using optical imaging, but these have significant limitations and motivate the development of additional modalities. Additionally, optical imaging is limited by the inherent diffraction properties of light. Mechanical and ion-conductance imaging approaches remedy limitations on spatial resolution but introduce additional challenges associated with the speed of acquisition of mechanical data. Future improvements to high-throughput biomechanical screening using these methods and the development of integrated platforms that combine them will open many new possibilities for biomechanical analysis and clinical diagnostics.

Acknowledgments

This work was supported by the Major International Joint Research Program of China (11120101002), the National 111 Project of China (B06024), the Key Grant Project of the Chinese Ministry of Education (313045), the National Natural Science Foundation of China (21105079), and the International Science and Technology Cooperation Program of China (2013DFG02930).

References

1. Liotta, L. A. & Kohn, E. C. "The Microenvironment of the Tumour-Host Interface." *Nature* 411(2001): 375–379.
2. Lutolf, M. P., Gilbert, P. M. & Blau, H. M. "Designing Materials to Direct Stem-Cell Fate." *Nature* 462(2009): 433–441.
3. Desprat, N., Supatto, W., Pouille, P.-A., Beaurepaire, E. & Farge, E. "Tissue Deformation Modulates Twist Expression to Determine Anterior Midgut Differentiation in Drosophila Embryos." *Developmental Cell* 15(2008): 470–477.
4. Pouille, P.-A., Ahmadi, P., Brunet, A.-C. & Farge, E. "Mechanical Signals Trigger Myosin II Redistribution and Mesoderm Invagination in Drosophila Embryos." *Science Signaling* 2(2009): ra16.
5. Fletcher, D. A. & Mullins, R. D. "Cell Mechanics and the Cytoskeleton." *Nature* 463(2010): 485–492.
6. Waite, J. H. & Broomell, C. C. "Changing Environments and Structure–Property Relationships in Marine Biomaterials." *The Journal of Experimental Biology* 215(2012): 873–883.
7. Mammoto, T. & Ingber, D. E. "Mechanical Control of Tissue and Organ Development." *Development* 137(2010): 1407–1420.
8. Huang, S. & Ingber, D. E. "Cell Tension, Matrix Mechanics, and Cancer Development." *Cancer Cell* 8(2005): 175–176.
9. Sakai, D., Kii, I., Nakagawa, K. et al. "Remodeling of Actin Cytoskeleton in Mouse Periosteal Cells under Mechanical Loading Induces Periosteal Cell Proliferation During Bone Formation." *PLos One* 6(2011): e24847.
10. Ozcivici, E., Luu, Y. K., Adler, B. et al. "Mechanical Signals as Anabolic Agents in Bone." *Nature Reviews. Rheumatology* 6(2010): 50–59.
11. Tschumperlin, D. J. & Drazen, J. M. "Chronic Effects of Mechanical Force on Airways." *Annual Review of Physiology* 68(2006): 563–583.
12. Moraes, C., Sun, Y. & Simmons, C. A. "(Micro) Managing the Mechanical Microenvironment." *Integrative Biology* 3(2011): 959–971.
13. Xu, F., Lu, T. J. & Seffen, K. A. "Biothermomechanics of Skin Tissues." *Journal of the Mechanics and Physics of Solids* 56(2008): 1852–1884.
14. Bhatia, V. A. & Troy, K. L. "A Portable Small-Scale Mechanical Loading and Testing Device: Validation and Application to a Mouse Tibia Loading Model." *Experimental Techniques* (2012). doi:10.1111/j.1747-1567.2012.00843.x.
15. Balguid, A., Mol, A., Van Vlimmeren, M. A. A., Baaijens, F. P. T. & Bouten, C. V. C. "Hypoxia Induces Near-Native Mechanical Properties in Engineered Heart Valve Tissue." *Circulation* 119(2009): 290–297.

16. Lanir, Y. & Fung, Y. C. "Two-Dimensional Mechanical Properties of Rabbit Skin—II. Experimental Results." *Journal of Biomechanics* 7(1974): 171–182.
17. Fung, Y. C. *Biomechanics: Mechanical Properties of Living Tissues*. (Springer, New York, 1993).
18. Fung, Y. C. *Biomechanics: Motion, Flow, Stress, and Growth*. (Springer, New York, 1990).
19. Fung, Y. C. *Biomechanics: Circulation*. (Springer, New York, 1997).
20. Zhao, R., Boudou, T., Wang, W. G., Chen, C. S. & Reich, D. H. "Decoupling Cell and Matrix Mechanics in Engineered Microtissues Using Magnetically Actuated Microcantilevers." *Advanced Materials* 25(2013): 1699–1705.
21. Bao, G. & Suresh, S. "Cell and Molecular Mechanics of Biological Materials." *Nature Material* 2(2003): 715–725.
22. Discher, D. E., Janmey, P. & Wang, Y.-L. "Tissue Cells Feel and Respond to the Stiffness of Their Substrate." *Science* 310(2005): 1139–1143.
23. Bercoff, J., Chaffai, S., Tanter, M. et al. "In Vivo Breast Tumor Detection Using Transient Elastography." *Ultrasound in Medicine and Biology* 29(2003): 1387–1396.
24. Mcknight, A. L., Kugel, J. L., Rossman, P. J. et al. "MR Elastography of Breast Cancer: Preliminary Results." *American Journal of Roentgenology* 178(2002): 1411–1417.
25. Cross, S. E., Jin, Y.-S., Rao, J. & Gimzewski, J. K. "Nanomechanical Analysis of Cells from Cancer Patients." *Nature Nanotechnology* 2(2007): 780–783.
26. Iyers, S., Gaikwad, R. M., Subba Rao, V., Woodworth, C. D. & Sokolov, I. "Atomic Force Microscopy Detects Differences in the Surface Brush of Normal and Cancerous Cells." *Nature Nanotechnology* 4(2009): 389–393.
27. Plodinec, M., Loparic, M., Monnier, C. A. et al. "The Nanomechanical Signature of Breast Cancer." *Nature Nanotechnology* 7(2012): 757–765.
28. Zhang, W., Wang, S., Lin, M. et al. "Advances in Experimental Approaches for Investigating Cell Aggregate Mechanics." *Acta Mechanica Solida Sinica* 25(2012): 473–482.
29. Tèorèok, P. & Kao, F. J. *Optical Imaging and Microscopy: Techniques and Advanced Systems*. (Springer-Verlag, New York, 2003).
30. Marquez, G., Wang, L. V., Lin, S.-P., Schwartz, J. A. & Thomsen, S. L. "Anisotropy in the Absorption and Scattering Spectra of Chicken Breast Tissue." *Applied Optics* 37(1998): 798–804.
31. Gurfinkel, M., Ke, S., Wen, X., Li, C. & Sevick-Muraca, E. M. "Near-Infrared Fluorescence Optical Imaging and Tomography." *Disease Markers* 19(2004): 107–121.
32. Hawrysz, D. J. & Sevick-Muraca, E. M. "Developments toward Diagnostic Breast Cancer Imaging Using near-Infrared Optical Measurements and Fluorescent Contrast Agents." *Neoplasia* 2(2000): 388.
33. Tuchin, V. V. *Tissue Optics: Light Scattering Methods and Instruments for Medical Diagnosis*, Vol. 13. (SPIE Press, Bellingham, 2007).
34. Yamamoto, J. A. "Shear-Stress-Sensing and Response Mechanisms in Vascular Endothelial Cells." *Interface Oral Health Science* 1(2010): 7.
35. Obi, S. & Shimizu, N. "Fluid Shear Stress Induces Arterial Differentiation of Endothelial Progenitor Cells." *Journal of Applied Physiology* 106(2009): 9
36. Yamamoto, K. & Ando, J. In 5th European Conference of the International Federation for Medical and Biological Engineering, 777–780 (Springer, New York, 2012).
37. Mingwan, L. *Fuxue Luo Foundation of Elasticity Theory*, Vol. 1. (Tsinghua University Press, Beijing, 1990).
38. Xu, B., Fan, X., Zhang, H., Wu, W. & Yang, G. "Application of Micropipette Aspiration and Half Space Model in the Study on Viscoelasticity of Rat Osteoblast." *Acta Biophysica Sinica* 14(1998): 360–366.
39. Gossett, D. R., Tse, H. T. K., Lee, S. A. et al. "Hydrodynamic Stretching of Single Cells for Large Population Mechanical Phenotyping." *Proceedings of the National Academy of Sciences of the United States of America* 109(2012): 7630–7635.
40. Guevorkian, K., Colbert, M.-J., Durth, M., Dufour, S. & Brochard-Wyart, F. "Aspiration of Biological Viscoelastic Drops." *Physical Review Letters* 104(2010): 218101.
41. Legant, W. R., Pathak, A., Yang, M. T. et al. "Microfabricated Tissue Gauges to Measure and Manipulate Forces from 3d Microtissues." *Proceedings of the National Academy of Sciences of the United States of America* 106(2009): 10097–10102.
42. Cappella, B. & Dietler, G. "Force-Distance Curves by Atomic Force Microscopy." *Surface Science Reports* 34(1999): 1–104.
43. Rotsch, C. & Radmacher, M. "Drug-Induced Changes of Cytoskeletal Structure and Mechanics in Fibroblasts: An Atomic Force Microscopy Study." *Biophysical Journal* 78(2000): 520–535.
44. Bearinger, J. P., Dugan, L. C., Wu, L. et al. "Chemical Tethering of Motile Bacteria to Silicon Surfaces." *BioTechniques* 46(2009): 209.
45. Cerf, A., Cau, J.-C., Vieu, C. & Dague, E. "Nanomechanical Properties of Dead or Alive Single-Patterned Bacteria." *Langmuir* 25(2009): 5731–5736.
46. Francius, G., Lebeer, S., Alsteens, D. et al. "Detection, Localization, and Conformational Analysis of Single Polysaccharide Molecules on Live Bacteria." *ACS Nano* 2(2008): 1921–1929.
47. Lulevich, V., Zink, T., Chen, H.-Y., Liu, F.-T. & Liu, G.-Y. "Cell Mechanics Using Atomic Force Microscopy-Based Single-Cell Compression." *Langmuir* 22(2006): 8151–8155.
48. Del Sol, R., Armstrong, I., Wright, C. & Dyson, P. "Characterization of Changes to the Cell Surface During the Life Cycle of Streptomyces Coelicolor: Atomic Force Microscopy of Living Cells." *Journal of Bacteriology* 189(2007): 2219–2225.
49. Lee, G. Y. H. & Lim, C. T. "Biomechanics Approaches to Studying Human Diseases." *Trends in Biotechnology* 25(2007): 111–118.
50. Lim, C. T., Zhou, E. H., Li, A., Vedula, S. R. K. & Fu, H. X. "Experimental Techniques for Single Cell and Single Molecule Biomechanics." *Materials Science and Engineering: C* 26(2006): 1278–1288.
51. Lekka, M., Laidler, P., Gil, D. et al. "Elasticity of Normal and Cancerous Human Bladder Cells Studied by Scanning Force Microscopy." *European Biophysics Journal* 28(1999): 312–316.

52. Li, Q. S., Lee, G. Y. H., Ong, C. N. & Lim, C. T. "Afm Indentation Study of Breast Cancer Cells." *Biochemical and Biophysical Research Communications* 374(2008): 609–613.
53. Müller, D. J. & Dufrêne, Y. F. "Atomic Force Microscopy: A Nanoscopic Window on the Cell Surface." *Trends in Cell Biology* 21(2011): 461–469.
54. Anariba, F., Anh, J. H., Jung, G.-E., Cho, N.-J. & Cho, S.-J. "Biophysical Applications of Scanning Ion Conductance Microscopy (SICM)." *Modern Physics Letters B* 26(2012): 1130003.
55. Bhargava, A., Wright, P. T. & Gorelik, J. "The Scanning Ion Conductance Microscope for Cellular Physiology." *American Journal of Physiology Heart and Circulatory Physiology* [illegible]
56. Chen, C.-C., Zhou, Y. & Baker, L. A. "Scanning Ion Conductance Microscopy." *Annual Review of Analytical Chemistry* 5(2012): 207–228.
57. Happel, P., Thatenhorst, D. & Dietzel, I. D. "Scanning Ion Conductance Microscopy for Studying Biological Samples." *Sensors (Basel, Switzerland)* 12(2012): 14983.
58. Rheinlaender, J., Geisse, N. A., Proksch, R. & Schäffer, T. E. "Comparison of Scanning Ion Conductance Microscopy with Atomic Force Microscopy for Cell Imaging." *Langmuir* 27(2010): 697–704.
59. Schäffer, T. E. "Nanomechanics of Molecules and Living Cells with Scanning Ion Conductance Microscopy." *Analytical Chemistry* 85(2013): 6988–6994.
60. Sánchez, D., Johnson, N., Li, C. et al. "Noncontact Measurement of the Local Mechanical Properties of Living Cells Using Pressure Applied Via a Pipette." *Biophysical Journal* 95(2008): 3017–3027.
61. Pellegrino, M., Orsini, P. & De Gregorio, F. "Use of Scanning Ion Conductance Microscopy to Guide and Redirect Neuronal Growth Cones." *Neuroscience Research* 64(2009): 290–296.
62. Rheinlaender, J. & Schäffer, T. E. "Mapping the Mechanical Stiffness of Live Cells with the Scanning Ion Conductance Microscope." *Soft Matter* 9(2013): 3230–3236.
63. Cook, S. M., Schäffer, T. E., Chynoweth, K. M. et al. "Practical Implementation of Dynamic Methods for Measuring Atomic Force Microscope Cantilever Spring Constants." *Nanotechnology* 17(2006): 2135–2145.
64. Liu, B.-C., Lu, X.-Y., Song, X. et al. "Scanning Ion Conductance Microscopy: A Nanotechnology for Biological Studies in Live Cells." *Frontiers in Physiology* 3(2012): 483.
65. Rheinlaender, J. & Schäffer, T. E. *Scanning Probe Microscopy of Functional Materials*, 433–460. (Springer Science+Business Media, LC, 2011).
66. Gorelik, J., Zhang, Y., Shevchuk, A. I. et al. "The Use of Scanning Ion Conductance Microscopy to Image A6 Cells." *Molecular and Cellular Endocrinology* 217(2004): 101–108.
67. Happel, P., Hoffmann, G., Mann, S. A. & Dietzel, I. D. "Monitoring Cell Movements and Volume Changes with Pulse-Mode Scanning Ion Conductance Microscopy." *Journal of Microscopy* 212(2003): 144–151.
68. Shevchuk, A. I., Gorelik, J., Harding, S. E., Klenerman, D. & Korchev, Y. E. "Simultaneous Measurement of Ca^{2+} and Cellular Dynamics: Combined Scanning Ion Conductance and Optical Microscopy to Study Contracting Cardiac Myocytes." *Biophysical Journal* 81(2001): 1759–1764.

Section VI

Tissue and Organ Imaging Methods

Multiscale Imaging and Modeling

15

Nonlinear and Poroelastic Biomechanical Imaging: Elastography beyond Young's Modulus

Paul E. Barbone, Assad A. Oberai, Jeffrey C. Bamber, Gearóid P. Berry, Jean-François Dord, Elizabete Rodrigues Ferreira, Sevan Goenezen, and Timothy J. Hall

CONTENTS

Recent developments in the elastography community have made it possible to measure, to unprecedented precision, deformation patterns within living tissues. By combining these measurements with faithful constitutive models of the tissue in an inverse problem framework, the local mechanical properties of soft and hard tissues may be inferred. Here the focus is on measuring the nonlinear elastic properties and some poroelastic properties of soft tissues. We motivate the biological interest in these properties, describe the in vivo measurement of tissue deformation and the relevant multiscale forward and inverse models of soft tissue deformations, and show some example applications.

15.1 Introduction and Background

Biomechanical imaging relies on imaging tissue while it is gently deformed by a set of externally applied forces. Through image processing, the displacement field everywhere in the region of interest is inferred. An inverse problem for the relevant mechanical properties is then solved, given the measured displacement fields, an assumed form of the tissue's constitutive equation, and the law of conservation of momentum. Hence, the goal of biomechanical imaging is to create a quantitative map of the mechanical properties of interest everywhere within the observed region. That is to say, our focus is less on imaging per se than it is to make densely sampled, quantitative, high-spatial-resolution mechanical measurements. From such measurements, of course, an image of tissue structure may be created, but that is a secondary goal.

The mechanical properties measurable by this approach depend upon the observable phenomena in the sample. For example, from observations of small quasi-static deformations, one may infer the linear elastic modulus. From transient deformations, one might infer viscoelastic parameters. The methods of observation and excitation of the tissue tend to determine the observable phenomena. Thus, magnetic resonance (MR) elastography[1–3] tends to be used to observe low-amplitude time-harmonic shear waves in the 100–500 Hz frequency range (see Chapters 4 and 5). Ultrasound may also be used to observe transient shear wave propagation. Typically, these low-amplitude waves are excited mechanically or by acoustic radiation force[4–6] and have frequency content also in the 100–500 Hz frequency range (see Chapters 7 and 8).

Thus far, we have applied biomechanical imaging to measure linear elastic isotropic behavior in 2-D[7–11] and in 3-D,[12,13] to linear elastic anisotropic properties in 3-D,[14] to linear viscoelastic properties in 2-D,[15] to nonlinear elastic properties in both 2-D and 3-D,[16–18] and to poroelastic properties.[19–23] In this chapter, the focus will be on inferring nonlinear elastic behavior from large-strain observations, and multiphase material properties from observations of strain relaxation. Such large deformations (up to 20%) and slow time scales (tens of seconds to 100 seconds) may be observed and measured using ultrasound imaging of quasi-static tissue deformations.

15.1.1 Background: Ultrasound Elastography

Biomechanical imaging has its origins in elastography,[24] which is the imaging of tissue structures based on contrast in their elastic properties (see Chapter 4). Excellent reviews of the subject are available, including Refs. 25–31. The emphasis of elastography has traditionally been imaging, with quantitation of linear elastic modulus a secondary priority. Over the years, elastography has seen an enormous number of potential applications in medicine and biological research. These include the detection, diagnosis, and treatment monitoring of diseases such as breast and prostate cancer,[32–37] local and diffuse coronary disease,[38–42] deep vein thrombosis,[43] liver cirrhosis,[6,44] and degenerative brain diseases.[45,46] Elastography-related methods have also been used to generate patient-specific anatomical models for surgical planning and image-guidance applications.[47–50] A much more recent and very exciting area of application for elastography and biomechanical imaging lies in mechanobiology and tissue engineering. Here, a detailed mechanical description of tissue or tissue-mimicking substrates is necessary in order to quantify the effect of the surrounding environment on cell function.[51–53]

15.1.2 Background: Nonlinear Elastic Properties of Breast Tissues

Most breast cancers in the United States are first detected by a patient or physician palpating a stiff lump.[54] Thus, it is widely recognized that cancer tends to alter the mechanical properties of breast tissue. Because ultrasound strain imaging (also known as quasi-static elastography) is sensitive to differences in tissue stiffness, it has often been applied to the problem of detection and diagnosis of breast cancer. Ultrasound strain imaging has been shown to aid the differentiation of benign from malignant solid breast masses.[34,55–62] Strain imaging is thought to display the relative stiffness of lesions compared to surrounding tissue. Stiffer areas deform less easily than their surroundings and tend to be depicted as dark, while softer areas deform more easily than their surroundings and tend to be depicted as light on strain images. Malignant masses typically appear dark and have high contrast with background breast tissue during deformation. Benign masses typically appear lighter and have lower contrast with background tissue with deformation, though strain image contrast alone is not a reliable criterion for classifying breast lesions.[34,57] In addition, malignant lesions tend to be larger on strain images than the corresponding ultrasound B-mode images,[34,57–59] which may be caused by the desmoplastic reaction commonly associated with many malignant tumors.[63]

At least as early as 2000, however, it has been recognized that strain contrast between a lesion and its background depends upon the total applied deformation of the tissue.[57,64–68] An equivalent phenomenon has been documented in shear wave elastography, where it is observed that shear wave speed increases substantially with overall compression.[69] The observed variation of material properties in vivo with overall applied strain is consistent with the fact that ex vivo measurement of the mechanical properties of breast tissues shows them to exhibit significant nonlinearity.[70–73] That is, the slope of the stress–strain curve is quite different at low strain than at high strain. Importantly, especially for the purposes of diagnosis, the nonlinear elastic properties differ from one tissue type to another. Thus, knowledge of the nonlinear parameter potentially enables the tissue to be unambiguously identified (Figure 15.1).

Several authors have attempted to exploit this phenomenon in a qualitative way to improve breast tumor

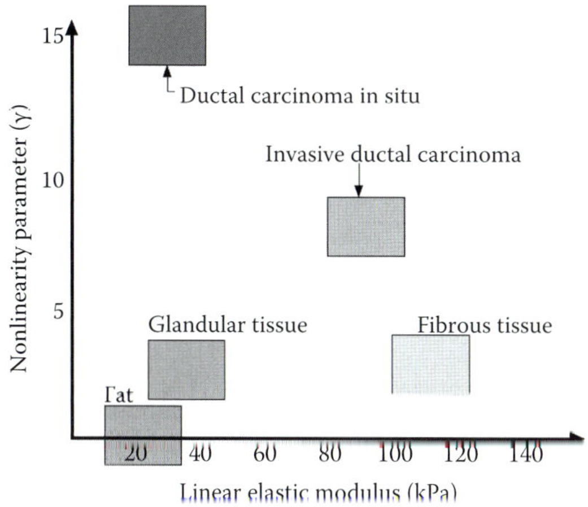

FIGURE 15.1
Nonlinear elastic stiffening parameter γ versus linear elastic modulus for several breast tissues. Note that malignant tissues tend to exhibit earlier onset of nonlinearity (i.e., larger γ) than benign tissues. The nonlinear parameter is estimated from a modified Blatz model, Equation 15.6, and the measurements reported in Refs. 70 and 71.

imaging.[64,67,68] Nitta and Shiina,[74] for instance, proposed creating what might be called "strain–slope" images, to capture the change of strain contrast with overall compression. The kinematic nonlinearities associated with large strains were accounted for in the work of Skovoroda[64]; however, the stress–strain behavior was assumed to be linear. More recent efforts, including those of our group, account for both geometric and material nonlinearity.[16–18,75,76]

15.1.2.1 Mammary Tumorigenesis Is Associated with Altered Collagen Orientation and Structure

It has long been known that tumors tend to be stiff. It is now recognized that the stiffness of the tumor environment is part of the malignant feedback cycle that leads to tumorigenesis.[77] Indeed, extracellular matrix stiffness can direct cell migration,[78–80] influence proliferation,[52,80,81] and determine differentiation.[82,83] These studies have focused on linear elastic modulus, which is related to density of collagen and its degree of cross-linking.

Malignancy is also associated with altered organization of stromal collagen.[52,84–87] In particular, different groups have noted that collagen fibers tend to be straighter in stiff malignant breast tumors than in either stiff benign tissues (e.g., fibroadenoma) or normal mammary tissue.[84–87] It is generally accepted that the mechanical response of a fiber network at small strain is dominated by the uncoiling of polymer fibers, while the large-strain stiffening occurs when the fibers become fully stretched and/or entangled with their neighbors. The straight, rodlike collagen fibers in a malignant tumor and surrounding stroma require relatively little strain to uncoil the collagen. As a result, one would expect early and strong onset of nonlinear stiffening in malignant tissues; this is precisely the behavior we see in macroscopic measurements of tumor tissue ex vivo; see Figure 15.1 and Refs. 70, 71, 73, and 88. Thus, while linear elastic modulus is related to collagen density, quantitative measures of macroscopic nonlinear mechanical properties reveal information related to morphology of collagen fibers at the microstructural level.

To summarize, we recognize that different breast tissues, including both benign and malignant breast tumors, tend to have different linear and nonlinear elastic properties. Several groups have documented a variety of mechanobiological effects, including the effect of the mechanical environment on cell growth, proliferation, motility, and differentiation. A working hypothesis in our group is that the linear elastic properties are thought to be associated with collagen density and degree of cross-linking, while the nonlinear elastic properties are derived from the microstructural organization of the collagen fibers.

15.1.3 Background: Multiphase Mechanics of Breast Tissues

Vascularized soft tissues, such as the breast, are recognized to be composed of both liquid and solid phases. At relatively rapid deformation rates, the phases tend to move in unison, so one may model the deformation as that of a single-phase material. Over longer time scales, however, the difference in motion of the different phases may become noticeable. Here, we briefly discuss the multiphase mechanics of breast tissue.

In addition to the elastic properties, vascularized soft tissue has several mechanical properties related to the fluid flow. Among these is the microvessel density (MVD). Typical values of MVD lie in the range of $10^2/\text{mm}^2$,[89] with the diameter of individual microvessels being as small as 5–10 μm. Thus, the microvasculature is well below the resolution of typical diagnostic scanners. Furthermore, the blood flow velocities in normal microvessels are very low (<5 mm/s) and are about an order of magnitude smaller in tumor microvessels.[90] Even very-high-frequency Doppler ultrasound is challenged to measure such low velocities.

Therefore, in considering the mechanics of breast tissue at the millimeter length scale, a continuum multiphase approach is both justified and warranted. In the first place, the microvasculature is treated as just that: a continuum-level quantity well below the resolution of our measurement system. Second, the multiphase model acknowledges the explicit link between the local elastic properties, the applied pressures, and the flow through the medium. Finally, the physical quantities defining the multiphase medium, in this case, the microvascular filtration coefficient, relate directly to quantities of histological significance (e.g., MVD, microvessel wall conductivity).

The biomechanics of the microcirculation has been well studied at both the microscopic and macroscopic levels; see, for example, the text by Fung[89] and the reviews in Refs. 91–96. Much of this work has been aimed at the microscopic level, though R.K. Jain and coworkers, in particular, have done some work in developing and experimentally validating *macroscopic models* of the mechanics of transport through solid tumors. (See the reviews in Refs. 90, 97, and 98 and the references therein.)

In the sections that follow, we will show that the microvascular filtration coefficient χ is the material parameter that governs observable tissue relaxation on the time scale of a few seconds. The microvascular filtration coefficient χ is closely related to intratumoral MVD, which itself is more strongly predictive of prognosis than tumor grade, size, estrogen-receptor positivity, or other prognostic markers.[99] Therefore, advances toward the noninvasive quantification of χ in vivo can have significant clinical impact.

15.1.4 Ultrasound Measurement of Soft-Tissue Deformation

Motion estimation from sequential images is a standard noncontact method to quantify deformation in experimental mechanics.[100] The technique is based on the assumption that any two image intensity fields I_0 and I_1 taken at different times differ from one another only by a displacement function $u(x)$ that varies only with position x (see Figure 15.2):

$$I_0(x) = I_1(x + u(x)). \tag{15.1}$$

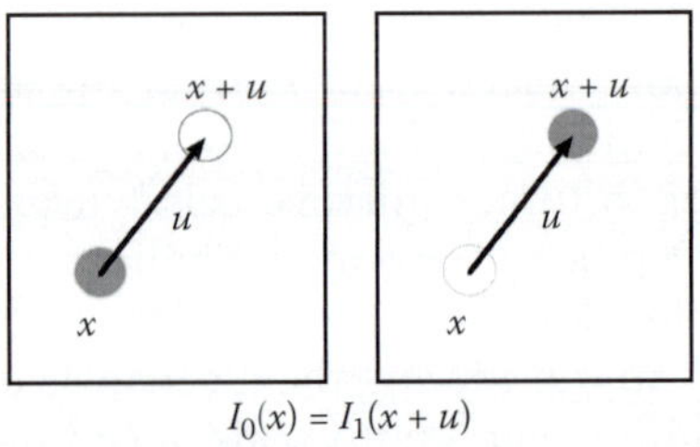

FIGURE 15.2
A particle (filled circle) imaged at point x_1 in the predeformation moves to point $x_2 = x_1 + u$ in the postdeformation image. Motion out of the imaging plane is neglected in 2-D images; 3-D vector displacements can be measured from 3-D images.

Equation 15.1 applies exactly only to a perfect imaging system with uniform sensitivity, delta point spread function (psf), and zero noise.

In using Equation 15.1 to find $u(x)$ from $I_0(x)$ and $I_1(x + u(x))$ in practice, therefore, one relaxes the equality in Equation 15.1 and seeks instead to maximize the "similarity" between the two images.

$$\Pi[u(x)] = S(I_0(x), I_1(x + u(x))). \tag{15.2}$$

Here, S is the chosen image similarity metric. The most commonly used similarity metrics are cross-correlation and sum-squared differences.

In principle, the "image" intensity field in Equation 15.1 could be any of a number of image data fields collected by ultrasound, including, for example, the radio-frequency (RF) image, its amplitude-detected envelope, or a log-compressed and time-gain compensated B-mode image. In practice, however, retaining phase information provides so much higher precision estimates of displacement that only RF images or their equivalent are used. The standard techniques are reviewed in Refs. 25–31 and 57, with more recent developments in displacement estimation described in Refs. 101–103.

Equation 15.1 gives a single displacement field from a pair of images. Given an image sequence with $M + 1$ images, M displacement fields may be obtained. Especially when one seeks to determine the nonlinear elastic parameters, it is important to recognize that these are Eulerian fields and need to be mapped back to a reference configuration to obtain the corresponding Lagrangian description of the deformation. With these fields in hand, we may consider formulating and solving an inverse problem for the material properties.

15.2 Inferring Elastic Nonlinearity of Breast Tissues with Ultrasound

In this section, we focus on cases where finite (i.e., not small) deformation has been observed and measured. The deformations of interest occur over a time period of a few seconds, and hence, inertia may be safely neglected. Following the observations of Fung,[89] the deformation is assumed to be quasi-elastic; that is, an elastic model may be used to represent pure loading, while a possibly different elastic model might be used to represent unloading. Therefore, restricting our attention to either loading or unloading only, we shall here focus on using a nonlinear elastic model for the tissue deformation and seek to determine spatial distribution of parameters related to the nonlinear elastic behavior of the tissue.

15.2.1 Formulation of the Inverse Problem

We formulate and solve the inverse problem for the material property distribution within the tissue as an optimization problem. That is, we seek the material property distribution that, when used in a forward model of the observed experiment, best predicts the observations. We therefore consider the following constrained minimization problem: find the spatial distribution of the elastic material parameters $\beta = (\beta_1, \beta_2, \ldots)$ that minimizes the objective function π:

$$\pi(\beta) = \frac{1}{2}\sum_{i=1}^{M}\int (u_i - \bar{u}_i)\cdot \boldsymbol{T}(u_i - \bar{u}_i)\,dx + \frac{\alpha}{2}\int \mathcal{R}(\beta)\,dx \qquad (15.3)$$

where each displacement field $\bar{u}_i$ is subject to the constraint imposed by the equations of motion:

$$\nabla \cdot \boldsymbol{P}_i = 0 \qquad (15.4)$$

and to appropriate boundary conditions. In Equation 15.3, u_i are the measured (Lagrangian) displacement fields obtained from minimizing Equation 15.2; $\bar{u}_i$ are the predicted displacements that are required to satisfy the constraint 15.4; and the index i runs from 1 to M, where M is the total number of independent measured displacement fields. In Equation 15.4, which is the equation of motion, $\boldsymbol{P}_i$ is the first Piola–Kirchhoff stress tensor, which relates to the predicted displacement field $\bar{u}_i$ according to a known constitutive function $\boldsymbol{P}_i = \boldsymbol{F}(\bar{u}_i, \beta)$, and β is the vector of material parameters whose spatial distribution is to be determined. The goal of the optimization problem is to find the spatial distribution of β that yields the displacements $\bar{u}_i$ that minimize the objective function π by reducing the first term in Equation 15.3. The uniqueness of this inverse problem for the nonlinear elastic parameter distributions, β, has recently been studied in 2-D.[104]

The tensor $\boldsymbol{T}$ in Equation 15.3 is a weighting tensor that allows those components of u_i with greater confidence to be weighted more strongly than those with lower confidence. In the context of maximum-likelihood estimation with Gaussian noise, $\boldsymbol{T}$ would be interpreted as an inverse covariance operator.

The second term in Equation 15.3 is a regularization or a smoothing term[105–108] that is added to control the effect of noise in the measurement. In this term, α is a regularization parameter, and $\mathcal{R}$ is a differential operator which determines the type of regularization.

15.2.2 Material Models: Phenomenological and Microstructural

15.2.2.1 Phenomenological Material Models

In an incompressible, elastic material, the second Piola–Kirchhoff stress $\boldsymbol{S}$ is related to the strain energy density $W(\boldsymbol{C},\beta)$ by

$$\boldsymbol{S} = -pJ\boldsymbol{C}^{-1} + 2\frac{\partial W}{\partial \boldsymbol{C}}. \qquad (15.5)$$

Here, p is the pressure, and W is the strain energy density function that, for an isotropic material, depends on the Cauchy–Green strain tensor $\boldsymbol{C} = \boldsymbol{F}^T\boldsymbol{F}$ through its invariants. The tensor $\boldsymbol{F}$ is the deformation gradient ($\boldsymbol{F} = \boldsymbol{I} + \nabla u$, in which $\boldsymbol{I}$ is the identity tensor), and $J = \det \boldsymbol{F}$ is its determinant. The number and physical interpretation of the parameters $\beta = (\beta_1, \beta_2, \ldots)$ depends upon the material model being used to describe the tissue deformation. We have considered two classes of nonlinear elastic models.

Phenomenological models are those that are proposed to capture the behavior observed but are agnostic to the origin of the behavior. The salient behavior to describe in the context of the mechanical response of breast tissue is the observation that breast tissues gradually stiffen with applied strain.[70–72,109] Several groups have modeled similar stiffening in other soft tissues by an exponential strain energy function (e.g., Refs. 109 and 110; see also the review focusing on arterial walls[111]).

In our work, we tend to focus on a modification of a model due to Blatz and coworkers.[110] The modification made here is to replace the appearance of I_1 with $J^{-2/3}I_1$.[18,112] This change renders the elastic energy contribution purely deviatoric, which is particularly appropriate for modeling incompressible materials. This model may also be seen as a special case of the Veronda–Westmann material model.[109] Thus, the strain energy function used is

$$W = \frac{\mu}{2\gamma}\left(e^{\gamma\left(J^{-\frac{2}{3}}I_1 - 3\right)} - 1\right). \qquad (15.6)$$

Here, μ is the shear modulus at zero strain, γ is a nonlinear parameter that determines the nonlinearity of the material response, and $I_1 = \mathrm{trace}(\boldsymbol{C})$ is the first principal invariant of the Cauchy–Green strain tensor. This strain energy function has several advantages in the inverse problem application. First, it is capable of modeling the observed exponential stiffening of soft tissues. Second, it contains only two material parameters to evaluate, μ and γ. Third, the material parameters have clear, distinct, physical interpretations: μ governs the small strain behavior independent of the value of γ, while γ controls the exponential stiffening, independent of μ. The respective roles of these two parameters may be most easily seen in uniaxial tension for small strains. In this case, the stress–strain relation may be approximated by

$$\sigma \approx 3\mu\varepsilon \exp[3\gamma\varepsilon^2] \quad \text{for } \varepsilon \ll 1. \tag{15.7}$$

Here, σ is the Cauchy stress, and $\varepsilon = \lambda - 1$ is the linearized strain measure, in which λ is the stretch ratio. The full nonlinear stress–strain behavior in uniaxial tension is given by the equation

$$\sigma = \mu\left(\lambda^2 - \frac{1}{\lambda}\right)e^{\gamma\left(\lambda^2+\frac{2}{\lambda}-3\right)}. \tag{15.8}$$

This phenomenological model was applied to approximate the deformation of a tissue-mimicking phantom created as described elsewhere.[113] The phantom was compressed to a maximum compression of 20%. Real-time ultrasound provided an image sequence of the deformation of a cross section of the phantom. Within the imaged cross section, there is a spherical inclusion with mechanical properties that differ from its surroundings. The displacement field was extracted from the ultrasound image sequence as described in Section 15.1.4, comparing pairwise frames that differed by an overall strain of roughly 0.005 (i.e., 0.5% strain.) Complete experimental details are reported elsewhere.[114]

Figure 15.3 shows the vertical component of normal incremental strain, ε_{yy}, with the load increment associated from frames 2–3 (Figure 15.3a), and 38–39 (Figure 15.3b), that is, near the beginning and near the end of the compression sequence. The strain increment at the smaller overall strain, Figure 15.3a, shows that the spherical inclusion behaves as if it is less compliant than the background and thus strains less than the background. Note, in particular, the range of the color bar. The strain increment at the larger overall strain, Figure 15.3b, shows that the inclusion and background behave similarly. That is, the inclusion no longer seems to be stiffer than the background. Again, note the range of the color bar. Thus, the data reveal a clear nonlinear phenomenon. Furthermore, while the inclusion apparently has a higher linear elastic modulus (i.e., tangent modulus at zero strain) than the background, the tangent modulus in the configuration of Figure 15.3b is roughly equal.

Figure 15.3c shows the reconstructed spatial distributions of μ(x) and γ(x) within the cross section of the

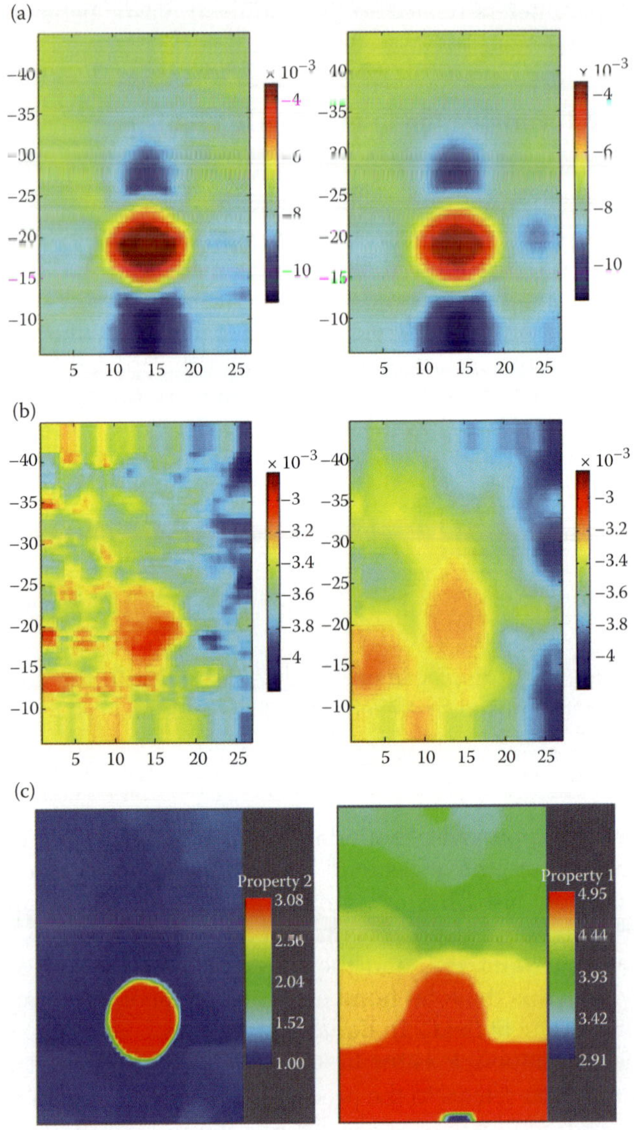

FIGURE 15.3
Vertical component of incremental normal strain in a tissue-mimicking phantom under vertical compression. Measured strain field on the left; reconstructed strain field from inverse problem solution on right. (a) Strain increment between (roughly) 1.0% and 1.5% strain. Note the (apparently stiff) inclusion with circular cross section. The similarity between data and model gives confidence that the model is accurately capturing the phenomenon of interest. (b) Strain increment between (roughly) 19.0% and 19.5% strain. Note the compressed color bar showing that the incremental strain field is roughly uniform within the phantom at this level of compression. This implies that the tangent moduli of the two materials are approximately equal at their respective overall strain levels. (c) Reconstructed spatial distributions of normalized shear modulus μ(x) (left) and elastic nonlinearity γ(x) (right) within the cross section of the phantom. The inverse problem was solved using the algorithms described elsewhere.[18] Independent mechanical tests of phantom material give $\mu_{incl}/\mu_{background} = 3.5$, with $4.3 < \gamma < 4.9$ within the inclusion and background.

phantom. The inverse problem was solved using algorithms described elsewhere,[18] assuming a plane stress model for the deformation. The results show clearly the inclusion with a circular cross section. The inclusion has a linear elastic shear modulus μ that is roughly three times stiffer than the background gel. Within the precision of the reconstruction, the elastic nonlinearity parameter, γ, is indistinguishable in the inclusion and the background. Both the inclusion and the background are found to have an elastic nonlinearity parameter $\gamma \approx$ 4.0–5.0. Independent mechanical testing of the materials used to create the phantom indicates that $\mu_{incl}/\mu_{background} \approx 3.5$, while γ ranged from 4.3 to 4.9 within the inclusion and background.[114]

Two generalizable observations may be made about these results. The first is that the contrast in shear modulus μ was slightly underestimated in the reconstruction. That is a systematic bias introduced by the regularization. In our experience, this bias tends to be about 20% in the contrast for contrast levels in the range of 1–10. The second is that while the materials involved in this particular study have nearly identical elastic nonlinearity parameters, γ, they nevertheless exhibit different nonlinear elastic behaviors. This is because the stiffer (at zero strain) inclusion tends to strain less than its surroundings, resulting in a highly heterogeneous strain field within the material. Therefore, at every loading increment, each of the two materials responds according to its own tangent modulus at its local level of overall strain. Hence, qualitative evaluations of nonlinear behavior may be a misleading indication of whether two materials actually have different elastic nonlinearity parameters.

15.2.2.2 Microstructural Material Models

In the previous example, we saw that parameters within a phenomenological model for nonlinear deformation can be accurately recovered by inverse problem methods. These parameters themselves, however, reveal nothing of the microstructural mechanisms that give rise to the resulting phenomena. In these situations where the mechanism is understood, however, a microstructural model can be used to infer directly the microstructural parameters that give rise to the observed macroscopic phenomenon. Here, we briefly describe how accounting for the tortuosity of collagen fibers at the microscale within soft tissue can predict the nonlinear stiffening observed at the macroscopic scale. This may be contrasted with the view that nonlinearity in polymer networks arises at the molecular scale.[115–118]

We use a homogenization method[119] to determine the effective macroscopic properties that arise from a periodic microstructure of wavy fibers. Both the individual and the background matrix are assumed to be neo-Hookean:

$$W = \frac{1}{2}\mu(I_1 - 3). \tag{15.9}$$

Thus, any observed nonlinear behavior arises from geometric nonlinearity during the deformation. The unit cell considered is shown in Figure 15.4.

The homogenization procedure proceeds roughly as follows. A macroscopic deformation gradient $\bar{\boldsymbol{F}}$ is applied. The total deformation gradient $\boldsymbol{F}$ is computed using periodic boundary conditions and the applied macroscopic deformation $\bar{\boldsymbol{F}}$. Thus, the equilibrium strain energy at each point within the unit cell is computed. For a given $\bar{\boldsymbol{F}}$, then, we may compute a macroscopic strain energy density:

$$\bar{W} = \frac{1}{V}\int_{\Omega} W(I_1)\,dV. \tag{15.10}$$

We further average over orientation to obtain an isotropic response.

This procedure yields the one-dimensional stress–stretch curves for the macroscopic response shown in Figure 15.5. These curves demonstrate that nonlinear stress–stretch response can result from linear constituents. The key to nonlinearity is the geometric rearrangement of the microstructure with increasing deformation. We note that this model predicts global nonlinear response whether the fibers or matrix are composed of polymers, amorphous solids, or polycrystaline solids. That is, the macroscopic behavior is determined almost entirely by the structure at this mesoscale, independent of the molecular scale. Similar observations are made elsewhere in the literature.[120]

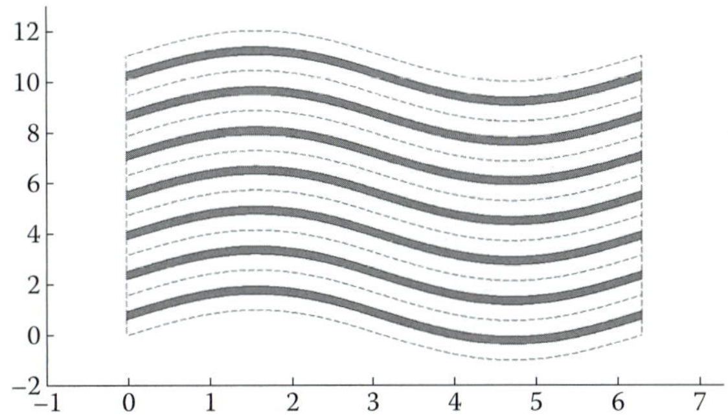

FIGURE 15.4
Seven adjacent unit cells used for homogenization calculation. Both fibers and background matrix are modeled with neo-Hookean strain energy functions and therefore individually exhibit linear behavior. The composite will stiffen under tension, however, when the fibers become fully elongated.

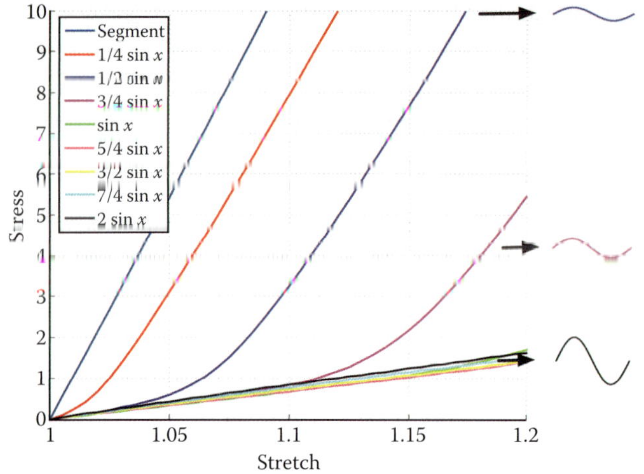

FIGURE 15.5
Effective macroscopic stress–stretch response predicted by homogenization. Microstructure is characterized by a collection of wavy fibers of the form $y = A \sin x$ with different peak amplitudes, A. We note that the wavier fibers have a delayed onset of nonlinear stiffening, while the microstructures with nearly straight fibers stiffen almost immediately.

15.3 Inferring Multiphase Mechanical Properties of Vascularized Soft Tissues with Ultrasound

15.3.1 Multiscale Multiphase Forward Model

Several multiphase models of tissue mechanics have been proposed to capture the link between fluid flow and tissue deformation. The primary application of such models has traditionally been to cartilage, starting with the "biphasic model,"[121] which may be regarded as a specialization of Biot's theory to incompressible phases. The biphasic model has several times been generalized, to include nonlinear deformation and to account for different rates of flow of solutes and solvent. Different "phases" in the model are distinguished by being physically different constituents.

When modeling the mechanics of vascularized soft tissues, however, one must account for different compartments within which the fluids may reside. This gives rise to different phases, which are, in fact, the same constituent (e.g., water or lymphatic fluid) but which currently reside in different locations within the tissue. Thus, water within the microvasculature is accounted for separately from water within the extravascular compartment, and this gives rise to two distinct phases in the model.

The simplest model that takes such an approach is a generalized poroelastic continuum model for fluid transport in solid tumors.[21,122] This model is described by the continuity and momentum equations:

$$\partial_t e - \nabla \cdot (K \nabla p) = \Omega(x,t) \tag{15.11}$$

$$\nabla \cdot \sigma = 0 \tag{15.12}$$

$$\mu(\nabla u + (\nabla u)^T) + \lambda e \boldsymbol{I} - p\boldsymbol{I} = \sigma \tag{15.13}$$

Darcy's law was used in writing Equation 15.11. The variables that appear are the dilatation of the solid, $e = \nabla \cdot \boldsymbol{u}$; the extravascular permeability, K; the displacement field $\boldsymbol{u}$ of the solid part of the tissue; the effective elastic Lamé constants of the tissue, λ and μ; the extravascular fluid pressure, p; and the mixture Cauchy stress, σ.

The "source" term, Ω, appearing on the right-hand side of Equation 15.11, accounts for fluid sources and sinks to the extravascular space. This term is given as[21,122]

$$\Omega = \frac{L_p S}{V}(p_c - p) - \frac{L_{pL} S_L}{V}(p - p_L) = -\chi p + \frac{L_p S}{V} p_c + \frac{L_{pL} S_L}{V} p_L. \tag{15.14}$$

In Equation 15.14, L_p is the "hydraulic conductivity" of the vessel walls (sometimes called the capillary permeability, or capillary water permeability); p_c is the effective vascular pressure (with osmotic adjustments to account for concentration gradients); p is again the pressure in the extravascular space; and S/V is the vessel surface area per unit volume of tissue. Quantities with subscript L correspond to values for the lymphatic system. This model may be thought of as a special case of a model with four phases: one fluid phase in each of three compartments and one solid phase. Here, the implicit assumption is made that flow through microvascular networks is sufficiently fast that p_c and p_L may be regarded as constant. In tumors with no functional lymphatics, $S_L/V \approx 0$, so Equation 15.14 is simplified accordingly. This mathematical model has since been validated experimentally in a mouse tail model[123] and here is applied to a phantom tumor model.

To appreciate the origin of Equations 15.11 and 15.14, we consider a representative volume element of tissue cartooned in Figure 15.6. Applying conservation of fluid mass over the extravascular space leads to

$$-\partial_t \int_{\Omega_e} \rho dV = \int_S \rho v_c \cdot n dS + \int_{S_L} \rho v_L \cdot n dS + \int_{S_{\text{box}}} \rho v \cdot n dS. \tag{15.15}$$

The left-hand side of Equation 15.15 represents the rate of change of fluid mass in the extravascular space, Ω_e, and $\boldsymbol{n}$ represents the unit outward normal vector. Noting that mass density ρ is constant for an incompressible fluid reduces the left-hand side to $-\rho\partial_t V(\Omega_e)$.

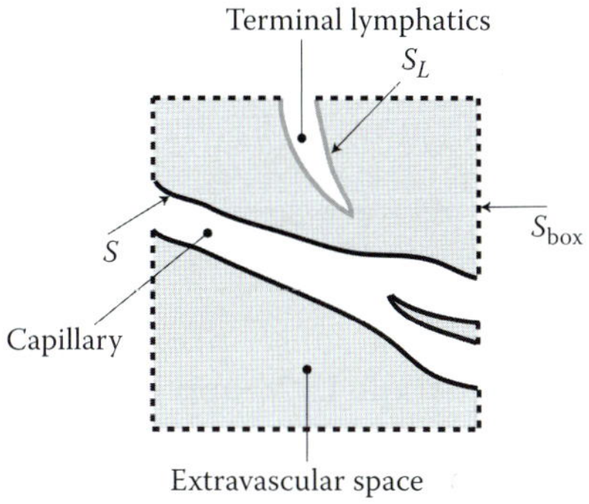

FIGURE 15.6
Cartoon of tissue structure used to derive Equations 15.11 and 15.14.

The three terms on the right-hand side of Equation 15.15 correspond, respectively, to flow out of Ω_e through the capillary walls (S), through the lymph vessel walls (S_L), and through the boundaries of the box S_{box}. For flows into the microvasculature, we use Starling's law[89]:

$$\rho v_c \cdot n = \rho L_p[(p - \hat{p}_c) - \sigma(\Pi - \Pi_c)],$$
$$\rho v_L \cdot n = \rho L_{pL}[(p - \hat{p}_L) - \sigma_L(\Pi - \Pi_L)]. \quad (15.16)$$

Here, Π, Π_c, and Π_L are the osmotic pressures of the extravascular space, capillaries, and lymphatic vessels, respectively. It is usual to make the assumption that $\Pi \approx \Pi_L$ because of the high permeability of the lymphatic vessels.

In order to obtain a continuum description of the phenomena considered here, we need to make the assumption that S_{box} is large enough (e.g., about 1 mm^3) so that averages over S_{box} are stationary. Equivalently, we consider an ensemble average of Equation 15.15 over all representative microstructural realizations. Thus, we replace

$$\int_S \rho v_c \cdot n\,dS = \rho L_p S[(p - \hat{p}_c) - \sigma(\Pi - \Pi_c)] \quad (15.17)$$

$$\int_{S_L} \rho v_L \cdot n\,dS = \rho L_{pL} S_L[(p - \hat{p}_L) - \sigma_L(\Pi - \Pi_L)] \quad (15.18)$$

where now, we use S and S_L to represent the (mean) areas of their respective surfaces. Further, all quantities in Equations 15.17 and 15.18 are understood to be spatial (or ensemble) averages over a representative volume element. Finally, we use Darcy's law, the divergence theorem, and ensemble average to write

$$\int_{S_{box}} \rho v \cdot n\,dS = -\rho \nabla \cdot K \nabla p. \quad (15.19)$$

Combining Equations 15.17 through 15.19 into Equation 15.15 and dividing by ρ and V yields

$$V^{-1}\partial_t V - \rho \nabla \cdot K \nabla p = \frac{L_p S}{V}[p_c - p] - \frac{L_{pL} S_L}{V}[p - p_L]. \quad (15.20)$$

In Equation 15.20, we have introduced $p_c = \hat{p} + \sigma(\Pi - \Pi_c)$, with p_L defined analogously.

Equation 15.20 shows one of the two explicit links between the fluid exchange and the solid deformation. That is the relation of specific volume change: $\partial_t e = \partial_t \nabla \cdot u = V^{-1}\partial_t V$. Making that substitution into Equation 15.20 yields Equation 15.11. Thus, we see that Equation 15.11 is a relation among quantities that are all averaged at the microstructural level. The second link between the solid deformation and the fluid flow is visible in the mixture momentum equation (Equation 15.12), showing how gradients in the fluid pressure can drive a solid deformation in addition to a fluid flow.

When a soft tissue is (quickly) compressed locally, it initially deforms as an incompressible elastic solid. The elevated internal pressure then drives fluid through the microvessels toward neighboring regions of uncompressed tissue. In the model considered here, this flow through the microvasculature is assumed to take place roughly instantaneously (in a time $\tau_0 < 1$ s), dropping the pressure in the microvasculature back to its equilibrium level. After this initial flow subsides, the extravascular pressure p remains elevated above p_c. The model just derived then predicts that fluid in the extravascular space flows toward the capillaries and toward the boundaries of the compressed region.

Equation 15.11 shows that the two different flow paths just described both contribute to the relaxation of the solid. To determine which effect is dominant, we consider a reference length a = 1 cm and, following the analysis in Ref. 21 and using reference parameter values therein, compute

$$\alpha = a\sqrt{\frac{L_p S}{K\,V}} \approx 40. \quad (15.21)$$

This indicates that the term containing K in Equation 15.11 may be neglected relative to the term containing $\chi = L_p S/V + L_p L S_L/V$. Note that Equation 15.11 is a reaction–diffusion equation. The parameter α in this context represents the Damkholer number. If the pressure varies over smaller length scales, however (for example, for tumor sizes $a < 0.5$ mm), the effect of K may become significant.

This analysis shows that the dominant parameter describing the relaxation rate of the tissue is

$$\chi = L_p S/V + L_{pL} S_L/V. \quad (15.22)$$

We call this quantity the "microvascular filtration coefficient." It controls the fluid transport between the microvasculature and the extravascular space. The larger the value of χ, the faster (more easily) fluid moves from one compartment to the other. Within tumors, $\chi \approx L_p S/V$.

The strain-relaxation time scale associated with this parameter has been estimated theoretically and confirmed experimentally to be on the order of 10 s.[90] This time scale is consistent with the predictions of Leiderman et al.[21] and correlates well with the relaxation times measured by elastography techniques in breast tumors.[124]

15.3.1.1 One-Dimensional Stress Relaxation in a Homogeneous Material

To gain a qualitative understanding of the predictions of this model, we consider uniaxial stress of a homogeneous cylinder. We restrict our attention to the special case that $K \approx 0$, motivated by Equation 15.21. For homogeneous uniaxial stress, we take $\sigma_{xx} = \sigma(t)$, with other stress components zero; hence, $\sigma = \sigma(t)\mathbf{e}_x \otimes \mathbf{e}_x$. The trace of Equation 15.13 then gives

$$\sigma(t) = -3p + (3\lambda + 2\mu)e. \quad (15.23)$$

On the other hand, the x component of Equation 15.13 gives

$$\sigma(t) = 2\mu\varepsilon_{xx} + \lambda e - p. \quad (15.24)$$

From this pair of equations, one may solve for p and e to obtain

$$e = 3\varepsilon_{xx} - \sigma/\mu \quad (15.25)$$

$$p = 3B\varepsilon_{xx} - \sigma/(1-2\nu) \quad (15.26)$$

Here, we introduced the bulk modulus $B = (\lambda + 2\mu/3)$ and Poisson's ratio $\nu = \lambda/(2(\lambda + \mu))$ of the solid frame.

In the special case $K \approx 0$, and $p_c = p_L = 0$, then Equation 15.11 simplifies to

$$\dot{e} + \chi p = 0. \quad (15.27)$$

Substituting Equations 15.25 and 15.26 into Equation 15.27 yields an effective constitutive equation of a standard linear solid:

$$\tau\dot{\sigma} + \sigma = 3\mu\tau\dot{\varepsilon} + E_\infty\varepsilon. \quad (15.28)$$

Here, we dropped the subscript on ε_{xx} with the understanding that Equation 15.28 applies to only uniaxial stress. Further, we introduced the time constant $\tau = \chi\mu(1 - 2\nu)$ and Young's modulus of the solid frame $E_\infty = 3B(1 - 2\nu)$. With the help of Equation 15.28, we can find the material response to an applied step in strain.

We now suppose the material is uniformly and instantly compressed with an applied strain magnitude, ε_0. Hence, the applied strain is $\varepsilon(t) = \varepsilon_0 H(t)$, where $H(t)$ is the Heaviside step function. Then solving Equation 15.28 yields

$$\sigma(t) = 3\mu\varepsilon_0 H(t) - (3\mu - E_\infty)\varepsilon_0 H(t)[1 - \exp(-t/\tau)]. \quad (15.29)$$

Equation 15.29 shows that at short times, $t \ll \tau$, the second term is negligible, and $\sigma \approx 3\mu\varepsilon$. That is, the material behaves as an incompressible elastic solid with shear modulus $\mu_0 = \mu$ and Poisson's ratio $\nu_0 = 1/2$. Hence, Young's modulus $E_0 = 3\mu$. At long times, $t \gg \tau$, the second term becomes constant, and $\sigma \approx E_\infty\varepsilon$. That is, the stress and strain are again related as in an elastic solid but now with Young's modulus E_∞.

This example demonstrates that the fluid redistribution in a homogeneous soft tissue by draining from the extravascular compartment into the microvascular network can give rise to a strain or a stress relaxation. The relaxation is monotonic, and its time scale may be related to a combination of intrinsic tissue material parameters: the product of the microvascular filtration coefficient χ and the elastic constant $\mu(1 - 2\nu)$. We shall see in the next section that the interpretation of similar relaxation phenomena in heterogeneous samples is less straightforward.

15.3.2 Application to Vascularized Tumor Phantom Model

A controlled laboratory model of fluid exchange between microvasculature and interstitium was created at the Institute of Cancer Research, by embedding a network of small (200 μm) permeable tubes into a biphasic hydrogel (see Figure 15.7). The network of tubes simulates the microvasculature, while the water-permeable hydrogel simulates the interstitial compartment. Observations clearly show a strain relaxation related to fluid exchange

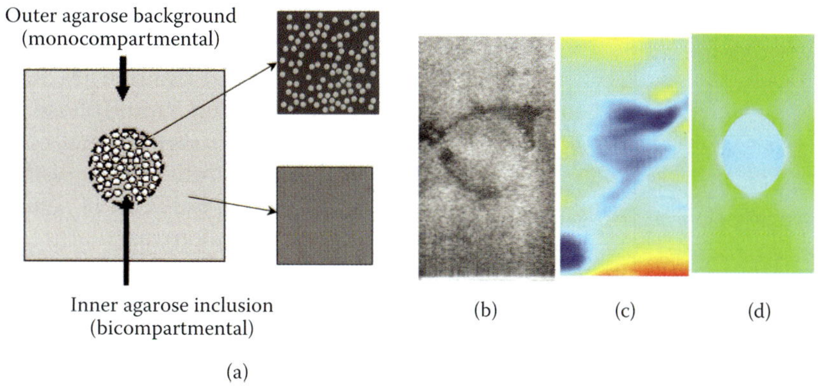

FIGURE 15.7
(a) Schematic of phantom cross section: 5 cm diameter inclusion in a [illegible] cm sample. The phantom is loaded under plane strain conditions. The inclusion is permeated by [illegible] pores of [illegible] μm diameter. The bottom is fixed, and the sides are traction free. A rigid compressor at the top applies a fixed degree of compression and then holds the specimen isometrically. (b) A B-mode ultrasound image of the phantom shows some damage around the cylinder. (c) Axial (vertical) strain distribution measured in the phantom immediately after an applied compression. Note the central region with vessels strains more as its effective modulus is lowered by the vessels. (d) This effect is captured in our macroscopic model as seen in this map of the axial strain predicted by the model.

between the compartments. Differential effects in the macroscopic responses of different regions of the phantom, which differ in the details of their microstructural arrangements, are also clearly visible in Figure 15.7.

In this experiment, the phantom was subjected to a sustained controlled compression. The compression drives fluid out of material, inducing a volumetric strain relaxation. The hypervascularization within the simulated tumor allows fluid to drain faster than in the surrounding medium. Based on the results of the homogeneous 1-D model above, one might expect simple monotonic strain relaxations in both the inclusion (simulated tumor) and its surroundings, with time constants corresponding to the local material properties. In fact, one observes the apparently counterintuitive behavior that after initially shrinking quickly, the inclusion subsequently swells! Strain histories after initial compression are shown in Figure 15.8, for representative regions within the inclusion and in the surroundings.

In fact, this behavior was predicted quantitatively using the model of Equations 15.11 and 15.12.[21] It may be understood qualitatively with the help of the insights gained in Section 15.3.1.1, and with the help of Figure 15.9. At short times immediately after compression, $t \ll \tau_{incl} \ll \tau_{gel}$, where τ_{incl} represents the relaxation time constant for the inclusion, while τ_{gel} represents the time constant for the surrounding gel. According to Equation 15.29, both phases will respond with their short time moduli, E_0. At this point, in a uniaxial stress model, $\varepsilon^{incl}/\varepsilon^{gel} = E_0^{gel}/E_0^{incl}$. At intermediate times, $t = O(\tau_{incl}) \ll \tau_{gel}$, the inclusion is now fully drained. It behaves effectively as an elastic solid, with $E^{incl} = E_\infty^{incl}$. The background gel, however, is only just beginning to relax and still exhibits its short time modulus, $E = E_0^{gel}$. Hence, in the uniaxial stress model, $\varepsilon^{incl}/\varepsilon^{gel} \approx E_0^{gel}/E_\infty^{incl}$. From this point forward, the gel continues to relax, thus reducing its effective modulus. The inclusion, therefore, is gradually less and less constrained by the gel and is allowed to expand. At long times, $\tau_{incl} \ll \tau_{gel} \ll t$, both phases are completely relaxed, with stress and strain related by their long time moduli $E = E_\infty$. Therefore, at this point, the uniaxial stress model would predict $\varepsilon^{incl}/\varepsilon^{gel} = E_\infty^{gel}/E_\infty^{incl}$.

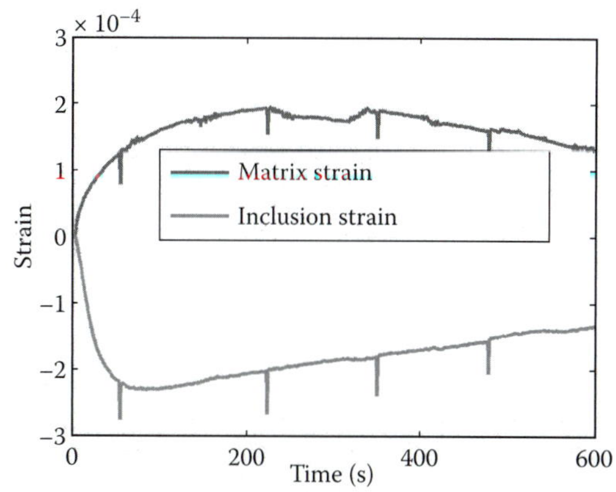

FIGURE 15.8
Measured strain histories in the microvascular phantom shown in Figure 15.7. Plotted is the relaxation strain history (normal strain in the direction of compression) in a representative region within the inclusion and in a representative region in the surrounding gel. The starting time is immediately after compression. The initial elastic strain is subtracted to emphasize the transient part of the strain history. The inclusion strain shows an initial shrinking followed by a slow swelling.

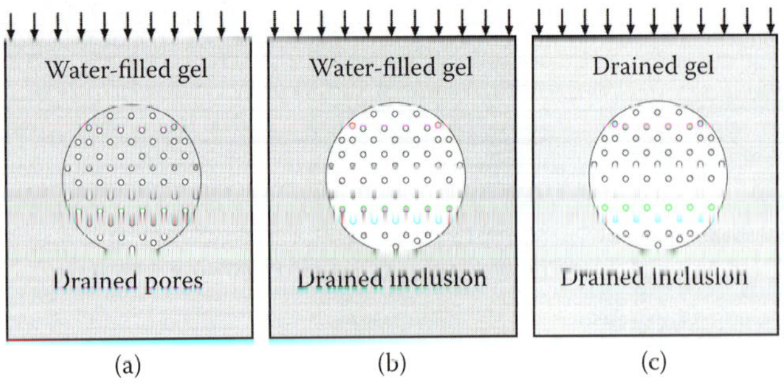

FIGURE 15.9
Schematic of vascularized phantom under compression: (a) Phantom initially compressed, $t \ll \tau_{incl} \ll \tau_{gel}$. Both phases are responding with their short time moduli, E_0, both phases relaxing as fluid drains. (b) For intermediate times, $t = O(\tau_{incl}) \ll \tau_{gel}$, the drained inclusion responds with E_∞, while the background gel still responds with its short time modulus, $E = E_0$. The inclusion is no longer relaxing but is constrained by surrounding gel. As liquid continues to drain from the gel, the gel relaxes (i.e., softens), thus allowing the inclusion to expand. (c) At long times, $\tau_{incl} \ll \tau_{gel} \ll t$, both phases eventually completely relax and respond with their long time moduli $E = E_\infty$.

15.4 Summary and Conclusions

Recent developments in the elastography community have made it possible to measure, to unprecedented precision, deformation patterns within living tissues. By combining these measurements with faithful constitutive models of the tissue in an inverse problem framework, the local mechanical properties of the soft tissue may be inferred. The current focus of much of the elastography community tends to be on linear elastic properties, while in biomechanical imaging, we seek to broaden that perspective to include any relevant mechanical phenomenon of interest. In particular, in this chapter, we have focused on nonlinear elastic properties relevant to large deformations of soft tissues and to multiphase mechanical properties, acknowledging the role of fluid motion in soft-tissue deformation.

We have discussed the fact that evidence for nonlinear elastic effects in soft-tissue deformation is abundant. The phenomenon can confound interpretation of strain images (quasi-static ultrasound elastography) and shear wave speed images (MR elastography and ultrasound radiation force elastography). Here, we have advocated both accounting for and quantifying the local nonlinear response of breast tissue. We have shown that quantifying the nonlinear elastic tissue parameter has potential to discriminate benign palpable breast masses from malignant palpable breast masses.

Soft tissues are often modeled as incompressible, with the acknowledgment that they have a large water content. Although the incompressible behavior of soft tissue is evident at short or early times, longer observation periods can reveal local volumetric changes as fluid moves from one region of the tissue into another. The effects of this fluid motion are observable in vivo.[22] Here, we have reviewed a multiphase model that accounts for fluid motion through several compartments, and demonstrated its validity as applied to a biphasic medium permeated by a network of semipermeable microtubes acting as the microvasculature. The determination of the microvascular parameters in this model through inverse techniques was demonstrated previously.[23]

In both the nonlinear elastic and linear multiphase models, the emergent phenomenon observable at the macroscopic scale is determined by the unresolvable microstructure. In the nonlinear elastic case, we described here how the emergent nonlinear stiffening often seen in soft tissues may be predicted by assuming a microstructure with stiff wavy fibers. In the multiphase case, we observed that time scales of deformation at the macroscopic scale are determined by fluid exchange between compartments at the microscale. In both situations, therefore, a multiscale model serves to bridge macroscopic-scale observations to microscale material properties.

This chapter was limited in scope to isotropic nonlinear elastic material behavior and isotropic linear multiphase material behavior. Clearly, opportunities for more detailed modeling exist where observations warrant their use. For example, we treated data in loading only for the case of nonlinear deformation and hence had no ability to assess hysteresis effects. These would be accessible through the techniques reviewed here within the pseudo-elastic model,[89] treating loading and unloading independently. The uniqueness of this inverse problem would follow that of nonlinear elasticity, which is understood in 2-D[104] but is an open question for 3-D. Anisotropy is another opportunity. Alignment of collagen fibers is a prognostic indicator in breast cancer[87] and would lead to an anisotropic response at the macroscopic level. Inverse elasticity with application to anisotropy has been examined[14,125] and requires observations of deformations along several axes. While the computational approach to this problem is straightforward (cf. Shore et al.[14]), the data requirements (e.g., conditions for uniqueness) of this inverse problem have not yet been settled. As a third example, it is known that different structural layers of soft tissues are separated by well-lubricated slip layers. Theoretically, such layers may be accommodated by modeling them as thin, compliant continua. In practice, however, they would be more accurately represented by allowing for discontinuous displacement fields within the medium. Neither the mathematical analysis nor the computational approach to this inverse problem has yet been addressed. Finally, an interesting and important opportunity related to the large deformation of soft tissues is the identification of a stress-free reference state. There is evidence to suggest that a connected and continuous stress-free reference state may not exist.[89] Notwithstanding the work treating homogeneous materials with connected reference configurations,[126–129] reconstructing a (possibly discontinuous) reference configuration within a material with unknown heterogeneous material properties is both completely open and highly relevant.

15.5 Prospects

In summary, the scope and detail of deformation measurements possible within soft tissues in vivo enables a host of material properties to be quantified with high spatial resolution. Reconstructing Young's modulus is only the beginning. Here, we described just two of several possible generalizations: nonlinear elasticity and multiphase phenomena. These quasi-static phenomena may be observed in vivo with ultrasound at clinical (i.e., millimeter) length scales. Opportunities to quantify other mechanical properties abound. At the clinical scale, these include identifying slipping fascia, quantifying viscoelastic contributions and decay exponents, and anisotropy resulting from aligned microstructures. At smaller scales, fundamental questions in mechanobiology may be addressed by similar techniques applied at the cellular scale,[130] or possibly even below. Accessibility to these scales and phenomena requires continued ingenious contributions from those pushing the frontiers of imaging technology. With such measurements, a great deal of applied mathematics and mechanics lies ahead before the full potential of biomechanical imaging can be realized.

Acknowledgments

The writing of this chapter and the work contained herein was partially supported by the National Science Foundation (NSF), the National Institutes of Health (NIH), and the US–Israeli Binational Science Foundation (BSF) (NSF grant nos. 1148124 and 1148111, NCIR01CA140271, BSF grant no. 2004400).

References

1. R. Muthupillai, D. J. Lomas, P. J. Rossman, J. F. Greenleaf, A. Manduca, and R. L. Ehman. Magnetic resonance elastography by direct visualization of propagating acoustic strain waves. *Science*, 269:1854–1857, 1995.
2. A. Manduca, T. E. Oliphant, M. A. Dresner, J. L. Mahowald, S. A. Kruse, E. Amromin, J. P. Felmlee, J. F. Greenleaf, and R. L. Ehman. Magnetic resonance elastography: Non-invasive mapping of tissue elasticity. *Medical Image Analysis*, 5(4):237–254, 2001.
3. T. E. Oliphant, A. Manduca, R. L. Ehman, and J. F. Greenleaf. Complex-valued stiffness reconstruction for magnetic resonance elastography by algebraic inversion of the differential equation. *Magnetic Resonance in Medicine*, 45(2):299–310, 2001.
4. L. Sandrin, M. Tanter, S. Catheline, and M. Fink. Shear modulus imaging with 2 D transient elastography. *Ultrasonics, Ferroelectrics and Frequency Control, IEEE Transactions on*, 49(4):426–435, 2002.
5. K. R. Nightingale, M. L. Palmeri, R. W. Nightingale, and G. E. Trahey. On the feasibility of remote palpation using acoustic radiation force. *Journal of the Acoustical Society of America*, 110(1):625–634, 2001.
6. K. R. Nightingale, R. W. Nightingale, D. Stutz, and G. E. Trahey. Acoustic radiation force impulse imaging of in vivo vastus medialis muscle under varying isometric load. *Ultrasonic Imaging*, 24(2):100–108, 2002.
7. P. E. Barbone, and J. C. Bamber. Quantitative elasticity imaging: What can and cannot be inferred from strain images. *Physics in Medicine and Biology*, 47:2147–2164, 2002.
8. P. E. Barbone, and N. Gokhale. Elastic modulus imaging: On the uniqueness and nonuniqueness of the elastography inverse problem in 2D. *Inverse Problems*, 20(1):283–296, 2004.
9. P. E. Barbone, and A. A. Oberai. Elastic modulus imaging: Some exact solutions of the compressible elastography inverse problem. *Physics in Medicine and Biology*, 52:1577–1593, 2007.
10. U. Albocher, A. A. Oberai, P. E. Barbone, and I. Harari. Adjoint-weighted equation for inverse problems of incompressible plane-stress elasticity. *Computer Methods in Applied Mechanics and Engineering*, 198(30–32):2412–2420, 2009.
11. R. A. Crescenti, J. C. Bamber, A. A. Oberai, P. E. Barbone, J. P. Richter, C. Rivas, N. L. Bush, and S. Webb. Quantitative ultrasonic elastography for gel dosimetry. *Ultrasound in Medicine and Biology*, 36(2):268–275, 2010.
12. T. G. Fisher, T. J. Hall, S. Panda, M. S. Richards, P. E. Barbone, J. Jiang, J. Resnick, and S. Barnes. Volumetric elasticity imaging with a 2-D CMUT array. *Ultrasound in Medicine and Biology*, 36(6):978–990, 2010.
13. M. S. Richards, P. E. Barbone, and A. A. Oberai. Quantitative three-dimensional elasticity imaging from quasi-static deformation: A phantom study. *Physics in Medicine and Biology*, 54:757–779, 2009.
14. S. W. Shore, P. E. Barbone, A. A. Oberai, and E. F. Morgan. Transversely isotropic elasticity imaging of cancellous bone. *Journal of Biomechanical Engineering*, 133:061002, 2011.
15. Y. Zhang, A. A. Oberai, P. E. Barbone, and I. Harari. Solution of the time-harmonic viscoelastic inverse problem with interior data in two dimensions. *International Journal for Numerical Methods in Engineering*, 92:1100–1116, 2012.

16. N. H. Gokhale, A. A. Oberai, and P. E. Barbone. Solution of the non-linear elasticity imaging inverse problem: The compressible case. *Inverse Problems*, 24(4):045010, 2008.
17. A. A. Oberai, N. H. Gokhale, S. Goenezen, P. E. Barbone, T. J. Hall, A. M. Sommer, and J. Jiang. Linear and nonlinear elasticity imaging of soft tissue in vivo: Demonstration of feasibility. *Physics in Medicine and Biology*, 54(5):1191, 2009.
18. S. Goenezen, P. Barbone, and A. A. Oberai. Solution of the nonlinear elasticity imaging inverse problem: The incompressible case. *Computer Methods in Applied Mechanics and Engineering*, 200(13):1406–1420, 2011.
19. G. P. Berry, J. C. Bamber, C. G. Armstrong, N. R. Miller, and P. E. Barbone. Towards an acoustic model-based poroelastic imaging method: I. Theoretical foundation. *Ultrasound in Medicine and Biology*, 32(4):547–567, 2006.
20. G. P. Berry, J. C. Bamber, N. R. Miller, P. E. Barbone, N. L. Bush, and C. G. Armstrong. Towards an acoustic model-based poroelastic imaging method: II. Experimental investigation. *Ultrasound in Medicine and Biology*, 32(12): 1869–1885, 2006.
21. R. Leiderman, P. E. Barbone, A. A. Oberai, and J. C. Bamber. Coupling between elastic strain and interstitial fluid flow: Ramifications for poroelastic imaging. *Physics in Medicine and Biology*, 51:6291–6313, 2006.
22. G. P. Berry, J. C. Bamber, P. S. Mortimer, N. L. Bush, N. R. Miller, and P. E. Barbone. The spatiotemporal strain response of oedematous and nonoedematous tissue to sustained compression in vivo. *Ultrasound in Medicine and Biology*, 34(4):617–629, 2008.
23. R. Leiderman, A. A. Oberai, and P. E. Barbone. Theory of reconstructing the spatial distribution of the filtration coefficient in vascularized soft tissues: Exact and approximate inverse solutions. *Comptes Rendus Mecanique*, 338:412–423, 2010.
24. J. Ophir, I. Cespedes, H. Ponnekanti, Y. Yazdi, and X. Li. Elastography—A quantitative method for imaging the elasticity of biological tissues. *Ultrasonic Imaging*, 13:111–134, 1991.
25. L. Gao, K. J. Parker, R. M. Lerner, and S. F. Levinson. Imaging of the elastic properties of tissue—A review. *Ultrasound in Medicine and Biology*, 22(8):959–977, 1996.
26. K. J. Parker, L. Gao, R. M. Lerner, and S. F. Levinson. Techniques for elastic imaging: A review. *IEEE Engineering in Medicine and Biology Magazine*, 15(6):52–59, 1996.
27. J. Ophir, S. K. Alam, B. Garra, F. Kallel, E. Konofagou, T. Krouskop, and T. Varghese. Elastography: Ultrasonic estimation and imaging of the elastic properties of tissues. *Proceedings of the Institution of Mechanical Engineers Part H-Journal of Engineering in Medicine*, 213(H3):203–233, 1999.
28. J. F. Greenleaf, M. Fatemi, and M. Insana. Selected methods for imaging elastic properties of biological tissues. *Annual Reviews in Biomedical Engineering*, 5(1):57–78, 2003.
29. K. J. Parker, L. S. Taylor, S. Gracewski, and D. J. Rubens. A unified view of imaging the elastic properties of tissue. *The Journal of the Acoustical Society of America*, 117:2705–2712, 2005.
30. K. J. Parker, M. M. Doyley, and D. J. Rubens. Imaging the elastic properties of tissue: The 20 year perspective. *Physics in Medicine and Biology*, 56:R1–R29, 2011.
31. M. M. Doyley. Model-based elastography: A survey of approaches to the inverse elasticity problem. *Physics in Medicine and Biology*, 57(3):R35–R73, 2012.
32. T. L. Chenevert, A. R. Skovoroda, M. O'Donnell, and S. Y. Emelianov. Elasticity reconstructive imaging by means of stimulated echo MRI. *Magnetic Resonance in Medicine*, 39(3):482–490, 1998.
33. C. R. Hill, M. Tristam, J. C. Bamber, M. Blaszczyk, D. Barbosa, D. Cosgrove, P. Simmonds. Ultrasonic Remote Palpation (URP): Use of shear elastic modulus to differentiate pathology. *Journal of Ultrasound in Medicine*, 7(Suppl.):S129, 1988.
34. B. S. Garra, I. Cespedes, J. Ophir, S. Spratt, R. A. Zuurbier, C. M. Magnant, and M. F. Pennanen. Elastography of breast lesions: Initial clinical results. *Radiology*, 202:79–86, 1997.
35. J. B. Weaver, E. E. W. Van Houten, M. I. Miga, F. E. Kennedy, and K. D. Paulsen. Magnetic resonance elastography using 3D gradient echo measurements of steady-state motion. *Medical Physics*, 28(8):1620–1628, 2001.
36. A. R. Skovoroda, S. Y. Emelianov, and M. O'Donnell. Tissue elasticity reconstruction based on ultrasonic displacement and strain images. *IEEE Transactions on Ultrasonics, Ferroelectrics, and Frequency Control*, 42:747–765, 1995.
37. C. Sumi, A. Suzuki, and K. Nakayama. Estimation of shear modulus distribution in soft tissue from strain distribution. *IEEE Transactions on Biomedical Engineering*, 42:193–202, 1995.
38. L. K. Ryan, and F. S. Foster. Ultrasonic measurement of differential displacement strain in a vascular model. *Ultrasonic Imaging*, 19(1):19–38, 1997.
39. C. L. de Korte, E. I. Cespedes, A. F. W. van der Steen, and C. T. Lancee. Intravascular elasticity imaging using ultrasound: Feasibility studies in phantoms. *Ultrasound in Medicine and Biology*, 23(5):735–746, 1997.
40. C. L. de Korte, A. F. W. van der Steen, E. I. Cespedes, and G. Pasterkamp. Intravascular ultrasound elastography in human arteries: Initial experience in vitro. *Ultrasound in Medicine and Biology*, 24(3):401–408, 1998.
41. C. L. de Korte, E. I. Cespedes, and A. F. W. van der Steen. Influence of catheter position on estimated strain in intravascular elastography. *IEEE Transactions on Ultrasonics, Ferroelectrics, and Frequency Control*, 46(3):616–625, 1999.
42. C. L. de Korte, A. F. van der Steen, E. I. Cepedes, G. Pasterkamp, S. G. Carlier, F. Mastik, A. H. Schoneveld, P. W. Serruys, and N. Bom. Characterization of plaque components and vulnerability with intravascular ultrasound elastography. *Physics in Medicine and Biology*, 45(6):1465–1475, 2000.
43. S. Y. Emelianov, X. Chen, M. O'Donnell, B. Knipp, D. Myers, T. W. Wakefield, and J. M. Rubin. Triplex ultrasound: Elasticity imaging to age deep venous thrombosis. *Ultrasound in Medicine and Biology*, 28(6):757–767, 2002.
44. K. R. Nightingale, M. S. Soo, R. W. Nightingale, and G. E. Trahey. Acoustic radiation force impulse imaging: In vivo demonstration of clinical feasibility. *Ultrasound in Medicine and Biology*, 28(2):227–235, 2002.

45. S. A. Kruse, G. H. Rose, K. J. Glaser, A. Manduca, J. P. Felmlee, C. R. Jack Jr., and R. L. Ehman. Magnetic resonance elastography of the brain. *Neuroimage*, 39(1):231–237, 2008.
46. M. C. Murphy, K. J. Glaser, A. Manduca, J. P. Felmlee, J. Huston III, and R. L. Ehman. Analysis of time reduction methods for magnetic resonance elastography of the brain. *Magnetic Resonance Imaging*, 28(10):1514–1524, 2010.
47. A. Samani, J. Bishop, M. J. Yaffe, and D. B. Plewes. Biomechanical 3-D finite element modeling of the human breast using MRI data. *IEEE Transactions on Medical Imaging*, 20(4):271–279, 2001.
48. M. I. Miga, D. W. Roberts, F. E. Kennedy, L. A. Platenik, A. Hartov, K. E. Lunn, and K. D. Paulsen. Modeling of retraction and resection for intraoperative updating of images. *Neurosurgery*, 49(1):75–84, 2001.
49. M. Ferrant, A. Nabavi, B. Macq, F. A. Jolesz, R. Kikinis, and S. K. Warfield. Registration of 3-D intraoperative MR images of the brain using a finite-element biomechanical model. *IEEE Transactions on Medical Imaging*, 20(12):1384–1397, 2001.
50. A. Hagemann, K. Rohr, H. S. Stiehl, J. M. Spetzger, and U. Gilsbach. Biomechanical modeling of the human head for physically based, nonrigid image registration. *IEEE Transactions on Medical Imaging*, 18(10):875–884, 1999.
51. R. J. Pelham, and Y.-L. Wang. Cell locomotion and focal adhesions are regulated by substrate flexibility. *Proceedings of the National Academy of Sciences of the United States of America*, 94(25):13661–13665, 1997.
52. M. J. Paszek, N. Zahir, K. R. Johnson, J. N. Lakins, G. I. Rozenberg, A. Gefen, C. A. Reinhart-King et al. Tensional homeostasis and the malignant phenotype. *Cancer Cell*, 8:241–254, 2005.
53. J. Solon, I. Levental, K. Sengupta, P. C. Georges, and P. A. Janmey. Fibroblast adaptation and stiffness matching to soft elastic substrates. *Biophysical Journal*, 93(12):4453–4461, 2007.
54. M. J. Reeves, P. A. Newcomb, P. L. Remington, and P. M. Marcus. Determinants of breast cancer detection among Wisconsin (United States) women. *Cancer-Causes-Control*, 6(2):103–111, 1995.
55. M. M. Doyley, J. C. Bamber, F. Fuechsel, and N. L. Bush. A freehand elastographic imaging approach for clinical breast imaging: System development and performance evaluation. *Ultrasound in Medicine and Biology*, 27(10):1347–1357, 2001.
56. K. M. Hiltawsky, M. Krüger, C. Starke, L. Heuser, H. Ermert, and A. Jensen. Freehand ultrasound elastography of breast lesions: Clinical results. *Ultrasound in Medicine and Biology*, 27(11):1461–1469, 2001.
57. T. J. Hall, Y. Zhu, and C. S. Spalding. In vivo real-time freehand palpation imaging. *Ultrasound in Medicine and Biology*, 29(3):427–435, 2003.
58. G. M. Giuseppetti, A. Martegani, B. D. Cioccio, and S. Baldassarre. Elastosonography in the diagnosis of the nodular breast lesions: Preliminary report. *Radiologia Medica*, 110:69–76, 2005.
59. A. Itoh, E. Ueno, E. Tohno, H. Kamma, H. Takahashi, T. Shiina, M. Yamakawa, and T. Matsumura. Breast disease: Clinical application of US elastography for diagnosis. *Radiology*, 239(2):341–350, 2006.
60. A. Thomas, T. Fischer, H. Frey, R. Ohlinger, S. Grunwald, J.-U. Blohmer, K.-J. Winzer et al. Real-time elastography an advanced method of ultrasound: First results in 108 patients with breast lesions. *Ultrasound in Obstetrics and Gynecology*, 28:335–340, 2006.
61. H. Zhi, B. Ou, B.-M. Luo, X. Feng, Y.-L. Wen, and H.-Y. Yang. Comparison of ultrasound elastography, mammography, and sonography in the diagnosis of solid breast lesions. *Journal of Ultrasound in Medicine*, 26:807–815, 2007.
62. E. S. Burnside, T. J. Hall, A. M. Sommer, G. K. Hesley, G. A. Sisney, W. E. Svensson, and N. J. Hangiandreou. Ultrasound strain imaging to improve the decision to biopsy solid breast masses. *Radiology*, 245(2):401–410, 2007.
63. C. A. Iacobuzio-Donahue, P. Argani, P. M. Hempen, J. Jones, and S. E. Kern. The desmoplastic response to infiltrating breast carcinoma: Gene expression at the site of primary invasion and implications for comparisons between tumor types. *Cancer Research*, 62(18):5351–5357, 2002.
64. A. R. Skovoroda, M. A. Lubinski, S. Y. Emelianov, and M. O'Donnell. Reconstructive elasticity imaging for large deformations. *IEEE Transactions on Ultrasonics, Ferroelectrics, and Frequency Control*, 46(3):523–535, 1999.
65. T. Varghese, J. Ophir, and T. A. Krouskop. Nonlinear stress–strain relationships in tissue and their effect on the contrast-to-noise ratio in elastograms. *Ultrasound in Medicine and Biology*, 26(5):839–851, 2000.
66. R. Q. Erkamp, S. Y. Emelianov, A. R. Skovoroda, X. Chen, and M. O'Donnell. Exploiting strain-hardening of tissue to increase contrast in elasticity imaging. *IEEE Ultrasonics Symposium*, 2:1833–1836, 2000.
67. R. Q. Erkamp, S. Y. Emelianov, A. R. Skovoroda, and M. O'Donnell. Nonlinear elasticity imaging. *IEEE Ultrasonics Symposium*, 2:1891–1894, 2002.
68. R. Q. Erkamp, S. Y. Emelianov, A. R. Skovoroda, and M. O'Donnell. Nonlinear elasticity imaging: Theory and phantom study. *Ultrasonics, Ferroelectrics and Frequency Control, IEEE Transactions on*, 51(5):532–539, 2004.
69. R. G. Barr. Sonographic breast elastography: A primer. *Journal of Ultrasound in Medicine*, 31(5):773–783, 2012.
70. T. A. Krouskop, T. M. Wheeler, F. Kallel, B. S. Garra, and T. Hall. Elastic moduli of breast and prostate tissues under compression. *Ultrasonic Imaging*, 20(4):260–274, 1998.
71. P. S. Wellman, E. P. Dalton, D. Krag, K. A. Kern, and R. D. Howe. Tactile imaging of breast masses: First clinical report. *Archives of Surgery*, 136(2):204–208, 2001.
72. A. Samani, and D. Plewes. A method to measure the hyperelastic parameters of ex vivo breast tissue samples. *Physics in Medicine and Biology*, 49(18):4395–4405, 2004.
73. J. J. O'Hagan, and A. Samani. Measurement of the hyperelastic properties of 44 pathological ex vivo breast tissue samples. *Physics in Medicine and Biology*, 54:2557–2569, 2009.
74. N. Nitta, and T. Shiina. A visualization of nonlinear elasticity property of tissues by ultrasound. *Electronics and Communications in Japan Part III-Fundamental Electronic Science*, 85(12):9–18, 2002.
75. H. Mehrabian, G. Campbell, and A. Samani. A constrained reconstruction technique of hyperelasticity parameters for breast cancer assessment. *Physics in Medicine and Biology*, 55:7489, 2010.

76. S. Goenezen, J.-F. Dord, Z. Sink, P. E. Barbone, J. Jiang, T. J. Hall, and A. A. Oberai. Linear and nonlinear elastic modulus imaging: An application to breast cancer diagnosis. *IEEE Transactions on Medical Imaging*, 31(8):1628–1637, 2012.
77. S. Huang, and D. E. Ingber. Cell tension, matrix mechanics, and cancer development. *Cancer Cell*, 8(3):175–176, 2005.
78. C.-M. Lo, H.-B. Wang, M. Dembo, and Y.-L. Wang. Cell movement is guided by the rigidity of the substrate. *Biophysical Journal*, 79(1):144–152, 2000.
79. S. R. Peyton, and A. J. Putnam. Extracellular matrix rigidity governs smooth muscle cell motility in a biphasic fashion. *Journal of Cellular Physiology*, 204(1):198–209, 2005.
80. T. A. Ulrich, E. M. de Juan Pardo, and S. Kumar. The mechanical rigidity of the extracellular matrix regulates the structure, motility, and proliferation of glioma cells. *Cancer Research*, 69(10):4167–4174, 2009.
81. J. Schrader, T. T. Gordon-Walker, R. L. Aucott, M. van Deemter, A. Quaas, S. Walsh, D. Benten, S. J. Forbes, R. G. Wells, and J. P. Iredale. Matrix stiffness modulates proliferation, chemotherapeutic response, and dormancy in hepatocellular carcinoma cells. *Hepatology*, 53(4): 1192–1205, 2011.
82. L. Kass, J. T. Erler, M. Dembo, and V. M. Weaver. Mammary epithelial cell: Influence of extracellular matrix composition and organization during development and tumorigenesis. *The International Journal of Biochemistry and Cell Biology*, 39(11):1987–1994, 2007.
83. A. J. Engler, S. Sen, H. L. Sweeney, and D. E. Discher. Matrix elasticity directs stem cell lineage specification. *Cell*, 126(4):677–689, 2006.
84. G. Falzon, S. Pearson, and R. Murison. Analysis of collagen fibre shape changes in breast cancer. *Physics in Medicine and Biology*, 53:6641, 2008.
85. J. I. Lopez, J. K. Mouw, and V. M. Weaver. Biomechanical regulation of cell orientation and fate. *Oncogene*, 27(55):6981, 2008.
86. P. P. Provenzano, D. R. Inman, K. W. Eliceiri, and P. J. Keely. Matrix density-induced mechanoregulation of breast cell phenotype, signaling and gene expression through a fak–erk linkage. *Oncogene*, 28(49):4326–4343, 2009.
87. M. W. Conklin, J. C. Eickhoff, K. M. Riching, C. A. Pehlke, K. W. Eliceiri, P. P. Provenzano, A. Friedl, and P. J. Keely. Aligned collagen is a prognostic signature for survival in human breast carcinoma. *The American Journal of Pathology*, 178(3):1221–1232, 2011.
88. A. Samani, J. Zubovits, and D. Plewes. Elastic moduli of normal and pathological human breast tissues: An inversion-technique-based investigation of 169 samples. *Physics in Medicine and Biology*, 52(6):1565–1576, 2007.
89. Y. C. Fung. *Biomechanics. Mechanical Properties of Living Tissues*, 2nd edition. Springer, New York, 1993.
90. R. K. Jain. Delivery of molecular medicine to solid tumors: Lessons from in vivo imaging of gene expression and function. *Journal of Controlled Release*, 74:7–25, 2001.
91. F. E. Curry. Mechanics and thermodynamics of transcapillary exchange. In *Handbook of Physiology*, American Physiological Society, Bethesda, MD, 4(Part 1):309–374, 1984.
92. C. C. Michel. Fluid movements through capillary walls. In *Handbook of Physiology. The Cardiovascular System. Microcirculation*, American Physiological Society, Bethesda, MD, 4:375–409, 1984.
93. J. B. Bassingthwaighte, and C. A. Goresky. Modeling in the analysis of solute and water exchange in the microvasculature. In *Handbook of Physiology*, Section, 2:549–626, 1984.
94. J. R. Levick. Capillary filtration–absorption balance reconsidered in light of dynamic extravascular factors. *Experimental Physiology*, 76(6):825–857, 1991
95. J. R. Levick. *An Introduction to Cardiovascular Physiology*. Arnold, London, 2003.
96. G. W. Schmid-Schönbein. Biomechanics of microcirculatory blood perfusion. *Annual Review of Biomedical Engineering*, 1(1):73–102, 1999.
97. R. K. Jain. Transport of molecules, particles, and cells in solid tumors. *Annual Review of Biomedical Engineering*, 1(1):241–263, 1999.
98. R. K. Jain. Delivery of molecular and cellular medicine to solid tumors. *Advanced Drug Delivery Reviews*, 64(Suppl.):353–365, 2012.
99. J. Folkman. Clinical applications of research on angiogenesis. *New England Journal of Medicine*, 333(26):1757–1763, 1995.
100. J. Westerweel. Fundamentals of digital particle image velocimetry. *Measurement Science and Technology*, 8(12):1379, 1997.
101. T. J. Hall, P. Barbone, A. A. Oberai, J. Jiang, J. F. Dord, S. Goenezen, and T. G. Fisher. Recent results in nonlinear strain and modulus imaging. *Current Medical Imaging Reviews*, 7(4):313, 2011.
102. M. S. Richards, and M. M. Doyley. Non-rigid image registration based strain estimator for intravascular ultrasound elastography. *Ultrasound in Medicine and Biology*, 39(3):515–533, 2013.
103. G. Treece, J. Lindop, L. Chen, J. Housden, R. Prager, and A. Gee. Real-time quasi-static ultrasound elastography. *Interface Focus*, 1(4):540–552, 2011.
104. E. R. Ferreira, A. A. Oberai, and P. E. Barbone. Uniqueness of the elastography inverse problem for incompressible nonlinear planar hyperelasticity. *Inverse Problems*, 28(6):065008, 2012.
105. A. N. Tikhonov. On the solution of incorrectly formulated problems and the regularization method. *Soviet Math Doklady*, 4:1035–1038, 1963.
106. A. N. Tikhonov. Regularization of incorrectly posed problems. *Soviet Math Doklady*, 4:1624–1627, 1963.
107. V. Isakov. *Inverse Problems for Partial Differential Equations*, 1st edition. Springer, New York, 1998.
108. C. R. Vogel. *Computational Methods for Inverse Problems*. SIAM, Philadelphia, PA, 2002.
109. D. R. Veronda, and R. A. Westmann. Mechanical characterization of skin–finite deformations. *Journal of Biomechanics*, 3(1):111–122, 1970.
110. P. J. Blatz, B. M. Chu, and H. Wayland. On the mechanical behavior of elastic animal tissue. *Journal of Rheology*, 13:83–102, 1969.
111. J. D. Humphrey. Mechanics of the arterial wall: Review and directions. *Critical Reviews in Biomedical Engineering*, 23(1–2):1–162, 1995.

112. G. A. Holzapfel. *Nonlinear Solid Mechanics: A Continuum Approach for Engineering*. John Wiley & Sons Ltd., Chichester, 2000.
113. T. Z. Pavan, E. L. Madsen, G. R. Frank, A. O. Carneiro, and T. J. Hall. Nonlinear elastic behavior of phantom materials for elastography. *Physics in Medicine and Biology*, 55(9):2679, 2010.
114. J.-F. Dord, S. Goenezen, A. A. Oberai, P. E. Barbone, J. Jiang, T. J. Hall, and T. Z. Pavan. Linear and nonlinear elastic imaging: A tissue-phantom study. In preparation, 2014.
115. E. M. Arruda, and M. C. Boyce. A three-dimensional constitutive model for the large stretch behavior of rubber elastic materials. *Journal of the Mechanics and Physics of Solids*, 41(2):389–412, 1993.
116. Y. Zhang, M. L. Dunn, E. S. Drexler, C. N. McCowan, A. J. Slifka, and D. D. Ivy. A microstructural hyperelastic model of pulmonary arteries under normo- and hypertensive conditions. *Annals of Biomedical Engineering*, 33(8):1042–1052, 2005.
117. C. Storm, J. J. Pastore, F. C. MacKintosh, T. C. Lubensky, and P. A. Janmey. Nonlinear elasticity in biological gels. *Nature*, 435(7039):191–194, 2005.
118. T. C. Gasser, R. W. Ogden, and G. A. Holzapfel. Hyperelastic modelling of arterial layers with distributed collagen fibre orientations. *Journal of the Royal Society Interface*, 3(6):15–35, 2006.
119. C. Miehe, J. Schröder, and M. Becker. Computational homogenization analysis in finite elasticity: Material and structural instabilities on the micro- and macro-scales of periodic composites and their interaction. *Computer Methods in Applied Mechanics and Engineering*, 191(44):4971–5005, 2002.
120. D. Vader, A. Kabla, D. Weitz, and L. Mahadevan. Strain-induced alignment in collagen gels. *PLoS One*, 4(6):e5902, 2009.
121. V. C. Mow, S. C. Kuei, W. M. Lai, and C. G. Armstrong. Biphasic creep and stress relaxation of articular cartilage in compression: Theory and experiments. *Journal of Biomechanical Engineering*, 102:73–84, 1980.
122. P. A. Netti, L. T. Baxter, Y. Boucher, R. Skalak, and R. K. Jain. Macro- and microscopic fluid transport in living tissues: Application to solid tumors. *AIChE Journal of Bioengineering, Food, and Natural Products*, 43(3):818–834, 1997.
123. A. Melody, M. A. Swartz, A. Kaipainen, P. A. Netti, C. Brekken, Y. Boucher, A. J. Grodzinsky, and R. K. Jain. Mechanics of interstitial-lymphatic fluid transport: Theoretical foundation and experimental validation. *Journal of Biomechanics*, 32:1297–1307, 1999.
124. Y. Qiu, M. Sridhar, J. K. Tsou, K. K. Lindfors, and M. F. Insana. Ultrasonic viscoelasticity imaging of nonpalpable breast tumors. 1: Preliminary Results. *Academic Radiology*, 15(12):1526–1533, 2008.
125. J.-L. Gennisson, S. Catheline, S. Chaffaï, and M. Fink. Transient elastography in anisotropic medium: Application to the measurement of slow and fast shear wave speeds in muscles. *The Journal of the Acoustical Society of America*, 114:536–541, 2003.
126. J. Lu, X. Zhou, and M. L. Raghavan. Inverse elastostatic stress analysis in pre-deformed biological structures: Demonstration using abdominal aortic aneurysms. *Journal of Biomechanics*, 40(3):693–696, 2007.
127. S. Govindjee, and P. A. Mihalic. Computational methods for inverse deformations in quasi-incompressible finite elasticity. *International Journal for Numerical Methods in Engineering*, 43(5):821–838, 1998.
128. J. Lu, X. Zhou, and M. L. Raghavan. Computational method of inverse elastostatics for anisotropic hyperelastic solids. *International Journal for Numerical Methods in Engineering*, 69(6):1239–1261, 2007.
129. S. Govindjee, and P. A. Mihalic. Computational methods for inverse finite elastostatics. *Computer methods in Applied Mechanics and Engineering*, 136(1):47–57, 1996.
130. E. P. Canović, D. T. Seidl, S. R. Polio, A. A. Oberai, P. E. Barbone, D. Stamenović, and M. L. Smith. Biomechanical imaging of cell stiffness and prestress with subcellular resolution. *Biomechanics and Modeling in Mechanobiology*, 13;665–678, 2014.

16

Quantitative Anatomy Using Design-Based Stereology

Peter R. Mouton

CONTENTS

In 1961, scientists from biology, geology, and material sciences convened a small meeting at a mountaintop retreat in Germany. Their purpose was to discuss an obscure problem common to all of their research: how to quantify the morphological features of arbitrary-shaped 3-D objects based on their 2-D appearance on cut surfaces. To describe this topic, they selected the term stereology, from the Greek stereos for "the study of objects in three dimensions," and a year later formed the International Society for Stereology (ISS). After more than a decade of debating the relative merits of different stereology approaches in their respective fields of study, the ISS categorically rejected older methods based on assumption- and model-based Euclidean geometry in favor of innovative new methods designed to avoid methodological bias. Today unbiased or design-based stereology is the sine qua non for the reliable quantification of structure in many fields of biological and biomedical research. This chapter examines key concepts of design-based stereology using systematic random sampling and objective geometry probes in comparison with biased methods. To exemplify an application to neurosciences (neurostereology), we review the results and discussion from our recently published study to quantify differences in postmortem brains from autistic and normal children. Continuing advances in computer-assisted microscopy, image segmentation, and whole slide scanning support the growing trend toward higher throughput and greater efficiency in computerized stereology, with the promise of fully automatic and accurate stereoanalysis of tissue sections and in vivo images in the near future.

16.1 Introduction

Biological morphometry is concerned with collecting quantitative data about the morphological characteristics of biological objects. In the field of morphometry, design-based stereology focuses on the first-order parameters of number (N), length (L), surface area (S) and volume (V), as well as second-order parameters like variation and spatial distribution. Like a Gallop poll, design-based stereology avoids systematic error (bias) by relying on assumption- and model-free sampling of a target population followed by data collection with objective probes. For instance, the Ford Motor Company might hire Gallop to survey the feelings of US car owners about Ford cars. The study design could involve asking car owners questions like, "How do you rate Fords against similar cars from other manufacturers?" After polling a random but representative sample of US car owners, the survey would provide Ford with a realistic idea about how all car owners in the United States feel about Ford cars. An alternative strategy would be to survey car owners outside a Ford production plant with questions like, "Fords are better cars than Chevrolets, don't you think?" Since the responses would likely favor Ford products, the conclusion might be well received by Ford but convey a biased view about Ford cars. In a closely similar way, reliable stereology studies are carefully designed, hence "design based," to avoid unverifiable assumptions, false models, and other factors that might skew morphological parameters from their expected values. Like a Gallop poll, the accuracy of all morphometric studies depends on how well the study excludes bias.

16.2 Design-Based Stereology

Morphometric analysis of biological tissue is concerned with quantifying parameters of scientific interest such as number (N, 0-D, no units); length (L, 1-D, units: μm, mm, cm…); surface area (S, 2-D, units: μm^2, mm^2, cm^2…); and volume (V, 3-D, units: μm^3, mm^3, cm^3…). Prior to the modern era of design-based stereology, morphometric studies attempted to quantify these parameters by relying on methodologically biased methods based on assumptions, models, and so-called correction factors (Wicksell 1925; Abercrombie 1946; for review, see Mouton 2002, 2011, 2013; West 2012). Because these older approaches were derived for Euclidean objects such as spheres and cylinders in known distributions, their applications to biological tissue introduced unmeasureable and nonremovable quantities of systematic error (bias). For example, a biased study design might quantify the volume of cells in tissue using the Euclidean formula for quantifying volume of a sphere ($V = 4\pi r^3/3$). The result will be biased by the degree of "nonsphericity" of the cells because the approach assumes that the cells are spheres. The problem with this approach is that biological structures are arbitrarily shaped, exist in unknown distributions in tissue, and vary under the influence of evolutionary and other forces. Rather than imposing Euclidean formulas on biological objects, stereologists developed bias-free probes and highly efficient systematic random sampling to quantify the same parameters (N, L, S, V), thereby eliminating the need for assumptions, models, and correction factors (Baddeley 1982, 1993, 1999; for reviews, see Gundersen 1984; Braendgaard and Gundersen 1986; Gundersen 1986; Howard and Reed 2005; Mouton and Gordon 2010; Mouton 2002, 2011, 2013; West 2012).

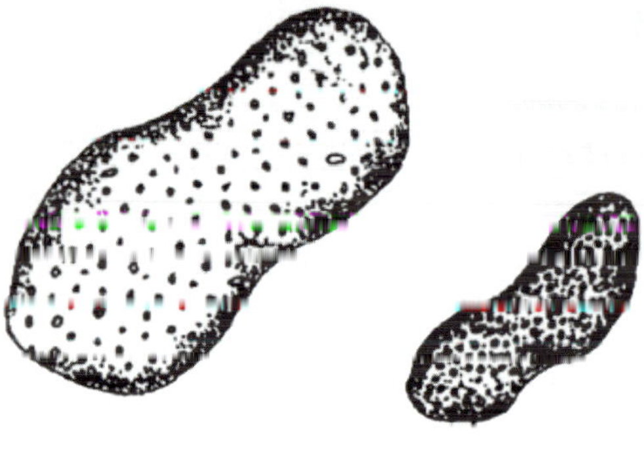

FIGURE 16.1
Shrinkage of reference space changing density without changing total number of contained objects.

16.2.1 The Reference Trap

Consider Monaco, a tiny country along the southern border of France with one of the highest population densities in the world. Monaco holds this distinction not because of its large population (about 36,400 people in total) but because all its citizens live in less than 1 square mile. If the population of Monaco were to double in size, the population density would fall by one-half, despite no change in the number of people. Thus, a country with less than 40,000 citizens has one of the highest population densities in the world. Assuming that density represents total values is called the reference trap. In morphometry, density estimates for number of objects per unit volume (N_v) represent the ratio of two parameters that vary independently: number (N) and the containing reference volume (V). The reference trap arises from the false assumption that shrinkage and/or expansion in the reference volume remains constant during tissue processing (fixation, freezing, dehydration) (Figure 16.1).

Factors such as aging, disease, and experimental lesions cause differential changes in reference volumes between groups (Haug 1984; Mouton 2011; West 2012). The reference trap applies not only to number density but also to length density (length per unit volume, L_v), surface density (surface per unit volume, S_v), and volume density (volume per unit volume, V_v). For this reason, the goal of design-based stereology studies is to quantify total parameters: N, L, S, and V.

16.2.2 Accuracy versus Precision

Design-based methods are accurate because as more sampling is carried out, the sample estimates progressively converge on the true or expected values for the population. As shown in Figure 16.2, biased (assumption- and model-based) methods are inaccurate because their central tendency does not converge on the true value (center of target).

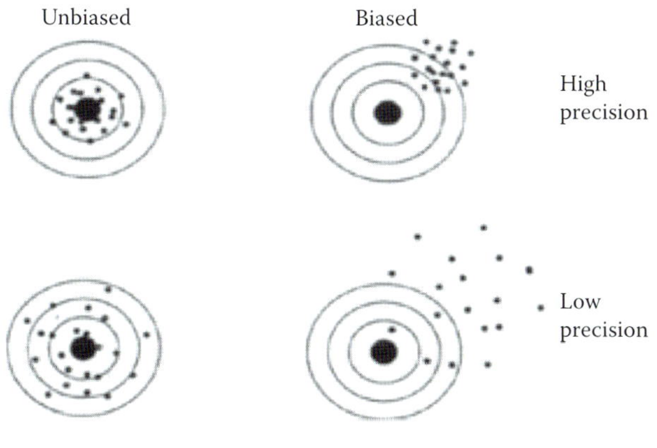

FIGURE 16.2
Schematic diagram of accuracy and precision for unbiased and biased methods.

Of further note is that no amount of sampling effort can produce accurate results for biased approaches (the upper and lower right targets). If the source of bias is quantifiable, it can be easily eliminated; otherwise, no correction factors are possible. A further advantage is that by optimizing the study design, appropriate precision can be achieved to find relevant differences between groups. Note that increasing sampling effort leads to greater precision for biased and unbiased methods.

While Figure 16.2 might seem to imply that the ideal study design would be an unbiased method with high precision, as shown by the target in the upper left, this turns out not to be the case in most studies. The adage "do more less well" was proposed by the esteemed Swiss stereologist Ewald Weibel to describe the most rational and efficient design for stereology studies (Gundersen and Osterby 1981). According to this notion, for the majority of biological studies, the most accurate, precise, and efficient approach uses the lowest amount of sampling effort (lower left target, Figure 16.2) to reveal significant differences between groups. This ideal level of precision is typically achieved by analyzing 100 to 200 probe locations spaced in a systematic random manner across eight to 12 sections through a well-defined reference space. When applied to about seven to eight cases (animals) per group, this efficient study design will usually reveal large to moderate group differences, if they exist.

16.3 Quantifying Reference Volume (V_{ref}) Using the Cavalieri Point Counting Method

About 400 years ago an Italian mathematician and student of Galileo Galilei, Buenaventura Cavalieri (1966), developed a straightforward design-based method to determine the volume of any arbitrary-shaped object. This method, known in Cavalieri's time as the theory of indivisibles (Figure 16.3, Equation 16.1), estimates region volume from a random set of equidistant slices through an object, with no assumptions about size, shape, or orientation of the region of interest (Cavalieri 1966).

$$\begin{aligned} V_{ref} &= k \cdot \Sigma V \\ &= k \cdot [(t \cdot A_1) + (t \cdot A_2) + (t \cdot A_3) + \ldots (t \cdot A_i)] \\ &= k \cdot \text{mean } t \cdot (A_1 \cdot A_2 \cdot A_3 \ldots A_i) \\ &= T \cdot \Sigma A \end{aligned} \tag{16.1}$$

where V = total volume of the reference space; k = sampling interval (every kth sampled section); ΣA = sum of the reference areas on the cut surface of each slab; mean t = average section thickness; and T = slab thickness (k · mean t).

Repeated volume estimation on individuals from the target population ensures that the sample estimate of mean V_{ref} converges on the true value. Numerous studies have empirically confirmed the accuracy of the Cavalieri method compared to the Archimedes water displacement method, the gold standard for volume measurement (Gundersen and Jensen 1987; Regeur and Pakkenberg 1989; Subbiah et al. 1996). Although Figure 16.3 shows the schematic reference volume sliced into three sections, for biological studies, eight to 12 sections per reference volume is optimal (Gundersen et al. 1999). Other requirements for the Cavalieri method for volume estimation include making the first slice at random to ensure equal

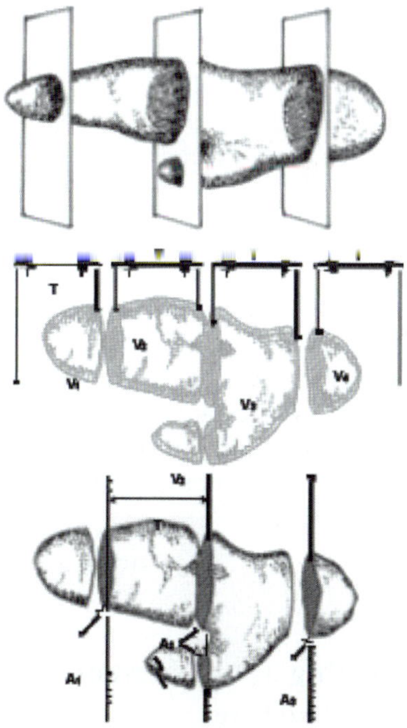

FIGURE 16.3
Sectioning a reference space for Cavalieri volume estimation.

sampling probability for all parts of the reference volume and that all slices are cut parallel to each other. If the reference volume consists of several parts, the most efficient approach is to embed them together prior to sectioning. Accuracy of the estimate is not affected by the orientation of the sections (coronal, axial, etc.); however, consistent orientation of sectioning leads to maximum efficiency.

16.3.1 Section Thickness

The mean section thickness, t, in Equation 16.1 refers to postprocessing thickness, that is, after fixation, sectioning, staining, and cover slipping is complete. Accurate determination of mean t is carried out with the thin focal plane obtained by high-magnification oil immersion objectives (60× to 100×). The instrument setting on the microtome, also known as block advance, is not used for the Cavalieri method, because less section thickness quantifies an intermediate volume, that is, after agonal changes and aldehyde fixation but before shrinkage from alcohol and other dehydrating reagents.

16.3.2 Area Estimation (Σ*A*) by Point Counting

Point counting provides an accurate and efficient method to estimate the sum of areas (ΣA) for estimation of the Cavalieri volume. To ensure random sampling, point counting requires random placement of the point grid on the reference area. As shown in Equation 16.2, adesign-based estimate of the sum of areas [$\Sigma A = [a(p)/M^2) \cdot \Sigma P$] is used to estimate V_{ref} using the Cavalieri point counting method,

$$V_{ref} = T \cdot \Sigma A$$
$$V_{ref} = T \cdot \Sigma P \cdot [a(p)/M^2] \quad (16.2)$$

where ΣP = number of points (P) falling within the reference area across all sections and $a(p)/M^2$ = area per point on point grid corrected for area magnification (M^2). The optimal spacing between points on the point grid allows a total of at least 150 and 200 point–object intersections on eight to 12 sections through the reference volume. For a detailed application of the Cavalieri point counting method for volume estimation, see Mouton et al. (1998).

16.4 Quantifying Total Number of Microstructures

Sectioning tissue into thin sections, a requirement for microscopic visualization, causes many objects (cells) to appear as profiles. As first shown by Delesse (1847), the number of 2-D profiles per unit area on a tissue section is not related to the true number of objects in the reference volume; that is, $N_A \neq N_V$. Almost a century ago, S.D. Wicksell introduced the corpuscle problem to explain Delesse's observation about the nonequality between the number of arbitrary-shaped objects and their profiles on thin tissue sections (Wicksell 1925). As shown in Figure 16.4, the corpuscle problem arises from three sources of bias—size, shape, and orientation.

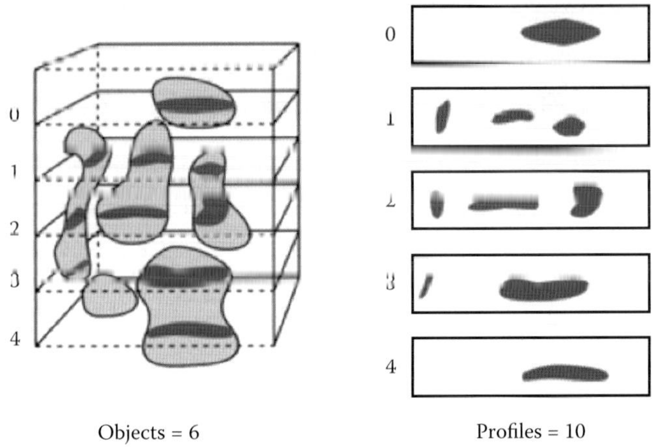

FIGURE 16.4
The corpuscle problem. For arbitrary-shaped 3-D objects sampled by a 2-D section, the number of profiles per unit area (N_A) does not equal the number of objects per unit volume (N_V).

The probability of a 3-D cell appearing as a cell profile on sections is not related to the number of cells but follows a multifactorial function related to the cell's size, complexity, and orientation of its long axis relative to the sectioning plane (Wicksell 1925). Considering the high biological variability of these factors, no known mathematical function can relate the number of profiles to the number of objects without assumptions and models that do not apply to biological objects, for example, "assume cells are spheres." Thus, counting cell profiles on a standard thin section leads to a biased estimate for cell number.

16.4.1 The Disector Principle

In 1984, an anonymous scientist using the nom de plume D.C. Sterio published the disector principle, the first and only solution to the corpuscle problem (Sterio 1984). A disector is a virtual 3-D volume used to quantify the numerical density (N_v), the total number of objects in a defined reference volume, without bias due to size, shape, or orientation of the objects. The disector method was initially proposed as two adjacent sections separated by a known distance, with one section as the reference section and the other section as the lookup section (physical disector). Objects are counted when they fall within an unbiased counting frame on the reference section, but only if they do not intersect the adjacent lookup section. Provided the distance between adjacent sections in the disector pair is less than the caliper height of smallest counting item (cell, nucleus, nucleolus), the disector method is unbiased for total object number (Sterio 1984; Colonnier and Beaulieu 1985; DeGroot and Bierman 1986).

In 1986, Gundersen introduced a simple and practical modification to the disector principle, the optical disector, which involves thin focal-plane optical scanning in the z-axis of relatively thick tissue sections (~30 to 50 μm). Like the physical disector, the optical disector uses the unbiased counting frame and counting rules to avoid bias from edge effects (Gundersen 1977). The optical disector requires objectives with relatively high resolution optics, such as 60× to 100× oil immersions with a numerical aperture of 1.3 or higher. The number of objects per unit volume (numerical density) is calculated for the optical disector using the formulas shown in Equations 16.3 and 16.4.

$$N_V = \Sigma Q^{-}/(\#\,\text{disectors} \cdot \text{volume of single disector}) \qquad (16.3)$$

where N_V = numerical density; ΣQ^{-} = number of objects counted; and volume of single disector = disector frame area · disector height.

As with the physical disector (Sterio 1984), the total number of objects (total N) for the optical disector is the product of N_V and the reference volume, V_{ref}.

$$\text{Total } N = N_V \cdot V_{ref}. \qquad (16.4)$$

The product of N_V and V_{ref} as shown in Equation 16.4 provides an unbiased estimate of total number, N. This two-stage approach is valid for calculation of absolute values for the other three first-order stereology parameters from their respective density estimates: surface area (total $S = S_V \cdot V_{ref}$); length (total $L = L_V \cdot V_{ref}$); and volume (total $V = V_V \cdot V_{ref}$). The cancellation of tissue deformation effects, however, requires estimation of both the density estimates, for example, N_V and V_{ref}, to be carried out on tissue processed in exactly the same manner.

16.4.2 The Fractionator

Gundersen's (1986) fractionator approach is a straightforward sampling scheme used with the disector principle to estimate first-order stereology parameters from stained tissue sections. Compared to the two-stage method (Equation 16.4), the fractionator approach is entirely unaffected by tissue deformation (shrinkage/expansion). In the early 1990s, Gundersen's group first introduced the optical fractionator as the most efficient method to date for quantifying total object number (West et al. 1991). The results for total N are calculated from the number of objects counted (ΣQ^{-}) and the reciprocals of known sampling fractions, as shown in Equation 16.5:

$$N = \Sigma Q^{-} \cdot 1/\text{SSF} \cdot 1/\text{ASF} \cdot 1/\text{TSF} \qquad (16.5)$$

where ΣQ^{-} = total number of objects counted; SSF = section sampling fraction = number of sections sampled ÷ total number of sections through the reference volume; ASF = area sampling fraction = total area sampled ÷ total reference area on all sampled sections; and TSF = thickness sampling fraction = height of the disector ÷ total section thickness.

When using a motorized stepping motor to count relatively rare events (about 2000 or fewer total objects of interest), reducing the x–y step distance is a practical though inefficient approach to avoid skipping the majority of objects. Decreasing the step distance, however, increases the number of disector probes, which in turn increases the time and effort to achieve a stable number estimate. The Rare Event Protocol (REP), a variation by Mouton (2011) on the fractionator scheme (Gundersen 1986; West et al. 1991), allows for accurate, precise, and efficient quantification of well-stained rare events in tissue sections. Briefly, at relatively low magnification (40×) the joystick control for a motorized stage is used to scan the entire section in a raster pattern until a rare event is located. The data collector switches to high magnification (63× to 100×) to confirm that the object comes into best focus at the top surface of the section; if so, the object's topmost point falls within the adjacent section, and the object is not counted. If the object comes into sharpest view after focusing into the z-axis, that, below the top plane of the sections, the object is counted. This procedure is repeated across all the sampled sections. The result for total number of rare events is calculated as the product of the number of objects counted and the reciprocal of the SSF, as shown in Equation 16.6.

$$\text{Total } N = \Sigma Q^{-} \cdot F_1 \qquad (16.6)$$

where ΣQ^{-} = the total number of objects with the topmost point within the sampled sections and F_1 = 1/section sampling fraction (SSF). Because the user scans the entire reference volume on each sampled section, the ASF and the TSF from Equation 16.5 are equal to 1. For example, if 150 objects are counting on a sample of sections that represents every fifth section, that is, one out of every five sections through the reference space, then the total count is 750 objects (150 · 1/1/5 = 150 · 5 = 750). The REP restores the sampling efficiency to the expected time for counting nonrare events using a stepping motor, that is, about 1 h per case.

16.5 Error Analysis for Stereology Studies

In contrast to widgets built on assembly lines, the morphological features of organic tissues and their contained microstructures vary under the control of evolutionary and other forces. This random variation (error) adds to the expected method error from sampling a fraction of the total number of sections, locations, and objects within the reference volume. Because biological variation and method error occur at random, the total observed error can be partitioned by subtraction of the calculated method error from the calculated total observed variation. The difference between these measures of variation is biological variation. Partitioning the observed variation into its component sources allows the sampling effort to be optimized for maximum efficiency, as outlined in the following section.

Partitioning variance in a pilot study allows sampling time and effort to be directed at the sampling levels contributing most error to the coefficient of variation (CV). From an initial analysis of $n = 2$–3 cases per group, the first step in partitioning error is to quantify the total observed error or CV (CV = standard deviation/mean). In contrast to CV and CE, BV cannot be directly estimated but can calculated from the difference of CV and CE.

$$CV^2 = BV^2 + CE^2 \quad (16.7)$$

where CV^2 = total observed variance (coefficient of variance); BV^2 = biological variance; and CE^2 = error variance (sampling variance) calculated using the quadratic approximation formula (Gundersen et al. 1999).

Partitioning the observed variation of stereology parameters provides a firm foundation for managing resources, for example, when preliminary statistics suggest a strong trend, for example, $P < .10$, toward significant group differences. The specific issue is whether the most efficient approach for reducing CV is decreasing the BV by analyzing more individuals from the population or reducing the mean CE by analyzing more locations within the cases already processed to sections, the least expensive option in terms of time, effort, and material resources. If error analysis shows mean CE to be small relative to BV, then the rational strategy for reducing the CV is to increase sampling within the already processed sections. If CE is already low relative to BV, however, the only choice is to reduce the BV by processing and analyzing more individuals from each group.

16.6 Computerized Stereology

The advent of low-cost, high-resolution digital imaging and computer controlled microscope stages in the past two decades led to the development of computerized stereology systems such as the Stereologer from the Stereology Resource Center (SRC, Tampa, FL). The Stereologer system uses a standard microscope equipped with low/high-power objectives (2× to 5×/63× to 100×), a motorized XYZ stepping motor to ensure objective tissue sampling, and moderate digital imaging resolution to allow unambiguous identification of objects of biological interest in stained sections. In practice, data collection using the Stereologer consists of two standardized steps. First, the user outlines the reference space (region of interest) on each stained tissue section using a low-power objective. In the second step, the user switches to high power, the motorized stage steps through the tissue in a systematic random manner, and the user clicks on intersections between unbiased sampling probes and the biological objects of interest. This process is repeated at about 100 to 200 systematic random locations in the x–y plane of eight to 12 sections through each reference space. Results with recommendations for optimizing sampling parameters in subsequent processes are automatically calculated. These software-controlled, microscope-based imaging systems have dramatically reduced the time and labor required previously through a combination of automatic procedures, from automatic section thickness measurement to automatic estimation of stereology parameters (Mouton et al. 2005; Elozory et al. 2012; Chaudhury et al. 2013). A typical study of total cell number and mean cell volume (MCV) requires about 45 min per case (animal, subject) as compared to weeks or months using previous manual methods.

16.7 Applied Stereology

The section provides a practical demonstration of design-based stereology concepts from our recent publication of abnormal findings in postmortem brains from autistic children (Courchesne et al. 2011c).

Numerous neuroimaging studies in autistic and age-matched children have reported early brain overgrowth in the prefrontal cortex (PFC), a brain region that mediates the majority of symptoms associated with autism including social interaction, communication skills, affective behavior, and cognitive function (Courchesne et al. 2001, 2003, 2007; Sparks et al. 2002; Redcay and Courchesne 2005; Hutsler et al. 2007; Dawson et al. 2007;

Amaral et al. 2008; Hazlett et al. 2011). One reason that the neural defects underlying autism remain a mystery is that the average age of postmortem cases analyzed in more than 50 studies of the autistic brain is 22 years (Courchesne et al. 2007; Amaral et al. 2008). The results from these studies of brains from older autism cases indicate reduced numbers of neurons in the cortex, amygdala, and cerebellum; proapoptotic and neuroimflammatory molecular signals; and reduction in cortical minicolumn size. The majority of MRI studies of the older autistic brain report cortical thinning, reduced size of the corpus callosum, reduced fractional anisotropy (FA) values in white matter fiber tracts, and either reduced brain size or absence of size differences as compared to normal (Courchesne et al. 2007, 2011a,b; Amaral et al. 2008). However, such findings in the adult autistic brain do not explain why infants and toddlers with autism show brain overgrowth. We designed a stereology study to test the hypothesis that early brain overgrowth in autism is the result of an excess numbers of neurons (Courchesne et al. 2001, 2003; Courchesne and Pierce 2005). To address this question, we used computerized stereology to quantify total neuron number (total *N*) and MCV in two subregions of the PFC: the dorsolateral (DL-PFC) and medial (M-PFC).

The extreme scarcity of young control and autistic postmortem cases is widely recognized and even more pronounced with regard to tissue suitable for modern stereological study of the entirety of DL-PFC and M-PFC. Material for this study included eight to 12 systematic random sampled sections stained with Nissl (cresyl violet) from $n = 13$ autopsied brains of males aged 2 to 16 years, including $n = 7$ brains from autistic males and $n = 6$ brains from age- and gender-matched controls. The cases represent all brains from young autism cases and all young male controls with systematically sectioned and stained sections through the whole frontal cortex available from the Autism Tissue Program (ATP), the Autism Brain Atlas set, the UCSD autism center of excellence (ACE), the National Institute of Child Health and Human Development (NICHD), and the University of Maryland Brain and Tissue Bank. Diagnostic classification of autism was based on the results of postmortem autism diagnostic interview-revised (ADI-R), the standard method for autism postmortem research; autism diagnostic observation schedule (ADOS) administered prior to death; or supporting clinical information available on the ATP Portal. For control cases, little information was available from brain and tissue bank records or the ATP Portal.

Landmarking of DL-PFC and M-PFC was carried out blind to diagnostic membership; age of case; and the purpose, literature, and theories behind this study to ensure completely bias-free anatomic decisions. Anatomical delineation of the M-PFC and DL-PFC regions throughout their rostrocaudal extent was based on previous definitions (Semendeferi et al. 1997; Carper and Courchesne 2005), which were based on overall gross anatomy. Tracking of sulci and reliable boundaries were further refined/verified based on cytoarchitectonic criteria, including layer IV granularity and, for instance, the presence of Betz cells indicating where the precentral cortex should be differentiated from DL-PFC (Semendeferi et al. 1997; Carper and Courchesne 2005). M-PFC included the entire medial surface of one hemisphere starting at the most rostral section and extending caudally until the paracentral lobule (PCL), replacing the PFC dorsally, the cingulate (CG) at mid-dorsoventral levels, and the subcallosal area most ventrally. The DL-PFC included the lateral portion of the superior frontal gyrus, the middle frontal gyrus, the inferior frontal gyrus, and the frontal operculum. The most rostral sections also included lateral portions of the frontopolar gyri. The dorsolateral surface was thus included starting rostrally and extending to a ventral termination at orbital PFC regions followed caudally by the insula/insular gyrus, and, within this gross definition, precentral and postcentral gyri were excluded. To simplify border delineation at each transition, a straight line was drawn normal to the curvature of the cortical surface. For further details on reference space boundaries for this study, see Courchesne et al. (2011c).

The stereological studies of total *N* and MCV within the DL-PFC and M-PFC were carried out with assistance from the Stereologer system by trained technicians blind to diagnostic membership; age of case; and the purpose, literature, and theories behind this study to effectively eliminate user bias. The optical fractionator method (West et al. 1991) was used to estimate the total number of neurons in the entire DL-PFC and M-PFC. The guard zone was >8 μm at the top and bottom of the section, and the optical disector height was 10 μm. The spacing between disectors was between 2.3 and 5 mm, depending on region size (typically 5 mm for lateral and 3.2 mm for medial, with a smaller distance used for some of the smaller brains to ensure sufficient sampling). The total neuron number was calculated as the product of the number of neurons counted within all disectors and the reciprocals of three sampling fractions: section sampling fraction, area sampling fraction, and thickness sampling fraction. Sampling was carried out at a minimum of 100 locations in the x–y axis across eight to 12 sections per region to a high sampling stringency (CE < 0.10) according to Gundersen et al. (1999).

Glial cells were distinguishable from neurons during the process of thin focal-plane scanning in the z-axis at high magnification (100× oil, numerical aperture of 1.4). Cells with noticeably more cytoplasm, a prominent nucleolus, a distinct nuclear membrane, and other characteristics not shared by either astrocytes or microglia were counted as neurons. Nissl-stained neurons with nucleoli touching the exclusion planes of the disector frame or in first focus within the guard zones were not counted. Neurons were counted when the topmost nucleolus first came into focus within the disector counting frame. About one of every 250 to 300 cells was difficult to categorize as a neuron or glial cell, which contributed to an insignificant percentage of uncertainty (<0.4% variability). Figure 16.5 shows the morphological differences between neurons and glial cells on 2-D images chosen to illustrate the most challenging cases.

The MCV of neurons was estimated using the rotator method of Jensen and Gundersen (1993) with assistance from the Stereologer software. After each cell was counted by the optical fractionator method, the data collector clicked on the cell, which caused a line to appear through the cell at a random angle. The user clicked at the top and bottom of the cell, causing three lines to appear in a systematic random orientation perpendicular to the random line. Finally, the user clicked at the boundary between the cell membrane and the three lines. Due to the random placement of the line probes, their average length was directly proportional to the MCV for each neuron. Since MCV estimates were, by necessity, carried out on tissue sectioned in a single (coronal) plane, that is, rather than planes selected at random across all possible orientations, a small orientation bias could be present in the MCV estimates. If present, however, this bias would be negligible with respect to the findings in this study.

Data were analyzed using SPSS software (SPSS Inc., Chicago, Illinois) to assess within-case, between-case, and main and interaction effects of region (DL-PFC, M-PFC) and diagnosis (autism, control). A multivariate analysis of covariance (MANCOVA) was done using diagnosis as the independent variable, age as a covariate, and brain data (i.e., deviation of brain weight from age norms, neuron counts, microglia counts, neuron volume) as the dependent variables. A full model with age, diagnosis, and their interaction was initially evaluated. Diagnosis remained a significant factor, while age and the interaction term were not significant. Pearson correlations were used to examine the relationship between neuron counts for the two regions, brain weight deviance, and age.

This study showed that at young but not older ages, the autistic brains were, on average, substantially heavier than controls. The seven young 2- to 16-year old autism cases exceeded normal age means by 212 g, while the two adult autism cases had brain weights slightly less than (−51 g or −3%) normal mean for age. As predicted by previous literature across age (Redcay and Courchesne 2005), brain weight strongly increased with age in controls ($r = 0.85$, $P = .008$, one-tailed) but did not change in autistic cases ($r = 0.4$, not significant). These autism and control brain weights reasonably represent the larger population of autistic and control cases reported in the literature (Redcay 2005).

Our stereology studies revealed 65% more neurons in DL-PFC in the autistic cases compared to the age- and gender-matched control group (see Figure 16.6a) [1.45 ± 0.280 billion vs. 0.88 ± 0.143 billion, $F(1,14) = 10.5$, $P = .006$] and 29% more neurons in M-PFC (Figure 16.6b) [0.356 ± 0.025 billion vs. 0.284 ± 0.035 billion, $F(1,14) = 11.2$, $P = .005$]. Significant group differences between autism and control remained after controlling for post-mortem interval. Overall, there were 65% more total neurons in PFC of autism cases as compared with controls (Figure 16.6c).

There was a strong and significant correlation between the number of DL-PFC neurons and the number of M-PFC neurons across autistic and control cases ($r = 0.59$, $P = .016$, data not shown). Neuron counts in DL-PFC among the youngest autistic cases were nearly twice control levels (Figure 16.6b). M-PFC displayed a nonsignificant decline in the autistic group ($r = -0.19$, ns). There was a significant age-related decline in total number of DL-PFC neuron numbers in the autistic cases ($r = -0.63$, $P = .033$, one-tailed) but not in controls ($r = -0.07$, ns).

There was a significant correlation between total number of prefrontal neurons and magnitude of deviation

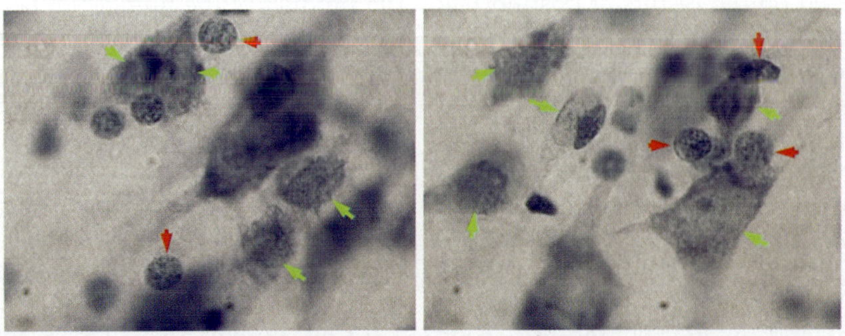

FIGURE 16.5
Morphological differences between neurons (green arrows) and glial cells (red arrows) on Nissl-stained sections.

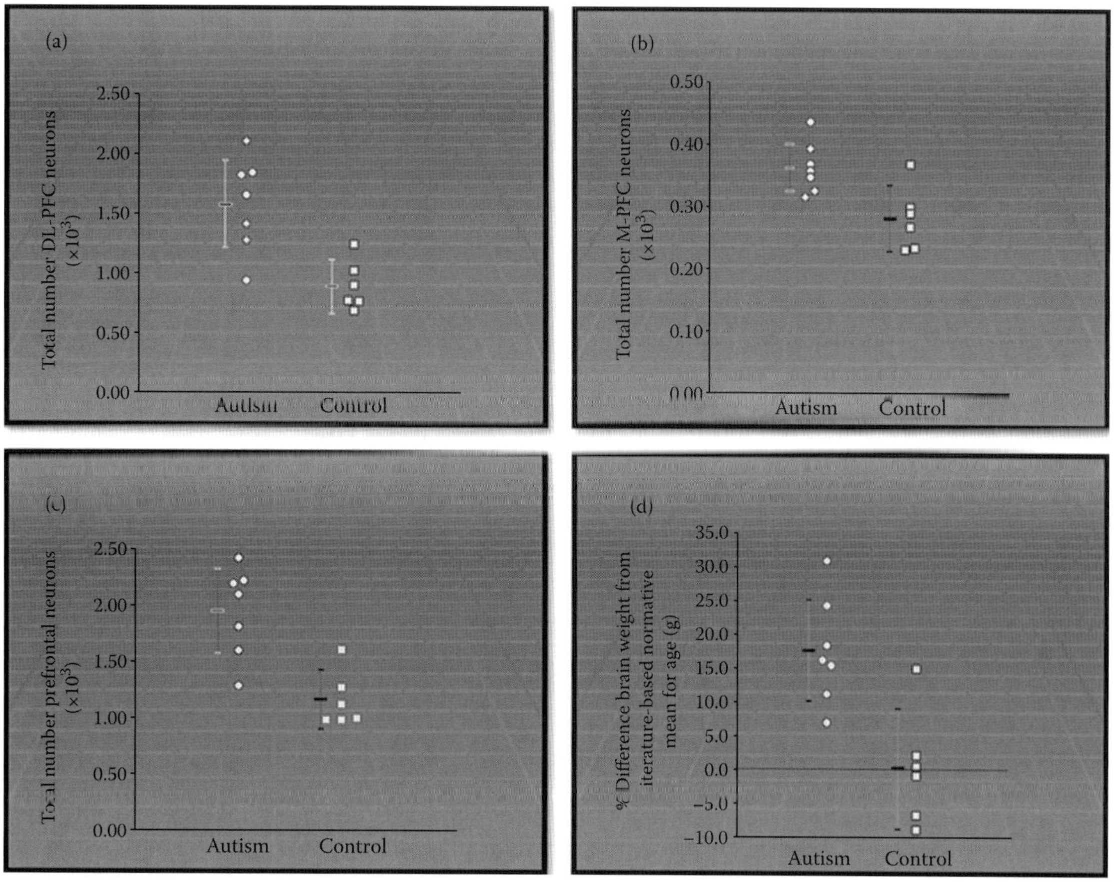

FIGURE 16.6
Results for autism cases (n = 7, male) and age-matched controls (n = 6, male) showing (a) 79% more DL-PFC neurons in autism cases ($P < .006$); (b) 29% higher total number of M-PFC neurons in autism cases ($P < .005$); (c) 67% more PFC neurons (DL + M) in autism cases ($P < .002$); and (d) 18% greater mean brain weight in autism cases compared to the literature-based mean for age ($P < .01$). For full details and results, see Courchesne et al. (2011c).

from normal mean brain weight for age ($r = 0.53$, $P = .035$, data not shown) driven primarily by DL-PFC ($r = 0.69$, $P = .003$) and, to a lesser degree, M-PFC effects ($r = 0.48$, $P = .057$). The higher total number of PFC neurons vis à-vis the larger postmortem brain weights further validates the blinded stereological analyses reported here for autism cases and controls.

In order to consider the question of the impact of neuron size on interpretations of brain tissue in autism, an analysis of covariance (ANCOVA) was performed using age as a covariate. Although mean neuron volume (MNV) was reduced slightly in autistic cases in both regions of interest, no statistically significant differences from normal were observed [DL-PFC, $F(1,12) = 0.34$. $P = .572$; M-PFC, $F(1,12) = 0.18$, $P = .6772$]. Examining the groups as a whole, there was no significant correlation between MNV and difference from age-matched average brain weight (DL-PFC, $r = -0.092$; M-PFC, $r = -0.127$, $P > .05$). There was also no significant correlation if the autism and control groups were considered as separate groups (all P values > .05). Likewise, there were no significant correlations between neuron volumes in either region of interest and neuron counts. Microglia counts in DL-PFC [$F(1,15) = 1.751$, ns] and M-PFC [$F(1,15) = 0.202$, ns] did not differ between groups and were not correlated with neuron counts. Further analyses revealed that null effects of microglia counts remained after controlling for PMI, autopsy brain weight, neuron numbers and neuron size (all P values > .05).

Normal brain growth and functional development in humans depend on a variety of foundational properties, one of which is establishing by around the time of birth a near-adult number of healthy neurons (Rabinowicz et al. 1996; Gohlke et al. 2007). Our results dovetail with this idea: The 2- and 3-year-old control cases in this study had a near-adult number of prefrontal neurons; thus, from age 2 years through young adulthood, neuron numbers were unchanged in controls. Therefore, one result of this study is that near-adult numbers of prefrontal neurons are indeed established during early brain development in humans. From this base of neurons, axon growth, synaptogenesis, and elaboration of connectivity occur in a gradual manner from the first postnatal years, when frontal circuits are sparse, on through 8 years of age, when connectivity has become highly elaborated (Conel 1939–1967; Huttenlocher 2002). Experience and learning guide axon growth and synapse selection during this period of time (Quartz and Sejnowski 1997).

These quantitative data reveal that autism deviates sharply from this normal pattern of brain development in two major ways. In autism at age 2 to 4 years, there is a substantial overabundance of prefrontal neurons, nearly twice the normal number. This major neuron excess suggests that molecular mechanisms that ordinarily regulate neuron numbers are globally abnormal in prefrontal regions of brains from autistic children. This overabundance of neurons must be due to defects in prenatal mechanisms that govern the number of neurons generated in the second trimester and/or removed by prenatal or perinatal apoptosis. It is important for future studies to assay neuron numbers in other cortical regions in autism using similarly rigorous stereological methods. An overabundance of prefrontal neurons will produce more, not fewer, connections (axons, dendrites, and synapses) during early development. By one calculation, the near doubling of neuron numbers could produce as many as four times the normal number of axons in the very young autistic brain (Ringo 1991). Autism, we argue based on the present data and other studies, begins as a disorder of both excess neuron numbers and aberrant overabundant connectivity, not disconnectivity.

These findings also point to a second critical neurodevelopmental abnormality in autism: the excess neuron numbers in the brains of autism cases appears to be removed across decades, unlike the normal brain, in which naturally occurring apoptosis removes excess rapidly across just a few months of prenatal and perinatal life. This abnormally slow age-related decline in prefrontal neuron number in autism must necessarily be accompanied by substantial removal and remodeling of axonal connections and synapses. The child with autism therefore faces years of neuronal apoptosis and substantial connectivity disassembly and, in the best of circumstances, a prolonged period of remodeling in an attempt to achieve improved circuit functioning. Postmortem studies of older children, adolescents, and adults with autism commonly point to degenerative and remodeling processes including microglia activation, age-related increases in haphazard neuron-to-neuron spatial organization, neuron loss, reduced minicolumn size, molecular and gene expression signals of proapoptosis, reduced long- but increased short-distance axons, synaptic remodeling, and neuroinflammation (Araghi-Niknam and Fatemi 2003; Vargas et al. 2005; Buxhoeveden et al. 2006; Schumann and Amaral 2006; Garbett et al. 2008; van Kooten et al. 2008; Morgan et al. 2010, 2012; Zikopoulos and Barbas 2010). These studies, we suggest, reflect not the origins of autism but also the response to the original neurodevelopmental defects consequent to pathologically overabundant neurons, axons, and synapses. In this light, it may be interesting to ask whether defects in synapse-relevant genes, several of which are risk factors for autism, would place an autistic child at a much greater disadvantage during circuit remodeling. In general, attention should be paid to the question of the genetics of recovery from autism as well as to the genetics that may cause autism in the first place.

Our stereology findings also support the view that early intervention is essential and beneficial to autistic children: removal and remodeling earlier in life when connectivity is still limited and malleable could preemptively reduce the number of abnormal connections and guide more functional connectivity among the excess neurons (Dawson et al. 2010). Conversely, interventions begun later would entail the much more difficult task of undoing and remodeling far more extensive networks of embedded dysfunctional connectivity; given the reduced neuroplasticity in older children and adults compared to early development, later interventions would not be expected to lead to optimal clinical outcomes. In testing the hypothesis of early overabundance of cortical neurons, these findings show a possible explanation for the behavioral, emotional, and cognitive disturbances in autism (Courchesne et al. 2001, 2003). This discovery opens new avenues for research into prenatal genetic and nongenetic causes of excess proliferation, the bases of abnormal connectivity, the reasons for neuroinflammation, proapoptosis and synaptic remodeling across the lifespan, recovery genetics, and how early identification and intervention may improve clinical outcome for affected babies and toddlers.

From its meager beginnings on a mountaintop retreat in Germany more than a half century ago, the field of stereology has successfully equipped biomedical scientists with accurate, precise, and efficient approaches for reliable morphometric studies of stained tissue sections and in vivo images. Current challenges include the need for enhanced image segmentation methods, the final obstacle to the realization of fully automatic computerized stereoanalysis that would accelerate progress on a wide variety of biomedical fronts. One exciting development in this arena is the recent demonstration that ensembles of segmentations (Chaudhury et al. 2013) in conjunction with design-based stereology can provide cancer pathologists with automatic data for predicting the malignancy potential from biopsy sections and cytology samples. With its team of multidisciplinary investigators, a common feature of many important breakthroughs, this study is prototypical for how a mathematically sound stereology approach can accelerate the journey toward greater longevity and improved health in the future.

References

Abercrombie, M. 1946. Estimation of nuclear population from microtome sections. *Anat Rec* 94:239–247.

Amaral, D., Schumann, C. M., Nordahl, C. W. 2008. Neuroanatomy of autism. *TINS* 31:137–145.

Araghi-Niknam, M., Fatemi, S. H. 2003. Levels of Bcl-2 and P53 are altered in superior frontal and cerebellar cortices of autistic subjects. *Cell Mol Neurobiol* 23:945–952.

Baddeley, A. J. 1982. Stochastic geometry: An introduction and reading list. *Int Statist Rev* 50:179–193.

Baddeley, A. J. 1993. Stereology and survey sampling theory. *Bull Intern Statis Inst, Proceedings 49th Session, Florence* 52(2):435–449.

Baddeley, A. J. 1999. Crash course in stochastic geometry, sampling and censoring. In: *Stochastic Geometry, Likelihood and Computation*, edited by O. E. Bandaroff-Nielsen, W. S. Kendall, and M. N. M. van Lieshout. Boca Raton, FL: Chapman & Hall/CRC, pp. 1–78.

Braendgaard, H., Gundersen, H. J. G. 1986. The impact of recent stereological advances on quantitative studies of the nervous system. *J Neurosci Methods* 18:39–78.

Buxhoeveden, D. P., Semendeferi, K., Buckwalter, J. et al. 2006. Reduced minicolumns in the frontal cortex of patients with autism. *Neuropathol Appl Neurobiol* 32(5):483–489.

Carper, R. A., Courchesne, E. 2005. Localized enlargement of the frontal cortex in early autism. *Biol Psychiatry* 57:126–133.

Cavalieri, B. 1966. *Geometria Indivisibilibus Continuorum. Bononi: Typis Clementis Ferronij, 1635.* Reprinted as Geometria degli Indivisibili. Torino: Unione Tipografico-Editrice Torinese.

Chaudhury, B., Phoulady, H. A., Goldgof, D., Hall, L. O., Mouton, P. R., Hakim, A., Siegel, E. M. 2013. An ensemble algorithm framework for automated stereology of cervical cancer. *Proceedings of the International Conference on Image Analysis and Processing (ICIAP)*, Berlin: Springer, Vol. 1, pp. 823–832.

Colonnier, M., Beaulieu, C. 1985. An empirical assessment of stereological formulae applied to the counting of synaptic disks in the cerebral cortex. *J Comp Neurol* 231:175–179.

Conel, J. L. 1939–1967. *The Postnatal Development of the Human Cerebral Cortex*, Vols. 1–8. Boston: Harvard University Press.

Courchesne, E., Karns, C., Davis, H. R. 2001. Unusual brain growth patterns in early life in patients with autistic disorder: An MRI study. *Neurology* 57:245–254.

Courchesne, E., Carper, R., Akshoomoff, N. 2003. Evidence of brain overgrowth in the first year of life in autism. *JAMA* 290(3):337–344.

Courchesne, E., Pierce, K. 2005. Brain overgrowth in autism during a critical time in development: Implications for frontal pyramidal neuron and interneuron development and connectivity. *Int J Dev Neurosci* 23:153–170.

Courchesne, E., Pierce, K., Schumann, C. M. et al. 2007. Mapping early brain development in autism. *Neuron* 56:399–413.

Courchesne, E., Webb, S. J., Schumann, C. M. 2011a. From toddlers to adults: The changing landscape of the brain in autism. In: *Autism Spectrum Disorders*, edited by D. G. Amaral, G. Dawson, and D. H. Geschwind. Oxford: Oxford University Press, 1456 pp.

Courchesne, E., Campbell, K., Solso, S. 2011b. Brain growth across the life span in autism: Age-specific changes in anatomical pathology. *Brain Res* 1380:138–145.

Courchesne, E., Mouton, P. R., Calhoun, M. E., Semendeferi, K., Ahrens-Barbeau, C., Carter, C., Pierce, K. 2011c. Neuron number and size in prefrontal cortex of children with autism. *JAMA* 306:2001–2010.

Dawson, G., Munson, J., Webb, S. J. et al. 2007. Rate of head growth decelerates and symptoms worsen in the second year of life in autism. *Biol Psychiatry* 61:458–464.

Dawson, G., Rogers, S., Munson, J. et al. 2010. Randomized, controlled trial of an intervention for toddlers with autism: The Early Start Denver Model. *Pediatrics* 125(1):17–23.

DeGroot, D. M. G., Bierman, E. P. B. 1986. A critical evaluation of methods for estimating the numerical density for synapses. *J Neurosci Meth* 18:79–101.

Delesse, M. A. 1847. Procede mecanique pour determiner la composition des roches. *C R Acad Sci Paris* 25:544–545.

Elozory, D. T., Bonam, O. P., Kramer, K., Goldgof, D., Hall, L., Mangual, O., Mouton, P. R. 2012. Automatic section thickness determination using an absolute gradient focus function. *J Microsc* 248(3):245–259.

Garbett, K., Ebert, P. J., Mitchell, A. et al. 2008. Immune transcriptome alterations in the temporal cortex of subjects with autism. *Neurobiol Dis* 30:303–311.

Gohlke, J. M., Griffith, W. C., Faustman, E. M. 2007. Computational models of neocortical neuronogenesis and programmed cell death in the developing mouse, monkey, and human. *Cereb Cortex* 17:2433–2442.

Gundersen, H. J. G. 1977. Notes on the estimation of the numerical density of arbitrary profiles: The edge effect. *J Microsc* 111:219–223.

Gundersen, H. J. G., Osterby, R. 1981. Optimizing sampling efficiency of stereological studies in biology: Or do more less well. *J Microsc* 121:65–73.

Gundersen, H. J. G. 1984. Stereology and sampling of biological surfaces. In: *Analysis of Organic and Biological Surfaces*, edited by P. Echlin. New York: Wiley Press, Vol. 71, pp. 477–506.

Gundersen, H. J. G. 1986. Stereology of arbitrary particles. A review of unbiased number and size estimators and the presentation of some new ones, in memory of William R. Thompson. *J Microsc* 143:3–45.

Haug, H., Kuhl, S., Mecke, E., Sass, N. L., Wasner, K. 1984. The significance of morphometric procedures in the investigation of age changes in cytoarchitectonic structures of human brain. *J Hirnforsch* 25:353–374.

Hazlett, H. C., Poe, M. D., Gerig, G. et al. 2011. Early brain overgrowth in autism associated with an increase in cortical surface area before age 2 years. *Arch Gen Psychiatry* 68(5):467–476.

Howard, C. V., Reed, M. G. 2005. *Unbiased Stereology*, 2nd ed. Oxford: BIOS Scientific Publishers.

Hutsler, J. J., Love, T., Zhang, H. 2007. Histological and magnetic resonance imaging assessment of cortical layering and thickness in autism spectrum disorders. *Biol Psychiatry* 61:449–457.

Huttenlocher, P. 2002. *Neural Plasticity: The Effects of Environment on the Development of Cerebral Cortex*. Boston: Harvard University Press.

Jensen, V. E. B., Gundersen, H. J. G. 1993. The rotator. *J Microsc* 170:35–44.

Morgan, J., Chana, G., Pardo, C. A. et al. 2010. Microglial activation and increased microglial density observed in the dorsolateral prefrontal cortex in autism. *Biol Psychiatry* 68:368–376.

Morgan, J. T., Chana, G., Abramson, I. et al. 2012. Abnormalities in microglial and neuronal spatial organization in the dorsolateral prefrontal cortex in autism. *Brain Res* 1456:72–81.

Mouton, P. R., Martin, L. J., Calhoun, M. E., Dal Forno, G., Troncoso, J. C., Price, D. L. 1998. Cognitive decline strongly correlates with cortical atrophy in Alzheimer's dementia. *Neurobiol Aging* 19:371–377.

Mouton, P. R. 2002. *Principles and Practices of Unbiased Stereology: An Introduction for Bioscientists*. Baltimore, MD: The Johns Hopkins University Press.

Mouton, P. R., Durgavich, J., Ingram, D. K. 2005. Automatic estimation of size parameters using verified computerized stereoanalysis. *Image Anal Stereol* 24:1–9.

Mouton, P. R., Gordon, M. 2010. Stereological and image analysis techniques for quantitative assessment of neurotoxicology. In: *Neurotoxicology*, 3rd edition, edited by G. J. Harry, and H. A. Tilson. London: Informa Press.

Mouton, P. R. 2011. *Unbiased Stereology: A Concise Guide*. Baltimore, MD: The Johns Hopkins University Press.

Mouton, P. R. 2013. *Neurostereology*, Boston: Wiley-Blackwell Press.

Quartz, S. R., Sejnowski, T. J. 1997. The neural basis of cognitive development: A constructivist manifesto. *Behav Brain Sci* 20(4):537–556.

Rabinowicz, T., deCourten-Myers, G. M., Petetot, J. M. et al. 1996. Human cortex development: Estimates of neuronal numbers indicate major loss late during gestation. *J Neuropathol Exp Neurol* 55(3):320–328.

Redcay, E., Courchesne, E. 2005. When is the brain enlarged in autism? A meta-analysis of all brain size reports. *Biol Psychiatry* 58:1–9.

Regeur, L., Pakkenberg, B. 1989. Optimizing sampling designs for volume measurements of components of human brain using a stereological method. *J Microsc* 155:113–121.

Ringo, J. L. 1991. Neuronal interconnections as a function of brain size. *Brain Behav Evol* 38:1–6.

Schumann, C. M., Amaral, D. G. F. 2006. Stereological analysis of amygdala neuron number in autism. *J Neurosci* 26(29):7674–7679.

Semendeferi, K., Damasio, H., Frank, R., VanHoesen, G. W. 1997. The evolution of the frontal lobes: A volumetric analysis based on three-dimensional reconstructions of magnetic resonance scans of human and ape brains. *J Hum Evol* 32(4):375–388.

Sparks, B. F., Friedman, S. D., Shaw, D. W. et al. 2002. Brain structural abnormalities in young children with autism spectrum disorder. *Neurology* 59:184–192.

Sterio, D. C. 1984. The unbiased estimation of number and sizes of arbitrary particles using the dissector. *J Microsc* 134:127–136.

Subbiah, P., Mouton, P. R., Fedor, H., McArthur, J. M., Glass, J. D. 1996. Stereological analysis of cerebral atrophy in human immunodeficiency virus-associated dementia. *J Neuropath Exp Neurol* 55:1032–1037.

van Kooten, I. A., Palmen, S. J., von Cappeln, P. et al. 2008. Neurons in the fusiform gyrus are fewer and smaller in autism. *Brain* 131:987–999.

Vargas, D. L., Nascimbene, C., Krishnan C. et al. 2005. Neuroglial activation and neuroinflammation in the brain of patients with autism. *Ann Neurol* 57(1):67–81.

West, M. J., Slomianka, L., Gundersen, H. J. 1991. Unbiased stereological estimation of the total number of neurons in the subdivisions of the rat hippocampus using the optical fractionator. *Anat Rec* 231:482–497.

West, M. J. 2012. *Basic Stereology for Biologists and Neuroscientists*. New York: Cold Spring Harbor Press.

Wicksell, S. D. 1925. The corpuscle problem. A mathematical study of a biometric problem. *Biometrika* 17:84–89.

Zikopoulos, B., Barbas, H. 2010. Changes in prefrontal axons may disrupt the network in autism. *J Neurosci* 30(44): 14595–14609.

17

Brain Shift Compensation via Intraoperative Imaging and Data Assimilation

Songbai Ji, Xiaoyao Fan, David W. Roberts, Alex Hartov, Timothy J. Schaewe, David A. Simon, and Keith D. Paulsen

CONTENTS

Image-guided neuronavigation based on coregistered preoperative magnetic resonance images has become the standard of care in many open cranial surgeries. Because the brain deforms even at the start of surgery, brain shift is widely recognized as the single most important factor that degrades navigational accuracy. Intraoperative imaging techniques such as volumetric true 3D ultrasound and stereovision capture subvolumes of the brain, but they, alone, are not sufficient to provide whole-brain image data for navigation. Using a biomechanical model to assimilate data from these intraoperative images and generate a model-updated magnetic resonance image volume (uMR) substantially increases their utility in image guidance, especially when uMR is integrated into a commercial navigational system. This chapter reports the important elements of the techniques required to produce sparse displacement data from intraoperative images and to generate uMR through model-based data inversion. Applications of these techniques are illustrated in actual patient cases to demonstrate the feasibility of the model-based brain shift compensation strategy for clinical deployment in the operating room.

17.1 Introduction

Image-guided neuronavigation based on coregistered preoperative magnetic resonance (pMR) images has become the standard-of-care in many open cranial procedures because of the improved surgical accuracy and patient outcomes that can be achieved. After coregistering the patient's head in the operating room (OR) with pMR to establish a spatial correspondence between anatomy in the physical and image spaces, localization of tissue of interest (e.g., tumor) in physical space is aided by navigation within the image space. Successful application of this image guidance technology has proven to be critical for brain tumor resections, where it has become increasingly evident that surgical accuracy impacts patient prognosis [1–3]. Clearly, successful application of image guidance technology depends on the level of registration accuracy or spatial correspondence that can be achieved between the anatomy of interest in the OR and its image data. Determining an accurate initial patient registration at the start of surgery is well established and is relatively easy and reliable using skull-implanted [4] or skin-affixed fiducials [5] (e.g., an accuracy of 1 mm or 3–5 mm in terms of fiducial registration error) or even without fiducials [6]. However, the brain often deforms even at the start of surgery, compromising registration between the patient in the OR and pMR. Because the magnitude of brain shift can be large (e.g., 20 mm or more at the cortical surface [7,8]) and the deformation pattern is dynamic (more significant as surgery progresses [9]) and complex (not necessarily correlated with gravity or equal in magnitude in the two hemispheres [10]), *maintaining* accurate patient registration throughout surgery is challenging.

To compensate for brain shift, intraoperative images such as volumetric true 3D ultrasound (3DUS [6,11–13]), stereovision [14–17], and laser range scanning [18–20], among others, have been used to observe brain shift during surgery. However, a limitation of these imaging techniques is that only a part of the brain (as opposed to the entire volume) is observable (e.g., a pyramid-shaped subvolume for 3DUS and exposed surface at the craniotomy for intraoperative stereovision [iSV]). A model-based data inversion scheme is attractive for brain shift compensation because it integrates displacements obtained from intraoperative images and estimates whole-brain deformation from which to generate a model-updated MR (uMR) image volume for subsequent image guidance. This chapter describes important aspects of a data inversion scheme for producing uMR views based on displacements derived from 3DUS and iSV. Applications of these techniques in actual patient cases are presented to illustrate the clinical feasibility of producing uMR during surgery for updated image guidance.

17.2 Image-Based Brain Shift Compensation

Intraoperative imaging is effective in compensating for brain shift during surgery. Intraoperative magnetic resonance (iMR; e.g., refs. [21–24]) captures whole-brain features and provides superior delineation of soft tissue. However, its substantial capital cost, intrusion on surgical workflow, and often long image acquisition times are obstacles to its gaining wide acceptance. Intraoperative ultrasound (iUS; e.g., refs. [6,11,25–30]), in contrast, offers real-time image acquisition and is easy to implement with low cost. However, it samples partial volumes of the brain and suffers from relatively low soft tissue contrast and greater operator dependence. Often, iUS is coregistered with pMR in order to improve the interpretation and understanding of intraoperatively imaged intracranial features. iSV (e.g., refs. [14–17,20,31]) and laser range scanning (e.g., refs. [18,19]) optically sample the exposed parenchymal surface to reconstruct texture-rich 3D surface profiles of the surgical field. A significant limitation of these intraoperative imaging techniques is that they sample either subvolumes of deep brain (iUS) or exposed surface profiles (iSV) but not the whole brain. Consequently, images generated by these techniques are typically not sufficient for direct image guidance. An attractive alternative is to assimilate parenchymal displacements obtained from these intraoperative images into a biomechanical model in order to generate full-volume uMR for subsequent neuronavigation.

17.3 Data Inversion in Model-Based Brain Shift Compensation

Biomechanical models are important adjuncts to intraoperative images that enable full-volume brain shift compensation for improved accuracy in neurosurgical image guidance (e.g., refs. [14,32–39]). A typical schematic of model-based brain shift compensation is illustrated in Figure 17.1, where the patient's head and intraoperative imaging sensors are continuously registered by a common tracking system. To improve model estimation accuracy, feature displacements detected from intraoperative images (e.g., iUS for features deep in the brain [34,35], and iSV [14,15,40] or laser range scanning [20,37] for exposed cortical surface features) can be used to guide the model solution, and the data inversion can be formulated in terms of known and unknown forcing conditions in which weighting factors are calculated to balance solution dependency on the forcing terms at observed (and unobserved) locations. An atlas-based inversion method can also be used to precompute basis solutions by varying the key determinants of brain shift (e.g., brain orientation, cerebrospinal fluid drainage, etc.) and estimating the actual intraoperative brain deformation via least-squares minimization between observed data and the precomputed atlas [37,38].

We have also developed a data-guided brain shift compensation strategy that iteratively estimates the right-hand-side forcing vector to minimize the misfit between model estimates and measured displacements (i.e., model–data misfit). The biomechanical model is based on consolidation theory [41], in which the brain is modeled as a biphasic porous medium. A detailed mathematical formulation has been developed previously [42,43] and is summarized by the following expressions:

$$\nabla \cdot G\nabla \mathbf{u} + \nabla \frac{G}{1-2\nu}(\nabla \cdot \mathbf{u}) - \alpha \nabla p = (\rho_t - \rho_f)\mathbf{g} \quad (17.1)$$

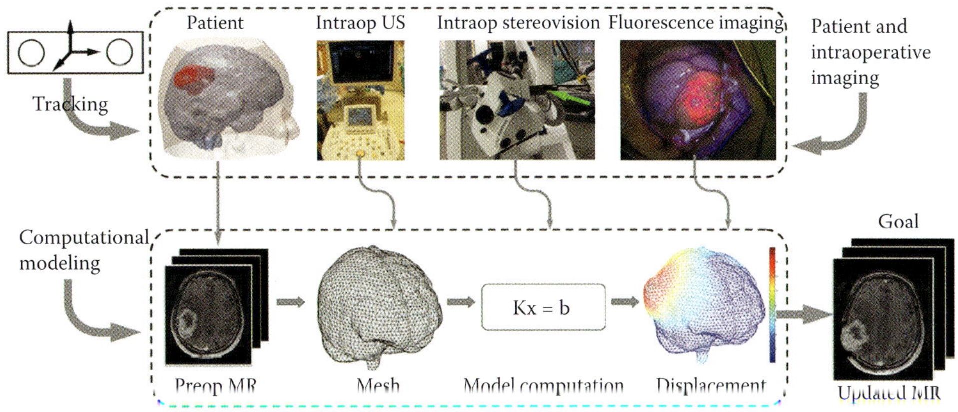

FIGURE 17.1
Schematic illustration of model-based brain shift compensation, where the patient's head and intraoperative imaging sensors are localized and coregistered through a common tracking system. Sparse displacement data are assimilated into a subject-specific biomechanical model to generate an updated MR volume for subsequent neuronavigation.

$$\alpha \frac{\partial}{\partial t}(\nabla \cdot \mathbf{u}) + \frac{1}{S}\frac{\partial p}{\partial t} - \nabla \cdot k\nabla p = \Psi \qquad (17.2)$$

where $\mathbf{u}$ and p represent the displacement and pore fluid pressure to be computed, respectively, and the corresponding material properties (and typical values for brain modeled as a homogeneous isotropic medium) are listed in Table 17.1 [43]. Discretizing these coupled equations through finite-element methods leads to the following linear algebraic system [42]:

$$\mathbf{Kx} = \mathbf{b} \qquad (17.3)$$

where **K** is the stiffness matrix that incorporates the discretized model equations, **x** is the solution vector of displacement and pore fluid pressure (4 degrees of freedom per node), and **b** is the forcing vector that includes forces and boundary conditions arising from gravitational effects and interactions between the parenchymal surface and the skull when a contact condition is used [44].

TABLE 17.1
Model Parameters and Their Typical Values in Brain

Parameter	Description	Value
G	Brain shear modulus	2100 Pa
N	Brain Poisson's ratio	0.46
ρ_t	Brain tissue density	1000 kg/m^3
$\rho_{f,s}$*	Surrounding fluid density for saturated tissue	1000 kg/m^3
$\rho_{f,u}$*	Surrounding fluid density for unsaturated tissue	1.0 kg/m^3
K	Hydraulic conductivity	1e^{-7} m^{3}s/kg
A	Ratio of fluid volume to change in solid volume	1
1/*S*	Fluid that can be forced into the tissue under constant volume	0
Ψ	Pressure source strength	0 Pa/s

*ρ_f in Equation 17.1 has either the value for $\rho_{f,s}$ or $\rho_{f,u}$ depending on whether the tissue is saturated or unsaturated with fluid.

Because incomplete displacement measurements are available, a generalized inversion using Lagrange multipliers can be used to minimize the difference (ε) between the model estimate (**x**) and measured data (**d**) through a least-squares optimization constrained by the model equations (Equation 17.3 [45]). Specifically, data error and forcing condition covariance models and a set of Lagrange multipliers (λ) are introduced to form an objective function, Ω [45]:

$$\Omega = (\mathbf{d} - \mathbf{Ax})^T \mathbf{W}_\varepsilon(\mathbf{d} - \mathbf{Ax}) + \mathbf{b}^T\mathbf{W}_b\mathbf{b} + \lambda^T(\mathbf{Kx} - \mathbf{b}) \qquad (17.4)$$

where **A** is a sampling matrix producing values of the computed solution at locations given by the measurement data. The matrices $\mathbf{W}_b$ and $\mathbf{W}_\varepsilon$ (inverses of the covariances of **b** and ε, respectively) describe the covariances of the forcing conditions between any two locations and the data error between any two measurement points, and are controlled by the standard deviation of the correlation length (*L*) and the measurement error (σ_ε), respectively [34,45]. To minimize the scalar objective function, Ω, its derivatives with respect to **x**, **b**, and λ are set to zero, generating the following set of adjoint equations after further mathematical simplification [45]:

$$\frac{\partial \Omega}{\partial \mathbf{x}} = \mathbf{K}^T\lambda - 2\mathbf{A}^T\mathbf{W}_\varepsilon\varepsilon = 0 \qquad (17.5)$$

$$\frac{\partial \Omega}{\partial \mathbf{b}} = 2\mathbf{W}_b\mathbf{b} - \lambda = 0 \qquad (17.6)$$

$$\frac{\partial \Omega}{\partial \lambda} = \mathbf{Kx} - \mathbf{b} = 0 \quad (17.7)$$

$$\varepsilon - (\mathbf{d} - \mathbf{Ax}) = 0 \quad (17.8)$$

These adjoint equations can be solved directly using the representer algorithm (REP) [34] or iteratively using the steepest gradient descent (SGD). When the number of sparse data points, N_{sps}, is large, SGD achieves acceptable solution accuracy and is more computationally efficient [46]. In addition, SGD allows brain–skull contact boundary conditions to be incorporated, whereas implementation of such constraints is not straightforward using the REP approach. Using data from both porcine brain retraction [35] and clinical patient cases [47], we have demonstrated significant improvement (~30%–40%) in the displacement accuracy achieved through data inversion relative to forcing the model directly with the displacement data. With more advanced contact boundary conditions between the brain and the skull, the accuracy in model estimation can be further improved at the tumor boundary (recovering about 85% of the deformation) as well as at the craniotomy (1.6 mm residual model–data misfit on average) due to the enhanced model flexibility [44]. Using iterative SGD coupled with a direct sparse solver, PARDISO, an average computational cost of approximately 2 min is achievable with sufficient accuracy (e.g., a mean model–data misfit reduced from 5.2 to 1.1 mm on average) when adequate sparse data (e.g., $N_{sprs} = 100$) are used [46]. These findings suggest that data inversion schemes based on locally measured sparse displacement data are capable of efficiently producing accurate whole-brain deformation intraoperatively for clinical applications.

17.4 Intraoperative Sparse Displacement Data

Apparently, sparse displacements determined from intraoperative images are important to the overall model-based compensation strategy because they are directly used to guide model deformation. Because iUS and iSV capture internal features and the exposed parenchymal surface, respectively, they are complementary and both can be assimilated for improved accuracy in uMR. In Sections 17.4.1 and 17.4.2, we illustrate the techniques we have developed to derive displacement fields from iUS and iSV for data assimilation.

17.4.1 Sparse Data from 3D iUS

Image rasterization is necessary to resample a 3D iUS image volume expressed in a pyramid-shaped coordinate system into a regular Cartesian coordinate system before image processing [12]. A trilinear interpolation algorithm operating in a parametric (as opposed to the physical) space offers improved accuracy and achieves a real-time computational performance that is comparable to the voxel nearest neighbor algorithm as well as the fastest pixel-based techniques (1.05 s per full volume with a 1 mm^3 voxel size). Rasterized 3D iUS allows direct intramodality registrations to extract feature displacement maps because of the large translational and rotational capture ranges (35.2 mm and 38.5°, respectively [13]) that are possible. An initial ("last-known-correct") intermodality registration is computed to establish correspondence between pMR and 3D iUS before dural opening, prior to the occurrence of significant brain shift. This step is important because it defines the spatial transformation required to transform sparse displacement data from the 3D iUS image space into the pMR image volume and compensates for any misalignment of tumor or other internal features between the two image modalities due to errors in patient registration and US transducer calibration. To compute feature displacements, a 3D iUS image volume acquired at a later surgical stage is transformed into the space of the 3D iUS image obtained before dural opening using the transformation from the tracking system. Because of errors in 3D iUS image acquisition and tracking, stationary features (e.g., parenchymal boundaries contralateral to the craniotomy/US transducer) may be misaligned, which could degrade the accuracy of the sparse data measurement. To correct for this misalignment, a rigid body registration can be performed between the two image volumes using only stationary features (e.g., by limiting voxels to the image volume contralateral to the US transducer). After realigning the two 3D iUS image volumes, residual feature misalignment would directly indicate tissue deformation, which can be derived through a B-spline nonrigid registration to generate displacement vectors so that the two 3D iUS image volumes align. Figure 17.2 presents a composite view of two 3D iUS image volumes after rigid registration using voxels in the contralateral regions, which clearly demonstrate feature misalignment or deformation near the tumor cavity. Displacement vectors in this area (Figure 17.2a and b) also demonstrate disparity in displacement magnitude and direction, suggesting the complex pattern of brain motion that occurs at this surgical stage (after tumor removal).

In total, one rigid registration (to correct for misalignment due to errors in tracking/transducer calibration) and one B-spline nonrigid registration (to extract feature displacements) are necessary to derive sparse displacements from 3D iUS. A combined computation time of 1–2 min is expected to generate sufficient sparse displacement vectors for use in the data assimilation [48] (e.g., $N = 100–200$) with a set of optimal parameters (e.g., number of grid nodes along each direction, down-sampling

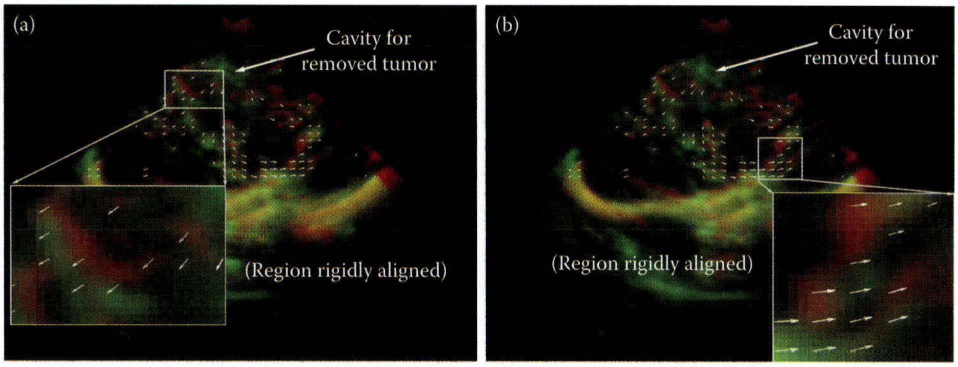

FIGURE 17.2
Composite 3DUS image for a patient (48-year-old female with lo- grade astrocytoma) after tumor resection (green) superimposed on the acquisition before dural opening (red). Two regional displacement maps (a and b) that show significantly different deformation patterns (insets) are magnified for comparison.

rate, and number of iterations). In practice, the rigid and nonrigid registrations can be executed simultaneously to further reduce computation time, in which case feature displacements due to rigid registration can be subtracted from those generated from the nonrigid registration to provide data guidance for the model.

17.4.2 Sparse Data from iSV

Stereovision is another important intraoperative imaging technique that captures a texture-rich profile of the exposed parenchymal surface during surgery. With a pair of stereoscopic images of the surgical scene, a complete 3D surface profile can be reconstructed based on a pinhole camera model using calibration parameters determined prior to surgery [15]. Similar to 3D iUS, an initial registration transformation between the pMR and iSV image modalities is needed to apply displacement maps computed from successive iSV acquisitions into the undeformed state. Registration methods based on either geometry (e.g., using iterative closest point [49]) or texture intensity can be employed. However, rigid registrations that depend solely upon geometry may not accurately capture surface deformation, especially when significant lateral shift occurs [16,50]. Incorporating texture intensity information into the registration can improve accuracy [16,20]. However, intensity-based methods may not be appropriate for registering stereo surfaces acquired before and after dural opening because of their substantial dissimilarity in appearance (i.e., dural surface vs. cortical surface). Instead of registering the stereo surfaces directly, the stereo surface obtained after dural opening is registered to a texture map generated from pMR in the region of the cortex sampled by iSV ([50]; see Figure 17.3a for a typical texture map generated from pMR in comparison with its corresponding stereo image after dural opening shown in Figure 17.3b). Common vascular features in both iSV and pMR texture maps allow registration using intensity-based algorithms (e.g., via maximization of mutual information). Because stereovision images can be coregistered with pMR using fiducial-based patient registration, the two 3D texture maps can be spatially merged to generate 2D projection images in the same coordinate system as long as no image "folding" occurs. For example, a local coordinate system can be established with its z-axis parallel to the average nodal normal direction of the texture surface from iSV to allow both images to be projected into the same coordinate system (Figure 17.3c). With proper image preprocessing (e.g., Gaussian smoothing), the two images can be registered with an intensity-based rigid algorithm to generate 2D displacement vectors (Figure 17.3d). Because the spatial correspondence between the textured cortical surface and its projection image can be inverted, full 3D displacements can be calculated by determining the 3D locations of the starting and ending points of each displacement vector.

Once the initial registration is obtained between iSV and pMR, subsequent registrations between iSV cortical surfaces acquired at different time points during surgery can be obtained by directly registering iSV surfaces using common features (e.g., vessels). Because the registration between iSV postdural opening and pMR is available, displacement maps can be expressed either relative to the initial undeformed pMR or to the previous deformed uMR at an earlier surgical stage. Figure 17.4 illustrates reconstructed cortical surfaces at three surgical stages: after dural opening, during, and after tumor resection (Figure 17.4a), where full 3D displacement vectors are computed either between two adjacent surgical stages (Figure 17.4b and c) or between the first and last surgical stages (Figure 17.4d).

High accuracy of iSV has been reported (e.g., ~1 mm in reconstruction [15] and 2 mm [16] or 2–3 pixels [20] in interframe registration between video frames), suggesting its clinical utility. However, reports on the computational

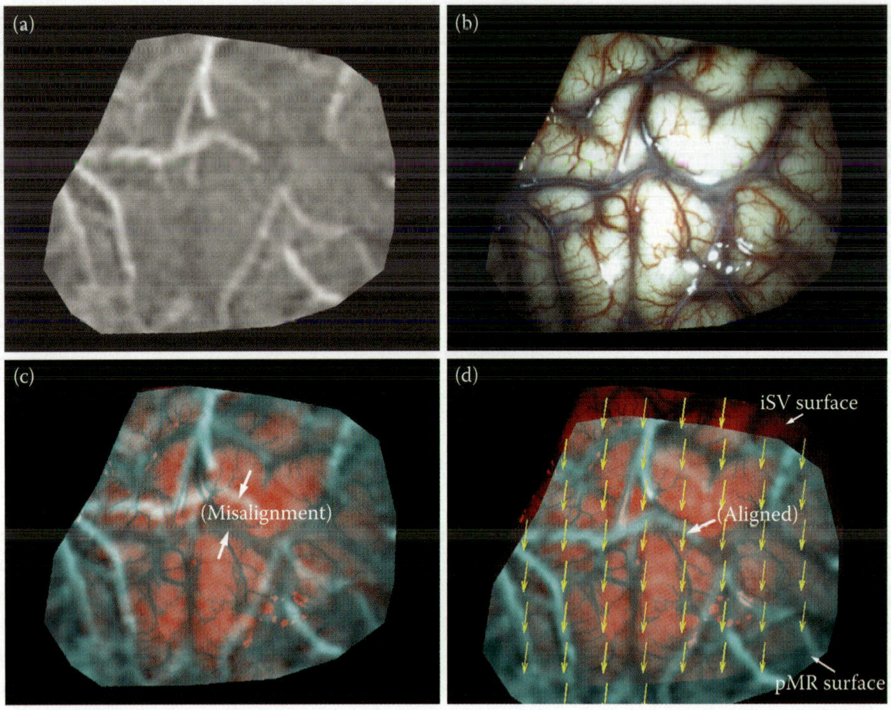

FIGURE 17.3
Texture-encoded map of the exposed cortical surface based on pMR (a) and its corresponding iSV image (b). The iSV projection image (red) is overlaid with its pMR counterpart (cyan) before (c) and after (d) registration. Apparent misalignment observed before registration (arrows in c) indicates significant lateral displacement, which is corrected after registration (arrow in d). Displacement vectors at sample locations are also shown (thin arrows in d). Patient was an 18-year-old female with epilepsy.

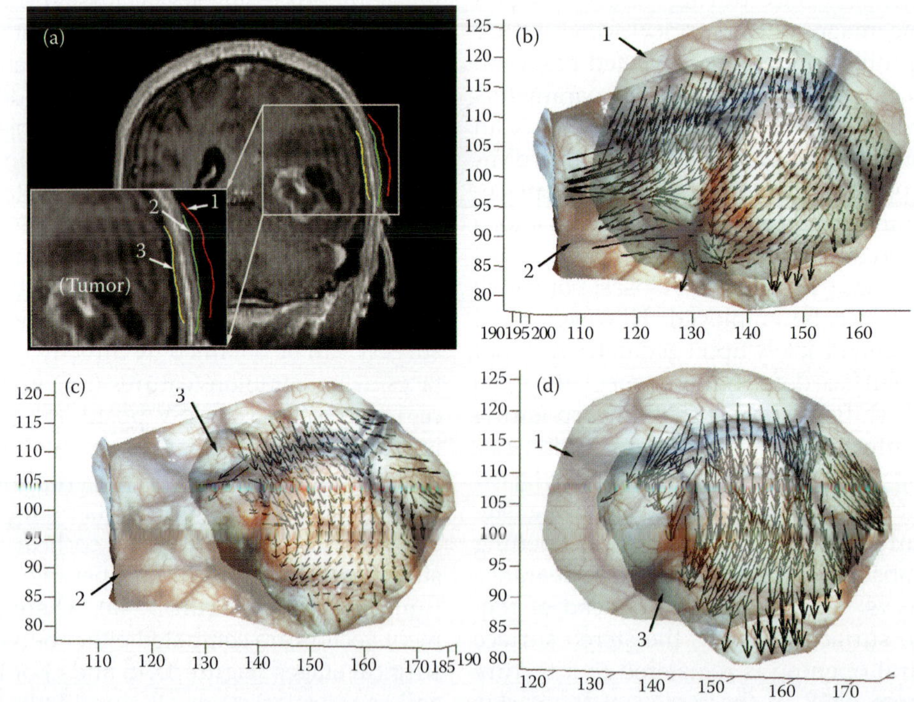

FIGURE 17.4
Cross sections of reconstructed cortical surfaces after dural opening (1), during resection (2), and after resection (3) on a coronal pMR image (a) for a 54-year-old male patient with glioblastoma multiforme. The resulting 3D displacements between the two surfaces are shown in b–d as indicated by the surface labels included.

efficiency of extracting cortical surface displacements are limited at present. Although a processing rate of 1 s per frame is indicated in ref. [20], the addition of semiautomatic vessel segmentation for feature extraction will likely compromise this efficiency [51]. The optical flow motion-tracking algorithm computes a dense displacement field without relying on landmark or feature segmentations; however, the computation time is significant (70 s for an image size of 1024 × 768 [31]). Because of the high resolution in iSV, down-sampling of the images or limiting the region of interest to the exposed cortical area (instead of the full size image) may be possible. Using these strategies, sparse displacement vectors can be generated within 10 s of registering the iSV images (e.g., with a down-sampling factor of 5 while maintaining an image resolution of 0.3 mm per pixel).

17.5 Illustration of Model-Estimated Brain Deformation

The goal of estimating whole-brain deformation is to generate uMR to allow subsequent neuronavigation. Here, we illustrate the results for a 66-year-old male with gliosarcoma, where both 3D iUS and iSV images were acquired after dural opening. Using sparse displacement data for internal features from iUS and surface features from iSV, the computational scheme described in Section 17.3 was employed to produce a whole-brain deformation field from which uMR was generated for updated image guidance. Figure 17.5a shows the alignment of internal features between iUS and uMR and between iSV and uMR at the cortical surface. Surface rendering of the uMR was also generated to visualize vessels on the parenchymal surface and is shown with the coregistered texture surface from iSV superimposed (Figure 17.5b), demonstrating accurate alignment of the vessels on the surface between the two image modalities. The alignments between intraoperative images and uMR are an encouraging qualitative indication of the accuracy of the model estimation process.

Ultimately, the accuracy of model-based brain shift compensation needs to be verified independently from the intraoperative images employed to generate the sparse displacement data that are incorporated into the model solution. With the ability to integrate uMR into the commercially available StealthStation® image-guided surgery platform now becoming available via the StealthLink® network communications interface (Medtronic, Inc., Louisville, Colorado) [52], we can verify the accuracy of uMR in maintaining patient registration during surgery using a tracked and coregistered surgical microscope. For example, Figure 17.6 compares uMR (top) and pMR (bottom) in image-to-patient registration for a patient case after dural opening when the microscope was carefully maneuvered to focus on the exposed cortical surface in the physical space in the OR. The accurate alignment between the focal point and the parenchymal boundary in uMR is a clear demonstration of effective compensation for brain deformation at this surgical stage. However, the initially accurate registration at the start of surgery (FRE of 1.3 mm for this patient) was lost with pMR because of significant parenchymal distention causing the focal point to be incorrectly localized on the skull. These illustrations provide important preliminary evidence that a model-based brain shift compensation strategy assimilating sparse information from intraoperative images is not only capable of generating uMR for more accurate image guidance but also feasible for routine clinical applications when integrated into a commercial navigational system.

(a) (b)

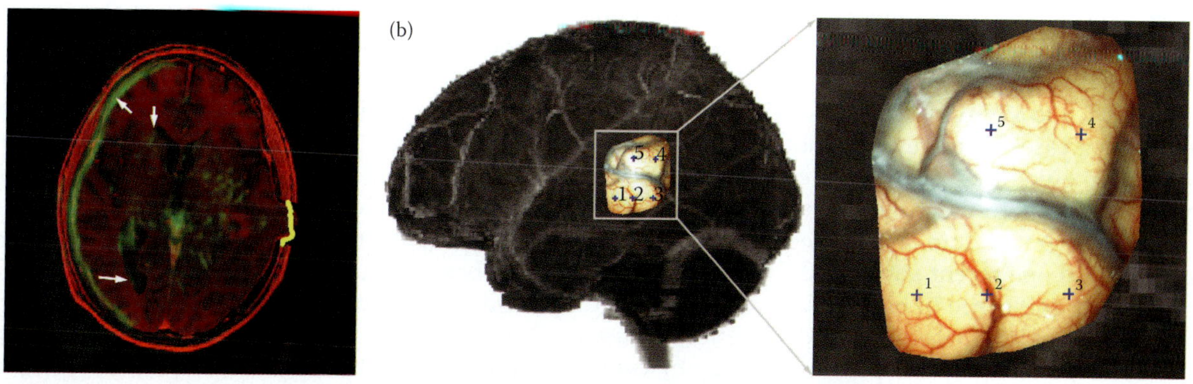

FIGURE 17.5
(a) Overlay of iUS (green) on a typical axial uMR image (red) and the corresponding cortical surface cross section obtained from iSV (thick line). (b) Surface rendering of uMR showing vessels on the parenchyma surface, which are well aligned with those captured from iSV. The alignments between intraoperative images (i.e., 3D iUS and iSV) and uMR indicate consistency between model estimates and intraoperative measurements. The patient was a 66-year-old male with gliosarcoma.

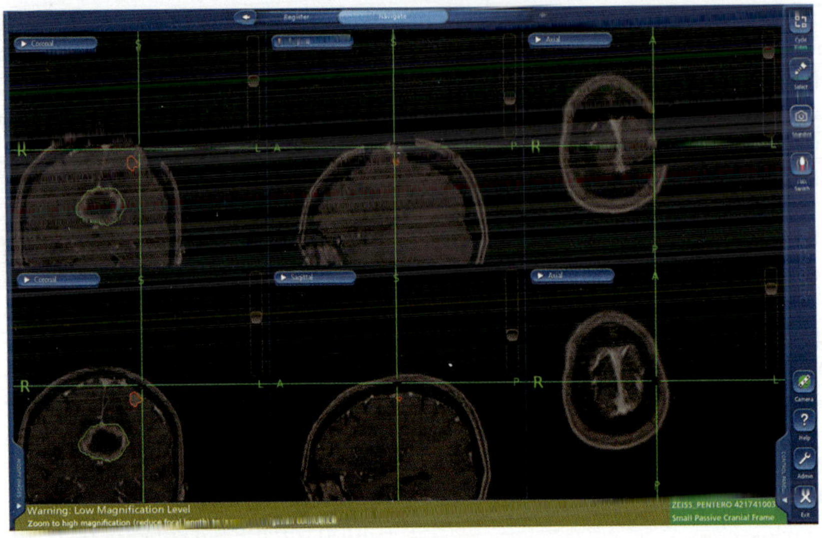

FIGURE 17.6
StealthStation® screen capture comparing uMR (top) and pMR (bottom) in three orthogonal views during actual surgery in a patient case. The surgical microscope was carefully focused on the exposed cortical surface at the craniotomy in the physical space in the OR after dural opening. The focal point (cross-hair) was significantly misaligned with the corresponding feature in pMR (bottom) but aligned well with the uMR equivalent (top) in all views, clearly demonstrating the loss of the initial patient registration in pMR that was maintained in uMR. Patient was a 65-year-old female with glioblastoma multiforme.

17.6 Summary and Conclusions

Portable, inexpensive intraoperative imaging techniques such as iUS and iSV continue to be valuable tools to compensate for brain deformation during open cranial neurosurgeries. Their utility is substantially increased when they are incorporated into a biomechanical model to estimate whole-brain deformation through data inversion. Because model-updated whole-brain MR image volumes are available, improved image guidance becomes possible at any point when intraoperative images are acquired. This chapter briefly describes important aspects of a data inversion scheme based on a biomechanical model as well as image registration techniques required to derive sparse displacement data from 3D iUS and iSV with illustrations from representative patient cases. The implementation of this model-based image guidance approach for clinical application becomes much more practical when uMR can be conveniently integrated into a commercial navigational system [52]. In one recent patient case, we have clearly verified the significance of brain shift even at the start of surgery (postdural opening), and the effectiveness of our intraoperative image updating strategy by comparing uMR with its pMR precursor in a side-by-side view in the neuronavigational system, when a tracked and coregistered microscope was focused on the actual parenchymal surface (Figure 17.6). In addition, realizing updated image guidance directly in the OR (instead of retrospectively) is a clear indication that the computational efficiency of our method is sufficient for clinical deployment.

Eventually, quantitative evaluation of the accuracy of updated image guidance is necessary to establish the effectiveness and robustness of the approach during surgery. Because the misfit between model solution and observed data indicates the congruence of model performance itself but not necessarily the degree of alignment with respect to intraoperative parenchymal anatomy, verification from an independent source of measurement is necessary. Although a surgical microscope or a tracked digitizing probe can be used to localize features on the exposed surface to quantify uMR registration accuracy (e.g., Figure 17.6), verification of the accuracy of internal features is challenging. In this case, intraoperative fluorescence imaging that highlights tumor tissue directly is an interesting approach [53]. However, this method of validation is only possible at later surgical stages when the tumor is exposed. Ideally, iMR captures whole-brain features during surgery and offers unsurpassed delineation of soft tissue and, therefore, could serve as the "gold standard" in assessing the accuracy of model-based brain shift compensation. This could especially be important toward later surgical operations when the model-based approach becomes increasingly challenging for brain shift compensation because of growing complexities in the tissue mechanical loading environment due to surgical instruments used (e.g., retractors), tissue incision and resection, and difficulty in tracking these events intraoperatively. Whether a more sophisticated biomechanical model (e.g., inclusion of dura septa [39], differentiation of normal and tumor tissue material properties, etc.) could facilitate a more realistic or accurate simulation of brain deformation is likely debatable as it introduces additional uncertainties in model parameters while measurement data are limited and computational cost is constrained. Although the choice of material properties of the brain itself may not influence the accuracy of the estimated deformation field when a set of sparse displacements on the stress-free cortical surface is known [54], the significance of the possible property differences between white and gray matter is unknown, and findings on the interregional variation in brain tissue properties are inconsistent at present [55]. In addition, the desired accuracy in brain shift compensation in targeted areas (e.g., within 1–2 mm for most surgical operations, comparable with typical MR image resolution) could also be poorly defined in practice especially for infiltrating tumors where no demarcation is present (e.g., high-grade gliomas). Regardless, the clinical efficacy of the model-based intraoperative image updating approach would be established when both its accuracy and efficiency are well characterized for practical applications either alone or in conjunction with intraoperative images (e.g., iMR).

Additionally, it may also be desirable to assess the clinical benefits of incorporating advanced multimodal neuroimaging increasingly becoming available into the model-based brain shift compensation strategy. Because of model-computed whole-brain deformation, it is possible to warp other coregistered image modalities to align with uMR intraoperatively (e.g., tractography from diffusion tensor imaging to serve as a structural substrate for functionally important neural pathways that could be critical for surgical decision making). Given the ever growing complexity in intraoperative imaging and information (e.g., uMR, intraoperative images such as iUS, iSV, iMR, fluorescence imaging, and warped neuroimaging, etc.), it is important to evaluate how best to utilize them in the OR for optimal surgical and patient outcome, which is not only a research question but also an engineering challenge that warrants further investigation.

Nevertheless, we have successfully demonstrated the use of intraoperative imaging techniques (3D iUS and iSV) coupled with a biomechanical data assimilation scheme for brain shift compensation through example patient cases. Our initial experience with integrating model-updated MR directly into a commercial navigational system suggests that this approach for brain shift characterization and compensation is feasible for surgical guidance in the OR.

Acknowledgments

Funding from the NIH grants R01 CA159324–01 and 1R21 NS078607 is acknowledged.

References

1. Lacroix, M., Abi-Said, D., Fourney, D.R., Gokaslan, Z.L., Shi, W., DeMonte, F., Lang, F.F. et al. A multivariate analysis of 416 patients with glioblastoma multiforme: Prognosis, extend of resection and survival. *J Neurosurg* **95** (2001) 190–198.
2. Laws, E.R., Shaffrey, M.E., Morris, A., Anderson, F.A.: Surgical management of intracranial gliomas—Does radical resection improve outcome? *Acta Neurochir* **85** (2002) 47–53.
3. Buckner, J.C.: Factors influencing survival in high-grade gliomas. *Semin Oncol* **30** (2003) 10–14.
4. Maurer, C.R., Fitzpatrick, J.M., Wang, M.Y., Galloway, R.L., Maciunas, R.J., Allen, G.S.: Registration of head volume images using implantable fiducial markers. *IEEE Tran Med Imag* **16**(4) (1997) 447–462.
5. Ammirati, M., Gross, J.D., Ammirati, G., Dugan, S.: Comparison of registration accuracy of skin-and bone-implanted fiducials for frameless stereotaxis of the brain: A prospective study. *Skull Base* **129**(3) (2002) 125–130.
6. Ji, S., Roberts, D.W., Hartov, A., Paulsen, K.D.: Intraoperative patient registration using volumetric true 3D ultrasound without fiducials. *Med Phys* **39**(12) (2012) 7540–7552.
7. Hill, D.L.G., Maurer, C.R., Maciunas, R.J., Barwise, J.A., Fitzpatrick, J.M., Wang, M.Y.: Measurement of intraoperative brain surface deformation under a craniotomy. *Neurosurgery* **43**(3) (1998) 514–526.
8. Roberts, D.W., Hartov, A., Kennedy, F.E., Miga, M.I., Paulsen, K.D.: Intraoperative brain shift and deformation: A quantitative analysis of cortical displacement in 28 cases. *Neurosurgery* **43**(4) (1998) 749–760.
9. Nabavi, A., Black, P.M., Gering, D.T., Westin, C.F., Mehta, V., Pergolizzi, R.S. Jr., Ferrant, M. et al. Serial intraoperative magnetic resonance imaging of brain shift. *Neurosurgery* **48**(4) (2001) 787–798.
10. Hartkens, T., Hill, D.L.G., Castellano-Smith, A.D., Hawkes, D.J. Jr., Maurer, C.R., Martin, M.J., Hall, W.A., Liu, H., Truwit, C.L.: Measurement and analysis of brain deformation during neurosurgery. *IEEE Trans Med Imaging* **22**(1) (2003) 82–92.

11. Ji, S., Fontaine, K., Hartov, A., Borsic, A., Roberts, D.W., Paulsen, K.D.: Coregistered volumetric true 3D ultrasonography in image-guided neurosurgery. In M.I. Miga, K.R. Cleary (Eds.) *Medical Imaging 2008: Visualization, Image-Guided Procedures, and Modeling*, Proceedings of SPIE, Vol. 6918 (SPIE, Bellingham, WA, 2008) 69180F.
12. Ji, S., Roberts, D.W., Hartov, A., Paulsen, K.D.: Real-time interpolation for true 3-Dimensional ultrasound image volumes. *J Ultrasound Med* **30** (2011) 241–250.
13. Ji, S., Roberts, D.W., Hartov, A., Paulsen, K.D.: Combining multiple volumetric true 3D ultrasound volumes through re-registration and rasterization. In G.-Z. Yang, D.J. Hawkes, D. Rueckert, A. Nobel, C. Taylor (Eds.) *MICCAI Part I, LNCS*, Vol. 5761 (Springer-Verlag, Berlin, Heidelberg, 2009) 795–802.
14. Skrinjar, O., Nabavi, A., Duncan, J.: Model-driven brain shift compensation. *Med Image Anal* **6**(4) (2002) 361–373.
15. Sun, H., Lunn, K.E., Farid, H., Wu, Z., Roberts, D.W., Hartov, A., Paulsen, K.D.: Stereopsis guided brain shift compensation. *IEEE Trans Med Imaging* **24**(8) (2005) 1039–1052.
16. Paul, P., Morandi, X., Jannin, P.: A surface registration method for quantification of intraoperative brain deformations in image-guided neurosurgery. *IEEE Trans Info Tech Biomed* **13**(6) (2009) 976–983.
17. DeLorenzo, C., Papademetris, X., Staib, L.H., Vives, K.P., Spencer, D.D., Duncan, J.S.: Image-guided intraoperative cortical deformation recovery using game theory: Application to neocortical epilepsy surgery. *IEEE Trans Med Imaging* **29**(2) (2010) 322–338.
18. Miga, M.I., Sinha, T.K., Cash, D.M., Galloway, R.L., Weil, R.J.: Cortical surface registration for image-guided neurosurgery using laser-range scanning. *IEEE Trans Med Imaging* **22**(8) (2003) 973–985.
19. Cao, A., Thompson, R.C., Dumpuri, P., Dawant, B.M., Galloway, R.L., Ding, S., Miga, M.I.: Laser range scanning for image-guided neurosurgery: Investigation of image-to-physical space registrations. *Med Phys* **35**(4) (2008) 1593–1605.
20. Ding, S., Miga, M.I., Pheiffer, T.S., Simpson, A.L., Thompson, R.C., Dawant, B.M.: Tracking of vessels in intra-operative microscope video sequences for cortical displacement estimation. *IEEE Trans Biomed Eng* **58**(7) (2011) 1985–1993.
21. Ferrant, M., Nabavi, A., Macq, B., Black, P.M., Joles, F.A., Kinkinis, R., Warfield, S.K.: Serial registration of intraoperative MR images of the brain. *Med Image Anal* **6** (2002) 337–359.
22. Hall, W.A., Liu, H., Martin, A.J., Pozza, C.H., Maxwell, R.E., Truwit, C.L.: Safety, efficacy, and functionality of high-field strength interventional magnetic resonance imaging for neurosurgery. *Neurosurgery* **46** (2000) 632–642.
23. Nimsky, C., Ganslandt, O., Hastreiter, P., Fahlbush, R.: Intraoperative compensation for brain shift. *Surg Neurol* **56** (2001) 357–365.
24. Wirtz, C.R., Bonsanto, M.M., Knauth, M., Tronnier, V.M., Albert, F.K., Staubert, A., Kunze, S.: Intraoperative magnetic resonance imaging to update interactive navigation in neurosurgery: Method and preliminary experience. *Comput Aided Surg* **2** (1997) 172–179.
25. Bonsanto, M.M., Staubert, A., Wirtz, C.R., Tronnier, V., Kunze, S.: Initial experience with an ultrasound-integrated single-rack neuronavigation system. *Acta Neurochir* **143**(11) (2001) 1127–1132.
26. Bucholz, R.D., Smith, K.R., Laycock, K.A., McDurmont, L.L.: Three-dimensional localization: From image-guided surgery to information-guided therapy. *Methods (Duluth)* **25**(2) (2001) 186–200.
27. Comeau, R.M., Sadikot, A.F., Fenster, A., Peters, T.M.: Intraoperative ultrasound for guidance and tissue shift correction in image-guided neurosurgery. *Med Phys* **27** (2000) 787–800.
28. Unsgaard, A., Rygh, O.M., Selbekk, T., Buller, T.B., Kolstad, F., Lindseth, F., Nagelhus Hernes, T.A.: Intraoperative 3D ultrasound in neurosurgery. *Acta Neurochir* **148** (2006) 235–253.
29. Rasmussen, I.A. Jr., Lindseth, F., Rygh, O.M., Berntsen, E.M., Selbekk, T., Xu, J., Nagelhus Hernes, T.A., Harg, E., Håberg, A., Unsgaard. G.: Functional neuronavigation combined with intra-operative 3D ultrasound: Initial experiences during surgical resections close to eloquent brain areas and future directions in automatic brain shift compensation of preoperative data. *Acta Neurochir (Wien)* **149**(4) (2007) 365–378.
30. Ji, S., Wu, Z., Hartov, A., Roberts, D.W., Paulsen, K.D.: Mutual-information-based patient registration using intraoperative ultrasound in image-guided neurosurgery. *Med Phys* **35**(10) (2008) 4612–4624.
31. Ji, S., Fan, X., Roberts, D.W., Paulsen, K.D.: Cortical surface strain estimation using stereovision. In G. Fichtinger, A. Martel, T. Peters (Eds.), *MICCAI, Part I, LNCS*, Vol 6891 (Springer, Berlin, 2011) 412–419.
32. Hu, J., Jin, X., Lee, J.B., Zhang, L., Chaudhary, V., Guthikonda, M., Yang, K.H., King, A.I.: Intraoperative brain shift prediction using a 3D inhomogeneous patient-specific finite element model. *J Neurosurg* **106** (2007) 164–169.
33. Wittek, A., Kikinis, R., Warfield, S.K., Miller, K.: Brain shift computation using a fully nonlinear biomechanical model. *Med Image Comput Comput Assist Inter* **8**(Pt 2) (2005) 583–590.
34. Lunn, K.E., Paulsen, K.D., Lynch, D.R., Roberts, D.W., Kennedy, F.E., Hartov, A.: Assimilating intraoperative data with brain shift modeling using the adjoint equations. *Med Image Anal* **9** (2005) 281–293.
35. Lunn, K.E., Paulsen, K.D., Liu, F., Kennedy, F.E., Hartov, A., Roberts, D.W.: Data-guided brain deformation modeling: Evaluation of a 3-D adjoint inversion method in porcine studies. *IEEE Trans Biomed Eng* **53**(10) (2006) 1893–1900.
36. Miller, K., Wittek, A., Joldes, G.: Biomechanics of the brain for computer-integrated surgery. *Acta Bioeng Biomech* **12**(2) (2010) 25–37.
37. Dumpuri, P., Thompson, R.C., Dawant, B.M., Cao, A., Miga, M.I.: An atlas-based method to compensate for brain shift: Preliminary results. *Med Image Anal* **11**(2) (2007) 128–145.

38. Dumpuri, P., Thompson, R.C., Cao, A., Ding, S., Garg, I., Dawant, B.M., Miga, M.I.: A fast efficient method to compensate for brain shift for tumor resection therapies measured between preoperative and postoperative tomograms. *IEEE Trans Biomed Eng* **57**(6) (2010) 1285–1296.
39. Chen, I., Coffey, A.M., Ding, S., Dumpuri, P., Dawant, B.M., Thompson, R.C., Miga, M.I.: Intraoperative brain shift compensation: Accounting for dural septa. *IEEE Trans Biomed Eng* **58**(3) (2011) 499–508.
40. DeLorenzo, C., Papademetris, X., Vives, K.P., Spencer, D.D., Duncan, J.S.: A comprehensive system for intraoperative 3D brain deformation recovery. In N. Ayache, S. Ourselin, A. Maeder (Eds.), *MICCAI Part II, LNCS*, Vol. 4792 (Springer-Verlag, Berlin, Heidelberg, 2007) 553–561.
41. Biot, M.: General theory of three-dimensional consolidation. *J Appl Phys* **12** (1941) 155–164.
42. Paulsen, K.D., Miga, M.I., Kennedy, F.E., Hoopes, P.J., Hartov, A., Roberts, D.W.: A computational model for tracking subsurface tissue deformation during stereotactic neurosurgery. *IEEE Trans Biomed Eng* **46** (1999) 213–225.
43. Miga, M.I.: *Development and Quantification of a 3D Brain Deformation Model for Model-Updated Image-Guided Stereotactic Neurosurgery* (Dartmouth College, Thayer School of Engineering, Hanover, NH, 1998).
44. Ji, S., Roberts, D.W., Hartov, A., Paulsen, K.D.: Brain-skull contact boundary condition in an inverse computational model. *Med Image Anal* **13**(4) (2009) 659–672.
45. Lynch, D.: *Numerical Partial Differential Equations for Environmental Scientists and Engineers* (Springer, Berlin, 2004).
46. Ji, S., Hartov, A., Roberts, D.W., Paulsen, K.D.: Data assimilation using a gradient descent method for estimation of intraoperative brain deformation. *Med Image Anal* **13**(5) (2009) 744–756.
47. Liu, F., Paulsen, K.D., Lunn, K., Sun, H., Hartov, A., Wu, Z., Roberts, D.W.: Comparative study of brain deformation estimation methods. In K.R. Cleary, R.L. Galloway Jr. (Eds) *Medical Imaging 2006: Visualization, Image-Guided Procedures, and Display*, Proceedings of SPIE Vol. 6141 (SPIE, Bellingham, WA, 2006) 61411D.
48. Ji, S., Fan, X., Roberts, D.W., Hartov, A., Paulsen, K.D.: Optimizing nonrigid registration performance between volumetric true 3D ultrasound images in image-guided neurosurgery. In D.R. Holmes III, K.H. Wong (Eds.) *Medical Imaging 2011: Visualization, Display and Image-Guided Procedures*, Proceedings of SPIE (Lake Buena Vista, Orlando, FL, 2011).
49. Besl, P.J., McKay, N.D.: A method for registration of 3-D shapes. *IEEE Trans Pattern Anal Mach Intell* **14**(2) (1992) 239–256.
50. Fan, X., Ji, S., Hartov, A., Roberts, D., Paulsen, K.: Registering stereovision surface with preoperative magnetic resonance images for brain shift compensation. In D.R. Holmes III, K.H. Wong (Eds.) *Medical Imaging 2012: Image-Guided Procedures, Robotic Interventions, and Modeling*, Proceedings of SPIE Vol. 8316 (SPIE, Bellingham, WA, 2012) 83161C.
51. Ding, S., Miga, M.I., Noble, J.H., Cao, A., Dumpuri, P., Thompson, R.C., Dawant, B.M.: Semiautomatic registration of pre- and post brain tumor resection laser range data: Method and validation. *IEEE Trans Biomed Eng* **56**(3) (2009) 770–780.
52. Schaewe, T.J., Fan, X., Ji, S., Hartov, A., Holton, L.H., Roberts, D.W., Paulsen, K.D., Simon, D.A.: Integration of intraoperative and model-updated images into an industry-standard neuronavigation system: Initial results. In D.R. Holmes, Z.R. Yaniv (Eds.) *Medical Imaging 2013: Visualization, Display and Image-Guided Procedures*, Proceedings of SPIE (Lake Buena Vista, Orlando, FL, 2013).
53. Valdes, P.A., Fan, X., Ji, S., Harris, B.T., Paulsen, K.D., Roberts, D.W.: Estimation of brain deformation for volumetric image updating in protoporphyrin IX fluorescence-guided resection. *Stereotact Funct Neurosurg* **88** (2010) 1–10.
54. Wittek, A., Hawkins, T., Miller, K. On the unimportance of constitutive models in computing brain deformation for image-guided surgery. *Biomech Model Mechanobiol* **8**(1) (2009) 77–84.
55. van Dommelen, J.A.W., van der Sande, T.P.J., Hrapko, M., Peters, G.W.M.: Mechanical properties of brain tissue by indentation: Interregional variation. *J Mech Behav Biomed Mater* **3** (2010) 158–166.

18

Noninvasive Determination of Material Properties for Biological Materials

Kent Butz, Deva Chan, Corey P. Neu, and Eric A. Nauman

CONTENTS

Biological materials exhibit nonlinear, heterogeneous, viscoelastic, and multiscale material properties, which are important to determine in situ or in vivo. Quantifying properties of tissues within complex systems of the body requires the noninvasive measurement of finite displacements and strains in conjunction with computational models of tissue mechanics. One of the most promising imaging methods for such a task is the use of magnetic resonance imaging (MRI). Combining MRI-based methods with numerical approximations of the tissue deformations can enable the use of noncontact and patient-specific modeling in a manner not currently possible with traditional mechanical testing or finite element techniques. Herein, we describe the application of these models to articular cartilage and the intervertebral disc.

18.1 Mechanical Modeling of Biomaterials

18.1.1 Computational Models

A number of numerical-based methods have been developed with the goal of more fully describing and characterizing the mechanics of soft tissue interactions in biomaterials and load bearing joints. The finite-element analysis has long been a popular means of estimating stress and other physical phenomena in biomechanics. While it was common in years past to use an idealized geometry approximating a joint or system, if proper care is not taken, these models have the risk of underestimating joint stresses due to the fact that perfect conformation between mating surfaces rarely occurs naturally within the body.[1,2] In recent years, image-based geometries have become the dominant basis for modeling. By this method, computerized tomography (CT) or magnetic resonance imaging (MRI) is used to create a digital replica of a given joint geometry for use in finite-element or mathematical analyses of the loading or contact response, thus providing a patient-specific representation of the anatomy.[3–12] Although typically used in the development of geometry for articular joints, MRI-based geometries have also been used in creating geometries of soft biological systems such as the brain[13,14] (see also Chapter 5).

A variety of constitutive models may be used in such modeling of joint mechanics. A linear elastic model, for example, provides a simple and straightforward relation between stress, strain, and material properties. Nonlinear models, in contrast, are more complex, yet far more descriptive of biomaterial behavior.[15,16] However, while the measurement of tissue mechanics ex vivo for compression/tension,[17,18] shear,[19] or poroelasticity[20] test conditions is common for determining relevant material parameters for these models, the results may not provide an accurate description of a living tissue's response under normal loading conditions within an intact joint. Consequently, a given simulation may pair a highly accurate and specific geometry with a very broad estimate of material properties in order to model behavior.[1] Such methods may provide an estimate of joint response within a statistical range found in normal individuals, but have little application to diagnose or monitor individual patients.

In the field of soft tissue biomechanics, biphasic and triphasic mixture theories are perhaps the most widely accepted models of mechanical behavior. The traditional biphasic model considers the loading response of cartilage to be the net result of the mechanical contributions of separate fluid and solid constituents.[21] In these models, the solid constituent is considered as the collagen/proteoglycan/bound water matrix and is typically assumed to behave as an incompressible, porous, hyperelastic solid. Frictional drag resulting from interstitial flow of unbound fluid is assumed to be a dominant factor contributing to a tissue's resistance to compression.[22–24]

The triphasic model is an extension of the biphasic model, which takes into account the effects of charge within the tissue.[25,26] This model is based on the concept that there is an interaction between fixed charges associated with proteoglycan content and the ions within the interstitial fluid that can affect fluid transport and diffusion throughout the interstitium.[27,28] Regions with a higher fixed charge density are observed to be more resistant to compression, thus resulting in changes to mechanical behavior.[22] Although differences may be observed between the more complex, charge-inclusive models and the simpler biphasic model over an extended period (10 s or more), within short loading cycles typical of gait (~1 s), the two models produce very similar results.[28]

18.1.2 Measuring Tissue Deformations In Vivo

While the preceding techniques use mathematical models to describe the mechanics of the cartilage system, certain imaging-based methods have been used in recent years to quantify deformations and mechanical properties of tissue noninvasively. This includes the use of ultrasound waves and a technique known as magnetic resonance elastography[29–31] (see also Chapters 4, 5, 7, and 8). Both of these methods rely on the imaging of transient shear waves to obtain a measure of the elastic response of an object and thus determine bulk material properties. While knowledge of the elastic behavior of a material can be valuable, little additional knowledge of the stresses of a system can be directly obtained by these methods. Alternatively, MR tagging techniques (see Chapters 2 and 3) may be used to directly track the deformation of a tissue throughout its volume and have found use in cardiovascular as well as cartilage systems.[32,33]

An alternative method whereby the deformation field of a biomaterial may be obtained is referred to as hyperelastic warping (see Chapter 12). In this method, differences in image intensities between a template (reference) image and a target (loaded) image are used to generate a body force that deforms a finite-element representation of the template such that it aligns with the target. To arrive at the final result, an optimization technique is used such that the strain energy function of the model is minimized.[34] While noninvasive, the technique does require some knowledge or estimate of the material properties of the system being analyzed, obtained either experimentally or from the literature.[35] Applications of this technique have included noninvasive measurement of strains in materials as diverse as ligament and the left ventricle.[36,37]

Direct measurement of displacements within tissue has been demonstrated recently using displacement-encoded MRI (see Chapter 6). Displacement-encoded MRI is a method that calculates the displacement of each pixel within an image set during low-frequency periodic motion. The technique has been used to noninvasively determine deformations across a range of applications, including the heart,[38] brain,[39] spine,[40] and joint cartilage.[41–43] Displacements calculated by this technique may then be used to directly calculate strain fields in the material associated with the deformation.

Despite the number of models and techniques that have been employed in soft tissue biomechanics, there has been a clear and significant gap in the ability to fully determine patient-specific, noninvasive estimates of soft tissue mechanics in vivo. In this work, a modeling technique for simultaneously evaluating stresses and estimating material properties is described utilizing displacement fields obtained by displacement-encoded MRI. Based on measured deformations in cartilage or other tissue, it is proposed that stress patterns and intrinsic properties may be estimated through computational analysis using any of the aforementioned biomechanical models or constitutive laws.

18.2 Noninvasive Evaluation of Linear and Nonlinear Constitutive Models in Articular Cartilage

Integrating MRI-based imaging methods with computational models for articular cartilage: One of the earliest studies in this field examined juvenile porcine tibiofemoral joints.[41] Juvenile joints were chosen because the cartilage layer is considerably thicker than typically found in adult joints, and displacement gradients are easier to observe. The joint was compressed in situ using a custom-built, MRI-safe pneumatic loading system and the contact surfaces identified. Once the geometry was defined, the two-dimensional (sagittal) displacement field was determined using displacement-encoding with stimulated echoes and a fast spin echo acquisition (DENSE-FSE; Figure 18.1).

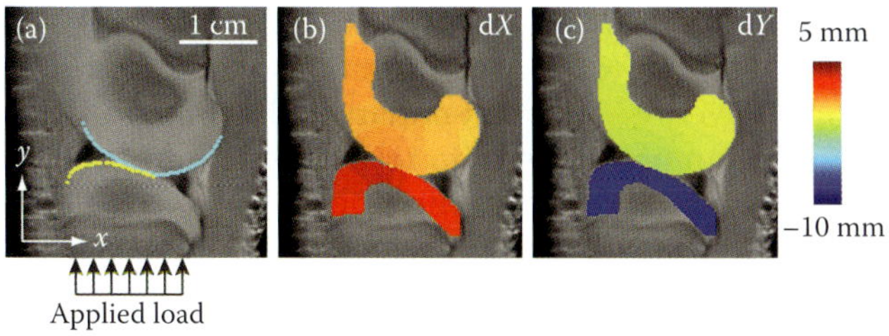

FIGURE 18.1
(a) Contact surfaces along the femur and tibia were identified and used to define geometric boundaries and boundary conditions for the computational models. Successive iterations of the material parameters in the constitutive model were input into the computational model. Convergence was assumed when the error between the measured force and the calculated force integrated across the boundaries was within 0.1. (b, c) For the juvenile porcine cartilage, displacements in the *x*- and *y*-directions, d*X* and d*Y*, respectively, were determined using DENSE-FSE.

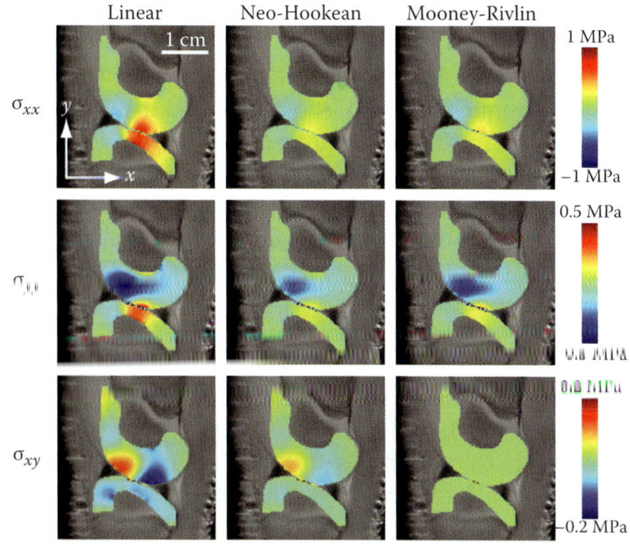

FIGURE 18.3
Stress fields predicted by three different constitutive laws using the same strain fields as inputs were surprisingly different. The nonlinear models (Neo-Hookean and Mooney-Rivlin) exhibited much more uniform distributions and did not possess the extremes observed when the linear elastic model was used.

Once the displacement gradients are established, it is a relatively straightforward matter to calculate the strain field (Figure 18.2), and the primary advantage of this particular method is realized. The stress field can be calculated by assuming initial values of the material parameters, comparing the calculated force across the joint to the measured applied load from the pneumatic actuator, and iterating until the two values approach each other. For this work, 0.1% error was assumed to define convergence.

One of the first insights gained using this model was that the choice of constitutive law dramatically affects the predicted stress distribution (Figure 18.3). The linear elastic model predicted the most extreme stress distributions with little mechanical penetration to the deeper regions of the cartilage. The nonlinear models (Neo-Hookean and Mooney–Rivlin) predicted much more uniform stress distributions throughout the cartilage and did not reveal peak stresses nearly as extreme as the linear model. While this may not be surprising to some, the primary question then is, which constitutive law is the best? There are a number of ways to grade these methods. We could look for the model that exhibits the lowest shear stress at the contact points, although accurate values may be difficult to obtain without very high resolution. It is also possible to examine the coefficient of variation (COV) in the predicted material properties under the assumption that the best constitutive law will also have the lowest COV. Perhaps a more instructive method would be to compress the joints at higher load and look for changes in structure that might indicate damage. This could be done within the MRI or histologically and the resulting maps overlaid on the stress distributions in an effort to determine the constitutive law that best predicts the injury pattern. It is likely that some combination of these methods will ultimately be required and should be an avenue of future research.

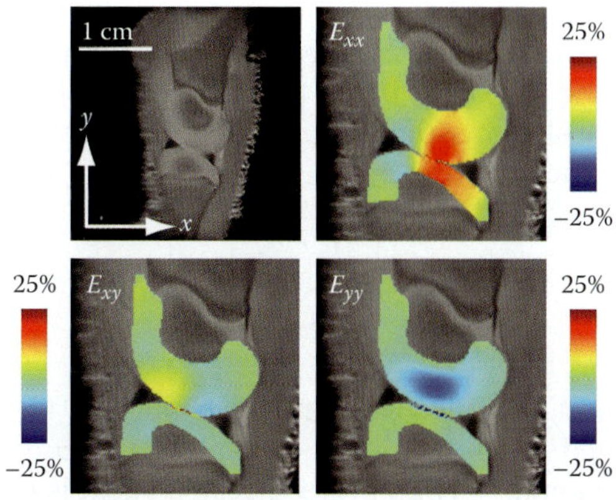

FIGURE 18.2
Displacement fields determined with DENSE-FSE in the juvenile porcine cartilage were used to calculate the Lagrangian strain fields in the *xx*-, *yy*-, and *xy*-directions.

18.3 Intervertebral Disk

Integrating MRI-based imaging methods with computational models for the intervertebral disk (IVD): While the integration of noninvasive imaging and material property determination in juvenile porcine cartilage was a useful first step, there were a number of unanswered questions. Perhaps the most important is, what happens when there are multiple tissue types that must be considered, especially those that exhibit heterogeneous properties? The IVD presents an ideal study system for the expansion of these techniques. The disk consists of two distinct material types that exhibit disparate location- and direction-dependent properties that together must satisfy known boundary conditions and displacements. Perhaps more importantly, it provides a relatively simple model of injury and provides an opportunity to investigate the extreme changes that occur when soft tissues are damaged.

For IVD experiments, New Zealand white rabbits were divided into two groups (healthy controls and experimentally ruptured disks).[40] At various time points, the disks were subjected to a cyclic compressive load of 30 N for a period of 1.5 s, with a total cycle length of 3.0 s (Figure 18.4). Displacements were then measured in the axial and radial directions of a coronal imaging plane taken through the center of the disk (Figure 18.5), and the deformation gradient and strains were then computed from the discrete displacement fields (Figure 18.6).

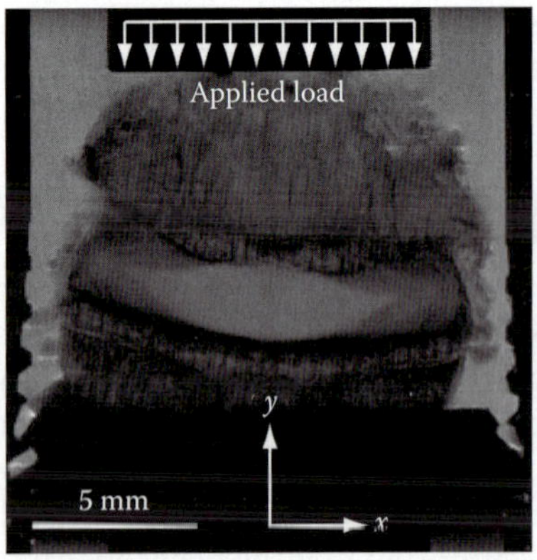

FIGURE 18.4
Two-dimensional displacements and strains were obtained with displacement-encoded MRI for rabbit IVDs subjected to a cyclic 30 N load in the axial (y) direction.

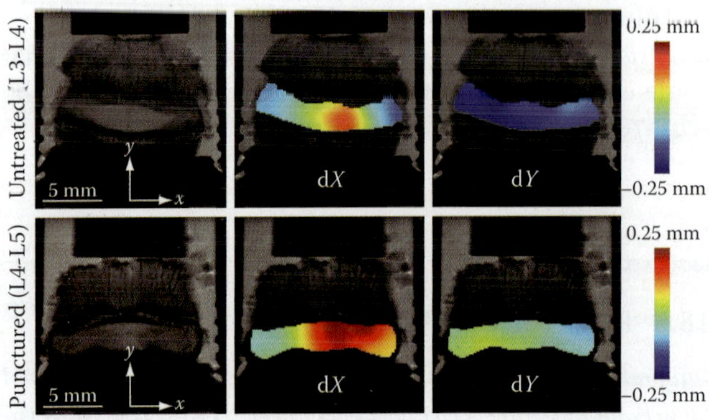

FIGURE 18.5
Displacements were measured within a coronal imaging plane oriented through the center of the disk for both punctured, degenerated disks as well as adjacent, unaltered disks, as shown in this representative pair.

In addition to strain, the same study[40] evaluated glycosaminoglycan (GAG) concentrations within the disks using delayed gadolinium enhanced MRI (dGEMRIC), providing an opportunity to relate the GAG concentration to material properties throughout the disk.

Stress distributions and material relations were determined in the IVD using a finite strain-based (FSB) computational model. MRI-based strains, describing the two-dimensional load–displacement response of a plane oriented through the center of the disk, were coupled with assumed material relations in this model. As such, a two-dimensional model with plane stress assumptions was used for the analysis. Importantly, the FSB model accounted for material composites, and the nucleus pulposus (NP) and annulus fibrosous (AF) were identified using a combination of histology and T_1 value maps acquired with MRI prior to exposure to the contrast agent (Figure 18.7).

Because both the NP and AF exhibit large strains and nonlinear stress–strain behaviors, both the NP and AF were modeled separately as incompressible Neo-Hookean materials,[44] represented by the strain energy density function,

$$W = \frac{c_1}{\beta}(J^{-2\beta} - 1) + c_1(I_1 - 3) - p(J - 1) \qquad (18.1)$$

where I_1 is the first invariant of the right Cauchy–Green tensor, $\boldsymbol{C}$, J is the determinant of the deformation gradient, c_1 and β are material parameters, and p represents a hydrostatic pressure term included to enforce incompressibility.[16] The second Piola–Kirchoff stress tensor, $\boldsymbol{S}$, was then determined by

$$\boldsymbol{S} = 2\left[c_1(\boldsymbol{I} - J^{-2\beta}\boldsymbol{C}^{-1}) - \frac{p}{2}(\det \boldsymbol{C})\boldsymbol{C}^{-1}\right], \qquad (18.2)$$

where $\boldsymbol{I}$ is the identity tensor.

Imposing the incompressibility assumption and neglecting the stress in the z-direction, consistent with the plane stress assumption, yields a relationship for the hydrostatic pressure:

$$p = \frac{2\left\{c_1\left[1 - (\boldsymbol{C}^{-1})_{33}\right]\right\}}{(\det \boldsymbol{C})(\boldsymbol{C}^{-1})_{33}} \qquad (18.3)$$

Material behavior during compression is known to be related to proteoglycan content and, subsequently, GAG content.[45,46] High proteoglycan content is typically accompanied by larger amounts of fluid associated with the surrounding matrix, and thus, a higher resistance to compression may be observed.[24,47] For example, a linear correlation between Gd-DTPA^{2-} and Young's modulus

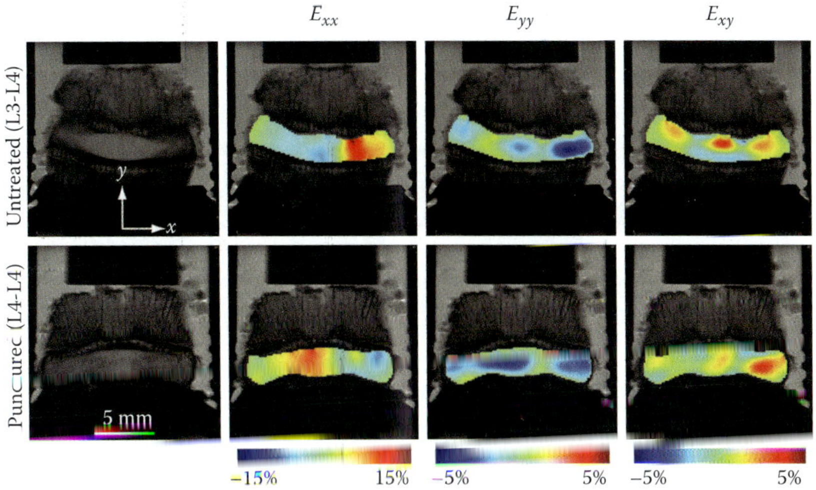

FIGURE 18.6
Strains were determined by dualMRI in paired samples of untreated (L3–L4) disks and adjacent (L4–L5) disks treated with an annular puncture to induce degeneration. In this representative pair, compressive strains in the loading (y) direction developed throughout the disk. In contrast, symmetric regions of tensile strain are observed to occur in the transverse (x) direction.

has been reported to occur in bovine articular cartilage.[46] In the model developed here, the Neo-Hookean material parameter c_1, which is linked to material stiffness, and proteoglycan content as measured by dGEMRIC were assumed to exhibit a similar linear relationship[46]:

$$c_1 = \gamma C \tag{18.4}$$

where γ is the local GAG content in units of μg/mL, and C is a constant to be calculated by the model. By these means, heterogeneous properties that are dependent upon biochemical composition may be estimated within the system.

Two sets of conditions were imposed upon the model. The first was an equilibrium condition for the entire system in the loading direction. The second condition applied to the system was the enforcement of a continuity condition across the boundaries separating the NP and AF. In essence, this condition states that, at any given point along the boundary between the NP and AF, forces normal to the boundary on either side must be equal to maintain equilibrium. This condition also allows for the determination of the relationship between material constants of the two regions relative to each other.

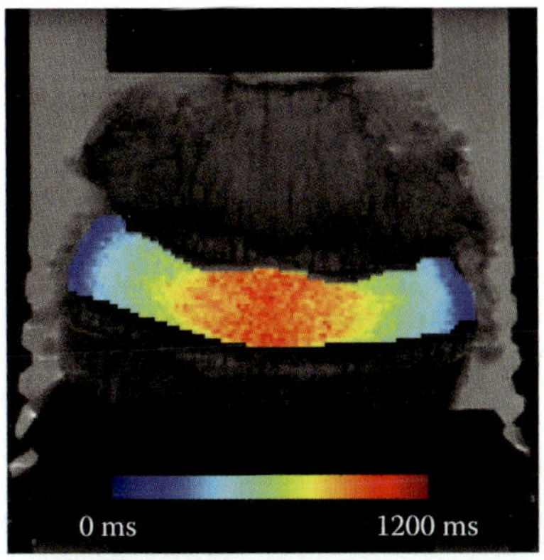

FIGURE 18.7
T_1 intensity was used to identify the nucleus pulposus and annulus fibrosus, with identifications confirmed by histology.

Heterogeneous stress fields were calculated for each of the IVDs. As expected, dominant compressive stresses were calculated in the direction of loading for each disk, while both tensile and compressive stresses were observed in the transverse direction (Figure 18.8).

A well-defined region with behavior similar to that expected of the NP was observed in the majority of untreated disks, characterized by a central region of compressive stress of the same shape and size in both the axial and radial directions (Figure 18.8). This region was observed to be less well defined or nonexistent in the punctured disks. These results corresponded well with observations of NP size and integrity made from histological images of the disks. Estimates for material constants within the NP of the untreated disks were higher than those of the punctured disks, a result that agrees well with direct measurements of healthy and degenerated NP in other studies.[48]

Taken together, the integration of noninvasive MRI sequences and computational models provided estimates for stresses in both the AF and NP that were a function of local biochemical composition, demonstrating the ability of the method to incorporate multiple heterogeneous materials. Stresses were within a

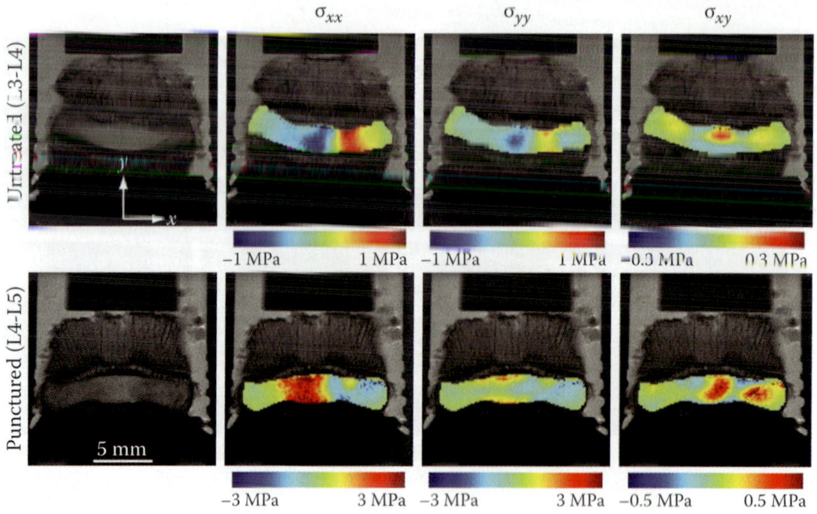

FIGURE 18.8
Stress fields were computed by a strain-based model as shown in these representative images. A large, defined central region of compressive stress was found to occur in both the loading and transverse directions of the untreated (L3–L4) disk, indicative of a healthy nucleus pulposus. A similar, but smaller, region was observed in the degenerated disk with compressive stresses more dispersed throughout the disk in the loading direction.

range reasonable for the magnitude of loading and size of the disks. Future work will ultimately be required to extend this method to one of clinical utility. Three-dimensional imaging data sets are currently time consuming, but will likely be a necessary next step, as will the incorporation of more complicated models of material parameter variation with GAG content. Interestingly, it may be possible to augment the current linear model using a combination of in vitro and in vivo experiments. Two additional features that will also be important and are currently under consideration are the addition of anisotropic constitutive laws and incorporation of mixture theory as a way to better understand the dynamic behavior of the skeletal tissue.

18.4 Conclusion

The methods described herein demonstrate the ability to noninvasively assess the mechanics of the articular cartilage layer and the IVD and, consequently, the physical health of skeletal tissues through visualization of the stress state during a known applied load. The techniques described here demonstrate the potential toward the analysis of musculoskeletal biomechanics in a clinical setting, where the evaluation of stress distributions and concentrations may lead to improved early diagnostic capabilities not possible with current techniques.

Acknowledgments

The authors gratefully acknowledge funding from NSF CMMI 1100554 (E.A.N. and C.P.N.) and NIH R01 AR063712 (C.P.N.)

References

1. Anderson, A. E., Ellis, B. J., Maas, S. A. & Weiss, J. A. Effects of idealized joint geometry on finite element predictions of cartilage contact stresses in the hip. *J Biomech* **43**, 1351–1357 (2010). doi:10.1016/j.jbiomech.2010.01.010.
2. Massey, C. J., van Donkelaar, C. C., Vresilovic, E., Zavaliangos, A. & Marcolongo, M. Effects of aging and degeneration on the human intervertebral disc during the diurnal cycle: A finite element study. *J Orthop Res* **30**, 122–128 (2012). doi:10.1002/jor.21475.
3. Pillai, R. P., Thoomukuntla, B., Ateshian, G. A. & Fischer, K. J. MRI-based modeling for evaluation of in vivo contact mechanics in the human wrist during active light grasp. *J Biomech* **40**, 2781–2787 (2007). doi:10.1016/j.jbiomech.2006.12.019.
4. Fitzpatrick, C. K., Baldwin, M. A. & Rullkoetter, P. J. Computationally efficient finite element evaluation of natural patellofemoral mechanics. *J Biomech Eng* **132**, 8 (2010).
5. Gislason, M. K., Stansfield, B. & Nash, D. H. Finite element model creation and stability considerations of complex biological articulation: The human wrist joint. *Med Eng Phys* **32**, 523–531 (2010).

6. Anderson, A. E., Peters, C. L., Tuttle, B. D. & Weiss, J. A. Subject-specific finite element model of the pelvis: Development, validation and sensitivity studies. *J Biomech Eng* **127**, 364–373 (2005).
7. Koo, S., Giori, N. J., Gold, G. E., Dyrby, C. O. & Andriacchi, T. P. Accuracy of 3D cartilage models generated from MR images is dependent on cartilage thickness: Laser scanner based validation of in vivo cartilage. *J Biomech Eng* **131**, 121004 (2009). doi:10.1115/1.4000087.
8. DeFrate, L. E., Sun, H., Gill, T. J., Rubash, H. E. & Li, G. In vivo tibiofemoral contact analysis using 3D MRI based knee models. *J Biomech* **37**, 1499–1504 (2004). doi:10.1016/j.jbiomech.2004.01.012.
9. Borotikar, B. S., Sipprell, W. H. III, Wible, E. F. & Sheehan, F. T. A methodology to accurately quantify patellofemoral cartilage contact kinematics by combining 3D image shape registration and cine-PC MRI velocity data. *J Biomech,* **45**, 1117–1122 (2012). doi:10.1016/j.jbiomech.2011.12.025.
10. Gold, G. E. et al. Weight-bearing MRI of patellofemoral joint cartilage contact area. *J Magn Reson Imaging* **20**, 526–530 (2004). doi:10.1002/jmri.20146.
11. Yang, N. H., Nayeb-Hashemi, H., Canavan, P. K. & Vaziri, A. Effect of frontal plane tibiofemoral angle on the stress and strain at the knee cartilage during the stance phase of gait. *J Orthop Res* **28**, 1539–1547 (2010). doi:10.1002/jor.211[illegible]4
12. Vaverka, M., Navrat, T. S., Vrbka, M., Florian, Z. & Fuis, V. Stress and strain analysis of the hip joint using FEM. *Technol Health Care* **14**, 271–279 (2006).
13. Chen, Y. & Ostoja-Starzewski, M. MRI-based finite element modeling of head trauma: Spherically focusing shear waves *Acta Mech* **213**, 155–167 (2010). doi:10.1007/s00707-009-0274-0.
14. Post, A., Hoshizaki, B. & Gilchrist, M. D. Finite element analysis of the effect of loading curve shape on brain injury predictors. *J Biomech* **45**, 679–683 (2012). doi:10.1016/j.jbiomech.2011.12.005.
15. Federico, S. & Herzog, W. Towards an analytical model of soft biological tissues. *J Biomech* **41**, 3309–3313 (2008)
16. Gallo, B., Ouyang, H., Shi, R. & Nauman, E. Correlations between tissue-level stresses and strains and cellular damage within the guinea pig spinal cord white matter. *J Biomech* **40**, 3029–3033 (2007). doi:10.1016/j.jbiomech.2007.03.014.
17. Julkunen, P. et al. Stress-relaxation of human patellar articular cartilage in unconfined compression: Prediction of mechanical response by tissue composition and structure. *J Biomech* **41**, 1978–1986 (2008).
18. Laksari, K., Shafieian, M. & Darvish, K. Constitutive model for brain tissue under finite compression. *J Biomech* **45**, 642–646 (2012). doi:10.1016/j.jbiomech.2011.12.023.
19. Wong, B. L. et al. Biomechanics of cartilage articulation: Effects of lubrication and degeneration on shear deformation. *Arthritis Rheum* **58**, 2065–2074 (2008).
20. Chin, H. C., Khayat, G. & Quinn, T. M. Improved characterization of cartilage mechanical properties using a combination of stress relaxation and creep. *J Biomech* **44**, 198–201 (2011).
21. Mow, V. C., Kuei, S. C., Lai, W. M. & Armstrong, C. G. Biphasic creep and stress relaxation of articular cartilage in compression? Theory and experiments. *J Biomech Eng* **102**, 73–84 (1980).
22. Nagel, T. & Kelly, D. J. The influence of fiber orientation on the equilibrium properties of neutral and charged biphasic tissues. *J Biomech Eng* **132**, 114506 (2010). doi:10.1115/1.4002589.
23. Kuo, J., Zhang, L., Bacro, T. & Yao, H. The region-dependent biphasic viscoelastic properties of human temporomandibular joint discs under confined compression. *J Biomech* **43**, 1316–1321 (2010). doi:10.1016/j.jbiomech.2010.01.020.
24. Mow, V. C., Ratcliffe, A. & Poole, A. R. Cartilage and diarthrodial joints as paradigms for hierarchical materials and structures. *Biomaterials* **13**, 67–97 (1992).
25. Lu, X. L., Miller, C., Chen, F. H., Guo, X. E. & Mow, V. C. The generalized triphasic correspondence principle for simultaneous determination of the mechanical properties and proteoglycan content of articular cartilage by indentation. *J Biomech* **40**, 2434–2441 (2007). doi:10.1016/j.jbiomech.2006.11.015.
26. Niu, H., Wang, Q., Zheng, Y. & Fan, Y. A new method for computing the uniaxial modulus of articular cartilages using modified inhomogeneous triphasic model. *Acta Mech Sin*, **26**, [illegible] (2010). doi:10.1007/s10409-009-0287-x.
27. Lu, X. L., Wan, L. Q., Guo, X. E. & Mow, V. C. A linearized formulation of triphasic mixture theory for articular cartilage, and its application to indentation analysis. *J Biomech* **43**, 673–679 (2010). doi:10.1016/j.jbiomech.2009.10.026.
28. Sun, D. D., Guo, X. E., Likhitpanichkul, M., Lai, W. M. & Mow, V. C. The influence of the fixed negative charges on mechanical and electrical behaviors of articular cartilage under unconfined compression. *J Biomech Eng* **126**, 6–16 (2004).
29. Bercoff, J., Tanter, M. & Fink, M. Supersonic shear imaging: a new technique for soft tissue elasticity mapping. *IEEE Trans Ultrason Ferroelectr Freq Control* **51**, 396–409 (2004).
30. Kwon, O. I. et al. Shear modulus decomposition algorithm in magnetic resonance elastography. *IEEE Trans Med Imaging* **28**, 1526–1533 (2009). doi:10.1109/tmi.2009.2019823.
31. Kiviranta, P. et al. Differences in acoustic properties of intact and degenerated human patellar cartilage during compression. *Ultrasound Med Biol* **35**, 1367–1375 (2009). doi:10.1016/j.ultrasmedbio.2009.03.003.
32. Neu, C. P., Hull, M. L. & Walton, J. H. Error optimization of a three-dimensional magnetic resonance imaging tagging-based cartilage deformation technique. *Magn Reson Med* **54**, 1290–1294 (2005). doi:10.1002/mrm.20669.
33. Shehata, M. L., Cheng, S., Osman, N. F., Bluemke, D. A. & Lima, J. A. Myocardial tissue tagging with cardiovascular magnetic resonance. *J Cardiovasc Magn Reson* **11**, 55 (2009). doi:10.1186/1532-429x-11-55.
34. Veress, A. I. et al. Measuring regional changes in the diastolic deformation of the left ventricle of SHR rats using microPET technology and hyperelastic warping. *Ann Biomed Eng* **36**, 1104–1117 (2008). doi:10.1007/s10439-008-9497-9.

35. Phatak, N. S. et al. Strain measurement in the left ventricle during systole with deformable image registration. *Med Image Anal* **13**, 354–361 (2009). doi:10.1016/j.media.2008.07.004.
36. Phatak, N. S. et al. Noninvasive determination of ligament strain with deformable image registration. *Ann Biomed Eng* **35**, 1175–1187 (2007). doi:10.1007/s10439-007-9287-9.
37. Veress, A. I., Gullberg, G. T. & Weiss, J. A. Measurement of strain in the left ventricle during diastole with cine-MRI and deformable image registration. *J Biomech Eng* **127**, 1195–1207 (2005).
38. Feng, L., Donnino, R., Babb, J., Axel, L. & Kim, D. Numerical and in vivo validation of fast cine displacement-encoded with stimulated echoes (DENSE) MRI for quantification of regional cardiac function. *Magn Reson Med* **62**, 682–690 (2009). doi:10.1002/mrm.22045.
39. Soellinger, M., Rutz, A. K., Kozerke, S. & Boesiger, P. 3D cine displacement-encoded MRI of pulsatile brain motion. *Magn Reson Med* **61**, 153–162 (2009). doi:10.1002/mrm.21802.
40. Chan, D. D. et al. Mechanical deformation and glycosaminoglycan content changes in a rabbit annular puncture disc degeneration model. *Spine (Phila Pa 1976)* **36**, 1438–1445 (2011). doi:10.1097/BRS.0b013e3181f8be52.
41. Chan, D. D., Neu, C. P. & Hull, M. L. In situ deformation of cartilage in cyclically loaded tibiofemoral joints by displacement-encoded MRI. *Osteoarthritis Cartilage* **17**, 1461–1468 (2009).
42. Chan, D. D., Neu, C. P. & Hull, M. L. Articular cartilage deformation determined in an intact tibiofemoral joint by displacement-encoded imaging. *Magn Reson Med* **61**, 989–993 (2009). doi:10.1002/mrm.21927.
43. Neu, C. P. & Walton, J. H. Displacement encoding for the measurement of cartilage deformation. *Magn Reson Med* **59**, 149–155 (2008). doi:10.1002/mrm.21464.
44. Butz, K. D., Chan, D. D., Nauman, E. A. & Neu, C. P. Stress distributions and material properties determined in articular cartilage from MRI-based finite strains. *J Biomech* **44**, 2667–2672 (2011). doi:10.1016/j.jbiomech.2011.08.005.
45. Nissi, M. J. et al. Estimation of mechanical properties of articular cartilage with MRI—dGEMRIC, T2 and T1 imaging in different species with variable stages of maturation. *Osteoarthritis Cartilage* **15**, 1141–1148 (2007). doi:10.1016/j.joca.2007.03.018.
46. Nieminen, M. T. et al. Prediction of biomechanical properties of articular cartilage with quantitative magnetic resonance imaging. *J Biomech* **37**, 321–328 (2004).
47. Samosky, J. T. et al. Spatially-localized correlation of dGEMRIC-measured GAG distribution and mechanical stiffness in the human tibial plateau. *J Orthop Res* **23**, 93–101 (2005). doi:10.1016/j.orthres.2004.05.008.
48. Johannessen, W. & Elliott, D. M. Effect of degeneration on the biphasic material properties of human nucleus pulposus in confined compression. *Spine (Phila Pa 1976)* **30**, E724–E729 (2005).

Section VII

Cellular and Molecular Imaging Methods

Microenvironments—Native ECM

19

Nonlinear Optical Microscopy in Biomechanics

Jessica C. Mansfield, James S. Bell, Julian Moger, and C. Peter Winlove

CONTENTS

The macroscopic mechanical properties of biological tissues such as cartilage and blood vessels are central to their functions and have been extensively investigated. However, knowledge of their mechanical properties on a microscopic scale is essential to fully understand the relationships between molecular structure and organization and mechanics, and thereby to assess the functional significance of molecular changes occurring during development, aging, and disease. Through its ability to simultaneously visualize cells and the principal fibrous proteins of the extracellular matrix and elastin in living tissues without the use of exogenous contrast agents, nonlinear microscopy provides an invaluable tool in this endeavor. We illustrate this using results from our recent investigations on the micromechanics of articular cartilage, small arteries and veins, and adipose tissue.

19.1 Introduction

The mechanical properties of tissues such as articular cartilage, the intervertebral disk, and blood vessels, known collectively as the connective tissues, are central to their functions. Their macroscopic mechanical properties have been extensively studied for at least 50 years (summarized in Fung[1]). A major outcome of this work has been a demonstration of the importance of changing mechanical properties in disease and with age and environment. The connective tissues have a low cell content, and their mechanical properties derive largely from the extracellular matrix. The extracellular matrix consists of a network of fibrous proteins immersed in a viscoelastic gel that is rich in proteoglycans. Proteoglycans are large, anionic molecules consisting of a protein core complexed to glycosaminoglycan side chains.[2] The fibrous proteins comprise the collagens, a family of rather rigid proteins with characteristic triple-helical domains, which are assembled into fibers (most commonly types I, II, and III collagens) or networks (e.g., type IV collagen) and elastin. Elastin is a unique protein possessing the unusual attribute for a protein of long-range elasticity and occurs in tissues in association with a number of other microfibrillar glycoproteins, forming tissue-specific networks of elastic fibers. The cells of the connective tissues frequently serve only the subsidiary role of repairing and maintaining the extracellular matrix.

Mechanical testing on macroscopic tissue samples, underpinned by histological assessment of tissue architecture, has provided a broad understanding of the roles of collagen and elastin networks and the proteoglycans in the mechanics of many tissues, particularly articular cartilage, the intervertebral disk, and large blood vessels.

This has informed the development of phenomenological or structurally based models of tissue mechanics, reviewed by Gribbon et al.[3] A vast body of biochemistry and cell and molecular biology research continues to reveal the additional complexity and heterogeneity in the composition of the extracellular matrix and demonstrates a rich synergy between cellular function and the composition of the extracellular matrix. Both lines of research promise insights into the normal processes of tissue growth and development as well the progression of disease. However, a deeper understanding of tissue mechanics is urgently required to realize these objectives. In particular, we need to

1. Understand the mechanical properties of each molecule of the extracellular matrix and the microscale mechanics of the structures it forms: this is necessary to determine the functional significance of changes, both at the molecular and the microscopic levels, in tissues and thereby to interpret the wealth of biochemical data in terms of causative or compensatory changes in matrix structure and organization.
2. Understand the mechanical interactions between cells and matrix: studies on cells in vitro have demonstrated a wide spectrum of responses to diverse mechanical forces. It is now essential to relate these observations to the forces communicated to cells in vivo through the extracellular matrix.

To achieve these objectives, it is necessary to investigate the structure and function (and dysfunction) of the composite components of living connective tissues on a subcellular scale. Optical microscopy is attractive because the diffraction limit of visible light provides subcellular spatial resolution, and its relatively noninvasive nature is compatible with living samples. However, traditional microscopy techniques, which derive contrast from linear optical processes, are limited to studies of surface structures or thin tissue sections because at greater depths, light scattering blurs the images. To some extent, this can be overcome with confocal microscopy; however, in highly scattering tissues, the signal is heavily attenuated since optical sectioning is achieved via a detection pinhole that rejects all light that appears not to originate from the focus. Inevitably, the required imaging depths can only be achieved by increasing the excitation power to a level that causes tissue damage. Moreover, other than a very limited number of tissue components that exhibit sufficiently strong autofluorescence, identification is derived from extrinsic fluorescent labeling, which limits application in living tissues.

During the past two decades, new optical microscopy techniques have emerged that use nonlinear light–matter interactions to generate signal contrast.[4–6] Due to several advantages over traditional methods, nonlinear optical (NLO) microscopy has become a powerful tool for studying living tissues. The nonlinear dependence on excitation intensity provides inherent ("confocal-like") 3-D sectioning, and the near-IR excitation provides both superior optical penetration and reduced photodamage. Moreover, techniques have emerged that allow quantitative label-free biochemical analysis within living tissues. These techniques derive chemically specific image contrast based on molecular vibrational resonances within a sample and are providing a potent new form of real-time bioimaging. The aim of this chapter is to illustrate the exciting opportunities provided by nonlinear microscopy and spectroscopy to pursue important, unanswered questions in tissue mechanics. We shall first review the available methodologies and then illustrate their application by summarizing some of the ongoing research in our laboratory. Finally, we shall point out further opportunities and challenges for future technical development.

19.2 Imaging Connective Tissue with NLO Microscopy

The immediate challenge for imaging the responses of tissues to mechanical loads is visualization of the fibrous proteins (elastin and collagen) and cellular bodies. Since the development of NLO imaging, it has become apparent that different techniques are best suited for different tissue components, giving rise to the instrumentation challenge of combining multiple techniques on a single platform.

19.2.1 Visualizing Elastin with Two-Photon Fluorescence

Two-photon fluorescence (TPF) of molecules is a nonlinear process involving the absorption of two photons with a combined energy that is sufficient to induce a molecular transition to an excited electronic state.[7] The absorption of multiple photons only occurs efficiently at very high light intensities as produced by ultrafast laser pulses, and because efficiency scales nonlinearly with the excitation intensity, it has an additional advantage of being confined to the focal volume of the microscope objective (where the photon flux is greatest) providing "confocal-like" optical sectioning. Moreover, by operating at twice the wavelength typically used for linear fluorescence, the excitation undergoes significantly less scattering, which extends depth penetration into tissues.

In connective tissues, elastin is readily imaged as it exhibits strong TPF, which provides better specificity and sensitivity than most classical histological or immunohistochemical techniques.[8,9] It has revealed extensive and potentially mechanically important networks of elastin in tissues such as cartilage and disk, where its presence has previously been overlooked.[10]

19.2.2 Visualizing Collagen with Second Harmonic Generation

Molecules that lack inversion symmetry are able to convert light to its second harmonic, i.e., emit light at twice the optical frequency of the excitation light. This process shares the benefit of other nonlinear techniques of high spatial resolution, and, as with TPF, pulsed excitation is required for efficiency. Because of its unusual molecular structure and its high degree of crystallinity, collagen is, by far, the strongest source of second harmonic generation (SHG) in animal tissue and provides a useful contrast mechanism for studying its distribution.[11–13]

19.2.3 Visualizing Cellular Components with Coherent Raman Scattering Microscopy

Coherent Raman scattering (CRS) microscopy has recently emerged as a powerful tool for providing image contrast of cellular structures based on the molecular vibrations of cellular lipids. CRS microscopy is performed by focusing two synchronized ultrafast pulse trains (pump and Stokes beams) into a sample with their optical frequency difference matched to a Raman active mode of a molecular species of interest. This leads to the coherent excitation of a nonlinear polarization that generates an optical signal with sufficient strength to be used as a contrast mechanism for bioimaging. As with TPF and SHG, CRS benefits from intrinsic "confocal-like" spatial resolution.

Two CRS processes are commonly used for biological microscopy: coherent anti-Stokes Raman scattering (CARS) and stimulated Raman scattering (SRS). CARS is a four-wave mixing (FWM) process in which the pump and Stokes beams, at frequencies ω_p and ω_S, respectively, interact within the sample. Matching the difference frequency ($\omega_p - \omega_S$) with a Raman active molecular vibration resonantly enhances the FWM signal, which is detected at the anti-Stokes frequency ($\omega_{as} = 2\omega_p - \omega_S$). However, the interaction of the same frequencies with the molecular electronic levels also generates an FWM signal that is spectrally indistinguishable from CARS. This nonresonant contribution originates solely from the third-order polarizability and is independent of the Raman resonance being probed; it reduces sensitivity and complicates spectral analysis.[14,15]

Recently, attention has turned to SRS as an alternative contrast mechanism for CRS microscopy.[16–19] SRS also utilizes pump and Stokes beams with the difference ($\omega_p - \omega_S$) tuned to a Raman resonance. However, in SRS, amplification of the Raman signal is achieved by virtue of stimulated excitation. The intensity of the Stokes beam, I_S, experiences a gain, ΔI_S (stimulated Raman gain), and the intensity of the pump beam, I_p, experiences a loss, ΔI_p (stimulated Raman loss). SRS can be viewed as the heterodyne detection of the NLO signal with the excitation field, and, in contrast to CARS, when ($\omega_p - \omega_S$) does not match a vibrational resonance, SRS cannot be detected. The spectral response of SRS matches that of spontaneous Raman scattering, which simplifies spectral analysis.

CRS microscopy can, in principle, be employed to image the distribution of any Raman active bond. However, it is particularly sensitive to lipid-rich structures because of the high density of CH bonds and has found niche application in this area for live cells and tissues.[20–23]

19.2.4 Multimodal NLO Imaging

TPF, SHG, and CRS provide contrast of elastin, collagen, and cell membranes, respectively, and combining the three modalities into a single "multimodal" platform provides an ideal tool for imaging the distribution and interaction of the principal components of connective tissues. For given excitation wavelengths, the different NLO processes generate signals that can be spectrally separated into independent image channels that are optically coregistered with submicron resolution.[24] There are several different optical setups and laser systems that can be successfully used for multimodal nonlinear microscopy. In our laboratory, we use a picosecond laser and OPO to excite coherent Raman and a femtosecond Ti:Sapphire laser system to excite TPF and SHG. Imaging is performed on modified confocal laser-scanning microscopes with a noninverted microscope being used for tensile loading and pressure myograph studies[25] and an inverted microscope being used for SRS studies.[26]

19.3 Tissue Micromechanics

In order to investigate the micromechanical properties of tissues, mechanical loading rigs that are compact enough to fit onto a microscope stage are required. We have constructed tensile and compressive loading rigs and a pressure myograph to fit onto the microscope stage (Figure 19.1). Each design incorporates a sensitive load cell to measure the applied force, and micrometers for accurate application of strain. This allows correlation of image data and macroscopic stress–strain behavior. In the tensile loading configuration, samples are glued

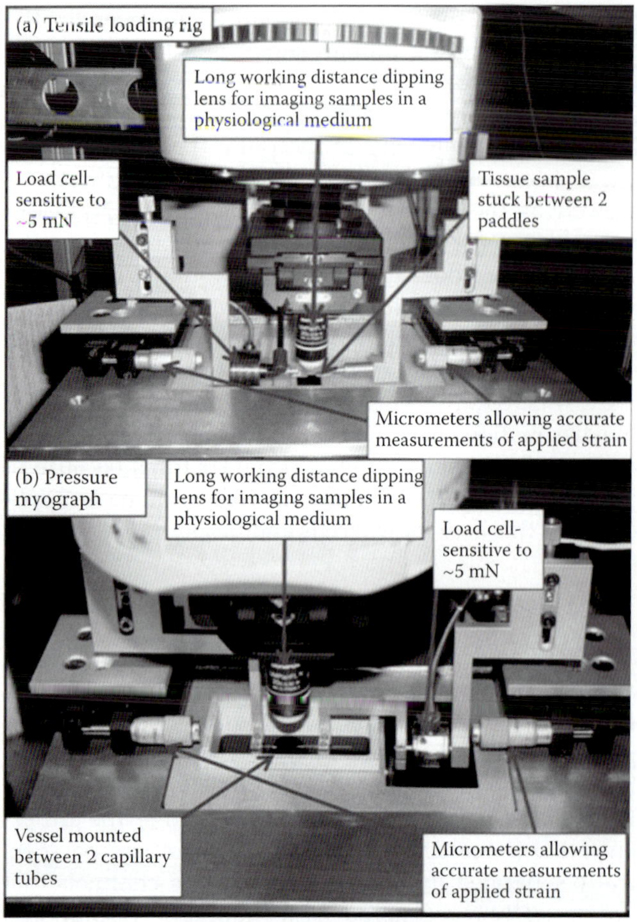

FIGURE 19.1
Mechanical loading rigs constructed on the stage of an Olympus BX51 microscope.

to the loading arms and immersed in physiological media. In the myograph configuration, small vessels are sutured to glass capillary tubes and perfused under physiological conditions.

In a typical experiment, 3D images of a set region of interest are acquired of the unloaded tissue and then under a series of progressive loads. In thick and highly scattering samples (e.g., cartilage and adipose tissue), all the signals are collected in the epi-direction, while in thinner samples (e.g., small resistance arteries), it is more efficient to collect the CARS and SHG signal in the forward direction. In order to measure tissue strain, grids or squares can be photobleached onto the samples to act as fiducial markers in the TPF image. Alternatively, distinctive patterns of cells and fibers can be used as intrinsic markers. From these image series, the overall strain over the image area can be investigated along with the strain on individual cells and fiber networks.

In Sections 19.3.1 to 19.3.3, we demonstrate the application of multimodal nonlinear microscopy to the study of three representative tissues: articular cartilage, small blood vessels, and adipose tissue.

19.3.1 Articular Cartilage

Articular cartilage covers the ends of the long bones and acts as a low friction bearing and shock absorber. Its macroscopic mechanical properties and their relationships to composition and morphology have been extensively investigated.[27–29] Cartilage mechanics is governed by interplay between the tensile properties of the type II collagen fibers that constitute approximately 20% of the dry weight of the tissue and the osmotic properties and hydraulic conductivity of the proteoglycan gel. However, recently, attention has turned toward the mechanical properties of the chondrocytes, and their responses to applied forces, either in isolation or model matrices or, in a few cases, in intact tissue.[30,31]

Figure 19.2 shows composite NLO images of the collagen matrix, the network of elastin fibers and the cells (chondrocytes), and their pericellular environment in fresh tissue explants.[32] The dense matrix of type II collagen fibers is clearly visualized in the SHG images, and although individual fibers are below the diffraction limit, the predominant fiber direction can be found from either polarization-sensitive measurements[33] or from Fourier transform analysis of the images.[34] TPF reveals

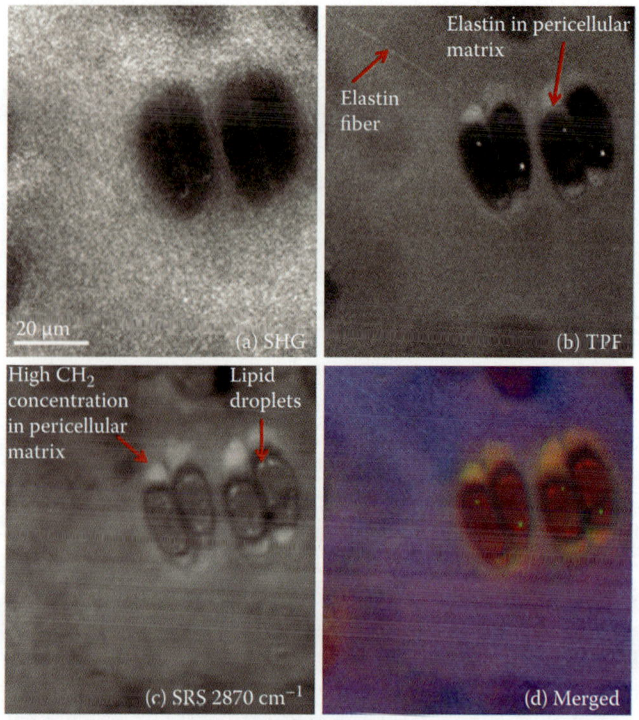

FIGURE 19.2
(a) SHG image of cartilage—here we can see the dense collagen matrix and the cell lacunae showing up as dark holes. (b) TPF image of the same area showing an elastin fiber and elastin fluorescence in the pericellular matrix. (c) SRS image of articular cartilage showing the chondrocytes within the lacunae, with lipid droplets and some increased lipid concentration in the pericellular matrix. (d) Merged image where blue = SHG, green = TPF, and red = SRS.

elastin fibers that previously had been overlooked as a significant component of the cartilage matrix in the superficial zone, as well as elastin fluorescence in the pericellular matrices of superficial chondrocytes.[10] CRS provides information on the morphology and cytoplasmic composition of the chondrocytes. SRS tuned to 2850 cm^{-1} (the symmetric CH_2 stretching mode in lipids) shows cell boundaries and intracellular lipid droplets as well as lipids in the pericellular matrix.[26]

Cartilage is exposed to a range of forces in vivo, and tensile loading of cartilage is useful for exploring the connectivity of the extracellular matrix fibers. Representative images of the superficial zone of articular cartilage under tensile load are shown in Figure 19.3. From these images, we are able to determine the reorientation and displacement of both matrix and cells. Interest in the elastin network has developed only recently, largely because of the improved visualization provided by TPF,[10] and a current question is whether elastin contributes to the mechanical properties of the superficial zone. Although it is not, at present, possible to determine intrafibrillar stresses and strains, the question has been approached by investigating the reorientation of the elastin network under tensile load. The elastin fibers are generally long and straight, and therefore, their orientation can be measured directly from 3D TPF images. It was established that fiber reorientation does not follow the strain field in the surrounding matrix, suggesting that the elastin network experiences different stresses to the rest of the local matrix (manuscript in preparation).

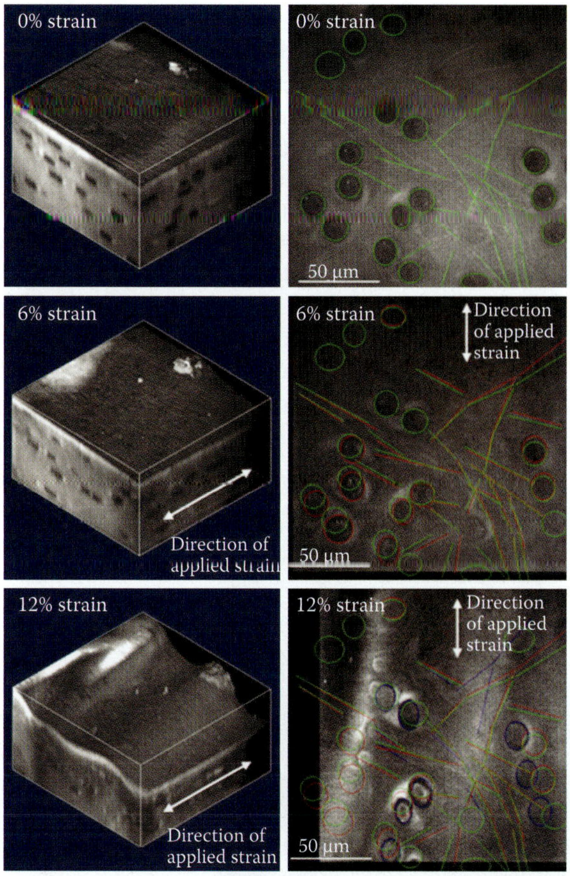

FIGURE 19.3
On the LHS are 3D reconstructions of image stacks taken at 0%, 6%, and 12% strain (volume reconstructed = 170 × 170 × 92 μm^3). RHS shows an image frame taken from approximately 12 μm below the articular surface. Overlays are included to show how individual cells and elastin fibers have reorientated between the progressive strains. At 12% strain, many of the original cells and elastin fibers had moved away from the field of view.

It is generally a nontrivial task to determine strain fields from sequences of 3D images, and the development of novel point recognition and tracking algorithms is required. A stochastic search algorithm was developed to find cells and determine their size and locations, complemented by a Bayesian point pattern-matching algorithm, to match patterns in the locations of cells in the image stacks before and after the strain is applied (see refs. 35 and 36 for further details). The pattern-matching algorithm assumes that the two point sets are related by a linear transformation consisting of a rotation, scaling, and translation. Deviations from this relation are then calculated and linearly interpolated across the image, yielding a map of the perturbation in strain. Ellipses fitted to the cells provide both a means of measuring strain on individual cells as well as statistics on the overall strains experienced by the cells (Figure 19.4a).

The local strain deviation from the mean (LSD) is shown in Figure 19.4b. A variation in the LSD with depth indicates the existence of a shear gradient that shows that the surface zone is stiffer than underlying zones.[37] This is consistent with an earlier observation from tensile-testing sequential tissue sections.[38,39]

The small-scale variations in the LSD represent variations in the local mechanical properties, which can be related to relative slippage of leaves of collagen fibers.

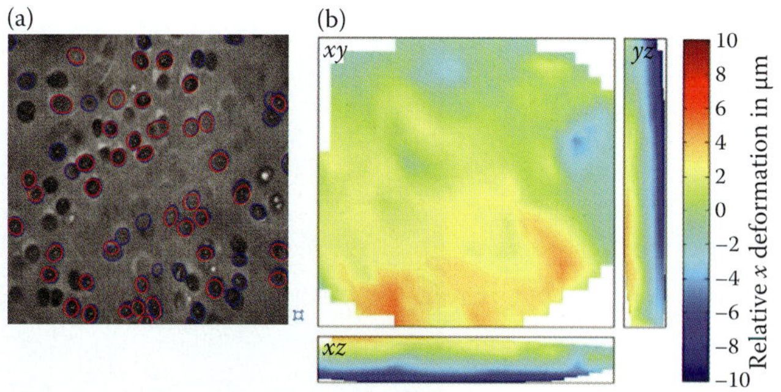

FIGURE 19.4
(a) TPF image of a sample of cartilage, taken at a depth of 29 μm from the surface. Blue ellipses correspond to automatically detected cells, and red ellipses correspond to tracked cells after application of strain. (b) LSD for the region in the left image. x–z and y–z cut-through passing through the center of the x–z image.

An important physiological implication of this observation is that there are corresponding variations in the mechanical stimuli received by different cells.

19.3.2 Small Blood Vessels

The mechanical properties of large arteries and veins and their relationships to the organization of cellular, elastin, and collagen content, both in health and disease, have been extensively characterized.[40] However, very little is known about small resistance arteries (vessels 100–400 μm in diameter),[40] which play a vital role in the distribution and regulation of blood flow and are thereby implicated in conditions ranging from hypertension to diabetes. The myograph described earlier has enabled us to determine the 3D structure of these small vessels and hence to explore the roles of the collagen and elastin networks in determining their passive mechanical properties. Current studies concern the interplay between matrix and smooth muscle in determining the mechanical response to changes in luminal pressure and vasoactive agents.

Figure 19.5 shows TPF and SHG images of a perfused microvessel from human adipose tissue, at different depths through the vessel cross section, at zero transmural pressure. In the section through the midpoint of the vessel, SHG reveals fibrous collagen only in the adventitial region. The TPF image shows two layers of elastin: an internal elastic lamina (similar to that observed in larger arteries) and an outer lamina (which is very different from the complex elastin network of large arteries). The TPF image of a tangential plane containing the internal elastic lamella reveals that the elastin fibers are highly ordered and longitudinally aligned in the unpressurized vessel. A tangential section through the adventitia shows the collagen network and also thinner fibers of elastin arranged in a more disordered, herringbone pattern.

The effect of increased lumenal pressure is shown in Figure 19.6. In these images, the vessel axis is oriented at a small angle to the image plane, which means that the lumen is visible at the bottom of the image and the adventitia at the top. At a physiological pressure of 30 mmHg, the fibers of the internal elastic lamella have reoriented to form a honeycomb lattice, and the wall has thinned. These data form the basis of structurally based finite-element models of wall mechanics, still under development, but the provisional interpretation is that, in contrast to large blood vessels, a component of the elastin network is responsible for supporting longitudinal forces arising from blood flow or the surrounding tissue.

19.3.3 Adipose Tissue

Adipose tissue provides a store of dietary fats at various locations throughout the body. Its macroscopic mechanical properties are pertinent to maintaining its integrity and to its additional functions in mechanical protection of internal organs. However, the current obesity epidemic has focused attention on a possible relationship between the mechanics of the tissue and its ability to store fats.[41–44] The adipocytes swell as they take up fat, and it has been suggested that if swelling is limited by constraints imposed by the pericellular matrix, dietary fats are deposited at exogenous sites.

Adipocytes store fats in large vacuoles that are readily visualized by CRS (see Figure 19.7b). Information on the

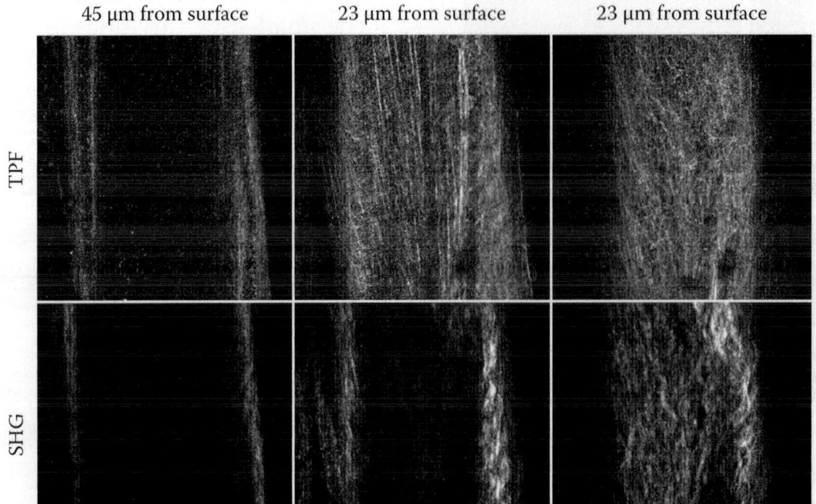

FIGURE 19.5
TPF and SHG images of a human resistance artery, taken at depths of 45, 23, and 15 μm from the outer surface.

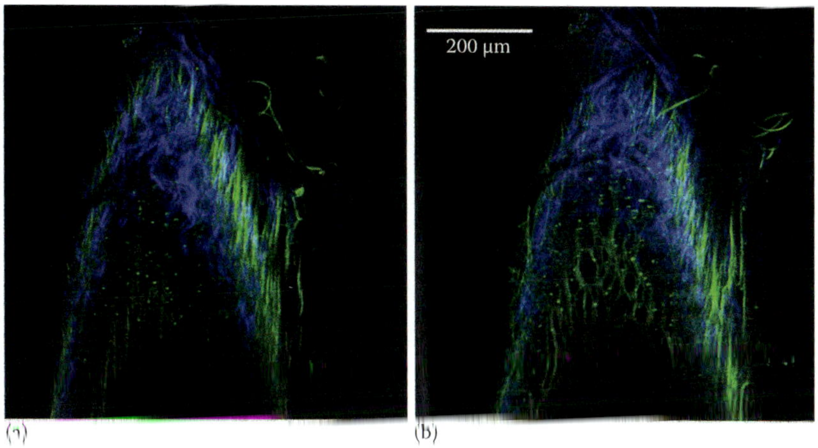

FIGURE 19.6
TPF (green) and SHG (blue) images of a 200 μm diameter vessel at neutral lumen pressure (a) and 30 mmHg lumen pressure (b).

lipid composition within the adipocytes can be obtained by acquiring CRS images over the range of Raman shifts spanning the CH vibrational region. A representative spectrum is shown in Figure 19.7a, where the spectral profile is indicative of unsaturated lipids.[45,46]

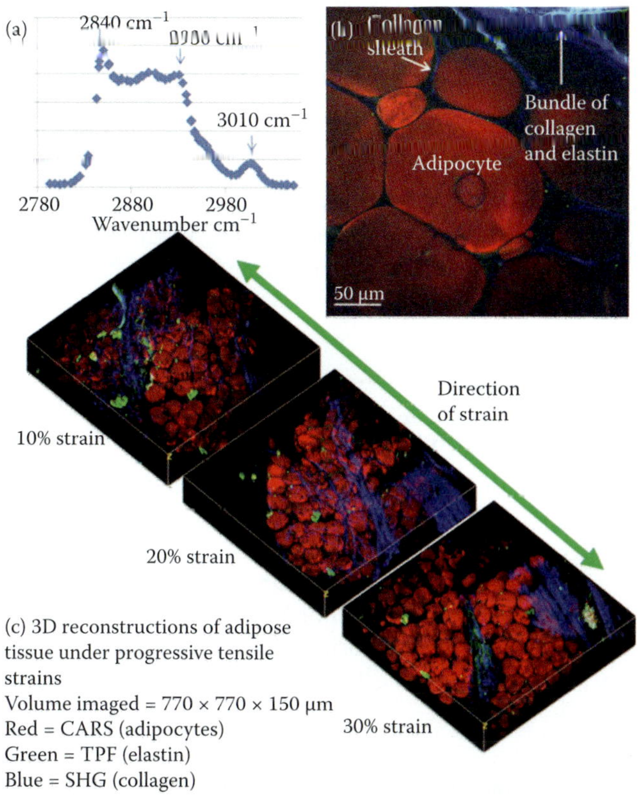

FIGURE 19.7
Adipose tissue imaged under progressive tensile strains. (a) SRS spectra taken of an adipocyte in the CH vibrational region. (b) Multimodal image of adipose tissue where red = CARS at 2840 cm^{-1} (showing lipids), green = TPF (showing elastin), and blue = SHG (showing collagen). (c) Images taken from a strip of adipose tissue imaged at progressive strains.

Adipocytes are surrounded by a pericellular matrix consisting of a basement membrane containing type IV collagen and an outer sheath containing fibrous collagens (types I, III, V, and VI), proteoglycans, and glycoproteins.[42,47] In addition, there are fiber bundles running through the tissue, which contain collagen types I and III along with elastin fibers. The type IV collagen in the basement membrane cannot be visualized with SHG; however, we are able to image the fibrous collagens in the sheath around the cells and in the bundles within the tissue (see Figure 19.7b).

The presumption is that the fibers provide whole-organ structure and mechanical support, while the pericellular matrix determines the behavior of individual cells. Micromechanical testing on the multiphoton microscope provides a means of testing this hypothesis.

We have investigated the microstructural response of adipose tissue to tensile loading. Images of a section of adipose tissue under incremental strain are shown in Figure 19.7c. There is little apparent deformation of individual adipocytes under strain, and the strain in the matrix is much less homogeneous than in cartilage or blood vessels. These unexpected observations raise difficulties in quantitative analysis and in developing models of whole tissue mechanics. However, the primary clinical interest in this tissue is in understanding the interrelationships between extracellular matrix mechanics and cell swelling; our ability to study intact, living tissue provides an unrivaled opportunity to achieve this, and such experiments are in progress.[25]

19.4 Conclusion

This chapter has illustrated the power of nonlinear microscopy to investigate tissue mechanics. Articular cartilage is probably the best-characterized connective tissue, both biochemically and biomechanically. Nevertheless, nonlinear microscopy has revealed novel details of micromechanics, particularly of its heterogeneity that, on one hand, poses a challenge to mechanical modelers, and, on the other, raises new questions concerning the role of mechanics in physiology and pathology. For blood vessels, nonlinear microscopy has provided the means of extending our understanding of the relationships between structure and mechanics down to the smallest components of the vascular tree. Adipose tissue is almost totally unexplored from a biomechanical perspective, and it is difficult to believe that the physiologically most important questions concerning mechanical interactions between cells and extracellular matrix could be tackled without the use of nonlinear microscopy.

These techniques would be applicable to almost all biological tissues where the mechanical properties are of interest (for example, lymphatics, ligaments, tendons, heart valves, and heart strings [chordae tendonae]). They would also be highly valuable to tissue engineering studies where the tissue-engineered constructs need to have mechanical properties as close as possible to native tissue on both the macroscopic and microscopic scale. Further technological developments that improve the signal-to-noise ratio and maximum imaging depth in tissue will allow nonlinear microscopy to be applied to even more biomechanical problems. Tissue autofluorescence is weak in comparison to many exogenous fluorophores, and this means that, when looking at rapid processes especially in in vivo applications, uses of exogenous dyes may still be preferable.[48] Currently, new microscope objectives specialized for nonlinear microscopy and the application of adaptive optics to nonlinear microscopy[49–51] are areas that are working to improve the image quality and depth penetration.

In this chapter, we have used spectral CRS imaging only to delineate the lipid-rich structures because of the strong signal generated by the high density of CH bonds in their hydrocarbon tails. However, SRS imaging has great potential in tapping all the information contained in the Raman spectrum. Elegant studies have been undertaken of cellular biochemistry using spontaneous Raman imaging,[52,53] and we have alluded above to the possibility of exploiting the advantages of CRS to follow changes in cellular metabolism associated with the application of mechanical stress to the tissue. It is also possible that, if appropriate markers can be identified in the Raman spectra of the proteins, spectral imaging can be used to map the distribution of cross-links, glycation sites, or other important features of the fibrous proteins.

A longer-term challenge in studying tissue micromechanics is to make direct measurements of fibril-level strains. A possible methodology is illustrated by recent work using spontaneous Raman spectroscopy to determine the molecular mechanisms of elasticity in matrix fibers. The collagen spectrum is reported to show shifts in peak positions and reorientation of peptide bonds when fibers are strained,[54] and elastin fibers show realignment of various bonds and changes in hydration water.[55] The additional sensitivity of CRS opens the possibility of repeating these measurements in the intact tissue as markers of fiber strain.

A final challenge for spectroscopic imaging is to explore the role of water in tissue mechanics. The extracellular matrix contains a large proportion of water whose redistribution under applied loads is a major determinant of tissue mechanics. This has led to the development of complex multiphasic or poroelastic models of tissue mechanics, and their experimental verification urgently requires measurements of pathways and rates of water movement on a microscopic scale. Experiments based on the spectroscopic monitoring of deuterium oxide are presently in progress in our laboratory that are designed to fulfill this need.

This chapter has focused on the potential of SHG, TPF, and coherent Raman techniques for investigating tissue mechanics. However, in addition to this, there are other nonlinear optical techniques that could be promising in this field such as transient absorption and photothermal lensing techniques.[56] These techniques are especially useful for imaging tissue components, which have a strong optical absorbance but do not exhibit fluorescence. Transient absorption techniques have proved useful for imaging skin pigments such as melanins,[57] whereas photothermal lensing has been applied to imaging red blood cells in the vasculature.[58–60]

Acknowledgments

A number of colleagues have contributed to the work summarized above, notably Ellen Green, Nadia Alkhouli, Aminat Adio, Kenton Arkill, and Kevyn White. The research on articular cartilage is funded by Arthritis Research UK (Grant No. 19432) and that on the microcirculation by the British Heart Foundation (Grant No. PG/11/17/28788).

References

1. Fung, Y. C. *Biomechanics Mechanical Properties of Living Tissues* (Springer-Verlag, London, 1981).
2. Comper, W. D. & Laurent, T. C. Physiological function of connective tissue polysaccharides. *Physiological Reviews* **58**, 255–315 (1978).

3. Gribbon, P. M., Maroudas, A., Parker, K. H. & Winlove, C. P. Water and solute transport in the extracellular matrix: Physical principles amd macromolecular determinants. In *Connective Tissue Biology: Integration and Reductionism* (eds. R. K. Reed & K. Rubin) 95–123 (Portland Press, London, 1997).
4. Zipfel, W. R. et al. Live tissue intrinsic emission microscopy using multiphoton-excited native fluorescence and second harmonic generation. *Proceedings of the National Academy of Sciences of the United States of America* **100**, 7075–7080 (2003).
5. Konig, K. Multiphoton microscopy in life sciences. *Journal of Microscopy-Oxford* **200**, 83–104 (2000).
6. Zipfel, W. R., Williams, R. M. & Webb, W. W. Nonlinear magic: Multiphoton microscopy in the biosciences. *Nature Biotechnology* **21**, 1369–1377 (2003).
7. Diaspro, A. & Sheppard, C. J. R. Two-photon microscopy: Basic principles and architectures. In *Confocal and Two-Photon Microscopy: Foundations, Applications and Advances* (ed. A. Diaspro) 39–73 (Wiley-Liss Inc., New York, 2002).
8. Zoumi, A., Yeh, A. & Tromberg, B. J. Imaging cells and extracellular matrix in vivo by using second-harmonic generation and two-photon excited fluorescence. *Proceedings of the National Academy of Sciences of the United States of America* **99**, 11014–11019 (2002).
9. Zoumi, A., Lu, X. A., Kassab, G. S. & Tromberg, B. J. Imaging coronary artery microstructure using second-harmonic and two-photon fluorescence microscopy. *Biophysical Journal* **87**, 2778–2786 (2004).
10. Mansfield, J. C. et al. The elastin network: Its relationship with collagen and cells in articular cartilage as visualized by multiphoton microscopy. *Journal of Anatomy* **215**, 682–691 (2009).
11. Roth, S. & Freund, I. Second harmonic generation in collagen. *Journal of Chemical Physics* **70**, 1637–1643 (1978).
12. Freund, I., Deutsch, M. & Sprecher, A. Connective-tissue polarity—Optical 2nd-harmonic microscopy, crossed-beam summation, and small-angle scattering in rat-tail tendon. *Biophysical Journal* **50**, 693–712 (1986).
13. Campagnola, P. J. et al. Three-dimensional high-resolution second harmonic generation imaging of endogenous structural proteins in biological tissues. *Biophysical Journal* **81**, 493–508 (2002).
14. Cheng, J. X. & Xie, X. S. Coherent anti-stokes Raman scattering microscopy: Instrumentation, theory, and applications. *Journal of Physical Chemistry B* **108**, 827–840 (2004).
15. Cheng, J. X., Volkmer, A. & Xie, X. S. Theoretical and experimental characterization of coherent anti-Stokes Raman scattering microscopy. *Journal of the Optical Society of America B-Optical Physics* **19**, 1363–1375 (2002).
16. Freudiger, C. W. et al. Label-free biomedical imaging with high sensitivity by stimulated Raman scattering microscopy. *Science* **322**, 1857–1861 (2008).
17. Nandakumar, P., Kovalev, A. & Volkmer, A. Vibrational imaging based on stimulated Raman scattering microscopy. *New Journal of Physics* **11**, 033026 (2009).
18. Ozeki, Y. & Itoh, K. Stimulated Raman scattering microscopy for live-cell imaging with high contrast and high sensitivity. *Laser Physics* **20**, 1114–1118 (2010).
19. Ozeki, Y. et al. High-speed molecular spectral imaging of tissue with stimulated Raman scattering. *Nature Photonics* **6**, 844–850 (2012).
20. Cheng, J. X., Jia, Y. K., Zheng, G. F. & Xie, X. S. Laser-scanning coherent anti-stokes Raman scattering microscopy and applications to cell biology. *Biophysical Journal* **83**, 502–509 (2002).
21. Nan, X. L., Cheng, J. X. & Xie, X. S. Vibrational imaging of lipid droplets in live fibroblast cells with coherent anti-Stokes Raman scattering microscopy. *Journal of Lipid Research* **44**, 2202–2208 (2003).
22. Wang, H. F., Fu, Y., Zickmund, P., Shi, R. Y. & Cheng, J. X. Coherent anti-Stokes Raman scattering imaging of axonal myelin in live spinal tissues. *Biophysical Journal* **89**, 581–591 (2005).
23. Slipchenko, M. N., Le, T. T., Chen, H. T. & Cheng, J. X. High-speed vibrational imaging and spectral analysis of lipid bodies by compound Raman microscopy. *Journal of Physical Chemistry B* **113**, 7681–7686 (2009).
24. Yue, S., Slipchenko, M. N. & Cheng, J.-X. Multimodal nonlinear optical microscopy. *Laser and Photonics Reviews* **5**, 496–512 (2011).
25. Alkhouli, N. et al. The mechanical properties of human adipose tissues and their relationships to the structure and composition of the extracellular matrix. *American Journal of Physiology, Endocrinology and Metabolism* **305**, 1427–1435 (2013).
26. Mansfield, J., Moger, J., Green, E., Moger, C. & Winlove, C. P. Chemically specific imaging and in-situ chemical analysis of articular cartilage with stimulated Raman scattering. *Journal of Biophotonics* **6**, 803–814 (2013).
27. Ratcliffe, A. & Mow, V. C. Articular cartilage. In *Extracellular Matrix*, Vol. 1 (ed. W. D. Comper) 235–302 (Harwood Academic Publishers, Amsterdam, 1996).
28. Urban, J. P. G. Solute transport in articular cartilage and interverbal disc. In *Connective Tissue Matrix*, Vol. Part 2 (ed. D. W. L. Hukins) 44–65 (Macmillan Press, Basingstoke, UK, 1990).
29. Roth, V., Mow, V. & Grodzinsky, A. J. Biophysical and electromechanical properties of articular cartilage. In *Skeletal Research: An Experimental Approach* (eds. D. J. Simmons & A. S. Kunin) 301–341 (Academic Press, New York, 1979).
30. Guilak, F., Jones, W. R., Ting-Beall, H. P. & Lee, G. M. The deformation behavior and mechanical properties of chondrocytes in articular cartilage. *Osteoarthritis and Cartilage* **7**, 59–70 (1999).
31. Chahine, N. O., Hung, C. T. & Ateshian, G. A. In-situ measurements of chondrocyte deformation under transient loading. *European Cells and Materials* **13**, 100–111 (2007).
32. Yeh, A. T. et al. Nonlinear optical microscopy of articular cartilage. *Osteoarthritis and Cartilage* **13**, 345–352 (2005).
33. Mansfield, J. C., Winlove, C. P., Moger, J. & Matcher, S. J. Collagen fiber arrangement in normal and diseased cartilage studied by polarization sensitive nonlinear microscopy. *Journal of Biomedical Optics* **13**, 044020 (2008).
34. Mansfield, J. C. & Winlove, C. P. A multi-modal multiphoton investigation of microstructure in the deep zone and calcified cartilage. *Journal of Anatomy* **220**, 405–416 (2012).

35. Bell, J. S., Christmas, J., Mansfield, J. C., Everson, R. M. & Winlove, C. P. Micromechanical response of articular cartilage to tensile load measured using nonlinear microscopy. *Acta Biomaterialia* **10**, 2574–2581 (2014).
36. Christmas, J., Everson, R. M., Bell, J. S. & Winlove, C. P. Inexact Bayesian point pattern matching for linear transformations. *Pattern Recognition* **47**, 3265–3275 (2014).
37. Bell, J. S., Winlove, C. P., Smith, C. W. & Dehghani, H. Modeling the steady-state deformation of the solid phase of articular cartilage. *Biomaterials* **30**, 6394–6401 (2009).
38. Kempson, G., Freeman, M. & Swanson, S. Tensile properties of articular cartilage. *Nature* **220**, 1127–1128 (1968).
39. Woo, S. L. Y., Akeson, W. & Jemmott, G. Measurements of nonhomogeneous, directional mechanical properties of articular cartilage in tension. *Journal of Biomechanics* **9**, 785–791 (1976).
40. Levick, J. R. *An Introduction to Cardiovascular Physiology* (Arnold, London, 2003).
41. Spencer, M. et al. Adipose tissue extracellular matrix and vascular abnormalities in obesity and insulin resistance. *Journal of Clinical Endocrinology and Metabolism* **96**, E1990–E1998 (2011).
42. Chun, T.-H. Peri-adipocyte ECM remodeling in obesity and adipose tissue fibrosis. *Adipocyte* **1**, 89–95 (2012).
43. Khan, T. et al. Metabolic dysregulation and adipose tissue fibrosis: Role of collagen VI. *Molecular and Cellular Biology* **29**, 1575–1591 (2009).
44. Divoux, A. et al. Fibrosis in human adipose tissue: Composition, distribution, and link with lipid metabolism and fat mass loss. *Diabetes* **59**, 2817–2825 (2010).
45. Wu, H. W. et al. In vivo lipidomics using single-cell Raman spectroscopy. *Proceedings of the National Academy of Sciences of the United States of America* **108**, 3809–3814 (2011).
46. Le, T. T., Duren, H. M., Slipchenko, M. N., Hu, C. D. & Cheng, J. X. Label-free quantitative analysis of lipid metabolism in living *Caenorhabditis elegans*. *Journal of Lipid Research* **51**, 672–677 (2010).
47. Comley, K. & Fleck, N. A. A micromechanical model for the Young's modulus of adipose tissue. *International Journal of Solids and Structures* **47**, 2982–2990 (2010).
48. Abusara, Z., Seerattan, R., Leumann, A., Thompson, R. & Herzog, W. A novel method for determining articular cartilage chondrocyte mechanics *in vivo*. *Journal of Biomechanics* **44**, 930–934 (2011).
49. Girkin, J. M., Poland, S. & Wright, A. J. Adaptive optics for deeper imaging of biological samples. *Current Opinion in Biotechnology* **20**, 106–110 (2009).
50. Wright, A. et al. Adaptive optics for enhanced signal in CARS microscopy. *Optics Express* **15**, 18209–18219 (2007).
51. Débarre, D. et al. Image-based adaptive optics for two-photon microscopy. *Optics Letters* **34**, 2495–2497 (2009).
52. van Manen, H.-J., Kraan, Y. M., Roos, D. & Otto, C. Intracellular chemical imaging of heme-containing enzymes involved in innate immunity using resonance Raman microscopy. *The Journal of Physical Chemistry B* **108**, 18762–18771 (2004).
53. van Manen, H.-J., Kraan, Y. M., Roos, D. & Otto, C. Single-cell Raman and fluorescence microscopy reveal the association of lipid bodies with phagosomes in leukocytes. *Proceedings of the National Academy of Sciences of the United States of America* **102**, 10159–10164 (2005).
54. Wang, Y. N., Galiotis, C. & Bader, D. L. Determination of molecular changes in soft tissues under strain using laser Raman microscopy. *Journal of Biomechanics* **33**, 483–486 (2000).
55. Green, E., Ellis, R. & Winlove, P. The molecular structure and physical properties of elastin fibers as revealed by Raman microspectroscopy. *Biopolymers* **89**, 931–940 (2008).
56. Ye, T., Fu, D. & Warren, W. S. Nonlinear absorption microscopy. *Photochemistry and Photobiology* **85**, 631–645 (2009).
57. Fu, D., Ye, T., Yurtsever, G., Warren, W. S. & Matthews, T. E. Two-color, two-photon, and excited-state absorption microscopy. *Journal of Biomedical Optics* **12**, 054004-054004-054008 (2007).
58. Fu, D. et al. High-resolution in vivo imaging of blood vessels without labeling. *Optics Letters* **32**, 2641–2643 (2007).
59. Moger, J. et al. Imaging cortical vasculature with stimulated Raman scattering and two-photon photothermal lensing microscopy. *Journal of Raman Spectroscopy* **43**, 668–674 (2012).
60. Lu, S., Min, W., Chong, S., Holtom, G. R. & Xie, X. S. Label-free imaging of heme proteins with two-photon excited photothermal lens microscopy. *Applied Physics Letters* **96**, 113701 (2010).

20

Collagen–Cell Interactions in Three-Dimensional Microenvironments

Sherry L. Voytik-Harbin and Bumsoo Han

CONTENTS

Scientists and engineers in academia and industry rely heavily on in vivo animal models and conventional two-dimensional (2D) cell culture for investigating various biological processes, chemical toxicity testing, as well as the discovery and development of next-generation drugs and medical therapies. While 2D culture platforms facilitate the study of molecular processes, these systems do not account for the complexity of biological signaling provided by the 3D extracellular microenvironment. This shortcoming has led to a substantial interest in the development of in vitro tissue systems that are human-cell based and recapitulate the complexities of 3D tissues, including the dynamic and spatiotemporal biochemical and physical signaling inherent to the in vivo microenvironment. This chapter discusses design considerations and approaches for development of 3D in vitro collagen-based tissue systems. In addition, an overview of state-of-the-art imaging tools and techniques for visualizing and quantifying multiscale cell–fluid–matrix interactions is provided.

20.1 Introduction

The transition from conventional 2D to emerging 3D cell and tissue culture formats is driving significant advances in basic and translational research related to cell biology (e.g., stem cell and tumor biology), tissue engineering, and drug development. This revolution in cell and tissue physiology is supported by a rapidly expanding body of compelling evidence showing that the interplay of biochemical and physical signaling inherent to 3D culture systems provides a more physiologically relevant context that supports the complex cellular processes observed in vivo (for reviews, see elsewhere[1–5]). In fact, the significant behavior differences observed between cells cultured in vitro using traditional 2D systems and their in vivo counterparts are largely owing to the dissociation of cells from their native 3D geometry and disruption of dynamic cell–fluid–matrix interactions inherent to the local tissue-specific microenvironment. Design principles for 3D in vitro systems, comprising individual or composite tissues, must continue to be inspired by knowledge of how cells sense and prioritize microenvironment cues in vivo, while emphasizing physicochemical parameters and mechanobiology principles.

Within tissues, the majority of cells reside within a 3D extracellular matrix (ECM), which represents primarily a network of type I collagen fibrils surrounded by an interstitial fluid containing numerous soluble protein and carbohydrate moieties. As shown in Figure 20.1, the collagen-fibril microstructure (defined by fibril density, interfibril branching [connectivity],

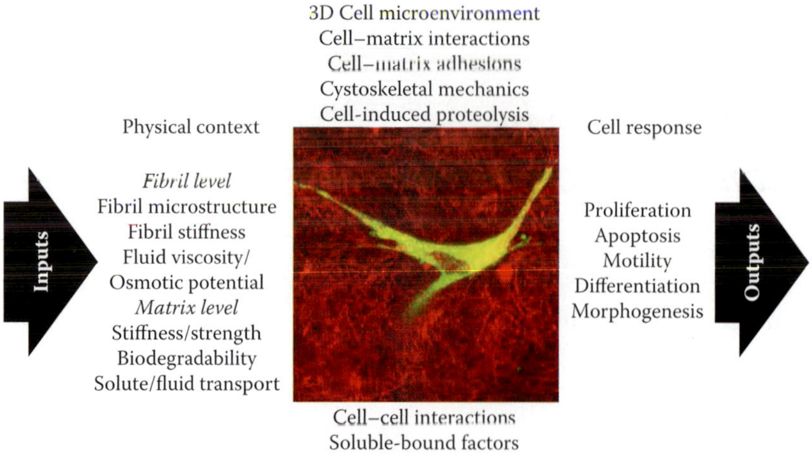

FIGURE 20.1
Fundamental cues of the 3D cell microenvironment including cell–matrix, cell–cell, and soluble/matrix-bound factor signaling. All cellular signaling occurs with the physical context established by the collagen-fibril matrix and its associated interstitial fluid to guide fundamental cell behavior.

and fibril diameter) as well as the viscosity and osmotic potential of the interstitial fluid provide the 3D physical context within which cells sense, integrate, and respond to various cues arising from other cells, soluble/bound factors (e.g., hormones, growth factors), and the matrix. These fibril-level features also work integrally to define matrix-level physicochemical properties including stiffness, biodegradability, and fluid and solute transport, which are all known to modulate fundamental cell behavior. A functional tissue microvasculature is also critical to the normal physiology of healthy tissue, and its compromise is implicated in the initiation and progression of many disease states. The microvasculature serves to provide convective transport through the tissue and a selective barrier that governs transport phenomena (nutrients, metabolic by-products, signaling molecules) between the blood, ECM, and cells.

Because of these complexities, advancements in design of 3D tissue systems require (i) defined strategies for precision tuning and quantification of relevant physical and biological cues; (ii) experiment development and optimization considering multiparametric interactions; (iii) incorporation of convective transport to overcome current size constraints imposed by diffusion limitations (approximately 100 μm); and (iv) new and adapted methods for visualizing and quantifying multiscale, spatiotemporal information relevant to interfacial phenomena between cells and their microenvironment. This chapter describes advances in the understanding of collagen assembly and approaches for precision-tuning the physical context of the collagen-fibril component of the 3D cell microenvironment. In addition, we summarize state-of-the-art imaging tools and techniques used for quantifying multiscale, dynamic cell–fluid–matrix interactions between cells and their 3D microenvironment.

20.2 Design of 3D In Vitro Collagen-Based Tissue Systems

20.2.1 Historic Perspective

Traditionally, the in vivo cell microenvironment has been mimicked by entrapping cells within reconstituted or self-assembled collagen-fibril matrices to create 3D tissue equivalents. In fact, one of the first studies addressing collagen–cell interactions in 3D was performed by Ehrmann and Gey[6] in 1956. Elsdale and Bard[7] later refined this approach to create hydrated collagen lattices or moldable collagen-fibril matrices within which cells could be embedded. Early work documented for a variety of cell types that 3D collagen matrices supported a more physiologically relevant cell morphology, phenotype, and function compared to 2D glass or plastic.[8–11] Since that time, numerous investigators have applied reconstituted collagen matrices to document how 3D geometric constraints and cell–matrix interactions affect cellular processes and tissue morphogenesis, including fibroblast motility and contraction,[12–15] tumor morphogenesis and invasion,[16–18] stem cell differentiation,[19] and angiogenesis and vasculogenesis.[20–23] It is now well established that, in addition to providing the structural framework, the collagen matrix imparts biological signaling capacity primarily through cell-surface integrin receptors. Integrins modulate the physical connectivity and mechanical force balance between the cell interior (cytoskeleton and nucleus) and the surrounding collagen fibrils. Collagen molecules also contain proteolytic motifs that contribute its metabolism (biodegradability) and cell-induced matrix remodeling.[24]

Despite the many advantages of using natural polymers, the more widespread use of collagen for creation of 3D in vitro tissue systems has been hindered by several persistent shortcomings. First, type I collagen formulations produced by individual research laboratories and commercial sources display significant variation in molecular composition and purity.[19,25,26] These variations arise since this natural protein is extracted from different animal tissues (e.g., pig skin, calf skin, and rat tail tendon) using a variety of different methods (e.g., pepsin digestion or acid solubilization). Furthermore, conventional monomer (single triple-helical molecules with amino- and carboxy-terminus telopeptides intact) and atelo-collagen (single triple-helical molecules devoid of telopeptides) formulations display significant source-to-source and lot-to-lot variation in polymerization capacity, long polymerization times, low mechanical integrity, and rapid biodegradation.[19,27] As a result, such formulations are difficult to handle and not conducive to homogenous cell entrapment. In addition, tissue constructs formed with conventional collagens often exhibit undesired cell-mediated contraction. Finally, since collagen cross-links determine tissue form and function and have been implicated in a number of pathological conditions, the inability to modulate these features when using traditional collagen formulations to design 3D cell microenvironments has been identified as a major shortcoming.[3,17] Collectively, these persistent limitations, along with a lack of collagen polymer standardization and comprehensive matrix characterization, make comparisons of results obtained between studies and laboratories difficult.[25,28]

20.2.2 Tunable Polymerizable Collagens for Creating Cellular Microenvironments In Vitro and In Vivo

While the type I collagen molecule primary sequence is identical across tissues, posttranscriptional modifications and formation of intermolecular cross-links contribute to diversification of collagen type I molecule building blocks, ECM collagen-fibril networks, and therefore tissue form and function.[29,30] In fact, cross-link chemistries appear to be more tissue-specific than species-specific.[31] In vivo, cross-link formation and specification involve (i) hydroxylation of select lysine residues by lysyl hydroxylase, which occurs intracellularly[32]; (ii) oxidative deamination of ε-amino groups on select lysine and hydroxylysine residues within the telopeptide regions by lysyl oxidase, as occurs within the extracellular space after collagen molecules are extruded from the cell[33]; (iii) reaction of resultant lysine aldehydes or hydroxylysine aldehydes with lysine or hydroxylysine groups within the central helical domain of adjacent molecules to form intermediate divalent cross-links of the aldol, hydroxy aldol, or ketoimine varieties; and (iv) conversion of a subset of divalent cross-links into more stable, mature trivalent cross-links. It is postulated that, during collagen assembly, lysyl oxidase binds to and catalyzes cross-link formation between prefibrillar aggregates of staggered collagen molecules (monomers) to create covalently cross-linked oligomers (e.g., at least to collagen molecules joined by a covalent cross-link).[33] In turn, these different oligomer precursors (dimers or trimers) direct the progressive molecular packing and assembly that eventually gives rise to tissue-specific fibril microstructure and matrix physical properties. In fact, cross-link chemistries derived from lysine aldehydes (allysine) route predominate in soft tissues including skin, cornea, and sclera. Alternatively, those derived via the hydroxylysine aldehyde (hydroxyallysine) route yield pyridinoline- and pyrrole-based cross-links, which are prominent in stiff tissues including bone, ligaments, and tendons.[34] Furthermore, the number and type of collagen cross-links also define tissue-specific collagen metabolism.

Recently, a collagen polymer engineering approach that is inspired by in vivo collagen-fibril assembly and incorporates natural intermolecular cross-link chemistries as oligomer building blocks has been described.[35] Here, tissue extraction methods were adapted to yield acid-soluble collagen composed primarily of oligomers.[19] Unlike conventional monomer and atelo-collagen preparations, oligomers comprise small aggregates of collagen molecules (e.g., trimers), which retain their tissue-specific, covalent intermolecular cross-links (Figure 20.2). Oligomers representing different cross-link chemistries can be obtained by applying similar extraction protocols to tissues that differ in their cross-link chemistry profile (e.g., skin, tendon, and bone). To support their application as robust and tunable matrix-forming polymers, collagens are subjected to an expanded set of standardization and characterization analyses[19,35,36] beyond what is routinely performed for research and medical-grade collagen products.[25,37] Molecular-level analyses include quantification of average polymer molecular weight and intermolecular cross-link composition, in addition to purity[19,35]

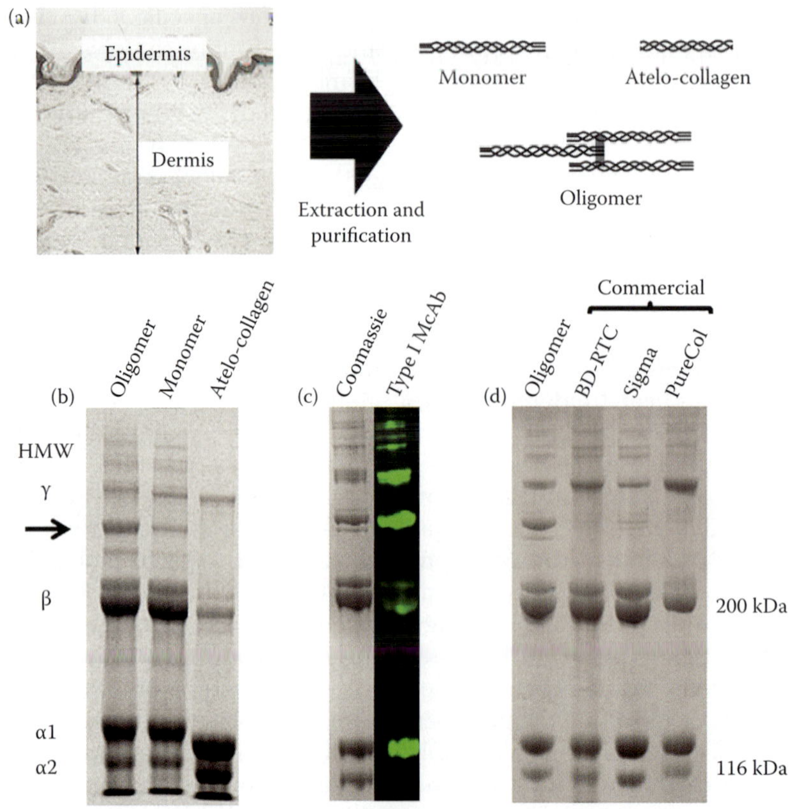

FIGURE 20.2
Molecular analysis of various laboratory-derived and commercial collagen formulations. (a) Different tissue extraction and purification methods can yield monomer, atelo-collagen, and oligomer formulations. (b, d) Sodium dodecyl sulfate-polyacrylamide gel electrophoresis (SDS-PAGE) (4%) and (c) Western blot (type I collagen monoclonal antibody) analyses are used to characterize and compare collagens. (From Kreger, S.T. et al., *Biopolymers* 93 (8), 690–707, 2010; Bailey, J.L. et al., *Biopolymers* 95 (2), 77–93, 2011.)

(Figure 20.2b–d). Polymerization potential is also measured and compared using quantitative descriptors of assembly kinetics and microstructure-mechanical properties of formed matrices (Figures 20.3 and 20.4). Tunability of collagen matrix physical properties is derived from established relationships between polymerization parameters (e.g., collagen concentration, pH, and presence of copolymers) and microstructure-mechanical properties of formed matrices (Table 20.1). In fact, an increase in collagen concentration is routinely used to increase matrix stiffness, which occurs as a result of increased fibril density. More recently, an increase in oligomer/monomer ratio (average polymer molecular) has been shown to increase matrix stiffness by increasing the extent of interfibril branching independent of fibril density. Collectively, these new collagen building blocks and the hierarchical approach to collagen polymer design and standardization support predictable and reproducible control and tuning of relevant fibril- and matrix-level physical properties for creation of 3D in vitro tissue systems.

Studies performed to date show that oligomer formulations offer several advantages over conventional collagens, including rapid polymerization times to support in vitro and in vivo cell entrapment, improved mechanical integrity due to increased interfibril branching, and broader range of tunable physicochemical attributes (e.g., microstructure-physicomechanical properties, biodegradability).[19,35,36] More importantly, oligomer formulations induce cellular processes and tissue morphogenesis events beyond what has been observed with conventional collagen formulations. More specifically, tuning fibril density, and therefore matrix stiffness, of oligomer matrices effectively induced different patterns of lineage-specific differentiation by embedded mesenchymal stem cells (MSCs).[19] While this work also documented that specific soluble factor cocktails were not sufficient to overcome matrix-induced MSC differentiation, it was apparent that both soluble factor and physical-based signaling could be used synergistically to guide the cell response. More recently, independent variation of collagen concentration (fibril density) and

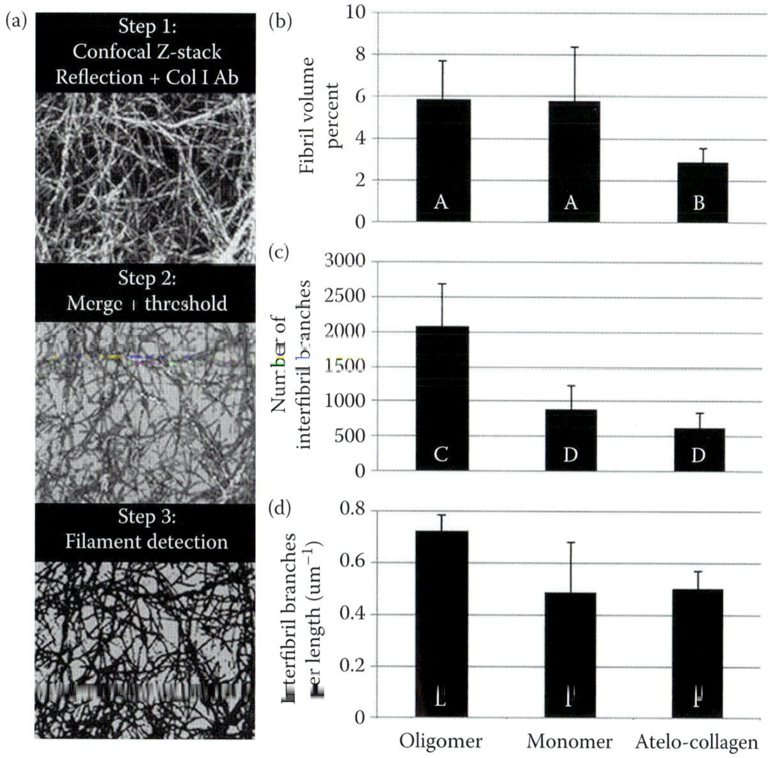

FIGURE 20.3
(a) Prediction and tuning of matrix microstructure for 3D in vitro tissue constructs. Microstructure analysis of collagen matrices involves collection of confocal image stacks in combined immunofluorescence and back-scattered (reflected) light modes, signal merging and image thresholding, and filament detection. (From Whittington, C.F. et al., *Microsc Microanal*, 2013.) Quantified microstructure parameters including (b), (c), and (d) for oligomer, monomer, and atelo-collagen matrices prepared at match concentration (1.5 mg/mL). Letters indicate statistically different groups for each parameter ($p < 0.05$). Such analyses are used for standardization and comparison of collagen polymerization capacity.

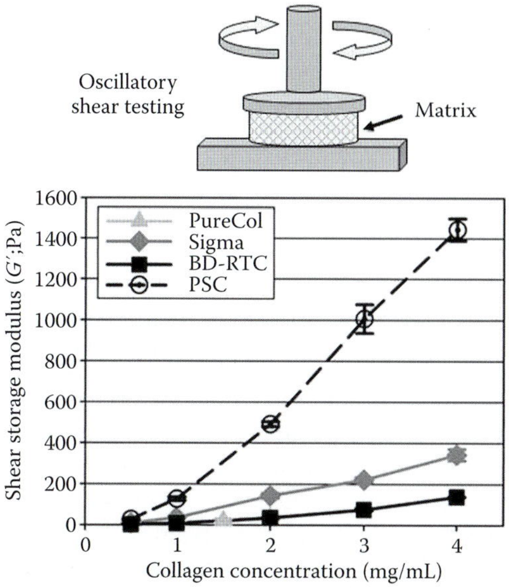

FIGURE 20.4
Prediction and tuning of matrix stiffness for 3D in vitro tissue constructs. Oscillatory shear testing of polymerized matrices produced at specified concentrations establishes G'-collagen concentration relationship for each collagen formulation. Such analyses are used for standardization and comparison of collagen polymerization capacity. (From Kreger, S.T. et al., *Biopolymers* 93 (8), 690–707, 2010.)

TABLE 20.1
Established Relationships between Collagen Polymerization Conditions and Fibril Microstructure and Mechanical Properties of Formed Collagen Matrices

Polymerization Parameter	Fibril-Level Microstructure	Matrix Stiffness	References
Collagen concentration	↑ Fibril density	↑ E_T; ↑ G'; ↑ E_C	19,35,38,39
Oligomer/monomer ratio	↑ Interfibril branching (connectivity)	↑ G'	23,35
Hyaluronic acid	↑ Interstitial fluid viscosity	↓ G'; ↑ E_C	40,41
pH	↑ Fibril length; ↓ Fibril diameter	↑ E_T	38,42
Type III collagen	↓ Fibril diameter	↓ G'	43
Temperature	↓ Fibril diameter; ↑ Fibril density	↑ G'	42,44

Note: G', shear storage modulus; E_T, tensile (Young's modulus); E_C, compressive modulus.

oligomer/monomer ratio (interfibril branching), two independent determinants of matrix stiffness, were found to guide the extent, architecture, and persistence of 3D vessel network formation by endothelial colony forming cells (ECFCs).[23,35] Interestingly, oligomer, but not monomer, matrices were sufficient to induced maturation and persistence (beyond 14 days) of ECFC vessel networks, as marked by basement membrane deposition, in absence of accessory cells.[23] It is also noteworthy that patterns of matrix-based vessel morphogenesis observed in vitro closely paralleled those observed for ECFC-seeded matrices implanted in vivo.[45] Based on these findings, as well as others, it is apparent that matrix–integrin–cytoskeletal signaling provides an important convergence point for biophysical and biochemical signaling pathways that drive cell fate. As such, bringing additional user specification and quantification of 3D cell microenvironment physical parameters, including convective and diffusive transport, provides a modernized approach to enhance the design and integration of physiologically relevant 3D in vitro tissue systems.

20.3 Multiscale Analysis of 3D Cell–Fluid–Matrix Interactions

20.3.1 Imaging of 3D Collagen Matrix Microstructure and Cell–Collagen Interactions

Advances in imaging technologies are providing useful tools for identifying the dynamic cell–fluid–matrix interactions and associated mechanotransduction mechanisms that regulate fundamental cell decision making. Confocal microscopy and multiphoton microscopy are now widely used for high-resolution imaging of optically thick specimens. Both imaging modalities offer the advantage of providing depth-resolved images and up to 4D information (x-, y-, z-, time), allowing identification of cellular features, including cell shape as well as intracellular and cell-surface molecules, and their spatiotemporal relationships to the collagen-fibril matrix.[46–48] Confocal laser-scanning microscopy limits out-of-focus signals by scanning the specimen and illuminating a single point along a plane with a pinpoint aperture. In contrast, multiphoton microscopy employs short pulses of longer wavelength light (near infrared), such that two or more photons are absorbed to excite a fluorophore. Excitation typically remains within the spot of focus since the probability of two or more photons simultaneously exciting a fluorophore outside the focal volume is extremely low. As a result, the lower energy excitation field and limited excitations of molecules outside the focal volume produce less photoinduced toxicity, greater imaging depths (up to 1000 μm versus 300 μm), and reduced photobleaching compared to confocal microscopy.

Both confocal microscopy and multiphoton microscopy provide imaging technologies that allow collection of 3D collagen microstructure details in absence of construct labeling or processing. Since collagen fibrils differ in their refractive index from their surroundings, laser-scanning confocal microscopy in reflection mode allows collection of reflected or back-scattered light from the collagen-fibril network.[49–52] Time-lapse confocal reflection imaging during collagen polymerization also provides useful information regarding kinetics and molecular mechanisms of fibril assembly.[53] On the other hand, the noncentrosymmetric, triple-helical structure inherent to collagen produces a nonlinear, second-order polarization of light.[54,55] Second-harmonic generation (SHG) produces scattered light that is polarized along the helix axis and half the wavelength of the incident multiphoton excitation. Since collagen possesses intrinsic fluorophores, its autofluorescence can also be visualized using confocal and multiphoton microscopy.[51] However, autofluorescence intensity is typically relatively low thereby compromising image quality and resolution.[51,56] Along with the collagen-fibril microstructure, the spatiotemporal distribution of cellular proteins and cell-associated ion fluxes (e.g., calcium) can be visualized within live cells using a wide range of vital dyes and fusion protein technology. For fixed specimens, collagen-specific antibodies[23] and probes[57] have been shown to improve signal-to-noise and image quality of collagen-fibril microstructure. Since these imaging modalities detect different collagen features, they can be combined to produce high-quality images to support comprehensive, quantitative matrix microstructure analyses.[23,58]

20.3.2 Measurement of Mechanical Load Transfer during Cell–Matrix–Fluid Interactions

It is now well established that the physical aspect of cell–collagen interactions and the associated mechanical force balance is a powerful modulator of cell fate. A number of in vitro systems consisting of collagen-based tissue equivalents have been specially designed and used to demonstrate how the balance of mechanical forces between cells and surrounding collagen fibrils modulates various functional tissue attributes including cell shape and alignment, morphology and composition of cell–matrix adhesions, ion fluxes, and cell motility. A number of outside-in and inside-out perturbation strategies have been established to perturb this delicate cell–matrix tensional force balance. Outside-in strategies include application of external mechanical loads, modulation of collagen matrix properties, and integrin-binding antagonists, while inside-out strategies target cytoskeletal organization and cell contractile machinery. Several integrated experimental setups and approaches have also been devised to gain information regarding the hierarchical and dynamic transmission of loads at multiple size scales extending from the tissue- to the molecular-level as illustrated in Figure 20.5.[59]

Digital image correlation (DIC) techniques have been instrumental for quantifying and comparing deformations and strains experienced by the cells and the matrix. DIC is a collective measurement technique where cells or tissues are continuously imaged using various imaging methods including optical microscopy,[60–64] magnetic resonance imaging,[65] and echocardiography.[66] These time-lapse images are cross-correlated to determine spatiotemporal deformation or deformation rate. DIC was initially developed to measure tissue or ECM level deformation[60–62] as shown in Figure 20.5a, but by using proper fluorescence dyes and nanoparticles under high-magnification microscopy, this technique can be applied to measure cellular and nuclear level deformation.[63,64,67] The DIC method is further extended to measure cellular traction forces.[68,69] In these studies, cells were cultured on a substrate containing fluorescence microbeads, and their movements were imaged to determine local deformation of the substrate caused by cellular contraction. This deformation was further analyzed to determine cellular traction force if elastic modulus of the substrate is known.

DIC is most widely used because of easy adaptation to fluorescence and phase contract microscopy techniques. However, it has been mainly used to study 2D environments (i.e., cells on a substrate or superficial deformation of tissues), and its application to 3D microenvironments (i.e., cells within a 3D collagen matrix) is still limited. Another technique limited to 2D contexts is atomic force microscopy (AFM). AFM involves probing the surface of biological samples including tissues, cells, and molecules by tapping with a sharp tip.[70,71] It can image the surface topology as well as directly measure force and displacement as shown in Figure 20.5b. The AFM tip may be functionalized with ligands to prove interactions with its corresponding cell surface receptors[72–75] and to quantify receptor–ligand adhesion strength.[76] The most notable advantage of AFM is its capacity to characterize both displacement and force

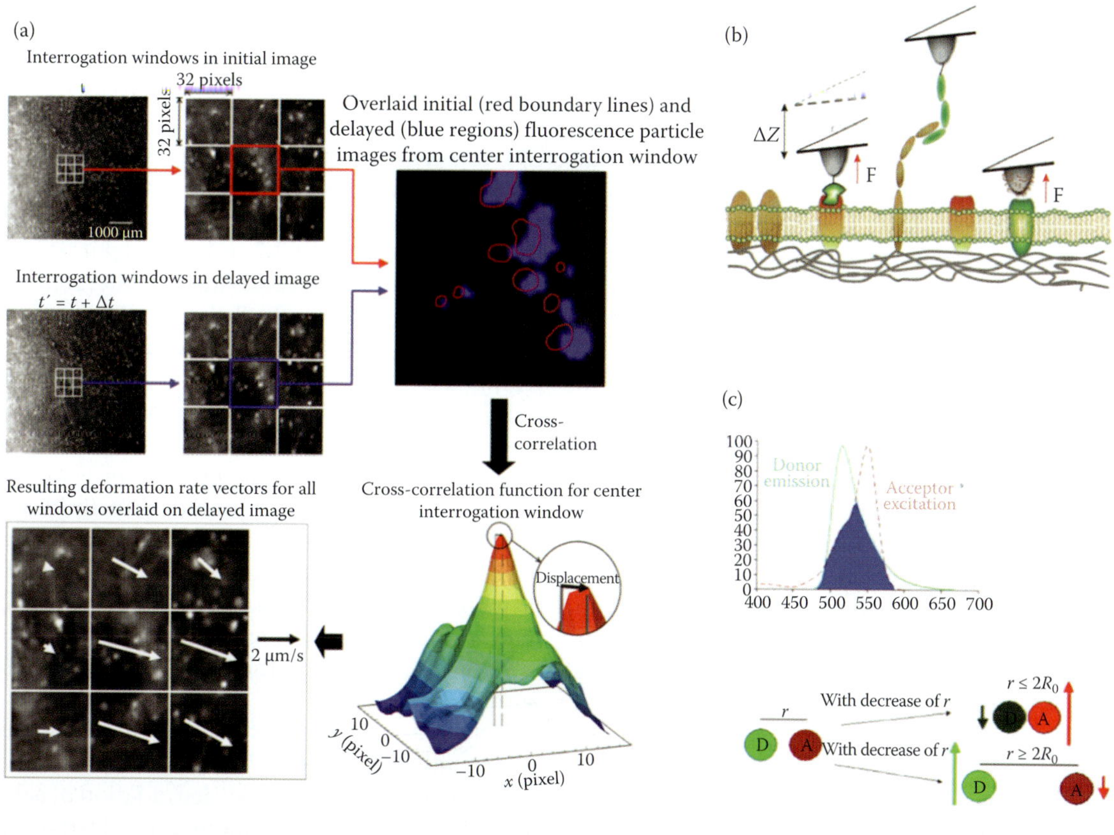

FIGURE 20.5
Schematic of currently available experimental methods to study cell–matrix interactions. (a) DIC. (From Nyberg, P. et al., *Front Biosci* 13, 6537–6553, 2008.) (b) AFM. (From Laplaca, M.C. and Prado, G.R., *J Biomech* 43 (1), 71–78, 2010.) (c) FRET. (From Yu, Z. & Morrison, B. III, *J Neurophysiol* 103 (1), 499–510, 2010.)

at the molecular level. Unfortunately, because measurements require interaction of the AFM tip with surfaces, its applicability is limited to 2D contexts consisting of cells seeded on top of substrates.

For measurement in 3D microenvironments, Roeder et al. developed and applied an incremental digital volume correlation algorithm[77] to time-lapse confocal images of collagen matrices during uniaxial tensile loading and quantified strain transmission in 3D from the tissue to collagen-fibril level.[78] Digital volume correlation was also applied to quantify cell-induced 3D local strains within fibroblast-populated matrices.[79] These studies documented that the collagen microstructure can be designed to regulate the transmission of externally applied loads from a tissue to cellular level as well as the shape and contractile capacity of fibroblasts during matrix remodeling.

A powerful tool for identification of molecular players involved in mechanotransduction is fluorescence resonance energy transfer (FRET). This technique has been applied to monitor intracellular molecular interactions and activities in live 3D tissue constructs with high spatiotemporal resolution.[80,81] FRET is a nonradiative energy transfer between two fluorescence molecules (from a donor to an acceptor). The extent of FRET is associated with the distance between these two molecules. Thus, change in the fluorescence signal can be used to estimate the distance between the donor and the acceptor (2 to 10 nm range comparable to the Foster radius) as illustrated in Figure 20.5c. For example, 3D time-lapse FRET was used to dissect how endothelial myosin light chain kinase activity is modulated during paracellular and transcellular migration of breast cancer cells in 3D vascularized collagen constructs.[82] More recently, a RhoA-specific FRET biosensor was used to show the preferential downregulation of RhoA activity at cell–cell contacts and its recruitment to cell–matrix adhesion sites.[83] The extremely high spatial resolution makes FRET an attractive technique to study cell–matrix interactions, but it has limitations of selecting dye pairs, needing to deal with weak fluorescence signals and photobleaching, and needing a two-photon setup to perform the measurements, which make its broader applications difficult.

In addition to direct mechanical interaction between cells and the matrix, indirect interactions occur via transport of interstitial fluid and solutes. The mass transport properties, also governed by matrix microstructure-mechanical properties, are responsible for nutrient exchange between cells embedded within the ECM as well as soluble factor availability and matrix remodeling.[1,84,85] Fluorescent recovery after photobleaching (FRAP) utilizes photobleaching of fluorescent tracers to measure intracellular or extracellular molecule mobility.[86] Fluorescent particles or proteins of various sizes have been uniformly distributed at dilute concentrations within 3D tissue constructs. A well-defined concentration gradient then is induced by laser-induced photobleaching, and relaxation of the tracer concentration profile is analyzed to yield the diffusion coefficient.[87,88] Alternatively, integrative optical imaging involves creation of a point source of fluorescent tracers via microinjection followed by image-based measurement of their temporal and spatial distribution.[36,89] FRAP also has been used to quantify regional interstitial flow velocities within a microfluidics-based tissue chip for purposes of defining how interstitial flow affects breast cancer cell invasion.[90]

20.3.3 Cell–Matrix Adhesions and Intracellular Fluid–Structure Interactions

During cell–fluid–matrix interactions, mechanical force and/or deformation is generated at the intracellular and/or the extracellular spaces and transmitted via the cell–matrix adhesion complex (CMAC) as illustrated in Figure 20.6a. The CMAC consists of integrins whose extracellular domains are bound to ECM proteins such as collagen and fibronectin, and intracellular domains are connected to actin cytoskeleton via various cytoskeletal proteins such as vinculin and talin. It is also connected to a plasma membrane. Moreover, as illustrated in Figure 20.6b, the molecular configuration of the CMAC varies depending on the structure and organization of ECM and cytoskeletal molecules.[91] Thus, the CMAC seems to play pivotal roles in mechanical cell–matrix interactions to sense extracellular mechanical environments and exert endogenous tension from the cytoskeleton on the ECM and remodel it.[92,93]

In addition to cell–fluid–matrix interactions, significance of intracellular fluid–structure interactions have been recognized in several recent works.[94,95] In these works, the cytoplasm is analyzed as a poroelastic material. Since mammalian cells have the fibrous porous cytoskeleton network whose void space is filled with the viscous fluidic cytosol, the cytoplasm is approximated as a porous structure saturated with fluid and bound by the plasma membrane. Thus, simultaneous knowledge of intracellular and extracellular deformation and/or forces is required to fully understand the cell–fluid matrix interactions and subsequent cell fate.

Since the integrity of the CMAC is strongly associated with various physiological processes and functions of cells/tissues, rupture or damage of the CMAC is usually associated with loss of cell/tissue functions and disease developments. For example, degradation of cell–matrix/cell–cell adhesion and increased plasma membrane permeability caused by tension on the CMACs have been observed in the development of traumatic brain and spinal cord injuries.[96–99] Moreover, the significance

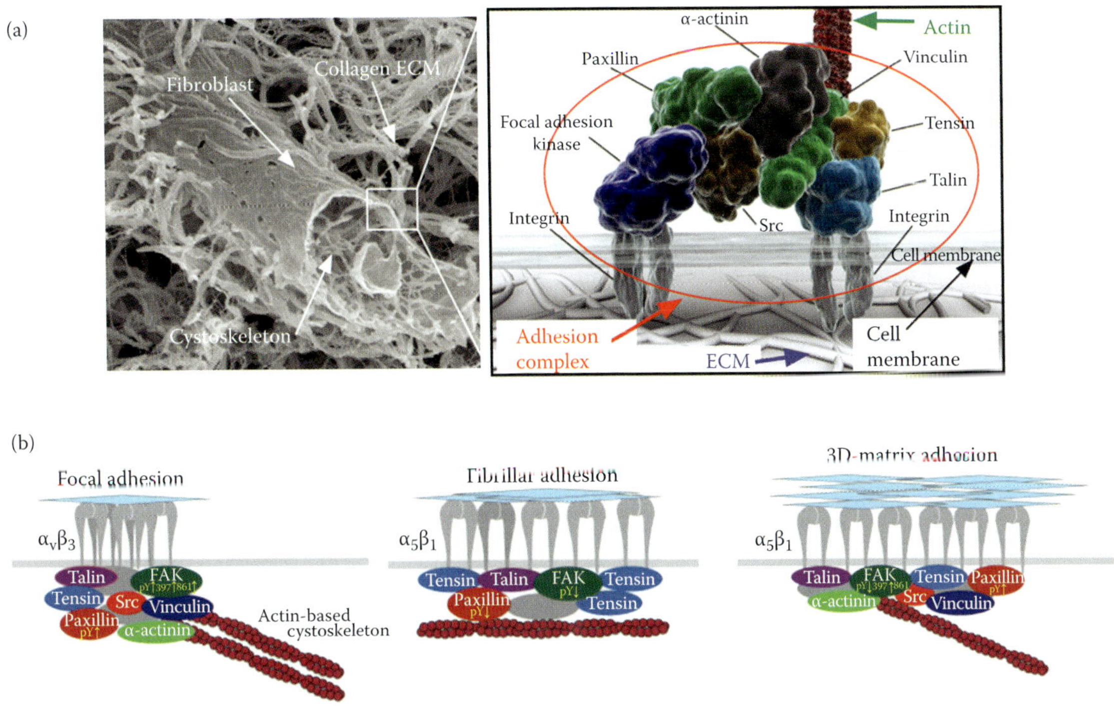

FIGURE 20.6
CMAC. (a) CryoSEM of fibroblasts in 3D collagen matrix and schematic diagram of CMAC. (b) Molecular structures of [illegible] configurations. (From [illegible], 1994.)

of the CMAC integrity is also recognized in cell/tissue engineering and processing technologies. For another example, one of the key enabling technologies for tissue engineering and regenerative medicine is cryopreservation, a long-term storage method of cells/tissues in a frozen state. Research on cryopreservation has been focused to understand and mitigate cell/tissue damage caused by subzero temperatures. However, recent studies have demonstrated that tissues experience freezing-induced spatiotemporal deformation that resulted from cell–fluid–matrix interactions.[61,62,100] These studies further showed that the cell–fluid–matrix interactions may lead to rupture of the CMACs and cellular detachment as shown in Figure 20.7.[101] In order to achieve

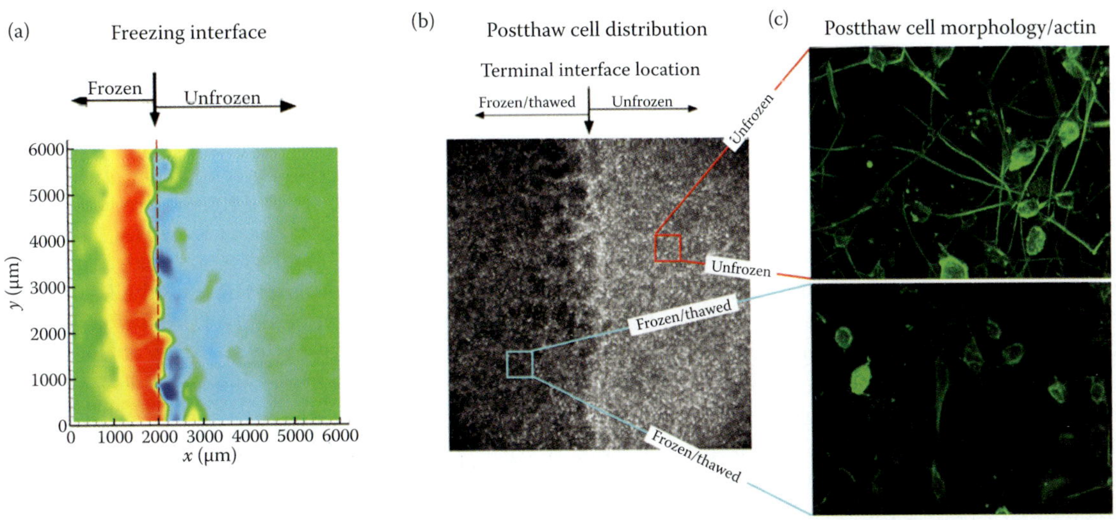

FIGURE 20.7
Freezing-induced deformation and its impact on the CMAC. (a) Spatiotemporal dilatation of tissues during freezing. Tissue was frozen from left to right. Red represents expansion and blue represents compression. (b) Cell density in the frozen/thawed region decreased significantly and implied cellular detachment. (c) Actin structures are damaged postthaw. (From Chouaib, S. et al., *Crit Rev Immunol* 30 (6), 529–545, 2010.)

reliable long-term cryopreservation of functional tissues, rupture of the CMAC needs to be minimized or would design freezing protocols only inducing deformation below the rupture threshold. However, quantitative understanding of its rupture mechanics has not been well established yet, especially when the mechanical loading rate is high. This mainly attributes to the difficulties associated with the necessity of knowledge of both intracellular and extracellular environments at multiple length scales—the ECM level is at microscale, the cellular level is at mesoscale, and the subcellular level is at nanoscale.

20.4 Conclusion

Understanding cellular behavior in more physiologically relevant 3D microenvironments can spark revolutionary change in (i) research of physiologic and pathologic processes, (ii) clinical translation of drugs and therapies, (iii) regenerative medicine (e.g., cellular therapies), and (iv) personalized medicine strategies. Although cell culture in 3D collagen matrices has been used for these purposes, recent developments are enabling precision control of the 3D microenvironment at multiple length scales including molecular scale by using a new type of collagen building blocks. This new level of user control can be harnessed to create 3D cell microenvironments with desired physicochemical properties and subsequent cell–fluid–matrix interactions. It is evident that we are in the early phases of this cell biology revolution, where the new perspective and capabilities offered by 3D culture platforms are working to narrow the gap between conventional 2D cell culture and in vivo animal models. Key future challenges that need to be addressed with respect to design and development of 3D tissue microenvironments include (1) integration of perfusion and/or functional vasculature to better capture in vivo tissue transport phenomena including matrix interstitial fluid flow as well as vasculature-associated convective and diffusive transport; (2) incorporation of physical gradients within a matrix as well as multiple tissue interfaces to better recapitulate the heterogeneous and anisotropic design and function of tissue and organ systems; (3) increased emphasis on quantitative biology and integration of predictive modeling of structure–function relationships to support design, scalability, and optimization of tissue/organ systems while effectively reducing the number of experiments; (4) continued adaptation and advancements of imaging tools (e.g., deep tissue microscopy) and biosensor techniques to support multiscale, spatiotemporal quantification of interfacial phenomena (e.g., molecular signaling) between individual cells and their microenvironment; and (5) validation of predictability between 3D in vitro models and human in vivo outcomes.

References

1. Griffith, L.G. and Swartz, M.A., Capturing complex 3D tissue physiology in vitro. *Nat Rev Mol Cell Biol* 7 (3), 211–224 (2006).
2. Lee, J., Cuddihy, M.J. and Kotov, N.A., Three-dimensional cell culture matrices: State of the art. *Tissue Eng Part B Rev* 14 (1), 61–86 (2008).
3. Yamada, K.M. and Cukierman, E., Modeling tissue morphogenesis and cancer in 3D. *Cell* 130 (4), 601–610 (2007).
4. Cukierman, E. and Bassi, D.E., Physico-mechanical aspects of extracellular matrix influences on tumorigenic behaviors. *Semin Cancer Biol* 20 (3), 139–145 (2010).
5. Baker, B.M. and Chen, C.S., Deconstructing the third dimension: How 3D culture microenvironments alter cellular cues. *J Cell Sci* 125 (Pt 13), 3015–3024 (2012).
6. Ehrmann, R.L. and Gey, G.O., The growth of cells on a transparent gel of reconstituted rat-tail collagen. *J Natl Cancer Inst* 16 (6), 1375–1403 (1956).
7. Elsdale, T. and Bard, J., Collagen substrata for studies on cell behavior. *J Cell Biol* 54 (3), 626–637 (1972).
8. Emerman, J.T. and Pitelka, D.R., Maintenance and induction of morphological differentiation in dissociated mammary epithelium on floating collagen membranes. In Vitro 13 (5), 316–328 (1977).
9. Emerman, J.T., Burwen, S.J. and Pitelka, D.R., Substrate properties influencing ultrastructural differentiation of mammary epithelial cells in culture. *Tissue Cell* 11 (1), 109–119 (1979).
10. Chambard, M., Gabrion, J. and Mauchamp, J., Influence of collagen gel on the orientation of epithelial cell polarity: Follicle formation from isolated thyroid cells and from preformed monolayers. *J Cell Biol* 91 (1), 157–166 (1981).
11. Bell, E., Ivarsson, B. and Merrill, C., Production of a tissue-like structure by contraction of collagen lattices by human fibroblasts of different proliferative potential in vitro. *Proc Natl Acad Sci U S A* 76 (3), 1274–1278 (1979).
12. Kolodney, M.S. and Wysolmerski, R.B., Isometric contraction by fibroblasts and endothelial cells in tissue culture: A quantitative study. *J Cell Biol* 117 (1), 73–82 (1992).
13. Wakatsuki, T., Kolodney, M.S., Zahalak, G.I. and Elson, E.L., Cell mechanics studied by a reconstituted model tissue. *Biophys J* 79 (5), 2353–2368 (2000).
14. Petroll, W.M. and Ma, L., Direct, dynamic assessment of cell–matrix interactions inside fibrillar collagen lattices. *Cell Motil Cytoskeleton* 55 (4), 254–264 (2003).
15. Grinnell, F. and Petroll, W.M., Cell motility and mechanics in three-dimensional collagen matrices. *Annu Rev Cell Dev Biol* 26, 335–361 (2010).
16. Hotary, K., Allen, E., Punturieri, A., Yana, I. and Weiss, S.J., Regulation of cell invasion and morphogenesis in a

three-dimensional type I collagen matrix by membrane-type matrix metalloproteinases 1, 2, and 3. *J Cell Biol* 149 (6), 1309–1323 (2000).

17. Sabeh, F., Shimizu-Hirota, R. and Weiss, S.J., Protease-dependent versus-independent cancer cell invasion programs: Three-dimensional amoeboid movement revisited. *J Cell Biol* 185 (1), 11–19 (2009).
18. Even-Ram, S. and Yamada, K.M., Cell migration in 3D matrix. *Curr Opin Cell Biol* 17 (5), 524–532 (2005).
19. Kreger, S.T. et al., Polymerization and matrix physical properties as important design considerations for soluble collagen formulations. *Biopolymers* 93 (8), 690–707 (2010).
20. Davis, G.E., Bayless, K.J. and Mavila, A., Molecular basis of endothelial cell morphogenesis in three-dimensional extracellular matrices. *Anat Rec* 268 (3), 252–275 (2002).
21. Koh, W., Stratman, A.N., Sacharidou, A. and Davis, G.E., In vitro three dimensional collagen matrix models of endothelial lumen formation during vasculogenesis and angiogenesis. *Methods Enzymol* 443, 83–101 (2008).
22. Sieminski, A.L., Hebbel, R.P. and Gooch, K.J., The relative magnitudes of endothelial force generation and matrix stiffness modulate capillary morphogenesis in vitro. *Exp Cell Res* 297 (2), 574–584 (2004).
23. Whittington, C.F., Yoder, M.C. and Voytik-Harbin, S.L., Collagen-polymer guidance of vessel network formation and stabilization by endothelial colony forming cells in vitro. *Macromol Biosci* 13, 1135–1149 (2013).
24. Perumal, S., Antipova, O. and Orgel, J.P., Collagen fibril architecture, domain organization, and triple-helical conformation govern its proteolysis. *Proc Natl Acad Sci U S A* 105 (8), 2824–2829 (2008).
25. Abraham, L.C., Zuena, E., Perez-Ramirez, B. and Kaplan, D.L., Guide to collagen characterization for biomaterial studies. *J Biomed Mater Res B Appl Biomater* 87 (1), 264–285 (2008).
26. Ruszczak, Z. and Friess, W., Collagen as a carrier for on-site delivery of antibacterial drugs. *Adv Drug Deliv Rev* 55 (12), 1679–1698 (2003).
27. Friess, W., Collagen—biomaterial for drug delivery. *Eur J Pharm Biopharm* 45 (2), 113–136 (1998).
28. Johnson, K.R., Leight, J.L. and Weaver, V.M., Demystifying the effects of a three dimensional microenvironment in tissue morphogenesis. *Methods Cell Biol* 83, 547–583 (2007).
29. Knott, L. and Bailey, A.J., Collagen cross-links in mineralizing tissues: A review of their chemistry, function, and clinical relevance. *Bone* 22 (3), 181–187 (1998).
30. Eyre, D.R. and Wu, J.J., Collagen cross-links. *Top Curr Chem* 247, 207–229 (2005).
31. Avery, N.C. and Bailey, A.J., Restraining cross-links responsible for the mechanical properties of collagen fibers: Natural and artificial. In *Collagen Structure and Mechanics*, edited by P. Fratzl (Springer, New York, 2008), pp. 81–110.
32. Henkel, W. and Glanville, R.W., Covalent crosslinking between molecules of type I and type III collagen. The involvement of the N-terminal, nonhelical regions of the alpha 1 (I) and alpha 1 (III) chains in the formation of intermolecular crosslinks. *Eur J Biochem* 122 (1), 205–213 (1982).
33. Cronlund, A.L., Smith, B.D. and Kagan, H.M., Binding of lysyl oxidase to fibrils of type I collagen. *Connect Tissue Res* 14 (2), 109–119 (1985).
34. Eyre, D.R., Paz, M.A. and Gallop, P.M., Cross-linking in collagen and elastin. *Annu Rev Biochem* 53, 717–748 (1984).
35. Bailey, J.L. et al., Collagen oligomers modulate physical and biological properties of three-dimensional self-assembled matrices. *Biopolymers* 95 (2), 77–93 (2011).
36. Whittington, C.F. et al., Oligomers modulate interfibril branching and mass transport properties of collagen matrices. *Microsc Microanal* 19, 1323–1333 (2013).
37. ASTM Standard F2212, 2002 (2011), Standard guide for characterization of type I collagen as starting material for surgical implants and substrates for tissue engineered medical products (TEMPs), ASTM International, West Conshohocken, PA, 2011, doi:10.1520/F2212-11, www.astm.org.
38. Roeder, B.A., Kokini, K., Sturgis, J.E., Robinson, J.P. and Voytik-Harbin, S.L., Tensile mechanical properties of three-dimensional type I collagen extracellular matrices with varied microstructure. *J Biomech Eng* 124 (2), 214–222 (2002).
39. Wood, G.C., The formation of fibrils from collagen solutions. 3. Effect of chondroitin sulphate and some other naturally occurring polyanions on the rate of formation. *Biochem J* 75, 605–612 (1960).
40. Kreger, S.T. and Voytik-Harbin, S.L., Hyaluronan concentration within a 3D collagen matrix modulates matrix viscoelasticity, but not fibroblast response. *Matrix Biol* 28 (6), 336–346 (2009).
41. Xin, X., Borzacchiello, A., Netti, P.A., Ambrosio, L. and Nicolais, L., Hyaluronic-acid-based semi-interpenetrating materials. *J Biomater Sci Polym Ed* 15 (9), 1223–1236 (2004).
42. Christiansen, D.L., Huang, E.K. and Silver, F.H., Assembly of type I collagen: Fusion of fibril subunits and the influence of fibril diameter on mechanical properties. *Matrix Biol* 19 (5), 409–420 (2000).
43. Stuart, K. and Panitch, A., Characterization of gels composed of blends of collagen I, collagen III, and chondroitin sulfate. *Biomacromolecules* 10 (1), 25–31 (2009).
44. Raub, C.B. et al., Noninvasive assessment of collagen gel microstructure and mechanics using multiphoton microscopy. *Biophys J* 92 (6), 2212–2222 (2007).
45. Critser, P.J., Kreger, S.T., Voytik-Harbin, S.L. and Yoder, M.C., Collagen matrix physical properties modulate endothelial colony forming cell-derived vessels in vivo. *Microvasc Res* 80 (1), 23–30 (2010).
46. Chen, X., Nadiarynkh, O., Plotnikov, S. and Campagnola, P.J., Second harmonic generation microscopy for quantitative analysis of collagen fibrillar structure. *Nat Protoc* 7 (4), 654–669 (2012).
47. Provenzano, P.P., Eliceiri, K.W. and Keely, P.J., Multiphoton microscopy and fluorescence lifetime imaging microscopy (FLIM) to monitor metastasis and the tumor microenvironment. *Clin Exp Metastasis* 26 (4), 357–370 (2009).
48. Georgakoudi, I., Rice, W.L., Hronik-Tupaj, M. and Kaplan, D.L., Optical spectroscopy and imaging for the noninvasive evaluation of engineered tissues. *Tissue Eng Part B Rev* 14 (4), 321–340 (2008).

49. Friedl, P. et al., Migration of highly aggressive MV3 melanoma cells in 3-dimensional collagen lattices results in local matrix reorganization and shedding of alpha2 and beta1 integrins and CD44. *Cancer Res* 57 (10), 2061–2070 (1997).
50. Voytik-Harbin, S.L., Roeder, B.A., Sturgis, J.E., Kokini, K. and Robinson, J.P., Simultaneous mechanical loading and confocal reflection microscopy for three-dimensional microbiomechanical analysis of biomaterials and tissue constructs. *Microsc Microanal* 9 (1), 74–85 (2003).
51. Voytik-Harbin, S.L., Rajwa, B. and Robinson, J.P., Three-dimensional imaging of extracellular matrix and extracellular matrix-cell interactions. *Methods Cell Biol* 63, 583–597 (2001).
52. Petroll, W.M., Cavanagh, H.D. and Jester, J.V., Dynamic three-dimensional visualization of collagen matrix remodeling and cytoskeletal organization in living corneal fibroblasts. *Scanning* 26 (1), 1–10 (2004).
53. Brightman, A.O. et al., Time-lapse confocal reflection microscopy of collagen fibrillogenesis and extracellular matrix assembly in vitro. *Biopolymers* 54 (3), 222–234 (2000).
54. Campagnola, P.J. et al., Three-dimensional high-resolution second-harmonic generation imaging of endogenous structural proteins in biological tissues. *Biophys J* 82 (1 Pt 1), 493–508 (2002).
55. Cox, G. et al., 3-dimensional imaging of collagen using second harmonic generation. *J Struct Biol* 141 (1), 53–62 (2003).
56. Zipfel, W.R. et al., Live tissue intrinsic emission microscopy using multiphoton-excited native fluorescence and second harmonic generation. *Proc Natl Acad Sci U S A* 100 (12), 7075–7080 (2003).
57. Boerboom, R.A. et al., High resolution imaging of collagen organisation and synthesis using a versatile collagen specific probe. *J Struct Biol* 159 (3), 392–399 (2007).
58. Raub, C.B. et al., Image correlation spectroscopy of multiphoton images correlates with collagen mechanical properties. *Biophys J* 94 (6), 2361–2373 (2008).
59. Nyberg, P., Salo, T. and Kalluri, R., Tumor microenvironment and angiogenesis. *Front Biosci* 13, 6537–6553 (2008).
60. Canal, C.E., Hung, C.T. and Ateshian, G.A., Two-dimensional strain fields on the cross-section of the bovine humeral head under contact loading. *J Biomech* 41 (15), 3145–3151 (2008).
61. Teo, K.Y., DeHoyos, T.O., Dutton, J.C., Grinnell, F. and Han, B., Effects of freezing-induced cell–fluid–matrix interactions on the cells and extracellular matrix of engineered tissues. *Biomaterials* 32 (23), 5380–5390 (2011).
62. Teo, K.Y., Dutton, J.C. and Han, B., Spatiotemporal measurement of freezing-induced deformation of engineered tissues. *J Biomech Eng* 132 (3), 031003 (2010).
63. Bieler, F.H. et al., Biaxial cell stimulation: A mechanical validation. *J Biomech* 42 (11), 1692–1696 (2009).
64. Ueki, Y., Sakamoto, N. and Sato, M., Direct measurement of shear strain in adherent vascular endothelial cells exposed to fluid shear stress. *Biochem Biophys Res Commun* 394 (1), 94–99 (2010).
65. Chan, D.D. et al., Mechanical deformation and glycosaminoglycan content changes in a rabbit annular puncture disc degeneration model. *Spine (Phila Pa 1976)* 36 (18), 1438–1445 (2011).
66. Grenne, B. et al., Mean strain throughout the heart cycle by longitudinal two-dimensional speckle-tracking echocardiography enables early prediction of infarct size. *J Am Soc Echocardiogr* 24 (10), 1118–1125 (2011).
67. Bell, B.J., Nauman, E. and Voytik-Harbin, S.L., Multiscale strain analysis of tissue equivalents using a custom-designed biaxial testing device. *Biophys J* 102 (6), 1303–1312 (2012).
68. Wang, J.H., Cell traction forces (CTFs) and CTF microscopy applications in musculoskeletal research. *Oper Tech Orthop* 20 (2), 106–109 (2010).
69. Bartalena, G. et al., A novel method for assessing adherent single-cell stiffness in tension: Design and testing of a substrate-based live cell functional imaging device. *Biomed Microdevices* 13 (2), 291–301 (2011).
70. Muller, D.J. and Dufrene, Y.F., Atomic force microscopy as a multifunctional molecular toolbox in nanobiotechnology. *Nat Nanotechnol* 3 (5), 261–269 (2008).
71. Muller, D.J., Helenius, J., Alsteens, D. and Dufrene, Y.F., Force probing surfaces of living cells to molecular resolution. *Nat Chem Biol* 5 (6), 383–390 (2009).
72. Fritz, J., Katopodis, A.G., Kolbinger, F. and Anselmetti, D., Force-mediated kinetics of single P-selectin/ligand complexes observed by atomic force microscopy. *Proc Natl Acad Sci U S A* 95 (21), 12283–12288 (1998).
73. Moy, V.T., Florin, E.L. and Gaub, H.E., Intermolecular forces and energies between ligands and receptors. *Science* 266 (5183), 257–259 (1994).
74. Baumgartner, W. et al., Cadherin interaction probed by atomic force microscopy. *Proc Natl Acad Sci U S A* 97 (8), 4005–4010 (2000).
75. Lee, G.U., Kidwell, D.A. and Colton, R.J., Sensing discrete streptavidin biotin interactions with atomic-force microscopy. *Langmuir* 10 (2), 354–357 (1994).
76. Evans, E.A. and Calderwood, D.A., Force and bond dynamics in cell adhesion. *Science* 316, 1148–1153 (2007).
77. Roeder, B.A., Kokini, K., Robinson, J.P. and Voytik-Harbin, S.L., Local, three-dimensional strain measurements within largely deformed extracellular matrix constructs. *J Biomech Eng* 126 (6), 699–708 (2004).
78. Roeder, B.A., Kokini, K. and Voytik-Harbin, S.L., Fibril microstructure affects strain transmission within collagen extracellular matrices. *J Biomech Eng* 131 (3), 031004 (2009).
79. Pizzo, A.M., Kokini, K., Vaughn, L.C., Waisner, B.Z. and Voytik-Harbin, S.L., Extracellular matrix (ECM) microstructural composition regulates local cell-ECM biomechanics and fundamental fibroblast behavior: A multidimensional perspective. *J Appl Physiol* 98 (5), 1909–1921 (2005).
80. Ballestrem, C. et al., Molecular mapping of tyrosine-phosphorylated proteins in focal adhesions using fluorescence resonance energy transfer. *J Cell Sci* 119 (Pt 5), 866–875 (2006).
81. Wang, Y. et al., Visualizing the mechanical activation of Src. *Nature* 434 (7036), 1040–1045 (2005).

82. Khuon, S. et al., Myosin light chain kinase mediates transcellular intravasation of breast cancer cells through the underlying endothelial cells: A three-dimensional FRET study. *J Cell Sci* 123 (Pt 3), 431–440 (2010).
83. Ponik, S.M., Trier, S.M., Wozniak, M.A., Eliceiri, K.W. and Keely, P.J., RhoA is down-regulated at cell–cell contacts via p190RhoGAP-B in response to tensional homeostasis. *Mol Biol Cell* 24 (11), 1688–1699 (2013).
84. Fisher, R.J. and Peattie, R.A., Controlling tissue microenvironments: Biomimetics, transport phenomena, and reacting systems. *Adv Biochem Eng Biotechnol* 103, 1–73 (2007).
85. Phelps, E.A. and Garcia, A.J., Update on therapeutic vascularization strategies. *Regen Med* 4 (1), 65–80 (2009).
86. Axelrod, D., Koppel, D.E., Schlessinger, J., Elson, E. and Webb, W.W., Mobility measurement by analysis of fluorescence photobleaching recovery kinetics. *Biophys J* 16 (9), 1055–1069 (1976).
87. Ramanujan, S. et al., Diffusion and convection in collagen gels: Implications for transport in the tumor interstitium. *Biophys J* 83 (3), 1650–1660 (2002).
88. Erikson, A., Andersen, H.N., Naess, S.N., Sikorski, P. and Davies Cde, L., Physical and chemical modifications of collagen gels: Impact on diffusion. *Biopolymers* 89 (2), 135–143 (2008).
89. Nicholson, C. and Tao, L., Hindered diffusion of high molecular weight compounds in brain extracellular microenvironment measured with integrative optical imaging. *Biophys J* 65 (6), 2277–2290 (1993).
90. Haessler, U., Teo, J.C., Foretay, D., Renaud, P. and Swartz, M.A., Migration dynamics of breast cancer cells in a tunable 3D interstitial flow chamber. *Integr Biol (Camb)* 4 (4), 401–409 (2012).
91. Berrier, A.L. and Yamada, K.M., Cell-matrix adhesion. *J Cell Physiol* 213 (3), 565–573 (2007).
92. Jean, C., Gravelle, P., Fournie, J.J. and Laurent, G., Influence of stress on extracellular matrix and integrin biology. *Oncogene* 30 (24), 2697–2706 (2011).
93. Rhee, S. and Grinnell, F., Fibroblast mechanics in 3D collagen matrices. *Adv Drug Deliv Rev* 59 (13), 1299–1305 (2007).
94. Mitchison, T.J., Charras, G.T. and Mahadevan, L., Implications of a poroelastic cytoplasm for the dynamics of animal cell shape. *Semin Cell Dev Biol* 19 (3), 215–223 (2008).
95. Moeendarbary, E. et al., The cytoplasm of living cells behaves as a poroelastic material. *Nat Mater* 12 (3), 253–261 (2013).
96. Cater, H.L., Sundstrom, L.E. and Morrison, B. III, Temporal development of hippocampal cell death is dependent on tissue strain but not strain rate. *J Biomech* 39 (15), 2810–2818 (2006).
97. Cullen, D.K., Simon, C.M. and LaPlaca, M.C., Strain rate-dependent induction of reactive astrogliosis and cell death in three-dimensional neuronal–astrocytic co-cultures. *Brain Res* 1158, 103–115 (2007).
98. Laplaca, M.C. and Prado, G.R., Neural mechanobiology and neuronal vulnerability to traumatic loading. *J Biomech* 43 (1), 71–78 (2010).
99. Yu, Z. and Morrison, B. III, Experimental mild traumatic brain injury induces functional alteration of the developing hippocampus. *J Neurophysiol* 103 (1), 499–510 (2010).
100. Han, B., Miller, J.D. and Jung, J.K., Freezing-induced fluid–matrix interaction in poroelastic material. *J Biomech Eng* 131 (2), 021002 (2009).
101. Chouaib, S. et al., Endothelial cells as key determinants of the tumor microenvironment: Interaction with tumor cells, extracellular matrix and immune killer cells. *Crit Rev Immunol* 30 (6), 529–545 (2010).

21

Integrated Modeling and Imaging for Quantifying the Mechanics of Cells in Three-Dimensional Culture

Guy M. Genin and Elliot L. Elson

CONTENTS

Tissue constructs synthesized by seeding natural extracellular matrix protein gels with living cells are important tools for the study of how cells interact mechanically with their neighborhood and their neighbors in natural, three-dimensional environments. Overcoming the challenge of quantifying cellular interactions from ensemble behaviors of entire tissue constructs has involved tight coupling of experimental, mathematical, and imaging techniques. This chapter summarizes the state of the art in such integrated approaches and presents some examples drawn from work in the authors' labs.

21.1 Introduction

Three-dimensional (3D) tissue constructs have played a central role in efforts to characterize the mechanics of cells that reside in solid tissues. Cells such as smooth muscle cells and fibroblasts from organs such as the heart, kidneys, and lungs serve in part to define the shape, tone, and structure of these organs. The structure of these tissues is believed to be important to the way that the cells themselves behave. Understanding many of the functions of these organs and the tissues that comprise them therefore requires an understanding of the mechanical properties and functioning of their cells, characterized in a natural 3D environment. Although much early work on cell mechanics derived from assays performed on cells that were cultured on two-dimensional (2D) substrata, a 2D substratum provides an unrealistic environment for many cell functions and indeed cell types. Increasing evidence indicates that cells in 2D culture differ substantially in structure and function from cells in a natural tissue (e.g., refs. 1–6). Even subcellular structures such as stress fibers and focal adhesions appear to be different in 2D culture than in 3D tissues.[2,7–9]

The logical course of action would seem to be to study the mechanics of cells within the natural tissues in which they reside. However, cells within natural tissues, especially adult tissues, are difficult to image, and the mechanical responses of such cells are often difficult to measure relative to the passive mechanics of the protein extracellular matrix (ECM) they synthesize and maintain. Engineered tissue constructs provide valuable degrees of flexibility such as the ability to construct a tissue model containing a single cell type or a defined mixture of cell types with specified cell density and ECM composition. This enables resolution of cell mechanics superior to that possible in natural tissues, and these tissue constructs have therefore proven to be useful models for studying cellular mechanics and function in a natural 3D environment.[2,10–14]

However, in each case, the application of these tissue constructs to the study of cellular mechanics requires overcoming several challenges related to integrating imaging, mechanical testing, and mechanical modeling. Deducing cellular behaviors in 3D requires models of cells, ECM, and their interactions, along with tightly integrated experiments that enable identification of model parameters through imaging. These methods and their application to some key questions in cellular biomechanics are the focus of this chapter.

The chapter begins with a brief summary of the role that imaging has played in the study of the mechanics of living cells that are cultured upon 2D substrata, and then describes several integrated experimental systems that have been developed to quantify cell mechanics in 3D. The chapter then describes mathematical models that have been developed to deduce mechanical responses of cells from some of these experimental systems, and the imaging tools developed to fit these models. Integrated into this are three examples of cellular behaviors deduced from tissue constructs, drawn from work in our own labs. The first is an estimate of the mechanical traction that a fibroblast cell exerts on a remodeled ECM scaffold. Contractile fibroblast cells are observed to exert mechanical stresses that are independent of the degree to which the cells are stretched. The second is a quantification of the way that contractile stresses relate to the density of stress fibers within the actin cytoskeleton. Contractile fibroblasts exert stress on a tissue construct in proportion to the time-varying density of stress fibers observed. The third is observations of the way that the actin cytoskeleton reacts to perturbations of a cell's mechanical environment. The degree of contractile force exerted by a cell in three dimensions is shown to be directly related to the average degree of polymerization of the actin cytoskeleton, and imaging studies reveal how repeatable ensemble behaviors arise from diverse responses of individual cells.

21.2 Integration of Mechanics and Imaging in Study of Cells Cultured in 2D

Advances in the study of the mechanics of cells cultured upon 2D substrata have historically involved advances in the ability to image mechanical structures that interact with living cells. We present here a brief summary of some of these efforts, and note that a broader picture can be found in an early review.[15]

One of the first protocols for measuring the mechanics of a living cell was the Cole experiment involving compression of an egg cell between two parallel plates.[16] The imaging challenges in this experiment were many, but particularly notable, was the strategy adopted for measuring the microgram forces needed to conduct the experiment. One end of a long, slender gold rod was placed against one of the parallel glass plates, while the other was pushed down gently. Cole photographed the deflection of the rod at specific levels of deflection and then applied beam theory to estimate the forces. The response to stretching or compression of a single cell held between two plates continues to provide important data about cellular viscoelasticity over timescale ranges pertinent to elastic and contractile responses,[17] and a similar approach was used to quantify for the first time the putative contributions of collagen, titin, microtubules, and intermediate filaments to the passive tension of individual cardiac muscle cells.[18] More generally, imaging the flexure of compliant beams has persisted as a prominent approach to quantifying cell mechanics in 2D, as described in the following. However, a weakness of all these approaches is that cells must be loaded in an unnatural environment.

The use of micropipettes (see Chapter 25), which dates to the 1950s,[19] requires integrated imaging and mechanical modeling that is somewhat more involved. Micropipette aspiration has been used extensively to study circulating cells such as erythrocytes and leukocytes, and to provide an excellent model of how erythrocytes with diameters of 7–8 μm could squeeze through capillaries with much smaller diameters.[20–23] These studies require imaging of the degree to which a living cell is drawn into a micropipette by a specified pressure differential. Interpreting such images through simple mechanical models has demonstrated that resistance of an erythrocyte to shear and bending is small compared to resistance to membrane area expansion. However, interpreting aspiration forces in terms of cytoskeletal and membrane responses remains an ongoing challenge.[24,25]

Long before their use for surface imaging, machines that are very similar to atomic force microscopes (AFMs) were developed to probe cell biomechanics. The very first AFM-like instrument of which we are aware was in fact invented specifically for this purpose.[26,27] At the core of this device is a long, slender cantilever whose deflection is measured by tracking the motion of a laser beam reflected from a mirror at the cantilever's end. The key to this measurement is the development of a sensitive quadrant detector. Early AFM-like devices were used to measure the area expansion modulus of erythrocytes,[28] study cytoskeletal proteins such as myosin II,[29,30] and quantify the mechanical effects of acute inflammatory processes.[31–33] As described in Chapters 26 and 36, AFM continues to be used to study the mechanical properties of adherent cells in a variety of contexts and, like micropipette aspiration, is attractive because the force transduced in such experiments arises directly from contact with a cell. However, again, like micropipette aspiration, the mode of loading of a cell by AFM is highly unnatural and must be interpreted through intricate mechanical models.[28,34,35]

A broad range of 2D systems for estimating cellular mechanics involve deducing mechanical forces from images of how cells deform their mechanical environment. Some of these and the models that they motivate are described in ref. 36. A notable early application of this to nonmuscle cells is an experiment that showed the ability of fibroblasts and other types of cells to exert contractile force through active wrinkling of a thin, soft silicone rubber layer.[37] A class of recent systems involves imaging the flexure of microposts upon which cells are cultured.[38] By culturing cells over defined patterns of microposts and imaging the resulting distributions of stress fibers, models relating stress fiber kinetics to mechanical forces have been derived and fit,[39] and many advances have been made in understanding the roles of topological and mechanical factors in determining the behavior and disposition of a cell in 2D.[40–47] Very intricate predictive models of cytoskeletal mechanics in 2D have been fit to the results of tests performed on flexible and structured 2D substrata.[39,48–53] As described in Chapter 24, imaging of cells on flexible substrata and associated models has elucidated further the mechanisms by which endothelial cells align in response to pulsatile stretch, and uncovered the ways that stress fibers gradually align in a direction of minimal substrate deformation.[54–58] Imaging of how cells displace fluorescent beads within elastic substrata has uncovered a range of dynamic responses of single cells and cell sheets.[40,59–63] A strength of this entire family of approaches is that the models required to infer cell forces from images taken in 2D can be quite straightforward and accurate: such models rely on the often linear elastic mechanics of the substratum, which can be known to great accuracy.

However, a challenge with all of these approaches is again that, for many types of cells, a 2D substratum is a poor mimic of a solid tissue, and the link between cellular mechanics and the environment cannot be fully captured on a 2D substratum. Therefore, despite the extensive challenges associated with imaging, systems to study cellular mechanics in a 3D scaffold that mimics a natural environment are essential. Section 21.3 describes several approaches.

21.3 Bird's Eye View of Systems for Measuring Cell Mechanics in 3D Tissue Constructs

All systems for quantifying the mechanics of cells in a 3D environment require tight coupling of imaging, modeling, and mechanical testing. The techniques of which we are aware can be categorized into three general approaches: (1) monitoring of cell-seeded collagen microspheres, (2) imaging of living cells within ECM scaffolds, and (3) mechanical testing of tissue constructs. These systems are all designed to present a 3D structure that cells can manipulate and remodel, and are all designed to work with imaging systems to produce the data needed to fit mathematical models for deducing cellular behavior from tissue construct behavior.

Although not a focus of this chapter, we begin by describing collagen microspheres. These continue to serve as one of the most important tools for characterizing and quantifying the ways that cells actively remodel ECM proteins.[64–67] The major imaging challenges in applying these approaches center on tracking the evolution of the cell population and tracking gross changes to the shape of the microspheres. These data, combined with creep data for the ECM, enable estimates of the tractions exerted by cells on the ECM over time.[64] The tool has persisted for 20 years as the most important laboratory model of development, and techniques for producing highly consistent microspheres continue to improve in part due to advances in microfluiduics.[68] These methods are especially well suited to determining cell mechanics over timescales that are associated with tissue remodeling (see, for example, refs. 64 and 69–73) and cellular remodeling.[74,75]

The remaining two approaches rely largely on tightly integrated mechanical, imaging, and biochemical studies. Tissue constructs containing cells and defined ECM are monitored while the tissue construct is perturbed mechanically through some combination of stretching of the tissue construct and active stressing of the construct by the cells within it. The same tissue construct is then monitored after a promoter or inhibitor of cellular contraction is applied, or after the cells are lysed using a detergent. From the results of these tests and associated imaging studies, the mechanics of the cells and the ECM can be estimated.

Early implementations of these approaches did not involve cellular imaging but nevertheless provided important data on cell biophysics (see review in ref. 36). For example, Delvoye et al. showed that contractile force developed by fibroblasts in collagen-based tissue constructs increased with the number of cells and with the application of an agonist (fetal calf serum), and decreased with the application of an actin cytoskeleton inhibitor (cytochalasin B).[76] However, extending this to the level of protein structures required imaging studies performed in conjunction with mechanical testing. For example, by applying a similar approach in conjunction with immunofluorescence imaging of the actin cytoskeleton, Kolodney and Wysolmerski[77] were able to provide greater detail on the role of cytoskeletal organization on these forces. Extending beyond these ensemble averages to estimates of how individual cells behave mechanically required additional mathematical and experimental techniques. A few of these coupled approaches are presented in Sections 21.4 and 21.5.

21.4 Imaging-Based Mathematical Models for Interpreting Cellular Biomechanics from Tensile Testing of Tissue Constructs

Interpreting the responses of a tissue construct to estimate the mechanics of the individual cells that comprise it requires a tightly integrated set of models and experiments, and calibration by unbiased stereological imaging techniques (cf. Chapter 16). A primary tool for this is a ring-shaped tissue construct that can loop over bars attached to stepper motors and force transducers (Figure 21.1). The specimens are formed in annular molds[10] from a suspension of cells in a solution of collagen kept in liquid form at low temperature. The solution is poured into the annular space (typically 2–3 mm wide) between the inner surface of the outer cylindrical mold and a central mandrel. When the mold is placed in a 37°C incubator, the collagen gels and the cells remodel and compress the collagen to form a tissue construct (typically reducing its volume to ~10% of its original value within 24 h). Removal of the construct from the mandrel yields a ring-shaped specimen that can be suspended between an isometric force transducer and a linear microstepping motor. The force transducer then measures active forces generated by cell contraction, and also passive forces developed when the tissue is stretched.

A typical test used to characterize these specimens is a viscoelastic relaxation test in which one end of the specimen is rapidly displaced to stretch the specimen a prescribed amount while the other end is held fixed. The force needed to sustain the specimen at its new length is monitored over time until it reaches a steady state. The test is repeated with additional increments of strain. Thereafter, the specimen is unloaded and allowed to relax viscoelastically to its original shape, a cytoskeletal inhibitor or cell lysing agent is applied, and the stretching protocol is repeated to quantify the mechanics of the

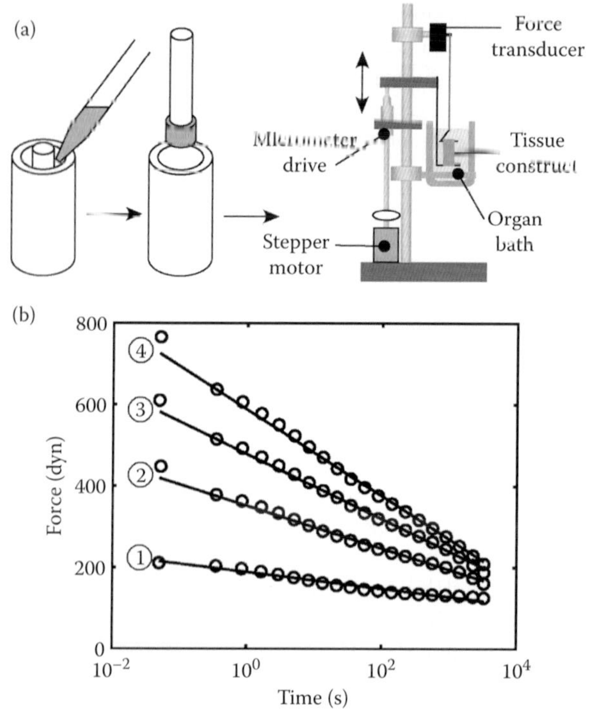

FIGURE 21.1
(a) Ring-shaped cell-seeded ECM scaffolds can be formed from cell solutions incubated in annular molds. After a few days of culture at 37°C, the ring-shaped tissue construct specimens can be placed onto a tensile testing apparatus. (b) Fibroblast populated matrix (FPM) specimens stretched a prescribed amount and then held isometrically while force was monitored show roughly logarithmic reductions in force over time. FPMs were preconditioned and then stretched to nominal strains of 2%, 8%, 14%, and 20%; the specimens were not unloaded between stretches. Tests were performed after 3 days of tissue culture. (Reprinted from Marquez JP et al., *Annals of Biomedical Engineering* 34(9): 1475–1482, 2006. With permission.)

ECM. Typical force relaxation data for a tissue construct containing fibroblast cells show an initial rise to a level that depends on strain ε and cell volume fraction f_c, and then follow a Fung-type (and hence logarithmic) decay with respect to time over an hour or more of isometric relaxation (Figure 21.1)[78]:

$$F(t, \varepsilon, f_c) \approx a_o(f_c) + a_1(f_c)\,\varepsilon + b_1(f_c)\,\varepsilon \ln(-t/t_o) \qquad (21.1)$$

where a_1, a_2, and b_1 are functions of the final cell volume fraction f_c only, t is time, and t_o is an arbitrary normalization constant. This expression is not valid at $t = 0$, but this does not cause difficulty because the first useful data point in a relaxation test is recorded a finite time after the completion of stretching. The fact that a_o varies only with f_c represents the observation that active contractile stresses exerted by fibroblasts in these tissue constructs are typically independent of strain and time.

Relating these data to the mechanics of individual cells requires multiscale homogenization models that incorporate the structure, organization, and composition of the tissue construct. These models, in turn, require imaging data to fit key parameters. As an example, we describe the Zahalak model[11] and some of its extensions. Although the Zahalak model is specific to the type of tissue construct described above, the approach is quite general, as are the imaging tools needed to make it work.

21.4.1 Estimation of Cell Volume Fractions

Zahalak estimated the response of a tissue construct containing a population of cells with defined orientation distribution and volume fraction, and ECM with prescribed (homogenous) mechanical properties. These contributions were taken to be in parallel, so that the stress at any material point is the sum of matrix (*m*) and cell (*c*) contributions:

$$\sigma_{ij}(t) = (1 - f_c)\sigma_{ij}^{(m)}(t) + f_c\sigma_{ij}^{(c)}(t) \tag{21.2}$$

where $\sigma_{ij}^{(m)}$ and $\sigma_{ij}^{(c)}$, the time-varying components of the engineering stress tensors, are weighted by cell and ECM volume fractions f_c and $f_m = 1 - f_c$. This assumption will be corrected in Section 21.5, but even the corrections involve f_c as a parameter of central importance.

Two approaches exist of which we are aware for estimating cell volume fractions. The first is immunofluorescence or live cell fluorescence assays, a technique that dates back to some of the earliest studies on the mechanics of cells in tissue constructs.[77] The basic idea is to insert a fluorescent label into the membrane, cytoskeleton, or cytoplasm of a cell, and then estimate the volume fraction of cells from a series of confocal microscopy slices. A logical approach is to reassemble these slices into a volume and attempt to reconstruct the fraction of the volume of cells, but this turns out to be inefficient and inaccurate compared to statistical sampling approaches. Instead, faster convergence to the true average volume fraction is attained by evaluating randomly slices using the Glagolev or Delesse principle to estimate volume fraction. These principles both rely on the observation that for randomly aligned, identical features, the expected value of the observed area fraction of a phase equals the volume fraction of that phase (see Chapter 16 for further details, and also refs. 79 and 80). In both cases, each of a series of fluorescent images is thresholded, and either the entire image (Delesse principle) or a random selected portion of it (Glagolev principle) is sampled to determine the ratio of fluorescent pixels to background pixels. Both of these ratios converge to the volume fraction of cells, with a few caveats. The first to note is that cell distributions within tissue constructs are rarely uniform, with cells often concentrated at the edges of the tissue construct. The second is that thresholding must be done with care, because the Delesse and Glagolev principles require infinitesimally thin optical slices, whereas confocal microscopy images contain fluorescence from a width that is usually large compared to the dimension of a pixel.

The second approach is to estimate the volumes of a few cells and then perform a total DNA quantification assay to determine the total number of cells.[78] After the total volume of a tissue construct is estimated, the tissue construct is centrifuged in capped tubes containing phosphate-buffered saline (PBS) and then transferred into a lysis buffer (e.g., 0.1% w/v sodium dodecyl sulfate in PBS) and sonicated. Then, a prescribed volume of solution is mixed with a fluorescent indicator of DNA (e.g., Hoechst 33258), and the total fluorescence of the solution is measured. By comparing these measurements to measurements of fluorescence obtained from known numbers of cells, an estimate of the cell number and hence cell volume fraction can be made.

21.4.2 Estimation of Cell Orientation Distributions

Zahalak idealized cells as thin contractile rods, appropriate for spindle-shaped cells in tissue constructs,[10] and performed statistical averaging of the passive and active contributions of the cells and the ECM to overall tissue construct mechanics. This statistical averaging requires optical estimation of the orientation distribution of cells, in the same way that many models for tissue mechanics build from models of collagen fiber geometry.[81–87]

Zahalak modeled each fibroblast with a linearized Hill element[88] (cf. Figure 21.2), including a parallel elastic component accounting for passive mechanical properties, a series elastic element to account for increased dynamic stiffness, and a damped contractile element, which embodies the active force generating and

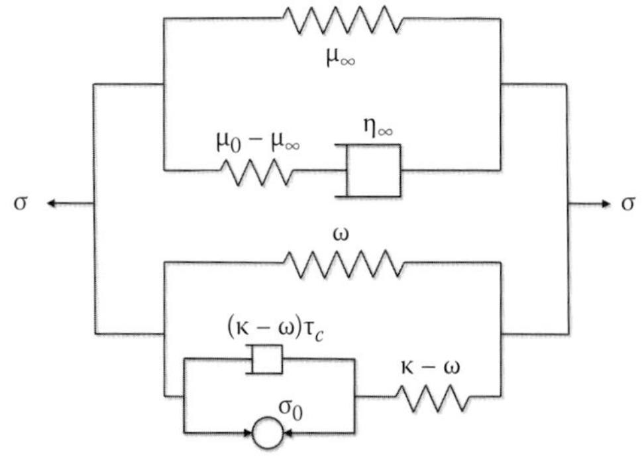

FIGURE 21.2
One-dimensional idealization of the Zahalak model, in which cells (Hill element, lower grouping) deform in parallel with the ECM (upper grouping, here represented as a linear viscoelastic solid).

quasi-viscous force–velocity response of the cell. The governing equation for such a cell is

$$\frac{dF}{dt}+\frac{F}{\tau_c}=\frac{F_o}{\tau_c}+l_o(k_s+k_p)\frac{d}{dt}\left(\frac{l}{l_o}\right)+\frac{l_o k_p}{\tau_c}\left(\frac{l}{l_o}-1\right) \qquad (21.3)$$

where F is the total force a cell exerts on the ECM; l and l_o are the current and initial cell length, respectively; k_s and k_p are the series and parallel elastic stiffness values of the cell, respectively; and $\tau_c = \eta_c/k_s$ is the cell time constant, in which η_c is the cell's effective viscosity. The cell-generated contractile force, F_o, can rise in response to activators such as fetal bovine serum or thrombin, and can fall in response to inhibitors such as Y27632. Taking N cells per unit volume, with cells directed along unit vectors $\mathbf{n}$ according to orientation distribution $p(\mathbf{n})$, yields

$$\sigma_{ij}^{(c)} = Nl_o\int_\Omega F(\Omega)n_i n_j p(\Omega)\,d\Omega \qquad (21.4)$$

where Ω is the surface of a unit sphere, $p(\Omega)d\Omega$ represents the probability of a cell having an axis parallel to the normal $\mathbf{n}$ to an infinitesimal patch on the surface of the unit sphere, and $\int_\Omega p(\Omega)d\Omega = 1$.

After some manipulation, an integral form of the constitutive law can be written as

$$\sigma_{ij}(t) = (1-f_c)\sigma_{ij}^{(m)}(t) + f_c\int_{-\infty}^{t}\exp\left(\frac{t-t'}{\tau_c}\right)\left[\frac{3}{\tau_c}\sigma_o A_{ij} + \left(\kappa\frac{\partial}{\partial t'}+\frac{\omega}{\tau_c}\right)B_{ijpq}\varepsilon_{pq}\right]dt' \qquad (21.5)$$

where $\kappa = Nl_o^2(k_s+k_p)$, $\omega = Nl_o^2 k_p$, and $\sigma_o = Nl_oF_o/3$ are tissue-level mechanical parameters associated with cell-level mechanical responses; the two "cell anisotropy tensors" $A_{ij} = \int_\Omega n_i n_j p(\Omega)d\Omega$ and $B_{ijpq} = \int_\Omega n_i n_j n_p n_q p(\Omega)d\Omega$ translate the anisotropy of the microscopic cell distribution into anisotropic mechanical behavior of the macroscopic tissue; and repeated indices imply summation. The parameter ω represents the macroscopic manifestation of cellular parallel elasticity ("slow" elasticity), and κ represents the combination of parallel and series elasticity ("fast" elasticity; Figure 21.2).

The key to estimating tissue level responses from models of individual cells is computing the anisotropy tensors **A** and **B**, which are based upon the orientation distribution $p(\mathbf{n})$ that must be estimated from imaging studies. Typical scanning confocal micrographs of fibroblasts within tissue constructs, following labeling with a fluorescent cytoplasmic dye (e.g., Cell Tracker), show that cells can present a range of orientation distributions from aligned to random (Figure 21.3). The planar projections of the orientation distribution can be estimated by hand from such images by assigning a dominant axis to each cell and binning the results (Figure 21.3). However, automated procedures exist as well based upon the Fourier transform of the images.[89] The foundation of such procedures is encapsulated by Figure 41.2 in Chapter 41: the power spectral density of the 2D discrete Fourier transform of an image containing a line presents a dominant axis that is rotated 90° from that line. Cells that have no preferred direction do not present such axes (cf. Figure 41.3 in Chapter 41). If a characteristic dimension of cells is known, a band-pass filter can be used to isolate information arising only from cell bodies, eliminating artifacts of differential brightness associated with cell staining (cf. Figure 41.4 in Chapter 41). By binning the amount of power in the band as a function of angle and normalizing, an orientation distribution arises. Rotating this by 90° yields an estimate

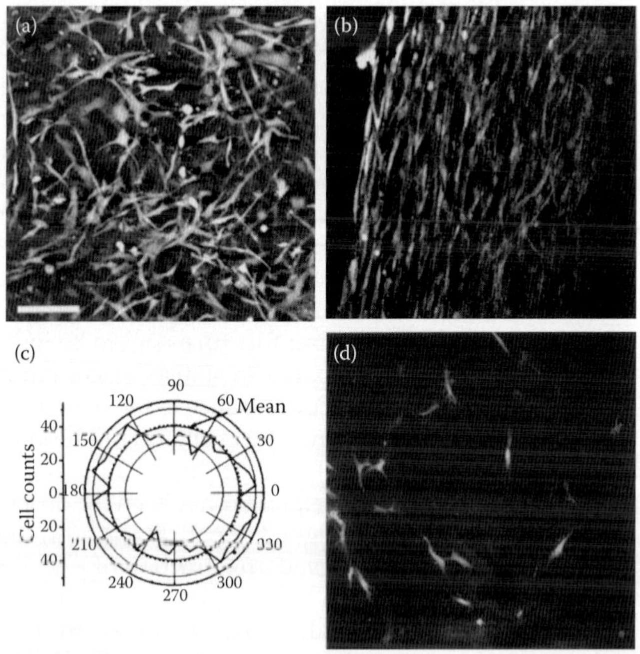

FIGURE 21.3
Optical sections of tissue constructs containing contractile fibroblast cells stained with a cytoplasmic dye (Cell Tracker). The tissue construct shown in (a) and (b) began with a cell concentration of 10^6 cells/mL of collagen solution, while that shown in (d) began with 10^6 cells/mL. (a) and (d) are the central region of the sample and (b) is near the edge. From these data, a polar histogram can be made for the orientation distribution (c). (Reprinted from Zahalak GI et al., *Biophysical Journal* 79(5): 2369–2381, 2000. With permission.)

of the function $p(\mathbf{n})$. Note that $p(\mathbf{n})$ is symmetric with respect to a 180° rotation.

21.5 Imaging and Interpreting the Mechanics of Cells and Cytoskeletal Proteins in 3D Culture

A series of technologies have emerged recently for the purpose of optically monitoring the actions of live cells in 3D. The Kaunas group has recently developed an innovative technology to enable stress fiber kinetics models such as those described in Chapter 24 to be tested on cyclically stretched cells in 3D culture.[90] The fundamental challenge is that collagen ECM scaffolds typically present important viscoelastic time constants in the 0.1–10 s range that is of the greatest interest for stress fiber kinetics.[91–93] A tissue construct can therefore not relax adequately following loadings in the 0.1–10 Hz range for a cyclic test to be performed. The strategy used by the Kaunas group was to coat a flexible elastic substratum with a cell-seeded collagen layer. By choosing a substratum that was stiff relative to the cell-seeded collagen layer and ensuring a well-bonded interface, the collagen layer could be forced to reverse direction at appropriate frequencies. By choosing a sufficiently thin layer, cells can be imaged to monitor stress fiber dynamics.

Technologies are just now becoming available for estimating mechanical tractions around individual cells within a cell-seeded ECM. The approach of Legant et al.[94] (see Chapter 22) is one in which cells are cultured within a well-defined hydrogel scaffold, and cell tractions are inferred from the motion of fluorescent particles embedded within the hydrogel. A central challenge that Legant et al. overcame is to design the hydrogel scaffold so that cells can remodel their environment and adopt realistic morphologies. The procedure is to estimate the displacement of the fluorescent beads associated with lysing cells and estimate a deformation gradient by taking the lysed state to be the reference configuration (note that this is a highly involved procedure; see Chapter 22 for details). A regularized inverse problem is then solved to estimate the cellular tractions that were most likely to have given rise to the observed displacement field. With this method, Legant et al.[94] obtained for the first time realistic estimates of tractions exerted by several different types of cells embedded within a 3D ECM scaffold.

The focus of this section are techniques to image cytoskeletal and cell shape changes resulting from mechanical loading in 3D. These techniques are all intertwined with corrections to Equation 21.2 to account for the fact that cells and the ECM in a tissue construct do not always share mechanical loads in parallel. The correction for this involves a "strain factor," S, defined as the relationship between the macroscopic strain applied to a tissue (resolved in a cell's axial direction) and the axial strain experienced by the cell, which is approximately constant for all cell orientations and is surprisingly insensitive to cell anisotropy.[78] In terms of this factor, Equation 21.5 can be rewritten as

$$\sigma_{ij}(t) = (1-f_c)\sigma_{ij}^{(m)}(t) + f_c \int_{-\infty}^{t} \exp\left(\frac{t-t'}{\tau_c}\right) \left[\frac{3}{\tau_c}\sigma_o A_{ij} + \left(\kappa\frac{\partial}{\partial t'} + \frac{\omega}{\tau_c}\right) B_{ijpq} S(t')\varepsilon_{pq}\right] dt' \tag{21.6}$$

where repeated indices imply summation. Note that the ECM response $\sigma_{ij}^{(m)}(t)$ must also be adjusted for the fact that cells and the ECM do not, in general, stretch the same amount. For example, if the ECM were represented by a Fung viscoelastic model, the response to a strain history $\varepsilon_{ij}(t)$ would have to be adapted as

$$\sigma_{ij}^{(m)}(t) = -p\delta_{ij} + \int_{-\infty}^{t} 2G(t-t')M(t')\left(\frac{\partial\sigma_{ij}^{(e)}\left(\varepsilon_{ij}(t')\right)}{\partial t'}\right) dt' \tag{21.7}$$

where the factor $M(t) = \varepsilon_{ij}^{(m)}(t)\varepsilon_{ij}^{-1}(t) = 1 + \left(1 - S(t)\right) f_c / (1 - f_c)$ provides the ratio between the average strain field $\varepsilon_{ij}^{(m)}(t)$ in the ECM to the average strain field $\varepsilon_{ij}(t)$ in the tissue construct as a whole, δ_{ij} is Kronecker's delta, p is the hydrostatic pressure, $G(t)$ is Fung's reduced relaxation function, and $\sigma_{ij}^{(e)}(t)$ is a function that defines the fast-loading constitutive response of the tissue. Equation 21.6 provides for straightforward calculation of the instantaneous and relaxed effective moduli of cells and the ECM because these responses depend upon tangent moduli at a particular strain level.[95] For this case, and limiting to the case of isotropic cells and the ECM, the strain factor for a thin tissue construct (thickness on the order of cell length) can be written in closed form using a self-consistent-type approach[96–98]:

$$S \approx \frac{1}{1 + 2\left(d/l_o\right)\left(E_c/E_m\right)} \equiv \frac{1}{1 + KY} \tag{21.8}$$

where E_c is the elastic modulus of the cells, E_m is the corresponding average tissue modulus, d is the (nominal, elliptical) cell width, l_o is the nominal cell length, $Y = dE_c/l_oE_m$, and K is a constant that equals 2 for long, slender elliptical cells. The ratio E_c/E_m can be expected to differ for the instantaneous and fully relaxed responses of the cell and the ECM, and S would then vary as a function of time. Determining K for other cell shapes requires numerical simulations of idealized, periodic microstructures of a tissue construct, but the value of K is relatively insensitive to this. For example, $K \approx 2.2$ for "rectangular" planar cells.[98] This also appears to be the case in 3D.[99]

These analyses underlie imaging studies to link mechanics of tissue constructs to mechanics of cells and related efforts to quantify cell shape changes in response to mechanical loading.

21.5.1 Interpreting the Mechanics of Cytoskeletal Proteins from Combined Mechanical and Imaging Studies of Tissue Constructs

For cell populations that are just beyond the percolation threshold (cf. refs. 100 and 101), simulations show that the strain factor is sufficiently close to 1 that the contributions of cells and the ECM can be treated as occurring in parallel. At low cell concentrations, cells behave as if fully isolated. In the case of very stiff cells, the axial strain experienced by the cells is small compared to that of the nearby ECM (Figure 21.4). At the percolation threshold (taking the dimensionless 2D cell concentration as $C = A_c N l_o^2 / d = f_c l_o / d$, in which A_c is the nominal cross-sectional area of a cell, this occurs near $C = 3$), the strain factor rises close to a value of 1, and the cells and the ECM deform nearly in registry.

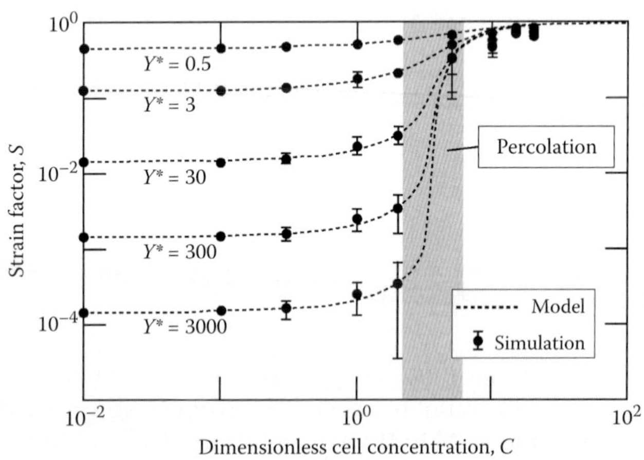

FIGURE 21.4
Strain factor indicates the degree that cells strain relative to the average strain tensor at a material point within a tissue construct. At the percolation threshold (here $C \approx 3$), cells and the ECM begin to stretch in registry ($S \approx 1$). (Adapted from Marquez JP et al., *Biophysical Journal* 88: 765–777, 2005.)

As an example of how this is useful, we present data from tissue constructs containing a population of contractile fibroblasts with final cell concentrations just above the percolation threshold. Mechanical relaxation data were obtained following relatively rapid (strain rate of 8 s^{-1}) stretching to 20% of a preconditioned tissue construct, and compared to data from identical tests performed on the decellularized (Triton X100-treated) ECM. Since $S \approx 1$, Equation 21.1 suggests that a subtraction of the ECM data from the tissue construct data should be a meaningful representation of the mechanical response of the cells (Figure 21.5).[102]

To interpret these data in the context of cytoskeletal mechanics, the mechanical studies must be combined with imaging studies. Fixing and staining tissue constructs with rhodamine phalloidin at prescribed times following stretch show an evolution of the stress fiber network over time (Figure 21.6).[102] This can be quantified by a Fourier space technique such as FABLE[103] (cf. Chapter 41) that integrates a band-pass-filtered power spectral density of an image to estimate the total length of stress fibers; images are preprocessed using median filtering to reduce artifacts. Results indicate that the total contribution of cells to tissue construct mechanics scales linearly with the total length of stress: the curve superimposed on data in Figure 21.6 is normalized data from Figure 21.5. This suggests that mechanical contributions of other protein structures such as microtubules

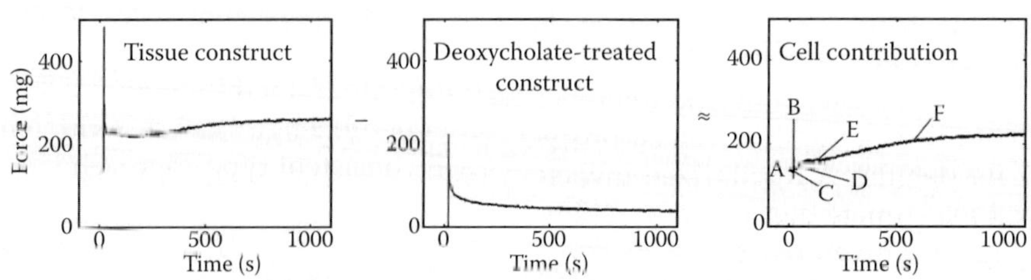

FIGURE 21.5
Isometric mechanical response of a tissue construct that is rapidly stretched and then held shows a characteristic rise followed by a reduction over time. The ECM contribution to this can be estimated by lysing cells with deoxycholate. Subtracting the second panel from the first provides a coarse qualitative estimate of the cellular contribution to the tissue construct response, provided that the cell population is above the percolation threshold. If the loading is sufficiently fast, the cell contribution shows a prestretch baseline force (A), a rise following stretch (B), a rapid drop shown below to result from cytoskeletal depolymerization (C), a rapid active response (D), a plateau (E), and a gradual active response (F). The final isometric force exceeds the prestretch value. This behavior is repeated upon subsequent loading cycles. (Adapted from Nekouzadeh A et al., *Journal of Biomechanics* 41(14): 2964–2971, 2008.)

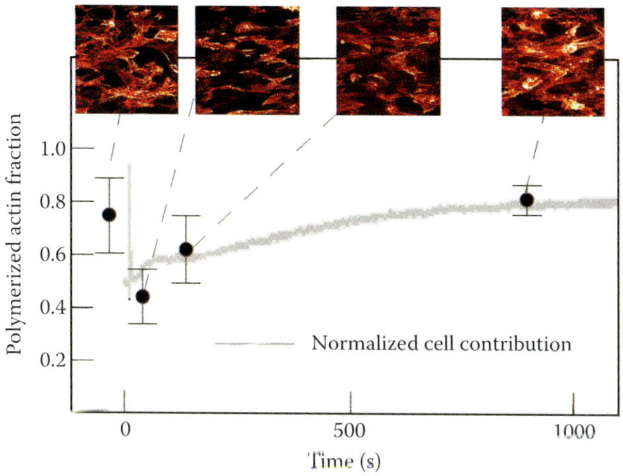

FIGURE 21.6
Fluidization and remodeling of the actin cytoskeleton in response to rapid external stressing, quantified by the density of F-actin at four time points: (1) after preconditioning but before the application of a rapid stretch; (2) at the end of the rapid reduction of force following a step stretch; (3) at the end of the rapid active response; and (4) at the end of the gradual active response. The trendline is the normalized cellular response curve from Figure 21.5. Error bars represent the range of data. The density of actin filaments scaled with the active cell force. (Adapted from Nekouzadeh A et al., *Journal of Biomechanics* 41(14): 2964–2971, 2008.)

either scale linearly with those of the actin cytoskeleton or are smaller in 3D tissue constructs than in 2D (cf. ref. 104).

These responses and their time courses can be further quantified using live cell imaging of the actin cytoskeleton.[105] Lee developed an apparatus to perform tensile tests on ring-shaped tissue constructs while simultaneously imaging fluorescently labeled cytoskeletal proteins over the objective of a confocal fluorescence microscope. Five to ten percent of chicken embryo fibroblast cells in each tissue construct were transfected with the F-actin reporter m-cherry LifeAct, a fluorescent probe that associates with actin filaments without affecting cellular dynamics. Looking at individual cells instead of populations of cells leads to much confusion, however. The ensemble behaviors shown in Figure 21.6 appear to be the result of a broad diversity of behaviors at the level of individual cells, with a range of "reinforcement" and "retraction" responses exhibited by cells and cytoskeletal proteins within a single tissue construct (Figure 21.7). Retraction responses occurred for all orientations of stress fibers and cellular protrusions relative to the stretch direction, while reinforcement responses, including extension of cellular processes and stress fiber formation, occurred predominantly in the

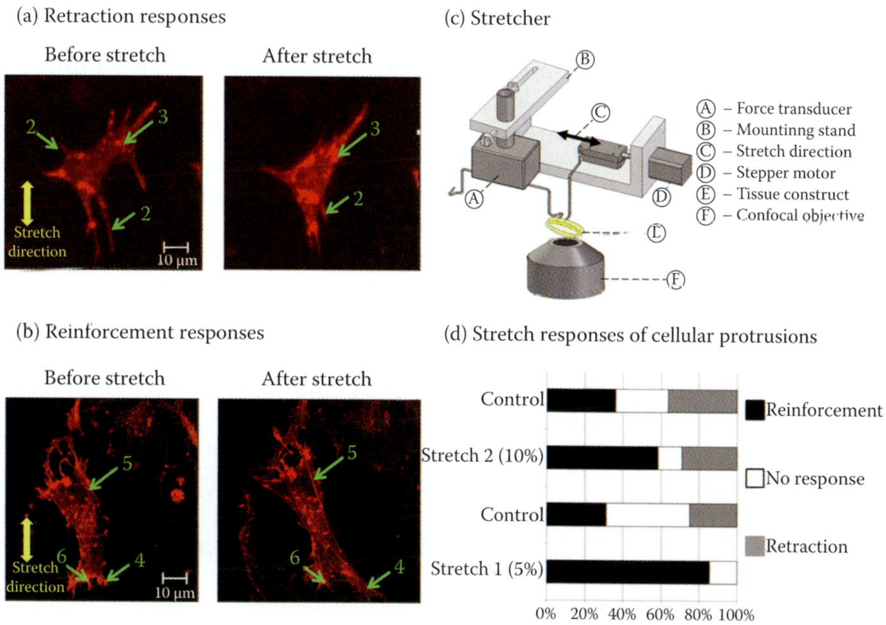

FIGURE 21.7
Preconditioned cells (a and b, left panels) within a tissue construct display two classes of responses when stretched and held for 30 min (a and b, right panels) using the device in (c). Retraction responses include transfer of actin from filopodium-like cellular protrusions (label 2) into F-actin reservoirs (label 3). Reinforcement responses include growth of cellular protrusions (label 4) and stress fibers (label 5), with actin likely drawn from F-actin reservoirs (label 6). Stretch increased the likelihood that cellular protrusions would extend, and decreased the likelihood that they would retract; however, only the former was significant statistically (d). (Reprinted from Lee SL et al., *PLoS One* 7(12): e45512, 2012. With permission.)

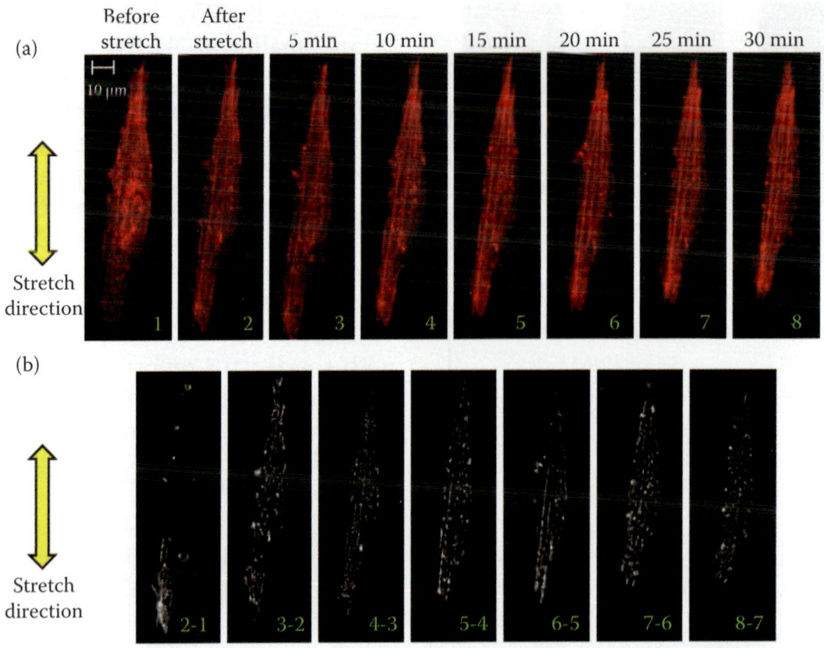

FIGURE 21.8
(a) Example of a time course of cytoskeletal remodeling (frame rate: one confocal image stack every 5 min). This cell exhibited partial fluidization followed by reinforcement and formation of actin stress fibers in response to mechanical stretch of 5%. (b) Time course of cytoskeletal remodeling is evident from subtraction images involving differences in image intensities of subsequent images. (Reprinted from Lee SL et al., *PLoS One* 7(12): e45512, 2012. With permission.)

stretch direction. A feature of cell kinetics in 3D are the F-actin clumps visible throughout the cytoskeleton that act as F-actin reservoirs for retraction and reinforcement responses during stretch.

Although several cells show a response of "fluidization" (cf. refs. 106 and 107) followed by reinforcement (e.g., Figure 21.8), others showed monotonic reinforcement or monotonic retraction. A convenient although imperfect metric by which these behaviors can be quantified is the FABLE measure of cytoskeletal polymerization (Chapter 41). Despite this measure being imperfect, it can be linked directly to mechanical and kinetic models and thus applied to help link images of mechanical responses to the physics underlying them. Lee et al. developed biomechanical models estimating how statistically varying levels of prestretch can combine with stress-dependent fiber growth and predict this range of responses within a single construct.

21.6 Concluding Remarks

Tissue constructs are an important platform for estimating the mechanical responses of cells in a natural 3D environment. Although qualitative observations of cell biophysics have been made for decades using tissue constructs, quantification of cellular responses requires tight integration of mechanical modeling with quantitative imaging. Key imaging needs include unbiased stereological measures such as volume fractions and spatial and orientation distributions of cells, along with observations of how changes to cellular cytoskeletons relate to measureable mechanical responses.

Key challenges that are, at present, unmet include technologies for accurately assessing spatial inhomogeneity in the mechanical properties of ECM and mathematical tools for linking imaging of spatially varying populations of cells to tissue construct response. An additional challenge is the variability of responses from cell to cell and tissue construct to tissue construct. Even with highly regular culture conditions, large variance can arise in the mechanics of tissue constructs. Although the nature of these variances is, at present, the source of much speculation, advances in imaging that can accurately quantify the cellular-level disposition of an entire tissue construct will be central to developing solutions to these challenges.

Acknowledgment

This work was funded in part by the National Institutes of Health grant R01HL109505.

References

1. Abbott A. Cell culture: Biology's new dimension. *Nature* 2003:424(6951): 870–872.
2. Lee SL, Nekouzadeh A, Butler B, Pryse KM, McConnaughey WB, Nathan AC, Legant WR, Schaefer PM, Pless RB, Elson EL et al. Physically-induced cytoskeleton remodeling of cells in three-dimensional culture. *PLoS One* 2012:7(12): e45512.
3. Cukierman E, Pankov R, Yamada KM. Cell interactions with three-dimensional matrices. *Current Opinion in Cell Biology* 2002:14(5): 633–640.
4. Paszek MJ, Zahir N, Johnson KR, Lakins JN, Rozenberg GI, Gefen A, Reinhart-King CA, Margulies SS, Dembo M, Boettiger D. Tensional homeostasis and the malignant phenotype. *Cancer Cell* 2005:8(3): 241–254.
5. Nelson CM, Bissell MJ. Modeling dynamic reciprocity: Engineering three-dimensional culture models of breast architecture, function, and neoplastic transformation. *Seminars in Cancer Biology* 2005:15: 342–352.
6. Vidi P-A, Bissell MJ, Lelièvre SA. Three-dimensional culture of human breast epithelial cells: The how and the why. In: *Epithelial Cell Culture Protocols*, edited by SH Randell and ML Fulcher. Totowa, NJ: Humana Press, 2013.
7. Fraley SI, Feng Y, Krishnamurthy R, Kim D-H, Celedon A, Longmore GD, Wirtz D. A distinctive role for focal adhesion proteins in three-dimensional cell motility. *Nature Cell Biology* 2010:12(6): 598–604.
8. Cukierman E, Pankov R, Stevens DR, Yamada KM. Taking cell–matrix adhesions to the third dimension. *Science* 2001:294(5547): 1708–1712.
9. Grinnell F. Fibroblast biology in three-dimensional collagen matrices. *Trends in Cell Biology* 2003:13(5): 264–269.
10. Wakatsuki T, Kolodney MS, Zahalak GI, Elson EL. Cell mechanics studied by a reconstituted model tissue. *Biophysical Journal* 2000:79(5): 2353–2368.
11. Zahalak GI, Wagenseil JE, Wakatsuki T, Elson EL. A cell-based constitutive relation for bio-artificial tissues. *Biophysical Journal* 2000:79(5): 2369–2381.
12. Lee PH, Vandenburgh H. Skeletal muscle atrophy in bioengineered skeletal muscle: A new model system. *Tissue Engineering Part A* 2013:19(19–20): 2147–2155.
13. Legant WR, Pathak A, Yang MT, Deshpande VS, McMeeking RM, Chen CS. Microfabricated tissue gauges to measure and manipulate forces from 3d microtissues. *Proceedings of the National Academy of Sciences of the United States of America* 2009:106(25): 10097–10102.
14. Asnes CF, Marquez JP, Elson EL, Wakatsuki T. Reconstitution of the frank-starling mechanism in engineered heart tissues. *Biophysical Journal* 2006:91(5): 1800–1810.
15. Elson EL. Cellular mechanics as an indicator of cytoskeletal structure and function. *Annual Review of Biophysics and Biophysical Chemistry* 1988:17: 397–430.
16. Cole KS. Surface forces of the Arbacia egg. *Journal of Cellular and Comparative Physiology* 1932:1(1): 1–9.
17. Thoumine O, Nerem RM, Girard PR. Oscillatory shear stress and hydrostatic pressure modulate cell–matrix attachment proteins in cultured endothelial cells. *In Vitro Cellular and Developmental Biology. Animal* 1995:31(1): 45–54.
18. Granzier HL, Irving TC. Passive tension in cardiac muscle: Contribution of collagen, titin, microtubules, and intermediate filaments. *Biophysical Journal* 1995:68(3): 1027–1044.
19. Mitchison JM, Swann MM. The mechanical properties of the cell surface. *Journal of Experimental Biology* 1954:31(3): 443–460.
20. Evans EA, Skalak R. *Mechanics and Thermodynamics of Biomembranes.* Boca Raton. CRC Press, 1980.
21. Evans E, Yeung A. Apparent viscosity and cortical tension of blood granulocytes determined by micropipet aspiration. *Biophysical Journal* 1989:56(1): 151–160.
22. Skalak R, Chien S, Schmid-Schonbein GW. Viscoelastic deformation of white cells: Theory and analysis. *Kroc Foundation Series* 1984:16: 3–18.
23. Elgsaeter A, Stokke BT, Mikkelsen A, Branton D. The molecular basis of erythrocyte shape. *Science* 1986:234 (4781): 1217–1223.
24. Chen Y, Liu B, Xu G, Shao J-Y. Validation, in-depth analysis, and modification of the micropipette aspiration technique. *Cellular and Molecular Bioengineering* 2009:2(3): 351–365.
25. Shao J-Y. Biomechanics of leukocyte and endothelial cell surface. *Current Topics in Membranes* 2009:64: 25–45.
26. Petersen NO, McConnaughey WB, Elson EL. Dependence of locally measured cellular deformability on position on the cell, temperature, and cytochalasin b. *Proceedings of the National Academy of Sciences of the United States of America* 1982:79(17): 5327–5331.
27. Zahalak GI, McConnaughey WB, Elson EL. Determination of cellular mechanical properties by cell poking, with an application to leukocytes. *Journal of Biomechanical Engineering* 1990:112(3): 283–294.
28. Daily B, Elson EL, Zahalak GI. Cell poking. Determination of the elastic area compressibility modulus of the erythrocyte membrane. *Biophysical Journal* 1984:45(4): 671–682.
29. Pasternak C, Elson EL. Lymphocyte mechanical response triggered by cross-linking surface receptors. *Journal of Cell Biology* 1985:100(3): 860–872.
30. Pasternak C, Spudich JA, Elson EL. Capping of surface receptors and concomitant cortical tension are generated by conventional myosin. *Nature* 1989:341(6242): 549–551.
31. Worthen GS, Schwab BD, Elson EL, Downey GP. Mechanics of stimulated neutrophils: Cell stiffening induces retention in capillaries. *Science* 1989:245(4914): 183–186.
32. Downey GP, Doherty DE, Schwab BD, Elson EL, Henson PM, Worthen GS. Retention of leukocytes in capillaries: Role of cell size and deformability. *Journal of Applied Physiology* 1990:69(5): 1767–1778.
33. Downey GP, Elson EL, Schwab BD, Erzurum SC, Young SK, Worthen GS. Biophysical properties and microfilament assembly in neutrophils: Modulation by cyclic amp. *Journal of Cell Biology* 1991:114(6): 1179–1190.

34. Radmacher M. Measuring the elastic properties of living cells by the atomic force microscope. *Methods in Cell Biology* 2002:68: 67–90.
35. Weafer P, Ronan W, Jarvis S, McGarry JP. Experimental and computational investigation of the role of stress fiber contractility in the resistance of osteoblasts to compression. *Bulletin of Mathematical Biology* 2013:75(8): 1284–1303.
36. Elson E, Genin G. The role of mechanics in actin stress fiber kinetics. *Experimental Cell Research* 2013:319(16): 2490–2500.
37. Harris AK, Wild P, Stopak D. Silicone rubber substrata: A new wrinkle in the study of cell locomotion. *Science* 1980:208(4440): 177–179.
38. Tan JL, Tien J, Pirone DM, Gray DS, Bhadriraju K, Chen CS. Cells lying on a bed of microneedles: An approach to isolate mechanical force. *Proceedings of the National Academy of Sciences of the United States of America* 2003:100(4): 1484–1489.
39. McGarry JP, Fu J, Yang MT, Chen CS, McMeeking RM, Evans AG, Deshpande VS. Simulation of the contractile response of cells on an array of micro-posts. *Philosophical Transactions of the Royal Society A: Mathematical, Physical and Engineering Sciences* 2009:367(1902): 3477–3497.
40. Chen C, Krishnan R, Zhou E, Ramachandran A, Tambe D, Rajendran K, Adam RM, Deng L, Fredberg JJ. Fluidization and resolidification of the human bladder smooth muscle cell in response to transient stretch. *PLoS One* 2010:5(8): e12035.
41. Lam RH, Sun Y, Chen W, Fu J. Elastomeric microposts integrated into microfluidics for flow-mediated endothelial mechanotransduction analysis. *Lab on a Chip* 2012:12(10): 1865–1873.
42. Polacheck WJ, Li R, Uzel SG, Kamm RD. Microfluidic platforms for mechanobiology. *Lab on a Chip* 2013:13(12): 2252–2267.
43. Han SJ, Bielawski KS, Ting LH, Rodriguez ML, Sniadecki NJ. Decoupling substrate stiffness, spread area, and micropost density: A close spatial relationship between traction forces and focal adhesions. *Biophysical Journal* 2012:103(4): 640–648.
44. Sniadecki NJ. Minireview: A tiny touch: Activation of cell signaling pathways with magnetic nanoparticles. *Endocrinology* 2010:151(2): 451–457.
45. Mann JM, Lam RH, Weng S, Sun Y, Fu J. A silicone-based stretchable micropost array membrane for monitoring live-cell subcellular cytoskeletal response. *Lab on a Chip* 2012:12(4): 731–740.
46. Lam RH, Weng S, Lu W, Fu J. Live-cell subcellular measurement of cell stiffness using a microengineered stretchable micropost array membrane. *Integrative Biology: Quantitative Biosciences from Nano to Macro* 2012:4(10): 1289–1298.
47. le Digabel J, Ghibaudo M, Trichet LA, Richert A, Ladoux B. Microfabricated substrates as a tool to study cell mechanotransduction. *Medical and Biological Engineering and Computing* 2010:48(10): 965–976.
48. Deshpande VS, McMeeking RM, Evans AG. A bio-chemo-mechanical model for cell contractility. *Proceedings of the National Academy of Sciences of the United States of America* 2006:103(38): 14015–14020.
49. Deshpande VS, McMeeking RM, Evans AG. A model for the contractility of the cytoskeleton including the effects of stress–fibre formation and dissociation. *Proceedings of the Royal Society A: Mathematical, Physical and Engineering Science* 2007:463(2079): 787–815.
50. Deshpande VS, Mrksich M, McMeeking RM, Evans AG. A bio-mechanical model for coupling cell contractility with focal adhesion formation. *Journal of the Mechanics and Physics of Solids* 2008:56(4): 1484–1510.
51. Pathak A, Deshpande VS, McMeeking RM, Evans AG. The simulation of stress fibre and focal adhesion development in cells on patterned substrates. *Journal of the Royal Society Interface* 2008:5(22): 507–524.
52. Ronan W, Deshpande VS, McMeeking RM, McGarry JP. Numerical investigation of the active role of the actin cytoskeleton in the compression resistance of cells. *Journal of the Mechanical Behavior of Biomedical Materials* 2012:14: 143–157.
53. Wei Z, Deshpande VS, McMeeking RM, Evans AG. Analysis and interpretation of stress fiber organization in cells subject to cyclic stretch. *Journal of Biomechanical Engineering* 2008:130(3): 031009.
54. Hsu HJ, Lee CF, Kaunas R. A dynamic stochastic model of frequency-dependent stress fiber alignment induced by cyclic stretch. *PLoS One* 2009:4(3): e4853.
55. Hsu HJ, Lee CF, Locke A, Vanderzyl SQ, Kaunas R. Stretch-induced stress fiber remodeling and the activations of jnk and erk depend on mechanical strain rate, but not fak. *PLoS One* 2010:5(8): e12470.
56. Kaunas R, Hsu H-J. A kinematic model of stretch-induced stress fiber turnover and reorientation. *Journal of Theoretical Biology* 2009:257(2): 320–330.
57. Kaunas R, Nguyen P, Usami S, Chien S. Cooperative effects of rho and mechanical stretch on stress fiber organization. *Proceedings of the National Academy of Sciences of the United States of America* 2005:102(44): 15895–15900.
58. Kaunas R, Usami S, Chien S. Regulation of stretch-induced jnk activation by stress fiber orientation. *Cellular Signalling* 2006:18(11): 1924–1931.
59. Butler JP, Tolic-Norrelykke IM, Fabry B, Fredberg JJ. Traction fields, moments, and strain energy that cells exert on their surroundings. *American Journal of Physiology. Cell Physiology* 2002:282(3): C595–C605.
60. Bursac P, Lenormand G, Fabry B, Oliver M, Weitz DA, Viasnoff V, Butler JP, Fredberg JJ. Cytoskeletal remodelling and slow dynamics in the living cell. *Nature Material* 2005:4(7): 557–561.
61. Trepat X, Wasserman MR, Angelini TE, Millet E, Weitz DA, Butler JP, Fredberg JJ. Physical forces during collective cell migration. *Nature Physics* 2009:5(6): 426–430.
62. Dembo M, Oliver T, Ishihara A, Jacobson K. Imaging the traction stresses exerted by locomoting cells with the elastic substratum method. *Biophysical Journal* 1996:70(4): 2008–2022.
63. Dembo M, Wang YL. Stresses at the cell-to-substrate interface during locomotion of fibroblasts. *Biophysical Journal* 1999:76(4): 2307–2316.

64. Barocas VH, Moon AG, Tranquillo RT. The fibroblast-populated collagen microsphere assay of cell traction force—Part 2: Measurement of the cell traction parameter. *Journal of Biomechanical Engineering* 1995:117(2): 161–170.
65. Barocas VH, Tranquillo RT. An anisotropic biphasic theory of tissue-equivalent mechanics: The interplay among cell traction, fibrillar network deformation, fibril alignment, and cell contact guidance. *Journal of Biomechanical Engineering* 1997:119(2): 137–145.
66. Moon A, Drubin DG. The adf/cofilin proteins: Stimulus-responsive modulators of actin dynamics. *Molecular Biology of the Cell* 1995:6(11): 1423–1431.
67. Knapp DM, Helou EF, Tranquillo RT. A fibrin or collagen gel assay for tissue cell chemotaxis: Assessment of fibroblast chemotaxis to grgdsp. *Experimental Cell Research* 1999:247(2): 543–553.
68. Hong S, Hsu H-J, Kaunas R, Kameoka J. Collagen microsphere production on a chip. *Lab on a Chip* 2012:12(18): 3277–3280.
69. Barocas VH, Tranquillo RT. A finite element solution for the anisotropic biphasic theory of tissue-equivalent mechanics: The effect of contact guidance on isometric cell traction measurement. *Journal of Biomechanical Engineering* 1997:119(3): 261–268.
70. Moon AG, Tranquillo RT. The fibroblast-populated collagen microsphere assay of cell traction force: Part 1 Continuum model. *AIChE Journal* 1993:39: 163–177.
71. Tranquillo RT. Self-organization of tissue-equivalents: The nature and role of contact guidance. *Biochemical Society Symposium* 1999:65: 27–42.
72. Tranquillo RT, Murray JD. Mechanistic model of wound contraction. *The Journal of Surgical Research* 1993:55(2): 233–247.
73. Tranquillo RT, Murray JD. Continuum model of fibroblast-driven wound contraction: Inflammation-mediation. *Journal of Theoretical Biology* 1992:158(2): 135–172.
74. Zemel A, Safran SA. Active self-polarization of contractile cells in asymmetrically shaped domains. *Physical Review. E, Statistical, Nonlinear, and Soft Matter Physics* 2007:76(2 Pt 1): 021905.
75. Zemel A, Bischofs IB, Safran SA. Active elasticity of gels with contractile cells. *Physical Review Letters* 2006:97(12): 128103.
76. Delvoye P, Wiliquet P, Levêque J-L, Nusgens BV, Lapière CM. Measurement of mechanical forces generated by skin fibroblasts embedded in a three-dimensional collagen gel. *Journal of Investigative Dermatology* 1991:97(5): 898–902.
77. Kolodney MS, Wysolmerski RB. Isometric contraction by fibroblasts and endothelial cells in tissue culture: A quantitative study. *The Journal of Cell Biology* 1992:117: 73–82.
78. Marquez JP, Genin GM, Pryse KM, Elson EL. Cellular and matrix contributions to tissue construct stiffness increase with cellular concentration. *Annals of Biomedical Engineering* 2006:34(9): 1475–1482.
79. Mouton PR. *Unbiased Stereology: A Concise Guide*. Baltimore: JHU Press, 2013.
80. Baddeley A, Vedel Jensen EB. *Stereology for Statisticians*. Boca Raton, FL: CRC Press, 2005.
81. Lanir Y. Constitutive equations for fibrous connective tissues. *Journal of Biomechanics* 1983:16(1): 1–12.
82. Lanir Y, Salant EL, Foux A. Physico-chemical and microstructural changes in collagen fiber bundles following stretch in-vitro. *Biorheology* 1988:25(4): 591–603.
83. Sacks MS. Incorporation of experimentally-derived fiber orientation into a structural constitutive model for planar collagenous tissues. *Journal of Biomechanical Engineering* 2003:125(2): 280–287.
84. Horowitz A, Lanir Y, Yin FC, Perl M, Sheinman I, Strumpf RK. Structural three-dimensional constitutive law for the passive myocardium. *Journal of Biomechanical Engineering* 1988:110(3): 200–207.
85. Hurschler C, Loitz-Ramage B, Vanderby R Jr. A structurally based stress–stretch relationship for tendon and ligament. *Journal of Biomechanical Engineering* 1997:119(4): 392–399.
86. Wren TA, Carter DR. A microstructural model for the tensile constitutive and failure behavior of soft skeletal connective tissues. *Journal of Biomechanical Engineering* 1998:120(1): 55–61.
87. Sander EA, Stylianopoulos T, Tranquillo RT, Barocas VH. Image-based multiscale modeling predicts tissue-level and network-level fiber reorganization in stretched cell-compacted collagen gels. *Proceedings of the National Academy of Sciences of the United States of America* 2009:106(42): 17675–17680.
88. Hill A. The heat of shortening and the dynamic constants of muscle. *Proceedings of the Royal Society of London. Series B, Biological Sciences* 1938:126(843): 136–195.
89. Marquez JP. Fourier analysis and automated measurement of cell and fiber angular orientation distributions. *International Journal of Solids and Structures* 2006:43(21): 6413–6423.
90. Tondon A, Kaunas R. The direction of stretch-induced cell and stress fiber orientation depends on collagen matrix stress. *PLoS One* 2014:9(2): e89592.
91. Nekouzadeh A, Genin GM. Adaptive quasi-linear viscoelastic modeling. In: *Computational Modeling in Tissue Engineering*, edited by L Geris. New York: Springer, 2013.
92. Pryse KM, Nekouzadeh A, Genin GM, Elson EL, Zahalak GI. Incremental mechanics of collagen gels: New experiments and a new viscoelastic model. *Annals of Biomedical Engineering* 2003:31(10): 1287–1296.
93. Nekouzadeh A, Pryse KM, Elson EL, Genin GM. A simplified approach to quasi-linear viscoelastic modeling. *Journal of Biomechanics* 2007:40(14): 3070–3078.
94. Legant WR, Miller JS, Blakely BL, Cohen DM, Genin GM, Chen CS. Measurement of mechanical tractions exerted by cells in three-dimensional matrices. *Nature Methods* 2010:7(12): 969–971.
95. Prager W. On the formulation of constitutive equations for living soft tissues. *Applied Math* 1969:27: 128–132.
96. Hill R. A self-consistent mechanics of composite materials. *Journal of the Mechanics and Physics of Solids* 1965:13: 213–222.
97. Budiansky B. On the elastic moduli of some heterogeneous materials. *Journal of the Mechanics and Physics of Solids* 1965:13: 223–227.
98. Marquez JP, Genin GM, Zahalak GI, Elson EL. Thin bio-artificial tissues in plane stress: The relationship between

cell and tissue strain, and an improved constitutive model. *Biophysical Journal* 2005:88: 765–777.

99. Marquez JP, Genin GM, Zahalak GI, Elson EL. The relationship between cell and tissue strain in three-dimensional bio-artificial tissues. *Biophysical Journal* 2005:88(2): 778–789.
100. Bug ALR, Safran SA, Grest GS, Webman I. Do interactions raise or lower a percolation threshold? *Physical Review Letters* 1985:55(18): 1896–1899.
101. Bug ALR, Safran SA, Webman I. Continuum percolation of rods. *Physical Review Letters* 1985:54(13): 1412–1415.
102. Nekouzadeh A, Pryse KM, Elson EL, Genin GM. Stretch-activated force shedding, force recovery, and cytoskeletal remodeling in contractile fibroblasts. *Journal of Biomechanics* 2008:41(14): 2964–2971.
103. Nekouzadeh A, Genin GM. Quantification of fibre polymerization through Fourier space image analysis. *Proceedings of the Royal Society A: Mathematical, Physical and Engineering Science* 2011:467(2132): 2310–2329.
104. Wang N, Naruse K, Stamenovic D, Fredberg JJ, Mijailovich SM, Tolic-Norrelykke IM, Polte T, Mannix R, Ingber DE. Mechanical behavior in living cells consistent with the tensegrity model. *Proceedings of the National Academy of Sciences of the United States of America* 2001:98(14): 7765–7770.
105. Lee CF, Haase C, Deguchi S, Kaunas R. Cyclic stretch-induced stress fiber dynamics—Dependence on strain rate, rho-kinase and mlck. *Biochemical and Biophysical Research Communications* 2010:401(3): 344–349.
106. Krishnan R, Park CY, Lin YC, Mead J, Jaspers RT, Trepat X, Lenormand G, et al. Reinforcement versus fluidization in cytoskeletal mechanoresponsiveness. *PLoS One* 2009:4(5): e5486.
107. Gavara N, Roca-Cusachs P, Sunyer R, Farre R, Navajas D. Mapping cell–matrix stresses during stretch reveals inelastic reorganization of the cytoskeleton. *Biophysical Journal* 2008:95(1): 464–471.

Section VIII

Cellular and Molecular Imaging Methods

Microenvironments—In Vitro Culture Systems

22

Measurement of Cellular Forces via Traction Force Microscopy

Wesley R. Legant

CONTENTS

Technologies capable of quantitatively measuring the forces exerted by single cells have enabled the field of mechanotransduction—the study of how mechanical forces are transduced into biochemical signals. Mechanical forces, which are exerted by nearly all eukaryotic cells, drive cell migration and maintain the mechanical tone of tissues. Moreover, the same forces feed back through biochemical signaling pathways to regulate cellular functions ranging from mitosis to stem cell differentiation and self-renewal. Here, I will review cell traction force microscopy, one of the most widespread methods for assessing the traction forces exerted by single cells and currently the only technique capable of providing vectorial information about the magnitude and direction of forces exerted in both 2D and 3D settings.

22.1 Introduction

As cells migrate through the extracellular space, they must exert or be subject to physical forces of some kind. During development, the collective forces exerted by many cells drive the reorganization events that generate distinct tissues and organs.[1–3] However, while it was known that physical forces must be driving these events, quantitatively measuring the forces generated by even an experimentally isolated single cell was technically restrictive. In 1980, Harris et al.[4] conclusively demonstrated that adherent cells exerted significant traction forces on their surroundings. He observed that, when cultured on a thin silicon membrane, isolated fibroblasts generated elastic wrinkles that disappeared and reappeared as the cells crawled to new areas. The magnitude and direction of the forces could be inferred from the size and direction of wrinkle propagation. Impressively, by combining this system with interference reflection microscopy, a correlation between focal adhesion development and wrinkle formation just distal to the leading edge was observed, indicating that forces transmitted through focal adhesions were critical for migration.

Since this seminal paper, numerous improvements in both material science and mathematical rigor have been made to what is now collectively termed traction force microscopy (TFM). Linearly elastic materials (polyacrylamide, polyethylene glycol, or polydimethylsiloxane [PDMS]) coupled to adhesive molecules were adopted in place of a thin silicon membrane, and fluorescent beads were utilized instead of wrinkles to report the substrate displacements.[5,6] Importantly, these changes replaced the generally nonlinear and mathematically complex description of wrinkle formation with a classical, linear, continuum mechanics model. With this as a starting point, TFM is now widely used in studies of cell mechanics and migration and has recently been extended into 2.5D (measuring the 3D forces exerted by cells cultured on planar substrates[7–10]) and fully 3D (measuring the 3D forces exerted by cells fully encapsulated within a hydrogel matrix[11]) settings (Figure 22.1). It has been shown that cells pull radially inward toward the center of mass, with forces concentrated around focal adhesion structures.[4,12–14] 2.5D TFM has revealed that cells also exert significant normal tractions, pulling up at the cell edges and pushing down near the nucleus.[8,10] These normal tractions can also result in rotational moments generated about focal adhesions.[9] Radially inward forces are conserved in fully 3D TFM, but force generation must be dynamically coupled to extracellular matrix (ECM) degradation to allow for cell spreading and migration.[11] By correlating readouts of cellular forces with changes in biochemistry and cell phenotype, it has been shown that cell traction forces not only drive migration but can also provide feedback to impact numerous cell functions such as differentiation and proliferation in both normal and pathologic settings.[15–18] The rest of this chapter will be devoted to discussing various approaches, which together define the current state of the art of TFM. A broad overview of different methods will be provided, and wherever possible, pros and cons of each approach will be discussed.

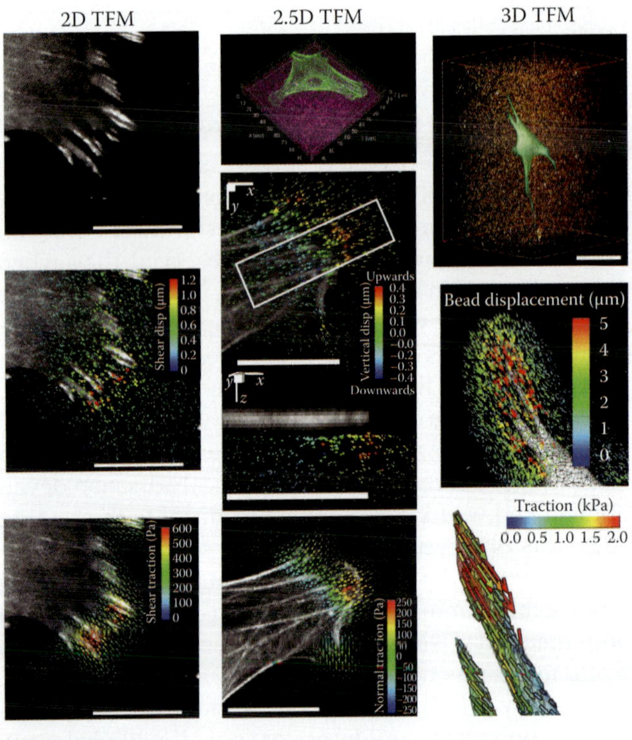

FIGURE 22.1
TFM in 2D, 2.5D, and 3D geometries. In both 2D and 2.5D TFM, cells are cultured on top of planar elastic substrates that are coated with adhesive ligands. For 2D TFM, the substrates are considered to be nearly incompressible, and displacements and tractions are only measured parallel to the planar surface. In 2.5D TFM, the geometry is unchanged, but normal displacements and tractions are also considered. 3D TFM fully encapsulates cells within proteolytically degradable hydrogels that permit both cell attachment and spreading into physiologically relevant 3D morphologies.

22.2 TFM—Methods

22.2.1 Measurement of the Displacement Field

The fundamental principles of TFM have remained constant since its initial development and are shared by all implementations outlined in the following. In TFM, the cell culture substrate deforms in response to cell-generated traction forces. These deformations are spatially and temporally mapped by monitoring the locations of embedded fluorescent beads. In order to measure cell forces, two images of the substrate are acquired—an image of the bead locations while the substrate is subject to cell traction forces (i.e., a "stressed" dataset) and a reference (stress-free) image of the bead locations after detaching the cell with trypsin or lysing the cell with a detergent (Figure 22.2a). The bead displacements are determined by comparing the reference and stressed images at any given time point. There are currently two main approaches to perform this

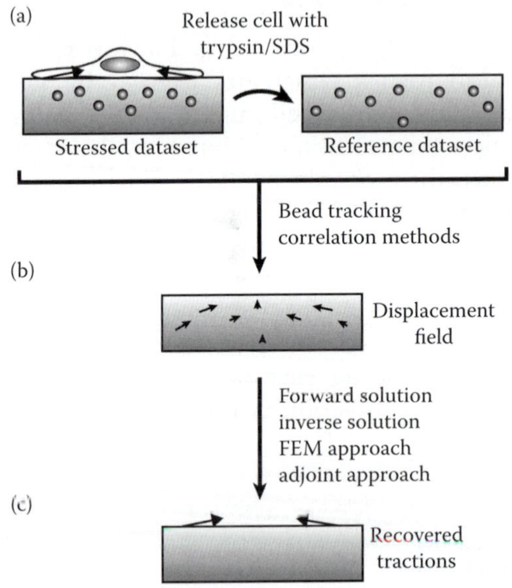

FIGURE 22.2
TFM flow chart. (a) Cells are cultured on elastic substrates containing fiducial markers. Two datasets are acquired, one subject to cell tractions (stressed dataset) and the other one after releasing the cell with trypsin or detergent (reference dataset). (b) Fiducials are tracked between the two datasets using particle tracking or image correlation methods to measure the displacement field. (c) Several different approaches can then be utilized to convert the displacement measurements into the source tractions that were applied by the cell.

comparison—localizing and tracking individual beads, and image correlation techniques (Figure 22.2b). The former generates estimates of bead centroids within an image by locating local maxima in pixel intensity. For small beads (i.e., 200 nm diameter), the image of the bead is defined by the optical diffraction limit; however, the true bead centroid can be identified with very high (potentially subnanometer) precision by utilizing single-particle localization algorithms.[19,20] This approach generates a list of discrete coordinates for bead centroids that must then be matched in the reference and stressed images. Several approaches have been developed for this purpose, including nearest neighbor analysis,[21,22] autoregression,[23] as well as more complex (and computationally demanding) pattern-matching algorithms.[11]

Alternatively, one may opt to forgo identifying and matching discrete bead centroids and instead locate regions, or windows, of each image that possess similar intensity distributions. This approach is often referred to as either particle imaging velocimetry or digital image correlation.[7,13,24,25] In the first step, both the reference and stressed images are partitioned into windows (~10 × 10 pixels). Within each of these windows, a cross-correlation function is calculated. The maximum of this function corresponds to the calculated shift between the image regions and is thus a measure of the displacement field. This method can be applied iteratively with smaller and smaller windows to increase the density of the displacement observations and has been shown to produce accuracies of ~140–280 nm.[25]

Both of these approaches have been applied in either two or three dimensions with minimal modifications. The question of which approach to use when is a matter of preference. Localizing the centroids of individual beads could potentially yield higher accuracy, but incorrect matches between beads in the reference and stressed images can contaminate the true displacement field. Thus, some filtering is often applied to identify and disregard any spurious matches. Correlation-based approaches are more robust against mismatches, but it is unclear whether they can obtain comparable accuracy. In either of these approaches, some care must be taken to correct for any drift (translational or rotational) or hydrogel swelling (e.g., due to changes in temperature), which could cause large-scale bead displacements independent of cell tractions.

22.2.2 Converting the Displacement Field to Tractions at the Interface

The next step in TFM is to convert the displacement field within the substrate into a mapping of the cellular traction forces at the interface (Figure 22.2c). Viewed another way, a relationship is sought to predict how the elastic volume will deform in response to some pattern of forces applied at the interface between the cell and the material. Once obtained, this relationship could be used to convert between material displacements and the cellular forces. Although the fundamental mechanics are the same, several different approaches now exist to accomplish this task. Here, an overview of the key equations and a discussion of the different methods will be provided.

22.2.3 Forward Approach

For an isotropic, linear elastic solid at constant temperature, the displacement, stress, and strain fields are related according to

$$\varepsilon_{ij} = \frac{1}{2}\left(\frac{\partial u_i}{\partial x_j} + \frac{\partial u_j}{\partial x_i}\right) \tag{22.1}$$

$$\sigma_{ij} = \frac{E}{1+v}\left(\varepsilon_{ij} + \frac{v}{1-2v}\varepsilon_{kk}\delta_{ij}\right) \tag{22.2}$$

where x is the position vector, u is the displacement vector, ε and σ are the strain and stress tensors, respectively, E is the Young's modulus, v is the Poisson's ratio, and i, j = 1,2,3 are the Cartesian basis vectors. These equations comprise a set of six coupled partial differential equations that relate the stress field to the displacement field within the hydrogel. Thus, given a displacement field within the hydrogel, Equations 22.1 and 22.2 can be used to compute the stress field within the material. The challenge arises in that the continuous displacement field is measured only at discrete locations (e.g., bead centroids); thus, some interpolation scheme is required to compute the differentials in Equation 22.1. A common approach is to use the bead centroids as nodes in a Delaunay triangulation scheme and then use standard trilinear shape functions to compute the strain field.[22] Higher-order shape functions, alternative geometric elements (e.g., quadrilateral), or local averaging schemes (such as the displacement–gradient technique or moving least squares[7,26]) may be used to improve accuracy. Upon obtaining a continuous approximation of the stress field within the solid, the tractions T at any surface may be computed according to

$$T_i = \sigma_{ij} n_j \tag{22.3}$$

where n is the outward normal to the surface at any given point.

The forward approach is a straightforward and computationally efficient method to measure cell traction forces. It can further be adopted for 2.5D and fully 3D traction force measurements with little modification. In addition, nonlinear, viscoelastic, or other material constitutive properties may be substituted without changing the general mathematical framework. However, there are trade-offs in using this approach. The first is that the stress field is not known immediately at the cell–ECM interface but rather is only measured at or below the layer of beads that is closest to this interface. In regions where the true interface lies some distance away from the bead locations, some method of extrapolation must be used to estimate the stress field at the interface. This may not contribute significant error in 2D or 2.5D TFM, where one can ensure that a large number of beads are very near the hydrogel surface, but for fully 3D TFM, the beads are randomly distributed in the hydrogel, and extrapolation of the stress field onto the complex surface of the cell is difficult. The second concern is that presumably the tractions at the interface, rather than the stress field in the bulk, are of key interest. However, only a small percentage of beads very close to the interface are utilized in the calculation of the local stress field via Equations 22.1 and 22.2. Thus, although bead displacements can be detected a significant distance away from the interface, information from these displacements is not used in the calculation of the traction field.

22.2.4 Inverse Approach

One technique to address these concerns is to formulate Equations 22.1 and 22.2 as an inverse problem. In this approach, the displacement field and the material stresses can be related via an integral equation:

$$u_i(x) = \int \sum_j G_{ij}(x, x') f_j(x') \mathrm{d}x' \quad (22.4)$$

where u is the displacement vector measured at position x, f is the force exerted at position x', and G_{ij} is a Green's function that relates the forces and displacements. This relationship is general for any geometry; however, the utilization of Green's functions requires that the stress and strain equations be linear, i.e., linearly elastic material properties, and subject to the small strain/small rotation approximation. A second difficulty arises in that analytic definitions for the Green's function G_{ij} are known only for a few geometries. Fortunately, the flat, planar hydrogel substrates commonly used for cell culture can be accurately modeled as a semi-infinite, homogeneous, elastic half-space for which G_{ij} can be defined by the Boussinesq–Cerruti solution.[5,6,12,27]

In order to solve Equation 22.4, both the displacement field and the forces are discretized. The discretization of the displacement field is achieved by the measurement of a finite number of bead displacements, and the discretization of the force field is typically accomplished using triangular elements and linear shape functions. Thus, the finite number of displacement points (observations) and the finite number of nodes at which the force field must be computed (variables) constitute a linear system of equations:

$$\bar{u} = G\bar{f} \quad (22.5)$$

that can be solved using matrix inversion. Unfortunately, Equation 22.4 is ill-posed, making the discretized version of G_{ij} in Equation 22.5 poorly conditioned. Because G_{ij} decays as ~$1/r$ with distance, the magnitude of the displacements decays slowly with distance from the location of force application, and Equation 22.4 constitutes a smoothing operation. This causes the displacements to be relatively insensitive to high-frequency variations in the forces. Upon inversion, this means that the calculated force field will be very sensitive to high-frequency fluctuations (i.e., noise) in the displacement measurements. Ill-posed inverse problems arise in many disciplines and are well investigated in numerical mathematics. Thus, many regularization schemes exist to stabilize the solution and produce reliable measurements of the forces.[28] Additionally, Equation 22.4 may be transformed and inverted in the Fourier domain,[13,29,30] thus dramatically reducing the computational expense. However, even with this approach, some presmoothing (effective regularization) of the displacement field is necessary to obtain a reasonable solution.[25]

The inverse approach utilizing Green's functions computes the forces based on all of the observed bead measurements. However, this comes at the cost of the need for regularization and introduces dependency of the computed forces on the choice of the regularization parameter and technique.[6,9,25] Moreover, the utilization of Green's functions assumes that the governing equations are linear. Using this method for nonlinear materials or large strains would introduce additional error into the measured forces.

This approach has been generalized to both 2.5D and fully 3D geometries[9,11,27] utilizing linearly elastic polyacrylamide or polyethylene glycol hydrogels. For cells fully encapsulated within a hydrogel, the interface between the cell and hydrogel is highly irregular. Thus, no analytic Green's function exists. Instead, a discretized Green's function can be generated numerically utilizing the finite-element method (FEM). Briefly, the cell and surrounding hydrogel volume are discretized using a 3D tetrahedral mesh. Unit tractions are then applied in each of the Cartesian directions at each facet on the interface between the cell and the hydrogel, and the induced displacement fields are recorded after solving the forward FEM problem for stress equilibrium. Once this function is obtained, the forces can be calculated utilizing the same numerical methods as for the 2D techniques.

22.2.5 Alternative Approaches

An alternative technique to either the forward or inverse approaches outlined above utilizes the classical FEM together with a displacement boundary condition to solve for stress equilibrium within the hydrogel.[8,31] In the first step, a finite-element model of the hydrogel geometry is generated. The displacement measurement within the hydrogel is extrapolated onto the hydrogel surface, and this is applied as a boundary condition within the context of the finite-element equations. Any geometry or constitutive equations can be modeled (i.e., 2D, 2.5D, or fully 3D with linear or nonlinear materials). However, as with the forward method, only a small fraction of the bead displacements are utilized in computing the displacement field at the surface, and, especially for complicated geometries, extrapolating this displacement field onto the surface of the cell can be challenging.

The inverse problem can also be solved by formulating Equation 22.4 in terms of the unknown cellular tractions and the adjoint state.[32,33] This results in a coupled set of partial differential equations that can be solved numerically for any geometry utilizing finite elements. Importantly, this method utilizes information from all observable bead displacements (an advantage of the inverse approach) while avoiding the requirements of utilizing Green's functions (i.e., the stress–strain equations need not be linear). Extensive numerical simulations have demonstrated that the adjoint approach can produce reliable estimates of fully 3D cellular tractions (at least in the context of a linearly elastic, homogeneous material model) with a lower computational cost compared with numerically generating a discretized Green's function.

22.2.6 Choice of TFM Method

Given the variety of methods capable of solving for cellular tractions in a variety of contexts, how does one go about choosing which approach to use? The forward approach is conceptually and computationally straightforward and produces a direct measurement of the strain field within the material. This is then coupled with an appropriate constitutive model for the surrounding material to produce the stress field. Importantly, this approach places no restrictions on the mechanical properties of the material and is thus applicable for both linear and nonlinear/discontinuous materials. However, in order to solve for surface tractions, the stress field must be extrapolated onto the hydrogel/cell interface. This extrapolation also arises in the displacement boundary condition-based FEM approaches. However, the error introduced by this process, especially in complex 3D geometries, has not been investigated thoroughly. Moreover, the surface tractions are computed solely from stress field and bead displacements within the immediate proximity of the interface. As such, only a small fraction of all accessible information is utilized in the computation of cellular tractions. In contrast, inverse approaches utilize information from all beads within the volume to compute the cellular tractions. However, the inverse problem itself is ill-posed and requires additional regularization. The use of regularization to stabilize the solution introduces additional dependency (i.e., on the method and value chosen for the regularization parameter) in the computed traction field. Moreover, the inverse problem is most commonly solved with the use of Green's functions, which require that the stress–strain equations be linear. Furthermore, analytical Green's functions only exist for a small subset of geometries (e.g., semi-infinite half spaces), and discretized versions must be generated numerically for more complex geometries. In theory, the adjoint approach offers perhaps the best compromise between forward techniques and Green's function-based inverse methods by utilizing information from all bead displacements while not explicitly requiring linearity. However, the ability of the finite-element software to converge on a solution has only been demonstrated for linearly elastic material models with relatively simple (nearly spherical) cellular geometries. Whether this approach will work for nonlinear materials, large strain datasets, or complex cellular geometries is an interesting avenue of investigation.

22.3 Resolution Metrics for TFM

Regardless of which approach is adopted, it is important to characterize the resolution and sensitivity of a given traction measurement under experimentally relevant conditions. There are many sources of error in TFM that can confound an accurate measurement. These include errors in determining the bead centroids and displacements and errors in the mechanical model used for the material properties and cell/substrate geometries. Moreover, the limited number of displacement measurements and the grid or mesh used for discretization will also determine the ultimate accuracy of the traction field.

Unfortunately, spatial resolution and traction sensitivity are inherently coupled in TFM. Qualitatively, one can imagine that a traction applied uniformly over a large area will introduce larger displacements in a greater number of beads than a traction of comparable magnitude exerted over a very small area. Because tractions have the units of force per unit area, the total force imparted by a widely distributed surface traction would be significantly higher than that imparted by a surface traction of identical magnitude, which is confined to a small area. Thus, any report of the sensitivity (i.e., what are the smallest magnitude tractions capable of being measured) must be accompanied by the length scale over which such tractions are applied. Similarly, any report of the resolution (i.e., what are the smallest spatial fluctuations capable of being measured) must also include the magnitude of the tractions at which this resolution is attainable. A full characterization would include a curve that plots the traction sensitivity as a function of the spatial frequency of the traction field.

Further complicating the analysis of resolution, traction measurements are vectors rather than scalars. The displacement field decays according to a different relationship depending on the direction in which the force is applied. Thus, the directions at which the surface tractions are oriented can also impact the true resolution and sensitivity of the technique. Fortunately, each of these aspects (traction sensitivity, spatial resolution, and the effect of traction orientation) can be readily assessed using simulated datasets (Figure 22.3).[8,9,11,25]

A loading of known magnitude and spatial localization can be applied to the surface of the simulated hydrogel volume. The forward problem is then solved to compute the displacement field that would be induced by such a loading. This displacement field can then be sampled at a similar distribution to experimentally measured bead locations. Experimentally relevant levels of noise are then overlaid onto these displacements, and the tractions are recovered according to the same method used for experimental measurements. This method could also be used to test the effect of any potential errors introduced by discretization or due to errors in inaccurately modeling the material properties. Thus far, two types of simulated loadings have been utilized to characterize TFM measurements—point-like tractions applied over circular areas of progressively decreasing magnitude and radius, and sinusoidally oscillating tractions of progressively decreasing magnitude and increasing spatial frequency. For fully 3D traction force measurements, it can be difficult to define a loading with controlled spatial frequency. Thus, simplified geometries such as a sphere or ellipsoid can be utilized to model the cell.[11,33]

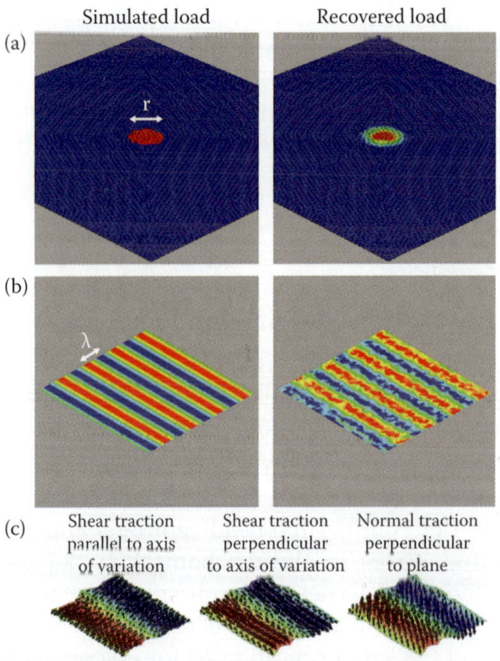

FIGURE 22.3
Resolution and sensitivity metrics for TFM. Simulated loadings can be used to assess the resolution and sensitivity of TFM. (a) Isolated loadings are simulated with controlled magnitude, direction, and area. These loadings can then be recovered under experimentally relevant conditions for bead density, regularization, and displacement noise. (b) Sinusoidally oscillating loadings can be simulated and recovered with controlled direction and wavelength. (c) Schematic of the different orientations possible for sinusoidally oscillating loadings.

22.4 Conclusion

TFM is now a well-established technique in studies of cell mechanics, migration, and mechanotransduction. It is one of the few techniques capable of providing 3D vectorial information about location, magnitude, and direction of cellular force generation. Many alternative techniques exist to assay cell forces including arrays of microfabricated pillars[34,35] and fluorescence-based molecular beacons.[36–40] In microfabricated pillar assays, each pillar acts as a spatially isolated force sensor. Computation of the cellular traction is directly measured by utilizing the observed degree of pillar bending and the known spring constant of the pillar. This approach avoids many of the mathematical difficulties contributed by the continuous substrate used in TFM. However, because forces can only be measured normal to the bending axis of the pillars, it is restricted to measuring 2D forces exerted by cells on planar surfaces. Alternatively, molecular beacons have recently been developed utilizing either fluorescence resonance energy transfer (FRET) or fluorescence quenching to indicate the degree of force application. In these systems, two fluorophores or one fluorophore and one quencher are placed at the opposite ends of an entropic molecular spring. As these molecules are subject to cellular forces, the spring section of the molecule stretches, and fluorescence intensity is modulated. Theoretically, such sensors can give very high spatial resolution (dictated primarily by the optical resolution of the microscope used for measurement) and, with a sufficient signal-to-noise ratio, could be readily incorporated into 3D materials. However, because these methods utilize changes in fluorescence intensity as the reporter for force application, they are inherently scalar in nature. In contrast to TFM, they do not report any information about which direction cellular forces are being applied.

To date, TFM has typically been applied to continuous, linearly elastic, homogeneous materials such as polyacrylamide, PDMS, or polyethylene glycol hydrogels. This is because the mechanics of such materials are easily modeled and incorporated into the algorithms for computing cell forces. Unfortunately, the vast majority of biological matrices consist of fibrillar matrices such as collagen or fibrin. These materials are known to be nonaffine, strain stiffening, and viscoelastic and can be plastically remodeled by encapsulated cells. Despite these challenges, understanding how cells exert forces within these mechanically complex materials will be critical for translating the wealth of knowledge acquired using TFM on elastic hydrogels into more clinically relevant scenarios. Many of the TFM methods outlined in Section 22.2 could be readily adapted to fibrillar or nonlinear materials through the use of an appropriate constitutive relation. Although not yet capable of reproducing the subcellular resolution provided by TFM on linearly elastic materials, useful information can be obtained by measuring the integrated strain energy produced by single cells[22] or by averaging the forces exerted by identically shaped micropatterned clusters of cells in collagen gels.[23] With future advances in material characterization and modeling, it should be possible to extend TFM into more mechanically complex and biologically relevant materials and ultimately into in vivo settings.

Acknowledgments

I thank Guy Genin, Christopher Chen, Colin Choi, and Colette Shen for helpful discussions and feedback on this manuscript.

References

1. Beloussov, L. V., Dorfman, J. G. & Cherdantzev, V. G. Mechanical stresses and morphological patterns in amphibian embryos. *J Embryol Exp Morph* **34**, 559–574 (1975).
2. Keller, R., Davidson, L. A. & Shook, D. R. How we are shaped: The biomechanics of gastrulation. *Differentiation* **71**, 171–205 (2003).
3. Thompson, D. A. W. *On Growth and Form* (Cambridge University Press, Cambridge, UK, 1917).
4. Harris, A. K., Wild, P. & Stopak, D. Silicone rubber substrata: A new wrinkle in the study of cell locomotion. *Science* **208**, 177–179 (1980).
5. Dembo, M., Oliver, T., Ishihara, A. & Jacobson, K. Imaging the traction stresses exerted by locomoting cells with the elastic substratum method. *Biophys J* **70**, 2008–2022 (1996).
6. Dembo, M. & Wang, Y. L. Stresses at the cell-to-substrate interface during locomotion of fibroblasts. *Biophys J* **76** 2307–2316 (1999).
7. Franck, C., Maskarinec, S. A., Tirrell, D. A. & Ravichandran, G. Three-dimensional traction force microscopy: A new tool for quantifying cell-matrix interactions. *PLoS One* **6**, e17833 (2011).
8. Hur, S. S., Zhao, Y. H., Li, Y. S., Botvinick, E. & Chien, S. Live cells exert 3-dimensional traction forces on their substrata. *Cell Mol Bioeng* **2**, 425–436 (2009).
9. Legant, W. R. et al. Multidimensional traction force microscopy reveals out-of-plane rotational moments about focal adhesions. *Proc Natl Acad Sci U S A* **110**, 881–886 (2013).
10. Maskarinec, S. A., Franck, C., Tirrell, D. A. & Ravichandran, G. Quantifying cellular traction forces in three dimensions. *Proc Natl Acad Sci U S A* **106**, 22108–22113 (2009).
11. Legant, W. R. et al. Measurement of mechanical tractions exerted by cells in three-dimensional matrices. *Nat Methods* **7**, 969–971 (2010).
12. Balaban, N. Q. et al. Force and focal adhesion assembly: A close relationship studied using elastic micropatterned substrates. *Nat Cell Biol* **3**, 466–472 (2001).
13. Butler, J. P., Tolic-Norrelykke, I. M., Fabry, B. & Fredberg, J. J. Traction fields, moments, and strain energy that cells exert on their surroundings. *Am J Physiol Cell Physiol* **282**, C595–C605 (2002).
14. Schwarz, U. S. et al. Calculation of forces at focal adhesions from elastic substrate data: The effect of localized force and the need for regularization. *Biophys J* **83**, 1380–1394 (2002). doi:10.1016/S0006-3495(02)73909-X.
15. Cohen, D. M. & Chen, C. S. Mechanical control of stem cell differentiation. In *StemBook* (eds. D. Melton & L. Girard) (2008). doi:10.3824/stembook.1.26.1.
16. DuFort, C. C., Paszek, M. J. & Weaver, V. M. Balancing forces: Architectural control of mechanotransduction. *Nat Rev Mol Cell Biol* **12**, 308–319 (2011). doi:10.1038/nrm3112.
17. Lauffenburger, D. A. & Horwitz, A. F. Cell migration: A physically integrated molecular process. *Cell* **84**, 359–369 (1996).
18. Yu, H., Mouw, J. K. & Weaver, V. M. Forcing form and function: Biomechanical regulation of tumor evolution. *Trends Cell Biol* **21**, 47–56 (2011). doi:10.1016/j.tcb.2010.08.015.
19. Gao, Y. X. & Kilfoil, M. L. Accurate detection and complete tracking of large populations of features in three dimensions. *Opt Express* **17**, 4685–4704 (2009).

20. Mortensen, K. I., Churchman, L. S., Spudich, J. A. & Flyvbjerg, H. Optimized localization analysis for single-molecule tracking and super-resolution microscopy. *Nat Methods* **7**, 377–381 (2010).
21. Hall, M. S., Long, R., Hui, C. Y. & Wu, M. M. Mapping three-dimensional stress and strain fields within a soft hydrogel using a fluorescence microscope. *Biophys J* **102**, 2241–2250 (2012).
22. Koch, T. M., Munster, S., Bonakdar, N., Butler, J. P. & Fabry, B. 3D traction forces in cancer cell invasion. *PLoS One* **7**, e33476 (2012).
23. Gjorevski, N. & Nelson, C. M. Mapping of mechanical strains and stresses around quiescent engineered three-dimensional epithelial tissues. *Biophys J* **103**, 152–162 (2012).
24. Pan, B., Qian, K. M., Xie, H. M. & Asundi, A. Two-dimensional digital image correlation for in-plane displacement and strain measurement: A review. *Meas Sci Technol* **20**, e062001 (2009).
25. Sabass, B., Gardel, M. L., Waterman, C. M. & Schwarz, U. S. High resolution traction force microscopy based on experimental and computational advances. *Biophys J* **94**, 207–220 (2008).
26. Belytschko, T., Lu, Y. Y. & Gu, L. Element-free galerkin methods. *Int J Numer Meth Eng* **37**, 229–256 (1994).
27. Delanoe-Ayari, H., Rieu, J. P. & Sano, M. 4D traction force microscopy reveals asymmetric cortical forces in migrating dictyostelium cells. *Phys Rev Lett* **105**, 248103 (2010).
28. Hansen, P. C. Regularization tools version 4.0 for MATLAB 7.3. *Numer Algorithms* **46**, 189–194 (2007).
29. Tambe, D. T. et al. Monolayer stress microscopy: Limitations, artifacts, and accuracy of recovered intercellular stresses. *PLoS One* **8**, e55172 (2013).
30. Trepat, X. et al. Physical forces during collective cell migration. *Nat Phys* **5**, 426–430 (2009).
31. Yang, Z. C., Lin, J. S., Chen, J. X. & Wang, J. H. C. Determining substrate displacement and cell traction fields—A new approach. *J Theor Biol* **242**, 607–616 (2006).
32. Vitale, G., Preziosi, L. & Ambrosi, D. Force traction microscopy: An inverse problem with pointwise observations. *J Math Anal Appl* **395**, 788–801 (2012).
33. Vitale, G., Preziosi, L. & Ambrosi, D. A numerical method for the inverse problem of cell traction in 3D. *Inverse Probl* **28**, 095013 (2012).
34. du Roure, O. et al. Force mapping in epithelial cell migration. *Proc Natl Acad Sci U S A* **102**, 2390–2395 (2005).
35. Tan, J. L. et al. Cells lying on a bed of microneedles: An approach to isolate mechanical force. *Proc Natl Acad Sci U S A* **100**, 1484–1489 (2003).
36. Grashoff, C. et al. Measuring mechanical tension across vinculin reveals regulation of focal adhesion dynamics. *Nature* **466**, 263–266 (2010).
37. Liu, Y., Yehl, K., Narui, Y. & Salaita, K. Tension sensing nanoparticles for mechano-imaging at the living/nonliving interface. *J Am Chem Soc* **135**, 5320–5323 (2013). doi:10.1021/ja401494e.
38. Morimatsu, M., Mekhdjian, A. H., Adhikari, A. S. & Dunn, A. R. Molecular tension sensors report forces generated by single integrin molecules in living cells. *Nano Lett* **13**, 3985–3989 (2013). doi:10.1021/nl4005145.
39. Wang, X. & Ha, T. Defining single molecular forces required to activate integrin and notch signaling. *Science* **340**, 991–994 (2013). doi:10.1126/science.1231041.
40. Wang, Y., Meng, F. & Sachs, F. Genetically encoded force sensors for measuring mechanical forces in proteins. *Commun Integr Biol* **4**, 385–390 (2011). doi:10.4161/cib.4.4.15505.

23

Cell Interactions in Wire (Fiber)-Based Structures and Scaffolds

Kevin Sheets, Puja Sharma, Brian Koons, and Amrinder Nain

CONTENTS

Recent advancements in micromanufacturing/nanomanufacturing processes have enabled the use of wire (hereafter called *fiber*)-based substrates for interrogating mechanobiology at single cell resolution. These circular cross-sectional, high aspect-ratio materials provide a unique approach in investigating cell-substrate interactions, which offer a closer semblance to interactions that may occur in the native cellular niche. Spatially confining a cell to a fiber causes it to simultaneously sense and respond to curvature and stiffness (i.e., structural and material properties), initiating changes to multiple critical cell behaviors including focal adhesion organization, cytoskeletal configuration, migration, and eventual fate. Here we provide an overview of recent developments in the design, fabrication, and use of fibrous scaffolds and structures and their contribution toward advancing our understanding of cellular mechanics with applications in tissue engineering and single cell disease models.

23.1 Introduction

Canonical *in vitro* studies in cell mechanobiology have often been performed on flat substrates.[1–3] Seeding cells on glass coverslips or gels of tunable elastic modulus have led to many critical and clinically relevant discoveries of cell–substrate interactions. However, the substrates from these studies have limited discernible topographical relation to *in vivo* architecture. With recent advancements in the ability to manufacture circular cross-sectional, high aspect-ratio fibers on the microscale/nanoscale, cellular investigations have revealed that several key cell behaviors are significantly altered when interacting with fiber structures.[4,5] Suspended fibers provide curvature constraints and gradients of structural stiffness that do not exist on flat surfaces. Cells interacting with individual fibers are also able to sense spacing and orientation between fibers as well as the hierarchical assembly of fibers in multiple layers (Figure 23.1).[6] Compared to flat substrates where the elastic modulus (N/m^2) plays a dominant role in cell behavior,[7–15] suspended fibers with structural stiffness (N/m) gradients directly affect cell shape, focal adhesion arrangement, and migration.[4,16,17] Suspended fiber networks are thus thought to elicit different cellular force responses (*inside-out*: ability to exert; *outside-in*: ability to withstand) due to curvature-induced focal adhesion clustering and wrapping.

In the body, cells migrate by attaching to and interacting with the fibrous extracellular matrix (ECM), primarily composed of fibrils ranging 30–70 nm in diameter which can bundle into 200 nm–1 μm fibers.[18–21] The ECM consists mainly of collagen, elastin, fibronectin,

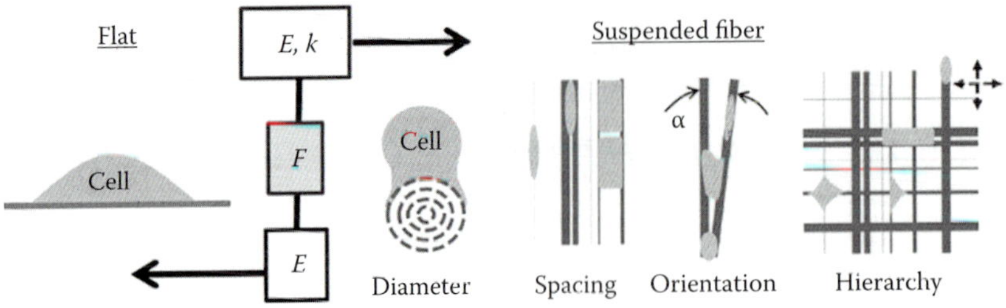

FIGURE 23.1
Fiber design space, which accounts for the substrate's elastic modulus (E: N/m^2), structural stiffness (k: N/m), diameter, interfiber spacing, orientation (α), and hierarchical assembly, compared to traditionally studied flat substrates. Forces (F) exerted and experienced by cells are currently thought to be different between the two systems. (From Sheets, K. et al., *J. Biomater. Tiss. Eng.*, 2013, 3, 353–368, Copyright American Scientific Publishers.)

laminin, glycosaminoglycan, and proteoglycans, but the exact composition and degree of alignment vary among different tissues.[22,23] The ECM delivers chemical and physical cues which significantly affect cell behaviors ranging from adhesion, migration, proliferation, and differentiation.[23–30] For this reason, *in vitro* culture conditions mimicking the ECM and replete with native mechanics/topography have tremendous potential for studying single-cell behavior for applications in tissue engineering-based therapies.[22,31–33]

Fiber scaffolds offer a unique platform to investigate cell–substrate interactions (Figure 23.2). Fibers possessing diameters closely resembling those of the assembled fibrous ECM proteins provide a curved surface for cell attachment and have mechanistic properties that can be tuned to investigate the biophysical influence on cell behaviors. Therefore, this work aims to elaborate upon and review the recent advancements in fabrication and use of fiber-based scaffolds for tissue engineering and single-cell–substrate interactions.

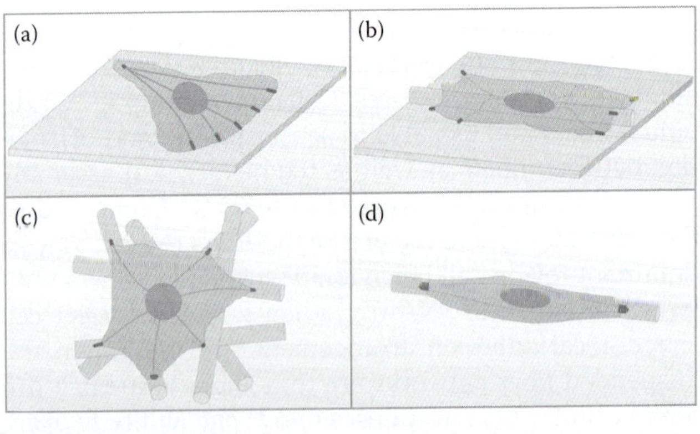

FIGURE 23.2
Illustration of substrates that cells interact with and their dimensionality. (a) Flat, featureless substrates such as cross-linkable gels with clear lamellipodia formation and strong stress fibers and focal adhesions but no directional preference. (b) Flat, featured substrates such as those created by soft lithography, which induce cell alignment. (c) Randomly oriented, unaligned fibrous meshes. (d) Single-fiber, single-cell interactions achieved by suspending a fiber across a substrate.

23.2 Fiber Fabrication

Several key factors impacting fiber use in cell studies include the ability to precisely control fiber diameter and deposition (both orientation and spacing) in both single and multiple layers. Advancements in manufacturing processes have enabled creation of smaller diameter fibers with higher degrees of alignment, thus making fiber manufacturing a powerful tool for investigating cell–substrate interactions (Table 23.1). In this regard, electrospinning is perhaps the most widely known and thoroughly studied method of forming polymeric nanofibers. In this process, polymer solution is pumped through a syringe to a needle where an electrical charge extrudes polymer fibers onto a collecting target.[34–38] The principle underlying this process was first discovered in the 1890s by Rayleigh but was not patented until 1934 by Formhals.[39–42] Taylor's research on electrically driven jets conducted in the late 1960s[43] paved the way for modern refinements that allow electrospinning to generate microscale/nanoscale fibers. With the realization that electrospinning could produce fibers with diameters on the order of those in native tissue, the bioengineering community has seen rapid growth in the use and improvement of the electrospinning technique to achieve higher degree of alignment and spatial organization.[35,36,44,45]

Some of the recent advancements in nanofiber fabrication include far-field electrospinning (FFES), near-field electrospinning (NFES), and nonelectrospinning direct drawing methods. FFES is widely used to fabricate continuous fibers of various materials (including engineering plastics, biopolymers, and ceramics).[46–50] To create substrates with aligned fibers, a high-speed rotating drum is commonly used as a collector in place of a stationary target.[51] Aligned polymer/ceramic nanofibers can also be achieved using patterned electrodes[52] or a wheel-like bobbin collector.[53,54] Modifications to the electric source include using biased AC potentials or using an auxiliary counterelectrode.[55,56] In all these approaches, due to the inherent electric instabilities of the

TABLE 23.1

Comparison of Fiber Manufacturing Techniques and Their Biological Applications

Method	Diameter Range	Uniqueness and Applications	Ref.
Electrospinning	80–1000 nm	Mass production; wide range of polymers, tissue engineering	35,36,44,57
Airbrushing	125–1950 nm	Target geometry variety, ease of use and low cost; tissue engineering	58
Rotary jet spinning	425–1600 nm	Mass production of aligned fibers; tissue engineering	59
STEP technique	70–5000 nm	Diameter, orientation, and spatial deposition control; single-cell–fiber interactions, tissue engineering	60,61–63

electrospinning process, a high degree of parallelism is often difficult to achieve along with the spacing between fibers in single and in multiple layers. On the other hand, NFES has demonstrated improved fiber patterning through reduction of applied voltage and the source-to-target distance.[59,64] However, precise multilayer fiber patterning on large surfaces using NFES has yet to be demonstrated. Since decreasing voltage enhances deposition capabilities, several approaches have removed the electric component entirely. For instance, direct drawing uses polymer wetted probe tips for precise fiber deposition.[60,65] Although some well-positioned fiber patterns have been demonstrated, this technique cannot operate continuously, and it is difficult to precisely deposit sub-250-nm-diameter fibers due to optical limitations.

Suspended fibers, which form a bridge across a gap, can be created by either hollowing out or forming trenches in the target substrate.[38,66] While less widely used, suspended air-gap nanofibers have been made with nonspinning techniques using micropillar arrays by brushing fibrinogen across thrombin-coated pillars[67] or melting polystyrene films over them.[68] Micropipette probes have also been used to directly draw fibers from solution droplets[65,69] or push and pull droplets across unique geometries.[70] Pseudo-dry spinning and nonelectrospinning spinneret-based tunable engineered parameters (STEP) fiber manufacturing platform has recently demonstrated the ability to deposit suspended fibers of a variety of polymers with control on fiber dimensions (diameter: sub-50 nm to microns, length: approximately in the centimeters range) and orientation (≤2.5°, submicron fiber-spacing deposition resolution in single and multiple layers; Figure 23.3).[61,62,69] With the ability to deposit nanofibers in a spatially controlled manner, the challenge then shifts from fabrication to characterization and eventual use in studying cell–substrate interactions.[71]

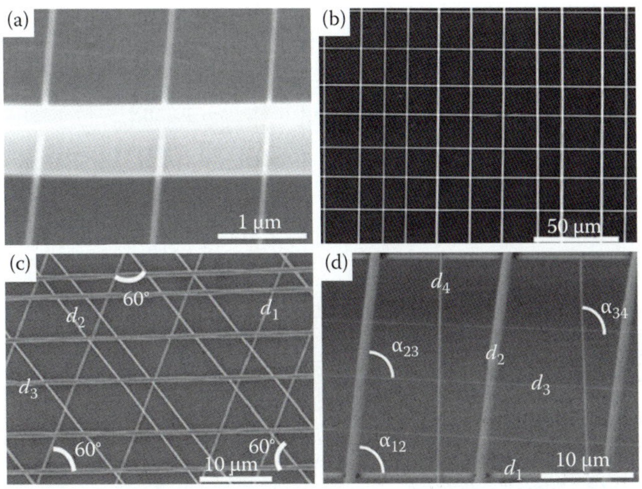

FIGURE 23.3
Fiber manufacturing using the STEP technique, which allows control of (a) multiple diameters (micro–nano) in the same scaffold, (b) spacing in a grid-like pattern, (c) orientation control between fibers, and (d) hierarchical assembly. (From Nain, A. S. and Wang, J., *Polym. J.* **45**, 695–700, 2013.)

23.3 Single-Cell–Fiber Interactions

23.3.1 Focal Adhesion Formation and Clustering

Cell attachment to a substrate is initialized when the dimer transmembrane protein integrin is activated upon binding of a suitable substrate ligand such as collagen or fibronectin.[72–75] Intracellular scaffolding proteins talin, kindlin, and paxillin are quickly recruited along with focal adhesion kinase (FAK) to stabilize activated integrin and link to cytoskeletal filamentous actin (F-actin), forming a nascent adhesion capable of sensing the mechanical state of the substrate.[76–79] As myosin-II applies tensile load to F-actin filaments, the nascent adhesion tugs the substrate and develops into a focal adhesion marked by the presence of α-actinin-bundled F-actin stress fibers.[80–82] The focal adhesion subsequently matures into a focal adhesion complex (FAC) after the recruitment of additional late-stage proteins vinculin and zyxin, which serve to further stabilize and strengthen the adhesion, thus in the process allowing it to apply even greater loads to the substrate.[83–85] Microtubules, kinesin-1, dynamin, and phosphatases facilitate the disassembly of FACs from the trailing edge of the cell with partial recycling to the leading edge of the cell.[25,86–88] Since the degree of maturation depends on the forces applied by the cell, the size and orientation of FACs reveal the spatiotemporal dynamics of cell contractility and its dependence on substrate

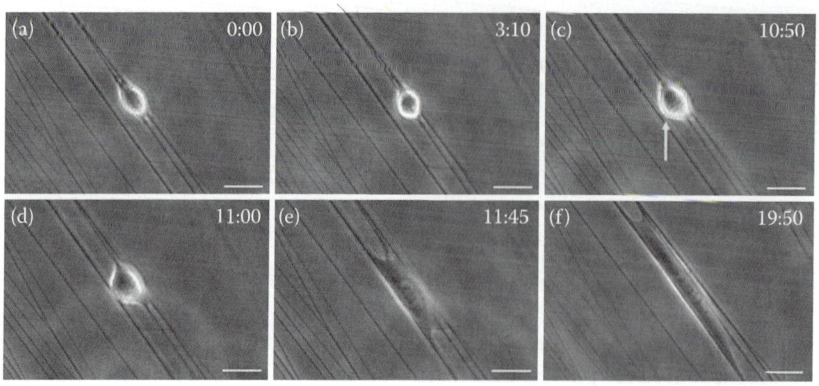

FIGURE 23.4
(a, b) C2C12 cell in spindle shape migrating back and forth on STEP fibers. (c) After nearly 11 h actively seeking a more stable configuration, the cell extends a protrusion (marked by the arrow) to a neighboring fiber. (d, e) In less than 1 h, the cell has spread into a more stable state. (f) The cell continues to spread, yet remains in the same configuration over 8 h later. Scale bar represents 20 μm.

properties. Imaging these structures allows one to visualize locations where cells are forming physical contacts and define force vectors along which cells apply contractile loads required to migrate. For these reasons, FACs are commonly used as indicators of both substrate stiffness and migration speed on flat surfaces.[26,89,90]

While FAC developmental mechanisms on suspended fibers likely remain unchanged, fiber topography forces cells to make several spatial adaptations which alters FAC size–migration speed correlations. Cells on suspended single fibers of 300–800 nm diameter are forced to attach within a spatially limited curved substrate, which causes the cell to form adhesions by elongating along the axis of the fiber.[4,91,92] In doing so, the cell arranges itself by partially wrapping around the fiber's curved surface, causing formation of classic spindle morphology with adhesions circumferentially clustered at the poles as opposed to the spread shapes with diffused adhesions seen throughout on flat surfaces.[93–95] These clusters have been characterized to be up to four times longer on suspended polystyrene fibers compared to flat polystyrene, leading to changes in migratory behavior.[4] On suspended parallel and intersecting fibers, cells are observed to spread between the fibers to form stable configurations (Figure 23.4 shown for parallel fibers).[4]

23.3.2 Migration

Cell migration is relevant to a myriad of physiological phenomena including developmental processes

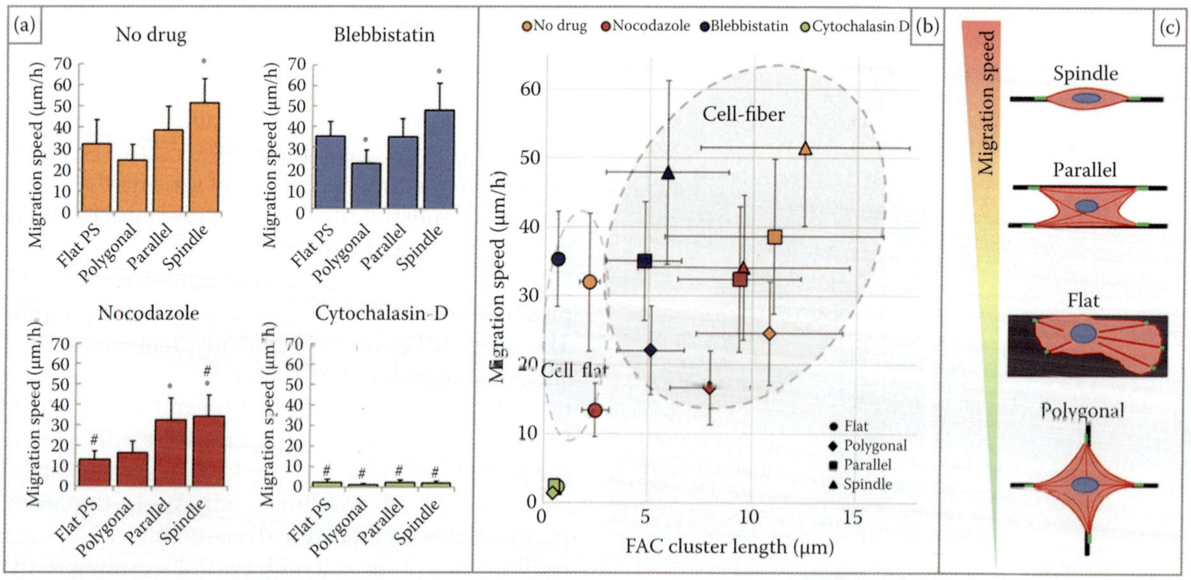

FIGURE 23.5
Cell behavioral response to STEP fiber substrates as opposed to flat substrates, including (a) migration as a function of shape and cytoskeletal knockdown agent, (b) adhesion-migration differences between cell-flat and cell-fiber substrates, and (c) schematic chart of speed for each shape seen on fibers. (From Sheets, K. et al., *Acta Biomater.* **9**, 7169–7177, 2013.)

and disease progression. While highly synchronized cell migration facilitates embryogenesis, nervous system development, wound healing, and inflammatory responses, unregulated cell migration is associated with cancer metastasis.[96–98]

Due to enhanced FAC alignment and polarization of contractile forces, cells have the capacity to nearly double their migration speed on fiber-based structures. On flat substrates, myoblast mobility follows a random walk motion with an average migration speed of ~30 μm/h (V_{flat}). On suspended fibers of 300–800 nm diameter, topography bias induces elongation and spindle formation with cells polarizing in the direction of increased structural stiffness.[16] Due to these effects, myoblast migratory speeds have been observed to be nearly 2 × V_{flat} (Figure 23.5).[4] Even with the administration of several cytoskeletal knockdown drugs known to affect FAC development, migration speeds remain higher on fibers compared to flat counterparts. This can be attributed to the longer FAC cluster lengths on fibrous scaffolds and also on the highly polarized cell shapes leading to directed cell migration.

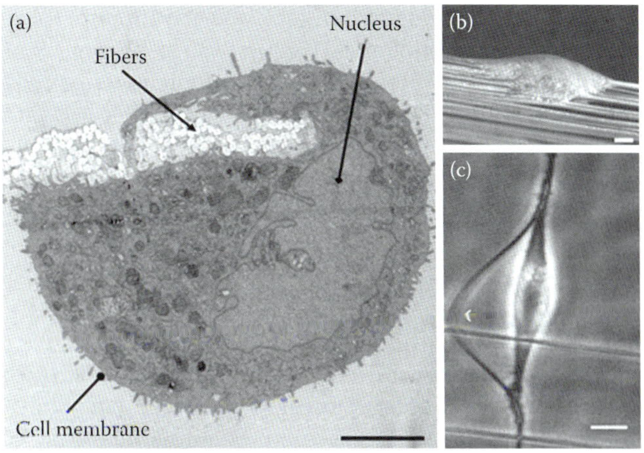

FIGURE 23.6
(a) Transmission electron microscopy image showing hepatocyte wrapping around STEP fiber bundles. (b) Scanning electron microscopy image showing DBTRG-05MG elongation along the STEP fiber axis. (c) Phase image of a C2C12 applying a load to the STEP suspended fiber sufficient to induce bucking (shown by arrow). Scale bar represents 4 μm.

23.3.3 Effect of Fiber Spacing and Hierarchy

Cell-fiber platforms can be characterized as pseudo-3D culture environments since cells wrap around the fiber and apply forces through focal adhesions clustered at the poles (Figure 23.6). Cells are constantly probing their immediate environment and, in doing so, form stable and elongated spindle morphologies on tightly spaced parallel fibers or on single fibers (Figure 23.7a–f, i). Cells

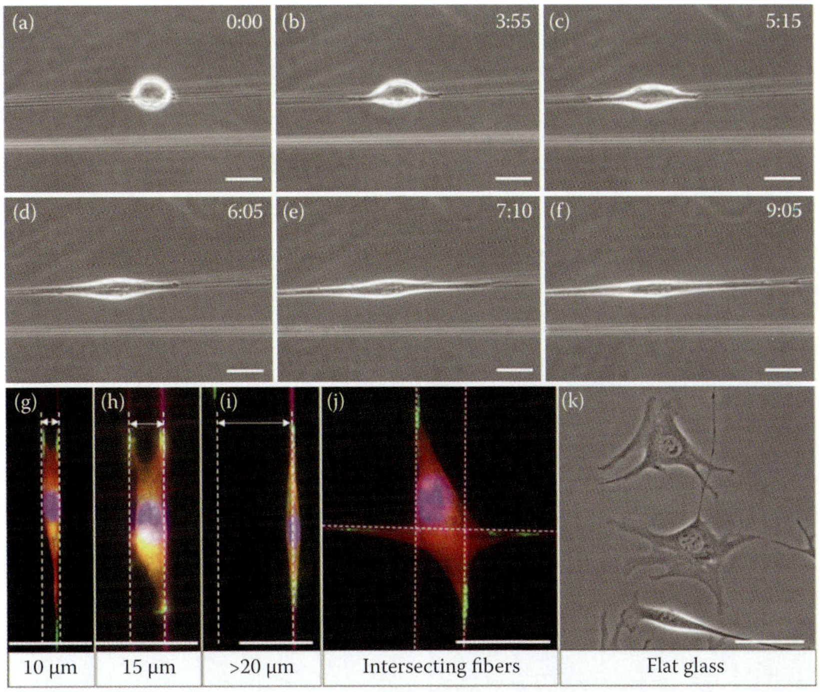

FIGURE 23.7
Effect of fiber spacing on cell behavior and morphology. (a–f) Time-lapse images showing spreading of a MSC on very tightly spaced STEP parallel fibers. Suspended substrate topography causes the cell to spread into a highly elongated spindle configuration. (g–j) Fluorescence images showing C2C12 cell configurations as the distance between STEP parallel fibers is increased. A gap size greater than 20 μm is often too large for the cell to spread between. (k) The aforementioned cell behaviors stand in contrast to those of cells on flat glass. Scale bar represents 50 μm.

are observed to bridge fiber gaps less than 20 μm, forming parallel morphologies (Figure 23.7g, h). The axial length of the cell between two parallel fibers is typically observed to be inversely related to fiber spacing. Furthermore, for intersecting fibers, the cell forms a polygonal morphology, which is typically nonmigratory (Figure 23.7j).[4,16]

23.3.4 Cell Response to Curvature and Structural Stiffness

Cell culture systems aim to capture the native stiffness of the tissue-cell model from the mechanistic viewpoint. Stiffness can be defined as either material or structural. Material stiffness is commonly referred to as Young's modulus and is geometry-independent with units of N/m^2. On the other hand, structural stiffness relates the size, organization, and shape of material to its ability to resist deformation and is represented by units of N/m. Cells sense and react to suspended fiber mechanical features via tension applied to focal adhesions, which is dependent upon both the structural stiffness and curvature.[16] These two parameters are coupled as it is possible to engineer suspended fibers having the same structural stiffness from different diameters and lengths. Therefore, experiments can be designed to reveal curvature (topographical) and stiffness (mechanical) effects independently on cell behavior.

Using this approach, it has recently been shown that at the same structural stiffness, focal adhesions were significantly longer on small-diameter (400 nm) than large-diameter (700–1200 nm) fibers with an observed decrease in migration speed.[16] The mechanism behind increased focal adhesion cluster length may perhaps originate from the need to maintain tensional homeostasis.

23.3.5 Cellular Force Generation

Forces have been known to directly affect cell behavior including migration, embryonic development, proliferation, differentiation, wound healing, and self-renewal.[3,9,11,83,99–121] Furthermore, mechanical forces have been shown to play a direct role in disease onset and progression including cancer.[122–124] Much of the current understanding of cell force generation has come from single-cell probe systems including microaspiration,[125,126] optical tweezers,[127–132] atomic force microscopy,[132–137] fluorescent bead embedded hydrogels,[104,122,138–141] and micropillars.[103,142–144] Suspended fibers provide an alternate route to study cellular force generation as it can be related to fiber structural stiffness and curvature. Using STEP platform, suspended fiber nanonets of structurally stiff fibers spaced orthogonally to low stiffness fibers and fused at intersections are used to investigate cell migratory and force modulation mechanisms. Cells attached to these nanonets are observed to deflect the low stiffness fibers, and using beam mechanics, the force required to cause these deflections can be calculated. Figure 23.8 demonstrates the use of this system in investigating both passive (inside-out) and active (outside-in) cell forces. In Figure 23.8a–c, a C2C12 cell is seen deflecting a fiber as a result of applying migratory forces to the scaffold with no external influence. However, the same scaffold can be used to investigate cell reactionary mechanisms to externally applied loads as shown in Figure 23.8d–f. These methods allow systematic investigation of the role of focal adhesions in force modulation

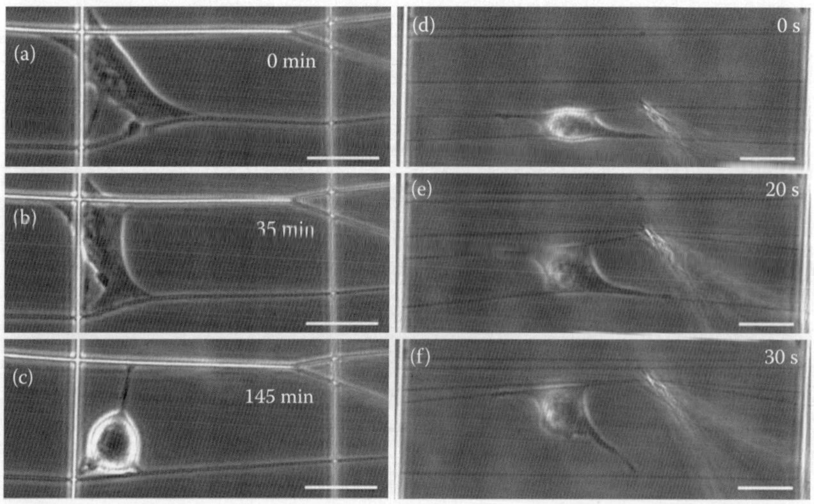

FIGURE 23.8
STEP based nanonets used to measure real-time cell forces. (a–c) Time lapse of cell forces exerted as a result of inside-out signaling. (d–f) Time lapse of cell forces as a result of outside-in manipulation. Scale bar represents 25 μm.

in the presence and absence of well-known pharmacological cytoskeletal-altering drugs.

Migratory force values from less than 1 to 100 nN have been calculated using different polymer fiber nanonets for glioma, mouse fibroblast, and myoblast cell lines (Nain et al., unpublished data). With further investigation, understanding how cell forces relate to biochemical and genetic differences in diseased cells will lead to improved disease prognosis.[145–148]

23.4 Population-Level Cell–Fiber Interactions

23.4.1 Effect on Differentiation and Proliferation

The clinical importance of the mechanistic environment in causing and supporting stem cell lineage specification is well established for both 2D and 3D environments.[149–157] Cell differentiation studies using mesenchymal stem cells (MSCs) on fiber networks have yielded bone, muscle, cartilage, tendon, and adipose cells depending on fiber mechanical properties, growth factors, and media used.[158] Even in the absence of chemical cues, neural stem cells (NSCs) have increased potential to differentiate into neuronal populations as opposed to astrocytes or oligodendrocytes due to fiber-induced elongated phenotypes.[159,160] Furthermore, electrospun poly(l-lactic acid)-collagen scaffolds have been used to drive cord blood stem cells toward osteoblasts.[161] Interestingly, fiber-based platforms have also shown increased potential to induce proliferation due to the ability to provide directional topographical cues.[162,163] Fibroblast metabolism doubled after 3 days of culture on fibrous electrospun polyurethane meshes vs. flat polyurethane, and when these fibers were aligned, increased collagen deposition was seen compared to randomly oriented fibers.[164] Osteoblasts have similarly been shown to proliferate on functionalized self-assembled nanofibers.[165] In each case, an enhanced cell function was achieved by providing the cell type with a fibrous substrate that more closely matched the topography and mechanics of its native tissue. Insights from these experiments apply directly to clinically relevant applications including nerve conduits,[166,167] wound dressings,[168,169] bone matrix scaffolds,[170,171] soft–hard tissue interface gradients,[172,173] blood vessel implants,[174,175] as well as targeted drug delivery platforms.[176,177] As additional details of cell–substrate interactions are uncovered with these novel fibrous substrates, tissue engineering therapies will continue to advance in sophistication and functionality.

23.4.2 Investigations into Tumor and Cancer Models

Many recent works have demonstrated that the tumor microenvironment strongly influences major processes in cancer including angiogenesis,[178,179] epithelial–mesenchymal transition,[180,181] metastatic potential,[182–184] aggressiveness,[185,186] and the response of cancer cells to anticancer agents.[187,188] While research in the last few decades has contributed significantly to a greater understanding of cancer biology, biochemistry, and genetics, the influence of biophysical factors on cancer behavior is relatively not well understood.[123,189,190] *In vitro* cancer models allow flexibility to tune the mechanistic properties of scaffolds to investigate these interactions.

The tumor microenvironment is composed of cancer cells, the ECM, fibroblasts, immune cells and cytokines.[191,192] Fibers serve as excellent platforms to closely represent the *in vivo* tumor environment. Focal adhesion patterns, protrusive behavior, and motility have been shown to be different when cells interact with fibrous scaffolds.[186] Furthermore, alignment and stiffness of nanofibers have been shown to influence cell motility and directionality in DBTRG-05MG and U251 glioma cells.[16,193] The current growth in investigations of cancer cells on nanofiber environments coupled with cancer biology, biochemistry, and genetics will contribute to better prognosis of cancer in the future.

23.5 Conclusions and Future Directions

Until recently, the study of cell–fiber interactions has been limited by difficulties in fabrication of controlled fibrous hierarchical assemblies, imaging techniques that comprehensively capture cell–fiber interactions, and microscale/nanoscale characterization of the cell–fiber interface. With the advancement of engineering platforms capable of producing fiber-based structures of ECM-mimicking topography, scale, and mechanics, single-cell analysis has revealed new interactions between a cell and its substrate. Cells are found to attach, arrange, migrate, and produce forces on fibers differently than on flat substrates. Such interactions stem from fiber suspension, structural stiffness, and substrate curvature.

Spatial and temporal cellular mechanosensing and cell adaptation capabilities to subtle changes in fiber curvature and structural stiffness are particularly challenging to study, and we believe them to be key contributors to eventual cell fate downstream. A number of studies in cell–fiber based interactions have shown promising results in advancing tissue engineering, stem cell research, and oncology. Knowledge of these fundamental interactions may be used in the future to control cell behavior and produce desired clinical outcomes such as directing cell motion with increased migration speed in wound healing models, slowing or thwarting migration in cancerous cells, or optimizing scaffold conditions to ensure proper colonization of tissue-engineered scaffolds. When coupled with computational and mathematical modeling, drug-eluting fibers can be used as powerful diagnostic platforms for drug efficacy prediction and testing. Furthermore, coupled with chemical cues, fibrous scaffolds can provide the unique opportunity to systematically study single and multicell (two, three, and more) aggregate behavior, which we believe to be critical in advancing mechanobiology. Overall, the development of fiber-based scaffolds has made significant contributions to the field of biotechnology and promises to advance studies in population level tissue engineering and single-cell disease models.

Acknowledgments

The authors acknowledge support from Virginia Tech's Institute for Critical Technology and Applied Science (ICTAS). Portions of this work were funded by the Bill and Andrea Waide Research Fund as well as the Jeffress Memorial Trust Fund.

References

1. Harris, A. K., Wild, P. & Stopak, D. Silicone rubber substrata: A new wrinkle in the study of cell locomotion. *Science* **208,** 177–179 (1980).
2. Lo, C. M., Wang, H. B., Dembo, M. & Wang, Y. L. Cell movement is guided by the rigidity of the substrate. *Biophys. J.* **79,** 144–152 (2000).
3. Engler, A. J., Sen, S., Sweeney, H. L. & Discher, D. E. Matrix elasticity directs stem cell lineage specification. *Cell* **126,** 677–689 (2006).
4. Sheets, K., Wunsch, S., Ng, C. & Nain, A. S. Shape-dependent cell migration and focal adhesion organization on suspended and aligned nanofiber scaffolds. *Acta Biomater.* **9,** 7169–7177 (2013).
5. Martin, L. Effect of extracellular function, matrix hepatocytes topology cultured on cell in a sandwich configuration and physiological responsiveness. *FASEB J.* **10,** 1471–1484 (1996).
6. Sheets, K., Wang, J., Meehan, S., Sharma, P., Ng, C., Khan, M., Koons, B., Behkam, B. & Nain, A. Cell-fiber interactions on aligned and suspended nanofiber scaffolds. *J. Biomater. Tiss. Eng.* **3,** 353–368 (2013).
7. Califano, J. P. & Reinhart-King, C. A. Substrate stiffness and cell area predict cellular traction stresses in single cells and cells in contact. *Cell. Mol. Bioeng.* **3,** 68–75 (2010).
8. Pathak, A. & Kumar, S. Independent regulation of tumor cell migration by matrix stiffness and confinement. *PNAS* **2012,** (2012).
9. Tee, S., Fu, J., Chen, C. S. & Janmey, P. A. Cell shape and substrate rigidity both regulate cell stiffness. *Biophys. J.* **100,** L25–L27 (2011).
10. Yeung, T. et al. Effects of substrate stiffness on cell morphology, cytoskeletal structure, and adhesion. *Cell Motil. Cytoskel.* **60,** 24–34 (2005).
11. Discher, D. E., Janmey, P. & Wang, Y. L. Tissue cells feel and respond to the stiffness of their substrate. *Science* **310,** 1139–1143 (2005).
12. Kong, H. J. et al. Non-viral gene delivery regulated by stiffness of cell adhesion substrates. *Nat. Mater.* **4,** 460–464 (2005).
13. Isenberg, B. C., DiMilla, P. A., Walker, M., Kim, S. & Wong, J. Y. Vascular smooth muscle cell durotaxis depends on substrate stiffness gradient strength. *Biophys. J.* **97,** 1313–1322 (2009).
14. Pelham, R. J. & Wang, Y. Cell locomotion and focal adhesions are regulated by substrate flexibility. *PNAS* **94,** 13661–13665 (1997).
15. Willits, R. K. & Skornia, S. L. Effect of collagen gel stiffness on neurite extension. *J. Biomat. Sci.-Polym. E.* **15,** 1521–1531 (2004).
16. Sharma, P., Sheets, K., Subbiah, E. & Nain, A. S. The mechanistic influence of aligned nanofiber networks on cell shape, migration and blebbing dynamics of glioma cells. *Integr. Biol.* **5,** 1036–1044 (2013).
17. Meehan, S. Structural stiffness gradient along a single nanofiber and associated single cell response (Master's thesis), *Virginia Tech,* Blacksburg, Virginia (2013).
18. Buehler, M. J. Nature designs tough collagen: Explaining the nanostructure of collagen fibrils. *PNAS* **103,** 12285–12290 (2006).
19. Ottani, V. et al. Collagen fibril arrangement and size distribution in monkey oral mucosa. *J. Anat.* **192,** 321–328 (1998).
20. Bozec, L., van der Heijden, G. & Horton, M. Collagen fibrils: Nanoscale ropes. *Biophys. J.* **92,** 70–75 (2007).
21. Mathur, A., Moore, S. W., Sheetz, M. P. & Hone, J. The role of feature curvature in contact guidance. *Acta Biomater.* **8,** 2595–2601 (2012).
22. Du, Y. et al. Synthetic sandwich culture of 3D hepatocyte monolayer. *Biomaterials* **29,** 290–301 (2008).
23. Rozario, T. & DeSimone, D. W. The extracellular matrix in development and morphogenesis: A dynamic view. *Dev. Biol.* **341,** 126–140 (2010).
24. Muiznieks, L. D. & Keeley, F. W. Molecular assembly and mechanical properties of the extracellular matrix: A fibrous protein perspective. *Acta Biochim. Biophys.* **1832,** 866–875 (2013).
25. Nagano, M., Hoshino, D., Koshikawa, N., Akizawa, T. & Seiki, M. Turnover of focal adhesions and cancer cell migration. *Int. J. Cell Biol.* **2012,** 310616 (2012).
26. Kim, M.-C. et al. Integrating focal adhesion dynamics, cytoskeleton remodeling, and actin motor activity for predicting cell migration on 3D curved surfaces of the extracellular matrix. *Integr. Biol.* **4,** 1386–1397 (2012).
27. Lock, J. G., Wehrle-Haller, B. & Strömblad, S. Cell-matrix adhesion complexes: Master control machinery of cell migration. *Semin. Cancer Biol.* **18,** 65–76 (2008).

28. Plotnikov, S. V., Pasapera, A. M., Sabass, B. & Waterman, C. M. Force fluctuations within focal adhesions mediate ECM-rigidity sensing to guide directed cell migration. *Cell* **151,** 1513–1527 (2012).
29. Ulrich, T. A., de Juan Pardo, E. M. & Kumar, S. The mechanical rigidity of the extracellular matrix regulates the structure, motility, and proliferation of glioma cells. *Cancer Res.* **69,** 4167–4174 (2009).
30. Ker, E. D. F. et al. Bioprinting of growth factors onto aligned sub-micron fibrous scaffolds for simultaneous control of cell differentiation and alignment. *Biomaterials* **32,** 8097–8107 (2011).
31. Shin, M., Ishii, O., Sueda, T. & Vacanti, J. P. Contractile cardiac grafts using a novel nanofibrous mesh. *Biomaterials* **25,** 3717–3723 (2004).
32. Bader, A. et al. Tissue engineering of heart valves—Human endothelial cell seeding of detergent acellularized porcine valves. *Eur. J. Cardio-Thorac.* **14,** 279–284 (1998).
33. Lutolf, M. P. & Hubbell, J. A. Synthetic biomaterials as instructive extracellular microenvironments for morphogenesis in tissue engineering. *Nat. Biotech.* **23,** 47–55 (2005).
34. Agarwal, S., Greiner, A. & Wendorff, J. H. Functional materials by electrospinning of polymers. *Prog. Polym. Sci.* **38,** 963–991 (2013).
35. Doshi, J. & Reneker, D. H. Electrospinning process and applications of electrospun fibers. *J. Electrostat.* **35,** 151–160 (1995).
36. Huang, Z.-M., Zhang, Y. Z., Kotaki, M. & Ramakrishna, S. A review on polymer nanofibers by electrospinning and their applications in nanocomposites. *Compos. Sci. Technol.* **63,** 2223–2253 (2003).
37. Matthews, J. A., Wnek, G. E., Simpson, D. G. & Bowlin, G. L. Electrospinning of collagen nanofibers. *Biomacromolecules* **3,** 232–238 (2002).
38. Orlova, Y., Magome, N., Liu, L., Chen, Y. & Agladze, K. Electrospun nanofibers as a tool for architecture control in engineered cardiac tissue. *Biomaterials* **32,** 5615–5624 (2011).
39. Formhals, A. Process and apparatus for preparing artificial threads. U.S. Patent No. 1,975,504 (1934).
40. Formhals, A. Method and apparatus for spinning. U.S. Patent No. 2,169,962 (1939).
41. Formhals, A. Artificial thread and method of producing same. U.S. Patent No. 2,187,306 (1940).
42. Zeleny, J. The electrical discharge from liquid points, and a hydrostatic method of measuring the electric intensity at their surfaces. *Phys. Rev.* **3,** 69–91 (1914).
43. Taylor, G. Electrically driven jets. *Proc. R. Soc. Lond. A Mat.* **313,** 453–475 (1969).
44. Li, W. J., Laurencin, C. T., Caterson, E. J., Tuan, R. S. & Ko, F. K. Electrospun nanofibrous structure: A novel scaffold for tissue engineering. *J. Biomed. Mater. Res.* **60,** 613 (2002).
45. Baker, B. M. & Mauck, R. L. The effect of nanofiber alignment on the maturation of engineered meniscus constructs. *Biomaterials* **28,** 1967 (2007).
46. Li, D., Wang, Y. & Xia, Y. Electrospinning of polymeric and ceramic nanofibers as uniaxially aligned arrays. *Nano Lett.* **3,** 1167–1171 (2003).
47. Li, D. & Xia, Y. Fabrication of titania nanofibers by electrospinning. *Nano Lett.* **3,** 555–560 (2003).
48. Larsen, G., Velarde-Ortiz, R., Minchow, K., Barrero, A. & Loscertales, I. G. A method for making inorganic and hybrid (organic/inorganic) fibers and vesicles with diameters in the submicrometer and micrometer range via sol–gel chemistry and electrically forced liquid jets. *J. Am. Chem. Soc.* **125,** 1154–1155 (2003).
49. Viswanathamurthi, P. et al. Preparation and morphology of niobium oxide fibres by electrospinning. *Chem. Phys. Lett.* **374,** 79–84 (2003).
50. Dai, H., Gong, J., Kim, H. & Lee, D. A novel method for preparing ultra fine alumina-borate oxide fibres via an electrospinning technique. *Nanotechnology* **13,** 674–677 (2002).
51. Katta, P., Alessandro, M., Ramsier, R. D. & Chase, G. G. Continuous electrospinning of aligned polymer nanofibers onto a wire drum collector. *Nano Lett.* **4,** 2215–2218 (2004).
52. Li, D., Ouyang, G., McCann, J. T. & Xia, Y. Collecting electrospun nanofibers with patterned electrodes. *Nano Lett.* **5,** 913–916 (2005).
53. Zussman, E., Theron, A. & Yarin, A. L. Formation of nanofiber crossbars in electrospinning. *Appl. Phys. Lett.* **82,** 973–975 (2003).
54. Theron, A., Zussman, E. & Yarin, A. Electrostatic field-assisted alignment of electrospun nanofibres. *Nanotechnology* **12,** 384–390 (2001).
55. Sarkar, S., Deevi, S. & Tepper, G. Biased AC electrospinning of aligned polymer nanofibers. *Macromol. Rap. Comm.* **28,** 1034–1039 (2007).
56. Carnell, L. S. et al. Aligned mats from electrospun single fibers. *Macromolecules* **41,** 5345–5349 (2008).
57. Sundaray, B. et al. Electrospinning of continuous aligned polymer fibers. *Appl. Phys. Lett.* **84,** 1222–1224 (2004).
58. Tutak, W. et al. The support of bone marrow stromal cell differentiation by airbrushed nanofiber scaffolds. *Biomaterials* **34,** 2389–2398 (2013).
59. Badrossamay, M. R., McIlwee, H. A., Goss, J. A. & Parker, K. K. Nanofiber assembly by rotary jet-spinning. *Nano Lett.* **10,** 2257–2261 (2010).
60. Nain, A. S., Wong, J. C., Amon, C. & Sitti, M. Drawing suspended polymer micro-/nanofibers using glass micropipettes. *Appl. Phys. Lett.* **89,** 183105 (2006).
61. Nain, A. S. et al. Dry spinning based spinneret based tunable engineered parameters (step) technique for controlled and aligned deposition of polymeric nanofibers. *Macromol. Rap. Comm.* **30,** 1406–1412 (2009).
62. Nain, A. S. & Wang, J. Polymeric nanofibers: Isodiametric design space and methodology for depositing aligned nanofiber arrays in single and multiple layers. *Polym. J.* **45,** 695–700 (2013).
63. Nain, A. S. et al. Control of cell behavior by aligned micro/nanofibrous biomaterial scaffolds fabricated by spinneret-based tunable engineered parameters (STEP) technique. *Small* **4,** 1153–1159 (2008).
64. Sun, D., Chang, C., Li, S. & Lin, L. Near-field electrospinning. *Nano Lett.* **6,** 839–842 (2006).
65. Harfenist, S. A. et al. Direct drawing of suspended filamentary micro- and nanostructures from liquid polymers. *Nano Lett.* **4,** 1931–1937 (2004).

66. Hong, J. K. & Madihally, S. V. Three-dimensional scaffold of electrosprayed fibers with large pore size for tissue regeneration. *Acta Biomater.* **6,** 4734–4742 (2010).
67. Pabba, S. et al. Biopolymerization-driven self-assembly of nanofiber air-bridges. *Soft Matter* **5,** 1378 (2009).
68. Rathfon, J. M., Grolman, J. M., Crosby, A. J. & Tew, G. N. Formation of oriented, suspended fibers by melting free standing polystyrene thin films. *Macromolecules* **42,** 6716–6722 (2009).
69. Nain, A. S., Amon, C. & Sitti, M. Proximal probes based nanorobotic drawing of polymer micro/nanofibers. *IEEE Trans. Nanotechnol.* **5,** 499–510 (2006).
70. Su, B. et al. Elaborate positioning of nanowire arrays contributed by highly adhesive superhydrophobic pillar-structured substrates. *Adv. Mat.* **24,** 559–564 (2012).
71. Su, B., Wu, Y. & Jiang, L. The art of aligning one-dimensional (1D) nanostructures. *Chem. Soc. Rev.* **41,** 7832–7856 (2012).
72. Bella, J. & Berman, H. M. Integrin–collagen complex: A metal–glutamate handshake. *Structure* **8,** R121–R126 (2000).
73. Li, F., Redick, S. D., Erickson, H. P. & Moy, V. T. Force measurements of the alpha5beta1 integrin-fibronectin interaction. *Biophys. J.* **84,** 1252–1262 (2003).
74. Humphries, M. J. Integrin activation: The link between ligand binding and signal transduction. *Curr. Opin. Cell Biol.* **8,** 632–640 (1996).
75. Plow, E. F., Haas, T. A., Zhang, L., Loftus, J. & Smith, J. W. Ligand binding to integrins. *J. Biol. Chem.* **275,** 21785–21788 (2000).
76. Digman, M. A., Brown, C. M., Horwitz, A. R., Mantulin, W. W. & Gratton, E. Paxillin dynamics measured during adhesion assembly and disassembly by correlation spectroscopy. *Biophys. J.* **94,** 2819–2831 (2008).
77. Zhang, X. et al. Talin depletion reveals independence of initial cell spreading from integrin activation and traction. *Nat. Cell Biol.* **10,** 1062–1068 (2008).
78. Wu, C.-C., Su, H.-W., Lee, C.-C., Tang, M.-J. & Su, F.-C. Quantitative measurement of changes in adhesion force involving focal adhesion kinase during cell attachment, spread, and migration. *Biochem. Biophys. Res. Comm.* **329,** 256–265 (2005).
79. Geiger, B., Spatz, J. P. & Bershadsky, A. D. Environmental sensing through focal adhesions. *Nat. Rev. Mol. Cell Biol.* **10,** 21–33 (2009).
80. Choi, C. K. et al. Actin and alpha-actinin orchestrate the assembly and maturation of nascent adhesions in a myosin II motor-independent manner. *Nat. Cell Biol.* **10,** 1039–1050 (2008).
81. Burridge, K. & Wittchen, E. S. The tension mounts: Stress fibers as force-generating mechanotransducers. *J. Cell Biol.* **200,** 9–19 (2013).
82. Roca-Cusachs, P. et al. Integrin-dependent force transmission to the extracellular matrix by α-actinin triggers adhesion maturation. *PNAS* **110,** E1361–E1370 (2013).
83. Gardel, M. L., Schneider, I. C., Aratyn-Schaus, Y. & Waterman, C. M. Mechanical integration of actin and adhesion dynamics in cell migration. *Ann. Rev. Cell Dev. Biol.* **26,** 315–333 (2010).
84. Galbraith, C. G., Yamada, K. M. & Sheetz, M. P. The relationship between force and focal complex development. *J. Cell Biol.* **159,** 695–705 (2002).
85. Möhl, C. et al. Becoming stable and strong: The interplay between vinculin exchange dynamics and adhesion strength during adhesion site maturation. *Cell Motil. Cytoskel.* **66,** 350–364 (2009).
86. Puklin-Faucher, E. & Sheetz, M. P. The mechanical integrin cycle. *J. Cell Sci.* **122,** 179–186 (2009).
87. Woods, A. J., White, D. P., Caswell, P. T. & Norman, J. C. PKD1/PKC[mu] promotes [alpha]v[beta]3 integrin recycling and delivery to nascent focal adhesions. *EMBO J.* **23,** 2531–2543 (2004).
88. Jones, M. C., Caswell, P. T. & Norman, J. C. Endocytic recycling pathways: Emerging regulators of cell migration. *Curr. Opin. Cell Biol.* **18,** 549–557 (2006).
89. Kim, D.-H. & Wirtz, D. Focal adhesion size uniquely predicts cell migration. *FASEB J.* **27,** 1351–1361 (2013).
90. Prager-Khoutorsky, M. et al. Fibroblast polarization is a matrix-rigidity-dependent process controlled by focal adhesion mechanosensing. *Nat. Cell Biol.* **13,** 1457–1465 (2011).
91. Andalib, M. N., Lee, J. S., Ha, L., Dzenis, Y. & Lim, J. Y. The role of RhoA kinase (ROCK) in cell alignment on nanofibers. *Acta Biomater.* **9,** 7737–7745 (2013).
92. Fuh, Y. K., Chen, S. Z. & He, Z. Y. Direct-write, highly aligned chitosan-poly(ethylene oxide) nanofiber patterns for cell morphology and spreading control. *Nanoscale Res. Lett.* **8,** 97 (2013).
93. Stricker, J., Aratyn-Schaus, Y., Oakes, P. W. & Gardel M. L. Spatiotemporal constraints on the force-dependent growth of focal adhesions. *Biophys. J.* **100,** 2883–2893 (2011).
94. Berginski, M. E., Vitriol, E. A., Hahn, K. M. & Gomez, S. M. High-resolution quantification of focal adhesion spatiotemporal dynamics in living cells. *PLOS One* **6,** e22025 (2011).
95. Hsu, Y.-M., Chen, C.-N., Chiu, J.-J., Chang, S.-H. & Wang, Y.-J. The effects of fiber size on MG63 cells cultured with collagen based matrices. *J. Biomed. Mater. Res. B: Appl. Biomater.* **91B,** 737–745 (2009).
96. Lauffenburger, D. A. & Horwitz, A. F. Cell migration: A physically integrated molecular process. *Cell* **84,** 359–369 (1996).
97. Friedl, P. & Gilmour, D. Collective cell migration in morphogenesis, regeneration and cancer. *Nat. Rev. Mol. Cell Biol.* **10,** 445–457 (2009).
98. Schäfer, M. & Werner, S. Cancer as an overhealing wound: An old hypothesis revisited. *Nat. Rev. Mol. Cell Biol.* **9,** 628–638 (2008).
99. Eyckmans, J., Boudou, T., Yu, X. & Chen, C. S. A hitchhiker's guide to mechanobiology. *Dev. Cell* **21,** 35–47 (2011).
100. Vuori, K. & Ruoslahti, E. Connections count in cell migration. *Nat. Cell Biol.* **1,** E85–E87 (1999).
101. Vicente-Manzanares, M., Choi, C. K. & Horwitz, A. R. Integrins in cell migration—The actin connection. *J. Cell Sci.* **122,** 199–206 (2009).

102. Campas, O., Mammoto, T., Hasso, S., Sperling, R. A., O'Connell, D., Bischof, A. G., Maas, R., Weitz, D. A., Mahadevan. L. & Ingber, D. E. Quantifying cell-generated mechanical forces within living embryonic tissues. *Nat. Meth.* **11**, 183–189 (2014).
103. Du Roure, O. et al. Force mapping in epithelial cell migration. *PNAS* **102,** 2390–2395 (2005).
104. Rape, A. D., Guo, W.-H. & Wang, Y.-L. The regulation of traction force in relation to cell shape and focal adhesions. *Biomaterials* **32,** 2043–2051 (2011).
105. Zhu, C., Bao, G. & Wang, N. Cell mechanics: Mechanical response, cell adhesion, and molecular deformation. *Annu. Rev. Biomed. Eng.* **2,** 189–226 (2000).
106. Pelham, R. J. & Wang, Y. L. High resolution detection of mechanical forces exerted by locomoting fibroblasts on the substrate. *Mol. Biol. Cell* **10,** 935–945 (1999).
107. Sheetz, M. P., Felsenfeld, D. P. & Galbraith, C. G. Cell migration: Regulation of force on extracellular-matrix–integrin complexes. *Trends Cell Biol.* **8,** 51–54 (1998).
108. Galbraith, C. G. & Sheetz, M. P. Forces on adhesive contacts affect cell function. *Curr. Opin. Cell Biol.* **10,** 566–571 (1998).
109. Galbraith, C. G. & Sheetz, M. P. Cell traction. *Curr. Protoc. Cell Biol.* **Chapter 12,** Unit 12.3 (2001).
110. Berry, M. F. et al. Mesenchymal stem cell injection after myocardial infarction improves myocardial compliance. *Am. J. Physiol.-Heart C.* **290,** H2196–H2203 (2006).
111. Richert, L., Engler, A. J., Discher, D. E. & Picart, C. Elasticity of native and cross-linked polyelectrolyte multilayer films. *Biomacromolecules* **5,** 1908–1916 (2004).
112. Chen, C. S. Mechanotransduction—A field pulling together? *J. Cell Sci.* **121,** 3285–3292 (2008).
113. Yang, M. T., Sniadecki, N. J. & Chen, C. S. Geometric considerations of micro- to nanoscale elastomeric post arrays to study cellular traction forces. *Adv. Mater.* **19,** 3119–3123 (2007).
114. Wozniak, M. A. & Chen, C. S. Mechanotransduction in development: A growing role for contractility. *Nat. Rev. Mol. Cell Biol.* **10,** 34–43 (2009).
115. Barone, V. & Heisenberg, C.-P. Cell adhesion in embryo morphogenesis. *Curr. Opin. Cell Biol.* **24,** 148–153 (2012).
116. Moore, K. A. et al. Control of basement membrane remodeling and epithelial branching morphogenesis in embryonic lung by Rho and cytoskeletal tension. *Dev. Dynam.* **232,** 268–281 (2005).
117. Levenberg, S. et al. Differentiation of human embryonic stem cells on three-dimensional polymer scaffolds. *PNAS* **100,** 12741–12746 (2003).
118. Engler, A. J. et al. Embryonic cardiomyocytes beat best on a matrix with heart-like elasticity: Scar-like rigidity inhibits beating. *J. Cell Sci.* **121,** 3794–3802 (2008).
119. Nikolić, D. L., Boettiger, A. N., Bar-Sagi, D., Carbeck, J. D. & Shvartsman, S. Y. Role of boundary conditions in an experimental model of epithelial wound healing. *Am. J. Phys. Cell Phys.* **291,** C68–C75 (2006).
120. Tremel, A. et al. Cell migration and proliferation during monolayer formation and wound healing. *Chem. Eng. Sci.* **64,** 247–253 (2009).
121. Ortega, S., Ittmann, M., Tsang, S. H., Ehrlich, M. & Basilico, C. Neuronal defects and delayed wound healing in mice lacking fibroblast growth factor 2. *PNAS* **95,** 5672–5677 (1998).
122. Koch, T. M., Münster, S., Bonakdar, N., Butler, J. P. & Fabry, B. 3D traction forces in cancer cell invasion. *PLOS One* **7,** e33476 (2012).
123. Kumar, S. & Weaver, V. M. Mechanics, malignancy, and metastasis: The force journey of a tumor cell. *Cancer Metast. Rev.* **28,** 113–127 (2009).
124. Suresh, S. Biomechanics and biophysics of cancer cells. *Acta Biomater.* **3,** 413–438 (2007).
125. Oh, M. J., Kuhr, F., Byfield, F. & Levitan, I. Micropipette aspiration of substrate-attached cells to estimate cell stiffness. *J. Vis. Exp.* **67,** 3886 (2012).
126. Stark, D. J., Killian, T. C. & Raphael, R. M. A microfabricated magnetic force transducer–microaspiration system for studying membrane mechanics. *Phys. Biol.* **8,** 56008 (2011).
127. Ashok, P. C. & Dholakia, K. Optical trapping for analytical biotechnology. *Curr. Opin. Biotechnol.* **1,** 16–21 (2011).
128. Guck, J. et al. The optical stretcher: A novel laser tool to micromanipulate cells. *Biophys. J.* **81,** 767–784 (2001).
129. Neuman, K. C. & Nagy, A. Single-molecule force spectroscopy: Optical tweezers, magnetic tweezers and atomic force microscopy. *Nat. Methods* **5,** 491–505 (2008).
130. Simmons, R. M., Finer, J. T., Chu, S. & Spudich, J. A. Quantitative measurements of force and displacement using an optical trap. *Biophys. J.* **70,** 1813–1822 (1996).
131. Li, Z. et al. Membrane tether formation from outer hair cells with optical tweezers. *Biophys. J.* **82,** 1386–1395 (2002).
132. Wu, Y., Sun, D. & Huang, W. Mechanical force characterization in manipulating live cells with optical tweezers. *J. Biomech.* **44,** 741–746 (2011).
133. Puech, P.-H., Poole, K., Knebel, D. & Muller, D. J. A new technical approach to quantify cell-cell adhesion forces by AFM. *Ultramicroscopy* **106,** 637–644 (2006).
134. Wu, X., Sun, Z., Meininger, G. A. & Muthuchamy, M. Application of atomic force microscopy measurements on cardiovascular cells. *Methods Mol. Biol.* **843,** 229–244 (2012).
135. Azeloglu, E. U. & Costa, K. D. Atomic force microscopy in mechanobiology: Measuring microelastic heterogeneity of living cells. *Methods Mol. Biol.* **736,** 303–329 (2011).
136. Alonso, J. L. & Goldmann, W. H. Feeling the forces: Atomic force microscopy in cell biology. *Life Sci.* **72,** 2553–2560 (2003).
137. Shahin, V. & Barrera, N. P. Providing unique insight into cell biology via atomic force microscopy. *Int. Rev. Cytol.* **265,** 227–252 (2008).
138. Legant, W. R. et al. Measurement of mechanical tractions exerted by cells in three-dimensional matrices. *Nat. Methods* **7,** 969–971 (2010).
139. Sunyer, R., Jin, A. J., Nossal, R. & Sackett, D. L. Fabrication of hydrogels with steep stiffness gradients for studying cell mechanical response. *PLOS One* **7,** e46107 (2012).
140. Underhill, G. H., Galie, P., Chen, C. S. & Bhatia, S. N. Bioengineering methods for analysis of cells in vitro. *Annu. Rev. Cell Dev. Biol.* **28,** 385–410 (2012).

141. Legant, W. R. et al. Multidimensional traction force microscopy reveals out-of-plane rotational moments about focal adhesions. *PNAS* **110,** 881–886 (2013).
142. Fu, J. et al. Mechanical regulation of cell function with geometrically modulated elastomeric substrates. *Nat. Methods* **7,** 733–736 (2010).
143. Han, S. J., Bielawski, K. S., Ting, L. H., Rodriguez, M. L. & Sniadecki, N. J. Decoupling substrate stiffness, spread area, and micropost density: A close spatial relationship between traction forces and focal adhesions. *Biophys. J.* **103,** 640–648 (2012).
144. Kajzar, A., Cesa, C. M., Kirchgessner, N., Hoffmann, B. & Merkel, R. Toward physiological conditions for cell analyses: Forces of heart muscle cells suspended between elastic micropillars. *Biophys. J.* **94,** 1854–1866 (2008).
145. Avvisato, C. L. et al. Mechanical force modulates global gene expression and beta-catenin signaling in colon cancer cells. *J. Cell Sci.* **120,** 2672–2682 (2007).
146. Resnick, N. & Gimbrone Jr., M. A. Hemodynamic forces are complex regulators of endothelial gene expression. *FASEB J.* **9,** 874–882 (1995).
147. Chan, M. W., Hinz, B. & McCulloch, C. A. Mechanical induction of gene expression in connective tissue cells. *Methods Cell Biol.* **98,** 178–205 (2010).
148. Tse, J. M. et al. Mechanical compression drives cancer cells toward invasive phenotype. *PNAS* **109,** 911–916 (2012).
149. Nikkhah, M., Edalat, F., Manoucheri, S. & Khademhosseini, A. Engineering microscale topographies to control the cell–substrate interface. *Biomaterials* **33,** 5230–5246 (2012).
150. Song, M. J., Dean, D. & Knothe Tate, M. L. Mechanical modulation of nascent stem cell lineage commitment in tissue engineering scaffolds. *Biomaterials* **34,** 5766–5775 (2013).
151. Jagielska, A. et al. Mechanical environment modulates biological properties of oligodendrocyte progenitor cells. *Stem Cells Dev.* **21,** 2905–2914 (2012).
152. Li, D. et al. Role of mechanical factors in fate decisions of stem cells. *Regen. Med.* **6,** 229–240 (2011).
153. Chang, H. & Knothe Tate, M. L. Structure-function relationships in the stem cell's mechanical world B: Emergent anisotropy of the cytoskeleton correlates to volume and shape changing stress exposure. *Mol. Cell. Biomech.* **8,** 297–318 (2011).
154. Clause, K. C., Liu, L. J. & Tobita, K. Directed stem cell differentiation: The role of physical forces. *Cell Commun. Adhes.* **17,** 48–54 (2010).
155. Ofek, G. et al. Mechanical characterization of differentiated human embryonic stem cells. *J. Biomech. Eng.* **131,** 61011 (2009).
156. Guilak, F. et al. Control of stem cell fate by physical interactions with the extracellular matrix. *Cell Stem Cell* **5,** 17–26 (2009).
157. Lund, A. W., Yener, B., Stegemann, J. P. & Plopper, G. E. The natural and engineered 3D microenvironment as a regulatory cue during stem cell fate determination. *Tissue Eng. Part B Rev.* **15,** 371–380 (2009).
158. Li, W.-J., Tuli, R., Huang, X., Laquerriere, P. & Tuan, R. S. Multilineage differentiation of human mesenchymal stem cells in a three-dimensional nanofibrous scaffold. *Biomaterials* **26,** 5158–5166 (2005).
159. Bakhru, S. et al. Direct and cell signaling-based, geometry-induced neuronal differentiation of neural stem cells. *Integr. Biol.* **3,** 1207–1214 (2011).
160. Christopherson, G. T., Song, H. & Mao, H.-Q. The influence of fiber diameter of electrospun substrates on neural stem cell differentiation and proliferation. *Biomaterials* **30,** 556–564 (2009).
161. Seyedjafari, E., Soleimani, M., Ghaemi, N. & Sarbolouki, M. N. Enhanced osteogenic differentiation of cord blood-derived unrestricted somatic stem cells on electrospun nanofibers. *J. Mater. Sci. Mater. Med.* **22,** 165–174 (2011).
162. Bao, C., Chen, W., Weir, M. D., Thein-Han, W. & Xu, H. H. Effects of electrospun submicron fibers in calcium phosphate cement scaffold on mechanical properties and osteogenic differentiation of umbilical cord stem cells. *Acta Biomater.* **7,** 4037–4044 (2011).
163. Borjigin, M. et al. Electrospun fiber membranes enable proliferation of genetically modified cells. *Int. J. Nanomed.* **8,** 855–864 (2013).
164. Lee, C. H. et al. Nanofiber alignment and direction of mechanical strain affect the ECM production of human ACL fibroblast. *Biomaterials* **26,** 1261–1270 (2005).
165. Horii, A., Wang, X., Gelain, F. & Zhang, S. Biological designer self-assembling peptide nanofiber scaffolds significantly enhance osteoblast proliferation, differentiation and 3-D migration. *PLOS One* **2,** e190 (2007).
166. Madduri, S., Papaloïzos, M. & Gander, B. Trophically and topographically functionalized silk fibroin nerve conduits for guided peripheral nerve regeneration. *Biomaterials* **31,** 2323–2334 (2010).
167. Bini, T. B. et al. Electrospun poly(L-lactide- co -glycolide) biodegradable polymer nanofibre tubes for peripheral nerve regeneration. *Nanotechnology* **15,** 1459 (2004).
168. Rho, K. S. et al. Electrospinning of collagen nanofibers: Effects on the behavior of normal human keratinocytes and early-stage wound healing. *Biomaterials* **27,** 1452–1461 (2006).
169. Katti, D. S., Robinson, K. W., Ko, F. K. & Laurencin, C. T. Bioresorbable nanofiber-based systems for wound healing and drug delivery: Optimization of fabrication parameters. *J. Biomed. Mater. Res. B Appl. Biomater.* **70B,** 286–296 (2004).
170. Yoshimoto, H., Shin, Y. M., Terai, H. & Vacanti, J. P. A biodegradable nanofiber scaffold by electrospinning and its potential for bone tissue engineering. *Biomaterials* **24,** 2077–2082 (2003).
171. Li, C., Vepari, C., Jin, H.-J., Kim, H. J. & Kaplan, D. L. Electrospun silk-BMP-2 scaffolds for bone tissue engineering. *Biomaterials* **27,** 3115–3124 (2006).
172. Seidi, A., Ramalingam, M., Elloumi-Hannachi, I., Ostrovidov, S. & Khademhosseini, A. Gradient biomaterials for soft-to-hard interface tissue engineering. *Acta Biomater.* **7,** 1441–1451 (2011).

173. Erisken, C., Kalyon, D. M. & Wang, H. Functionally graded electrospun polycaprolactone and β-tricalcium phosphate nanocomposites for tissue engineering applications. *Biomaterials* **29,** 4065–4073 (2008).
174. Vaz, C. M., van Tuijl, S., Bouten, C. V. C. & Baaijens, F. P. T. Design of scaffolds for blood vessel tissue engineering using a multi-layering electrospinning technique. *Acta Biomater.* **1,** 575–582 (2005).
175. Xu, C. Y., Inai, R., Kotaki, M. & Ramakrishna, S. Aligned biodegradable nanofibrous structure: A potential scaffold for blood vessel engineering. *Biomaterials* **25,** 877–886 (2004).
176. Zeng, J. et al. Biodegradable electrospun fibers for drug delivery. *J. Control. Release* **92,** 227–231 (2003).
177. Kretlow, J. D., Klouda, L. & Mikos, A. G. Injectable matrices and scaffolds for drug delivery in tissue engineering. *Adv. Drug Deliv. Rev.* **59,** 263–273 (2007).
178. Nyberg, P., Salo, T. & Kalluri, R. Tumor microenvironment and angiogenesis. *Front. Biosci.* **13,** 6537–6553 (2008).
179. Watnick, R. S. The role of the tumor microenvironment in regulating angiogenesis. *Cold Spring Harb. Perspect. Med.* **2,** a006676 (2012).
180. Talbot, L. J., Bhattacharya, S. D. & Kuo, P. C. Epithelial-mesenchymal transition, the tumor microenvironment, and metastatic behavior of epithelial malignancies. *Int. J. Biochem. Mol. Biol.* **3,** 117–136 (2012).
181. Lee, K. & Nelson, C. M. New insights into the regulation of epithelial-mesenchymal transition and tissue fibrosis. *Int. Rev. Cell Mol. Biol.* **294,** 171–221 (2012).
182. Wirtz, D., Konstantopoulos, K. & Searson, P. C. The physics of cancer: The role of physical interactions and mechanical forces in metastasis. *Nat. Rev. Cancer* **11,** 512–522 (2011).
183. Park, J. E. et al. Hypoxic tumor cell modulates its microenvironment to enhance angiogenic and metastatic potential by secretion of proteins and exosomes. *Mol. Cell Proteomics* **9,** 1085–1099 (2010).
184. Robinson, B. D. et al. Tumor microenvironment of metastasis in human breast carcinoma: A potential prognostic marker linked to hematogenous dissemination. *Clin. Cancer Res.* **15,** 2433–2441 (2009).
185. Koontongkaew, S. The tumor microenvironment contribution to development, growth, invasion and metastasis of head and neck squamous cell carcinomas. *J. Cancer* **4,** 66–83 (2013).
186. Fraley, S. I. et al. A distinctive role for focal adhesion proteins in three-dimensional cell motility. *Nat. Cell Biol.* **12,** 598–604 (2010).
187. Straussman, R. et al. Tumour micro-environment elicits innate resistance to RAF inhibitors through HGF secretion. *Nature* **487,** 500–504 (2012).
188. Sun, Y. et al. Treatment-induced damage to the tumor microenvironment promotes prostate cancer therapy resistance through WNT16B. *Nat. Med.* **18,** 1359–1368 (2012).
189. Davies, P. C., Demetrius, L. & Tuszynski, J. A. Cancer as a dynamical phase transition. *Theor. Biol. Med. Model* **8,** 30 (2011).
190. Baker, E. L., Srivastava, J., Yu, D., Bonnecaze, R. T. & Zaman, M. H. Cancer cell migration: Integrated roles of matrix mechanics and transforming potential. *PLOS One* **6,** e20355 (2011).
191. Korkaya, H., Liu, S. & Wicha, M. S. Breast cancer stem cells, cytokine networks, and the tumor microenvironment. *J. Clin. Invest.* **121,** 3804–3809 (2011).
192. Mbeunkui, F. & Johann Jr., D. J. Cancer and the tumor microenvironment: A review of an essential relationship. *Cancer Chemother. Pharmacol.* **63,** 571–582 (2009).
193. Johnson, J. et al. Quantitative analysis of complex glioma cell migration on electrospun polycaprolactone using time-lapse microscopy. *Tissue Eng. Part C Methods* **15,** 531–540 (2009).

24

Mechanical Stretch Assays in Cell Culture Systems

Abhishek Tondon, Candice Haase, and Roland Kaunas

CONTENTS

This chapter will focus on various techniques for imaging stretched cells and review several studies that have imaged and quantified stretch-induced reorganization of specific cellular structures. In addition, the use of imaging to estimate stretch-induced force development will be presented. Finally, perspectives for future research and development in the area of imaging stretch mechanobiology will be discussed.

Imaging the effects of dynamic stretching on cells and their ECM has revealed a great deal regarding how forces applied at the macroscale result in reorganization of the ECM, resident cells, and subcellular structures. These structural changes are driven by the forces the cells perceive at their cell–matrix adhesions. Creative imaging techniques have been developed to estimate these forces to elucidate cellular mechanics at the microscopic level.

24.1 Introduction

Mechanical forces applied to soft tissues result in substantial deformation of the extracellular matrix (ECM) and the cells that reside within them. Respiration involves periodic inflation of the lungs, while the cardiac cycle involves cyclic distention of the heart and arteries. Other tissues, such as skeletal muscle and connective tissues, are also periodically distended during limb movements. Experiments studying the responses of the ECM and cells from these tissues to stretching in cell culture have revealed that these deformations have profound effects on cell structure, tissue organization, and cell function.

24.2 Instrumentation

24.2.1 Stretching Cells on 2-D Substrates

The vast majority of stretch devices are based on stretching cell cultures on flat elastomeric sheets. Typically, the elastomer used is silicone rubber consisting of polydimethyl-siloxane (PDMS). The elastomer surface chemistry is modified to accommodate cell adhesion using various ECM proteins, including fibronectin and collagen type I, or short synthetic peptides containing the integrin binding sequences of these matrix proteins.[1] Methods for deforming these elastomers include vacuum suction, indentation, and longitudinal pulling.

In 1985, Banes et al.[2] introduced and later commercialized the Flexcell® system, which consists of a computer-controlled pump that applies suction to conventional 6-well plates containing silicone rubber bottoms. In the original version of the device, the strain pattern was heterogeneous, with cells located at the periphery of the well subjected to high, anisotropic strains and cells at the center subjected to low, equibiaxial strain.[3] The addition of a loading post increases the uniformity of the strain field over the post, with equibiaxial strain produced with a circular loading post and uniaxial strain with an oblong loading post. The StageFlexer® is a version of the system that can be mounted on the stage of an upright microscope for live cell imaging throughout the stretch cycle (Figure 24.1). Variations on this approach have also been developed.[4]

As an alternative to the vacuum approach, the loading post can be pushed against the elastomer sheet to produce a homogenous equibiaxial strain,[5–7] although uniaxial strain can also be produced by modifying the post geometry.[8,9] As with the Flexcell system with a loading post, continuous observations of cultured cells is possible since the stationary indenter allows the membrane to remain in the same plane while being stretched. A DC motor-cam is typically used to drive the cyclic load, which allows control of both stretch magnitude and frequency, but the stretch waveform is limited to sinusoidal function,[7] so a stepper motor can be used instead.[9]

The simplest manifestation of longitudinal pulling involves anchoring one end of a rectangular elastomeric sheet and pulling on the other end with a motorized actuator.[10] Due to the Poisson effect, the sheet contracts in the lateral direction, which results in the alignment of cells at an oblique angle (θ) relative to the direction of pulling[11] that depends on the Poisson ratio (υ) of the elastomer as described by the equation $\theta = \cos^{-1}\sqrt{\frac{\upsilon}{1+\upsilon}}$ 12. Various stretch devices have been developed that limit the extent of lateral contraction,[11–13] including the commercially available STREX® system.[14] While a microscope stage-mounted version of the STREX system is available, custom-built systems using linear actuators can also be used (Figure 24.2). This system provides

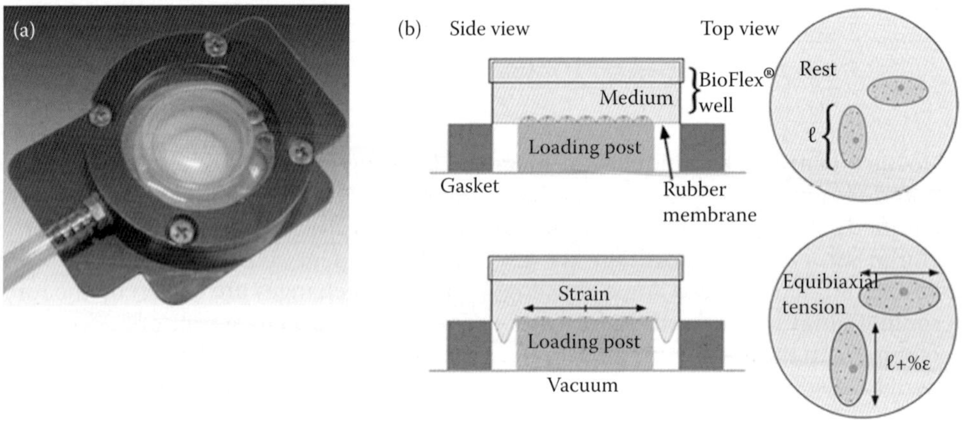

FIGURE 24.1
StageFlexer® system. Image (a) and schematic diagram (b) illustrating how the application of a vacuum generates equibiaxial strain to cells cultured in the area above an annular loading post.

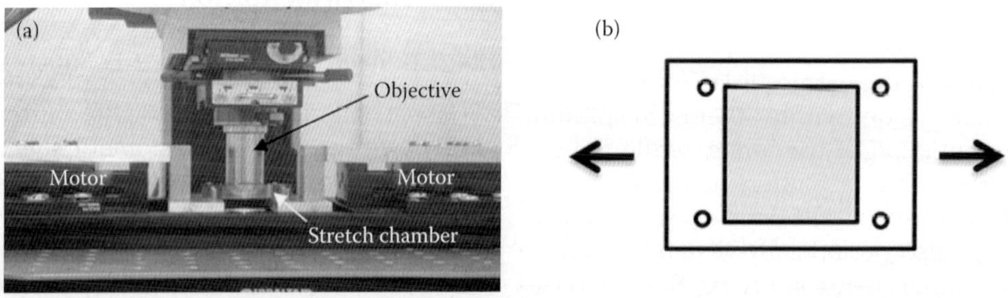

FIGURE 24.2
Uniaxial stretch device. (a) The device is mounted on the stage of an upright confocal microscope with a water dipping objective inserted into the center of a STREX uniaxial stretch chamber. (b) Schematic top view of the STREX chamber. The outer walls consist of thick silicone rubber, while the inner region (gray) is a very thin silicone rubber sheet. The entire chamber is stretched axisymmetrically using two linear motors connected to the four holes of the chamber. The side walls constrain lateral contraction of the inner sheet during longitudinal pulling.

much flexibility in the temporal waveform of a stretch cycle. For example, we have used this system to show that cells are much more sensitive to the rate of lengthening than to the rate of shortening of the stretch cycle.[15]

24.2.2 Stretching Cells in 3-D Matrices

Cells in 3-D matrices are exposed to a much different environment than cells attached to flat substrates.[16] Unlike cells on flat substrates, cells in 3-D matrices have no apical–basolateral polarity, interact with soft networks of discrete fibrils, and may be sterically hindered from spreading.[17] Furthermore, cells in 3-D collagen matrices align parallel, not perpendicular, to the direction of stretch, as do the collagen fibrils themselves.[18,19] These observations are consistent with the circumferential alignment of smooth muscle cells and collagen fibrils parallel to the principal direction of cyclic stretch in arteries. Thus, when studying the effects of stretch on cells that reside within tissues, it is appropriate to use a system in which the cells are cultured in a 3-D matrix.

Fibrillar collagen hydrogels are a commonly used material for stretching cells in a 3-D matrix. Various approaches have been used to handle these very soft gels. A rigid material (e.g., hydrophilic porous polyethylene) can be used as an interface between the gel and the actuators used to stretch the gel by allowing the gel to intercalate within the pores before gelation.[20] One example is the Flexcell Tissue Train® culture system, which applies a uniaxial strain to the matrix to result in alignment of F-actin, fibroblasts, and collagen fibrils parallel to the direction of stretch.[18,21] A biaxial configuration provides control over strains in the x- and y-directions of a cruciform-shaped collagen gel.[20] Alternatively, the gel can be cast in a hoop shape and pulled in opposite directions using two hooks.[19] A relatively flat collagen gel can also be adhered onto a silicone rubber membrane using an adhesive intermediate.[22]

24.2.3 Live Imaging

Most images of stretch-induced changes in cell structures are derived from fixed cells, which provide little information regarding the spatiotemporal dynamics. Time lapse imaging of live cells can be performed to image dynamic changes in cells or fluorescently labeled subcellular structures, including actin stress fibers and cell–matrix adhesions.[23–25] Stretching often causes cells to move out of field unless the strain is axisymmetric about the field of view or the stretch device itself is very small.[26] Further, the flexibility of the elastomer can result in small changes in position in both the xy-plane and z-direction after several stretch cycles. Consequently, images are typically captured in the unstretched state with the cyclic stretch temporarily stopped to reposition the sample for each image capture (Figure 24.3).

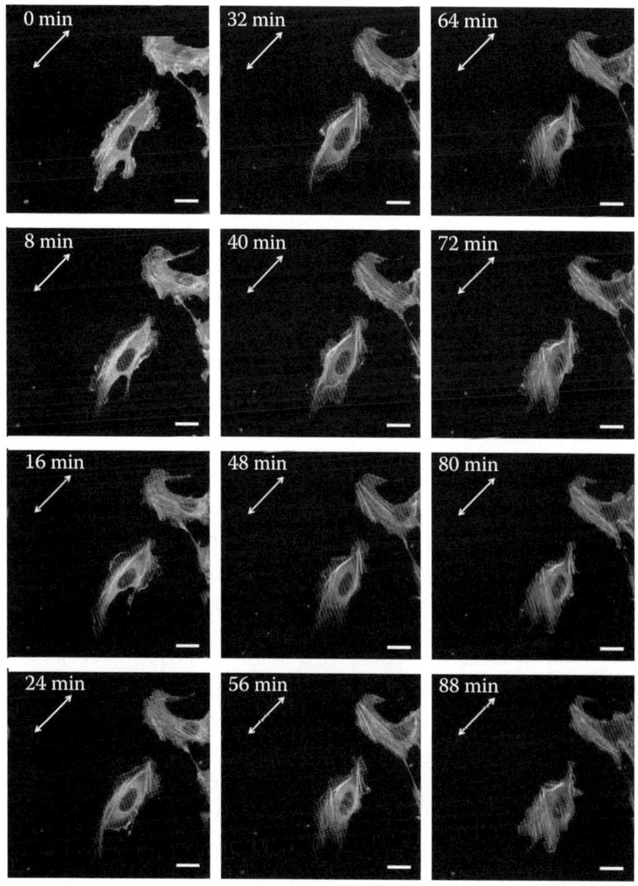

FIGURE 24.3
Dynamic realignment of stress fibers in response to cyclic stretch. Time-lapse images were captured at 8 min intervals of U2OS cells expressing green fluorescent protein (GFP)-actin subjected to 10% cyclic uniaxial stretch at 1 Hz. Bars, 25 μm.

24.3 Imaging and Quantification of Cellular Structures

24.3.1 Cell Morphology

Uniaxial cyclic stretch induces cell alignment in most adherent cell types, with the extent of alignment dependent on stretch amplitude, frequency, and spatial pattern.[8,11,27,28] Cell alignment is typically imaged using differential interference contrast (DIC) or phase contrast methods,[24] though fluorescently stained cells generally provide greater image contrast (Figure 24.4a). Although manual outlining is reliable, it is time-consuming and prone to individual bias. With a high-contrast fluorescent signal, image thresholding can be used to automatically

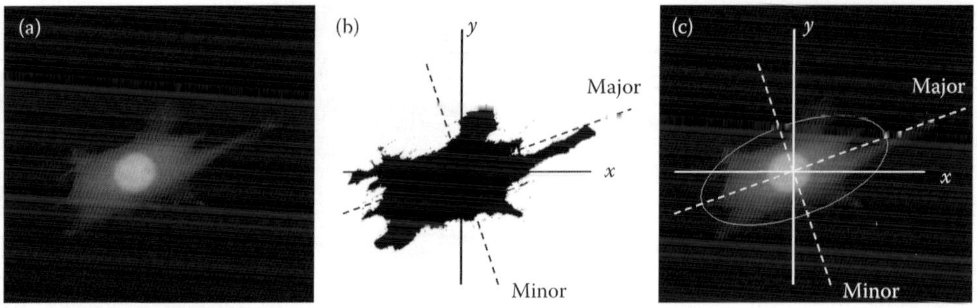

FIGURE 24.4
Quantifying cell shape. The image of a cell expressing GFP (a) is binarized (b) to allow automated measurement of the second moments and principal axes of the cell area. An ellipse with equivalent principal axes and major/minor moments is then fit to the cell shape (c).

extract the cell shape using image processing software, such as ImageJ (Figure 24.4b). Typically, the software detects individual cells in a binarized image and fits each shape to an ellipse (Figure 24.4c). The ellipse is determined from the second moments of inertia of the original image with respect to the centroidal x- and y-axes:

$$I_x = \sum_{i=1}^{n} (y_i - \bar{y})^2,\ I_y = \sum_{i=1}^{n} (x_i - \bar{x})^2,\ I_{xy} = \sum_{i=1}^{n} (x_i - \bar{x})(y_i - \bar{y}) \qquad (24.1)$$

where the centroid is located at $(\bar{x}, \bar{y})$ and (x_i, y_i) is the location of the ith point in the cell. Using the concept of Mohr's circle, the principle moments about the major and minor axes are

$$I_{\text{minor,major}} = \frac{I_x + I_y}{2} \pm \sqrt{\left(\frac{I_x + I_y}{2}\right)^2 + I_{xy}^2} \qquad (24.2)$$

and the orientation of the major principal axis can be calculated as

$$\theta = \left| 90° - \frac{1}{2}\arctan\left(\frac{2I_{xy}}{I_y - I_x}\right) \right| \qquad (24.3)$$

The fit ellipse has the same principle moments and axes as the original cell shape. The elongation of the cell can be quantified as either the ratio of the principal second moments or the ratio of the major and minor radii of the ellipse. Fast Fourier transform (FFT) methods have also been used to automatically quantify cell alignment, though these tend to yield low alignment scores for even images having highly aligned cells since the frequency spectrum has components in directions other than the cell orientation.[29]

Changes in cell shape are dependent on the dynamic movement of cellular structures such as lamellipodia, which are also affected by stretch.[9] Katsumi et al.[30] demonstrated that a step increase in equibiaxial stretch causes a transient retraction of lamellipodia in all directions, while uniaxial stretch only causes retraction of lamellipodia located along the sides of the cell perpendicular to the direction of stretching. Using fluorescence resonance energy transfer (FRET) microscopy with cells expressing GFP-labeled RacGTPase and Alexa-p21 binding domain of PAK1 (Alexa-PBD), FRET emission was reduced in retracting areas, suggesting that Rac activity is decreased at the edges that are subjected to increased tangential tension because of the stretch. Time lapse imaging of cells expressing GFP-labeled actin subjected to cyclic equibiaxial stretching shows sustained

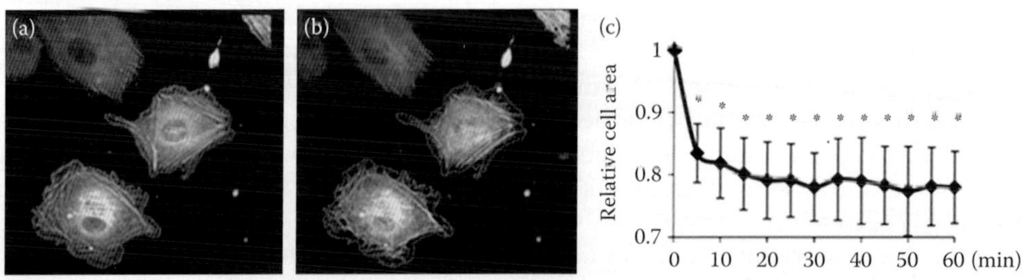

FIGURE 24.5
Lamellipodial retraction in response to cyclic equibiaxial stretch. U2OS cells expressing GFP-actin imaged immediately before (a) and 5 min after (b) applying 10% equibiaxial stretch at 1 Hz. The outlines of the cells prior to stretching (red) are shown for reference. Bar = 20 µm. (c) Ratio of cell areas (after/before stretch) for the indicated durations of stretch (mean ± S.D.; $n = 6$). (From Hsu, H. J. et al., *PLoS One* **5**, e12470, 2010. doi:10.1371/journal.pone.0012470.)

lamellipodial retraction[31] (Figure 24.5). Thus, cells can detect the spatiotemporal characteristics of the stretch pattern to regulate intracellular signaling that regulates cell membrane protrusions and, ultimately, cell shape.

24.3.2 Stress Fibers

Cell shape is highly dependent on the organization of actin stress fibers. Stress fibers align in response to cyclic stretch to an extent that depends on stretch magnitude and frequency[8,31] (Figure 24.6). Time-lapse images (cf. Figure 24.3) illustrate that stress fibers in cells expressing fluorescently labeled actin gradually align perpendicular to the stretch direction through stress fiber turnover and rotation.[23,25] Automated edge-detection algorithms have been used to quantify stress fiber orientation distributions.[31–33] These algorithms involve the convolution of a small section of an image (e.g., 20 × 20 pixels) with filter matrices (e.g., Sobel kernels) in order to quantify intensity gradients in the horizontal (G_x) and vertical (G_y) directions, from which the direction of the lowest brightness gradient can be calculated[8,33] (Figure 24.7). Areas lacking stress fibers are excluded by skipping subregions with average pixel brightness below a threshold value. While a mean stress fiber orientation angle can be calculated by vectorially summing all the orientation angles, a distribution should be recorded for each cell, which also contains information on the extent of alignment within an individual cell, with the lengths normalized so that each cell's distribution contributes equally to the distribution representing a group of cells (Figure 24.6). The circular variance can be computed as a metric quantifying the uniformity in the orientations by vectorially summing each orientation vector component, normalizing the result by the total number of vectors (N), and subtracting the value from unity (Equation 24.4)[34]:

$$\text{circular variance} = 1 - \frac{1}{N}\sqrt{\left(\sum_{i=1}^{N}\sin 2\theta^{i}\right)^{2} + \left(\sum_{i=1}^{N}\cos 2\theta^{i}\right)^{2}} \tag{24.4}$$

The values range from zero (perfect alignment) to one (random distribution).

FFT has also been used to quantify the alignment of stress fibers and microtubules.[35] In a side-by-side comparison, Kemeny and Clyne[32] demonstrated that edge detection methods are more sensitive than FFT

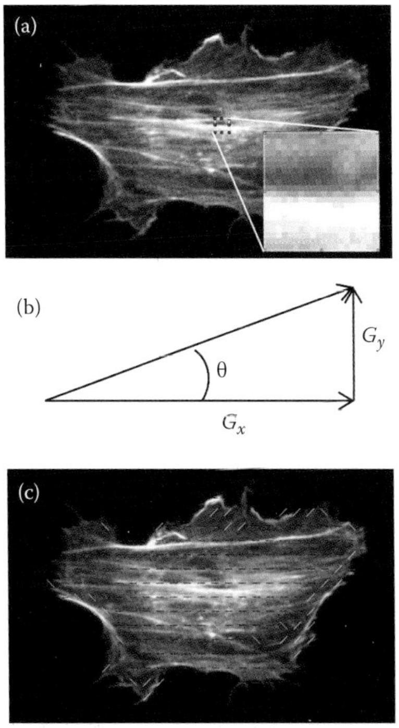

FIGURE 24.7
Quantifying stress fiber alignment. Small subregions of an image of GFP-labeled actin stress fibers (a) can be analyzed with an edge detection algorithm to determine the direction of the lowest pixel brightness gradient from gradients in the x- and y-directions (b). This calculation is then repeated in all subregions containing stress fibers (c).

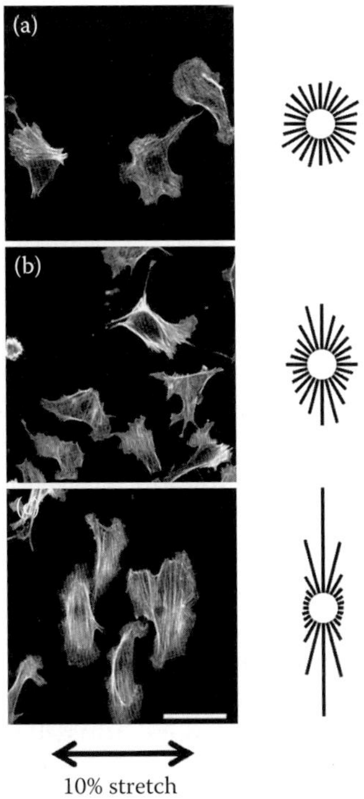

FIGURE 24.6
Extent of stress fiber alignment depends on the frequency of cyclic uniaxial stretch. Bovine aortic ECs were subjected to 4 h of 10% cyclic uniaxial stretch at frequencies of 1 (a), 0.1 (b), and 0.01 Hz (c), fixed, and stained for F-actin. The distributions of stress fiber orientations were determined using an edge detection algorithm, and the results from multiple cells (n = 50 cells) are summarized as angular histograms. (From Hsu, H. J. et al., *PLoS One* **4**, e4853, 2009. doi:10.1371/journal.pone.0004853.)

methods. The accuracy of any of these automated image analyses depends highly on image contrast, which in turn depends on magnification, fluorescence intensity, and photobleaching.[33]

24.3.3 Focal Adhesions

Cyclic stretch induces significant changes in cell–matrix adhesion-associated components, including vinculin and paxillin.[36] Goldyn et al.[24] followed individual focal adhesions in time-lapse images using the Manual Tracking feature in ImageJ to show that cyclic stretch can cause sliding of focal adhesions, which was subsequently shown to allow the rotation of associated stress fibers perpendicular to the stretch direction.[25] Na et al.[4] developed a stretch device mounted within an imaging system combining confocal and atomic force microscopy (AFM) to show that stretch-induced increases in focal adhesion area correspond with simultaneous increases in cell stiffness. While AFM was used to quantify mechanical properties, AFM in this system could potentially be used to detect stretch-induced changes in cellular features below the resolution limit of light microscopy.

24.3.4 Caveolae

Caveolae are invaginations in the plasma membrane consisting of caveolin, cholesterol, and glycosphingolipids that may act as membrane-mediated sensors and regulators of the plasma membrane tension.[37] Though too small to be resolved by light microscopy,[38] FRET biosensors provide a method to detect nanoscale strains in living cells. Gervasio et al.[39] stretched myoblasts that were incubated with a mixture of Alexa-Fluor-555-conjugated cholera toxin (FRET donor) and Alexa-Fluor-647-conjugated cholera toxin (FRET acceptor) to randomly label glycosphingolipid molecules. Reduction in FRET efficiency was observed on stretching that recovered upon stretch release, indicating a reversible increase in average spacing between glycosphingolipids within caveolae. Using total internal reflection fluorescence (TIRF), Sinha et al.[40] demonstrated that the number of caveolae along the basal surface of cells expressing Enhanced Green Fluorescent Protein (EGFP)-tagged caveolin1 decreased upon stretching to an extent proportional to the magnitude of stretch.

24.4 Measuring Forces Generated by Stretch

As discussed in Chapter 22, powerful tools are being developed to quantify the localized force production by cells on their adjacent substrate.[41] Recently, devices have been developed to measures changes in these forces in cells subjected to stretch.[22,42] Gavara et al.[22] stretched cells adhering to the top of a collagen gel supported on an elastic silicone rubber membrane. The locations of fluorescent beads on the surface of the gel were imaged at different stages of a stretch cycle. This was repeated after trypsinization to remove the cells. Cell traction forces increased upon stretching the cells, but release of the stretch causes the forces to transiently drop below the baseline levels. These observations are consistent with those reported by Trepat et al.[43] who demonstrated that transient stretching rapidly attenuates cell stiffness followed by slow recovery. This behavior was attributed to fluidization and resolidification of the cytoskeleton.

While collagen networks provide a more physiological environment than the polyacrylamide commonly used for traction microscopy, collagen has nonlinear elastic properties and is susceptible to remodeling by proteolytic degradation, cell contractile forces, and applied stretching forces.[44] Another approach, developed by Krishnan et al.,[42] involves using a punch-indentation system to stretch cells seeded on an elastic polyacrylamide gel. The indenter is mounted coaxial to the objective lens of an inverted microscope and lowered manually to produce a predetermined strain in the underlying substrate. The indentation amplitude and shape of the punch determine the strain pattern produced. An annular punch produces a biaxial strain, whereas a punch composed of two parallel plates produces uniaxial strain. Using this system, Krishnan et al. reported that a stretch and release causes rapid fluidization followed by slow resolidification of the cell cytoskeleton. When subjecting cells to periodic stretching, Krishnan et al.[45] demonstrated that cellular reorientation in response to stretch involves rapid attenuation of traction forces followed by recovery of the traction forces perpendicular to the direction of the imposed stretch. Moreover, reorientation in cell shape systematically lagged reorientation of the traction field.

Recently, Mann et al.[46] designed a relatively simple device for estimating stretch-induced traction forces. The device consists of a microfabricated array of silicone elastomeric microposts integrated onto a stretchable elastomeric membrane. The deformation of the membrane does not propagate into the posts, thus allowing local cell-derived forces to be estimated with simple beam theory.

24.5 Conclusion

While a number of studies have involved imaging cells stretched within 3-D matrices, the vast majority of studies involve imaging cells stretched on relatively stiff and flat, elastic substrates due to the relative ease of the experiments. A major challenge for the future of this field is to expand our understanding of cellular responses to stretch within the context of living tissues. This will require the development of advanced imaging techniques, such as nonlinear multiphoton microscopy,[20] to making imaging cells deep within these tissues practical.

Engineered tissues are often subjected to stretch to mimic physiological loads. Strains are expected to be heterogeneous in a system as simple as cells in a collagen gel.[47] The strains produced within more complex scaffolds consisting of multiple components with different mechanical properties may be very complex, ultimately affecting the way resident cells respond. New methods must be developed to characterize these spatially heterogeneous strains in complex scaffolds in order to understand how these strains are transferred to cells and their intracellular structures.

Quantitative information regarding the spatiotemporal reorganization of cellular structures in response to stretch is necessary to develop mechanical models of cells that accurately predict how cells will remodel in response to different patterns of stretch.[15,25] These models predict the involvement of specific proteins, including myosin motors and focal adhesion proteins, in regulating cellular responses to stretch.[48] The roles of other cytoskeleton-associated proteins, such as zyxin, p130Cas, and α-actinin, in stretch-induced cell remodeling are only beginning to be revealed.[49] Creative methods combining multiple imaging modalities may be necessary to delve deeper. For example, the application of forces at the scale relevant to individual ligand–receptor bonds and the use of TIRF and AFM have revealed that the stretching of single talin rods in focal adhesions exposes cryptic binding sites for vinculin.[50] Improvements in imaging spatial and temporal resolution are needed to elucidate the roles of individual molecules in the dynamic and complex events occurring within cells as they adapt to their mechanical environment.

References

1. Boateng, S. Y. et al. RGD and YIGSR synthetic peptides facilitate cellular adhesion identical to that of laminin and fibronectin but alter the physiology of neonatal cardiac myocytes. *Am J Physiol Cell Physiol* **288**, C30–C38 (2005). doi:10.1152/ajpcell.00199.2004.
2. Banes, A. J., Gilbert, J., Taylor, D. & Monbureau, O. A new vacuum-operated stress-providing instrument that applies static or variable duration cyclic tension or compression to cells in vitro. *J Cell Sci* **75**, 35–42 (1985).
3. Gilbert, J. A., Weinhold, P. S., Banes, A. J., Link, G. W. & Jones, G. L. Strain profiles for circular cell culture plates containing flexible surfaces employed to mechanically deform cells in vitro. *J Biomech* **27**, 1169–1177 (1994).
4. Na, S. et al. Time-dependent changes in smooth muscle cell stiffness and focal adhesion area in response to cyclic equibiaxial stretch. *Ann Biomed Eng* **36**, 369–380 (2008). doi:10.1007/s10439-008-9438-7.
5. Schaffer, J. L. et al. Device for the application of a dynamic biaxially uniform and isotropic strain to a flexible cell culture membrane. *J Orthop Res* **12**, 709–719 (1994). doi:10.1002/jor.1100120514.
6. Hung, C. T. & Williams, J. L. A method for inducing equibiaxial and uniform strains in elastomeric membranes used as cell substrates. *J Biomech* **27**, 227–232 (1994). doi:10.1016/0021-9290(94)90212-7 [pii].
7. Sotoudeh, M., Jalali, S., Usami, S., Shyy, J. Y. & Chien, S. A strain device imposing dynamic and uniform equi-biaxial strain to cultured cells. *Ann Biomed Eng* **26**, 181–189 (1998).
8. Kaunas, R., Nguyen, P., Usami, S. & Chien, S. Cooperative effects of Rho and mechanical stretch on stress fiber organization. *Proc Natl Acad Sci U S A* **102**, 15895–15900 (2005). doi:10.1073/pnas.0506041102 [pii].
9. Huang, L., Mathieu, P. S. & Helmke, B. P. A stretching device for high-resolution live-cell imaging. *Ann Biomed Eng* **38**, 1728–1740 (2010). doi:10.1007/s10439-010-9968-7.
10. Buck, R. C. Reorientation response of cells to repeated stretch and recoil of the substratum. *Exp Cell Res* **127**, 470–474 (1980).
11. Wang, J. H., Goldschmidt-Clermont, P., Wille, J. & Yin, F. C. Specificity of endothelial cell reorientation in response to cyclic mechanical stretching. *J Biomech* **34**, 1563–1572 (2001).
12. Costa, K. D., Hucker, W. J. & Yin, F. C. Buckling of actin stress fibers: A new wrinkle in the cytoskeletal tapestry. *Cell Motil Cytoskeleton* **52**, 266–274 (2002). doi:10.1002/cm.10056.
13. Naruse, K., Yamada, T. & Sokabe, M. Involvement of SA channels in orienting response of cultured endothelial cells to cyclic stretch. *Am J Physiol* **274**, H1532–H1538 (1998).
14. Morioka, M. et al. Microtubule dynamics regulate cyclic stretch-induced cell alignment in human airway smooth muscle cells. *PLoS One* **6**, e26384 (2011). doi:10.1371/journal.pone.0026384.
15. Tondon, A., Hsu, H. J. & Kaunas, R. Dependence of cyclic stretch-induced stress fiber reorientation on stretch waveform. *J Biomech* **45**, 728–735 (2012). doi:10.1016/j.jbiomech.2011.11.012.
16. Cukierman, E., Pankov, R., Stevens, D. R. & Yamada, K. M. Taking cell-matrix adhesions to the third dimension. *Science* **294**, 1708–1712 (2001). doi:10.1126/science.1064829.
17. Baker, B. M. & Chen, C. S. Deconstructing the third dimension: How 3D culture microenvironments alter cellular cues. *J Cell Sci* **125**, 3015–3024 (2012). doi:10.1242/jcs.079509.
18. Nieponice, A., Maul, T. M., Cumer, J. M., Soletti, L. & Vorp, D. A. Mechanical stimulation induces morphological and phenotypic changes in bone marrow-derived

progenitor cells within a three-dimensional fibrin matrix. *J Biomed Mater Res A* **81**, 523–530 (2007). doi:10.1002/jbm.a.31041

19. Nekouzadeh, A., Pryse, K. M., Elson, E. L. & Genin, G. M. Stretch-activated force shedding, force recovery, and cytoskeletal remodeling in contractile fibroblasts. *J Biomech* **41**, 2964–2971 (2008). doi:10.1016/j.jbiomech.2008.07.033.
20. Hu, J. J., Humphrey, J. D. & Yeh, A. T. Characterization of engineered tissue development under biaxial stretch using nonlinear optical microscopy. *Tissue Eng Part A* **15**, 1553–1564 (2009). doi:10.1089/ten.tea.2008.0287.
21. Henshaw, D. R., Attia, E., Bhargava, M. & Hannafin, J. A. Canine ACL fibroblast integrin expression and cell alignment in response to cyclic tensile strain in three-dimensional collagen gels. *J Orthop Res* **24**, 481–490 (2006). doi:10.1002/jor.20050.
22. Gavara, N., Roca-Cusachs, P., Sunyer, R., Farre, R. & Navajas, D. Mapping cell-matrix stresses during stretch reveals inelastic reorganization of the cytoskeleton. *Biophys J* **95**, 464–471 (2008). doi:10.1529/biophysj.107.124180 [pii].
23. Lee, C. F., Haase, C., Deguchi, S. & Kaunas, R. Cyclic stretch-induced stress fiber dynamics—Dependence on strain rate, Rho-kinase and MLCK. *Biochem Biophys Res Commun* **401**, 344–349 (2010). doi:10.1016/j.bbrc.2010.09.046.
24. Goldyn, A. M., Rioja, B. A., Spatz, J. P., Ballestrem, C. & Kemkemer, R. Force-induced cell polarisation is linked to RhoA driven microtubule-independent focal-adhesion sliding. *J Cell Sci* **122**, 3644–3651 (2009). doi:10.1242/jcs.054866.
25. Chen, B., Kemkemer, R., Deibler, M., Spatz, J. & Gao, H. Cyclic stretch induces cell reorientation on substrates by destabilizing catch bonds in focal adhesions. *PLoS One* **7**, e48346 (2012). doi:10.1371/journal.pone.0048346.
26. Caille, N., Tardy, Y. & Meister, J. J. Assessment of strain field in endothelial cells subjected to uniaxial deformation of their substrate. *Ann Biomed Eng* **26**, 409–416 (1998). doi:10.1114/1.132.
27. Kaunas, R., Usami, S. & Chien, S. Regulation of stretch-induced JNK activation by stress fiber orientation. *Cell Signal* **18**, 1924–1931 (2006). doi:10.1016/j.cellsig.2006.02.008.
28. Jungbauer, S., Gao, H., Spatz, J. P. & Kemkemer, R. Two characteristic regimes in frequency-dependent dynamic reorientation of fibroblasts on cyclically stretched substrates. *Biophys J* **95**, 3470–3478 (2008). doi:10.1529/iophysj.107.128611.
29. Xu, F., Beyazoglu, T., Hefner, E., Gurkan, U. A. & Demirci, U. Automated and adaptable quantification of cellular alignment from microscopic images for tissue engineering applications. *Tissue Eng Part C Methods* **17**, 641–649 (2011). doi:10.1089/ten.tec.2011.0038.
30. Katsumi, A. et al. Effects of cell tension on the small GTPase Rac. *J Cell Biol* **158**, 153–164 (2002). doi:10.1083/jcb.200201105 [pii].
31. Hsu, H. J., Lee, C. F., Locke, A., Vanderzyl, S. Q. & Kaunas, R. Stretch-induced stress fiber remodeling and the activations of JNK and ERK depend on mechanical strain rate, but not FAK. *PLoS One* **5**, e12470 (2010). doi:10.1371/journal.pone.0012470.
32. Kemeny, S. F. & Clyne, A. M. A simplified implementation of edge detection in MATLAB is faster and more sensitive than fast Fourier transform for actin fiber alignment quantification. *Microsc Microanal* **17**, 156–166 (2011). doi:10.1017/S143192761100002X [pii].
33. Yoshigi, M., Clark, E. B. & Yost, H. J. Quantification of stretch-induced cytoskeletal remodeling in vascular endothelial cells by image processing. *Cytometry A* **55**, 109–118 (2003). doi:10.1002/cyto.a.10076.
34. Hsu, H. J., Lee, C. F. & Kaunas, R. A dynamic stochastic model of frequency-dependent stress fiber alignment induced by cyclic stretch. *PLoS One* **4**, e4853 (2009). doi:10.1371/journal.pone.0004853.
35. Goldyn, A. M., Kaiser, P., Spatz, J. P., Ballestrem, C. & Kemkemer, R. The kinetics of force-induced cell reorganization depend on microtubules and actin. *Cytoskeleton* **67**, 241–250 (2010). doi:10.1002/cm.20439.
36. Cunningham, J. J., Linderman, J. J. & Mooney, D. J. Externally applied cyclic strain regulates localization of focal contact components in cultured smooth muscle cells. *Ann Biomed Eng* **30**, 927–935 (2002).
37. Sens, P. & Turner, M. S. Budded membrane microdomains as tension regulators. *Phys Rev E Stat Nonlin Soft Matter Phys* **73**, 031918 (2006).
38. Dulhunty, A. F. & Franzini-Armstrong, C. The relative contributions of the folds and caveolae to the surface membrane of frog skeletal muscle fibres at different sarcomere lengths. *J Physiol* **250**, 513–539 (1975).
39. Gervasio, O. L., Phillips, W. D., Cole, L. & Allen, D. G. Caveolae respond to cell stretch and contribute to stretch-induced signaling. *J Cell Sci* **124**, 3581–3590 (2011). doi:10.1242/jcs.084376 [pii].
40. Sinha, B. et al. Cells respond to mechanical stress by rapid disassembly of caveolae. *Cell* **144**, 402–413 (2011). doi:10.1016/j.cell.2010.12.031 [pii].
41. Butler, J. P., Tolic-Norrelykke, I. M., Fabry, B. & Fredberg, J. J. Traction fields, moments, and strain energy that cells exert on their surroundings. *Am J Physiol Cell Physiol* **282**, C595–C605 (2002). doi:10.1152/ajpcell.00270.2001.
42. Krishnan, R. et al. Reinforcement versus fluidization in cytoskeletal mechanoresponsiveness. *PLoS One* **4**, e5486 (2009). doi:10.1371/journal.pone.0005486.
43. Trepat, X. et al. Universal physical responses to stretch in the living cell. *Nature* **447**, 592–595 (2007). doi:10.1038/nature05824.
44. Vader, D., Kabla, A., Weitz, D. & Mahadevan, L. Strain-induced alignment in collagen gels. *PLoS One* **4**, e5902 (2009). doi:10.1371/journal.pone.0005902.
45. Krishnan, R. et al. Fluidization, resolidification, and reorientation of the endothelial cell in response to slow tidal stretches. *Am J Physiol Cell Physiol* **303**, C368–C375 (2012). doi:10.1152/ajpcell.00074.2012.

46. Mann, J. M., Lam, R. H., Weng, S., Sun, Y. & Fu, J. A silicone-based stretchable micropost array membrane for monitoring live-cell subcellular cytoskeletal response. *Lab Chip* **12**, 731–740 (2012). doi:10.1039/c2lc20896b.
47. Stevenson, M. D. et al. Pericellular conditions regulate extent of cell-mediated compaction of collagen gels. *Biophys J* **99**, 19–28 (2010). doi:10.1016/j.bpj.2010.03.041.
48. Kaunas, R., Hsu, H. J. & Deguchi, S. Sarcomeric model of stretch-induced stress fiber reorganization. *Cell Health Cytoskelet* **3**, 13–22 (2011).
49. Hoffman, L. M., Jensen, C. C., Chaturvedi, A., Yoshigi, M. & Beckerle, M. C. Stretch-induced actin remodeling requires targeting of zyxin to stress fibers and recruitment of actin regulators. *Mol Biol Cell* **23**, 1846–1859 (2012). doi:10.1091/mbc.E11-12-1057.
50. del Rio, A. et al. Stretching single talin rod molecules activates vinculin binding. *Science* **323**, 638–641 (2009). doi:10.1126/science.1162912.

Section IX

Cellular and Molecular Imaging Methods

Cellular and Subcellular Biophysics

25

Micropipette Aspiration and Subcellular Biophysics

Baoyu Liu and Jin-Yu Shao

CONTENTS

Subcellular structures adopt unique biophysical properties to suit their specific functions. Quantification of such properties is an indispensable part of understanding the structure–function relationship. In this chapter, we focus on the quantitative techniques with cellular and subcellular capacity that are based on micropipette aspiration.[1] These techniques all include as an essential component a micrometer-size pipette with controlled inner suction pressure. They are widely used for easy manipulation of single live cells and their versatile applicability. Here, we review major techniques of this type, describing their technical details, unique capabilities, and key applications.

25.1 Introduction

Quantitative physical understanding of biological processes is the primary goal of biophysics. For studies of microscale subjects, one of the principal challenges is to perform direct experimental measurements on the subjects of interest. One school of method developed to meet this demand includes ultrasensitive mechanical techniques that integrate micromanipulation and minuscule force application and measurement. Several of these techniques are based on micropipette aspiration.[1] The first micropipette-aspiration technique dates back decades earlier[2] than the well-known atomic force microscopy (AFM)[3] or the optical trap (OT).[4] In 1954, Mitchison and Swann designed a "cell elastimeter," which used micropipette suction to measure surface properties of individual sea urchin egg cells.[2,5] Ten years later, the technique was improved by Rand and Burton to study mechanical properties of red blood cell (RBC) membranes.[6,7] In the 1970s and 1980s, the technique was refined and applied in a myriad of studies in quantifying material constants of subcellular structures or the whole cell. Meanwhile, researchers operationally modified the technique and started using it to quantify adhesion forces or strength of cell–cell interactions such as wheat germ agglutinin (WGA)- or antibody-mediated RBC–RBC binding[8,9] and killer T cell–target cell binding.[10]

In the majority of these early studies, the theme of force application was to aspire a portion of a cell to seal a micropipette and observe the resultant cell projection inside the pipette; mechanical force *per se* was not of particular interest. In only a few studies mentioned earlier,[8–10] the technique was used to measure force. At that time, the force-measuring capacity was fairly limited with ambiguity for force calculation, a low degree of force resolution, and significant uncertainty in force measurement due to nonspecific cell-pipette adhesion; furthermore, the experimental procedure was quite cumbersome in practice. Two key developments[11,12] in the 1990s significantly changed this picture and greatly expanded the technique's general applicability. In 1995, Evans et al.[11] invented the biomembrane force probe (BFP). The BFP uses a micropipette to aspire a soft RBC to form an ultrasensitive force transducer, capable of measuring a wide range of forces from 0.01 to 1000 pN. In the year following the invention of BFP, Shao and Hochmuth[12] introduced the modern version of the micropipette aspiration technique (MAT). In the MAT, instead of sealing the pipette opening, a closely fitting spherical object is placed inside the pipette to act as the force transducer, capable of measuring forces from a few piconewtons to hundreds of nanonewtons. As the force transducer in the BFP and the MAT can be easily coupled with proteins, force application can be directed to specific cellular/subcellular/molecular targets, which opens up a broad spectrum of new applications.[13–40] Since the inception of BFP and MAT, many technical improvements and new functional capabilities have been added,[38–42] making the micropipette aspiration-based techniques powerful tools in the field of biophysics.

25.2 The "Old Fashioned" Whole-Cell Separation

The design principle of a micropipette aspiration-based technique had experienced little change for nearly four decades since the introduction of the "cell elastimeter" in the 1950s. It was an "extremely simple"[2] and effective

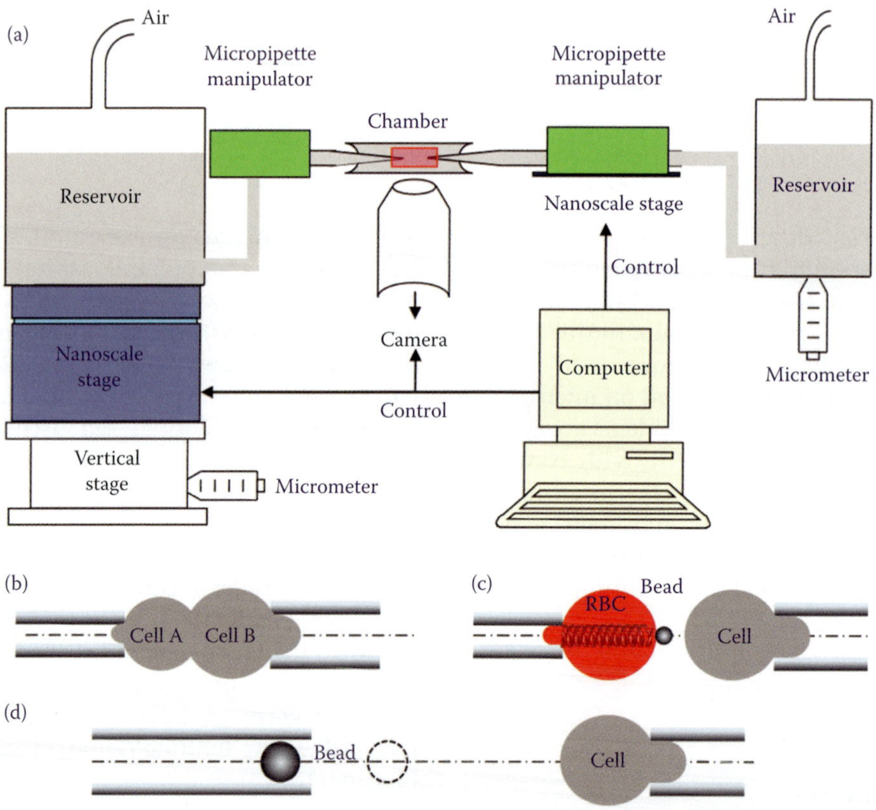

FIGURE 25.1
(a) Micropipette aspiration system. Two micropipettes are inserted into the experimental chamber. Each is controlled by a manual 3D hydraulic micromanipulator connected to a water reservoir. The right manipulator and the left reservoir sit on a horizontal (one-axis) and a vertical piezo stage, respectively, the movement of which is controlled by a computer. (b) Schematic of the SPT. Two cells are brought together to allow conjugation to form and then separated by retraction of one of the pipettes. By gradually increasing the suction pressure, a critical force can be determined that is just enough for the separation. (c) Schematic of the BFP. A glass bead is mounted on a pipette-aspirated RBC to form an ultrasoft spring. The stiffness can be adjusted in situ by adjusting the suction pressure. (d) Schematic of the MAT. The original MAT has a closely fitting bead inside the pipette to act as the force transducer. Placing the bead outside the pipette (position indicated by the dashed circle) enhanced the force measuring capability (to femtonewton level). In all three experimental setups (b, c, and d), the cell on the right can also be replaced by a bead or other substrates coupled with subjects of interest.

design (Figure 25.1a). A glass micropipette, which connects to a water reservoir, holds a cell at its opening with a gentle hydrostatic pressure. By changing the water height with a manometer, different aspiration pressures (with ~0.1 Pa resolution) can be applied to the cell, generating responses in accordance to the cell's intrinsic mechanical properties. The process is observed under a microscope and recorded.

In early applications of the technique, the primary readout was the cell projection length inside the pipette. Under a common light microscope, the tracking resolution of the leading edge of the cell projection is about 25 nm. To convert the simple experimental readout into mechanical properties, theoretical modeling needs to be carried out. Usually, a sophisticated model was formulated specifically for the material parameter of interest, and fitting the model to experimental data yields the material constant.

The micropipette manipulation system provided an economic and convenient means of manipulating single cells under the microscope. With developments like adding three-dimensional (3D) micromanipulators to control the pipette movement, the micropipette manipulation system gained more popularity in single-cell research. Owing to a wide range of pipette size (from ~1 μm up), virtually all types of cells can be manipulated and investigated. The micropipette-based techniques have been successfully applied to the measurement of a variety of material properties of cellular and subcellular structures such as Young's moduli of sea urchin egg surface (~10^3 pN/μm^2)[6,43]; endothelial cells (~4×10^3 pN/μm^2)[44] and chondrocytes (6.5×10^3 pN/μm^2)[45]; RBC membrane moduli for area dilation (0.3 pN/μm), shear (7×10^{-6} pN/μm), and bending (1.8×10^{-19} pN μm)[46–51]; membrane viscosity of RBC (3×10^{-4} pN s/μm)[49] and lipid vesicles (~10^{-6} pN s/μm)[52,53]; shear modulus of bovine endothelial cells (~10^{-4} pN/μm)[54]; and cortical tension (3.5×10^{-5} pN/μm) and cell interior viscosity (~10^2 pN s/μm^2) of human granulocytes.[14,55] These biophysical measurements not only provided valuable material information but also, together with biochemical evidence, helped depict a much clearer picture of the interrelationship among cell deformability, membrane composition, and microstructure.

25.3 Step-Pressure Technique

As cell adhesion research started gathering momentum in the beginning of the 1980s,[56] there was a great need to understand the biophysical nature of cell adhesion. Evans and Leung[9] pioneered the studies of cell–cell adhesion by quantitating cellular adhesion strength with the step-pressure technique (SPT), which was modified and established by Sung et al.[10] a couple of years later. The SPT was based on a typical micropipette manipulation system with two micropipettes for simultaneous manipulation of two cells (Figure 25.1b). In the procedure described by Sung et al., one pipette firmly holds one cell with a large suction pressure, while the other smaller pipette gently aspires the other cell with a smaller pressure. The two cells are brought together by moving the smaller pipette to allow conjugation to form between the cell pair for a defined time period. This is followed by retraction of the small pipette. If strong conjugation was developed during contact, the gently aspirated cell would be pulled out of the pipette. In this case, the procedure is repeated multiple times by step-increasing the suction pressure inside the small pipette until it can hold the cell just tightly enough to completely separate the cell–cell conjugation. The threshold suction pressure was referred to as the critical separation pressure (P_c). The corresponding critical separation force (F_c) or stress (S_c) can be calculated as[10]

$$F_c = 2\pi P_c R_p^2 \quad \text{or} \quad S_c = \frac{2P_c R_p^2}{R_c^2} \tag{25.1}$$

where R_p and R_c are the radii of the smaller pipette and the circular contact area, respectively.

The SPT was the first micropipette-aspiration-based technique used to measure mechanical force.[8–10] Evans and Leung[9] used it to measure the adhesion strength between two RBCs mediated by WGA adhesion in 1984. It was found that the separation force increased approximately tenfold from the start to completion of separation as the contact area diminished. The maximal separation force, which has a strong dependence on WGA concentration, was on the order of several nanonewtons. In 1986, Sung et al.[10] applied the SPT to the junctional avidity between a killer T cell and its target. For human T-cell clone F1 (HLA-DRw6 restricted) and its specific target cell JY (a B lymphoblastoid cell line), the critical separation force and stress were found to be ~6 nN and ~1.5×10^3 pN/μm^2, respectively. They also showed in a following study that interleukin-1 treatment of endothelial cells enhanced their adhesiveness with HL-60 cells (a human blood cell line) via increased expression of E-selectin.[57] Theoretical modeling revealed that the adhesive energy density (defined as the energy per unit conjugation area) increases as the cell–cell conjugation area decreases, and the separation force is concentrated on the conjugation edge.[58] A slightly different version of the SPT, where the firmly aspirated cell is replaced by a planar substrate coated with specific ligands, was adapted to measure the adhesion strengths between phorbol 12-myristate-13-acetate stimulated T lymphocytes and intercellular adhesion molecule-1 (ICAM-1),[59] between Jurkat cells and lymphocyte function-associated antigen-3 (LFA-3),[60] between human ligament fibroblasts and laminin,[61] and between osteoblasts and titanium surfaces.[62] Using adhesion strength as the readout, Sung et al.[63] investigated the role of various cytoskeletal proteins on the adhesion of anterior cruciate ligament (ACL) and medial collateral ligament (MCL) to a fibronectin-coated surface, shedding insight on the ligament-healing process.

The SPT was a novel technique in quantifying adhesion forces between live cells. Other techniques such as the AFM and OT were not readily applicable to studying live cell–cell interactions. Furthermore, the AFM cantilever obstructs the microscopic view during single cell characterization (AFM); the OT usually could not apply sufficiently large forces required for total cell separation. However, a few drawbacks of the SPT also exist. First, force calculation was not straightforward as the calculation formula relies on how the cell body is modeled mechanically. Consequently, there were several formulae (including Equation 25.1) that differ a great deal.[8,9,58,59] Besides, the pressure inside the cell was usually neglected for simplicity. Second, it was assumed that the nonspecific adhesion between the pipette wall and the aspirated cell is negligible, which is an issue that is difficult to assess. All these factors increase the uncertainty in force calculation and lower the force resolution of the technique, making any measurement of very small forces (e.g., piconewton forces) unreliable. As far as data interpretation was concerned, because adhesion was mediated by many receptor-ligand bonds (in hundreds) of sometimes many types, it was difficult if not impossible to pinpoint the underlying molecular mechanism. Moreover, the SPT studies did not consider the possible effect of the separation speed, which was shown later to greatly change the bond strength.[25] Nevertheless, the SPT represented a significant technical advance in its time and made much important contribution to the cell adhesion research.

25.4 Biomembrane Force Probe

When the molecular era of cell adhesion research came of age,[56] force techniques of single bond sensitivity emerged and quickly gained popularity. To study molecular adhesion and probe ultrastructures at the biological membrane surface, Evans et al.[11] developed an ingenious micropipette-based technique called the biomembrane force probe (BFP) in 1995. The BFP uses a micropipette to aspirate a flexible RBC to create an ultrasoft force transducer or "spring" (Figure 25.1c). The spring constant (k_p) is calculated, based on membrane mechanics, from the aspiration pressure and geometries of the setup as[35]

$$k_p = \frac{\pi R_p P_p}{(1 - R_p/R_o)\ln\left[4R_o^2/(R_p R_c)\right]} \tag{25.2}$$

where P_p, R_p, R_o, and R_c are the aspiration pressure and the radii of the probe holding pipette, the outside portion of the RBC, and the RBC–probe contact area, respectively. The spring displacement is monitored by tracking the black edge of a glass bead (probe) mounted on the RBC. To improve tracking efficiency, the BFP tracks a narrow rectangular strip across the bead–RBC interface; with a high-speed digital camera, the displacement can be tracked and recorded online at 1600 frames per second (fps) with a spatial resolution of 3 nm.[42] As the spring constant can be tuned in situ from ~0.1 to 3 pN/nm by adjusting the aspiration pressure,[64] the BFP can easily apply forces from ~0.5 pN to more than 1000 pN.

25.4.1 Dynamic Force Spectroscopy

Using the BFP, the Evans group has pioneered the method of dynamic force spectroscopy (DFS).[25,26,28,30,64,65] DFS is a single-molecule technique that maps the energy landscape of a receptor–ligand bond. Due to its finite lifetime, a molecular bond spontaneously dissociates at any level of constant force if held for sufficient time. However, when forced to rupture at faster than spontaneous rates, molecular bonds resist dissociation and exhibit strength manifested by characteristic rupture forces. Mechanical force lowers energy barriers along the force-driven dissociation pathway, causing faster bond separation. Because of exponential dependence of kinetic rates on the barrier energies, DFS experiments are most revealing when done over many logs of force loading rates (force/time), delineating an inner picture of molecular interaction difficult or even impossible to probe by other means.

The BFP is ideal for such application as it has a superb capacity of applying piconewton forces spanning a vast range of loading rates from 10^{-1} to 10^5 pN/s.[26] By controlling receptor–ligand interaction at infinitely low concentration (decorating the probe bead with a controlled small number of molecules of interest), the BFP is capable of probing single molecular bonds one at a time. In a typical DFS experiment, the ligand-coupled probe bead is kept in place while a cell or another bead (called target) bearing the corresponding receptor is computer-driven to approach, contact, and retract from the probe bead in automated repetitive cycles. The contact is kept brief to prevent multibond formation. Due to the stochastic nature of single bond behavior, bond strength or rupture force at any given loading rate is collected from multiple rupture events and presented as a distribution histogram, from which most probable rupture forces are derived. The signature graph of the DFS is obtained by plotting rupture forces against the logarithm of loading rates. Fitting theoretical models to such data yields the dissociation rate and the characteristic length of each energy barrier.[26,64]

The BFP has been used to explore energy landscapes of a variety of molecular pairs, including (strept) avidin–biotin,[25] lipid anchoring in membrane bilayers,[34,66] selectins and their ligands,[67,68] β_2 integrin and ICAM-1,[33] and cell surface receptor dissociation from the cytoskeleton.[35] For receptor–ligand interactions with a single energy barrier, there is a simple linear relationship in the rupture force vs. log(loading rate) graph; on the other hand, multiple linear segments suggest multiple energy barriers along the force-driven dissociation pathway. DFS analysis of single bonds has brought to the attention of the biophysics community the quantitative relationship between the bond strength and the force loading rate, which has far-reaching implications in cell–cell adhesion and other biological phenomena involving molecular interactions in a dynamic force environment.

25.4.2 Direct Measurement of the Bond Lifetime

The derivation of bond lifetime from the DFS relies on modeling of molecular bonds. However, due to limited knowledge of bond nature, it is possible that current theoretical modeling cannot account for all essential features pertaining to bond lifetime determination. Thus, direct measurement of bond lifetime also has its practical appeal. In this regard, the BFP has been used as a platform to implement a couple of assays to measure the lifetime of single molecular bonds.[39,42]

The thermal fluctuation assay was designed to measure the bond lifetime in a force-free environment.[39] This assay utilizes thermal fluctuation of the probe bead and the fact that an additional linkage to the target via a molecular bond reduces the magnitude of the probe bead's thermal fluctuation. The experimental procedure resembles that of a DFS experiment with the exception that the target retraction stops at the point where the impinging force just vanishes. At such a close proximity between the probe and the target, receptors and ligands on opposing surfaces bind and dissociate driven by thermal fluctuation. The bond lifetime is defined as the period from reduction to resumption of bead fluctuation. In addition to the bond lifetime (the reciprocal of the off-rate), one can also measure the on-rate from waiting times in between lifetime events. Yet another variation of the DFS assay is the force clamp assay,[38] which directly measures the bond lifetime at different clamped force levels. Here, the target retraction stops when the loading force reaches a preset value and the force is clamped until spontaneous bond dissociation, and then the bond lifetime is measured as the force clamp period. These BFP assays have been applied to receptor–ligand interactions such as P- and L-selectin, P-selectin and P-selectin glycoprotein ligand-1 (PSGL-1),[39] OT1 TCR and its specific ligands,[23] and LFA-1 and ICAM-1.[24,38] One important finding was that the lifetime of LFA-1 on the cell surface binding to ICAM-1 was prolonged by force application, a counterintuitive phenomenon called *catch bond*.[69] This demonstrates unexpected complexity of molecular interactions as it cannot be explained with the commonly accepted model of molecular bond dissociation assuming that force shortens the bond lifetime as it lowers the energy barrier of dissociation.

25.5 Micropipette Aspiration Technique

Unlike the BFP that is based on membrane mechanics, the MAT has its foundation in fluid mechanics.[12,37] The MAT uses a closely fitting latex bead in a micropipette as the force transducer (Figure 25.1d). The pressure in the bead-containing pipette generates fluid flow around the bead. If no force is applied, the bead moves freely at a certain velocity V_f; when an external force is applied after the bead adheres to the target (another bead or a cell) at the opposing side, the bead moves at a slower velocity (V). The force (F) can be calculated by[12]

$$F = \pi R_p^2 \Delta p \left(1 - \frac{4}{3}\varepsilon\right)\left(1 - \frac{V}{V_f}\right) \tag{25.3}$$

where R_p, Δp, and $\varepsilon = (R_p - R_b)/R_p$ are the pipette radius, the aspiration pressure, and the relative pipette/bead radius difference (R_b is the bead radius), respectively. The process of theoretical derivation for Equation 25.3 has been updated to improve accuracy if higher orders of ε are to be included.[40] With a vertical nanostage controlling the water reservoir height, the MAT has a pressure resolution of 0.001 Pa (pN/μm^2), corresponding to a theoretical force resolution of 0.05 pN for a force transducer of typical size (8 μm in diameter). The force range of the MAT is from a few piconewtons to ~100 nN.

In a typical MAT experiment, a ligand-coated bead is moved by a positive lung pressure (against a preset suction pressure) to the pipette opening to gently touch the target. After a short contact, the positive pressure is released, leaving the suction pressure to generate a point pulling force on the target via receptor–ligand bonds. If no adhesion occurs, the bead moves back freely; in the case of adhesion, the bead either stops or moves back at a slower velocity, which is followed by free motion after bond rupture. With a two-dimensional (2D) tracking algorithm,[70] the bead movement can be monitored at a high resolution of 1–2 nm, allowing precise measurement of not only the bead velocity but substrate deformation as well. Using the MAT, direct cell–cell interaction can be readily studied by replacing the transducer bead with a live round cell.[71]

25.5.1 Technical Improvements and Analyses

The formula for the force calculation of the MAT was derived from the lubrication theory.[12] The theory assumes that the pipette–bead gap is much smaller than the pipette radius ($\varepsilon \ll 1$), and the bead moves concentrically in the pipette. For the former concern, it is important to measure the radii of the pipette and the bead precisely to guarantee its validity and minimize uncertainty in using Equation 25.3. Due to optical diffraction, direct measurement under a light microscope comes with a large error. To overcome this problem, Shao and Xu[41] calibrated the pipette diameter (directly measured by differential interference contrast [DIC] microscopy) against that by electron microscopy and showed that the actual pipette diameter is the DIC measurement divided by a correction coefficient (1.120). They also provided a simple method for measuring the bead diameter, which makes use of a bead triplet configuration. Assuming that direct bright field measurements of three center-to-center distance are l_1, l_2, and l_3 (each representing the center-to-center distance between the other two beads), the diameter of bead #1 is given by $l_2 l_3 / l_1$.

The second assumption of concentric bead movement could become invalid in experiments. For example, as bonds can form at any local spot in the bead-substrate contact area (on the order of 1 μm²), it cannot be guaranteed that the point force aligns perfectly with the pipette axis. This could cause deviation of the bead center from the pipette axis. To evaluate the off-axis effect on force calculation, Chen at al.[40] performed computer simulations with a 3D nonsymmetrical model. Varying the distance between the bead center and the pipette axis from 0% to 80% of the maximal gap ($R_p - R_b$) causes only a minor deviation (less than 1%) of simulated force from that calculated from Equation 25.3. The same study further looked at the possible influence of the presence of the substrate outside the pipette opening, which was not considered in the lubrication theory. It was found that, for a typical sized substrate (e.g., a 10 μm live cell), Equation 25.3 has more than 95% accuracy as long as the substrate is positioned more than 2 μm away from the pipette opening.

In the original design of the MAT, only a constant suction pressure and thus a constant force can be applied in each approach–contact–retraction cycle, which has limited general applicability of the MAT. Chen et al.[40] replaced the original manual manometer with a motorized piezo stage such that varying forces can be applied. As the stage movement is programmable, essentially any pattern of force signal is achievable. The range of force loading rate that can be applied with the modified MAT is from several to tens of thousands of piconewtons per second.

25.5.2 Membrane Tether Extraction

The MAT has played an important role in quantifying tether extraction from leukocyte and endothelial cells.[12–15,17–19,40,41,72] Tethers are membranous nanotubes extruded from the cell surface, usually as a result of point force pulling. Tether extraction is a critical phenomenon in the transmigration process of leukocytes during infection.[73,74] Traveling leukocytes are first captured from the blood stream and roll on the endothelium as mediated by selectins (L-selectin on neutrophils and P- and E-selectin on the endothelium). Tethers from leukocytes and endothelial cells stabilize rolling by lowering the disruptive hydrodynamic force on selectin bonds and increasing the duration of the leukocyte–endothelium interaction. Therefore, it is imperative to quantify how fast tethers grow under specified forces. One early and simple model (constitutive equation) of tether extraction is[41,75]

$$F = F_0 + 2\pi\mu_{\mathrm{eff}}\, U_{\mathrm{t}} \tag{25.4}$$

where F_0, μ_{eff}, and U_{t} are the minimal force required to extract a tether (mainly determined by membrane–cytoskeleton adhesion), the effective viscosity (mainly determined by interbilayer slip and membrane slip over cytoskeleton), and the growth velocity of tether length, respectively. The first application of the MAT was to study tether extraction from human neutrophils.[12] Using neutrophils interacting with beads coated with antibodies for L-selectin, the β_2 integrins, or CD45, it was found that F_0 is ~45 pN. For cells like neutrophils with a large membrane reservoir (stored in microvilli), F_0 has a simple relation to the adhesion energy per unit area (γ_{t}) of the membrane–cytoskeleton adhesion ($\gamma_{\mathrm{t}} = F_0^2/(8\pi^2 B)$, where B is the membrane bending modulus).[75] Then γ_{t} can be estimated to be ~130 pN/μm (B ~ 0.2 pN μm).[76] The effective viscosity μ_{eff} for human neutrophils was later determined to be ~1.8 pN μm/s.[41] Xu and Shao[19] showed that human CD4+ T lymphocytes have almost identical tether extraction parameters (F_0 and μ_{eff} are ~45 pN and 1.6 pN μm/s, respectively),[19] suggesting a common mechanical characteristic of tether extraction for leukocytes.

In parallel with the leukocyte studies, the F–U_t relationships for human endothelial cells were also obtained.[13,18] For three endothelial cell types—umbilical vein (HUVEC) and adult or neonatal dermal microvascular endothelial cells—F_0 and μ_{eff} are all ~50 pN and ~0.5 pN μm/s, respectively. In these experiments, tethers were extracted from cells with beads coupled with antibodies for cell surface receptors. The similar F_0 values for leukocytes and endothelial cells predicted extraction of simultaneous tethers (one from an endothelial cell and the other from a leukocyte connected by selectin bonds in between) during leukocyte rolling on the endothelium of which one cell is used as the force transducer of the MAT. Moreover, in simultaneous tether extraction, endothelial cells should contribute more to the composite tether length as their effective viscosity is much smaller compared with leukocytes. Indeed, simultaneous tether extraction was demonstrated by fluorescent membrane labeling with a neutrophil acting as the force transducer interacting with an endothelial cell.[17] As expected, F_0 (46 pN) is similar and μ_{eff} (0.39 pN μm/s) is smaller than that of single tether extraction from either cell type. Simulation studies showed that simultaneous tether extraction is more effective in stabilizing rolling than single tether extraction.[17,77] In addition to single tethers, it is also possible to extract double tethers in parallel from either leukocytes or endothelial cells, which was shown to have F_0 and μ_{eff} double of those of single tether extraction.[15,19]

Tether extraction is an intrinsic membrane protrusion process in response to point force application that is dictated by cell membrane property and membrane–cytoskeleton interactions. Consistent with this view, the obtained material parameters (F_0 and μ_{eff}) do not depend on what cell surface molecules used to extract tethers[12,13,15,18,19,41] or whether endothelial cells are attached to a substrate or suspended in solution.[13,15] Furthermore, pharmaceutical treatments (e.g., latrunculin A or cytochalasin D) that disrupt cytoskeletal integrity decrease the effective viscosity approximately fourfold for neutrophils[41] and approximately tenfold for endothelial cells.[13] In comparison, stimulatory agents (e.g., IL8, PMA, TNF-α, or IL-1β) increase the threshold force twofold and decrease the effective viscosity threefold for neutrophils[19,41] but did not affect those of endothelial cells.[13,15] All the aforementioned tether extraction experiments were performed at room temperature. Compared with room temperature experiments, pulling tethers from neutrophils at 37°C produces a twofold lower effective viscosity (no significant effect on the threshold force).[14] It is thus possible that tethers are extracted more easily at our body temperature for both leukocytes and endothelial cells, which is advantageous to their function of rolling stabilization.

The aforementioned tether studies were conducted with the MAT of constant force capability where one F–U_t relationship requires tether extraction measurements at multiple forces at different suction pressures and linear regression of the experimental data. To improve efficacy of tether extraction experiments, Chen et al.[72] used the motorized stage to control the suction pressure to apply changing forces with the MAT. In this manner, a single pulling cycle with a linearly ramping pressure yields a complete F–U_t curve. With this capability, it was possible to examine tether extraction data from a new perspective. For HUVECs, F_0 increased from ~40 to 80 pN and μ_{eff} decreased from ~1.6 to 0.2 pN μm/s as U_t increased from 2 to 60 μm/s. The shear thinning phenomenon (μ_{eff} decreasing with rising U_t; due to the small range of U_t, shear thinning was not obvious in previous MAT experiments) agrees with tether extraction experiments by the BFP.[32] Using the programmable stage, it was also possible to measure forces at negative U_t. To capture all characteristics of tether extraction, Chen et al. proposed a new phenomenological constitutive equation:

$$F = \alpha(U_t - U_0)^\beta \tag{25.5}$$

where α and β are two constants and U_0 is the tether velocity when $F = 0$. This new equation is more consistent with the available data for tether extraction than the power law relationship proposed by Heinrich et al.,[32] as it not only explains the shear thinning phenomenon but also accounts for the fact that a finite force exists at $U_t = 0$ and U_t may be negative.

Besides tether extraction, the MAT has seen many other applications of quantifying the mechanical properties of human neutrophils, including the resistance of squeezing through small capillaries,[16] the anchoring strength of surface receptors to membrane and cytoskeleton,[20] microvillar length,[22] cortical tension,[14,41] and surface protrusion.[14] These mechanical characterizations of leukocytes and endothelial cells have contributed to a better understanding of the complex leukocyte–endothelium interaction and its important role in the leukocyte transmigration process.

25.5.3 Femtonewton Force Application

In all published findings with the MAT, only piconewton or higher levels of forces have been applied. In 2012, the Shao group proposed a novel modification to the MAT to enable femtonewton force application.[37] The idea was to move the transducer bead to the outside of the aspirating micropipette. In this configuration, the inward fluid flow generated by the suction pressure inside the micropipette should impose a hydrodynamic force of a

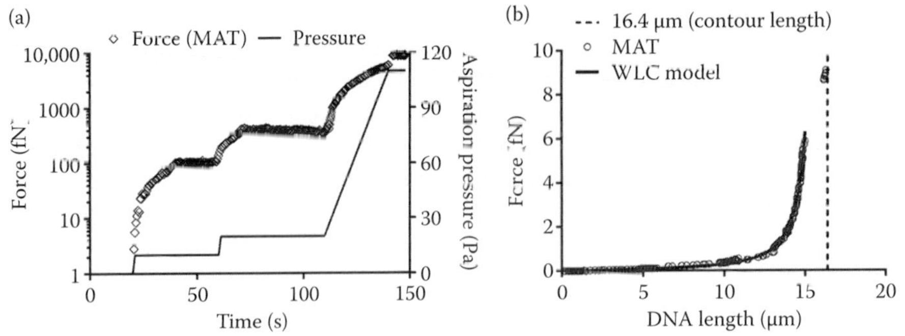

FIGURE 25.2
Stretching two DNA molecules in parallel with the MAT. Two beads were connected by lambda DNA molecules, one of which was used as the transducer on the left and the other was held stationary on the right (cf. Figure 25.1d). The transducer bead was initially ~20 μm from the left pipette opening. An aspiration pressure (panel a, solid line, right ordinate) pulled the transducer to the left, applying forces on the DNAs at the same time (panel a, open circles, left ordinate). At ~140 s, the force jumped to a new equilibrium, likely indicating the dissociation of one DNA molecule. The WLC model (panel b, solid line) was fitted to the force vs. DNA length data prior to the force jump (panel b, open circles), yielding a persistent length of 28.7 nm—half of that for a single DNA, suggesting that two DNA molecules in parallel were stretched. After the force jump, the DNA length was close to the contour length of a single lambda DNA (indicated with the vertical dashed line in panel b), further supporting the interpretation that one DNA dissociated at the force jump and the other DNA was still stretched after the force jump.

much smaller magnitude than that of the original MAT. In principle, force can be as small as one wishes simply by increasing the distance between the pipette opening and the bead (D_{bp}). The concept has been verified by finite-element simulations based on a 2D axisymmetric model. For example, if R_p, R_b, and D_{bp} are all equal to 2 μm, the force on the bead is ~100 fN for a suction pressure of 0.2 Pa; increasing D_{bp} to 6 μm corresponds to a force of 30 fN. In reality, deviation from an axisymmetric configuration is inevitable even for experienced experimenters. To evaluate the effects of a nonsymmetrical configuration on force calculation, a 3D finite-element model was developed (unpublished data). There are two major factors of deviation, namely, the offset of the bead center from the pipette axis (O_{bp}) and the tilt angle of the pipette (θ_p). 3D simulations showed that as long as O_{bp} is less than 25% of R_b and θ_p is smaller than 10°, the relative error of force calculated by 2D vs. 3D models is within 5%.

Recently, Liu et al. experimentally verified the femtonewton capability of the MAT by applying it to stretching single DNA molecules (unpublished data). In this experiment, linearized Lambda DNA molecules were anchored between two beads. One of the beads was used as the force transducer and the other was held stationary (cf. Figure 25.1d). By controlling the suction pressure using the motorized vertical stage, forces of varying magnitude were applied to the transducer, which in turn extended the DNA molecules in between the two beads (Figure 25.2). According to the worm-like-chain (WLC) model,[78] the force–DNA extension relationship follows:

$$\frac{F\xi}{k_BT} = \frac{1}{4}\left[\left(1-\frac{x}{l_0}\right)^{2} - 1\right] + \left(\frac{x}{l_0}\right) \quad (25.6)$$

where ξ, k_B, T, x, and l_0 are the persistent length, the Boltzmann constant, the absolute temperature, the DNA extension, and the contour length, respectively.

One example of stretching two DNA molecules is shown in Figure 25.2. There was a sudden force jump at around 140 s, after which the bead-to-bead distance (16.3 μm) was close to the contour length of a Lambda DNA (16.5 μm), consistent with the data in the literature.[78] Furthermore, fitting the WLC model to the data prior to the force jump yielded a persistent length of 28 nm. This is half of that for a single DNA strand, strongly indicating that two DNA molecules were pulled in parallel initially and the force jump was likely caused by dissociation of one of the two DNA molecules, ending with one DNA molecule stretched to the contour length at ~9 pN. Most importantly, this example illustrates the MAT's capability of applying femtonewton forces (Figure 25.2a, open circles). For instance, as the DNA extension was fluctuating around an equilibrium length of 5.28 μm from 40 to 60 s, force was ~105 fN with a small standard deviation of 2 fN. With this technical update, the MAT can now apply a wide range of forces (spanning seven logs from several femtonewtons to hundreds of nanonewtons) with just a single experimental setup, which is unprecedented to our knowledge.

25.6 Conclusions

Micropipette aspiration-based techniques have made significant contributions in the field of cellular, subcellular, and molecular biophysics. We believe that this trend will continue in the future as researchers strive to improve the existing techniques and invent new or better alternatives. Two major obstacles of using such techniques are a low degree of automation and a low throughput nature of the experiments. We have made great progress pertaining to the first aspect as more high-tech automated components are incorporated into the manipulation system, and many sophisticated software tools are developed to control tedious and complicated experimental procedures and data analysis.[38,39,41,79,80] However, one cell/molecule at a time is still the common strategy in such experiments, confining the techniques largely as a laboratory usage. One possible breakthrough could involve novel designs that will successfully marry unique features of micropipette techniques with high-throughput microfluidic devices, which should promise exciting technological advances and possible solutions of many important biological questions.

Acknowledgments

This work was supported by grants from the National Institutes of Health and the American Heart Association.

References

1. Shao, J. Y., Xu, G. & Guo, P. Quantifying cell-adhesion strength with micropipette manipulation: Principle and application. *Front Biosci* 9, 2183–2191 (2004).
2. Mitchison, J. M. & Swann, M. M. The mechanical properties of the cell surface: I. The cell elastimeter. *J Exp Biol* 31, 443–460 (1954).
3. Binnig, G., Quate, C. F. & Gerber, C. Atomic force microscope. *Phys Rev Lett* 56, 930–933 (1986).
4. Ashkin, A., Dziedzic, J. M., Bjorkholm, J. E. & Chu, S. Observation of a single-beam gradient force optical trap for dielectric particles. *Opt Lett* 11, 288 (1986).
5. Mitchison, J. M. & Swann, M. M. The mechanical properties of the cell surface: II. The unfertilized sea-urchin egg. *J Exp Biol* 31, 461–472 (1954).
6. Rand, R. P. Mechanical properties of the red cell membrane: II. Viscoelastic breakdown of the membrane. *Biophys J* 4, 303–316 (1964).
7. Rand, R. P. & Burton, A. C. Mechanical properties of the red cell membrane: I. Membrane stiffness and intracellular pressure. *Biophys J* 4, 115–135 (1964).
8. Tha, S. P. & Goldsmith, H. L. Interaction forces between red cells agglutinated by antibody. III. Micromanipulation. *Biophys J* 53, 677–687 (1988).
9. Evans, E. & Leung, A. Adhesivity and rigidity of erythrocyte membrane in relation to wheat germ agglutinin binding. *J Cell Biol* 98, 1201–1208 (1984).
10. Sung, K. L., Sung, L. A., Crimmins, M., Burakoff, S. J. & Chien, S. Determination of junction avidity of cytolytic T cell and target cell. *Science* 234, 1405–1408 (1986).
11. Evans, E., Ritchie, K. & Merkel, R. Sensitive force technique to probe molecular adhesion and structural linkages at biological interfaces. *Biophys J* 68, 2580–2587 (1995).
12. Shao, J. Y. & Hochmuth, R. M. Micropipette suction for measuring piconewton forces of adhesion and tether formation from neutrophil membranes. *Biophys J* 71, 2892–2901 (1996).
13. Chen, Y., Girdhar, G. & Shao, J. Y. Single membrane tether extraction from adult and neonatal dermal microvascular endothelial cells. *Am J Physiol Cell Physiol* 292, C1272–C1279 (2007).
14. Liu, B., Goergen, C. J. & Shao, J. Y. Effect of temperature on tether extraction, surface protrusion, and cortical tension of human neutrophils. *Biophys J* 93, 2923–2933 (2007).
15. Girdhar, G., Chen, Y. & Shao, J. Y. Double-tether extraction from human umbilical vein and dermal microvascular endothelial cells. *Biophys J* 92, 1035–1045 (2007).
16. Shao, J. Y. & Hochmuth, R. M. The resistance to flow of individual human neutrophils in glass capillary tubes with diameters between 4.65 and 7.75 mu m. *Microcirculation-London* 4, 61–74 (1997).
17. Girdhar, G. & Shao, J. Y. Simultaneous tether extraction from endothelial cells and leukocytes: Observation, mechanics, and significance. *Biophys J* 93, 4041–4052 (2007).
18. Girdhar, G. & Shao, J. Y. Membrane tether extraction from human umbilical vein endothelial cells and its implication in leukocyte rolling. *Biophys J* 87, 3561–3568 (2004).
19. Xu, G. & Shao, J. Y. Double tether extraction from human neutrophils and its comparison with CD4(+) T-lymphocytes. *Biophys J* 88, 661–669 (2005).
20. Shao, J. Y. & Hochmuth, R. M. Mechanical anchoring strength of L-selectin, beta(2) integrins, and CD45 to neutrophil cytoskeleton and membrane. *Biophys J* 77, 587–596 (1999).
21. Hochmuth, R. M., Shao, J. Y., Dai, J. W. & Sheetz, M. P. Deformation and flow of membrane into tethers extracted from neuronal growth cones. *Biophys J* 70, 358–369 (1996).
22. Shao, J. Y., Ting-Beall, H. P. & Hochmuth, R. M. Static and dynamic lengths of neutrophil microvilli. *Proc Natl Acad Sci U S A* 95, 6797–6802 (1998).
23. Huang, J. et al. The kinetics of two-dimensional TCR and pMHC interactions determine T-cell responsiveness. *Nature* 464, 932–936 (2010).
24. Chen, W., Lou, J., Evans, E. A. & Zhu, C. Observing force-regulated conformational changes and ligand dissociation from a single integrin on cells. *J Cell Biol* 199, 497–512 (2012).
25. Merkel, R., Nassoy, P., Leung, A., Ritchie, K. & Evans, E. Energy landscapes of receptor-ligand bonds explored with dynamic force spectroscopy. *Nature* 397, 50–53 (1999).
26. Evans, E. Looking inside molecular bonds at biological interfaces with dynamic force spectroscopy. *Biophys Chem* 82, 83–97 (1999).

27. Evans, E. & Kinoshita, K. In: *Methods in Cell Biology*, edited by W. Yu-Li & E. D. Dennis, Vol. 83, pp. 373–396. Academic Press, Waltham, MA (2007).
28. Evans, E. A. & Calderwood, D. A. Forces and bond dynamics in cell adhesion. *Science* 316, 1148–1153 (2007).
29. Simon, S. I. et al. Dynamics of neutrophil membrane compliance and microstructure probed with a micropipet-based piconewton force transducer. *Ann Biomed Eng* 35, 595–604 (2007).
30. Evans, E. & Ritchie, K. Dynamic strength of molecular adhesion bonds. *Biophys J* 72, 1541–1555 (1997).
31. Kinoshita, K., Leung, A., Simon, S. & Evans, E. Long-lived, high-strength states of ICAM-1 bonds to beta2 integrin, II: Lifetimes of LFA-1 bonds under force in leukocyte signaling. *Biophys J* 98, 1467–1475 (2010).
32. Heinrich, V., Leung, A. & Evans, E. Nano- to microscale dynamics of P-selectin detachment from leukocyte interfaces. II. Tether flow terminated by P-selectin dissociation from PSGL-1. *Biophys J* 88, 2299–2308 (2005).
33. Evans, E., Kinoshita, K., Simon, S. & Leung, A. Long-lived, high-strength states of ICAM-1 bonds to beta2 integrin, I: Lifetimes of bonds to recombinant alphaL-beta2 under force. *Biophys J* 98, 1458–1466 (2010).
34. Evans, E., Heinrich, V., Ludwig, F. & Rawicz, W. Dynamic tension spectroscopy and strength of biomembranes. *Biophys J* 85, 2342–2350 (2003).
35. Evans, E., Heinrich, V., Leung, A. & Kinoshita, K. Nano- to microscale dynamics of P-selectin detachment from leukocyte interfaces. I. Membrane separation from the cytoskeleton. *Biophys J* 88, 2288–2298 (2005).
36. Liu, B. Y., Yu, Y., Yao, D. K. & Shao, J. Y. A direct micropipette-based calibration method for atomic force microscope cantilevers. *Rev Sci Instrum* 80, 065109 (2009).
37. Shao, J. Y. Finite element analysis of imposing femtonewton forces with micropipette aspiration. *Ann Biomed Eng* 30, 546–554 (2002).
38. Chen, W., Lou, J. & Zhu, C. Forcing switch from short- to intermediate- and long-lived states of the alphaA domain generates LFA-1/ICAM-1 catch bonds. *J Biol Chem* 285, 35967–35978 (2010).
39. Chen, W., Evans, E. A., McEver, R. P. & Zhu, C. Monitoring receptor-ligand interactions between surfaces by thermal fluctuations. *Biophys J* 94, 694–701 (2008).
40. Chen, Y., Liu, B., Xu, G. & Shao, J. Y. Validation, in-depth analysis, and modification of the micropipette aspiration technique. *Cell Mol Bioeng* 2, 351–365 (2009).
41. Shao, J. Y. & Xu, J. A modified micropipette aspiration technique and its application to tether formation from human neutrophils. *J Biomech Eng* 124, 388–396 (2002).
42. Chen, W., Zarnitsyna, V. I., Sarangapani, K. K., Huang, J. & Zhu, C. Measuring receptor-ligand binding kinetics on cell surfaces: From adhesion frequency to thermal fluctuation methods. *Cell Mol Bioeng* 1, 276–288 (2008).
43. Hiramoto, Y. Mechanical properties of sea urchin eggs. I. Surface force and elastic modulus of the cell membrane. *Exp Cell Res* 32, 59–75 (1963).
44. Theret, D. P., Levesque, M. J., Sato, M., Nerem, R. M. & Wheeler, L. T. The application of a homogeneous half-space model in the analysis of endothelial cell micropipette measurements. *J Biomech Eng* 110, 190–199 (1988).
45. Jones, W. R. et al. Alterations in the Young's modulus and volumetric properties of chondrocytes isolated from normal and osteoarthritic human cartilage. *J Biomech* 32, 119–127 (1999).
46. Evans, E. A. Bending elastic modulus of red blood cell membrane derived from buckling instability in micropipet aspiration tests. *Biophys J* 43, 27–30 (1983).
47. Evans, E. A., Waugh, R. & Melnik, L. Elastic area compressibility modulus of red cell membrane. *Biophys J* 16, 585–595 (1976).
48. Hochmuth, R. M. & Evans, E. A. Extensional flow of erythrocyte membrane from cell body to elastic tether. I. Analysis. *Biophys J* 39, 71–81 (1982).
49. Hochmuth, R. M., Wiles, H. C., Evans, E. A. & McCown, J. T. Extensional flow of erythrocyte membrane from cell body to elastic tether. II. Experiment. *Biophys J* 39, 83–89 (1982).
50. Chien, S., Sung, K. L., Skalak, R., Usami, S. & Tozeren, A. Theoretical and experimental studies on viscoelastic properties of erythrocyte membrane. *Biophys J* 24, 463–487 (1978).
51. Evans, E. A. New membrane concept applied to the analysis of fluid shear- and micropipette-deformed red blood cells. *Biophys J* 13, 941–954 (1973).
52. Waugh, R. E. Surface viscosity measurements from large bilayer vesicle tether formation. II. Experiments. *Biophys J* 38, 29–37 (1982).
53. Waugh, R. E. Surface viscosity measurements from large bilayer vesicle tether formation. I. Analysis. *Biophys J* 38, 19–27 (1982).
54. Sato, M., Levesque, M. J. & Nerem, R. M. An application of the micropipette technique to the measurement of the mechanical properties of cultured bovine aortic endothelial cells. *J Biomech Eng* 109, 27–34 (1987).
55. Evans, E. & Yeung, A. Apparent viscosity and cortical tension of blood granulocytes determined by micropipet aspiration. *Biophys J* 56, 151–160 (1989).
56. Horwitz, A. R. The origins of the molecular era of adhesion research. *Nat Rev Mol Cell Biol* 13, 805–811 (2012).
57. Sung, K. L. P., Saldivar, E. & Phillips, L. Interleukin-1β induces differential adhesiveness on human endothelial cell surfaces. *Biochem Biophys Res Commun* 202, 866–872 (1994).
58. Tozeren, A., Sung, K. L. & Chien, S. Theoretical and experimental studies on cross-bridge migration during cell disaggregation. *Biophys J* 55, 479–487 (1989).
59. Tözeren, A. et al. Micromanipulation of adhesion of phorbol 12-myristate-13-acetate-stimulated T lymphocytes to planar membranes containing intercellular adhesion molecule-1. *Biophys J* 63, 247–258 (1992).
60. Tözeren, A. et al. Micromanipulation of adhesion of a Jurkat cell to a planar bilayer membrane containing lymphocyte function-associated antigen 3 molecules. *J Cell Biol* 116, 997–1006 (1992).
61. Sung, K. L. P., Steele, L. L., Whittermore, D., Hagan, J. & Akeson, W. H. Adhesiveness of human ligament fibroblasts to laminin. *J Orthop Res* 13, 166–173 (1995).
62. Nugiel, D. J., Wood, D. J. & Sung, K. L. Quantification of adhesiveness of osteoblasts to titanium surfaces in vitro by the micropipette aspiration technique. *Tissue Eng* 2, 127–140 (1996).

63. Sung, K. L. et al. The differential adhesion forces of anterior cruciate and medial collateral ligament fibroblasts: Effects of tropomodulin, talin, vinculin, and alpha-actinin. *Proc Natl Acad Sci U S A* 93, 9182–9187 (1996).
64. Evans, E. Probing the relation between force—lifetime—and chemistry in single molecular bonds. *Annu Rev Biophys Biomol Struct* 30, 105–128 (2001).
65. Evans, E. Energy landscapes of biomolecular adhesion and receptor anchoring at interfaces explored with dynamic force spectroscopy. *Faraday Discuss* 111, 1–16 (1998).
66. Evans, E. & Ludwig, F. Dynamic strengths of molecular anchoring and material cohesion in fluid biomembranes. *J Phys: Condens Matter* 12, A315–A320 (2000).
67. Evans, E., Leung, A., Heinrich, V. & Zhu, C. Mechanical switching and coupling between two dissociation pathways in a P-selectin adhesion bond. *Proc Natl Acad Sci U S A* 101, 11281–11286 (2004).
68. Evans, E., Leung, A., Hammer, D. & Simon, S. Chemically distinct transition states govern rapid dissociation of single L-selectin bonds under force. *Proc Natl Acad Sci U S A* 98, 3784–3789 (2001).
68. Marshall, B. T. et al. Direct observation of catch bonds involving cell-adhesion molecules. *Nature* 423, 190–193 (2003).
70. Gelles, J., Schnapp, B. J. & Sheetz, M. P. Tracking kinesin-driven movements with nanometre-scale precision. *Nature* 331, 450–453 (1988).
71. Levin, J. D., Ting-Beall, H. P. & Hochmuth, R. M. Correlating the kinetics of cytokine-induced E-selectin adhesion and expression on endothelial cells. *Biophys J* 80, 656–667 (2001).
72. Chen, Y., Yao, D. K. & Shao, J. Y. The constitutive equation for membrane tether extraction. *Ann Biomed Eng* 38, 3756–3765 (2010).
73. Ley, K., Laudanna, C., Cybulsky, M. I. & Nourshargh, S. Getting to the site of inflammation: The leukocyte adhesion cascade updated. *Nat Rev Immunol* 7, 678–689 (2007).
74. Sundd, P., Gutierrez, E., Koltsova, E. K., Kuwano, Y., Fukuda, S., Pospieszalska, M. K., Groisman, A. & Ley, K. "Slings" enable neutrophil rolling at high shear. *Nature* 488, 399–403 (2012).
75. Hochmuth, R. & Marcus, W. Membrane tethers formed from blood cells with available area and determination of their adhesion energy. *Biophys J* 82, 2964–2969 (2002).
76. Hwang, W. C. & Waugh, R. E. Energy of dissociation of lipid bilayer from the membrane skeleton of red blood cells. *Biophys J* 72, 2669–2678 (1997).
77. Yu, Y. & Shao, J.-Y. Simultaneous tether extraction contributes to neutrophil rolling stabilization: A model study. *Biophys J* 92, 418–429 (2007).
78. Bustamante, C., Marko, J. F., Siggia, E. D. & Smith, S. Entropic elasticity of lambda-phage DNA. *Science* 265, 1599–1600 (1994).
79. Shojaei-Baghini, E., Zheng, Y. & Sun, Y. Automated micropipette aspiration of single cells. *Ann Biomed Eng* 41(6), 1208–1216 (2013).
80. Zhang, X. et al. Controlled aspiration and positioning of biological cells in a micropipette. *IEEE Trans Biomed Eng* 59, 1032–1040 (2012).

26

Scanning Probe Investigation of Cellular and Subcellular Biomechanics

Sirimuvva Tadepalli, Keng-Ku Liu, and Srikanth Singamaneni

CONTENTS

The introduction of the atomic force microscope (AFM) is often regarded as one of the most important milestones in fields of nanoscience, nanotechnology, and surface science. AFMs enabled new possibilities in imaging and probing surface, interfacial, mechanical, and chemical properties of soft and biological materials with unprecedented spatial and temporal resolution. Specific interactions, which form the fundamental character of most biological materials, have been probed using bioconjugated AFM cantilevers as force transducers. These investigations provide new insights that are often lost and inaccessible in conventional ensemble measurements. In this chapter, we briefly review AFM-based surface force spectroscopy (SFS) as applied to cellular and subcellular components. Following a brief introduction to SFS, we highlight several examples from recent literature, where AFMs provide new insight into receptor–ligand, cell–cell, and cell–surface interactions. We also briefly discuss recent reports that demonstrate the application of AFM-based nanomechanical measurements of individual cells for cancer detection.

26.1 Introduction

Devices and methods to probe the mechanics of cellular and subcellular components have been a longstanding interest. Even before the invention of atomic force microscope (AFM), devices such as "cell poker" were specially designed and employed to probe the mechanics of cells growing on a culture dish.[1,2] The introduction of AFM has completely changed the landscape of soft matter imaging and probing. Surface force spectroscopy (SFS), based on AFM, is a powerful method to probe the nanomechanical and adhesive properties of surfaces, such as elastic modulus, adhesion, chemical binding, intermolecular/intramolecular forces, selective interactions, chemical composition, relaxation times, and resilience. Modified SFS techniques are also extremely useful for electrical and thermal characterization of materials. SFS is a form of force spectroscopy (also known as static force spectroscopy) that is limited to probing surfaces (limited to the closest subsurfaces), as opposed to more complicated force spectroscopic measurements, such as deforming and indenting structures, or molecular chain pulling.

A so-called "pulling-off" version of SFS is widely utilized for the investigation of protein unfolding, segment flexibility, molecular weight, brush stretching, and other tensile-related mechanical properties of individual molecules or surface structures and aggregates, and typically requires special tip modification with selective binding groups. Detailed discussion of this subject can be found elsewhere.[3–6] An important variation of force spectroscopy is dynamic force measurement, which involves the oscillation of a cantilever at a few kilohertz (much smaller than the resonance frequency of the cantilever) to obtain a simultaneous map of topography, relative stiffness, adhesion, and electrostatic properties much faster than conventional, static force spectroscopy.

This chapter provides a brief overview of SFS based on atomic force microscopy and its application to probing cellular and subcellular biomechanics. Rather than providing an extensive literature review, we discuss general principles involved in such measurements and highlight specific examples to illustrate the application of SFS in probing cell mechanics, cell–cell, and cell–surface adhesion.

26.2 Introduction to Force Spectroscopy

As a surface-based technique, SFS based on AFM is well suited to study the mechanical and adhesion properties of materials spanning a broad range of stiffness (glassy polymers [a few gigapascals] to highly swollen gels [a few tens of pascals]). SFS in different modes is particularly useful in characterizing the material properties of biological structures exhibiting multiscale hierarchical structure, where the contributions of individual components and their interaction can be quantified and spatially mapped. Furthermore, SFS can be used to spatially map the mechanical properties of complex biological structures under controlled environment (e.g., temperature, solvent, and pH).

Force spectroscopy is a tedious multistep process, which should be done with great care to ensure accurate results. Without care, force spectroscopy measurements can be very inaccurate and misleading. Therefore, it is important to fully understand the process and the sources of error. Furthermore, like many experimental methods, practice and experience with reference samples with known properties are invaluable for successful probing and analysis. This section is focused on a brief discussion of SFS measurement techniques. Specifically, the main emphasis is on obtaining quantitative data, analyzing the data, and avoiding common pitfalls associated with this method, with coverage on basics and capabilities. A comprehensive, in-depth review of force spectroscopy including contact theories and complex force curve interpretation can be found in some reviews.[7,8]

26.2.1 Force–Distance Curves

A single force curve, commonly called a force–distance curve (FDC), is a plot of tip-sample force versus piezoelement movement (Figure 26.1a).[9,10] Figure 26.1a and b

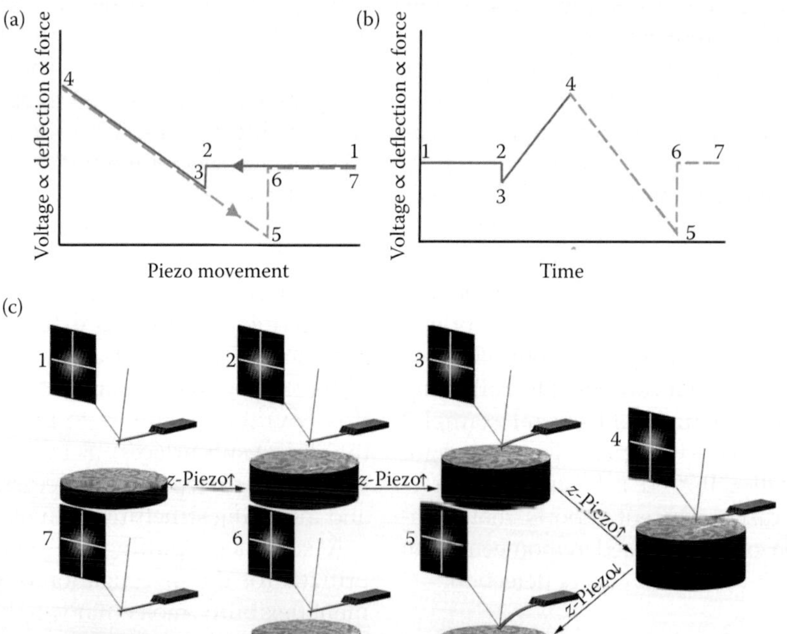

FIGURE 26.1
(a) An ideal FDC as explained in the text. (b) The deflection data from (a) plotted with respect to time. (c) A schematic explaining the different regions of the force curve. Note that all the numbers in a–c correspond to each other. It is also important to carefully note that in the schematic, the laser spot, cantilever deflection, and sample height (piezoelement position) correspond to their positions on the force curve. (Reprinted from McConney, M.E. et al., *Polym. Rev.*, 50 (3), 235–286, 2010. With permission.)

shows the plotting of deflection data with displacement and with time, respectively. Figure 26.1c is a schematic explaining the particular regions of force curves in terms of piezoelement movements, cantilever deflection, and surface deformation.

In Figure 26.1a, an ideal FDC is plotted in the conventional trace–retrace manner. The x-axis is the distance between the tip and the surface and/or a measure of the interaction between the sample and the probe. The larger the piezoelement position value is, the farther the probe is from the surface. First, the vertical piezoelement is moved in the extension direction, which is depicted in the solid line in Figure 26.1a and b. In the curve, line 1–2 is called the extension zero-line, which corresponds to the region when the sample is not in contact with the tip but is moving toward the probe. Line 2–3 corresponds to the "jump to contact" region (also known as the "snap-to" region), when the probe is initially attracted to the sample surface, thereby bending the cantilever downward. In this region, the surface might be also slightly deformed toward the tip. As the piezoelement moves the sample surface closer to the tip, the cantilever passes the zero deflection and is further bent up. This is an unstable region, like the snap-from region, which is not monitored, and in the minimum point, the balance of vertical forces is reached and the tip rests onto a slightly deformed local region (the contact point).

The deflection of the cantilever, when in physical contact with the sample surface, is indicated by the line 3–4. This region is linear for purely elastic deformation with a slope directly related to surface stiffness. For infinitely stiff substrates utilized for sensitivity calibrations (see Figure 26.1), the slope is 1 due to the fact that the cantilever deflection is exactly equal to the piezoelement displacement. In the case of time-dependent surface deformation (viscoelasticity phenomenon), nonuniform deformation, or plastic deformation, this region becomes highly nonlinear. Such an instrumentational factor as piezoelement hysteresis in this region, which can be critical for stiff materials probed with very high forces, is usually barely noticeable and not generally very critical for soft materials.

Point 4 indicates the end of the piezoelement extension sequence and the beginning of the retraction sequence. Ideally, the lines 3–4 and 4–5 will partially overlap and have the same slope during extension and retraction. Generally, line 5–6 represents the force of adhesion, "pulling forces," or the "snap-from-contact" region. It is vertical in ideal cases but can display complex shapes in special cases (e.g., "saw-toothed" unfolding events). Line 6–7 is a region where the cantilever is once again free from contact with the surface and rests in the neutral position.

It is important to note that the applied force is only indirectly measured by the AFM via the cantilever deflection. The microscope itself only monitors the movement of the laser spot on the photodiode, which is related to the cantilever deflection via the photodiode sensitivity. The photodiode sensitivity, expressed in units of nm/V, must be calibrated for each set of measurements. Cantilever deflection is directly related to the applied force via the cantilever spring constant, which is expressed in units of N/m or nN/nm. The cantilever spring constant must also be calibrated for each set of force measurements. The elastic modulus of the sample can be deduced from the FDCs using one of the various contact mechanics models. Three popular models that are frequently employed for data analysis are the Hertzian, Sneddons, and Johnson–Kendall–Roberts (JKR) models, each more suitable than the others for different conditions. In general, the elastic modulus of the sample being elastically indented is related to the indentation depth, h, by

$$E \propto ch^{\alpha} \tag{26.1}$$

where E is the elastic modulus, c is a coefficient that is a function of the cantilever spring constant and tip radius, and α is a constant that depends on the contact model chosen for analysis.

26.2.2 Force Mapping Mode

Force mapping (sometimes referred to as force volume) is a spatial map of FDCs collected across the selected surface area by pixel-to-pixel motion. This FDC matrix can be used for sampling statistics, as well as for relating surface features to mechanical and chemical (adhesion) properties. FDCs are obtained by monitoring the applied force while extending the vertical piezoelement to diminish the probe-sample distance until the cantilever is deflected by a set amount (trigger), followed by retraction of the vertical piezoelement. Unless otherwise stated, for the following discussion, the convention of a sample-scanning AFM is used, where the vertical piezoactuator moves the sample with respect to the tip, as opposed to tip-scanning AFM instruments.

26.3 Probing Cell Mechanics

26.3.1 Single-Molecule Force Spectroscopy on Cell Surface

As discussed above, AFM has piconewton force sensitivity and nanometer positional accuracy, which make it an indispensible tool for exploring the forces and the dynamics of the interaction between individual ligands and receptors, either on isolated molecules or on cellular surfaces. The adhesive forces between the AFM tip and the live cells can be measured by modifying the tip with specific biomolecules.[11–13] Single-molecule force spectroscopy lies on the principle of modifying the AFM tip by attaching a biomolecule to study the adhesion on a cell surface or to transfer molecules. In order to avoid the tip sample repulsive forces, the entire process needs to be done either in vacuum or in the presence of water.

Numerous functionalization strategies have been developed to attach biomolecules to AFM tips for single-molecule force spectroscopy measurements. Some of the common functionalization methods include (i) gold-coated tip with biotinylated bovine serum albumin (BBSA) modified with streptavidin and attached to biotinylated protein; (ii) gold-coated tip with alkanethiol terminated ethylene glycol modified with NTA-alkanethiol and attached to histidine-tagged protein; and (iii) silicon tip modified with silanes and attached with hetero-bifunctional polyethylene glycol (PEG) using 2-pyridyldithiopropionyl (PDP) and *N*-hydroxysuccinimide (NHS) chemistry (Figure 26.2).[12]

One particularly promising approach for binding biomolecules to an AFM tip, which has consistently yielded reliable results, employs PEG as a spacer, which is terminated by NHS on one end and a maleimide group on the other end. Michael addition between the maleimide group and a thiol group (either naturally present on the biomolecule or introduced by mutagenesis) of the biomolecule is used to bind the biomolecule to the tip. A recent report by Zimmermann et al.[14] provides a detailed protocol for this immobilization procedure. The PEG spacer chains (typically a few tens of nanometers long), which are commonly employed for immobilizing biomolecules, space the biomolecules from the surface of the tip. This arrangement preserves the natural conformation of the immobilized biomolecule, significantly reducing nonspecific adsorption onto the tip, providing necessary conformation freedom, and more importantly

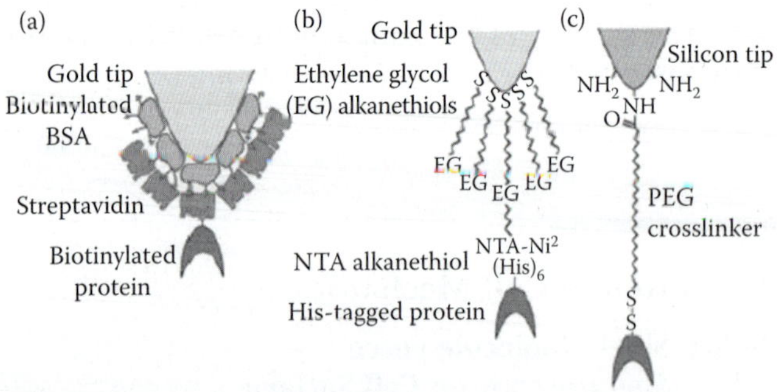

FIGURE 26.2
Strategies commonly used for modifying AFM tips for single molecular recognition studies; (a) physisorption of proteins such as biotinylated BSA; (b) chemisorption of alkanethiols on gold; and (c) covalent coupling of silanes on silicon oxide. (Reprinted from Hinterdorfer, P., Dufrêne, Y., *Nat. Methods*, 3 (5), 347, 2006. With permission.)

serves as an internal control for force measurements. Extension of the PEG (unraveling the polymer chain) during the retraction of the AFM tip results in a characteristic distance in the FDC indicating force curves with specific interaction events among other unwanted non-specific interactions.

In a typical SFS measurement, as the modified tip approaches the surface, the cantilever may bend owing to the presence of repulsive forces like electrostatic, hydrostatic, or steric forces. As the tip is brought in contact with the substrate, the binding takes place. When the tip is retracted from the surface, the curve shows a hysteresis corresponding to the adhesion "pull-off" force during which the unbinding takes place. In molecular recognition studies, this value represents the unbinding forces between complementary receptors and ligand molecules. However, the interpretation of the unbinding forces is complicated by the presence of nonspecific binding.

The change in the slope of the curve during retraction, which is due to a decreased effective spring constant, indicates specific binding (Figure 26.3a). This indicates that the cantilever is relaxed during the beginning of the retraction, and as the retraction process proceeds, the cantilever, protein, and ligand are stretched. In contrast to this, nonspecific binding exhibits a constant slope during retraction as shown in Figure 26.3b.[15]

The loading rate plays a critical role in determining the unbinding forces. The unbinding force usually scales with the logarithmic of the loading rate. The receptor–ligand bonds have limited lifetimes, and the lifetime is shortened when a force is applied to a bond owing to thermal activation. The spontaneous dissociation of the bond $\tau(0)$ determines the lowest time in which dissociation can take place without resistance. When a bond is detached at time spans greater than $\tau(0)$, the resistance to detachment makes the unbinding force measurable. The force acting on the complex lowers the dissociation barrier and lowers the bond lifetime[12]:

$$\tau(F) = \tau(0)\exp\left(\frac{-x_\beta F}{k_B T}\right) \tag{26.2}$$

where x_β is the thermally averaged projection of the energy barrier along the direction of the force, $\tau(F)$ is the bond lifetime under force F, k_B is the Boltzmann constant, and T is the absolute temperature. Dynamic rupture force spectroscopy is a mode in which the unbinding forces of the receptor–ligand complex are measured as a function of the actual loading rate. This approach is useful in assessing the kinetic parameters of the unbinding process like the height of the energy barrier and the number of barriers.

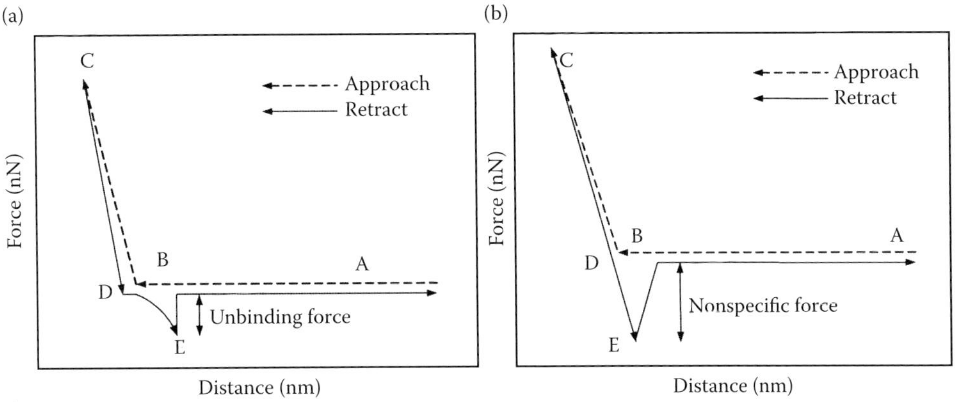

FIGURE 26.3
(a) A schematic diagram of a typical specific force measurement curve, using an AFM to measure the force required to separate individual intermolecular protein–ligand (protein) interaction. Approach (dash line): as the piezoscanner moves toward the AFM tip at a constant velocity from A to B, and then where the AFM tip comes into contact with the surface at point B. As the scanner continues approaching the tip, the cantilever bends upward until it reaches point C. The retract (solid line): occurs when the tip reaches point C, at which the piezoscanner moves away from the AFM tip and the cantilever then begins to retract. During the retraction process, the tip adheres to the surface due to the interaction between the tip and the sample and results in a change in the slope of the curve (points C–D to D–E). As the scanner continues to retract, the cantilever is bent downward until it reaches point E. The tip sample starts to break from point E. Finally, the cantilever then returns to its original equilibrium state at point A. (b) A schematic diagram of a typical nonspecific force measurement curve. (A) A change in the slope of the curve (points C–D and D–E) during the retraction process. The curve of (B) retains the same slope (points C–E) during the process of retraction. (Reprinted from Lee, C. et al., *Micron*, 38 (5), 446–461, 2007. With permission.)

We will briefly discuss a few examples of SFS on proteins and subcellular components. In the first example, fibronectin was investigated with SFS on a gold-coated substrate under water.[16] The stretching range for fibronectin molecules was selected to be around 250 nm. The experiment was repeated under different speeds from 0.01 to 10 nm/ms to investigate the rate dependence of the unfolding events. Sawtooth-like FDCs were observed with a rupture force of 145 pN and a contour length of 285 nm. Analysis of the unfolding events for fibronectin domain showed that this length corresponds to a domain length composed of 90 amino acids. Proteins with different fibronectin domains and polyproteins composed of individual domains were investigated as well in order to confirm the suggestions (Figure 26.4).[17] It was found that multidomain molecules showed several histogram peaks for the force and length that are in good agreement with theoretical models. The worm-like chain (WLC) analysis produced the persistent length of 0.4–2.5 nm depending on the domain sequence.

In a related study, Meadows et al.[18] conducted SFS for the same proteins deposited on a mica or glass surface and dried. The experiment was done in a water or phosphate-buffered saline (PBS) environment with cantilevers having spring constants from 30 to 70 pN/nm. The Z-range was from 60 to 300 nm, and the pulling speed was 1 nm/ms. The authors found that the isolated molecules have a tendency to denature and the contour length tends to be 12 nm. Aggregation of proteins in densely packed films on some surface areas did not affect their unfolding behavior.

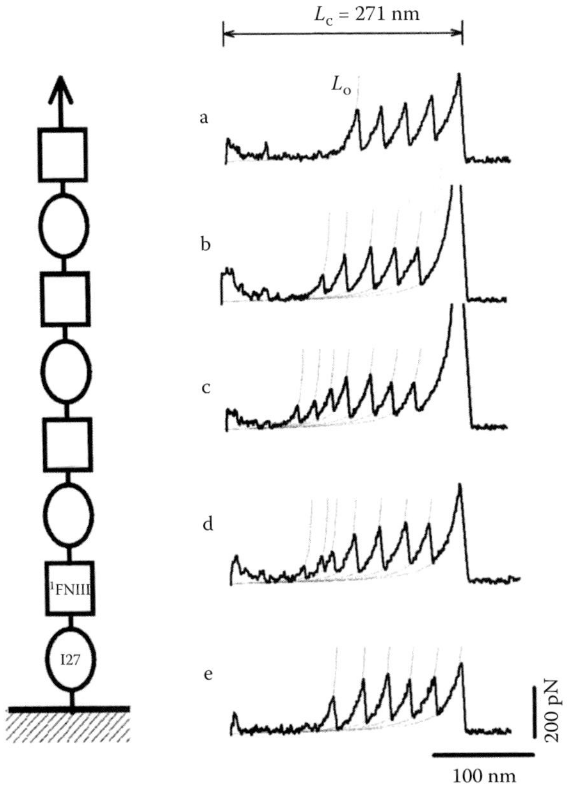

FIGURE 26.4
FNIII (fibronectin) FDC, domain model, and fitting (thin lines) by WLC model. (a–e) Force extension relationship for an (I27-FNII) polyprotein. All contour lengths are around 280 nm, which is close to the expected value. (Reprinted from Oberhauser, A.F. et al., *J. Mol. Biol.*, 319 (2), 433–447, 2002. With permission.)

Gutsmann et al.[19] conducted extensive SFS experiments for collagen molecules. They used AFM cantilevers with a spring constant of ~25 mN/m and different pulling speeds of 7 and 14 nm/ms. It was found that the distribution of rupture events (contour lengths) possesses two peaks. The authors explained them as the presence of two different domains, and it was noticed that AFM imaging shows a domain with a size close to the size of a single peak on the FDC data.

Another important protein, spectrin is a cytoskeletal protein composed of two subunits, α- and β-chain, which form laterally associated heterodimers. Spectrin is on the intracellular side of the plasma membrane of many cell types playing an important role in the maintenance of the mechanical integrity of the plasma membrane and cytoskeletal structure. Repeat units of this protein fold into triple-helical coiled coils composed of 106 amino acid residues. The SFS approach was employed by Rief et al.[20] to mechanically unfold these repeats, and the force required to unfold was found to be 25 to 35 pN. The stretching of the protein was performed at two different speeds: 0.08 and 0.8 μm/s. As discussed above, it was observed that protein unfolding is a rate-dependent process. Pulling at a higher speed (0.8 μm/s) required higher unfolding forces compared to a lower speed (0.08 μm/s). Monte Carlo simulations of unfolding events exhibited excellent quantitative agreement with the observed force–displacement curves. This example demonstrates that AFM force measurements can also provide insight into the dynamic nature of the mechanical behavior of the protein molecules.

SFS was employed to probe the interaction between $\alpha_5\beta_1$ integrin and its ligand, fibronectin (FN), as it plays a vital role in cell differentiation, proliferation, and migration between a cell and the extracellular matrix (ECM). Studies on this system showed that the dynamic rupture force of $\alpha_5\beta_1$/FN7-10 complex revealed two regimes of loading corresponding to the inner and outer activation energies of the complex.[21] These regimes correspond to the fast- and slow-loading rates, respectively, establishing the Arrhenius relationship of the bond lifetime as shown in Equation 26.2. Work done on these interactions on vascular smooth muscle cells (VSMCs) suggested that with the increase in contact time, the force required to dislodge the fibronectin-coated bead on VSMCs increased indicating that the number of adhesion bonds increased with contact time.[22] Further, research showed that histamine treatment increased the bond strength of fibronectin interactions.[23]

26.3.2 Cellular Mechanics

Living cells and their condition are regulated by the mechanical environment on the cells. So, the deformation characteristics provide important information about their structural and biological functions. The resistance of a cell to the deformation from the mechanical environment is dependent on the elastic modulus of the cell, which is three orders of magnitude lower when compared to engineering materials as shown in Figure 26.5a. The variability in the mechanical rigidity of the cell arises from the presence of integral protein molecules and cholesterol in the hydrophilic head and the hydrophobic tails as shown in Figure 26.5b.[24]

There are various models that describe the reason how external forces influence the live cells. They include disruption of the signal generated by the cell in the presence of external forces, alteration in the native physiological state of the cell by disruption in the normal functioning causing a diseased state, and generation of delusionary signals, which might affect the cell functioning. Discher et al.[25] proposed that the behavior of the cell and the mechanical properties depend on the substrate on which it is placed. SFS was used to demonstrate that tip indentation caused changes in elasticity of osteoblasts during the cell cycle.[26] SFS can also be used to determine the turgor pressure of bacteria by relating it to the depth of indentationof the AFM tip.[27,28]

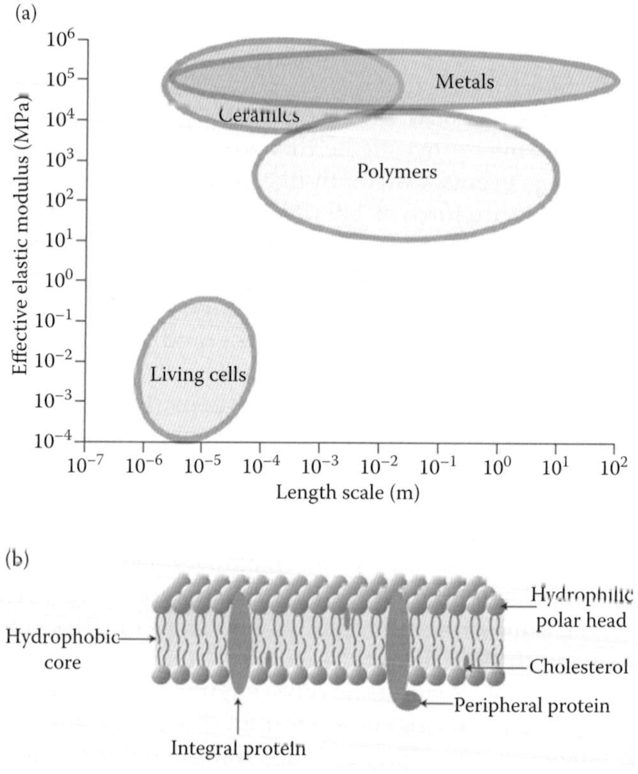

FIGURE 26.5
(a) Approximate elastic modulus when compared to engineering materials. (b) Schematic of the membrane of the cell with the phospholipid layer. (Reprinted from Bao, G., Suresh, S., *Nat. Mater.*, 2 (11), 715–725, 2003. With permission.)

When an AFM tip approaches the surface of the cell, the indentation can be considered to be in a homogenously elastic surface by a rigid probe and can be assumed to follow the Hertz model[29]:

$$F = kd = \frac{\pi E\varphi(\delta)}{(1-v^2)} \tag{26.3}$$

where E is the elastic modulus φ, (δ) is a function of the tip dimension, and ν is the Poisson's ratio.

When the cell surface is approached by the AFM tip, the tip velocity is an important factor determining the force–distance plot of cells.[30] With the increase in the tip velocity, the hysteresis loss increases in the FDCs of cardiac apical cells, as shown in Figure 26.6, caused by viscous losses. Similar curves have been generated for endothelial cells and skeletal muscle cells. It is to be noted that the force retraced to zero after the cycle, demonstrating that there is no permanent deformation of the cell surface.

The mechanical behavior of the cells is an important factor that determines the condition of the cell. As cancer is a highly complex biochemical phenomenon, the change in stiffness of the cells in a diseased state, which can be probed using an AFM, proves to be a useful "marker" to predict the state of a live cell.

The stiffness of the cell shows a direct correlation with the state of the cell. However, this behavior is different in a single cell and in live tissue. Several studies indicate that single isolated cancer cells have lower elastic modulus (E) than healthy cells, and the skewness of the elastic modulus distribution constitutes a reliable fingerprint of the diseased population.[31,32] On the contrary, several studies have indicated that tumor cells are stiffer than healthy cells, and continuous increase in stiffness is a promoter of cancer.[33–35]

Although a lot of work has been done in this direction, there are certain limitations that pose difficulties in interpreting these results in a clinical background.

i. A single cell behaves mechanically different when compared to a group of cells or tissue due to the absence of interactions.
ii. The modulus of elasticity of the cell is influenced by the substrate on which it is placed substantially causing error in the interpretation.

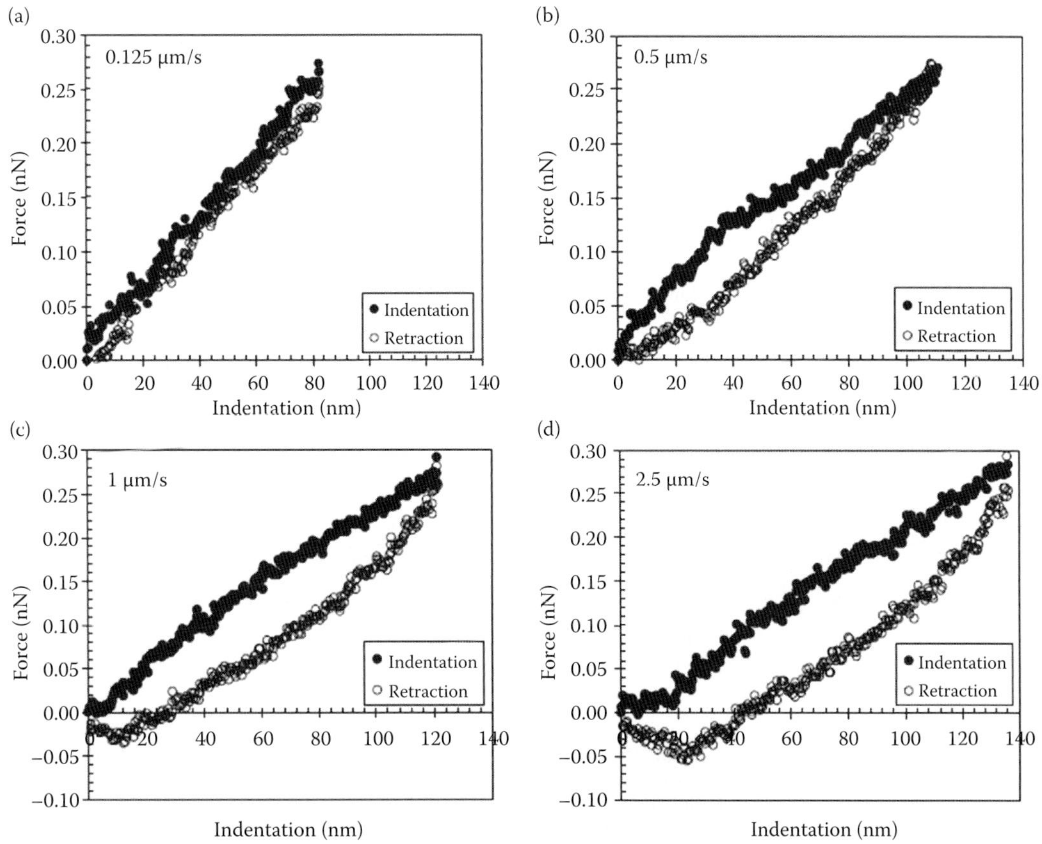

FIGURE 26.6
FDCs on the cardiac apical surface for different tip velocities (a) 0.125 μm/s, (b) 0.5 μm/s, (c) 1 μm/s, and (d) 2.5 μm/s showing that an increase in tip velocity increases the hysteresis of the curves. (Reprinted from Mathura, A. et al., *J. Biomech.*, 34 (12), 1545–1553, 2001. With permission.)

iii. The modulus of elasticity of the cell varies along the surface at the nucleus and in the ECM.
iv. Studies suggest that tumor (cultured) cells are softer than their healthy counterparts.[36]

The error in the interpretation due to the presence of the substrate can be overcome by introducing a correction factor to the Sneddon's model[37]:

$$F = \frac{8E\tan(\theta)\delta^2}{3\pi} \times K_{BECC} \tag{26.4}$$

where F is the applied force, δ is the indentation, θ is the half opening angle of the cone, and K_{BECC} is the correction factor introduced by the bottom effect cone correction model.

To understand the stiffness paradox described in this section, variation of E across the cell surface needs to be understood using a smaller length scale. A recent work on nanomechanical properties of human breast biopsies using indentation-type AFM suggests that stiffness of the cell across the cell surface varies significantly. The normal tissue showed a sharp peak of stiffness distribution,

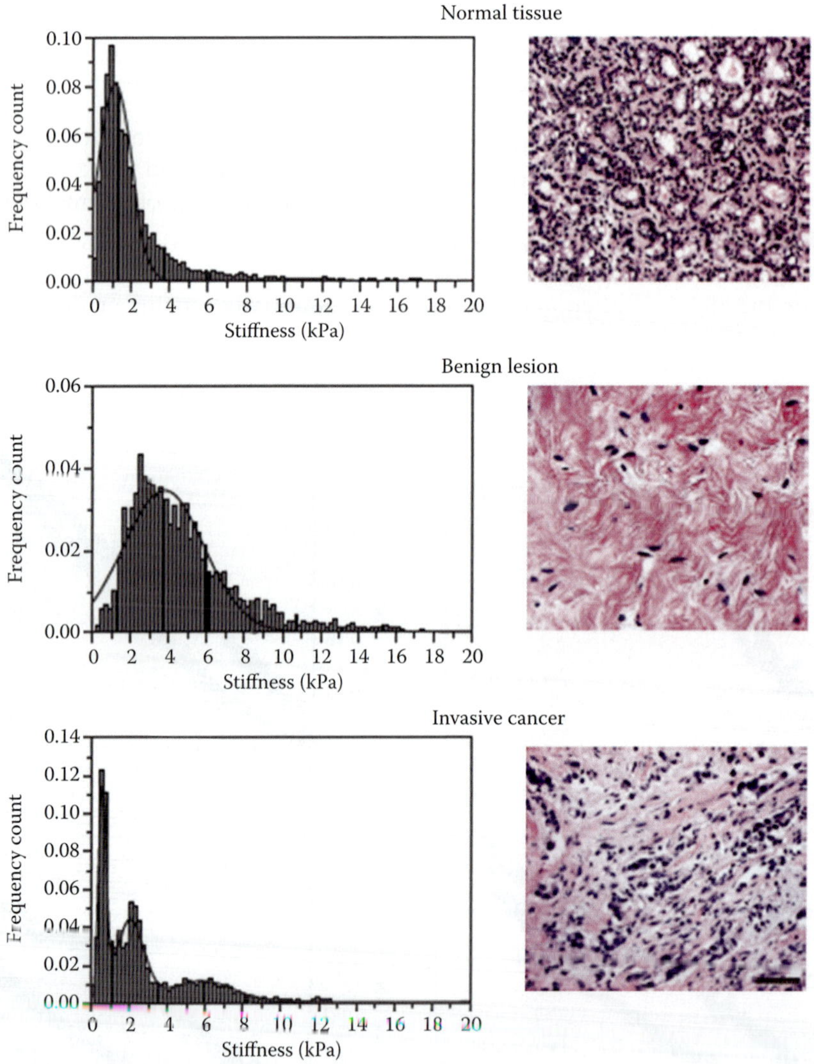

FIGURE 26.7
Top left: Stiffness distribution for normal mammary gland tissue is unimodal. Top right: Post-AFM hematoxylin and eosin (H&E) stained section reveals the terminal ductal lobular unit of a normal mammary gland fenced by interstitial connective tissue. Middle left: Biopsy-wide histogram for a benign lesion reveals a unimodal but broader stiffness distribution with an increase in stiffness compared with the healthy biopsy. Middle right: H&E-stained section reveals extensive fibrotic stroma interspersed with fibroblasts typical for fibroadenoma. Bottom left: Heterogeneous stiffness distribution with a characteristic soft peak for malignant tumor tissue is consistent with histopathology (bottom right), revealing an invasive breast carcinoma with infiltrating nests of cancer cells that have evoked a dense fibrous tissue response. Scale bar to all images, 50 μm. (Reprinted from Plodinec, M. et al., *Nat. Nanotechnol.*, 7 (11), 757–765, 2012. With permission.)

and the benign lesion cells broadened the peak with the progress of cancer. Invasive cancer showed a bimodal peak of stiffness distribution as shown in Figure 26.7. This work suggested that cancer progression is not only restricted to ECM stiffening but also associated with a corresponding softening of tumor epithelial cells when compared to normal mammary epithelial cells.[38] This distinctive softening of tumor epithelial cells probably went unnoticed because the colloidal probe used cannot provide sufficient resolution to distinguish individual tissue components.

Although these studies indicate that AFM is a reliable tool to distinguish normal, benign tumor and invasive tumor, the application of a 2D characterization technique to a 3D tissue environment requires further investigation. The issues pertaining to differences in cell mechanical behavior in vivo and in vitro need to be addressed. An understanding of the correlation between the biochemical changes that occur during cancer progression and the mechanical properties is essential.

26.4 Nanosurgical Manipulation of Living Cells

In addition to studying the mechanical properties, AFM force measurements can be done to study the mechanical response of the cell upon indentation. The mechanical response can be in the form of changes in intracellular calcium reactions or cytoskeletal disruption.[39]

Cell manipulation is possible with AFM by altering the force on the cantilever beam by controlling the voltage on the piezodevice. To insert a solid material, the size and shape of the tip should be improved for certainty of insertion and minimization of damage to a living cell. The shape of the tip is an important factor in determining the relationship between the force applied and the indentation of the cell surface. From classical elastic theory, the force applied on the surface is given by

$$F_{\text{conical}} = C_{\text{O}}\,(\Delta z)^2 \tag{26.5a}$$

$$F_{\text{parabola}} = C_{\text{P}}\,(\Delta z)^{3/2} \tag{26.5b}$$

$$F_{\text{cylindrical}} = C_{\text{Y}}\,(\Delta z) \tag{26.5c}$$

where F is the applied force and Δz is the indentation in the cell surface.[29] The fine-tuning of the shape and size of the AFM tip by focus-ion beam etching and other lithography processes makes it possible to study the behavior of the in vitro cell in the presence of an applied force. The length of the nanoneedle should be greater than the height of the living cells in order to avoid tapering and disturbances in the force analysis.[40]

The force distance curves of the indentation can be divided into two regions:

i. *Cell elastic deformation*: The force increased with the increase in the depth of penetration into the surface. The increase in force depends on the shape of the tip as shown in Equations 26.5a to 26.5c.
ii. *Cell rupture*: The force drops because of the cell rupture after the force reaches a certain critical value. The critical value of the force depends on the shape of the tip and the type of cell.

The plasma membrane on the surface of the cell is soft. The tip either deforms the membrane or penetrates the membrane to image the harder structures inside the cell. Obataya et al.[41] studied the effect of the shape of the tip on the tip penetration process. It was suggested that the probability of penetration, applied force, and indentation depends on the shape of the edge of the nanoneedle used (Table 26.1).[41]

Force response depends on the condition of the cells. For a living cell, for low deformations, the FDCs are completely retraceable, and the cell behaves elastically. Beyond a particular amount of deformation, cell rupturing occurs, which causes the cell fluid to ooze out, rupturing the cell membrane. The complexity in the force response of a live cell is evident from the multiple peaks as shown in Figure 26.8a. The typical range of deformation that causes cells to burst out is about

TABLE 26.1
Properties of Indentation and Penetration for Nanoneedles

Nanoneedle	Diameter (nm)	Shape of the Edge	Probability of Penetration (%)	Force of Indentation	Depth of Indentation
A	200	Cone	70	0.78 ± 0.25	1200 ± 490
B	200	Cylinder	92	0.65 ± 0.28	610 ± 270
C	800	Pyramid	21	1.9 ± 0.92	2800 ± 900
D	800	Prism	56	0.67 ± 0.37	1900 ± 810

Source: Obataya, I. et al., *Biosens. Bioelectron.*, 20 (8), 1652–1655, 2005.

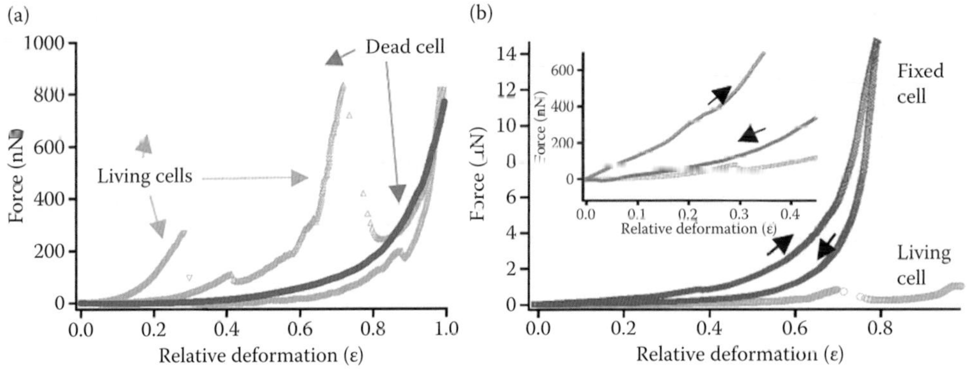

FIGURE 26.8
(a) Force response comparison of living and dead cells. (b) Force response comparison of live cells in fixed and free state. (Reprinted from Lulevich, V. et al., *Langmuir*, 22 (19), 8151–8155, 2006. With permission.)

30% deformation with about 0.1 to 0.3 μN force.[42] In contrast to live cells, dead cells exhibit smooth FDCs demonstrating the absence of the vital components of the cell. However, fixed cells behave differently and do not exhibit any peaks even at very high deformations as shown in Figure 26.8b.

The force response of the cells is dependent on a number of factors and can be studied using an AFM probe. Improvement in the stiffness of cardiac myocytes due to aging[43] and changes in the stiffness of the tissue cells due to the presence of a substrate[25] are a couple of them. The relationship between the variation in stiffness and the diseased condition is discussed in Section 26.3.

26.5 Probing Cell–Surface Adhesion

Cell adhesion is the binding between a cell and a surface; the surface could be another cell, ECM, or synthetic surface (e.g., organic or inorganic substrate). Cell adhesion is mediated by specific interactions between the cell adhesion molecules (CAMs) and the molecules to which the receptors bind (ligands). The CAMs recognize and interact with the neighboring cell adhesion reactors or interact with the proteins of the ECM. Binding is generally heterotypic, but if the CAMs involve cadherins, it can be homotypic.[44]

In the process of cell adhesion, live cells are subjected to exert mechanical forces supported by cell adhesion receptors such as cadherins, integrins, selectins, immunoglobulins, and proteoglycans.[44] Cell–cell and cell–surface adhesions are mediated by transmembrane receptor molecules, which recognize selective ligands and may cooperate to enhance the interaction. The ability of these cell adhesive interactions is critical to their function. The unbinding rate at which the CAMs and the ligands dissociate depends on the force that is applied to pull them apart[45]; the off-rate is expected to rise exponentially with increasing force.

Cell adhesion often involves the cooperation of multiple individual binding sites forming clusters, leading to a dynamic behavior different from that observed on single-molecule interactions discussed in Section 26.3.1. Thus, the dynamic response of multiple bonds under parallel loading is critical to understand cell adhesion.[46] Adhesions between cell–cell and cell–ECM have been measured using various mechanical methods, such as micropipette manipulation[47] and induction of hydrodynamic stress.[48] AFM-based force spectroscopy is a useful approach to measure cell–cell and cell–ECM adhesive force.

Cell adhesion analysis with AFM generally requires functionalizing the AFM cantilevers with molecules of interest.[49–51] Figure 26.9 shows a different experimental setup for cell adhesion measurements. Figure 26.9a depicts the interaction between a cell bound at the tip of a cantilever interacting with a cell immobilized on a substrate. Cell–surface interactions can be measured by both immobilizing the cell on the cantilever or a substrate and mating it with a substrate or a cantilever with desired surface functionality (Figure 26.9b and c).

During the force measurement of adhesive interaction, stable cantilever functionalization is an important

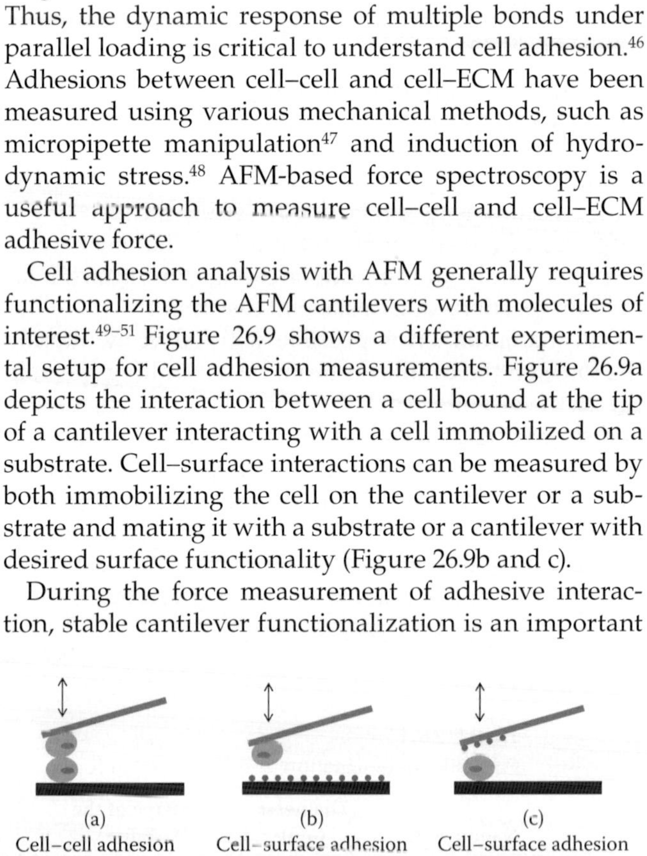

FIGURE 26.9
Different experimental setup for cell adhesion measurement. (a) Cantilever immobilized with single cell and approach to a second cell immobilized on the substrate. (b) Cantilever immobilized with single cell and approach to adhesive substrate. (c) Cantilever immobilized with ligand and approach to a single cell immobilized substrate.

step to obtain reproducible force measurement. The cell attached to the cantilever has to be stronger compared to the adhesive force between the substrate to the cell to prevent the cell from detaching during the adhesive force measurement process. Hence, cell attachment to a cantilever is a key step for performing reliable and reproducible force measurements. There are three general methods to attach the cell to a cantilever. The first one is to attach a suspended cell to a functionalized cantilever. The second involves attaching an adherent cell to an ECM-functionalized cantilever. In the third method, cells are directly cultivated on a cantilever. To attach a cell to the cantilever surface, typically a tip-less (commercially available) cantilever is engaged on a preselected cell surface followed by gently lifting the cantilever, resulting in attachment of the cell to the cantilever. Subsequently, the bound cell can be brought in contact with a target cell, and force measurements can be performed to probe the cell–cell adhesion. A cantilever functionalized with an ECM can be used to test higher adhesion forces, such as well-adhered cells. The cells can also be directly cultured on the cantilever, which provide extremely strong cell attachment. However, in this case, usually more than one cell attaches to the cantilever complicating the force experiment.

As mentioned earlier, the cell–cell adhesion is mediated by specific interactions between a cell and adjacent cells. Cell–cell adhesion is a homophilic binding; cells interact with adjacent cells by using specific and controllable adhesion mechanisms. During the cell–cell adhesion process, the cells recognize each other, and the interaction results in the formation of cell adhesion contact. For measuring the cell–cell adhesion force using AFM, the deflection of the cantilever during the cell–cell separation is monitored. Recently, Beckmann et al.[52] demonstrated that teneurins select for homophilic interaction patterns (same teneurins) and function as cell–cell adhesion receptors. They use a combination of deletions and swaps of teneurin-1 and teneurin-2 domains to identify which teneurin domains facilitate molecular recognition between cells and strength cell–cell adhesion. Figure 26.10 illustrates the single cell force spectroscopy (SCFS) experiments.[52] In period i, the probe moves close to the target. As the distance between cells is close to a point, the cells form adhesive contact, and the force is increased (period ii). After this contact, both cells were separated (period iii). The FDC records the maximum adhesion force (F_{max}) between the probe and the target cell. AFM-based SCFS has been used to quantitatively measure the contribution of specific cell surface receptors to cell adhesion. Beckmann et al.[52] introduced the combination of SCFS with cell biological and genetic recombination approaches to quantify the mechanisms of how teneurins establish homophilic interactions and strengthen cell–cell adhesion by using AFM-based SCFS. Retraction FDCs can be separated into different phases, which show a characteristic and complex pattern, such as maximum force peak and several small force steps. The diversity in the mechanochemistry of cell adhesion bonds reflects how mechanical force applied to a bond between molecules alters activation energy barriers along kinetic pathways, or switches pathways, which lead to dissociation.[52] Cell–cell adhesion forces have been measured using SCFS for

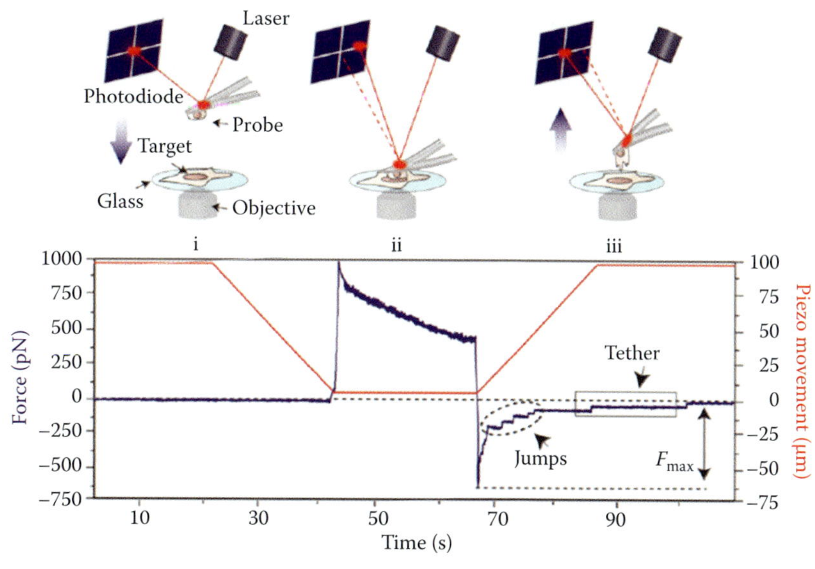

FIGURE 26.10
SCFS experiments. A single cell ("probe") immobilized on an AFM cantilever is approached into contact with a second cell ("target"). After a predefined contact time, the probe cell was retracted (i–iii) and adhesive forces were detected recording the cantilever deflection over the distance traveled by the cantilever. (Reprinted from Beckmann, J. et al., *Nano Lett.*, 13 (6), 2937–2946, 2013. With permission.)

various cells.[53–56] The cell–surface adhesion is mediated by specific interactions between a cell and other surfaces instead of cells, such as an ECM. The cell–surface adhesion is heterotypic binding, which use integrins as CAMs. Integrins are CAMs that mediate the adhesion between a cell and the surface. The cell–surface adhesion force can be measured by using AFM-based SCFS according to procedures described above.

26.6 Conclusion

To summarize, let us quickly recall Richard Feynman's[57] remarks: "What biology needs is not more math, but to see better at the atomic level." Owing to its high resolution, ability to image and probe under native conditions (hydrated state, controlled pH, and temperature), and relatively simple sample preparation, AFM, after its introduction, was immediately embraced by the biophysical and biochemical research communities as an invaluable tool for the imaging and probing of biological materials. In particular, surface force measurements of various forms have proven to be crucial in probing native properties of biological materials under in vivo conditions.[58–61] In addition to accurate AFM imaging, SFS provided a deep understanding and novel insights into the mechanics of biological structures at different length scales (from tissues to individual protein molecules). In this chapter, we briefly reviewed the general principle of SFS with particular emphasis on single-molecule force spectroscopy and SCFS. AFM-based force measurements continue to be revisited and refined making them more reliable and broadly applicable. We believe that force spectroscopy will continue to play an important role in understanding cellular and subcellular processes under biologically relevant conditions. An emerging trend is the integration of AFM-based SFS measurements (and more generally AFM) with complementary spectroscopic tools such as fluorescence and Raman scattering spectroscopy. These developments are expected to open new windows into cell biology and a better understanding of the cellular and subcellular processes at the nanoscale.

Acknowledgment

We acknowledge the financial support from the National Science Foundation under CBET-1254399 (NSF CAREER award).

References

1. McConnaughey, W.B., Nils, O.P. (1980) Cell poker: An apparatus for stress-strain measurements on living cells. *Rev. Sci. Instrum.*, **51** (5), 575–580.
2. McConnaughey, W.B., Nils, O.P., Elson, E.L. (1982) Dependence of locally measured cellular deformability on position on the cell, temperature, and cytochalasin B. *Proc. Natl. Acad. Sci. U.S.A.*, **79** (17), 5327–5331.
3. Zhang, W., Zhang, X. (2003) Single molecule mechanochemistry of macromolecules. *Prog. Polym. Sci.*, **28** (8), 1271–1295.
4. Kühner, F., Gaub, E.H. (2006) Modelling cantilever-based force spectroscopy with polymers. *Polymer*, **47** (7), 2555–2563.
5. Zou, S., Korczagin, I., Hempenius, M.A., Schöherr, H., Vancso, G.J. (2006) Single molecule force spectroscopy of smart poly(ferrocenylsilane) macromolecules: Towards highly controlled redox-driven single chain motors. *Polymer*, **47** (7), 2483–2892.
6. Al-Maawali, S., Bemis, J.E., Akhremitchev, B.B., Liu, H., Walker, G.C. (2005) Single-molecule AFM study of polystyrene grafted at gold surfaces. *J. Adhes.*, **81** (10–11), 999–1016.
7. Butt, H.-J., Cappella, B., Kappl, M. (2005) Force measurements with the atomic force microscope: Technique, interpretation, and applications. *Surf. Sci. Rep.*, **59** (1–6), 1–152.
8. Cappella, B., Dietler, G. (1999) Force–distance curves by atomic force microscopy. *Surf. Sci. Rep.*, **34** (1–3), 1–104.
9. Tsukruk, V.V., Singamaneni, S. (2012) *Scanning Probe Microscopy of Soft Matter: Fundamentals and Practices*. Wiley-VCH: Weinheim, Germany.
10. McConney, M.E., Singamaneni, S., Tsukruk, V.V. (2010) Probing soft matter with the atomic force microscopies: Imaging and force spectroscopy. *Polym. Rev.*, **50** (3), 235–286.
11. Marszalek, P., Dufrêne, Y. (2012) Stretching single polysaccharides and proteins using atomic force microscopy. *Chem. Soc. Rev.*, **41** (9), 3523–3534.
12. Hinterdorfer, P., Dufrêne, Y. (2006) Detection and localization of single molecular recognition events using atomic force microscopy. *Nat. Methods*, **3** (5), 347.
13. Gerber, C., Lang, H. (2006) How the doors to the nanoworld were opened. *Nat. Nanotechnol.*, **1** (1), 3–5.
14. Zimmermann, J.L., Nicolaus, T., Neuert, G., Blank, K. (2010) Thiol-based, site-specific and covalent immobilization of biomolecules for single-molecule experiments. *Nat. Protoc.*, **5** (5), 975–985.
15. Lee, C., Wang, Y., Huang, L., Lin, S. (2007) Atomic force microscopy: Determination of unbinding force, off rate and energy barrier for protein-ligand interaction. *Micron*, **38** (5), 446–461.
16. Li, H., Oberhauser, A., Redick, S., Carrion-Vazquez, M., Erickson, H., Fernandez, J. (2001) Multiple conformations of PEVK proteins detected by single-molecule techniques. *Proc. Natl. Acad. Sci. U.S.A.*, **98** (19), 10682–10686.

17. Oberhauser, A.F., Badilla-Fernandez, C., Carrion-Vazquez, M., Fernandez, J.M. (2002) The mechanical hierarchies of fibronectin observed with single-molecule AFM. *J. Mol. Biol.*, **319** (2), 433–447.
18. Meadows, P., Bemis, J., Walker, G. (2003) Single-molecule force spectroscopy of isolated and aggregated fibronectin proteins on negatively charged surfaces in aqueous liquids. *Langmuir*, **19** (23), 9566–9572.
19. Gutsmann, T., Fantner, G., Kindt, J., Venturoni, M., Danielsen, S., Hansma, P. (2004) Force spectroscopy of collagen fibers to investigate their mechanical properties and structural organization. *Biophys. J.*, **86** (5), 3186–3193.
20. Rief, M., Pascual, J., Saraste, M., Gaub, H.E. (1999) Single molecule force spectroscopy of spectrin repeats: Low unfolding forces in helix bundles. *J. Mol. Biol.*, **286** (12), 553–561.
21. Li, F., Redick, S., Erikson, H., Moy, V. (2003) Force measurements of the α5β1 integrin-fibronectin interaction. *Biophys. J.*, **84** (2), 1252–1262.
22. Zhe, S., Martinez-Lemus, L., Trache, A., Trzeciakowski, J., Davis, G., Pohl, U., Meininger, G.A. (2005) Mechanical properties of the interaction between fibronectin and α5β1-integrin on vascular smooth muscle cells studied using atomic force microscopy. *Am. J. Physiol.-Heart C*, **289** (6), H2526–H2535.
23. Trache, A., Trzeciakowski, J., Gardiner, L., Sun, Z., Muthuchamy, M., Guo, M., Yuan S.Y., Meininger, G.A. (2005) Histamine effects on endothelial cell fibronectin interaction studied by atomic force microscopy. *Biophys. J.*, **89** (4), 2888–2898.
24. Bao, G., Suresh, S. (2003) Cell and molecular mechanics of biological materials. *Nat. Mater.*, **2** (11), 715–725.
25. Discher, D., Janmey, P., Wang, Y. (2005) Tissue cells feel and respond to the stiffness of their substrate. *Science*, **310** (5751), 1139–1143.
26. Kelly, G., Campbell, V., Jarvis, S., Prendergast, P. (2008) AFM indentation on Osteoblasts shows that elasticity changes during the cell cycle. *J. Biomech.*, **41** (1), S21.
27. Arnoldi, M., Fritz, M., Bauerlein, E., Radmacher, M., Sackmann, E., Boulbitch, A. (2000) Bacterial turgor pressure can be measured by atomic force microscopy. *Phys. Rev. E.*, **62** (1), 1034–1044.
28. Yao, X., Walter, J., Burke, S., Stewart, S., Jericho, M.H., Pink, D., Hunter, R., Beveridge, T.J. (2002) Atomic force microscopy and theoretical considerations of surface properties and turgor pressures of bacteria. *Colloids Surf. B*, **23** (2–3), 213–230.
29. Weisenhorn, A., Khorsandi, M., Kasas, S., Gotzos, V., Butt, H. (1993) Deformation and height anomaly of soft surfaces studied with an AFM. *Nanotechnology*, **4** (2), 106.
30. Mathura, A., Collinswortha, A., William, M., Krausb, W., Truskeya, G. (2001) Endothelial, cardiac muscle and skeletal muscle exhibit different viscous and elastic properties as determined by atomic force microscopy. *J. Biomech.*, **34** (12), 1545–1553.
31. Cross, S., Jin, Y., Rao, J., Gimzewsk, J. (2007) Nanomechanical analysis of cells from cancer patients. *Nat. Nanotechnol.*, **2** (12), 780–783.
32. Suresh, S. (2007) Biomechanics and biophysics of cancer cells. *Acta Biomater.*, **3** (4), 413–438.
33. Levental, K., Yu, H., Kass, L., Lakins, J., Egeblad, M., Erler, J.T., Fong, S.F. et al. (2009) Matrix crosslinking forces tumour progression by enhancing integrin signaling. *Cell*, **139** (5), 891–906.
34. Sedwick, C., Weaver, V. (2011) Overcoming cancer's stiff resistance. *J. Cell. Biol.*, **193** (5), 802–803.
35. Lopez, J., Kang, I., You, W., McDonald, D., Weaver, V. (2011) In-situ force mapping of mammary gland transformation. *Integr. Biol.*, **3** (9), 910–921.
36. Guck, J., Schinkinger, S., Lincoln, B., Wottawah, F., Ebert, S., Romeyke, M. (2005) Optical deformability as an inherent cell marker for testing malignant transformation and metastatic competence. *Biophys. J.*, **88** (5), 3689–3698.
37. Gavara, N., Chadwick, R. (2012) Determination of the elastic moduli of thin samples and adherent cells using conical atomic force microscope tips. *Nat. Nanotechnol.*, **7** (11), 733–736.
38. Plodinec, M., Loparic, M., Monnier, C., Obermann, E., Dallenbach, Z., Oertle, P., Hyotyla, J.T. et al. (2012) The nanomechanical signature of breast cancer. *Nat. Nanotechnol.*, **7** (11), 757–765.
39. Charras, G., Horton, M. (2002) Single cell mechanotransduction and its modulation analyzed by atomic force microscope indentation. *Biophys. J.*, **82** (6), 2970–2981.
40. Obataya, I., Nakamura, C., Han, S., Nakamura, N., Miyake, J. (2005) Nanoscale operation of a living cell using an atomic force microscope with a nanoneedle. *Nano Lett.*, **5** (1), 27–30.
41. Obataya, I., Nakamuraa, C., Han, S., Nakamuraa, N., Miyakea, J. (2005) Mechanical sensing of the penetration of various nanoneedles into a living cell using atomic force microscopy. *Biosens. Bioelectron.*, **20** (8), 1652–1655.
42. Lulevich, V., Zink, T., Chen, H., Liu, F., Liu, G. (2006) Cell mechanics using atomic force microscopy-based single-cell compression. *Langmuir*, **22** (19), 8151–8155.
43. Lieber, S., Aubry, N., Pain, J., Diaz, G., Kim, S., Vatner, S. (2004) Aging increases stiffness of cardiac myocytes measured by atomic force microscopy nanoindentation. *Am. J. Physiol.-Heart C.*, **287** (2), 645–651.
44. Gumbiner, B. (1996) Cell adhesion: The molecular basis of tissue architecture and morphogenesis. *Cell*, **84** (3), 345–357.
45. Bell, G. (1978) Models for the specific adhesion of cells to cells. *Science*, **200** (4342), 618–627.
46. Rico, F., Roca-Cusachs, P., Sunyer, R., Farré, R., Navajas, D. (2007) Cell dynamic adhesion and elastic properties probed with cylindrical atomic force microscopy cantilever tips. *J. Mol. Recognit.*, **20** (6), 459–466.
47. Evans, E.A. (1985) Detailed mechanics of membrane-membrane adhesion and separation. II. Discrete kinetically trapped molecular cross-bridges. *Biophys. J.*, **48** (1), 185–192.
48. Chen, S., Springer, T. (1999) An automatic braking system that stabilizes leukocyte rolling by an increase in selectin bond number with shear. *J. Cell Biol.*, **144** (1), 185–200.

49. Krieg, M., Helenius, J., Heisenberg, C., Muller, D. (2008) A bond for a lifetime: Employing membrane nanotubes from living cells to determine receptor-ligand kinetics. *Angew. Chem. Int. Ed.*, **47** (50), 9775–9777.
50. Fritz, J., Katopodis, A., Kolbinger, F., Anselmetti, D. (1998) Force-mediated kinetics of single P-selectin/ligand complexes observed by atomic force microscopy. *Proc. Natl. Acad. Sci. U.S.A.*, **95** (21), 12283–12288.
51. Müller, D., Dufrêne, Y. (2011) Atomic force microscopy: A nanoscopic window on the cell surface. *Trends Cell Biol.*, **21** (8), 461–469.
52. Beckmann, J., Schubert, R., Ehrismann, R., Müller, D. (2013) Deciphering teneurin domains that facilitate cellular recognition, cell–cell adhesion, and neurite outgrowth using atomic force microscopy-based single-cell force spectroscopy. *Nano Lett.*, **13** (6), 2937–2946.
53. Thie, M., Röspel, R., Dettmann, W., Benoit, M., Ludwig, M., Gaub, H.E., Denker, H.W. (1998) Interactions between trophoblast and uterine epithelium: Monitoring of adhesive forces. *Hum. Reprod.*, **13** (11), 3211–3219.
54. Zhang, X., Wojcikiewicz, E., Moy, V.T. (2002) Force spectroscopy of the leukocyte function-associated antigen-1/intercellular adhesion molecule-1 interaction. *Biophys. J.*, **83** (4), 2270–2279.
55. Zhang, X., Wojcikiewicz, E., Moy, V.T. (2006) Dynamic adhesion of T lymphocytes to endothelial cells revealed by atomic force microscopy. *Exp. Biol. Med.*, **231** (8), 1306–1312.
56. Alon, R., Feigelson, S.W., Manevich, E., Rose, D.M., Schmitz, J., Overby, D.R., Winter, E. et al. (2005) α4β1-dependent adhesion strengthening under mechanical strain is regulated by paxillin association with the α4-cytoplasmic domain. *J. Cell. Biol.*, **171** (6), 1073–1084.
57. Feynman, R.P. (1999) *The Pleasure of Finding Things Out: The Best Short Works of Richard P. Feynman*. Perseus Books: New York.
58. Kienberger, F., Ebner, A., Gruber, H.J., Hinterdorfer, P. (2006) Molecular recognition imaging and force spectroscopy of single biomolecules. *Acc. Chem. Res.*, **39** (1), 29–36.
59. Engel, A., Muller, D.J. (2000) Observing single biomolecules at work with the atomic force microscope. *Nat. Struct. Biol.*, **7** (2000), 715–718.
60. Hansma, H.G., Kim, K.J., Laney, D.E. (1997) Properties of biomolecules measured from atomic force microscope images: A review. *J. Struct. Biol.*, **119** (2), 99–108.
61. Fotiadis, D., Scheuring, S., Muller, S.A., Engel, A., Muller, D.J. (2002) Imaging and manipulation of biological structures with the AFM. *Micron*, **33** (4) 385–397.

27

Fluorescence-Based Tools for Quantifying Adhesion Dynamics

Haguy Wolfenson

CONTENTS

Integrin adhesions are multimolecular assemblies that mediate the connection between cells and the extracellular matrix. These structures can be very stable in stationary cells, maintaining a firm grip on the microenvironment, but are highly dynamic in order to support the movements of motile cells. The dynamic characteristics of adhesions are presumably regulated by the complex interactions between multiple adhesion-resident proteins. This chapter describes several fluorescence-based tools that are used to study adhesion dynamics both at the structural and molecular levels. The challenge in the future would be to use these (and other) tools to accurately quantify adhesion dynamics and correlate them to the molecular composition of the adhesions under different conditions. This would allow building a dynamic systems-level model to explain adhesion regulation in different cell types.

27.1 Introduction

27.1.1 Structural Organization of Integrin Adhesions

The physical connection between cells and the extracellular matrix (ECM) is mediated via multimolecular complexes known as "integrin adhesions."[1,2] These structures are located at cells' ventral surfaces when plated on 2-dimensional (2D) matrices[3] (see Figure 27.1) and in cell bodies and protrusions when embedded in 3-dimensional (3D) environments.[4] Integrin adhesions not only serve as local anchors for the cells but also play active roles in chemical and mechanical sensing of the ECM and are thus important in maintaining cellular homeostasis.[5]

At the core of these adhesions, members of the transmembrane integrin family bind to ECM proteins at their extracellular end and to the actin cytoskeleton, via various adapter proteins, at their intracellular end.[6,7] Many signaling proteins are also recruited to the adhesions,[8,9] and together with the adaptors, they make up the adhesion "plaque," located at the interface between the integrins and actin. Beyond these plaque proteins, the network of molecules that interact with integrin adhesions (collectively known as the "adhesome," with ~180 molecules known to date[10]) is even more complex and includes cytoskeletal elements, lipids, ion channels, and other transmembrane receptors.[7]

27.1.2 Lifecycle of Integrin Adhesions

Integrin adhesions are dynamic structures that appear in different sizes and forms, depending mainly on their

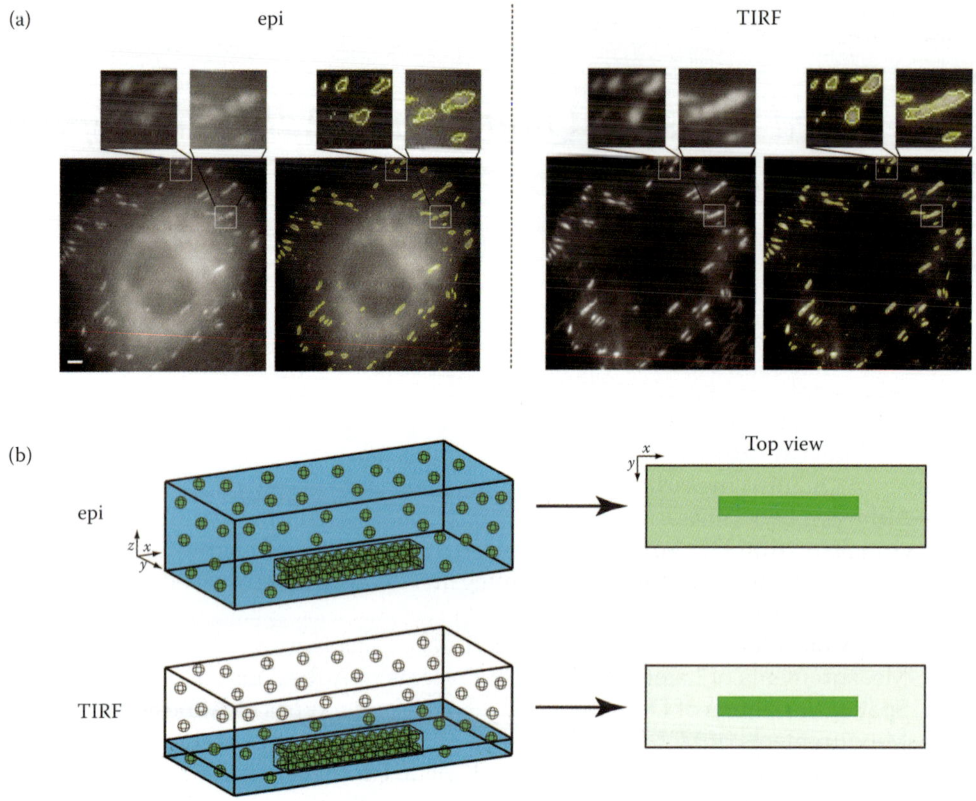

FIGURE 27.1
Imaging integrin adhesions. (a) Epifluorescence and TIRF images of the same HeLa cell expressing the plaque protein paxillin labeled with YFP. Bar: 4 μm. The right image in each case shows the adhesions that were detected using the "focal adhesion analysis system."[11] Note the differences in detection of both focal adhesions (right insets in each case) and focal complexes (left insets). (b) Illustration of epifluorescence vs. TIRF imaging of an adhesion (inner 3D rectangle). Since in epifluorescence imaging, the cell is illuminated throughout the entire *z*-axis (blue), fluorophores (circles) emit fluorescence (green) to a large degree in the cytoplasm, resulting in a high background signal compared to the adhesion. In contrast, in TIRF imaging, only fluorophores located near the coverslip are illuminated, resulting in a high signal-to-background ratio of fluorophores that are in the adhesion.

cellular location and the motility of the cell. The current classification of integrin adhesions includes four major classes: nascent adhesions, focal complexes, focal adhesions, and fibrillar adhesions (listed from less to more stable).[12,13] The distinction between the different forms is done by their size, distance from the cell edge, and specific protein repertoire. In general, less motile cells display larger and more stable adhesions.

Both in stationary and motile cells, the cell edge (known as the "lamellipodium") constantly undergoes protrusion/retraction cycles to probe the underlying matrix.[14] Under the right conditions, integrin molecules located at the protruding lamellipodium establish initial contacts with ECM proteins and cluster together to form nascent adhesions. These are small, dot-like structures that are barely visible using ordinary fluorescence microscopy and are relatively short-lived (~1 to 2 min).[15] At early times, they contain very few integrins and plaque proteins, but actin filaments begin to polymerize from them.[16] During the continued protrusion of the lamellipodium forward, the nascent adhesions typically remain in place and then either disassemble or begin to grow into the slightly larger focal complexes (typically ~1 μm diameter)[15] following force-induced stimuli that originate from myosin-II pulling on the actin filaments connected to the adhesions.[17] At this stage, the focal complexes are located at the boundary between the lamellipodium and the lamellum, and continue to grow in a centripetal manner (toward the cell center) into the more elongated focal adhesions (typically ~1 μm wide, ~3 to 5 μm long).[1] The growth process is force-dependent and is accompanied by bundling of the actin filaments into large stress fibers, primarily by α-actinin and myosin-II.[15,18] When the focal adhesions cease to grow, they typically reach steady state in which their size, location, and orientation do not change significantly for up to several tens of minutes (especially in nonmotile cells).[15,19] Under certain conditions, focal adhesions can next transform into fibrillar adhesions, which are much more elongated and stable, and are typically located around the cell center, where they are associated with fibronectin fibrils.[20]

27.1.3 Challenges in Quantifying Adhesion Dynamics

The major roles played by adhesions, both in regulating the steady state of stationary cells and in supporting cell migration, highlight the importance of quantitatively characterizing their dynamics. In particular, the processes of assembly, growth, stability, and disassembly of the adhesions are important to study since they are tightly connected to the motility of the cell. In motile cells, the adhesions are generally smaller since they undergo rapid assembly/disassembly cycles to support the movement of the cell.[21] But even in such cells, the adhesions can be stable on the order of several minutes (especially focal adhesions). Importantly, this structural stability may be misleading, since even under steady-state conditions, there is constant exchange of plaque proteins in and out of the adhesions. Presumably, this turnover is regulated by the available binding sites of each protein within the adhesions,[22] which in turn may vary significantly, since, on average, each adhesome molecule has ~8 to 10 potential binding partners.[7] Moreover, the gradual transition between the different adhesion forms involves changes in the repertoire of the plaque proteins and the available binding sites.[23,24]

Despite much experimental work that has been done over the years, considerable information is still missing about what regulates, at the molecular level, the transitions between the different adhesion forms, as well as the disassembly of the adhesions. Thus, the major challenge in characterizing integrin adhesions is to correlate the structural dynamics of the adhesions with their molecular dynamics (i.e., the turnover of proteins within them). This is particularly challenging due to the high complexity of the adhesions and the number of different molecules that are involved. As outlined below, the tools for quantifying both the structural and molecular dynamics of integrin adhesions are now widely available. What is required is a wide set of experiments, under different conditions and for different cell types, which would link among protein composition, adhesion dynamics, and the state of the cell. Analyses of such quantitative data can help identify the central players that regulate adhesion dynamics and how their activity is altered under different conditions.

27.2 Tools to Study Adhesion Dynamics

27.2.1 Quantifying Structural Dynamics of Adhesions by Time-Lapse Imaging

Since integrin adhesions interchange between different forms over time, gaining knowledge about their structural dynamics requires performing long-term time-lapse imaging of cells expressing adhesion proteins labeled with fluorescent proteins (FPs). While it is very common to transfect cells with DNA plasmids for expression of fluorescently labeled proteins, one should be cautious that overexpression of certain plaque proteins does not alter the adhesions themselves or the morphology of the cells. For instance, overexpression of Src or focal adhesion kinase (FAK), which reside in adhesions and regulate their size and dynamics, can result in altered adhesions compared to wild-type cells.[25] Typically, in such experiments, the cells are spread on a glass coverslip coated with an ECM protein. The choice of ECM proteins used for such studies could affect the adhesions too, since different ECM components activate different members of the integrin family. It was shown, for instance, that collagen-coated surfaces lead to higher adhesive strength of fibroblasts than fibronectin.[26]

The time-lapse movies should be long enough to cover adhesion assembly, growth, (possibly) merging with other adhesions, and disassembly. For motile cells, which usually display shorter adhesion lifetimes, movies of 20–30 min can prove sufficient, but for nonmotile cells, movies of 1 h or longer may be essential. The lifetimes of the adhesions also affect the required imaging frame rate—the more stable the adhesions, the lower the frame rate. Usually, frame rates of 4–12 frames per minute should suffice.

27.2.1.1 Imaging Techniques for Tracking Integrin Adhesions in Live Cells

In principle, imaging integrin adhesions over time can be done using a standard epifluorescence microscope; however, in cases of some adhesion plaque proteins, other imaging methods would be required. The reason is that plaque proteins can reside both in the cytoplasm and in adhesions. While many plaque proteins are very highly concentrated in the adhesions, and therefore display high contrast between their fluorescent signal in the adhesion and its surroundings, other proteins do not display such high density in adhesions. In such cases, it is required to reduce the fluorescent signal from the cytoplasm. Two very effective imaging methods to do so are confocal microscopy and total internal reflection fluorescence (TIRF) microscopy, both of which involve collection of the fluorescent signal from a limited layer in the *z*-axis. Since the concentration of plaque proteins is higher inside adhesions compared to the cytoplasm (to different degrees), this will increase substantially the signal-to-background ratio for fluorophores that are located within the adhesions (see Figure 27.1). TIRF is the more effective method since its collection depth in the *z*-axis is approximately fourfold less than in a typical confocal microscope. Also, since TIRF involves fluorescence excitation within an evanescent field of only

~150 nm close to the coverslip (formed by the internal reflection of an incident light), the overall illumination intensity is reduced significantly compared to a confocal or epifluorescence setup, resulting in less photobleaching and less photodamage to the cells.[27]

27.2.1.2 Image Processing and Analysis

After acquisition of the time-lapse movies, the first step in tracking adhesions over time involves filtering the background. This is typically done by applying a high-pass filter to the images, which subtracts the mean intensity from a region around each pixel for a region larger than a typical adhesion but smaller than the cell.[28,29] Once high-pass filtration is applied to all the frames of a movie, the next step is to perform segmentation for each image in order to define the contours of individual adhesions (see Figure 27.1, for example). The segmentation is done by assigning certain area and intensity thresholds to the images that define the lower limit of what is considered an adhesion (both thresholds should be determined empirically based on the typical intensity and sizes of adhesions in the cells used). A very useful algorithm for performing image segmentation is the "Water" algorithm,[28] in which a pixel above the defined intensity threshold is assigned to a "patch" (adhesion) if its neighboring pixels are also above this threshold, and if together they make up a patch bigger than the defined area threshold. If the patch is too small, and there isn't an adjacent patch that it could be merged with, this patch is excluded. The segmented patches are then being tracked over time based on their overlap between frames, while in each frame, a patch can either be "born," persist, merge with other patches, or "die." In this way, various characteristics of each adhesion (patch) are monitored, including adhesion area, center of mass, axial ratio, and intensity.

Notably, Berginski and Gomez have recently introduced the "focal adhesion analysis system,"[11] a robust web-based tool that automatically analyzes time-lapse data from users based on the algorithms described above. Using this tool, it is now possible to analyze large data sets and obtain highly quantitative information about adhesion structural dynamics, including adhesion assembly, growth, and disassembly, in different regions of the cell.

27.2.2 Quantifying Molecular Dynamics of Adhesions

Unlike studies of adhesion structural dynamics, where the adhesion is viewed as a single entity, quantifying the dynamics of adhesion proteins requires tracking the movements of subgroups of molecules within adhesions. Importantly, plaque proteins can theoretically both move laterally in the membrane plane and exchange between the adhesion and the cytoplasm. Several methods have been used over the years to track adhesion molecular dynamics, which proved to be complementary both in terms of lateral mobility versus exchange and in terms of timescales, as outlined in the following.

27.2.2.1 Measurements of Lateral Mobility of Proteins in Adhesions

Single particle tracking (SPT) is a particularly useful method for studying the lateral mobility of proteins in the membrane plane.[30] In such studies, proteins are sparsely labeled, allowing their detection as single particles that are then being tracked over time using high-speed imaging. The advantage of SPT is that besides quantitative analysis of protein kinetics, it can also provide important nanoscale spatial information,[31] which is missing when using ensemble-average techniques such as fluorescence recovery after photobleaching (FRAP) or fluorescence correlation methods (see Sections 27.2.2.2 and 27.2.2.3). Also, by measuring the mean square displacement (MSD) of a particle's trajectory over time, one can obtain information on the nature of the movement (i.e., pure diffusion, subdiffusion, or superdiffusion[32]).

SPT studies of transmembrane proteins such as integrins are usually much simpler to perform compared to nonintegral membrane proteins that can also reside in the cytoplasm (such as adhesion plaque proteins). The reason is that the fast cytoplasmic diffusion that a protein undergoes once it exits the adhesion into the cytoplasm makes it extremely difficult to continue its tracking. Moreover, whereas integrins can be externally labeled with nonbleaching fluorescent particles,[33] tracking FP-labeled intracellular proteins restricts the duration of the measurements due to continuous photobleaching of the FPs over time. Recently, Rossier et al.[34] overcame this difficulty by performing SPT studies using photoactivatable FPs (PAFPs), which unlike traditional FPs cannot be excited until they undergo an activation step by a burst of light in a specific wavelength.[35] By using continuous photoactivation and excitation, they controlled the density of fluorescence-emitting PAFPs[36] and were thus able to track multiple trajectories of both integrins and the plaque protein talin over a course of several minutes. In this way, they showed that activated integrins are mostly immobile in focal adhesions, but also that a subset of them undergo inactivation and free diffusion. Furthermore, they showed that talin is mainly stationary in adhesions, but that it also exhibits slow rearward flow.[34] Notably, the notion that plaque proteins display rearward movements in adhesions was reported by Hu et al. using another particle tracking technique: speckle microscopy.[37] The authors showed that different classes of plaque proteins exhibited varying degrees of correlated motions with the rearward-flowing actin in adhesions. The flow velocities were in the range of ~100 to 250 nm/min, in agreement with Rossier et al.[34]

27.2.2.2 Spatial Variations of Dynamics between Different Types of Adhesions

In SPT or FRAP experiments, one would like to restrict the level of photobleaching prior to starting the measurement. Therefore, knowledge on the history of the adhesion being studied is usually missing. Recently, several image correlation spectroscopy (ICS)[38] techniques were developed that allow taking long movies and subsequently measuring protein kinetics at different time points during adhesions' lifetimes. The basis for these techniques was fluorescence correlation spectroscopy (FCS), which is an ensemble-average technique that has proven to be very powerful in studying diffusive behavior and concentrations of proteins in cells and solutions[39] (see Chapter 40). The original FCS studies involved illumination of very small volumes, but the ICS techniques rely on analysis of the spatiotemporal correlation of intensity fluctuations within whole images,[40] as well as between frames of a movie.[41] The advantages of these techniques are that protein kinetics can be studied over whole cells, and that a range of diffusion coefficients can be measured with detailed spatial information.

Digman et al.[42,43] recently used a combination of such correlation microscopy techniques to study the dynamics of adhesion proteins in and out of adhesions. Like with the previous techniques, these studies demonstrated slow movements of plaque proteins in focal adhesions, but also that the binding or unbinding of adhesion proteins depends on the adhesion assembly/disassembly status, location in the cell, and apparent translocation. They also showed that in adhesions that undergo "sliding," plaque proteins enter as monomers at the assembling end of focal adhesions, but exit as large aggregates from the disassembling end. While on and off rate constants are difficult to obtain from such fluorescence correlation measurements, they do provide important information on the spatial differences in protein kinetics, including those within single adhesions.

27.2.2.3 Measurements of Molecular Exchange of Plaque Proteins in Adhesions

Probably the most popular method for measuring plaque protein dynamics is FRAP. In this method, the fluorophores in a region of interest (ROI) are bleached by a brief pulse of an intense laser beam, after which the fluorescence recovery is monitored by an attenuated beam, as bleached molecules are replaced by nonbleached ones (for a detailed FRAP tutorial, see reference 44). The advantage of using FRAP is that it can measure the exchange between adhesion-bound and cytoplasmic proteins. However, unlike the previously mentioned techniques, FRAP experiments require the adhesions to be at steady state. The reason is that to calculate what fraction of the protein's population is recovering and the time constant for its recovery, one has to know the prebleach intensity; in adhesions that are not at steady state (for instance, focal complexes), this intensity usually displays large fluctuations that preclude such calculations.

Typical FRAP experiments are performed in a confocal setup, and therefore, when the laser beam is focused at the membrane plane, the fluorescence is collected both from the membrane and from the adjacent cytoplasm. Thus, fluorescence can recover by three different mechanisms: diffusion of cytoplasmic proteins, diffusion/flow of membrane-bound proteins, and exchange between membrane-bound and cytoplasmic proteins. In many cases of cytoplasmic proteins that interact with the membrane (e.g., ras proteins[45]), the recovery of fluorescence due to cytoplasmic diffusion is very rapid and therefore can be neglected when analyzing the FRAP curves. However, in the case of adhesion plaque proteins, this is usually not the case: high time-resolution FRAP experiments on different plaque proteins revealed that the recovery curves are biphasic (Figure 27.2), and that the first phase is due to cytoplasmic diffusion that is attenuated approximately fourfold compared to typical diffusion, and therefore cannot be neglected.[22] This was shown using a method called FRAP beam-size analysis, which can distinguish between diffusion and exchange mechanisms (see Figure 27.2 for details). Since fluorescence in these experiments was collected only from ~600 nm in the *z*-axis (the confocal volume penetration depth), this attenuated diffusion rate is attributed to a juxtamembrane cytoplasmic subpopulation next to the adhesions.[21] The second (slower) phase of the FRAP curves is dominated by exchange between adhesion-bound and cytoplasmic molecules, as indicated by both beam-size analysis (Figure 27.2) and complete bleaching of focal adhesions so that any recovery has to come from out of the adhesions.[22]

27.2.2.4 Bringing It All Together—Extracting Kinetic Parameters

The results depicted above show that plaque proteins that are attached to the adhesion can either be stably bound and undergo slow centripetal flow, or exit their binding sites and exchange with other molecules of the same kind from the nearby cytoplasm. The flow is typically much slower than the exchange process, as indicated by comparing the time constant of flow to the $T_{1/2}$ of the second phase of the FRAP curves (dominated by exchange). Doing so for paxillin and vinculin (flow

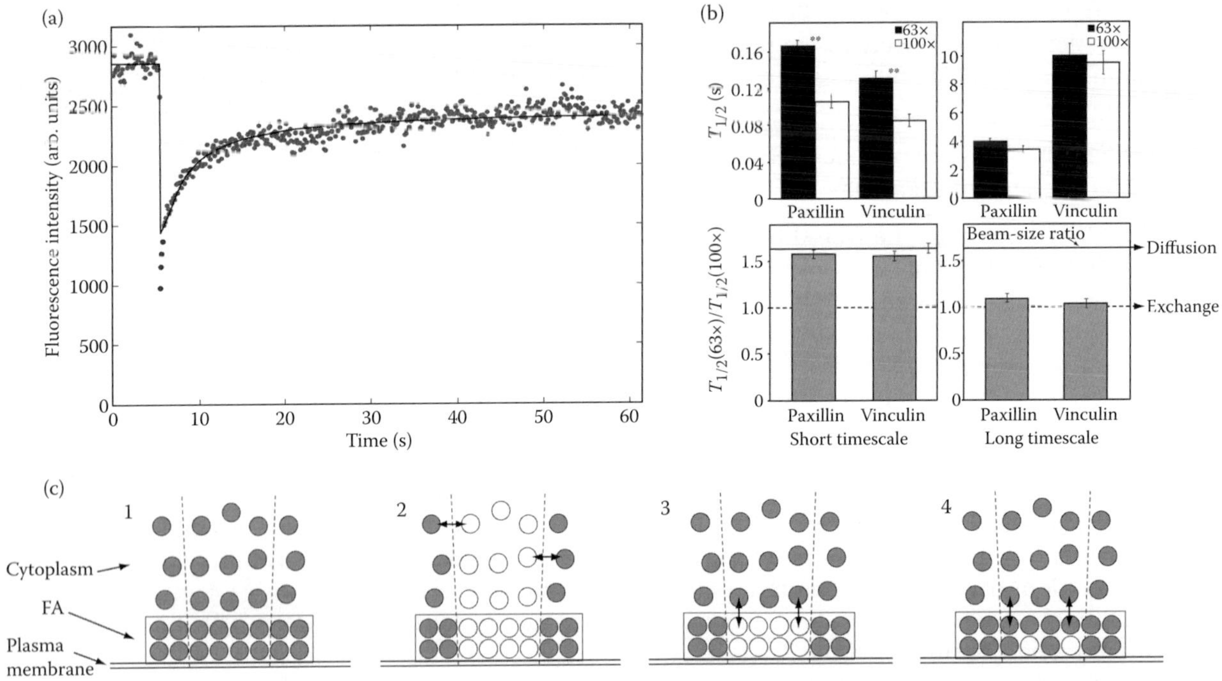

FIGURE 27.2
FRAP measurements on plaque proteins in focal adhesions. (a) Typical FRAP curve of YFP-paxillin in HeLa cells. Note the biphasic recovery (solid line is the fit of only the second step of recovery). (b) FRAP beam-size analysis to determine the modes of recovery of the two phases of FRAP for paxillin and vinculin. (Adapted from Wolfenson, H. et al., *PLoS ONE* 4, e4304, 2009.) For each protein, measurements on a short timescale (4 s for the first recovery phase) and a long timescale (60 s for the second phase) were each performed twice with two different beam sizes, using 63× (larger beam area) and 100× (smaller beam area) objectives. Mean $T_{1/2}$ ± SEM values are displayed for each case (upper boxes), as well as the ratio between $T_{1/2}(63\times)$ and $T_{1/2}(100\times)$ (lower boxes). If the recovery of a specific protein is by pure diffusion, $T_{1/2}$ would be proportional to the square radius of the laser beam, and therefore, the ratio between the recovery times from the two measurements would be equal to the ratio between the areas of the respective beams. However, if recovery is only by exchange, the recovery time does not depend on the size of the beam, and therefore, the ratio between the recovery times equals 1. Notably, on the short timescale, diffusion is dominant, as both proteins display a $T_{1/2}(63\times)/T_{1/2}(100\times)$ ratio of ~1.6, which is the ratio between the beam areas. On the long timescale, this ratio is close to 1 for both proteins, indicating that exchange is the dominant process. (c) Model for protein dynamics underlying the FRAP curves of adhesion plaque proteins. (1) Side view of a focal adhesion and its adjacent cytoplasm before the bleach (area between the dashed lines is the laser beam); (2) immediately after the brief pulse of an intense laser beam, fluorescent proteins in the membrane and the cytoplasm are bleached, and fluorescence begins to recover in the cytoplasm due to diffusion of proteins there; (3) after the cytoplasmic fluorescence is fully recovered, exchange becomes the dominant process, until (4) the FRAP curve plateaus as most bleached molecules are replaced by nonbleached ones in the adhesion (incomplete recovery; see panel a).

rates of ~100 nm/min,[37] $T_{1/2}$ values of 4 and 10 s, respectively[22]), while taking into account the beam size used in the FRAP studies (Gaussian beam radius = 0.77 μm), yields a flow time constant of ~450 s, approximately 50- to 100-fold longer than the exchange time constants. Notably, in many cases, the fluorescence recovery of plaque proteins is not complete; rather, it goes up to 60%–80% of the prebleach intensity, indicating that ~20% to 40% of the population inside the adhesions is immobile on the FRAP experimental timescale.[22] This seemingly immobile population may correspond to the slow flow of proteins discussed above.[34,37]

To accurately quantify the turnover rates of specific proteins using the FRAP data, one should fit the FRAP curves to a mathematical expression derived from a model that describes the dynamic mechanisms underlying fluorescence recovery in adhesions. In such a model, the plaque protein's population is divided into cytoplasmic and adhesion-bound subpopulations. The kinetics of these subpopulations are described by the following differential equations:

$$\frac{\partial N_M}{\partial t} = D_M \nabla^2 N_M - k_{off} N_M + k_{on} N_C \quad (27.1)$$

$$\frac{\partial N_C}{\partial t} = D_C \nabla^2 N_C + k_{off} N_M - k_{on} N_C \quad (27.2)$$

where N_M is the number of membrane-bound proteins with diffusion coefficient D_M, N_C is the number of proteins in the juxtamembrane cytoplasmic pool with diffusion coefficient D_C, and k_{on} and k_{off} are the rate constants for entering and exiting the adhesion, respectively. Importantly, since exchange between membrane-bound and cytoplasmic proteins involves transition from a 2D surface to a 3D volume, it is more convenient to express the steady-state situation with protein numbers rather than concentrations, namely, $k_{off}N_M = k_{on}N_C$. Thus, N_C is the number of molecules available for exchange from the cytoplasmic layer in a single average step to the membrane during the FRAP time interval (see reference 46 for details).

Originally in the derivation of the above model,[22] the kinetics of the two subpopulations were not coupled, and also $k_{on}N_C$ was treated as a constant. Therefore, only the values of k_{off} and D_C could be extracted from the fit (similar to other reports[47]). In a recent, more complex derivation of the model, the two subpopulations are coupled, and therefore, the value of k_{on} can also be determined (see reference 46 for a detailed description and derivation of this model, as well as the MATLAB® code for analyzing the FRAP data). However, since in this case there are more fitting parameters (and therefore more degrees of freedom), some of the parameters must be measured in independent experiments and entered as a fixed input while fitting the FRAP data. Recently, the on and off rates for eight different plaque proteins under different conditions were extracted in this manner, showing high correlation between adhesion molecular and structural dynamics.[48]

27.3 Conclusions and Future Prospects

With the advent of commercial automated fluorescence microscopes in the last couple of decades, performing dynamic measurements such as those discussed above has become a much simpler task than before. Still, *quantitative analysis* of such measurements often requires developing the right mathematical model that describes the system being studied. Specifically in the case of focal adhesions, we now have a variety of tools and models to accurately characterize the molecular dynamics of plaque proteins and link those to the state of the adhesion. The importance of obtaining the accurate kinetic parameters lies in the fact that the on and off rates of a specific protein reflect its binding affinity to the adhesion, which is presumably regulated by the adhesion's protein composition. This affinity may change, as evident by the different time constants reported in the literature for the same protein in different cell types (Table 27.1). Therefore, analysis of adhesion composition

TABLE 27.1

Reported Values in the Literature of Half-Time to Recovery ($T_{1/2}$) and Mobile Fraction of Vinculin from FRAP Measurements Performed in Different Cell Types and under Different Conditions

$T_{1/2}$ (s)	Mobile Fraction (%)	Cell Type	Fibronectin Concentration/Comments	Reference No.
83	57	Vinculin null mouse embryonic fibroblasts	20 μg/mL	49
20	85	PtK1 cells	5 μg/mL	50
9	81	PtK1 cells	10 μg/mL	50
36	87	PtK1 cells	30 μg/mL	50
20	55	NIH-3T3 cells	10 μg/mL	51
52	N.A.	Vinculin null mouse embryonic fibroblasts	20 μg/mL	51
17	78	NIH 3T3 cells	10 μg/mL	52
13	83	NIH 3T3 cells	10 μg/mL Measurements were taken under treatment with the Rac1 inhibitor NSC23766	52
57	52	Normal human epidermal keratinocytes	2.5 μg/cm² Stationary cells	53
36	70	Normal human epidermal keratinocytes	2.5 μg/cm² Mobile cells	53
40	63	HeLa cells	20 μg/mL	54
10	88	HeLa cells	20 μg/mL Measurements were taken under treatment with the actomyosin-contractility inhibitor blebbistatin	54
20–120	60–80	Vinculin null mouse embryonic fibroblasts	50 μg/mL The range of values corresponds to different traction-forces applied by the adhesions to the matrix	55

under different conditions (e.g., under different force regimes[54,55]) in parallel to accurate measurements of protein kinetics would help build a *dynamic systems-level model* of the integrin adhesome.

Some limitations still exist, for instance, in quantifying the molecular dynamics of focal complexes (since they are not at steady state) or of sliding adhesions (as demonstrated by the ICS studies). Importantly, other novel tools such as three-dimensional SPT,[56] which enables tracking the trajectories of particles also in the *z*-axis, or fluorescence loss after photoconversion,[57] which could allow distinguishing between on and off rates under nonsteady-state conditions, are expected to help overcome such drawbacks in the future. Another very intriguing prospect for the future is using specific knockout of a variety of plaque proteins to study their role in regulating adhesion dynamics. New technologies such as CRISPR[58] now allow knocking out specific genes in cell lines in a relatively straightforward manner. Knocking out plaque proteins one by one will allow identifying the key molecules and the mechanisms that regulate adhesion dynamics.

Acknowledgments

I acknowledge the support from the FP7 Marie Curie Actions of the European Commission via the International Outgoing Fellowship (FP7-PEOPLE-2012-IOF-332045). I also thank Thomas Iskratsch and Simon Moore for helpful comments on the text.

References

1. Zamir, E. and Geiger, B. Molecular complexity and dynamics of cell-matrix adhesions. *J. Cell Sci.* **114**, 3583–3590 (2001).
2. Burridge, K. and Chrzanowska-Wodnicka, M. Focal adhesions, contractility, and signaling. *Annu. Rev. Cell Dev. Biol.* **12**, 463–518 (1996).
3. Abercrombie, M. and Dunn, G. A. Adhesions of fibroblasts to substratum during contact inhibition observed by interference reflection microscopy. *Exp. Cell Res.* **92**, 57–62 (1975).
4. Cukierman, E., Pankov, R., Stevens, D. R. and Yamada, K. M. Taking cell-matrix adhesions to the third dimension. *Science* **294**, 1708–1712 (2001).
5. Geiger, B., Bershadsky, A., Pankov, R. and Yamada, K. M. Transmembrane crosstalk between the extracellular matrix and the cytoskeleton. *Nat. Rev. Mol. Cell Biol.* **2**, 793–805 (2001).
6. Hynes, R. O. Integrins: A family of cell surface receptors. *Cell* **48**, 549–554 (1987).
7. Zaidel-Bar, R., Itzkovitz, S., Ma'ayan, A., Iyengar, R. and Geiger, B. Functional atlas of the integrin adhesome. *Nat. Cell Biol.* **9**, 858–867 (2007).
8. Avizienyte, E. and Frame, M. C. Src and FAK signalling controls adhesion fate and the epithelial-to-mesenchymal transition. *Curr. Opin. Cell Biol.* **17**, 542–547 (2005).
9. Dupont, S. *et al.* Role of YAP/TAZ in mechanotransduction. *Nature* **474**, 179–183 (2011).
10. Zaidel-Bar, R. and Geiger, B. The switchable integrin adhesome. *J. Cell Sci.* **123**, 1385–1388 (2010).
11. Berginski, M. E. and Gomez, S. M. The Focal Adhesion Analysis Server: A web tool for analyzing focal adhesion dynamics. *F1000Research.* [v1; ref status: indexed, http://f1000r.es/ycl, (2013). doi: 10.12688/f1000research.2-68.v1.
12. Vicente-Manzanares, M. and Horwitz, A. R. Adhesion dynamics at a glance. *J. Cell Sci.* **124**, 3923–3927 (2011).
13. Wolfenson, H., Lavelin, I. and Geiger, B. Dynamic regulation of the structure and functions of integrin adhesions. *Dev. Cell* **24**, 447–458 (2013).
14. Giannone, G. *et al.* Periodic lamellipodial contractions correlate with rearward actin waves. *Cell* **116**, 431–443 (2004).
15. Choi, C. K. *et al.* Actin and alpha-actinin orchestrate the assembly and maturation of nascent adhesions in a myosin II motor-independent manner. *Nat. Cell Biol.* **10**, 1039–1050 (2008).
16. Yu, C. H., Law, J. B., Suryana, M., Low, H. Y. and Sheetz, M. P. Early integrin binding to Arg-Gly-Asp peptide activates actin polymerization and contractile movement that stimulates outward translocation. *Proc. Natl. Acad. Sci. U.S.A.* **108**, 20585–20590 (2011).
17. Roca-Cusachs, P., Gauthier, N. C., Del Rio, A. and Sheetz, M. P. Clustering of alpha(5)beta(1) integrins determines adhesion strength whereas alpha(v)beta(3) and talin enable mechanotransduction. *Proc. Natl. Acad. Sci. U.S.A.* **106**, 16245–16250 (2009).
18. Riveline, D. *et al.* Focal contacts as mechanosensors: Externally applied local mechanical force induces growth of focal contacts by an mDia1-dependent and ROCK-independent mechanism. *J. Cell Biol.* **153**, 1175–1186 (2001).
19. Zaidel-Bar, R., Milo, R., Kam, Z. and Geiger, B. A paxillin tyrosine phosphorylation switch regulates the assembly and form of cell-matrix adhesions. *J. Cell Sci.* **120**, 137–148 (2007).
20. Pankov, R. *et al.* Integrin dynamics and matrix assembly: Tensin-dependent translocation of alpha(5)beta(1) integrins promotes early fibronectin fibrillogenesis. *J. Cell Biol.* **148**, 1075–1090 (2000).
21. Ridley, A. J. *et al.* Cell migration: Integrating signals from front to back. *Science* **302**, 1704–1709 (2003).
22. Wolfenson, H. *et al.* A role for the juxtamembrane cytoplasm in the molecular dynamics of focal adhesions. *PLoS One* **4**, e4304 (2009).
23. Zamir, E. *et al.* Dynamics and segregation of cell-matrix adhesions in cultured fibroblasts. *Nat. Cell Biol.* **2**, 191–196 (2000).
24. Zamir, E., Geiger, B. and Kam, Z. Quantitative multicolor compositional imaging resolves molecular domains in cell-matrix adhesions. *PLoS One* **3**, e1901 (2008).

25. Webb, D. J. *et al.* FAK-Src signalling through paxillin, ERK and MLCK regulates adhesion disassembly. *Nat. Cell Biol.* **6**, 154–161 (2004).
26. Yamamoto, A., Mishima, S., Maruyama, N. and Sumita, M. Quantitative evaluation of cell attachment to glass, polystyrene, and fibronectin- or collagen-coated polystyrene by measurement of cell adhesive shear force and cell detachment energy. *J Biomed. Mater. Res.* **50**, 114–124 (2000).
27. Axelrod, D. Total internal reflection fluorescence microscopy in cell biology. *Traffic* **2**, 764–774 (2001).
28. Zamir, E. *et al.* Molecular diversity of cell-matrix adhesions. *J. Cell Sci.* **112 (Pt 11)**, 1655–1669 (1999).
29. Berginski, M. E., Vitriol, E. A., Hahn, K. M. and Gomez, S. M. High-resolution quantification of focal adhesion spatiotemporal dynamics in living cells. *PLoS One* **6**, e22025 (2011).
30. Saxton, M. J. and Jacobson, K. Single-particle tracking: Applications to membrane dynamics. *Annu. Rev. Biophys. Biomol. Struct.* **26**, 373–399 (1997).
31. Jaqaman, K. *et al.* Cytoskeletal control of CD36 diffusion promotes its receptor and signaling function. *Cell* **146**, 593–606 (2011).
32. Lubelski, A., Sokolov, I. M. and Klafter, J. Nonergodicity mimics inhomogeneity in single particle tracking. *Phys. Rev. Lett.* **100**, 250602 (2008).
33. Chen, H., Titushkin, I., Stroscio, M. and Cho, M. Altered membrane dynamics of quantum dot-conjugated integrins during osteogenic differentiation of human bone marrow derived progenitor cells. *Biophys. J.* **92**, 1399–1408 (2007).
34. Rossier, O. *et al.* Integrins beta1 and beta3 exhibit distinct dynamic nanoscale organizations inside focal adhesions. *Nat. Cell Biol.* **14**, 1057–1067 (2012).
35. Lippincott-Schwartz, J. and Patterson, G. H. Photoactivatable fluorescent proteins for diffraction-limited and super-resolution imaging. *Trends Cell Biol.* **19**, 555–565 (2009).
36. Manley, S. *et al.* High-density mapping of single-molecule trajectories with photoactivated localization microscopy. *Nat. Methods* **5**, 155–157 (2008).
37. Hu, K., Ji, L., Applegate, K. T., Danuser, G. and Waterman-Storer, C. M. Differential transmission of actin motion within focal adhesions. *Science* **315**, 111–115 (2007).
38. Wiseman, P. W., Squier, J. A., Ellisman, M. H. and Wilson, K. R. Two-photon image correlation spectroscopy and image cross-correlation spectroscopy. *J. Microsc.* **200**, 14–25 (2000).
39. Elson, E. L. Fluorescence correlation spectroscopy: Past, present, future. *Biophys. J.* **101**, 2855–2870 (2011).
40. Digman, M. A. *et al.* Measuring fast dynamics in solutions and cells with a laser scanning microscope. *Biophys. J.* **89**, 1317–1327 (2005).
41. Hebert, B., Costantino, S. and Wiseman, P. W. Spatiotemporal image correlation spectroscopy (STICS) theory, verification, and application to protein velocity mapping in living CHO cells. *Biophys. J.* **88**, 3601–3614 (2005).
42. Digman, M. A., Brown, C. M., Horwitz, A. R., Mantulin, W. W. and Gratton, E. Paxillin dynamics measured during adhesion assembly and disassembly by correlation spectroscopy. *Biophys. J.* **94**, 2819–2831 (2008).
43. Digman, M. A., Wiseman, P. W., Horwitz, A. R. and Gratton, E. Detecting protein complexes in living cells from laser scanning confocal image sequences by the cross correlation raster image spectroscopy method. *Biophys. J.* **96**, 707–716 (2009).
44. Carisey, A., Stroud, M., Tsang, R. and Ballestrem, C. Fluorescence recovery after photobleaching. *Methods Mol. Biol.* **769**, 387–402 (2011).
45. Eisenberg, S. *et al.* Raft protein clustering alters N-Ras membrane interactions and activation pattern. *Mol. Cell Biol.* **31**, 3938–3952 (2011).
46. Berkovich, R. *et al.* Accurate quantification of diffusion and binding kinetics of non-integral membrane proteins by FRAP. *Traffic* **12**, 1648–1657 (2011).
47. Lele, T. P. *et al.* Mechanical forces alter zyxin unbinding kinetics within focal adhesions of living cells. *J. Cell Physiol.* **207**, 187–194 (2006).
48. Lavelin, I. *et al.* Differential effect of actomyosin relaxation on the dynamic properties of focal adhesion proteins. *PLoS One* **8**, e73549 (2013).
49. Cohen, D. M., Kutscher, B., Chen, H., Murphy, D. B. and Craig, S. W. A conformational switch in vinculin drives formation and dynamics of a talin-vinculin complex at focal adhesions. *J. Biol. Chem.* **281**, 16006–16015 (2006).
50. Gupton, S. and Waterman-Storer, C. Spatiotemporal feedback between actomyosin and focal-adhesion systems optimizes rapid cell migration. *Cell* **125**, 1361–1374 (2006).
51. Humphries, J. D. *et al.* Vinculin controls focal adhesion formation by direct interactions with talin and actin. *J. Cell Biol.* **179**, 1043–1057 (2007).
52. Deakin, N. O., Ballestrem, C. and Turner, C. E. Paxillin and Hic-5 interaction with vinculin is differentially regulated by Rac1 and RhoA. *PLoS One* **7**, e37990 (2012).
53. Mohl, C. *et al.* Becoming stable and strong: The interplay between vinculin exchange dynamics and adhesion strength during adhesion site maturation. *Cell Motil. Cytoskeleton* **66**, 350–364 (2009).
54. Wolfenson, H., Bershadsky, A., Henis, Y. I. and Geiger, B. Actomyosin-generated tension controls the molecular kinetics of focal adhesions. *J. Cell Sci.* **124**, 1425–1432 (2011).
55. Dumbauld, D. W. *et al.* How vinculin regulates force transmission. *Proc. Natl. Acad. Sci. U.S.A.* **110**, 9788–9793 (2013).
56. Toprak, E., Balci, H., Blehm, B. H. and Selvin, P. R. Three-dimensional particle tracking via bifocal imaging. *Nano Lett.* **7**, 2043–2045 (2007).
57. Ehrlicher, A. J., Nakamura, F., Hartwig, J. H., Weitz, D. A. and Stossel, T. P. Mechanical strain in actin networks regulates FilGAP and integrin binding to filamin A. *Nature* **478**, 260–263 (2011).
58. Mali, P. *et al.* RNA-guided human genome engineering via Cas9. *Science* **339**, 823–826 (2013).

28

Nanoscale Mechanical Testing of FIB-Isolated Biological Specimens

Ines Jimenez-Palomar, Russell J. Bailey, and Asa H. Barber

CONTENTS

Focused ion beam (FIB) microscopy provides unprecedented opportunities in isolating discrete volumes of biological material for further experimental studies. While FIB has been historically exploited for preparing relatively thin biological samples for high-resolution transmission electron microscopy, more recent works have applied a range of small-scale mechanical testing techniques, such as nanoindentation and atomic force microscopy, to evaluate the mechanical properties of these FIB-prepared samples. The overall structure–mechanical function of biological materials can therefore be assessed by isolated discrete volumes of materials from a variety of spatial locations and subsequent mechanical evaluation of these volumes. The aim of this chapter is to review the development of FIB and its application in understanding the mechanical properties of complex biological materials. Particular attention is given to mineralized tissues where distinct composite structures that provide clear mechanical functions exist. Emphasis on the use of FIB to prepare samples free from artifacts that would dictate resultant mechanical properties is made, including efforts to understand the hydration state of mineralized tissue. Finally, the ability to mechanically test discrete volumes of biological material while monitoring using in situ microscopy is highlighted as a progression in the field, especially when specific deformation and failure mechanisms of a region within a biological material are to be explored.

28.1 Advantages of Isolating Discrete Volumes in Complex Materials

Biological materials such as wood, bone, and shell are inherently complex structures. In many cases, these biological materials can be considered as reinforced composites with a hierarchical structure extended across a range of length scales. As with man-made composite materials, these natural composites are optimized for a specific mechanical function such as load bearing or protection, as well as performing a number of biological roles. In contrast to man-made composites, biological materials have the ability to grow and remodel depending on the need to adapt to its environment and ultimately heal or self-repair (Fratzl and Weinkamer 2007). The hierarchical structure in biological materials is therefore highly complex so that each specific structure contributes both to overall mechanical performance as well as to being adapted to local needs (Jeronimidis 2000).

The difficulties in describing the mechanics of biological materials originate from the relative complexity found in these structures and applying current solid mechanics theories commonly developed for structurally simpler man-made composites. A range of issues must therefore be considered in biological materials such as the mechanical properties of individual components, the interaction between components during mechanical loading, and design strategies driving the assembly of components in order to create structural complexity across all of the different hierarchical length scales. Understanding the mechanical properties of a particular length scale in order to describe the performance of the whole biological material therefore requires techniques that are able to both isolate a discrete volume of material and subsequently probe the mechanical performance of the isolated unit. This chapter describes recent efforts in experimental procedures to isolate discrete volumes of biological material using focused ion beam (FIB) microscopy and provide subsequent mechanical testing to evaluate the mechanical properties of the selected volume. While our current efforts are focused on the understanding of mineralized tissues, the approach to isolating biological material volumes is flexible and can be used for a wide variety of biological systems.

28.2 FIB Microscopy

28.2.1 Historical Development

FIB microscopy is a powerful tool that was initially implemented as early as the 1960s for the fine polishing of material surfaces, such as those of optical lenses (Schroeder et al. 1966) and, at a much smaller scale, for the thinning of transmission electron microscope (TEM) samples (Abrahams et al. 1968; Abrahams and Buiocchi 1974; Yoshida and Yamada 1984). These early attempts used Ar^+ ions set at shallow angles of incidence (1°–10°) from two Penning ionization sources to impinge on a sample's surface. The momentum of the impinging ions caused removal of atoms and fragments of the sample in a process known as sputtering. This ion system allowed the milling of samples for the production of an electron transparent section at thicknesses of the order of 100 nm. Further work was carried out to improve the production of ions in the 1970s by using the first field emission ion sources for FIB applications, especially maskless doping of surfaces in the semiconductor industry (Escovitz et al. 1975; Orloff and Swanson 1975, 1978). The original field emission ion sources used gas phase field ionization, whereby the neutral gas of the element to be ionized was fed to the tip of a sharp electrode. A large positive potential was applied to the electrode, and the high field strength, due to the sharpened tip, caused tunneling of electrons from the gas atoms and repulsion of the generated ions. Seliger and Fleming (1974) further demonstrated a FIB system that could achieve a spot size of 3.5 μm at a beam energy of 60 keV, which would be suitable for maskless doping.

Concurrently, interest in FIB systems grew as an alternative to electron beam lithography (EBL) of resist layers in microfabrication, and a critical breakthrough was made in the development of liquid metal ion sources (LMISs), which are commonly used in modern FIB microscopy. The LMIS is capable of producing ion beams with greater brightness and a higher current density than the standard field ionization for both increased sputtering rates and smaller ion beam spot sizes (Clampitt et al. 1975; Broers 1979). LMISs utilize the principle of field evaporation from the liquid metal surface, whereby the application of a strong electrical field reduces the potential barrier and causes atom evaporation followed by ionization in the field at the source tip. LMISs commonly use a crucible or needle point source, coated with the ion source liquid, which is fed to the tip from a reservoir (Orloff 1993; Giannuzzi and Stevie 1999). Field evaporation occurs upon application of an extracting voltage to the sharp source needle, which results in a large electric field of the order of 1×10^8 V cm^{-1}. The liquid metal in the presence of the extraction field distorts according to a Taylor cone, where the shape of the cone is achieved from a balance between the surface tension holding the liquid together and the electrostatic forces that cause repulsion within the liquid (Taylor 1964). Taylor cones importantly stabilize the liquid at a half angle of 49.3° to give a further increase in field strength during the field evaporation process for a high ion emission flux. The majority of LMIS FIB microscopes use gallium as the metal ion source, and improved ion optics have led to gallium ion beams with probe sizes ranging from initial sub-50 nm probe diameters to typically reported ion beam resolutions of 5 nm (Gamo and Namba 1991; Swanson 1994). Further use of a bias electrode around the source needle of the FIB has resulted in stabilization of the emission of ions from the LMIS (Kawakatsu 1981) to provide a variable ion beam current while maintaining a constant beam energy (Orloff and Swanson 1981). These developments have therefore resulted in control of gallium ion beam energy, flux, and spot size for the precision removal of any material through sputtering processes.

28.2.2 Use of FIB in Micromachining and Nanomachining

One of the major applications of the FIB, especially with regard to the high brightness beams produced by LMIS, has been micromachining and nanomachining. The ability to sputter material accurately, reproducibly, and with high-resolution means has historically promoted the use of FIB microscopy within modern semiconductor and other electronics fields (Prewett and Mair 1991; Orloff et al. 2003; Giannuzzi and Stevie 2004), high-performance optics (Harriott et al. 1987), and specimen preparation for high-resolution imaging techniques (Prenitzer et al. 1998a,b; Giannuzzi and Stevie 1999; Langford and Clinton 2004). Specifically, Harriot et al. (1987) demonstrated FIB applications in sputtering material to produce optical cavities used in laser systems. However, the mill times were relatively long, of the order of 20 min, for a single mirror surface, and so they surmised that FIB would only be suitable for unique, high-value devices or for prototyping and research.

Considerable applications of FIB were found in the preparation of samples with thicknesses of the order of 100 nm for high-resolution TEM. Previous methods of specimen preparation involved the use of microtomes, polishing discs, and polishing ion systems, which were relatively crude and lacked controlled selection of the sample to be evaluated. With the use of FIB microscopy, specific sites of interest were shown to be selected, exposed, and examined in a number of ways, depending on the final technique to be used. Specifically, Giannuzzi and Stevie (1999) described two distinct methods of preparing specimens for subsequent TEM evaluations. The first "trench" method is best used to thin a sample that has already been produced mechanically and requires the use of the FIB to mill parallel trenches into opposite edges of the sample film to give a thin foil in selected regions within a mechanically prepared sample. The "lift-out" method is a technique that has been developed for use with small dual beam (SDB) systems, which consist of a FIB with an additional scanning electron microscope (SEM) for in situ electron image capability. The advantage of the SDB system lies in the SEM allowing direct observation of the progress of the FIB, with little or no damage to the sample. The "lift-out" method uses the FIB to mill parallel trenches normal to the surface of the sample. This bulk material around the site of interest is removed rapidly using high beam currents, leaving a thin section, containing the site of interest, between the trenches. The sample section is then thinned further using lower beam currents and tilted into the FIB for a final polishing step (Prenitzer et al. 1998a,b). The final TEM foil is then tilted relative to the FIB so that it can be detached from the side walls, and the sample attached to a micromanipulator before the final cut is made across the base of the sample using FIB to provide detachment of the sample. The TEM specimen is then lifted from the bulk material by the manipulator and deposited onto a TEM grid ready for transfer to the TEM. Further application of the FIB in an SDB system is the development of "slice and view" or 3D FIB tomography. This three-dimensional technique uses the FIB to prepare sequential cross sections of a sample that can be imaged using the in situ SEM as shown in Figure 28.1. The series of images obtained by SEM can then be aligned and reconstituted into a three-dimensional model of the sample. Slice and view has been used in evaluating synthetic materials such as, for example, visualization of the internal structure of semiconductors and other solid devices (Sugiyama and Sigesato 2004) and characterization of the deformation and damage due to nanoindentation of multilayered composites (Singh et al. 2010).

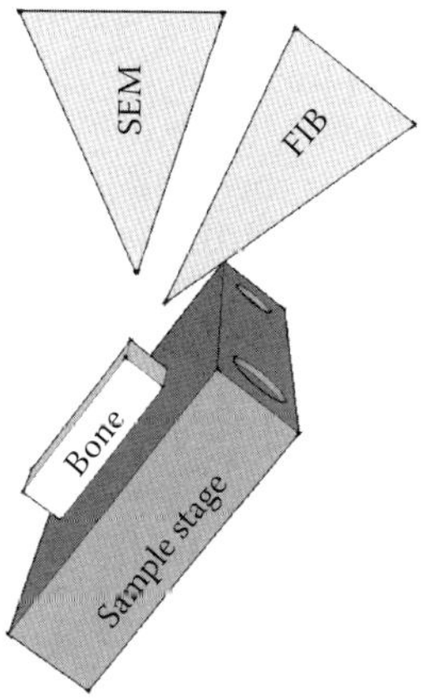

FIGURE 28.1
Schematic of the SDB system. The SEM beam allows top-down imaging of the sample, whereas the sample is typically tilted to present a surface parallel to the FIB set at a 52° angle for subsequent sample patterning.

FIB has more recently been applied to the studies of biological materials through the use of sample thinning for subsequent analysis using TEM and 3D FIB tomography (Drobne 2013). For example, whole cells and tissues have been reconstructed fully in 3D through application of a FIB section and SEM observation (Kamino et al. 2004; Heymann et al. 2006). In 2006, Heymann et al. demonstrated the abilities of 3D FIB tomography by imaging a number of biological samples including cross sections as well as 3D reconstructions of embedded, critical-point dried and cryogenically frozen yeast cells. The study also demonstrated the reconstruction of a section of embedded tumor tissue, lift-out of embedded yeast cells for TEM, and the production of cylindrical specimens for electron tomography (Heymann et al. 2006). Further biological work was shown in 2009, where Schroeder-Reiter et al. (2009) demonstrated the potential of 3D FIB tomography for imaging the structural elements in critical-point dried chromosomes. Indeed, the use of sample drying for FIB preparation has been described in detail for a range of biological materials (Drobne 2013). While sample drying is suitable for structural investigations, removal of water from a biological sample is treated with caution here due to its clearly expected impact on the mechanical properties of the sample.

28.2.3 Evaluation and Control of FIB-Induced Sample Damage

The isolation of discrete sample volumes using FIB raises issues of the ion beam inducing damage during sample preparation. This FIB-induced damage will become critical in modifying the mechanical properties of the isolated sample volume and therefore requires understanding of ion beam interactions. Many studies have evaluated the interaction between gallium ions and silicon samples, especially as the FIB has a historical basis in the semiconductor industry. Specifically, FIB-induced damage in silicon is greatly dependent on the accelerating voltage applied to the ion beam and the relative angle of the incident beam to the sample's surface (Kempshall et al. 2002; Brezna et al. 2003; Rubanov and Munroe 2004; Kolibal et al. 2011). Initial investigations into damage in biological samples have described artifacts from FIB preparation, including morphological changes of "melt-like" effects and "curtaining" of the exposed surface (Drobne et al. 2007). Similar morphological changes have been additionally observed in synthetic polymers exposed to FIB (Brostow et al. 2007; Wong et al. 2010). Consequently, methods to remove artifacts from FIB processing have been explored, with the most popular being the use of cryopreparation techniques to reduce the effects of sample heating during exposure to the FIB (Hayles et al. 2007). However, cryopreparation is useful for preserving the structure for observation at the corresponding low temperature, whereas taking the sample back up to room temperature to explore mechanical properties will additionally introduce preparation artifacts, especially in biological materials where amorphous solid water forms into ice crystals of increased volume.

Recent work has explored FIB-induced damage of soft materials by taking an amorphous synthetic polymer (polycarbonate) and exposing the polymer surface to a range of FIB energies, varied using ion beam current and applied accelerating voltage (Bailey et al. 2013), and at the standard grazing angle configuration as shown in Figure 28.1. The work noted that FIB currents above 7 nA produced thermal damage at the surface of the sample but was useful in sputtering away materials from the sample rapidly, indicating that high beam currents should be used for initial isolation of a discrete sample volume, whereas finer, and less evasive, FIB milling should be performed with lower beam currents. Critically, the work examined the mechanical properties of the polymer surface using phase contrast atomic force microscopy (AFM). Figure 28.2 shows a plot of the phase shift, indicative of the mechanical properties of the sample, and the root-mean-square (RMS) roughness of the polymer surface when varying the voltage applied to the impinging FIB.

Figure 28.2 highlights the phase shift of the unexposed polymer surface, when the ion beam energy is 0 keV, and an initial drop of the phase shift with low FIB energy. The recovery of the phase shift only occurs when the FIB energy exceeds 15 keV, suggesting that the ion beam causes damage in the sample at relatively low beam currents. Further simulation work supported these experimental observations by modeling the gallium ion interactions with sample atoms (Bailey et al. 2013). A mechanism was found that highlighted how gallium ions with low-energy travel scatter by a small distance within the sample but do not have sufficient energy to sputter significant volumes of the material. Increasing the ion

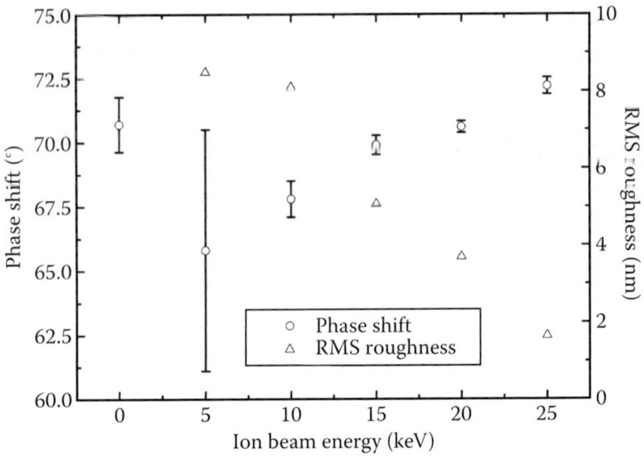

FIGURE 28.2
Plot of the phase shift, indicative of the mechanical properties of the sample surface, and the RMS roughness of the sample against the ion beam energy as a function of the applied voltage. (Reprinted from Bailey, R. J., R. Geurts, D. J. Stokes, F. d. Jong and A. H. Barber, Evaluating focused ion beam induced damage in soft materials, *Micron*, 50, 51–56, Copyright 2013, with permission from Elsevier.)

beam energy causes an increase in the scattering distance traveled within the sample, but these higher energy ions are able to sputter away the volumes in which they are contained within, thus providing a sample without any embedded gallium and resulting in mechanical properties of the sample that are unmodified by FIB exposure. FIBs should therefore be operated at relatively low ion beam currents below 7 nA but at higher beam energies above 15 keV in order to produce samples with mechanical properties that are unmodified by exposure to the FIB.

28.3 FIB Preparation of Biological Samples

The standard method for producing discrete volumes of material in biological specimens with both mineralized and protein phases was previously fully described by Jimenez-Palomar et al. (2012) for rat bone femur. In summary, the bone sample was firstly dehydrated by submerging it in a series of water/ethanol solutions, summarized in Table 28.1, prior to FIB milling. The dehydrated

TABLE 28.1
Dehydration of Bone Using Increasing Concentrations of Ethanol in Water

% Ethanol	Time (min)
70%	Storage
90%	60
95%	30
100%	30
100%	30

sample was then coated with a continuous thin layer (<10 nm) of gold and fixed to the sample stage of a dual beam microscope (Quanta 3D FEG, FEI, USA/EU) using a two-part epoxy glue (Poxipol, Arg.). The gold coating is required to avoid the sample from becoming electrically charged under the ion beam, which causes subsequent repulsion of the FIB from the charged sample surface. Sample dehydration is also required as the FIB operates in a vacuum. FIB milling to produce a discrete sample volume from the bulk bone material was performed by a succession of FIB milling steps summarized in Figure 28.3.

In order to successfully produce discrete beam volumes of biological structures using a FIB, a stepped scaling-down approach is used as shown in Table 28.2. Initially flat edges of the biological structure are produced using a high current ion beam of 65 nA and an accelerating voltage of 30 kV. As the material surrounding the desired structure is removed, the current is reduced in a series of steps to 0.1 nA as shown in Figure 28.3. The smaller ion beam currents avoid observable ion beam damage in the sample due to thermal effects. In addition, the FIB milling is always performed parallel to the produced sample faces at a grazing angle to reduce the embedding of the impinging gallium ions from the FIB within the discrete beam volumes produced. These operating parameters are consistent with the minimization of sample damage during FIB exposure. The corresponding relationship among the ion beam conditions, the FIB preparation steps in Figure 28.3, and the dimensions of beams produced at each preparation step are summarized in Table 28.2.

TABLE 28.2
FIB Conditions Used to Define the Discrete Beam Volume Size as Shown in Figure 28.3

Current (nA)	Dimension (μm) $x \times y \times z$	Time (min)
65	100 × 20 × 2	20
65	100 × 20 × 2	20
15	100 × 10 × 1	10
1	5 × 10 × 1	10
0.1	Polish around the beam	5–10

Note: The steps are chronological from top (high current) to bottom (low current).

A common discrete volume geometry prepared using a FIB is the rectangular beam. To achieve this beam geometry, a short column between each of the FIB-produced beams is used in order to prevent the redeposition of milled material and gallium ions onto the neighboring beams. This sputtered material redeposits onto the short columns instead of the sample beams, which results in more regular rectangular beam geometry produced from the FIB process as shown in Figure 28.4.

The beams in the bulk rat bone sample were noted as being dehydrated, and the need to subsequently rehydrate after the FIB milling process was thus required. Initial efforts to rehydrate by placing the sample into water were unsuccessful as the surface tension of the liquid applied capillary forces sufficient to fail the FIB-prepared beams in Figure 28.4. A less evasive rehydration is therefore used where the sample was removed from the dual beam chamber and placed in a closed vessel containing a high vapor concentration of Hank's

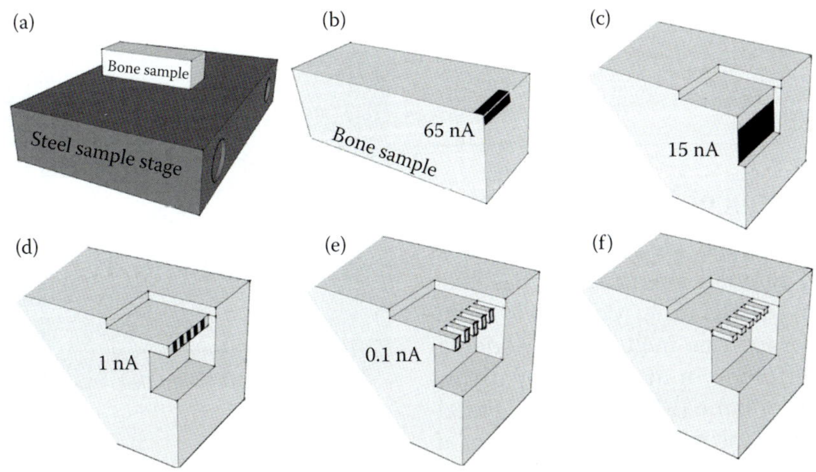

FIGURE 28.3
Schematic diagram showing the FIB preparation steps required to produce a series of beams within a bulk material. The black rectangles mark the FIB milled area that sputters away sample. (a) The sample, such as bone, is first mounted on a steel sample stage. In (b), two initial edges of the sample at a corner are milled to a flat surface using a FIB current of 65 nA. (c) A further relatively large sample volume is sputtered from the bulk using a FIB current of 15 nA, followed by (d) isolation of beams using a FIB current of 1 nA. Further fine milling in (e) to shape the beams toward more regular geometries is achieved using a FIB current of 0.1 nA to give (f) final samples. All cuts performed with a FIB acceleration voltage of 30 kV. (Reprinted from Jimenez-Palomar, I., A. Shipov, R. Shahar and A. H. Barber, Influence of SEM vacuum on bone micromechanics using in situ AFM, *Journal of the Mechanical Behavior of Biomedical Materials*, 5, 149–155, Copyright 2012, with permission from Elsevier.)

FIGURE 28.4
(a) Scanning electron micrograph of a series of beams fabricated from the bulk rat bone sample and (b) higher magnification backscattered electron micrograph showing an individual rat bone beam. (Reprinted from Jimenez-Palomar, I., A. Shipov, R. Shahar and A. H. Barber, Influence of SEM vacuum on bone micromechanics using in situ AFM, *Journal of the Mechanical Behavior of Biomedical Materials*, 5, 149–155, Copyright 2012, with permission from Elsevier.)

buffer solution for 2 h. Literature shows that bone samples have been observed to require only 45 min for rehydration (Utku et al. 2008), indicating that 2 h is sufficient for hydration, although adjustments to these times may be required depending on the biological sample examined. We note that the literature presented on dehydration and rehydration is a persistent problem in mineralized tissue preparation. Sample dehydration after preparation using a FIB is particularly problematic for further mechanical tests as "structural" water has been shown to influence the mechanical properties of mineralized tissue (see discussion in Jimenez-Palomar et al. 2012).

28.4 Mechanical Testing of Discrete Volumes

The process of measuring the mechanical properties of biological materials at small length scales has been commonly employed in the literature. Methods to evaluate local mechanical properties include nanoindentation, which has been effective in making comparisons between different areas of bone (Rho et al. 1997; Fratzl-Zelman et al. 2009), variations in bone types (Rho et al. 1997), and the effect of disease (Guo and Goldstein 2000; Fratzl-Zelman et al. 2009; Maïmoun et al. 2012) on the mechanical properties of bone at smaller hierarchical levels. However, these results have shown a significant deviation from results for larger length scale testing (Choi et al. 1990; Barengolts et al. 1993; Kasra et al. 1997; Cory et al. 2010). A more recent study on nanoindentation as a technique to test the mechanical properties of bone has revealed the strong effect of the hydration state, probe geometry, and the assumptions of the analysis methods on the results (Rodriguez-Florez et al. 2013).

Further determination of mechanical properties of constituents in biological materials has been more recently available with the advent of more sophisticated and higher force resolution, which is able to elucidate constituent mechanics directly such as AFM tensile testing of individual mineralized collagen fibrils (Hang and Barber 2011). Mechanical testing of the discrete volumes of biological materials selected using a FIB is therefore a logical development from the uncertainties in deformation volume when using nanoindentation of bulk specimens while allowing to probe the mechanical properties of the "unit cell" of the biological material. However, mechanical testing of FIB-prepared discrete volumes is generally based on similar low load and high force resolution techniques. Previous literature has highlighted the production of beams from the bulk biological materials using a FIB, with examples of triangular cross-sectional beams in enamel and square cross-sectional beams in limpet teeth shown in Figure 28.5.

Direct mechanical testing of the beams as produced in Figure 28.5 has been performed using either nanoindentation (Chan et al. 2009) or AFM (Lu and Barber 2012) to apply a load to the end of the FIB-fabricated beam. Both techniques are able to apply a force and record the displacement of the beam under the applied force. The resultant force–displacement curves can be further converted to stress–strain curves for the determination of the inherent material's mechanical properties within the discrete volume tested using the established beam bending theory. Figure 28.6 shows an example of a

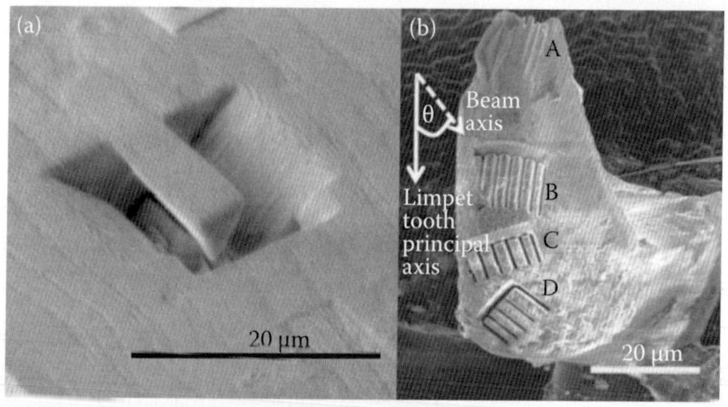

FIGURE 28.5
Scanning electron micrographs showing (a) a FIB-prepared beam with a triangular cross section in enamel and (b) series of square cross-sectional beams patterned at increasing orientations, labeled A to D in the image, along the long axis of the limpet tooth. (Reprinted from Chan, Y. L., A. H. W. Ngan and N. M. King, Use of focused ion beam milling for investigating the mechanical properties of biological tissues: A study of human primary molars, *Journal of the Mechanical Behavior of Biomedical Materials*, 2, 375–383, Copyright 2009, with permission from Elsevier; Adapted from Lu, D. and A. H. Barber, *Journal of the Royal Society Interface* 9: 1318–1324, 2012.)

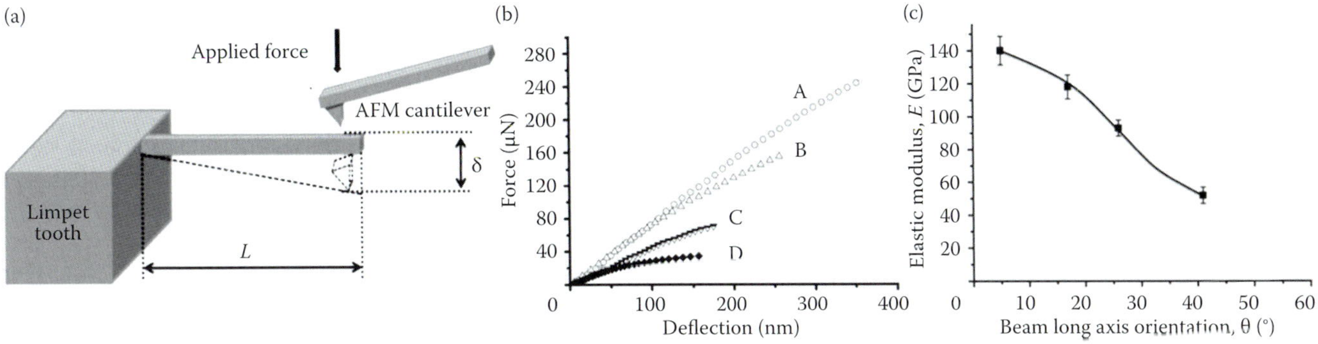

FIGURE 28.6
(a) Schematic showing the force applied to the end of the FIB-prepared beam using AFM; (b) the resultant beam deflection with applied force for beams located in regions A, B, C, and D in Figure 28.5, and (c) the calculated elastic modulus plotted as a function of beam orientation as shown in Figure 28.5. (Adapted from Lu, D. and A. H. Barber, *Journal of the Royal Society Interface* 9: 1318–1324, 2012.)

force–displacement curve for bending the beams shown for the limpet tooth sample in Figure 28.5.

The force F applied to the end of the beam provides a corresponding deflection δ, the plot gradient of which can be used to calculate the elastic modulus E of the beam using the established bending theory (Lu and Barber 2012):

$$\frac{F}{\delta} = \frac{Etw^3}{4L^3} \quad (28.1)$$

where t, w, and L are the thickness, width, and length of the beam, respectively. The strength of the triangular cross-sectional beams has additionally used the beam bending theory to give the flexural strength σ using (Chan et al. 2009)

$$\sigma = \frac{3F_{max}L}{2wt^3} \quad (28.2)$$

where F_{max} is the maximum force applied to fail the beam. These methods have therefore been suitable in elucidating local mechanical properties of biological materials, with the scale of the testing defined by the volume selected by the FIB down to nanometer resolution.

The application of force to FIB-prepared samples has recently driven the requirement to observe mechanical deformation using a suitable imaging technique that is able to resolve these discrete volumes. In situ mechanical testing is therefore required to monitor external loading on the discrete volume and has been applied to the observation of beam bending in rat bone femur. A generic configuration integrates the mechanical testing device within the imaging techniques, with Figure 28.7 showing the result of integrating a custom-built AFM into the chamber of an SEM, where the AFM provides the application of force while allowing access for the electron beam of the SEM. This configuration is a powerful synergy between the samples prepared using a FIB having dimensions in the microscale to nanoscale and the corresponding SEM imaging able to resolve such dimensions.

These in situ mechanical testing techniques are a powerful addition to understanding material failure as, in principle, the propagation of failure events such as crack propagation or plastic deformation in a biological material can be observed directly using the

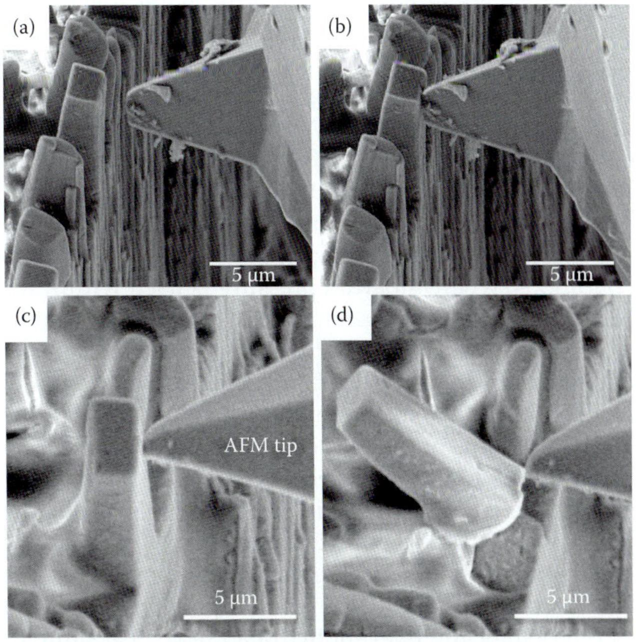

FIGURE 28.7
Scanning electron micrographs showing a beam bending configuration of FIB-prepared rat bone beams during in situ mechanical testing using AFM. Beam bending (a) before and (b) during testing can be observed with the SEM. (Reprinted from Jimenez-Palomar, I., A. Shipov, R. Shahar and A. H. Barber, Influence of SEM vacuum on bone micromechanics using in situ AFM, *Journal of the Mechanical Behavior of Biomedical Materials*, 5, 149–155, Copyright 2012, with permission from Elsevier.) (c) AFM tip contact with the beam can be used to apply increasing force until (d) failure of the beam is observed with the SEM.

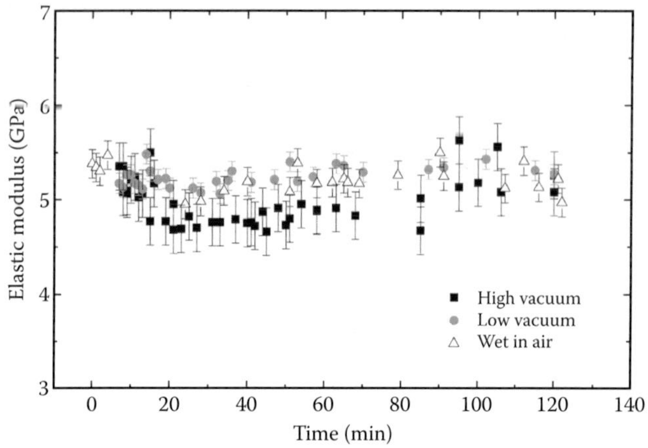

FIGURE 28.8
Plot of the elastic modulus of rat bone beams against time of exposure within high- and low-vacuum SEM conditions, baselined against bending tests for samples wet in air. (Reprinted from Jimenez-Palomar, I., A. Shipov, R. Shahar and A. H. Barber, Influence of SEM vacuum on bone micromechanics using in situ AFM, *Journal of the Mechanical Behavior of Biomedical Materials*, 5, 149–155, Copyright 2012, with permission from Elsevier.)

SEM and can be related to the corresponding force–displacement behavior recorded by nanoindentation or AFM. A potential limit to in situ mechanical testing is the requirement of using electron microscopy for direct observation of a sample with dimensions typically produced when prepared using a FIB, yet electron microscopy uses at least partial vacuums when operated. A recent study explored the effect of SEM vacuum on the mechanical properties of rat bone material in a range of environmental conditions ranging from hydrated samples through to SEM high-vacuum and low-vacuum conditions, where a partial pressure is used within the SEM chamber using water vapor (Jimenez-Palomar et al. 2012). The elastic modulus for rat bone beams, calculated from in situ AFM bending tests, in a range of environmental conditions is shown in Figure 28.8. The elastic modulus of the rat bone beams in low and high vacuum was found to be generally similar to wet-in-air environments up to a 2 h timeframe, with perhaps the low-vacuum testing showing the closest comparable mechanical behavior to the wet conditions across the timeframe considered. This result is perhaps significant as physiologically relevant hydrated biological samples can thus be mechanically studied within the SEM.

28.5 Conclusions

This chapter describes the justification in using a FIB to isolate discrete volumes from complex biological materials in order to assess local mechanical properties. The evaluation of local behavior in biological materials is expected to be important in understanding the contribution of mechanic at spatially selected volumes to the overall mechanical performance of the system. The use of a FIB requires understanding the interactions between the ion beam and the sample, with a comprehensive summary of FIB conditions highlighted so that samples free from ion beam artifacts can be produced following the experimental procedures described. The corresponding mechanical testing of the FIB-prepared biological samples is reviewed to show how local mechanical properties, including elastic modulus and strength, for the discrete volumes can be assessed. Finally, more recent work showing in situ mechanical testing where the deformation of the sample can be observed using electron microscopy is highlighted, as well as the ability to mechanical deform samples in a hydrated state. This strategy is expected to become more relevant in future work where in situ mechanics and concurrent observation of material deformation can be collected for a comprehensive description of local material behavior. Such an approach is generic and will progress with the development of future techniques. Mechanical testing is reaching maturity for studying FIB-prepared samples, but imaging methods are continuing to expand. "Super"-resolution optical microscopy is perhaps the next logical development for in situ mechanical testing and shows distinct advantages in observing biological materials in a fully hydrated state. This optical microscopy development would require corresponding developments in a FIB for preparing soft biological materials and retaining their mechanical properties for subsequent testing.

References

Abrahams, M. S. and C. J. Buiocchi (1974). "Cross-sectional specimens for transmission electron-microscopy." *Journal of Applied Physics* **45**(8): 3315–3316.

Abrahams, M. S., C. J. Buiocchi and M. D. Coutts (1968). "Improvements to the ALBA machine for thinning specimens for electron microscopy." *Review of Scientific Instruments* **39**(12): 1944–1945.

Bailey, R. J., R. Geurts, D. J. Stokes, F. d. Jong and A. H. Barber (2013). "Evaluating focused ion beam induced damage in soft materials." *Micron* **50**: 51–56.

Barengolts, E. I., D. J. Curry, M. S. Bapna and S. C. Kukreja (1993). "Effects of endurance exercise on bone mass and mechanical properties in intact and ovariectomized rats." *Journal of Bone and Mineral Research* **8**(8): 937–942.

Brezna, W., H. Wanzenbock, A. Lugstein, E. Bertagnolli, E. Gornik and J. Smoliner (2003). "Focussed ion beam induced damage in silicon studied by scanning capacitance microscopy." *Semiconductor Science and Technology* **18**: 195–198.

Broers, A. N. (1979). "High-resolution systems for microfabrication." *Physics Today* **32**(11): 38.

Brostow, W., B. P. Gorman and O. Olea-Mejia (2007). "Focused ion beam milling and scanning electron microscopy characterization of polymer + metal hybrids." *Materials Letters* **61**(6): 1333–1336.

Chan, Y. L., A. H. W. Ngan and N. M. King (2009). "Use of focused ion beam milling for investigating the mechanical properties of biological tissues: A study of human primary molars." *Journal of the Mechanical Behavior of Biomedical Materials* **2**: 375–383.

Choi, K., J. L. Kuhn, M. J. Ciarelli and S. A. Goldstein (1990). "The elastic moduli of human subchondral, trabecular, and cortical bone tissue and the size-dependency of cortical bone modulus." *Journal Biomechanics* **23**(11): 1102–1113.

Clampitt, R., K. L. Aitken and D. K. Jefferies (1975). "Intense field-emission ion source of liquid metals." *Journal of Vacuum Science and Technology* **12**(6): 1208.

Cory, E., A. Nazarian, V. Entezari, V. Vartanians, R. Müller and B. D. Snyder (2010). "Compressive axial mechanical properties of rat bone as functions of bone volume fraction, apparent density and micro-ct based mineral density." *Journal of Biomechanics* **43**(5): 953–960.

Drobne, D. (2013). "3D imaging of cells and tissues by focused ion beam/scanning electron microscopy (FIB/SEM)." *Methods in Molecular Biology* **950**: 275–292.

Drobne, D., M. Milani, V. Leser and F. Tatti (2007). "Surface damage induced by FIB milling and imaging of biological samples is controllable." *Microscopy Research and Technique* **70**: 895–903.

Escovitz, W. H., T. R. Fox and R. Leviselli (1975). "Scanning transmission ion microscope with a field ion source." *Proceedings of the National Academy of Sciences of the United States of America* **72**(5): 1826–1828.

Fratzl, P. and R. Weinkamer (2007). "Nature's hierarchical materials." *Progress in Materials Science* **52**(8): 1263–1334.

Fratzl-Zelman, N., P. Roschger, A. Gourrier, M. Weber, B. M. Misof, N. Loveridge, J. Reeve, K. Klaushofer and P. Fratzl (2009). "Combination of nanoindentation and quantitative backscattered electron imaging revealed altered bone material properties associated with femoral neck fragility." *Calcified Tissue International* **85**(4): 335–343.

Gamo, K. and S. Namba (1991). "Recent advance of focused ion beam technology in maskless deposition and patterning." *Nuclear Instruments and Methods in Physics Research Section B—Beam Interactions with Materials and Atoms* **59**: 190–196.

Giannuzzi, L. A. and F. A. Stevie (1999). "A review of focused ion beam milling techniques for TEM specimen preparation." *Micron* **30**(3): 197–204.

Giannuzzi, L. A. and F. A. Stevie (2004). *Introduction to Focused Ion Beams: Instrumentation, Theory, Techniques and Practice*. New York, Springer.

Guo, X. E. and S. A. Goldstein (2000). "Vertebral trabecular bone microscopic tissue elastic modulus and hardness do not change in ovariectomized rats." *Journal of Orthopaedic Research* **18**(2): 333–336.

Hang, F. and A. H. Barber (2011). "Nano-mechanical properties of individual mineralized collagen fibrils from bone tissue." *Journal of the Royal Society, Interface* **8**(57): 500–505.

Harriott, L. R., R. E. Scotti, K. D. Cummings and A. F. Ambrose (1987). "Micromachining of optical structures with focused ion-beams." *Journal of Vacuum Science and Technology* **5**(1): 207–210.

Hayles, M. F., D. J. Stokes, D. Phifer and K. C. Findlay (2007). "A technique for improved focused ion beam milling of cryo-prepared life science specimens." *Journal of Microscopy-Oxford* **226**: 263–269.

Heymann, J. A. W., M. Hayles, I. Gestmann, L. A. Gianuzzu, B. Lich and S. Subramaniam (2006). "Site-specific 3D imaging of cells and tissues with a dual beam microscope." *Journal of Structural Biology* **155**(1): 63–73.

Jeronimidis, G. (2000). "Structure–property relationships in biological materials." In *Structural Biological Materials: Design and Structure Property Relationships in Structural Biological Materials*, M. Elices (ed.). Amsterdam, Pergamon: 3–29.

Jimenez-Palomar, I., A. Shipov, R. Shahar and A. H. Barber (2012). "Influence of SEM vacuum on bone micromechanics using in situ AFM." *Journal of the Mechanical Behavior of Biomedical Materials* **5**: 149–155.

Kamino, T., T. Yaguchi, T. Ishitani and M. Osumi (2004). "Application of a FIB-STEM system for 3D observation of a resin-embedded yeast cell." *Journal of Electron Microscopy* **53**(5): 563–566.

Kasra, M., C. M. Vanin, N. J. MacLusky, R. F. Casper and M. D. Grynpas (1997). "Effects of different estrogen and progestin regimens on the mechanical properties of rat femur." *Journal of Orthopaedic Research* **15**: 118–123.

Kawakatsu, M. K. H. (1981). "Field-emission liquid-metal ion source and triode ion gun." *Journal of Applied Physics* **52**(4): 2642–2645.

Kempshall, B. W., L. A. Giannuzzi, B. I. Prenitzer, F. A. Stevie and S. X. Da (2002). "Comparative evaluation of protective coatings and focused ion beam chemical vapor deposition processes." *Journal of Vacuum Science and Technology B* **20**: 286–290.

Kolibal, M., T. Matlocha, T. Vystavel and T. Sikola (2011). "Low energy focused ion beam milling of silicon and germanium nanostructures." *Nanotechnology* **22**: 105304.

Langford, R. M. and C. Clinton (2004). "In situ lift-out using a FIB-SEM system." *Micron* **35**(7): 607–611.

Lu, D. and A. H. Barber (2012). "Optimized nanoscale composite behaviour in limpet teeth." *Journal of the Royal Society, Interface* **9**: 1318–1324.

Maïmoun, L., T. C. Brennan-Speranza, R. Rizzoli and P. Ammann (2012). "Effects of ovariectomy on the changes in microarchitecture and material level properties in response to hind leg disuse in female rats." *Bone* **51**(3): 586–591.

Orloff, J. (1993). "High-resolution focused ion beams." *Review of Scientific Instruments* **64**(5): 1105–1130.

Orloff, J. and L. W. Swanson (1975). "Study of a field-ionization source for microprobe applications." *Journal of Vacuum Science and Technology* **12**(6): 1209–1213.

Orloff, J. and L. W. Swanson (1978). "Fine-focus ion beams with field ionization." *Journal of Vacuum Science and Technology* **15**(3): 845–848.

Orloff, J. and L. W. Swanson (1981). "Optical column design with liquid metal ion sources." *Journal of Vacuum Science and Technology* **19**(4): 1149–1152.

Orloff, J., M. W. Utlaut and L. Swanson (2003). *High Resolution Focused Ion Beams: FIB and Its Applications*. X. New York, London, Kluwer Academic/Plenum.

Prenitzer, B. I., L. A. Giannuzzi, S. R. Brown, R. B. Irwin, T. L. Shofner and F. A. Stevie (1998a). "The influence of incident ion range on the efficiency of TEM and SEM specimen preparation of focused ion beam milling." In *Electron Microscopy 1998, ICEM14, Symposium K Institute of Physics*, H. A. C. Benavides and M. J. Yacaman (eds.). Bristol, Institute of Physics, **III**: 711–712.

Prenitzer, B. I., L. A. Giannuzzi, K. Newman, S. R. Brown, T. L. Shofner, R. B. Irwin and F. A. Stevie (1998b). "Transmission electron microscope specimen preparation of Zn powders using the focused ion beam lift-out technique." *Metallurgical and Materials Transactions A* **29**(9): 2399.

Prewett, P. D. and G. L. R. Mair (1991). *Focused Ion Beams from Liquid Metal Ion Sources*. XI. Baldock, UK, Research Studies Press.

Rho, J. Y., T. Y. Tsui and G. M. Pharr (1997). "Elastic properties of human cortical and trabecular lamellar bone measured by nanoindentation." *Biomaterials* **18**: 1325–1330.

Rodriguez-Florez, N., M. L. Oyen and S. J. Shefelbine (2013). "Insight into differences in nanoindentation properties of bone." *Journal of the Mechanical Behavior of Biomedical Materials* **18**: 90–99.

Rubanov, S. and P. R. Munroe (2004). "FIB-induced damage in silicon." *Journal of Microscopy-Oxford* **214**: 213–221.

Schroeder, J. B., S. Bashkin and J. F. Nester (1966). "Ionic polishing of optical surfaces." *Applied Optics* **5**(6): 1031–1034.

Schroeder-Reiter, E., F. Perez-Willard, U. Zeile and G. Wanner (2009). "Focused ion beam (FIB) combined with high resolution scanning electron microscopy: A promising tool for 3D analysis of chromosome architecture." *Journal of Structural Biology* **165**(2): 97–106.

Seliger, R. L. and W. P. Fleming (1974). "Focused ion beams in microfabrication." *Journal of Applied Physics* **45**(3): 1416–1422.

Singh, D. R. P., N. Chawla and Y. L. Shen (2010). "Focused Ion Beam (FIB) tomography of nanoindentation damage in nanoscale metal/ceramic multilayers." *Materials Characterization* **61**(4): 481–488.

Sugiyama, M. and G. Sigesato (2004). "A review of focused ion beam technology and its applications in transmission electron microscopy." *Journal of Electron Microscopy* **53**(5): 527–536.

Swanson, L. W. (1994). "Use of the liquid metal ion source for focused beam applications." *Applied Surface Science* **76–77**(1–4): 80–88.

Taylor, G. (1964). "Disintegration of water drops in an electric field." *Proceedings of the Royal Society of London. Series A, Containing Papers of a Mathematical and Physical Sciences* **280**(1382): 383–397.

Utku, F. S., E. Klein, H. Saybasili, C. A. Yucesoy and S. Weiner (2008). "Probing the role of water in lamellar bone by dehydration in the environmental scanning electron microscope." *Journal of Structural Biology* **162**(3): 361–367.

Wong, K. C., C. M. Haslauer, N. Anantharamaiah, B. Pourdeyhimi, A. D. Batchelor and D. P. Griffis (2010). "Focused ion beam characterization of bicomponent polymer fibers." *Microscopy and Microanalysis* **16**(3): 282–290.

Yoshida, K. and T. Yamada (1984). "Ion thinning apparatus for the preparation of transmission electron-microscopy specimens using new-type ion guns." *Review of Scientific Instruments* **55**(4): 551–557.

29

Imaging Cellular Mechanotransduction Using FRET-Based Biosensors

Qiaoqiao Wan, Yu-Hui Lai, and Sungsoo Na

CONTENTS

Mechanical force plays an important role in regulating fundamental cell functions and has been implicated in various diseases. On the molecular level, however, the mechanism by which how cells sense and transduce mechanical signals has been largely elusive. The measurement and visualization of such activities at the subcellular or molecular level require tools of high spatial and temporal resolution and sensitivity. Recent advancements in live cell imaging and genetically encoded biosensors based on fluorescence resonance energy transfer (FRET) offer greater understanding of the dynamic molecular activities in living cells under mechanical force at the high spatial and temporal resolutions. This chapter will discuss signaling proteins involved in mechanotransduction. Emphasis will be placed on the structure and usage of the FRET-based biosensors and new findings on mechanotransduction revealed by the FRET-based biosensors.

29.1 Introduction

Mechanotransduction is the process by which cells sense and respond to their physical microenvironments and convert them into biochemical signaling that affects their various functions including growth, survival, differentiation, and migration.[1–3] Changes in physical microenvironments, such as magnitude and frequency of mechanical forces or stiffness and geometry of extracellular matrices, result in altered mechanotransduction signaling and consequently lead to, or at least contribute to, physiological or pathophysiological outcomes of tissue and organ development and functions.[4–6] Moderate, physiological mechanical stress, for example, decreases the proteolytic activity of chondrocytes in the articular cartilage, whereas excessive mechanical stress increases the activity.[7,8] In bone, application of mechanical loading accelerates bone formation and fracture healing, whereas removal of loading due to microgravity or bed rest leads to bone loss.[9,10] In the cardiovascular system, laminar flow has an atheroprotective effect, while disturbed flow can develop atherosclerotic regions.[11] In addition, stiffness of the extracellular matrix is shown to control the differentiation of mesenchymal stem cells toward the neuronal cell lineage on soft matrix and the osteogenic cell lineage on the rigid matrix.[12] Thus, understanding the molecular mechanism of mechanotransduction has important implications in human health and disease.

Studies of the cellular and molecular mechanics revealed various mechanisms of mechanotransduction.[2,13] These include opening of mechanosensitive ion channels[14,15] or signaling activation through integrins.[16,17] Integrins are cell surface receptors that mediate cell adhesion to the extracellular matrix. Under mechanical stimulation, integrins undergo conformational change by binding to extracellular matrix proteins and form an activated state, which promotes the linkage of integrins to the cytoskeleton and focal adhesion proteins to initiate intracellular signal transduction pathways.[1,18] Interestingly, recent studies demonstrate that cytoskeletal structures such as microtubules or actin filaments can transmit extracellular force via integrins to induce rapid signal transduction.[19–23] These studies have utilized fluorescent proteins that allow real-time visualization of the activities of proteins involved in mechanotransduction at the subcellular resolutions. Although traditional biochemical assays have been extensively used for study of mechanotransduction, their measurements are based on the large population of cells and thus cannot provide information on transient and rapid dynamics of molecular activities within a single-cell level. In this chapter, we will focus on the use of fluorescent proteins, particularly fluorescence resonance energy transfer (FRET) imaging, for the study of mechanotransduction.

29.2 Fluorescent Proteins and Their Application in FRET-Based Biosensors

Green fluorescent protein (GFP) was first purified from jellyfish in the 1960s.[24] About three decades later, it was shown that the gene of the GFP can be fused to a target molecule to visualize and track the dynamic localization of the molecule.[25,26] At the same time, Tsien's group first demonstrated that GFP can produce distinct colors, suggesting the possibility of simultaneous assessment of multiple gene expression or protein trafficking.[27] Since then, a significant progress has been made in developing a variety of GFP variants, including the most popular FRET pair, cyan fluorescent protein (CFP) and yellow fluorescent protein (YFP).[28] In addition to the discovery of GFP, red fluorescent protein (DsRed), spectrally distinct from GFP variants, was isolated from the nonbioluminescent reef corals.[29] Although the discovery of DsRed opened a new avenue in the simultaneous visualization of multiple signaling events in a single cell, its application to live cell imaging was hindered by a slow and incomplete maturation and obligate tetramerization. To overcome these problems, Tsien's group developed monomeric red fluorescent proteins.[30,31] Currently, many fluorescent proteins are available with different emission spectrums ranging from blue to far-red colors for live cell imaging and FRET applications.[32–35]

FRET is a phenomenon that occurs when two fluorescent proteins, namely, a donor and an acceptor, are in proximity (~10 nm), and the emission spectrum of the donor overlaps the excitation spectrum of the acceptor.[36] Several different structures of FRET-based biosensors are available in terms of their design strategies, including intramolecular and intermolecular FRET biosensors. Intramolecular FRET-based biosensors are advantageous in live cell imaging because the donor and acceptor fluorescent proteins are fused together, and thus, their ratio within the cell is constant during experiment, allowing the use of a simple ratiometric imaging approach to monitor and analyze FRET changes.[37] Intramolecular FRET-based biosensors consist of a pair of fluorescent proteins fused together with a target molecule and a specific binding domain (Figure 29.1). When the molecule of interest is at its rest (inactive) state, donor and acceptor fluorescent proteins (CFP and YFP, respectively, in the example in Figure 29.1) are far apart, and excitation of CFP only results in the emission of CFP. When the molecule is activated, the high affinity and interaction between the molecule and the binding domain cause conformational change of the biosensor, resulting in FRET from CFP to YFP and an increase in YFP emission. Therefore, the molecular activity can be represented by the FRET changes (e.g., the emission ratio of CFP to YFP or YFP to CFP) of the FRET-based biosensor.[38]

29.3 Mechanotransduction Studies Using FRET-Based Biosensors

Numerous studies have shown that various signaling molecules can be activated or inhibited by mechanical stimuli. However, it is unclear about how cells perceive

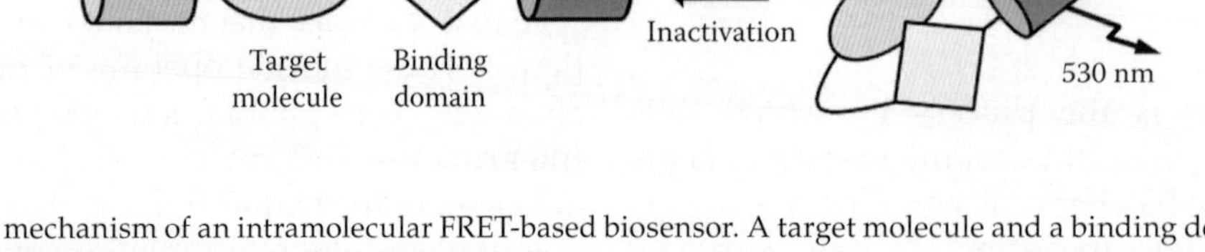

FIGURE 29.1
Structure and activation mechanism of an intramolecular FRET-based biosensor. A target molecule and a binding domain are connected by a flexible linker, and a pair of fluorescent proteins (e.g., CFP and YFP) serves as a donor and an acceptor. The biosensor changes its conformation upon activation of the molecule, resulting in the changes in FRET between CFP and YFP.

changes in mechanical stimuli and upregulate or downregulate signaling molecules at the subcellular level that can consequently lead to selective cell functions. Recent developments in live cell imaging and FRET-based biosensors have significantly advanced our understanding of the molecular mechanism in mechanotransduction. In this section, we will discuss several FRET-based biosensors that have been used for the study in mechanotransduction.

29.3.1 Src

Src kinase regulates many cellular functions including cell differentiation, proliferation, migration, and apoptosis.[39] It consists of phosphotyrosine binding domains (SH3 and SH2) and tyrosine kinase domains.[40] The activation of Src is modulated by the binding of those domains, such as the binding of Src homology 2 (SH2) and kinase domains or SH3 and the linker segment between the SH2 and the kinase domain. Src is also shown to be activated by integrins, which play an important role in mechanotransduction. When integrins are activated, they interact with the SH3 domain of Src, resulting in the conformational change and activation of Src.[41] Tsien's group developed a genetically encoded Src biosensor.[42] It consists of fusions of the SH2 domain, a phosphorylation substrate specific to SH2, and a pair of CFP and YFP. Upon activation of Src, SH2 forms an intramolecular complex with the substrate domain, giving rise to a conformational change of the Src biosensor, which alters the FRET between CFP and YFP. They further developed the Src biosensor with higher specificity and a better defined response mechanism by replacing the original substrate with an Src substrate derived from p130Cas.[43] Using this Src biosensor and a laser tweezer that controls the movement of fibronectin-coated beads attached to the cell surface, Wang et al.[43] demonstrated that Src at the plasma membrane is activated by mechanical force with a wave propagation, the direction of which is opposite to that of the force. Wang's group further enhanced the sensitivity of the Src biosensor by replacing CFP and YFP with enhanced CFP (ECFP) and YPet, respectively.[44] We have used this improved Src biosensor to measure the mechanical activation of Src in the cytoplasm.[21] We have found that local mechanical force at the cell surface can be converted into rapid (<300 ms) Src activation in regions of the cytoplasm that are distant from the site of force application. This rapid mechanochemical conversion by mechanical force has been found to colocalize with regions of microtubule deformation by force[21] (Figure 29.2). We used this Src

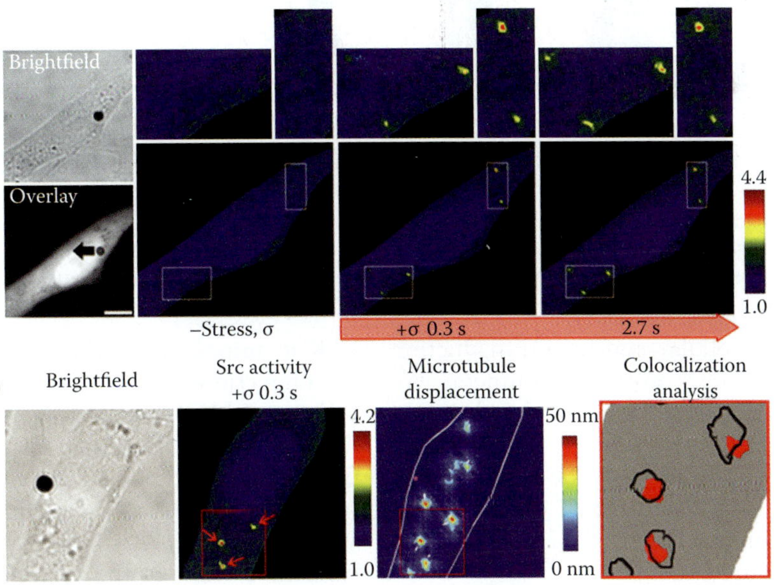

FIGURE 29.2
Rapid Src activation by mechanical force. (a) A FRET-based cytosolic Src biosensor was transfected into the smooth muscle cells. A 4.5 μm RGD-coated ferromagnetic bead (a black circle in the Brightfield image) was attached to the apical surface of the cell. The bead was subjected to a vertical magnetic field that applies a mechanical stress (σ) to the cell. The stress application induced rapid changes (<0.3 s) in FRET of the Src biosensor at discrete, distant sites in the cytoplasm, indicating rapid Src activation. Images are scaled, and regions of large FRET changes (strong Src activity) are shown in red. The black arrow indicates bead movement direction. (b) The cell was cotransfected with an Src biosensor and mCherry-tubulin. A step function stress (17.5 Pa) was first applied for 3 s via an RGD-coated bead, and FRET changes were recorded. Then the microtubule deformation map was acquired when an oscillatory stress was applied for 30 s (0.3 Hz; peak stress = 24.5 Pa, equivalent to a constant stress of 17.5 Pa). In this representative cell, strong Src activation sites coincide with large deformation sites (>15 nm) of microtubules in the same cell at the same focal plane (1 μm above cell base). Red arrows point to strong Src activation sites. In the colocalization analysis panel, red represents strong Src activation, and black lines represent large microtubule displacements. Scale bars, 10 μm. (Adapted from Na, S. et al., *Proc Natl Acad Sci U S A* **105**, 6626–6631. Copyright 2008 National Academy of Sciences, U.S.A.)

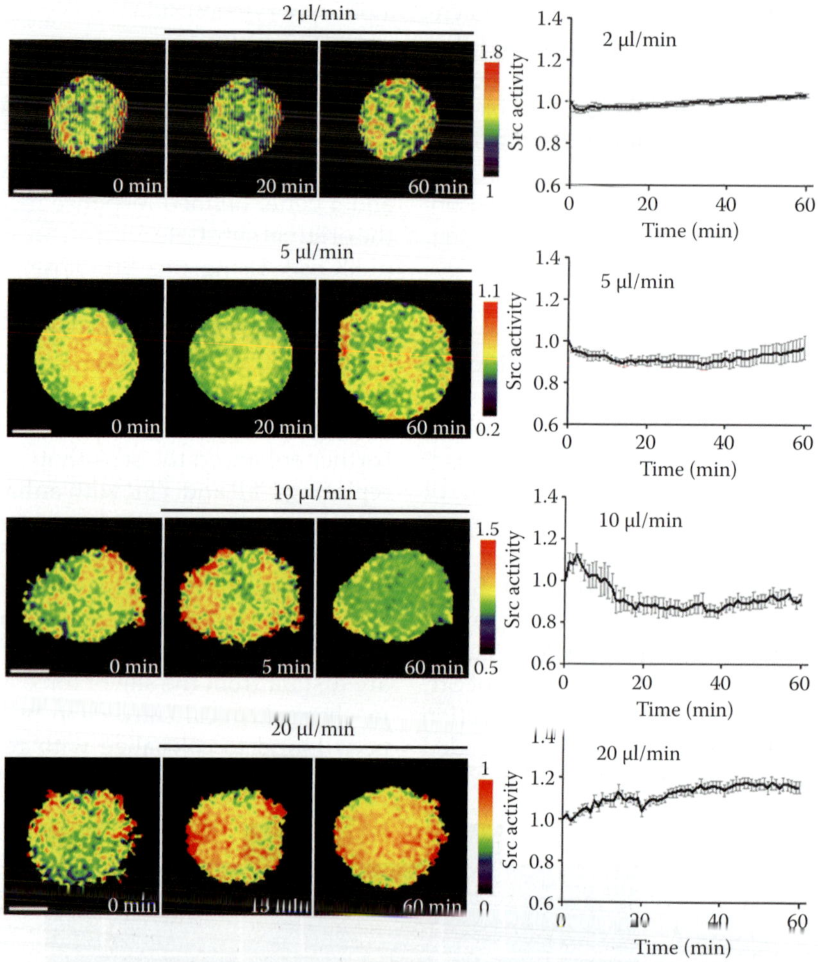

FIGURE 29.3
Src activity is mechanical loading-magnitude dependent under interstitial fluid flow. CFP–YFP Src biosensor was transfected into C28/I2 human chondrocytes. After transfection, the cells were transferred and seeded in the collagen-conjugated agarose gel. During imaging, dynamic, pulsed fluid flow was applied to the cells/gel construct in the chamber at 0.5 Hz. Scale bars, 10 μm.

biosensor to visualize cellular mechanotransduction in a three-dimensional (3D) microenvironment. We have recently developed a 3D cell culture chamber that allows for monitoring spatiotemporal dynamics of Src activity under interstitial fluid flow. The results show that flow-induced Src activity in C28/I2 chondrocytes is spatiotemporally dynamic and is dependent on the magnitude of force (Figure 29.3).

29.3.2 Focal Adhesion Kinase

Focal adhesion kinase (FAK) plays an important role in cell adhesion, migration, differentiation, and survival.[45–48] Because FAK has been implicated in integrin-mediated signaling pathways,[49,50] it is expected that FAK can be regulated by mechanical force.[1] For example, fluid shear stress activates FAK in endothelial cells.[51] Application of cyclic strain increases FAK activity in osteoblasts[52] and cardiac myocytes.[53] Upon activation, FAK is autophosphorylated at tyrosine 397, leading to its interaction with SH2 domain of Src, which in turn activates FAK.[54] Wang et al.[55] developed a FRET-based FAK biosensor. It consists of the SH2 domain derived from c-Src, a substrate derived from FAK containing tyrosine 397, and a pair of fluorescent proteins (ECFP and YPet). When FAK is activated, it phosphorylates the tyrosine 397 site of the substrate in the FAK biosensor, which subsequently binds to the SH2 domain of the biosensor, resulting in the conformational change of the biosensor and alteration of FRET between ECFP and YPet.[55] Because it is known that FAK is localized at a detergent-resistant membrane (DRM),[56] Wang et al. also developed a FAK biosensor specific to the DRM by fusing a DRM-targeting motif derived from Lyn. Using the Src and FAK biosensors targeted to DRM regions, Wang et al. showed that Src activity is required for the soluble factor-induced FAK activation, whereas FAK activity is required for Src activation in integrin-mediated cell adhesion process, suggesting that soluble factor-induced signaling and integrin-mediated signaling use a distinct signaling mechanism.[55]

29.3.3 Rho Family GTPases

The Rho family GTPases act as a molecular switch that regulates fundamental processes including morphogenesis, polarity, movement, and cell division.[57] They present an active (GTP-bound) and inactive (GDP-bound) states, which are controlled by guanine nucleotide exchange factors (GEFs) and GTPase-activating proteins (GAPs). RhoA, Rac1, and Cdc42 are most well studied members of the GTP-binding proteins and are known to play distinctive roles in regulating actin cytoskeletal dynamics of stress fibers, lamellipodia, and filopodia, respectively.[58] Accumulating evidence indicates that they respond to mechanical loading in various cell types, including osteoblasts,[59] smooth muscle cells,[23] endothelial cells,[60] and mesenchymal progenitor cells.[61] They are also involved in bone resorption by regulating osteoclast functions.[62–64] Hahn et al.[65,66] developed Rac1 and RhoA biosensors utilizing FRET. Using the Rac1 biosensor, Tzima in Schwartz's group showed that flow induced Rac1 activation is localized at the downstream edge of the endothelial cell[67] (Figure 29.4). Collins in Tzima's group later showed using the RhoA biosensor and a magnetic tweezer that application of local force to endothelial cells induces cell stiffening and global RhoA activation, which are mediated by platelet endothelial cell adhesion molecule-1 (PECAM-1).[68] Matsuda's group also independently developed FRET-based Rho family GTPase biosensors.[69,70] These biosensors consist of one of the GTPases, RhoA, Rac1, or Cdc42, a binding domain of an effector protein for the specific GTPase (protein kinase N for RhoA, and p21 protein-activated kinase 1 for Rac1 or Cdc42), and a pair of CFP and YFP. Using these

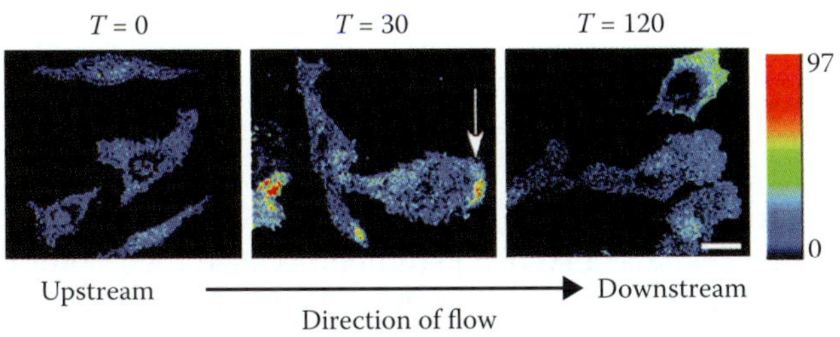

FIGURE 29.4
Spatiotemporal activity of Rac1 in response to fluid flow. Bovine aortic endothelial cells were transfected with GFP fused wild-type Rac1 and Alexa568-Rac1 binding domain. Rac1 activation leads to binding of these two molecules and subsequently results in FRET increase from GFP to Alexa568. Cells were subjected to shear stress for 120 min. Rac1 activity increases at the downstream edges of cells. Scale bar, 20 μm. (Reprinted by permission from Macmillan Publishers Ltd: *EMBO Journal*,[67] copyright 2002.)

biosensors and a magnetic twisting cytometer, we have shown that localized mechanical force induces rapid (<0.3 s) Rac1 activation, which is independent of Src activation.[23] We have also shown that chondrocytes exhibit shear stress magnitude-dependent RhoA activity. High shear stress (10 or 20 dyn/cm^2) increases RhoA activity, whereas intermediate shear stress (5 dyn/cm^2) decreases the activity. Furthermore, these two-level activities of RhoA by shear stress are shown to be closely linked to the shear stress-induced alterations in actin cytoskeleton and intracellular tension[71] (Figure 29.5). In addition to fluid flow, RhoA activity and associated actin cytoskeletal organization were also shown to be decreased by simulated unloading (e.g., clinorotation) in osteoblasts.[72] We have also shown that Rho family GTPases are distinctively activated under flow. Importantly, inhibition of Rac1 and Cdc42 blocks flow-induced β-catenin signaling activity, which is essential for bone formation and osteoblast proliferation and differentiation.[73]

29.3.4 Calcium

Calcium (Ca^{2+}) signaling is known to affect a diverse range of cell functions including gene transcription, muscle contraction, and cell proliferation.[74] Its activity is highly heterogeneous within the cell, and the spatiotemporal distribution of Ca^{2+} concentration is determined by the balance between extracellular Ca^{2+} influx and Ca^{2+} release from intracellular stores such as endoplasmic reticulum or sarcoplasmic reticulum.[75,76] A calcium biosensor based on FRET has been developed by Tsien's group to measure the spatiotemporal distribution of Ca^{2+} signaling.[77] It consists of calmodulin, the calmodulin-binding peptide, and two fluorescent proteins (blue and green or cyan and yellow) that serve as a FRET pair. Wang's group has recently improved the sensitivity of the biosensor by replacing the original FRET pairs with ECFP and YPet.[44] They have shown that the substrate with lower stiffness (1 kPa) significantly decreases the magnitude and frequency of the Ca^{2+} oscillation in human mesenchymal stem cells.[78] This improved Ca^{2+} biosensor has also been used in endothelial cells under flow. The results show that high shear stress (65 dyn/cm^2) induced two distinct phases of calcium signaling with different sources and mechanisms: an early phase due to calcium influx across the plasma membrane and a late phase due to calcium release from endoplasmic reticulum.[79]

29.3.5 Cytoskeletal Structures and Membrane Receptors

It has been well established that cytoskeletal structures and focal adhesion proteins are essential in mechanotransduction. Grashoff et al.[80] developed a FRET-based

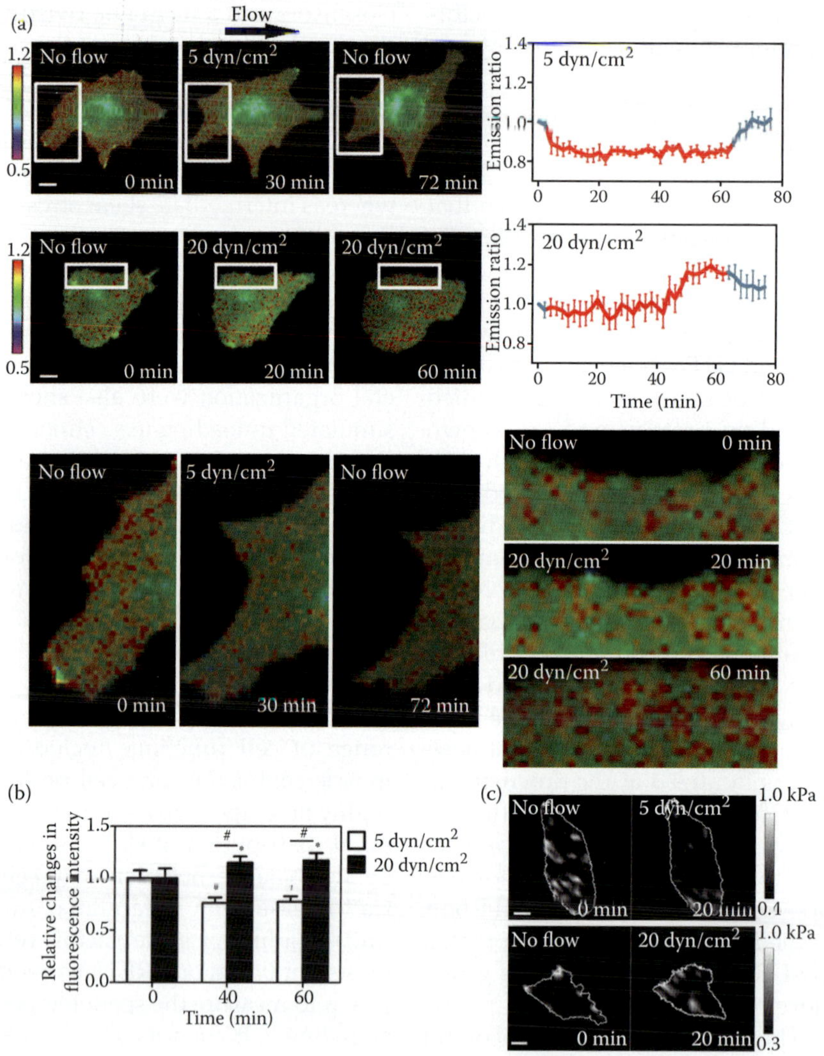

FIGURE 29.5
Shear stress-induced dynamics of RhoA activity is linked to actin cytoskeleton remodeling and intracellular tension. (a) C28/I2 chondrocytes exhibit shear stress magnitude-dependent RhoA activity. Cells that transfected a RhoA biosensor were subjected to shear stress for 1 h. The white boxes are enlarged for better visualization. In the time courses, red and blue colors represent flow and no-flow conditions, respectively. (b) Relative changes in fluorescent intensity of actin cytoskeleton under 5 and 20 dyn/cm^2 of shear stress. (c) Changes in intracellular tension revealed by a dynamic traction map in response to 5 and 20 dyn/cm^2 of shear stress. Scale bars, 10 μm. (Wan, Q., Kim, S.J., Yokota, H. and Na, S.: Differential activation and inhibition of RhoA by fluid flow induced shear stress in chondrocytes. *Cell Biol Int* 37. 568–576. 2013. Copyright Wiley-VCH Verlag GmbH & Co. KGaA. Reproduced with permission.)

biosensor that can detect tensional force across vinculin, one of the focal adhesion proteins important in cellular mechanosensing and mechanotransduction.[80] They found that tension across vinculin in stable focal adhesions is ~2.5 pN, and activation and recruitment of vinculin and force transmission across vinculin are regulated independently. Meng and Sachs[81] developed a similar biosensor that can measure the force across the cytoskeleton, specifically actinin. With this biosensor, stress across actinin has been shown to be upregulated by flow (10 dyn/cm^2), which is not dependent on the intracellular Ca^{2+} level.[82] In addition to cytoskeletal structures, membrane proteins (e.g., G protein-coupled receptors) have also been studied because their conformational changes play an important role in transducing extracellular stimuli into intracellular signals.[83] For example, Zhang et al.[84] developed a FRET-based biosensor that can detect conformational change of the parathyroid hormone type 1 receptor (PTH1R), which is important in bone growth. They showed that fluid shear stress can induce ligand-independent PTH1R conformation in osteoblasts, suggesting a crucial role of the mechanical force in bone formation via G protein-coupled receptors.[84]

29.4 Conclusion

Revealing the molecular mechanism of cellular mechanotransduction requires imaging and biomechanical loading systems with high spatial and temporal resolution and sensitivity. The FRET-based biosensors have been successfully used to provide a detailed understanding of the molecular mechanism and a visualization of the molecular activity at the subcellular resolution *in vitro*. Studies in mechanotransduction using FRET-based biosensors rely on the development of the biosensors that can probe the activity of signaling proteins of interest. The development and improvements of the FRET biosensors, however, require several design strategies, for example, dynamic range and target specificity of the biosensor and minimal perturbation of endogenous signaling activity by the biosensor. The development of mechanical loading systems and cell culture systems will also be required to mimic the physiologically or pathologically relevant microenvironment of specific cell types *in vivo*. At the current stage, however, there are still many challenges in using the FRET imaging technique *in vivo*, including the delivery, expression, and imaging of FRET-based biosensors in specific cell types. Further improvements of FRET imaging in cellular mechanotransduction studies *in vivo* require the advanced molecular delivery systems that allow the incorporation of FRET-based biosensors into specific cells so that the cells of interest can be visualized and monitored in live organs; the new optical and imaging devices that provide higher resolution and improved quality of imaging *in vivo*; and the novel FRET-based biosensors with longer emission wavelengths that are able to enhance light penetration by minimizing the light absorption by live tissues. The excellent articles on general information and strategies for FRET imaging and the development of FRET-based biosensors have been published.[33,37,85–87] It is expected that current developments in molecular biology and tissue engineering as well as the integration of FRET imaging and mechanical loading will accelerate our understanding of cellular mechanotransduction and its linkage to physiological or pathological outcomes.

References

1. Schwartz, M.A. Integrins and extracellular matrix in mechanotransduction. *Cold Spring Harb Perspect Biol* **2**, a005066 (2010).
2. Wang, N., Tytell, J.D. and Ingber, D.E. Mechanotransduction at a distance: Mechanically coupling the extracellular matrix with the nucleus. *Nat Rev Mol Cell Biol* **10**, 75–82 (2009).
3. Ingber, D.E. Cellular mechanotransduction: Putting all the pieces together again. *FASEB J* **20**, 811–827 (2006).
4. Farge, E. Mechanotransduction in development. *Curr Top Dev Biol* **95**, 243–265 (2011).
5. Jaalouk, D.E. and Lammerding, J. Mechanotransduction gone awry. *Nat Rev Mol Cell Biol* **10**, 63–73 (2009).
6. Ingber, D.E. Mechanobiology and diseases of mechanotransduction. *Ann Med* **35**, 564–577 (2003).
7. Yokota, H., Goldring, M.B. and Sun, H.B. CITED2-mediated regulation of MMP-1 and MMP-13 in human chondrocytes under flow shear. *J Biol Chem* **278**, 47275–47280 (2003).
8. Yokota, H., Leong, D.J. and Sun, H.B. Mechanical loading: Bone remodeling and cartilage maintenance. *Curr Osteoporos Rep* **9**, 237–242 (2011).
9. Rubin, C.T. and Lanyon, L.E. Regulation of bone formation by applied dynamic loads. *J Bone Joint Surg Am* **66**, 397–402 (1984).
10. Lang, T. et al. Cortical and trabecular bone mineral loss from the spine and hip in long-duration spaceflight. *J Bone Miner Res* **19**, 1006–1012 (2004).
11. Hahn, C. and Schwartz, M.A. Mechanotransduction in vascular physiology and atherogenesis. *Nat Rev Mol Cell Biol* **10**, 53–62 (2009).
12. Engler, A.J., Sen, S., Sweeney, H.L. and Discher, D.E. Matrix elasticity directs stem cell lineage specification. *Cell* **126**, 677–689 (2006).
13. Vogel, V. and Sheetz, M. Local force and geometry sensing regulate cell functions. *Nat Rev Mol Cell Biol* **7**, 265–275 (2006).
14. Martinac, B. Mechanosensitive ion channels: Molecules of mechanotransduction. *J Cell Sci* **117**, 2449–2460 (2004).
15. Sukharev, S. and Anishkin, A. Mechanosensitive channels: What can we learn from "simple" model systems? *Trends Neurosci* **27**, 345–351 (2004).
16. Maniotis, A.J., Chen, C.S. and Ingber, D.E. Demonstration of mechanical connections between integrins, cytoskeletal filaments, and nucleoplasm that stabilize nuclear structure. *Proc Natl Acad Sci U S A* **94**, 849–854 (1997).
17. Wang, N., Butler, J.P. and Ingber, D.E. Mechanotransduction across the cell surface and through the cytoskeleton. *Science* **260**, 1124–1127 (1993).
18. Hynes, R.O. Integrins: Bidirectional, allosteric signaling machines. *Cell* **110**, 673–687 (2002).
19. Hu, S. et al. Intracellular stress tomography reveals stress focusing and structural anisotropy in cytoskeleton of living cells. *Am J Physiol Cell Physiol* **285**, C1082–C1090 (2003).
20. Matthews, B.D. et al. Ultra-rapid activation of TRPV4 ion channels by mechanical forces applied to cell surface beta1 integrins. *Integr Biol (Camb)* **2**, 435–442 (2010).
21. Na, S. et al. Rapid signal transduction in living cells is a unique feature of mechanotransduction. *Proc Natl Acad Sci U S A* **105**, 6626–6631 (2008).
22. Hayakawa, K., Tatsumi, H. and Sokabe, M. Actin stress fibers transmit and focus force to activate mechanosensitive channels. *J Cell Sci* **121**, 496–503 (2008).
23. Poh, Y.C. et al. Rapid activation of Rac GTPase in living cells by force is independent of Src. *PLoS One* **4**, e7886 (2009).

24. Shimomura, O., Johnson, F.H. and Saiga, Y. Extraction, purification and properties of aequorin, a bioluminescent protein from the luminous hydromedusan, Aequorea. *J Cell Comp Physiol* **59**, 223–239 (1962).
25. Chalfie, M., Tu, Y., Euskirchen, G., Ward, W.W. and Prasher, D.C. Green fluorescent protein as a marker for gene expression. *Science* **263**, 802–805 (1994).
26. Inouye, S. and Tsuji, F.I. Aequorea green fluorescent protein. Expression of the gene and fluorescence characteristics of the recombinant protein. *FEBS Lett* **341**, 277–280 (1994).
27. Heim, R., Prasher, D.C. and Tsien, R.Y. Wavelength mutations and posttranslational autoxidation of green fluorescent protein. *Proc Natl Acad Sci U S A* **91**, 12501–12504 (1994).
28. Tsien, R.Y. The green fluorescent protein. *Annu Rev Biochem* **67**, 509–544 (1998).
29. Matz, M.V. et al. Fluorescent proteins from nonbioluminescent Anthozoa species. *Nat Biotechnol* **17**, 969–973 (1999).
30. Campbell, R.E. et al. A monomeric red fluorescent protein. *Proc Natl Acad Sci U S A* **99**, 7877–7882 (2002).
31. Shaner, N.C. et al. Improved monomeric red, orange and yellow fluorescent proteins derived from Discosoma sp. red fluorescent protein. *Nat Biotechnol* **22**, 1567–1572 (2004).
32. Tsien, R.Y. Building and breeding molecules to spy on cells and tumors. *FEBS Lett* **579**, 927–932 (2005).
33. Tsien, R.Y. Constructing and exploiting the fluorescent protein paintbox (Nobel Lecture). *Angew Chem Int Ed Engl* **48**, 5612–5626 (2009).
34. Shaner, N.C., Steinbach, P.A. and Tsien, R.Y. A guide to choosing fluorescent proteins. *Nat Methods* **2**, 905–909 (2005).
35. Miyawaki, A., Nagai, T. and Mizuno, H. Engineering fluorescent proteins. *Adv Biochem Eng Biotechnol* **95**, 1–15 (2005).
36. Truong, K. and Ikura, M. The use of FRET imaging microscopy to detect protein–protein interactions and protein conformational changes in vivo. *Curr Opin Struct Biol* **11**, 573–578 (2001).
37. Zhang, J., Campbell, R.E., Ting, A.Y. and Tsien, R.Y. Creating new fluorescent probes for cell biology. *Nat Rev Mol Cell Biol* **3**, 906–918 (2002).
38. Wang, Y., Shyy, J.Y. and Chien, S. Fluorescence proteins, live-cell imaging, and mechanobiology: Seeing is believing. *Annu Rev Biomed Eng* **10**, 1–38 (2008).
39. Thomas, S.M. and Brugge, J.S. Cellular functions regulated by Src family kinases. *Annu Rev Cell Dev Biol* **13**, 513–609 (1997).
40. Boggon, T.J. and Eck, M.J. Structure and regulation of Src family kinases. *Oncogene* **23**, 7918–7927 (2004).
41. Arias-Salgado, E.G. et al. Src kinase activation by direct interaction with the integrin beta cytoplasmic domain. *Proc Natl Acad Sci U S A* **100**, 13298–13302 (2003).
42. Ting, A.Y., Kain, K.H., Klemke, R.L. and Tsien, R.Y. Genetically encoded fluorescent reporters of protein tyrosine kinase activities in living cells. *Proc Natl Acad Sci U S A* **98**, 15003–15008 (2001).
43. Wang, Y. et al. Visualizing the mechanical activation of Src. *Nature* **434**, 1040–1045 (2005).
44. Ouyang, M., Sun, J., Chien, S. and Wang, Y. Determination of hierarchical relationship of Src and Rac at subcellular locations with FRET biosensors. *Proc Natl Acad Sci U S A* **105**, 14353–14358 (2008).
45. Mitra, S.K., Hanson, D.A. and Schlaepfer, D.D. Focal adhesion kinase: In command and control of cell motility. *Nat Rev Mol Cell Biol* **6**, 56–68 (2005).
46. Michael, K.E., Dumbauld, D.W., Burns, K.L., Hanks, S.K. and Garcia, A.J. Focal adhesion kinase modulates cell adhesion strengthening via integrin activation. *Mol Biol Cell* **20**, 2508–2519 (2009).
47. Frisch, S.M., Vuori, K., Ruoslahti, E. and Chan-Hui, P.Y. Control of adhesion-dependent cell survival by focal adhesion kinase. *J Cell Biol* **134**, 793–799 (1996).
48. Leucht, P., Kim, J.B., Currey, J.A., Brunski, J. and Helms, J.A. FAK-Mediated mechanotransduction in skeletal regeneration. *PLoS One* **2**, e390 (2007).
49. Guan, J.L. Role of focal adhesion kinase in integrin signaling. *Int J Biochem Cell Biol* **29**, 1085–1096 (1997).
50. Renshaw, M.W., Price, L.S. and Schwartz, M.A. Focal adhesion kinase mediates the integrin signaling requirement for growth factor activation of MAP kinase. *J Cell Biol* **147**, 611–618 (1999).
51. Li, S. et al. Fluid shear stress activation of focal adhesion kinase. Linking to mitogen-activated protein kinases. *J Biol Chem* **272**, 30455–30462 (1997).
52. Boutahar, N., Guignandon, A., Vico, L. and Lafage-Proust, M.H. Mechanical strain on osteoblasts activates autophosphorylation of focal adhesion kinase and proline-rich tyrosine kinase 2 tyrosine sites involved in ERK activation. *J Biol Chem* **279**, 30588–30599 (2004).
53. Torsoni, A.S., Constancio, S.S., Nadruz, W. Jr., Hanks, S.K. and Franchini, K.G. Focal adhesion kinase is activated and mediates the early hypertrophic response to stretch in cardiac myocytes. *Circ Res* **93**, 140–147 (2003).
54. Mitra, S.K. and Schlaepfer, D.D. Integrin-regulated FAK-Src signaling in normal and cancer cells. *Curr Opin Cell Biol* **18**, 516–523 (2006).
55. Seong, J. et al. Detection of focal adhesion kinase activation at membrane microdomains by fluorescence resonance energy transfer. *Nat Commun* **2**, 406 (2011).
56. Park, E.K. et al. Cholesterol depletion induces anoikis-like apoptosis via FAK down-regulation and caveolae internalization. *J Pathol* **218**, 337–349 (2009).
57. Etienne-Manneville, S. and Hall, A. Rho GTPases in cell biology. *Nature* **420**, 629–635 (2002).
58. Nobes, C.D. and Hall, A. Rho, rac, and cdc42 GTPases regulate the assembly of multimolecular focal complexes associated with actin stress fibers, lamellipodia, and filopodia. *Cell* **81**, 53–62 (1995).
59. Hamamura, K. et al. RhoA-mediated signaling in mechanotransduction of osteoblasts. *Connect Tissue Res* **53**, 398–406 (2012).
60. Tzima, E. Role of small GTPases in endothelial cytoskeletal dynamics and the shear stress response. *Circ Res* **98**, 176–185 (2006).
61. Arnsdorf, E.J., Tummala, P., Kwon, R.Y. and Jacobs, C.R. Mechanically induced osteogenic differentiation—the role of RhoA, ROCKII and cytoskeletal dynamics. *J Cell Sci* **122**, 546–553 (2009).

62. Chellaiah, M.A., Biswas, R.S., Rittling, S.R., Denhardt, D.T. and Hruska, K.A. Rho-dependent Rho kinase activation increases CD44 surface expression and bone resorption in osteoclasts. *J Biol Chem* **278**, 29086–29097 (2003).
63. Ito, Y. et al. Cdc42 regulates bone modeling and remodeling in mice by modulating RANKL/M-CSF signaling and osteoclast polarization. *J Clin Invest* **120**, 1981–1993 (2010).
64. Razzouk, S., Lieberherr, M. and Cournot, G. Rac-GTPase, osteoclast cytoskeleton and bone resorption. *Eur J Cell Biol* **78**, 249–255 (1999).
65. Kraynov, V.S. et al. Localized Rac activation dynamics visualized in living cells. *Science* **290**, 333–337 (2000).
66. Pertz, O., Hodgson, L., Klemke, R.L. and Hahn, K.M. Spatiotemporal dynamics of RhoA activity in migrating cells. *Nature* **440**, 1069–1072 (2006).
67. Tzima, E. et al. Activation of Rac1 by shear stress in endothelial cells mediates both cytoskeletal reorganization and effects on gene expression. *EMBO J* **21**, 6791–6800 (2002).
68. Collins, C. et al. Localized tensional forces on PECAM-1 elicit a global mechanotransduction response via the integrin-RhoA pathway. *Curr Biol* **22**, 2087–2094 (2012).
69. Itoh, R.E. et al. Activation of rac and cdc42 video imaged by fluorescent resonance energy transfer-based single-molecule probes in the membrane of living cells. *Mol Cell Biol* **22**, 6582–6591 (2002).
70. Yoshizaki, H. et al. Activity of Rho-family GTPases during cell division as visualized with FRET-based probes. *J Cell Biol* **162**, 223–232 (2003).
71. Wan, Q., Kim, S.J., Yokota, H. and Na, S. Differential activation and inhibition of RhoA by fluid flow induced shear stress in chondrocytes. *Cell Biol Int* **37**, 568–576 (2013).
72. Wan, Q., Cho, E., Yokota, H. and Na, S. RhoA GTPase interacts with beta-catenin signaling in clinorotated osteoblasts. *J Bone Miner Metab* **31**, 520–532 (2013).
73. Wan, Q., Cho, E., Yokota, H. and Na, S. Rac1 and Cdc42 GTPases regulate shear stress-driven beta-catenin signaling in osteoblasts. *Biochem Biophys Res Commun* **433**, 502–507 (2013).
74. Berridge, M.J., Lipp, P. and Bootman, M.D. The versatility and universality of calcium signalling. *Nat Rev Mol Cell Biol* **1**, 11–21 (2000).
75. Clapham, D.E. Calcium signaling. *Cell* **80**, 259–268 (1995).
76. Berridge, M.J., Bootman, M.D. and Roderick, H.L. Calcium signalling: Dynamics, homeostasis and remodelling. *Nat Rev Mol Cell Biol* **4**, 517–529 (2003).
77. Miyawaki, A. et al. Fluorescent indicators for Ca2+ based on green fluorescent proteins and calmodulin. *Nature* **388**, 882–887 (1997).
78. Kim, T.J. et al. Substrate rigidity regulates Ca2+ oscillation via RhoA pathway in stem cells. *J Cell Physiol* **218**, 285–293 (2009).
79. Liu, B., Lu, S., Zheng, S., Jiang, Z. and Wang, Y. Two distinct phases of calcium signalling under flow. *Cardiovasc Res* **91**, 124–133 (2011).
80. Grashoff, C. et al. Measuring mechanical tension across vinculin reveals regulation of focal adhesion dynamics. *Nature* **466**, 263–266 (2010).
81. Meng, F. and Sachs, F. Visualizing dynamic cytoplasmic forces with a compliance-matched FRET sensor. *J Cell Sci* **124**, 261–269 (2011).
82. Rahimzadeh, J. et al. Real-time observation of flow-induced cytoskeletal stress in living cells. *Am J Physiol Cell Physiol* **301**, C646–C652 (2011).
83. Venkatakrishnan, A.J. et al. Molecular signatures of G-protein-coupled receptors. *Nature* **494**, 185–194 (2013).
84. Zhang, Y.L., Frangos, J.A. and Chachisvilis, M. Mechanical stimulus alters conformation of type 1 parathyroid hormone receptor in bone cells. *Am J Physiol Cell Physiol* **296**, C1391–C1399 (2009).
85. Aoki, K., Kamioka, Y. and Matsuda, M. Fluorescence resonance energy transfer imaging of cell signaling from in vitro to in vivo: Basis of biosensor construction, live imaging, and image processing. *Dev Growth Differ* **55**, 515–522 (2013).
86. Giepmans, B.N., Adams, S.R., Ellisman, M.H. and Tsien, R.Y. The fluorescent toolbox for assessing protein location and function. *Science* **312**, 217–224 (2006).
87. Aoki, K. and Matsuda, M. Visualization of small GTPase activity with fluorescence resonance energy transfer-based biosensors. *Nat Protoc* **4**, 1623–1631 (2009).

30

Intracellular Particle Tracking Microrheology

Michelle R. Dawson, Yiider Tseng, Jerry S.H. Lee, and Kathleen M. McAndrews

CONTENTS

Intracellular particle tracking microrheology offers a method of directly quantifying the mechanical properties of live cells from the displacements of particles embedded in the cytoplasmic region. Given the heterogeneities of architecture within a cell and between cells, the microenvironment-dependent mean square displacements (MSDs) can be used individually to estimate the local rheological properties or can be ensemble-averaged to determine the overall cellular mechanical properties depending on the biological phenomenon being examined. As such, intracellular particle tracking microrheology has been used to map the rheological properties of individual cells in more physiologically relevant conditions and monitor mechanical responses of cells to extracellular chemical and physical stimuli over long time periods (hours or days). In addition, changes in the intracellular mechanical properties, which contribute to cell shape, structure, and function, are dynamically altered by minute mechanical forces from both the extracellular environment and intracellular actomyosin contraction, and can be monitored by intracellular particle tracking microrheology. This chapter provides general information regarding particle tracking microrheology and highlights recent advancements in this field.

30.1 Introduction

Mechanical stress can be generated from cellular processes such as cell migration,[1–3] stem cell differentiation,[4] protein unfolding,[5] and cancerous transformation.[6] It also plays a role in various physiological and pathological events, such as embryonic development.[7,8] organogenesis,[9] tissue regeneration,[10] and tumor growth.[11] During these events, cells respond to the mechanical stress not only with simultaneous cytoplasmic deformation but also with dynamic cytoskeleton remodeling to continually dissipate stress. Together, how the cytoplasmic region of a cell responds to external forces, i.e., the intracellular rheological property, is of interest and is essential to thoroughly understand those cellular events.

The molecular mechanisms for cells to dynamically regulate the intracellular rheological properties to respond to the physical shear stress and chemical cues from external environments are conserved across all cell types and species.[12] It requires the reorganization of the cytoskeleton, a complex network that acts as a scaffold to support the cell's structure from the cell cortex to the nucleus.[13] The cytoskeleton contains three filamentous structures: actin (F-actin), microtubules, and intermediate filaments. Among them, the rheological properties of actin-cytoskeleton are most well studied. Intracellular mechanics have previously been related mainly to the organization of F-actin,[14,15] which controls cell shape, structure, and force generation.[16] Chemical and physical stimuli from the extracellular environment alter cell mechanics through Rho GTPases, including RhoA, Rac1, and Cdc42, which act as the molecular switches to interact with downstream effector molecules and trigger signaling pathways to manifest actin cytoskeleton reorganization.[17–19]

The actin-cytoskeleton contains a multitude of regulatory proteins, molecular motors, and actin-binding proteins (ABPs) that together mediate the organization of actin-cytoskeleton. The actin stress fibers link the actin-cytoskeleton to the extracellular matrix via focal adhesions. Motor protein myosin binds to F-actin and utilizes ATP to generate contraction force that propagates tension along actin filaments.[20] Distinct ABPs individually possess specific actin-binding activities, such as severing and elongating to regulate the filamentous length of actin filaments, branching, cross-linking and bundling to formulate the architecture of the actin network, and sequestering and nucleation to control the amount of available globular actin (G-actin) for filament formation in the cytoplasmic pool. Together, the functions and activities of motor proteins and ABPs provide dynamics and structural diversity to support a broad range of rheological properties and architecture to the actin network.

30.2 Fluid Mechanics and Cell Rheology

Rheology is the study of deformation of soft matter, which is used to explain the mechanical behavior of complex fluids, including polymeric solutions, colloidal suspensions, and biological gels. Complex fluids are viscoelastic in nature since they act as elastic solids (resisting permanent deformation) at short times and viscous fluids (deforming irreversibly) at longer times.[21] A rheometer is a tool to characterize the time- or frequency-dependent viscoelastic properties of complex fluids from the mechanical deformations that result from the application of controlled stress or strain.[22] Bulk-fluid rheometers have previously been used to characterize the mechanical behavior of reconstituted solutions of cytoskeletal proteins[23]; however, these tools are not applicable to living cells with respect to their small size and soft cytoplasm that requires the application of extremely small forces (in the piconewton range) and nanoscale deformations.[24] Quantitative microscopic techniques, such as atomic force microscopy (AFM), micropipette aspiration, magnetic twisting cytometry (MTC), and multiple particle tracking microrheology (MPTM), have been developed in recent years to characterize the rheological properties of the cell cytoskeleton in microscopic domains for their rheological response during intracellular processes.

30.2.1 Limitations of Current Methods for Measuring Cell Mechanics

Current methods for measuring cellular rheology include AFM, micropipette aspiration, and MTC, as shown in Figure 30.1. Utilizing AFM, cell mechanical properties are determined from the deflections of a cantilever tip on a cell surface. AFM results depend heavily on the accuracy of the spring constant of the tip and the interaction of the tip with the cell surface, which in turn may change the mechanical properties of the cell for subsequent measurements.[25] Moreover, it is only capable of measuring surface properties that may reflect larger actin stress fibers or membrane properties. Micropipette aspiration involves deforming a cell by aspirating it into a micropipette and measuring the geometric changes of the whole cell as bulk cellular rheology. In MTC, ligand-coated magnetic beads are attached to the surface of the cell, and magnetic pulses are applied, which deform the cell. Rheological properties measured with MTC vary depending on the ligand chemistry and density on the probe beads,[26] which, given heterogeneous molecule expression across the cell, could lead to results that are inconsistent or unable to be compared between experiments. All of the aforementioned methods are also unable to simultaneously

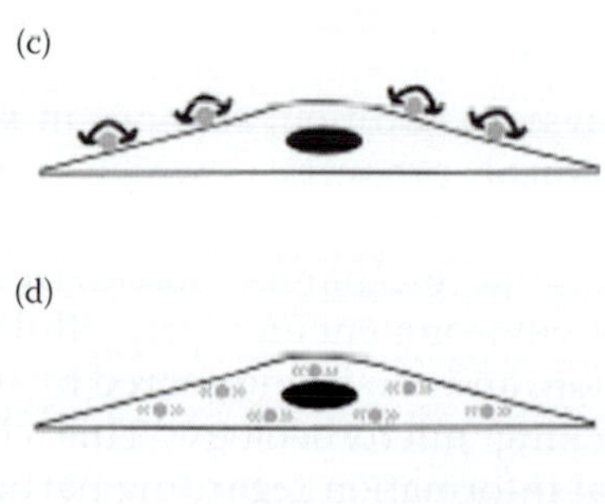

FIGURE 30.1
Current methods for measuring cell rheology. (a) Atomic force microscopy. (b) Micropipette aspiration. (c) Magnetic twisting cytometry. (d) Intracellular particle tracking nanorheology.

measure multiple points across a cell and as such are time-consuming and produce relatively low throughput. In addition, as mechanical properties are probed from the extracellular surface, this limits local measurements of intracellular microrheology and analysis of cytoskeletal heterogeneity. Finally, applications of AFM and micropipette aspiration are limited to those in which the probe can directly contact the cell and thus prevents the measurement of cells in three-dimensional environments or cells subjected to flow.

30.2.2 Intracellular Particle Tracking Microrheology

In order to better assess intracellular cytoskeletal heterogeneity and avoid the difficulties associated with other extracellular rheological measurement methods, MPTM was developed.[3] MPTM can be used to determine the mechanical properties of complex fluids with the accuracy of traditional bulk-fluid techniques, such as strain-controlled cone and plate rheometry.[27] MPTM can also be used to probe the local mechanical properties of complex fluids within microscopic domains, which is not possible with bulk-fluid rheological technique. For the original approach, fluorescent submicron particles were microinjected into the cytoplasmic region of individual cells. This method was improved in efficiency by the development of ballistic intracellular particle tracking nanorheology to massively inject submicron particles into a million cells at once,[28] thereby making it a powerful technique for quickly characterizing the rheological properties of samples that could not be achieved before.

Using MPTM, the displacements of the submicron particles, which result from the interaction between the particles and its surrounding cytoskeletal network, are used to determine the rheological properties of the cell cytoplasm (Figure 30.1d). It offers unique capacities compared to other methods for measuring the rheological properties of cells. Embedding particles in the cytoplasm allows for highly localized rheological measurements without the use of external probes.[29] Cytoplasmic viscoelasticity can be analyzed with this method in a variety of different conditions, including 3D culture,[6,30] shear,[28] and *in vivo*.[31] Moreover, the contributions of the plasma membrane and nucleus to the mechanical properties of the cell can be distinguished from cytoskeletal contributions.[32] Rheological measurements for a single cell can be acquired under a minute so that the short-term response of cells to specific factors can be analyzed.[1,33] The technique has also been used to determine the viscoelasticity of concentrated solutions of DNA,[34] cystic fibrotic sputum,[35] and actin filament solutions[36–39] in addition to the cell cytoplasm.[1,2,28–30,32,33]

The displacements of fluorescent particles embedded in the filamentous network (Figure 30.2) are captured

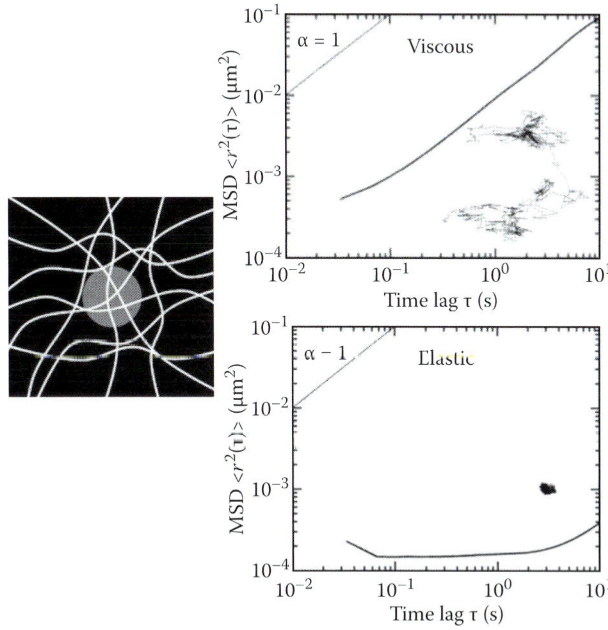

FIGURE 30.2
Brownian motion and MSDs of nanoparticles embedded in the cytoplasm. Schematic of particle embedded in the cytoplasm (left). Example traces and MSD plots of cytoplasm with viscous (top right) and elastic (bottom right) character.

using a fluorescence microscope with high magnification and a fast-speed camera for good spatial and temporal resolution. High resolution of particle displacements is obtained by tracking the intensity weighted centroids of the particles in the plane of focus of the objective. Particle trajectories in the focal plane are 2D projections of 3D displacements; thus, 2D tracking assumes that the fluid is locally isotropic. In 2D, the mean square displacement (MSD) of the particle is obtained by the following equation:

$$\text{MSD} =< r^2(\tau) >=< [x(t+\tau) - x(t)]^2 + [x(t+\tau) - y(t)]^2 \quad (30.1)$$

The MSD of the particle is related to the local diffusivity of the network, which is determined as

$$D = \frac{\langle \Delta r^2(\tau) \rangle}{4\tau} \quad (30.2)$$

In a purely viscous fluid, such as water or glycerol, the thermal fluctuation-driven particle motion is only hindered by viscous drag with the Stokes–Einstein equation:

$$D = \frac{k_B T}{6\pi a \eta} \quad (30.3)$$

TABLE 30.1
Reported Values of the Elastic Modulus at 1 Hz and Shear Viscosity Measured by Intracellular Particle Tracking Microrheology for Different Cell Types

Cell Type and Condition	Average Elasticity (dyn/cm²)	Average Viscosity (Poise)	Reference
Swiss 3T3 fibroblast	50 ± 20	10 ± 3	56
Swiss 3T3 fibroblast treated with LPA	120 ± 30	95 ± 20	56
Swiss 3T3 fibroblast under shear flow	600 ± 50	300 ± 40	28
Swiss 3T3 fibroblast treated with PDGF	190 ± 30	24 ± 8	2
Mouse embryonic fibroblast	140 ± 30	18 ± 2	57
HUVEC in 3D peptide matrix	55 ± 4	14 ± 1	30
HUVEC in 3D peptide matrix treated with VEGF	40 ± 3	18 ± 1	30
HUVEC in 3D fibronectin matrix	58 ± 6	NA	58
C. elegans embryo	Negligible	10 ± 1	31
Human IMR90 fetal lung fibroblasts	800 ± 100	24 ± 3	48
hiPSC	NA	15 ± 1	48
Mouse mesenchymal stem cell (mMSC)	42 ± 29	29 ± 5.6	1
mMSC treated with 4T1 TCM for 1 h	412 ± 40	330 ± 53	1
Low passage human MSC (hMSC)	3.6 ± 0.9	9 ± 3	33
Low passage hMSC treated with MCF7 TCM for 1 h	194 ± 72.1	118 ± 11	33
SKOV3	3.5 ± 0.5	4.4 ± 1.1	Present study

where D is the diffusion coefficient, k_B is Boltzmann's constant, T is the temperature, a is the particle radius, and η is the viscosity of the fluid. Therefore, the particle tracking data can be used to extract the viscosity of the fluid surrounding the particle.

However, in the complex filamentous network space when the particle size is greater than the pore size of the meshwork, which is the distance between filaments aligned in any direction, particle motion is affected by not only the viscosity of the fluid but also the elasticity of the filamentous network. The creep compliance (Γ), which measures the deformability of the cytoplasm, can be directly calculated from the MSD:

$$\Gamma(\tau) = \left(\frac{3\pi a}{2k_BT}\right) MSD(\tau) \tag{30.4}$$

In order to facilitate analysis of additional viscoelastic properties of the cytoplasm, the MSD is transferred to the frequency domain. The frequency-dependent form of the Stokes–Einstein equation is used to determine the viscoelastic properties of the fluid:

$$G^*(\omega) = \frac{2k_BT}{3\pi a\left\langle \Delta r^2\left(\frac{1}{\omega}\right)\right\rangle \Gamma[(1+\alpha(\omega))(1+\beta(\omega)/2)]} \tag{30.5}$$

where G^* is the frequency-dependent complex shear modulus, Γ is the gamma function (different from the creep compliance, Γ), α is the first derivative, and β is the second derivative of the MSD curve. The complex shear modulus can be further divided into the in-phase component, or elastic (storage) modulus (G'), and the out-of-phase component, or viscous (loss) modulus (G''):

$$G'(\omega) = |G^*(\omega)|\cos\left(\frac{\pi\alpha(\omega)}{2}\right)$$
$$G''(\omega) = |G^*(\omega)|\cos\left(\frac{\pi\alpha(\omega)}{2}\right) \tag{30.6}$$

Cells behave more like an elastic solid at high frequencies, or low timescales, with G' being greater than G''. Typical values of G' at 1 Hz are shown in Table 30.1. The shear viscosity can be estimated as the product of the plateau modulus and the relaxation time. The plateau modulus is the value of the elastic modulus at intermediate frequencies where it reaches a quasi plateau value. Relaxation time is the inverse of frequency at which elastic and viscous moduli are equal.

30.2.3 Limitations of Intracellular Particle Tracking Microrheology

Since intracellular particle tracking is used to characterize the microrheology of the fluid surrounding tracked particles, this technique is limited by a particle diameter, which must be small enough for particles to be embedded in the cytosol (100–200 nm) but large enough to be imaged with fluorescence microscopy (~60 nm). This particle size range is also ideal for vesicle-mediated endocytosis, which results in active transport along microtubules via kinesin or dynein motors.[40] Since the intracellular rheology is related to the random motion of microparticles, not motor-driven transport, this issue can be circumvented by introducing the particles into the cytoplasm using physical methods, such as ballistic bombardment and microinjection. In addition, the motion of injected nanoparticles is not affected by myosin-induced contraction forces in the cytoplasm.[32] Before the intensity of fluorescence can be enhanced, intracellular particle tracking microrheology may be unable to probe the rheological properties in densely cross-linked regions of the actin cortex, such as the lamellipodia, which have overall thicknesses on the length scale of a particle diameter (100–200 nm).[15] Also, for the cytoskeletal filaments, which have a pore size smaller than the diameter of the probe particles, it would not be effective for the particles to penetrate into those networks unless the structure forms after the particles have been located there. Furthermore, relying strictly on Brownian motion for quantification of rheological properties typically only allows for analysis of materials within a certain range of properties, though most cell types fall within this range. In the lower limiting case, particles must stay in the field of view for sufficient time to track their motion; thus, particle tracking microrheology cannot measure the rheological properties of low viscosity fluids, such as water, without acquisition of expensive high-end high-speed cameras. Since the elastic modulus measured by this method is dependent on thermal energy (k_bT, approximately constant at a physiological temperature of 37°C), the relatively small magnitude of this energy does not allow for probing high elastic moduli without measuring particle motion over longer intervals of time. Taking this limit in applied energy into account with current typical tracking resolution of 5 nm, one may probe elastic moduli of up to ~1000 Pa[41]; consequently, for materials with elastic moduli greater than ~1000 Pa, other methods may be more ideal. In addition, as particle displacements approach the resolution limit, care must be taken to ensure that vibration and drift do not occur, as these can alter tracking results.

30.3 Intracellular Particle Tracking Microrheology Applications

30.3.1 Model Systems of Actin Filaments

Actin solutions can be used to model the complex viscoelastic properties of the cytoplasm. Diffusing wave spectroscopy (DWS) has been used to determine the thermally excited motion of microspheres in actin solutions.[42] The mechanical properties of actin networks are highly dependent on the rate of exchange of dynamic cross-links between actin filaments by α-actinin.[43] MPTM studies of actin solutions with fascin, an F-actin bundling protein, demonstrated that increasing actin or fascin concentration increases and skews the distribution of MSDs.[44] Numerous other studies have highlighted the important role that ABPs play in determining the rheology of actin networks.[36,37,45–47]

30.3.2 Heterogeneity Analysis of Single-Cell Mechanics

With MPTM, the mechanical properties of the cytoplasm are determined from the transport patterns of hundreds of individual nanoparticle probes. Each particle probes the rheology of the surrounding cytoplasm, which is heavily affected by differences in the degree of actin cross-linking and the local microstructure. Micromechanical mapping of Swiss 3T3 fibroblasts revealed that the cytoplasm is significantly more heterogeneous than reconstituted actin filament networks.[29] In addition, the perinuclear region is typically more compliant than the lamella.[29] Cross-linking actin with α-actinin increases the stiffness and heterogeneity of the cytoplasm,[29] which is in agreement with actin solution studies.[43] Statistical analysis of the distributions of particles' MSDs at specific timescales is used for heterogeneity analysis of the cytoplasmic region for different cell types or the same cell type under different treatments. Local mechanical properties from particle tracking analysis can also be traced back to specific regions of the cell, which can be used to determine how differences in cytoskeletal mechanics affect cell function.

30.3.3 Single-Cell Particle Tracking Rheology

Intracellular particle tracking microrheology has been used for a wide variety of cell types and conditions, as shown in Table 30.1. This technique has revealed that the cytoplasm of induced pluripotent stem cells (iPSCs) closely mimics their parental cells with a disassembled actin network.[48] This method has also been used to determine the rheological properties of a developing *Caenorhabditis elegans* embryo.[31] The cytoplasm of these embryos is highly viscous without measurable elasticity, in contrast to fully differentiated cells, which are viscoelastic.[31]

Intracellular particle tracking microrheology has also been used to elucidate how fibroblasts spatially coordinate movement.[2] Rac1 and Cdc42 activation coordinate stiffening in the leading lamella and perinuclear region during migration.[2] Although depolymerizing microtubules has no effect on overall cell rheology, it prevents mechanical polarization and cell migration.[2] In addition, cancerous transformation potential has been associated with cytoplasmic stiffening in response to matrix stiffening,[49] which is correlated with increased migration.[50] In contrast, treatment of human umbilical vein endothelial cells (HUVECs) embedded in a 3D peptide-based hydrogel with vascular endothelial growth factor (VEGF), which increases endothelial cell migration, decreases the elasticity of the cytoplasm.[30] This increase in elasticity can be reversed by inhibition of ROCK.[30] ROCK also plays an important role in stiffening of fibroblasts in response to flow.[28] Tumor conditioned media (TCM), which is a cocktail of promigratory factors that mimic factors secreted by tumors, induces stiffening and migration in both human and murine mesenchymal stem cells (MSCs).[1,33] AFM, micropipette aspiration, and MPTM have been used to characterize the viscosity of human MSCs (Table 30.2).[33,51,52] While AFM and micropipette aspiration viscosity measurements are similar, the viscosity measured with ballistic intracellular particle tracking nanorheology is significantly different, presumably because it measures the local viscoelasticity of the cytoplasm, while the other methods mainly probe large stress fibers and include contributions of the nucleus and the plasma membrane.

Cytoplasmic stiffness has been associated with metastatic potential in both breast cancer[53] and ovarian cancer.[54] The cytoplasm of OVCAR3 human ovarian cancer cells is viscoelastic, whereas OSE10 normal human ovarian epithelial cells are primarily viscous.[55] SKOV3 human metastatic ovarian cancer cells have a higher α value, or slope of the MSD curve, than both OSE10 and OVCAR3 cells (Figure 30.3b), suggesting less restricted particle movement in the cytoplasm. A distribution of MSDs exists in particles embedded in the cytoplasm of SKOV3 cells (Figure 30.3c), similar to previous reports with other cell lines.[2,28,29,31–33,48] The slope of the time-dependent creep compliance curve is higher than OSE10 and OVCAR3 cells (Figure 30.4a), indicating that SKOV3 cells have a more compliant cytoplasm in agreement with previous AFM measurements.[54] The viscous modulus (G'') dominates over the elastic modulus (G') at all frequencies. These results suggest that the cytoskeletal network of SKOV3 is highly deformable, which may be important for squeezing through blood vessels in order to metastasize.[3]

TABLE 30.2
Viscosity of Human MSCs Measured by Common Rheology Methods

Intracellular Rheology Measurement Method	Viscosity (Pa·s)	Reference
Micropipette aspiration	2700 ± 1600	52
AFM	3591 ± 630	51
BIPTN	0.9 ± 0.3	33

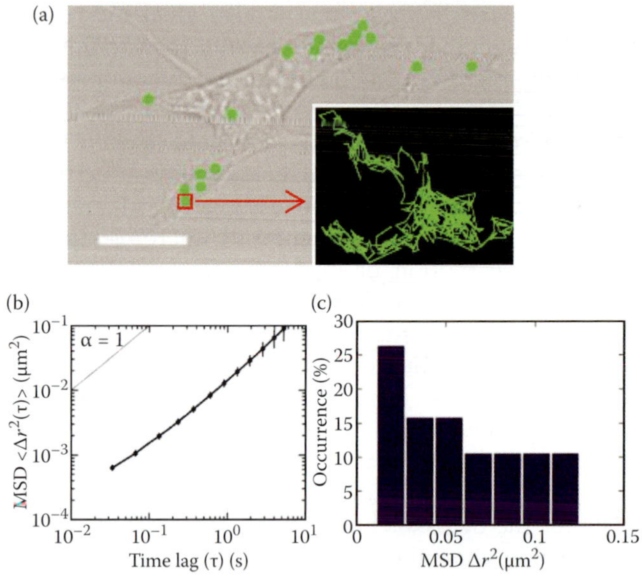

FIGURE 30.3
Brownian motion of nanoparticles embedded in the cytoplasm of SKOV3 cells. Phase-contrast micrograph of SKOV3 cell ballistically injected with 100 nm green particles and trace of a single particle's motion (inset) (a). Time-dependent ensemble average MSD of particles embedded in the cytoplasm of SKOV3 cells (b). Histogram showing the distribution of particle MSDs within a single SKOV3 cell for timescale $\tau = 1$ s (c).

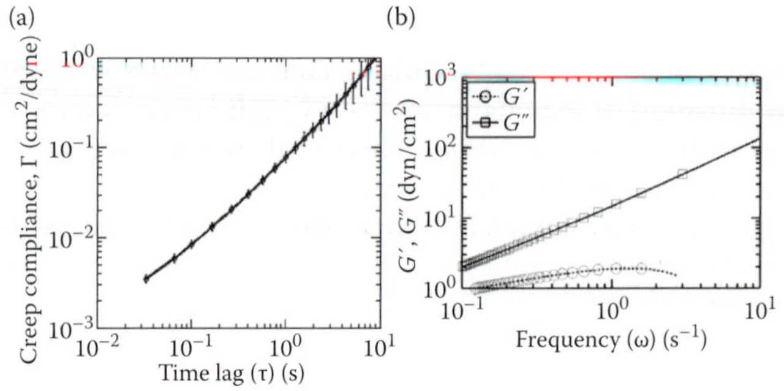

FIGURE 30.4
Creep compliance and viscous and elastic moduli for SKOV3 cells. Average creep compliance (Γ) of SKOV3 cells (a). Frequency-dependent viscous (G', dashed line) and elastic (G'', solid line) moduli of SKOV3 cells (b). Properties determined from ensemble-averaged MSD in Figure 30.2b.

30.4 Conclusion

Intracellular particle tracking microrheology is a powerful tool for measuring the mechanical properties of cells in physiologically relevant conditions. This method allows for the measurement of local rheological properties, which cannot be measured with any other current method. In addition, intracellular particle tracking microrheology can be high throughput and can measure changes in the intracellular rheological properties at multiple timescales. The corresponding changes in intracellular viscoelasticity can elucidate a wide variety of critical cellular processes including differentiation, migration, cancerous transformation, and response to extracellular stimuli. Future applications for this technique include *in vivo* three-dimensional microrheology and screening of drugs for mechanical-related mechanisms.

Acknowledgments

The authors thank the funding provided by the National Science Foundation (M.D.—1066585) and the National Cancer Institute (Y.T.—NIH/NCI U54CA143868).

References

1. McGrail, D. J., Ghosh, D., Quach, N. D. and Dawson, M. R. Differential mechanical response of mesenchymal stem cells and fibroblasts to tumor-secreted soluble factors. *PLoS One* **7,** e33248 (2012).
2. Kole, T. P., Tseng, Y., Jiang, I., Katz, J. L. and Wirtz, D. Intracellular mechanics of migrating fibroblasts. *Biophys. J.* **16,** 328–38 (2005).
3. Balzer, E. M. et al. Physical confinement alters tumor cell adhesion and migration phenotypes. *FASEB J.* **26,** 4045–56 (2012).
4. Engler, A. J., Sen, S., Sweeney, H. L. and Discher, D. E. Matrix elasticity directs stem cell lineage specification. *Cell* **126,** 677–89 (2006).
5. Homouz, D., Perham, M., Samiotakis, A., Cheung, M. S. and Wittung-Stafshede, P. Crowded, cell-like environment induces shape changes in aspherical protein. *Proc. Natl. Acad. Sci. U. S. A.* **105,** 11754–9 (2008).
6. Baker, E. L., Bonnecaze, R. T. and Zaman, M. H. Extracellular matrix stiffness and architecture govern intracellular rheology in cancer. *Biophys. J.* **97,** 1013–21 (2009).
7. Ingber, D. E. Mechanical control of tissue morphogenesis during embryological development. *Int. J. Dev. Biol.* **50,** 255–66 (2006).
8. Keller, P. J., Schmidt, A. D., Wittbrodt, J. and Stelzer, E. H. K. Reconstruction of zebrafish early embryonic development by light sheet microscopy. *Science* **322,** 1065–9 (2008).
9. Mammoto, T. and Ingber, D. E. Mechanical control of tissue and organ development. *Development* **137,** 1407–20 (2010).
10. Vacanti, J., Langer, R., Upton, J. and Marler, J. Transplantation of cells in matrices for tissue regeneration. *Adv. Drug Deliv. Rev.* **33,** 165–82 (1998).
11. Suresh, S. Biomechanics and biophysics of cancer cells. *Acta Mater.* **55,** 3989–4014 (2007).
12. Orr, A., Helmke, B., Blackman, B. and Schwartz, M. Mechanisms of mechanotransduction. *Dev. Cell* **10,** 11–20 (2006).
13. Alberts, B. et al. *Molecular Biology of the Cell* (Garland Science Taylor and Francis Group, New York, 2002).
14. Atilgan, E., Wirtz, D. and Sun, S. X. Mechanics and dynamics of actin-driven thin membrane protrusions. *Biophys. J.* **90,** 65–76 (2006).
15. Atilgan, E., Wirtz, D. and Sun, S. X. Morphology of the lamellipodium and organization of actin filaments at the leading edge of crawling cells. *Biophys. J.* **89,** 3589–602 (2005).
16. Stricker, J., Falzone, T. and Gardel, M. L. Mechanics of the F-actin cytoskeleton. *J. Biomech.* **43,** 9–14 (2010).
17. McBeath, R., Pirone, D. M., Nelson, C. M., Bhadriraju, K. and Chen, C. S. Cell shape, cytoskeletal tension, and RhoA regulate stem cell lineage commitment. *Dev. Cell* **6,** 483–95 (2004).
18. Treiser, M. D. et al. Cytoskeleton-based forecasting of stem cell lineage fates. *Proc. Natl. Acad. Sci. U. S. A.* **107,** 610–5 (2010).
19. Kilian, K. A., Bugarija, B., Lahn, B. T. and Mrksich, M. Geometric cues for directing the differentiation of mesenchymal stem cells. *Proc. Natl. Acad. Sci. U. S. A.* **107,** 4872–7 (2010).
20. Pollard, T. D. and Weihing, R. R. Actin and myosin and cell movement. *CRC Crit. Rev. Biochem.* **2,** 1–65 (1974).
21. Waigh, T. A. Microrheology of complex fluids. *Rep. Prog. Phys.* **68,** 685–742 (2005).
22. Larson, R. G. *The Structure and Rheology of Complex Fluids* (Oxford University Press, New York, 1999).
23. Tseng, Y., Fedorov, E., McCaffery, J. M., Almo, S. C. and Wirtz, D. Micromechanics and ultrastructure of actin filament networks crosslinked by human fascin: A comparison with alpha-actinin. *J. Mol. Biol.* **310,** 351–66 (2001).
24. Kollmannsberger, P. and Fabry, B. Linear and nonlinear rheology of living cells. *Annu. Rev. Mater. Res.* **41,** 75–97 (2011).
25. Kirmizis, D. and Logothetidis, S. Atomic force microscopy probing in the measurement of cell mechanics. *Int. J. Nanomed.* **5,** 137–45 (2010).
26. Puig-de-Morales, M. et al. Cytoskeletal mechanics in adherent human airway smooth muscle cells: Probe specificity and scaling of protein–protein dynamics. *Am. J. Physiol. Cell Physiol.* **287,** C643–54 (2004).
27. Mason, T. G. and Weitz, D. A. Optical measurements of frequency-dependent linear viscoelastic moduli of complex fluids. *Phys. Rev. Lett.* **74,** 1250–3 (1995).
28. Lee, J. S. H. et al. Ballistic intracellular nanorheology reveals ROCK-hard cytoplasmic stiffening response to fluid flow. *J. Cell Sci.* **119,** 1760–8 (2006).
29. Tseng, Y., Kole, T. P. and Wirtz, D. Micromechanical mapping of live cells by multiple-particle tracking microrheology. *Biophys. J.* **83,** 3162–76 (2002).

30. Panorchan, P., Lee, J. S. H., Kole, T. P., Tseng, Y. and Wirtz, D. Microrheology and ROCK signaling of human endothelial cells embedded in a 3D matrix. *Biophys. J.* **91,** 3499–507 (2006).
31. Daniels, B. R., Masi, B. C. and Wirtz, D. Probing single-cell micromechanics in vivo: The microrheology of C. elegans developing embryos. *Biophys. J.* **90,** 4712–9 (2006).
32. Hale, C. M., Sun, S. X. and Wirtz, D. Resolving the role of actomyosin contractility in cell microrheology. *PLoS One* **4,** e7054 (2009).
33. McGrail, D. J., McAndrews, K. M. and Dawson, M. R. Biomechanical analysis predicts decreased human mesenchymal stem cell function before molecular differences. *Exp. Cell Res.* **319,** 684–96 (2013).
34. Mason, T. G., Dhople, A. and Wirtz, D. Linear viscoelastic moduli of concentrated DNA solutions. *Macromolecules* **9297,** 3600–3 (1998).
35. Dawson, M., Wirtz, D. and Hanes, J. Enhanced viscoelasticity of human cystic fibrotic sputum correlates with increasing microheterogeneity in particle transport. *J. Biol. Chem.* **278,** 50393–401 (2003).
36. Tseng, Y., An, K. M., Esue, O. and Wirtz, D. The bimodal role of filamin in controlling the architecture and mechanics of F-actin networks. *J. Biol. Chem.* **279,** 1819–26 (2004).
37. Tseng, Y., Schafer, B. W., Almo, S. C. and Wirtz, D. Functional synergy of actin filament cross-linking proteins. *J. Biol. Chem.* **277,** 25609–16 (2002).
38. Tseng, Y., An, K. M. and Wirtz, D. Microheterogeneity controls the rate of gelation of actin filament networks. *J. Biol. Chem.* **277,** 18143–50 (2002).
39. Tseng, Y. and Wirtz, D. Mechanics and multiple-particle tracking microheterogeneity of alpha-actinin-cross-linked actin filament networks. *Biophys. J.* **81,** 1643–56 (2001).
40. Hoffman, B. D., Massiera, G., Van Citters, K. M. and Crocker, J. C. The consensus mechanics of cultured mammalian cells. *Proc. Natl. Acad. Sci. U. S. A.* **103,** 10259–64 (2006).
41. Gardel, M. L., Valentine, M. T. and Weitz, D. A. Microrheology. In *Microscale Diagnostic Tech* (Breuer, K. ed.) 1–49 (Springer, Berlin, Heidelberg, 2005). doi: 10.1007/3-540-26449-3_1.
42. Palmer, A., Cha, B. and Wirtz, D. Structure and dynamics of actin filament solutions in the presence of latrunculin A. *J. Polym. Sci. Part B Polym. Phys.* **36,** 3007–15 (1998).
43. Xu, J., Wirtz, D. and Pollard, T. D. Dynamic cross-linking by alpha-actinin determines the mechanical properties of actin filament networks. *J. Biol. Chem.* **273,** 9570–6 (1998).
44. Apgar, J. et al. Multiple-particle tracking measurements of heterogeneities in solutions of actin filaments and actin bundles. *Biophys. J.* **79,** 1095–106 (2000).
45. Gardel, M., Valentine, M., Crocker, J., Bausch, A. and Weitz, D. Microrheology of entangled F-actin solutions. *Phys. Rev. Lett.* **91,** 158302 (2003)
46. Gardel, M. L. et al. Elastic behavior of cross-linked and bundled actin networks. *Science* **304,** 1301–5 (2004).
47. Tseng, Y. et al. How actin crosslinking and bundling proteins cooperate to generate an enhanced cell mechanical response. *Biochem. Biophys. Res. Commun.* **334,** 183–92 (2005).
48. Daniels, B. R. et al. Differences in the microrheology of human embryonic stem cells and human induced pluripotent stem cells. *Biophys. J.* **99,** 3563–70 (2010).
49. Baker, E. L., Lu, J., Yu, D., Bonnecaze, R. T. and Zaman, M. H. Cancer cell stiffness: Integrated roles of three-dimensional matrix stiffness and transforming potential. *Biophys. J.* **99,** 2048–57 (2010).
50. Baker, E. L., Srivastava, J., Yu, D., Bonnecaze, R. T. and Zaman, M. H. Cancer cell migration: Integrated roles of matrix mechanics and transforming potential. *PLoS One* **6,** e20355 (2011).
51. Darling, E. M., Topel, M., Zauscher, S., Vail, T. P. and Guilak, F. Viscoelastic properties of human mesenchymally-derived stem cells and primary osteoblasts, chondrocytes, and adipocytes. *J. Biomech.* **41,** 454–64 (2008).
52. Tan, S. C. W. et al. Viscoelastic behaviour of human mesenchymal stem cells. *BMC Cell Biol.* **9,** 40 (2008).
53. Cross, S. E., Jin, Y.-S., Rao, J. and Gimzewski, J. K. Nanomechanical analysis of cells from cancer patients. *Nat. Nanotechnol.* **2,** 780–3 (2007).
54. Swaminathan, V. et al. Mechanical stiffness grades metastatic potential in patient tumor cells and in cancer cell lines. *Cancer Res.* **71,** 5075–80 (2011).
55. Wu, P.-H. et al. High-throughput ballistic injection nanorheology to measure cell mechanics. *Nat. Protoc.* **7,** 155–70 (2012).
56. Kole, T. P., Tseng, Y., Huang, L., Katz, J. L. and Wirtz, D. Rho kinase regulates the intracellular micromechanical response of adherent cells to Rho activation. *Mol. Biol. Cell* **15,** 3475–84 (2004).
57. Lee, J. S. H. et al. Nuclear lamin A/C deficiency induces defects in cell mechanics, polarization, and migration. *Biophys. J.* **93,** 2542–52 (2007).
58. Zhou, X., Rowe, R. and Hiraoka, N. Fibronectin fibrillogenesis regulates three-dimensional neovessel formation. *Genes Dev.* **22,** 1231–43 (2008).

Section X

Cellular and Molecular Imaging Methods

Nuclear Biophysics

31

Probing Chromatin Structure and Dynamics Using Fluorescence Anisotropy Imaging

Ekta Makhija, K. Venkatesan Iyer, Shefali Talwar, and G.V. Shivashankar

CONTENTS

In this chapter, we describe a high-resolution fluorescence anisotropy imaging (FAI) based method to study chromatin structure and dynamics. The rotational mobility of fluorescent tagged core histone proteins in live cells detected using FAI provides a measure of the local chromatin structure. The anisotropy-based compaction maps were validated by labeling known epigenetic markers of higher-order chromatin structure. Time lapse FAI combined with image correlation analysis was used to study (a) differential dynamics between heterochromatin and euchromatin regions, (b) impact of cell geometric constraints on chromatin dynamics, and (c) compaction states of chromatin assembly in undifferentiated and differentiated cells. Numerical simulation of image correlation provides a quantitative measure of the relaxation timescales associated with chromatin dynamics in these cases. Taken together, this technique provides us with a handle to analyze the spatiotemporal organization of chromatin in live cells.

31.1 Introduction

Chromatin of eukaryotic cells is packaged in the nucleus with the help of various histone and nonhistone proteins to create heterogeneously compacted regions, which provides spatial regulation of gene expression [1,2]. The relative chromatin packaging provides an insight into the state of the nucleus, i.e., its transcriptional activity [3], the phase of cell cycle [4,5], and lineage commitment [6]. The chromatin organization in interphase cells has been shown to be dynamic, thus facilitating temporal regulation of gene expression [7–9]. These dynamics are regulated by various processes: binding/unbinding of core and linker histones, turnover of histone tail modifications, interaction with chromatin remodeling complexes, and Brownian motion of chromosomes [10]. These different processes contribute to chromatin remodeling at different length and timescales: from a few milliseconds for nucleosomal arrays [11], seconds for heterochromatin binding proteins [12,13], minutes for linker histones [14] and RNA polymerase II [15], to hours for core histones [16] and lamin proteins [17].

Various imaging techniques have been used to study chromatin structure and dynamics [18]. The most commonly used techniques are electron microscopy [19], immunostaining of various chromatin binding proteins, and fluorescent labeling of the DNA in fixed cells. While these methods account for the chromatin compaction state, they lack information about its dynamics. In such fixed samples, biochemical methods like chromosome conformation capture and its variants have also been used to generate 3D maps of intergenomic distances [17]. Confocal microscopy combined with fluorescence recovery after photobleaching (FRAP) to measure translational diffusion [20,21] of fluorescently tagged proteins is commonly used to study chromatin dynamics. Further, the higher-order packaging of chromatin has also been studied using fluorescence resonance energy transfer (FRET) [22,23]. However, techniques to measure the local DNA compaction as well as spatial heterogeneities and dynamics associated with packaging chromatin in living cells are limited.

In this chapter, we use the rotational mobility of fluorophore-bound proteins, commonly used in biochemical assays [24,25] to implement live cell fluorescence anisotropy imaging (FAI) to probe chromatin assembly and dynamics. Rotational mobility of the fluorophore depends on its shape and size and the viscosity of the solvent [26]. The effect of rotational diffusion of fluorophore on anisotropy r is given by the Perrin equation:

$$\frac{r_0}{r} = 1 + \frac{\tau}{\theta} \quad (31.1)$$

where r_0 is the fundamental anisotropy of randomly oriented fluorophores in fixed configuration, τ is the fluorescence lifetime, and θ is the rotational correlation time. FA is calculated as

$$r = \frac{I_{II} - I_{\perp}}{I_{II} + 2I_{\perp}} \quad (31.2)$$

which is a measure of the emission depolarization owing to the rotational diffusion of the fluorophore (Figure 31.1a). We utilized this concept in the imaging mode to visualize chromatin compaction profile and its dynamics by labeling EGFP-tagged core histone proteins (H2B-EGFP) in live cells [21,27–33]. The anisotropy values H2B-EGFP represent the local chromatin compaction, as both interaction of DNA in nucleosomes and packaging of nucleosomes to higher-order structure affect its rotational mobility. Thus, a lower anisotropy value (higher depolarization) would represent loosely packed chromatin and vice versa. Further,

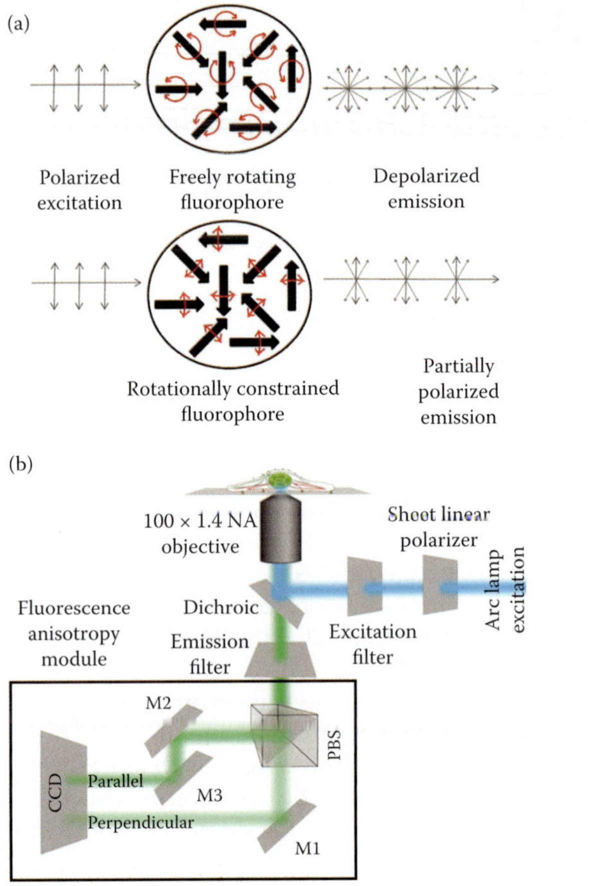

FIGURE 31.1
FA imaging. (a) Schematic depicting the principle of FA; emission depolarization is proportional to rotational mobility of fluorophore molecules. (b) Schematic of the experimental setup.

in this chapter, we describe a quantitative analysis of FAI to explore the changes in chromatin structure and dynamics upon application of geometric constraints and in stem cells.

31.2 Materials and Methods

31.2.1 Cell Culture

Stable transfections were carried out in NIH3T3 with fusion plasmid for core histone H2B tagged with EGFP. Embryonic stem (ES) cells and primary mouse embryonic fibroblasts (PMEFs) were obtained from H2B-EGFP transgenic mice, B6.Cg-Tg (HIST1H2BB/EGFP)1Pa/J (Jackson Laboratories, Bar Harbor, Maine). NIH3T3 and PMEF cells were cultured in DMEM supplemented with 10% fetal bovine serum and 1% penicillin–streptomycin. ES cells were cultured in K/O DMEM supplemented with 15% knockout fetal bovine serum,

1 mM sodium pyruvate (Sigma), 0.1 mM nonessential amino acids, 2 mM L-glutamine, 0.1 mM b-mercaptoethanol (Sigma), and 500 U/mL leukemia inhibitory factor (LIF) (Chemicon) and penicillin–streptomycin. All cell culture reagents were from GIBCO Invitrogen unless mentioned otherwise.

31.2.2 Fluorescence Anisotropy Imaging

Fluorescence anisotropy (FA) images were generated from intensity images acquired in parallel and perpendicular polarization on an inverted microscope (Eclipse Ti, Nikon Instruments Inc., USA, 100×, 1.4NA objective). Linearly polarized light was generated by passing the arc lamp beam through a sheet polarizer (New Focus, USA) and an excitation filter (HQ470/40x, Chroma Tech, USA). The emission from the sample plane was collected using an emission filter (HQ 525/50m, Chroma Tech, USA) and was split into parallel and perpendicular polarizations using the Roper Dual View polarization module (Photometrics, Tucson, Arizona). The parallel and perpendicular images were acquired on two halves of an Andor iXon 897 BV EMCCD camera (Andor Technology, USA). Time lapse images were captured using NIS Elements (Nikon Instruments Inc., USA). Schematic of the setup is shown in Figure 31.1b. A custom-written program in MATLAB (Mathworks, USA) was used to compute anisotropy from the split image on a pixel-by-pixel basis after performing background subtraction and 3 × 3 smoothing of the original image. For each pixel, anisotropy was calculated as $r = \frac{I_{II} - g{*}I_{\perp}}{I_{II} + 2g{*}I_{\perp}}$, where g (g-factor) is the ratio of the sensitivity of the experimental setup for parallel and perpendicular channels [26]. To calculate the g-factor, anisotropy of fluorescein isothiocyanate solution in water ($r = 0.02$) was used in the above formula.

31.2.3 Image Analysis

The parallel and perpendicular images were aligned for translational and rotational shifts between the two channels using MATLAB. The perpendicular emission intensity image was translated by Δx and Δy up to ±5 pixels and rotated by $\Delta\theta$ up to ±5°. Correlation was calculated in each case with the parallel image. The Δx, Δy, and $\Delta\theta$ values, which gave maximum correlation, were used to translate the original perpendicular images to obtain the best alignment. The complete temporal sequence was corrected for any translation and rotation with respect to the first time frame. Pearson correlation coefficient (PCC) was calculated using the "corr" function in MATLAB and then averaged over all image frames with time lag ranging from 1 to 30 min. Before calculating the PCC between anisotropy images, a 5 pixel border was removed from the nucleus periphery to avoid the artifacts that may arise from edge effects.

31.2.4 Immunofluorescence Staining

Cells were grown on glass bottom dishes and were fixed with 4% paraformaldehyde (PFA) for 15 min prior to permeabilization with 0.5% Triton X-100 for 15 min. Triton X-100 was washed off twice with 1× phosphate buffered saline (PBS), followed by blocking in 1% bovine serum albumin (BSA) for 30 min. Primary antibody diluted in 1% BSA was incubated for 1 h followed by washing with 1% BSA and incubation in secondary antibody diluted 1:300 in 1% BSA for 1 h. Cells were then washed with 1× PBS and stored in PBS or mounted using prolong antifade mounting reagent (Life Technologies, USA). The following antibodies were used: anti-Histone H3 (acetyl K9) (Abcam—ab10812), anti-H3 (trimethyl K9) (Abcam—ab8898), anti-RNA polymerase II CTD repeat YSPTSPS (phospho S5) (Abcam—ab5131), and Anti-HP1α (Abcam—ab77256). Various Alexa-647 labeled secondary antibodies (Life Technologies, Invitrogen) were used.

31.2.5 Microcontact Printing

Polydimethyl sulfate (PDMS) and a cross-linker were mixed in 10:1 ratio, poured over silicon wafers with 1600 μm² triangle and circular patterns, and heated at 80°C for 2 h. The stamps were then peeled, coated with labeled fibronectin, and stamped on untreated plastic bottom dishes (Ibidi). Dishes were then treated with 0.2% pluronic acid for 30 min, washed twice with PBS, once with DMEM, and then 40,000 cells seeded per 35 mm dish.

31.3 Results and Discussion

31.3.1 Core Histone Anisotropy Measures Chromatin Compaction

Chromatin is known to be organized into highly compact heterochromatin forming distinct nodes and relatively decompact euchromatin regions in fibroblasts. FAI of core histone H2B in PMEFs was characterized as a measure of chromatin compaction by comparing with well-established epigenetic markers. FA was measured in a 10 × 10 pixels² region around the heterochromatin regions and was compared with the measurement in a

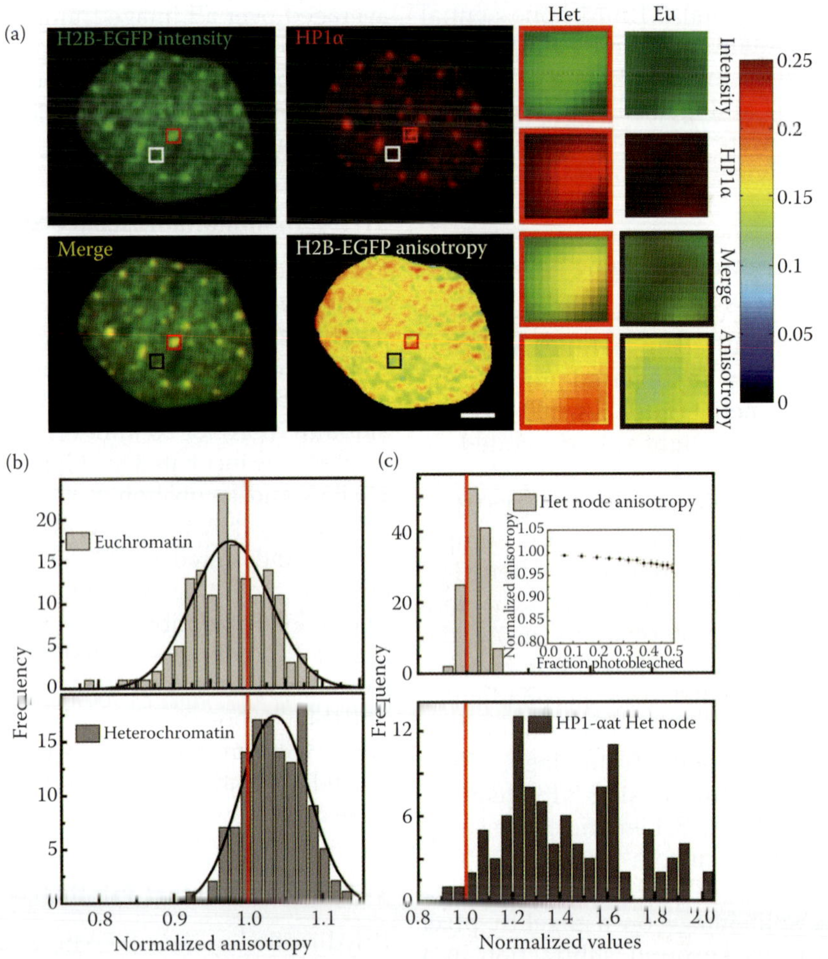

FIGURE 31.2
H2B-EGFP anisotropy represents chromatin structure. (a) Images showing H2B-EGFP intensity, HP1α antibody staining, their merging, and H2B-EGFP anisotropy for a PMEF cell nucleus. The collage on the right shows enlarged images of heterochromatin (red ROI) and euchromatin (black ROI). (b) Histograms showing the distribution of anisotropy values (normalized to the mean nuclear anisotropy) in the heterochromatin and euchromatin region. (c) Histograms showing the distribution of anisotropy and HP1α intensities (normalized to their respective mean nuclear values) in the heterochromatin region of the nucleus. The red line indicates the mean nuclear value. Inset shows anisotropy as a function of photobleaching, illustrating that anisotropy does not depend on intensity. Scale bar, 5 μm.

similar sized euchromatin region (Figure 31.2a) to generate a histogram of anisotropy values (Figure 31.2b). The values were normalized to the mean nuclear anisotropy to represent the difference in chromatin compaction between the euchromatin and heterochromatin. The heterochromatin nodes, observed as bright regions in the H2B-EGFP intensity image, were observed to correspond to high anisotropy. But all high anisotropy regions may not correspond to heterochromatin nodes (Figure 31.2a).

To further validate these observations, known markers of chromatin compaction were visualized in regions of high anisotropy. Toward this, images were acquired for anisotropy followed by fixation by 4% PFA and immunostaining for heterochromatin binding protein (HP1α). The anisotropy at heterochromatin nodes as well as the HP1α intensity values were significantly larger than their respective mean nuclear values (Figure 31.2c). However, anisotropy was found to be independent of fluorescence intensity, as verified by photobleaching (Figure 31.2c, inset). This further negated the presence of homoFRET between the EGFP-tagged core histones.

Next, anisotropy in heterochromatin regions was compared with various immunolabeled transcriptional markers. Heterochromatin regions with high anisotropy showed enrichment of the transcriptionally repressive marker H3K9Me3 (Figure 31.3a) and very less activation marker H3K9Ac (Figure 31.3b), and depletion of actively transcribing RNA Pol. II (Figure 31.3c). Anisotropy of H2B-EGFP and intensity of these transcriptional markers were measured in multiple 10 × 10 pixels^2 regions

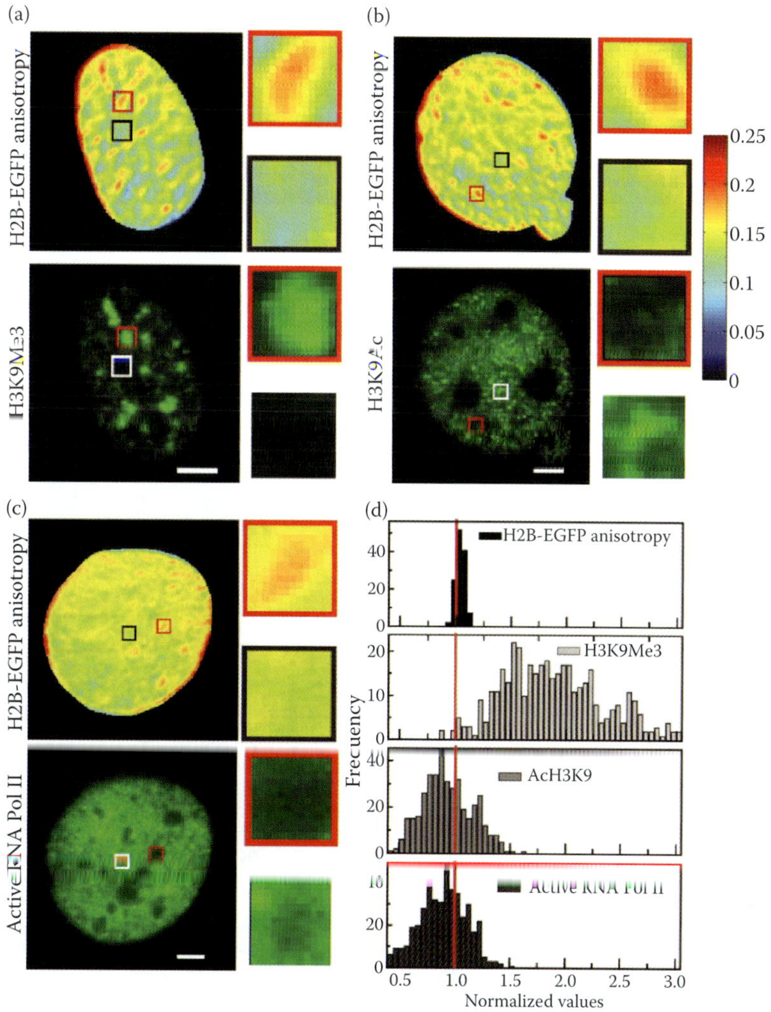

FIGURE 31.3
Anisotropy of H2B-EGFP to visualize transcriptional state of chromatin. (a) Images showing anisotropy of H2B-EGFP and immunofluorescence staining for H3K9Me3. The collage shows enlarged anisotropy and H3K9Me3 images of heterochromatin (red ROI) and euchromatin (black ROI) regions. (b) Similar images for H2B-EGFP anisotropy and H3K9Ac. (c) Images of H2B-EGFP anisotropy and active RNA Pol. II. (d) Histograms showing the distribution of anisotropy and immunofluorescence staining values (normalized to their respective mean nuclear values). The red line indicates the mean value of the nucleus. Scale bar, 5 μm.

around heterochromatin nodes in several cells, and normalized histograms were generated to quantify these observations (Figure 31.3d). In Section 31.3.2, we develop methods to visualize dynamics of chromatin in live cells.

31.3.2 Time Lapse Anisotropy Imaging to Study Chromatin Dynamics

Time lapse FAI of cells expressing H2B-EGFP was carried out to determine the temporal changes in anisotropy values. As a first step, we describe numerical approaches to probe image correlation using a simulated image (Figure 31.4a) and a series of decorrelated images generated by exchanging a pair of randomly chosen pixels in every iteration (Figure 31.4b). To quantify this decorrelation of image sequences, PCC of each image with respect to the first image was calculated and plotted. It was observed that the PCC curve is sensitive to the rate of change (no. of pixel pairs exchanged per iteration) and addition of random noise (Figure 31.4c). Typical PCC vs. time lag plot for an anisotropy image sequence is shown in Figure 31.4d. Such a curve can be fitted with a single exponential decay equation with three parameters—*noise*(η), *time constant*(τ), and *drop*(α). Simulated exponential curves with varying η, τ, and α are shown in Figure 31.4e–g, respectively. The drop in correlation at time lag $t = 1$, which is a measure of decorrelation between consecutive anisotropy images, was termed as the *noise*(η). It was observed

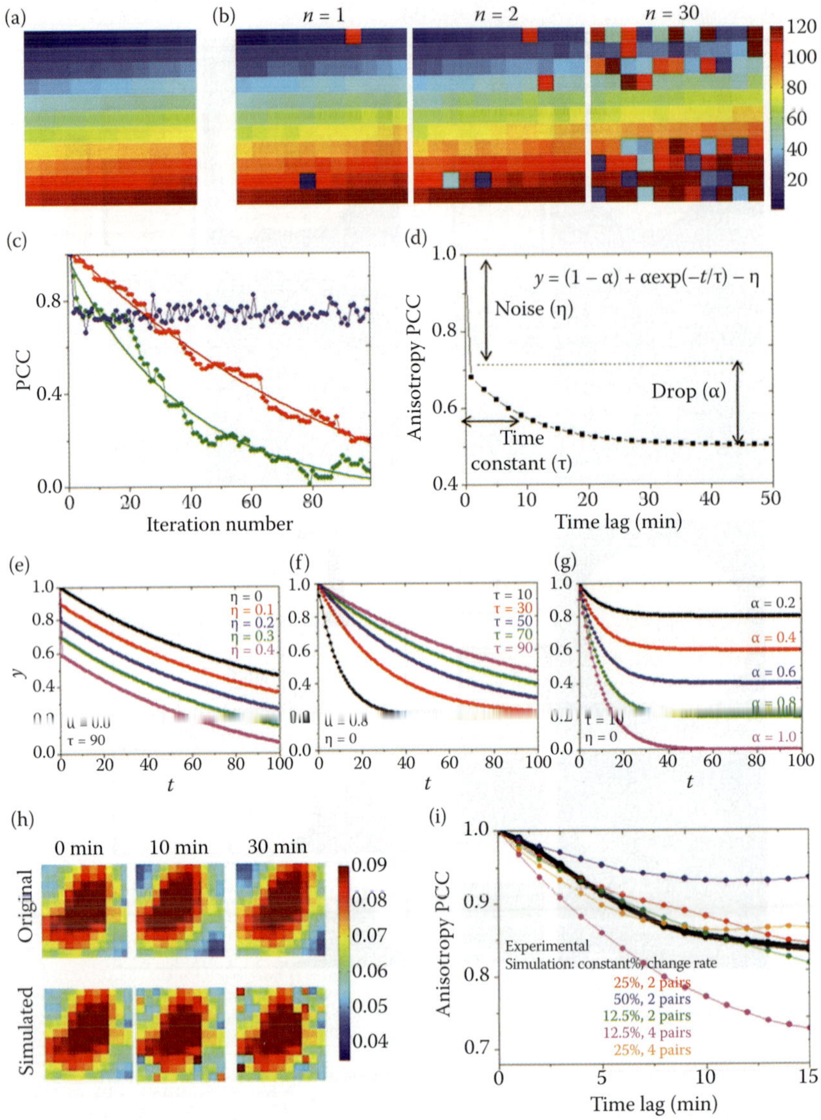

FIGURE 31.4
Characterization of image dynamics using PCC. (a) An 11 × 11 pixel increasing numbered image with values ranging from 1 to 121. (b) Simulations generated from this image by exchanging pixel pairs, as an example of image decorrelation. (c) PCC vs. iteration curve, which indicates the correlation with respect to initial image, for 1 pixel pair exchange (red), 2 pixel pairs exchange (green), and addition of random noise without any pixel pair exchange (blue). (d) A typical PCC curve for H2B-EGFP anisotropy time series fitted with the exponentially decaying equation having three parameters, *noise*(η), *time constant*(τ), and *drop*(α). Exponential decay curves with varying *noise* (e), *time constant* (f), and *drop* (g). (h) Simulation of dynamics in an 11 × 11 pixels² ROI (1.76 × 1.76 μm²) cropped from H2B-EGFP anisotropy image. (i) PCC curves for the original ROI compared with various simulations.

that the PCC does not drop to zero at long timescales but stabilizes after some time. This stable PCC value ($PCC_{t=\infty}$) is a measure of the fraction of anisotropy image that remains unchanged. It was used to define the parameter *drop*(α) = 1 − $PCC_{t=\infty}$ − *noise*(η). The *time constant*(τ) is obtained by fitting a single exponential decay equation to the PCC curve after removing the *noise*(η). τ is a measure of time taken for the anisotropy image to decorrelate to $PCC_{t=\infty}$. In further experiments, the parameter *drop rate* = α/τ was used to quantify chromatin dynamics.

Next, the above parameters were used to understand the PCC curves obtained from anisotropy images of the nucleus at different time points and compared with those obtained by simulation. The top panel in Figure 31.4h is a time series of a region of 11 × 11 pixels² from a nucleus anisotropy image. Using the experimental image at $t = 0$, simulations were carried out by varying either the number of randomly chosen pairs of pixels to be exchanged or the fraction of image accessible for pixel exchange (bottom panel, Figure 31.4h). These simulation parameters affect the *time*

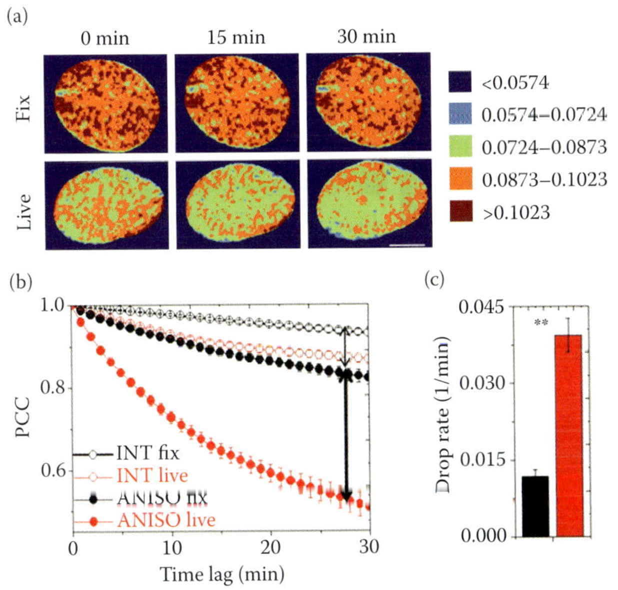

FIGURE 31.5
Anisotropy PCC can be used to measure chromatin dynamics. (a) Time series of anisotropy profiles for fixed and live cell nuclei. Scale bar, 10 μm. (b) Comparison of intensity and anisotropy PCC curves for fixed and live cells. (c) Comparison of drop rates for fixed and live cell anisotropy curves. Error bars show standard error. $^{**}p = 3.1E{-}8$.

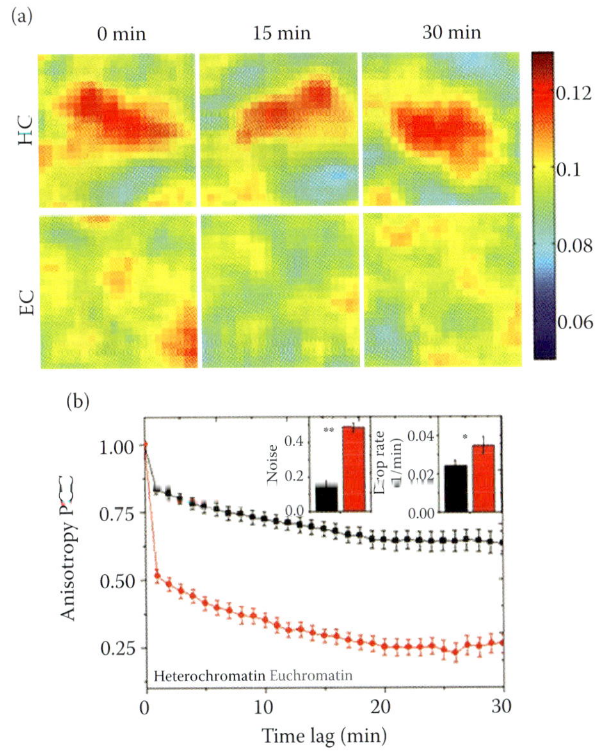

FIGURE 31.6
Relative chromatin dynamics at local sites. (a) Time series of anisotropy profile at 21 × 21 pixels (3.5 × 3.5 μm^2) heterochromatin and euchromatin regions. (b) Comparison of anisotropy PCC curves for heterochromatin and euchromatin regions. Insets show drop rate and noise for the two curves. Error bars show standard error. $^{*}p = 0.06$; $^{**}p = 1.5E{-}12$.

constant(τ) and the *drop*(α) of the PCC curve, respectively. By comparing the original and simulated PCC curves (Figure 31.4i, black and green curves, respectively), it is estimated that anisotropy changes is 3% ((2*2)/(11*11))*100 of area per minute and 88% of the total area is dynamic.

The sensitivity in FAI was further confirmed by comparing the anisotropy PCC curves of PFA fixed cells with those of live cells (Figure 31.5a and b). As expected, the drop rate for fixed cells (0.01) was significantly lower than for live cells (0.04), indicating reduced dynamics in chromatin compaction (Figure 31.5c). It can be noted that the intensity PCC curves are much less sensitive than the anisotropy PCC curves (Figure 31.5b) suggesting that FAI captures subtle changes in chromatin dynamics that cannot be observed by intensity profile. In the next three sections, we show the utility of anisotropy PCC to probe chromatin remodeling in different functional context.

31.3.3 Time Lapse FAI Reveals Distinct Dynamics between Heterochromatin and Euchromatin Assembly

The local variations in chromatin dynamics were probed in either heterochromatin or euchromatin (21 × 21 pixels^2) regions (Figure 31.6a), and anisotropy PCC curves were plotted. It was observed that euchromatin regions have significantly higher noise ($\eta = 0.5$) than heterochromatin ($\eta = 0.2$, Figure 31.6b, inset 1), which implies rapid loss of structural information in the loosely packaged euchromatin region. Consistent with this, the drop rate (a measure of the time to reach steady state during polymer relaxation) is slightly higher in the euchromatin region (0.035) when compared to heterochromatin (0.025, Figure 31.6b, inset 2).

31.3.4 Cell Geometric Constraints Affect Chromatin Dynamics

Previous reports indicate that the cellular geometry impinges on cytoskeleton and chromatin architecture resulting in altered genetic programs [34]. To quantitatively analyze the change in chromatin dynamics arising due to cell geometric constraints, we map anisotropy PCC curves in cells plated on micropatterned triangle and circle shapes. Figure 31.7a shows a representative image of single cells labeled with actin (lifeact RFP) and H2B-EGFP plated on triangular and circular micropatterns, and Figure 31.7b shows time lapse anisotropy images. The anisotropy PCC curves were plotted (Figure 31.7c), and it was observed that the circular cells have a significantly higher drop rate

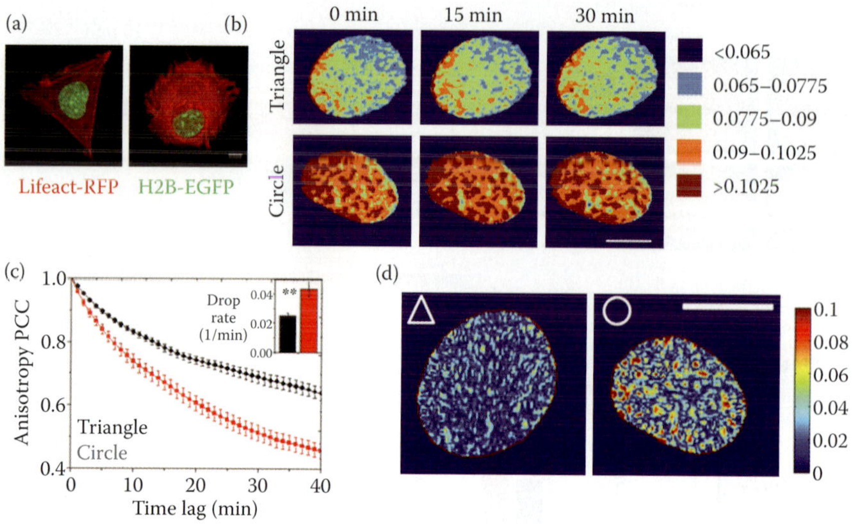

FIGURE 31.7
Effect of constrained cellular geometry on chromatin dynamics. (a) Images of NIH3T3 cells expressing lifeact-RFP and H2B-EGFP on 1600 μm^2 triangular and circular fibronectin micropatterns. (b) Time series of anisotropy profile for triangular and circular cells. (c) Comparison of anisotropy PCC curves for triangular and circular cells. Inset shows the drop rates for the two curves. Error bars show standard error. $^{**}p = 2.4E{-}4$. (d) Difference image of anisotropy |5–0 min| for triangular and circular cells. Red color at the periphery shows the boundary of the nucleus. Scale bar, 10 μm.

(0.04) than triangular ones (0.02), which implies faster chromatin remodeling (Figure 31.7c inset). Normalized difference images (5–0 min) of anisotropy for the two geometries show a higher difference in the case of circle cells (Figure 31.7d) and also confirm the faster chromatin dynamics in cells on circular geometry.

31.3.5 FAI Captures Changes in Chromatin Dynamics in Distinct Cellular Differentiation States

FRAP studies of chromatin binding proteins reveal a hyperdynamic nuclear structure in stem cells [21,33,35]. We therefore applied FAI to quantitatively measure the differences in chromatin compaction dynamics between ES and PMEF cells. Time lapse anisotropy imaging was recorded for 1 h, and anisotropy PCC was plotted (Figure 31.8a and b). Stem cells exhibit a significantly dynamic chromatin structure (drop rate = 0.12) as compared to PMEFs (drop rate = 0.06) (Figure 31.8b inset). Further, to assess local variations in chromatin dynamics, anisotropy PCC curves were plotted by dividing the nucleus into 11 × 11 pixels^2 regions, and region-wise drop rates were measured (Figure 31.8c). The region with a lower drop rate is less dynamic and vice versa as seen in the color-coded images (Figure 31.8c). The region-wise drop rates in ES cells were, on average, higher compared to PMEFs, suggesting that the chromatin structure in stem cells undergoes rapid conformational changes.

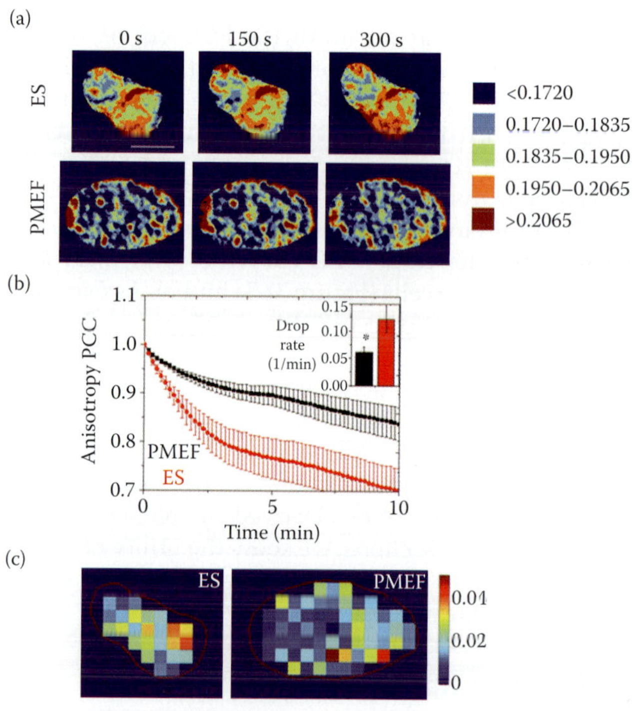

FIGURE 31.8
Chromatin dynamics of stem cells vs. differentiated cells. (a) Typical H2B-EGFP anisotropy images of ES and PMEF cells. (b) Comparison of anisotropy PCC curves for ES and PMEF cells. Inset shows the drop rates for the two curves. $^{*}p = 0.03$. (c) Comparison of local anisotropy drop rates in ES and PMEF cells. Box size is 10 × 10 pixels^2 (1.6 × 1.6 μm^2). Red color at the periphery shows the boundary of the nucleus. Scale bar, 10 μm.

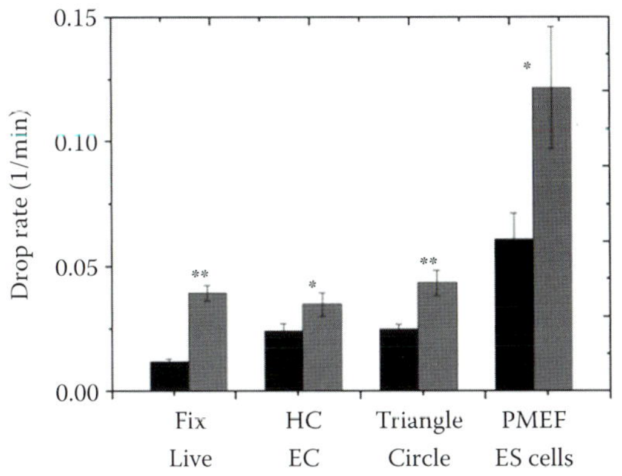

FIGURE 31.9
Comparison of chromatin dynamics in different conditions. PCC curve drop rates for various conditions. *$p < 0.05$, **$p < 0.005$. The drop rates in these different conditions were calculated by fitting the PCC data with single exponential decay.

31.4 Conclusion

In this chapter, we have established FAI as a method to measure differential chromatin compaction in living cells. The compaction maps obtained by FAI were correlated with structural, epigenetic, and transcriptional markers. Using PCC curves derived from time-lapse FAI, we demonstrated the dependence of rate and extent of chromatin remodeling on the structural variation in packaging (Figure 31.9). Further, using FAI, we quantify changes in chromatin dynamics in distinct cellular geometries or differentiation states (Figure 31.9). Taken together, time lapse FAI provides a sensitive measure to capture subtle chromatin dynamics in various cellular functional states.

Acknowledgment

We thank the Mechanobiology Institute, National University Singapore, for funding.

References

1. Li, G. and D. Reinberg, Chromatin higher-order structures and gene regulation. *Curr Opin Genet Dev*, 2011. **21**(2): pp. 175–86.
2. Moindrot, B., P. Bouvet, and F. Mongelard, Chromatin structure and organization: The relation with gene expression during development and disease. *Subcell Biochem*, 2013. **61**: pp. 373–96.
3. Hubner, M.R., M.A. Eckersley-Maslin, and D.L. Spector, Chromatin organization and transcriptional regulation. *Curr Opin Genet Dev*, 2013. **23**(2): pp. 89–95.
4. Pederson, T., Chromatin structure and the cell cycle. *Proc Natl Acad Sci U S A*, 1972. **69**(8): pp. 2224–8.
5. Essers, J. et al., Dynamics of relative chromosome position during the cell cycle. *Mol Biol Cell*, 2005. **16**(2): pp. 769–75.
6. Meshorer, E. and T. Misteli, Chromatin in pluripotent embryonic stem cells and differentiation. *Nat Rev Mol Cell Biol*, 2006. **7**(7): pp. 540–6.
7. Belmont, A., Dynamics of chromatin, proteins, and bodies within the cell nucleus. *Curr Opin Cell Biol*, 2003. **15**(3): pp. 304–10.
8. Cheung, A.Y. and A.S. Reddy, Nuclear architecture and dynamics: Territories, nuclear bodies, and nucleocytoplasmic trafficking. *Plant Physiol*, 2012. **158**(1): pp. 23–5.
9. Hemmerich, P., L. Schmiedeberg, and S. Diekmann, Dynamic as well as stable protein interactions contribute to genome function and maintenance. *Chromosome Res*, 2011. **19**(1): pp. 131–51.
10. Marshall, W.F. et al., Interphase chromosomes undergo constrained diffusional motion in living cells. *Curr Biol*, 1997. **7**(12): pp. 930–9.
11. Poirier, M.G. et al., Dynamics and function of compact nucleosome arrays. *Nat Struct Mol Biol*, 2009. **16**(9): pp. 938–44.
12. Festenstein, R. et al., Modulation of heterochromatin protein 1 dynamics in primary mammalian cells. *Science*, 2003. **299**(5607): pp. 719–21.
13. Cheutin, T. et al., Maintenance of stable heterochromatin domains by dynamic HP1 binding. *Science*, 2003. **299**(5607): pp. 721–5.
14. Misteli, T. et al., Dynamic binding of histone H1 to chromatin in living cells. *Nature*, 2000. **408**(6814): pp. 877–81.
15. Kimura, H., K. Sugaya, and P.R. Cook, The transcription cycle of RNA polymerase II in living cells. *J Cell Biol*, 2002. **159**(5): pp. 777–82.
16. Kimura, H. and P.R. Cook, Kinetics of core histones in living human cells: Little exchange of H3 and H4 and some rapid exchange of H2B. *J Cell Biol*, 2001. **153**(7): pp. 1341–53.
17. van Steensel, B. and J. Dekker, Genomics tools for unraveling chromosome architecture. *Nat Biotechnol*, 2010. **28**(10): pp. 1089–95.
18. Sajan, S.A. and R.D. Hawkins, Methods for identifying higher-order chromatin structure. *Annu Rev Genomics Hum Genet*, 2012. **13**: pp. 59–82.
19. Daban, J.R., Electron microscopy and atomic force microscopy studies of chromatin and metaphase chromosome structure. *Micron*, 2011. **42**(8): pp. 733–50.
20. Mueller, F. et al., Monitoring dynamic binding of chromatin proteins *in vivo* by fluorescence recovery after photobleaching. *Methods Mol Biol*, 2012. **833**: pp. 153–76.
21. Bhattacharya, D. et al., Spatio-temporal plasticity in chromatin organization in mouse cell differentiation and during Drosophila embryogenesis. *Biophys J*, 2009. **96**(9): pp. 3832–9.
22. Lleres, D. et al., Quantitative analysis of chromatin compaction in living cells using FLIM-FRET. *J Cell Biol*, 2009. **187**(4): pp. 481–96.

23. Sasaki, K. et al., Real-time imaging of histone H4 hyperacetylation in living cells. *Proc Natl Acad Sci U S A*, 2009. **106**(38): pp. 16257–62.
24. Zhang, C., S.K. Nordeen, and D.J. Shapiro, Fluorescence anisotropy microplate assay to investigate the interaction of full-length steroid receptor coactivator-1a with steroid receptors. *Methods Mol Biol*, 2013. **977**: pp. 339–51.
25. Audran, E. et al., A general framework to characterize inhibitors of calmodulin: Use of calmodulin inhibitors to study the interaction between calmodulin and its calmodulin binding domains. *Biochim Biophys Acta*, 2013. **1833**(7): pp. 1720–31.
26. Lakowicz, J., *Principles of Fluorescence Spectroscopy*. 1st ed. New York: Springer, 2006.
27. Banerjee, B., D. Bhattacharya, and G.V. Shivashankar, Chromatin structure exhibits spatio-temporal heterogeneity within the cell nucleus. *Biophys J*, 2006. **91**(6): pp. 2297–303.
28. Hameed, F.M., M. Rao, and G.V. Shivashankar, Dynamics of passive and active particles in the cell nucleus. *PLoS One*, 2012. **7**(10): p. e45843.
29. Iyer, K.V. et al., Mechanical activation of cells induces chromatin remodeling preceding MKL nuclear transport. *Biophys J*, 2012. **103**(7): pp. 1416–28.
30. Maharana, S. et al., Dynamic organization of transcription compartments is dependent on functional nuclear architecture. *Biophys J*, 2012. **103**(5): pp. 851–9.
31. Rao, J. et al., Trichostatin-A induces differential changes in histone protein dynamics and expression in HeLa cells. *Biochem Biophys Res Commun*, 2007. **363**(2): pp. 263–8.
32. Sinha, D.K. et al., Probing the dynamic organization of transcription compartments and gene loci within the nucleus of living cells. *Biophys J*, 2008. **95**(11): pp. 5432–8.
33. Talwar, S. et al., Correlated spatio-temporal fluctuations in chromatin compaction States characterize stem cells. *Biophys J*, 2013. **104**(3): pp. 553–64.
34. Kilian, K.A. et al., Geometric cues for directing the differentiation of mesenchymal stem cells. *Proc Natl Acad Sci U S A*, 2010. **107**(11): pp. 4872–7.
35. Meshorer, E. et al., Hyperdynamic plasticity of chromatin proteins in pluripotent embryonic stem cells. *Dev Cell*, 2006. **10**(1): pp. 105–16.

32

Nuclear Imaging in Mechanobiology

Irena L. Ivanovska, Joe Swift, Jerome Irianto, Kyle Spinler, Jae-Won Shin, Amnon Buxboim, and Dennis E. Discher

CONTENTS

All of the different cells within an individual share the same genome, whereas epigenetic variations underlie the plasticity in gene expression from the stem cells that differentiate during development, repair, and aging. Beyond differences in covalent and noncovalent modifications of chromatin between different cell types, there are also lineage-relevant variations in the structure proteins of the nuclear envelope called lamins that contribute to chromatin organization and even transcription factor localization. Here we focus on imaging these various molecular factors with the relevant optical microscopy in the context of multiple methods for physically manipulating nuclei.

32.1 Introduction

The nucleus is generally the largest single organelle of a eukaryotic cell and is the cell's defining feature. Within the nuclei of a given organism, the DNA is essentially identical, but the many differentiated cell types possess epigenetic differences that fix the fates of progenitors and stem cells. DNA methylation, histone isoforms, and histone modifications collectively regulate gene expression but so do nuclear envelope proteins that also change in development. For example, lamin-A/C is largely absent from human embryonic stem cells (hESCs)[1] and low in blood cells and neuronal tissues,[2] but spliced isoforms accumulate at the nuclear periphery of mesenchymal tissue cells (such as bone and fibroblasts) in aged, normal humans.[3] We hypothesized that such epigenetic plasticity in normal development and aging would be mechanically measurable and meaningful. Methodologies were therefore sought that could lend physical insight into nuclear physiology. For example, blood capillaries at 2–3 μm in diameter are similar in size or smaller than nuclei, which means that nuclei in cells ranging from white blood cells to circulating metastatic cancer cells must be able to deform and flow in order to access peripheral tissues.

This chapter will describe a number of techniques that combine imaging with physical manipulation to provide insight into mechanobiological processes. The nucleus is perhaps the key mechanosensing organelle, and we present methods for applying controlled deformations: micropipette aspiration, which we use to measure nuclear compliance; and rheometry, which provides insight into nuclear robustness and osmotic stressing, where we describe imaging methods to determine the extent of chromatin condensation. Finally, we discuss the challenges of quantitative imaging, particularly in tracking the location of low-abundance transcription factors.

32.2 Imaging in Micropipette Aspiration

32.2.1 Basic Experimental Method

This method has several important features: (1) the length scale of deformation is similar in length scale to nuclear subdomains or "nuclear territories,"[4] (2) the change in nuclear shape under stress can be measured over a wide range of time, and (3) the deformation modes of the different nuclear subcomponents labeled fluorescently (with DNA dyes or green fluorescent protein [GFP]-constructs) can be visualized by fluorescence microscopy. With a nucleus that is isolated from a cell (see Section 32.3.1), aspiration reveals nuclear responses without the effects of physical links between the nucleus and different components of the cytoskeleton. Solution conditions are readily changed, including both ionic strength and osmotic strength that control nuclear volume and molecular interactions.[5] However, measurements made on nuclei within intact cells can, in principle, provide more relevant in situ insight into the physical behavior of the nucleus provided that the cytoskeleton and other cell components do not interfere. This has been achieved when necessary by depolymerizing filamentous actin in cells with Latrunculin immediately prior to aspiration.[6] Whether the nucleus is isolated or not, it is partially aspirated into a micropipette with a preset pressure, and the increase in the projection length of the nucleus is measured as a function of time (Figure 32.1a).

Micropipettes were prepared from a 1-mm-diameter glass capillary by pulling with a micropipette puller, following standard electrophysiology methods. Typically, the inner diameter was between 1 and 5 μm, and was constant over at least 10 μm in order to allow visualization of the aspirated projection of nucleus. This was achieved by

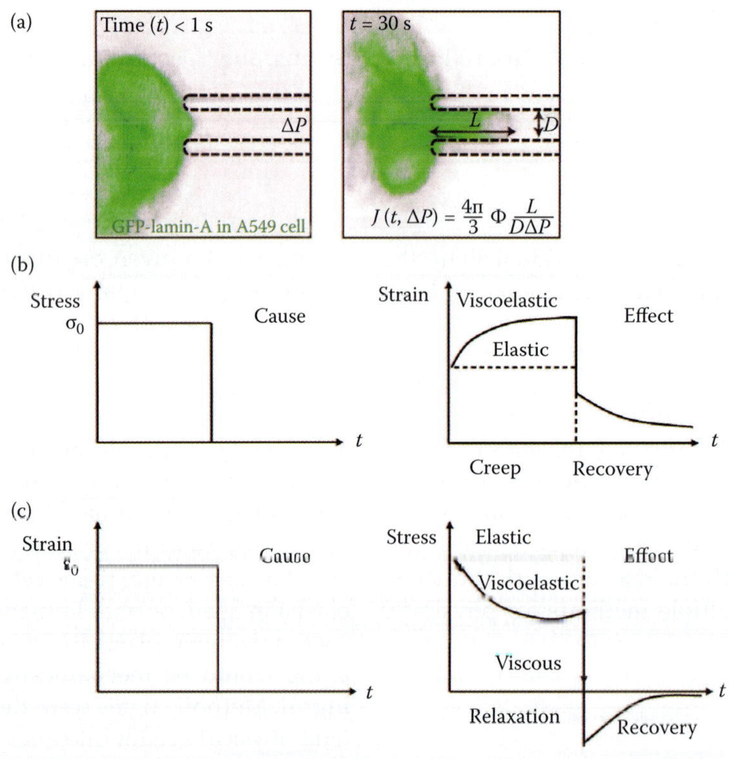

FIGURE 32.1
(a) A549 nucleus (visualized by GFP-lamin-A expression) subjected to deformation by micropipette aspiration. The equation allows calculation of nuclear compliance, *J*. (b) Mechanical response to constant pressure effects a change in strain, and release of the stress leads to recovery except for a residual plastic strain. (c) Mechanical response to constant strain effects a change in stress, and release of the strain leads to recovery except for a residual plastic stress.

optimizing the pulling rate. Micropipettes were fractured and forged to create a flat tip, necessary to ensure symmetric aspiration and flow of the nucleus into the micropipette. The micropipette inner and/or outer surface was then passivated to minimize cell or nuclear adhesion and friction. This was achieved by immersing into albumin or silanizing solutions. The micropipette was then backfilled with physiological buffer using a syringe with a suitably narrow gauge needle. The micropipette was mounted on a micromanipulator connected to a pressure-controlled system in which the applied pressure (negative relative to atmospheric) was measured by a calibrated pressure transducer or manometer. Using the micromanipulators, the micropipette was then positioned horizontally within the focal plane of the microscope and close to the nucleus to be aspirated. Following slight aspiration of the nucleus, the micropipette was raised above the coverslip, which the cell or nucleus had settled upon in order to avoid friction between the nucleus and the supportive surface during the aspiration. We typically found that negative pressures from −1 to −10 kPa were required for aspiration of the nucleus. Higher pressures were avoided to prevent rupture of the nucleus. In the case of measurements of a nucleus within a cell, repetitive aspiration of the cell could be used to mechanically disrupt the cell wall and thus isolate the nucleus.[7]

32.2.2 Mathematics of Physical Responses

The DNA in each of the 46 human chromosomes is a massive macromolecule (1–10 cm long) that is condensed as chromatin into a microns-diameter nucleus and would be expected—as with most polymers—to exhibit complex flow behavior (i.e., complex rheology). While some past studies of nuclear deformation have assumed fully recoverable elastic behavior,[8] others have suggested viscoelastic behavior[7] or shown more complex power law rheology.[5,6]

For small deformations, most solid materials can be described by Hooke's law of linear elasticity in which E is Young's elastic modulus for the material. The one-dimensional relationship between the applied stress σ and caused strain ε is given by

$$\sigma = E\varepsilon \tag{32.1}$$

The inverse of the modulus E is called compliance:

$$J = \frac{1}{E} \tag{32.2}$$

A purely viscous fluid under shear obeys the following relationship between the stress and strain rate:

$$\sigma = \eta \frac{d\varepsilon}{dt} \tag{32.3}$$

where η is the viscosity. Real materials often deviate from pure elastic or viscous behavior and exhibit a more complex and time-dependent stress versus deformation response. If they yield or break after some period of stress and thereafter do not recover their deformation, then they are referred to as "plastic."

Creep is defined as a progressive deformation of a material held under constant stress. If a sudden stress is applied to a material (Figure 32.1b), then it can be described with

$$\sigma = \sigma_0 H(t) \tag{32.4}$$

where $H(t) = 0$ for $t < 0$ and $H(t) = 1$ for $t > 0$. The compliance might then prove time-dependent and is called the creep compliance:

$$J(t) = \frac{\varepsilon(t)}{\sigma_0} \tag{32.5}$$

Creep curves can capture material behaviors over many decades of time, but if the load is released, the material might begin to recover. If the recovery never reaches zero strain, then the remaining strain is an indication of the plasticity of the material. Linear viscoelastic materials are those for which the creep compliance is independent of the stress.

On the other hand, if the strain is suddenly imposed and held constant, then one can present the strain history as a step function (Figure 32.1c):

$$\varepsilon = \varepsilon_0 H(t) \tag{32.6}$$

A decrease in stress in a material held under constant strain is called relaxation and can be described with relaxation modulus:

$$E(t) = \frac{\sigma(t)}{\varepsilon_0} \tag{32.7}$$

Creep and relaxation function can be experimentally obtained, and their mathematical descriptions involve attempts to capture a solid–liquid duality.

Differential constitutive models relate the stress and the strain in linear differential equations with constant coefficients by connecting the basic elements—elastic (springs) and viscous (dashpots)—in different ways. A classical Maxwell model consists of a spring and dashpot connected in series with a constitutive equation that reads

$$\frac{d\varepsilon}{dt} = \frac{1}{E}\frac{d\sigma}{dt} + \frac{\sigma}{\eta} \tag{32.8}$$

A spring and dashpot connected in parallel (Kelvin–Voigt model) has the form

$$\sigma = E\varepsilon = \eta \frac{d\varepsilon}{dt} \tag{32.9}$$

For elements in series, the stresses coincide with the total strain being a sum of the elemental strains. For elements in parallel, the strains coincide and the total stress is a sum of stresses in the individual elements. This approach can be used to design more complicated models by combining different Maxwell and Voigt elements in parallel and/or in series to capture the complex rheological behavior of the biological materials.

32.2.3 Complexity in Fractional Differential Models

More complex rheological properties may be expressed in terms of fractional derivatives. An element might obey an equation of the form

$$\sigma(t) = E\tau_r^{\beta} \frac{d^{\beta}\varepsilon(t)}{dt^{\beta}} \tag{32.10}$$

where E is the stiffness, τ_r is the relaxation time constant, and β is a material-dependent parameter that can be used to obtain the relaxation function $E(t)$. In such a case, the relaxation function is a power law in time that may be approximated by

$$E(t) = Bt^{-n} \tag{32.11}$$

Then using the Laplace transform, it can be shown that the creep function is also a power law:

$$J(t) = At^{n} \tag{32.12}$$

For $n = 0$, the compliance is independent of time, which is the case for purely elastic response, and for $n = 1$, the response is purely viscous. For any n's that are between these extremes, the viscoelastic complexity of the material can be captured. Power-law models describe adequately the creep and relaxation function of many materials with a small number of adjustable parameters and over many decades of time.

32.2.4 Nuclear Creep in Micropipette Aspiration

For the particular case of micropipette aspiration of a nucleus that is much larger than the micropipette diameter, the creep of a nucleus into the micropipette can be measured as

$$J(t) = \frac{2\pi\Phi}{3} \frac{1}{P} \frac{\Delta L(t)}{R_p} \tag{32.13}$$

where Φ is a geometry-dependent numerical prefactor (~2.1), P is the constant pressure applied by the micropipette, and the strain is given by ΔL, which is the dynamic aspirated length normalized to the pipette radius R_p. This equation for $J(t)$ provides the instantaneous measurable creep; nuclei could aspirate independent of time or exhibit a more complex power-law response.

Micropipette measurements performed on both stem cells and differentiated cells over timescales of about 100 s or more tend to exhibit a power-law creep compliance with creep exponents of $n \sim 0.2$–0.6.[6] After approximately 10 s, the deformation proves irreversible, and this provides clear evidence of the plasticity of the nucleus, that is, a permanent rearrangement of the nucleus and its chromatin. In general, the nuclear compliance of fully differentiated and aged cells (fibroblast and epithelial cells) is demonstrably lower than that in younger cells, including both human hematopoietic stem cells (HSCs) and pluripotent hESCs.[2,6,9,10] The latter nuclei, over several days in differentiation media, exhibit a six-fold increase in stiffness. Visualization of the chromatin in differentiated cells with GFP-histones shows that the chromatin is pinned at places to the nuclear envelope at what some have called lamina-associated domains (LADs) that are repress genes.[2] Seeing chromatin extend and flow into the aspirating micropipette can be a useful way to measure the interaction strengths at LADs but will require further advances in gene-specific labeling for live cell imaging.

32.3 Imaging Nuclei after Bulk Stressing in Rheometer

32.3.1 Isolating Nuclei

Gene-specific labeling and imaging of fixed cells is well established, but applying controlled stresses to nuclei continues to be developed. Nuclei from many cell types can be isolated by both single-cell mechanical extraction[7] and bulk methods.[5,8,11] Extremely fragile nuclei are more difficult to isolate, and since the nuclear envelope breaks down during mitosis, the methods apply only to mechanically stable interphase nuclei. Isolation also takes advantage of the nucleus' relative rigidity and its tenuous connections to the rest of the cell.

One detailed protocol for bulk isolation, described in detail previously[12] and shown to yield nuclei suited to micromechanical characterization,[5] is briefly described in the following. It has the advantage that further biochemical analyses such as proteomic profiling can be applied to the nuclei,[13,14] perhaps even under stressed conditions.[2,15] In general, the cell's plasma membrane is disrupted mechanically with hypotonic swelling or chemically with digitonin or other surfactants that perturb the plasma membrane but not nuclear membranes. The cell is then opened with mechanical homogenization, and cellular contents are separated from the nuclei by ultracentrifugation through a sucrose gradient.

The bulk nuclear isolation protocol requires preparation of a number of solutions. The following can be preprepared and kept as stocks: "10 TKMC" (500 mM Tris, 250 mM KCl, 25 mM $MgCl_2$, 30 mM $CaCl_2$; adjusted to pH 8); "2.3 STKMC" (in 0.1× 10 TKMC, 2.3 M sucrose, adjusted to pH 7.6 at 4°C); "5 STKMC" (in 0.5× 10 TKMC, 1.25 M sucrose, adjusted to pH 7.6 at 4°C); 10 mM 2-[4-(2-hydroxyethyl)piperazin-1-yl]ethanesulfonic acid (HEPES), pH 7.5. The following should be prepared on the day of use and cooled to 4°C: 200 μL 5 STKMC, 2.5 μL protease inhibitor cocktail (PIC); 3 mL 2.3 STKMC, 7.5 μL PIC; "2.3 STKMC–BSA" (900 μL of 2.3 STKMC with bovine serum albumin [BSA] to 0.2 mg/mL); 10 mL of 10 mM HEPES, pH 7.5 with 1 mM DTT (dithiothreitol).

Cells were harvested at near-confluency (e.g., A549 or U251 cells): 10^5 cells minimum, 10^8 maximum. Typically, a T75 flask was washed twice with cold phosphate buffered saline (PBS, 5 mL), then once with cold HEPES/DTT buffer (5 mL). Cells were harvested by scraping and suspended in 1 mL HEPES/DTT buffer. Suspended cells were then cooled on ice for 10 min, before lysing with 25 strokes in a Dounce homogenizer (avoiding air bubbles). An amount of 180 μL 5 STKMC was added to 900 μL of homogenized cells and cooled on ice for a further 10 min. A further 2 mL 2.3 STKMC was added, ensuring that the solution was well mixed, bringing the overall sucrose concentration to 1.6 M. An amount of 200 μL 2.3 STKMC–BSA was pipetted into each of the four ultracentrifuge tubes (polycarbonate 8 × 34 mm^2), followed by a 750 μL layer of cell lysate solution (creating a visible interface between the two solutions). The tubes were ultracentrifuged for 1 h at 50,000 rpm in a TLS 55 rotor at 4°C, resulting in a pellet of nuclear material and other cell material localized at the interface between sucrose solutions. The supernatant was removed and the pellet resuspended.

Nuclei were resuspended into a variety of buffers, dependent on the subsequent experiments, although adding BSA was found to help break up clumps and facilitate later manipulation. Nuclei were counted using a hemocytometer. Subsequent to isolation, we found it important to assess the quality of the nuclei by comparison with nuclei in intact cells and to make sure that the nuclear envelope remained intact while avoiding excessive coseparation of membranes such as the endoplasmic reticulum or cytoskeletal structures. DNA stains such as Hoechst dyes or, better, GFP-lamins allowed fluorescence visualization for assessment of nuclear morphology.

For imaging, isolated nuclei were resuspended in 50 μL 10 mM Tris with 10,000:1 diluted Hoescht stain and incubated at 37°C for 10 min. The sample was then centrifuged to pellet the nuclei and the supernatant subsequently removed. Nuclei were then resuspended in 50 μL 10 mM Tris with 500:1 diluted phalloidin–rhodamine stain, incubated again at 37°C for 30 min. Nuclei were pelleted again by centrifugation and resuspended in 50 μL 10 mM Tris before examination by fluorescence microscopy.

Last, we note that the concentrations of salts can dramatically affect nuclear mechanics since divalent and, to a lesser extent, monovalent cations affect chromatin condensation.[5,16] With changes in salt, nuclear volume can change significantly and might stress the nuclear lamina. It is also unclear exactly what effect salt concentrations have on protein–protein interactions within the nucleus. Nuclei isolated in buffers with low salt tend to swell and resemble closely the contours and size of nuclei inside live cells. To accurately approximate nuclear mechanics inside the cell, it is usually desirable to mimic the intracellular ion concentrations, but the difficulty is that the ion concentrations inside the cytoplasm and within organelles have not been reproducibly determined, and values for ions such as calcium range from submillimolar to millimolar and can vary greatly as a function of disease.[17] It is also difficult to predict the shift in nuclear salt homeostasis after isolation from the cell. Some studies have deliberately examined mechanics of isolated nuclei at extreme salt conditions to determine the maximum possible range of mechanical responses. This strategy proves effective since the chromatin condensation and dilation appear to shift the load within the nucleus from the chromatin to the lamina.[5]

32.3.2 Stressing, Fixing, and Imaging Nuclei

The consequences of shear stresses on nuclei can be analyzed by placing isolated nuclei in a cone and plate rheometer (Figures 32.2 through 32.4). The following method was used to ascertain the role of nuclear lamin in providing stiffness and physical robustness to nuclei.[2] A range of cell types were surveyed, with increasing ratios of A-type to B-type lamin: U251 (glioblastoma) < A549 (lung cancer epithelial) < A549 with lamin-A overexpression (A549-OE). Samples were prepared for imaging by immunofluorescence microscopy, but sufficient cellular material was available for mass spectrometry proteomics.

Nuclei were isolated from U251, A549, and A549-OE cells as described above and suspended in nuclei buffer

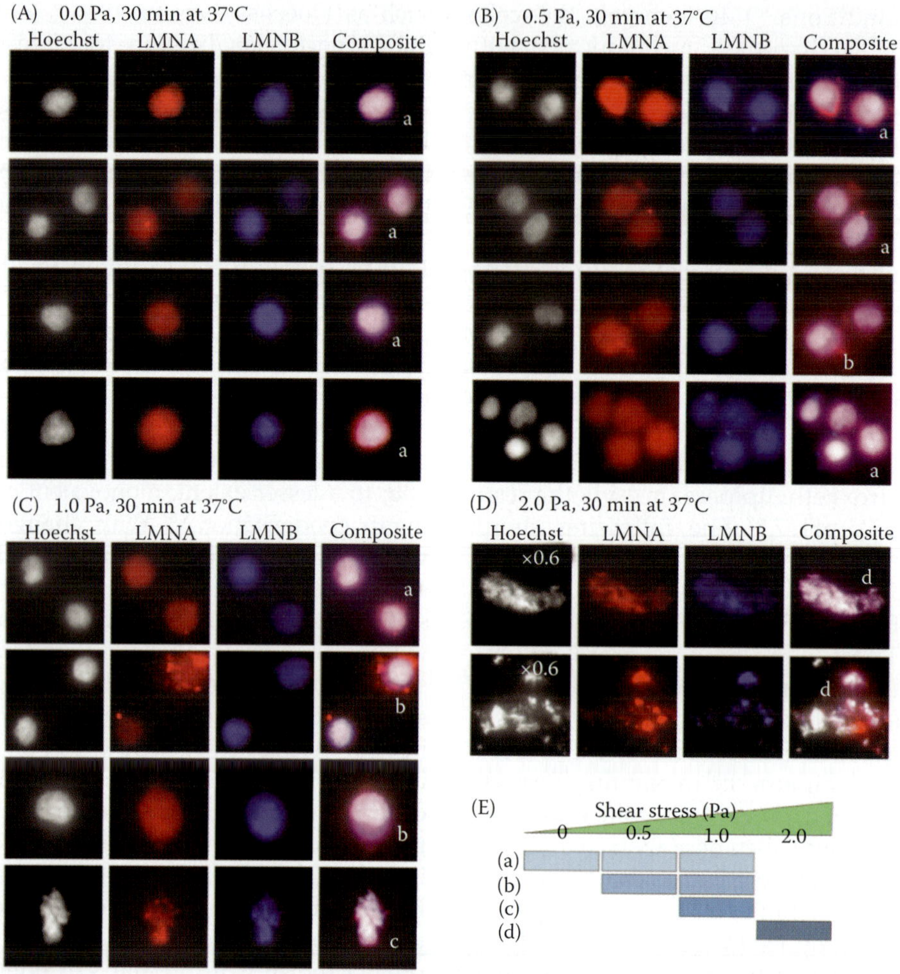

FIGURE 32.2
Nuclei isolated from U251 cells, subjected to increasing shear stress in a cone and plate rheometer (30 min shear at 37°C, 0 to 2 Pa, A–D). Following separation and enrichment (see methods description), nuclei were resuspended in 0.5 mM $MgCl_2$, 0.2 M sucrose, 10 mM HEPES at pH 7.6. (E) Images were characterized by inspection, with occurrences of the following phenomena noted: (a) distinct nuclei with colocalization of DNA (Hoechst stain), lamin-A, and lamin-B. (b) Nuclei with sloughing of the nuclear membrane, a partial separation of DNA from lamins A and B. (c) Relaxation of chromatin packing. (d) Formation of larger aggregates of nuclear material.

(0.2 M sucrose, 10 mM HEPES, 1 mM $MgCl_2$, 0.1% PIC) at a concentration of about 250 nuclei/μL. Nuclei samples were loaded onto a cone and plate rheometer (Bohlin Gemini) with the stage heated to 37°C. Shear stress was varied between 0.5 and 5 Pa (using a 10 μm gap, with the computer-controlled rheometer setting the rotation speed as required), and shear stress load duration was varied up to 40 min. Appropriate unstressed control samples were also prepared. Following shear stress, nuclei were fixed with 4% paraformaldehyde for 10 min and stained for immunofluorescence imaging using repeat cycles of spin-down (in a bench-top centrifuge) and resuspension in order to wash the nuclei. Primary antibodies used were lamin-A/C mouse monoclonal sc-7292 and lamin-B goat polyclonal sc-6217 (Santa Cruz), with the addition of Hoechst to image the chromatin.

The resulting images exhibit the following features. At low shear stress or with high lamin-A/C relative to lamin-B, nuclei are distinct and exhibit sharp, colocalized lamina components and chromatin. With increasing shear stress, the nuclear membrane sloughs away with partial separation of chromatin and the lamina, and we observe a relaxation of chromatin packing. At the highest stresses, large aggregates of nuclear material form. Examples of these characteristic features are shown in Figures 32.2 through 32.4. In general, we find that a high level of lamin-A/C in the nucleus (relative to lamin-B) confers robustness to shear stress, consistent with a polymer physics model of a stiffer and more stable nucleus.[2]

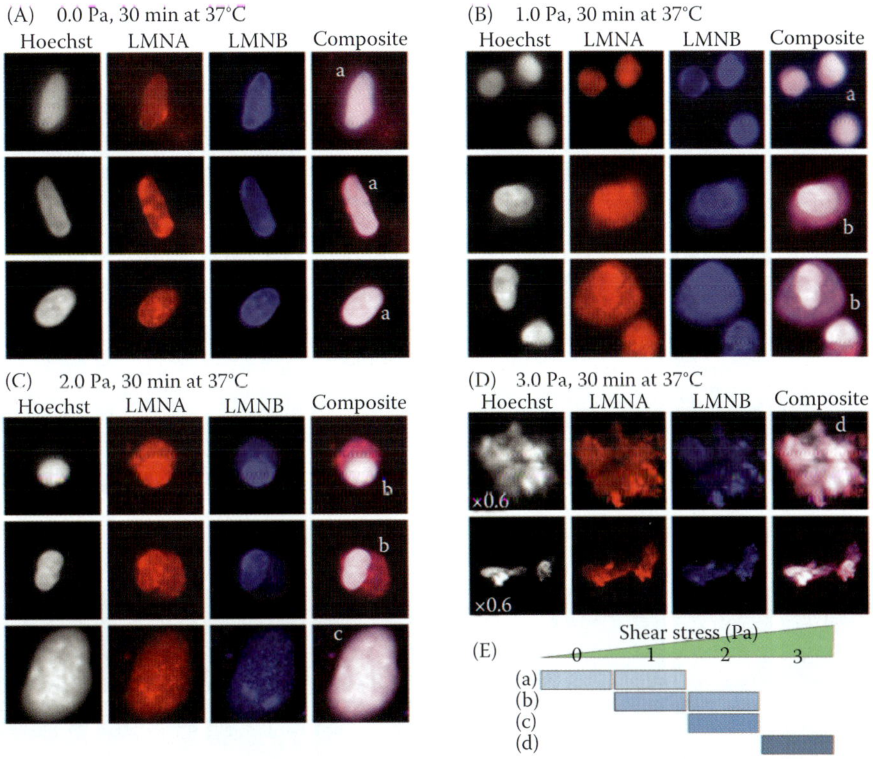

FIGURE 32.3
Nuclei isolated from A549 cells, subjected to shear stress in a rheometer (nuclei suspended in 0.5 mM $MgCl_2$, 0.2 M sucrose, 10 mM HEPES at pH 7.6; 30 min shear at 37°C, 0 to 3 Pa, A–D). (E) Nuclei deformations were characterized by visual inspection (see Figure 32.3).

32.4 Confocal Microscopy Imaging of Stressed Nuclei

32.4.1 Confocal Basics and Pitfalls

For typical confocal laser scanning microscopy, the laser beam passes through an excitation pinhole and is focused by an objective lens into a small focal volume within the sample. The fluorescence signal emitted by the sample (at higher wavelength) comes back through the same objective and passes through an emission pinhole to a photodetection device, such as a photomultiplier tube. The pinholes physically block out-of-focus light, and the intensity recorded from a focal volume represents a "voxel" (the volume equivalent of a pixel) in the resulting image. Resolution limits of voxel height and size dictate the accuracy of key nuclear properties such as nuclear volume, which dictates the space for chromatin and the factors that regulate its expression.

Trade-offs in imaging balance the purpose of imaging and the characteristics of the specimen. The optical resolution limit must be balanced against focal or working distance. An objective with higher numerical aperture (NA), such as a 63× oil immersion lens with an NA of 1.4, will provide high optical resolution but at the cost of reduced focal depth. Monolayers of cells seeded onto a glass coverslip will be ideal for high-resolution imaging of the nucleus, but the specimens must be close to the objective. The objective's NA, with the laser wavelength (λ) and the refractive index of the substance at the focal place (η), determines the optical resolution limit, which can be calculated as[18,19]

$$R_a = \frac{0.61 \times \lambda}{\text{NA}} \quad (32.14)$$

$$R_l = \frac{1.4 \times \eta \times \lambda}{\text{NA}^2} \quad (32.15)$$

where R_a is the axial resolution limit and R_l is the lateral resolution limit. A second trade-off is between specimen preparation/preservation and image quality, including the resolution and the signal-to-noise ratio. Greater spatial resolution, image averaging, and laser intensity will generally provide better image quality. However, a greater level of laser exposure will induce greater damage to the fluorophores (photobleaching) through the sample depth and, in the case of live cell imaging, on the

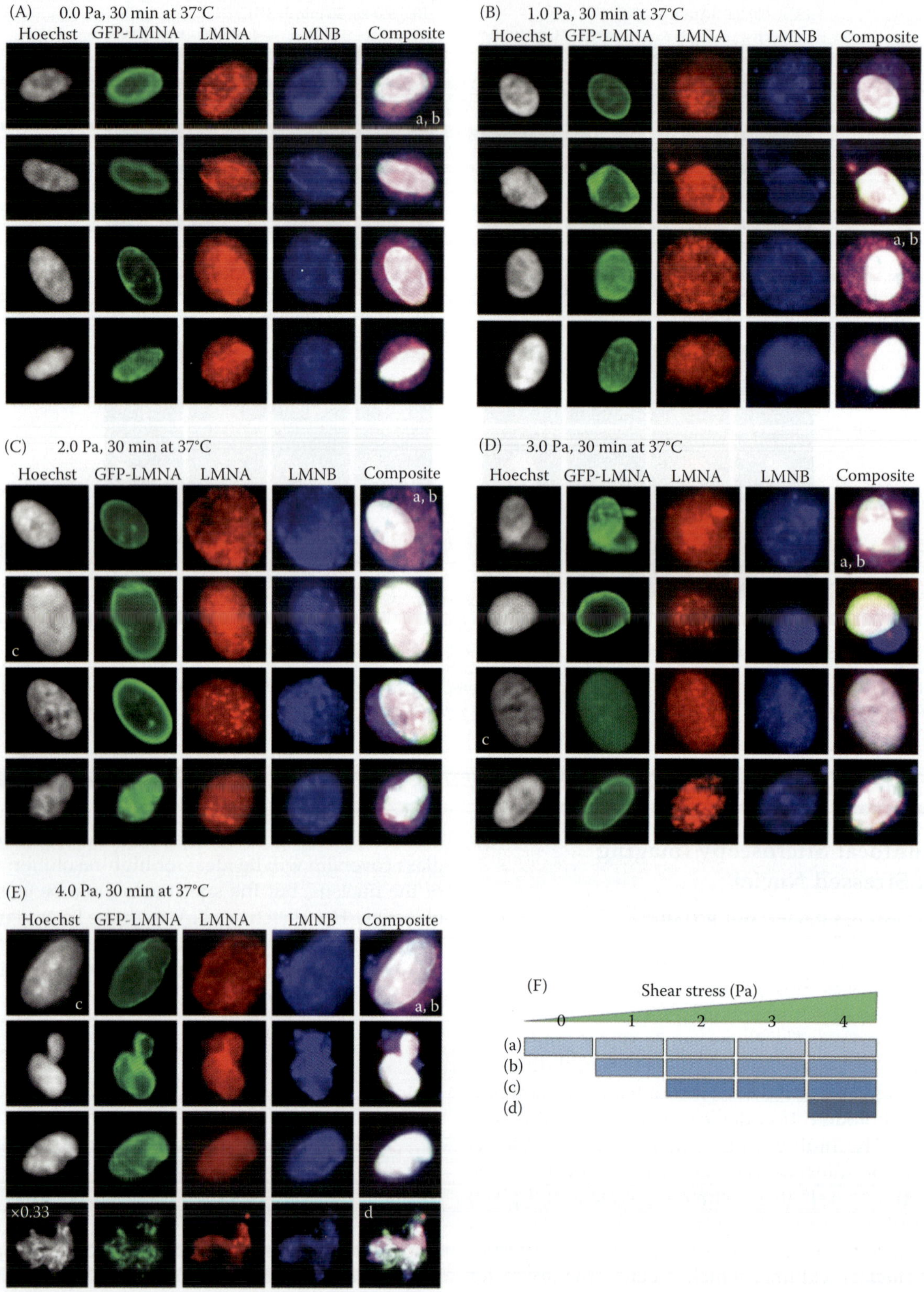

FIGURE 32.4
Nuclei isolated from A549 cells with GFP-lamin-A overexpression, subjected to shear stress in a rheometer (nuclei suspended in 0.5 mM $MgCl_2$, 0.2 M sucrose, 10 mM HEPES at pH 7.6; 30 min shear at 37°C, 0 to 4 Pa, A–E). (F) Nuclei deformations were characterized by visual inspection (see Figure 32.3).

specimen itself (phototoxicity). Hence, optimal imaging settings should allow the user to acquire images of the specimen that fulfill the aim of the imaging while keeping the laser exposure to a minimum. The following parameters should therefore be considered: voxel size, pinhole size, and scanning time.

The resolution of the image produced by confocal microscopy depends on the axial and lateral dimensions of the voxels. Image resolution is always limited by the optical resolution limit, and so lowering the voxel size to below the optical resolution limit will not provide more information on the sample but could increase laser exposure and induce further damage to the sample. Increasing the pinhole size will provide brighter signals, but it will also allow the detection of out-of-focus signals from around the focal point, hence decreasing the axial and lateral resolution. Ideally, the pinhole size should be adjusted to only allow the signal from the focused volume to pass through. Scanning speed, line, and frame averaging also play a part in image quality. Increasing the scan speed will lessen the time required for a full scan of the specimen. However, it also brings down the sampling time for each voxel, consequently reducing the number of photons gathered and the signal-to-noise ratio. In contrast, reducing the scan speed will improve the image quality, but higher sampling time means longer laser exposure. The use of frame averaging allows an entire section of the image to be repeatedly scanned and the average intensity taken for each voxel, thereby reducing noise. This may also be achieved a line at a time (line averaging), which is more appropriate for making time-resolved measurements. Although averaging improves the signal-to-noise ratio, it also increases laser exposure. Furthermore, when imaging a series of lateral sections, it is recommended to start the section from the bottom of the sample in order to avoid unnecessary photobleaching as the laser travels through the specimen.

Beyond imaging settings, sample preparation is also important. In case of live cell or tissue imaging, the culture media should not contain phenol red, as it is known to autofluoresce, and the culture environment should be maintained, including temperature and CO_2 levels. If CO_2 levels are not maintained, the culture media should not contain sodium bicarbonate. Following fixation, specimens are often mounted with commercially available "mounting media" that allows for both long-term specimen storage and inhibition of photobleaching. However, mounting can induce morphological artifacts to the specimens, such as flattening of the nucleus. If nuclear morphology is essential to maintain for a study, mounting is not recommended or at least must be compared to unmounted samples. For example, imaging of osmotically stressed nuclei that aims to measure changes in nuclear volume[20] must be done with careful account of all the factors above.

32.4.2 Modeling of Osmotic Stressing of Nuclei

Given a set of images such as from nuclei in cells that are imaged under different osmotic conditions, one would like to quantify changes and relate mathematically to an appropriate model. In recent studies,[20] the extent of local chromatin heterogeneity was determined using a chromatin condensation parameter, which reflects the edge density (ρ) in the image of a nucleus (Figure 32.5). In modeling this, we first consider the osmotic pressure of a polymer solution. In the semidilute regime where the volume occupied by the individual chains just covers the total sample volume, the osmotic pressure of a polymer solution (Π) is dependent only on monomer concentration, c_m[20,21]:

$$\frac{\Pi}{k_b T} \propto c_m^{9/4} \tag{32.16}$$

where k_b is the Boltzmann constant and T is the absolute temperature.

Treating a polymer as a chain of monomer subunits, the polymer concentration (c_p), which we approximate as proportional to the chromatin condensation parameter (ρ), relates to the monomer concentration by the degree of polymerization, N:

$$Nc_p = c_m \tag{32.17}$$

$$\rho \propto c_p \tag{32.18}$$

Confinement within the nucleus adds a "basal" osmotic pressure, Π_b:

$$\Pi = \Pi_b + B\rho^{9/4} \tag{32.19}$$

$$B \propto k_b T N^{9/4} \tag{32.20}$$

In addition, chromatin entanglements and cross-linking lead to a resistance to dilational stretching (as well as shearing) of the chromatin. Modifying the semidilute polymer equation with a term compensating for elastic deformation of the polymer, $\frac{C}{\rho^d}$:

$$\Pi = \Pi_b + B\rho^{9/4} - \frac{C}{\rho^d} \tag{32.21}$$

This last term could be justifiably replaced by a sum of terms with $d = 1, 2, 3$ as is appropriate for wormlike chain elasticity, but this will generally introduce more fitting factors. We therefore use the parsimonious model above in fitting the data of chromatin condensation parameter versus osmolality, with four fit parameters: Π_b, B, C, and d.

Curve fitting in Figure 32.5b and c was performed with Igor Pro (version 6.22A, WaveMetrics, Portland, Oregon) using the chromatin condensation parameter

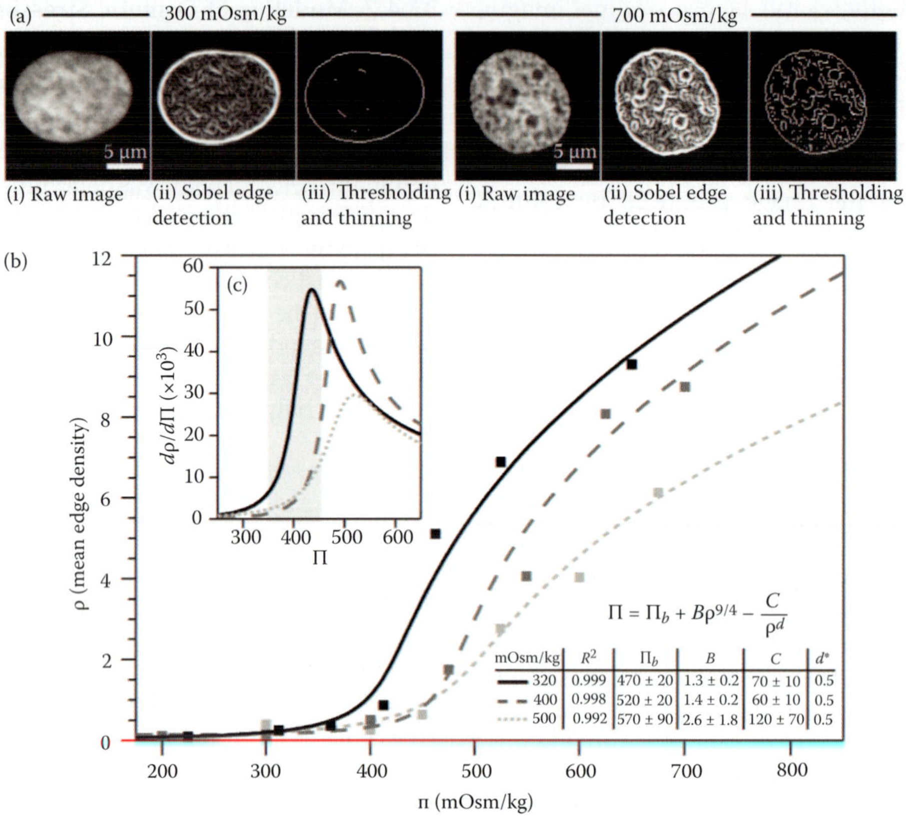

FIGURE 32.5
(a) Images of chondrocyte nuclei at osmolalities of 300 and 700 mOsm/kg, demonstrating how image processing methods can be used to quantify the extent of chromatin condensation. (b) Quantified chromatin condensation parameter fit to a semidilute polymer-gel model with an additional term compensating for chain elasticity. d is constrained to 0.5 to give the best fit for the datasets. Data are binned within the osmolality range of reversible density changes. (c) Calculated from the model and, for cells precultured 320 mOsm/kg, a large change in density is observed within the range of osmolalities associated with in vivo articular cartilage. (From Irianto, J. et al., *Biophys. J.* 104, 759–769, 2013.)

data binned in pairs within the osmolality range of reversible nuclear deformation and the cell volume data, respectively. As with the micropipette aspiration and the imaging after shear stress, this example of imaging nuclear features followed by physics-based modeling of extracted features illustrates an overall aim of obtaining imaging results that seem physically plausible. For some processes such as imaging of transcription factors that shuttle in and out of nuclei, the field remains very much in the experimental phase.

32.5 Mechanosensitive Transcription Factors in Nuclei

32.5.1 Immunostaining for Low Abundance Factors

Immunofluorescence is often used to study the localization and distribution of transcription factors that control gene expression, but transcription factors are often very low in abundance compared to structural proteins such as lamin-A/C and thus present some distinct challenges. Nonetheless, when transcription factor imaging works well on single cells, localization can be compared to other cellular or nuclear factors under different physical perturbations. Immunofluorescence is always limited by the availability, quality, and specificity of antibodies, and this remains an issue. If imaging results can be proven specific, transcription factor localization from the cytoplasm into the nucleus (or the reverse) as a function of physical perturbations to cells can provide important insight into mechanisms of differential gene expression in development and differentiation. Mechanotransduction signals might thus propagate from outside the cell to the nucleus where transcription is regulated, and translocation has been reported to depend on matrix elasticity for various lineage-specific transcription factors in mesenchymal stem cells (MSCs)[22] as well as the Hippo pathway transcription factor yes-associated protein 1 (YAP1)[23] and at least one of the retinoic acid (RA) regulated transcription factors (RARs) implicated in development and differentiation.[2]

In general, the immunostaining procedure involves (1) fixation of the cells; (2) permeabilization to allow antibody access; (3) "blocking" to minimize nonspecific antibody binding; (4) primary antibody staining; and (5) fluorescent secondary antibody staining. The low abundance of transcription factors can result in a very low fluorescent signal, and so high exposure times are needed for imaging. An important control is to check for and rule out nonspecific binding of the secondary antibody alone under each set of imaging conditions.

To image YAP1 and RARG in MSCs, cultured cells were first rinsed with PBS, fixed with 4% paraformaldehyde for 20 min, and then washed twice with PBS. Following fixation, cells were permeabilized with 0.5% Triton-X in PBS for 20 min, then washed with PBS, and treated with 5% BSA (blocking solution) for 1 h. Following two additional PBS washes, samples were incubated overnight with fresh primary antibodies at 1:300 dilution (or as indicated by the manufacturer) in 2% BSA solution with gentle agitation at 4°C. Cells were then washed twice in PBS and incubated with the corresponding secondary antibodies at 1:500 dilution (or as indicated by the manufacturer) for 30–60 min. Cells were imaged directly or mounted with mounting media with an antifading agent for later imaging.

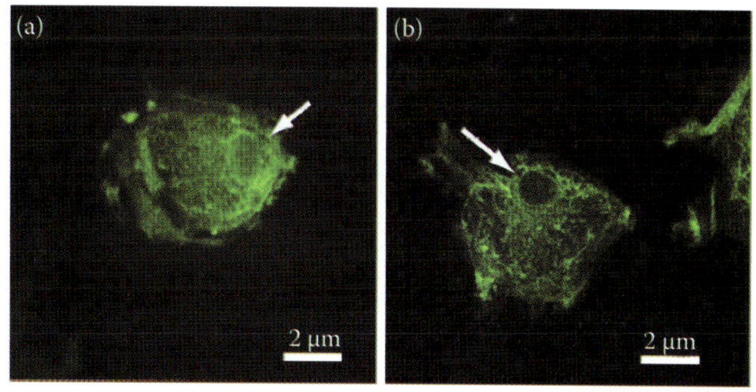

FIGURE 32.6
Immunofluorescent images of YAP1 distribution in MSCs cultured on soft gels in conventional fluorescence microscopy (a) and confocal microscopy (b), taken with the same objective. The confocal image taken at the midsection of the cell clearly shows that YAP1 is excluded from the nucleus compared with the left image, where signals from above and below the nucleus are collected.

32.5.2 Imaging and the Many Controls

Conventional fluorescence microscopy (epifluorescence) typically provides good signal-to-noise, and although highly detailed images can be produced, the nuclear location of an immunostained factor can be difficult to be certain of as the fluorescent signal can originate from out-of-focus fluorophores in the cytoplasm above or below the nucleus. Confocal microscopy can eliminate such signals and allow one to collect a series of images at different z-positions, for three-dimensional reconstruction. Figure 32.6 shows the difference between an epifluorescence image and a confocal image with YAP1 distributed inside and outside the nuclei of MSCs cultured on a soft substrate. Because confocal microscopy rejects out-of-plane fluorescence, the overall signal is reduced. This can be problematic for low abundance factors that may already have a low signal. In order to increase the signal, high excitation power is needed, leading to extensive photobleaching of the fluorophore. Additional studies of YAP1 in MSCs that were engineered to overexpress lamin-A/C suggested that a fraction of YAP1 is enriched at the nuclear envelope, which raises additional questions about thorough approaches to molecular colocalization.

32.5.3 Colocalization Imaging, GFP Variant Fusion Proteins, and Spectral Controls

In order to collect information about spatial colocalization, imaging more than one protein in the same cell is common but must be done with care, especially when one of the proteins is in low abundance as a transcription factor. For illustration, one can express wild-type or mutated proteins that have been tagged with fluorescent constructs (e.g., green or red fluorescent protein, GFP or RFP); Figure 32.7 shows an example of lung carcinoma A549 cells cotransfected with both GFP- and RFP-lamin-A. Both proteins should appear nuclear, but because of the random nature of transfection, some cells express only one of the proteins, and so such cells should only appear green or red and not be visible in both color channels. Thus, when two or more fluorescently tagged proteins and/or immunostained factors are to be imaged

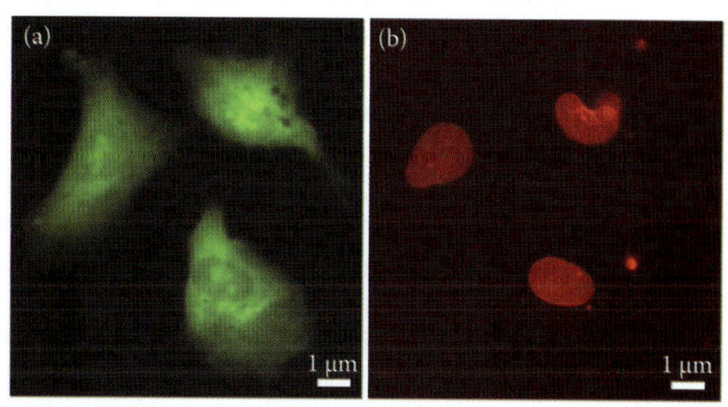

FIGURE 32.7
A549 cells cotransfected with both GFP (a) and RFP-lamin-A (b). GFP is known to have a cryptic nuclear localization signal (NLS), and so both proteins should appear nuclear as well as cytoplasmic. However, because of the random nature of transfection, some cells express only one of the proteins, and so such cells should only appear green or red and not be visible in both color channels.

in the same specimen, it becomes especially important to check for compatibility between wavelengths of the light source and excitation/emission spectra of the fluorophores. Many fluorophores have broad spectra, and if the excitation/emission ranges of two fluorophores overlap significantly, cross-talk between the two channels can be expected (referred to as "bleed-through").

When combining two fluorophores with overlapping excitation or emission spectra, it is important to ensure that each can be individually excited by different laser lines. If there is any overlap, sequential scanning with single laser lines should be performed instead of simultaneous scanning. Additionally, two fluorophores might have partially overlapping excitation ranges, and the excitation wavelength of one of the fluorophores also excites the second fluorophore but with lower efficiency. In this case, single-label control specimens must be prepared, and the signal measured in the main channel in addition to all other channels in which bleed-through may occur. Laser power, gains, and offsets should be adjusted through all the channels. With ideal settings, unused channels should appear dark in order to avoid channel cross-talk affecting the final multilabeled, compositional image. The fraction of the bleed-through signal recorded with different settings through the channels should be taken into account when quantitative analysis of colocalization is performed.

32.6 Conclusion

Epigenetic variations contribute to the plasticity in gene expression from stem cells that commit during both development and aging, and it seems increasingly likely that stresses and strains on nuclei contribute in part to epigenetic regulation. Methods are thus emerging as findings show that these mechanobiological epigenetic processes can range from lineage-relevant variations in the nuclear envelope structure proteins called lamins (that protect DNA) to chromatin organization and transcription factor localization. Any in-depth study of mechanobiology must try to link microscopic-scale processes to molecular events and, reciprocally, ought to clarify how molecular events influence broader processes. Mechanically perturbed systems should be increasingly studied at the molecular level, particularly with powerful microscopy methods, but, at the same time, these observations should also be combined with physical descriptions and models of the broader system, as attempted herein.

One of the future challenges for the study of mechanobiology will be to track broad changes in nuclear morphology to a specific epigenetic program. Methods such as "chromosome painting" with fluorescence in situ hybridization (FISH) probes[24] and, more recently, "Hi-C" spatial mapping[4] have shown that certain features maintain discrete positions within the nucleus, but due to technical complexities, methods have focused on static, fixed systems rather than dynamic responses to stimuli. Spatial relationships are thought to be important factors in determining gene activity, and studying how they progress with time will be crucial. Recent work by Talwar et al.[25] has characterized differences in chromatin compaction states between mouse embryonic stem and fibroblast cells, and emerging technologies such as TALE-mediated genome visualization (TGV) might allow the study of the spatial organization of specific nuclear features as live cells respond to their environment.[26] Integrating such approaches into the types of visualized mechanical experiments here could yield exciting new insights into some of the most basic responses in mechanobiology.

Acknowledgments

We greatly appreciate support from the US National Institutes of Health (NIH R01HL062352; P01DK032094; R01EB007049; NCATS-8UL1TR000003), the US National Science Foundation (grant 1200834, Penn's Materials Research Science and Engineering Center, and Nano Science and Engineering Center-Nano Bio Interface Center), the Human Frontier Science Program (I.L.I., D.E.D.), and the Stem Cell Xenograft Core (A. Secreto, J. Glover, and G. Danet-Desnoyers) and Microarray Cores (P30-DK090969). We thank David Gilbert (Florida State University, Department of Biological Science), Harald Herrmann (Heidelberg, DKFZ), and Howard Worman (Columbia University, Department of Medicine) for protein constructs.

References

1. Constantinescu, D., Gray, H. L., Sammak, P. J., Schatten, G. P. and Csoka, A. B. Lamin A/C expression is a marker of mouse and human embryonic stem cell differentiation. *Stem Cells* **24**, 177–185 (2006).
2. Swift, J. et al. Nuclear lamin-A scales with tissue stiffness and enhances matrix-directed differentiation. *Science* **341**, 1240104 (2013).
3. Scaffidi, P. and Misteli, T. Lamin A-dependent nuclear defects in human aging. *Science* **312**, 1059–1063 (2006).
4. Lieberman-Aiden, E. et al. Comprehensive mapping of long-range interactions reveals folding principles of the human genome. *Science* **326**, 289–293 (2009).

5. Dahl, K. N., Engler, A. J., Pajerowski, J. D. and Discher, D. E. Power-law rheology of isolated nuclei with deformation mapping of nuclear substructures. *Biophys. J.* **89**, 2855–2864 (2005).
6. Pajerowski, J. D., Dahl, K. N., Zhong, F. L., Sammak, P. J. and Discher, D. E. Physical plasticity of the nucleus in stem cell differentiation. *Proc. Natl. Acad. Sci. USA* **104**, 15619–15624 (2007).
7. Guilak, F. The deformation behavior and viscoelastic properties of chondrocytes in articular cartilage. *Biorheology* **37**, 27–44 (2000).
8. Deguchi, S., Maeda, K., Ohashi, T. and Sato, M. Flow-induced hardening of endothelial nucleus as an intracellular stress-bearing organelle. *J. Biomech.* **38**, 1751–1759 (2005).
9. Dahl, K. N. et al. Distinct structural and mechanical properties of the nuclear lamina in Hutchinson–Gilford progeria syndrome. *Proc. Natl. Acad. Sci. USA* **103**, 10271–10276 (2006).
10. Shin, J.-W., Spinler, K. R., Swift, J., Chasis, J. A., Mohandas, N. and Discher, D. E. Lamins regulate cell trafficking and lineage maturation of adult human hematopoietic cells. *Proc. Natl. Acad. Sci. USA* **110**, 18892–18897 (2013).
11. Caille, N., Thoumine, O., Tardy, Y. and Meister, J. J. Contribution of the nucleus to the mechanical properties of endothelial cells. *J. Biomech.* **35**, 177–187 (2002).
12. Dean, D. A. and Kasamatsu, H. Signal- and energy-dependent nuclear transport of SV40 Vp3 by isolated nuclei. Establishment of a filtration assay for nuclear protein import. *J. Biol. Chem.* **269**, 4910–4916 (1994).
13. Andersen, J. S. et al. Nucleolar proteome dynamics. *Nature* **433**, 77–83 (2005).
14. Black, B. E., Brock, M. A., Bedard, S., Woods, V. L. and Cleveland, D. W. An epigenetic mark generated by the incorporation of CENP-A into centromeric nucleosomes. *Proc. Natl. Acad. Sci. USA* **104**, 5008–5013 (2007).
15. Johnson, C., Tang, H., Carag, C., Speicher, D. and Discher, D. Forced unfolding of proteins within cells. *Science* **317**, 663–666 (2007).
16. Aaronson, R. P. and Woo, E. Organization in the cell nucleus: Divalent cations modulate the distribution of condensed and diffuse chromatin. *J. Cell Biol.* **90**, 181–186 (1981).
17. Dobi, A. and Agoston, D. Submillimolar levels of calcium regulates DNA structure at the dinucleotide repeat (TG/AC)n. *Proc. Natl. Acad. Sci. USA* **95**, 5981–5986 (1998).
18. Corle, T. R., Chou, C. H. and Kino, G. S. Depth response of confocal optical microscopes. *Optics Lett.* **11**, 770–772 (1986).
19. Pawley, J. B. *Handbook of Biological Confocal Microscopy*, 2nd edn. (Plenum Press, NY, 2005).
20. Irianto, J. et al. Osmotic challenge drives rapid and reversible chromatin condensation in chondrocytes. *Biophys. J.* **104**, 759–769 (2013).
21. Strobl, G. *The Physics of Polymers, Concepts for Understanding Their Structure and Behavior*. (Springer, Berlin Heidelberg, 2007).
22. Engler, A. J., Sen, S., Sweeney, H. L. and Discher, D. E. Matrix elasticity directs stem cell lineage specification. *Cell* **126**, 677–689 (2006).
23. Dupont, S., Morsut, L., Aragona, M., Enzo, E., Giulitti, S., Cordenonsi, M., Zanconato, F., Le Digabel, J., Forcato, M., Bicciato, S., Elvassore, N. and Piccolo, S. Role of YAP/TAZ in mechanotransduction. *Nature* **474**, 179–183 (2011).
24. Cremer, T. and Cremer, C. Chromosome territories, nuclear architecture and gene regulation in mammalian cells. *Nat. Rev. Genet.* **2**, 292–301 (2001).
25. Talwar, S., Kumar, A., Rao, M., Menon, G. I. and Shivashankar, G. V. Correlated spatio-temporal fluctuations in chromatin compaction states characterize stem cells. *Biophys. J.* **104**, 553–564 (2013).
26. Miyanari, Y., Ziegler-Birling, C. and Torres-Padilla, M.-E. Live visualization of chromatin dynamics with fluorescent TALEs. *Nat. Struct. Mol. Biol.* **20**, 1321–1324 (2013).

33

Intranuclear Measurement of Deformation in Single Cells

Jonathan T. Henderson and Corey P. Neu

CONTENTS

Nuclear mechanics play a pivotal role in cell survival and gene expression. Mechanical stimulation of cells and nuclei regulates specific mechanotransduction pathways that aid in cell homeostasis. This chapter describes current and emerging methodologies that are used to study nuclear mechanics at the single cell level. We highlight a hybrid method based on optical microscopy and hyperelastic warping that was developed to measure intranuclear strain within individual nuclei maintained in cultured tissue explants. The combined use of imaging and postprocessing methods allows for the study of nuclear mechanics and mechanobiology in single cells, permits simultaneous study of newly synthesized RNAs, and overcomes limitations in competing factors like spatial and temporal resolution of acquisition systems. We discuss the implementation of methods to measure nuclear mechanics, as well as their potential to increase understanding of diseases involved in compromised nuclear structures.

33.1 Introduction

Diseases such as osteoarthritis, laminopathies, and specific cancers are shown to have compromised tissue, cellular, and nuclear mechanics, potentially interrupting normal mechanobiological processes and providing an origin point for disease progression.[1–3] Osteoarthritis of cartilage and articulating joints has a multifactorial etiology resulting in part from a homeostatic imbalance in the tissue. Cartilage degradation causes an altered micromechanical environment, increasing the expression of degrading enzymes that exacerbate the disease state.[4] Laminopathies specifically compromise the rigidity of the nuclear envelope and are linked to premature aging and muscular dystrophy. Cancer cells actively alter their nuclear stiffness, facilitating cell mobility and metastasis,[3,5] and pluripotent cells alter their nuclear structure and stiffness prior to cell lineage commitment.[6] To better understand healthy and diseased or damaged tissues, and how mechanical forces applied to tissues and cells influence individual cell nuclei, it is important to measure the mechanics (e.g., strain) in the nucleus and quantify how nuclear mechanics regulate gene expression.

The nucleus is largely occupied by chromatin, dynamic structures that are influenced by diverse chemical and mechanical signals.[7,8] Chromatin itself is a hierarchical structure composed of DNA molecules wrapped around histones to form nucleosomes, which are then further packed together to form chromatin structures. Within the nucleus, chromatin is often classified as

TABLE 33.1
Common Techniques for Measurement of Nuclear Mechanics

Measurement Technique	Description	Spatial Resolution	Bulk or Local Measurement	Isolated or Embedded (2D/3D) Cells	References
Fluorescence anisotropy	Local compaction	High	Local	Isolated/2D	11–13
Aspect ratio	Morphological change	Low	Bulk	Isolated/2D/3D	14,15
Volume change	Morphological change	Low	Bulk	Isolated/2D/3D	16,17
Micropipette aspiration	Material properties	Medium	Bulk	Isolated	18,19
Nanoindentation	Material properties	High	Local (surface)	Isolated	20,21
Texture correlation	Nuclear strain	High	Local	Isolated/2D/3D	22,23
Hyperelastic warping	Intranuclear strain	High	Local	Isolated/2D/3D	24

heterochromatin, i.e., tightly packed structures that are thought to be areas of relatively silenced gene expression, or euchromatin, i.e., lower density structures that exhibit higher levels of gene expression. Gene expression is affected by nuclear receptors, secondary messengers, and other chemical signaling cascades triggered from chemical binding to cellular receptors. Recently, direct regulation of gene expression by mechanical force has been hypothesized to work in parallel with chemical signaling, with transcriptional activity and accessibility regulated through physical changes in DNA regions.[9,10]

Several methodologies have been developed to measure nuclear mechanics and help identify the role of the nuclear structure in the mechanoregulation of gene expression. Methods to measure nuclear mechanics are numerous and include fluorescence anisotropy, micropipette aspiration or nanoindentation, changes in nuclear aspect ratios and/or volume, and texture correlation or deformable image registration (Table 33.1). Understanding the strengths and limitations of each method better define their best applications for intended experimental needs and objectives. This chapter will cover different methodologies that are used to investigate nuclear mechanics, with special attention paid to the new implementation of deformable image registration using hyperelastic warping and concurrent quantification of nascent RNA expression in single cells.

33.2 Deformed Nucleus

33.2.1 Connecting the Nucleus to the Cell and Extracellular Matrix

The nucleus is a distinct structure inside the cell that is linked through cytoskeletal components to the extracellular matrix (ECM). Among different tissue and cell types, the structures connecting the nucleus to the extracellular environment can be very unique. In this chapter, we will largely focus on the primary cell type (chondrocyte) found in cartilage, a load-bearing tissue with significant relevance to human disease (e.g., osteoarthritis).[25,26] In the chondrocyte, there are multiple structural elements that connect the nucleus to the ECM (Figure 33.1). The cell is anchored to the ECM by cell membrane proteins (e.g., integrins and CD44), which bind to collagen and proteoglycans. These anchoring transmembrane proteins are associated with intracellular cytoskeletal components such as actin, intermediate filaments, and microtubules. The microtubules interact with the endoplasmic reticulum (ER), which is continuous with the outer nuclear membrane. The microtubules, along with other cytoskeleton components, are also connected to the nucleus through a group of nuclear transmembrane proteins called the linker of nucleoskeleton and cytoskeleton complex (LINC). These structural connections from the ECM to the nucleus provide a mechanism for stress and strain to directly transfer to the nucleus when applied at relatively distant tissue surfaces.

33.2.2 Nuclear Structures: Form and Function

Within the nucleus, there are many specific structural elements that help to regulate gene expression. DNA nucleotides that code for genes make up the majority of material in the nucleus. Nuclear function is regulated by transcription factors,[27] nuclear pores,[28] and the nucleolus,[29] about which the referenced reviews provide more in-depth detail. Within the nucleus, distinct euchromatin and heterochromatin regions are observed, in addition to the interchromatin space, where nuclear components are located, such as the nucleolus, which synthesizes rRNA.[29,30] Like a shell, the nuclear membrane is made up of a network of lamins that provide structural rigidity to the nucleus. Distributed in the nuclear membrane, nuclear pores regulate the transport of molecules in and

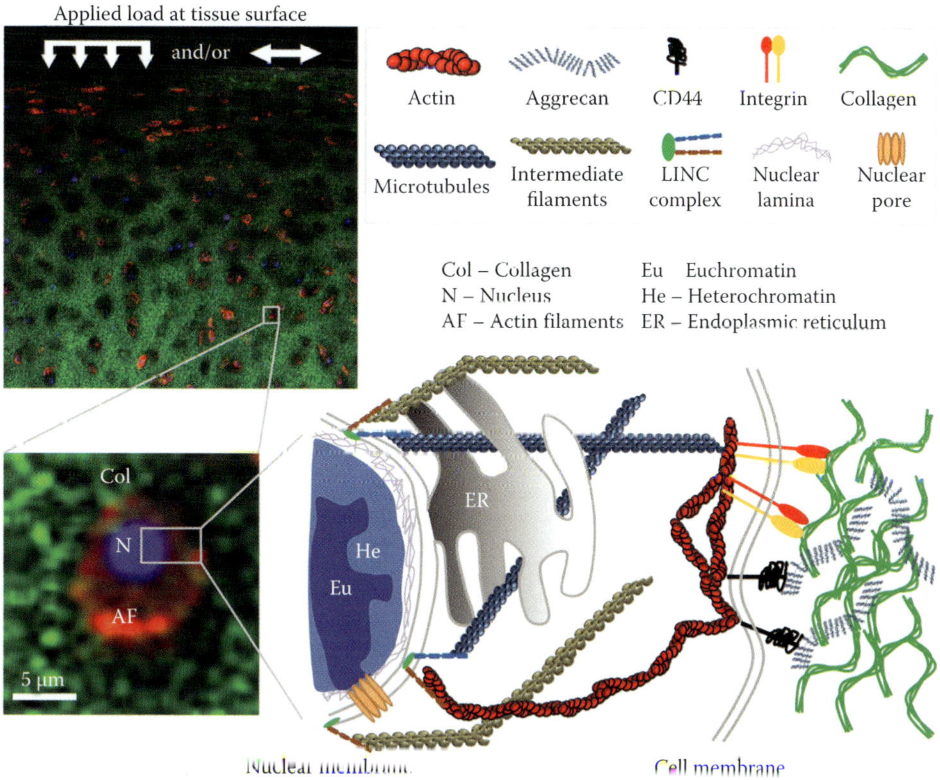

FIGURE 33.1
Structural components transfer strain from the tissue and ECM to the cell nucleus. Strain is transferred from an applied load (e.g., compression or shear) at the tissue surface to the cells and nuclei maintained alive within the interior. Articular cartilage is shown here as a model hierarchical system that normally undergoes many thousands of loading cycles during daily activities like walking. Multiple structural components, including surface receptors, cytoskeletal proteins, and the nuclear lamina, connect the ECM to the nucleus.

out of the nucleus (e.g., RNA; Figure 33.1). The integration of the nuclear components and global structures provides the biophysical conditions for the nucleus to maintain the integrity of the DNA and regulate cellular biosynthesis by modulating the shape, size, and internal patterns of deformation (e.g., displacements and strain).

Local mechanical deformations within the nucleus are thought to alter early transcriptional activities for gene expression. One proposed mechanism is that intranuclear mechanical deformation alters the structure of the chromatin domains thereby modulating the gene expression in the cell.[31]

33.3 Measuring Nuclear Mechanics and Mechanobiology

33.3.1 Nuclear Imaging, Deformation, and Measurement Techniques

Imaging of intranuclear mechanics in single cells is challenging. Several contemporary methods attempt to meet this need through the use of high-resolution imaging (e.g., superresolution microscopy, scanning electron microscopy), although visualization of cell and subcellular structural morphology does not necessarily provide information describing how these structures may deform under applied loading or during active cytoskeletal or chromatin remodeling. Typically, imaging involves visualization of chemically fixed cells, preventing an analysis of deformation and associated mechanotransduction activity. Widefield and confocal microscopy have visualized nuclei in living cells before and after applied deformations,[22] and methods have been used to mechanically deform the nuclei of isolated or tissue/matrix-embedded cells[32] (Table 33.1 and Figure 33.2). A commonality among the deformation and measurement techniques is that the nuclei are fluorescently imaged in predeformation and postdeformation states, so that motion in a current configuration can be measured from a reference configuration.

Considering the variety of methods available to measure nuclear mechanics (Table 33.1), we focus on three common ways to measure predeformation and postdeformation changes of the nucleus (Figures 33.2 and

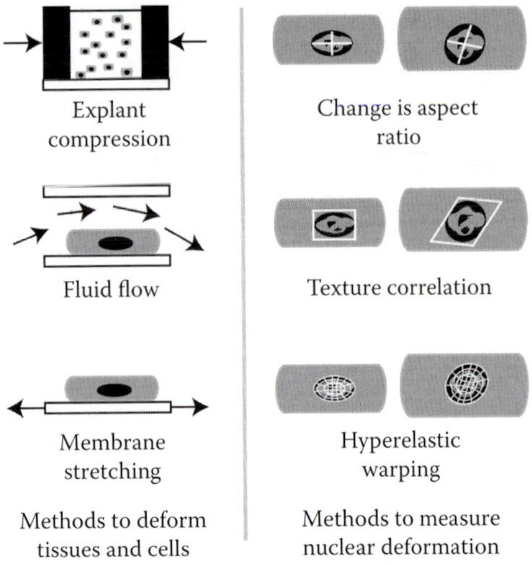

FIGURE 33.2
Common methods to measure mechanical deformation in the nucleus. Flow and stretch methods deform tissues and isolated cells in 2D (monolayer) or 3D (embedded) native or engineered microenvironments. Simple measures of nuclear shape and morphology changes are documented by changes in the aspect ratio. Texture correlation and hyperelastic warping methods provide details of internal strains with different levels of error as described in Figure 33.3.

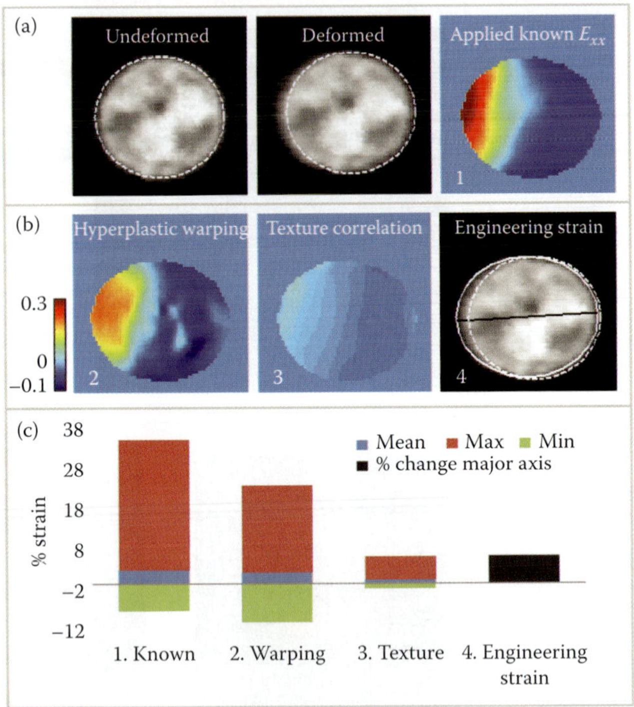

FIGURE 33.3
Simple comparison of three techniques to quantify nuclear deformation. A finite-element model was used to apply a known deformation (strain) to a nucleus and create the deformed nucleus from the undeformed z-stack images (a). Knowing the applied intranuclear strain (a.1), three measurement techniques, hyperelastic warping (b.2), texture correlation (b.3), and engineering strain in the direction of the major axis (b.4), were used to measure the strain fields for comparison between the techniques. Results from the hyperelastic warping method best described the known deformation, with decreased sensitivity observed from texture correlation and engineering strain measurements (c).

33.3), each with distinct strengths and weaknesses. First, changes in bulk morphology of the nucleus (e.g., aspect ratios, nuclear volume, or major/minor axes) enable an average mechanical response of nuclei after mechanical perturbation, often using automated or semiautomated algorithms. While bulk measurements can be acquired from large numbers of cells, important spatial information from within individual nuclei is not captured or described. Second, texture correlation measures intranuclear strain fields by tracking intrinsic high contrast regions, revealed by spatial distributions of fluorescent markers, in the images depicting undeformed and deformed nuclei.[22] Unfortunately, the small volume of the nucleus, coupled with large subset sizes required to help minimize error,[22] limits the ability to reliably quantify intranuclear strain. Third, deformable image registration (e.g. hyperelastic warping) permits intranuclear strain measurements by iteratively deforming a three-dimensional (3D) finite-element mesh representing the undeformed nucleus until it matches the deformed nucleus.[24] A 3D intranuclear strain field is calculated from the displaced nodes of the registered nuclear mesh.

Bulk morphology of the nucleus, texture correlation, and hyperelastic warping increasingly predict known nuclear deformations applied in simulations (Figure 33.3).

Compared to the known deformation applied, texture correlation underestimates the magnitude of the strain field, and the (bulk) engineering strain measured by changes in width fails to capture the strain distribution or magnitudes. In contrast, hyperelastic warping best approximated the known magnitude and spatial distribution of the applied known strain field, with decreased error (i.e., higher precision and minimal bias).[24]

33.3.2 Design Criteria to Combine Nuclear Mechanics and Mechanobiology

Limitations in the spatial and temporal resolution of image data, and the impact of these limitations on image quality (e.g., signal-to-noise ratio [SNR]), must be considered when designing techniques to study nuclear mechanics and mechanobiology in single cells. Trade-offs in spatial and temporal resolutions may be accepted for specific applications, noting that the error of intranuclear deformation depends largely on the ability to track the motion of structures on the organelle interior with a sufficient SNR. Methods that analyze long-term responses ranging from minutes to days may provide insufficient temporal resolution to capture direct relationships between nuclear structure and gene expression.[33–36] For example, to capture nuclear mechanics and mechanobiology in a single cell, the measurement technique needs to be fast enough to capture a possible rapid gene expression response to the applied load. The technique also needs to have a high enough spatial resolution to image the nuclear shape changes due to the applied load. Spatial resolution is often sacrificed to improve temporal resolution, and vice versa, when designing techniques for the study of single cells.[36]

33.4 Hybrid Technique—Microscopy and Hyperelastic Warping

A hybrid confocal and hyperelastic warping methodology has recently been demonstrated to measure and calculate intranuclear mechanics and biophysical activities.[24] Combining the high spatial resolution of optical (e.g., confocal or multiphoton) microscopy with hyperelastic warping to measure intranuclear strain maps has provided detailed intranuclear strains compared to bulk measurement of nuclear deformation (Table 33.1). Optical microscopy also enables imaging deep within tissues to calculate nuclear mechanics of cells maintained in native tissue cultures. Z-stack images capture the full volume of the nucleus in the undeformed and deformed state. The volume of the nucleus is converted into a finite-element mesh that is used in an iterative hyperelastic warping-based method, where the undeformed nuclear mesh is optimally warped until it matches the deformed nucleus, to measure 3D intranuclear deformations (Figure 33.4). The intranuclear strain map can be spatially and temporally compared to any fluorescent indicator of gene expression that can be imaged concurrently with the nuclei stain in a second imaging channel, such as nascent RNA synthesis (Figure 33.4).

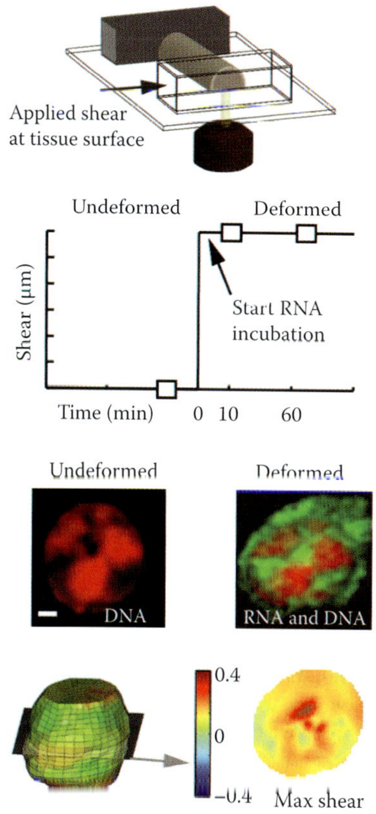

FIGURE 33.4
Experimental overview for methods to measure single-cell intranuclear mechanics and RNA synthesis. For articular cartilage, a tissue explant is created as a half cylinder from a cored osteochondral plug and placed in an electronically controlled deformation device that allows for single cells to be imaged within the deforming tissue using optical (e.g., confocal, multiphoton) microscopy. Immediately after shearing, newly synthesizing RNAs are labeled using a click chemistry-based incubation. The deformed nucleus is imaged at 10 min and then at the end of the incubation time (60 min), and the data are used for the measurement of intranuclear strain using hyperelastic warping. Images of the nucleus are taken from the center z-slice (red = DNA stain, green = RNA stain) (Scale bar = 1 μm).

33.4.1 Strain Transfer: Tissue to Nucleus

Strain transfer to the nucleus from an applied mechanical load at a distant tissue surface can be measured as amplification and attenuation in local subnuclear regions by microscopy and hyperelastic warping. When cartilage explants undergo shear loading, there is higher tissue strain in the superficial zone that diminishes in the deep zone[37] (Figure 33.5). Likewise, compressed cartilage explants influence nuclei more near the articulating surface (superficial zone) and progressively less in the middle and deep zones.[38] These bulk nuclear volumetric strain results demonstrate a depth-dependent strain transfer from the applied tissues strain to the cell. However, more detailed intranuclear strains are resolved by using hyperelastic warping that shows amplification and attenuation of strain that the bulk measurement techniques do not reveal. The combined use of optical microscopy and hyperelastic warping provides a new method that gives 3D local nuclear deformations that were previously immeasurable by bulk measurement techniques.

33.4.2 Measuring Single-Cell 3D Intranuclear Time-Dependent Strain Fields

Single-cell intranuclear strain fields show complex strain patterns with time-varying magnitudes that are spatially correlated to DNA intensity. Confocal z-stack images of chondrocyte nuclei taken before and at 10 and

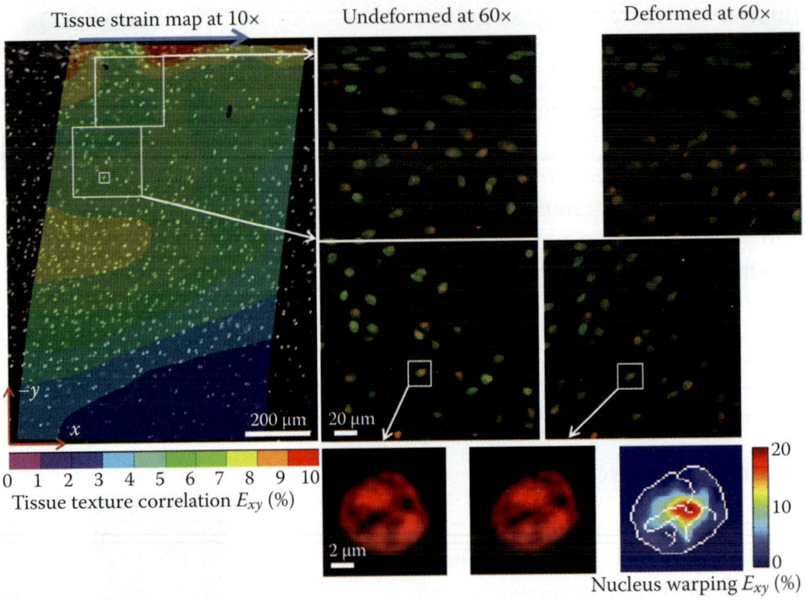

FIGURE 33.5
Strain transfer from the tissue surface to intranuclear regions. A shear load was applied at the articular cartilage surface (blue arrow), and texture correlation was used to measure the E_{xy} strain field for the tissue scale deformation (10× objective) using cells as fiducial markers. Hyperelastic warping was used to measure the strain fields of the nucleus highlighted by the smaller boxes in images acquired using 10× and 60× objectives. Depending on the intranuclear region, the measured tissue scale strain (5%–6%; i.e., tissue texture correlation E_{xy} [%] strain field) was amplified or attenuated when compared to the nucleus warping E_{xy} (%) strain field.

60 min after a shear load is applied to the articular surface of a cartilage explant show similar strain field patterns within a cell, although the magnitudes of the patterns at the two time points differ cell to cell.[24] The intranuclear strain field patterns are spatially complex, with brighter DNA intensity regions in specific cell populations studied typically corresponding to tensile strain regions. A current challenge remains to define mechanisms of strain transfer, including specific cytoskeletal elements responsible for load transmission *in situ*. Also, the role of directionally concentrated strains in the physical regulation of gene expression is not yet known. However, the hybrid methodology does provide high spatial resolution strain maps of the intranuclear space that will allow for future investigation of these questions.

33.4.3 Detecting Nascent RNA Synthesis in Single Cells

Combining measurements of intranuclear mechanics with biosynthesis assays, e.g., detection of nascent RNAs, facilitates analysis of mechanoregulated gene expression within the nuclei of single cells. Nascent RNA detection has been measured in cells embedded in their native ECM[24] using a commercially available global RNA detection kit (Click-iT® RNA Alexa Fluor® 488 Imaging Kit, Invitrogen, Carlsbad, California).[39,40] In previous studies, the incubation for nascent RNA detection began immediately after the deformation and lasted for 60 min, and the sample was subsequently fixed for fluorescent labeling (Figure 33.4). In the cell population studied, the RNA was located in the lower intensity DNA regions, i.e., the interchromatin space, where the DNA is less densely packed and thought to be highly active for transcription.[24,30,41] The spatial location of RNA was typically correlated with the more compressive nuclear regions. However, the spatial correlation of mechanics and biosynthesis should be interpreted with caution because it was observed that RNA was also transported outside of the nucleus during the incubation period. In a control study, with no tissue scale deformation applied, there were fewer nuclei (26%) with detectable RNA synthesis compared to the deformed nuclei results (59%). The mechanism for why the nuclei in the mechanically strained tissue had more detectable RNA nuclei has not been explained, but it is nevertheless important to employ methods that allow for direct spatial correlation of intranuclear strain fields and RNA synthesis in nuclei to help elucidate pathways for gene expression in single cells.

33.4.4 Specificity and Experimental Tradeoffs

There is still a need to increase the specificity of hybrid microscopy and hyperelastic warping methods to reveal biophysical and biosynthesis actions at even smaller spatial and temporal scales. Labeling of genes related to disease pathogenesis or tissue regeneration may allow for the discovery of local mechanical factors that can alter clinically relevant gene transcription. One challenge is the development of high (e.g., super) resolution imaging to better reveal spatial patterns while also not sacrificing temporal resolution of the desired response. For example, an experimental protocol that captures time-consuming high-resolution images of cell nuclei and fast calcium signaling is affected by competing spatiotemporal factors. Calcium fluxes can occur on the order of seconds to milliseconds, a time duration that is faster than the time required for image acquisition depicting nuclei at high resolution. In contrast, biosynthesis processes like RNA transcription can occur over longer time durations and may be more appropriately combined with higher-resolution imaging to simultaneously provide intranuclear strain fields (Figure 33.4 and Table 33.1). However, the localization of specific gene activity or gene products is not often resolved in living single cells,[36] and it is likely that biotechnology advances for gene expression profiling, coupled with hybrid microscopy, will only improve our understanding of mechanotransduction overall.

33.5 Conclusion

This chapter has demonstrated that a hybrid method combining optical microscopy and hyperelastic warping provides detailed intranuclear strain patterns and represents a balanced approach to meet the competing acquisition factors of spatial and temporal resolutions. The measurement of nascent RNA synthesis was a fundamental step to show that intranuclear mechanics and gene expression-related events can be measured within the same nucleus maintained in a native 3D microenvironment. As research objectives turn toward targeting specific genes or profiling of mRNA expression in single cells, new approaches will be developed and will be easily included for spatial correlation of strain fields measured by hyperelastic warping.

Acknowledgments

The authors gratefully acknowledge funding from the Andrews Fellowship and Hugh W. and Edna M. Donnan Fellowship (JTH).

References

1. Goldman, R. D., Gruenbaum, Y., Moir, R. D., Shumaker, D. K. and Spann, T. P. Nuclear lamins: Building blocks of nuclear architecture. *Genes and Development* **16**, 533–547 (2002). doi:10.1101/gad.960502.
2. Gerace, L. and Huber, M. D. Nuclear lamina at the crossroads of the cytoplasm and nucleus. *Journal of Structural Biology* **177**, 24–31 (2012). doi:10.1016/j.jsb.2011.11.007.
3. Li, Q. S. and Lim, C. T. Structure–mechanical property changes in nucleus arising from breast cancer. *Cellular and Biomolecular Mechanics and Mechanobiology* **4**, 465–475 (2011). doi:10.1007/8415_2010_19.
4. Aigner, T., Kurz, B., Fukui, N. and Sandell, L. Roles of chondrocytes in the pathogenesis of osteoarthritis. *Current Opinion in Rheumatology* **14**, 578–584 (2002). doi:10.1097/01.bor.0000022355.07263.81.
5. Hou, H. W. et al. Deformability study of breast cancer cells using microfluidics. *Biomedical Microdevices* **11**, 557–564 (2009). doi:10.1007/s10544-008-9262-8.
6. Chalut, K. J. et al. Chromatin decondensation and nuclear softening accompany Nanog downregulation in embryonic stem cells. *Biophysical Journal* **103**, 2060–2070 (2012). doi:10.1016/j.bpj.2012.10.015.
7. Chuang, C. H. and Belmont, A. S. Moving chromatin within the interphase nucleus-controlled transitions? *Seminars in Cell and Developmental Biology* **18**, 698–706 (2007). doi:10.1016/j.semcdb.2007.08.012.
8. Trinkle-Mulcahy, L. and Lamond, A. I. Nuclear functions in space and time: Gene expression in a dynamic, constrained environment. *FEBS Letters* **582**, 1960–1970 (2008). doi:10.1016/j.febslet.2008.04.029.
9. Martins, R. P., Finan, J. D., Guilak, F. and Lee, D. A. Mechanical regulation of nuclear structure and function. *Annual Review of Biomedical Engineering* **14**, 431–455 (2012). doi:10.1146/annurev-bioeng-071910.
10. Teperino, R., Lempradl, A. and Pospisilik, J. A. Bridging epigenomics and complex disease: The basics. *Cellular and Molecular Life Sciences* **70**, 1609–1621 (2013). doi:10.1007/s00018-013-1299-z.
11. Banerjee, B., Bhattacharya, D. and Shivashankar, G. V. Chromatin structure exhibits spatio-temporal heterogeneity within the cell nucleus. *Biophysical Journal* **91**, 2297–2303 (2006). doi:10.1529/biophysj.105.079525.
12. Bhattacharya, D., Talwar, S., Mazumder, A. and Shivashankar, G. V. Spatio-temporal plasticity in chromatin organization in mouse cell differentiation and during drosophila embryogenesis. *Biophysical Journal* **96**, 3832–3839 (2009). doi:10.1016/j.bpj.2008.11.075.
13. Talwar, S., Kumar, A., Rao, M., Menon, G. I. and Shivashankar, G. V. Correlated spatio-temporal fluctuations in chromatin compaction states characterize stem cells. *Biophysical Journal* **104**, 553–564 (2013).
14. Haudenschild, D. R. et al. Vimentin contributes to changes in chondrocyte stiffness in osteoarthritis. *Journal of Orthopaedic Research* **29**, 20–25 (2011). doi:10.1002/jor.21198.

15. Knight, M. M. et al. Cell and nucleus deformation in compressed chondrocyte-alginate constructs: Temporal changes and calculation of cell modulus. *Biochimica et Biophysica Acta-General Subjects* **1570**, 1–8 (2002).
16. Guilak, F., Ratcliffe, A. and Mow, V. C. Chondrocyte deformation and local tissue strain in articular-cartilage—A confocal microscopy study. *Journal of Orthopaedic Research* **13**, 410–421 (1995).
17. Abusara, Z., Seerattan, R., Leumann, A., Thompson, R. and Herzog, W. A novel method for determining articular cartilage chondrocyte mechanics in vivo. *Journal of Biomechanics* **44**, 930–934 (2011). doi:10.1016/j.jbiomech.2010.11.031.
18. Guilak, F., Alexopoulos, L. G., Haider, M. A., Ting-Beall, H. P. and Setton, L. A. Zonal uniformity in mechanical properties of the chondrocyte pericellular matrix: Micropipette aspiration of canine chondrons isolated by cartilage homogenization. *Annals of Biomedical Engineering* **33**, 1312–1318 (2005). doi:10.1007/s10439-005-4479-7.
19. Bader, D. L., Ohashi, T., Knight, M. M., Lee, D. A. and Sato, M. Deformation properties of articular chondrocytes: A critique of three separate techniques. *Biorheology* **39**, 69–78 (2002).
20. Darling, E. M. Force scanning: A rapid, high-resolution approach for spatial mechanical property mapping. *Nanotechnology* **22**, 175707 (2011).
21. Darling, E. M., Wilusz, R. E., Bolognesi, M. P., Zauscher, S. and Guilak, F. Spatial mapping of the biomechanical properties of the pericellular matrix of articular cartilage measured in situ via atomic force microscopy. *Biophysical Journal* **98**, 2848–2856 (2010). doi:10.1016/j.bpj.2010.03.037.
22. Lim, K. Y., Henderson, J. T. and Neu, C. P. Cell and tissue deformation measurements: Texture correlation with third-order approximation of displacement gradients. *Journal of Biomechanics* **46**, 2490–2496 (2013). doi:10.1016/j.jbiomech.2013.07.035.
23. Gilchrist, C. L., Witvoet-Braam, S. W., Guilak, F. and Setton, L. A. Measurement of intracellular strain on deformable substrates with texture correlation. *Journal of Biomechanics* **40**, 786–794 (2007). doi:10.1016/j.jbiomech.2006.03.013.
24. Henderson, J. T., Shannon, G., Veress, A. I. and Neu, C. P. Direct measurement of intranuclear strain distributions and RNA synthesis in single cells embedded within native tissue. *Biophysical Journal* **105**, 2252–2261 (2013).
25. Neu, C. P., Khalafi, A., Komvopoulos, K., Schmid, T. and Reddi, A. H. Mechanotransduction of articular cartilage superficial zone protein by TGF-b signaling. *Arthritis and Rheumatism* **56**, 3706–3714 (2007).
26. Neu, C. P., Reddi, A. H., Komvopoulos, K., Schmid, T. M. and Di Cesare, P. E. Increased friction coefficient and superficial zone protein expression in patients with advanced osteoarthritis. *Arthritis and Rheumatism* **62**, 2680–2687 (2010). doi:10.1002/art.27577.
27. Latchman, D. S. Transcription factors: An overview. *International Journal of Biochemistry and Cell Biology* **29**, 1305–1312 (1997). doi:10.1016/s1357-2725(97)00085-x.
28. Gorlich, D. and Kutay, U. Transport between the cell nucleus and the cytoplasm. *Annual Review of Cell and Developmental Biology* **15**, 607–660 (1999). doi:10.1146/annurev.cellbio.15.1.607.
29. Raska, I., Shaw, P. J. and Cmarko, D. New insights into nucleolar architecture and activity. *International Review of Cytology—A Survey of Cell Biology* **255**, 177–235 (2006).
30. Cremer, T. et al. Chromosome territories, interchromatin domain compartment, and nuclear matrix: An integrated view of the functional nuclear architecture. *Critical Reviews in Eukaryotic Gene Expression* **10**, 179–212 (2000).
31. Mammoto, A., Mammoto, T. and Ingber, D. E. Mechanosensitive mechanisms in transcriptional regulation. *Journal of Cell Science* **125**, 3061–3073 (2012).
32. Guilak, F. Compression-induced changes in the shape and volume of the chondrocyte nucleus. *Journal of Biomechanics* **28**, 1529–1541 (1995).
33. Campbell, J. J., Blain, E. J., Chowdhury, T. T. and Knight, M. M. Loading alters actin dynamics and up-regulates cofilin gene expression in chondrocytes. *Biochemical and Biophysical Research Communications* **361**, 329–334 (2007). doi:10.1016/j.bbrc.2007.06.185.
34. Chan, P. S. et al. Gene expression profile of mechanically impacted bovine articular cartilage explants. *Journal of Orthopaedic Research* **23**, 1146–1151 (2005). doi:10.1016/j.orthres.2005.01.016.
35. Dehne, T. et al. Gene expression profiling of primary human articular chondrocytes in high-density micromasses reveals patterns of recovery, maintenance, re- and dedifferentiation. *Gene* **462**, 8–17 (2010). doi:10.1016/j.gene.2010.04.006.
36. Mousoulis, C., Xu, X., Reiter, D. A. and Neu, C. P. Single cell spectroscopy: Noninvasive measures of small-scale structure and function. *Methods* **64**, 119–128 (2013).
37. Buckley, M. R., Bergou, A. J., Fouchard, J., Bonassar, L. J. and Cohen, I. High-resolution spatial mapping of shear properties in cartilage. *Journal of Biomechanics* **43**, 796–800 (2010). doi:10.1016/j.jbiomech.2009.10.012.
38. Choi, J. B. et al. Zonal changes in the three-dimensional morphology of the chondron under compression: The relationship among cellular, pericellular, and extracellular deformation in articular cartilage. *Journal of Biomechanics* **40**, 2596–2603 (2007).
39. Jao, C. Y. and Salic, A. Exploring RNA transcription and turnover in vivo by using click chemistry. *Proceedings of the National Academy of Sciences of the United States of America* **105**, 15,779–15,784 (2008).
40. Paredes, E. and Das, S. R. Click chemistry for rapid labeling and ligation of RNA. *Chembiochemistry* **12**, 125–131 (2011).
41. Politz, J. C., Tuft, R. A., Pederson, T. and Singer, R. H. Movement of nuclear poly(A) RNA throughout the interchromatin space in living cells. *Current Biology* **9**, 285–291 (1999). doi:10.1016/s0960-9822(99)80136-5.

Section XI

Cellular and Molecular Imaging Methods

Channels

34

Fluorescence Methods for Monitoring Mechanosensitive Channels

Alexander Macmillan, Charles G. Cranfield, and Boris Martinac

CONTENTS

Over the last 10 to 15 years, fluorescence methods have become frequently used for studies of the conformational changes of various membrane proteins including ion channels that function as gated membrane pores in numerous physiological processes essential for the normal function of biological cells. In combination with molecular biological methods that allow introduction of fluorescent labeling *in vivo* and *in vitro*, and functional electrophysiological voltage- and patch-clamp recording techniques, fluorescence spectroscopy methods, such as *Förster resonance energy transfer* (FRET), have made ion channel proteins accessible to studies of protein structure and function relationship. Here we focus on the use of these methods in studies of structural dynamics of bacterial mechanosensitive channels, which to date are among the best studied membrane channels functioning in mechanosensory transduction in living cells.

34.1 Introduction

Human perception depends on the senses of vision, hearing, touch, smell, and taste without which the outside world would apparently cease to exist. Sensations of touch and hearing are manifestations of mechanical stimuli resulting from a mechanical contact and air pressure acting on touch receptors and hair cells of the inner ear, respectively. In bacteria, mechanical stimuli originate from osmotic forces acting on the microbes.

Among cellular receptors that play a prominent role in converting mechanical stimuli into a cascade of intracellular signals contributing to changes in electrical, chemical, or osmotic activity in biological cells are mechanosensitive (MS) ion channels, which, next to voltage- and ligand-gated ion channels, form an additional class of membrane channels. Mechanical force acting on the channels in the form of membrane tension gates the MS channels, open and closed, regulating the trafficking of ions and organic solutes across the cellular membrane and thus affecting the cell activity.

Since their discovery in 1984 (Guharay and Sachs 1984; Brehm et al. 1984), MS channels have been documented in living cells belonging to all of the three major domains of life: Archaea, Eucarya, and Bacteria (Martinac 2004). A significant step forward in studies of MS channels was the cloning (Sukharev et al. 1994; Levina et al. 1999) and solving of the crystal structure (Chang et al. 1998; Bass et al. 2002) of MscL and MscS, the bacterial MS channels of large and small conductance, respectively. This has greatly contributed to a better understanding of the basic physical principles underlying mechanosensory transduction in living cells. Over the last 25 years, a multidisciplinary approach that has been employed in studies of these channels has brought an unprecedented wealth of information and new knowledge about their structure and function. Consequently, MS channels are, to date, among the best characterized biological force-sensing systems coupling molecular dynamics to cellular mechanics. This chapter focuses on fluorescence spectroscopy methods in the study of the structure and function of MS channels.

34.2 Preparative Methods for Fluorescence Spectroscopy of MS Channels

34.2.1 Purification and Liposome Reconstitution of MS Channel Proteins

Much of what we know about the basic physical principles underlying the conversion of mechanical force into membrane tension causing conformational changes of MS channels comes from studies of the bacterial *mechanosensitive channel of large conductance* (MscL) and the *mechanososensitive channel of small conductance* (MscS) reconstituted into liposomes. Isolating MS channels in liposomes removes any ambiguity that would arise from the presence of other membrane-bound proteins and simplifies the fluorescence labeling of proteins. The production of both MscL and MscS channels for liposome reconstitution requires protein expression in *Escherichia coli* as fusion proteins with a selection tag, allowing a simple and rapid isolation of these proteins from bacterial membrane preparations using commonly available commercial purification supplies (Martinac et al. 2010). Two affinity expression tag systems are commonly used. A gene construct used to produce a fusion with six histidine amino acids (6×His-tag) attached to either the amino or carboxyl terminal of the protein relies on charged affinity allowing the fusion protein to bind to nickel, or cobalt, chelated to sepharose beads. Purification of MS proteins is achieved by washing the beads with a buffer to remove nonspecifically bound impurities and final elution of MS proteins with a solution containing high concentration of imidazole as a charged competitor. Affinity histidine tags can either be noncleavable or cleavable with thrombin. Alternatively, enzymatic affinity based on a fusion product with glutathione-S-transferase (GST) may also be used to achieve higher purity at lower yields. This involves specific tight binding to sepharose beads coupled to the GST target, glutathione. The MS protein, which is studied, is cleaved off GST with thrombin via a target site that has been inserted between the two parts of the fusion products: the channel and the GST. It should be noted, however, that thrombin cleavage will leave a portion of its recognition sequence on the finally purified protein, which, in the case of MscL and MscS, does not affect the channel function.

Different GST or 6×His-tagged constructs of MscL and MscS have been used for liposome reconstitution. For structural studies using Förster resonance energy transfer (FRET) spectroscopy (see Section 34.3.1), usually 6×His-tagged proteins are sufficient. The amount of protein required for reconstitution depends on the protein-to-lipid ratio (w/w for patch clamp experiments or mol/mol for structural FRET experiments), the reconstitution method, and the type of experiment. For MscL reconstitution, we have also found that the lipid-to-protein ratio depends on which construct is being used. Particularly for MscS, we have found that the lipids commonly used for electrophysiological measurements, such as 1,2-dioleoyl-sn-glycero-3-phosphatidylcholine or *E. coli* polar lipids, cannot be used for MscS structural determination through spectroscopy, since they promote two-dimensional aggregation, regardless of the expression and purification conditions.

The most frequently used method for liposome reconstitution is the so-called dehydration/rehydration (D/R) method (Delcour et al. 1989). It should be noted that the number of active channels per liposome patch varies with the protein-to-lipid ratio and is higher when the GST-MscL constructs are reconstituted compared to 6×His-tagged MscL constructs. Another reconstitution method that has proven particularly useful for reconstitution of MscS is a sucrose reconstitution method (Battle et al. 2009). When using this method less protein is required and a higher percentage of liposome patches with active channels is available for patch clamp experiments. On average, 5 μg of any of the MS channels described here is sufficient for several liposome preparations using the D/R method for patch clamping and fluorescence imaging.

34.2.2 Fluorescent Labeling of MS Channels

Careful deliberation needs to be undertaken when setting out to fluorescently label proteins such as stretch activated channels (Martinac and Cranfield 2012). Ideally, the best fluorescent label is one that has a high fluorescence quantum yield, that has an emission wavelength far removed from any endogenous fluorescence (i.e., autofluorescence), that is attached to 100% of the desired stretch activated protein target, that is bound uniquely to the protein target and with no nonspecific binding to other proteins, and whose labeling does not affect the structure or function of the target protein. Given that there is no such "ideal" fluorescent label, compromises in fluorescent labeling will have to be made. Here we discuss the differing fluorescent labeling options available and, depending on the questions being asked, which labeling techniques are best suited for stretch activated channel experiments.

34.2.2.1 Fluorescent Proteins

Fluorescent protein labeling of stretch activated channels using fluorescent proteins such as *green fluorescent protein* (GFP) has a long history. MscL–GFP conjugates have been used by a number of researchers to visualize the location of the protein expressed in bacteria (Norman et al. 2005; Wahome et al. 2009). Likewise, MscS analogues in plant cells were identified by conjugating GFPs (Haswell et al. 2008). The advantage of fluorescent protein labeling is that, in most cases, it ensures 100% binding of the fluorophore to its target stretch activated channel. In fact, as many stretch activated channels are homomeric in nature, there is actually a greater than 100% binding efficiency. The stretch activated channel MscL, for instance, is a homopentameric channel, so one would expect 500% fluorescent protein binding. This can be a very useful property if the oligomerization of a homomeric stretch activated protein is unknown, as single-molecule photobleaching steps can then be used to determine just how many monomers are needed to create a functioning membrane-bound protein. This has been accomplished recently, with Coste et al. (2012) demonstrating that the MS channel *MmPiezo1* homo-oligomerizes to form at least a tetramer. The major drawback regarding fluorescent protein labeling is in regard to maintaining the function and the structure of stretch activated channels. As GFP and other fluorescent proteins are large proteins themselves, the position of their incorporation into MS channels is limited, usually to the C or N terminal regions, in order to maintain channel structure and function. This makes them typically unsuitable for *intraprotein* FRET imaging.

34.2.2.2 Organic Fluorophore Labeling

Instead of labeling proteins *in vivo*, as with fluorescent proteins mentioned above, organic dyes can be used to label proteins *in situ* or, more commonly, *in vitro*. Unlike fluorescent proteins, the labeling efficiency will rarely, if ever, be 100%, but the quantum yields for selected fluorophores are typically much better with a greater selection of wavelengths available. In terms of labeling stretch activated channels, the most common organic fluorescent dyes used have been the Alexa Fluor® probes (Molecular Probes, Invitrogen; Haugland et al. 2005). The popularity of these fluorescent dyes is due to their high quantum yield and good photostability. MscL proteins were tagged with *Alexa Fluor 488* by Grage et al. (2011) to investigate clustering of the protein in liposomes. They bound the fluorophore to their protein using a dye with a maleimide group attached, which can then be covalently attached to the thiol group of any available cysteine amino acid within proteins. As the MscL has no native cysteine residues, mutated forms of the protein with prepositioned cysteine groups need to be made. The placement of the cysteine residues was manipulated by Corry et al. (2010) and Wang et al. (2014) to determine the size of the pore in the open state of MscL. Using Alexa Fluor 488 in conjunction with Alexa Fluor 568, they used FRET measurements to determine the distance between their various mutated cysteine groups in the protein. FRET measurements using *fluorescence lifetime imaging* (FLIM) between these same two Alexa fluorophores were undertaken by Nomura et al. (2012) to demonstrate how MscL clusters not only with itself in lipid membranes but also with MscS. The aggregation state of MscS mutants with themselves was also monitored over time using the FRET pair of *fluorescein* and *tetramethylrhodamine* by Vásquez et al. (2007).

As well as the thiol groups of cysteines, the NH_2 groups of lysine or arginine residues can also be used to attach fluorophores via acylation chemistries. This method is not generally preferred when investigating the structure and function of proteins, however, because the abundance of lysine residues in most proteins makes specific residue targeting problematic (Haugland et al. 2005).

34.2.2.3 Biarsenical Dyes

Biarsenical dyes *FlAsH* and *ReAsH* attach to tetracysteine motifs mutated into proteins in the form of CC–XX–CC, where XX is any amino acid other than cysteine (Adams et al. 2002). The advantage of these fluorophores is that they will traverse cell membranes to form attachments uniquely to the mutated protein of choice, and they are essentially nonfluorescent until conjugated to the tetracysteine motif. Romantsov et al. (2008, 2010) used FlAsH to look at the localization of their osmoregulated transporter *ProP* in *E. coli* and demonstrated that the response of this transporter is dependent on *cardiolipin* lipids present in the bacterial membrane. Using the same labeling technique, they have also shown that, similar to ProP, MscS concentrated at the poles depending on the cellular cardiolipin content (Romantsov et al. 2010).

34.2.2.4 Lanthanide Metal Labeling

Lanthanide group metals have also been used to label MS channels (Becker et al. 2004). Terbium or europium ions have the advantage over other fluorophores in that they exhibit a longer fluorescence lifetime than other fluorophores; however, as lanthanides have a poor extinction coefficient, they are typically linked to chelators to improve their light absorption profile (Selvin and Hearst 1994). This same chelate, though, can then be conjugated with maleimides to make them thiol reactive and so enable the lanthanide chelates to selectively bind to select residues in proteins and peptides (Chen and Selvin 1999).

34.3 Fluorescence Spectroscopy Techniques

34.3.1 Förster Resonance Energy Transfer

The recent development of superresolution microscopy techniques allows us to readily achieve resolutions of tens of nanometers; however, FRET allows us to look at interactions between 1 and 10 nm, which exceeds the resolution afforded by advanced microscopy techniques. Biological processes are often driven by the interaction and conformational changes of proteins (Shih et al. 2000), so it is imperative that these processes can be monitored and understood, and FRET provides a very useful tool (Hink et al. 2002; Sekar and Periasamy 2003; Vogel et al. 2006). The popularity of the technique is evident by the number of papers and reviews (Jares-Erijman and Jovin 2003) covering the topic, of which there are too many too list here.

34.3.2 FRET of MS Channels

FRET has been utilized to study conformational changes involved during the gating of MscL channels (Figure 34.1; Corry et al. 2005, 2010) using acceptor photobleaching. In this study, the protein diameter was found to increase by 16 Å. The measurement was performed on an ensemble of Alexa Fluor-labeled cysteine mutants reconstituted into artificial liposomes before and after acceptor photobleaching. The FRET technique can also be applied at the single molecule level (Roy et al. 2008), which allows us to look at individual processes rather than the ensemble "average." This is of particular benefit in measuring conformational changes due to the intrinsic inhomogeneity of the process. Single-molecule FRET (smFRET) was demonstrated by Harms et al. (2003) in measuring the conformational changes of the gramicidin ion channel, which also exhibits mechanosensitivity dependent on the bilayer thickness (Martinac and Hamill 2002). The conformational changes were determined in a lipid bilayer formed at a patch-clamp micropipette tip under a buffer solution (Harms et al. 2003). Recently, the pore diameter of MscL in the open state was measured by smFRET using total internal reflection fluorescence microscopy (Wang et al. 2014). The determined diameter of the open pore of 2.8 nm is in good agreement with the results of the electron paramagnetic resonance (Perozo et al. 2002) and ensemble FRET studies of MscL (Corry et al. 2010). FLIM provides a very sensitive method for measuring FRET through measurement of the donor lifetime and is free of some

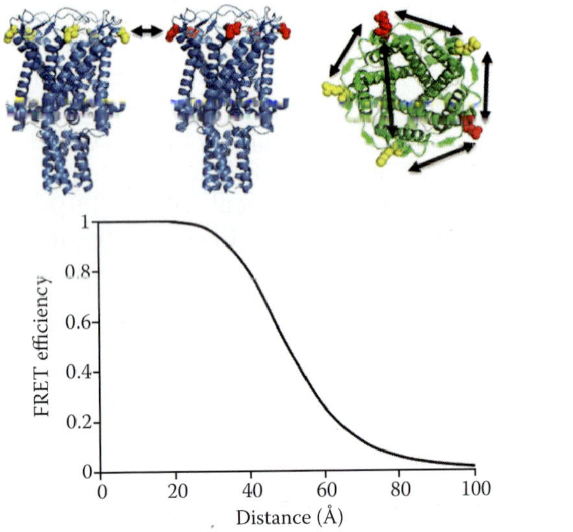

FIGURE 34.1
Schematic showing the use of FRET to study MS channels. The top right figure demonstrates how interprotein FRET can be used to measure clustering of MS channels *in vitro* (top left figure). In this case, two separate populations of MscL protein can be labeled, and FRET can be expected to occur when two proteins are within ~10 nm from each other. The top right figure is an example of intraprotein FRET. In this example, a specific loci of each protein monomer can be labeled with a corresponding FRET pair and the distance between each loci evaluated (see Corry et al. 2010). The bottom plot demonstrates the relationship between *distance* and *FRET efficiency*, which is dependent on the *Förster distance* (R_0) of the fluorophores used (for a good description of this process, see Corry et al. 2013).

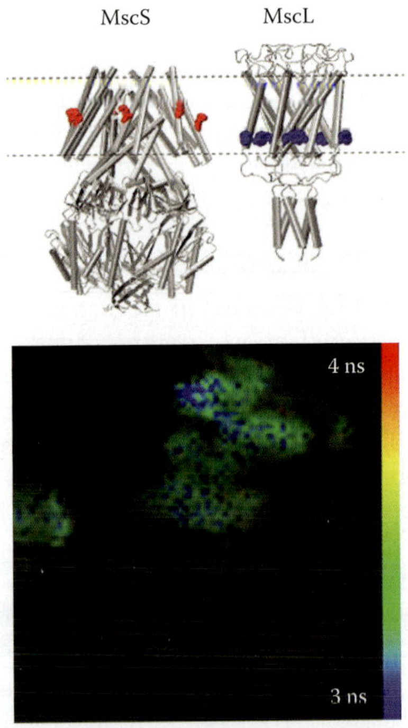

FIGURE 34.2
Investigation of MS channel clustering using FLIM. Regions colored blue are patches exhibiting a shorter fluorescence lifetime as a result of quenching of the donor fluorophore due to FRET. (From Nomura, T. et al., *Proceedings of the National Academy of Sciences* 109 (22):8770–8775, 2012. With permission.)

of the drawbacks that affect steady-state measurement. The application of the technique was demonstrated in measuring the clustering of MscL and MscS (Figure 34.2; Nomura et al. 2012).

34.3.3 Patch Fluorimetry

A recent report by Nomura et al. (2012) demonstrated that the actual membrane tension required for MS channel opening patch clamp experiments can be determined quite readily using fluorescently tagged lipids. When pressure is applied, the membrane tension is a function of the applied pressure and the curvature radius of the patch membrane. Using Laplace' law,

$$T = P\frac{r}{2} \tag{34.1}$$

where T is the membrane tension, P is the membrane pressure, and r is the radius of the membrane curvature of the threshold tension for channel opening that can be readily calculated, so long as the patch membrane is visible under the microscope (Sachs 2010; Sokabe et al. 1991). Using patch pipettes that had been specially curved to sit parallel to glass coverslips, Nomura et al. reconstituted MscS and MscL in lipids containing 0.1% *rhodamine* conjugated to *phosphatidylethanolamine*. In this way, they were able to visualize the patch membrane using the confocal microscope and calculate the radius of curvature when pressure was applied (Figure 34.3). Membrane tension calculations using this method are considered to be a far more accurate indicator of mechanosensation thresholds than pressure measurements alone.

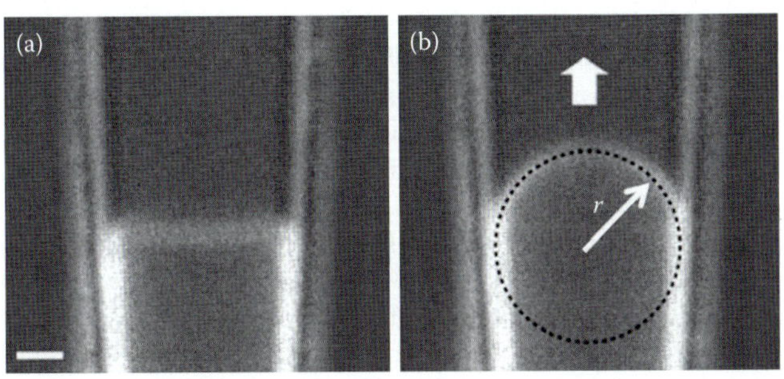

FIGURE 34.3
Patch fluorimetry. Fluorescently labeled lipids enable a calculation of membrane tension in the patch pipette using Laplace's law (Equation 34.1). (a) A glass micropipette containing a fluorescently labelled liposome patch with no suction (negative pressure) applied to the pipette. (b) The same liposome patch shown in (a) after suction was applied (white arrow) to the patch pipette. (From Nomura, T. et al., *Proceedings of the National Academy of Sciences* 109 (22):8770–8775, 2012. With permission.)

34.4 Conclusions

The examples of MS channels MscL and MscS we described in this chapter demonstrate the requirements for the use of FRET for the study of the structural dynamics of ion channels. Besides the channel-forming peptide gramicidin, MscL and MscS are, to date, the only MS channels examined using this fluorescence technique. The FRET technique enables detailed structural information to be obtained for these MS channels including the major advancement in determining the diameter of the open channel pore of MscL. Although the use of FRET requires cysteine scanning mutagenesis for fluorophore labeling and careful analysis of the orientations, geometries, and number of fluorescent probes, this experimental approach provides an excellent tool for studying the structures of membrane proteins.

There are a variety of advanced FRET-based fluorescence microscopy techniques that can be employed for studies of molecular dynamics of ion channels, which include *pulsed interleaved excitation* and *alternating laser excitation*. They can provide detailed and rapid information on FRET stoichiometry (Lee et al. 2005, 2007; Müller et al. 2005). Recently, the use of FLIM-FRET analysis has also been greatly simplified in biological systems by the development of *phasor analysis*, which provides a model-free interpretation of the data (Hinde et al. 2013). Furthermore, the light transparency of the Casper strain of zebra fish (White et al. 2008) offers new possibilities for FRET studies of MS ion channels. Although no MS channels have currently been identified in the zebra fish at the molecular level, they have been implicated in the mechanosensitivity of hair cells in the developing fish (Kindt et al. 2012). A future challenge will be to adapt various FRET techniques to study MS channels *in vivo* not only in embryos but also in the adult zebra fish.

Acknowledgments

We thank Dr. Philip Gottlieb for his comments and critical reading of the manuscript. We also thank the National Health and Medical Research Council of Australia for funding support.

References

Adams, S. R., R. E. Campbell, L. A. Gross et al. 2002. New biarsenical ligands and tetracysteine motifs for protein labeling in vitro and in vivo: Synthesis and biological applications. *Journal of the American Chemical Society* 124 (21):6063–76.

Bass, R. B., P. Strop, M. Barclay, and D. C. Rees. 2002. Crystal structure of *Escherichia coli* MscS, a voltage-modulated and mechanosensitive channel. *Science* 298 (5598):1582–7.

Battle, A. R., E. Petrov, P. Pal, and B. Martinac. 2009. Rapid and improved reconstitution of bacterial mechanosensitive ion channel proteins MscS and MscL into liposomes using a modified sucrose method. *FEBS Letters* 583 (2):407–12.

Becker, C. F. W., D. Clayton, G. Shapovalov, H. A. Lester, and G. G. Kochendoerfer. 2004. On-resin assembly of a linkerless lanthanide(III)-based luminescence label and its application to the total synthesis of site-specifically labeled mechanosensitive channels. *Bioconjugate Chemistry* 15 (5):1118–24.

Brehm, P., R. Kullberg, and F. Moody-Corbett. 1984. Properties of non-junctional acetylcholine receptor channels on innervated muscle of *Xenopus laevis*. *The Journal of Physiology* 350 (1):631–48.

Chang, G., R. H. Spencer, A. T. Lee, M. T. Barclay, and D. C. Rees. 1998. Structure of the MscL homolog from *Mycobacterium tuberculosis*: A gated mechanosensitive ion channel. *Science* 282 (5397):2220.

Chen, J., and P. R. Selvin. 1999. Thiol-reactive luminescent chelates of terbium and europium. *Bioconjugate Chemistry* 10 (2):311–5.

Corry, B., A. C. Hurst, P. Pal, T. Nomura, P. Rigby, and B. Martinac. 2010. An improved open-channel structure of MscL determined from FRET confocal microscopy and simulation. *The Journal of General Physiology* 136 (4):483.

Corry, B., C. G. Cranfield, and B. Martinac. 2013. Structure and physiological role of ion channels studied by fluorescence spectroscopy. In *Encyclopedia of Analytical Chemistry*, R. A. Meyers (ed.). John Wiley & Sons, Ltd., West Sussex, UK.

Coste, B., B. Xiao, J. S. Santos et al. 2012. Piezo proteins are pore-forming subunits of mechanically activated channels. *Nature* 483 (7388):176–81.

Delcour, A. H., B. Martinac, J. Adler, and C. Kung. 1989. Modified reconstitution method used in patch-clamp studies of *Escherichia coli* ion channels. *Biophysical Journal* 56 (3):631–6.

Grage, S. L., A. M. Keleshian, T. Turdzeladze et al. 2011. Bilayer-mediated clustering and functional interaction of MscL channels. *Biophysical Journal* 100 (5):1252–1260. (doi:10. 1016/j.bpj. 2011.01.023).

Guharay, F., and F. Sachs. 1984. Stretch-activated single ion channel currents in tissue-cultured embryonic chick skeletal muscle. *The Journal of Physiology* 352 (1):685–701.

Harms, G. S., G. Orr, M. Montal, B. D. Thrall, S. D. Colson, and H. P. Lu. 2003. Probing conformational changes of gramicidin ion channels by single-molecule patch-clamp fluorescence microscopy. *Biophysical Journal* 85:1826–38.

Haswell, E. S., R. Peyronnet, H. Barbier-Brygoo, E. M. Meyerowitz, and J.-M. Frachisse. 2008. Two MscS homologs provide mechanosensitive channel activities in the Arabidopsis root. *Current Biology* 18 (10):730–4.

Haugland, R. P., M. T. Z. Spence, and I. D. Johnson. 2005. In *The Handbook: A Guide to Fluorescent Probes and Labeling Technologies*. Molecular Probes, Eugene, OR.

Hinde, E., M. A. Digman, K. M. Hahn, and E. Gratton. 2013. Millisecond spatiotemporal dynamics of FRET biosensors by the pair correlation function and the phasor approach to FLIM. *Proceedings of the National Academy of Sciences of the United States of America* 110:135–40.

Hink, M. A., T. Bisseling, and A. Visser. 2002. Imaging protein-protein interactions in living cells. *Plant Molecular Biology* 50 (6):871–83.

Jares-Erijman, E. A., and T. M. Jovin. 2003. FRET imaging. *Nature Biotechnology* 21 (11):1387–95.

Kindt, K. S., G. Finch, and T. Nicolson. 2012. Kinocilia mediate mechanosensitivity in developing zebrafish hair cells. *Developmental Cell* 23 (2):329–41.

Lee, N. K., A. N. Kapanidis, H. R. Koh et al. 2007. Three-color alternating-laser excitation of single molecules: Monitoring multiple interactions and distances. *Biophysical Journal* 92:303–12.

Lee, N. K., A. N. Kapanidis, Y. Wang et al. 2005. Accurate FRET measurements within single diffusing biomolecules using alternating-laser excitation. *Biophysical Journal* 88:2939–53.

Levina, N., S. Totemeyer, N. R. Stokes, P. Louis, M. A. Jones, and I. R. Booth. 1999. Protection of *Escherichia coli* cells against extreme turgor by activation of MscS and MscL mechanosensitive channels: Identification of genes required for MscS activity. *European Molecular Biology Organization Journal* 18 (7):1730–7.

Martinac, B. 2004. Mechanosensitive ion channels: Molecules of mechanotransduction. *Journal of Cell Science* 117 (12):2449–60.

Martinac, B., and C. Cranfield. 2012. Shining a light on the structural dynamics of ion channels using Förster resonance energy transfer (FRET). *IPSI BgD Transactions on Advanced Research* 8 (1):19–25.

Martinac, B., and O. P. Hamill. 2002. Gramicidin A channels switch between stretch activation and stretch inactivation depending on bilayer thickness. *Proceedings of the National Academy of Sciences of the United States of America* 99 (7). 4308–12.

Martinac, B., P. R. Rohde, A. R. Battle et al. 2010. Studying mechanosensitive ion channels using liposomes. In *Methods in Molecular Biology*, V. Weissig (ed.). Humana Press, Springer, New York.

Müller, B. K., E. Zaychikov, C. Bräuchle, and D. C. Lamb. 2005. Pulsed interleaved excitation. *Biophysical Journal* 89:3508–22.

Nomura, T., C. G. Cranfield, E. Deplazes et al. 2012. Differential effects of lipids and lyso-lipids on the mechanosensitivity of the mechanosensitive channels MscL and MscS. *Proceedings of the National Academy of Sciences of the United States of America* 109 (22):8770–5.

Norman, C., Z.-W. Liu, P. Rigby, A. Raso, Y. Petrov, and B. Martinac. 2005. Visualisation of the mechanosensitive channel of large conductance in bacteria using confocal microscopy. *European Biophysics Journal* 34 (5):396–402.

Perozo, E., D. M. Cortes, P. Sompornpisut, A. Kloda, and B. Martinac. 2002. Open channel structure of MscL and the gating mechanism of mechanosensitive channels. *Nature* 418:942–8.

Romantsov, T., A. R. Battle, J. L. Hendel, B. Martinac, and J. M. Wood. 2010. Protein localization in *Escherichia coli* cells: Comparison of the cytoplasmic membrane proteins ProP, LacY, ProW, AqpZ, MscS, and MscL. *Journal of Bacteriology* 192 (4):912.

Romantsov, T., L. Stalker, D. E. Culham, and J. M. Wood. 2008. Cardiolipin controls the osmotic stress response and the subcellular location of Transporter ProP in *Escherichia coli*. *Journal of Biological Chemistry* 283 (18):12314–23.

Roy, R., S. Hohng, and T. Ha. 2008. A practical guide to single molecule FRET. *Nature Methods* 5 (6):507–16.

Sachs, F. 2010. Stretch-activated ion channels: What are they? *Physiology* 25 (1):50–6.

Sekar, R. B., and A. Periasamy. 2003. Fluorescence resonance energy transfer (FRET) microscopy imaging of live cell protein localizations. *Journal of Cell Biology* 160 (5):629–33.

Selvin, P. R., and J. E. Hearst. 1994. Luminescence energy transfer using a terbium chelate: Improvements on fluorescence energy transfer. *Proceedings of the National Academy of Sciences of the United States of America* 91 (21):10024–8.

Shih, W. M., Z. Gryczynski, J. R. Lakowicz, and J. A. Spudich. 2000. A FRET-based sensor reveals large ATP hydrolysis-induced conformational changes and three distinct states of the molecular motor myosin. *Cell* 102 (5):683–94.

Sokabe, M., F. Sachs, and Z. Q. Jing. 1991. Quantitative video microscopy of patch clamped membranes stress, strain, capacitance, and stretch channel activation. *Biophysical Journal* 59 (3):722–8.

Sukharev, S. I., P. Blount, B. Martinac, F. R. Blattner, and C. Kung. 1994. A large-conductance mechanosensitive channel in *E. coli* encoded by mscL alone. *Nature* 368 (6468):265–8.

Vásquez, V., D. M. Cortes, H. Furukawa, and E. Perozo. 2007. An optimized purification and reconstitution method for the MscS channel: Strategies for spectroscopical analysis. *Biochemistry* 46 (23):6766–73.

Vogel, S. S., C. Thaler, and S. V. Koushik. 2006. Fanciful FRET. *Science's STKE: Signal Transduction* 2006 331:re2, doi: 10.1126/stke.3312006re2.

Wahome, P. G., A. E. Cowan, B. Setlow, and P. Setlow. 2009. Levels and localization of mechanosensitive channel proteins in *Bacillus subtilis*. *Archives of Microbiology* 191 (5):403–14.

Wang, W., Y. Liu, H. A. DeBerg et al. 2014. Single molecule FRET reveals pore size and opening mechanism of MscL. *eLife* 3:e01834.

White, R. M., A. Sessa, C. Burke et al. 2008. Transparent adult zebrafish as a tool for in vivo transplantation analysis. *Cell Stem Cell* 2 (2):183–9.

35

Channel Activation and Mechanotransduction

Henry J. Donahue, Randall L. Duncan, and Damian C. Genetos

CONTENTS

Cells are exquisitely sensitive and adaptive to changes in environment. Within the mammalian skeleton, routine activity creates an array of biophysical forces on bone cells. Cells that perceive and respond to these forces initiate intracellular signaling cascades designed to alter cell metabolism or phenotype to maintain or alter bone mass and architecture. Perceiving an altered biophysical environment and altering cell function is termed mechanotransduction. Understanding the mechanotransduction processes that enable cells to perceive an altered extracellular environment not only is critical for a systems-level biologic approach but may also provide the foundation for treating diseases of bone loss, such as senile osteoporosis. Osteocytes may play a crucial role in mechanotransduction because of their localization and plenitude within bone matrix; further, dendritic processes enable intercellular connectivity to neighboring osteocytes, osteoblasts, osteoclasts, and stem cells present within the marrow and periosteum. In vitro models for mechanical loading have begun to identify cellular responses to biophysical forces like fluid flow; chemical antagonists, siRNA, or cells from knockout animals have enabled investigators to delineate the contribution of specific proteins or signaling cascades to a particular response, increases in cytosolic calcium levels, or induction of a particular candidate gene. One of the earliest measurable responses to a biophysical force is activation of an ion channel in the plasma membrane. Within this chapter, we review evidence for ion channel activation within bone, including common optical imaging methods, the role of these ion channels in skeletal mechanotransduction in vitro, and in vivo evidence either corroborating or refuting in vitro findings.

35.1 Introduction

Mechanotransduction is the integrative process whereby an extracellular biophysical stimulus is perceived by cells and translated into an intracellular electrical or biochemical response. It is a process critical in development and homeostasis, such that aberrant mechanosensing or mechanoresponsiveness produces pathological phenotypes such as cancer, atherosclerosis, and osteoporosis.[1–4] Providing a physiological basis for senses like touch, hearing, and proprioception in electrically excitable cells,[5] mechanotransduction is now recognized to occur in cells traditionally considered nonelectrically excitable. Within the skeleton, mechanotransduction is crucial for increased bone formation to adapt to increased mechanical forces in order to prevent fracture. Conversely, when insufficient mechanical forces are applied during limb immobilization or microgravity, bone is lost. Similarly, endothelial cells that line the inner surfaces of blood vessels are constantly exposed to hemodynamic forces and, in response to altered shear stress, release vasodilators to modulate vascular smooth muscle tone.[6,7]

Diverse biophysical stimuli are constantly imparted upon cells. For example, during mechanical loading associated with locomotion, cells within the skeleton are exposed to substrate strain from the bending of bones, shear stress generated by inhomogeneous interstitial pressure, and electric fields generated by streaming potentials or piezoelectric effects. Dynamic and static loading of articular cartilage causes tension, compression, and shear forces, which are translated to resident chondrocytes in the forms of strain, osmotic changes, shear stress, and streaming potentials.[8] Cells within a given tissue may be exposed to stimuli that other cell types may not experience or respond to (e.g., a renal epithelial cell is exposed to changes in osmolality that skeletal muscle will likely not experience), indicating that these diverse biophysical stimuli are perceived by the cell through a diverse array of protein-, carbohydrate-, and lipid-based mechanosensors. For example, physical stress causes strain upon the extracellular matrix substrate that is translated through integrin engagement to activate focal adhesion signaling. The same physical stress can cause tension within the lipid bilayer of the plasma membrane to alter the open probability of mechanosensitive ion channels (MSCs). Thus, a plethora of putative mechanosensory apparatuses has been implicated in translating a biophysical signal into an intracellular response. This chapter focuses on mechanosensory systems with clearly defined ion or chemical channel activity and will not address mechanotransduction machinery such as integrins and focal adhesions, the glycocalyx, lipid bilayer, or metabotropic receptors that are addressed in other reviews.[9–16] Further, while mechanotransduction occurs in almost every cell, we will focus primarily on ion channels activated within osteoblasts and osteocytes.

35.2 Ion Channels and Intracellular Calcium Signaling

Typically, ion channel activation in response to a biophysical signal causes an increase in cytosolic calcium (Ca_i^{2+}) that is dependent on both entry from the extracellular environment and release from intracellular stores.[17] This increase in Ca_i^{2+} acts as a second messenger to allosterically activate or inhibit proteins such as transcription factors and kinases[18] or to induce release of factors from the cell to stimulate surrounding cells to amplify and further propagate the initial signal. Fine-tuning of the Ca_i^{2+} amplitude, duration, and frequency by intracellular calcium-binding proteins encodes specific cellular responses.[19,20] Within osteoblasts, Ca_i^{2+} transients play crucial roles in the release of paracrine factors like ATP and PGE_2,[21] kinase activation,[22] transcription factor translocation,[23] cytoskeletal organization,[24] proliferation,[25] and differentiation.[26,27] Whether the Ca_i^{2+} transient involves extracellular entry or intracellular release may determine the cellular response, as inhibition of calcium entry through a membrane-bound ion channel is implicated in PGE_2[28] or TGF-β release,[14] whereas G protein-induced intracellular calcium release is required for changes in gene expression.[23] Further, there is evidence that the sensitivity and dynamics of Ca_i^{2+} transient in response to mechanical stimulation are a function of the degree of osteogenic differentiation: osteocytes are more sensitive to fluid flow than osteoblasts, especially under low-level mechanical stimulation.[29] While cellular responses to biophysical signals do not have to involve changes in Ca_i^{2+}, this is a critical aspect to consider in mechanotransduction especially with regard to the role of ion channels.

35.3 Mechanosensitive Ion Channels

MSCs are directly responsive to changes in membrane tension, either increasing or decreasing their open channel probability. Altered open channel probability is a response to tension transmitted to the channel by surrounding lipids within the plasmalemma or cytoskeletal elements with plasmalemmal insertion[30] or direct physical interaction with the channel.[31]

MSCs are expressed in nearly all organisms. A subset of channels with mechanosensitive gating have been cloned,[30] including the MscL from bacteria,[32] the NMDA channel,[33] and a Ca^{2+}-activated K^+ channel in smooth muscles,[34] among others.[30] A mechanosensitive cation-selective channel (MSCC) has been demonstrated in osteoblasts and osteocytes,[35] as well as a high-conductance K^+-selective channel.[36] The kinetics of the MSCC are altered by chronic, intermittent strain,[37] which suggests that the MSCC can be sensitized for greater activation by additional mechanical loading. MSCC inhibition with trivalent lanthanides like Gd^{3+} modulates numerous early osteoblastic and osteocytic responses to biophysical forces,[38] including Ca_i^{2+} transient,[17,39] NO release,[38] and changes in gene expression.[38] MSCC channel activity is required for induction of VEGF and M-CSF, which are crucial initiators of angiogenesis and osteoclastogenesis, respectively.[40]

MSC channel activity may involve the actin cytoskeleton directly opening the MSC channel pore.[30] Proper cytoskeletal architecture has been shown to be a critical determinant of a cell's capacity to respond to biophysical forces (reviewed in Ref. 41). Parathyroid hormone (PTH), which exerts an anabolic effect on the skeleton when administered intermittently, alters osteoblast and osteocyte morphology through disruption of F-actin stress fibers into monomeric G-actin.[42–44] Thus, cytoskeletal-driven alterations in MSCC activity may be achieved by PTH. Zhang et al.[45] demonstrated that the MSCC channel open probability was significantly increased under pipette pressure with PTH pretreatment, and that the synergistic effect of pressure and PTH on the channel open probability was eliminated when the cytoskeleton was stabilized with phalloidin. This observation extended to Ca_i^{2+} transients in response to shear stress: cotreatment with 50 nM bPTH(1-34) significantly increased the percentage of cells responding to unidirectional 10 dyn/cm^2 shear stress (62% control vs. 82% +bPTH) as well as per-cell amplitude increase (185% control vs. 295% +bPTH). However, inhibiting actin reorganization with phalloidin reduced both the percentage and the amplitude of cells responding to flow. These and other data suggest the structural control of MSCC channel activity by the cell, not by phosphorylation changes induced by PTH activation of second messengers. In turn, this suggests a mechanism whereby PTH can prime osteoblasts to respond to mechanical stimulation through alterations in cellular structure (MSCC) and ion channel phosphorylation (L-type VSCC).[45]

35.4 Voltage-Sensitive Channels

The gating of voltage-sensitive channels depends upon the depolarization of the cell membrane. They can also be activated indirectly in response to mechanical loading, likely through membrane depolarization caused by a mechanosensitive channel or a ligand-activated ion channel like a purinoceptor (described in Section 35.5) that sufficiently depolarizes the cell to open voltage-activated channels.[46] Two types of voltage-sensitive Ca^{2+} channels (VSCCs) are present in bone.[47] Long-lasting (L-type) VSCCs are activated by high-voltage depolarizations, are slow to inactivate, and are sensitive to dihydropyridines. These channels are expressed in preosteoblasts, osteoblasts,[48] and osteosarcoma cells[35] and appear to regulate proliferative capacity and differentiation state.[25,26,49] Transient (T-type) VSCCs are activated by lower levels of depolarization, are rapidly inactivated, and are found in preosteoblasts, osteoblasts, and osteocytes,[47] although the function of these channels has yet to be determined.

In vitro and in vivo evidence demonstrates the importance of L-type VSCCs in mechanotransduction within the skeleton. In vitro pharmacologic inhibition of L-type VSCCs decreases substrate strain-induced Ca_i^{2+} transients and osteocalcin secretion in osteoblasts,[50] and, similarly, fluid shear stress (FSS)-induced ATP release,[21] Ca_i^{2+} increase,[39] and ERK1/2 phosphorylation.[22] The effect of PTH on the skeleton may involve modulation of the L-type VSCC function as well. PTH enhances FSS-induced Ca_i^{2+} increases, but the synergistic effect of shear and PTH was eliminated when the L-type VSCC was blocked.[51] Similar findings were reported using osteocytic cells challenged by altered osmolality.[52] In vivo inhibition of the L-type VSCC with verapamil or nifedipine, which work through distinct but separate mechanisms, decreased endosteal bone formation by 50%–75% in rats or mice exposed to tibial bending or ulnar loading.[53,54] Of note, the identity of the MSCC may be revealed by the L-type VSCC[55]: heterologously expressed $Ca_{V1.2}$ (the pore-forming subunit of the L-type VSCC) in HEK293 cells increased normalized Ba^{2+} current and open channel probability in response to shear stress (10 mL/min for 30 s), positive pipette pressure (+5 to +40 mmHg), or negative pipette pressure (–20 to –40 mmHg).[56]

T-type VSCCs are expressed in preosteoblasts, osteoblasts, and osteocytes. While osteoblasts predominantly express L-type channels, the expression of the L-VSCCs is not found in osteocytes, leaving T type VSCCs as the only VSCC expressed in osteocytes.[57] Knockout of this channel in mice significantly alters the skeletal phenotype with a reduced mineral density and bone mineral content of both cortical and trabecular bone observed in long bones.[58] Preliminary data show that when the knockout mice were unloaded via tail suspension, followed by reloading for a period of time equal to the unloading, the null mice failed to regain bone mass lost during suspension, suggesting that T-type VSCC activity supports mechanically induced bone formation in vivo. This concept is supported by in vitro data showing that primary osteocytes isolated from TVSCC null animals fail to respond to fluid flow-induced shear with an increase in Ca_i^{2+} or a release of ATP compared to wild-type primary osteocytes.[59]

Osteoblasts and osteocytes also express voltage-sensitive potassium channels, whose currents are characterized by inactivating, noninactivating, or developing and are primarily outwardly conducting.[60] The large-conductance Ca^{2+}-activated K^+ channel hSlo physically associates with focal adhesion kinase ppFAK125 and increased open probability in response to hypotonic challenge.[61] Similarly, the two-pore domain K^+ channel TREK-1 is expressed in osteoblastic cells and demonstrates mechanosensitivity.[62] Further characterization of

K^+ channels under mechanical stimulation is required to delineate their role in skeletal mechanotransduction.

35.5 Purinoceptors

Purinoceptors, or purinergic receptors, are a class of membrane-spanning receptors activated by extracellular nucleotides and are functionally characterized as ionotropic P2X or metabotropic P2Y receptors. The ionotropic P2X receptors consist of seven subunits ($P2X_{1-7}R$) and generally demonstrate high permeability to Ca^{2+} and somewhat less permeability to Na^+. Heterologously expressed P2XR can generate heterotrimers or homotrimers,[63,64] and a variety of compositions impart functional and biochemical diversity with respect to current, ionic permeability, desensitization, and agonist and antagonist selectivity.[63,65]

P2X receptors are activated in a variety of tissues in response to altered mechanical loading. Within visceral organs such as the bladder, ureter, and colon, distension provokes ATP release, which then acts on P2XR on subepithelial nerve plexus, which transmits the nociceptive stimulus to pain centers within the brain.[66–68] Within renal epithelial cells, FSS induces ATP release, which then functions in an autocrine fashion through a P2XR to increase Ca_i^{2+}.[69]

In the skeleton, $P2X_7$ receptors are activated in response to mechanical loading in vitro and in vivo, and play a critical role in load-induced bone formation. In vitro activation of $P2X_7R$ in response to FSS is implicated in pore formation,[70] prostaglandin E_2 release,[70] ERK1/2 phosphorylation,[22,71] and NF-κB translocation.[9] Mice lacking $P2X_7R$ demonstrate reduced periosteal bone formation and increased trabecular bone resorption,[2] mimicking disuse-induced bone loss[11] and suggesting a role for $P2X_7R$ in skeletal mechanotransduction. Indeed, axial loading of the right ulna demonstrated a pronounced reduction in bone formation parameters (rMS/BS, rMAR, and rBFR/BS) in $P2X_7R^{-/-}$ mice compared to wild-type controls.[70] Similarly, Viecilli et al.[72] observed altered root and bone resorption in the molars of wild-type versus $P2X_7R^{-/-}$ mice in response to external stress. Collectively, these data demonstrate a crucial role for the $P2X_7R$ in skeletal mechanotransduction.

35.6 Transient Receptor Potential Channels

The transient receptor potential (TRP) ion channel superfamily includes candidate mechanosensors in a variety of organ systems. TRP channels are classified on the basis of sequence homology into one of six subfamilies: TRPC (canonical), TRPM (melastatin), TRPV (vanilloid), TRPP (polycystin), TRPA (ankyrin), and TRPML (mucolipin).[73,74] TRP channels are broadly expressed by nonexcitable and excitable cells and varying cation selectivity. For example, TRPV1–4 are demonstrate generally nonselective for cations Ca^{2+}, Na^+, and Mg^{2+}, whereas TRPV5 and TRPV6 are highly Ca^{2+}-selective.[74,75]

TRP channels are implicated in mechanotransduction because of their activation by biophysical stressors (e.g., osmolality, FSS, pH, heat, and pressure) in endothelial cells and renal epithelial cells.[73,76] In skeletal cells, TRP channels participate in the maintenance of extracellular Ca^{2+} balance as well as intracellular Ca^{2+} signaling.[74] Osteoblasts express TRPV1, TRPV4, and low levels of TRPV6, while osteoclasts express TRPV1, TRPV2, TRPV4, TRPV5, and TRPV6.[74,77] Activating mutations in TRPV4 result in spectrum of autosomal-dominant skeletal dysplasias characterized by decreased bone density and irregular endochondral ossification (reviewed in Ref. 78). TRPV4 expression increases during osteogenic differentiation, and primary osteoblasts derived from $Trpv4^{-/-}$ mice demonstrate reduced Ca_i^{2+} in response to FSS compared to wild-type animals.[77] In a parallel fashion, TRPV4 deficiency in vivo prevents immobilization-induced reductions in bone formation and increases in resorption.[79] In osteoclasts, TRPV5 is critically involved in RANKL-induced Ca_i^{2+} mobilization[80] and transcellular Ca^{2+} transport from the resorption lacuna to the basolateral membrane.[81] Recent data suggest that the TRPV4 channel is also important in osteoclastogenesis and has been shown to regulate osteoclast activity.[82–84] However, no evidence has been produced to date that suggests a mechanosensory role for either TRPV4 or TRPV5 in osteoclasts. Chondrocytes within articular cartilage express TRPV4, which enables intracellular Ca^{2+} transients and regulatory volume decrease in response to hypo-osmolality.[85]

35.7 Gap Junctions, Hemichannels, and Pannexins

Gap junctions provide a specialized form of intercellular communication by allowing the direct transfer of low molecular weight substances from the cytosol of one cell to the cytosol of another. They are composed of two connexons (or hemichannels), one from each of the paired cells, with each connexon composed of six connexins. While gap junctions are well characterized as to their role in, for example, electrochemical coupling in cardiac muscle, it is only more recently appreciated that a hemichannel can function independently of another hemichannel and, in doing so, form a channel-like conduit from the cytosol to the pericellular space.[86]

An array of studies suggests a role for hemichannels in skeletal mechanotransduction. Jiang and Cherian[87] first demonstrated that FSS upon osteocytic MLO-Y4 cells activates Cx43 hemichannels to promote uptake of extracellular dye and PGE_2 release; similar results were reported by Genetos et al.,[88] although their data indicate that ATP, not PGE_2, is released through hemichannels, as activation of purinoceptors with exogenous ATP was able to induce PGE_2 release in Cx43 siRNA transfected osteocytes. Activation of hemichannels in response to a mechanical stimulus is also observed in astrocytes,[89,90] glioma cells,[91] and chondrocytes.[92] Hemichannels are not thought to be directly mechanosensitive themselves, but instead appear to require integrin activation for hemichannel opening.[93] Due to space limitations, we cannot include a discussion of gap junction intercellular communication in bone mechanotransduction. For reviews of this area of research, see refs. 94–97 and references therein.

Studies such as those described above have generally relied upon the use of pharmacologic antagonists, such as 18α-glycyrrhetinic acid, in sparsely seeded nonjunctional cells, in order to implicate hemichannel function versus gap junctional intercellular communication in promoting dye uptake and release of ATP or PGE_2. Connexin paralogs, termed pannexins 1, 2, and 3, are also inhibited by such pharmacologic agents, thereby introducing confusion as to the role of connexins or pannexins in the controlled release of signaling molecules.[98] In contrast to connexins, pannexins do not form junctional complexes in appositional membranes. Pannexin 1 forms a directly mechanosensitive, large-conductance channel in nonjunctional membranes of cells, as heterologously expressed pannexin 1 in human oocytes expressed characteristics of a mechanosensitive channel: 40 mbar negative pressure increased channel open probability from both a totally closed channel and a channel already open but in a subconductance level.[99]

Recently, Thi et al.[98] provided evidence that functions previously attributed to connexon hemichannels might instead be performed by pannexins: extracellular dye uptake was consistent in osteoblastic, osteocytic, and primary osteoblasts lacking Cx43, but was not observed in osteoblasts from mice lacking $P2X_7R$. Similarly, using chemical antagonists that are more potent for pannexins than connexins, they observed similar reductions in ATP-induced dye uptake in both wild-type and Cx43-deficient osteoblasts, and loss of ATP-induced dye uptake in $P2X_7R$-deficient osteoblasts. Thus, their model suggests that a $P2X_7R$–Panx1 complex is responsible for dye uptake and PGE_2 release but does not consider the role of pericellular ATP responsible for activating the $P2X_7R$, which presumably would still require the involvement of a Cx43 hemichannel based upon the findings by Genetos et al.[89] Nonetheless, these data demonstrate the potential role and interplay of connexins and pannexins in mechanotransduction.

35.8 Future Directions and Concluding Remarks

Despite tremendous advances in understanding how osteoblastic and osteocytic cells may initially perceive changes in their biophysical environment, a great deal of work remains to be done in order to formulate a cohesive model for mechanotransduction in the skeleton. The vast majority of in vitro studies in the literature and highlighted above were performed under two-dimensional conditions that do not closely mimic the native three-dimensional environment in which an osteoblast or osteocyte is found. Continued development of tissue perfusion bioreactors will enable investigators to confirm mechanotransduction machinery and events in a more native environment. Doing so, however, generates new problems to overcome, such as impaired access for patch clamping or penetration depth of cells in a three-dimensional bioreactor. This prevents direct analysis of calcium currents or transients, for example, that is more readily achieved in a two-dimensional parallel-plate chamber. Antagonists may still be added to perfusion media within a bioreactor, resulting in reduced calcium signaling, but it is unclear in the absence of path clamping whether this is due to direct block of the calcium channel or reduced calcium-induced calcium release.

Improvements in fluorometric indicator dyes concomitant with advances in microscopy have increased our understanding of the spatiotemporal dynamics of calcium signaling in cells in response to biophysical stimuli. Fura-2 has high affinity for Ca^{2+} (K_D = 140 nM) and is useful for measuring smaller changes in Ca_i^{2+}, but releases calcium more slowly. Thus, changes in Fura-2 fluorescence may not reveal rapid fluctuations in Ca_i^{2+} and cannot provide as accurate kinetics of Ca_i^{2+} transients as do dyes with lower Ca^{2+} affinity, such as Fluo-4.[100] This is demonstrated in Figure 35.1 in osteoblasts loaded with Fura-2 and Fluo-4; small perturbations in Ca_i^{2+} are apparent by monitoring changes in Fluo-4 intensity (arrowheads in Figure 35.1a; trace in Figure 35.1b) that are not apparent by monitoring Fura-2 fluorescence (trace in Figure 35.1c). The ability to visualize such small, localized Ca_i^{2+}

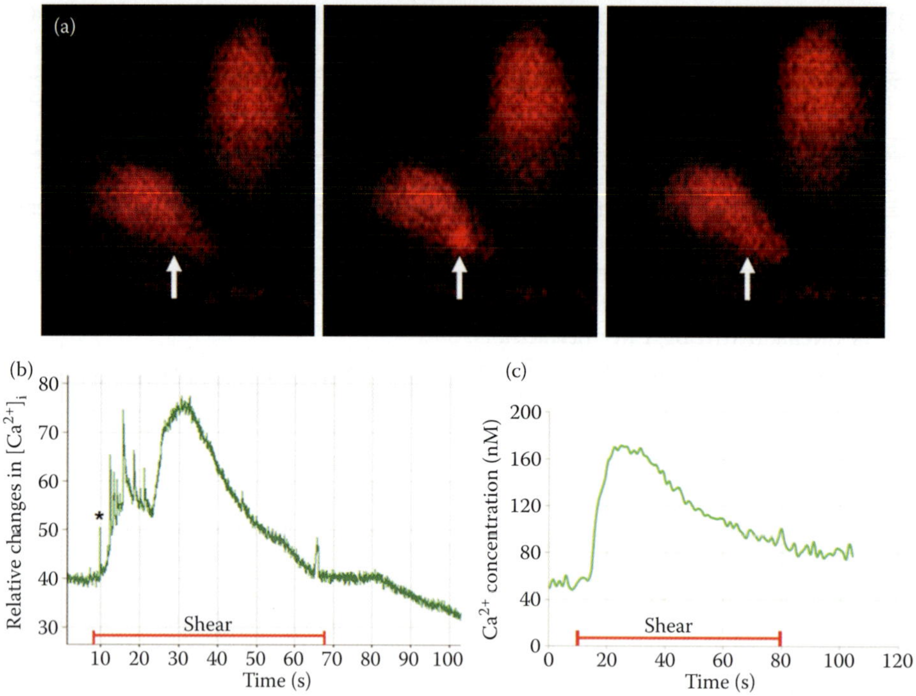

FIGURE 35.1
Activation of L-type voltage-sensitive calcium channels in MC3T3-E1 osteoblasts with fluid shear. MC3T3-E1 osteoblasts were grown on type 1 collagen coated slides until 50% confluent, loaded with 5 μM Fluo-4 for 30 min, then mounted on a Zeiss LSM510 DUO high-speed confocal microscope and subjected to 12 dyn/cm^2 fluid shear (time 0). (a) Three images, taken every 25 ms, demonstrate that within 10 s of the onset of fluid shear, brief, localized increases in Ca_i^{2+} were observed (arrow, center panel) that were not observed in the images taken 25 ms before (left panel) and after (right panel) the increase. (b) Relative changes in Ca_i^{2+} in a 2 × 2 μm region above the arrow over time indicate that the calcium response is biphasic with early responses observed as spikes (* denotes the calcium transient shown in a) that result in an accumulation of calcium prior to an overall Ca_i^{2+} response induced by intracellular calcium release. Inhibition of the L-type voltage-sensitive calcium channel with 5 μM nifedipine abolished the rapid transient Ca_i^{2+} spikes and reduced the overall Ca_i^{2+} response to shear (data not shown). (c) MC3T3-E1 cells loaded with 5 μM Fura-2AM and subjected to fluid shear in the same manner as shown in (a). Images were recorded at 340 and 380 nm for 433 and 100 ms, respectively. The ratio of 340:380 recorded emission was compared to that of a standard curve, resulting in the calculated concentration of intracellular Ca^{2+}. Thus, more than 500 ms is required for acquisition of each data point.

transients may be critical for identifying a mechanistic role for Ca_i^{2+} microdomains in whole-cell Ca_i^{2+} increases or localized secretion of ATP-filled vesicles, for example.

Mechanotransduction, and the cellular and phenotypic responses that result, is a critical process during the development and life of an organism. Here, we focused upon the role of ion channels in mechanotransduction within the skeleton. Ion channels are rapidly activated (<1 min) in response to biophysical stimuli and thus are likely to play a role early in mechanotransduction by translating an extracellular signal into a cellular response, or in transmitting a signal (i.e., Ca_i^{2+} signal) to a neighboring cell.[101] In addition to the classes of ion channels described above, there are many more that may be involved in mechanotransduction in osteoblastic and osteoclastic cells, but were omitted because their role in mechanotransduction is nascent or in the early stages of discovery. These include annexins,[102] the epithelial sodium channel, ENaC,[103] and stretch-activated potassium channels.[104] Additionally, other ion channels have been omitted from this chapter because of space limitations and the breadth of current reviews on their function in the skeleton, such as polycystins.[105,106]

Clearly, additional information is required to integrate ion channel activation and rapid Ca_i^{2+} transients with other cellular processes activated by mechanotransduction, such as reorganization of the actin cytoskeleton and nuclear shuttling of transcription factors (discussed in greater detail in refs. 41 and 103). The actin cytoskeleton regulates MSCC channel activity through an unknown mechanism,[45] and the adapter protein AHNAK physically links the actin cytoskeleton to regulatory β_2 subunits of L-type VSCCs.[107] Are these ion channels equally important in mechanotransduction as a cell differentiates from a mesenchymal stem cell into an osteoblast and then into an osteocyte? Limited data suggest that this is not true, as osteocytes express a different repertoire of voltage-sensitive calcium channels than osteoblasts.[57] Nonetheless, a model (Figure 35.2) has emerged wherein osteocytes embedded within the

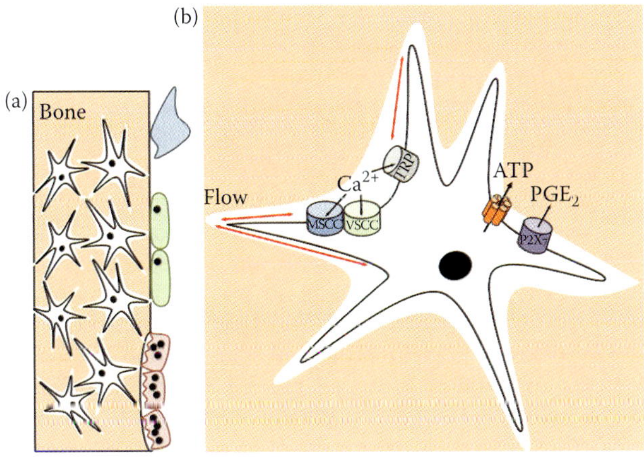

FIGURE 35.2
Mechanistic model of ion channel activation in skeletal mechanotransduction. (a) Osteocytes embedded within a bone are capable of interacting with mesenchymal stem cells (blue), osteoblasts (green), and osteoclasts (red). (b) A biophysical signal such as fluid flow-induced shear stress across an osteocyte activates ion channels such as mechanosensitive cation channel, voltage-sensitive cation channel, or TRP channel to enable entry of extracellular Ca^{2+}. Increases in Ca_i^{2+} then mediate ATP release through vesicular or gap junction hemichannel or pannexin channels; secreted ATP then binds to $P2X_7$ purinoceptors to cause release of osteotropic factors like PGE_2.

mineralized matrix of bone perceive a biophysical signal, such as fluid flow-induced shear stress, which activates mechanosensitive and voltage-sensitive anion or cation channel activity to ultimately produce a skeletally anabolic or catabolic response. While this model awaits experimental validation, it serves as a useful paradigm in directing future research.

Acknowledgments

This work was supported by NIH/NIA R01 AG13087 and the Michael and Myrtle Baker Orthopaedic Research Endowment (to HJD); by NIH/NIAMS R01 AR043222 (RLD); and by NIH/NIAMS R03 AR0575447 and R01 AR064255 (DCG).

References

1. Genetos DC, Jacobs CR. *Studies in Mechanobiology, Tissue Engineering and Biomaterials*. Berlin: Springer, 2011.
2. Ke HZ, Qi H, Weidema AF, Zhang Q, Panupinthu N, Crawford DT et al. Deletion of the P2X7 nucleotide receptor reveals its regulatory roles in bone formation and resorption. *Mol Endocrinol*. 2003;17(7):1356–67.
3. Jaalouk DE, Lammerding J. Mechanotransduction gone awry. *Nat Rev Mol Cell Biol*. 2009;10(1):63–73.
4. Sukharev S, Blount P, Martinac B, Kung C. Mechanosensitive channels of *Escherichia coli*: The MscL gene, protein, and activities. *Annu Rev Physiol*. 1997;59:633–57.
5. Hill JK, Gillespie PG. Seeing touch: Moving closer to the worm mechanotransduction complex. *Curr Biol*. 2003;13(24):R967–9.
6. Vanhoutte PM. The endothelium—modulator of vascular smooth-muscle tone. *N Engl J Med*. 1988;319(8):512–3.
7. Paniagua OA, Bryant MB, Panza JA. Role of endothelial nitric oxide in shear stress-induced vasodilation of human microvasculature: Diminished activity in hypertensive and hypercholesterolemic patients. *Circulation*. 2001;103(13):1752–8.
8. Guilak F, Mow VC. The mechanical environment of the chondrocyte: A biphasic finite element model of cell–matrix interactions in articular cartilage. *J Biomech*. 2000;33(12):1663–73.
9. Genetos DC, Karin NJ, Geist DJ, Donahue HJ, Duncan RL. Purinergic signaling is required for fluid shear stress-induced NF-κB translocation in osteoblasts. *Exp Cell Res*. 2011;317(6):737–44. PMCID: PMC3049820.
10. Klein-Nulend J, Bacabac RG, Bakker AD. Mechanical loading and how it affects bone cells: The role of the osteocyte cytoskeleton in maintaining our skeleton. *Eur Cell Mater*. 2011;24:278–91.
11. Orriss I, Syberg S, Wang N, Robaye B, Gartland A, Jorgensen N et al. Bone phenotypes of P2 receptor knockout mice. *Front Biosci (Schol Ed)*. 2011;3:1038–46.
12. Jacobs CR, Temiyasathit S, Castillo AB. Osteocyte mechanobiology and pericellular mechanics. *Annu Rev Biomed Eng*. 2010;12:369–400.
13. Williams BO, Insogna KL. Where Wnts went: The exploding field of Lrp5 and Lrp6 signaling in bone. *J Bone Miner Res*. 2009;24(2):171–8. PMCID: PMC3276354.
14. Sakai K, Mohtai M, Iwamoto Y. Fluid shear stress increases transforming growth factor beta 1 expression in human osteoblast-like cells: Modulation by cation channel blockades. *Calcif Tissue Int*. 1998;63(6):515–20.
15. Zaragoza C, Márquez S, Saura M. Endothelial mechanosensors of shear stress as regulators of atherogenesis. *Curr Opin Lipidol*. 2012;23(5):446–52.
16. Ghazi A, Ajouz B, Besnard M. Mechanosensitive ion channels and their mode of activation. *Biochimie*. 1998;80(5–6):357–62.
17. Hung C, Allen F, Pollack S, Brighton C. Intracellular Ca^{2+} stores and extracellular Ca^{2+} are required in the real-time Ca^{2+} response of bone cells experiencing fluid flow. *J Biomech*. 1996;29(11):1411–7.
18. Salazar C, Höfer T. Allosteric regulation of the transcription factor NFAT1 by multiple phosphorylation sites: A mathematical analysis. *J Mol Biol*. 2003;327(1):31–45.
19. Dolmetsch R, Lewis R, Goodnow C, Healy J. Differential activation of transcription factors induced by Ca^{2+} response amplitude and duration. *Nature*. 1997;386(6627):855–8.
20. Dolmetsch R, Xu K, Lewis R. Calcium oscillations increase the efficiency and specificity of gene expression. *Nature*. 1998;392(6679):933–6.

21. Genetos DC, Geist DJ, Liu D, Donahue HJ, Duncan RL. Fluid shear-induced ATP secretion mediates prostaglandin release in MC3T3-E1 osteoblasts. *J Bone Miner Res*. 2005;20(1):41–9. PMCID: PMC2929123.
22. Liu D, Genetos DC, Shao Y, Geist DJ, Li J, Ke HZ et al. Activation of extracellular-signal regulated kinase (ERK1/2) by fluid shear is Ca(2+)- and ATP-dependent in MC3T3-E1 osteoblasts. *Bone*. 2008;42(4):644–52. PMCID: PMC2937351.
23. Chen NX, Geist DJ, Genetos DC, Pavalko FM, Duncan RL. Fluid shear-induced NFkappaB translocation in osteoblasts is mediated by intracellular calcium release. *Bone*. 2003;33(3):399–410.
24. Chen NX, Ryder KD, Pavalko FM, Turner CH, Burr DB, Qiu J et al. Ca(2+) regulates fluid shear-induced cytoskeletal reorganization and gene expression in osteoblasts. *Am J Physiol Cell Physiol*. 2000;278(5):C989–97.
25. Wen L, Wang Y, Wang H, Kong L, Zhang L, Chen X et al. L-type calcium channels play a crucial role in the proliferation and osteogenic differentiation of bone marrow mesenchymal stem cells. *Biochem Biophys Res Commun*. 2012;424(3):439–45.
26. Guggino SE, Lajeunesse D, Wagner JA, Snyder SH. Bone remodeling signaled by a dihydropyridine- and phenylalkylamine-sensitive calcium channel. *Proc Natl Acad Sci USA*. 1989;86(8):2957–60. PMCID: PMC287039.
27. Barradas AMC, Fernandes HAM, Groen N, Chai YC, Schrooten J, van de Peppel J et al. A calcium-induced signaling cascade leading to osteogenic differentiation of human bone marrow-derived mesenchymal stromal cells. *Biomaterials*. 2012;33(11):3205–15.
28. Ajubi NE, Klein-Nulend J, Nijweide PJ, Vrijheid-Lammers T, Alblas MJ, Burger EH. Pulsating fluid flow increases prostaglandin production by cultured chicken osteocytes—A cytoskeleton-dependent process. *Biochem Biophys Res Commun*. 1996;225(1):62–8.
29. Lu XL, Huo B, Chiang V, Guo XE. Osteocytic network is more responsive in calcium signaling than osteoblastic network under fluid flow. *J Bone Miner Res*. 2012;27(3):563–74. PMCID: PMC3343217.
30. Sachs F, Morris CE. Mechanosensitive ion channels in nonspecialized cells. *Rev Physiol Biochem Pharmacol*. 1998; 132:1–77.
31. Martinac B. The ion channels to cytoskeleton connection as potential mechanism of mechanosensitivity. *Biochim Biophys Acta*. 2014;1838:682–91.
32. Sukharev SI, Blount P, Martinac B, Blattner FR, Kung C. A large-conductance mechanosensitive channel in *E. coli* encoded by mscL alone. *Nature*. 1994;368(6468):265–8.
33. Paoletti P, Ascher P. Mechanosensitivity of NMDA receptors in cultured mouse central neurons. *Neuron*. 1994;13(3):645–55.
34. Dopico AM, Kirber MT, Singer JJ, Walsh JV. Membrane stretch directly activates large conductance Ca(2+)-activated K^+ channels in mesenteric artery smooth muscle cells. *Am J Hypertens*. 1994;7(1):82–9.
35. Duncan R, Misler S. Voltage-activated and stretch-activated Ba^{2+} conducting channels in an osteoblast-like cell line (UMR 106). *FEBS Lett*. 1989;251(1–2):17–21.
36. Davidson R, Tatakis D, Auerbach A. Multiple forms of mechanosensitive ion channels in osteoblast-like cells. *Pflugers Arch*. 1990;416(6):646–51.
37. Duncan RL, Hruska KA. Chronic, intermittent loading alters mechanosensitive channel characteristics in osteoblast-like cells. *Am J Physiol*. 1994;267(6 Pt 2):F909–16.
38. Rawlinson S, Pitsillides A, Lanyon L. Involvement of different ion channels in osteoblasts' and osteocytes' early response to mechanical strain. *Bone*. 1996;19(6):609–14.
39. You J, Reilly GC, Zhen X, Yellowley CE, Chen Q, Donahue HJ et al. Osteopontin gene regulation by oscillatory fluid flow via intracellular calcium mobilization and activation of mitogen-activated protein kinase in MC3T3-E1 osteoblasts. *J Biol Chem*. 2001;276(16):13365–71.
40. Motokawa M, Kaku M, Tohma Y, Kawata T, Fujita T, Kohno S et al. Effects of cyclic tensile forces on the expression of vascular endothelial growth factor (VEGF) and macrophage-colony-stimulating factor (M-CSF) in murine osteoblastic MC3T3-E1 cells. *J Dent Res*. 2005;84(5):422–7.
41. Bidwell JP, Pavalko FM. The load-bearing mechanosome revisited. *Clin Rev Bone Miner Metab*. 2010;8(4):213–23. PMCID: PMC3071518.
42. Miller S, Wolf A, Arnaud C. Bone cells in culture: Morphologic transformation by hormones. *Science*. 1976;192(4246):1340–3.
43. Lomri A, Marie P. Changes in cytoskeletal proteins in response to parathyroid hormone and 1,25-dihydroxyvitamin D in human osteoblastic cells. *Bone Miner*. 1990;10(1):1–12.
44. Lomri A, Marie P. Effect of parathyroid hormone and forskolin on cytoskeletal protein synthesis in cultured mouse osteoblastic cells. *Biochim Biophys Acta*. 1988;970(3):333–42.
45. Zhang J, Ryder KD, Bethel JA, Ramirez R, Duncan RL. PTH-induced actin depolymerization increases mechanosensitive channel activity to enhance mechanically stimulated Ca^{2+} signaling in osteoblasts. *J Bone Miner Res*. 2006;21(11):1729–37.
46. Hur E-M, Park T-J, Kim K-T. Coupling of L-type voltage-sensitive calcium channels to P2X2 purinoceptors in PC-12 cells. *Am J Physiol Cell Physiol*. 2001;280(5):C1121–9.
47. Duncan R, Akanbi K, Farach-Carson M. Calcium signals and calcium channels in osteoblastic cells. *Semin Nephrol*. 1998;18(2):178–90.
48. Chesnoy-Marchais D, Fritsch J. Voltage-gated sodium and calcium currents in rat osteoblasts. *J Physiol*. 1988;398:291–311.
49. Duriez J, Flautre B, Blary M, Hardouin P. Effects of the calcium channel blocker nifedipine on epiphyseal growth plate and bone turnover: A study in rabbit. *Calcif Tissue Int*. 1993;52(2):120–4.
50. Walker LM, Publicover SJ, Preston MR, Said Ahmed MA, Haj El AJ. Calcium-channel activation and matrix protein upregulation in bone cells in response to mechanical strain. *J Cell Biochem*. 2000;79(4):648–61.
51. Ryder KD, Duncan RL. Parathyroid hormone enhances fluid shear-induced $[Ca^{2+}]$i signaling in osteoblastic cells through activation of mechanosensitive and voltage-sensitive Ca^{2+} channels. *J Bone Miner Res*. 2001;16(2):240–8.

52. Miyauchi A, Notoya K, Mikuni-Takagaki Y, Takagi Y, Goto M, Miki Y, et al. Parathyroid hormone-activated volume-sensitive calcium influx pathways in mechanically loaded osteocytes. *J Biol Chem*. 2000;275(5):3335–42.
53. Li J, Duncan RL, Burr DB, Turner CH. L-type calcium channels mediate mechanically induced bone formation in vivo. *J Bone Miner Res*. 2002;17(10):1795–800.
54. Li J, Duncan RL, Burr DB, Gattone VH, Turner CH. Parathyroid hormone enhances mechanically induced bone formation, possibly involving L-type voltage-sensitive calcium channels. *Endocrinology*. 2003;144(4):1226–33.
55. Duncan RL, Kizer N, Barry EL, Friedman PA, Hruska KA. Antisense oligodeoxynucleotide inhibition of a swelling-activated cation channel in osteoblast-like osteosarcoma cells. *Proc Natl Acad Sci USA*. 1996;93(5):1864–9. PMCID: PMC39873.
56. Lyford G, Strege P, Shepard A, Ou Y, Ermilov L, Miller S et al. alpha(1C) (Ca(V)1.2) L-type calcium channel mediates mechanosensitive calcium regulation. *Am J Physiol Cell Physiol*. 2002;283(3):C1001–8.
57. Shao Y, Alicknavitch M, Farach-Carson M. Expression of voltage sensitive calcium channel (VSCC) L-type Cav1.2 (alpha1C) and T-type Cav3.2 (alpha1H) subunits during mouse bone development. *Dev Dyn*. 2005;234(1):54–62.
58. Kronbergs A, Shao Y, Campbell KP, Duncan RL, Farach-Carson MC. Assessment of changes in thoracic and lumbar vertebrae in T type Cav3.2 (a1H) voltage sensitive calcium channel knock out mice. *J Bone Miner Res*. 2008;23(Suppl 1). Available at http://onlinelibrary.wiley.com/doi/10.1002/jbmr.5650231306/full. Accessed July 1, 2014.
59. Kronbergs A, Campbell K, Burrd DB, Farach-Carson MC, Duncan RL. Transactions of the 56th Annual Meeting of the Orthopaedic Research Society, New Orleans, LA, 2010, p. 0680.
60. Gu Y, Preston MR, Haj El AJ, Howl JD, Publicover SJ. Three types of K(+) currents in murine osteocyte-like cells (MLO-Y4). *Bone*. 2001;28(1):29–37.
61. Rezzonico R, Cayatte C, Bourget-Ponzio I, Romey G, Belhacene N, Loubat A et al. Focal adhesion kinase pp125FAK interacts with the large conductance calcium-activated hSlo potassium channel in human osteoblasts: Potential role in mechanotransduction. *J Bone Miner Res*. 2003;18(10):1863–71.
62. Hughes S, Magnay J, Foreman M, Publicover SJ, Dobson JP, Haj El AJ. Expression of the mechanosensitive 2PK+ channel TREK-1 in human osteoblasts. *J Cell Physiol*. 2006;206(3):738–48.
63. Nicke A. Homotrimeric complexes are the dominant assembly state of native P2X7 subunits. *Biochem Biophys Res Commun*. 2008;377(3):803–8.
64. Li M, Kawate T, Silberberg SD, Swartz KJ. Pore opening mechanism in trimeric P2X receptor channels. *Nat Comms*. Nature Publishing Group; 2010;1(4):1–7.
65. North R. Molecular physiology of P2X receptors. *Physiol Rev*. 2002;82(4):1013–67.
66. Burnstock G. Release of vasoactive substances from endothelial cells by shear stress and purinergic mechanosensory transduction. *J Anat*. 1999;194(Pt 3):335–42. PMCID: PMC1467933.
67. Wynn G, Rong W, Xiang Z, Burnstock G. Purinergic mechanisms contribute to mechanosensory transduction in the rat colorectum. *Gastroenterology*. 2003;125(5):1398–409.
68. Wynn G, Bei M, Huai-Zhen R, Burnstock G. Purinergic component of mechanosensory transduction is increased in a rat model of colitis. *Am J Physiol Gastrointest Liver Physiol*. 2004;287:G647–57.
69. Rodat-Despoix L, Hao J, Dandonneau M, Delmas P. Shear stress-induced Ca(2+) mobilization in MDCK cells is ATP dependent, no matter the primary cilium. *Cell Calcium*. 2013;53(5–6):327–37.
70. Li J, Liu D, Ke HZ, Duncan RL, Turner CH. The P2X7 nucleotide receptor mediates skeletal mechanotransduction. *J Biol Chem*. 2005;280(52):42952–9.
71. Okumura H, Shiba D, Kubo T, Yokoyama T. P2X7 receptor as sensitive flow sensor for ERK activation in osteoblasts. *Biochem Biophys Res Commun*. 2008;372(3):486–90.
72. Viecilli RF, Katona TR, Chen J, Hartsfield JK, Roberts WE. Orthodontic mechanotransduction and the role of the P2X7 receptor. *Am J Orthod Dentofacial Orthop*. 2009;135(6):694.e1–16; discussion 694–5.
73. Yin J, Kuebler WM. Mechanotransduction by TRP channels: General concepts and specific role in the vasculature. *Cell Biochem Biophys*. 2010;56(1):1–18.
74. Lieben L, Carmeliet G. The involvement of TRP channels in bone homeostasis. *Front Endocrinol (Lausanne)*. 2012;3:99. PMCID: PMC3422722.
75. Berrout J, Jin M, Mamenko M, Zaika O, Pochynyuk O, O'Neil RG. Function of transient receptor potential cation channel subfamily V member 4 (TRPV4) as a mechanical transducer in flow-sensitive segments of renal collecting duct system. *J Biol Chem*. 2012;287(12):8782–91. PMCID: PMC3308780.
76. Christensen AP, Corey DP. TRP channels in mechanosensation: Direct or indirect activation? *Nat Rev Neurosci*. 2007;8(7):510–21.
77. Suzuki T, Notomi T, Miyajima D, Mizoguchi F, Hayata T, Nakamoto T et al. Osteoblastic differentiation enhances expression of TRPV4 that is required for calcium oscillation induced by mechanical force. *Bone*. 2013;54(1):172–8.
78. Kang SS, Shin SH, Auh C-K, Chun J. Human skeletal dysplasia caused by a constitutive activated transient receptor potential vanilloid 4 (TRPV4) cation channel mutation. *Exp Mol Med*. 2012;44(12):707–22. PMCID: PMC3538978.
79. Mizoguchi F, Mizuno A, Hayata T, Nakashima K, Heller S, Ushida T et al. Transient receptor potential vanilloid 4 deficiency suppresses unloading-induced bone loss. *J Cell Physiol*. 2008;216(1):47–53.
80. Chamoux E, Bisson M, Payet MD, Roux S. TRPV-5 mediates a receptor activator of NF-kappaB (RANK) ligand-induced increase in cytosolic Ca^{2+} in human osteoclasts and down-regulates bone resorption. *J Biol Chem*. 2010;285(33):25354–62. PMCID: PMC2919098.
81. van der Eerden BCJ, Hoenderop JGJ, de Vries TJ, Schoenmaker T, Buurman CJ, Uitterlinden AG et al. The epithelial Ca^{2+} channel TRPV5 is essential for proper osteoclastic bone resorption. *Proc Natl Acad Sci USA*. 2005;102(48):17507–12. PMCID: PMC1297662.

82. Kato K, Morita I. Acidosis environment promotes osteoclast formation by acting on the last phase of preosteoclast differentiation: A study to elucidate the action points of acidosis and search for putative target molecules. *Eur J Pharmacol*. 2011;663(1–3):27–39.
83. Masuyama R, Vriens J, Voets T, Karashima Y, Owsianik G, Vennekens R et al. TRPV4-mediated calcium influx regulates terminal differentiation of osteoclasts. *Cell Metab*. 2008;8(3):257–65.
84. Masuyama R, Mizuno A, Komori H, Kajiya H, Uekawa A, Kitaura H et al. Calcium/calmodulin-signaling supports TRPV4 activation in osteoclasts and regulates bone mass. *J Bone Miner Res*. 2012;27(8):1708–21.
85. Phan MN, Leddy HA, Votta BJ, Kumar S, Levy DS, Lipshutz DB et al. Functional characterization of TRPV4 as an osmotically sensitive ion channel in porcine articular chondrocytes. *Arthritis Rheum*. 2009;60(10):3028–37. PMCID: PMC2846816.
86. Contreras JE, Sánchez HA, Eugenin EA, Speidel D, Theis M, Willecke K et al. Metabolic inhibition induces opening of unapposed connexin 43 gap junction hemichannels and reduces gap junctional communication in cortical astrocytes in culture. *Proc Natl Acad Sci USA*. 2002;99(1):495–500. PMCID: PMC117588.
87. Jiang JX, Cherian PP. Hemichannels formed by connexin 43 play an important role in the release of prostaglandin E(2) by osteocytes in response to mechanical strain. *Cell Commun Adhes*. 2003;10(4–6):259–64.
88. Genetos DC, Kephart CJ, Zhang Y, Yellowley CE, Donahue HJ. Oscillating fluid flow activation of gap junction hemichannels induces ATP release from MLO-Y4 osteocytes. *J Cell Physiol*. 2007;212(1):207–14. PMCID: PMC2929812.
89. Stout C, Costantin J, Naus C, Charles A. Intercellular calcium signaling in astrocytes via ATP release through connexin hemichannels. *J Biol Chem*. 2002;277(12):10482–8.
90. Verderio C, Bruzzone S, Zocchi E, Fedele E, Schenk U, De Flora A et al. Evidence of a role for cyclic ADP-ribose in calcium signalling and neurotransmitter release in cultured astrocytes. *J Neurochem*. 2001;78(3):646–57.
91. Cotrina M, Lin J, Alves-Rodrigues A, Liu S, Li J, Azmi-Ghadimi H et al. Connexins regulate calcium signaling by controlling ATP release. *Proc Natl Acad Sci USA*. 1998;95(26):15735–40.
92. Graff R, Lazarowski E, Banes A, Lee G. ATP release by mechanically loaded porcine chondrons in pellet culture. *Arthritis Rheum*. 2000;43(7):1571–9.
93. Batra N, Kar R, Jiang JX. Gap junctions and hemichannels in signal transmission, function and development of bone. *Biochim Biophys Acta*. 2012;1818(8):1909–18. PMCID: PMC3440861.
94. Lloyd SAJ, Donahue HJ. Gap junctions and biophysical regulation of bone cells. *Clinic Rev Bone Miner Metab*. 2010;8(4):189–200. PMCID: PMC3677790.
95. Plotkin LI, Bellido T. Beyond gap junctions: Connexin43 and bone cell signaling. *Bone*. 2013;52(1):157–66. PMCID: PMC3513515.
96. Jiang JX, Siller-Jackson AJ, Burra S. Roles of gap junctions and hemichannels in bone cell functions and in signal transmission of mechanical stress. *Front Biosci*. 2007;12:1450–62. PMCID: PMC1797155.
97. Civitelli R. Cell-cell communication in the osteoblast/osteocyte lineage. *Arch Biochem Biophys*. 2008;473(2):188–92. PMCID: PMC2441851.
98. Thi MM, Islam S, Suadicani SO, Spray DC. Connexin43 and pannexin1 channels in osteoblasts: Who is the "hemichannel?" *J Membr Biol*. 2012;245(7):401–9. PMCID: PMC3427001.
99. Bao L, Locovei S, Dahl G. Pannexin membrane channels are mechanosensitive conduits for ATP. *FEBS Lett*. 2004;572(1–3):65–8.
100. Bruton JD, Cheng AJ, Westerblad H. Methods to detect Ca(2+) in living cells. *Adv Exp Med Biol*. 2012;740:27–43.
101. Duncan RL, Turner CH. Mechanotransduction and the functional response of bone to mechanical strain. *Calcif Tissue Int*. 1995;57(5):344–58.
102. Haut Donahue TL, Genetos DC, Jacobs CR, Donahue HJ, Yellowley CE. Annexin V disruption impairs mechanically induced calcium signaling in osteoblastic cells. *Bone*. 2004;35(3):656–63.
103. Pavalko FM, Norvell SM, Burr DB, Turner CH, Duncan RL, Bidwell JP. A model for mechanotransduction in bone cells: The load-bearing mechanosomes. *J Cell Biochem*. 2003;88(1):104–12.
104. Chen X, Macica CM, Ng KW, Broadus AE. Stretch-induced PTH-related protein gene expression in osteoblasts. *J Bone Miner Res*. 2005;20(8):1454–61.
105. Hoey DA, Chen JC, Jacobs CR. The primary cilium as a novel extracellular sensor in bone. *Front Endocrinol (Lausanne)*. 2012;3:75. PMCID: PMC3374377.
106. Temiyasathit S, Tang WJ, Leucht P, Anderson CT, Monica SD, Castillo AB et al. Mechanosensing by the primary cilium: Deletion of Kif3A reduces bone formation due to loading. *PLoS One*. 2012;7(3):e33368. PMCID: PMC3299788.
107. Shao Y, Czymmek KJ, Jones PA, Fomin VP, Akanbi K, Duncan RL et al. Dynamic interactions between L-type voltage-sensitive calcium channel Cav1.2 subunits and ahnak in osteoblastic cells. *Am J Physiol Cell Physiol*. 2009;296(5):C1067–78. PMCID: PMC2681378.

Section XII

Cellular and Molecular Imaging Methods

Surfaces and Single Molecule Measurements

36

Characterization of Intermolecular and Intramolecular Interactions with the Atomic Force Microscope

Gil U. Lee, Krasimir Ivanov, Devrim Kilinc, Elena Martines, Agata Blasiak, Peng Li, and Michael J. Higgins

CONTENTS

The atomic force microscope (AFM) is a powerful nanoscale instrument that is capable of simultaneously imaging biomolecules and measuring the forces between biomolecules attached to the probe and the surface. In this review, we provide an introduction to the experimental and theoretical considerations that are involved in studying single molecular interactions with the AFM. We start by defining the performance of the AFM and types of biomolecular interactions that have been characterized with it. We then describe the surface chemistries that are most commonly used to immobilize biomolecules to AFM surfaces and introduce the theoretical framework for the analysis of the nonequilibrium kinetics of single molecule interactions. Specific molecular interactions for a model cell adhesion molecule are analyzed to introduce how the technique is implemented. We conclude with a brief consideration of the emerging applications of single molecule measurements.

36.1 Introduction

Recent advances in biomolecular science have provided us with a detailed map of the molecular networks that control cellular behavior, and it is clear that these networks are composed of a large number of glycoproteins that sense, signal, and respond to a cell's environment.[1] The response of these networks is determined by the interaction of two or more proteins that can be described in terms of chemical bonds and significant conformational changes. Biomedical research is now increasingly focused on characterizing and modifying the chemical and physical nature of specific molecular interactions that have been linked to specific diseases. In the simplest case, the thermodynamics and kinetics of these biomolecular reactions can be described in terms of the association (k_{on}) and dissociation (k_{off}) rate constants.[2] These parameters are now routinely measured using synthetic surfaces and in vitro cell cultures.[3,4] For simple ligand–receptor interactions, these biomolecular interactions have magnitudes that fall in specific ranges, i.e., 10^5 M^{-1} min^{-1} < k_{on} < 10^{10} M^{-1} min^{-1} and 10^{-3} min^{-1} < k_{off} < 10^1 min^{-1}. However, there is a growing appreciation of the fact that biological systems exist in a nonequilibrium state and that mechanical changes associated with nonequilibrium conditions are not detected using equilibrium binding assays.[5]

The atomic force microscopy (AFM) is a powerful nanoscale technique that is capable of imaging biomolecules and cells with subnanometer resolution and simultaneously characterizing their mechanical properties.[6,7] The development of microfabricated cantilevers, very sensitive displacement sensors, and piezoelectric transducers has made it possible to use the AFM to measure forces in a range of 1 to 1000 pN while controlling displacements with subnanometer resolution. This gives the AFM a force and temporal resolution that is not as high as optical or magnetic tweezers; nevertheless, it has the unique capacity to image and measure force–distance curves over a range of relevant displacements. AFM is also now commonly combined with high-resolution, optical microscopy techniques to allow cells to be simultaneously studied across length scales ranging from 1 nm to 100 μm. Thus, it is possible to use the AFM to image biological macromolecules and cells, and measure intermolecular and intramolecular interactions under biologically relevant conditions.

There are two main approaches to measure single-molecule intermolecular and intramolecular interactions with the AFM (Figure 36.1). Single-molecule intermolecular force measurements were first made with the AFM by physically adsorbing or covalently linking model molecular receptors to the probe and a surface. These measurements were used to demonstrate that biomolecular interactions result in adhesive forces that were related to the speed at which bonds were loaded and to the energy of the bond by classical Boltzmann behavior.[8–10] *Catch-bonds* have also been identified and are believed to be associated with more complex intermolecular interactions.[11] Intramolecular forces within single molecules have also been identified in the force–distance behavior. The intramolecular forces can be broadly classified in terms of the level of force that is applied to a molecule. At low levels of force, $0.001 < f < 0.1$ pN, entropic forces dominate the behavior of the molecules, and these forces fall outside a level that can be easily measured with the AFM. At higher forces, i.e., $1 < f < 300$ pN, intermolecular and intramolecular conformational changes that take place in the molecule dominate. At high levels of forces, $f > 500$ pN, covalent bonds begin to break and dominate the force behavior of the molecule. Table 36.1 identifies some of the intermolecular and intramolecular interactions that have been measured with the AFM to provide insight into this technique.

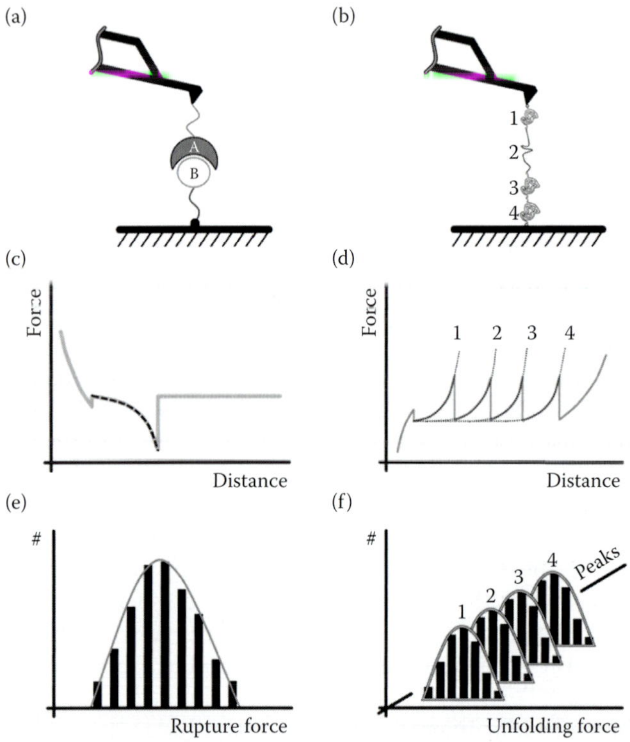

FIGURE 36.1
Single-molecule force measurements with the AFM. (a) Intermolecular force measurements between a biomolecule A and B attached to the probe and surface, respectively. (b) Intramolecular force measurements made within four subdomains of a biomolecule attached to the probe and the surface. (c) Force–distance curve exhibiting the intermolecular stretching of the tethered molecules and ultimate rupture of the A–B bond. (d) Force–distance curves exhibiting the unfolding of the four subunits of the biomolecule. (e) Typical rupture force histogram measure for an intermolecular interaction at a single loading rate. (f) Typical rupture force histogram measure for an intramolecular interaction measured at increasing loading rates.

It is our goal in this chapter to introduce the reader to the application of AFM to single-molecule force measurements. We will start by reviewing the physical scaling laws that define the performance of the instrument and then focus on strategies for immobilizing specific receptors on the probe, which is an important consideration in the design of a single-molecule AFM study. We will

TABLE 36.1
Selection of Intermolecular and Intramolecular Force Measurements Made with AFM

Intermolecular Interactions	
Streptavidin–biotin	Florin et al.,[12] Lee et al.,[13] Moy et al.[14]
Complementary strands of DNA	Grange et al.,[15] Strunz et al.,[16] Lee et al.[13]
Monoclonal antibodies	Preiner et al.,[17] Sulchek et al.,[18] Neuert et al.,[19] Schwesinger et al.,[20] Raab et al.,[21] Ros et al.,[22] Hinterdorfer et al.,[23] Dammer et al.[24]
Fibrin	Chtcheglova et al.[25]
Actomyosin bond	Colombini et al.[26]
DNA binding proteins/peptides	Eckel et al.,[27] Kuhner et al.[28]
NTA-His6	Verbelen et al.[29]
Cell adhesion molecules	Zhang et al.,[30] Wieland et al.[31]
Aptamer complex	Jiang et al.[32]
Cadherins	Baumgartner et al.[33]
Poly(ethylene glycol)	Oesterhelt et al.[34]
Single-stranded DNA	Rief et al.,[35] Lee et al.[13]
Polysaccharides	Rief et al.[36]
Beta-amino acids	Acevedo-Velez et al.[37]
Titin	Rief et al.[38]
Bacteriorhodopsin	Janovjak et al.,[39] Muller et al.[40]
Ubiquitin	Li et al.[8]

then provide an overview of the theoretical treatment of the experimental results necessary to determine k_{off}, k_{on}, and the bond rupture length, which is the key parameter describing nonequilibrium kinetics. These tools are then used to analyze results from the force measurements made for the homophilic binding interaction of apCAM, a cell adhesion molecule (CAM) found on the neuron growth cones of *Aplysia*. We conclude by providing a brief discussion of emerging application of AFM in the measurement of mechanotransduction events in living cells.

36.2 Experimental Methods

36.2.1 Atomic Force Microscopy

Commercial AFMs are powerful instruments capable of characterizing the structure and mechanical properties of a biomolecule at the nanoscale. The key elements of the AFM include a microfabricated cantilever with an integrated probe, a cantilever displacement detector for measuring the force, a piezoelectric transducer to move the sample relative to the probe, and a feedback loop (Figure 36.2). The physical performance of the AFM is typically determined by the microfabricated cantilever that is used to sense the force between the probe and the surface, either through the direct detection of the deflection of the cantilever or through the change in the resonance of the cantilever as it comes into intermittent contact with the surface. The laser deflection system, first introduced by Meyer and Amer,[41,42] is the most commonly used force detection system in the AFM setups and relies on a four-quadrant position sensitive detector that can be used to quantify the deflection of the cantilever. Images of the surface are generated by rastering the probe over the sample while maintaining a constant static force or change in the amplitude of the cantilever oscillation at a specific frequency. It is critical to use low imaging forces when studying biomolecules and living cells as these materials are easily destroyed by both normal and lateral forces.[43] Force measurements are typically made by either cycling the tip sample at a predefined distance or moving the sample until a specific compressive or tensile load is achieved.

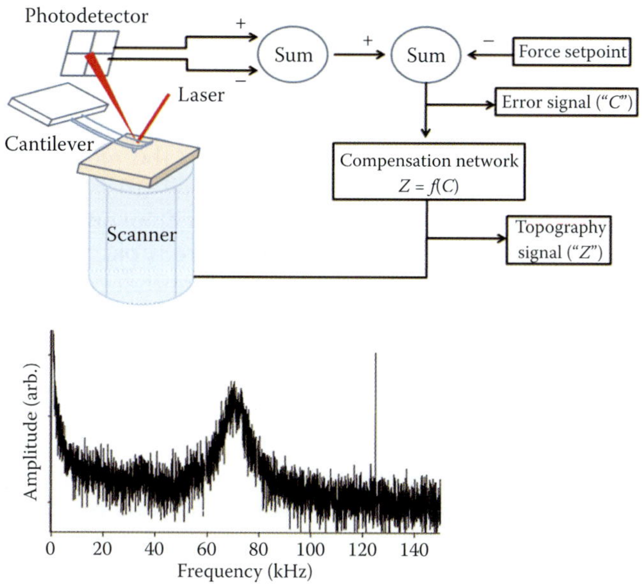

FIGURE 36.2
AFM. (a) Schematic diagram of the main components of the feedback system of an AFM. (b) Dynamic response of a 90 μm long, 35 μm wide, and 1 μm thick silicon microfabricated cantilever (NSC36C, Mikromasch, Madrid, Spain) in water. The observed resonance frequency of the cantilever is 77.723 kHz and Q is 6.2.

The force, spatial, and temporal resolution of the AFM instrument is ultimately limited by the physical properties of the microfabricated force sensor and integrated probe.[44] The spring constant (k) of a linear cantilever scales as $k = \frac{Ew}{4}\left(\frac{t}{l}\right)^3$, where E is the elastic modulus, w is the width, t is the thickness, and l is the length of the cantilever. The resonance frequency, f, of a linear cantilever scales as $f \propto \sqrt{\frac{E}{\rho}\frac{t}{l^2}}$, where ρ is the density of the cantilever material. The thermal and piezo-activated frequency spectra of a silicon cantilever in pure water that has a nominal spring constant of 4.5 N/m and nominal resonance frequency of 150 kHz in air can be measured. A single peak is observed at approximately 77 kHz in the amplitude–frequency spectrum produced by thermal fluctuations, which is characteristic of the highly damped cantilever motion (Figure 36.2).

Optimum conditions for imaging and measuring biological interactions are produced by soft cantilevers with high resonance frequencies, which provide high levels of force and temporal resolution. Microfabrication of thin, long cantilevers from stiff, lightweight materials, such as silicon and silicon oxynitride, has made it possible to create cantilevers with $0.01 < k < 1$ N/m and $1 < f < 100$ kHz in aqueous buffers. The development of small, highly sensitive displacement sensors has made it possible to operate AFMs near the theoretical limits of force, temporal, and spatial resolution.

36.2.2 Surface Chemistries for Immobilization of Biomolecules

A key step in the design of a single-molecule force measurement is the selection of a chemistry to immobilize the biomolecules on the probe and surface. The selection of a specific surface chemistry should be based on four criteria, i.e., the need to orient the biomolecules, the need for steric mobility, the level of nonspecific forces resulting from the surface chemistry, and the strength of the intermolecular interaction.[45,46]

There are three chemistries that are commonly used to immobilize biomolecules on the AFM probe and surface. Perhaps the simplest approach is to spontaneously adsorb the biomolecules on the surfaces. It has been found that the physical adsorption of a protein on an inorganic surface in a buffer at its isoelectric point often results in a strong interaction as a result of many weak bonds, i.e., hydrogen or van der Waals bonds, or one or more covalent bonds forming between the protein and the surface. This allows intramolecular interactions to be measured within a protein if it can be irreversibly adsorbed on a surface and picked up with the AFM probe (Figure 36.3a). Although this approach is simple to implement, there are a limited number of proteins in which spontaneous adsorption produces stable structures that can result in reproducible force measurements. Rather, it is likely that the bonds between the biomolecules and surfaces will result in force curves that cannot be easily interpreted. There is also a high probability that proteins will experience a significant change in their structure on adsorption and that interpretation of the forces will be complicated by nonspecific interactions.

The two most common covalent chemical approaches used to immobilize biomolecules on a surface involve the use of silane and thiol self-assembled monolayers (SAMs). The advantage of using SAMs is that a highly ordered and uniform monolayer can be formed on the surface through the spontaneous adsorption of a silane or thiol compound on oxide or gold surfaces, respectively.[47,48] The silanization approach is well developed since highly polished silicon wafers and silicon AFM probes are readily available. A protein-R′-silane monolayer can be formed on a silica surface, where the R′ group may be a short-chain alkane or a poly(ethylene glycol) (PEG) polymer linker that provides steric

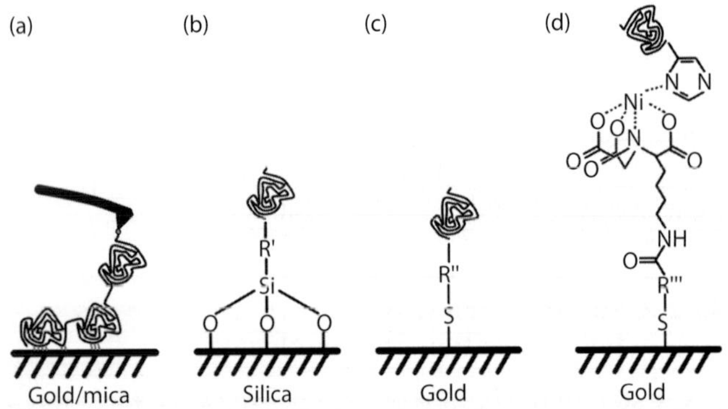

FIGURE 36.3
Surface chemistries for the immobilization of proteins on AFM probe and surface. (a) Physical adsorption, (b) silanization, (c) gold-thiol SAM, and (d) NTA-Ni-His complex.

mobility to the immobilized biomolecules (Figure 36.3b). The surfaces are typically formed in a three-step process: First, the surfaces are carefully cleaned to create a high density of silanol groups on the surface. Second, a silane monolayer is formed by reacting a commercially available R′-trialkoxy-silane with the silica surface followed by heating to drive the dehydrolysis reaction that results in the covalent immobilization of silanes. This is followed by reaction with specific nucleophilic groups on the proteins with the silane monolayer that has been activated by a cross-linker. Commercially available heterobifunctional cross-linkers link lysine, thiol, or carboxyl functional groups on biomolecules with surfaces, avoiding self-similar reactions. The advantage of using covalent chemistries to immobilize the biomolecules is that the force measurements are more easily interpreted since the covalent bonds are stronger than the specific intermolecular and intramolecular interactions that are being studied.

Gold-coated cantilevers and surfaces are also commercially available and allow alkane-thiol SAMs to be used to link the biomolecules to the surface. An R″-thiol protein monolayer can be immobilized on a gold surface (Figure 36.3c). The use of mixed films of thiols functionalized with hydroxyl, carboxyl, amide, and amino groups makes it possible to covalently link a biomolecule to a hydrophilic SAM using heterobifunctional cross-linkers. The advantage of using thiol SAMs is that the biomolecules maintain their functionality while nonspecific interactions between the biomolecules and the surface are minimized by the highly ordered alkane monolayers.

For many single measurements, it is desirable to orient the biomolecules on the surface in order to maximize the probability of specific interactions. In some cases, it may be possible to use a unique functional group on a protein, such as a thiol, to orient the molecules. However, when using heterobifunctional cross-linkers, it is likely that they will react with more than one active group on the biomolecule, e.g., there are typically multiple lysine groups on a protein surface that can react with a heterobifunctional cross-linker. The nitrilotriacetic acid (NTA)–polyhistidine (His_n) bond is a means for orienting a molecule at a surface because the His tag can be easily inserted in the protein structure.[29] The NTA-Ni-His approach has been used for affinity purification of recombinant proteins, but more recently, it has been utilized for the immobilization of ligands and receptors on AFM tips and substrates. This technique relies on the formation of a metal chelate complex between a NTA molecule and several histidine amino acids situated on the target protein (Figure 36.3d). The Ni^{2+} ion has six coordination sites, four of which are occupied by the NTA molecule and two by the His tag. The advantage of this technique is that, aside from the attached protein being fully functional, the protein can be oriented *via* its histidine tag. A potential disadvantage of this technique is that it may result in a bond that is not as strong as the specific molecular interaction of interest.

36.3 Theoretical Analysis of Single-Molecule Force Measurements

36.3.1 Introduction to Nonequilibrium Kinetics of Biomolecular Interactions

The kinetics of biomolecular interactions are determined by transition state theory and are characterized by the off rate, k_{off}, and the on rate, k_{on}. The off rate for a second-order reaction with a single transition state with activation energy E_a may be defined in terms of the Arrhenius or Eyring equations, which have the form[49]

$$k_{off} = A'\exp\left(\frac{-E_a}{k_BT}\right) = \frac{k_BT}{h}\exp\left(\frac{-\Delta G}{k_BT}\right) \quad (36.1)$$

Eyring's equation replaces the Arrhenius pre-exponential factor A' with the constant k_BT/h, where k_B is Boltzmann's constant, h is the Plank constant, and T is the temperature; and the activation energy, E_a, is replaced by the Gibbs free energy difference between the initial and the final states, ΔG. Eyring's equation was derived for gas phase reactions, but the form has been broadly used as it includes the chemical potential, ΔG, instead of the activation energy.

Kramers developed the means to analyze the nonequilibrium behavior of reaction kinetics from transition state theory. Kramers' equation for off-rates is

$$k_{off}(F) = \frac{\omega_{min}\omega_{ts}}{2\pi\gamma}\exp\left(\frac{-E_a + Fx}{k_BT}\right) = k_{off,0}\exp\left(\frac{Fx}{k_BT}\right) \quad (36.2)$$

where ω_{min} and ω_{ts} are the angular frequencies of the energy minimum and transition state, respectively, γ is the damping relaxation rate, $k_{off,0}$ is the equilibrium off-rate, and x is the distance between the energy minimum and the transition state along the direction of the applied force, F. Bell used Kramers' theory to develop a simplified expression for the equilibrium bond lifetime of a bond, $\tau = 1/k_{off}$, in terms of $k_{off,0}$, x, and F.[50] We can see from Equation 36.2 that by increasing the applied force, the bond lifetime will decrease exponentially with a scaling factor that is determined by the thermal activation force k_BT/x. This analysis of a *slip bond* is based on the assumption that there is a single transition state and that the distance x remains constant with the increase in the loading rate.

The application of Kramers' theory to single-molecule force measurements allows the key nonequilibrium parameters, i.e., $k_{off,0}$ and x, to be determined by measuring the change in k_{off} as a function of applied force. Evans and Ritchie[10] showed that this can be achieved by either measuring the bond rupture force distribution as a function of the rate at which the external force is applied to the bond, r_F, or measuring the bond lifetime as a function of applied force.[10] The first approach has been widely applied in atomic force microscopy measurements as force measurements can be easily made at specific loading rates. We will demonstrate how this approach is implemented in Section 36.4. The second approach is also readily implemented if a force clamp can be applied to a bond, i.e., the half-life of the bond is determined by fitting a distribution on bond lifetimes across a range of loads. We will not analyze the force clamp technique but will present results of measurements in Section 36.4.

In order to relate the nonequilibrium kinetics of a biomolecular interaction to loading dynamics, Evans and Ritchie showed that the most probable adhesive force, F^*, and the loading rate at which the forces are measured, r_F, are related through the equation

$$F^* = \frac{k_BT}{x}\ln\left(\frac{r_F}{k_{off,0}k_BT}\right) \quad (36.3)$$

where F^* is determined from a histogram of rupture forces at a specific loading rate, and $r_F = \frac{k_c k_t s}{(k_c + k_t)}$ with k_c being the spring constant of the cantilever, k_t being the instantaneous spring constant of the tether, and s being the probe velocity.[10] The key nonequilibrium parameters can be determined by characterizing the increase in unbinding forces over a range of loading rates. In this case, the standard deviation in measured unbinding forces, σ_F, is proportional to the thermal activation energy x/k_BT. This analysis is based on the assumption that the transition state distance x remains constant with the increase in the loading rate.

36.3.2 Formation of Multiple Bonds

The formation of multiple distinct bonds can take place within a single intermolecular interaction or when multiple biomolecules on the AFM probe bind to their complementary partners on a surface. For example, CAMs are composed of up to seven distinct domains, and these molecules are known to form homophilic assemblies at points of contact between cells.[51,52] Williams[53] developed models of multiple bond formations in which the load was distributed equally across all bonds, i.e., parallel bond configuration, or through correlated binding, i.e., zipper configuration. Assuming that the system is composed of multiple uncorrelated single bonds that break at once, the multiple bond model could be formulated as

$$r_F = k_{off}\frac{k_BT}{x}\left[\sum_{n=1}^{N}\frac{1}{n^2}\exp\left(\frac{Fx}{nk_BT}\right)\right] \quad (36.4)$$

where N is the total number of bonds.

36.3.3 Association Rate Constant

The association rate, k_{on} ($M^{-1}\,s^{-1}$), for the reaction of the molecules on the tip and surface has been detected through the bond force,[54] and it has been empirically observed that the association rate is determined by a first-order reaction, which suggests that the biomolecules on one surface is in significant excess. The half-life of the reaction, τ, is determined by fitting the probability of binding, P_b, for the time, t, that the molecules on the probe and the surface are in contact:

$$P_b = P_{max}\left[1-\exp\left(\frac{-(t-t_0)}{\tau}\right)\right] \quad (36.5)$$

where t_0 is an observed lag time of binding, and P_{max} is the maximum observable binding probability. According to the first-order reaction kinetic theory, $k_{on}=\frac{1}{\tau C_{eff}}$, where C_{eff} is the effective concentration of the reactant on the probe (mol/m^3). The effective concentration of the reactants has been estimated by calculating the density of molecules generated by a half-sphere of radius equal to the PEG length. The k_{on} measured by AFM is strongly influenced by the immobilized chemistry used to link the biomolecule to the surfaces and is unlikely to be equivalent to k_{on} measured by other techniques, such as surface plasmon resonance or homogenous kinetic assays.

36.4 Analysis of apCAM–apCAM Binding with Force Spectroscopy

CAMs, such as integrin, cadherin, and the immunoglobulin superfamily CAMs, are of primary importance in cell–cell and cell–substrate interactions, signaling pathways, and other crucial biological processes. The *Aplysia* CAM (apCAM) is the analog of neuronal CAM (NCAM), both in structure and function, and is preferentially expressed in the growth cones of the axons of the giant sea slug *Aplysia*.[55] We have recently completed several studies of the nonequilibrium binding kinetic of apCAM homophilic interactions due to the important role that CAMs play in neural development, regeneration, and synaptic plasticity, including learning and memory consolidation.[56,57]

The extracellular portion of apCAM consists of five N-terminal Ig domains followed by two fibronectin (Fn) type III domains, and in this study, a recombinant protein was produced with a C-terminus His6-tag. apCAM was immobilized on the gold-coated AFM cantilevers and surfaces using mixed SAMs of COOH-PEG-SH (MW 3.4 kDa PEG) and CH_3O-PEG-SH (MW 2.0 kDa PEG).[56] In initial studies, apCAM was immobilized on a nitrilotriacetic acid group conjugated to COOH-PEG to allow the extracellular portion of apCAM to be oriented on the surface; however, this approach was abandoned after it was found that the apCAM monolayers lost activity during force measurements. To ensure the strong binding of apCAM to the surfaces, it was directly linked to N-hydroxysuccinimide activated COOH-PEG.[56] This chemistry resulted in the random orientation of covalently immobilized apCAM.

To characterize the nonequilibrium binding kinetics of apCAM–apCAM interactions, adhesion forces were measured with the AFM at tip velocities from 70 to 700 nm s^{-1}. Representative unbinding force histogram, force–distance curves, and rupture distance histogram for apCAM monolayers formed on a 100% COOH-PEG SAM for a probe velocity of 140 nm s^{-1} are provided (Figure 36.4). The unbinding force data in Figure 36.4a were fit with two Gaussian curves, with a weak interaction centered at 75 ± 25 pN and a stronger interaction centered at 181 ± 42 pN. These two adhesion forces could be attributed either to a single homophilic apCAM interaction composed of two distinct bonds or multiple apCAM interactions. The molecular origin of this distinct force distribution cannot be determined from the form of the force histogram (Figure 36.4a).

Force–distance curves for weak and strong bonds can be distinguished (black and red lines, respectively, in Figure 36.4b). As the distance between the tip and the surface increases from 0 to 20 nm, the force applied to the bonds increased nonlinearly until the intermolecular interaction ruptured. The stretching of simple polymer bonds with the AFM has been found to be described by the freely jointed chain (FJC) model at low forces.[18] The contour length of the FJC model was 22 ± 3 nm if we assume that the stretching region has a Kuhn length of 0.7 ± 0.2 nm (Figure 36.4c). This Kuhn length is a characteristic of linear PEG polymers and was selected because our previous study demonstrated that this polymer dominates the apCAM–apCAM interaction.[56] The observed contour length is approximately half the length previously measured for the apCAM-PEG assembly, which leads us to conclude that the measurements presented are the result of two sets of apCAM–apCAM

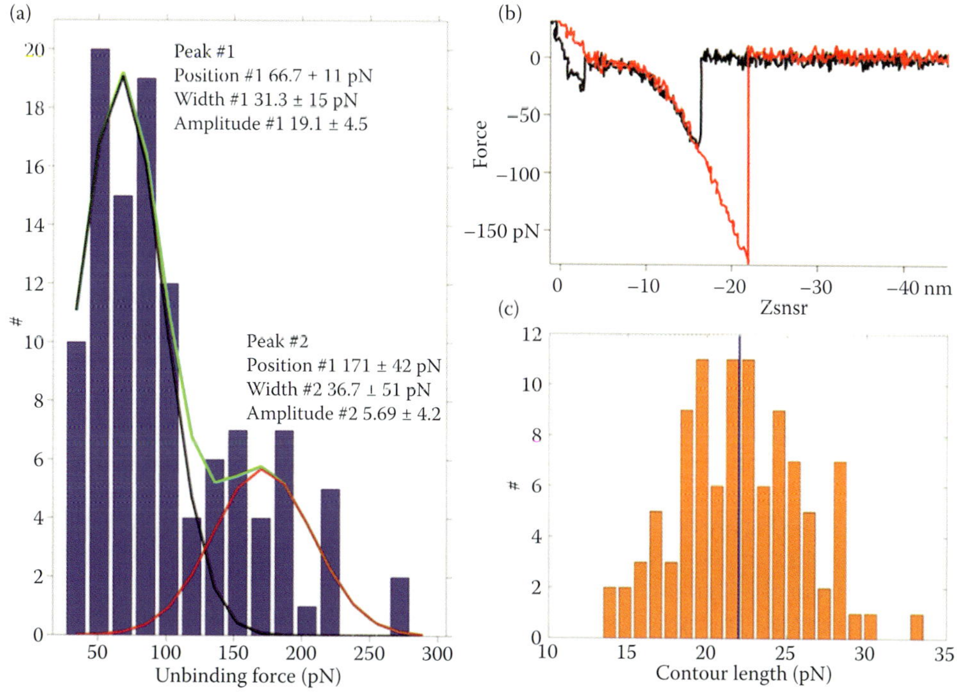

FIGURE 36.4
Analysis of force [illegible] apCAM functionalized cantilevers and gold substrates. Single-molecule force measurements were made at a tip velocity of 140 nm s^{-1} over a 5 × 5 μm area with a 70 pN maximum force in 0.2 μm filtered phosphate-buffered saline at room temperature. The spring constants of the cantilevers were determined by thermal fluctuation in air.[58] (a) The force histogram had two distinct peaks centered at 75 ± 25 and 181 ± 42 pN. The bin size in the force histograms was set at two times the root mean square noise at the different loading rates and ranged from 11 to 26 pN. (b) Force–distance curves showing unbinding events at 100 and 200 pN. (c) Histogram of the contour length distribution at 140 nm s^{-1} tip velocity, showing a mean value of 22 nm.

interactions. This is consistent with the dense apCAM monolayers that were used in the measurements presented, i.e., previous studies of single-molecule interactions used COOH-PEG to M-PEG ratios of 1:200 with M-PEG.[56] This leads us to conclude that the two peaks observed (Figure 36.4a) are the result of the coherent rupture of a pair of higher-order apCAM interactions. Previous studies of NCAM interactions indicate that both *cis* and *trans* homophilic bonds drive the formation of higher-order NCAM assemblies that play a central role in neural development, regeneration, and synaptic plasticity (Figure 36.5).[51,59]

Homophilic binding of NCAM is suggested to be based on zipper formation, which was supported with crystallographic evidence.[59] Similar to apCAM, the extracellular portion of the NCAM molecule consists

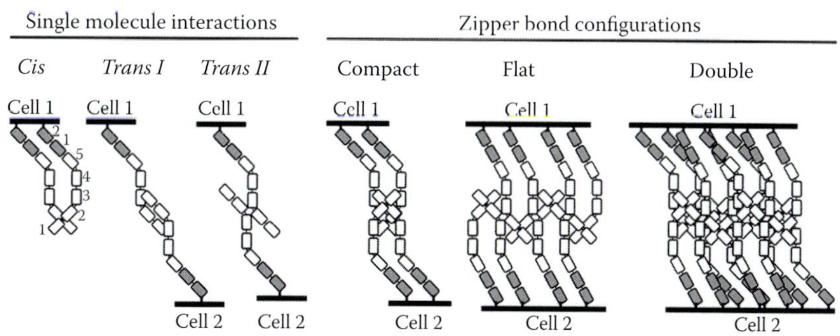

FIGURE 36.5
Cis, trans, and zipper bond types proposed for the homophilic NCAM bond. The extracellular portion of the NCAM molecule consists of two Fn type III domains (gray) and five Ig domains (white) extending from the cell membrane. *Cis* interactions form between molecules on the same cell through Ig1–Ig2 interactions. *Trans* interactions form between molecules on opposing cells through Ig1–Ig3 (*trans I*) or through Ig2–Ig3 (*trans II*) interactions. Zipper bonds form with increasing surface density of the NCAM molecule. Compact zipper is a combination of *cis* and *trans I* bonds, flat zipper is a combination of *cis* and *trans II* bonds, and double zipper is a combination of compact and flat zipper bonds.

of two Fn type III domains and two Ig domains (gray and white shapes in Figure 36.5, respectively). Cis bonds form between two NCAM molecules of the same cell through two interactions between Ig1 and Ig 2 domains of each molecule. Trans bonds form between two NCAM molecules on two opposing cells. Two types of trans bonds are proposed: one that is based on Ig2–Ig3 interactions (trans I) and one that is based on Ig1–Ig3 interactions (trans II). Depending on the surface density and mobility of the NCAM molecules on respective cell membranes, several distinct zipper bond configurations may occur based on the cis and trans bonds. Compact zipper forms as a combination of cis and trans I bonds; flat zipper forms as a combination of cis and trans II bonds. Finally, double zipper configuration occurs as a combination of compact and flat zippers and contains cis, trans I, and trans II bonds (Figure 36.5).

The nonequilibrium kinetics of the apCAM interaction can be determined from the dynamic force spectra data (such as that presented in Figure 36.6) using Equation 36.3. The most probable rupture force, F^*, at a specific loading rate, r_F, is determined from a Gaussian fit of the force histograms. The transition distance, x, is determined from the slope, and the dissociation rate, $k_{off,0}$, is determined from the intercept at $r_F = 0$ of the F^* vs. $\ln(r_F)$ curve. Figure 36.6 presents the F^* vs. $\ln(r_F)$ curve for single homophilic apCAM interaction previously described[56] as well as the high-density apCAM measurements presented above. The single homophilic apCAM interaction was found to be characterized by weak and strong slip bonds associated with specific Ig domain interactions. The weak bond had an x of 0.1 ± 0.06 nm and a dissociation rate k_{off} of 0.72 ± 1 s^{-1}, while the strong bond had an x of 0.46 ± 0.88 nm and a dissociation rate k_{off} of 0.05 ± 0.88 s^{-1}. The strong bond results proved to have a significant level of uncertainty as a result of the relatively low slope of the F^* vs. $\ln(r_F)$ curve. This measurement error is inherent to the force ramp mode of bond characterization of single-molecule interactions that can be overcome by using force clamp measurements.[57] The results of the high-density apCAM measurements shown in Figure 36.6 were strikingly different from the single-molecule measurements as they do not appear to be slip bonds, i.e., F^* does not appear to have strong r_F dependence. We attribute the *nonslip* properties of these bonds to the formation of higher-order apCAM interactions, which could play an important role in neural development, regeneration, and synaptic plasticity.

Association rate measurements. The results of the apCAM homophilic bond association rate, k_{on}, measurements are presented (Figure 36.7). The data were collected as a binding probability function of the dwell time and fit to Equation 36.5, yielding a contact time τ of 319 ms. Assuming a PEG linker length of 24 nm and an apCAM length of 20 nm, the effective volume was calculated to be 2.29×10^{-5} μm^3, and the effective concentration was calculated to be 4.37×10^4 molecules μm^{-3}. This results in an association rate of 1.37×10^5 M^{-1} s^{-1}, which is approximately one order of magnitude larger than the 2×10^4 M^{-1} s^{-1} on rate measured for single apCAM–apCAM interactions.[56]

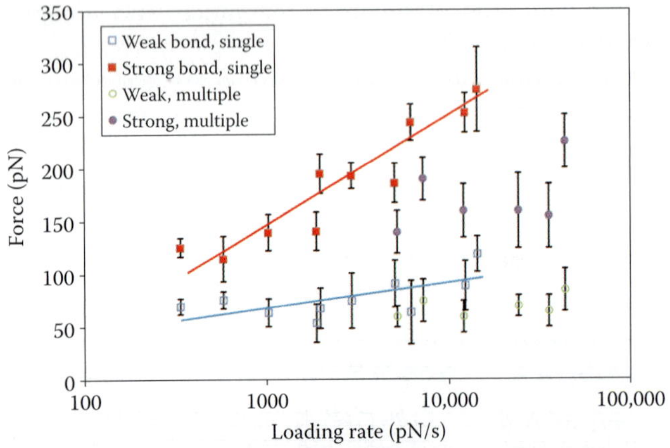

FIGURE 36.6
Analysis of the nonequilibrium binding kinetics of the homophilic apCAM intermolecular interaction. Force spectra were measured for approximately 100 apCAM interactions for probe velocities that range between 70 and 700 nm s^{-1}. The unbinding force was determined by fitting the force histograms with two Gaussian distributions, as shown in Figure 36.4a, and the error bars represent the width of the Gaussian fit. The unbinding force-probe velocity data were fit with the Bell–Evans model (solid lines) to extract values of x and k_{off} for two uncorrelated bonds.

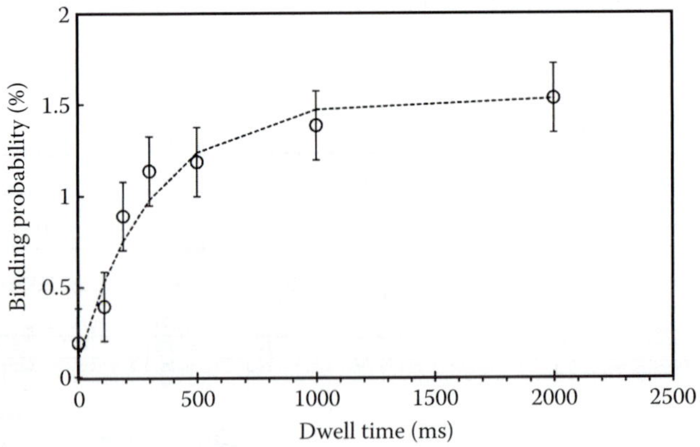

FIGURE 36.7
Analysis of the on-rate of the homophilic apCAM interaction. The binding probability, P_b, was measured as a function of the dwell time, t, of the probe on the surface. The data were fitted to Equation 36.5 to determine the maximum probability of interaction, P_{max} = 1.5%, and the half-life of the reaction, τ = 319 ms. These measurements were performed with a probe velocity of 700 nm s^{-1}.

36.5 Conclusions

The AFM has been used to characterize a range of biomolecular interactions, and many of these interactions have been found to have *slip-bond* behavior, i.e., the bonds appear to be stiffer the higher the rate at which force is applied to them. The molecular details of these interactions have been determined using site-directed mutagenesis as well as all-atomic molecular dynamics simulations, which have not been described in this review. These advanced techniques provide us with the means to define the molecular nature of more complex interactions, e.g., higher-order interactions such as the double apCAM–apCAM interaction or catch–bond interaction, in which intramolecular interactions may lead to the adsorption of energy or changes in conformation. AFM measurements promise to provide a means to understand the more complex molecular mechanisms of cellular interactions in the coming years. Ideally, these measurements will be made with cultured cells, although single-molecule force measurements on live cells remain challenging.[7] Clearly, improvements in biomolecular surface chemistry will play an important role in these developments. Combining AFM with recent advances in high-resolution, live cell imaging also promises to allow us to study the chemical mechanisms of mechanotransduction.

Acknowledgments

The authors thank Fernando Mancilla Perez, Ying Xiong, and Jay Zhong for their thoughtful discussions. This work was supported by Nanoremedies Programme, which is funded by the Irish Higher Education Administration (HEA) Programme for Research in Third Level Institutions (PRTLI5) and the European Regional Development Fund. The authors also acknowledge the support of the Science Foundation Ireland (Grant Nos. 05/W.1/B31, 07/RP1/B1376, and 08/IN1/B2072), IRCSET Postdoctoral Fellowship Scheme (EMPOWER), and the Marie Curie Intra-European Fellowship Scheme (Fp7-People-2010-IEF-273420).

References

1. Aloy, P. & Russell, R. B. Structural systems biology: Modelling protein interactions. *Nat. Rev. Mol. Cell Biol.* **7**, 188–197 (2006).
2. Kueng, A., Kranz, C. & Mizaikoff, B. Scanning probe microscopy with integrated biosensors. *Sens. Lett.* **1**, 2–15 (2003). doi:10.1166/sl.2003.001.
3. Moi, N. J. & Fischer, M. J. E. Surface plasmon resonance: Methods and protocols. *Methods Mol. Biol.* **627**, 15–54 (2010).
4. Shoemaker, B. A. & Panchenko, A. R. Deciphering protein–protein interactions. Part I. Experimental techniques and databases. *PLoS Comput. Biol.* **3**, e42 (2007). doi:10.1371/journal.pcbi.0030042.
5. Hoffman, D. B., Grashoff, C. & Schwartz, M. A. Dynamic molecular processes mediate cellular mechanotransduction. *Nature* **475**, 316–323 (2011).
6. Hinterdorfer, P., Garcia Parajo, M. F. & Dufrene, Y. F. Single-molecule imaging of cell surfaces using near field nanoscopy. *Acc Chem Res* **45**, 327–336 (2012). doi:10.1021/ar2001167.
7. Dufrene, Y. F. et al. Five challenges to bringing single-molecule force spectroscopy into living cells. *Nat. Methods* **8**, 123–127 (2011). doi:10.1038/nmeth0211-123.
8. Li, J. Y., Fernandez, J. M. & Berne, B. J. Water's role in the force-induced unfolding of ubiquitin. *Proc. Natl. Acad. Sci. U. S. A.* **107**, 19284–19289 (2010). doi:10.1073/pnas.1013159107.
9. Merkel, R., Nassoy, P., Leung, A., Ritchie, K. & Evans, E. Energy landscapes of receptor-ligand bonds explored with dynamic force spectroscopy. *Nature* **397**, 50–53 (1999).
10. Evans, E. & Ritchie, K. Dynamic strength of molecular adhesion bonds. *Biophys. J.* **72**, 1541–1555 (1997).
11. Marshall, B. T. et al. Direct observation of catch bonds involving cell-adhesion molecules. *Nature* **423**, 190–193 (2003). doi:10.1038/nature01605
12. Florin, E. L., Moy, V. T. & Gaub, H. E. Adhesion forces between individual ligand-receptor pairs. *Science* **264**, 415–417 (1994). doi:10.1126/science.8153628.
13. Lee, G. U., Chrisey, L. A. & Colton, R. J. Direct measurement of the forces between complementary strands of DNA. *Science* **266**, 771–773 (1994). doi:10.1126/science.7973628.
14. Moy, V. T., Florin, E. L. & Gaub, H. E. Intermolecular forces and energies between ligands and receptors. *Science* **266**, 257–259 (1994). doi:10.1126/science.7939660.
15. Grange, W., Strunz, T., Schumakovitch, I., Guntherodt, H. J. & Hegner, M. Molecular recognition and adhesion of individual DNA strands studied by dynamic force microscopy. *Single Mol.* **2**, 75–78 (2001). doi:10.1002/1438-5171(200107)2:2<75::aid-simo75>3.3.co;2-#.
16. Strunz, T., Oroszlan, K., Schafer, R. & Guntherodt, H. J. Dynamic force spectroscopy of single DNA molecules. *Proc. Natl. Acad. Sci. U. S. A.* **96**, 11277–11282 (1999). doi:10.1073/pnas.96.20.11277.
17. Preiner, J. et al. Imaging and detection of single molecule recognition events on organic semiconductor surfaces. *Nano Lett.* **9**, 571–575 (2009). doi:10.1021/nl802721g.
18. Sulchek, T. A. et al. Dynamic force spectroscopy of parallel individual Mucin1-antibody bonds. *Proc. Natl. Acad. Sci. U. S. A.* **102**, 16638–16643 (2005). doi:10.1073/pnas.0505208102.
19. Neuert, G., Albrecht, C., Pamir, E. & Gaub, H. E. Dynamic force spectroscopy of the digoxigenin-antibody

complex. *FEBS Lett.* **580**, 505–509 (2006). doi:10.1016/j.febslet.2005.12.052.

20. Schwesinger, F. et al. Unbinding forces of single antibody-antigen complexes correlate with their thermal dissociation rates. *Proc. Natl. Acad. Sci. U. S. A.* **97**, 9972–9977 (2000). doi:10.1073/pnas.97.18.9972.
21. Raab, A. et al. Antibody recognition imaging by force microscopy. *Nat. Biotechnol.* **17**, 902–905 (1999).
22. Ros, R. et al. Antigen binding forces of individually addressed single-chain Fv antibody molecules. *Proc. Natl. Acad. Sci. U. S. A.* **95**, 7402–7405 (1998). doi:10.1073/pnas.95.13.7402.
23. Hinterdorfer, P., Baumgartner, W., Gruber, H. J., Schilcher, K. & Schindler, H. Detection and localization of individual antibody-antigen recognition events by atomic force microscopy. *Proc. Natl. Acad. Sci. U. S. A.* **93**, 3477–3481 (1996). doi:10.1073/pnas.93.8.3477.
24. Dammer, U. et al. Binding strength between cell-adhesion proteoglycans measured by atomic-force microscopy. *Science* **267**, 1173–1175 (1995). doi:10.1126/science.7855599.
25. Chtcheglova, L. A., Haeberli, A. & Dietler, G. Force spectroscopy of the fibrin(ogen)—fibrinogen interaction. *Biopolymers* **89**, 292–301 (2008). doi:10.1002/bip.20910.
26. Colombini, B., Bagni, M. A., Romano, G. & Cecchi, G. Characterization of actomyosin bond properties in intact skeletal muscle by force spectroscopy. *Proc. Natl. Acad. Sci. U. S. A.* **104**, 9284–9289 (2007). doi:10.1073/pnas.0611070104.
27. Eckel, R., Ros, R., Decker, B., Mattay, J. & Anselmetti, D. Supramolecular chemistry at the single-molecule level. *Angew. Chem.-Int. Ed.* **44**, 484–488 (2005). doi:10.1002/anie.200461382.
28. Kuhner, F. et al. LexA-DNA bond strength by single molecule force spectroscopy. *Biophys. J.* **87**, 2683–2690 (2004). doi:10.1529/biophysj.104.048868.
29. Verbelen, C., Gruber, H. J. & Dufrene, Y. F. The NTA-HiS(6) bond is strong enough for AFM single-molecular recognition studies. *J. Mol. Recognit.* **20**, 490–494 (2007). doi:10.1002/jmr.833.
30. Zhang, X. H., Craig, S. E., Kirby, H., Humphries, M. J. & Moy, V. T. Molecular basis for the dynamic strength of the integrin alpha(4)beta(1)/VCAM-1 interaction. *Biophys. J.* **87**, 3470–3478 (2004). doi:10.1529/biophysj.104.045690.
31. Wieland, J. A., Gewirth, A. A. & Leckband, D. E. Single molecule adhesion measurements reveal two homophilic neural cell adhesion molecule bonds with mechanically distinct properties. *J. Biol. Chem.* **280**, 41037–41046 (2005). doi:10.1071/jbc.M503975200.
32. Jiang, Y. X. et al. Specific aptamer–protein interaction studied by atomic force microscopy. *Anal. Chem.* **75**, 2112–2116 (2003). doi:10.1021/ac026182s.
33. Baumgartner, W. et al. Cadherin interaction probed by atomic force microscopy. *Proc. Natl. Acad. Sci. U. S. A.* **97**, 4005–4010 (2000). doi:10.1073/pnas.070052697.
34. Oesterhelt, F., Rief, M. & Gaub, H. E. Single molecule force spectroscopy by AFM indicates helical structure of poly(ethylene-glycol) in water. *New J. Phys.* **1**, 6.1–6.11 (1999). doi:10.1088/1367-2630/1/1/006.
35. Rief, M., Clausen-Schaumann, H. & Gaub, H. E. Sequence-dependent mechanics of single DNA molecules. *Nat. Struct. Biol.* **6**, 346–349 (1999).
36. Rief, M., Oesterhelt, F., Heymann, B. & Gaub, H. E. Single molecule force spectroscopy on polysaccharides by atomic force microscopy. *Science* **275**, 1295–1297 (1997). doi:10.1126/science.275.5304.1295.
38. Rief, M., Gautel, M., Oesterhelt, F., Fernandez, J. M. & Gaub, H. E. Reversible unfolding of individual titin immunoglobulin domains by AFM. *Science* **276**, 1109–1112 (1997). doi:10.1126/science.276.5315.1109.
37. Acevedo-Velez, C., Andre, G., Dufrene, Y. F., Gellman, S. H. & Abbott, N. L. Single-molecule force spectroscopy of beta-peptides that display well-defined three-dimensional chemical patterns. *J. Am. Chem. Soc.* **133**, 3981–3988 (2011). doi:10.1021/ja1089183.
39. Janovjak, H., Kessler, M., Oesterhelt, D., Gaub, H. & Muller, D. J. Unfolding pathways of native bacteriorhodopsin depend on temperature. *EMBO J.* **22**, 5220–5229 (2003). doi:10.1093/emboj/cdg509.
40. Muller, D. J., Baumeister, W. & Engel, A. Controlled unzipping of a bacterial surface layer with atomic force microscopy. *Proc. Natl. Acad. Sci. U. S. A.* **96**, 13170–13174 (1999). doi:10.1073/pnas.96.23.13170.
41. Meyer, G. & Amer, N. M. Novel optical approach to atomic force microscopy. *Appl. Phys. Lett.* **53**, 1045–1047 (1988). doi:10.1063/1.100061.
42. Alexander, S. et al. An atomic-resolution atomic-force microscope implemented using an optical-lever. *J. Appl. Phys.* **65**, 164–167 (1989). doi:10.1063/1.342563.
43. Green, J. B. D., Novoradovsky, A. & Lee, G. U. Effect of mechanical contact on the molecular recognition of biomolecules. *Langmuir* **15**, 238–243 (1999). doi:10.1021/la980934p.
44. Walters, D. A. et al. Short cantilevers for atomic force microscopy. *Rev. Sci. Instrum.* **67**, 3583–3590 (1996). doi:10.1063/1.1147177.
45. Hinterdorfer, P. & Dufrene, Y. F. Detection and localization of single molecular recognition events using atomic force microscopy. *Nat. Methods* **3**, 347–355 (2006). doi:10.1038/nmeth871.
46. Lee, G. U. et al. Chemically-specific probes for the atomic force microscope. *Isr. J. Chem.* **36**, 81–87 (1996).
47. Sagiv, J. Organized monolayers by adsorption. 1. Formation and structure of oleophobic mixed monolayers on solid-surfaces. *J. Am. Chem. Soc.* **102**, 92–98 (1980). doi:10.1021/ja00521a016.
48. Nuzzo, R. G. & Allara, D. L. Adsorption of bifunctional organic disulfides on gold surfaces. *J. Am. Chem. Soc.* **105**, 4481–4483 (1983). doi:10.1021/ja00351a063.
49. Zambelli, S. Chemical kinetics and diffusion approach: The history of the Klein-Kramers equation. *Arch. Hist. Exact Sci.* **64**, 395–428 (2010). doi:10.1007/s00407-010-0059-9.
50. Bell, G. I. Models for specific adhesion of cells to cells. *Science* **200**, 618–627 (1978). doi:10.1126/science.347575.
51. Atkins, A. R. et al. Association between the first two immunoglobulin-like domains of the neural cell adhesion molecule N-CAM. *FEBS Lett.* **451**, 162–168 (1999). doi:10.1016/s0014-5793(99)00554-2.

52. Aricescu, A. R. & Jones, E. Y. Immunoglobulin superfamily cell adhesion molecules: Zippers and signals. *Curr. Opin. Cell Biol.* **19**, 543–550 (2007). doi:10.1016/j.ceb.2007.09.010.
53. Williams, P. M. Analytical descriptions of dynamic force spectroscopy: Behaviour of multiple connections. *Anal. Chim. Acta* **479**, 107–115 (2003). doi:10.1016/s0003-2670(02)01569-6.
54. Rankl, C. et al. Multiple receptors involved in human rhinovirus attachment to live cells. *Proc. Natl. Acad. Sci. U. S. A.* **105**, 17778–17783 (2008). doi:10.1073/pnas.0806451105.
55. Mayford, M., Barzilai, A., Keller, F., Schacher, S. & Kandel, E. R. Modulation of an NCAM-related adhesion molecule with long-term synaptic plasticity in aplysia. *Science* **256**, 638–644 (1992). doi:10.1126/science.1585176.
56. Martines, E. et al. Single-molecule force spectroscopy of the aplysia cell adhesion molecule reveals two homophilic bonds. *Biophys. J.* **103**, 649–657 (2012). doi:10.1016/j.bpj.2012.07.004.
57. Kilinc, D., Blasiak, A., O'Mahony, J. J., Suter, D. M. & Lee, G. U. Magnetic tweezers-based force clamp reveals mechanically distinct apCAM domain interactions. *Biophys. J.* **103**, 1120–1129 (2012). doi:10.1016/j.bpj.2012.08.025.
58. Hutter, J. L. & Bechhoefer, J. Calibration of atomic-force microscope tips. *Rev. Sci. Instrum.* **64**, 1868–1873 (1993). doi:10.1063/1.1143970.
59. Soroka, V. et al. Structure and interactions of NCAM Ig1-2-3 suggest a novel zipper mechanism for homophilic adhesion. *Structure* **11**, 1291–1301 (2003). doi:10.1016/j.str.2003.09.006.

37

Characterization of Biomolecular Interactions with the Surface Forces Apparatus

Marina Ruths, Carlos Drummond, and Jacob N. Israelachvili

CONTENTS

Biological macromolecules and surfaces interact through different nonspecific and specific forces. These interactions, which can be highly complex, affect the organization and morphology of biological systems at different length scales. Since biological systems are not at thermodynamic equilibrium, the interactions are best described as processes that evolve in time and space. Static and dynamic forces between surfaces functionalized with biomolecules can be directly measured with the surface force apparatus. The general principles of such measurements are described, and examples are shown of measured forces and film thickness transitions associated with hemifusion of supported bilayers, interactions between end-adsorbed glycoproteins and between ligands and receptors, and forces generated due to polymerization of confined actin.

37.1 Interactions in Biological Systems

Most chemical and colloidal interactions are considered as occurring within a single macromolecule (e.g., a folding protein), between a molecule and its surrounding medium (solvent effects), or among pairs of molecules, particles, or surfaces (binding, aggregation, and adsorption). However, many biological interactions are many-body interactions that progress in space and time, and specific interactions also exist that are not described by these general concepts. The interaction forces between biological molecules and surfaces are thus much more complex than those between nonbiological surfaces.

37.1.1 Biological Interactions Are Processes

The complexity of interactions in biological systems is due to a number of factors: (i) the simultaneous involvement of many different molecules and different noncovalent forces—van der Waals, electrostatic, solvation (hydration, hydrophobic), steric, entropic, and specific; and (ii) the flexibility of biological macromolecules and fluidity of membranes. Biological interactions are better thought of as processes that evolve in space and time and, under physiological conditions, involve a continuous input of energy.[1,2] Such systems are, therefore, not at thermodynamic equilibrium or even tending toward equilibrium.

A good example is the saltatory motion of white blood cells along the endothelium surfaces of blood capillaries (Figure 37.1a), which follows a chaotic, but not random, stick–slip type motion (Figure 37.1b).[1] This behavior has been successfully modeled in terms of the thermally controlled binding and unbinding of specific ligand–receptor (L–R) bonds on the two surfaces.[3,4] Interestingly, the theoretical model used is conceptually similar to models used to analyze stick–slip friction, sound generation (e.g., the sound of a squeaking door or a violin), seismic movements, and earthquakes.

37.1.2 Intermolecular and Interparticle (Surface) Forces

Interactions between macroscopic bodies arise from the interactions between the constituent atoms and molecules of each body across the gap separating them. Figure 37.2 is a schematic representation of common intermolecular and interparticle forces or pair potentials (interaction energies as a function of separation distance).[1,5,6] However, the interactions or processes occurring between biomolecular assemblies such as those shown in Figure 37.1 can be far more complicated than these potentials because they are not additive in space, usually act at different times, and may be strongly dependent on approach and separation rates, as indicated schematically with arrows in Figure 37.2.

Theoretical models of interactions typically describe the energy between two flat surfaces (plates), whereas the force between two spheres, a sphere and a flat surface, or two crossed cylinders (the geometry in the surface forces apparatus [SFA] described in Section 37.2) is more easily accessed experimentally.[5,7] The latter two geometries are equivalent at small separation distances. The interaction energy per unit area between two flat surfaces, $E(D)_{\text{flat}}$, and the force between a sphere and a flat surface, $F(D)_{\text{curved}}$, are related through the Derjaguin approximation[8]:

$$F(D)_{\text{curved}} = 2\pi R E(D)_{\text{flat}} \tag{37.1}$$

where R is the radius of curvature and D is the distance between the opposing surfaces. SFA data are commonly presented as $F(D)/R$ vs. D to facilitate comparison (i) between experiments with different values of R and

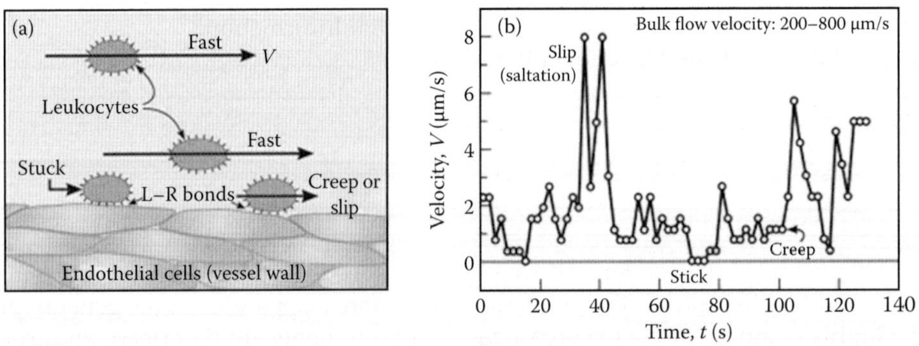

FIGURE 37.1
Biological interactions are processes. Leukocyte rolling along the endothelium in vitro (a) involves (b) sticking (V = 0), rolling or creep ($V \approx 1$ μm/s), stick–slip or saltation ($V \approx 5$ μm/s), and free or bulk flow (V = 200–800 μm/s). Each rolling process involves many mobile molecules, known as selectins, on both surfaces that interact via L–R bonds that in turn form and break in a complex but well-orchestrated way. (a: Reprinted from *Intermolecular and Surface Forces*, 3rd ed., Israelachvili, J. N., Copyright 2011, with permission from Elsevier; b: Reprinted from *Biophys. J.*, **66**, Goetz, D. J., et al., 2202–2209, Copyright 1994, with permission from Elsevier.)

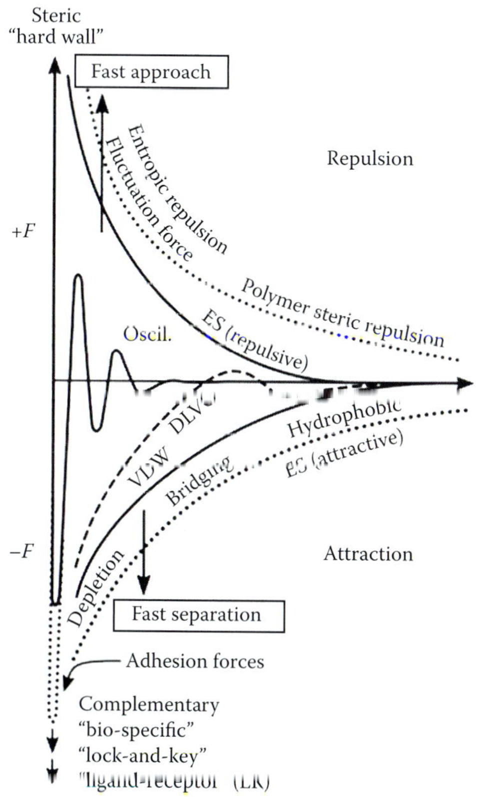

FIGURE 37.2
Intermolecular and surface forces. Schematic illustration of different types of pair potentials or force laws between complex macromolecules or biomembrane surfaces. Biological molecules and membrane surfaces often interact via four or more of these attractive and repulsive forces either simultaneously or in some systematic order in space and/or in time. Oscillatory forces due to solvent structure occur between smooth hard surfaces and are rarely, if ever, encountered in biological systems where the surfaces are heterogeneous and dynamically rough (fluctuating) on the scale of a water molecule. (Reprinted from *Intermolecular and Surface Forces*, 3rd ed., Israelachvili, J. N., Copyright 2011, with permission from Elsevier.)

(ii) between experimental force data and theories for the interaction energy according to Equation 37.1 (cf. Figure 37.2). Since within a given geometry the force is the gradient of the energy, i.e., the derivative with respect to separation distance, $F(D) = -dE(D)/dD$, one can also obtain $F(D)_{flat}$ from $E(D)_{flat}$ and $E(D)_{curved}$ from $F(D)_{curved}$ as needed.

37.1.2.1 van der Waals Forces

van der Waals forces ("vdW" in Figure 37.2) is a collective name for three different interactions responsible for the short-range attraction (cohesion) between atoms and molecules. These three interactions occur both across vacuum and across a medium and consist of electrostatic forces between permanent dipoles, between permanent and induced dipoles, and between induced dipoles (London dispersion or fluctuation interaction). Except for highly polar materials such as water, the London interaction gives the largest contribution (70%–100%) to the van der Waals attraction.[5]

The interaction energy of the van der Waals attraction between atoms or molecules depends on their intermolecular separation distance r as

$$E(r) = -C_{vdW}/r^6 \tag{37.2}$$

where C_{vdW} is a materials-dependent constant. The van der Waals interaction energy between two macroscopic bodies separated by a distance D is dependent on their geometry and is always attractive between two bodies of the same material. For a sphere interacting with a flat surface, or two crossed cylinders with $R >> D$, the corresponding force is given by[5]

$$F(D) = -A_H R/(6D^2) \tag{37.3}$$

where A_H, the Hamaker constant, depends on the dielectric properties of the interacting materials. In a medium, the van der Waals forces are significantly lower than in vacuum, since the attractive interaction between two solute molecules in a medium (solvent) involves displacement and reorientation of solvent molecules. Typically, Hamaker constants for interactions in a medium are an order of magnitude lower than for interactions between the same surfaces in vacuum. Between macroscopic surfaces in liquids, van der Waals forces become important at $D < 10$–15 nm and may at these distances start to dominate interactions of different origin that are observed at larger separations (see below).

37.1.2.2 Electrostatic Double Layer Forces and the Derjaguin–Landau–Verwey–Overbeek (DLVO) Theory

When in contact with water, surfaces may become charged, either by dissolution of ions that become solvated and are distributed into the solution due to entropy, or by preferential adsorption of ions from the solution. This surface charge is balanced by a decaying concentration of counterions within a thin layer of solution close to the surface. The surface charge and this "atmosphere of counterions" near the surface (in excess of the bulk concentration) form a "diffuse double layer."[5] At low ionic strength (i.e., in dilute solution), a measure of its thickness is the Debye (screening) length, κ^{-1}:

$$\kappa^{-1} = \sqrt{\varepsilon_0 \varepsilon_r k_B T \Big/ \left(e^2 N_A \sum_i z_i^2 c_i\right)} \tag{37.4}$$

where e is the electron charge, N_A is Avogadro's constant, z_i is the ionic charge, c_i is the solution concentration, ε_0 is the vacuum permittivity, ε_r is the relative permittivity of the solvent, and the sum is carried over all ionic species present. The Debye length is a property of the solution, not of the charged surface, and is large in dilute solution (for example, 30 nm in 0.1 mM 1:1 electrolyte and about 1 μm in pure water). Interactions due to this diffuse double layer are thus expected to be long range at low ionic strength, but short range at high electrolyte concentration or high or low pH.

As the separation between two surfaces is decreased, the counterions stay between the surfaces because of their attraction to the surface charges; therefore, their concentration is increased, and repulsion arises due to the increased osmotic pressure in the confined thin film. The so-called "electrostatic double-layer force" ("ES repulsive" in Figure 37.2 for two identical charged surfaces, and "ES attractive" for surfaces of opposite charge) is thus of entropic origin. At large separation distances D (i.e., at weak overlap between the regions with excess ion concentration and thus weak overlap of the potentials), the interaction energy between two similarly charged surfaces decays exponentially, with the decay length being the Debye length, κ^{-1}:

$$E(D) = \text{constant} \times e^{-\kappa D} \tag{37.5}$$

where the constant depends on the geometry of the interacting surfaces, on the surface charge density, and on the solution conditions (temperature and ionic strength). The constant is found by solving the Poisson–Boltzmann equation under different constraints and approximations described in detail in textbooks on colloid science, where solutions for different geometries are provided.[5,9] The equations above are accurate only at D larger than about one Debye length. At smaller D, it can be shown that the interaction energy depends on whether the surfaces remain at constant potential (as assumed above), or at constant charge density (which would result in a larger repulsion), or somewhere between these limits.

All surfaces experience van der Waals forces, and these become important at small separation distances where they can overcome the repulsive double-layer forces because of their different distance dependence. This is the theoretical prediction behind the so-called Derjaguin–Landau–Verwey–Overbeek (DLVO) theory.[5,10,11] If the van der Waals forces and the double-layer force are assumed to be additive, the total force–distance curve can show several maxima and minima ("DLVO" in Figure 37.2), where the outer or "secondary" minimum is typically not more than a few k_BT deep and is responsible for the (reversible) flocculation of colloidal particles from an aqueous suspension. The DLVO theory also predicts an inner or "primary" minimum at smaller separation distances where interacting particles experience irreversible coagulation.

In practice, there are also forces other than van der Waals and double-layer forces at small separation distances, and a strongly confined thin film can have more features that cause additional repulsive barriers or stronger attraction than the one in the DLVO theory, as discussed in the following sections. Thermal fluctuations of flexible surfaces and adsorption of macromolecules may cause interaction forces that entirely dominate the properties of a system.

37.1.2.3 Hydrogen Bonding, Hydration, and Hydrophobic Forces

In many aqueous systems, the interactions between molecules, particles, and surfaces are not well described by the DLVO theory. Additional forces can arise through hydrogen bonding, which is an interaction between the permanent dipoles formed by certain functional groups (e.g., hydroxyl –OH or carboxyl –COOH). Because of the strong polarity of the O–H bond and the small size of the hydrogen atom, a strong interaction (up to ~10 k_BT/bond) can occur between electronegative atoms and hydrogen, such as between the oxygen atom of a water molecule and a hydrogen atom from another.[5] Such a hydrogen bonding transient "network" structure is the reason for many of the unusual properties of water, for example, its high boiling point compared to molecules of the same size that interact only through van der Waals forces. Hydrogen bonding is also important for the folding of proteins and the interaction of lipid headgroups with water.[12]

Other forces, whose origin is being investigated and debated, can also be observed in aqueous systems. Between hydrophilic surfaces, monotonically repulsive, exponentially decaying forces called structural or hydration forces may act at short range (D < 3–5 nm) and keep the particles or surfaces from coagulating in the manner expected by the DLVO theory.[13] This interaction is related to the energy needed to dehydrate (remove water from) surfaces with strongly hydrogen-bonding groups such as hydrated ions or hydroxyl groups, which modify the hydrogen-bonding network of liquid water adjacent to them.[14] At D < 1.5 nm, this hydration force may become oscillatory with an oscillation period corresponding to the diameter of a water molecule.[15,16]

Between hydrophobic surfaces in water, a monotonically attractive force can be found at longer range and with a much larger adhesion (attractive force upon separation from contact) than that expected from van der Waals forces ("hydrophobic" in Figure 37.2).[17–21] This force appears to be quite insensitive to the electrolyte concentration at D < 10 nm; it has been suggested that it is related to a depleted region of water with different density around hydrophobic surfaces.[22,23] This hydrophobic force has many important consequences such as the low miscibility of water and oil molecules, nonwetting of water on hydrophobic surfaces, and micelle formation, and it is believed to play a role in the rapid folding of proteins and in the adhesion and fusion of biological membranes and cells.

37.1.2.4 Thermal Fluctuation Forces

Between surfaces that are soft or fluid-like (as is common in biological systems), short-range repulsive forces called thermal fluctuation forces (also called entropic or steric forces) may occur. They are very effective at stabilizing the van der Waals attraction at some small but finite separation (Figure 37.2). These arise from thermal fluctuations, which make these surfaces dynamically rough, even though they may be perfectly smooth on a time average. Surfaces that fall into this category are fluid-like amphiphilic surfaces of micelles, bilayers, emulsions, and soap films, but also solid colloidal particle surfaces that are coated with fluid-like surfactant monolayers, as occurs in lubricating oils, paints, and toners.

This phenomenon can reduce the adhesion energy or force by up to three orders of magnitude, and it is mainly for this reason that fluid-like micelles and bilayers, biological membranes, emulsion droplets, or gas bubbles adhere to each other only very weakly. It has been shown[24,25] that these repulsive forces have an entropic origin arising from the osmotic repulsion between exposed thermally mobile surface groups once these overlap in a liquid. These phenomena include undulating and peristaltic forces between membranes or bilayers and, on the molecular scale, protrusion and head-group overlap forces where the interactions are also influenced by hydration forces.

37.1.2.5 Polymer-Mediated Forces

If large molecules protrude from surfaces, they may cause a long-range steric repulsion due to compression (molecular chain elasticity) and osmotic effects as the surfaces come close together. This is commonly observed in systems containing adsorbed polymers or macromolecules (Figure 37.2).[5,26] However, if the adsorbed amount of polymer is low, the interaction may have an attractive component due to bridging, i.e., adsorption of a polymer chain to two opposing surfaces. In the case of nonadsorbing polymer, an attractive depletion force can arise at a small separation distance between two surfaces as a result of the osmotic pressure of the solution. In cases where a chain end group or block has different chemical properties from the rest of the chain, it may become preferentially adsorbed, which gives rise to structures called polymer "mushroom" and "brush" layers,[27] with distinctive, repulsive interactions.[26] Contributions from polymer-mediated interactions will be discussed in more detail in the examples of interactions in biological model systems in Section 37.3.

37.1.3 Synergy and Complexity of Biological Interactions

Biological cells have clearly developed means for controlling different interaction forces over different distances and timescales. For example, a transient change in the local pH or ion concentration at a membrane surface will modify the *long-range* electrostatic double-layer interaction but may not affect the *short-range* L–R interaction. A change in the receptor–binding-site affinity could alter the short-range force but not the long-range force. Moreover, a change in the local membrane fluidity will alter the lateral diffusion rate of a ligand, thereby preventing or enhancing its probability of capture by a receptor.

The opening up or exposure of hydrophobic groups can be triggered by a number of stress-inducing effects, such as a change in the conformation of a protein, a local change in the curvature of a membrane, calcium-ion binding to one side of a membrane, transmembrane electric fields, flow stresses, and changes in the transmembrane osmotic pressure. There is now ample evidence that the exposure of hydrophobic "pockets" leads to adhesion and fusion, both of proteins and lipid bilayers. Moreover, the stresses that induce these pockets can occur well away from the pockets and, therefore, from the adhesion or fusion site. For example, an osmotic pressure change triggered in a vesicle by a particular protein will cause water to flow into the vesicle interior. This will cause the vesicle to swell, increasing the tension of the entire membrane that might result in fusion (e.g., with another membrane) in a completely different place on the vesicle surface.

Many of these simultaneously occurring interactions can be observed and their individual contribution quantified through force measurements on carefully constructed model systems using the SFA, which is described in Section 37.2. Experimental data on model biological systems are shown subsequently, together with comparisons with theoretical models.

37.2 Surface Forces Apparatus (SFA)

The SFA[7,28,29] is an instrument for sensitive measurements of normal and lateral forces between two macroscopic surfaces in contact or separated by a thin film. The surface separation distance can be measured and independently controlled to 0.1 nm, and the force resolution in the direction normal to the surfaces is 10 nN. Time- and rate-dependent effects in the measured forces can be ascribed to phenomena in the thin films or adsorbed layers confined between the surfaces.

37.2.1 Basics of the Technique

In SFA experiments, interaction forces are measured between two surfaces across air or a confined liquid film. The most commonly used surface substrates are back-silvered, molecularly smooth muscovite mica sheets glued to half-cylindrical fused-silica disks. These half-cylindrical surfaces are mounted in the SFA in a crossed-cylinder configuration (cf. Figure 37.3). At surface separation distances $D \ll R$, this is equivalent to a sphere-on-flat geometry, which presents several advantages over that of two parallel plates: alignment issues and edge effects are avoided, and different points of contact between the two surfaces can be easily investigated by displacing the disks laterally (for example, to check the repeatability of measurements at different regions of a pair of surfaces, or to move away from contamination or wear debris). In addition, the normal force $F(D)$ measured between cylindrical surfaces is related to the interaction energy between flat surfaces $E(D)$ according to the Derjaguin approximation (Equation 37.1).

Several different SFA setups have been developed for the measurements of normal and lateral forces. Early designs[18] were refined for measurements in liquids and vapors (the Mk II).[28] More recent models are easier to assemble and clean (the Mk IV)[30] or have improved distance controls and many different attachments (the Mk III/SFA3 model[29] in Figure 37.3 and the SFA2000[31]). The newer designs are also more user-friendly and have improved stability against thermal and mechanical drifts.

37.2.1.1 Distance, Force, and Adhesion Measurements

In the SFA, interaction forces normal to the surfaces are determined from the deflection of a spring of known spring constant, k, supporting the lower surface (Figure 37.3). The spring deflection as a function of separation distance or film thickness, D, is found by monitoring the change in distance with multiple beam interferometry (MBI)[32,33] as the surfaces are moved toward or away from one another using motorized stages or a piezoelectric actuator. Any deviation in the measured distance from the one expected from a calibration of the movement at large separation distances (where no forces act between the surfaces) implies a deflection of the spring, ΔD. The change in force when moving from one distance to another is $\Delta F = k\Delta D$, according to Hooke's law.

Regions of a force vs. distance curve where the gradient of the force, dF/dD, exceeds the spring constant k are inaccessible to the technique due to mechanical instability of the setup. When reaching such a region, the surfaces will spontaneously jump from one stable distance to the next. This is commonly encountered on approach and separation in adhesive systems, where parts of the force curve cannot be measured if a weak spring is used,

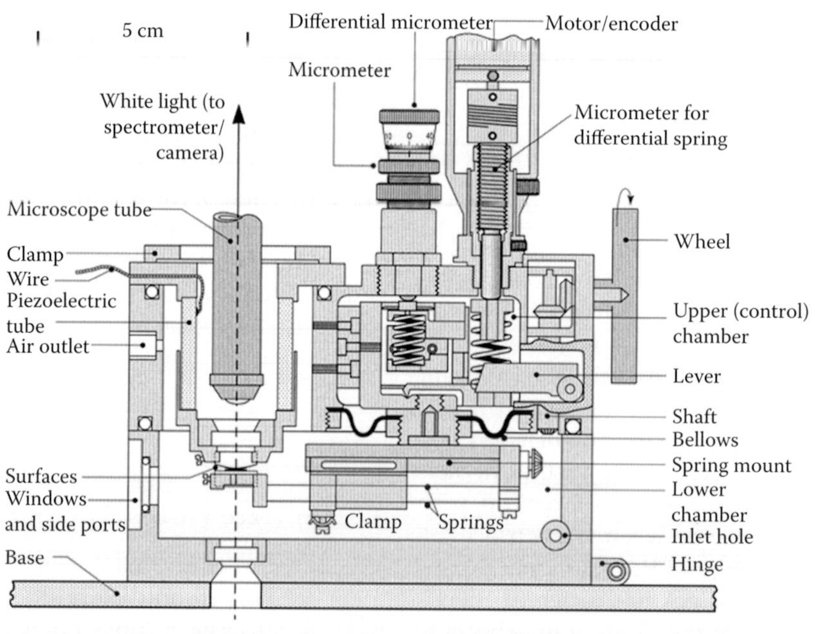

FIGURE 37.3
Surface Forces Apparatus (SFA). Schematic drawing of the SFA3 (Mk III) configured for measurements of forces normal to the surfaces.[29] The lower surface is supported on a "force-measuring" double cantilever spring with movable clamping, and the upper surface is mounted on a holder containing a piezoelectric tube for fine control of the separation distance between the surfaces, D. The mechanical distance control system (a system of weak and stiff springs reducing the movement of the manual and motor-driven micrometers) is separated from the main chamber of the instrument by Teflon bellows. The path of the light used for MBI is indicated by the dashed line. (Reprinted from Ruths, M., Israelachvili, J.N., In Springer Handbook of Nanotechnology, 3rd ed., pp. 857–922, 2010. With permission of Springer.)

but the "pull-off" or maximum adhesion force on separation can be accurately determined. The choice of spring stiffness is therefore of importance for the force sensitivity and for the detection of different regions of the force curve. Several SFA designs incorporate springs whose length can be changed during the experiment (cf. Figure 37.3), or use force feedback to control the force applied to the surfaces independently of the displacement.[7]

37.2.1.2 Control and Optical Imaging of Film Thickness and Contact Area

MBI[32,33] is used to determine D with subnanometer resolution (0.1 nm) and to measure the refractive index of confined films to 0.01. MBI is a particularly useful feature of the standard SFA setup, since it allows real-time in situ monitoring of the geometry of the interacting surfaces and of the region of contact. Parameters such as the film thickness (including the thickness of individual molecular layers) and the diameter or radius of the contact area (to a lateral resolution of about 1 μm) can be readily measured. The occurrence of protrusions or regions of different refractive index can also be detected,[34–36] which may indicate accumulation or depletion of material in some region of the contact.

The surfaces, any material confined between them, and the semitransparent, highly reflective silver layers on their back sides form a built-in Fabry–Perot interferometer. White light is passed through the lower surface, and the wavelengths that interfere constructively after several reflections between the silver mirrors emerge through the top surface. This light is focused onto the slit of a grating spectrometer. The resulting pattern, fringes of equal chromatic order (FECO, Figure 37.4),[32,33] maps the shape of the space between the silver layers and depends on the optical path length through the materials in the interferometer (the thicknesses and refractive indexes of the mica sheets and any material confined between them). A measurement of the interfering wavelengths, for example, from surfaces in mica–mica contact in dry N_2 gas (Figure 37.4a), is used as the basis for calculation of distances (D) between the front sides of the mica surfaces (and thus of film thicknesses of confined materials), since an increase in optical path causes the fringes to shift to larger wavelengths with respect to this value (Figure 37.4b). Relationships between the measured wavelengths, refractive indexes, and D for a symmetric interferometer (two mica sheets of equal thickness)[32] have later been extended to a variety of arrangements such as asymmetric, absorbing, anisotropic, or more complicated multilayer systems.[33]

Since the contact geometry in the SFA closely resembles a sphere-on-flat contact and the surfaces can be very smooth (atomically smooth in the case of bare mica surfaces), a single-asperity contact can be formed whose size,

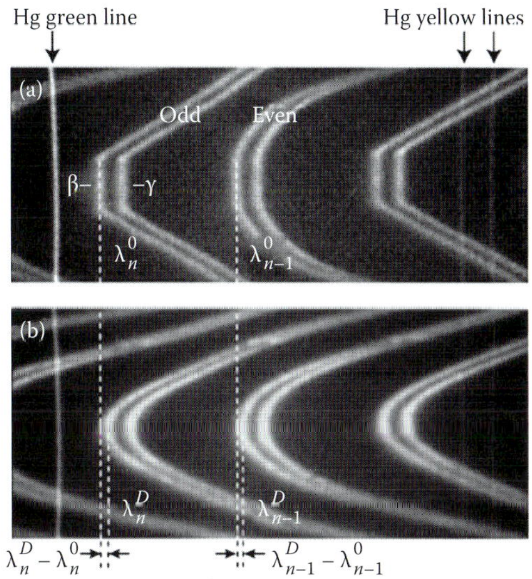

FIGURE 37.4
FECO. The horizontal direction is wavelength (increasing toward the right) and the vertical direction is position along the plane of contact. Hg spectral lines are used for calibration of the wavelengths. Each fringe appears as a doublet (with β and γ components) because of the birefringence of the mica and represents an odd or even order of interference, n. (a) When surfaces are in adhesive contact or pressed together, the glue layers behind the substrates deform and flatten so that the surfaces come in contact over a region of tens to hundreds of micrometers (flat part of fringes). Wavelengths λ_n^0 measured in this situation (for example, in mica–mica or monolayer–monolayer contact) define $D = 0$ in a particular experiment. (b) Surfaces separated a distance D that can be calculated based on the new wavelengths λ_n^D compared to λ_n^0. (From Israelachvili, J. et al., *Rep. Progr. Phys.* **73**, 036601, 2010. © IOP Publishing. With permission.)

position, and deformation are easily monitored using MBI. If the surfaces adhere or if the load is sufficiently high, the glue layers under the mica sheets deform elastically,[18,37] and a flat, circular contact area of uniform film thickness is formed. The corresponding FECO (Figure 37.4a) shows a characteristic flattening indicating a region of uniform film thickness. The diameter of the contact area can be measured directly (to about 1 μm) from this flattened region of the FECO (observations can be made in two orthogonal directions by rotating the light entering the spectrometer with a dove prism, not shown here). Surfaces with a radius of curvature $R \approx 2$ cm typically give rise to contacts with a diameter of a few micrometers to a few hundreds of micrometers, depending on the applied force and strength of adhesion, and the resulting maximum pressure in the middle of the contact is typically not more than a few tens of megapascals (a few hundred atm). Because of the curved surfaces in the SFA, the pressure in the direction normal to the surfaces is not a constant over the flattened contact area but varies in a manner that can be described by contact mechanics models,[5,37] and the mechanical properties of confined films, including protein and cell layers,[35,38–41] between them can be studied.

37.2.2 Surface Modifications for Biological Studies

Muscovite mica is the most commonly used substrate in SFA experiments, because of its transparency and the relative ease by which large areas of atomically smooth, step free sheets of uniform thickness (typically 2–5 μm) can be cleaved from larger crystals.[28,32] The compressibility of mica is quite low, and it is chemically inert, i.e., in its native state, it does not expose functional groups to which chemical reactions can occur. On the one hand, this implies that its properties are well known from one experiment to the next. On the other hand, a substantial effort is needed to modify mica for different purposes by physisorption or chemical reactions. The substrate plays a large role in the surface phenomena investigated by SFA, and unmodified (bare) mica is not necessarily representative of many surfaces of biomedical or engineering interest.

A large number of procedures have been developed for the purpose of modifying or replacing the mica surfaces with, for example, organic thin films, metals, or metal oxides to get surfaces with different chemical properties and surface energies.[5,18] Since mica becomes negatively charged in aqueous solution (due to desorption of K^+ ions), positively charged species (for example, cationic molecules) are attracted to its surface, but in many cases, the high surface energy of mica alone causes physisorption of uncharged amphiphiles and macromolecules. Surfactant, lipid, polymer, or protein layers can be deposited from solution (either before the surfaces are mounted in the instrument or by adsorption in situ) or by Langmuir Blodgett (LB) deposition and subsequent transfer into the instrument chamber.[5,18,42] Bilayers of surfactant or lipid, which expose their hydrophilic groups to the surroundings, need to be kept and transferred underwater to avoid removal of the outer monolayer. Details on surface preparation protocols are given in the literature referred to in Section 37.3.

37.3 Forces Measured in Biological Model Systems

As mentioned previously, interactions in biological systems can be very complex because they occur between multicomponent self-assembled structures, over different length scales, and, commonly, as processes or sequences where certain requirements in terms of energy input or molecular conformation and orientation must be met. In the following, examples are shown of how direct measurements of forces can be used to understand and control the interactions in models that closely mimic certain aspects of biological systems such as cell membranes and surfaces in synovial joints.

37.3.1 Supported Bilayer Adhesion versus Hemifusion

Hemifusion and fusion of amphiphilic bilayers in solution (vesicles) or supported on solid surfaces were originally thought to involve direct (molecular) contact between the surfaces, requiring them to overcome short-range repulsive electrostatic double layer and/or hydration forces before fusion would occur. Forces that increase the attraction and adhesion in a system, including bridging of charged bilayers by divalent ions,[43] were thought to be necessary to induce fusion, whereas additional repulsive forces were expected to prevent it.[44] However, experiments have shown that fusion can be initiated both in adhesive and repulsive systems at distances larger than direct molecular contact,[45–49] and it is now recognized that bilayer stress, deformation, and the related exposure of the hydrophobic regions of the molecules (either locally or throughout the bilayer) play an important role. For example, van der Waals attraction and other forces may cause significant bilayer adhesion, but fusion of fully developed bilayers (in equilibrium with excess amphiphile above its critical micelle or aggregation concentration, CMC) is unlikely since their hydrophobic regions are well shielded from the aqueous environment. Stress (osmotic or electric field-induced) or defects (for example, domains in multicomponent systems or inserted proteins in real biological membranes) are required.

In model systems, this can be induced in different manners, e.g., by depositing the second (outer) layer of a bilayer at a reduced packing density,[50] by depleting (thinning) it by lowering the concentration of a contacting solution of the amphiphile, by inducing other components into the bilayer to cause packing mismatches and local defects, or by changing the conformation of the molecules to expose hydrophobic moieties.[21] Figures 37.5 through 37.7 show examples of how FECO images and forces measured with the SFA can give information on events at the molecular scale during such processes and experimental conditions.[45,49,51]

37.3.1.1 *Imaging of Local Deformations and Molecular Rearrangement*

Local deformations can be directly observed on the FECO as illustrated in Figure 37.5, where hemifusion of partially depleted bilayers of L-α-dimyristoylphosphatidylcholine (DMPC, an uncharged zwitterionic lipid) in the fluid state developed from the initial breakthrough at the center of the contact (appearing as a "blip" pointing to the left on the flat part of the FECO, toward smaller wavelengths, i.e., toward lower film thickness) at a hydrated bilayer separation of about 1 nm and spread rapidly over the area of contact.[48] Hemifusion of DMPC bilayers at lower temperature, where the hydrocarbon chains are

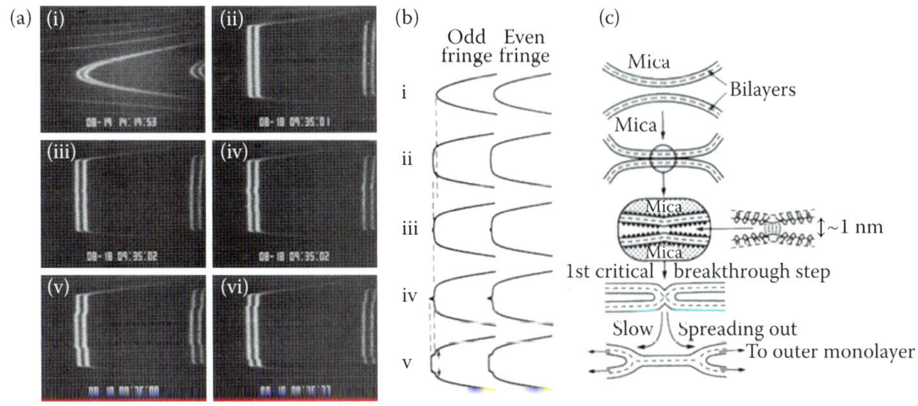

FIGURE 37.5
FECO imaging of bilayer hemifusion. (a) FECO showing hemifusion of two partially depleted DMPC bilayers in the fluid state. (i) Separated, undeformed surfaces. (ii) Surfaces in contact at a pressure of about 50 atm, flattened over an area with diameter ~95 μm because of deformation of the glue layer behind the mica substrates (cf. Figure 37.4). The distance between the outer surfaces of the bilayers is about 1 nm. (iii–v) Breakthrough and rapid expansion of the hemifused area over a lateral distance of ~50 μm within about 6 s (the numbers at the bottom of the images indicate the month and date and the time in hours, minutes, and seconds). (vi) Spreading continues at a slower rate. (From Helm, C. A. et al., *Science* **246**, 919–922, 1989. Reprinted with permission of AAAS.) (b) Shapes of fringes (FECO). (c) Shapes of surfaces. (b and c) Intermediate stages in hemifusion showing the molecular rearrangement, with schematics i–v corresponding to FECO images i–v on the left. After the critical breakthrough in step iii, lipids slide out from the fused outer monolayers until a single hemifused bilayer remains at step v. (Reprinted with permission from Helm, C. A. et al., *Biochemistry* **31**, 1794–1805. Copyright 1992 American Chemical Society.)

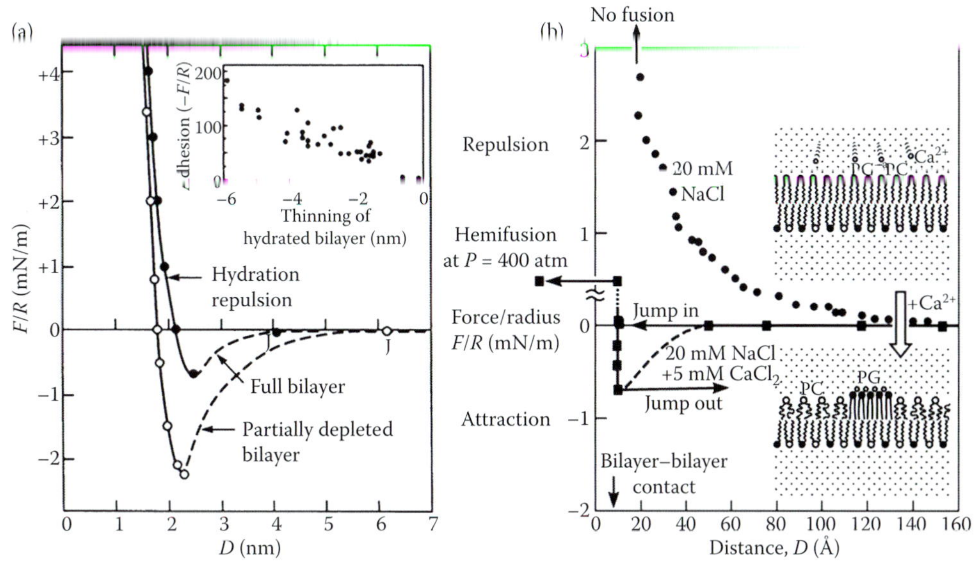

FIGURE 37.6
Bilayer hemifusion due to hydrophobic interaction. (a) Forces between uncharged, fluid DLPC monolayers on solid DPPE monolayers at 22°C. $D = 0$ is at nominal dehydrated bilayer–bilayer contact. (●) van der Waals attraction caused fully developed DLPC surfaces to jump into adhesive contact from "J" at $D = 4.2$ nm. After the jump-in, the bilayers were held apart by hydration repulsion, and no fusion was observed even at very high pressures of 40 atm ($F/R \approx 1000$ mN/m). (○) Forces between depleted DLPC monolayers (on DPPE) in contact with partially saturated solution, where the DLPC head group areas in the monolayers were about 15% larger than at equilibrium. A stronger attraction due to hydrophobic interaction and a jump in from "J" at $D = 6.2$ nm were observed. At a pressure of 3 atm, the bilayers fused into one (not shown). Insert: Depth and relative position of the adhesive minimum in forces between increasingly depleted (thinned) bilayers, compared to the fully developed hydrated bilayers shown with (●) symbols in the main graph. (From Helm, C. A. et al., *Science* **246**, 919–922, 1989. Reprinted with permission of AAAS.) (b) Forces between mixed anionic/uncharged DLPG⁻/DLPC bilayers in 20 mM NaCl at pH 5.6 with (■) and without (●) added 5 mM $CaCl_2$. In NaCl solution (without calcium), the force was dominated by double-layer repulsion and was similar to the force between other pure (one-component) anionic bilayers. No adhesion was observed nor could fusion be induced between such bilayers. In contrast, the force in the presence of calcium was completely different from the forces between either pure DLPC or pure DLPG⁻ bilayers. With the mixed lipid bilayers, calcium-induced phase separation (clustering) of the negatively charged PG⁻ lipids separates the charged from the uncharged lipids, as shown in the lower inset. The stressed DLPC regions become more hydrophobic and act as adhesion and fusion sites with other, similarly stressed bilayers. (Reprinted with permission from Leckband, D. E. et al., *Biochemistry* **32**, 1127–1140. Copyright 1993 American Chemical Society.)

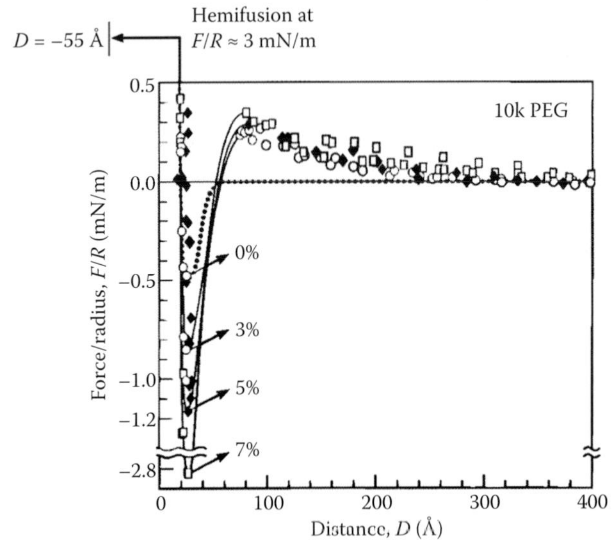

FIGURE 37.7
Osmotic stress due to depletion attraction. Depletion force and hemifusion between two DMPC-on-DPPE bilayers mediated by free PEG (molecular weight 1×10^4 g/mol) in concentrated aqueous solutions. $D = 0$ at the contact between nominally dehydrated bilayers. In the absence of PEG, the bilayers interact only through van der Waals attraction. An enhanced attraction due to depletion forces is observed with increasing PEG concentration. The enhanced adhesion at contact is well described by the osmotic pressure if the bulk concentration is replaced by the *activity* of PEG in aqueous solution at this concentration. The long-range repulsion is due to a weak double-layer force and/or to the long-range repulsive tail of the depletion force (depletion stabilization). Hemifusion occurred at a pressure of ~0.3 atm. (Reprinted with permission from Kuhl, T. et al., *Langmuir* 12, 3003–3014. Copyright 1996 American Chemical Society.)

in the solid state, occurred in a similar fashion but much slower, over a period of hours.[49] DMPC bilayers are uncharged and attract one another weakly through van der Waals forces and, in the case of partially depleted ones, through hydrophobic interaction (in a manner similar to the systems shown in Figure 37.6a). In the case of bilayer systems that interact through DLVO forces, such as CTAB (hexadecyltrimethylammonium bromide, a cationic surfactant), and adhere more strongly than the DMPC bilayers, the hemifusion started not from the midpoint of the contact but from the edges.[48]

37.3.1.2 Interactions between Uncharged and Charged Bilayers

Figure 37.6 shows examples of forces measured between uncharged and charged lipid bilayers. The uncharged zwitterionic system in Figure 37.6a consisted of a bilayer on each mica surface of an LB-deposited fluid-like L-α-dilauroylphosphatidylcholine (DLPC) monolayer on top of a solid L-α-dipalmitoylphosphatidylethanolamine (DPPE) monolayer.[48,49] The forces were first measured between fully developed DLPC layers in the presence of a saturated solution of monomers (to prevent thinning of the DLPC monolayer). The interactions were well described by van der Waals forces. When the DLPC portion of the bilayer was depleted by diluting the surrounding monomer solution, a stronger attraction, a combination of van der Waals and hydrophobic interaction, was observed. In the latter case, hemifusion could be induced at elevated pressures, and the strength of the adhesion increased as the degree of depletion was increased (insert in Figure 37.6a).

The influence of the fluidity of uncharged bilayers was investigated further by supporting them not directly on the mica (as in Figures 37.5a and 37.6a) but on a polyethyleneimine layer.[52] In this configuration, hemifusion occurred at a lower pressure than on mica, and complete fusion (which was not observed on mica) could be observed above the chain transition temperature.

Figure 37.6b shows hydrophobic and electrostatic interactions between mixed lipid bilayers containing 75% uncharged DLPC and 25% charged (anionic) dilauroylphosphatidyl glycerol ($DLPG^-$) at pH 5.6.[51] The top half of Figure 37.6b shows the bilayer structure and measured forces in monovalent electrolyte (NaCl), and the lower half shows the effects of adding Ca^{2+} ions to the solution, which caused adhesion followed by easy hemifusion. The effects of calcium on the mixed lipid bilayers (in the absence or presence of NaCl) were quite different from those induced on pure, single-component charged or uncharged bilayers, which are well described by DLVO theory and hydration forces. Ca^{2+} did not induce fusion of unstressed single-component bilayers but could give an increased attraction due to "ion correlation forces" (caused by density fluctuations of ions in the double layer).[5,53] With mixed bilayers, lateral phase separation or clustering occurred where PG^- lipids were pulled close together (condensed) by calcium binding, as indicated in the lower schematic insert in Figure 37.6b. The condensed PG–calcium domains resulted in a tensile stress on the remaining PC-rich regions, which became more hydrophobic and thereby acted as adhesion and/or fusion sites, and the interaction forces became dominated by a hydrophobic attraction. In this system, fusion did not occur unless a number of favorable factors were present simultaneously: mixed rather than single-component bilayers, divalent counterions, and fluid membranes that allowed molecular rearrangements to occur.

With this type of fusion mechanism, fusion will occur at a different location from the calcium-binding sites. In addition, in a living system, unlike the model system shown in Figure 37.6b, the calcium ions that trigger the fusion event are likely to be released as a transient spurt, so that after some (short) time, they will redisperse in the surrounding solution, bringing the system back into its quiescent state, i.e., one or a few fusion events will occur, rather than a continuous buildup of fusion sites.

The above example applies to lipid bilayers but is relevant, in principle, to other biological systems. The role of calcium, and to a lesser extent magnesium, in triggering membrane fusion is generally thought to occur more commonly via their effect on protein structure, which causes them to expose hydrophobic domains, rather than by inducing structural changes in the lipids. The membrane fusion channels opened up by soluble *N*-ethylmaleimide-sensitive-factor attachment protein receptor (SNARE) proteins at vesicle–presynaptic junctions when calcium is released into the junction may be an example of this type of mechanism involving both lipids and proteins. SFA studies of the interaction forces and kinetics associated with the assembly of SNARE proteins between lipid bilayers have revealed a reproducible assembly and disassembly on subsequent approaches.[54–56]

37.3.1.3 Osmotic Stress-Induced Adhesion and Destabilization

Figure 37.7 shows the interaction between supported bilayers of DMPC on DPPE in aqueous solution of 0.5 mM KNO_3 saturated with DMPC, without and with polyethylene glycol (PEG) with a molecular weight of 1×10^4 g/mol.[57–59] In the absence of PEG, the bilayers interacted through van der Waals forces (dotted line "0%" in Figure 37.7), similarly to the "full bilayers" in Figure 37.6a. As PEG was introduced, a stronger adhesion and a weak, long-range repulsion were found. The magnitudes of these forces increased with increasing PEG concentration, but the hard-wall film thickness remained at the same distance as without PEG. This excluded adsorption or bridging interaction, since the adsorbed polymer would have been detected as an increased thickness. Hydrophobic attraction due to thinning of the lipid could also be excluded, since this would have resulted in smaller thickness. This system is an example of polymer-mediated osmotic forces (depletion forces), which first affect the adhesion between lipid bilayers[60,61] and then may cause their thinning, destabilization, and fusion.

In systems containing sufficiently high concentrations of nonadsorbing polymers or large aggregates of self-assembled molecules (e.g., micelles), such an attractive force may arise at small surface separations (smaller than the coil or aggregate diameter) from the difference in osmotic pressure of the solution remaining in the gap between the surfaces and the surrounding (bulk) solution containing a high concentration of polymer.[60–67] When the surfaces are brought closer together than the average diameter of the polymer coils ($2R_g$ or $2R_F$, where R_g and R_F are the radius of gyration and the Flory radius, respectively), the remaining coils between the surfaces are expelled from the gap, and the surfaces are then pulled closer together to compensate for the local change in the number concentration of solute (polymer).

Between two flat surfaces at a separation below $2R_g$, the depletion force per area is the osmotic pressure of the solution, i.e., $\pi_{osm} = -\rho kT$, where ρ is the number density of polymer coils. For flat surfaces, the depletion force is thus independent of the separation distance D and acts only at separations below $2R_g$. Integrating the force per area (i.e., the pressure) between flat plates from $2R_g$ to D to get the corresponding interaction energy, and using the Derjaguin approximation (Equation 37.1), one obtains the depletion force between a sphere and a flat surface (or between two interacting crossed cylinders)[5,26]:

$$\frac{F(D)}{R} = -2\pi\rho k_B T(2R_g - D), \quad \text{for } D < 2R_g \qquad (37.6)$$

The adhesive minima in Figure 37.7 were found to be of the magnitude expected for depletion attraction.[57,58]

With PEG present, hemifusion could be induced at a fairly low pressure compared to the systems in Figures 37.5 and 37.6. The fusion event has been proposed to be induced because of a local destabilization of the DMPC layer. PEG is known to stabilize bilayers by producing a condensation of headgroups.[68,69] In the attractive regime of the force curve in Figure 37.7, where the PEG has been expelled from the gap between the surfaces, this stabilizing effect is absent within the contact region. The DMPC outer monolayers, which were in a packing state between the liquid expanded and liquid condensed, thinned within the contact area and became more hydrophobic, which resulted in hemifusion.

Figure 37.7 also shows a repulsive force barrier whose magnitude increased with increasing concentration. Several possible reasons have been discussed,[57–59] including demixing free energy, DLVO forces, or "depletion stabilization"[65,66] due to compression of polymer coils, i.e., a weakly repulsive force at a separation distance of about twice the polymer coil diameter. Albeit found over a different distance regime, depletion interactions are conceptually similar to solvation (oscillatory) forces indicated in Figure 37.2, where alternating repulsive barriers and attractive minima can be observed as the separation distance is decreased such that only discrete numbers of semiordered solvent layers can be accommodated between the surfaces. Careful studies revealed that PEG can chelate K^+ ions present in the solution and thus alter the charge distribution near the surface, so that the long-range repulsion in Figure 37.7 might be of electrostatic (double-layer) origin.[59]

Further investigations have revealed similar depletion forces between cationic bilayers in electrolyte solutions containing PEG or cationic polyelectrolyte, where DLVO interactions are also present.[70]

37.3.2 Biopolymer Brush-Like and Bridging Interactions

As seen in the example of PEG-induced depletion attraction above, the presence of macromolecules can radically change the type and magnitude of interaction forces in a system. If a macromolecule contains an end group or a block of segments that is different from the rest of the chain, which is very common among biopolymers, this group may preferentially adsorb on a surface due to electrostatic interaction, chemical bonding, or different solubility in the solvent. End-adsorbed macromolecules are attached to the surface in only one point, and the extension of the chain, which arises from a balance among elastic energy, intramolecular and intermolecular interactions, and conformational entropy,[26,27] is dependent on the grafting density, i.e., on the grafting distance s between end groups adsorbed on the surface.[27,71,72]

As the grafting density is increased, the chains avoid overlapping one another and will therefore extend far into the solution. In order of increasing overlap (decreasing grafting distance s), systems with different stretching are called "pancake," "mushroom," and "brush" layers.[27,72] In the mushroom regime (low coverage Γ), the molecules are not stretched and interact only with the opposing bare surface or with opposing unstretched polymer mushrooms. In this regime, the interaction forces between two polymer coated surfaces (in a sphere-on-flat or crossed cylinder geometry) can be approximated by an exponentially decaying function.[73] Such interactions have been measured between low molecular weight PEG chains with a lipid end group inserted in lipid bilayers on mica at low grafting density.[74,75]

At high Γ, in the brush regime, the repulsive interaction does not decay exponentially for monodisperse systems but has a characteristic S-shape.[26,74] Typically, there is very little interdigitation between two interacting brush layers (especially for long chains), so that their interaction forces in a good solvent show very little hysteresis between the compression (approach) and separation, and the interaction forces when compressing two brush layers can be based on those obtained when compressing one brush layer against a wall. Several increasingly advanced models have been developed to describe the segment density of densely grafted (strongly stretched) chains in the direction normal to a substrate, and the resulting range and magnitude of the interactions between two brushes or one brush and a bare surface.[26,27,71,72,76,77]

Loops formed by chains where both ends adsorb to the same surface can also be considered a type of a brush layer.[78] An example of such a system is lubricin,[79,80] a glycoprotein found in the synovial fluid in mammalian joints. Lubricin has a molecular weight of 2.3×10^5 g/mol, an extended (contour) length of 200 ± 50 nm, and a chain diameter of a few nanometers (Figure 37.8). It consists

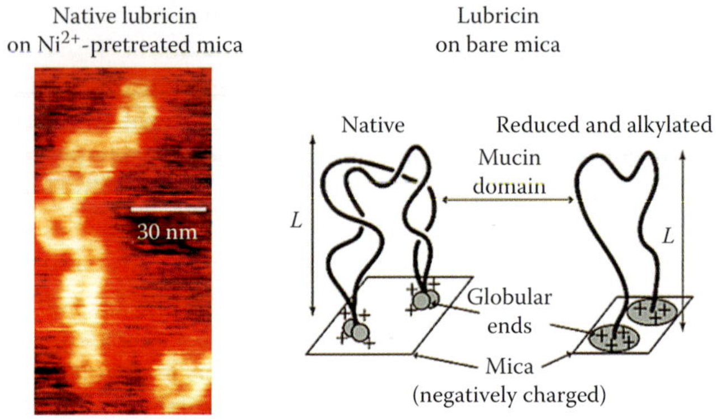

FIGURE 37.8
Adsorbed glycoprotein (lubricin). Atomic force microscopy (AFM) image and suggested "loop" conformations of dimeric LUB and monomeric R-LUB adsorbed nonspecifically to negatively charged surfaces such as mica. The height, L, of the loop and the surface coverage, Γ, are similar in both cases, which implies that the average distance between neighboring molecules is larger for LUB than for R-LUB. (Reprinted with permission from Zappone, B. et al., *Langmuir* **24**, 1495–1508. Copyright 2008 American Chemical Society.)

of an extended backbone with glycosylated side chains, which are strongly hydrated, and hydrophobic globular end groups that also contain anionic and cationic moieties. Lubricin can adsorb through its end groups on hydrophilic (charged and uncharged) or hydrophobic surfaces. It prevents cell proliferation and has been indicated as a friction- and wear-controlling compound on cartilage. Force measurements discussed below suggested that on mica, lubricin adsorbs in a brush-like loop configuration as shown in Figure 37.8.

The interaction forces between mica surfaces modified by lubricin are given in Figure 37.9. Figure 37.9a shows the forces measured between two mica surfaces covered with lubricin adsorbed for 3 h from 290 μg/mL (physiological concentration) and 64 μg/mL solution in phosphate buffered saline (PBS) at pH 7.2. Independently of the adsorption concentration, two different force curves could be measured at a given position. The force curves were monotonically repulsive with almost no hysteresis and had a shape characteristic of end-adsorbed polymers in a brush-like conformation. This indicates the formation of irreversibly adsorbed layers that prevent further adsorption and adhesion with other surfaces. After rinsing with PBS, only the shorter-ranged force was observed. Further experiments[80] attributed the longer force range to dimeric lubricin (LUB) with a "tail"-like (single-end adsorbed) conformation and the shorter range to LUB and monomeric R-LUB in loop-like conformations as illustrated schematically in Figure 37.8.

The forces in Figure 37.9a are well described by a simple model for the interactions of monodisperse brush systems, the Alexander–de Gennes model,[27,71,72] which

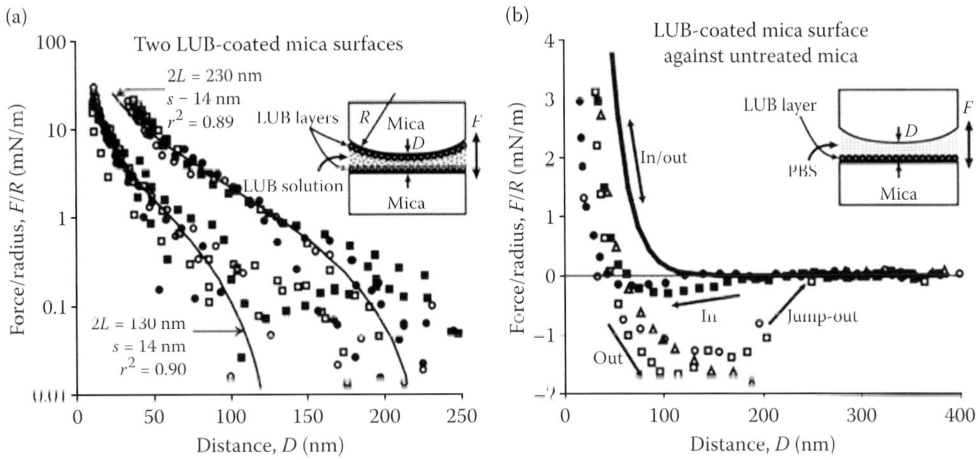

FIGURE 37.9
Polymer brush-like and bridging forces. (a) Forces between two lubricin-coated mica surfaces in PBS, shown on semilog scales. Filled symbols indicate approach and open symbols show data on separation of the surfaces. $D = 0$ at mica–mica contact in dry N_2 gas. The two data ranges correspond to different molecular conformations described in the text. The solid curves are fits of the Alexander–de Gennes model, Equation 37.7, with values of the brush height, L, and average grafting distance, s, given in the figure. The noise level in the force measurements was about 0.1 mN/m. (b) Forces between one lubricin-coated and one bare (uncoated) mica surface in PBS, shown on linear scales. The solid curve indicates the force measured in (a). An arrow indicates the jump-out from an "adhesive well" caused by bridging of lubricin from one surface to the other. (Reprinted from *Biophys. J.* **92**, Zappone, B., et al. 1693–1708, Copyright 2007, with permission from Elsevier.)

assumes that the segment density profile of the brush is uniform, i.e., a step function. In this approximation, scaling analysis gives the force between a sphere and flat surface (or between two crossed cylinders) covered with polymer brushes as[26,27]

$$\frac{F(D)}{R} = \frac{16\pi k_B T L_0}{35 s^3}\left[7\left(\frac{2L_0}{D}\right)^{5/4} + 5\left(\frac{D}{2L_0}\right)^{7/4} - 12\right] \quad \text{for } D < 2L_0, \tag{37.7}$$

where s is the grafting distance and L_0 is the height of one brush layer ($L_0 = R_F^{5/3} s^{-2/3}$). The first term arises from the osmotic pressure and the second from chain elasticity. Comparisons of predicted forces from this model to experimental data on monodisperse polymer brushes have supported its prediction of the brush height increasing linearly with the polymer length (number of monomers N) for a fixed grafting density. In many cases, such as in Figure 37.9a, this and more sophisticated models[76,77] also describe the force between two confined brushes well, especially at larger compression (in the region dominated by osmotic interactions).[74,81–84]

The main differences in the interactions between polymer brushes and adsorbed homopolymer layers are the reduction or absence of hysteresis in the force–distance curve on approach and separation of brush layers, and the lack of interpenetration and entanglement due to the strong stretching of the chains. Bridging is also avoided with polymer brush layers, which makes them a useful surface modification strategy for many applications (e.g., for colloid stabilization or lubrication). Figure 37.9b shows an example of bridging of molecules from a lubricin-covered surface to bare (uncoated) mica in PBS. Instead of the repulsion seen in Figure 37.9a, attraction is observed both on approach (in) and separation (out). The long range of the latter suggests a lower limit for the extended length of the bridging molecules of about 200 nm, which agrees well with the expected value.

Bridging occurs when segments from a chain adsorbed on one surface are able to reach over to another surface and adsorb on it. This requires available adsorption sites on the opposing surface, either due to low adsorption density as in Figure 37.9b (a low adsorbed amount also facilitates the diffusion of chains through an opposing layer, if present) or rearrangements on compression. Once attached to the opposing surface, the bridging polymer chain would gain conformational entropy if the two surfaces came closer together, and the result is an attractive force acting between the two surfaces. Depending on the extension of the polymer, this attraction may be long range, typically much longer than the van der Waals attraction or the depletion interaction.[26] In addition, in a poor solvent, there are attractive segment–segment interactions. These may affect the range of the attraction, since segment–segment attraction is possible as soon as there is some overlap between the adsorbed layers (chains crossing the midplane between the surfaces), whereas bridging requires chains to reach the opposing surface. At small separation distances and restricted equilibrium, repulsion is observed due to the osmotic interaction of the adsorbed, highly confined polymer.

37.3.3 Specific Interactions: Ligand–Receptor (L–R)

In addition to the nonspecific forces described in the previous sections, biological systems may interact through specific interactions commonly referred to as lock-and-key[85] or L–R interactions, which were first proposed to describe the action of enzymes[85] and immunological responses.[86] Early on, it was recognized that such interactions occur between different active sites on complex, often asymmetrical molecules, and that the exact geometry and orientation of the interacting moieties play important roles. These interactions are short-ranged and typically involve molecular rearrangements or changes in conformation, which manifest themselves as forces that depend on contact time and approach rate as well as on the mobility of molecules along bilayers or membranes.

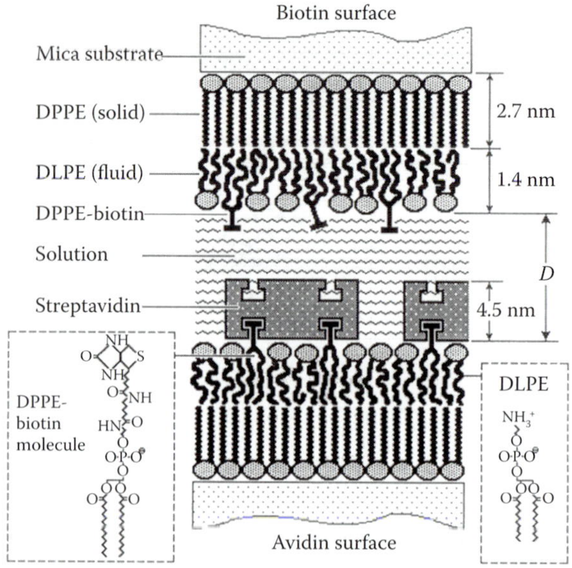

FIGURE 37.10
Supported bilayers with ligands and receptors. Schematic illustration of the ligand biotin and the receptor streptavidin ("avidin") incorporated in LB-deposited supported bilayers. Streptavidin has four receptor sites, two on either side of the molecule, and form an about 4.5 nm thick single layer that is easily compressed to 3.5 nm. The solution contains electrolyte and DLPE at its CMC. $D = 0$ at nominally dehydrated bilayer–bilayer contact. (Reprinted from Helm, C. A. et al., *Proc. Natl. Acad. Sci., USA* **88**, 8169–8173, 1991.)

37.3.3.1 Bilayer-Bound L–R Interactions

Although L–R interactions are physical bonds, they can be very strong. The bond energy in one of the most studied systems, biotin (the ligand, with a molecular weight of 244 g/mol) and streptavidin (the receptor, a protein with molecular weight ~6×10^4 g/mol), is about 35 k_BT/bond, and the binding site is located in a 1.5 nm deep recess on the streptavidin surface. The unbinding rate is very low. The SFA has been used to study the long- and short-range forces between these molecules in the bilayer-bound configuration shown in Figure 37.10.[87–89] Bilayers carrying biotin were prepared by LB deposition of an outer fluid monolayer of *N*-[6-(biotinylamino)hexanoyl]dipalmitoyl-L-α-phosphatidylethanolamine (biotin-X-DPPE) in L-α-dilauroylphosphatidylethanolamine (DLPE) on a solid monolayer of DPPE. Streptavidin was then adsorbed to one of the biotin-carrying bilayer surfaces from aqueous solution. Force measurements were performed under a wide range of experimental conditions with different ratios of biotin-X-DPPE to DLPE (0.5 or 5%) in the outer monolayer, in phosphate buffer solution containing DLPE to prevent thinning of the bilayer (as in the section above) and varying concentrations of electrolytes, and at temperatures above and below the freezing point of the hydrocarbon chains in DLPE (about 30°C) to change the lateral mobility of the lipids and thus that of the attached ligands and receptors.[87–89]

The interaction between two biotin-covered bilayers (without streptavidin, like the upper surface in Figure 37.10) was purely repulsive without any adhesion and well described by DLVO forces between two charged surfaces in electrolyte solutions.[87–89] DLPE is uncharged and interacts through van der Waals forces.[48,90] The charges are located on anionic phosphate groups of the biotinylated molecules, which dissociate (pK ~3) to different degrees depending on the solution conditions. Two streptavidin-covered surfaces (like the lower surface in Figure 37.10) are hydrophilic but uncharged, interact through van der Waals forces, and first come in adhesive contact at a film thickness corresponding to two streptavidin layers.[87–89] On further compression to overcome a low repulsive barrier to compress the film and laterally part the streptavidin (which can diffuse laterally because of the fluidity of the DLPE monolayer), the surfaces jumped in to a film thickness corresponding to a single layer of streptavidin. The adhesion at this film thickness was also well described by van der Waals interactions. There are thus no unusual or specific forces occurring in symmetric systems of two biotin or two streptavidin-functionalized surfaces.

The forces in an asymmetric system of biotin and streptavidin (the configuration in Figure 37.10) were quite different from those in symmetric systems and are shown in Figure 37.11.[87–89] Detailed experiments at different electrolyte concentrations (which regulate the DLVO forces), pH (which controls the dissociation of charged groups, even enabling charge reversal of streptavidin, which has an isoelectric point of pI ~ 6.0–6.5), and temperature (affecting the lateral diffusion) showed that the long-range interactions are dominated by DLVO forces, hydrophobic attraction, and nonequilibrium steric repulsion due to time-dependent reorientation and lateral diffusion of biotin. Specific interactions give rise

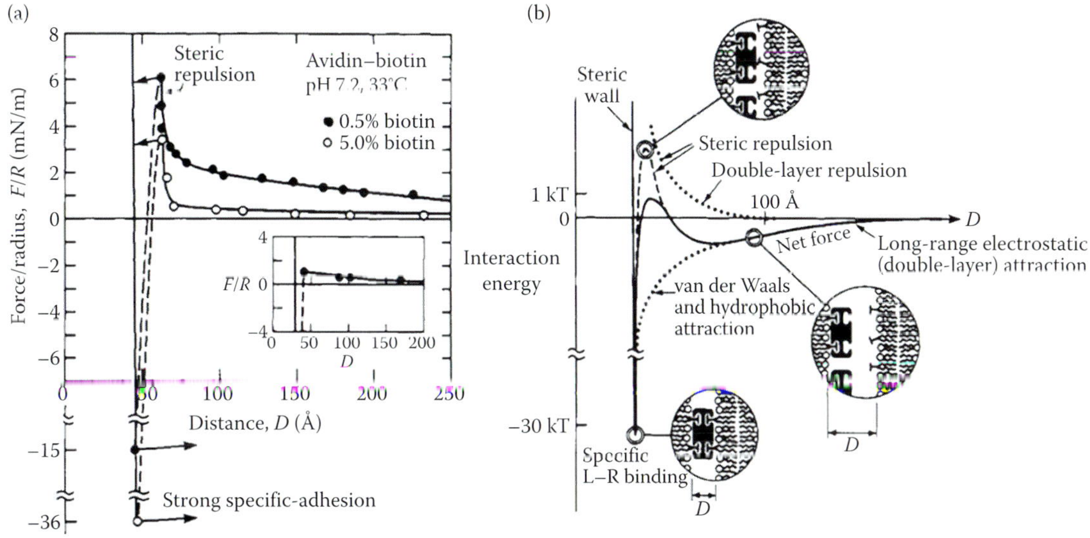

FIGURE 37.11
Bilayer-bound L–R interactions. (a) Forces measured in the system shown in Figure 37.10. The solution contained DLPE at its CMC and 0.3 mM NaCl in 0.1 mM phosphate buffer (pH 7.2). A long-range double-layer repulsion was followed by an approach-rate and contact-time-dependent steric repulsion between biotin and streptavidin. The specific interaction is short ranged. At constant streptavidin coverage, the adhesion of 0.5% biotin was stronger than expected based on the value for 5% biotin because of diffusion of biotinylated lipid into the contact region. Insert: Equilibrium interactions on a very slow approach, where the steric barrier did not occur as the lipids had time to diffuse laterally. $D = 0$ at nominally dehydrated bilayer–bilayer contact. (b) Schematic illustration of the interactions governing L–R interactions at different separation distances. The double-layer interaction can be repulsive as in (a) or attractive depending on the charge of the surfaces at different pH (cf. Figure 37.2). (Reprinted with permission from Leckband, D.E., et al., *Biochemistry* **33**, 4611–4624. Copyright 1994 American Chemical Society.)

to a strong adhesion only in the last 0.5 nm gap between the outer surface of the streptavidin layer and the opposing biotinylated bilayer surface.[89]

Under certain combinations of solution conditions, temperature, and slow approach rate, only a very weak repulsion was measured on approach, followed by the jump into final contact (insert in Figure 37.11a). On the other hand, at 25°C (below the chain melting temperature), the steric repulsion was always present at all approach rates, and the adhesion force in contact was only $F/R = -5$ mN/m, emphasizing the importance of molecular mobility for the ligands to "find" and interact with the sites on the receptors.[89] This important issue will be discussed further in later sections.

Reversible separation of the biotin from the binding sites on streptavidin was not possible because of the strong interaction in contact, which was almost two orders of magnitude larger than the van der Waals attraction (cf. the adhesion of "full bilayer" in Figure 37.6a). Instead, biotin-X-DPPE molecules attached to opposing streptavidin were pulled out from the monolayers, and a second approach at the same location showed larger film thickness in contact and a weak adhesion due only to van der Waals attraction. The adhesion energy thus reflected the lipid–lipid interaction in the monolayer and was only a lower limit of the strength of the specific L–R interaction. After a few hours, the surfaces reannealed due to lateral diffusion, and the original high adhesion could be measured on a new approach at the same location.[89]

37.3.3.2 Tethered L–R Interactions

The results described demonstrate the importance of molecular flexibility and mobility in the formation of specific bonds. In many real biological systems, ligands are not immobilized on the surface of a macromolecule or on a solid-supported surface as shown in Figure 37.10, but are attached to flexible or semiflexible tethers. A system with ligands tethered to a surface by PEG chains, as illustrated schematically in Figure 37.12, thus represents a more realistic model of systems that involve molecular recognition and specific interaction.[91–94]

Bilayers were formed by LB deposition in a manner similar to the systems in Section 37.3.3.1. In Figure 37.12, the biotin-functionalized surface was a monolayer of biotin-PEG-DSPE in DSPE (distearoylphosphatidylethanolamine) on a solid monolayer of DPPE. The streptavidin-covered surface was formed by deposition of biotin-X-DHPE (X being a 6-carbon spacer and DHPE being 1,2-dihexadecanoyl-*sn*-glycero-3-phosphoethanolamine) in DLPE on DPPE, followed by immersion in streptavidin solution, and removal of excess streptavidin by rinsing.[91]

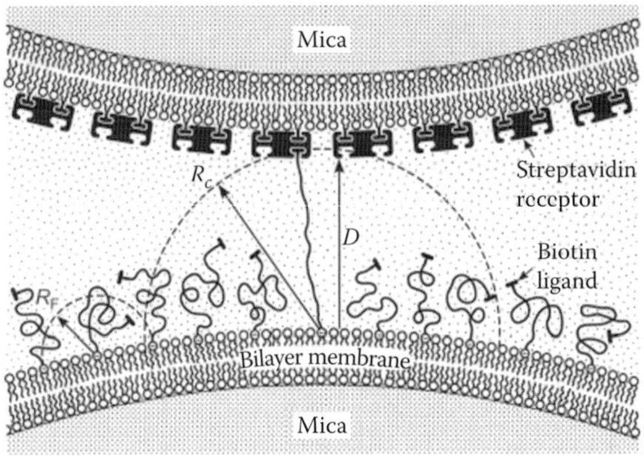

FIGURE 37.12
Tethered L–R system. Schematic illustration of the streptavidin and biotin-PEG molecular configurations. The biotin-PEG coverage was 950 Å^2 per molecule or 4.5 mol% and corresponded to the so-called "weakly overlapping mushroom regime." Here and in Figure 37.13a, *D* corresponds to the distance from the outer edge of the streptavidin layer to the outer lipid head group surface on the opposing surface. The lipid head groups of biotin-PEG were negatively charged and the biotin molecule itself was neutral; the streptavidin protein was negatively charged at pH 7.2. (Reprinted from *Intermolecular and Surface Forces*, 3rd ed. Israelachvili, J. N., Copyright 2011, with permission from Elsevier.)

The interaction forces shown in Figure 37.13a were measured in phosphate buffer containing 0.5 mM NaCl at 25°C, i.e., below the chain melting temperatures of the lipids, which minimized their lateral diffusion. As mentioned in Section 37.3.3.1, the interaction forces between streptavidin and biotin at this temperature and in the absence of PEG consist of double-layer repulsion, steric repulsion, and a weak adhesion, as seen as a solid line in Figure 37.13a. With PEG tethers, there is a weak double-layer repulsion (no steric repulsion) followed by a long-range attraction with a long jump-in on approach (open symbols, points B to C in Figure 37.13a) to a compressed film thickness corresponding to confined PEG chains between the surfaces. Further compression of the PEG chains gives steric and osmotic repulsion. On separation (filled symbols), there was a stronger adhesion than in the case without tethers. In this case, as well as with the system in Figure 37.10a, the separation was essentially irreversible and caused pull-out of lipids from the monolayer once the PEG chains had become stretched, seen as a flat region of the adhesive part of the force curve (cf. Figure 37.13b). On subsequent approaches, the surfaces came in contact at a slightly larger film thickness and lower adhesion. When using longer PEG tethers, the jump-in distances and the flat parts of the curve on separation were longer.[92]

Clearly, the tethers allowed the ligands to interact with the receptors over a much larger distance range as compared to the system in Figure 37.10 (about 15 nm vs. 0.5 nm). This is possible because the PEG chains sample all possible conformations including ones extending well beyond the average "mushroom" height shown in Figure 37.12. The calculated full extended length of these PEG chains (molecular weight 2×10^3 g/mol) was 16 nm. At D = 13 nm in Figure 37.13, a small fraction of the chains are sufficiently extended so that their biotin ends are within 0.5 nm of the opposing streptavidin where binding can occur. Bridging, even of a small number of chains, results in an attractive force (cf. Figure 37.9b)

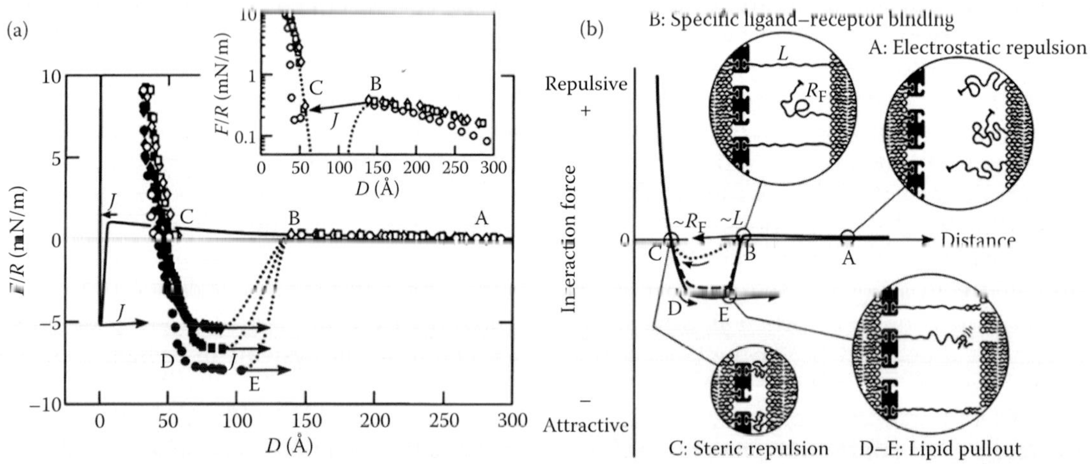

FIGURE 37.13
Tethered L–R interactions. (a) Forces between streptavidin and biotin-PEG. The solid curve shows the interaction between streptavidin with 5% biotin without the PEG tether.[89] Symbols: first approach (○) and separation (●); second approach (□) and separation (■); third approach (◇) and separation (◆). Arrows *J* denote inward and outward jumps on approach and separation, respectively. The insert shows the weakly repulsive, electrostatic double-layer force. (b) Schematic construction of the "ideal" tethered L–R interaction potential, with notation as in (a). (From Wong, J. Y. et al., *Science* **275**, 820–822, 1997. Reprinted with permission of AAAS.)

that brings the surfaces closer, and other, less extended chains can then bridge over to cause the large jump-in indicated in Figure 37.13a. Experiments with longer PEG tethers showed that the jump-in distance scaled linearly with the molecular weight, i.e., with the number of PEG monomers *N*, which is consistent with a dynamic sampling of conformations including highly extended ones (whereas the average coil diameter indicated in Figure 37.12 scales as $N^{3/5}$).[92] The total interaction energy of the system is determined by the distance over which the force acted, which is significantly different from the case of the bilayer-bound system.

Figure 37.13b shows a calculated force curve (shown as a solid line) and schematic illustrations of the relevant parameters used in the calculations. The details of the calculation are given in ref. 93. As in Figure 37.11, the interaction at large *D* is well described by electrostatic double-layer repulsion (A). At B, a small fraction of the biotin-functionalized PEG chains at their near-full extension are able to "lock" into the biotin-binding site on the opposing streptavidin-functionalized surface. On approach, the minimum of the attractive potential (dotted curve) results from the polymer coil elastic restoring force and is located at a separation distance corresponding closely to the Flory radius R_F of the PEG. At smaller distances, $D < R_F$, polymer steric repulsion (C) leads to a larger confined film thickness than in the case of bilayer-bound ligands. On separation (D and E), the effective separation potential (solid curve) is determined by the strength of lipid pullout or L–R bond breakage (whichever is the weakest—here the lipid pullout, dashed curve) and the polymer coil interaction (dotted curve), and acts over a range of the fully extended tether length, *L*.[91–93]

The probability of a successful binding of a tethered ligand L to a receptor R at a large distance away (the capture distance, R_c) increases with the waiting time and depends more on the dynamics of the tether (polymer dynamics) than the L–R binding energy. As in the reverse case of separation or pull-out, capture depends on a statistically rare event. The statistics of the whole process involves knowledge of single-molecule dynamics, the dynamic spatial distribution and 3D diffusion of the molecules (many-molecule dynamics), the shapes and relative motion of the two surfaces (macroscopic structure and dynamics), and the L–R binding energies.

In addition to the L–R systems above, other protein systems and also DNA base pairs have been studied with the SFA and typically show complex interactions where DLVO forces, steric repulsion, hydrophobic interactions, and orientation- and rate-dependent forces can be identified.[54,55,95–100] The concept of tethered functional groups has been explored further in studies of acid-based and hydrophobic interactions.[101] A system with a very different interaction mechanism will be discussed in Section 37.3.4.

37.3.4 Force Generation by Polymerization of Confined Actin

A recent example of interactions not explained by the general pair potentials in Figure 37.2 are the forces generated through polymerization of confined actin.[102,103] Actin is a protein with different functions as mechanical support in the cell cytoskeleton, as a scaffold for contraction-generating proteins in muscle cells,[104] and in cell motility. Under physiological conditions, G-actin, a monomeric globular protein (molecular weight ~4.2 × 10^4 g/mol), is transformed by ATP into F-actin, a polymeric filament containing many G-actin monomers. The polymerization and depolymerization are very rapid. F-actin forms single-stranded, dynamic, helical structures.

SFA experiments on end-tethered actin filaments (Figure 37.14) showed several unusual features. After an initial nucleation stage where a solution of 2–3 mg/mL G-actin in Tris buffer (pH 8) containing 0.2 mM ATP and 0.2 mM dithiothreitol was confined to a film thickness of 50 nm between two mica surfaces, actin filaments became bound to the surface at one end and grew in length (Figure 37.14a). The polymerization/depolymerization process continued at the free end, generating forces that pushed the surfaces apart and showed fluctuations. The filament density on the surfaces was estimated to be 0.5 × 10^4/μm^2 to 1 × 10^4/μm^2, and the force generated by each filament was 5–11 pN, in agreement with the values estimated by others using other techniques.[102,103,105,106]

Figure 37.14b shows that the polymerization process generated opposing elastic forces that were larger than the compressive force, i.e., on decompression (separation), a repulsive force was present that was stronger than that on compression (approach). In contrast, typical hysteretic forces (seen, for example, as a result of compression-induced rearrangements of adsorbed homopolymer layers) and viscous forces give a lower force on separation than on compression. The lateral shear viscosity (not shown) was also larger on decompression and also showed fluctuations, indicating that filaments are oriented in all directions. Refractive index measurements did not show a detectable change in the concentration of actin within the contact region during compression and decompression.

The rapid polymerization and depolymerization generated fluctuations in force ΔF (and thus in space, ΔD) with time, which were seen as a saw-tooth shape superimposed on a monotonically repulsive background force (Figure 37.14b). The frequency of the fluctuations was typically 0.2–1.0 Hz, and their presence suggested a collective behavior of the filaments, since forces generated by random elongation and shrinking of individual filaments should "cancel out." The fluctuations continued as long as the solution contained ATP. When all the ATP had been consumed, the fluctuations could be resumed by adding ATP to the solution. The addition of 1 μg/mL

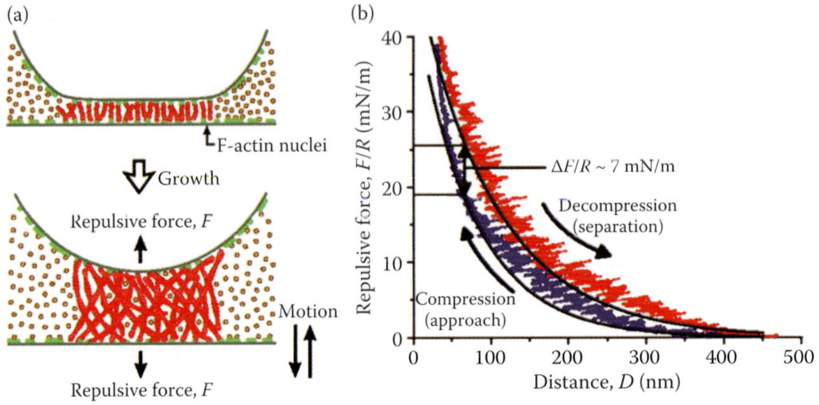

FIGURE 37.14
Force generation by actin polymerization. (a) Polymerization of G-actin monomers to F-actin filaments anchored to one mica surface was nucleated by pressure and confinement for about 10 min at $D = 50$ nm ($D = 0$ at mica–mica contact). As the surfaces were separated at a speed of 10 nm/s, the filaments grew rapidly and generated repulsive, fluctuating forces. Only the filaments giving rise to forces normal to the surfaces are shown. (b) The fluctuations in the force (and thus in distance) were superimposed on an elastic (exponentially decaying) background force (black curves). Each compression/decompression cycle showed an unusual hysteresis where on decompression, the repulsive force was stronger than on approach. (From Greene, G. W. et al., *Proc. Nat. Acad. Sci. USA* **106**, 445–449, 2009.)

gesolin, a capping agent, halted the fluctuations, resulting in a monotonically repulsive force. However, the unusual hysteresis with the stronger repulsion on decompression was still observed in the presence of gesolin. This was thus not a result of the polymerization/depolymerization process but ascribed to a stiffening process of the actin filaments to bending, where a different binding state was introduced by compression.[102,103]

37.4 Combination of SFA with Other Techniques: Local Structural Information

The data gathered in a conventional SFA experiment are the result of an average response of the confined film to compression and shear. Information on the molecular-level film structure during these processes would add significantly to the interpretation of the measured forces. Obtaining such information is very challenging because of the nature of the confined region: the response arises from a relatively small number of molecules spread in a thin layer, which implies that any characteristic spectroscopic signal from this region will be of low intensity. The confined film is surrounded by thicker layers of materials that will give rise to spectroscopic signals of their own, typically much stronger than those from the film. Despite these experimental difficulties, interesting observations on film structure and molecular alignment have been made by combining SFA with several different techniques, as briefly described in the following. Further development of techniques suitable for studying the molecular properties of highly confined films can be expected to significantly improve our understanding of confined thin film structure and adhesion.

Extensions of standard MBI have been used to obtain structural information on the orientation and intermolecular interactions of optically active molecules in confined films.[107,108] Since the light adsorption of optically active (dye) molecules is enhanced by the multiple reflections in the optical cavity, very thin films can be studied, and local information in the contact area can be obtained.

Combining the SFA with other optical techniques has been limited by the presence of the reflective silver layers needed for the standard MBI approach. These layers strongly reduce the illumination of confined films, limiting the performance of spectrometric techniques. This has been successfully overcome by replacing the silver layers with multilayer dielectric coatings,[109] with transparency to different wavelengths, so that information on the structure of ultrathin films can be obtained. For example, the SFA has been combined with fluorescence correlation spectroscopy to measure the molecular diffusion coefficient in thin films within spots of submicron size, obtaining spatially resolved measurements.[109] A drawback is that fluorescent molecules have to be added to the film to be investigated. A reduction in the diffusion coefficient by about two orders of magnitude was found with confinement of films of simple liquids, and the diffusion coefficient was found to decrease when going from the edges toward the center of the contact region. Results have also been reported from combining the SFA with confocal Raman spectroscopy,[109] which avoids the problem of the bulk contributing to the scattered signal. Using multilayer reflective coatings, the geometry of the contact area could be monitored simultaneously with the Raman signal, showing the effects of shear and confinement on the orientations of molecules within the confined film.

The SFA has been combined with x-ray diffraction measurements (XSFA).[110,111] Although this technique has thus far been limited to films thicker than 500 nm, the shear-induced alignment of liquid crystal molecules in confined films has been shown. Because of the size of the x-ray beam, the results represent an average over the contact area. Active research in this area is aiming at reducing the investigated region and film thickness, recently indicating that structural information on confined molecular films can be obtained with x-ray reflectivity.[112]

The application of electric fields is highly relevant for studies of biological systems and has also been implemented in the SFA,[113–117] either by depositing conducting layers on top of the mica or by using the reflective silver layers on the backsides as electrodes. Thus far, studies have concerned polymer and polyelectrolyte films, colloidal suspensions, and liquid crystals, but this technique development is expected to be extendable to biological systems as well.

37.5 Nonequilibrium, Rate, and Time-Dependent Interactions

In a classic paper in *Science*, in 1978, Bell[118] quantitatively analyzed the role of thermal fluctuations on the measured force needed to unbind or separate two adhering molecules (or a lipid or protein from a membrane surface). He showed that even as the interaction potential is a well-defined function of space, the separation force is not, but depends on the length or size of the "bond," the temperature, the pulling speed, and the waiting time: the slower the pulling speed or the longer the waiting time, the lower the separation or "adhesion" force. Later refinements to this approach have included viscous dissipation, which further affects the lifetimes of bonds.[119] Such forces have been measured using both the atomic force microscope and SFA techniques, mainly between ligands and their receptors,[120] but also in the unfolding of proteins and nucleic acids. The important conceptual message here is that separation or unbinding is determined by a statistically rare event, i.e., not only a particular binding energy but also, for example, a particular required combination of molecular position and orientation.

Since separation is determined by a number of factors that are likely to be different for different types of binding interactions, when a number of bonds occur in series, they can break in one place at one pulling speed, but break at another place at a different pulling speed. The quantitative measurement and analysis of multiple bonds in biological systems, including bonds in series and bonds in parallel, remain an important but unsolved problem for both experimentalists and theoreticians.

Equally interesting and subtle is the reverse process of "binding," "capture," or "recognition," which again is often determined by a statistically rare event. This is illustrated in Figure 37.15 for the case

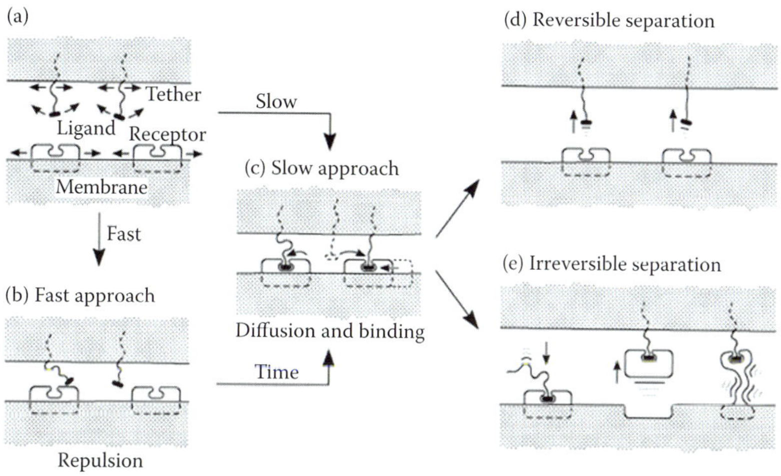

FIGURE 37.15
Nonequilibrium, rate- and time-dependent interactions. Examples of the complexity of biological interactions involving molecular binding, transfer, and unfolding reactions. On fast approach (a)→(b), there is an additional kinetic force barrier because the ligand does not have time to find the binding site until some finite time after the surfaces have been close together (b)→(c). On separation, depending on the relative bond energies, bond lengths, and the rate of separation, the original L–R bonds may break (d, reversible separation), or lipid molecules may be pulled out of the membrane, or the receptor molecules may be pulled out, or the protein may unfold (e, irreversible separation). Other scenarios are also possible, including statistical combinations of the above. Biological systems typically involve many different bonds acting (binding and unbinding) "in parallel" and/or "in series," both of which are very difficult to model. (Reprinted from *Intermolecular and Surface Forces*, 3rd ed. Israelachvili, J. N., Copyright 2011, with permission from Elsevier.)

of a surface consisting of tethered ligands approaching or moving past a surface covered with receptors (cf. Figure 37.1).

When considering a repetitive or cyclic process, such as the approach *and* separation of two molecules, or the continuous flow of cells past each other (cf. Figure 37.1a), subtle and unintuitive things can happen. Examples of some possible scenarios are illustrated in Figure 37.15, where two membranes—one exposing tethered ligands and the other with receptor proteins—come together and then separate. Such processes are rarely reversible, either energetically or structurally, and many quite different scenarios can arise depending on the relative values of the bond energies and bond lengths of the L–R bond, the lipid–membrane "bond," the protein–membrane "bond," and the internal bonds holding the protein receptor together. In addition, the tether length, the temperature, the fluidity of the membrane lipids, and particularly the approach and separation rates all play important roles. Thus, in Figure 37.15, we find that, on fast approach, there may be no binding (a)→(b), whereas on slow approach, there is (a)→(c) or (b)→(c). On separation, again depending on the factors mentioned above, the bonds could break at their original sites (d), but they could also break elsewhere including the unraveling of the protein as often occurs in AFM pulling experiments. One interesting possibility, which no doubt occurs in some biological interactions, is the transfer of a molecule, via binding and pull-out, from one vesicle or cell membrane to another as the two pass each other, as exemplified by the first two diagrams of Figure 37.15e.

37.6 Conclusion

The SFA technique can be used to directly measure interaction forces between biological model surfaces of different degrees of complexity. Typically, several forces of different origin appear superposed on one another in the total interaction potential, and differentiating between them requires careful construction of model systems of increasing complexity, as well as precise control of the experimental conditions. Some of these forces occur also in simpler colloidal system, whereas others, such as bilayer fusion events, L–R interaction, and protein polymerization, are not described by common pair potentials and show dependence on molecular orientation, temperature, pressure, contact or waiting time, or approach and separation rate.

Acknowledgments

MR gratefully acknowledges support through a CAREER award from the National Science Foundation (CMMI-0645065). JNI's contribution to this research was supported by the U.S. Department of Energy, Office of Basic Energy Sciences, Division of Materials Sciences and Engineering under Award # DE-FG02-87ER-45331 (development of instrumentation and surface force measurements described in the review). This research was supported by the National Institutes of Health under award number R01 DE 018468 (experiments on biological molecules and surfaces).

References

1. Leckband, D., Israelachvili, J. Intermolecular forces in biology. *Quart. Rev. Biophys.* **34**, 105–267 (2001).
2. Israelachvili, J. Differences between non-specific and bio-specific, and between equilibrium and non-equilibrium, interactions in biological systems. *Quart. Rev. Biophys.* **38**, 331–337 (2005).
3. Goetz, D. J., El-Sabban, M. E., Pauli, B. U., Hammer, D. A. Dynamics of neutrophil rolling over stimulated endothelium in vitro. *Biophys. J.* **66**, 2202–2209 (1994).
4. Hammer, D. A., Tirrell, M. Biological adhesion at interfaces. *Annu. Rev. Mater. Sci.* **26**, 651–691 (1996).
5. Israelachvili, J. N. *Intermolecular and Surface Forces*, 3rd ed. (Academic Press, Amsterdam, 2011), and references therein.
6. Ruths, M. Surface forces, surface tension, and adhesion. In *Encyclopedia of Tribology* (eds. Wang, Q. J., Chung, Y. W.) 3435–3443 (Springer, New York, 2013).
7. Drummond, C., Ruths, M. Surface forces apparatus. In *Encyclopedia of Nanotechnology* (ed. Bhushan, B.) 2574–2582 (Springer, Dordrecht, 2012).
8. Derjaguin, B. V. Untersuchungen über die Reibung und Adhäsion, IV. Theorie des Anhaftens kleiner Teilchen. *Kolloid Z.* **69**, 115–164 (1934).
9. Hunter, R. J. *Foundations of Colloid Science*, 2nd ed. (Oxford Univ. Press, New York, 2001).
10. Derjaguin, B. V., Landau, L. Theory of the stability of strongly charged lyophobic sols and of the adhesion of strongly charged particles in solution of electrolytes. *Acta Physicochim. URSS* **14**, 633–662 (1941).
11. Verwey, E. J. W., Overbeek, J. T. G. *Theory of the Stability of Lyophobic Colloids* (Elsevier, Amsterdam, 1948).
12. Tareste, D., Pincet, F., Perez, E., Rickling, S., Mioskowski, C., Lebeau, L. Energy of hydrogen bonds probed by the adhesion of functionalized lipid bilayers. *Biophys. J.* **83**, 3675–3681 (2002).
13. van Olphen, H. *An Introduction to Clay Colloid Chemistry*, 2nd ed. (Wiley, New York, 1977).

14. Stanley, H. E., Teixeira, J. Interpretation of the unusual behavior of H_2O and D_2O at low temperatures: Tests of a percolation model. *J. Chem. Phys.* **73**, 3404–3422 (1980).
15. Pashley, R. M. DLVO and hydration forces between mica surfaces in Li^+, Na^+, K^+, and Cs^+ electrolyte solutions: A correlation of double-layer and hydration forces with surface cation exchange properties. *J. Colloid Interface Sci.* **83**, 531–546 (1981).
16. Pashley, R. M. Hydration forces between mica surfaces in electrolyte solution. *Adv. Colloid Interface Sci.* **16**, 57–62 (1982).
17. Israelachvili, J., Pashley, R. The hydrophobic interaction is long range, decaying exponentially with distance. *Nature* **300**, 341–342 (1982).
18. Christenson, H. K., Claesson, P. M. Direct measurements of the force between hydrophobic surfaces in water. *Adv. Colloid Interface Sci.* **91**, 391–436 (2001).
19. Meyer, E. E., Lin, Q., Hassenkam, T., Oroudjev, E., Israelachvili, J. Origin of the long-range attraction between surfactant-covered surfaces. *Proc. Natl. Acad. Sci. USA* **102**, 6839–6842 (2005).
20. Ruths, M., Israelachvili, J. N. Surface forces and nanorheology in molecularly thin films. In *Springer Handbook of Nanotechnology* (ed. Bhushan, B.) Chapter 29, 857–922 (Springer, Berlin & Heidelberg, 2010), and references therein.
21. Donaldson, S. H. Jr., Lee, C. T. Jr., Chmelka, B. F., Israelachvili, J. N. General hydrophobic interaction potential for surfactant/lipid bilayers from direct force measurements between light-modulated bilayers. *Prod. Natl. Acad. Sci. USA* **108**, 15699–15704 (2011).
22. Leung, K., Luzar, A. Dynamics of capillary evaporation. II. Free energy barriers. *J. Chem. Phys.* **113**, 5845–5852 (2000).
23. Mao, M., Zhang, J., Yoon, R.-H., Ducker, W. A. Is there a thin film of air at the interface between water and smooth hydrophobic solids? *Langmuir* **20**, 1843–1849 (2004). Erratum, *Langmuir* **20**, 4310 (2004).
24. Israelachvili, J., Wennerström, H. Entropic forces between amphiphilic surfaces in liquids. *J. Phys. Chem.* **96**, 520–531 (1992).
25. Pabst, G., Katsaras, J., Raghunathan, V. A. Enhancement of steric repulsion with temperature in oriented lipid multilayers. *Phys. Rev. Lett.* **88**, 128101 (2002).
26. Ruths, M. Characterization of molecularly thin polymer layers with the surface forces apparatus (SFA). In *Functional Polymer Films: Assembly, Functionalization and Characterization* (eds. Knoll, W., Advincula, R.) Chapter 23, 745–769 (Wiley-VCH, Weinheim, 2011), and references therein.
27. de Gennes, P. G. Polymers at a interface: A simplified view. *Adv. Colloid Interface Sci.* **27**, 189–209 (1987).
28. Israelachvili, J. N., Adams, G. E. Measurements of forces between two mica surfaces in aqueous electrolyte solutions in the range 0–100 nm. *J. Chem. Soc. Faraday Trans. I* **74**, 975–1001 (1978).
29. Israelachvili, J. N., McGuiggan, P. M. Adhesion and short-range forces between surfaces. Part 1: New apparatus for surface force measurements. *J. Mater. Res.* **5**, 2223–2231 (1990).
30. Parker, J. L., Christenson, H. K., Ninham, B. W. Device for measuring the force and separation between two surfaces down to molecular separations. *Rev. Sci. Instrum.* **60**, 3135–3138 (1989).
31. Israelachvili, J., Min, Y., Akbulut, M., Alig, A., Carver, C., Greene, W., Kristiansen, K., Meyer, E., Pesika, N., Rosenberg, K., Zeng, H. Recent advances in the surface forces apparatus (SFA) technique. *Rep. Progr. Phys.* **73**, 036601 (2010), and references therein.
32. Israelachvili, J. N. Thin film studies using multiple-beam interferometry. *J. Colloid Interface Sci.* **44**, 259–272 (1973).
33. Heuberger, M. The extended surface forces apparatus. Part I. Fast spectral correlation interferometry. *Rev. Sci Instrum.* **72**, 1700–1707 (2001) and references therein.
34. Palkar, S. A. Lenhoff, A. M. Diffusion and adsorption of proteins in a model pore in the surface forces apparatus. *Colloids Surf. A* **110**, 119–127 (1996).
35. Gauthier-Manuel, B., Gallinet, J. P. Study of the behavior of biological molecules in a thin gap by refractive index and force measurements with an automatic surface force apparatus. *Biochemie* **80**, 391–399 (1998).
36. Gallinet, J.-P., Gauthier-Manuel, B. Protein behavior in a thin gap: Progress obtained from simultaneous measurements of refractive index and interaction with an automated surface force apparatus. *Curr. Top. Colloid Interface Sci.* **5**, 291–306 (2002).
37. Horn, R. G., Israelachvili, J. N., Pribac, F. Measurements of the deformation and adhesion of solids in contact. *J. Colloid Interface Sci.* **115**, 480–492 (1987).
38. Suda, H., Sasaki, Y. C., Oishi, N., Hiraoka, N., Sutoh, K. Elasticity of mutant myosin subfragment-1 arranged on a functional silver surface. *Biochem. Biophys. Res. Comm.* **261**, 276–282 (1999).
39. Martin-Molina, A., Moreno-Flores, S., Perez, E., Pum, D., Sleytr, U. B., Toca-Herrera, J. L. Structure, surface interactions, and compressibility of bacterial S-layers through scanning force microscopy and the surface force apparatus. *Biophys. J.* **90**, 1821–1829 (2006).
40. Banquy, X., Rabanel, J.-M., Hildgen, P., Giasson, S. Direct measurement of mechanical and adhesive properties of living cells using surface forces apparatus. *Austr. J. Chem.* **60**, 638–645 (2007).
41. Curry, J. E., Heo, C. H., Mainer, R. M. Directly measuring adhesive and elastic properties of bacteria using a surface force apparatus. *ACS Symp. Ser.* **984**(Microbial Surfaces), 217–229 (2008).
42. Leckband, D. The surface force apparatus—A tool for probing molecular protein interactions. *Nature* **376**, 617–618 (1995).
43. Portis, A., Newton, C., Pangborn, W., Papahadjopoulos, D. Studies on the mechanism of membrane fusion: Evidence for an intermembrane Ca^{2+}–phospholipid complex, synergism with Mg^{2+}, and inhibition by spectrin. *Biochemistry* **18**, 780–790 (1979).
44. Rand, R. P., Parsegian, V. A. Hydration forces between phospholipid bilayers. *Biochim. Biophys. Acta* **988**, 351–376 (1989).
45. Horn, R. G. Direct measurement of the force between two lipid bilayers and observation of their fusion. *Biochim. Biophys. Acta* **778**, 224–228 (1984).

46. Ohki, S. A mechanism of divalent ion-induced phosphatidylserine membrane fusion. *Biochim. Biophys. Acta* **689**, 1–11 (1982).
47. Maezawa, S., Yoshimura, T., Hong, K., Düzgünes, N., Papahadjopoulos, D. Mechanism of protein-induced membrane fusion: Fusion of phospholipid vesicles by clathrin associated with its membrane binding and conformational change. *Biochemistry* **28**, 1422–1428 (1989).
48. Helm, C. A., Israelachvili, J. N., McGuiggan, P. M. Molecular mechanisms and forces involved in the adhesion and fusion of amphiphilic bilayers. *Science* **246**, 919–922 (1989).
49. Helm, C. A., Israelachvili, J. N., McGuiggan, P. M. Role of hydrophobic forces in bilayer adhesion and fusion. *Biochemistry* **31**, 1794–1805 (1992) and references therein.
50. Benz, M., Gutsmann, T., Chen, N., Tadmor, R., Israelachvili, J. Correlation of AFM and SFA measurements concerning the stability of supported lipid bilayers. *Biophys. J.* **86**, 870–879 (2004).
51. Leckband, D. E., Helm, C. A., Israelachvili, J. Role of calcium in the adhesion and fusion of bilayers. *Biochemistry* **32**, 1127–1140 (1993).
52. Wong, J., Park, C. K., Seitz, M., Israelachvili, J. N. Polymer-cushioned bilayers. II. An investigation of interaction forces and fusion using the surface forces apparatus. *Biophys. J.* **77**, 1458–1468 (1999).
53. Marra, J. Direct measurement of the interaction between phosphatidylglycerol bilayers in aqueous electrolyte solutions. *Biophys. J.* **50**, 815–825 (1986).
54. Li, F., Pincet, F., Perez, E., Eng, W. S., Melia, T. J., Rothman, J. E., Tareste, D. Energetics and dynamics of SNAREpin folding across lipid bilayers. *Nat. Struct. Mol. Biol.* **14**, 890–896 (2007).
55. Perez, E., Li, F., Tareste, D., Pincet, F. The surface forces apparatus to reveal the energetic of biomolecules assembly. Application to DNA bases pairing and SNARE fusion proteins folding. *Cell. Mol. Bioeng.* **1**, 240–246 (2008).
56. Li, F., Pincet, F., Perez, E., Giraudo, G., Tareste, D., Rothman, J. E. Complexin activates and clamps SNAREpins by a common mechanism involving an intermediate energetic state. *Nat. Struct. Mol. Biol.* **18**, 941–946 (2011).
57. Kuhl, T., Guo, Y., Alderfer, J. L., Berman, A. D., Leckband, D., Israelachvili, J., Hui, S. W. Direct measurement of polyethylene glycol induced depletion attraction between lipid bilayers. *Langmuir* **12**, 3003–3014 (1996).
58. Kuhl, T. L., Berman, A. D., Hui, S. W., Israelachvili, J. N. Part 1. Direct measurement of depletion attraction and thin film viscosity between lipid bilayers in aqueous polyethylene glycol solutions. *Macromolecules* **31**, 8250–8257 (1998).
59. Kuhl, T. L., Berman, A. D., Hui, S. W., Israelachvili, J. N. Part 2. Crossover from depletion attraction to adsorption: Polyethylene glycol induced electrostatic repulsion between lipid bilayers. *Macromolecules* **31**, 8250–8257 (1998).
60. Evans, E., Needham, D. Attraction between lipid bilayer membranes in concentrated solutions of nonadsorbing polymer: Comparison of mean-field theory with measurements of adhesion energy. *Macromolecules* **21**, 1822–1831 (1988).
61. Evans, E. A. Force between surfaces that confined a polymer solution: Derivation from self-consistent field theories. *Macromolecules* **22**, 2277–2286 (1989).
62. Asakura, S., Oosawa, F. Interaction between particles suspended in solutions of macromolecules. *J. Polym. Sci.* **33**, 183–192 (1958).
63. Joanny, J. F., Leibler, L., de Gennes, P. G. Effects of polymer solutions on colloid stability. *J. Polym. Sci. Polym. Phys.* **17**, 1073–1084 (1979).
64. Vincent, B., Luckham, P. F., Waite, F. A. The effect of free polymer on the stability of sterically stabilized dispersions. *J. Colloid Interface Sci.* **73**, 508–521 (1980).
65. Feigin, R. I., Napper, D. H. Stabilization of colloids by free polymer. *J. Colloid Interface Sci.* **74**, 567–521 (1980).
66. Feigin, R. I., Napper, D. H. Depletion stabilization and depletion flocculation. *J. Colloid Interface Sci.* **75**, 525–541 (1980).
67. Fleer, G. J., Tuinier, R. Concentration and solvency effects on polymer depletion and the resulting pair interaction of colloidal particles in a solution of non-adsorbing polymer. *Polymer Prepr.* **46**, 366 (2005).
68. Maggio, B., Lucy, J. A. Interactions of water-soluble fusogens with phospholipids in monolayers. *FEBS Lett.* **94**, 301–304 (1978).
69. Yamazaki, M., Ohshika, M., Kashiwagi, N., Asano, T. Phase transitions of phospholipid vesicles under osmotic stress and in the presence of ethylene glycol. *Biophys. Chem.* **43**, 29–37 (1992), and references therein.
70. Anderson, T. H., Donaldson, S. H., Zeng, H., Israelachvili, J. N. Direct measurement of double-layer, van der Waals, and polymer depletion attraction forces between supported cationic bilayers. *Langmuir* **26**, 14458–14465 (2010).
71. Alexander, S. Adsorption of chain molecules with a polar head—A scaling description. *J. Phys. (Paris)* **38**, 983–987 (1977).
72. de Gennes, P. G. Conformations of polymers attached to an interface. *Macromolecules* **13**, 1069–1075 (1980).
73. Dolan, A. K., Edwards, S. F. Theory of the stabilization of colloids by adsorbed polymer. *Proc. R. Soc. London A* **337**, 509–516 (1974).
74. Kuhl, T. L., Leckband, D. E., Lasic, D. D., Israelachvili, J. N. Modulation of interaction forces between bilayers exposing short-chained ethylene oxide headgroups. *Biophys. J.* **66**, 1479–1488 (1994).
75. Li, F., Pincet, F. Confinement free energy of surfaces bearing end-grafted polymers in the mushroom regime and local measurement of the polymer density. *Langmuir* **23**, 12541–12548 (2007).
76. Milner, S. T., Witten, T. A., Cates, M. E. Theory of the grafted polymer brush. *Macromolecules* **21**, 2610–2619 (1988).
77. Zhulina, E. B., Borisov, O. V., Priamitsyn, V. A., Birstein, T. M. Theory of steric stabilization of colloid dispersions by grafted polymers. *J. Colloid Interface Sci.* **137**, 495–511 (1990).
78. Milner, S. T., Witten, T. A. Bridging attraction by telechelic polymers. *Macromolecules* **25**, 5495–5503 (1992).
79. Zappone, B., Ruths, M., Greene, G. W., Jay, G. D., Israelachvili, J. N. Adsorption, lubrication, and wear of lubricin on model surfaces: Polymer brush-like behavior of a glycoprotein. *Biophys. J.* **92**, 1693–1708 (2007).

80. Zappone, B., Greene, G. W., Oroudjev, E., Jay, G. D., Israelachvili, J. N. Molecular aspects of boundary lubrication by human lubricin: Effects of disulfide bonds and enzymatic digestion. *Langmuir* **24**, 1495–1508 (2008).
81. Taunton, H. J., Toprakcioglu, C., Fetters, L. J., Klein, J. Interactions between surfaces bearing end-adsorbed chains in a good solvent. *Macromolecules* **23**, 571–580 (1990).
82. Hadziioannou, G., Patel, S., Granick, S., Tirrell, M. Forces between surfaces of block copolymers on mica. *J. Am. Chem. Soc.* **108**, 2869–2876 (1986).
83. Watanabe, H., Tirrell, M. Measurement of forces in symmetric and asymmetric interactions between diblock copolymer layers adsorbed on mica. *Macromolecules* **26**, 6455–6466 (1993).
84. Liao, W.-P., Kuhl, T. L. Steric forces of tethered polymer chains as a function of grafting density: Studies with a single diblock molecular weight. *Macromolecules* **45**, 5766–5772 (2012).
85. Fisher, E. Einfluss der Configuration auf die Wirkung der Enzyme. *Ber. Dtsch. Chem. Ges.* **27**, 2985–2992 (1894).
86. Ehrlich, P. On immunity with special reference to cell life. *Proc. R. Soc. London* **66**, 424–448 (1900).
87. Helm, C. A., Knoll, W., Israelachvili, J. N. Measurement of ligand–receptor interactions. *Proc. Natl. Acad. Sci. USA* **88**, 8169–8173 (1991).
88. Leckband, D. E., Israelachvili, J. N., Schmitt, F.-J., Knoll, W. Long-range attraction and molecular rearrangements in receptor-ligand interactions. *Science* **255**, 1419–1421 (1992).
89. Leckband, D. E., Schmitt, F.-J., Israelachvili, J. N., Knoll, W. Direct force measurements of specific and nonspecific protein interactions. *Biochemistry* **33**, 4611–4624 (1994).
90. Marra, J., Israelachvili, J. N. Direct measurements of forces between phosphatidylcholine and phosphatidylethanolamine bilayers in aqueous electrolyte solution. *Biochemistry* **24**, 4608–4618 (1985).
91. Wong, J. Y., Kuhl, T. L., Israelachvili, J. N., Mullah, N., Zalipsky, S. Direct measurement of a tethered ligand–receptor interaction potential. *Science* **275**, 820–822 (1997).
92. Wong, J. Y., Kuhl, T. L. Dynamics of membrane adhesion: The role of polyethylene glycol spacers, ligand-receptor bond strength, and rupture pathway. *Langmuir* **24**, 1225–1231 (2008).
93. Jeppesen, C., Wong, J. Y., Kuhl, T. L., Israelachvili, J. N., Mullah, N., Zalipsky, S., Marques, C. M. Impact of polymer tether length on multiple ligand–receptor bond formation. *Science* **293**, 465–468 (2001).
94. Kekicheff, P. From heterocoagulation to biorecognition: Preparation of surfaces for direct measurement of their interactions in a surface force apparatus. *Macromol. Symp.* **335**, 24–42 (2014).
95. Leckband, D., Chen, Y. L., Israelachvili, J., Wickman, H. H., Fletcher, M., Zimmerman, R. Measurements of conformational changes during adhesion of lipid and protein (polylysine and S-layer) surfaces. *Biotechnol. Bioeng.* **42**, 167–177 (1993).
96. Hartley, P., Bailey, A. I., Luckham, P. F., Batts, G. Interactions between adsorbed blood clotting factors probed at the nanometer scale using the surface force apparatus. *Nanobiology* **3**, 167–175 (1994).
97. Bayas, M. V., Kearney, A., Avramovic, A., van der Merwe, P. A., Leckband, D. E. Impact of salt bridges on the equilibrium binding and adhesion of human CD2 and CD58. *J. Biol. Chem.* **282**, 5589–5596 (2007).
98. Rosenberg, K. J., Ross, J. L., Feinstein, H. E., Feinstein, S. C., Israelachvili, J. N. Complementary dimerization of microtubule-associated tau protein: Implications for microtubule bundling and tau-mediated pathogenesis. *Prod. Natl. Acad. Sci. USA* **105**, 7445–7450 (2008).
99. Zappone, B., Thurner, P. J., Adams, J., Fantner, G. E., Hansma, P. K. Effect of Ca^{2+} ions on the adhesion and mechanical properties of adsorbed layers of human osteopontin. *Biophys. J.* **95**, 2939–2950 (2008).
100. Ramos, S., Campos-Teran, J., Mas-Oliva, J., Nylander, T., Castillo, R. Forces between hydrophilic surfaces adsorbed with apolipoprotein AII alpha helices. *Langmuir* **24**, 8568–8575 (2008).
101. Valtiner, M., Donaldson, S. H. Jr., Gebbie, M. A., Israelachvili, J. N. Hydrophobic forces, electrostatic steering, and acid-base bridging between atomically smooth self-assembled monolayers and end-functionalized PEGolated lipid bilayers. *J. Am. Chem. Soc.* **134**, 1746–1753 (2012).
102. Greene, G. W., Anderson, T. H., Zeng, H., Zappone, B., Israelachvili, J. N. Force amplification response of actin filaments under confined compression. *Proc. Nat. Acad. Sci. USA* **106**, 445–449 (2009).
103. Banquy, X., Greene, W. G., Zappone, B., Kolomeisky, A. B., Israelachvili, J. N. Dynamics of force generation by confined actin filaments. *Soft Matter* **9**, 2389–2392 (2013).
104. Suda, H., Sasaki, N., Sasaki, Y. C., Goto, K. Force generation by recombinant myosin heads trapped between two functionalized surfaces. *Eur. Biophys. J.* **33**, 469–476 (2004).
105. Mogilner, A., Oster, G. Force generation by actin polymerization II: The elastic ratchet and tethered filaments. *Biophys. J.* **84**, 1591–1605 (2003).
106. Upadhyaya, A., Chabot, J. R., Andreeva, A., Samadani, A., van Oudenaarden, A. Probing polymerization forces by using actin-propelled lipid vesicles. *Proc. Natl. Acad. Sci. USA* **100**, 4521–4526 (2003).
107. Mächtle, P., Müller, C., Helm, C. A. A thin absorbing layer at the center of a Fabry–Perot interferometer. *J. Phys. II* **4**, 481–500 (1994).
108. Alig, A. R. G., Gourdon, D., Israelachvili, J. Properties of confined and sheared rhodamine B films studied by SFA-FECO spectroscopy. *J. Phys. Chem. B* **111**, 95–106 (2007).
109. Bae, S. C., Wong, J. S., Kim, M., Jiang, S., Hong, L., Granick, S. Using light to study boundary lubrication: Spectroscopic study of confined films. *Phil. Trans. R. Soc. A* **366**, 1443–1454 (2008) and references therein.
110. Idziak, S. H. J., Koltover, I., Israelachvili, J. N., Safinya, C. R. Structure in a confined smectic liquid crystal with competing surface and sample elasticities. *Phys. Rev. Lett.* **76**, 1477–1480 (1996).
111. Golan, Y., Seitz, M., Luo, C., Martin-Herranz, A., Yasa, M., Li, Y., Safinya, C. R., Israelachvili, J. The x-ray surface forces apparatus for simultaneous x-ray diffraction and direct normal and lateral force measurements. *Rev. Sci. Instrum.* **73**, 2486–2488 (2002).

112. Seeck, O. H., Kim, H., Lee, D. R., Shu, D., Kaendler, I. D., Basu, J. K., Sinha, S. K. Observation of thickness quantization in liquid films confined to molecular dimension. *Europhys. Lett.* **60**, 376–382 (2002).
113. Cho, Y.-K., Watanabe, H., Granick, S. Dielectric response of polymer films confined between mica surfaces. *J. Chem. Phys.* **110**, 9688–9696 (1999).
114. Cho, Y.-K., Granick, S. A surface forces platform for dielectric measurements. *J. Chem. Phys.* **119**, 547–554 (2003).
115. Zeng, H., Tian, Y., Anderson, T. H., Tirrell, M., Israelachvili, J. N. New SFA techniques for studying surface forces and thin film patterns induced by electric fields. *Langmuir* **24**, 1173–1182 (2008).
116. Valtiner, M., Banquy, X., Kristiansen, K., Greene, G. W., Israelachvili, J. N. The electrochemical surface forces apparatus: The effects of surface roughness, electrostatic surface potentials, and anodic oxide growth on interaction forces, and friction between dissimilar surfaces in aqueous solutions. *Langmuir* **28**, 13080–13093 (2012).
117. Drummond, C. Electric-field-induced friction reduction and control. *Phys. Rev. Lett.* **109**, 154302 (2012).
118. Bell, G. I. Models for the specific adhesion of cells to cells. *Science* **200**, 618–627 (1978).
119. Evans, E., Ritchie, K. Dynamic strength of molecular adhesion bonds. *Biophys. J.* **72**, 1541–1555 (1997).
120. Moore, N. W., Mulder, D. J., Kuhl, T. L. Adhesion from tethered ligand–receptor bonds with microsecond lifetimes. *Langmuir* **24**, 1212–1218 (2008).

38

Magnetic Tweezers Force Spectroscopy

Eric A. Galburt

CONTENTS

The magnetic tweezers has contributed to a unique set of discoveries about both the fundamental physics of nucleic acids and the mechanisms through which enzymes manipulate these polymers. Since its invention more than 20 years ago,[1] it has been used to study a wide range of topics including the linear and torsional elasticity of DNA and the activities of protein translocases, topoisomerases, polymerases, and helicases. Here, the basic setup of a magnetic tweezers microscope is covered in addition to specific modifications and variations in the technique. Selected examples from the literature are provided throughout in an attempt to survey of the diversity of applications of this technique.

38.1 Introduction

Since the inception of magnetic tweezers (MT),[1] a handful of groups have consistently pushed the methodology forward. Several outstanding comprehensive reviews of MT have been published by the innovators of the technique and are must reads for anyone contemplating building an MT or performing MT experiments.[2–5]

A typical MT setup is made up of an inverted microscope equipped with magnets that have been mounted to rotation and translational stages. The magnetic field gradient emanating from the magnets results in a force being exerted on magnetized superparamagnetic particles that have been tethered to the surface of a flow cell by the molecule of interest. The microscope allows for the visualization and three-dimensional tracking of beads, while the moveable magnets allow for the control of both the force and torque on the molecule. The third crucial component of the modern MT setup is the bead tracking software that allows for the localization of both magnetic and reference beads with high precision.

38.2 MT Microscope

The tweezers described here was heavily based on microscopes in the Seidel laboratory at the Biotechnology Center TU-Dresden, Germany. It consists of an inverted microscope built around an oil-immersion objective lens with magnification between 25× and 100× (typically 40×) depending on the desired field of view and the desired tracking resolution (Figure 38.1a and b). The objective is mounted on a piezoelectric stage that allows the height of the objective to be computer-controlled over a distance of around 100 μm with nanometer precision. The piezo stage is required for generating calibrated stacks of bead images at different distances from the objective that are then used by the tracking algorithms. A mirror is mounted below the objective stage to reflect the image toward the tube lens after which it eventually arrives at the charge-coupled device camera. The light source is a light-emitting diode (LED) that is mounted above the objective on a stage that can be moved *X* and *Y* micrometers to center the light over the objective lens. Laser illumination is not recommended as the coherent beam leads to spurious interference patterns that can interfere with bead tracking algorithms.[6] Two 5 mm cube NdFeB magnets are held in a collar that allows one to specify the distance between the magnets. The magnets can be encircled by soft iron, which focuses the magnetic field to produce higher forces.[6] A spacing of 1 mm results in the ability to apply constant forces up to ~10 pN with 1 μm superparamagnetic beads. The magnet collar is mounted on a hollow shaft, which is connected to the rotary stage that is used to rotate the magnets to apply torque. The rotary stage is mounted on a translation stage that may be used to change magnet height and control the magnitude of the pulling force. The entire magnet assembly (collar, shaft, translational, and rotary stages) is mounted on an X–Y translational stage to allow for alignment so that the gap between the magnets is centered on the light path (from LED to objective lens).

38.3 Sample Preparation—Making a Tether

To perform experiments on a single molecule with the MT, the molecule of choice must be fixed between the glass coverslip and a magnetic bead. In this section, the preparation of the beads, flow cell, and molecule is described.

38.3.1 Flow Cell

A flow cell consists of two glass coverslips with a melted parafilm spacer (Figure 38.1c). One cover glass has a hole drilled at each end to allow for buffer exchange, and the spacer is cut with a channel that connects the holes. The sandwich is then heated on a hot plate (125°C) until the parafilm becomes clear and has fully melted. The volume of a flow cell made in this way is typically

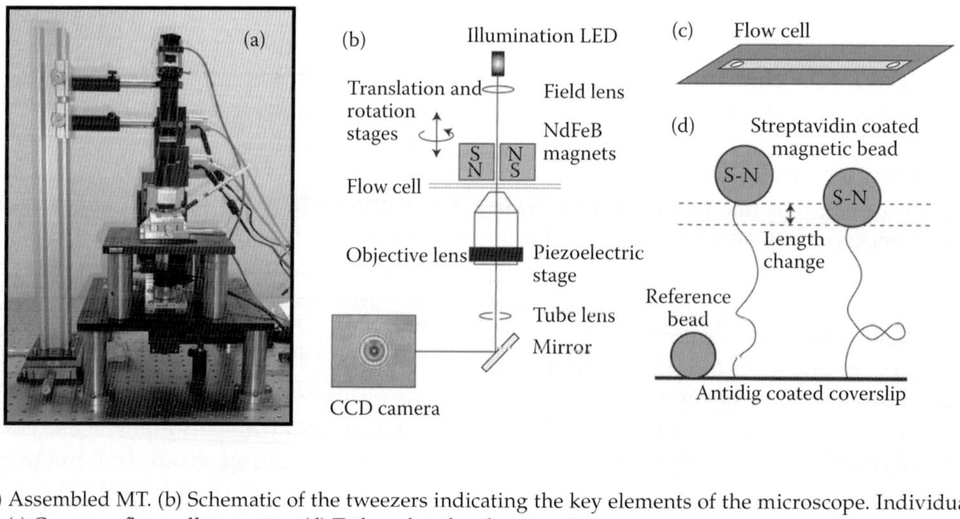

FIGURE 38.1
MT instrument. (a) Assembled MT. (b) Schematic of the tweezers indicating the key elements of the microscope. Individual components are not drawn to scale. (c) Common flow cell geometry. (d) Tethered molecule geometry.

between 15 and 30 μL. Before sandwiching, the glass is cleaned and treated to varying degrees depending on the desired surface coating. These treatments allow for the specific attachment of molecules of interest to the bottom cover glass while excluding nonspecific binding.

The two most common coating techniques involve either biotinylated polyethylene glycol (PEG) or antibody-coated surfaces. In the PEG technique,[7] the glass is cleaned via a series of washes (10% alconox, tap H_2O, acetone, KOH, milliQ H_2O) and amino-modified with amino silane in a 5% acetic acid/methanol solution (20 min) to create a reactive surface. A mixture of biotinylated and unbiotinylated mPEG-SC (succinimidyl carbonate, Laysan Bio) is introduced into the cell and incubated overnight in a humid chamber. After subsequent incubation with streptavidin, the flow cell is ready for the specific binding of biotinylated molecules. Biotinylated, nonmagnetic beads may be introduced into the chamber and used as reference beads.

In the antibody technique,[8] the glass is cleaned via sonication with isopropanol (15 min) and washed three times in water. After drying the glass with nitrogen, it is coated with a 0.1% wt/vol nitrocellulose solution in amyl acetate (this solution should be stored at 1% wt/vol and diluted at the time of use). After sandwiching, the flow cell is rinsed with phosphate-buffered saline (PBS) and incubated with polystyrene beads that serve as reference beads during data acquisition. A 1:1000 dilution of carboxylated polystyrene beads (Spherotech #CP-15-10) incubated for 3–5 min results in one to three reference beads per field of view with a 40× objective. After flowing out reference beads with PBS, a 100 μg/mL solution of antibody (i.e., antidigoxigenin) is incubated for at least 30 min. Last, a 2–10 mg/mL solution of bovine serum albumin (BSA) is added to the cell for passivation of nonspecific binding sites and incubated until the time of use (at least an hour and up to several days).

38.3.2 Printed Array Flow Cells

It is also possible to increase the throughput of MT measurements by using microcontact printing techniques to generate an array of attachment points (i.e., antibodies) on a flow cell.[9] The printing technique increases the number of observable molecules in a given frame by an order of magnitude as compared to the density one can generate by stochastic binding.[9] This strategy coupled with quadrant interpolation-based tracking described in Section 38.4.2 allows for the tracking of over 100 tethers in a single experiment at nanometer spatial resolution.

38.3.3 Magnetic Beads

Coated superparamagnetic beads (i.e., streptavidin-coated) may be passivated and equilibrated to a buffer of choice by a series of washing steps with buffer supplemented with 1 mg/mL BSA. Alternatively, an antibody of choice may be cross-linked to protein G-coupled beads or directly to surface carboxylated beads with 1-ethyl-3-[3-dimethylaminopropyl]carbodiimide hydrochloride (EDC). The inclusion of 0.1% detergent (i.e., Pluronic F-127 or Tween 20) prevents irreversible bead clumping. These washes may be conveniently performed using an external magnet to pellet the beads while exchanging wash buffers. A typical wash consists of three 1 mL washes of 100 μL of magnetic beads in PBS, 0.1% detergent, 1 mg/mL BSA followed by the resuspension of the beads in 100 μL of wash buffer for storage at 4°C.

38.3.4 Tethered Molecule

With a prepared flow cell and selected magnetic beads, one only requires that the molecule of interest be specifically modified on each end with an appropriate label. An important note is that while single labels are fine for pulling experiments (see below, Section 38.7.2), any experiment probing torsional degrees of freedom must have multiple attachment points to both the surface of the flow cell and the magnetic beads to prevent torsional relaxation through the swiveling of single bonds. This can be accomplished by creating multivalent handles that attach to the bead and the surface at multiple points. In practice, 1 kb pair multivalent DNA handles may be amplified by polymerase chain reaction (PCR) in the presence of either biotin- or digoxigenin-labeled nucleotides (dUTP). These handles contain unique restriction sites for the enzymatic digestion and subsequent ligation to a digested DNA construct of study (Figure 38.2).[10,11] A recently described procedure generates similar tethers but avoids digestion and ligation completely by performing PCR with labeled 400 bp megaprimers.[12]

The addition of the molecule to the flow cell can be performed in one of two ways. The molecule may be incubated to either the flow cell or the beads first, followed by either the addition of bead-bound molecules to the

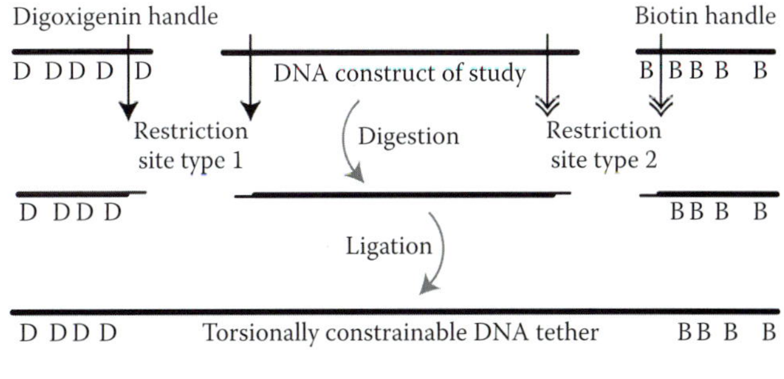

FIGURE 38.2
Construction of a torsionally constrained DNA tether. The products of three individual PCRs (digoxigenin handle, DNA sequence of study, and biotin handle) are digested with restriction enzymes that generate compatible sticky ends that can then be ligated together.

flow cell or the incubation of flow cell-bound molecules with beads. Concentrations and incubation times vary and are best determined empirically for specific systems. As a guideline, typical protocols add 50 pM molecule to 10 μL of beads, which are then diluted 10× with buffer and added to the flow cell. After a 10 min incubation, the unbound beads are slowly washed out with buffer (10 μL/min). By fine-tuning this protocol, one can generate a field of view with five to ten magnetic beads that are tethered to the surface by a single molecule of interest.

38.4 Video-Based Optical Bead Tracking

38.4.1 Three-Dimensional Tracking

The subpixel X–Y position of the bead can be estimated using tracking algorithms that find the centroid of the bead image by considering the linear X and Y intensity profiles near a visually estimated center of the image.[5,13] Cross-correlations with the mirror images of these profiles result in an offset that allows for a corrected estimate of the center of the bead. The local intensity profile near the estimate may then be fit with a second-order polynomial to obtain a subpixel resolution center position.

Tracking in Z is based on the acquisition of a calibration stack of bead images generated at known spacings. Images of the bead at different heights relative to the objective lens are acquired by using the piezoelectric stage to step the objective lens through several microns at intervals of 100 nm. These images are then used as references to determine how the bead position changes during an experiment where the objective lens is now fixed. With slightly convergent illumination, the phase of the ring pattern shifts as a function of the height of the bead. By measuring this phase relative to interpolated phases of the calibration stack, the Z-position may be determined.[4–6] Importantly, one must include a multiplicative factor in this comparison. Theoretically, when the objective moves a distance Δx during acquisition of the calibration images, the focus inside the flow cell changes by $\Delta x(n_{water}/n_{oil})$, where n_{water} and n_{oil} are indices of refraction.[14] However, this approximation can lead to errors as large as 10% with high numerical aperture objectives, and it is recommended to empirically determine the correction factor for a given setup. This can be done by comparing the distance the objective must be moved to focus on the two flow cell surfaces when it is filled with either water or immersion oil.[15]

Once the calibration stack is acquired, it remains to compare experimental bead positions to the stack to generate Z-positions. A variety of algorithms for the determination of bead positions to 1 nm in X–Y and better than 10 nm in Z (depending on the magnification of the objective lens) at bandwidths of 1 s have been reported.[5,6,15]

38.4.2 Quadrant Interpolation

A recent improvement in tracking called quadrant interpolation from Cees Dekker's group allows for low-magnification (<50×) images to be tracked at resolutions previously only obtained with higher magnification.[15] The method is based on the interpolation of the bead image on a circular grid to reduce pixel bias to more accurately determine the X–Y position of the bead. Furthermore, the increased lateral precision leads to a similar increase in the precision of the Z-tracking algorithms. This technique allows increased experimental throughput by increasing the field of view and tracking more tethers concurrently, and the LABVIEW code is freely available for download from the Dekker website.

38.4.3 Millisecond Bead Tracking

The recent use of a superluminescent diode coupled with a high-speed camera has allowed for the collection of MT data at rates greater than 10,000 Hz.[16] This significant increase in time resolution has the power to reveal intermediate states with lifetimes on the millisecond timescale.

38.5 Force Calibration

The magnetic field gradient results in a force that can be determined empirically for different magnet heights.[11] A bead that is bound to the surface via a single tether and is being pulled upward by a force can be modeled as an inverted pendulum, where the gravitational field is replaced by the magnetic field (Figure 38.3). A straightforward analysis equates the transverse spring constant (κ) generated from the transverse restoring force of the magnets (F_m/L), where L is the end-to-end distance of the tether at a given magnet height. By equating the average thermal energy and the average energy stored in a Hookean spring and rearranging terms, one can determine the magnetic pulling force: $F_m = (k_B T)(L)/\langle \Delta x^2 \rangle$. One should use the fluctuations of the bead in the direction of the magnetic field lines where the bead is constrained by the field as opposed to the orthogonal direction where the bead can pivot around the tether in the direction of the measurement and introduce errors.[17] By measuring the end-to-end distance of the tether and the Brownian fluctuations of the bead, one arrives at an estimate of the force for a given magnet height. In the far field, where the bead is farther away from the magnets

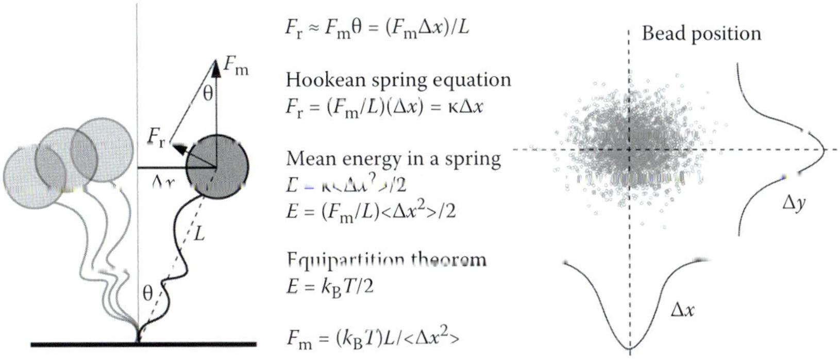

FIGURE 38.3
Force calibration. The Brownian fluctuations of the tethered bead may be analyzed as an inverted pendulum. For small angles (θ), the lateral spring constant of the trap generated by the magnets and the tether can be approximated as F_m/L. By comparing the average energy in a Hookean spring with the available thermal energy from the equipartition theorem, the pulling force of the magnets is determined.

than the magnet–magnet spacing, the force varies exponentially with magnet height. Force measurements at a series of magnet heights may be fit exponentially to determine a continuous force–calibration curve.

38.6 Variations in Experimental (Magnetic) Geometry

38.6.1 Freely Orbiting MT

Different magnet geometries lead to MT that are specialized for specific types of measurements. The freely orbiting MT (FOMT) uses a cylindrical magnet instead of a pair of cubes to flip the direction of the magnetic field from horizontal to vertical and allow for free rotation of the molecule while still applying a pulling force[18] (Figure 38.4b). After proper alignment of the bead with the magnet, the magnetic bead freely orbits the tether position and traces out a circle in the X–Y plane. This technique works because the magnetic beads possess a preferred magnetization axis that causes the bead to align to the field instead of simply reorienting its magnetization vector. This arrangement allows for rotation of the molecule that can be followed simply by tracking the bead position in X and Y. Since this angular movement is only constrained by the molecule's torsional stiffness, it permits a measurement of the torsional stiffness of the tether itself. Buffeted by Brownian motion, a torsionally stiff molecule will exhibit small excursions around its orbit compared to a compliant one. Analogously to the increase in DNA stiffness as a function of force described by the worm-like chain, the increase in torsional stiffness with increasing force has been quantitated using this technique.[18] Furthermore, changes in twist of the molecule caused by protein binding may be directly monitored. The power of this technique has been demonstrated by tracking twist

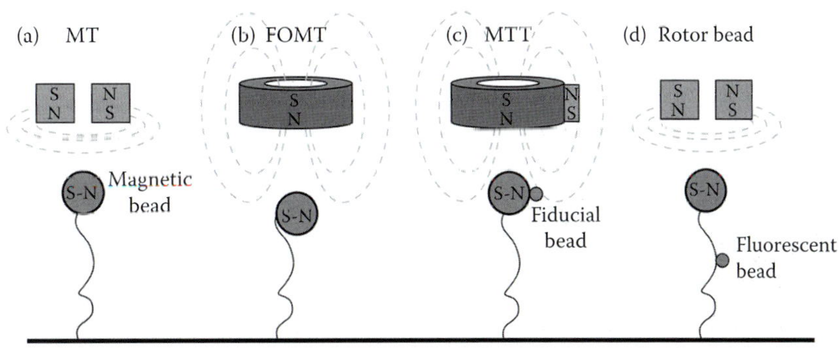

FIGURE 38.4
Variations on magnet geometry. Four different magnetic geometries lead to different types of magnetic traps. (a) Standard MT. (b) Rotor bead assay requires a second bead attached to the side of the DNA and allows for twist to be directly monitored. (c) FOMT uses a cylindrical magnet to create a trap in which the bead is free to rotate around the tether point. (d) MTT represents a cross between the MT and the FOMT where the bead is not completely free to rotate as in the FOMT but is not completely constrained to a particular angle as in the MT either. The result is a geometry that allows for the measurement of torques.

during the formation of the RecA filament on dsDNA and ultimately determining the effect of filament formation on torsional stiffness.[18]

38.6.2 Magnetic Torque Tweezers

A further modification adds two small cube magnets to one side of the pair of the cylindrical magnet to create an asymmetry in the magnetic field creating a magnetic torque tweezers (MTT)[8] (Figure 38.4c). This leads to the bead having a preferred angular position in contrast to both the freely orbiting geometry described above and the essentially fixed angular position found in conventional tweezers. Instead of rotational trap stiffness on the order of 10^4 pN nm rad^{-1}, this geometry generates trap stiffness of around 300 pN nm rad^{-1}. In this case, the beads rotational position will lag behind that of the magnets by an amount that depends on the restoring torque of the DNA. The addition of a small nonmagnetic fiducial bead to the surface of the magnetic bead allows for angular tracking. Using the measured rotational trap stiffness (k_{rot}), determined by monitoring angular fluctuations of the bead, and by comparing the bead angle to the angle of magnet rotation, experimental measures of torque are possible: $\tau_{DNA} = k_{rot}(\theta - \theta_0)$.

38.6.3 Rotor Beads

An interesting experimental geometry that does not depend on changes to magnet geometry also gives one the ability to measure twist and torque. The rotor-bead assay, originally used in an optical tweezers setup,[19] consists of modifications to the DNA tether that result in the binding of a small nonmagnetic fluorescent bead to an internal site in the tether (Figure 38.4d). This geometry allows for measurements of twist by monitoring changes in the X–Y position of the rotor bead as a function of experimental parameters. Important examples of this technique include mechanistic studies of DNA gyrase[20] as well as experiments used to show that DNA overwinds when stretched.[21] This technique has also recently been modified and used to study the sequence dependence of cooperative transitions of supercoiled DNA including strand separation and Z DNA formation in both dynamic and static torque measurements.[22,23]

38.6.4 Electromagnets

Multipole electromagnets allow for real-time control of both the direction and magnitude of the magnetic field gradient.[5] As the field is limited only by the current through the poles, nanonewton forces may be obtained. These instruments have found application principally in cell biology applications to study the response of cells to externally and internally applied forces.[24–26]

38.6.5 High-Force MT

Recently, by using low-carbon-steel focusing tips that surround two NdFeB cube magnets, forces in the nanonewton range have been reached with 4.5 μm magnetic beads making permanent-magnet MT useful in microrheology experiments.[27]

38.6.6 Total Internal Reflection Fluorescent Microscopy

One weakness of molecular manipulation techniques including MT is that the signal (i.e., change in extension or rotation) may originate anywhere along the molecule. However, the MT setup is well suited for the inclusion of objective-based total internal reflection fluorescent microscopy (TIRFM) (Figure 38.5). This powerful combination instrument allows for concurrent mechanical manipulation and single-molecule fluorescent microscopy allowing for the localization of mechanical changes to specifically labeled base pairs. This technique also has potential for the concurrent observation of fluorescently labeled protein binding and mechanical changes to the DNA. This strategy has been used to measure the Forster resonant energy transfer (FRET) between two dyes as a function of force,[28] to monitor fluorescent polarization in order to assay for rotation of the phi-29 viral DNA packaging motor,[29] to study the force and superhelicity dependence of B–Z transitions in DNA,[30] and to follow the mechanical unfolding of teleomeric G-quadruplex DNA.[31]

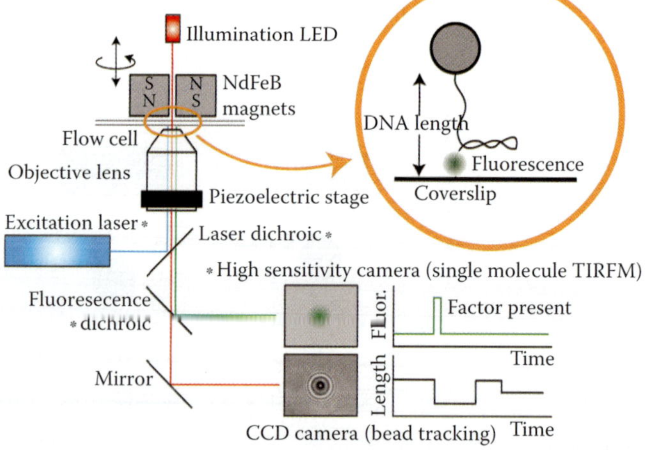

FIGURE 38.5
TIRFM–tweezers combination. The addition of an excitation laser and detection pathways to the standard MT setup allows for the monitoring of single-molecule fluorescence concurrent with the application of force or torque.

38.7 Molecular Manipulation

Although the MT may be used to manipulate any molecule, the majority of MT work consists of experiments manipulating nucleic acids, in particular, double-stranded DNA. The following sections will summarize the molecular manipulations possible with MT using dsDNA as an example.

38.7.1 Stretching

To measure how the length of a molecule responds to force, the end-to-end length of the molecule is measured by tracking the Z-position of the bead at different forces. A plot of extension versus force can then be used to assess the stiffness of the molecule (Figure 38.6a). Whereas a simple Hookean spring has a linear force–extension plot, polymers typically exhibit nonlinear behavior. In the case of dsDNA, this nonlinearity is well described by the worm-like chain model from polymer physics.[32–34] The nonextensible worm-like-chain (WLC) model (valid at forces less than 5 pN) has two parameters: the contour and persistence lengths. The contour length for a given DNA construct can be calculated as the average B-form base-pair spacing of 3.4 Å times the length of the DNA in base pairs. The persistence length describes the entropic elasticity of the molecule and represents the length over which bending the molecule costs an energy k_BT.[33,34] For example, the difference in persistence length between dsDNA (~45 nm) and single-stranded DNA (~3 nm) quantitates the relative rigidity of dsDNA compared to ssDNA.

At forces between 5 and 60 pN, an additional term must be added to the model that describes the enthalpic stretching of the molecule past its contour length.[34,35] At 60 pN, the DNA undergoes a phase change known as the overstretch transition where the force is maintained constant and the molecule extends to approximately twice its contour length.[36]

One important caveat to the worm-like chain model is that the polymer is assumed to not self-interact. In the case of ssDNA, which may base-pair with itself, models of this interaction may be included to fit experimental pulling data, which permit estimates of average interaction energies.[37]

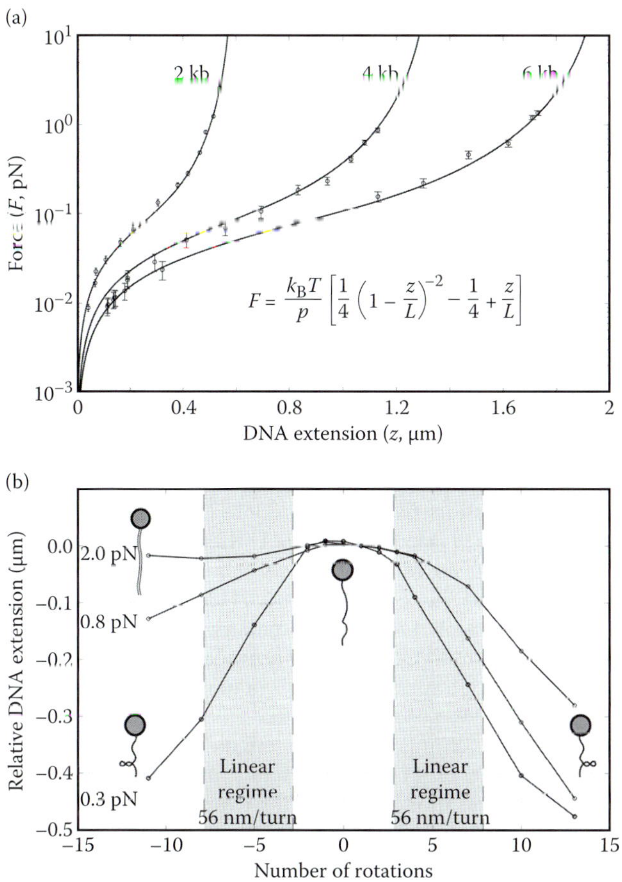

FIGURE 38.6
Pulling and twisting DNA. (a) Force–extension curves of three lengths (2, 4, and 6 kb) of DNA collected on MT are shown along with fits using the displayed worm-like chain model. (b) Rotation–extension curves for a 2 kb piece of DNA at three forces (2, 0.8, and 0.3 pN).

38.7.2 Twisting

An important advantage of the MT approach as compared to other single-molecule manipulations is the ability to introduce and measure changes in molecular rotation. In the case of the double-helical DNA, this degree of freedom is of utmost importance for understanding both how DNA is packaged and how the information it contains is maintained and accessed. Turning the magnets introduces torque on torsionally constrained molecules (molecules that cannot relax turn through the swiveling of single covalent bonds, i.e., un-nicked dsDNA that is attached at multiple points to both the magnetic bead and the glass surface; see Section 38.3.4).

The linking number (Lk, the number of times the individual strands of the DNA cross each other) of a torsionally constrained DNA molecule is conserved and can be written as the sum of writhe (Wr) and twist (Tw).[38–40] Twist counts the crossings of individual strands around the axis of the DNA, while writhe counts axis–axis crossings. A full turn of the magnets increments the linking number by exactly 1. Extra links may be distributed through the molecule as either twist or writhe depending on the force and buffer conditions at which the tether is being held.[41] Changes in writhe cost less energy in buffers with moderate levels of salt; however, they require that the end-to-end distance of the molecule change significantly (i.e., at 0.3 pN, $\Delta z \sim \pm 50$ nm/writhe). Therefore, at low forces (F < 0.3 pN), the linking number changes of either sign partition into writhes and decrease the end-to-end distance. At high forces (F > 2 pN), negative links are absorbed by a reduction of twist, while positive links are still absorbed by an increase in writhe. Plots of DNA extension versus magnet turns illustrate the relationship between force, twist, and writhe (Figure 38.6b). The characteristic asymmetry in these plots is a consequence of the chirality of DNA. It also allows for a measurement of the balance of energies between DNA compaction and DNA unwinding.[41,42] Intermediate forces (0.3 pN < F < 2.0 pN) result in rotation–extension plots, where removal of links leads to variable amounts of DNA unwinding and smaller changes in extension, while positive links still lead to the generation of positive writhes.

38.8 Conclusions and Future Directions

The MT technique allows for unique experiments to further our knowledge of the polymer physics of DNA. It also empowers research into the detailed molecular mechanisms of biologically important enzymes that manipulate the ubiquitous polymer. The ability to simply exert both force and torque along with the relative ease of construction and the availability of algorithms and software for data analysis makes MT a natural choice for single-molecule manipulation. Use of this technique is becoming more and more widespread, and in the near future, the complexity of molecular systems to which it is applied will increase. Furthermore, to deal with increasing system complexity, combination techniques such as MT/fluorescent microscopy will become more necessary. The continued development of these combination techniques will result in the ability to track multiple observables concurrently and to dissect the mechanisms of larger and more complex molecular systems.

References

1. Smith, S. B., Finzi, L. & Bustamante, C. Direct mechanical measurements of the elasticity of single DNA molecules by using magnetic beads. *Science* **258,** 1122–1126 (1992).
2. Lionnet, T. et al. Magnetic trap construction. *Cold Spring Harb Protoc* **2012,** 133–138 (2012).
3. De Vlaminck, I. & Dekker, C. Recent advances in magnetic tweezers. *Annu Rev Biophys* **41,** 453–472 (2012).
4. Lionnet, T. et al. Single-molecule studies using magnetic traps. *Cold Spring Harb Protoc* **1,** 34–49 (2012). doi:10.1101/pdb.top067488.
5. Gosse, C. & Croquette, V. Magnetic tweezers: Micromanipulation and force measurement at the molecular level. *Biophys J* **82,** 3314–3329 (2002).
6. Seol, Y. & Neuman, K. C. Magnetic tweezers for single-molecule manipulation. *Methods Mol Biol* **783,** 265–293 (2011).
7. Selvin, P. R. & Ha, T. *Single-Molecule Techniques*. Cold Spring Harbor Laboratory Press, Woodbury, New York (2008).
8. Lipfert, J., Kerssemakers, J. W. J., Jager, T. & Dekker, N. H. Magnetic torque tweezers: Measuring torsional stiffness in DNA and RecA-DNA filaments. *Nat Meth* **7,** 977–980 (2010).
9. De Vlaminck, I. et al. Highly parallel magnetic tweezers by targeted DNA tethering. *Nano Lett* **11,** 5489–5493 (2011).
10. Revyakin, A., Ebright, R. H. & Strick, T. R. Single-molecule DNA nanomanipulation: Improved resolution through use of shorter DNA fragments. *Nat Meth* **2,** 127–138 (2005).
11. Strick, T. R., Allemand, J. F., Bensimon, D. & Croquette, V. Behavior of supercoiled DNA. *Biophys J* **74,** 2016–2028 (1998).
12. Paik, D. H., Roskens, V. A. & Perkins, T. T. Torsionally constrained DNA for single-molecule assays: An efficient, ligation-free method. *Nucleic Acids Res* **41,** e179 (2013). doi:10.1093/nar/gkt699.
13. Gelles, J., Schnapp, B. J. & Sheetz, M. P. Tracking kinesin-driven movements with nanometre-scale precision. *Nature* **331,** 450–453 (1988).
14. Neuman, K. C., Abbondanzieri, E. A. & Block, S. M. Measurement of the effective focal shift in an optical trap. *Optics Lett* **30,** 1318–1320 (2005).
15. van Loenhout, M. T. J., Kerssemakers, J. W. J., De Vlaminck, I. & Dekker, C. Non-bias-limited tracking of spherical particles, enabling nanometer resolution at low magnification. *Biophys J* **102,** 2362–2371 (2012).
16. Lansdorp, B. M., Tabrizi, S. J., Dittmore, A. & Saleh, O. A. A high-speed magnetic tweezer beyond 10,000 frames per second. *Rev Sci Instrum* **84,** 044301 (2013).
17. te Velthuis, A. J. W., Kerssemakers, J. W. J., Lipfert, J. & Dekker, N. H. Quantitative guidelines for force calibration through spectral analysis of magnetic tweezers data. *Biophys J* **99,** 1292–1302 (2010).
18. Lipfert, J., Wiggin, M., Kerssemakers, J. W. J., Pedaci, F. & Dekker, N. H. Freely orbiting magnetic tweezers to directly monitor changes in the twist of nucleic acids. *Nat Commun* **2,** 1–9 (2011). doi:10.1038/ncomms1450.
19. Bryant, Z. et al. Structural transitions and elasticity from torque measurements on DNA. *Nature* **424,** 338–341 (2003).
20. Gore, J. et al. Mechanochemical analysis of DNA gyrase using rotor bead tracking. *Nature* **439,** 100–104 (2006).
21. Gore, J. et al. DNA overwinds when stretched. *Nature* **442,** 836–839 (2006).

22. Oberstrass, F. C., Fernandes, L. E. & Bryant, Z. Torque measurements reveal sequence-specific cooperative transitions in supercoiled DNA. *Proc Natl Acad Sci USA* **109,** 6106–6111 (2012). doi:10.1073/pnas.1113532109.
23. Oberstrass, F. C., Fernandes, L. E., Lebel, P. & Bryant, Z. Torque spectroscopy of DNA: Base-pair stability, boundary effects, backbending, and breathing dynamics. *Phys Rev Lett* **110,** 178103 (2013).
24. Kollmannsberger, P. & Fabry, B. High-force magnetic tweezers with force feedback for biological applications. *Rev Sci Instrum* **78,** 114301 (2007).
25. de Vries, A. H. B., Krenn, B. E., van Driel, R., Subramaniam, V. & Kanger, J. S. Direct observation of nanomechanical properties of chromatin in living cells. *Nano Lett* **7,** 1424–1427 (2007).
26. Bausch, A. R., Möller, W. & Sackmann, E. Measurement of local viscoelasticity and forces in living cells by magnetic tweezers. *Biophys J* **76,** 573–579 (1999).
27. Lin, J. & Valentine, M. T. High-force NdFeB-based magnetic tweezers device optimized for microrheology experiments. *Rev Sci Instrum* **83,** 053905 (2012).
28. Shroff, H. et al. Biocompatible force sensor with optical readout and dimensions of 6 nm 3. *Nano Lett* **5,** 1509–1514 (2005).
29. Hugel, T. et al. Experimental test of connector rotation during DNA packaging into bacteriophage phi29 capsids. *PLoS Biol* **5,** e59 (2007).
30. Lee, M., Kim, S. H. & Hong, S.-C. Minute negative superhelicity is sufficient to induce the B-Z transition in the presence of low tension. *Proc Natl Acad Sci USA* **107,** 4985–4990 (2010).
31. Long, X., Parks, J. W., Bagshaw, C. R. & Stone, M. D. Mechanical unfolding of human telomere G-quadruplex DNA probed by integrated fluorescence and magnetic tweezers spectroscopy. *Nucleic Acids Res* **41**, 2746–2755 (2013). doi:10.1093/nar/gks1341.
32. Bustamante, C., Marko, J. F., Siggia, E. D. & Smith, S. Entropic elasticity of lambda-phage DNA. *Science* **265,** 1599–1600 (1994).
33. Marko, J. F. & Siggia, E. D. Stretching DNA. *Macromolecules* **28,** 8759–8770 (1995).
34. Odijk, T. Stiff chains and filaments under tension. *Macromolecules* **28,** 7016–7018 (1995).
35. Baumann, C., Smith, S., Bloomfield, V. & Bustamante, C. Ionic effects on the elasticity of single DNA molecules. *Proc Natl Acad Sci USA* **94**, 6185–6190 (1996).
36. Smith, S., Cui, Y. & Bustamante, C. Overstretching B-DNA: The elastic response of individual double-stranded and single-stranded DNA molecules. *Science* **271**, 795–799 (1995).
37. Dessinges, M.-N. et al. Stretching single stranded DNA, a model polyelectrolyte. *Phys Rev Lett* **89,** 248102 (2002).
38. Fuller, F. B. The writhing number of a space curve. *Proc Natl Acad Sci USA* **68,** 815–819 (1971).
39. Crick, F. H. Linking numbers and nucleosomes. *Proc Natl Acad Sci USA* **73**, 2639–2643 (1976).
40. Fuller, F. B. Decomposition of the linking number of a closed ribbon: A problem from molecular biology. *Proc Natl Acad Sci USA* **75,** 3557–3561 (1978).
41. Tempestini, A. et al. Magnetic tweezers measurements of the nanomechanical stability of DNA against denaturation at various conditions of pH and ionic strength. *Nucleic Acids Res* **41,** 2009–2019 (2013).
42. Salerno, D. et al. Single-molecule study of the DNA denaturation phase transition in the force-torsion space. *Phys Rev Lett* **109,** 118303 (2012).

39

Analytical Electron Microscopy of Bone and Mineralized Tissue

Michal Klosowski, Alexandra E. Porter, Sandra J. Shefelbine, and David W. McComb

CONTENTS

In this chapter, the use of transmission electron microscopy (TEM) techniques for investigation of mineralized tissues is reviewed. The high-resolution imaging capabilities of modern TEM instruments are yielding new insights into the nanostructure of materials such as bone and dentine that improve our understanding of the complex multiscale structures. Analytical techniques such as electron energy-loss spectroscopy and energy dispersive x-ray analysis in the TEM are providing insights into chemistry and bonding in mineralized tissues with molecular-scale spatial resolution. Dual beam focused ion beam instruments have revolutionized sample preparation; TEM samples can be prepared from site-specific regions of interest such as cracks and interfaces. In combination, these methods can shed light on the exact relationship between the mineral and organic phases in mineralized tissues and ultimately provide unique insights in the complex relationship between chemistry, structure, and bonding with bone-related pathologies.

39.1 Introduction to Transmission Electron Microscopy

The need to characterize the structure and chemistry of mineralized tissues, at the nanometer and subnanometer scale, stems, at least in part, from the increasing number of bone-related pathologies in our aging society. Moreover, while the macrostructure and microstructure of mineralized tissues have become well known, the inherent complexity of such specimens makes it very difficult to investigate and understand their nanostructure. Transmission electron microscopy (TEM) and associated spectroscopies have been used to great effect to relate the ultrastructure and nanostructure of bone and other mineralized tissues to the mechanical and structural properties.

TEM has been used to image and analyze bone since the early 1950s. The early studies prepared the bone by blending slivers of bone in distilled water and evaporating the water from the suspension to observe bone crystals [1]. These studies established that bone crystals were "tablet"-shaped [2] and were found inside collagen fibrils [3]. Later studies maintained the local organization of the tissue by microtoming thin sections and observed that the collagen fibrils were highly aligned and the mineral crystals were "needle"-shaped apatite particles [4,5]. The needle-like morphology of bone crystals was the dominant view in most subsequent studies that used thin cut specimens until tilted images (early tomography) indicated indeed the tablet structure [6,7]. Electron microscopists should take heed of this historical lesson. New preparation and analysis methods provided different perspectives on the bone structure, revealing features that were previously not apparent. Here, we outline briefly the physics of TEM and the general areas that it has been used in studying bone. We also detail the current preparation and analysis methods for imaging and analyzing bone using TEM.

39.1.1 Basics of TEM

In electron microscopy, a sample is illuminated with an electron beam created by heated tungsten, LaB_6 filaments, or more sophisticated field emission system. In conventional TEM, the incident beam is parallelized with a condenser lens and transmitted through a thin specimen producing an image (Figure 39.1). Depending on sample properties, incident beam electrons can be directly transmitted through a thin layer or interact with the material in elastic or inelastic scattering events. These events can be used to determine not only the structure of a given specimen but also its chemistry. In scanning TEM (STEM), the electron beam is focused and scanned over the specimen.

39.1.2 TEM of Bone Ultrastructure

Most TEM of bone has examined the ultrastructural organization of the molecular and mineral components of bone, which are primarily collagen and hydroxyapatite. A collagen fibril imaged by TEM has a distinctive banding pattern with a 64–70 nm spacing due to the quarter-staggered arrangement of collagen (Figure 39.2a) [8]. In a mineralized fibril (Figure 39.2b), the dark regions in

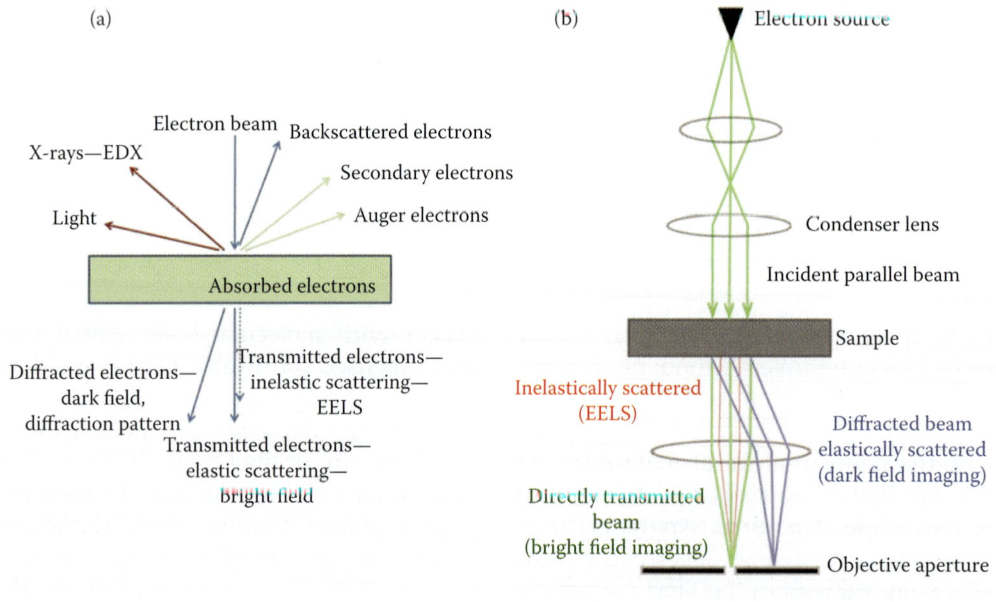

FIGURE 39.1
(a) Schematic of electron–sample interactions used in electron microscopy techniques. (b) Schematic of standard TEM.

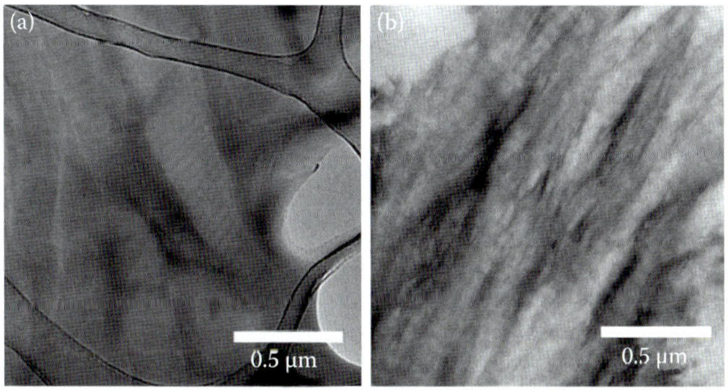

FIGURE 39.2
(a) Banding in a demineralized collagen fibril of a 14-week-old turkey tendon. Bright field (BF) TEM renders the overlap region of the demineralized fibril dark. (b) Mineralized fibril of a 14-week-old turkey tendon. BF TEM renders the crystals dark. Both samples were prepared by freeze substitution (FS) without staining and microtome sectioning.

the banding are due to the high electron scattering of the dense mineral crystals in the gap; the lighter regions are the overlap of the collagen molecules. Crystals are found both within the gap region of the collagen fibril (intrafibrillar) and the coating outside of the fibril (extrafibrillar). In early TEM studies, it was assumed that the majority of the mineral was intrafibrillar, but now it is generally accepted that nearly 70% of the mineral content of bone is extrafibrillar [9–11]. This extrafibrillar arrangement is particularly striking when fibrils are viewed in cross section. In this view, the mineral within the fibril has such low contrast that it is not visible. TEM images can indicate the size of individual crystals (approximately 15–200 nm length × 10–80 nm width × 2–7 nm thickness) [2,12,13] (unlike x-ray methods that measure an average crystal size in bulk). The crystallographic *c*-axis is typically aligned with the long axis of the fibril [14,15].

The arrangement of the mineralized fibers into a lamellar structure is characteristic of osteonal bone in which "rings" of bone form around a central vessel. Studies have used TEM to examine the fibril orientation within each lamella and proposed a plywood-like structure with alternating orientation of the fibers [16,17]. A recent study using lower magnification TEM revealed that within a lamellar layer, fibers actually have a heterogeneous woven structure [18]. Previous studies with higher magnification were probably examining a single fiber bundle within lamella. Particularly with the high resolution afforded by TEM, one should be wary of drawing conclusions about structure/function on a very small and perhaps not representative region examined. This sampling problem can be overcome by complementing analysis with bulk techniques such as x-ray diffraction or small-angle x-ray scattering.

Using analytical techniques such as electron energy loss spectroscopy (EELS), compositional mapping at the nanometer level is possible [19]. EELS has been used to examine compositional structure across a mineralized fibril in dentin [20] and intracellular and extracellular composition of vesicles, particularly examining calcium transport in mineralization [21,22] and resorption [23] processes.

39.1.3 Imaging Bone Cell Function

One advantage of TEM is that, with appropriate preparation techniques, bone cells can be imaged and probed for composition and structure at a much higher resolution than afforded by light microscopy and with spatial resolution that is not possible with other compositional methods such as x-ray diffraction, Fourier transform infrared, or Raman. For decades, TEM has been used to examine the mineralization process in bone or in bone cells *in vitro*. Landis showed that early mineral is found primarily in the gap regions of the fibrils [24] as well as on the surface of the fibrils [25], suggesting a matrix-mediated mineralization process regulated by noncollagenous proteins. Others have proposed that mineralization is mediated by vesicles that are emitted from the cell containing calcium and phosphate ions [26–29]. A recently proposed theory, supported by EELS of the mitochondria and associated vesicles, suggests that amorphous calcium phosphate (ACP) from the mitochondria is delivered in vesicles to the matrix where it then crystalizes [21]. As preparation techniques and analysis methods improve, it is likely that TEM will allow us to determine the structure–function relationship within a cell and its interaction with the surrounding matrix. Few other imaging modalities allow the probing of composition while preserving structure at sufficient spatial resolution to decipher the components of the specific parts of a cell.

39.1.4 Other Applications of TEM in Bone

TEM is often utilized to study the interface between an implant and native bone in order to characterize nano-osseous integration. In such studies, it is critical that all tissue types be preserved and the interface structure be maintained. The implant material (titanium, cobalt chrome, etc.) often precludes microtoming. Studies have found that the interface structure is best preserved if the implant is removed after embedding the tissues [30]. Others have used ion beam milling [31,32] to preserve both implant and bone at the interface. Volume renderings of the interfaces are possible with electron tomography (ET) [33] and compositional characterization with energy dispersive x-ray (EDX) analysis [34]. Because careful sample preparation techniques (discussed in Section 39.2) can preserve chemistry, structure, and cellular details, TEM allows complicated interface properties to be well characterized at high spatial resolution.

TEM has been used in the paleontological and anthropological contexts to understand the process of fossilization of bone, particularly changes that occur in apatite crystals [35] and osteocytes [36]. Such studies help to understand the preservation state of the bone and the recrystallization processes that have occurred. Analytical TEM methods have been applied to dinosaur bones [37–39] in order to look for original composition in extinct organisms subjected to millions of years of geochemical influences. The destructive nature of TEM sample preparation and the chemical changes that occur over millions of years make TEM of paleontological samples challenging. Nonetheless, TEM offers a modality with a potential for uncovering nanostructural changes in bone evolution.

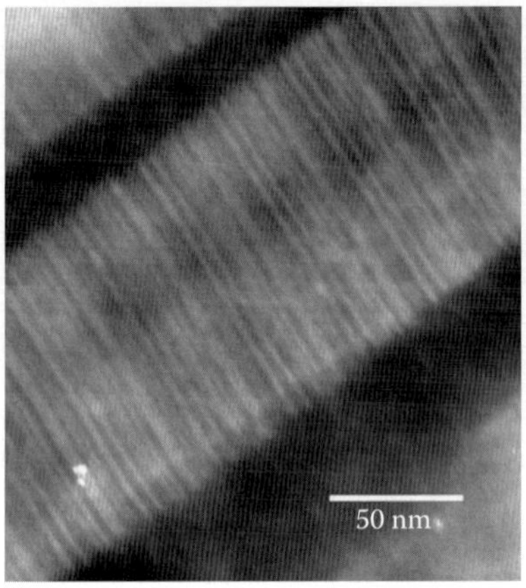

FIGURE 39.3
Mouse tail collagen fibril bulk stained with osmium tetroxide and uranyl acetate. Acquired in STEM DF mode at 300 kV. The uranyl acetate strongly stains negatively charged carboxylate groups, which generates contrast from the bands.

39.2 Biological Sample Preparation of Hard Tissues

Imaging and analysis of bone in its native state are challenging as the tissue contains liquids that cannot be observed directly in the TEM, which is operated under high vacuum. A number of preparation protocols exist for preparing mineralized cells and tissues for examination by TEM; however, simultaneously preserving the chemistry and structure of the mineral and the cells during preparation is challenging. For this reason, a number of complementary sample preparation routes are typically employed, which are designed to optimize preservation of the mineral and cell structure and chemistry. A further limitation is that the bone can be susceptible to damage under the electron beam, which can alter the structure and chemistry of the specimens. These factors can all create artifacts, such as crystallization of the mineral phase, and make the results liable to misinterpretation. Artifacts introduced during TEM sample preparation can be extremely subtle, and several techniques must frequently be used in parallel.

39.2.1 Chemical Fixation

The most common method for preparing bone for TEM observation is chemical fixation followed by dehydration and embedding in a plastic resin. This method preserves the cell structure and also the organization of the collagen fibrils and their banding pattern. During fixation, the cell structure is fixed by cross-linking the amino acid groups using aldehydes (i.e., glutaraldehyde and formaldehyde). The contrast from the cell membranes and internal cell organelles can be also increased using heavy metal stains, such as osmium tetroxide and uranyl acetate. For example, osmium tetroxide, which is a strong oxidant, and therefore acts as a stain and also a secondary fixative, increases contrast by incorporating into unsaturated carbon bonds in the lipids (Figure 39.3). After fixation, the tissue is dehydrated in graded concentrations of ethanol dilutions before embedding in a polymer. Ultrathin sections of ~70 nm thickness are prepared for TEM observation and analysis by ultramicrotomy (sections are cut mechanically using a diamond knife). Alternatively, sections can be milled and placed on a TEM grid using a focused ion beam (FIB) milling [20].

Chemical fixation methods preserve cell and collagen organization and periodic banding structure; however, this method introduces a high percentage of water to the tissue and will transform ACP to crystalline calcium phosphate. In addition, the chemical fixation and dehydration as part of the sample preparation can cause shrinkage and collapse of delicate structures. For this reason, while these methods are useful to study bone ultrastructure, other methods must be employed for more detailed analysis of the bone chemistry and cell structure.

39.2.2 Cryofixation

Cryofixation is an alternative method for preparing bone for TEM imaging and analysis. The advantage of cryofixation methods is that all structure and chemistry can be preserved simultaneously [40]. This method freezes specimens at a very low temperature, typically using liquid propane or liquid nitrogen. The specimen is frozen within milliseconds, which is fast enough to capture biological events. For example, fibrillogenesis of collagen was captured first in chick embryo tendon prepared by rapid freezing [41] and is not usually observed using chemical fixation. Cryofixation is the best method to preserve the chemistry of the sample, and several groups have studied bone mineralization by transferring and imaging frozen hydrated bone into the TEM [42]. However, the limitation of cryofixation at ambient atmosphere is that it is limited to small sample sizes and problems can be encountered with ice crystal damage, which may disrupt fine structures of the cell, for example. Chemical analysis of frozen hydrated bone is not yet possible in the TEM as the samples damage under the electron beam, although application of low-loss electron energy loss techniques may have potential for mapping the distribution of calcium, phosphorus, and nitrogen within the tissues.

39.2.3 Anhydrous Fixation

Bone contains the mineral components of several solid phases of calcium phosphate. ACP is a precursor phase, which later converts to crystalline bone apatite [42–44]. ACP is thermodynamically unstable in aqueous environments and quickly converts to a more stable, crystalline phase of carbonated hydroxyapatite [45] in a process, which is temperature and pH dependent [46]. ACP has been reported to be stable in ethanol and acetone. Immersion in 50% (v/v) acetone solution delays the transformation of ACP to hydroxyapatite (HA) compared to immersion in water [45]. ACP processed by chemical fixation significantly converts to poorly crystalline HA, as demonstrated by x-ray and electron diffraction. The aqueous solution has been reported to change the solid phase of calcium phosphate [24]. Therefore, chemical fixation in solution would lead to poor analysis of the calcium phosphate phases in bones/calcified tissues. An alternative room temperature preparation technique, the anhydrous processing method, does not significantly change crystallinity or chemistry of bone apatite, ACP, or brucite (observed by electron diffraction). Samples are dehydrated directly in ethylene glycol before embedding in resin. The anhydrous preparation procedure does not fix the organic components of the tissue, so the cellular structure is not preserved well. This method has been used for various high-resolution TEM studies of bone and HA implants [47,48] to preserve the chemistry and crystallinity of the mineral phase of bone. Full details of anhydrous sample preparation procedures for preparation of bone and bone–implant interfaces are provided in ref. [48].

A limitation of the anhydrous room temperature method of preparation is that it does not preserve the distribution of diffusible ions, such as calcium. In order to simultaneously preserve the mineral and organic phases in bone and analyze their architecture and chemistry in the TEM, high-pressure freezing (HPF) and freeze substitution (FS) must be employed.

39.2.4 High-Pressure Freezing

HPF has an advantage over other freezing techniques as it can increase the depth of a sample that becomes well frozen without ice damage. At higher pressures (i.e., 210 MPa), the melting point of water is minimized to −22°C and water can be supercooled to −92°C [49] (Figure 39.4). This maximizes the supercooling window at this pressure. Water therefore can be vitrified with a lower cooling rate. A cooling rate of −500°C/s is sufficient to freeze biological samples at 210 MPa [50]; thus, thicker specimens can be vitrified. In addition, under high pressure, water crystallizes as ice II and III, which are denser and have a lower growth rate of crystals than ice I that forms at ambient atmosphere. Thus, vitrification of samples can be achieved with thickness of up to 50 μm when cooling is applied from both sides [51,52]. This method has been successfully employed in our own studies to prepare bone nodules for TEM analysis and study intracellular and extracellular transport of calcium ions [21].

Frozen samples (either by high pressure freezing or cryofixation) can be transferred directly into the TEM and imaged at liquid nitrogen/helium temperatures (e.g., refs. [43] and [53]). Alternatively, specimens may have the ice removed using a solvent such as acetone and methanol at a low temperature (−90°C). This substitution process does not cause osmotic stress in the frozen samples [54]. Following HPF and FS, the specimens are embedded in resin and processed using a procedure similar to the chemical fixation protocol. Freezing combined with FS is a hybrid method combining

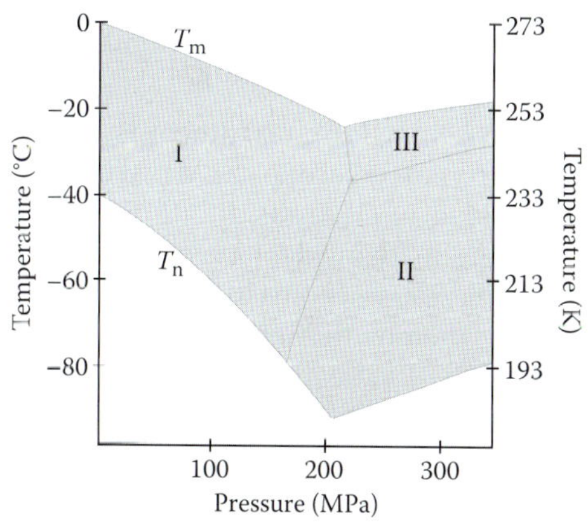

FIGURE 39.4
Part of water phase diagram showing the melting point (T_m) and the homogeneous nucleation temperature (T_n) as a function of pressure. At 210 MPa, the melting temperature is at a minimum as well as the nucleation temperature. The possibility of supercooling is maximized. At high pressure (i.e., above 200 MPa), water crystallizes as ice II and III, rather than ice I.

preservation and resin embedding. Chemical agents that are generally used in conventional fixation may be added during substitution at low temperature, around −90°C, to fix cells and enhance contrast from cell organelles. The fixation process takes place during the heating phase of the procedure [55]. Osmium tetroxide starts reacting with the lipid membrane at −70°C [56], and the glutaraldehyde starts cross-linking proteins at −40°C [57]. FS with suitable solvents has been reported to be sufficient to preserve the position and chemistry of the diffusible ions within the sample to study cell-mediated mechanisms of bone mineralization [21].

HPF followed by FS procedure will induce a significantly lower amount of conversion to crystalline calcium phosphate since only 1% water is added; however, this method is not completely anhydrous and so cannot be used to study crystallinity or the crystal structure (although the amount of crystalline calcium phosphate converted will be very low). Therefore, this HPF/FS must be combined with the anhydrous preparation protocol to study the crystallinity of the mineral.

39.2.5 Ultramicrotome Cutting

Once the bone tissue has been resin embedded, the final stage of processing is to ultramicrotome thin sections of the bone samples. Typically, a 45° diamond knife is used to cut 70 nm thick sections onto TEM grids into a water bath. The sections must be collected immediately and dried on filter paper to ensure that the bone mineral has not dissolved the ultramicrotomy process [47]. The challenge with ultramicrotoming bone lies in matching the hardness of the embedding resin with the hardness of the bone to ensure that the interface does not damage during sectioning. For this reason, choice of embedding resin is crucial.

39.2.6 Focused Ion Beam Milling

A focussed ion beam (FIB) instrument is analogous to a scanning electron microscope (SEM), with a source of heavy ions instead of electrons. In a FIB instrument, the high momentum of the incident ion beam, as it is scanned across the sample, causes atoms in the target material to be sputtered away, thereby milling the surface of the sample [58]. Control of the ion beam allows precision milling of thin sections of a sample. Secondary electrons (SEs) generated during the ion collision cascade can be used for imaging. This allows a region of interest to be selected for milling with a precision of 10–20 nm, which is a major advantage compared with ultramicrotomy.

FIB milling is an alternative route for preparation of TEM samples that offers several advantages over ultramicrotomy. The technique has been extensively used in nonbiological material preparation [58,59], and its use for biological systems is increasing. The first data on FIB sample preparation for TEM of tooth dentine were published by Hayashi et al. [60]. Subsequently, high-quality TEM sections of human dentine were obtained and investigated by Hoshi et al. [61], who used energy-filtering TEM on FIB milled sections. Volkert et al. [62] and Nalla et al. [63] concluded that the FIB technique better preserved the original structure of the specimen. TEM observation of dentinal tubules prepared by FIB milling and ultramicrotomy suggested that tubules sectioned with the microtome seemed to have been only partially filled, implying that some mineral may have been lost during the slicing process [63]. More recently, Giannuzzi and Stevie [58] prepared very high-quality TEM sections of bone–implant interfaces using the FIB.

In a detailed comparison of FIB and ultramicrotomy, Jantou et al. [64] showed that it is possible to prepare high-quality TEM sections of ivory dentine, for high-resolution imaging and analysis, using the FIB milling technique. It was demonstrated that FIB has a significant advantage over ultramicrotomy in that demineralization, dehydration, embedding, and section flotation can be avoided, eliminating potential mineral dissolution. The characteristic periodic contrast of collagen fibrils is clearly seen in FIB sections without the need for any chemical staining. Comparing transverse and longitudinal sections of ivory dentin prepared by both preparation methods emphasized their structural and chemical differences. Therefore, with the FIB milling sample preparation technique, it is now possible to investigate the exact relationship between the organic and inorganic phase of mineralized tissues without the chemical complications arising from the sample preparation procedure.

39.3 TEM Imaging

39.3.1 Conventional TEM Imaging of Bone

There are two imaging modes in the TEM: bright-field (BF) and dark-field (DF) imaging. BF imaging uses an objective aperture that selects the directly transmitted beams, DF imaging uses the diffracted beams (Figure 39.1b). Contrast is defined as the difference in intensity between two adjacent areas. For collagen and cells, contrast is predominantly caused by mass and thickness (defined as mass–thickness contrast). When electrons pass through the specimen, they are scattered to different angles depending on the nature of the variations within the samples. The scattering cross section is a function of atomic number (Z), density (ρ), and thickness (t). The regions that contain heavy atoms (higher Z) scatter more electrons than the regions that contain lighter atoms. The thicker regions also scatter

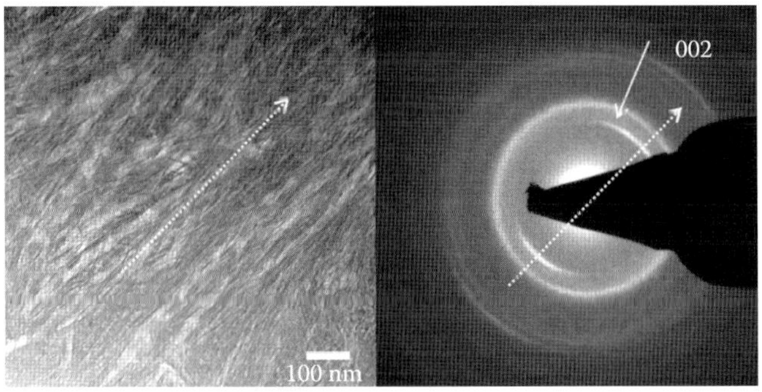

FIGURE 39.6
BF image of mouse bone and corresponding diffraction pattern (002 crystal plane of hydroxyapatite) indicating the alignment of the crystals (dotted arrow).

more electrons than the thinner regions when they both have the same average Z. In BF imaging, the regions that scatter more electrons (with high mass–thickness) have lower intensity profiles than the regions with lower mass–thickness; therefore, thick regions appear dark (Figure 39.5). For biological samples, a mass–thickness contrast is typically enhanced by selectively staining cell organelles with heavy elements (e.g., osmium and uranium). Specimens can be stained *en bloc* before embedding and/or later, directly after sectioning, by brief exposure of the sections to solutions of the heavy metal stains.

39.3.2 Electron Diffraction

The electron beam is partly scattered when it hits the sample. Atoms scattering in phase give rise to constructive interference of the electrons, resulting in a diffracted beam. This is only possible at specific angles for monochromatic incident waves. In 1912, W. Bragg showed that constructive interference occurs when the path difference between two interfering waves is an integer of the incident wavelength, satisfying what is now known as Bragg's law:

$$n\lambda = 2d \sin(\theta_B) \quad (39.1)$$

with n the order of diffraction, λ the wavelength, d the distance between two atomic planes, and θ_B the semiangle between the incident wave vector and the diffracted wave vector.

Electron diffraction patterns can be used to characterize the mineral phases present in bone, the degree of atomic order within the mineral, and the orientation of the apatite crystal with respect to the collagen fibrils (e.g., ref. [21]). DF imaging of bone can also be applied to map the distribution of different mineral phases within the sample by imaging using a selected diffracted beam.

39.3.3 Scanning Transmission Electron Microscopy

Scanning transmission electron microscopy (STEM) operates by converging the incident beam to form a probe that is rastered across the sample. Two pairs of scan coils are placed between the second condenser lens and the upper objective pole piece to pivot the beam about the front focal plane of the upper objective pole pieces. A set of electron detectors is placed after the specimen to collect electrons that are scattered to different angles; this allows us to select electrons to form an image. The BF detector collects the direct beam, and the annular detector (ADF) collects electrons scattered through a semiangle of 10–50 mrad. The high angular annular DF (HAADF) detector only collects electrons scattered by Rutherford scattering to high semiangles of >50 mrad (~3°). At these high angles, the contrast depends mainly on the atomic number of the sample (contrast Z^2). As the detector collection angles are reduced, the contribution of mass–thickness (ρt) increases. Thus, it is possible for the user to control which electron/specimen interactions contribute to an image. Our work has shown that it is possible to generate contrast from the bands within the collagen fibrils without the use of heavy metal staining by controlling the detector collection angles to optimize mass–thickness contrast [20]. The advantage of using unstained samples over stained ones is that the staining procedure can alter the chemistry of the mineral phase as water is introduced to the sample; in addition, the stain reduces the spatial resolution that is achievable from substructures within the collagen fibrils, and stained sections are not suitable for conducting subsequent analysis of the tissues by spectroscopic techniques as the stain interferes with the analysis.

39.3.4 Electron Tomography

Electron tomography (ET) is a technique for obtaining a three-dimensional image from a TEM specimen. Data collection for ET involves collecting a series of images while tilting the specimen around a single axis at regular intervals. A three-dimensional structure can then be reconstructed using the series of images collected to give high-resolution information about the material being analyzed. ET techniques have successfully been used to describe the spatial relationship between collagen and mineral [58]. A BF ET study of turkey tendon has suggested that three-dimensional fibrils are composed of two-dimensional arrays of collagen. Gaps exist between adjacent arrays, which are believed to be the sites for the mineral crystals to grow into after the mineral has nucleated in the gap zone [58]. More recently, HAADF-STEM tomography techniques have successfully been applied to study mineralization at interfaces between bone and Ti6Al4V alloys [31]. HAADF-STEM has a distinct advantage over other imaging methods for generating 3-D ET reconstructions of bone. The reason for this is that thicker samples can be studied as the effects of chromatic aberration are eliminated. In addition, the HAADF detector collects electrons that are not Bragg scattered. As such, HAADF images show little or no diffraction effects, and their intensity is approximately proportional to Z^2. This imaging technique proves ideal for tomographic reconstruction as it generates strong contrast that has a fully monotonic relationship with thickness. Clearly, there exists great potential to apply STEM tomography techniques to further our understanding of collagen fibril mineralization.

39.3.5 Radiation Damage of Bone in TEM

Depending on experimental conditions, several different electron beam damage events may occur. Both elastic and inelastic interactions between the electron beam and a specimen may cause damage to the specimen, resulting in structural and chemical changes. There are two main damage mechanisms: radiolysis and knock-on damage.

39.3.5.1 Radiolysis

Radiolysis occurs when the incident electron beam causes the breakage of chemical bonds in organic compounds such as polymers and biological samples. The breakage of the primary polymer chains changes the structure of material, resulting in mass loss. The breaking-off of side chain molecules leaves reactive free radicals, which may react and cross-link to form a new structure. In the TEM, mass loss of the specimen can be seen as thinning of the specimens and eventually the drilling of a hole through the specimen. Radiolytic damage can also lead to the loss of the crystallinity of a polymer. This damage mechanism can be minimized by operating at a high accelerating voltage (e.g., 300 kV) as this will decrease the probability of electron interaction and energy transfer to the sample. Also, electrons accelerated through a high potential have high velocity; as a consequence, the electron spends less time in a given volume. Therefore, operating at a high voltage minimizes electronic interactions between incident electrons and the sample. However, it also lowers the probability of elastic scattering, resulting in reduction of image contrast.

Radiolysis is particularly visible when bone sections are examined using the STEM probe and can result in "hole drilling." The electron beam excites oxygen atoms bound to the structure of bone. Once the oxygen atoms are released, either as free radicals or O_2 molecules, they form a "bubble" inside the sample. The growing "bubble" may finally burst from the surface of the thin section, creating a void in the sample. This effect may be minimized by reducing electron dose or coating the sample with a thin carbon layer.

Specimen heating is also a problem in the TEM, particularly for specimens that have a low thermal conductivity, such as organic materials. When incident electrons collide with electrons in the specimen, they can transfer energy to the specimen, and most of the transferred energy ends up as heat in the specimen. Heating damage can be minimized by using a high accelerating voltage and by reducing the specimen thickness.

As for many other organic materials, bone is susceptible to radiolytic damage as it is a poor heat and charge conductor. Therefore, illuminating a thin section of bone gradually heats the sample, which leads to bond excitation and finally to breakdown of the sample structure. Prolonged exposure to electron beam radiation may cause amorphization not only of organic chains but also of the apatite crystals. These changes can lead to formation of new compounds: apatite can reorganize into CaO and phosphate can transform into phosphide once the total electron fluence exceeds above 10^8 electrons/nm^2 [65].

39.3.5.2 Knock-On Damage

Knock-on damage occurs when electron beam transfer energy is sufficient to knock out an atom from its atomic site leading to the generation of a vacancy; this affects the crystalline perfection of the materials. The threshold energy required for this event to occur is dependent on the materials being studied. In TEM, knock-on damage is not expected to occur at an accelerating voltage of 60 kV, unless a specimen contains hydrogen atoms. This damage mechanism occurs at higher voltages. To minimize knock-on damage, operation should be carried out using an incident beam with kinetic energy lower than the threshold displacement energy of the material. If a higher energy is used, then the electron dose (electrons/nm^2) should be minimized. For the atoms on the specimen surface, the threshold energy for displacement is lower, as the surface atoms are free to leave the specimen and do not have to be displaced into interstitial sites. This mechanism is called sputtering.

39.4 Analytical Methods

The ability to investigate the chemical composition and bonding environment in a localized region of a specimen allows the analytical electron microscope to provide deep and novel insights into the relationship between chemistry and structure in mineralized tissues. EELS and EDX analysis provide such information.

39.4.1 Electron Energy-Loss Spectroscopy

The fast primary electrons that interact with a thin sample in the TEM can be scattered elastically or inelastically. In electron energy-loss spectroscopy (EELS), the inelastic electrons, which have lost energy due to interactions with the specimen, can be analyzed by the electron energy-loss spectrometer, giving detailed information on the local chemistry, electronic structure, and bonding of the sample [66–68]. The energy-loss spectrum (Figure 39.6a) can be divided into three main regions: the zero-loss peak (ZLP; at 0 eV), the low-loss region (typically up to 50 eV), and the core loss region (from 50 eV onward). The ZLP is usually the most intense feature on the spectrum. As the name implies, it contains electrons that have lost no measurable amount of energy. The full-width at half maximum (FWHM) of the ZLP is a measure of the energy resolution of the spectrum. The energy resolution mainly depends on the energy spread of the electron beam, which, in turn, depends on the electron source. A Schottky field emission gun (FEG) source gives a typical energy spread of 0.6–0.8 eV while a cold FEG gives 0.3–0.5 eV. Use of a monochromator can improve the energy resolution to 0.2 eV or better.

The low-loss region is dominated by peaks associated with collective oscillation of the valence electron density, commonly referred to as plasmons, but may also contain single electron excitations from the valence band to low-energy unoccupied electronic states above the Fermi level. The low-loss region can be used to determine the relative thickness (t/λ, where λ is the inelastic mean free path) of a sample [19].

At higher energy losses, the incident electrons excite atomic electrons from inner (or core) shells to unoccupied energy levels [19,67,69]. This leads to the appearance of ionization "edges" superimposed on the exponentially decreasing background. The background on an EELS spectrum arises from contributions from plural-scattering events and from the tail of core-loss edges lying at lower energies. The threshold energy of a particular edge is determined by the energy required to promote an inner-shell electron to an unoccupied state above the Fermi level within an atom. The EELS technique can therefore be used to give compositional information as the presence of an edge on the spectrum can be used to identify the type of atoms in the sample. Moreover, the intensity of a particular edge depends on the number of atoms present and may be used for quantitative analysis [19].

The fine structure observed on core-loss edges is termed the electron energy-loss near-edge structure (ELNES) [19]. Provided the energy resolution is sufficient, analysis of the ELNES can provide information on the bonding environments, oxidation states, chemistry, and electronic structure with high spatial resolution [66]. The ELNES region can extend up to 30 eV above the edge threshold, although the effect of final state broadening decreases the usefulness of the information further from the threshold. Fine structure beyond the ELNES region is termed the extended energy-loss fine structure (EXELFS). This region is usually weak and may be obscured by other edges lying at higher energies [69].

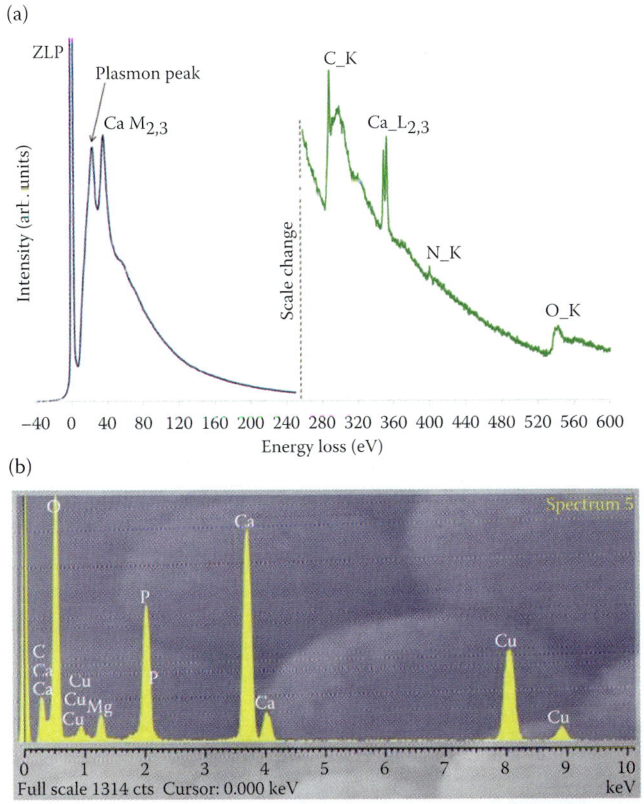

FIGURE 39.6
(a) Typical EELS spectrum of ivory dentine showing the intense ZLP and the low-loss region (containing the plasmon peak), up to 50 eV, followed by the core-loss region at higher energies. (b) Typical EDX spectrum of ivory dentine. (From Jantou, V, "Analytical electron microscopy of mineralized dentine," Thesis PhD and DIC Department of Materials, Imperial College London, 2008.)

39.4.2 EDX Analysis

When a fast primary electron loses energy due to an inelastic scattering event, the atom/ion in the sample is placed in an excited state. This species will return to its normal equilibrium state by emission of energy in the form of a phonon, electron, or photon. Two types of x-ray photons may be emitted when an electron interacts with a sample: characteristic x-rays and bremsstrahlung x-rays [70,71]. Bremsstrahlung x-rays are emitted when electrons are decelerated in the specimen, when being scattered through a large angle following interaction with the nucleus of an atom. The energy of these x-rays is not species specific. Characteristic x-rays are emitted from a specimen upon inelastic interaction with the incident electrons. The energy of these x-rays depends on the atomic species and on the energy difference between its initial and final states. In EDX analysis, the number of x-rays detected with a given energy is plotted over a range of energies to produce a spectrum (Figure 39.6b). The spectrum shows the characteristic x-ray energies as peaks that can be used to identify the atomic species within the sample. Although the bremsstrahlung x-rays contribute to the total x-ray signal, their energies are continuous, and therefore, they are considered part of the background.

39.4.3 STEM-EELS-EDX for the Study of Mineralized Tissues

EELS/EDX when combined with HAADF-STEM allows spectroscopy to be performed with high spatial resolution. Using a spectrum imaging (SI) approach is an especially powerful method of acquisition. In this mode, the incident electron beam is scanned across the specimen, and EELS/EDX spectra (or both) are collected at every point (x,y) [72]. The result is a 3-D dataset that can be postprocessed in a number of ways to obtain line profiles, filtered images, thickness maps, etc.

39.4.3.1 Zero-Loss Filtering and Thickness Mapping

The low-loss part of the energy-loss spectrum can be used for zero-loss filtering and thickness mapping. Conventional BF TEM images contain both elastic and inelastic contributions, and the latter increases with sample thickness (Figure 39.7). Using an energy-selecting slit in the spectrometer, it is possible to use only the electrons in the ZLP to form an image (Figure 39.7). Removing the majority of the inelastically scattered electrons significantly improves the image contrast (although the overall image intensity is lower) and resolution due to the reduction in chromatic aberrations. Such images are called zero-loss filtered or elastic images [67]. This technique is referred to as energy filtered TEM (EFTEM). A thickness map can be obtained by acquiring both an unfiltered and zero-loss filtered image from the same region of interest. Unfiltered images are acquired with the energy-selecting slit removed so that both elastic and inelastic electrons fall onto the spectrometer detector. By processing the intensity of both images at each pixel, a map of the relative thickness (t/λ) can be obtained (Figure 39.7).

EFTEM uses only some of the scattered electrons to form the image, and many electrons do not contribute. An alternative approach is to operate the microscope in STEM mode and to acquire an EELS spectrum at every pixel in the STEM image. This STEM-EELS spectrum image can then be postprocessed to obtain zero-loss filtered images and thickness maps. This has significant advantages over the traditional EFTEM approach as it allows significantly more user intervention in the calculation procedures to ensure that any energy drift in the spectrum image is compensated for by alignment of the ZLP, to monitor the quality of the fitting and subtraction of the ZLP, and to correct for other instrumentation artifacts such as gain nonuniformities and variations in beam current.

39.4.3.2 Chemical Mapping

The ability to obtain two-dimensional distributions of elements in a specimen can be enormously useful in material characterization, and EELS performed either in TEM or STEM mode allows this to be achieved. In EFTEM, elemental maps are generated by imaging with electrons that have lost energies corresponding to the characteristic inner-shell ionization edges of interest. In order to generate such a map, the "three-window method" is used with two "preedge" images collected in order to fit the background before the ionization edge (see Figure 39.6) and one "postedge" image [19,20,66,68,73]. Thus, each EFTEM map obtained (Figure 39.8) is actually the result of three image acquisitions, and consequently, the specimen is exposed to a significant total electron dose [73].

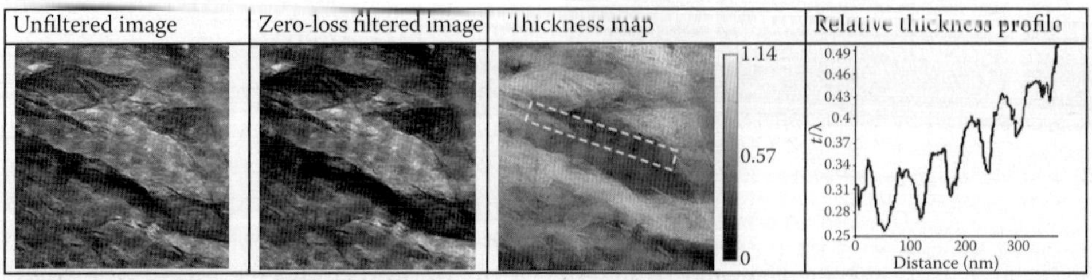

FIGURE 39.7
Comparison of unfiltered and zero-loss filtered images acquired in EFTEM mode from mineralized collagen. The thickness map and corresponding profiles of t/λ are also shown. (From V. Jantou-Morris et al., *Biomaterials*, vol. 31, no. 19, pp. 5275–5286, 2010; V. Jantou, "Analytical electron microscopy of mineralized dentine," Thesis PhD and DIC—Department of Materials, Imperial College London, 2008.)

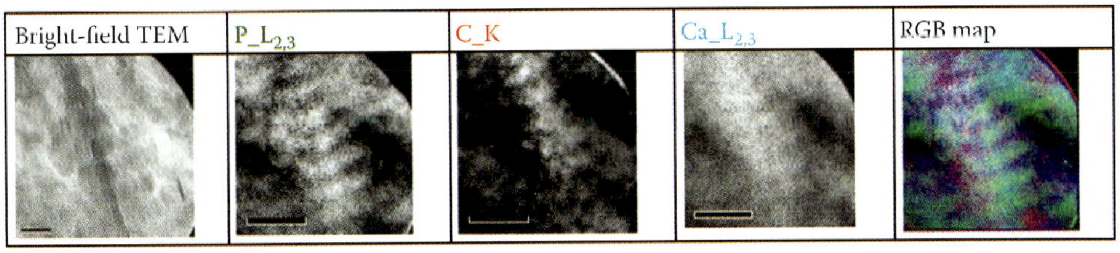

FIGURE 39.8
EFTEM maps of phosphorus ($P\_L_{2,3}$), carbon (C_K), and calcium ($Ca\_L_{2,3}$) obtained using the three-window method. The RGB map shows the combined data in false color to emphasize the spatial correlation between the maps. (From V. Jantou-Morris et al., *Biomaterials*, vol. 31, no. 19, pp. 5275–5286, 2010; V. Jantou, "Analytical electron microscopy of mineralized dentine," Thesis PhD and DIC—Department of Materials, Imperial College London, 2008.)

As discussed previously, elemental maps of this nature can also be directly obtained by utilizing the STEM SI approach. Since data are acquired over the full desired spectral range at every pixel, this allows post-processing to obtain elemental maps. This is illustrated in Figure 39.9 where intensity profiles for a range of elements were extracted from a STEM-EELS spectrum image dataset recorded from a longitudinal section of mineralized collagen. As well as lower total electron dose, the additional advantage of this approach compared with EFTEM imaging is that the processing steps can be easily modified to deal with complexities such as overlapping edges or instrument artifacts.

39.4.3.3 Mapping Chemical Bonding

The use of ELNES as a probe of local chemical environment has proved extremely powerful in the study of inorganic materials. ELNES has been used to gain information on coordination environment, oxidation state, and chemical bonding. The use of ELNES for the study of organic materials has been somewhat limited

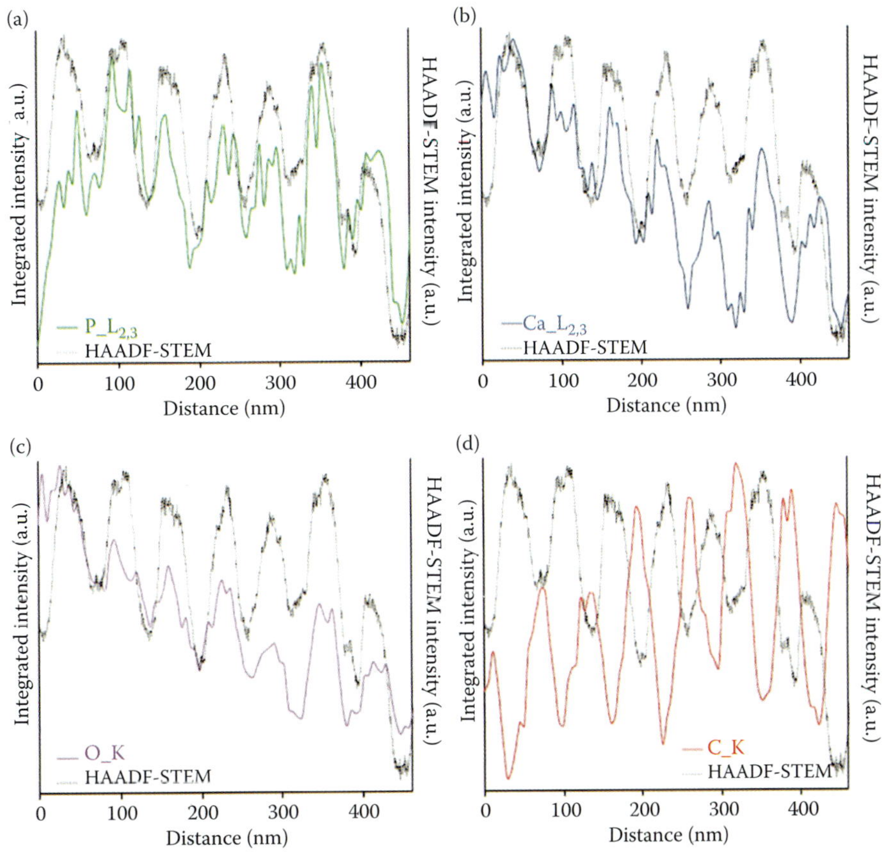

FIGURE 39.9
EELS signal intensity profiles across collagen banding: (a) $P\_L_{2,3}$, (b) $Ca\_L_{2,3}$, (c) O_K, (d) C_K. Also shown is the intensity profile from the HAADF image for comparison. (From V. Jantou-Morris et al., *Biomaterials*, vol. 31, no. 19, pp. 5275–5286, 2010.)

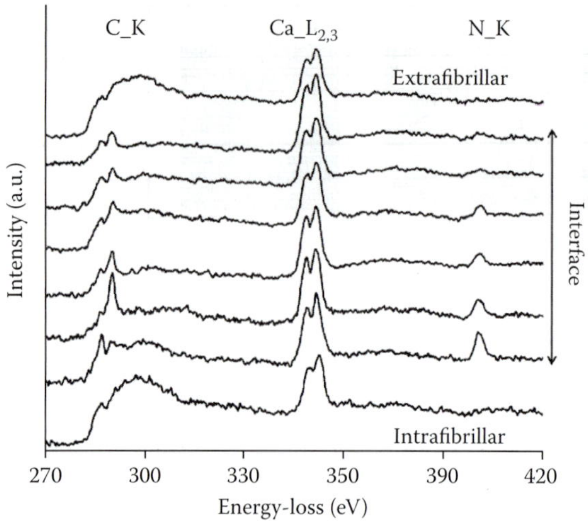

FIGURE 39.10
ELNES data acquired across the interface between extrafibrillar mineral and collagen in a transverse section of ivory dentin. The sharp features on the carbon K-edge can be associated with molecular units that are associated with the interfacial bonding. (From V. Jantou-Morris et al., *Biomaterials*, vol. 31, no. 19, pp. 5275–5286, 2010.)

by electron beam damage. However, the use of the x-ray equivalent technique, x-ray absorption near-edge structure (XANES), has been shown to be extremely powerful for the study of functional groups in organic polymers. With the development of electron monochromators, it is now possible to achieve the energy resolution to identify functional groups in organic polymers and biomaterials with nanometer-scale spatial resolution (Figure 39.10), and we can anticipate that, with further improvements in detector technology, this area of research will develop significantly in the next few years [20,73].

39.5 Conclusions

There is a clear need to characterize the structure and chemistry of mineralized tissues at the nanometer and subnanometer scale in order to improve our understanding of the inherent complexity of such materials. TEM/STEM when combined with analytical techniques such as EELS and EDX has already been used to great effect to relate the ultrastructure and nanostructure of bone and other mineralized tissues to the mechanical and structural properties. These analytical methods have the potential to cast yet more light on the exact origin of the collagen banding and the precise relationship between the mineral and organic phases of mineralized tissues, and indeed on the actual mechanism of mineralization [21]. The dual beam FIB instruments have had major impact on sample preparation. Preparation of site-specific samples associated with regions of interest such as defects, cracks, interfaces, etc., is likely to have a major impact on the correlation of properties across multiple length scales. The development of electron monochromators, aberration-corrected electron optics, and new detectors for imaging and spectroscopy has already had a major impact on the use of analytical TEM in the physical sciences. It is in the application of these technologies to the study of complex interfaces in bone and mineralized tissues that we can anticipate many exciting new insights in the future. Ultimately, this may then lead to a better understanding of how common skeletal disorders, such as osteoporosis, affect the structure and chemistry of bone at the nanometer scale. In turn, this may allow the development of effective new treatments.

References

1. Robinson, R. A. and F. W. Bishop, "Methods of preparing bone and tooth samples for viewing in the electron microscope," *Science*, vol. 111, no. 2894, pp. 655–657, Jun. 1950.
2. Robinson, R. A., "An electron-microscopic study of the crystalline inorganic component of bone and its relationship to the organic matrix," *J. Bone Joint Surg. Am.*, vol. 34-A, no. 2, pp. 389–435; passim, Apr. 1952.
3. Sheldon, H. and R. A. Robinson, "Electron microscope studies of crystal-collagen relationships in bone. IV. The occurrence of crystals within collagen fibrils," *J. Biophys. Biochem. Cytol.*, vol. 3, no. 6, pp. 1011–1016, Nov. 1957.
4. Engstrom, A. and H. Fernandez-Moran, "Ultrastructural organization of bone," *Nature*, vol. 178, no. 4531, pp. 494–495, Sep. 1956.
5. Fernandez-Moran, H. and A. Engstrom, "Electron microscopy and x-ray diffraction of bone," *Biochim. Biophys. Acta*, vol. 23, no. 2, pp. 260–264, Feb. 1957.
6. Bocciarelli, P. D. S., "Morphology of crystallites in bone," *Calcif. Tissue Res.*, vol. 5, no. 1, pp. 261–269, Dec. 1970.
7. Johansen, E. and H. F. Parks, "Electron microscopic observations on the three-dimensional morphology of apatite crystallites of human dentine and bone," *J. Biophys. Biochem. Cytol.*, vol. 7, no. 4, pp. 743–746, Jul. 1960.
8. Petruska, J. A. and A. J. Hodge, "A subunit model for the tropocollagen macromolecule," *Proc. Natl. Acad. Sci. U. S. A.*, vol. 51, no. 5, pp. 871–876, May 1964.
9. Alexander, B., T. L. Daulton, G. M. Genin, J. Lipner, J. D. Pasteris, B. Wopenka, and S. Thomopoulos, "The nanometre-scale physiology of bone: steric modelling and scanning transmission electron microscopy of collagen-mineral structure," *J. R. Soc. Interface R. Soc.*, vol. 9, no. 73, pp. 1774–1786, Aug. 2012.

10. McNally, E., F. Nan, G. A. Botton, and H. P. Schwarcz, "Scanning transmission electron microscopic tomography of cortical bone using Z-contrast imaging," *Micron*, vol. 49, pp. 46–53, Jun. 2013.
11. McNally, E. A., H. P. Schwarcz, G. A. Botton, and A. L. Arsenault, "A model for the ultrastructure of bone based on electron microscopy of ion-milled sections," *PLoS One*, vol. 7, no. 1, p. e29258, 2012.
12. Rubin, M. A., I. Jasiuk, J. Taylor, J. Rubin, T. Ganey, and R. P. Apkarian, "TEM analysis of the nanostructure of normal and osteoporotic human trabecular bone," *Bone*, vol. 33, no. 3, pp. 270–282, Sep. 2003.
13. Moradian-Oldak, J., S. Weiner, L. Addadi, W. J. Landis, and W. Traub, "Electron imaging and diffraction study of individual crystals of bone, mineralized tendon and synthetic carbonate apatite," *Connect. Tissue Res.*, vol. 25, no. 3–4, pp. 219–228, 1991.
14. Posner, A. S. "Crystal chemistry of bone mineral," *Physiol. Rev.*, vol. 49, no. 4, pp. 760–792, Oct. 1969.
15. Engström, A. and R. Zetterström, "Studies on the ultrastructure of bone," *Exp. Cell Res.*, vol. 2, no. 2, pp. 268–274, 1951.
16. Weiner, S., T. Arad, and W. Traub, "Crystal organization in rat bone lamellae," *FEBS Lett.*, vol. 285, no. 1, pp. 49–54, Jul. 1991.
17. Giraud-Guille, M. M., "Twisted plywood architecture of collagen fibrils in human compact bone osteons," *Calcif. Tissue Int.*, vol. 42, no. 3, pp. 167–180, Mar. 1988.
18. Marotti, G., M. Ferretti, and C. Palumbo, "The problem of bone lamellation: An attempt to explain different proposed models," *J. Morphol.*, vol. 274, no. 5, pp. 543–550, May 2013.
19. Egerton, R. F., *Electron Energy-Loss Spectroscopy in the Electron Microscope*, 3rd ed. NY: Plenum, 1996.
20. Jantou-Morris, V., M. A. Horton, and D. W. McComb, "The nano-morphological relationships between apatite crystals and collagen fibrils in ivory dentine," *Biomaterials*, vol. 31, no. 19, pp. 5275–5286, Jul. 2010.
21. Boonrungsiman, S., E. Gentleman, R. Carzaniga, N. D. Evans, D. W. McComb, A. E. Porter, and M. M. Stevens, "The role of intracellular calcium phosphate in osteoblast-mediated bone apatite formation," *Proc. Natl. Acad. Sci. U. S. A.*, vol. 109, no. 35, pp. 14170–14175, Aug. 2012.
22. Bordat, C., O. Bouet, and G. Cournot, "Calcium distribution in high-pressure frozen bone cells by electron energy loss spectroscopy and electron spectroscopic imaging," *Histochem. Cell Biol.*, vol. 109, no. 2, pp. 167–174, Feb. 1998.
23. Kawahara, I., M. Koide, O. Tadokoro, N. Udagawa, H. Nakamura, N. Takahashi, and H. Ozawa, "The relationship between calcium accumulation in osteoclast mitochondrial granules and bone resorption," *Bone*, vol. 45, no. 5, pp. 980–986, Nov. 2009.
24. Landis, W. J. and M. J. Glimcher, "Electron diffraction and electron probe microanalysis of the mineral phase of bone tissue prepared by anhydrous techniques," *J. Ultrastruct. Res.*, vol. 63, no. 2, pp. 188–223, May 1978.
25. Landis, W. J., K. J. Hodgens, M. J. Song, J. Arena, S. Kiyonaga, M. Marko, C. Owen, and B. F. McEwen, "Mineralization of collagen may occur on fibril surfaces: Evidence from conventional and high-voltage electron microscopy and three-dimensional imaging," *J. Struct. Biol.*, vol. 117, no. 1, pp. 24–35, Aug. 1996.
26. Anderson, H. C., "Matrix vesicles and calcification," *Curr. Rheumatol. Rep.*, vol. 5, no. 3, pp. 222–226, Jun. 2003.
27. Gay, C. V., H. Schraer, and T. E. Hargest Jr., "Ultrastructure of matrix vesicles and mineral in unfixed embryonic bone," *Metab. Bone Dis. Relat. Res.*, vol. 1, no. 2, pp. 105–108, 1978.
28. Morris, D. C., H. K. Väänänen, and H. C. Anderson, "Matrix vesicle calcification in rat epiphyseal growth plate cartilage prepared anhydrously for electron microscopy," *Metab. Bone Dis. Relat. Res.*, vol. 5, no. 3, pp. 131–137, 1983–1984.
29. Cecil, R. N. A. and H. Clarke Anderson, "Freeze-fracture studies of matrix vesicle calcification in epiphyseal growth plate," *Metab. Bone Dis. Relat. Res.*, vol. 1, no. 2, pp. 89–95, 1978.
30. Thomsen, P. and L. E. Ericson, "Light and transmission electron microscopy used to study the tissue morphology close to implants," *Biomaterials*, vol. 6, no. 6, pp. 421–424, Nov. 1985.
31. Grandfield, K., A. Palmquist, and H. Engqvist, "Three-dimensional structure of laser-modified Ti6Al4V and bone interface revealed with STEM tomography," *Ultramicroscopy*, vol. 127, pp. 48–52, Apr. 2013.
32. Tracy, B. M. and R. H. Doremus, "Direct electron microscopy studies of the bone-hydroxylapatite interface," *J. Biomed. Mater. Res.*, vol. 18, no. 7, pp. 719–726, Sep. 1984.
33. Grandfield, K., A. Palmquist, and H. Engqvist, "High-resolution three-dimensional probes of biomaterials and their interfaces," *Philos. Transact. A Math. Phys. Eng. Sci.*, vol. 370, no. 1963, pp. 1337–1351, Mar. 2012.
34. Grandfield, K., S. Gustafsson, and A. Palmquist, "Where bone meets implant: The characterization of nano-osseointegration," *Nanoscale*, vol. 5, no. 10, pp. 4302–4308, May 2013.
35. Reiche, I., C. Vignaud, and M. Menu, "The crystallinity of ancient bone and dentine: New insights by transmission electron microscopy," *Archaeometry*, vol. 44, no. 3, pp. 447–459, 2002.
36. Bell, L. S., M. Kayser, and C. Jones, "The mineralized osteocyte: A living fossil," *Am. J. Phys. Anthropol.*, vol. 137, no. 4, pp. 449–456, Dec. 2008.
37. Zheng, W. and M. H. Schweitzer, "Chemical analyses of fossil bone," *Methods Mol. Biol.*, vol. 915, pp. 153–172, 2012.
38. Lindgren, J., P. Uvdal, A. Engdahl, A. H. Lee, C. Alwmark, K.-E. Bergquist, E. Nilsson et al., "Microspectroscopic evidence of cretaceous bone proteins," *PLoS One*, vol. 6, no. 4, p. e19445, 2011.
39. Schweitzer, M. H., Z. Suo, R. Avci, J. M. Asara, M. A. Allen, F. T. Arce, and J. R. Horner, "Analyses of soft tissue from Tyrannosaurus rex suggest the presence of protein," *Science*, vol. 316, no. 5822, pp. 277–280, Apr. 2007.
40. McDonald, K. L., M. Morphew, P. Verkade, and T. Müller-Reichert, "Recent advances in high-pressure freezing: Equipment- and specimen-loading methods," *Methods Mol. Biol.*, vol. 369, pp. 143–173, 2007.

41. Mallein-Gérin, F. and R. Garrone, "Tendon collagen fibrillogenesis is a multistep assembly process as revealed by quick-freezing and freeze-substitution," *Biol. Cell Auspices Eur. Cell Biol. Organ.*, vol. 69, no. 1, pp. 9–16, 1990.
42. Mahamid, J., A. Sharir, D. Gur, E. Zelzer, L. Addadi, and S. Weiner, "Bone mineralization proceeds through intracellular calcium phosphate loaded vesicles: A cryo-electron microscopy study," *J. Struct. Biol.*, vol. 174, no. 3, pp. 527–535, Jun. 2011.
43. Mahamid, J., A. Sharir, L. Addadi, and S. Weiner, "Amorphous calcium phosphate is a major component of the forming fin bones of zebrafish: Indications for an amorphous precursor phase," *Proc. Natl. Acad. Sci. U. S. A.*, vol. 105, no. 35, pp. 12748–12753, Sep. 2008.
44. Posner, A. S., F. Betts, and N. C. Blumenthal, "Properties of nucleating systems," *Metab. Bone Dis. Relat. Res.*, vol. 1, no. 2, pp. 179–183, 1978.
45. Boskey, A. L. and A. S. Posner, "Conversion of amorphous calcium phosphate to microcrystalline hydroxyapatite. A pH-dependent, solution-mediated, solid–solid conversion," *J. Phys. Chem.*, vol. 77, no. 19, pp. 2313–2317, Sep. 1973.
46. Combes, C. and C. Rey, "Amorphous calcium phosphates: Synthesis, properties and uses in biomaterials," *Acta Biomater.*, vol. 6, no. 9, pp. 3362–3378, Sep. 2010.
47. Porter, A. E., N. Patel, J. N. Skepper, S. M. Best, and W. Bonfield, "Comparison of in vivo dissolution processes in hydroxyapatite and silicon-substituted hydroxyapatite bioceramics," *Biomaterials*, vol. 24, no. 25, pp. 4609–4620, Nov. 2003.
48. Porter, A. E., N. Patel, J. N. Skepper, S. M. Best, and W. Bonfield, "Effect of sintered silicate-substituted hydroxyapatite on remodelling processes at the bone–implant interface," *Biomaterials*, vol. 25, no. 16, pp. 3303–3314, Jul. 2004.
49. Kanno, H., R. J. Speedy, and C. A. Angell, "Supercooling of water to -92{degrees}C under pressure," *Science*, vol. 189, no. 4206, pp. 880–881, Sep. 1975.
50. Riehle, U., *Über die Vitrifizierung verdünnter wässriger Lösungen*. Zurich, Juris, 1968.
51. Dahl, R. and L. A. Staehelin, "High-pressure freezing for the preservation of biological structure: Theory and practice," *J. Electron Microsc. Tech.*, vol. 13, no. 3, pp. 165–174, Nov. 1989.
52. Moor, H., "Theory and practice of high pressure freezing," in *Cryotechniques in Biological Electron Microscopy*, P. D. R. A. Steinbrecht, and D. K. Zierold, Eds. Berlin, Heidelberg: Springer, 1987, pp. 175–191.
53. Nudelman, F., K. Pieterse, A. George, P. H. H. Bomans, H. Friedrich, L. J. Brylka, P. A. J. Hilbers, G. de With, and N. A. J. M. Sommerdijk, "The role of collagen in bone apatite formation in the presence of hydroxyapatite nucleation inhibitors," *Nat. Mater.*, vol. 9, no. 12, pp. 1004–1009, Dec. 2010.
54. Studer, D., H. Hennecke, and M. Müller, "High-pressure freezing of soybean nodules leads to an improved preservation of ultrastructure," *Planta*, vol. 188, no. 2, pp. 155–163, Sep. 1992.
55. Studer, D., B. M. Humbel, and M. Chiquet, "Electron microscopy of high pressure frozen samples: Bridging the gap between cellular ultrastructure and atomic resolution," *Histochem. Cell Biol.*, vol. 130, no. 5, pp. 877–889, Nov. 2008.
56. White, D. L., S. B. Andrews, J. W. Faller, and R. J. Barrnett, "The chemical nature of osmium tetroxide fixation and staining of membranes by x-ray photoelectron spectroscopy," *Biochim. Biophys. Acta*, vol. 436, no. 3, pp. 577–592, Jul. 1976.
57. Hodge, A. J. and J. A. Petruska, "Recent studies with the electron microscope on ordered aggregates of the tropocollagen molecule," in *Aspects of Protein Structure*, G. N. Ramachandran, Ed. New York: Academic Press, 1963, pp. 289–300.
58. Giannuzzi, L. A. and F. A. Stevie, *Introduction to Focused Ion Beams: Instrumentation, Theory, Techniques and Practice*. New York: Springer, 2005.
59. Orloff, J., L. Swanson, and M. Utlaut, *High Resolution Focused Ion Beams: FIB and Its Applications—The Physics of Liquid Metal Ion*. New York: Springer, 2002.
60. Hayashi, Y., T. Yaguchi, K. Ito, and T. Kamino, "High-resolution electron microscopy of human enamel sections prepared with focused ion beam system," *Scanning*, vol. 20, no. 3, pp. 234–235, Apr. 1998.
61. Hoshi, K., S. Ejiri, W. Probst, V. Seybold, T. Kamino, T. Yaguchi, N. Yamahira, and H. Ozawa, "Observation of human dentine by focused ion beam and energy-filtering transmission electron microscopy," *J. Microsc.*, vol. 201, Pt 1, pp. 44–49, Jan. 2001.
62. Volkert, C. A., S. Busch, B. Heiland, and G. Dehm, "Transmission electron microscopy of fluorapatite–gelatine composite particles prepared using focused ion beam milling," *J. Microsc.*, vol. 214, Pt 3, pp. 208–212, Jun. 2004.
63. Nalla, R. K., A. E. Porter, C. Daraio, A. M. Minor, V. Radmilovic, E. A. Stach, A. P. Tomsia, and R. O. Ritchie, "Ultrastructural examination of dentin using focused ion-beam cross-sectioning and transmission electron microscopy," *Micron Oxf. Engl. 1993*, vol. 36, no. 7–8, pp. 672–680, 2005.
64. Jantou, V., M. Turmaine, G. D. West, M. A. Horton, and D. W. McComb, "Focused ion beam milling and ultramicrotomy of mineralised ivory dentine for analytical transmission electron microscopy," *Micron Oxf. Engl. 1993*, vol. 40, no. 4, pp. 495–501, Jun. 2009.
65. Eddisford, P., A. Brown, and R. Brydson, "Identifying and quantifying the mechanism of electron beam induced damage and recovery in hydroxyapatite," *J. Phys. Conf. Ser.*, vol. 126, no. 1, p. 012008, Aug. 2008.
66. Garvie, L. A. J., A. J. Craven, and R. Brydson, "Use of electron-energy loss near-edge fine structure in the study of minerals," *Am. Mineral.*, vol. 79, no. 5–6, pp. 411–425, Jun. 1994.
67. Brydson, R., *Electron Energy Loss Spectroscopy*. London: Taylor & Francis, 2006.
68. Keast, V. J., A. J. Scott, R. Brydson, D. B. Williams, and J. Bruley, "Electron energy-loss near-edge structure—a tool for the investigation of electronic structure on the nanometre scale," *J. Microsc.*, vol. 203, no. 2, pp. 135–175, 2001.

69. Williams, D. B. and C. B. Carter, *Transmission Electron Microscopy: A Textbook for Materials Science. 3, 3.* New York [u.a.]: Plenum Press, 2009, pp. 751–754.
70. Goldstein, J., *Scanning Electron Microscopy and X-Ray Microanalysis*. New York: Kluwer Academic/Plenum Publishers, 2003.
71. Goodhew, P. J., F. J. Humphreys, and R. Beanland, *Electron Microscopy and Analysis*. London; New York: Taylor & Francis, 2001.
72. Jeanguillaume, C. and C. Colliex, "Spectrum-image: The next step in EELS digital acquisition and processing," *Ultramicroscopy*, vol. 28, no. 1–4, pp. 252–257, Apr. 1989.
73. Jantou, V., "Analytical electron microscopy of mineralized dentine," Thesis PhD and DIC—Department of Materials, Imperial College, London, 2008.

40

Optical Microscopy Methods for Measuring Structure and Dynamic Processes of Cells and Tissues

Elliot L. Elson

CONTENTS

Fluorescence correlation spectroscopy (FCS) and, more broadly, optical microscopy methods that harness information from diffusing particles are important tools for the study of structural and dynamic processes in cells and tissues. This chapter describes three such methods. The first is FCS, which measures dynamic processes via fluorescence fluctuations in steady-state systems, and the second is fluorescence photobleaching recovery, which measures the relaxation of macroscopic concentration gradients produced by localized photobleaching. These closely related methods supply information about rates of diffusion, convection, chemical reaction, and other dynamic processes. They have been extensively used to measure dynamic molecular behavior both in purified systems and in cells. The third is single particle tracking, which allows the observation of different mechanisms of molecular transport in cells and in very small subcellular domains. These measurements indicate heterogeneity due to localized structures that influence particle transport. Single particle tracking can also be used to determine the viscoelastic properties of highly localized regions of cells and polymeric systems.

40.1 Introduction

Cell and tissue mechanics is typically thought of mainly in terms of mechanical properties such as stiffness or viscoelasticity or dynamic functions such as force generation or locomotion. These properties and functions are often interpreted using measurements at the whole cell or multicellular tissue level, e.g., measurements of stiffness and viscoelasticity on the scale of many micrometers up to centimeters. Measurements on a finer spatial scale (<1 μm) can supply complementary information about the structures and mechanical functions of cells and tissues. This chapter describes methods based on optical microscopy that measures dynamic processes at the subcellular level and at a spatial scale limited only by the optical resolution limit, i.e., dimensions <0.5 μm for visible light. These methods yield transport coefficients (diffusion coefficients and velocities of convective flow) of specific fluorescent molecules, extents of molecular aggregation, rates and extents of interactions of specified molecular species, and highly local values of viscoelastic parameters.

One subject area of interest for cell mechanics is the structure and dynamics of the actin cytoskeleton and the interaction of the cytoskeleton with the cell membrane. Although the methods described in Sections 40.2 and 40.3 have a much wider range of application, their application to aspects of the structure and function of actin filaments in cells can be illustrated. Fluorescence correlation spectroscopy (FCS) and fluorescence photobleaching recovery (FPR, also known as fluorescence recovery after photobleaching [FRAP]) are closely related and versatile methods that can yield diffusion coefficients and convective velocities of fluorescent molecules, as well as extents and rates of aggregation and molecular interaction. Among many other applications,[1] FCS and its extensions have been used to study the structure and dynamic properties of actin filament gels,[2] the dependence between the diffusion of protein and lipid probes and the density of actin filaments on the cell membrane,[3] and segmental dynamics of actin filaments.[4] FPR has also seen many and diverse applications,[5,6] including the study of the dynamics of actin and actin-binding proteins in lamellipodia of motile cells (e.g., refs. 7 and 8). Single particle tracking (SPT) yields transport rates of individual molecules or particles.[9] It has been essential to the unraveling of the processes and structures that limit the rate of translational diffusion on cell surfaces.[10] Particle tracking microrheology can yield viscoelastic properties of a gel or solution by measuring the highly localized thermal motions of microscopic particles embedded within a much larger sample volume.[11] The approach has been applied to in vitro actin gels and also to living cells.[12]

The methods to be described here share several features and advantages that make them especially useful for studies of animal cells. Their high spatial resolution (<1 μm), limited only by the optical diffraction limit, enables measurements in different regions of a cell, e.g., nucleus versus cytoplasm or lamellipodium versus cell body. They can measure processes that have characteristic times over a very wide range. For example, FCS can measure diffusion or chemical reaction kinetics over the range from microseconds to seconds, whereas FPR can measure similar processes from tens of milliseconds to many minutes. Furthermore, since these approaches are based on fluorescence detection for the most part, they have high molecular sensitivity down to the single molecule level and high chemical selectivity enabled by the ability to detect specifically fluorescence-labeled molecules even within a background of unlabeled molecules at high concentration.

40.2 Fluorescence Correlation Spectroscopy

In the simplest possible FCS experiment, a laser beam is focused to form a tiny illuminated subvolume (the "detection volume," radius ~300 nm) of a dilute solution (~0.1 nM) of a fluorescent compound. Any time a fluorescent molecule diffuses across the detection volume, the laser excites its fluorescence. A light detector, e.g., a photomultiplier or an avalanche photodiode, detects the fluorescent pulses, which are recorded in a computer. The duration of an emitted pulse of light depends on how rapidly the fluorescent molecule diffuses across the excitation volume, which, in turn, depends on the size of the volume and the rate of diffusion. If the size of the detection volume is known, the rate of diffusion can be computed from the fluorescence fluctuation lifetimes. The fluctuations are, however, stochastic; they differ from one another. Therefore, they must be analyzed statistically. This is most conveniently done by calculating a fluorescence fluctuation autocorrelation function.[13] Suppose that the fluorescence at time t is $F(t)$ and that the mean fluorescence is $F(t)$. Then, the fluorescence fluctuation at time t is $\delta F(t) = F(t) - F(t)$. The correlation function is[14]

$$G(\tau) = \frac{\delta F(t)\delta F(t+\tau)}{F(t)^2} = \frac{\int_0^\infty \delta F(t)\delta F(t+\tau)\mathrm{d}t}{\left(\int_0^\infty F(t)\mathrm{d}t\right)^2}. \quad (40.1)$$

An FCS measurement requires both an experimentally measured correlation function $\hat{G}(\tau)$ determined from the record of fluorescence fluctuations and a theoretical $G(\tau)$ expressed in terms of the diffusion coefficient D. A solution of the diffusion equation supplies $c(r, t)$, the concentration of the fluorophore at position r and time t. The fluorescence is simply

$$F(t) = q\int I(r)c(r,t)\mathrm{d}^3r. \quad (40.2)$$

Here $I(r)$ is the intensity profile of the laser-illuminated detection volume, typically taken to be a Gaussian:

$$I(r) = I_0\exp\left[-2\frac{(x^2+y^2)}{w^2}\right]\exp\left[-\frac{2z^2}{w_z}\right]. \quad (40.3)$$

(Due to the nature of conventional microscope optics, the detection volume is a circular Gaussian in the focal plane with radius w, while the Gaussian radius in the z direction has a different value, w_z.[14]) Solving the diffusion equation then leads to an expression for the correlation function for simple diffusion in a three-dimensional space[13,15]:

$$G(\tau) = \frac{G(0)}{\left(1+\frac{\tau}{\tau_D}\right)\left(1+\frac{\tau}{\tau_{D_z}}\right)^{\frac{1}{2}}}. \quad (40.4)$$

The characteristic diffusion times are $\tau_D = \frac{w^2}{4D}$ and $\tau_{D_z} = \frac{w_z}{4D}$.

The zero-time amplitude simply follows from the fact that the number of fluorescent molecules in the detection volume is governed by a Poisson distribution. For a single component system, $G(0) = \frac{1}{\langle N \rangle}$, where $\langle N \rangle$ is the average number of fluorophores in the detection volume.[1,13] For this three-dimensional confocal excitation intensity profile, the effective observation volume is defined as $V_{obs} = \pi^{3/2} w^2 w_z$.[16] It is noteworthy that the correlation times, τ_D and τ_{D_z}, determine the diffusion coefficient, but the amplitude $G(0)$ is a thermodynamic quantity that provides a direct estimate of the concentration of fluorophores in the system.

Finally, to determine the diffusion coefficient, one fits the theoretical expression of the correlation function, Equation 40.4, to the experimental correlation function:

$$\hat{G}(\tau) = \frac{\lim_{N\to\infty} \frac{1}{N} \sum_{i=1}^{N} \delta\hat{F}(idt)\delta\hat{F}(idt+\tau)}{\left(\lim_{N\to\infty} \hat{F}(t)\right)^2}. \quad (40.5)$$

The fluorescence intensity is recorded as a sequence of photon counts measured in time intervals or "bins" of duration dt. Hence, the experimentally measured fluorescence at a time idt is $\hat{F}(idt)$, which is the number of photon counts measured in the ith time bin. Commercial instruments for performing FCS measurements are available from a number of suppliers.

FCS can measure various dynamic processes in addition to diffusion, including convective flow, chemical reaction kinetics, and photophysical processes such as triplet decay. Table 40.1 indicates some of these. Figure 40.1 illustrates some characteristic correlation decay patterns for systems undergoing both diffusion and convection. As expected, the correlation decay observed for rapid convection in the presence of slow diffusion is sharper than the inverse case of fast diffusion and slow convection. The analysis of the kinetics of chemical reaction systems by FCS is complicated by the coupling of the reaction to diffusion in the open detection volume. As an example, consider the simple second-order reaction: $A + B \underset{k_b}{\overset{k_f}{\rightleftarrows}} C$, and suppose that C has a different fluorescence than $A + B$. The kinetics of the reaction can be measured conventionally by displacing the system (slightly) from equilibrium, e.g., by a temperature jump, and then observing the fluorescence change over time as the system relaxes back to equilibrium. Typically, this kind of measurement is performed in a closed system so that $-\Delta C = \Delta A = \Delta B$. Then, the kinetics of the chemical relaxation can be described in terms of a single differential equation $\frac{d\Delta C_c}{dt} = -\tau_{chem}^{-1}\Delta C_c$ yielding $\Delta C_c(t) = \Delta C_c(0)\exp\left(-\frac{t}{\tau_{chem}}\right)$, where $(\tau_{chem})^{-1} = k_f(\bar{C}_A + \bar{C}_B) + k_b$ and $\bar{C}_x$ is the equilibrium concentration of component x. In contrast, in an

TABLE 40.1

Some of the Processes That Can Be Measured by FCS

Process	$\frac{\partial \delta c_i}{\partial t}$	$\frac{G(\tau)}{G(0)}$	Parameter	Reference
Diffusion	$D_i \nabla^2 \delta c_i$	$\left(1+\frac{\tau}{\tau_{Di}}\right)^{-1}$	$\left(\tau_{Di} = \frac{w^2}{4D_i}\right)$	17
Convection	$-V_i \nabla \delta c_i$	$\text{Exp}\left[-\left(\frac{\tau}{\tau_{V_i}}\right)^2\right]$	$\tau_{V_i} = \frac{w}{V_i}$	27
Diffusion + convection	$D_i \nabla^2 \delta c_i - V_i \nabla \delta c_i$	$\frac{\text{Exp}\left[\frac{-\left(\frac{\tau}{\tau_{V_i}}\right)^2}{1+\frac{\tau}{\tau_{Di}}}\right]}{1+\frac{\tau}{\tau_{Di}}}$	As above	27
Chemical reaction	$\sum_{j=1}^{M} T_{ij}\delta c_j$	$\sum_{s=1}^{N} A_s \text{Exp}\left[-\frac{\tau}{\tau_s}\right]$	T_{ij}, A_s, and τ_s are functions of the rate constants and equilibrium concentrations	17

Source: Reprinted from Elson, E.L., *Biophys J*, 101(12): 2855–70, 2011. With permission.

Note: For simplicity, we suppose a 2D system. There are M chemical components with concentration c_i and N chemical reactions. For the ith component, the diffusion coefficient and convection velocity are D_i and V_i, respectively. For coupled diffusion and reaction typical for an FCS measurement in an open chemical reaction system (Equation 40.3), the relaxation modes are often complicated functions mixing diffusion (and convection) and chemical reactions.

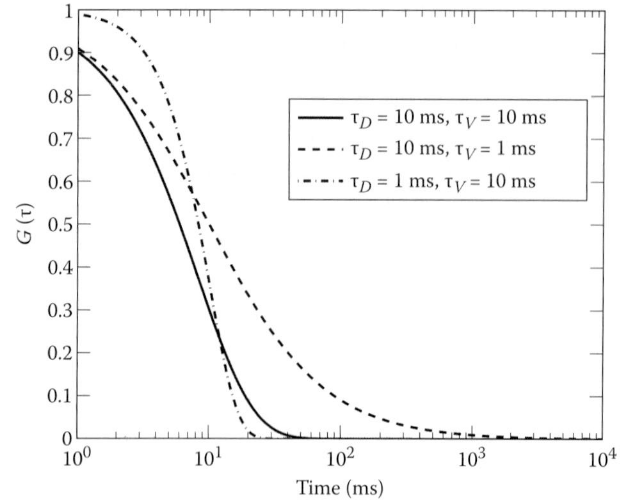

FIGURE 40.1
Correlation function for a system that experiences both diffusion and convection. For simplicity, we consider a two-dimensional system. Suppose that the system experiences both isotropic diffusion and convection along the x-axis at constant velocity V_x. This system is described by the following equation $\frac{\partial \delta c}{\partial t} = D\nabla^2 \delta c - V_x \frac{d\delta c}{\partial x}$, and the correlation function is $G(\tau) = \frac{G(0)}{1+\frac{\tau}{\tau_D}} \exp\left[-\frac{\left(\frac{\tau}{\tau_V}\right)^2}{1+\frac{\tau}{\tau_D}}\right]$. The characteristic times are $\tau_V = \frac{w}{V}$ and $\tau_D = \frac{w^2}{4D}$ with w the radius of the Gaussian excitation intensity profile in the focal plane.

FCS measurement, the detection volume is open; the components A, B, and C can diffuse independently into and out of the volume. As a result, three coupled reaction–diffusion equations are required to describe the kinetics of the system, and the fluctuations of the concentrations of all three components must be accounted. More generally, a system of N components will require N coupled equations.[13] Because the concentration fluctuations are small, all the kinetic equations can be linearized, and so the chemical kinetic contributions to the FCS correlation function are sums of exponential terms. Note that for direct measurement of chemical kinetics by FCS, it is necessary not only that the chemical reactions cause changes in fluorescence but also that the chemical reactions be fast compared to diffusion, in effect, $\tau_{chem} \ll \tau_D$. Otherwise, the correlation decay is dominated by diffusion, and the chemical kinetic contribution is small.

In general, a good fit of a theoretical to an experimental correlation function is insufficient to decide with confidence that the chosen theoretical model adequately describes the experimental system. Typically, one would need additional evidence. For example, to distinguish between diffusion and convection, one may vary the laser beam radius, w, noting that τ_D and τ_v vary as w^2 and w, respectively. It is also useful, in studying the kinetics of second- and higher-order reactions, to vary the reactant concentrations and thereby the chemical relaxation times of the system as illustrated for the simple bimolecular reaction above.[13,17]

FCS can be useful to study isomerization reactions that are sufficiently rapid and cause a substantial change in fluorescence. Conformational fluctuations of proteins and nucleic acids provide especially interesting examples. Intestinal fatty acid binding protein (IFAB) is a small (15 kDa) protein formed from two β-sheets each composed of five β-strands. The fatty acid binds in a cavity formed between the two β-sheets. FCS measurements detected fluctuations of the fluorescence of fluorescein bound inside the cavity that could be attributed to conformational motions of the protein.[18] These fluctuations were not seen from fluorescein bound outside the cavity. Other kinds of conformational fluctuations could be observed from the unfolded form of the protein to which pairs of rhodamine fluorophores were attached at some distance one from the other.[19] FCS detected fluorescence fluctuations due to self-quenching of the rhodamines when they occasionally came into contact as the conformation of the unfolded polypeptide chain fluctuated randomly. Conformational fluctuations due to segmental motions have shown interesting differences between single- and double-stranded DNA.[20] The behavior of double-stranded DNA can be accounted for without accounting for hydrodynamic interactions among the polymer segments, while these interactions are necessary to account for the behavior of the single-stranded conformations. This difference is presumed to result from the fact that double-stranded DNA is in an extended conformation due to molecular stiffness (greater persistence length), while the single-stranded molecule is in a more compact conformation. FCS has also been used to study conformational fluctuations of DNA hairpin helices. A fluorophore and a quencher were attached to the 5′ and 3′ ends of the hairpin oligonucleotides. Upon formation of the hairpin, the two ends were brought together, and the fluorescence was quenched. When the helix unwound, the two ends separated and fluorescence increased.[21] FCS measurements provided estimates of the rate constants and their temperature dependence for helix formation and dissolution based on a simple one-step (two-state) model.[21,22] Subsequent work showed that the two-state model was insufficient and that a three-state model was required to account for measurements that revealed the existence of a previously unsuspected intermediate.[23]

Recent years have seen the rapid development of studies at the single molecule level in which chemical reaction kinetics, conformational transformations, and other processes have been studied in individual molecules one at a time.[24] Although FCS detects many molecules during a single measurement, in two important respects, it also is a single molecule approach. The very dilute solutions to which FCS is applied are essentially ideal. The fluorescent molecules do not significantly interact with one another, and so each of these molecules correlates only with itself and not with any of the other fluorescent molecules. Hence, FCS is a bridge between conventional methods based on ensembles of many molecules and the new single molecule approaches. More important, the statistical analysis of fluctuations is at the heart of both FCS and single molecule measurements since both address mesoscopic molecular systems.[25] Hence, even though studies of conformational fluctuations of proteins and nucleic acids by FCS encounter many molecules, they are conceptually related to corresponding single molecule studies.[26]

An often encountered subject of study is the aggregation or polymerization of molecules in a cell-free solution, in a cytoplasm, and on the cell surface. FCS provides several approaches to measuring these phenomena. The diffusion coefficient diminishes as the size of a particle or molecular complex increases. For a spherical particle, however, the translational diffusion coefficient varies inversely as the radius of the particle. Hence, the diffusion coefficient varies only as $M^{-1/3}$, where M is the molecular weight of the particle or the number of subunits in the spherical aggregate. Measurements based on fluctuation amplitudes yield a more sensitive approach. Consider a system in which $nA \underset{k_b}{\overset{k_f}{\rightleftarrows}} A_n$. The FCS correlation function amplitude supplies a sensitive indicator of whether the fluorophores are present as A or as A_n. Each A monomer has a fluorescent label with a "molecular" brightness q. (Brightness is the number of photons emitted per second by a fluorophore when it is at the center of the Gaussian laser excitation profile.) As has been shown, for a one component system, $G(0) = \frac{1}{\langle N \rangle}$, where $\langle N \rangle$ is the average number of independently diffusing fluorophores in the detection region. The mean fluorescence, i.e., the average number of photon counts per second emitted from the detection region, is $\langle F \rangle$. Hence, the effective brightness (averaged over the entire Gaussian detection volume) is defined as $B = G(0)\langle F \rangle$. (Note that by this definition, B is one half the "molecular" brightness q of the same particle at the center of the excitation profile.) In the absence of optical interactions among its constituents, the brightness of a polymer or aggregate containing n monomers is n times the brightness of a single monomer. Thus, the brightness of an aggregate varies as n and so is a much more sensitive indicator of the degree of aggregation than the diffusion coefficient, which varies as $n^{-1/3}$.[27]

This example is oversimplified in supposing that the fluorophore population will be constituted either entirely of A or entirely of A_n. More realistically, there also will be intermediate species A_m with $0 < m < n$. Hence, the more general problem is to determine the set $\{N_j, q_j\}$ of all species j in the system. In principle, this is accomplished using the photon count histogram, $P(k)$, which is the probability of measuring k photons in one time bin. The number of molecules of each species in the detection volume and the photon counts emitted by each fluorophore are both governed by Poisson distributions. For a single species, it is straightforward to express $P(n)$ in terms of N and q. Neglecting detector (shot) noise,

$$\begin{aligned} P(k) &= \sum_{N=0}^{\infty} P(N)P(k \mid N) \\ &= \sum_{N=0}^{\infty} \frac{\langle N \rangle^N \exp(-\langle N \rangle)}{N!} \frac{\langle Nq \rangle^k \exp(-\langle Nq \rangle)}{k!}. \end{aligned} \tag{40.6}$$

The compound distribution for a system containing several different species of different brightness values can be computed either by convolution of the one-component distributions[28] or by adding generating functions of the one-component distributions.[29] $P(n)$ can be used to calculate moments of the distribution,[30] or $P(n)$ can be fitted directly to a model that specifies the values of N_j and q_j.[28,31] Among many other applications, these approaches have been used to study the formation of oligomeric complexes in cells[32] and the clustering of EGF receptors in the absence of epidermal growth factor (EGF).[33]

Although interactions among proteins can be detected via changes of diffusion and brightness, fluorescence cross-correlation spectroscopy (FCCS) provides a direct indicator of interactions among molecules labeled with distinguishable fluorophores.[34] Suppose that molecules of species A and B are labeled with red and green fluorophores, respectively. The cross-correlation $G_{AB}(\tau) = \langle \delta F_A(t)\delta F_B(t+\tau)\rangle = \langle \delta F_{red}(t)\delta F_{green}(t+\tau)\rangle$ is measured by detecting the red and green fluorescence in two different channels and then cross-correlating the two fluctuation records. If the A and B molecules diffuse independently, $G_{AB}(0) = 0$. If A and B are linked and diffuse together, however, their fluctuations are correlated, and so $G_{AB}(0) \neq 0$, and the magnitude of $G_{AB}(0)$ is related to the extent of molecular linkage. The first practical demonstration of FCCS was the separation by action of nuclease cleavage of a polynucleotide to which fluorophores of different colors were attached at each end.[35,36] A number of examples of FCCS to measure molecular linkages in living cells have been reviewed.[37]

Recently, fluorescence fluctuation methods have been exploited in an imaging mode using fluorescence microscopy especially to measure aggregation of fluorescence-labeled proteins on cell membranes. This allows correlation of fluorescence fluctuations over an area of the cell surface at a particular time ("image correlation spectroscopy"). Just as with temporal correlation spectroscopy, a few highly fluorescent aggregates of fluorescent "monomers" will give a larger fluctuation signal than many fluorescent monomers at the same total fluorescence. This principle has been used to measure membrane protein clusters such as the fusion of virus particles into cell membranes as well as the clustering of proteins in coated pits.[38] Image correlation techniques have been extended to include both spatial and temporal correlation and cross-correlation.[39]

The optical diffraction limit sets the minimum size of the detection region in conventional FCS in the range of 200–300 nm (for visible light). There are, however, advantages to shrinking these small dimensions even further, e.g., to measure highly localized diffusion of molecules that are closely confined by impermeable barriers. A smaller detection region also diminishes the diffusion correlation time, which can be useful for measuring slow-moving molecules. Moreover, since FCS measurements require that there be a relatively small number of molecules in the detection region, shrinking this region enables measurements at higher concentrations. Several approaches have been introduced to overcome the optical resolution limit in FCS measurements.[40] Of these, two approaches are of particular utility. Total internal reflection microscopy (TIRFM), which uses an evanescent field to illuminate a thin layer of the sample immediately above its interface with a glass surface, has been applied to both FCS and FPR measurements[41] and is still in wide use, e.g., refs. 42 and 43. While TIRFM diminishes the dimensions of the detection region along the optical axis, it does not, however, reduce its size in the focal plane. In contrast, stimulated emission depletion (STED) microscopy allows close confinement of the detection/excitation region by using a ring-shaped laser beam to deplete the fluorescence that is emitted from the outer regions of the conventional diffraction-limited Gaussian laser profile.[44] An interesting study has used STED microscopy to observe the dynamics of membrane lipids within very small areas.[45]

40.3 Fluorescence Photobleaching Recovery (Fluorescence Recovery after Photobleaching)

FPR and FCS measurements are similar both in theory and in experimental implementation. In an FPR experiment, after the fluorescence is measured in a small laser-illuminated region, a fraction of the fluorescent molecules in the region are irreversibly photobleached by increasing the intensity of the laser by a factor of 10^3–10^4 for a brief period of time. After the brief bleach interval, the laser intensity is reduced to its original intensity to monitor the intensity emitted from the detection region. The photobleaching pulse should be so brief that redistribution of the fluorescent molecules by diffusion or chemical reaction is minimal. After this pulse, fluorescence increases from its diminished level as the bleached molecules are replaced by unbleached fluorescent molecules that diffuse into the detection region (or are produced by chemical reactions). The same confocal microscope setup can be used for both FCS and FPR measurements except that an FPR measurement requires a method for rapidly increasing and decreasing the laser intensity at the detection region.[46] This could be done using a beam splitter and a mechanical shutter or an electro-optic element, like an acousto-optic tunable filter (AOTF) or an acousto-optic modulator (AOM). The theoretical analyses of both FCS and FPR measurements are based on a similar approach. A diffusion equation or a reaction–diffusion equation specifies the concentrations of the relevant components as a function of position and time.[13,47] The correlation function, $G(\tau)$, for FCS and the fluorescence recovery, $\Delta F(t)$, for FPR are calculated by integrating the concentration over the laser excitation intensity profile. The differences between the two approaches arise from differences in the fluorescence intensities measured and from the different weightings of recovery modes.[48–50] For simple diffusion, it is straightforward to show that at low bleaching intensities, $G(\tau)$ is proportional to $\Delta F(t)$.[48]

The original platform for FPR experiments was a confocal fluorescence microscope (also set up for FCS measurements) equipped with a fast shutter, beam splitters, and electronics for shutter control, data registration, and analysis.[46] Currently, it is more likely that commercial laser scanning confocal microscope (LSCM) systems are used for this purpose. An advantage of the custom-build instruments is that they can have very rapid photobleaching intervals (<1 ms), limited only by the intensity of the laser and the speed of the shutter/switch. In contrast, commercial LSCM systems also enjoy advantages such as immediate availability of the instrument and the ability to control the size and shape of the photobleached region of the cell.[51] Since the LSCM systems are used in an imaging mode, however, only relatively slow recovery processes, with characteristic times in the range of 100 ms or longer, can be measured. For faster diffusion, it is difficult to obtain sufficient time points to characterize the recovery, and it may be necessary to account for redistribution of the fluorophore during the bleach interval.[52] If the system is spatially homogeneous so that high spatial resolution is unnecessary, then one can enlarge the bleached area, thereby slowing the recovery to a range more suitable for using a LSCM.

At a practical level, FCS and FPR serve complementary applications. FCS is suitable to measure systems in either equilibrium or nonequilibrium steady states that have very low concentrations of fluorophore and without perturbing the steady state. The requirement for low concentrations of fluorophore arises from the fact that the relative fluctuation amplitude in an FCS experiment is determined by Poisson statistics. The amplitude of the FCS correlation function can be approximated as $G(0) = \frac{\langle (\delta N)^2 \rangle}{\langle N \rangle^2} = \frac{\langle (N - \langle N \rangle)^2 \rangle}{\langle N \rangle^2}$, where N is the number of fluorophores in the detection region. For a Poisson distribution, $\langle (N - \langle N \rangle)^2 \rangle = \langle N \rangle$. Then $G(0) = \frac{1}{\langle N \rangle}$. Hence, $G(0)$ diminishes as $\langle N \rangle$ increases. Furthermore, since FCS is based on stochastic fluctuations, it is necessary to calculate $G(\tau)$ from a long record of fluorescence fluctuations to ensure accuracy. In contrast, FPR measures the relaxation of a macroscopic, i.e., nonstochastic, fluorescence transient. Hence, FPR yields the desired rate coefficient with an accuracy limited only by the accuracy of the measurement of the individual transient. For measurements on living animal cells, the short measurement time needed for FPR is an especially desirable quality because cells are relatively unstable targets that can cause spurious fluorescence fluctuations by motions associated with locomotion and other dynamic processes. Furthermore, in contrast to FCS, the strength of an FPR signal *increases* with the concentration of the fluorophore, and also the macroscopic transients typically have larger amplitudes than the spontaneous fluctuations measured by FCS. Another important practical difference between FCS and FPR is that only the confocal detection volume is important for the former. For FPR, however, while the fluorescence is detected from the confocal volume, the region photobleached is not controlled by the confocal aperture and so is much larger along the optic axis (x-axis) of the microscope, cf. ref. 53. This distinction is irrelevant to the many applications of FPR to two-dimensional membrane systems. For applications in three dimensions, both the detection and photobleaching volumes can be closely confined using multiphoton excitation.[54,55]

An important application of both FCS and FPR is to measure the kinetics and extent of interactions between fluorescence-labeled molecules and other cellular molecules or structures (see also Chapter 27). As pointed out above, this can be accomplished directly if the reaction causes a change in fluorescence, e.g., quenching or enhancement, and the reaction is fast compared to diffusion. Even without a change in fluorescence, however, the effects of the reaction on the diffusion can yield useful information about the reaction kinetics. Consider again the reaction $A + B \underset{k_b}{\overset{k_f}{\rightleftarrows}} C$. The equilibrium constant is $K = \frac{k_f}{k_b} = \frac{\bar{C}_C}{\bar{C}_A \bar{C}_B}$, where $\bar{C}_X$ is the equilibrium concentration of X. Because the full analysis even of this simple reaction is complicated by the coupling of diffusion and chemical reaction,[13] it is helpful to examine some simplified cases.[56,57] Suppose, for example, that component A is nonfluorescent and diffuses slowly, that fluorescent component B diffuses rapidly, and that C has the same fluorescence as B (the reaction causes no change in fluorescence) and diffuses slowly at the same rate as A. Sprague et al.[57] define three different regimes. In the "diffusion-dominant" regime, the reaction equilibrium is either far to the left so that component B is, in effect, the only fluorescent species in the system or the reaction is so slow that the ratio of B and C does not change during the period of fluorescent recovery. In the first instance, only the diffusion of component B is detected, and in the second, both B and C diffuse and contribute to the fluorescence fluctuations independently of each other. Then, the fluorescence recovery for a single component can be represented either for a Gaussian excitation intensity profile[47]:

$$\Delta F_k(t) = Q \sum_{n=0}^{\infty} \left[\frac{(-K)^n}{n!} \right] \left[1 + n \left(1 + \frac{2t}{\tau_D} \right) \right]^{-1}. \quad (40.7)$$

or a uniform circular profile of radius w[58]:

$$\Delta F(t) = \exp\left(-\frac{2\tau_D}{t} \right) \left[I_0\left(\frac{2\tau_D}{t} \right) + I_1\left(\frac{2\tau_D}{t} \right) \right]. \quad (40.8)$$

where I_0 and I_1 are modified Bessel functions. Note that retaining only the $n = 1$ term for $K \ll 1$ in Equation 40.7 matches the FCS correlation function for simple diffusion. The Gaussian intensity profile is used for conventional "spot" photobleaching measurements,[46,47] while the uniform disk profile can be used in photobleaching with an LSCM.[51,59] When there are two or more fluorescent components that diffuse independently, the measured recovery curve is a superposition of the recovery curves of each of the components.

When the reaction process is much faster than diffusion ($\tau_D \gg \tau_{chem}$), then, as a B molecule diffuses across the detection area, it binds to and dissociates from A many times. Therefore, it appears to have a single "effective" diffusion coefficient, $D_e = D_C f_C + D_B f_B$, where D_B and D_C are the diffusion coefficients of B and C and $f_C = \frac{C_C}{C_B + C_C} = \frac{KC_A}{1 + KC_A}$ and $f_B = 1 - f_C = \frac{C_B}{C_B + C_C} = \frac{1}{1 + KC_A}$.[56] More simply, if $D_C \to 0$, then $D_e = \frac{D_B}{1 + \frac{k_f C_A}{k_b}}$.[57] Therefore, the diffusion of component B is slowed to the extent that it is bound to immobile component C, and the single apparent correlation time is $\tau_D = \frac{w^2}{4D_e}$.

Finally, in a commonly observed situation, diffusion is fast compared to the chemical reaction. Consider a system in which the rate limiting step of the chemical reaction is the dissociation of B from C, while the formation of C from A and B is fast. Then, after the photobleach, the equilibrium concentration of B is restored by diffusion so rapidly that it is not seen in the fluorescence recovery $\Delta F(t)$, but bleached B molecules are retained in complex with A to form nonfluorescent C molecules. Over time, the bleached (dark) B molecules dissociate from A and are replaced by unbleached fluorescent B molecules, reconstituting fluorescent C molecules. Thus, the fluorescence recovery is measured as $\Delta F(t) = 1 - C_C \exp(-k_b t)$[57] to yield k_b. An experimental study by FPR of the motion of glucocorticoid receptors in the nuclei of living cells nicely illustrates the practical utility of this formulation.[57]

40.4 Single Particle Tracking

High-speed digital imaging methods make it possible to track the trajectory of individual fluorescent molecules (or larger particles). As described in Chapter 30, these measurements provide a high-resolution analysis of the mechanism of motion, e.g., diffusive or convective, and the effects of barriers and interactions and have been especially useful in studies of dynamic processes on cell membranes. It is useful to characterize the motion in terms of the mean square displacement (MSD) of the diffusing particle: $\rho(t) = \langle (r(t) - r(0))^2 \rangle$. For simple diffusion in two dimensions, it is well known that $\rho(t) = 4Dt$, where D is the diffusion coefficient. For a particle that is moving deterministically at velocity V, $\rho(t) = (Vt)^2$. Therefore, a particle that is undergoing both diffusion and convection, e.g., a particle in a flowing liquid, $\rho(t) = 4Dt + (Vt)^2$. It is also possible, however, that, if a particle encounters barriers or interactions that retard its diffusion, its MSD will scale as t^α here $\alpha < 1$. Then, the three kinds of motion will display distinctly different plots of $\rho(t)$ vs. t. Simple diffusion is linear, whereas a significant contribution of convection or of retardation by barriers causes positive or negative curvature, respectively, as shown in Figure 40.2. This approach has been used to distinguish between convective and diffusive motion on cell surfaces, e.g., ref. 9. Furthermore, diffusion of proteins on cell surface membranes is surprisingly complicated. For some time, it has been known that many membrane proteins diffuse more slowly over a range ≥1 μm than would be expected from the apparent viscosity of the lipid bilayer in which they are embedded, and some appear to be immobile on the timescale of the measurement.[60] SPT measurements have revealed a range of diffusion styles and a mosaic of membrane regions displaying either fast or slow diffusion,[61] and computational modeling has clarified the effects of different kinds of interactions on diffusion as measured by SPT.[62,63] High-speed nanometer-scale SPT measurements have led to the view that the cell surface membrane is partitioned into submicron compartments within which proteins diffuse rapidly limited by the viscosity of the lipid bilayer.[10] To move greater distances, however, the proteins must hop from one compartment to another in a much slower process.

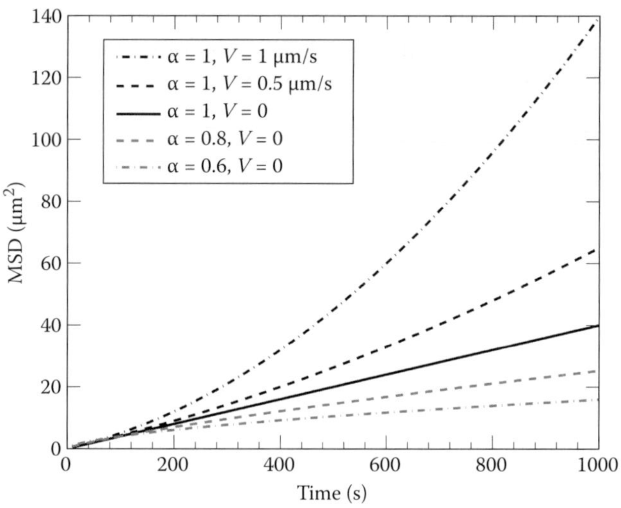

FIGURE 40.2
MSD versus time for simple diffusion (solid black line), diffusion plus convection (dashed and dot-dashed black lines), and obstructed (anomalous) diffusion (dashed and dot-dashed gray lines). The MSD as a function of time is given by $\mathrm{MSD}(t) = 4Dt^\alpha + (Vt)^2$. For all curves, $D = 1\ \mu m^2 s^{-1}$.

40.4.1 Microrheology

The rate and extent of spontaneous thermal motion of a particle in a viscoelastic liquid solution or solid matrix are limited by the viscous and elastic resistance imposed by the particle's environment. Microrheology determines the viscous and elastic parameters of a solution or matrix by tracking the motion of the particle.

This is most simply demonstrated for a three-dimensional fluid system in which the MSD in three dimensions is $\rho(t) = 6Dt$. The Stokes–Einstein equation for the diffusion coefficient of a spherical particle of radius a is $D = \frac{k_B T}{\zeta} = \frac{k_B T}{6\pi\eta a}$, where k_B, ζ, and η are the Boltzmann's constant, the frictional coefficient of the particle, and the viscosity of the fluid medium, respectively. Hence, the viscosity is $\eta = \frac{k_B 1t}{\pi a \rho(t)}$. Conversely, a particle embedded in a viscoelastic solid matrix could be represented as a tethered bead, i.e., a harmonic oscillator, in a viscous medium. For simplicity, this model is considered in one dimension as embodied in the following Langevin equation: $\zeta\, x = -kx + f(t)$, where k is the spring constant and $f(t)$ is a random force with the properties $\langle f(t) \rangle = 0$ and $\langle f(t) f(t') \rangle = 2\zeta k_B T \delta(t - t')$.[64] Then, the MSD is

$$\langle x(t)^2 \rangle = \frac{2k_B T}{k}\left[1 - \exp\left(-\frac{t}{\tau}\right)\right]. \tag{40.9}$$

The relaxation time $\tau = \frac{\zeta}{k}$. Thus, measuring the long-time limit of the MSD yields k, the elastic parameter of the system. Measuring the rate of relaxation to that limit in response to a perturbation, e.g., an abrupt shear force, yields η, the viscous parameter of the system.

A microscopic bead bound to one end of a DNA molecule that is tethered at its other end to a substrate provides one interesting application of the harmonic oscillator model.[65] Tracking the bead by optical microscopy yields a projection of its trajectory on the focal plane. The spatial range of this bead motion can be used to indicate the interaction of the *Lac* repressor protein with its binding sites on DNA. The protein has two binding sites for two specific and separated regions of DNA in the lac operon. For both sites to be bound to the repressor, the DNA must form a loop, which decreases the range of motion of the tethered bead. Hence, the range of motion indicates DNA looping and therefore binding the repressor. Even though DNA is a relatively stiff polymer, a molecule with a contour length that is long compared to its persistence length behaves as a statistical coil with a defined mean squared end-to-end distance.[64] The statistical force that restores conformational fluctuations of the DNA molecule to its mean state can be represented as resulting from a simple harmonic potential.[64] One can also interpret the dynamics of the DNA molecule in terms of a coupled bead and spring model that either does not[66] or does[67] account for hydrodynamic interactions among the polymer segments.[65]

Mason and Weitz have derived a more general relationship in terms of the Laplace transforms of the position, $\tilde{r}(s)$, the complex modulus, $\tilde{G}(s)$, and the frequency-dependent viscosity, $\tilde{\eta}(s)$, respectively (neglecting an inertial term):

$$\langle \Delta\tilde{r}(s)^2 \rangle = \frac{k_B T}{\pi a s \tilde{G}(s)} = \frac{k_B T}{\pi a s^2 \tilde{\eta}(s)} \tag{40.10}$$

in which a is the radius of the tracked particle.[11] SPT can provide $\langle \Delta r(t)^2 \rangle$ with which to evaluate the viscoelastic parameters.[68] This approach has been used extensively to study the viscoelastic properties of cytoskeletal filaments such as actin and microtubules, both in purified systems and in cells as summarized in useful reviews.[12,69,70]

40.5 Summary

FCS, which measures dynamic processes via fluorescence fluctuations in steady-state systems, and FPR, which measures the relaxation of macroscopic concentration gradients produced by localized photobleaching, are closely related methods that supply information about rates of diffusion, convection, chemical reaction, and other dynamic processes. They have been extensively used to measure dynamic molecular behavior both in purified systems and in cells. SPT allows the observation of different mechanisms of molecular transport in cells and in very small subcellular domains. These measurements indicate heterogeneity due to localized structures that influence particle transport. SPT can also be used to determine the viscoelastic properties of highly localized regions of cells and polymeric systems.

References

1. Elson EL. Fluorescence correlation spectroscopy: Past, present, future. *Biophys J* 2011:101(12): 2855–2870.
2. Qian H, Elson EL, Frieden C. Studies on the structure of actin gels using time correlation spectroscopy of fluorescent beads. *Biophys J* 1992:63(4): 1000–1010.
3. Heinemann F, Vogel SK, Schwille P. Lateral membrane diffusion modulated by a minimal actin cortex. *Biophys J* 2013:104(7): 1465–1475.
4. Bernheim-Groswasser A, Shusterman R, Krichevsky O. Fluorescence correlation spectroscopy analysis of segmental dynamics in actin filaments. *J Chem Phys* 2006:125(8): 084903-1–084903-11.
5. Rayan G, Guet J-E, Taulier N, Pincet F, Urbach W. Recent applications of fluorescence recovery after photobleaching (frap) to membrane bio-macromolecules. *Sensors* 2010:10(6): 5927–5948.

6. Lippincott-Schwartz J, Snapp E, Kenworthy A. Studying protein dynamics in living cells. *Nat Rev Mol Cell Biol* 2001:2(6): 444–456.
7. Lai FP, Szczodrak M, Block J, Faix J, Breitsprecher D, Mannherz HG, Stradal TE, Dunn GA, Small JV, Rottner K. Arp2/3 complex interactions and actin network turnover in lamellipodia. *EMBO J* 2008:27(7): 982–992.
8. Smith MB, Kiuchi T, Watanabe N, Vavylonis D. Distributed actin turnover in the lamellipodium and frap kinetics. *Biophys J* 2013:104(1): 247–257.
9. Sheetz MP, Turney S, Qian H, Elson EL. Nanometre-level analysis demonstrates that lipid flow does not drive membrane glycoprotein movements. *Nature* 1989:340(6231): 284–288.
10. Kusumi A, Nakada C, Ritchie K, Murase K, Suzuki K, Murakoshi H, Kasai RS, Kondo J, Fujiwara T. Paradigm shift of the plasma membrane concept from the two-dimensional continuum fluid to the partitioned fluid: High-speed single-molecule tracking of membrane molecules. *Annu Rev Biophys Biomol Struct* 2005:34: 351–378.
11. Mason TG, Weitz DA. Optical measurements of frequency-dependent linear viscoelastic moduli of complex fluids. *Phys Rev Lett* 1995:74: 1250–1253.
12. Wirtz D. Particle-tracking microrheology of living cells: Principles and applications. *Annu Rev Biophys* 2009:38: 301–326.
13. Elson E, Magde D. Fluorescence correlation spectroscopy. I. Conceptual basis and theory. *Biopolymers* 1974:13: 1–27.
14. Lee SL, Nekouzadeh A, Butler B, Pryse KM, McConnaughey WB, Nathan AC, Legant WR et al. Physically-induced cytoskeleton remodeling of cells in three-dimensional culture. *PLoS One* 2012:7(12): e45512.
15. Aragon SR, Pecora R. Fluorescence correlation spectroscopy as a probe of molecular-dynamics. *J Chem Phys* 1976:64(4): 1791–1803.
16. Ruttinger S, Buschmann V, Kramer B, Erdmann R, Macdonald R, Koberling F. Comparison and accuracy of methods to determine the confocal volume for quantitative fluorescence correlation spectroscopy. *J Microsc* 2008:232(2): 343–352.
17. Magde D, Elson EL, Webb WW. Fluorescence correlation spectroscopy. Ii. An experimental realization. *Biopolymers* 1974:13(1): 29–61.
18. Chattopadhyay K, Saffarian S, Elson EL, Frieden C. Measurement of microsecond dynamic motion in the intestinal fatty acid binding protein by using fluorescence correlation spectroscopy. *Proc Natl Acad Sci U S A* 2002:99(22): 14171–14176.
19. Chattopadhyay K, Elson EL, Frieden C. The kinetics of conformational fluctuations in an unfolded protein measured by fluorescence methods. *Proc Natl Acad Sci U S A* 2005:102(7): 2385–2389.
20. Shusterman R, Alon S, Gavrinyov T, Krichevsky O. Monomer dynamics in double- and single-stranded DNA polymers. *Phys Rev Lett* 2004:92(4): 048303.
21. Bonnet G, Krichevsky O, Libchaber A. Kinetics of conformational fluctuations in DNA hairpin-loops. *Proc Natl Acad Sci U S A* 1998:95(15): 8602–8606.
22. Bonnet G, Tyagi S, Libchaber A, Kramer FR. Thermodynamic basis of the enhanced specificity of structured DNA probes. *Proc Natl Acad Sci U S A* 1999:96(11): 6171–6176.
23. Jung J, Ihly R, Scott E, Yu M, Van Orden A. Probing the complete folding trajectory of a DNA hairpin using dual beam fluorescence fluctuation spectroscopy. *J Phys Chem B* 2008:112(1): 127–133.
24. Tinnefeld P, Sauer M. Branching out of single-molecule fluorescence spectroscopy: Challenges for chemistry and influence on biology. *Angew Chem Int Ed Engl* 2005:44(18): 2642–2671.
25. Qian H, Saffarian S, Elson EL. Concentration fluctuations in a mesoscopic oscillating chemical reaction system. *Proc Natl Acad Sci U S A* 2002:99(16): 10376–10381.
26. Michalet X, Weiss S, Jager M. Single-molecule fluorescence studies of protein folding and conformational dynamics. *Chem Rev* 2006:106(5): 1785–1813.
27. Magde D, Webb WW, Elson EL. Fluorescence correlation spectroscopy iii. Uniform translation and laminar flow. *Biopolymers* 1978:17: 361–376.
28. Muller JD, Chen Y, Gratton E. Resolving heterogeneity on the single molecular level with the photon-counting histogram. *Biophys J* 2000:78(1): 474–486.
29. Kask P, Palo K, Ullmann D, Gall K. Fluorescence-intensity distribution analysis and its application in biomolecular detection technology. *Proc Natl Acad Sci U S A* 1999:96(24): 13756–61.
30. Qian H, Elson EL. On the analysis of high order moments of fluorescence fluctuations. *Biophys J* 1990:57(2): 375–380.
31. Kask P, Palo K, Fay N, Brand L, Mets U, Ullmann D, Jungmann J, Pschorr J, Gall K. Two-dimensional fluorescence intensity distribution analysis: Theory and applications. *Biophys J* 2000:78(4). 1703–1713.
32. Chen Y, Wei LN, Muller JD. Probing protein oligomerization in living cells with fluorescence fluctuation spectroscopy. *Proc Natl Acad Sci U S A* 2003:100(26): 15492–15497.
33. Saffarian S, Li Y, Elson EL, Pike LJ. Oligomerization of the egf receptor investigated by live cell fluorescence intensity distribution analysis. *Biophys J* 2007:93(3): 1021–1031.
34. Ries J, Petrasek Z, Garcia-Saez AJ, Schwille P. A comprehensive framework for fluorescence cross-correlation spectroscopy. *New J Phys* 2010:12: 113009.
35. Schwille P, Meyer-Almes FJ, Rigler R. Dual-color fluorescence cross-correlation spectroscopy for multicomponent diffusional analysis in solution. *Biophys J* 1997:72(4): 1878–1886.
36. Kettling U, Koltermann A, Schwille P, Eigen M. Real-time enzyme kinetics monitored by dual-color fluorescence cross-correlation spectroscopy. *Proc Natl Acad Sci U S A* 1998:95(4): 1416–1420.
37. Bacia K, Kim SA, Schwille P. Fluorescence cross-correlation spectroscopy in living cells. *Nat Methods* 2006:3(2): 83–89.
38. Petersen NO, Brown C, Kaminski A, Rocheleau J, Srivastava M, Wiseman PW. Analysis of membrane protein cluster densities and sizes in situ by image correlation spectroscopy. *Faraday Discuss* 1998:111: 289–305.

39. Kolin DL, Wiseman PW. Advances in image correlation spectroscopy: Measuring number densities, aggregation states, and dynamics of fluorescently labeled macromolecules in cells. *Cell Biochem Biophys* 2007:49(3). 141–164.
40. Wenger J, Gérard D, Lenne P-F, Rigneault H, Bonod N, Popov E, Marguet D, Nelep C, Ebbesen T. Biophotonics applications of nanometric apertures. *Int J Mater Prod Technol* 2009:34(4): 488–506.
41. Thompson NL, Burghardt TP, Axelrod D. Measuring surface dynamics of biomolecules by total internal reflection fluorescence with photobleaching recovery or correlation spectroscopy. *Biophys J* 1981:33(3): 435–454.
42. Lieto AM, Cush RC, Thompson NL. Ligand-receptor kinetics measured by total internal reflection with fluorescence correlation spectroscopy. *Biophys J* 2003:85(5): 3294–3302.
43. Ohsugi Y, Saito K, Tamura M, Kinjo M. Lateral mobility of membrane-binding proteins in living cells measured by total internal reflection fluorescence correlation spectroscopy. *Biophys J* 2006:91(9): 3456–3464.
44. Hell SW, Wichmann J. Breaking the diffraction resolution limit by stimulated emission: Stimulated-emission-depletion fluorescence microscopy. *Opt Lett* 1994:19(11): 780–782.
45. Eggeling C, Ringemann C, Medda R, Schwarzmann G, Sandhoff K, Polyakova S, Belov VN et al. Direct observation of the nanoscale dynamics of membrane lipids in a living cell. *Nature* 2009:457(7233): 1159–1162.
46. Koppel DE, Axelrod D, Schlessinger J, Elson EL, Webb WW. Dynamics of fluorescence marker concentration as a probe of mobility. *Biophys J* 1976:16(11): 1315–1329.
47. Axelrod D, Koppel DE, Schlessinger J, Elson E, Webb WW. Mobility measurement by analysis of fluorescence photobleaching recovery kinetics. *Biophys J* 1976:16(9): 1055–1069.
48. Elson EL. Fluorescence correlation spectroscopy and photobleaching recovery. *Ann Rev Phys Chem* 1985:36: 379–406.
49. Icenogle RD, Elson EL. Fluorescence correlation spectroscopy and photobleaching recovery of multiple binding reactions. II. Fpr and fcs measurements at low and high DNA concentrations. *Biopolymers* 1983:22(8): 1949–1966.
50. Icenogle RD, Elson EL. Fluorescence correlation spectroscopy and photobleaching recovery of multiple binding reactions. I. Theory and fcs measurements. *Biopolymers* 1983:22(8): 1919–1948.
51. Lippincott-Schwartz J, Altan-Bonnet N, Patterson GH. Photobleaching and photoactivation: Following protein dynamics in living cells. *Nat Cell Biol* 2003:Suppl: S7–S14.
52. Kang M, Day CA, Drake K, Kenworthy AK, DiBenedetto E. A generalization of theory for two-dimensional fluorescence recovery after photobleaching applicable to confocal laser scanning microscopes. *Biophys J* 2009:97(5): 1501–1511.
53. Qian H, Elson EL. Analysis of confocal laser-microscope optics for 3-fluorescence correlation spectroscopy. *Appl Opt* 1991:30: 1185–1195.
54. Mazza D, Braeckmans K, Cella F, Testa I, Vercauteren D, Demeester J, De Smedt SS, Diaspro A. A new frap/frapa method for three-dimensional diffusion measurements based on multiphoton excitation microscopy. *Biophys J* 2008:95(7): 3457–3469.
55. Mazza D, Cella F, Vicidomini G, Krol S, Diaspro A. Role of three-dimensional bleach distribution in confocal and two-photon fluorescence recovery after photobleaching experiments. *Appl Opt* 2007.46(30): 7401–7411.
56. Elson EL, Reidler JA. Analysis of cell surface interactions by measurements of lateral mobility. *J Supramol Struct* 1979:12(4): 481–489.
57. Sprague BL, Pego RL, Stavreva DA, McNally JG. Analysis of binding reactions by fluorescence recovery after photobleaching. *Biophys J* 2004:86(6): 3473–3495.
58. Soumpasis DM. Theoretical analysis of fluorescence photobleaching recovery experiments. *Biophys J* 1983:41(1): 95–97.
59. Stavreva DA, McNally JG, Allis CD, Carl W. Fluorescence recovery after photobleaching (frap) methods for visualizing protein dynamics in living mammalian cell nuclei. *Methods in Enzymology* 2004:375: 443–455.
60. Schlessinger J, Koppel DE, Axelrod D, Jacobson K, Webb WW, Elson EL. Lateral transport on cell membranes: Mobility of concanavalin a receptors on myoblasts. *Proc Natl Acad Sci U S A* 1976:73(7): 2409–2413.
61. Simson R, Yang B, Moore SE, Doherty P, Walsh FS, Jacobson KA. Structural mosaicism on the submicron scale in the plasma membrane. *Biophys J* 1998:74(1): 297–308.
62. Saxton MJ. Single-particle tracking: The distribution of diffusion coefficients. *Biophys J* 1997:72(4): 1744–1753.
63. Saxton MJ, Jacobson K. Single-particle tracking: Applications to membrane dynamics. *Annu Rev Biophys Biomol Struct* 1997:26: 373–399.
64. Doi M, Edwards SF. *The Theory of Polymer Dynamics*. Oxford: Clarendon Press, 1986.
65. Qian H, Elson EL. Quantitative study of polymer conformation and dynamics by single particle tracking. *Biophys J* 1999:76(3): 1598–1605.
66. Rouse PE. A theory of the linear viscoelastic properties of dilute solutions of coiling polymers. *J Chem Phys* 1953:21(7): 1272–1280.
67. Zimm BH. Dynamics of polymer molecules in dilute solution: Viscoelasticity, flow birefringence and dielectric loss. *J Chem Phys* 1956:24(2): 269–278.
68. Mason TG, Ganesan K, van Zanten JH, Wirtz D, Kuo SC. Particle tracking microrheology of complex fluids. *Phys Rev Lett* 1997:79: 3282–3285.
69. Cicuta P, Donald AM. Microrheology: A review of the method and applications. *Soft Matter* 2007:3(12): 1449–1455.
70. Kasza KE, Rowat AC, Liu J, Angelini TE, Brangwynne CP, Koenderink GH, Weitz DA. The cell as a material. *Curr Opin Cell Biol* 2007:19(1): 101–107.

41

FABLE and Image Processing

Ali Nekouzadeh and Niloufar Ghoreishi

CONTENTS

Quantification of the total length of fibers in an image is a necessity in a wide variety of biological applications. In an image, fibers usually appear as curved or straight lines with different lengths, thicknesses, orientations, and curvatures. Here, we present an automated approach based upon Fourier space analysis for quantifying the total length of the fibers in an image. The approach is termed FABLE (Fourier-space automated band-pass length estimation) and integrates the power spectral densities of grayscale images within a band-pass filter to provide a metric for the total length of fibers. Using mathematical analysis and computer simulations, it is shown that FABLE can provide an accurate metric, termed fibrosity, for the comparison of the total length of the fibers in different images. The limitations of the technique for application to biological images are discussed along with possible mitigations.

41.1 Introduction

The density and orientation of fibers in a material usually affect the mechanical properties of the material and are often considered as characteristic features. High-resolution images of the tissue or material are usually used in combinations with image processing algorithms for quantifying the density and orientation of the fibers.[1,2] Fibers appear as lines with different lengths, thicknesses, orientations, and curvatures in the image. The total length of these lines is a metric to quantify the amount of tissue fibers and is important in a broad range of biological and mechanical applications including trabecular bone microfracture,[3] microcracking in fiber-reinforced concrete,[4–6] crack growth within metals and plastics,[7,8] nondestructive testing of the welds, and total internal reflection microscopy.[9,10]

Distribution of the fiber orientation can be determined using Fourier space image processing techniques.[1,11] These techniques are based on the fact that the spatial Fourier transform of a fiber is extended in a direction associated with the fiber direction. The angular variation of the power spectrum in radial segments of the Fourier space is proportional to the orientation distribution of fibers in an image. The exact amplitude of the power spectrum is not important in these techniques, and a normalized spectrum may be used.

Determining the total fiber length using image processing techniques is more complex and requires careful evaluation of the methodology and its underlying assumptions. The most appropriate techniques to estimate the total length of curved fibers in an image are the edge detection techniques. The edge detection techniques are based on radon transform and its discrete implementation, the Hough transform.[12] The radon transform is used for detection of thin and straight lines; a modified form of Hough transform was used to detect thicker lines as well.[13] Hough transform of an image is the number of pixels that are located along any possible line in an image. The number of pixels along a particular line in the image is stored in an accumulator cell associated with that line. Each point in the Hough space defines a line in the image, and its value is the number of pixels along that line (stored in the accumulator). A high value point in the Hough space indicates many pixels along the associated line and therefore suggests a line along that direction in the image. Although in this method the value of a point in Hough space is proportional to the net length of the line,[14] application of the method is limited to images with a few straight lines and no other features. The method is not accurate in measuring the total length of curved lines or lines with different thicknesses. Additionally, the accuracy of the method is compromised if an image contains features other than thin lines (e.g., noise or 2D bodies).

In this chapter, we present and discuss a novel automated method developed for estimating the total length of the lines within an image.[15] We abbreviated this method as FABLE (Fourier-space automated band-pass length estimation). This method is based on the Fourier transform of the image and estimates the total length of the fibers regardless of their thickness, orientation, and curvature. The application scope and the limitations of the method will be discussed, and possible mitigations will be suggested. The exact amplitudes of the Fourier spectrum are required in this method (unlike the fiber orientation method that can use a normalized Fourier spectrum), and the image may require some preprocessing for more accurate results. Using a series of images with defined fiber dimensions and orientation distributions, we examine the accuracy of FABLE to show that it can provide statistically meaningful information about fiber length changes.

The typical application of this method involves quantifying changes in the actin cytoskeletons of cells in response to various conditions that promote their remodeling. Cytoskeletons can be imaged after fixing the cell with formalin, and then staining the cytoskeleton-associated proteins using immunofluorescence techniques,[16] or by inducing cells to express fluorescently labeled actin monomers.[17] Measurement of fiber length is currently achieved through manual estimation.[10] We explain the application of this method in remodeling of myofibroblast, in which the total length of stress fibers within a myofibroblast can change in response to biochemical[18–20] and mechanical[21,22] stimuli. Stress fibers are made from actin filaments and form a mechanical structure within the myofibroblast; their interaction with the nonmuscle myosins is the underlying mechanism of the cell contraction and contracture. The contractile force of a myofibroblast is directly proportional to the total length of its stress fibers.[21] The orientation distribution of the stress fibers is also a determinant of the mechanical functioning of myofibroblasts.[23–27]

41.2 FABLE Theory

41.2.1 Fourier Transform of a Fiber

A fiber may be modeled, in an ideal case, by a thin and long white rectangle on a black background. Therefore, a fiber of length a and thickness b that is located in the center of the image at (0,0) may be defined by function $f(x,y)$ as

$$f(x,y) = \begin{cases} 1, & -a/2 < x < a/2 \text{ and } -b/2 < y < b/2 \\ 0, & \text{otherwise} \end{cases} \tag{41.1}$$

The 2D Fourier transform of $f(x,y)$ is

$$F(u,v) = \int_{-\infty}^{\infty}\int_{-\infty}^{\infty} f(x,y)e^{-2\pi j(ux+vy)}\,\mathrm{d}x\,\mathrm{d}y = \frac{\sin(\pi ua)}{\pi u}\frac{\sin(\pi vb)}{\pi v}, \tag{41.2}$$

where $j = \sqrt{-1}$. Translating the fiber from (0,0) to (x_0,y_0) changes its Fourier transform to

$$F^*(u,v) = e^{-2\pi j(ux_0+vy_0)}\frac{\sin(\pi ua)}{\pi u}\frac{\sin(\pi vb)}{\pi v}. \tag{41.3}$$

Comparing $F(u,v)$ and $F^*(u,v)$ shows that the amplitudes of the Fourier transforms of fibers located at different locations of an image are equal, but the phases are not. Figure 41.1 shows an ideal fiber defined by Equation 41.1 and its Fourier transform power spectrum.

41.2.2 Effect of Fiber Orientation on Its Fourier Transform

The fiber introduced by Equation 41.1 is located horizontally along the x direction. Assume that $g(x,y)$ defines a fiber that is rotated by θ degrees with respect to the horizontal orientation. $g(x,y)$ can be written in terms of $f(x,y)$ as

$$g(x,y) = f(x^*, y^*) \tag{41.4}$$

where

$$\begin{bmatrix} x^* \\ y^* \end{bmatrix} = \begin{bmatrix} \cos\theta & \sin\theta \\ -\sin\theta & \cos\theta \end{bmatrix} \begin{bmatrix} x \\ y \end{bmatrix} \tag{41.5}$$

And the Fourier transform of the rotated fiber is

$$\begin{aligned} G(u,v) &= \int_{-\infty}^{\infty}\int_{-\infty}^{\infty} g(x,y) e^{-2\pi j(ux+vy)}\, dx\, dy \\ &= \int_{-\infty}^{\infty}\int_{-\infty}^{\infty} f(x^*,y^*) e^{-2\pi j\left([u\ v]\begin{bmatrix} \cos\theta & -\sin\theta \\ \sin\theta & \cos\theta \end{bmatrix}\begin{bmatrix} x^* \\ y^* \end{bmatrix}\right)} dx^*\, dy^* \\ &= F(u^*, v^*) \end{aligned} \tag{41.6}$$

where

$$\begin{bmatrix} u^* \\ v^* \end{bmatrix} = \begin{bmatrix} \cos\theta & \sin\theta \\ -\sin\theta & \cos\theta \end{bmatrix} \begin{bmatrix} u \\ v \end{bmatrix}. \tag{41.7}$$

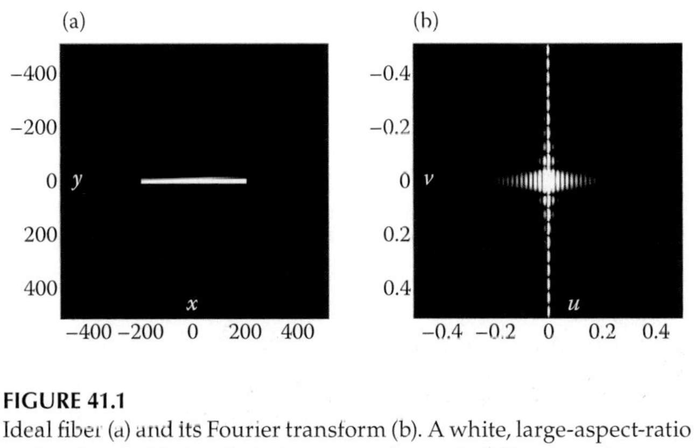

FIGURE 41.1
Ideal fiber (a) and its Fourier transform (b). A white, large-aspect-ratio rectangle on a black substrate represents an ideal fiber in an image. u is the frequency of x and v is the frequency of y. Unit of both x and y is pixel and unit of both u and v is pixel^{-1}.

It means that if a fiber (or, in general, an image) rotates by θ degrees, its Fourier transform rotates by $-\theta$ degrees. Figure 41.2 shows the effect of fiber orientation on its Fourier transform.

41.2.3 Fourier Transform Power Scales with the Number of Identical Fibers

An image of several nonoverlapping fibers (on a black background) may be considered as the sum of the images of the individual fibers. Fourier transform is a linear transform, and therefore, the Fourier transform of the image of several fibers is the sum of Fourier transforms of all individual fibers:

$$\begin{aligned} F_{\text{tot}}(u,v) &= \sum_{i=1}^{n} e^{-2\pi j(ux_i+vy_i)} \frac{\sin(\pi ua)}{\pi u}\frac{\sin(\pi vb)}{\pi v} \\ &= \frac{\sin(\pi ua)}{\pi u}\frac{\sin(\pi vb)}{\pi v} \sum_{i=1}^{n} \cos(2\pi ux_i + 2\pi vy_i) \\ &\quad - j\sin(2\pi ux_i + 2\pi vy_i), \end{aligned} \tag{41.8}$$

where x_i and y_i are the center of the ith fiber in the image, and n is the total number of fibers. The power of the Fourier transform (power spectrum) is the square of its amplitude:

$$\begin{aligned} F_{\text{tot}}(u,v)^2 &= \left(\frac{\sin(\pi ua)}{\pi u}\frac{\sin(\pi vb)}{\pi v}\right)^2 \\ &\quad \left[\left(\sum_{i=1}^{n} \cos(2\pi ux_i + 2\pi vy_i)\right)^2 \right. \\ &\quad \left. + \left(\sum_{i=1}^{n} \sin(2\pi ux_i + 2\pi vy_i)\right)^2\right]. \end{aligned} \tag{41.9}$$

If fibers are distributed randomly, the expectation value of the power spectrum is

$$\begin{aligned} E\left[F_{\text{tot}}(u,v)^2\right] &= \left(\frac{\sin(\pi ua)}{\pi u}\frac{\sin(\pi vb)}{\pi v}\right)^2 \\ &\quad \left(E\left[\left(\sum_{i=1}^{n} \cos(2\pi ux_i + 2\pi vy_i)\right)^2\right] \right. \\ &\quad \left. + E\left[\left(\sum_{i=1}^{n} \sin(2\pi ux_i + 2\pi vy_i)\right)^2\right]\right). \end{aligned} \tag{41.10}$$

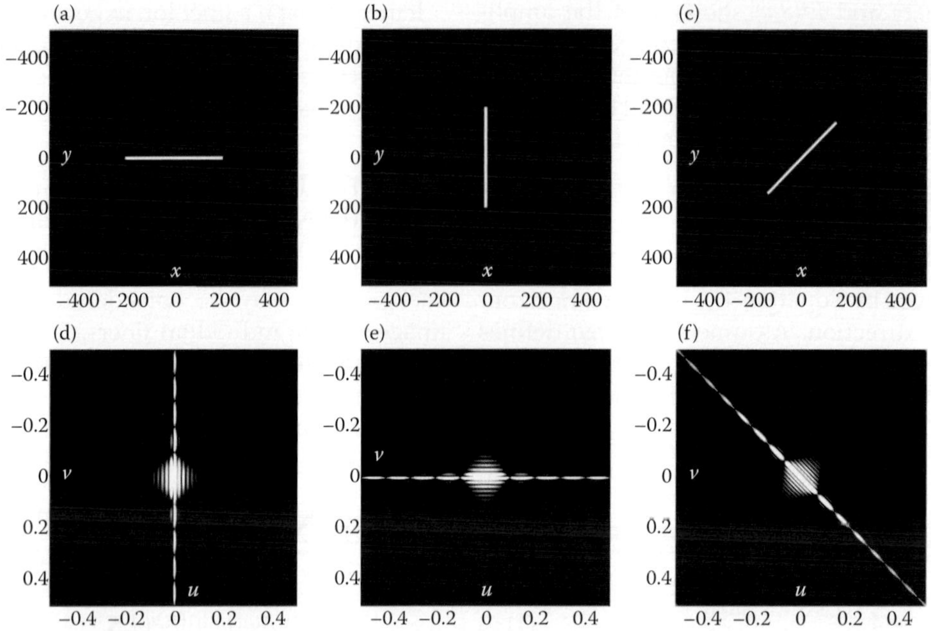

FIGURE 41.2
Horizontal (a), vertical (b), and oblique (c) fibers and their Fourier transforms (d, e, and f, respectively). A θ rotation of the fiber results to a −θ rotation of its Fourier transform.

This can be simplified using the following expansion:

$$E\left[\left(\sum_{i=1}^{n}\cos(\alpha_i)\right)^2\right]$$

$$= E\left[\sum_{i=1}^{n}\cos^2(\alpha_i)\right] + E\left[\sum_{i=1}^{n}\sum_{\substack{j=1\\ j\neq i}}^{n}\cos(\alpha_i)\cos(\alpha_j)\right]$$

$$= \sum_{i=1}^{n}E[\cos^2(\alpha_i)] + \sum_{i=1}^{n}\sum_{\substack{j=1\\ j\neq i}}^{n}E[\cos(\alpha_i)\cos(\alpha_j)]$$

$$= \frac{n}{2} + 0. \qquad (41.11)$$

This relation is valid for both the sine and cosine series in Equation 41.10. Therefore,

$$E\left[F_{tot}(u,v)^2\right] = n\left(\frac{\sin(\pi ua)}{\pi u}\frac{\sin(\pi vb)}{\pi v}\right)^2. \qquad (41.12)$$

If either of u or v is sufficiently large that the random variable $2\pi ux_i + 2\pi vy_i$ is distributed almost uniformly over all angles, the expectation value of the power spectrum is almost equal to the power spectrum itself:

$$F_{tot}(u,v)^2 \approx E\left[F_{tot}(u,v)^2\right] - n\left(\frac{\sin(\pi ua)}{\pi u}\frac{\sin(\pi vb)}{\pi v}\right)^2, \qquad (41.13)$$

indicating that the square of the Fourier power spectrum scales with the number of fibers within the image.

41.2.4 Relative Numbers of Fibers in Images Can Be Estimated Using a Circular Band-Pass Filter

The majority of the power in a Fourier transform is distributed in the center of the spectrum (where both spatial frequencies u and v are small). The center of the power spectrum includes information of all the objects in the image, including fibrous and bulky features. Farther from the center of the transform, information about the fibers persists, while information about the bulky objects decays (Figure 41.3). Additionally, Equation 41.13 may not hold where both frequencies u and v are small (in the vicinity of the center of the Fourier transform). Therefore, the low frequencies have to be filtered out for estimating the total length of the fibers. A fiber rotation leads to a rotation in its Fourier transform and, therefore, alters the amplitude of the Fourier transform at a particular frequency pair (u,v). However, the integral of the amplitudes along any circular path centered at the origin of the frequency domain remains unchanged by any rotation of the fibers.

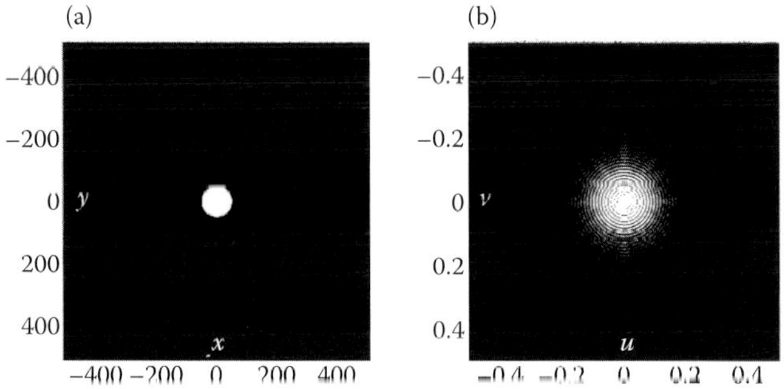

FIGURE 41.3
Bulky object (a) and its Fourier transform (b). The area of the circle in (a) is equal to the area of fiber in Figure 41.1a. There is no significant power in high frequencies of the Fourier transform.

To quantify the total number of fibers independent of their angular distribution, we considered the Fourier transform powers in a circular band-pass filter located at high frequencies (Figure 41.4). The high frequencies at the corners of the Fourier transform are filtered out by the circular band-pass filter to avoid providing an undesired bias to fibers oriented at 45° with respect to the image axes. The scaling in Equation 41.13 holds for arbitrary locations and orientations of fibers within an image, under two conditions: (1) the locations and orientations must be independent variables, and (2) at least one of the frequencies must be sufficiently large that $2\pi ux_i + 2\pi vy_i$ maps uniformly over all angles for the range of frequencies within the circular band-pass filter, independent of the exact distribution of x_i and y_i. If $|ux_i|+|vy_i| \gg 1$, then the latter requirement is satisfied.

For pixilated images, a pixel may be considered as the unit of length, and the number of pixels is the measure of length. In this system, pixel^{-1} is the unit of frequency. The fast Fourier transform (FFT) of a 1024 × 1024 image is itself a 1024 × 1024 image, with u and v ranging from

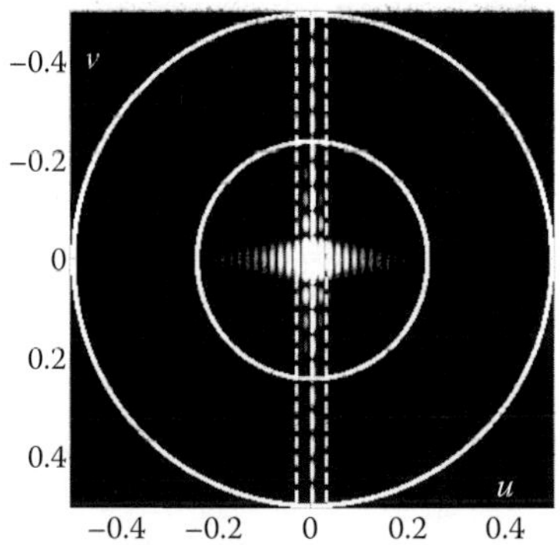

FIGURE 41.4
Circular filter used to compute the fibrosity. The upper band of the filter (the exterior circle) is usually the maximum frequency, but the lower band of the filter (the interior circle) may vary. Fibrosity is the sum of all Fourier powers between the two circles. For a single fiber, the majority of the powers between the two circles are concentrated within a small area (between the two dashed lines).

−511/1024 pixel^{-1} to +512/1024 pixel^{-1}. In order to satisfy the uniform distribution of the phase angles, we may assume that $|ux_i|+|vy_i| \geq 50$. For a 1024 × 1024 image, a band-pass filter with a lower cutoff frequency (radius) greater than f_L = 50/1024 pixel^{-1} generates a measure proportional to the number of fibers in an image. We term this measure the "fibrosity," Φ, of the image:

$$\Phi \equiv \iint_{\text{circular filter}} F_{\text{tot}}(u,v)^2 \, dv \, du. \tag{41.14}$$

41.2.5 Fibrosity Is Proportional to Fiber Length, but Relatively Insensitive to Fiber Width

The fibrosity measure in Equation 41.14 is also appropriate for images that contain fibers with different dimensions; it estimates the total length of all fibers regardless of the length and thickness of individual fibers. Fibrosity varies in proportion to fiber length but is relatively insensitive to changes in fiber width. For a rectangular fiber, fibrosity Φ_f is

$$\Phi_f = \iint_{\text{ring region}} \left(\frac{\sin(\pi ua)}{\pi u} \frac{\sin(\pi vb)}{\pi v} \right)^2 dv \, du. \tag{41.15}$$

For a fiber that is extended horizontally in the x direction ($a \gg b$), the power spectrum is extended vertically toward v frequencies. For example, in the Fourier transform of Figure 41.1, the power spectrum is concentrated within the range $-u_c < u < u_c$ (parallel dashed lines) in the horizontal direction but extends vertically all the way to the maximum v frequency. Approximately 95% of the power lies within $\pm u_c$ for a cutoff frequency of u_c = 50/1024 pixel^{-1}. For a band-pass filter ($B(u,v)=1, f_L \leq \sqrt{u^2+v^2} \leq f_U; B(u,v)=0$ otherwise), the power is mostly concentrated within two almost rectangular regions at the top and bottom of the spectrum (Figure 41.4). Therefore, Equation 41.15 is approximately

$$\Phi_f \approx \left(2a^2 \int_0^{u_c} \left(\frac{\sin(\pi ua)}{\pi ua} \right)^2 du \right) \left(2b^2 \int_{f_L}^{f_U} \left(\frac{\sin(\pi vb)}{\pi vb} \right)^2 dv \right) \equiv (I_u)(I_v). \tag{41.16}$$

where I_u and I_v are defined as

$$I_u = 2a^2 \int_0^{u_c} \left(\frac{\sin(\pi ua)}{\pi ua} \right)^2 du$$
$$I_v = 2b^2 \int_{f_L}^{f_U} \left(\frac{\sin(\pi vb)}{\pi vb} \right)^2 dv. \tag{41.17}$$

The sensitivity of the fibrosity to fiber length and fiber thickness was assessed by comparing the fibrosity Φ_f of a fiber with dimensions $a \times b$ to the fibrosity Φ_α of an elongated or shortened fiber with dimensions $\alpha a \times b$, and the fibrosity Φ_β of a thickened or narrowed fiber with dimensions $a \times \beta b$. Defining $i_u(u)$ such that $I_u(u) = 2a^2 \int_0^{u_c} i_u(u)\,\mathrm{d}u$, and choosing a common u_c such that Equation 41.16 is an acceptable approximation for both u_c and αu_c:

$$\Phi_\alpha \approx \left(2\alpha^2 a^2 \int_0^{u_c} i_u(\alpha u)\,\mathrm{d}u\right)(I_v)$$
$$= \left(2\alpha a^2 \int_0^{\alpha u_c} i_u(u^*)\,\mathrm{d}u^*\right)(I_v) = (\alpha I_u)(I_v) = \alpha\Phi_f \quad (41.18)$$

where $u^* = \alpha u$. Therefore, scaling the length of a fiber scales the fibrosity.

Now let us assume that the fiber width is scaled by β. Defining $i_v(v)$ such that $I_v(v) = 2b^2 \int_{f_L}^{f_U} i_v(v)\,\mathrm{d}v$:

$$\Phi_\beta \approx (I_u)\left(2\beta^2 b^2 \int_{f_L}^{f_U} i_v(\beta v)\,\mathrm{d}v\right) = (I_u)\left(2\beta b^2 \int_{\beta f_L}^{\beta f_U} i_v(v^*)\,\mathrm{d}v^*\right) \quad (41.19)$$

where $v^* = \beta v$. Unlike the case of the u term, scaling the integral limits of the v term changes the integral in Equation 41.19 in an oscillatory manner due to the oscillatory nature of the sinc^2 function in $i_v(v)$ (Figure 41.5): scaling the limits can increase or decrease $I_v(\beta v)$ relative to $I_v(v)$.

The oscillations are smaller for larger fiber widths and for higher filter cutoff frequencies, because both move the limits of integration toward right in Figure 41.5, where the oscillations are less significant. As an estimate of the effect of width changes on Φ, we integrate over an averaged function of sinc^2, $i_v(v) = 1/(2\pi^2 v^2)$ (solid gray curve, Figure 41.5). Φ_f is then approximately

$$\Phi_f = (I_u)(I_v) \approx (I_u)\left(2b^2 \int_{f_L}^{f_U} \frac{1}{2\pi^2 b^2 v^2}\,\mathrm{d}v\right)$$
$$= (I_u)\left(\frac{1}{\pi^2}\left(\frac{1}{f_L} - \frac{1}{f_U}\right)\right), \quad (41.20)$$

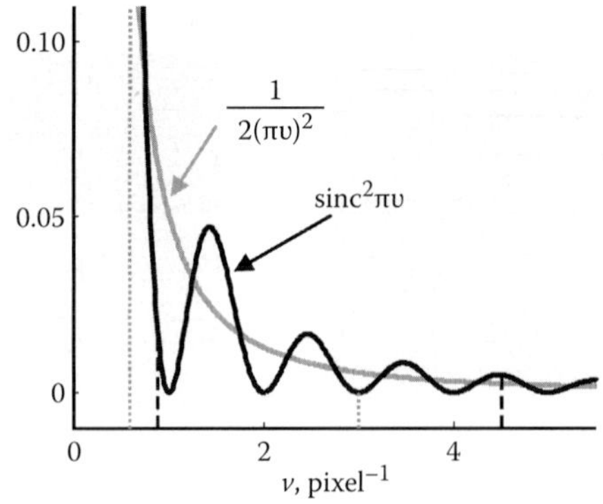

FIGURE 41.5
Oscillatory nature of the sinc^2 function (black curve) and its average (gray curve). Scaling the fiber thickness scales the integral limits (e.g., from gray dotted line to black dashed line).

and Φ_β is approximately

$$\Phi_\beta \approx (I_u)\left(2\beta b^2 \int_{\beta f_L}^{\beta f_U} \frac{1}{2\pi^2 b^2 v^2}\,\mathrm{d}v\right)$$
$$= (I_u)\left(\frac{\beta}{\pi^2}\left(\frac{1}{\beta f_L} - \frac{1}{\beta f_U}\right)\right) = \Phi_f. \quad (41.21)$$

Therefore, for long, slender fibers, Φ does not change on average with the fiber width but increases in direct proportion to the fiber length. Computer simulations showed that this would be valid for $a/b > 5$.

Equation 41.16 would be a close approximation of Equation 41.15 if the inner arc of the semirectangular regions of integration (Figure 41.4) is approximately less than $\pi/4$ radians. The enclosed angle by this arc, ω, can be calculated in terms of u_c and f_L as

$$\sin\frac{\omega}{2} = \frac{u_c}{f_L}. \quad (41.22)$$

About 95% of the power of the sinc^2 function is within its first three domes. Therefore, we may choose u_c so that the circular filter covers the first three domes of the u frequencies for all the fibers:

$$u_C a_{\min} > 3 \quad (41.23)$$

where $a_{\min}$ is the minimum length of the fibers of interests in the image. Combining Equations 41.22 and 41.23, we conclude the following condition for the lower frequency of the circular filter:

$$f_L > \frac{3}{a_{\min}\sin\frac{\pi}{8}} \quad (41.24)$$

41.3 Simulation Results

To assess the accuracy of the FABLE in predicting the total length of fibers in an image, a series of test images was generated with different fiber characteristics. FABLE could provide accurate estimates for the relative fibrosity, regardless of the fiber dimensions, directions, and curvature. The relative fibrosity is defined as the fibrosity of an image divided by the fibrosity of a reference image, and is used for comparison between the total fiber length in different images.

41.3.1 Fibrosity Is Proportional to the Total Length of the Fibers in an Image and Does Not Depend on Position or Length of Individual Fibers

To show that fibrosity scales with the total length of the fibers, 15 images were generated at three different fiber concentrations (five images at each concentration): 30, 60, and 90 fibers per image (Figure 41.6a–c). Fibers are identical with a dimension of 200 × 11 pixels and are located at randomly selected positions. The upper and lower bands of the filter are set to f_U = 0.5 pixel^{-1} and f_L = 0.1 pixel^{-1}. The measured relative fibrosities are 0.34, 0.67, and 1.0, respectively (Figure 41.6d). The standard deviations of relative fibrosity among images of each group are 0.009, 0.012, and 0.013, respectively. Fibrosity scales proportionally to the number of fibers, and so it is proportional to the total length of the image fibers. The small standard deviations among the images with the same number of fibers indicate that fibrosity is not sensitive to the fiber locations. Note that in Section 41.2, we proved that the expectation value of the fibrosity for randomly located fibers scales with the number of fibers.

To show that fibrosity does not depend on the individual fiber lengths, we generated 15 images with three different fiber lengths but the same total length of fibers. Each of the five images of the first group consists of 60 fibers with dimensions of 200 × 11 pixels (Figure 41.7a). Each of the five images of the second group consists of 40 fibers with dimensions of 300 × 11 pixels (Figure 41.7b), and each of the five images of the third group consists of 30 fibers with dimensions of 400 × 11 pixels (Figure 41.7c). The upper and lower frequencies of the circular filter are f_U = 0.5 pixel^{-1} and f_L = 0.1 pixel^{-1}. The averages of the relative fibrosities are 1.03, 1.03, and 1.00, and the standard deviations are 0.02, 0.01, and 0.01 among the three groups (Figure 41.7d). In all images of the three

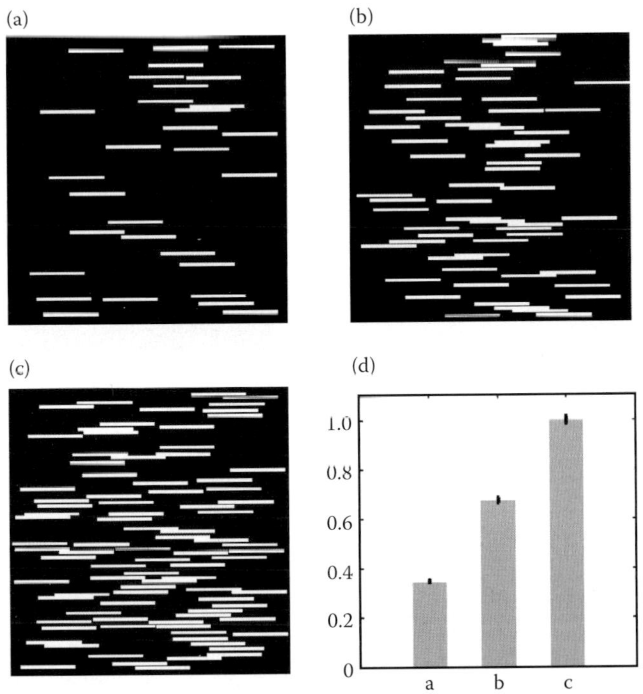

FIGURE 41.6
Three sample images with three different fiber densities (a–c) and the FABLE prediction of the relative fibrosities for the three different fiber densities (d). The fibrosities are measured relative to the highest density (c). Vertical lines on the bars represent the standard deviations of five images at each density.

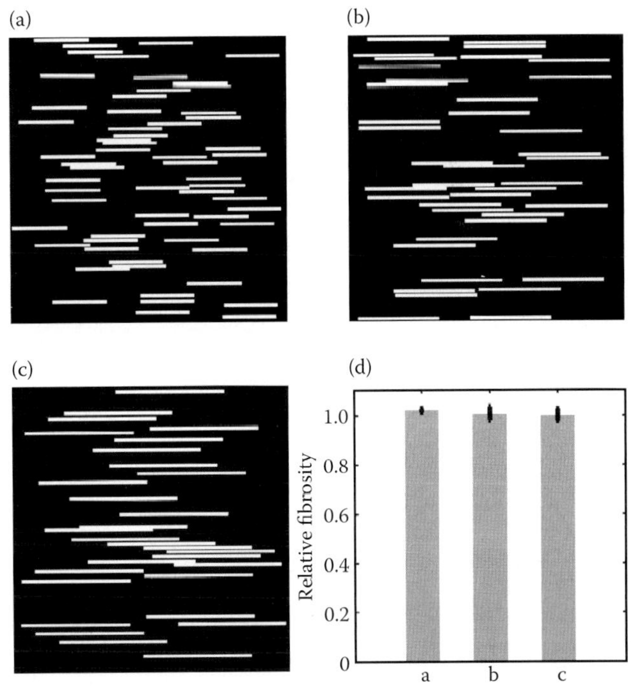

FIGURE 41.7
Three sample images with the same total fiber length but different individual fiber length (a–c) and the FABLE prediction of the relative fibrosities for the three types of fibers (d). The fibrosities are measured relative to the images with the longest fibers (c). Vertical lines on the bars represent the standard deviations of five images in each group.

groups, the total length of the fibers is 12,000 pixels. Fibrosity is almost equal in all images, which means that fibrosity is insensitive to the length of individual fibers.

41.3.2 Fibrosity Has Negligible Dependence on the Fiber Thickness, Fiber Curvature, and Fiber Orientation

Here we evaluate dependence of fibrosity to the shape of the fibers, in particular, to the fiber thickness, fiber curvature, and fiber orientation. To evaluate the dependence of fibrosity to the fiber thickness, we generated a series of images with variable thicknesses ranging from 5 to 80 pixels (five images for each fiber thickness). All the fiber lengths are 200 pixels. The fibrosities were computed and are presented in Figure 41.8 in terms of fiber aspect ratios. Three sample images at the aspect ratios of 0.045, 0.225, and 0.405 are shown in Figure 41.8a,b, and c, respectively. The upper band of the circular filter is $f_U = 0.5$ pixel^{-1} and the lower band is $f_L = 0.25$ pixel^{-1}. Up to an aspect ratio of about 0.2 ($b < a/5$), the fibrosity is almost independent of the fiber aspect ratio; for aspect ratios higher than 0.2, fibrosity increases gradually by increasing the fiber thickness (Figure 41.8d). When the fiber thickness increases beyond a certain limit, its width will contribute to the Fourier transform powers within the circular filter. Therefore, FABLE overestimates the total length of the fibers. It is noteworthy that usually the aspect ratio of a fiber is expected to be far smaller than 0.2. Fibrosity has a slight oscillatory variation with respect to the fiber aspect ratio (Figure 41.8d). These oscillations are caused by the oscillatory nature of the sinc^2 function and depend on the upper and lower bands of the circular filter (Figure 41.5). We discussed the theoretical basis of these oscillations in Section 41.2.

As we discussed in Section 41.2, rotating a fiber by θ degrees results in a rotation of −θ degrees in its Fourier transform and should not affect the fibrosity. Therefore, we expect the fibrosity of an image of several fibers to be independent of the fiber directions. To evaluate that, we generated 30 images in three different groups: in the first group, fibers are distributed within ±20° of the horizontal direction (Figure 41.9a); in the second group, they are distributed within ±20° of the vertical direction (Figure 41.9b); and in the third group, they are distributed uniformly in all directions (Figure 41.9c). The upper band of the circular filter is $f_U = 0.5$ pixel^{-1} and the lower band is $f_L = 0.25$ pixel^{-1}. The images of fibers that are distributed in all directions result to about 5% lower fibrosity compared with the images of the fibers that are distributed within a limited range of vertical

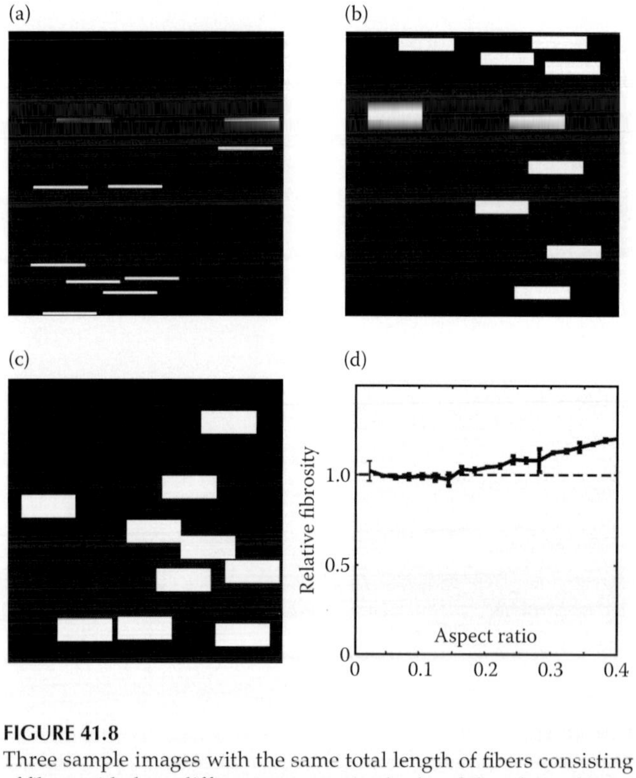

FIGURE 41.8
Three sample images with the same total length of fibers consisting of fibers with three different aspect ratios (a–c) and the relative fibrosity in terms of the fiber aspect ratio (d). The aspect ratios of fibers in (a), (b), and (c) are 0.045, 0.225, and 0.405, respectively. Vertical lines on the curve represent the standard deviations of the images at each aspect ratio.

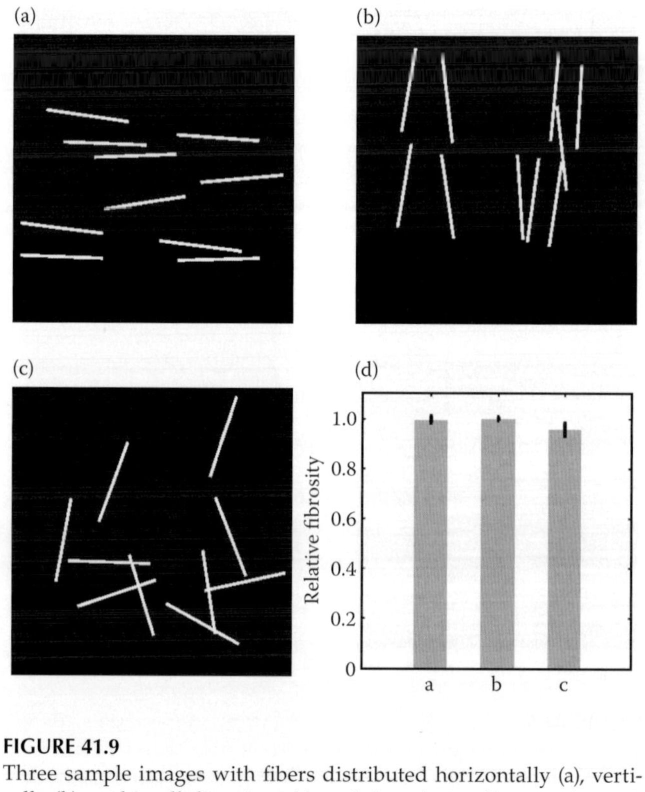

FIGURE 41.9
Three sample images with fibers distributed horizontally (a), vertically (b), and in all directions (c), and the relative fibrosity for each fiber distribution (d). Vertical lines on the bars represent the standard deviations of ten images in each group.

or horizontal direction (Figure 41.9d). The resultant 5% error may be considered negligible for many practical applications of FABLE.

To determine whether this reduction is a consequence of having different directions in one image or is caused by the individual fiber direction, we computed the fibrosity of a single fiber at different directions (Figure 41.10). Clearly, the fibrosity varies with fiber direction and decreases significantly when the fiber forms an angle of 45° with respect to the horizontal direction. Variation of fibrosity by the fiber direction depends on the bandwidth of the circular filter and worsens by increasing the lower frequency of the circular filter (Figure 41.10, panels a, b, and c). Variation of fibrosity by the fiber direction is a result of discretized images and Fourier transforms. When a pixel on the perimeter of the fiber rotates by and an angle of θ about the origin of the image, it falls between 4 pixels. However, as the image is discretized, the pixel that is the closest to the rotated location is considered the rotated pixel. This shift in the location of the rotated pixel deviates the FFT from the theoretical continuous Fourier transform of the continuous image and leads to erroneous variations in the fibrosity for different fiber directions.

FABLE also generates negligible deviations in the fibrosity of images with the same total length of the fiber and different fiber curvatures (Figure 41.11). The fibrosities of two groups of images with two different sinusoidally curved fibers are computed and compared with the fibrosity of images with straight horizontal fibers (sample images of Figure 42.11a–c). The curved fibers are defined as $y_i = C(\lambda, a)\sin(\pi x_i/\lambda)$, where x_i and y_i are the Cartesian coordinates of pixel i, λ is a constant, and $C(\lambda,a)$ is the maximum deflection from the straight line. λ varies between 0 and the fiber length a; $C(\lambda,a)$ varies between $a/2$ (infinite curvature, $\lambda = 0$) and 0 (no curvature, $\lambda = a$). The upper band of the circular filter is $f_U = 0.5$ pixel^{-1} and the lower band is $f_L = 0.25$ pixel^{-1}. Fibrosity is relatively insensitive to moderate fiber curvature where $\lambda/a \geq 0.5$ (Figure 41.11d). For very small λ, fibers fold at their midpoints, with the resulting overlap of the fiber's two halves halving the fibrosity.

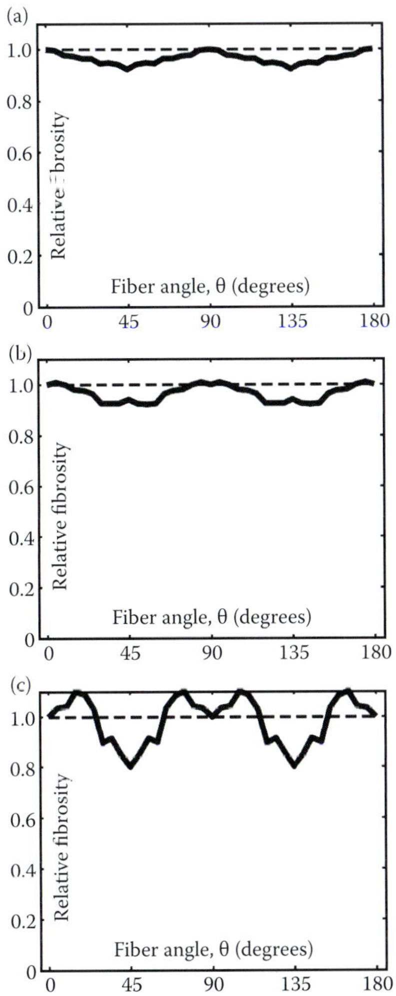

FIGURE 41.10
Variations of the single fiber fibrosity with respect to the fiber angle (relative to the horizontal direction). The upper filter frequency is 0.5 pixel^{-1}, and the lower frequency is 0.05 pixel^{-1} in (a), 0.15 pixel^{-1} in (b), and 0.25 pixel^{-1} in (c).

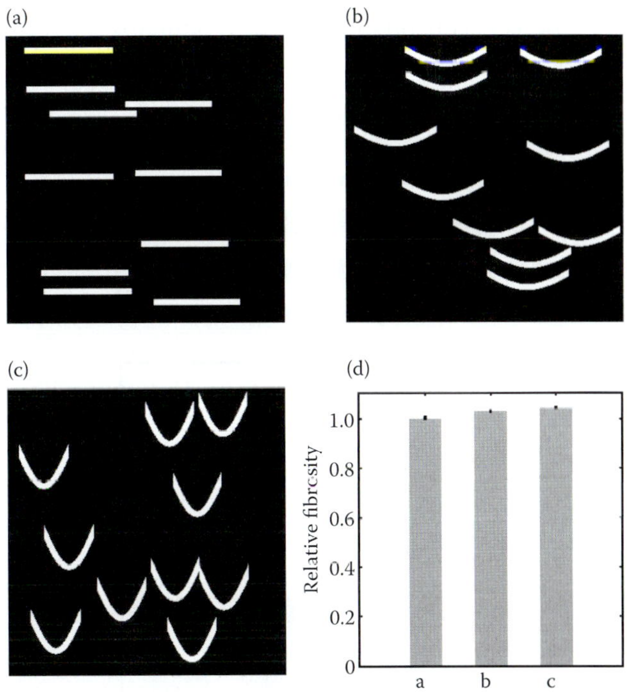

FIGURE 41.11
Three sample images with fibers of different curvature (a–c) and the relative fibrosities for the images with different fiber curvatures (d). Vertical lines on the bars represent the standard deviations of five images in each group. Lengths of all individual fibers are equal regardless of their curvatures.

41.4 Limitations

41.4.1 Changing the Image Sharpness Affects the Fibrosity

One of the limitations of the FABLE is its sensitivity to image sharpness. In particular, on the edges of the fiber, if the transition from substrate luminance to fiber luminance occurs abruptly (a sharp image), the measured fibrosity is higher than when the transition occurs over a few pixels. Figure 41.12 compares the relative fibrosities of three fibers with the same dimensions and different transitions. The upper band of the circular filter is $f_U = 0.5$ pixel^{-1} and the lower band is $f_L = 0.1$ pixel^{-1}. Although the fiber length is the same in all images, the fibrosities are significantly different because of different transition gradients (Figure 41.12b). This means that the application FABLE is limited to images that have the same level of sharpness. Increasing the lower frequency of the circular filter increases the sensitivity of FABLE to the fiber transition gradients. Fibers with smoother transitions have significantly lower powers at higher frequencies compared with the fibers with sharp transitions.

41.4.2 Fibrosity Is Sensitive to Image Luminance

Scaling the image luminance scales the Fourier transform and consequently the fibrosity. Similarly, increasing the luminance of the fibers by a constant value increases the fibrosity. To compare these effects, we

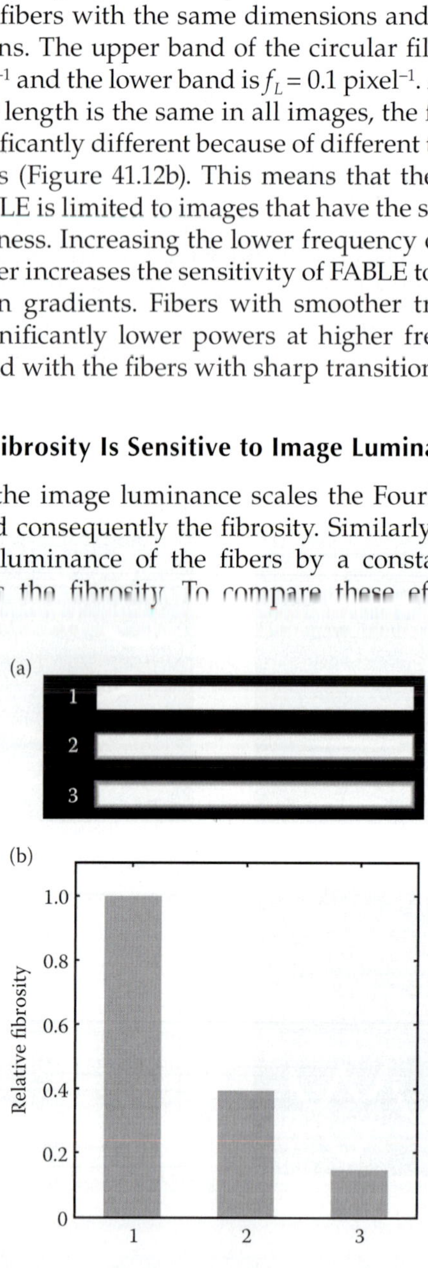

FIGURE 41.12
Three fibers with three different transition gradients (a) and their fibrosities (b). Fiber 1 transitions in one pixel, while fibers 2 and 3 transition uniformly within three and five pixels.

generated three groups of images (five in each group). Each image consists of 60 fibers (200 × 11 pixels). In the first group, the fiber luminance is scaled by a random variable distributed uniformly between 0.2 and 0.7 (sample image in Figure 41.13a). This random variable is distributed uniformly between 0.2 and 1.0 in the second group for more contrast (sample image in Figure 41.13b), and between 0.5 and 1.0 in the third group for more brightness (sample image in Figure 41.13c). The average fibrosities of these three groups of images are shown in Figure 41.13d. Fibrosity is higher in images that the average luminance of their fibers is higher. To mitigate this artifact, the luminance of the images may be adjusted prior to the application of the FABLE method. Let us define the average image luminance as the average of the brightness levels of all the pixels of above a certain threshold. The brightness level of each image is scaled so that the average image luminance is the same in all images. The computed fibrosities for the adjusted images are shown in Figure 41.13e. Scaling all images to the same level of average luminance brings the computed fibrosities very close to each other (within 10%) and mitigates this limitation of the FABLE.

41.4.3 Image Noise Introduces an Artifact

Image noises can significantly affect the fibrosity. Image noises are usually small features with sharp transitions (e.g., salt-and-pepper noises). Therefore, they contribute to the high-frequency powers of the Fourier spectrum and are artifacts of the FABLE technique, especially if the transitions at fiber edges are smooth. Median filters are powerful image processing tools to remove salt-and-pepper noises while keeping the fiber transitions intact. However, the noise dimensions should be smaller than the fiber width for appropriate filtering. The fibrous image of Figure 41.14a is contaminated with fine salt-and-pepper noises (Figure 41.14b) and is filtered by a median filter (Figure 41.14c). Each dimension of the noise is less than 5 pixels, the width of the fiber is 15 pixels, and the bandwidth of the median filter is 7 pixels. Figure 41.14d compares the fibrosities of the three images. The fine salt-and-pepper noises of Figure 41.14b increase the image fibrosity by more than 50%. However, these noises can be removed almost completely without affecting the transition gradients of the fibers (Figure 41.14c). The filtered image restores the initial fibrosity of the image (Figure 41.14d).

41.4.4 FABLE Is Suitable for Relative Comparison of Total Fiber Lengths in Similar Images

Because of the previously explained limitations, computing the total length of fibers in an image depends on many factors like the fiber and image luminance and

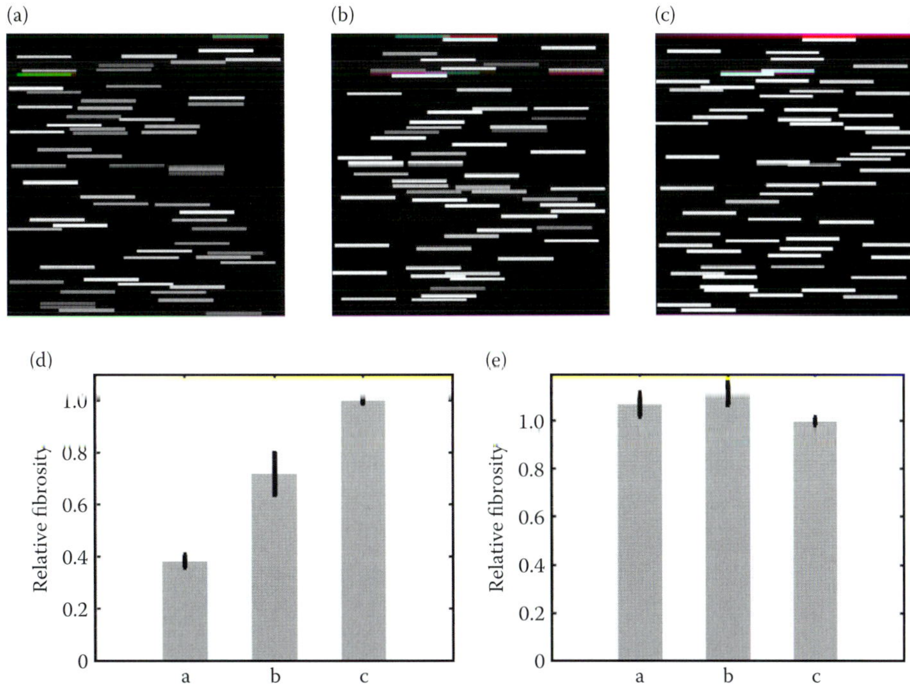

FIGURE 41.13
Three sample images with three different brightness distributions (a–c), the relative fibrosities for the three brightness distributions (d), and the relative fibrosities after brightness adjustment (e).

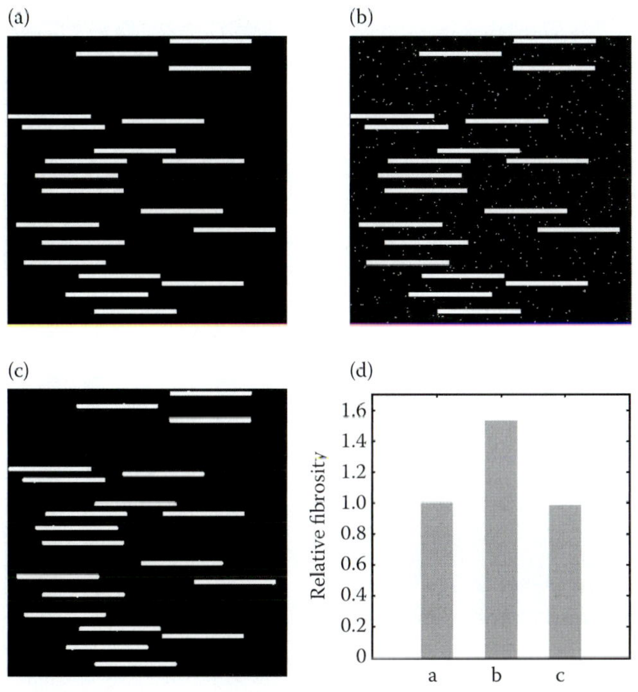

FIGURE 41.14
Control fibrous image (a) contaminated with salt-and-pepper noises (b), filtered with a median filter (c), and the fibrosities of control, noisy, and filtered image (d).

the transition gradient of the fibers. These features are difficult to measure or approximate with automated image processing techniques. Therefore, in practice, the application of FABLE is limited to prediction of the relative fibrosity. It is assumed implicitly that all images are taken from similar fibrous media under the same conditions (e.g., same ambient light, same magnification, same photo sensor, same focus, etc.).

41.5 Application of FABLE for Monitoring the Formation of Stress Fibers

Stress fibers of the myofibroblasts depolymerize in response to a sudden stretch and then repolymerize over time to regain their prestretch density.[21] Stress fibers are the backbone of the force-generating mechanism of myofibroblasts. Density of stress fibers is a characteristic property of the myofibroblasts as the capability of a fibroblast cell in generating tensile force is controlled by its stress fiber network. Stress fibers appear as a few pixel wide fibers in the images of actin cytoskeleton, and the FABLE technique can be used to monitor the variation of these fibers in response to a sudden stretch. Figure 41.15 shows the actin cytoskeleton of the myofibroblast before

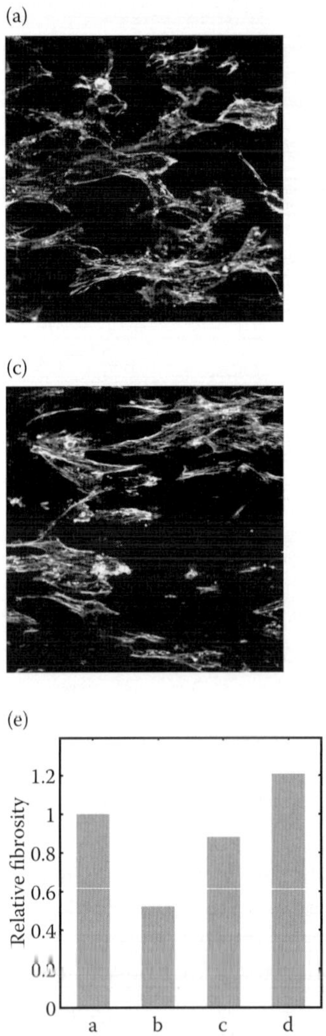

FIGURE 41.15
Four images of myofibroblast actin cytoskeleton at different stage of stress fiber depolymerization and repolymerization (a–d) and the FABLE predictions for the relative fibrosities of each image (e).

the stretch (Figure 41.15a), right after a sudden stretch (Figure 41.15b), 1 min after the stretch (Figure 41.15c), and 20 min after the stretch (Figure 41.15d). Variation of stress fibers among these images is quantified using FABLE (Figure 41.15e).

41.6 Discussion

In this chapter, the ability and accuracy of FABLE in providing a metric, termed fibrosity, for comparison of the total length of the fiber in fibrous images was explained and reviewed. Fibrosity is proportional to the total length of the fibers in an image regardless of the distribution of the length, thickness, and curvature of the individual fibers (Figure 41.16). The coefficient of this proportionality is not a constant number and depends on several image parameters like image brightness, image sharpness (especially at the transition of the fibers), and image noises. Broader analytical studies and simulations are required to determine all the image parameters that affect this coefficient of proportionality and to formulate how the fibrosity is affected by those parameters. Authors propose these analyses as the future advancement of the FABLE technique. In this chapter, lack of these analyses is discussed as artifacts and limitation of the technique (see Section 41.4). The effects of these artifacts would be minimal for comparison of image fibrosities if images were recorded under the same level of lumination, with the same resolution, with the same sensitivity level of the camera, and possibly noise free.

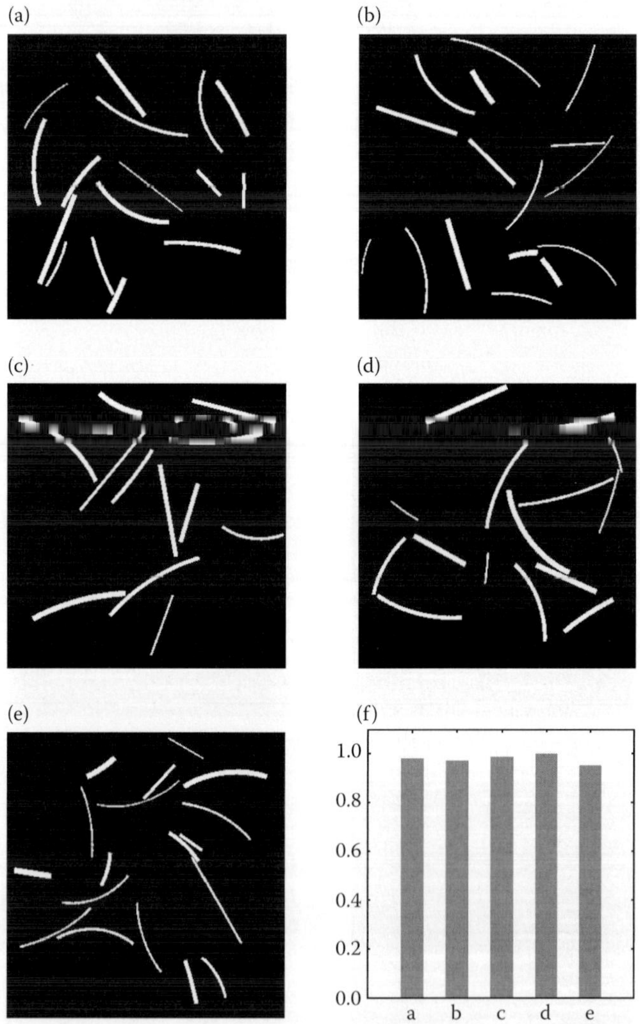

FIGURE 41.16
Five images with the same total length of fibers (a–e) and their relative fibrosities (f). The length, thickness, curvature, and orientation of fibers vary within each image and between images. The relative fibrosity is almost equal (within ±3%) for all images regardless of different length, thickness, curvature, and orientation of fibers.

References

1. Marquez, J. P. Fourier analysis and automated measurement of cell and fiber angular orientation distributions. *Int J Solids Struct* **43**, 6413–6423 (2006).
2. Nekouzadeh, A., Pryse, K. M., Elson, E. L. & Genin, G. M. A simplified approach to quasi-linear viscoelastic modelling. *J Biomech* **40**, 3070–3078 (2007).
3. Lee, T. C., Arthur, T. L., Gibson, L. J. & Hayes, W. C. Sequential labelling of microdamage in bone using chelating agents. *J Orthop Res* **18**, 322–325 (2000).
4. Farhat, F. A., Nicolaides, D., Kanellopoulos, A. & Karihaloo, B. L. High performance fibre-reinforced cementitious composite (CARDIFRC)—Performance and application to retrofitting. *Eng Fract Mech* **74**, 151–167 (2007).
5. Nicolaides, D., Kanellopoulos, A. & Karihaloo, B. L. Effect of fiber distribution on the fatigue performance and autogenous shrinkage of CARDIFRC. In *Measuring, Monitoring and Modeling Concrete Properties* (ed. M. S. Konsta-Gdoutos) 3–16 (Springer, Netherlands, 2006).
6. Stähli, P., Custer, R. & van Mier, J. On flow properties, fibre distribution, fibre orientation and flexural behaviour of FRC. *Mater Struct* **41**, 189–196 (2008).
7. Anstis, G. R., Chantikul, P., Lawn, B. R. & Marshall, D. B. A critical evaluation of indentation techniques for measuring fracture toughness: I, direct crack measurements. *Eng Fract Mech* **4**, 175–179 (1972).
8. Saxena, A. & Hudak, S. J. Review and extension of compliance information for common crack growth specimens. *Int J Fract* **14**, 453–468 (1978).
9. Cooper, J. A. & Sept, D. New insights into mechanism and regulation of actin capping protein. *Int Rev Cell Mol Biol* **267**, 183–206 (2008).
10. Kim, K. et al. Structure/function analysis of the interaction of phosphatidylinositol 4,5-bisphosphate with actin-capping protein: Implications for how capping protein binds the actin filament. *J Biol Chem* **282**, 5871–5879 (2007).
11. Sander, E. A. & Barocas, V. H. Comparison of 2D fiber network orientation measurement methods. *J Biomed Mater Res Part A* **88**, 322–331 (2009).
12. Hart, P. E. How the Hough transform was invented. *IEEE Signal Proc Mag* **26**, 18–22 (2009).
13. Zhang, Q. P. & Couloigner, I. Accurate centerline detection and line width estimation of thick lines using the radon transform. *IEEE T Image Process* **16**, 310–316 (2007).
14. Akhtar, M. W. & Atiquzzaman, M. Determination of line length using Hough transform. *Electron Lett* **28**, 94–96 (1992).
15. Nekouzadeh, A. & Genin, G. M. Quantification of fibre polymerization through Fourier space image analysis. *Proc R Soc A* **467**, 2310–2329 (2011).
16. Cooper, J. A., Loftus, D. J., Frieden, C., Bryan, J. & Elson, E. L. Localization and mobility of gelsolin in cells. *J Cell Biol* **106**, 1229–1240 (1988).
17. Colombelli, J. et al. Mechanosensing in actin stress fibers revealed by a close correlation between force and protein localization. *J Cell Sci* **122**, 1665 (2009).
18. Wakatsuki, T., Kolodney, M. S., Zahalak, G. I. & Elson, E. L. Cell mechanics studied by a reconstituted model tissue. *Biophys J* **79**, 2353–2368 (2000).
19. Marquez, J. P., Genin, G. M., Zahalak, G. I. & Elson, E. L. Thin bio-artificial tissues in plane stress: The relationship between cell and tissue strain, and an improved constitutive model. *Biophys J* **88**, 765–777 (2005).
20. Marquez, J. P., Genin, G. M., Zahalak, G. I. & Elson, E. L. The relationship between cell and tissue strain in three-dimensional bio-artificial tissues. *Biophys J* **88**, 778–789 (2005).
21. Nekouzadeh, A., Pryse, K. M., Elson, E. L. & Genin, G. M. Stretch-activated force shedding, force recovery, and cytoskeletal remodeling in contractile fibroblasts. *J Biomech* **41**, 2964–2971 (2008).
22. Lee, S. L. et al. Physically-induced cytoskeleton remodeling of cells in three-dimensional culture. *PLoS One* **7**, e45512 (2012).
23. Deshpande, V. S., McMeeking, R. M. & Evans, A. G. A bio-chemo-mechanical model for cell contractility. *Proc Natl Acad Sci USA* **103**, 14015 (2006).
24. Deshpande, V. S., McMeeking, R. M. & Evans, A. G. A model for the contractility of the cytoskeleton including the effects of stress-fibre formation and dissociation. *Proc R Soc A* **463**, 787 (2007).
25. Kaunas, R., Nguyen, P., Usami, S. & Chien, S. Cooperative effects of Rho and mechanical stretch on stress fiber organization. *Proc Natl Acad Sci USA* **102**, 15895–15900 (2005).
26. McGarry, J. P. et al. Simulation of the contractile response of cells on an array of micro-posts. *Philosophical Transactions of the Royal Society A: Mathematical, Physical and Engineering Sciences* **367**, 3477 (2009).
27. Nicolas, A., Besser, A. & Safran, S. A. Dynamics of cellular focal adhesions on deformable substrates: Consequences for cell force microscopy. *Biophys J* **95**, 527–539 (2008).

Index

Page numbers followed by f and t indicate figures and tables, respectively.

E

G

H

J

K

L

M

N

O

S

T

U